FOURTH EDITION
ESSENTIAL
CELL BIOLOGY

FOURTH EDITION
ESSENTIAL
CELL BIOLOGY

GS Garland Science
Taylor & Francis Group

ALBERTS · BRAY · HOPKIN · JOHNSON · LEWIS · RAFF · ROBERTS · WALTER

Garland Science
Vice President: Denise Schanck
Senior Editor: Michael Morales
Production Editor and Layout: Emma Jeffcock of EJ Publishing Services
Illustrator: Nigel Orme
Developmental Editor: Monica Toledo
Editorial Assistants: Lamia Harik and Alina Yurova
Copy Editor: Jo Clayton
Book Design: Matthew McClements, Blink Studio, Ltd.
Cover Illustration: Jose Ortega
Authors Album Cover: Photography, Christophe Carlinet; Design, Nigel Orme
Indexer: Bill Johncocks

Essential Cell Biology Website
Artistic and Scientific Direction: Peter Walter
Narrated by: Julie Theriot
Producer: Michael Morales

About the Authors

Bruce Alberts received his PhD from Harvard University and is the Chancellor's Leadership Chair in Biochemistry and Biophysics for Science and Education, University of California, San Francisco. He was the editor-in-chief of *Science* magazine from 2008–2013, and for twelve years he served as President of the U.S. National Academy of Sciences (1993–2005).

Dennis Bray received his PhD from Massachusetts Institute of Technology and is currently an active emeritus professor at the University of Cambridge.

Karen Hopkin received her PhD in biochemistry from the Albert Einstein College of Medicine and is a science writer in Somerville, Massachusetts. She is a contributor to *Scientific American's* daily podcast, *60-Second Science*, and to E. O. Wilson's digital biology textbook, *Life on Earth*.

Alexander Johnson received his PhD from Harvard University and is Professor of Microbiology and Immunology at the University of California, San Francisco.

Julian Lewis received his DPhil from the University of Oxford and is an Emeritus Scientist at the London Research Institute of Cancer Research UK.

Martin Raff received his MD from McGill University and is at the Medical Research Council Laboratory for Molecular Cell Biology and Cell Biology Unit at University College London.

Keith Roberts received his PhD from the University of Cambridge and was Deputy Director of the John Innes Centre, Norwich. He is currently Emeritus Professor at the University of East Anglia.

Peter Walter received his PhD from The Rockefeller University in New York and is Professor of the Department of Biochemistry and Biophysics at the University of California, San Francisco, and an Investigator of the Howard Hughes Medical Institute.

ISBNs: 978-0-8153-4454-4 (hardcover); 978-0-8153-4455-1 (softcover).

Published by Garland Science, Taylor & Francis Group, LLC, an informa business, 711 Third Avenue, New York, NY 10017, USA, and 3 Park Square, Milton Park, Abingdon, OX14 4RN, UK.

Printed in the United States of America

15 14 13 12 11 10 9 8 7 6 5 4 3 2 1

Library of Congress Cataloging-in-Publication Data

Alberts, Bruce.
  Essential cell biology / Bruce Alberts [and seven others].
-- Fourth edition.
    pages cm.
  ISBN 978-0-8153-4454-4 (hardback)
  1. Cytology. 2. Molecular biology. 3. Biochemistry. I. Title.
QH581.2.E78 2013
571.6--dc23
                                        2013025976

Garland Science
Taylor & Francis Group

Visit our website at http://www.garlandscience.com

Preface

In our world there is no form of matter more astonishing than the living cell: tiny, fragile, marvelously intricate, continually made afresh, yet preserving in its DNA a record of information dating back more than three billion years, to a time when our planet had barely cooled from the hot materials of the nascent solar system. Ceaselessly re-engineered and diversified by evolution, extraordinarily versatile and adaptable, the cell retains a complex core of self-replicating chemical machinery that is shared and endlessly repeated by every living organism on the face of the Earth—in every animal, every leaf, every bacterium in a piece of cheese, every yeast in a vat of wine.

Curiosity, if nothing else, should drive us to study cell biology; we need to understand cell biology to understand ourselves. But there are practical reasons, too, why cell biology should be a part of everyone's education. We are made of cells, we feed on cells, and our world is made habitable by cells. The challenge for scientists is to deepen our knowledge of cells and find new ways to apply it. All of us, as citizens, need to know something of the subject to grapple with the modern world, from our own health affairs to the great public issues of environmental change, biomedical technologies, agriculture, and epidemic disease.

Cell biology is a big subject, and it has links with almost every other branch of science. The study of cell biology therefore provides a great scientific education. However, as the science advances, it becomes increasingly easy to become lost in detail, distracted by an overload of information and technical terminology. In this book we therefore focus on providing a digestible, straightforward, and engaging account of only the essential principles. We seek to explain, in a way that can be understood even by a reader approaching biology for the first time, how the living cell works: to show how the molecules of the cell—especially the protein, DNA, and RNA molecules—cooperate to create this remarkable system that feeds, responds to stimuli, moves, grows, divides, and duplicates itself.

The need for a clear account of the essentials of cell biology became apparent to us while we were writing *Molecular Biology of the Cell (MBoC)*, now in its fifth edition. *MBoC* is a large book aimed at advanced undergraduates and graduate students specializing in the life sciences or medicine. Many students and educated lay people who require an introductory account of cell biology would find *MBoC* too detailed for their needs. *Essential Cell Biology (ECB)*, in contrast, is designed to provide the fundamentals of cell biology that are required by anyone to understand both the biomedical and the broader biological issues that affect our lives.

This fourth edition has been extensively revised. We have brought every part of the book up to date, with new material on regulatory RNAs, induced pluripotent stem cells, cell suicide and reprogramming, the human genome, and even Neanderthal DNA. In response to student feedback, we have improved our discussions of photosynthesis and DNA repair. We have added many new figures and have updated our coverage of many exciting new experimental techniques—including RNAi, optogenetics, the applications of new DNA sequencing technologies, and the use of mutant organisms to probe the defects underlying human disease. At the same time, our "How We Know" sections continue to present experimental data and design, illustrating with specific examples how biologists tackle important questions and how their experimental results shape future ideas.

As before, the diagrams in *ECB* emphasize central concepts and are stripped of unnecessary details. The key terms introduced in each chapter are highlighted when they first appear and are collected together at the end of the book in a large, illustrated glossary.

A central feature of the book is the many questions that are presented in the text margins and at the end of each chapter. These are designed to provoke students to think carefully about what they have read, encouraging them to pause and test their understanding. Many questions challenge the student to place the newly acquired information in a broader biological context, and some have more than one valid answer. Others invite speculation. Answers to all the questions are given at the end of the book; in many cases these provide a commentary or an alternative perspective on material presented in the main text.

For those who want to develop their active grasp of cell biology further, we recommend *Molecular Biology of the Cell, Fifth Edition: A Problems Approach*, by John Wilson and Tim Hunt. Though written as a companion to *MBoC*, this book contains questions at all levels of difficulty and contains a goldmine of thought-provoking problems for teachers and students. We have drawn upon it for some of the questions in *ECB*, and we are very grateful to its authors.

The explosion of new imaging and computer technologies continues to provide fresh and spectacular views of the inner workings of living cells. We have captured some of this excitement in the new *Essential Cell Biology* website, located at *www.garlandscience.com/ECB4-students*. This site, which is freely available to anyone in the world with an interest in cell biology, contains over 150 video clips, animations, molecular structures, and high-resolution micrographs—all designed to complement the material in individual book chapters. One cannot watch cells crawling, dividing, segregating their chromosomes, or rearranging their surface without a sense of wonder at the molecular mechanisms that underlie these processes. For a vivid sense of the marvel that science reveals, it is hard to match the narrated movie of DNA replication. These resources have been carefully designed to make the learning of cell biology both easier and more rewarding.

Those who seek references for further reading will find them on the *ECB* student and instructor websites. But for the very latest reviews in the current literature, we suggest the use of web-based search engines, such as PubMed (*www.ncbi.nlm.nih.gov*) or Google Scholar (*scholar.google.com*).

As with *MBoC*, each chapter of *ECB* is the product of a communal effort, with individual drafts circulating from one author to another. In addition, many people have helped us, and these are credited in the Acknowledgments that follow. Despite our best efforts, it is inevitable that there will be errors in the book. We encourage readers who find them to let us know at science@garland.com, so that we can correct these errors in the next printing.

Acknowledgments

The authors acknowledge the many contributions of professors and students from around the world in the creation of this fourth edition. In particular, we are grateful to the students who participated in our focus groups; they provided invaluable feedback about their experiences using the book and our multimedia, and many of their suggestions were implemented in this edition.

We would also like to thank the professors who helped organize the student focus groups at their schools: Nancy W. Kleckner at Bates College, Kate Wright and Dina Newman at Rochester Institute of Technology, David L. Gard at University of Utah, and Chris Brandl and Derek McLachlin at University of Western Ontario. We greatly appreciate their hospitality and the opportunity to learn from their students.

We also received detailed reviews from many instructors who used the third edition, and we would like to thank them for their contributions: Devavani Chatterjea, Macalester College; Frank Hauser, University of Copenhagen; Alan Jones, University of North Carolina at Chapel Hill; Eugene Mesco, Savannah State University; M. Scott Shell, University of California Santa Barbara; Grith Lykke Sørensen, University of Southern Denmark; Marta Bechtel, James Madison University; David Bourgaize, Whittier College; John Stephen Horton, Union College; Sieirn Lim, Nanyang Technological University; Satoru Kenneth Nishimoto, University of Tennessee Health Science Center; Maureen Peters, Oberlin College; Johanna Rees, University of Cambridge; Gregg Whitworth, Grinnell College; Karl Fath, Queens College, City University of New York; Barbara Frank, Idaho State University; Sarah Lundin-Schiller, Austin Peay State University; Marianna Patrauchan, Oklahoma State University; Ellen Rosenberg, University of British Columbia; Leslie Kate Wright, Rochester Institute of Technology; Steven H. Denison, Eckerd College; David Featherstone, University of Illinois at Chicago; Andor Kiss, Miami University; Julie Lively, Sewanee, The University of the South; Matthew Rainbow, Antelope Valley College; Juliet Spencer, University of San Francisco; Christoph Winkler, National University of Singapore; Richard Bird, Auburn University; David Burgess, Boston College; Elisabeth Cox, State University of New York, College at Geneseo; David L. Gard, University of Utah; Beatrice Holton, University of Wisconsin Oshkosh; Glenn H. Kageyama, California State Polytechnic University, Pomona; Jane R. Dunlevy, University of North Dakota; Matthias Falk, Lehigh University. We also want to thank James Hadfield of Cancer Research UK Cambridge Institute for his review of the methods chapter.

Special thanks go to David Morgan, a coauthor of *MBoC*, for his help on the signaling and cell division chapters.

We are very grateful, too, to the readers who alerted us to errors they had found in the previous edition.

Many staff at Garland Science contributed to the creation of this book and made our work on it a pleasure. First of all, we owe a special debt to Michael Morales, our editor, who coordinated the whole enterprise. He organized the initial reviewing and the focus groups, worked closely with the authors on their chapters, urged us on when we fell behind, and played a major part in the design, assembly, and production of *Essential Cell Biology* student website. Monica Toledo managed the flow of chapters through the book development and production process, and oversaw the writing of the accompanying question bank. Lamia Harik gave editorial assistance. Nigel Orme took original drawings created by author Keith Roberts and redrew them on a computer, or occasionally by hand, with great skill and flair. To Matt McClements goes the credit for the graphic design of the book and the creation of the chapter-opener sculptures. As in previous editions, Emma Jeffcock did a brilliant job in laying out the whole book and meticulously incorporating our endless corrections. Adam Sendroff and Lucy Brodie gathered user feedback and launched the book into the wide world. Denise Schanck, the Vice President of Garland Science, attended all of our writing retreats and orchestrated everything with great taste and diplomacy. We give our thanks to everyone in this long list.

Last but not least, we are grateful, yet again, to our colleagues and our families for their unflagging tolerance and support.

Resources for Instructors and Students

The teaching and learning resources for instructors and students are available online. The instructor's resources are password protected and available only to qualified instructors. The student resources are available to everyone. We hope these resources will enhance student learning, and make it easier for instructors to prepare dynamic lectures and activities for the classroom.

INSTRUCTOR RESOURCES

Instructor Resources are available on the Garland Science Instructor's Resource Site, located at *www.garlandscience.com/instructors.* The website provides access not only to the teaching resources for this book but also to all other Garland Science textbooks. Qualified instructors can obtain access to the site from their sales representative or by emailing science@garland.com.

Art of Essential Cell Biology, Fourth Edition

The images from the book are available in two convenient formats: PowerPoint® and JPEG. They have been optimized for display on a computer. Figures are searchable by figure number, figure name, or by keywords used in the figure legend from the book.

Figure-Integrated Lecture Outlines

The section headings, concept headings, and figures from the text have been integrated into PowerPoint presentations. These will be useful for instructors who would like a head start creating lectures for their course. Like all of our PowerPoint presentations, the lecture outlines can be customized. For example, the content of these presentations can be combined with videos and questions from the book or "Question Bank," in order to create unique lectures that facilitate interactive learning.

Animations and Videos

The 130+ animations and videos that are available to students are also available on the Instructor's Resource site in two formats. The WMV-formatted movies are created for instructors who wish to use the movies in PowerPoint presentations on Windows® computers; the QuickTime-formatted movies are for use in PowerPoint for Apple computers or Keynote® presentations. The movies can easily be downloaded to your computer using the "download" button on the movie preview page.

Question Bank

Written by Linda Huang, University of Massachusetts, Boston, and Cheryl D. Vaughan, Harvard University Division of Continuing Education, the revised and expanded question bank includes a variety of question formats: multiple choice, fill-in-the-blank, true-false, matching, essay, and challenging "thought" questions. There are approximately 60–70 questions per chapter, and a large number of the multiple-choice questions will be suitable for use with personal response systems (that is, clickers). The *Question Bank* was created with the philosophy that a good exam should do much more than simply test students' ability to memorize information; it should require them to reflect upon and integrate information as a part of a sound understanding. It provides a comprehensive sampling of questions that can be used either directly or as inspiration for instructors to write their own test questions.

References

Adapted from the detailed references of *Molecular Biology of the Cell*, and organized by the table of contents for *Essential Cell Biology*, the "References" provide a rich compendium of journal and review articles for reference and reading assignments. The "References" PDF document is available on both the instructor and student websites.

Medical Topics Guide

This document highlights medically relevant topics covered throughout the book, and will be particularly useful for instructors with a large number of premedical, health science, or nursing students.

Media Guide

This document overviews the multimedia available for students and instructors and contains the text of the voice-over narration for all of the movies.

Blackboard® and LMS Integration

The movies, book images, and student assessments that accompany the book can be integrated into Blackboard or other learning management systems. These resources are bundled into a "Common Cartridge" that facilitates bulk uploading of textbook resources into Blackboard and other learning management systems. The LMS Common Cartridge can be obtained on a DVD from your sales representative or by emailing science@garland.com.

STUDENT RESOURCES

The resources for students are available on the *Essential Cell Biology* Student Website, located at *www.garland science.com/ECB4-students*.

Animations and Videos

There are over 130 movies, covering a wide range of cell biology topics, which review key concepts in the book and illuminate the cellular microcosm.

Student Self-Assessments

The website contains a variety of self-assessment tools to help students.

- Each chapter has a multiple-choice quiz to test basic reading comprehension.

- There are also a number of media assessments that require students to respond to specific questions about movies on the website or figures in the book.

- Additional concept questions complement the questions available in the book.

- "Challenge" questions are included that provide a more experimental perspective or require a greater depth of conceptual understanding.

Cell Explorer

This application teaches cell morphology through interactive micrographs that highlight important cellular structures.

Flashcards

Each chapter contains a set of flashcards, built into the website, that allow students to review key terms from the text.

Glossary

The complete glossary from the book is available on the website and can be searched or browsed.

References

A set of references is available for each chapter for further reading and exploration.

Contents and Special Features

Detailed Contents

CHAPTER **ONE**

1

Cells: The Fundamental Units of Life

What does it mean to be living? Petunias, people, and pond scum are all alive; stones, sand, and summer breezes are not. But what are the fundamental properties that characterize living things and distinguish them from nonliving matter?

The answer begins with a basic fact that is taken for granted now, but marked a revolution in thinking when first established 175 years ago. All living things (or *organisms*) are built from **cells**: small, membrane-enclosed units filled with a concentrated aqueous solution of chemicals and endowed with the extraordinary ability to create copies of themselves by growing and then dividing in two. The simplest forms of life are solitary cells. Higher organisms, including ourselves, are communities of cells derived by growth and division from a single founder cell. Every animal or plant is a vast colony of individual cells, each of which performs a specialized function that is regulated by intricate systems of cell-to-cell communication.

Cells, therefore, are the fundamental units of life. Thus it is to *cell biology*—the study of cells and their structure, function, and behavior—that we must look for an answer to the question of what life is and how it works. With a deeper understanding of cells, we can begin to tackle the grand historical problems of life on Earth: its mysterious origins, its stunning diversity produced by billions of years of evolution, and its invasion of every conceivable habitat. At the same time, cell biology can provide us with answers to the questions we have about ourselves: Where did we come from? How do we develop from a single fertilized egg cell? How is each of us similar to—yet different from—everyone else on Earth? Why do we get sick, grow old, and die?

In this chapter, we begin by looking at the great variety of forms that cells can show, and we take a preliminary glimpse at the chemical machinery that all cells have in common. We then consider how cells are made visible under the microscope and what we see when we peer inside them. Finally, we discuss how we can exploit the similarities of living things to achieve a coherent understanding of all forms of life on Earth—from the tiniest bacterium to the mightiest oak.

UNITY AND DIVERSITY OF CELLS

Cell biologists often speak of "the cell" without specifying any particular cell. But cells are not all alike; in fact, they can be wildly different. Biologists estimate that there may be up to 100 million distinct species of living things on our planet. Before delving deeper into cell biology, we must take stock: What does a bacterium have in common with a butterfly? What do the cells of a rose have in common with those of a dolphin? And in what ways do the plethora of cell types within an individual multicellular organism differ?

Cells Vary Enormously in Appearance and Function

Let us begin with size. A bacterial cell—say a *Lactobacillus* in a piece of cheese—is a few **micrometers**, or µm, in length. That's about 25 times smaller than the width of a human hair. A frog egg—which is also a single cell—has a diameter of about 1 millimeter. If we scaled them up to make the *Lactobacillus* the size of a person, the frog egg would be half a mile high.

Cells vary just as widely in their shape (**Figure 1–1**). A typical nerve cell in your brain, for example, is enormously extended; it sends out its electrical signals along a fine protrusion that is 10,000 times longer than it is thick, and it receives signals from other nerve cells through a mass of shorter processes that sprout from its body like the branches of a tree (see Figure 1–1A). A *Paramecium* in a drop of pond water is shaped like a submarine and is covered with thousands of *cilia*—hairlike extensions whose sinuous beating sweeps the cell forward, rotating as it goes (Figure 1–1B). A cell in the surface layer of a plant is squat and immobile, surrounded

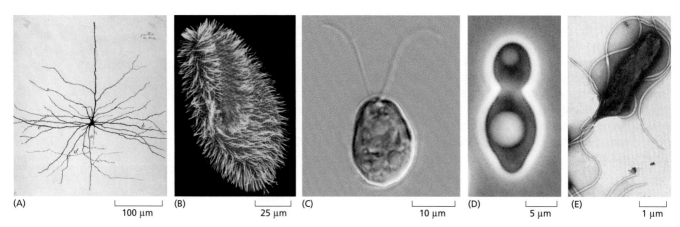

(A)   100 µm   (B)   25 µm   (C)   10 µm   (D)   5 µm   (E)   1 µm

Figure 1–1 **Cells come in a variety of shapes and sizes.** Note the very different scales of these micrographs. (A) Drawing of a single nerve cell from a mammalian brain. This cell has a huge branching tree of processes, through which it receives signals from as many as 100,000 other nerve cells. (B) *Paramecium.* This protozoan—a single giant cell—swims by means of the beating cilia that cover its surface. (C) *Chlamydomonas.* This type of single-celled green algae is found all over the world—in soil, fresh water, oceans, and even in the snow at the top of mountains. The cell makes its food like plants do—via photosynthesis—and it pulls itself through the water using its paired flagella to do the breaststroke. (D) *Saccharomyces cerevisiae.* This yeast cell, used in baking bread, reproduces itself by a process called budding. (E) *Helicobacter pylori.* This bacterium—a causative agent of stomach ulcers—uses a handful of whiplike flagella to propel itself through the stomach lining. (A, copyright Herederos de Santiago Ramón y Cajal, 1899; B, courtesy of Anne Fleury, Michel Laurent, and André Adoutte; C, courtesy of Brian Piasecki; E, courtesy of Yutaka Tsutsumi.)

by a rigid box of cellulose with an outer waterproof coating of wax. A neutrophil or a macrophage in the body of an animal, by contrast, crawls through tissues, constantly pouring itself into new shapes, as it searches for and engulfs debris, foreign microorganisms, and dead or dying cells. And so on.

Cells are also enormously diverse in their chemical requirements. Some require oxygen to live; for others this gas is deadly. Some cells consume little more than air, sunlight, and water as their raw materials; others need a complex mixture of molecules produced by other cells.

These differences in size, shape, and chemical requirements often reflect differences in cell function. Some cells are specialized factories for the production of particular substances, such as hormones, starch, fat, latex, or pigments. Others are engines, like muscle cells that burn fuel to do mechanical work. Still others are electricity generators, like the modified muscle cells in the electric eel.

Some modifications specialize a cell so much that they spoil its chances of leaving any descendants. Such specialization would be senseless for a cell that lived a solitary life. In a multicellular organism, however, there is a division of labor among cells, allowing some cells to become specialized to an extreme degree for particular tasks and leaving them dependent on their fellow cells for many basic requirements. Even the most basic need of all, that of passing on the genetic instructions of the organism to the next generation, is delegated to specialists—the egg and the sperm.

Living Cells All Have a Similar Basic Chemistry

Despite the extraordinary diversity of plants and animals, people have recognized from time immemorial that these organisms have something in common, something that entitles them all to be called living things. But while it seemed easy enough to recognize life, it was remarkably difficult to say in what sense all living things were alike. Textbooks had to settle for defining life in abstract general terms related to growth, reproduction, and an ability to respond to the environment.

The discoveries of biochemists and molecular biologists have provided an elegant solution to this awkward situation. Although the cells of all living things are infinitely varied when viewed from the outside, they are fundamentally similar inside. We now know that cells resemble one another to an astonishing degree in the details of their chemistry. They are composed of the same sorts of molecules, which participate in the same types of chemical reactions (discussed in Chapter 2). In all organisms, genetic information—in the form of *genes*—is carried in DNA molecules. This information is written in the same chemical code, constructed out of the same chemical building blocks, interpreted by essentially the same chemical machinery, and replicated in the same way when an organism reproduces. Thus, in every cell, the long **DNA** polymer chains are made from the same set of four monomers, called *nucleotides*, strung together in different sequences like the letters of an alphabet to convey information. In every cell, the information encoded in the DNA is read out, or *transcribed*, into a chemically related set of polymers called **RNA**. A subset of these RNA molecules is in turn *translated* into yet another type of polymer called a **protein**. This flow of information—from DNA to RNA to protein—is so fundamental to life that it is referred to as the *central dogma* (**Figure 1–2**).

The appearance and behavior of a cell are dictated largely by its protein molecules, which serve as structural supports, chemical catalysts,

QUESTION 1–1

"Life" is easy to recognize but difficult to define. According to one popular biology text, living things:
1. Are highly organized compared to natural inanimate objects.
2. Display homeostasis, maintaining a relatively constant internal environment.
3. Reproduce themselves.
4. Grow and develop from simple beginnings.
5. Take energy and matter from the environment and transform it.
6. Respond to stimuli.
7. Show adaptation to their environment.
Score a person, a vacuum cleaner, and a potato with respect to these characteristics.

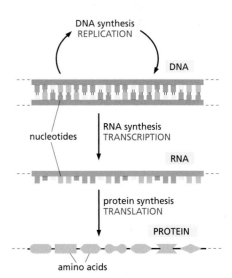

Figure 1–2 In all living cells, genetic information flows from DNA to RNA (transcription) and from RNA to protein (translation)—a sequence known as the central dogma. The sequence of nucleotides in a particular segment of DNA (a gene) is transcribed into an RNA molecule, which can then be translated into the linear sequence of amino acids of a protein. Only a small part of the gene, RNA, and protein are shown.

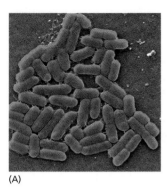
(A)

(B)

(C)

(D)

Figure 1–3 All living organisms are constructed from cells. A colony of bacteria, a butterfly, a rose, and a dolphin are all made of cells that have a fundamentally similar chemistry and operate according to the same basic principles. (A, courtesy of Janice Carr; C, courtesy of the John Innes Foundation; D, courtesy of Jonathan Gordon, IFAW.)

molecular motors, and so on. Proteins are built from *amino acids*, and all organisms use the same set of 20 amino acids to make their proteins. But the amino acids are linked in different sequences, giving each type of protein molecule a different three-dimensional shape, or *conformation*, just as different sequences of letters spell different words. In this way, the same basic biochemical machinery has served to generate the whole gamut of life on Earth (**Figure 1–3**). A more detailed discussion of the structure and function of proteins, RNA, and DNA is presented in Chapters 4 through 8.

If cells are the fundamental unit of living matter, then nothing less than a cell can truly be called living. Viruses, for example, are compact packages of genetic information—in the form of DNA or RNA—encased in protein but they have no ability to reproduce themselves by their own efforts. Instead, they get themselves copied by parasitizing the reproductive machinery of the cells that they invade. Thus, viruses are chemical zombies: they are inert and inactive outside their host cells, but they can exert a malign control over a cell once they gain entry.

All Present-Day Cells Have Apparently Evolved from the Same Ancestral Cell

A cell reproduces by replicating its DNA and then dividing in two, passing a copy of the genetic instructions encoded in its DNA to each of its daughter cells. That is why daughter cells resemble the parent cell. However, the copying is not always perfect, and the instructions are occasionally corrupted by *mutations* that change the DNA. For this reason, daughter cells do not always match the parent cell exactly.

Mutations can create offspring that are changed for the worse (in that they are less able to survive and reproduce), changed for the better (in that they are better able to survive and reproduce), or changed in a neutral way (in that they are genetically different but equally viable). The struggle for survival eliminates the first, favors the second, and tolerates the third. The genes of the next generation will be the genes of the survivors.

On occasion, the pattern of descent may be complicated by sexual reproduction, in which two cells of the same species fuse, pooling their DNA. The genetic cards are then shuffled, re-dealt, and distributed in new combinations to the next generation, to be tested again for their ability to promote survival and reproduction.

These simple principles of genetic change and selection, applied repeatedly over billions of cell generations, are the basis of **evolution**—the process by which living species become gradually modified and adapted to their environment in more and more sophisticated ways. Evolution offers a startling but compelling explanation of why present-day cells are so similar in their fundamentals: they have all inherited their genetic instructions from the same common ancestor. It is estimated that this ancestral cell existed between 3.5 and 3.8 billion years ago, and we must

QUESTION 1–2

Mutations are mistakes in the DNA that change the genetic plan from the previous generation. Imagine a shoe factory. Would you expect mistakes (i.e., unintentional changes) in copying the shoe design to lead to improvements in the shoes produced? Explain your answer.

suppose that it contained a prototype of the universal machinery of all life on Earth today. Through a very long process of mutation and natural selection, the descendants of this ancestral cell have gradually diverged to fill every habitat on Earth with organisms that exploit the potential of the machinery in an endless variety of ways.

Genes Provide the Instructions for Cell Form, Function, and Complex Behavior

A cell's **genome**—that is, the entire sequence of nucleotides in an organism's DNA—provides a genetic program that instructs the cell how to behave. For the cells of plant and animal embryos, the genome directs the growth and development of an adult organism with hundreds of different cell types. Within an individual plant or animal, these cells can be extraordinarily varied, as we discuss in Chapter 20. Fat cells, skin cells, bone cells, and nerve cells seem as dissimilar as any cells could be. Yet all these *differentiated cell types* are generated during embryonic development from a single fertilized egg cell, and all contain identical copies of the DNA of the species. Their varied characters stem from the way that individual cells use their genetic instructions. Different cells *express* different genes: that is, they use their genes to produce some proteins and not others, depending on their internal state and on cues that they and their ancestor cells have received from their surroundings—mainly signals from other cells in the organism.

The DNA, therefore, is not just a shopping list specifying the molecules that every cell must make, and a cell is not just an assembly of all the items on the list. Each cell is capable of carrying out a variety of biological tasks, depending on its environment and its history, and it selectively uses the information encoded in its DNA to guide its activities. Later in this book, we will see in detail how DNA defines both the parts list of the cell and the rules that decide when and where these parts are to be made.

CELLS UNDER THE MICROSCOPE

Today, we have the technology to decipher the underlying principles that govern the structure and activity of the cell. But cell biology started without these tools. The earliest cell biologists began by simply looking at tissues and cells, and later breaking them open or slicing them up, attempting to view their contents. What they saw was to them profoundly baffling—a collection of tiny and scarcely visible objects whose relationship to the properties of living matter seemed an impenetrable mystery. Nevertheless, this type of visual investigation was the first step toward understanding cells, and it remains essential in the study of cell biology.

Cells were not made visible until the seventeenth century, when the **microscope** was invented. For hundreds of years afterward, all that was known about cells was discovered using this instrument. *Light microscopes* use visible light to illuminate specimens, and they allowed biologists to see for the first time the intricate structure that underpins all living things.

Although these instruments now incorporate many sophisticated improvements, the properties of light itself set a limit to the fineness of detail they reveal. *Electron microscopes*, invented in the 1930s, go beyond this limit by using beams of electrons instead of beams of light as the source of illumination, greatly extending our ability to see the fine details of cells and even making some of the larger molecules visible individually. These and other forms of microscopy remain vital tools in the modern cell biology laboratory, where they continue to reveal new and sometimes surprising details about the way cells are built and how they operate.

The Invention of the Light Microscope Led to the Discovery of Cells

The development of the light microscope depended on advances in the production of glass lenses. By the seventeenth century, lenses were powerful enough to make out details invisible to the naked eye. Using an instrument equipped with such a lens, Robert Hooke examined a piece of cork and in 1665 reported to the Royal Society of London that the cork was composed of a mass of minute chambers. He called these chambers "cells," based on their resemblance to the simple rooms occupied by monks in a monastery. The name stuck, even though the structures Hooke described were actually the cell walls that remained after the living plant cells inside them had died. Later, Hooke and his Dutch contemporary Antoni van Leeuwenhoek were able to observe living cells, seeing for the first time a world teeming with motile microscopic organisms.

For almost 200 years, such instruments—the first light microscopes—remained exotic devices, available only to a few wealthy individuals. It was not until the nineteenth century that microscopes began to be widely used to look at cells. The emergence of cell biology as a distinct science was a gradual process to which many individuals contributed, but its official birth is generally said to have been signaled by two publications: one by the botanist Matthias Schleiden in 1838 and the other by the zoologist Theodor Schwann in 1839. In these papers, Schleiden and Schwann documented the results of a systematic investigation of plant and animal tissues with the light microscope, showing that cells were the universal building blocks of all living tissues. Their work, and that of other nineteenth-century microscopists, slowly led to the realization that all living cells are formed by the growth and division of existing cells—a principle sometimes referred to as the *cell theory* (**Figure 1–4**). The implication that

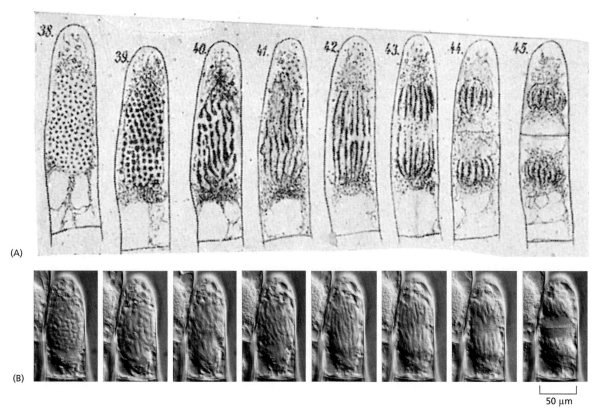

(A)

(B)

50 μm

Figure 1–4 New cells form by growth and division of existing cells. (A) In 1880, Eduard Strasburger drew a living plant cell (a hair cell from a *Tradescantia* flower), which he observed dividing into two daughter cells over a period of 2.5 hours. (B) A comparable living plant cell photographed recently through a modern light microscope. (B, courtesy of Peter Hepler.)

living organisms do not arise spontaneously but can be generated only from existing organisms was hotly contested, but it was finally confirmed in the 1860s by an elegant set of experiments performed by Louis Pasteur.

The principle that cells are generated only from preexisting cells and inherit their characteristics from them underlies all of biology and gives the subject a unique flavor: in biology, questions about the present are inescapably linked to questions about the past. To understand why present-day cells and organisms behave as they do, we need to understand their history, all the way back to the misty origins of the first cells on Earth. Charles Darwin provided the key insight that makes this history comprehensible. His theory of evolution, published in 1859, explains how random variation and natural selection gave rise to diversity among organisms that share a common ancestry. When combined with the cell theory, the theory of evolution leads us to view all life, from its beginnings to the present day, as one vast family tree of individual cells. Although this book is primarily about how cells work today, we will encounter the theme of evolution again and again.

Light Microscopes Allow Examination of Cells and Some of Their Components

If you cut a very thin slice from a suitable plant or animal tissue and view it using a light microscope, you will see that the tissue is divided into thousands of small cells. These may be either closely packed or separated from one another by an *extracellular matrix*, a dense material often made of protein fibers embedded in a polysaccharide gel (**Figure 1–5**). Each cell is typically about 5–20 µm in diameter. If you have taken care of your specimen so that its cells remain alive, you will be able to see particles moving around inside individual cells. And if you watch patiently, you may even see a cell slowly change shape and divide into two (see Figure 1–4 and a speeded-up video of cell division in a frog embryo in Movie 1.1).

To see the internal structure of a cell is difficult, not only because the parts are small, but also because they are transparent and mostly colorless. One way around the problem is to stain cells with dyes that color particular components differently (see Figure 1–5). Alternatively, one can exploit the fact that cell components differ slightly from one another in

QUESTION 1–3

You have embarked on an ambitious research project: to create life in a test tube. You boil up a rich mixture of yeast extract and amino acids in a flask along with a sprinkling of the inorganic salts known to be essential for life. You seal the flask and allow it to cool. After several months, the liquid is as clear as ever, and there are no signs of life. A friend suggests that excluding the air was a mistake, since most life as we know it requires oxygen. You repeat the experiment, but this time you leave the flask open to the atmosphere. To your great delight, the liquid becomes cloudy after a few days and under the microscope you see beautiful small cells that are clearly growing and dividing. Does this experiment prove that you managed to generate a novel life-form? How might you redesign your experiment to allow air into the flask, yet eliminate the possibility that contamination is the explanation for the results? (For a ready-made answer, look up the classic experiments of Louis Pasteur.)

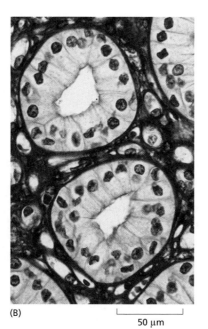

(A)   50 µm   (B)   50 µm

Figure 1–5 Cells form tissues in plants and animals. (A) Cells in the root tip of a fern. The nuclei are stained *red*, and each cell is surrounded by a thin cell wall (*light blue*). (B) Cells in the urine-collecting ducts of the kidney. Each duct appears in this cross section as a ring of closely packed cells (with nuclei stained *red*). The ring is surrounded by extracellular matrix, stained *purple*. (A, courtesy of James Mauseth; B, from P.R. Wheater et al., Functional Histology, 2nd ed. Edinburgh: Churchill Livingstone, 1987. With permission from Elsevier.)

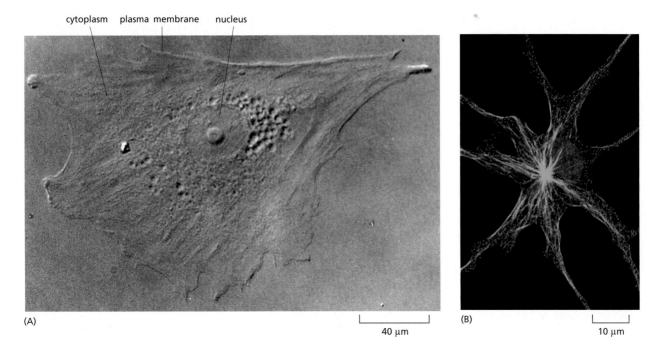

cytoplasm   plasma membrane   nucleus

(A)   (B)

40 μm   10 μm

Figure 1–6 Some of the internal structures of a living cell can be seen with a light microscope. (A) A cell taken from human skin and grown in culture was photographed through a light microscope using interference-contrast optics (see Panel 1–1, pp. 10–11). The nucleus is especially prominent. (B) A pigment cell from a frog, stained with fluorescent dyes and viewed with a confocal fluorescence microscope (see Panel 1–1). The nucleus is shown in *purple*, the pigment granules in *red*, and the microtubules—a class of filaments built from protein molecules in the cytoplasm—in *green*. (A, courtesy of Casey Cunningham; B, courtesy of Stephen Rogers and the Imaging Technology Group of the Beckman Institute, University of Illinois, Urbana.)

refractive index, just as glass differs in refractive index from water, causing light rays to be deflected as they pass from the one medium into the other. The small differences in refractive index can be made visible by specialized optical techniques, and the resulting images can be enhanced further by electronic processing.

The cell thus revealed has a distinct anatomy (**Figure 1–6A**). It has a sharply defined boundary, indicating the presence of an enclosing membrane. A large, round structure, the *nucleus*, is prominent in the middle of the cell. Around the nucleus and filling the cell's interior is the **cytoplasm**, a transparent substance crammed with what seems at first to be a jumble of miscellaneous objects. With a good light microscope, one can begin to distinguish and classify some of the specific components in the cytoplasm, but structures smaller than about 0.2 μm—about half the wavelength of visible light—cannot normally be resolved; points closer than this are not distinguishable and appear as a single blur.

In recent years, however, new types of **fluorescence microscopes** have been developed that use sophisticated methods of illumination and electronic image processing to see fluorescently labeled cell components in much finer detail (**Figure 1–6B**). The most recent super-resolution fluorescence microscopes, for example, can push the limits of resolution down even further, to about 20 nanometers (nm). That is the size of a single **ribosome**, a large macromolecular complex composed of 80–90 individual proteins and RNA molecules.

The Fine Structure of a Cell Is Revealed by Electron Microscopy

For the highest magnification and best resolution, one must turn to an **electron microscope**, which can reveal details down to a few nanometers. Cell samples for the electron microscope require painstaking preparation. Even for light microscopy, a tissue often has to be *fixed* (that is, preserved by pickling in a reactive chemical solution), supported by *embedding* in a solid wax or resin, cut or *sectioned* into thin slices, and *stained* before it is viewed. For electron microscopy, similar procedures are required, but the sections have to be much thinner and there is no possibility of looking at living, wet cells.

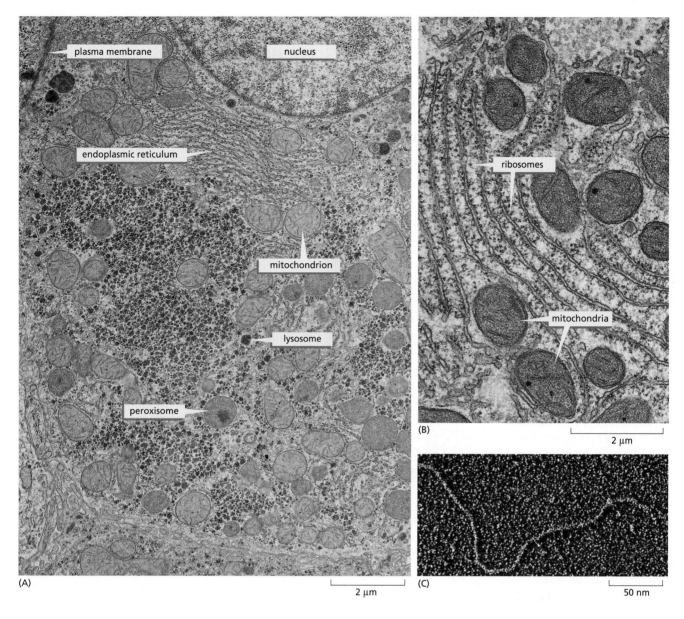

(A)

(B)

(C)

Figure 1–7 **The fine structure of a cell can be seen in a transmission electron microscope.** (A) Thin section of a liver cell showing the enormous amount of detail that is visible. Some of the components to be discussed later in the chapter are labeled; they are identifiable by their size and shape. (B) A small region of the cytoplasm at higher magnification. The smallest structures that are clearly visible are the ribosomes, each of which is made of 80–90 or so individual large molecules. (C) Portion of a long, threadlike DNA molecule isolated from a cell and viewed by electron microscopy. (A and B, courtesy of Daniel S. Friend; C, courtesy of Mei Lie Wong.)

When thin sections are cut, stained, and placed in the electron microscope, much of the jumble of cell components becomes sharply resolved into distinct **organelles**—separate, recognizable substructures with specialized functions that are often only hazily defined with a light microscope. A delicate membrane, only about 5 nm thick, is visible enclosing the cell, and similar membranes form the boundary of many of the organelles inside (**Figure 1–7A, B**). The membrane that separates the interior of the cell from its external environment is called the **plasma membrane**, while the membranes surrounding organelles are called *internal membranes*. All of these membranes are only two molecules thick (as discussed in Chapter 11). With an electron microscope, even individual large molecules can be seen (**Figure 1–7C**).

The type of electron microscope used to look at thin sections of tissue is known as a *transmission electron microscope*. This is, in principle, similar to a light microscope, except that it transmits a beam of electrons rather than a beam of light through the sample. Another type of electron microscope—the *scanning electron microscope*—scatters electrons off the surface of the sample and so is used to look at the surface detail of cells and other structures. A survey of the principal types of microscopy used to examine cells is given in **Panel 1–1** (pp. 10–11).

THE LIGHT MICROSCOPE

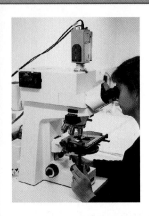

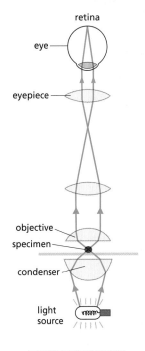

retina

eye

eyepiece

objective
specimen

condenser

light
source

the light path in a
light microscope

The light microscope allows us to magnify cells up to 1000 times and to resolve details as small as 0.2 μm (a limitation imposed by the wavelike nature of light, not by the quality of the lenses). Three things are required for viewing cells in a light microscope. First, a bright light must be focused onto the specimen by lenses in the condenser. Second, the specimen must be carefully prepared to allow light to pass through it. Third, an appropriate set of lenses (objective and eyepiece) must be arranged to focus an image of the specimen in the eye.

FLUORESCENCE MICROSCOPY

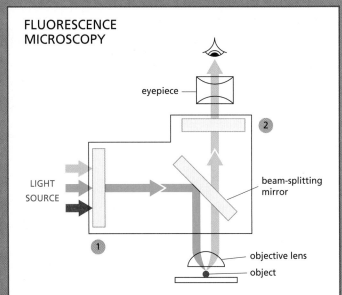

eyepiece

2

beam-splitting
mirror

LIGHT
SOURCE

1

objective lens
object

Fluorescent dyes used for staining cells are detected with the aid of a *fluorescence microscope*. This is similar to an ordinary light microscope except that the illuminating light is passed through two sets of filters. The first (1) filters the light before it reaches the specimen, passing only those wavelengths that excite the particular fluorescent dye. The second (2) blocks out this light and passes only those wavelengths emitted when the dye fluoresces. Dyed objects show up in bright color on a dark background.

LOOKING AT LIVING CELLS

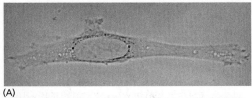

(A)

(B)

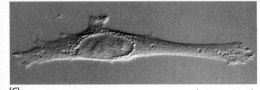

(C)

50 μm

The same unstained, living animal cell (fibroblast) in culture viewed with (A) straightforward (bright-field) optics; (B) phase-contrast optics; (C) interference-contrast optics. The two latter systems exploit differences in the way light travels through regions of the cell with differing refractive indexes. All three images can be obtained on the same microscope simply by interchanging optical components.

FIXED SAMPLES

Most tissues are neither small enough nor transparent enough to examine directly in the microscope. Typically, therefore, they are chemically fixed and cut into very thin slices, or *sections*, that can be mounted on a glass microscope slide and subsequently stained to reveal different components of the cells. A stained section of a plant root tip is shown here (D). (Courtesy of Catherine Kidner.)

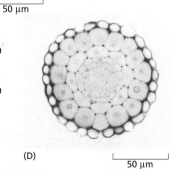

(D)

50 μm

FLUORESCENT PROBES

Dividing nuclei in a fly embryo seen with a fluorescence microscope after staining with specific fluorescent dyes.

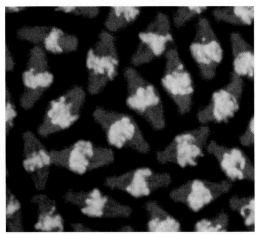

Fluorescent dyes absorb light at one wavelength and emit it at another, longer wavelength. Some such dyes bind specifically to particular molecules in cells and can reveal their location when examined with a fluorescence microscope. An example is the stain for DNA shown here (*green*). Other dyes can be coupled to antibody molecules, which then serve as highly specific and versatile staining reagents that bind selectively to particular large molecules, allowing us to see their distribution in the cell. In the example shown, a microtubule protein in the mitotic spindle is stained *red* with a fluorescent antibody. (Courtesy of William Sullivan.)

CONFOCAL MICROSCOPY

A confocal microscope is a specialized type of fluorescence microscope that builds up an image by scanning the specimen with a laser beam. The beam is focused onto a single point at a specific depth in the specimen, and a pinhole aperture in the detector allows only fluorescence emitted from this same point to be included in the image. Scanning the beam across the specimen generates a sharp image of the plane of focus—an *optical* section. A series of optical sections at different depths allows a three-dimensional image to be constructed. An intact insect embryo is shown here stained with a fluorescent probe for actin filaments. (A) Conventional fluorescence microscopy gives a blurry image due to the presence of fluorescent structures above and below the plane of focus. (B) Confocal microscopy provides an optical section showing the individual cells clearly. (Courtesy of Richard Warn and Peter Shaw.)

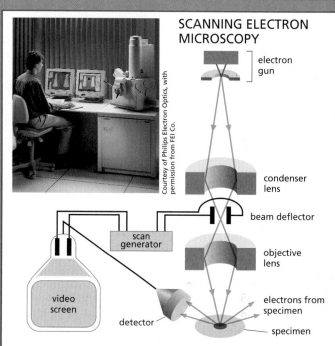

(A)   (B) |_____|
                                            10 μm

TRANSMISSION ELECTRON MICROSCOPY

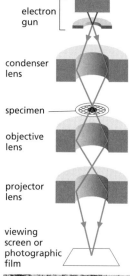

electron gun

condenser lens

specimen

objective lens

projector lens

viewing screen or photographic film

Courtesy of Philips Electron Optics, with permission from FEI Co.

The electron micrograph below shows a small region of a cell in a piece of testis. The tissue has been chemically fixed, embedded in plastic, and cut into very thin sections that have then been stained with salts of uranium and lead. (Courtesy of Daniel S. Friend.)

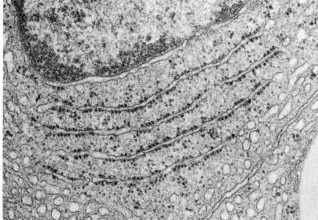

|_____|
0.5 μm

The transmission electron microscope (TEM) is in principle similar to a light microscope, but it uses a beam of electrons instead of a beam of light, and magnetic coils to focus the beam instead of glass lenses. The specimen, which is placed in a vacuum, must be very thin. Contrast is usually introduced by staining the specimen with electron-dense heavy metals that locally absorb or scatter electrons, removing them from the beam as it passes through the specimen. The TEM has a useful magnification of up to a million-fold and can resolve details as small as about 1 nm in biological specimens.

SCANNING ELECTRON MICROSCOPY

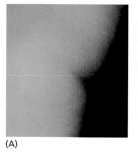

Courtesy of Philips Electron Optics, with permission from FEI Co.

electron gun

condenser lens

beam deflector

scan generator

objective lens

video screen

detector

electrons from specimen

specimen

In the scanning electron microscope (SEM), the specimen, which has been coated with a very thin film of a heavy metal, is scanned by a beam of electrons brought to a focus on the specimen by magnetic coils that act as lenses. The quantity of electrons scattered or emitted as the beam bombards each successive point on the surface of the specimen is measured by the detector, and is used to control the intensity of successive points in an image built up on a video screen. The microscope creates striking images of three-dimensional objects with great depth of focus and can resolve details down to somewhere between 3 nm and 20 nm, depending on the instrument.

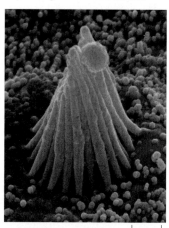

|_____|
5 μm

Scanning electron micrograph of stereocilia projecting from a hair cell in the inner ear (*left*). For comparison, the same structure is shown by light microscopy, at the limit of its resolution (*above*). (Courtesy of Richard Jacobs and James Hudspeth.)

|_____|
1 μm

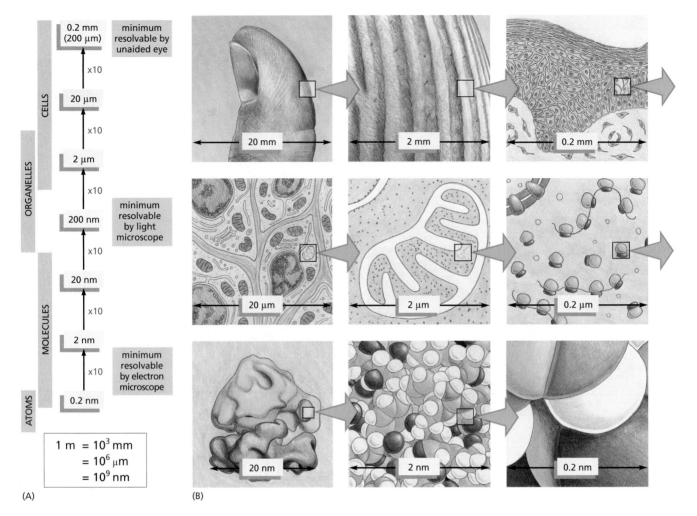

Figure 1–8 How big is a cell and its components? (A) The sizes of cells and of their component parts, plus the units in which they are measured. (B) Drawings to convey a sense of scale between living cells and atoms. Each panel shows an image that is magnified by a factor of 10 compared to its predecessor—producing an imaginary progression from a thumb, to skin, to skin cells, to a mitochondrion, to a ribosome, and ultimately to a cluster of atoms forming part of one of the many protein molecules in our bodies. Note that ribosomes are present inside mitochondria (as shown here), as well as in the cytoplasm. Details of molecular structure, as shown in the last two panels, are beyond the power of the electron microscope.

Even the most powerful electron microscopes, however, cannot visualize the individual atoms that make up biological molecules (**Figure 1–8**). To study the cell's key components in atomic detail, biologists have developed even more sophisticated tools. A technique called X-ray crystallography, for example, is used to determine the precise three-dimensional structure of protein molecules (discussed in Chapter 4).

THE PROKARYOTIC CELL

Of all the types of cells revealed by the microscope, *bacteria* have the simplest structure and come closest to showing us life stripped down to its essentials. Indeed, a bacterium contains essentially no organelles—not even a nucleus to hold its DNA. This property—the presence or absence of a nucleus—is used as the basis for a simple but fundamental classification of all living things. Organisms whose cells have a nucleus are called **eukaryotes** (from the Greek words *eu*, meaning "well" or "truly," and *karyon*, a "kernel" or "nucleus"). Organisms whose cells do not have a nucleus are called **prokaryotes** (from *pro*, meaning "before"). The terms

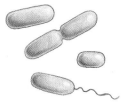
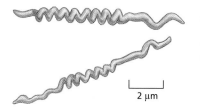

2 μm

spherical cells,
e.g., *Streptococcus*

rod-shaped cells,
e.g., *Escherichia coli*,
Salmonella

spiral cells,
e.g., *Treponema pallidum*

Figure 1–9 Bacteria come in different shapes and sizes. Typical spherical, rodlike, and spiral-shaped bacteria are drawn to scale. The spiral cells shown are the organisms that cause syphilis.

"bacterium" and "prokaryote" are often used interchangeably, although we will see that the category of prokaryotes also includes another class of cells, the *archaea* (singular archaeon), which are so remotely related to bacteria that they are given a separate name.

Prokaryotes are typically spherical, rodlike, or corkscrew-shaped (**Figure 1–9**). They are also small—generally just a few micrometers long, although there are some giant species as much as 100 times longer than this. Prokaryotes often have a tough protective coat, or cell wall, surrounding the plasma membrane, which encloses a single compartment containing the cytoplasm and the DNA. In the electron microscope, the cell interior typically appears as a matrix of varying texture, without any obvious organized internal structure (**Figure 1–10**). The cells reproduce quickly by dividing in two. Under optimum conditions, when food is plentiful, many prokaryotic cells can duplicate themselves in as little as 20 minutes. In 11 hours, by repeated divisions, a single prokaryote can give rise to more than 8 billion progeny (which exceeds the total number of humans presently on Earth). Thanks to their large numbers, rapid growth rates, and ability to exchange bits of genetic material by a process akin to sex, populations of prokaryotic cells can evolve fast, rapidly acquiring the ability to use a new food source or to resist being killed by a new antibiotic.

Prokaryotes Are the Most Diverse and Numerous Cells on Earth

Most prokaryotes live as single-celled organisms, although some join together to form chains, clusters, or other organized multicellular structures. In shape and structure, prokaryotes may seem simple and limited, but in terms of chemistry, they are the most diverse and inventive class of cells. Members of this class exploit an enormous range of habitats, from hot puddles of volcanic mud to the interiors of other living cells, and they vastly outnumber all eukaryotic organisms on Earth. Some are aerobic, using oxygen to oxidize food molecules; some are strictly anaerobic and are killed by the slightest exposure to oxygen. As we discuss later in this

> **QUESTION 1–4**
>
> A bacterium weighs about 10^{-12} g and can divide every 20 minutes. If a single bacterial cell carried on dividing at this rate, how long would it take before the mass of bacteria would equal that of the Earth (6×10^{24} kg)? Contrast your result with the fact that bacteria originated at least 3.5 billion years ago and have been dividing ever since. Explain the apparent paradox. (The number of cells N in a culture at time t is described by the equation $N = N_0 \times 2^{t/G}$, where N_0 is the number of cells at zero time and G is the population doubling time.)

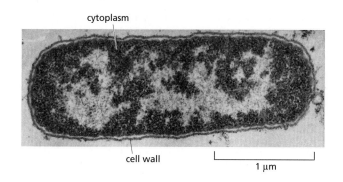

cytoplasm

cell wall

1 μm

Figure 1–10 The bacterium *Escherichia coli (E. coli)* has served as an important model organism. An electron micrograph of a longitudinal section is shown here; the cell's DNA is concentrated in the lightly stained region. (Courtesy of E. Kellenberger.)

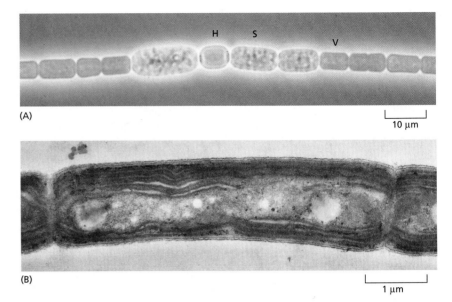

(A)

10 μm

(B)

1 μm

Figure 1–11 Some bacteria are photosynthetic. (A) *Anabaena cylindrica* forms long, multicellular filaments. This light micrograph shows specialized cells that either fix nitrogen (that is, capture N_2 from the atmosphere and incorporate it into organic compounds; labeled *H*), fix CO_2 through photosynthesis (labeled *V*), or become resistant spores (labeled *S*). (B) An electron micrograph of a related species, *Phormidium laminosum*, shows the intracellular membranes where photosynthesis occurs. These micrographs illustrate that even some prokaryotes can form simple multicellular organisms. (A, courtesy of David Adams; B, courtesy of D.P. Hill and C.J. Howe.)

chapter, *mitochondria*—the organelles that generate energy in eukaryotic cells—are thought to have evolved from aerobic bacteria that took to living inside the anaerobic ancestors of today's eukaryotic cells. Thus our own oxygen-based metabolism can be regarded as a product of the activities of bacterial cells.

Virtually any organic, carbon-containing material—from wood to petroleum—can be used as food by one sort of bacterium or another. Even more remarkably, some prokaryotes can live entirely on inorganic substances: they can get their carbon from CO_2 in the atmosphere, their nitrogen from atmospheric N_2, and their oxygen, hydrogen, sulfur, and phosphorus from air, water, and inorganic minerals. Some of these prokaryotic cells, like plant cells, perform *photosynthesis*, using energy from sunlight to produce organic molecules from CO_2 (**Figure 1–11**); others derive energy from the chemical reactivity of inorganic substances in the environment (**Figure 1–12**). In either case, such prokaryotes play a unique and fundamental part in the economy of life on Earth: other living things depend on the organic compounds that these cells generate from inorganic materials.

Plants, too, can capture energy from sunlight and carbon from atmospheric CO_2. But plants unaided by bacteria cannot capture N_2 from the atmosphere, and in a sense even plants depend on bacteria for photosynthesis. It is almost certain that the organelles in the plant cell that

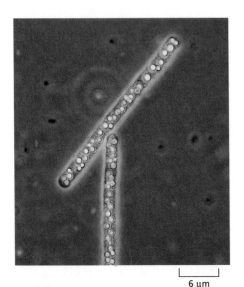

6 μm

Figure 1–12 A sulfur bacterium gets its energy from H_2S. *Beggiatoa*, a prokaryote that lives in sulfurous environments, oxidizes H_2S to produce sulfur and can fix carbon even in the dark. In this light micrograph, yellow deposits of sulfur can be seen inside both of the cells. (Courtesy of Ralph W. Wolfe.)

perform photosynthesis—the *chloroplasts*—have evolved from photosynthetic bacteria that long ago found a home inside the cytoplasm of a plant cell ancestor.

The World of Prokaryotes Is Divided into Two Domains: Bacteria and Archaea

Traditionally, all prokaryotes have been classified together in one large group. But molecular studies reveal that there is a gulf within the class of prokaryotes, dividing it into two distinct *domains* called the **bacteria** and the **archaea**. Remarkably, at a molecular level, the members of these two domains differ as much from one another as either does from the eukaryotes. Most of the prokaryotes familiar from everyday life—the species that live in the soil or make us ill—are bacteria. Archaea are found not only in these habitats, but also in environments that are too hostile for most other cells: concentrated brine, the hot acid of volcanic springs, the airless depths of marine sediments, the sludge of sewage treatment plants, pools beneath the frozen surface of Antarctica, and in the acidic, oxygen-free environment of a cow's stomach where they break down cellulose and generate methane gas. Many of these extreme environments resemble the harsh conditions that must have existed on the primitive Earth, where living things first evolved before the atmosphere became rich in oxygen.

THE EUKARYOTIC CELL

Eukaryotic cells, in general, are bigger and more elaborate than bacteria and archaea. Some live independent lives as single-celled organisms, such as amoebae and yeasts (**Figure 1–13**); others live in multicellular assemblies. All of the more complex multicellular organisms—including plants, animals, and fungi—are formed from eukaryotic cells.

By definition, all eukaryotic cells have a nucleus. But possession of a nucleus goes hand-in-hand with possession of a variety of other organelles, most of which are membrane-enclosed and common to all eukaryotic organisms. In this section, we take a look at the main organelles found in eukaryotic cells from the point of view of their functions, and we consider how they came to serve the roles they have in the life of the eukaryotic cell.

The Nucleus Is the Information Store of the Cell

The **nucleus** is usually the most prominent organelle in a eukaryotic cell (**Figure 1–14**). It is enclosed within two concentric membranes that form the *nuclear envelope*, and it contains molecules of DNA—extremely long polymers that encode the genetic information of the organism. In the light microscope, these giant DNA molecules become visible as individual **chromosomes** when they become more compact before a cell divides into two daughter cells (**Figure 1–15**). DNA also carries the genetic information in prokaryotic cells; these cells lack a distinct nucleus not because they lack DNA, but because they do not keep their DNA inside a nuclear envelope, segregated from the rest of the cell contents.

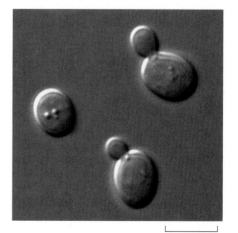

10 μm

Figure 1–13 Yeasts are simple free-living eukaryotes. The cells shown in this micrograph belong to the species of yeast, *Saccharomyces cerevisiae*, used to make dough rise and turn malted barley juice into beer. As can be seen in this image, the cells reproduce by growing a bud and then dividing asymmetrically into a large mother cell and a small daughter cell; for this reason, they are called budding yeast.

Figure 1–14 The nucleus contains most of the DNA in a eukaryotic cell. (A) This drawing of a typical animal cell shows its extensive system of membrane-enclosed organelles. The nucleus is colored *brown*, the nuclear envelope is *green*, and the cytoplasm (the interior of the cell outside the nucleus) is *white*. (B) An electron micrograph of the nucleus in a mammalian cell. Individual chromosomes are not visible because at this stage of the cell's growth its DNA molecules are dispersed as fine threads throughout the nucleus. (B, courtesy of Daniel S. Friend.)

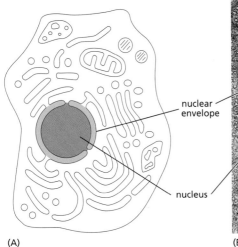

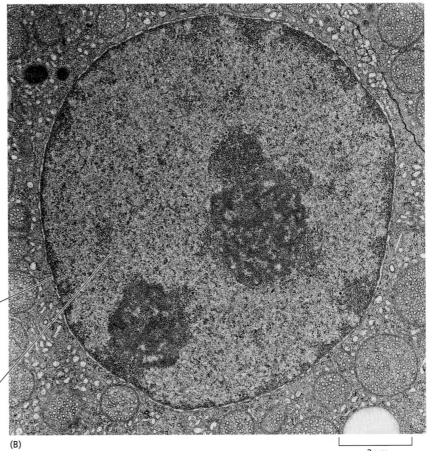

(A)

(B)

2 μm

Mitochondria Generate Usable Energy from Food to Power the Cell

Mitochondria are present in essentially all eukaryotic cells, and they are among the most conspicuous organelles in the cytoplasm (see Figure 1–7B). In a fluorescence microscope, they appear as worm-shaped structures that often form branching networks (**Figure 1–16**). When seen with an electron microscope, individual mitochondria are found to be enclosed in two separate membranes, with the inner membrane formed into folds that project into the interior of the organelle (**Figure 1–17**).

Microscopic examination by itself, however, gives little indication of what mitochondria do. Their function was discovered by breaking open cells and then spinning the soup of cell fragments in a centrifuge; this

Figure 1–15 Chromosomes become visible when a cell is about to divide. As a eukaryotic cell prepares to divide, its DNA molecules become progressively more compacted (condensed), forming wormlike chromosomes that can be distinguished in the light microscope. The photographs show three successive steps in this process in a cultured cell from a newt's lung; note that in the last micrograph on the right, the nuclear envelope has broken down. (Courtesy of Conly L. Rieder.)

nucleus   nuclear envelope   condensed chromosomes

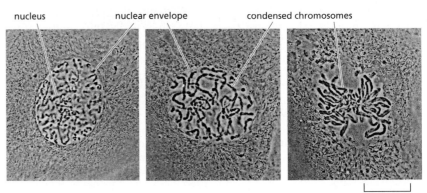

25 μm

Figure 1–16 **Mitochondria can be variable in shape and size.** This budding yeast cell, which contains a green fluorescent protein in its mitochondria, was viewed in a super-resolution confocal fluorescence microscope. In this three-dimensional image, the mitochondria are seen to form complex branched networks. (From A. Egner et al., *Proc. Natl Acad. Sci. USA* 99:3370–3375, 2002. With permission from the National Academy of Sciences.)

10 μm

separates the organelles according to their size and density. Purified mitochondria were then tested to see what chemical processes they could perform. This revealed that mitochondria are generators of chemical energy for the cell. They harness the energy from the oxidation of food molecules, such as sugars, to produce *adenosine triphosphate*, or *ATP*— the basic chemical fuel that powers most of the cell's activities. Because the mitochondrion consumes oxygen and releases carbon dioxide in the course of this activity, the entire process is called *cellular respiration*— essentially, breathing on a cellular level. Without mitochondria, animals, fungi, and plants would be unable to use oxygen to extract the energy they need from the food molecules that nourish them. The process of cellular respiration is considered in detail in Chapter 14.

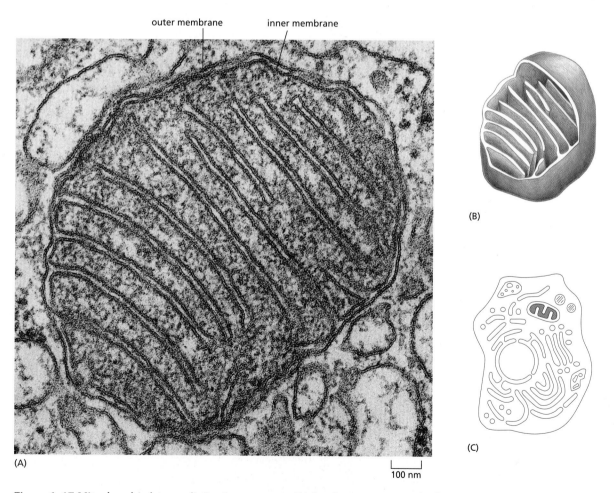

outer membrane   inner membrane

(A)

(B)

(C)

100 nm

Figure 1–17 **Mitochondria have a distinctive structure.** (A) An electron micrograph of a cross section of a mitochondrion reveals the extensive infolding of the inner membrane. (B) This three-dimensional representation of the arrangement of the mitochondrial membranes shows the smooth outer membrane (*gray*) and the highly convoluted inner membrane (*red*). The inner membrane contains most of the proteins responsible for cellular respiration—one of the mitochondrion's main functions—and it is highly folded to provide a large surface area for this activity. (C) In this schematic cell, the interior space of the mitochondrion is colored *orange*. (A, courtesy of Daniel S. Friend.)

Figure 1–18 Mitochondria most likely evolved from engulfed bacteria. It is virtually certain that mitochondria originate from bacteria that were engulfed by an ancestral pre-eukaryotic cell and survived inside it, living in symbiosis with their host. Note that the double membrane of present-day mitochondria is thought to have been derived from the plasma membrane and outer membrane of the engulfed bacterium.

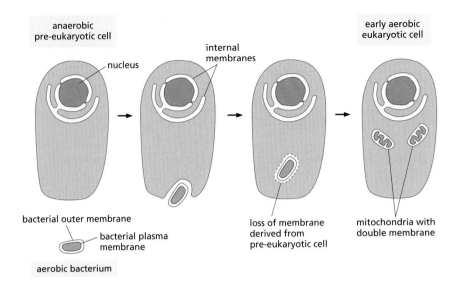

Mitochondria contain their own DNA and reproduce by dividing in two. Because they resemble bacteria in so many ways, they are thought to have been derived from bacteria that were engulfed by some ancestor of present-day eukaryotic cells (**Figure 1–18**). This evidently created a *symbiotic* relationship in which the host eukaryote and the engulfed bacterium helped one another to survive and reproduce.

Chloroplasts Capture Energy from Sunlight

Chloroplasts are large, green organelles that are found only in the cells of plants and algae, not in the cells of animals or fungi. These organelles have an even more complex structure than mitochondria: in addition to their two surrounding membranes, they possess internal stacks of membranes containing the green pigment *chlorophyll* (**Figure 1–19**).

Chloroplasts carry out **photosynthesis**—trapping the energy of sunlight in their chlorophyll molecules and using this energy to drive the manufacture of energy-rich sugar molecules. In the process, they release

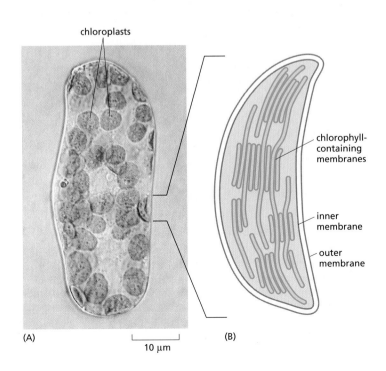

Figure 1–19 Chloroplasts in plant cells capture the energy of sunlight. (A) A single cell isolated from a leaf of a flowering plant, seen in the light microscope, showing many green chloroplasts. (B) A drawing of one of the chloroplasts, showing the inner and outer membranes, as well as the highly folded system of internal membranes containing the green chlorophyll molecules that absorb light energy. (A, courtesy of Preeti Dahiya.)

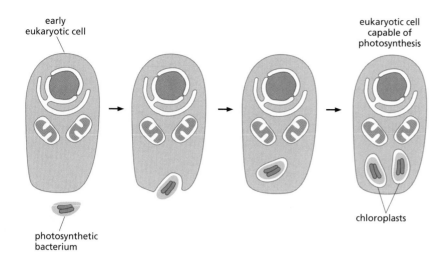

early
eukaryotic cell

eukaryotic cell
capable of
photosynthesis

chloroplasts

photosynthetic
bacterium

Figure 1–20 Chloroplasts almost certainly evolved from engulfed photosynthetic bacteria. The bacteria are thought to have been taken up by early eukaryotic cells that already contained mitochondria.

oxygen as a molecular by-product. Plant cells can then extract this stored chemical energy when they need it, by oxidizing these sugars in their mitochondria, just as animal cells do. Chloroplasts thus enable plants to get their energy directly from sunlight. And they allow plants to produce the food molecules—and the oxygen—that mitochondria use to generate chemical energy in the form of ATP. How these organelles work together is discussed in Chapter 14.

Like mitochondria, chloroplasts contain their own DNA, reproduce by dividing in two, and are thought to have evolved from bacteria—in this case, from photosynthetic bacteria that were engulfed by an early eukaryotic cell (**Figure 1–20**).

Internal Membranes Create Intracellular Compartments with Different Functions

Nuclei, mitochondria, and chloroplasts are not the only membrane-enclosed organelles inside eukaryotic cells. The cytoplasm contains a profusion of other organelles that are surrounded by single membranes (see Figure 1–7A). Most of these structures are involved with the cell's ability to import raw materials and to export both the useful substances and waste products that are produced by the cell.

The *endoplasmic reticulum* (*ER*) is an irregular maze of interconnected spaces enclosed by a membrane (**Figure 1–21**). It is the site where most cell-membrane components, as well as materials destined for export from the cell, are made. This organelle is enormously enlarged in cells that are specialized for the secretion of proteins. Stacks of flattened, membrane-enclosed sacs constitute the *Golgi apparatus* (**Figure 1–22**), which modifies and packages molecules made in the ER that are destined to be either secreted from the cell or transported to another cell compartment. *Lysosomes* are small, irregularly shaped organelles in which intracellular digestion occurs, releasing nutrients from ingested food particles and breaking down unwanted molecules for either recycling within the cell or excretion from the cell. Indeed, many of the large and small molecules within the cell are constantly being broken down and remade. *Peroxisomes* are small, membrane-enclosed vesicles that provide a safe environment for a variety of reactions in which hydrogen peroxide is used to inactivate toxic molecules. Membranes also form many different types of small *transport vesicles* that ferry materials between one membrane-enclosed organelle and another. All of these membrane-enclosed organelles are sketched in **Figure 1–23A**.

Figure 1–21 **The endoplasmic reticulum produces many of the components of a eukaryotic cell.** (A) Schematic diagram of an animal cell shows the endoplasmic reticulum (ER) in *green*. (B) Electron micrograph of a thin section of a mammalian pancreatic cell shows a small part of the ER, of which there are vast amounts in this cell type, which is specialized for protein secretion. Note that the ER is continuous with the membranes of the nuclear envelope. The black particles studding the particular region of the ER shown here are ribosomes, structures that translate RNAs into proteins. Because of its appearance, ribosome-coated ER is often called "rough ER" to distinguish it from the "smooth ER," which does not have ribosomes bound to it. (B, courtesy of Lelio Orci.)

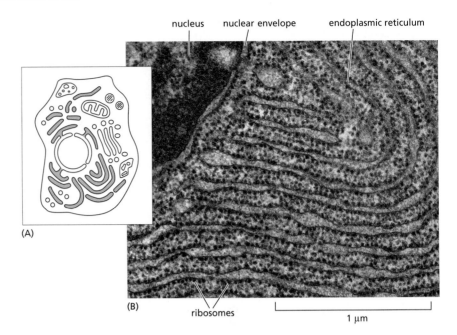

A continual exchange of materials takes place between the endoplasmic reticulum, the Golgi apparatus, the lysosomes, and the outside of the cell. The exchange is mediated by transport vesicles that pinch off from the membrane of one organelle and fuse with another, like tiny soap bubbles budding from and rejoining larger bubbles. At the surface of the cell, for example, portions of the plasma membrane tuck inward and pinch off to form vesicles that carry material captured from the external medium into the cell—a process called *endocytosis* (**Figure 1–24**). Animal cells can

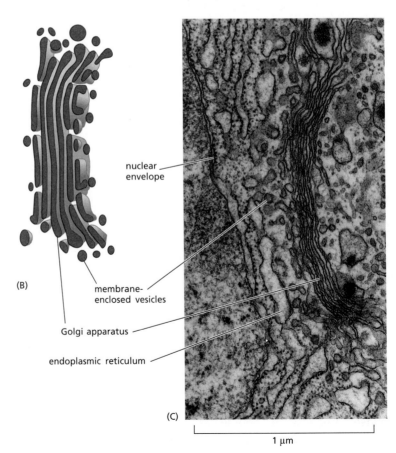

Figure 1–22 **The Golgi apparatus is composed of a stack of flattened discs.** (A) Schematic diagram of an animal cell with the Golgi apparatus colored *red*. (B) More realistic drawing of the Golgi apparatus. Some of the vesicles seen nearby have pinched off from the Golgi stack; others are destined to fuse with it. Only one stack is shown here, but several can be present in a cell. (C) Electron micrograph that shows the Golgi apparatus from a typical animal cell. (C, courtesy of Brij J. Gupta.)

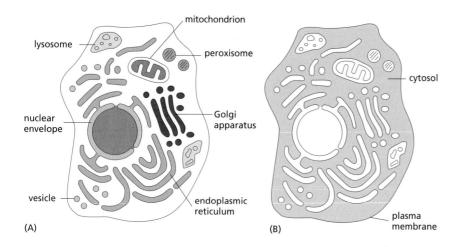

Figure 1–23 **Membrane-enclosed organelles are distributed throughout the eukaryotic cell cytoplasm.** (A) The membrane-enclosed organelles, shown in different colors, are each specialized to perform a different function. (B) The cytoplasm that fills the space outside of these organelles is called the cytosol (colored *blue*).

engulf very large particles, or even entire foreign cells, by endocytosis. In the reverse process, called *exocytosis*, vesicles from inside the cell fuse with the plasma membrane and release their contents into the external medium (see Figure 1–24); most of the hormones and signal molecules that allow cells to communicate with one another are secreted from cells by exocytosis. How membrane-enclosed organelles move proteins and other molecules from place to place inside the cell is discussed in detail in Chapter 15.

The Cytosol Is a Concentrated Aqueous Gel of Large and Small Molecules

If we were to strip the plasma membrane from a eukaryotic cell and then remove all of its membrane-enclosed organelles, including the nucleus, endoplasmic reticulum, Golgi apparatus, mitochondria, chloroplasts, and so on, we would be left with the **cytosol** (see Figure 1–23B). In other words, the cytosol is the part of the cytoplasm that is not contained within intracellular membranes. In most cells, the cytosol is the largest single compartment. It contains a host of large and small molecules, crowded together so closely that it behaves more like a water-based gel than a liquid solution (**Figure 1–25**). The cytosol is the site of many chemical reactions that are fundamental to the cell's existence. The early steps in the breakdown of nutrient molecules take place in the cytosol, for example, and it is here that most proteins are made by ribosomes.

The Cytoskeleton Is Responsible for Directed Cell Movements

The cytoplasm is not just a structureless soup of chemicals and organelles. Using an electron microscope, one can see that in eukaryotic cells the cytosol is criss-crossed by long, fine filaments. Frequently, the filaments are seen to be anchored at one end to the plasma membrane or to radiate out from a central site adjacent to the nucleus. This system of protein filaments, called the **cytoskeleton**, is composed of three major filament types (**Figure 1–26**). The thinnest of these filaments are the *actin filaments*; they are abundant in all eukaryotic cells but occur in especially large numbers inside muscle cells, where they serve as a central part of the machinery responsible for muscle contraction. The thickest filaments in the cytosol are called *microtubules*, because they have the form of minute hollow tubes. In dividing cells, they become reorganized into a spectacular array that helps pull the duplicated chromosomes in opposite

IMPORT BY ENDOCYTOSIS

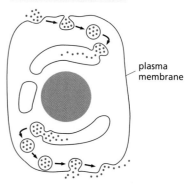

plasma membrane

EXPORT BY EXOCYTOSIS

Figure 1–24 **Eukaryotic cells engage in continual endocytosis and exocytosis.** They import extracellular materials by endocytosis and secrete intracellular materials by exocytosis.

Figure 1–25 The cytoplasm is stuffed with organelles and a host of large and small molecules. This schematic drawing, which extends across two pages and is based on the known sizes and concentrations of molecules in the cytosol, shows how crowded the cytoplasm is. Proteins are *blue*, membrane lipids are *yellow*, and ribosomes and DNA are *pink*. The panorama begins on the far left at the plasma membrane, moves through the endoplasmic reticulum, Golgi apparatus, and a mitochondrion, and ends on the far right in the nucleus. (Courtesy of D. Goodsell.)

directions and distribute them equally to the two daughter cells (**Figure 1–27**). Intermediate in thickness between actin filaments and microtubules are the *intermediate filaments*, which serve to strengthen the cell. These three types of filaments, together with other proteins that attach to them, form a system of girders, ropes, and motors that gives the cell its mechanical strength, controls its shape, and drives and guides its movements (Movie 1.2 and Movie 1.3).

Because the cytoskeleton governs the internal organization of the cell as well as its external features, it is as necessary to a plant cell—boxed in by a tough wall of extracellular matrix—as it is to an animal cell that freely bends, stretches, swims, or crawls. In a plant cell, for example, organelles such as mitochondria are driven in a constant stream around the cell interior along cytoskeletal tracks (Movie 1.4). And animal cells and plant cells alike depend on the cytoskeleton to separate their internal components into two daughter cells during cell division (see Figure 1–27).

The cytoskeleton's role in cell division may be its most ancient function. Even bacteria contain proteins that are distantly related to those of eukaryotic actin filaments and microtubules, forming filaments that play a part in prokaryotic cell division. We examine the cytoskeleton in detail in Chapter 17, discuss its role in cell division in Chapter 18, and review how it responds to signals from outside the cell in Chapter 16.

The Cytoplasm Is Far from Static

The cell interior is in constant motion. The cytoskeleton is a dynamic jungle of protein ropes that are continually being strung together and taken apart; its filaments can assemble and then disappear in a matter of minutes. *Motor proteins* use the energy stored in molecules of ATP to trundle along these tracks and cables, carrying organelles and proteins throughout the cytoplasm, and racing across the width of the cell in seconds. In addition, the large and small molecules that fill every free space in the cell are swept to and fro by random thermal motion, constantly colliding with one another and with other structures in the cell's crowded cytoplasm (Movie 1–5).

QUESTION 1–5

Suggest a reason why it would be advantageous for eukaryotic cells to evolve elaborate internal membrane systems that allow them to import substances from the outside, as shown in Figure 1–24.

Figure 1–26 The cytoskeleton is a network of protein filaments that crisscrosses the cytoplasm of eukaryotic cells. The three major types of filaments can be detected using different fluorescent stains. Shown here are (A) actin filaments, (B) microtubules, and (C) intermediate filaments. (A, courtesy of Simon Barry and Chris D'Lacey; B, courtesy of Nancy Kedersha; C, courtesy of Clive Lloyd.)

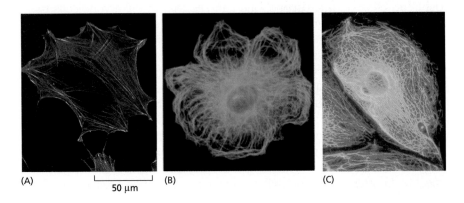

(A)   50 µm   (B)   (C)

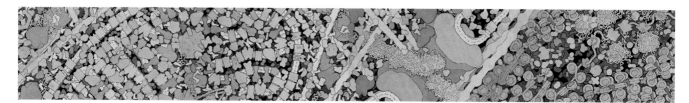

Of course, neither the bustling nature of the cell's interior nor the details of cell structure were appreciated when scientists first peered at cells in a microscope; our knowledge of cell structure accumulated slowly. A few of the key discoveries are listed in **Table 1–1**. In addition, **Panel 1–2** summarizes the differences between animal, plant, and bacterial cells.

Eukaryotic Cells May Have Originated as Predators

Eukaryotic cells are typically 10 times the length and 1000 times the volume of prokaryotic cells, although there is huge size variation within each category. They also possess a whole collection of features—a cytoskeleton, mitochondria, and other organelles—that set them apart from bacteria and archaea.

When and how eukaryotes evolved these systems remains something of a mystery. Although eukaryotes, bacteria, and archaea must have diverged from one another very early in the history of life on Earth (discussed in Chapter 14), the eukaryotes did not acquire all of their distinctive features at the same time (**Figure 1–28**). According to one theory, the ancestral eukaryotic cell was a predator that fed by capturing other cells. Such a way of life requires a large size, a flexible membrane, and a cytoskeleton to help the cell move and eat. The nuclear compartment may have evolved to keep the DNA segregated from this physical and chemical hurly-burly, so as to allow more delicate and complex control of the way the cell reads out its genetic information.

Such a primitive cell, with a nucleus and cytoskeleton, was most likely the sort of cell that engulfed the free-living, oxygen-consuming bacteria that were the likely ancestors of the mitochondria (see Figure 1–18). This partnership is thought to have been established 1.5 billion years ago, when the Earth's atmosphere first became rich in oxygen. A subset of

QUESTION 1–6

Discuss the relative advantages and disadvantages of light and electron microscopy. How could you best visualize (a) a living skin cell, (b) a yeast mitochondrion, (c) a bacterium, and (d) a microtubule?

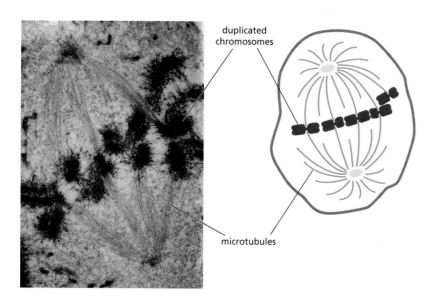

duplicated chromosomes

microtubules

Figure 1–27 Microtubules help distribute the chromosomes in a dividing cell. When a cell divides, its nuclear envelope breaks down and its DNA condenses into visible chromosomes, each of which has duplicated to form a pair of conjoined chromosomes that will ultimately be pulled apart into separate cells by microtubules. In the transmission electron micrograph (*left*), the microtubules are seen to radiate from foci at opposite ends of the dividing cell. (Photomicrograph courtesy of Conly L. Rieder.)

TABLE 1–1 HISTORICAL LANDMARKS IN DETERMINING CELL STRUCTURE

1665	Hooke uses a primitive microscope to describe small chambers in sections of cork that he calls "cells."
1674	Leeuwenhoek reports his discovery of protozoa. Nine years later, he sees bacteria for the first time.
1833	Brown publishes his microscopic observations of orchids, clearly describing the cell nucleus.
1839	Schleiden and Schwann propose the cell theory, stating that the nucleated cell is the universal building block of plant and animal tissues.
1857	Kölliker describes mitochondria in muscle cells.
1879	Flemming describes with great clarity chromosome behavior during mitosis in animal cells.
1881	Cajal and other histologists develop staining methods that reveal the structure of nerve cells and the organization of neural tissue.
1898	Golgi first sees and describes the Golgi apparatus by staining cells with silver nitrate.
1902	Boveri links chromosomes and heredity by observing chromosome behavior during sexual reproduction.
1952	Palade, Porter, and Sjöstrand develop methods of electron microscopy that enable many intracellular structures to be seen for the first time. In one of the first applications of these techniques, Huxley shows that muscle contains arrays of protein filaments—the first evidence of a cytoskeleton.
1957	Robertson describes the bilayer structure of the cell membrane, seen for the first time in the electron microscope.
1960	Kendrew describes the first detailed protein structure (sperm whale myoglobin) to a resolution of 0.2 nm using X-ray crystallography. Perutz proposes a lower-resolution structure for hemoglobin.
1965	Christian de Duve and his colleagues use a cell-fractionation technique to separate peroxisomes, mitochondria, and lysosomes from a preparation of rat liver.
1968	Petran and collaborators make the first confocal microscope.
1970	Frye and Edidin use fluorescent antibodies to show that plasma membrane molecules can diffuse in the plane of the membrane, indicating that cell membranes are fluid.
1974	Lazarides and Weber use fluorescent antibodies to stain the cytoskeleton.
1994	Chalfie and collaborators introduce green fluorescent protein (GFP) as a marker to follow the behavior of proteins in living cells.

these cells later acquired chloroplasts by engulfing photosynthetic bacteria (see Figure 1–20). The likely history of these endosymbiotic events is illustrated in Figure 1–28.

That single-celled eukaryotes can prey upon and swallow other cells is borne out by the behavior of many of the free-living, actively motile

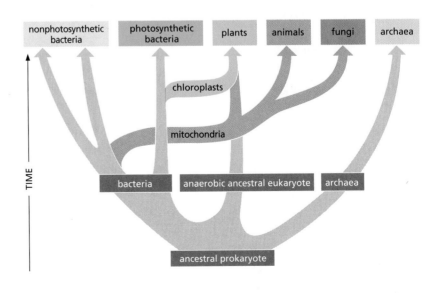

Figure 1–28 Where did eukaryotes come from? The eukaryotic, bacterial, and archaean lineages diverged from one another very early in the evolution of life on Earth. Some time later, eukaryotes are thought to have acquired mitochondria; later still, a subset of eukaryotes acquired chloroplasts. Mitochondria are essentially the same in plants, animals, and fungi, and therefore were presumably acquired before these lines diverged.

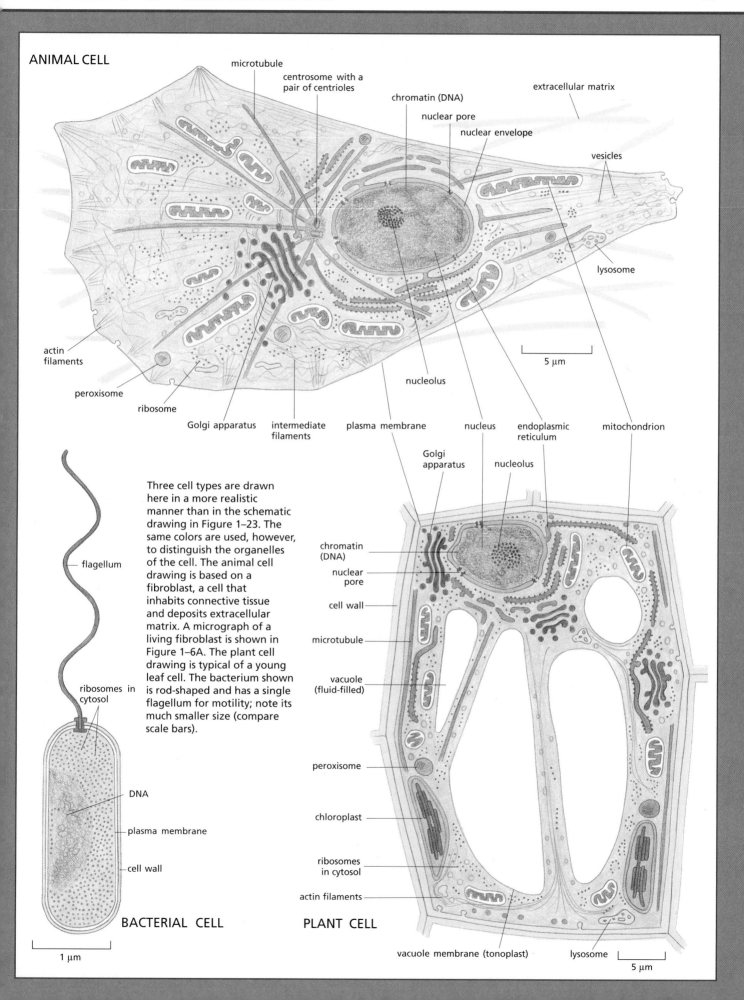

ANIMAL CELL

microtubule

centrosome with a pair of centrioles

chromatin (DNA)

nuclear pore

nuclear envelope

extracellular matrix

vesicles

lysosome

actin filaments

peroxisome

ribosome

Golgi apparatus

intermediate filaments

plasma membrane

nucleolus

nucleus

endoplasmic reticulum

mitochondrion

5 µm

flagellum

ribosomes in cytosol

DNA

plasma membrane

cell wall

BACTERIAL CELL

1 µm

Three cell types are drawn here in a more realistic manner than in the schematic drawing in Figure 1–23. The same colors are used, however, to distinguish the organelles of the cell. The animal cell drawing is based on a fibroblast, a cell that inhabits connective tissue and deposits extracellular matrix. A micrograph of a living fibroblast is shown in Figure 1–6A. The plant cell drawing is typical of a young leaf cell. The bacterium shown is rod-shaped and has a single flagellum for motility; note its much smaller size (compare scale bars).

Golgi apparatus

nucleolus

chromatin (DNA)

nuclear pore

cell wall

microtubule

vacuole (fluid-filled)

peroxisome

chloroplast

ribosomes in cytosol

actin filaments

vacuole membrane (tonoplast)

lysosome

PLANT CELL

5 µm

Figure 1–29 One protozoan eats another.
(A) The scanning electron micrograph shows *Didinium* on its own, with its circumferential rings of beating cilia and its "snout" at the top. (B) *Didinium* is seen ingesting another ciliated protozoan, a *Paramecium*. (Courtesy of D. Barlow.)

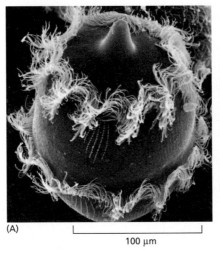

(A)

100 μm

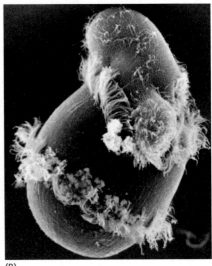

(B)

microorganisms called **protozoans**. *Didinium*, for example, is a large, carnivorous protozoan with a diameter of about 150 μm—roughly 10 times that of the average human cell. It has a globular body encircled by two fringes of cilia, and its front end is flattened except for a single protrusion rather like a snout (**Figure 1–29A**). *Didinium* swims at high speed by means of its beating cilia. When it encounters a suitable prey, usually another type of protozoan, it releases numerous small, paralyzing darts from its snout region. *Didinium* then attaches to and devours the other cell, inverting like a hollow ball to engulf its victim, which can be almost as large as itself (**Figure 1–29B**).

Not all protozoans are predators. They can be photosynthetic or carnivorous, motile or sedentary. Their anatomy is often elaborate and includes such structures as sensory bristles, photoreceptors, beating cilia, stalk-like appendages, mouthparts, stinging darts, and musclelike contractile bundles (**Figure 1–30**). Although they are single cells, protozoans can be as intricate and versatile as many multicellular organisms. Much remains to be learned about fundamental cell biology from studies of these fascinating life-forms.

MODEL ORGANISMS

All cells are thought to be descended from a common ancestor, whose fundamental properties have been conserved through evolution. Thus knowledge gained from the study of one organism contributes to our understanding of others, including ourselves. But certain organisms are easier than others to study in the laboratory. Some reproduce rapidly and are convenient for genetic manipulations; others are multicellular but transparent, so that one can directly watch the development of all their internal tissues and organs. For reasons such as these, large communities of biologists have become dedicated to studying different aspects of the biology of a few chosen species, pooling their knowledge to gain a deeper understanding than could be achieved if their efforts were spread over many different species. Although the roster of these representative organisms is continually expanding, a few stand out in terms of the breadth and depth of information that has been accumulated about them over the years—knowledge that contributes to our understanding of how all cells work. In this section, we examine some of these **model organisms** and review the benefits that each offers to the study of cell biology and, in many cases, to the promotion of human health.

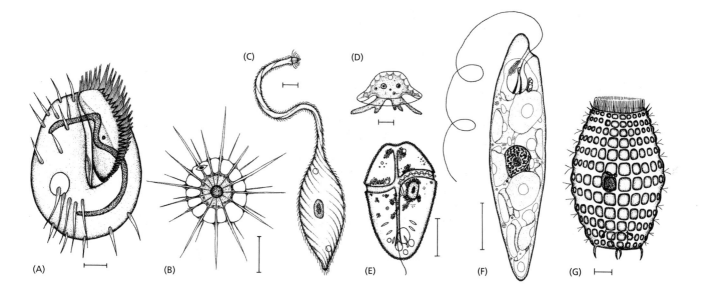

Figure 1–30 An assortment of protozoans illustrates the enormous variety within this class of single-celled microorganisms. These drawings are done to different scales, but in each case the scale bar represents 10 μm. The organisms in (A), (C), and (G) are ciliates; (B) is a heliozoan; (D) is an amoeba; (E) is a dinoflagellate; and (F) is a euglenoid. To see the latter in action, watch Movie 1.6. (From M.A. Sleigh, The Biology of Protozoa. London: Edward Arnold, 1973. With permission from Edward Arnold.)

Molecular Biologists Have Focused on *E. coli*

In molecular terms, we understand the workings of the bacterium *Escherichia coli—E. coli* for short—more thoroughly than those of any other living organism (see Figure 1–10). This small, rod-shaped cell normally lives in the gut of humans and other vertebrates, but it also grows happily and reproduces rapidly in a simple nutrient broth in a culture bottle.

Most of our knowledge of the fundamental mechanisms of life—including how cells replicate their DNA and how they decode these genetic instructions to make proteins—has come from studies of *E. coli*. Subsequent research has confirmed that these basic processes occur in essentially the same way in our own cells as they do in *E. coli*.

Brewer's Yeast Is a Simple Eukaryotic Cell

We tend to be preoccupied with eukaryotes because we are eukaryotes ourselves. But human cells are complicated and reproduce relatively slowly. To get a handle on the fundamental biology of eukaryotic cells, it is often advantageous to study a simpler cell that reproduces more rapidly. A popular choice has been the budding yeast *Saccharomyces cerevisiae* (**Figure 1–31**)—the same microorganism that is used for brewing beer and baking bread.

S. cerevisiae is a small, single-celled fungus that is at least as closely related to animals as it is to plants. Like other fungi, it has a rigid cell wall, is relatively immobile, and possesses mitochondria but not chloroplasts. When nutrients are plentiful, *S. cerevisiae* reproduces almost as rapidly as a bacterium. Yet it carries out all the basic tasks that every eukaryotic cell must perform. Genetic and biochemical studies in yeast have been crucial to understanding many basic mechanisms in eukaryotic cells, including the cell-division cycle—the chain of events by which the nucleus and all the other components of a cell are duplicated and parceled out to create two daughter cells. The machinery that governs cell division has been

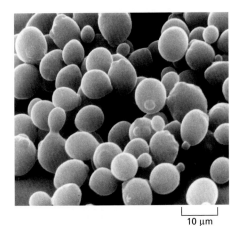

Figure 1–31 The yeast *Saccharomyces cerevisiae* is a model eukaryote. In this scanning electron micrograph, a few yeast cells are seen in the process of dividing, which they do by budding. Another micrograph of the same species is shown in Figure 1–13. (Courtesy of Ira Herskowitz and Eric Schabatach.)

10 μm

so well conserved over the course of evolution that many of its components can function interchangeably in yeast and human cells (see **How We Know**, pp. 30–31). Darwin himself would no doubt have been stunned by this dramatic example of evolutionary conservation.

Arabidopsis Has Been Chosen as a Model Plant

The large multicellular organisms that we see around us—both plants and animals—seem fantastically varied, but they are much closer to one another in their evolutionary origins, and more similar in their basic cell biology, than the great host of microscopic single-celled organisms. Whereas bacteria, archaea, and eukaryotes separated from each other more than 3 billion years ago, plants, animals, and fungi diverged only about 1.5 billion years ago, and the different species of flowering plants less than 200 million years ago.

The close evolutionary relationship among all flowering plants means that we can gain insight into their cell and molecular biology by focusing on just a few convenient species for detailed analysis. Out of the several hundred thousand species of flowering plants on Earth today, molecular biologists have focused their efforts on a small weed, the common wall cress *Arabidopsis thaliana* (**Figure 1–32**), which can be grown indoors in large numbers: one plant can produce thousands of offspring within 8–10 weeks. Because genes found in *Arabidopsis* have counterparts in agricultural species, studying this simple weed provides insights into the development and physiology of the crop plants upon which our lives depend, as well as into the evolution of all the other plant species that dominate nearly every ecosystem on Earth.

Model Animals Include Flies, Fish, Worms, and Mice

Multicellular animals account for the majority of all named species of living organisms, and the majority of animal species are insects. It is fitting, therefore, that an insect, the small fruit fly *Drosophila melanogaster* (**Figure 1–33**), should occupy a central place in biological research. In fact, the foundations of classical genetics were built to a large extent on studies of this insect. More than 80 years ago, genetic analysis of the fruit fly provided definitive proof that genes—the units of heredity—are carried on chromosomes. In more recent times, *Drosophila*, more than any other organism, has shown us how the genetic instructions encoded in DNA molecules direct the development of a fertilized egg cell (or *zygote*) into an adult multicellular organism containing vast numbers of different cell types organized in a precise and predictable way. *Drosophila* mutants with body parts strangely misplaced or oddly patterned have provided the key to identifying and characterizing the genes that are needed to make a properly structured adult body, with gut, wings, legs, eyes, and all the other bits and pieces in their correct places. These genes—which are copied and passed on to every cell in the body—define how each cell will behave in its social interactions with its sisters and cousins, thus controlling the structures that the cells can create. Moreover, the genes

Figure 1–32 *Arabidopsis thaliana*, the common wall cress, is a model plant. This small weed has become the favorite organism of plant molecular and developmental biologists. (Courtesy of Toni Hayden and the John Innes Centre.)

1 cm

Figure 1–33 ***Drosophila melanogaster* is a favorite among developmental biologists and geneticists.** Molecular genetic studies on this small fly have provided a key to the understanding of how all animals develop. (Courtesy of E.B. Lewis.)

1 mm

responsible for the development of *Drosophila* have turned out to be amazingly similar to those of humans—far more similar than one would suspect from outward appearances. Thus the fly serves as a valuable model for studying human development and disease.

Another widely studied organism is the nematode worm *Caenorhabditis elegans* (**Figure 1–34**), a harmless relative of the eelworms that attack the roots of crops. Smaller and simpler than *Drosophila*, this creature develops with clockwork precision from a fertilized egg cell into an adult that has exactly 959 body cells (plus a variable number of egg and sperm cells)—an unusual degree of regularity for an animal. We now have a minutely detailed description of the sequence of events by which this occurs—as the cells divide, move, and become specialized according to strict and predictable rules. And a wealth of mutants are available for testing how the worm's genes direct this developmental ballet. Some 70% of human genes have some counterpart in the worm, and *C. elegans*, like *Drosophila*, has proved to be a valuable model for many of the developmental processes that occur in our own bodies. Studies of nematode development, for example, have led to a detailed molecular understanding of *apoptosis*, a form of programmed cell death by which surplus cells are disposed of in all animals—a topic of great importance for cancer research (discussed in Chapters 18 and 20).

Another organism that is providing molecular insights into developmental processes, particularly in vertebrates, is the *zebrafish*. Because this

0.2 mm

Figure 1–34 ***Caenorhabditis elegans* is a small nematode worm that normally lives in the soil.** Most individuals are hermaphrodites, producing both sperm and eggs (the latter of which can be seen along the underside of the animal). *C. elegans* was the first multicellular organism to have its complete genome sequenced. (Courtesy of Maria Gallegos.)

LIFE'S COMMON MECHANISMS

All living things are made of cells, and all cells—as we have discussed in this chapter—are fundamentally similar inside: they store their genetic instructions in DNA molecules, which direct the production of RNA molecules, which in turn direct the production of proteins. It is largely the proteins that carry out the cell's chemical reactions, give the cell its shape, and control its behavior. But how deep do these similarities between cells—and the organisms they comprise—really run? Are parts from one organism interchangeable with parts from another? Would an enzyme that breaks down glucose in a bacterium be able to digest the same sugar if it were placed inside a yeast cell or a cell from a lobster or a human? What about the molecular machines that copy and interpret genetic information? Are they functionally equivalent from one organism to another? Insights have come from many sources, but the most stunning and dramatic answer came from experiments performed on humble yeast cells. These studies, which shocked the biological community, focused on one of the most fundamental processes of life—cell division.

Division and discovery

All cells come from other cells, and the only way to make a new cell is through division of a preexisting one. To reproduce, a parent cell must execute an orderly sequence of reactions, through which it duplicates its contents and divides in two. This critical process of duplication and division—known as the *cell-division cycle*, or *cell cycle* for short—is complex and carefully controlled. Defects in any of the proteins involved can be devastating to the cell.

Fortunately for biologists, this acute reliance on crucial proteins makes them easy to identify and study. If a protein is essential for a given process, a mutation that results in an abnormal protein—or in no protein at all—can prevent the cell from carrying out the process. By isolating organisms that are defective in their cell-division cycle, scientists have worked backward to discover the proteins that control progress through the cycle.

The study of cell-cycle mutants has been particularly successful in yeasts. Yeasts are unicellular fungi and are popular organisms for such genetic studies. They are eukaryotes, like us, but they are small, simple, rapidly reproducing, and easy to manipulate genetically. Yeast mutants that are defective in their ability to complete cell division have led to the discovery of many genes that control the cell-division cycle—the so-called *Cdc* genes—and have provided a detailed understanding of how these genes, and the proteins they encode, actually work.

Paul Nurse and his colleagues used this approach to identify *Cdc* genes in the yeast *Schizosaccharomyces pombe*, which is named after the African beer from which it was first isolated. *S. pombe* is a rod-shaped cell, which grows by elongation at its ends and divides by fission into two, through the formation of a partition in the center of the rod. The researchers found that one of the *Cdc* genes they had identified, called *Cdc2*, was required to trigger several key events in the cell-division cycle. When that gene was inactivated by a mutation, the yeast cells would not divide. And when the cells were provided with a normal copy of the gene, their ability to reproduce was restored.

It's obvious that replacing a faulty *Cdc2* gene in *S. pombe* with a functioning *Cdc2* gene from the same yeast should repair the damage and enable the cell to divide normally. But what about using a similar cell-division gene from a different organism? That's the question the Nurse team tackled next.

Next of kin

Saccharomyces cerevisiae is another kind of yeast and is one of a handful of model organisms biologists have chosen to study to expand their understanding of how cells work. Also used to brew beer, *S. cerevisiae* divides by forming a small bud that grows steadily until it separates from the mother cell (see Figures 1–13 and 1–31). Although *S. cerevisiae* and *S. pombe* differ in their style of division, both rely on a complex network of interacting proteins to get the job done. But could the proteins from one type of yeast substitute for those of the other?

To find out, Nurse and his colleagues prepared DNA from healthy *S. cerevisiae*, and they introduced this DNA into *S. pombe* cells that contained a mutation in the *Cdc2* gene that kept the cells from dividing when the temperature was elevated. And they found that some of the mutant *S. pombe* cells regained the ability to proliferate when warm. If spread onto a culture plate containing a growth medium, the rescued cells could divide again and again to form visible colonies, each containing millions of individual yeast cells (**Figure 1–35**). Upon closer examination, the researchers discovered that these "rescued" yeast cells had received a fragment of DNA that contained the *S. cerevisiae* version of *Cdc2*—a gene that had been discovered in pioneering studies of the cell cycle by Lee Hartwell and colleagues.

The result was exciting, but perhaps not all that surprising. After all, how different can one yeast be from another? A more demanding test would be to use DNA from a more distant relative. So Nurse's team repeated the experiment, this time using human DNA. And the results were the same. The human equivalent of the

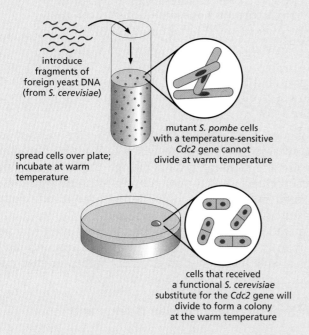

introduce
fragments of
foreign yeast DNA
(from *S. cerevisiae*)

mutant *S. pombe* cells
with a temperature-sensitive
Cdc2 gene cannot
divide at warm temperature

spread cells over plate;
incubate at warm
temperature

cells that received
a functional *S. cerevisiae*
substitute for the *Cdc2* gene will
divide to form a colony
at the warm temperature

Figure 1–35 *S. pombe* mutants defective in a cell-cycle gene can be rescued by the equivalent gene from *S. cerevisiae*. DNA is collected from *S. cerevisiae* and broken into large fragments, which are introduced into a culture of mutant *S. pombe* cells dividing at room temperature. We discuss how DNA can be manipulated and transferred into different cell types in Chapter 10. These yeast cells are then spread onto a plate containing a suitable growth medium and are incubated at a warm temperature, at which the mutant Cdc2 protein is inactive. The rare cells that survive and proliferate on these plates have been rescued by incorporation of a foreign gene that allows them to divide normally at the higher temperature.

S. pombe Cdc2 gene could rescue the mutant yeast cells, allowing them to divide normally.

Gene reading

This result was much more surprising—even to Nurse. The ancestors of yeast and humans diverged some 1.5 billion years ago. So it was hard to believe that these

two organisms would orchestrate cell division in such a similar way. But the results clearly showed that the human and yeast proteins are functionally equivalent. Indeed, Nurse and colleagues demonstrated that the proteins are almost exactly the same size and consist of amino acids strung together in a very similar order; the human Cdc2 protein is identical to the *S. pombe* Cdc2 protein in 63% of its amino acids and is identical to the equivalent protein from *S. cerevisiae* in 58% of its amino acids (**Figure 1–36**). Together with Tim Hunt, who discovered a different cell-cycle protein called cyclin, Nurse and Hartwell shared a 2001 Nobel Prize for their studies of key regulators of the cell cycle.

The Nurse experiments showed that proteins from very different eukaryotes can be functionally interchangeable and suggested that the cell cycle is controlled in a similar fashion in every eukaryotic organism alive today. Apparently, the proteins that orchestrate the cycle in eukaryotes are so fundamentally important that they have been conserved almost unchanged over more than a billion years of eukaryotic evolution.

The same experiment also highlights another, even more basic, point. The mutant yeast cells were rescued, not by direct injection of the human protein, but by introduction of a piece of human DNA. Thus the yeast cells could read and use this information correctly, indicating that, in eukaryotes, the molecular machinery for reading the information encoded in DNA is also similar from cell to cell and from organism to organism. A yeast cell has all the equipment it needs to interpret the instructions encoded in a human gene and to use that information to direct the production of a fully functional human protein.

The story of Cdc2 is just one of thousands of examples of how research in yeast cells has provided critical insights into human biology. Although it may sound paradoxical, the shortest, most efficient path to improving human health will often begin with detailed studies of the biology of simple organisms such as brewer's or baker's yeast.

human	···FGLARAFGIPIRVYTHEVVTLWYRSPEVLLGSARYSTPVDIWSIGTIFAELATKLPLFHGDSEIDQLFRIPRALGTPNNEVWPEVESLQDYKNTFP···
S. pombe	···FGLARSFGVPLRNYTHEIVTLWYRAPEVLLGSRHYSTGVDIWSVGCIFAENIRRSPLFPGDSEIDEIFKIPQVLGTPNEEVWPGVTLLQDYKSTFP···
S. cerevisiae	···FGLARAFGVPLRAYTHEIVTLWYRAPEVLLGGKQYSTGVDTWSIGCIFAEHCNRLPIFSGDSEIDQIFKIPRVLGTPNEAIWPDIVYLPDFKPSFP···

Figure 1–36 The cell-division-cycle proteins from yeasts and human are very similar in their amino acid sequences. Identities between the amino acid sequences of a region of the human Cdc2 protein and a similar region of the equivalent proteins in *S. pombe* and *S. cerevisiae* are indicated by *green* shading. Each amino acid is represented by a single letter.

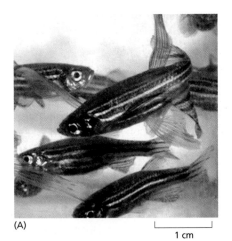

(A)

|_____|_____|
1 cm

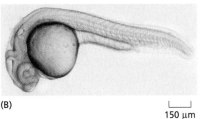

(B)

|___|
150 μm

Figure 1–37 Zebrafish are popular models for studies of vertebrate development. (A) These small, hardy, tropical fish are a staple in many home aquaria. But they are also ideal for developmental studies, as their transparent embryos (B) make it easy to observe cells moving and changing their characters in the living organism as it develops. (A, courtesy of Steve Baskauf; B, from M. Rhinn et al., *Neural Dev.* 4:12, 2009. With permission from BioMed Central Ltd.)

creature is transparent for the first 2 weeks of its life, it provides an ideal system in which to observe how cells behave during development in a living animal (**Figure 1–37**).

Mammals are among the most complex of animals, and the mouse has long been used as the model organism in which to study mammalian genetics, development, immunology, and cell biology. Thanks to modern molecular biological techniques, it is now possible to breed mice with deliberately engineered mutations in any specific gene, or with artificially constructed genes introduced into them. In this way, one can test what a given gene is required for and how it functions. Almost every human gene has a counterpart in the mouse, with a similar DNA sequence and function. Thus, this animal has proven an excellent model for studying genes that are important in both human health and disease.

Biologists Also Directly Study Human Beings and Their Cells

Humans are not mice—or fish or flies or worms or yeast—and so we also study human beings themselves. Like bacteria or yeast, our individual cells can be harvested and grown in culture, where we can study their biology and more closely examine the genes that govern their functions. Given the appropriate surroundings, most human cells—indeed, most cells from animals or plants—will survive, proliferate, and even express specialized properties in a culture dish. Experiments using such cultured cells are sometimes said to be carried out *in vitro* (literally, "in glass") to contrast them with experiments on intact organisms, which are said to be carried out *in vivo* (literally, "in the living").

Although not true for all types of cells, many types of cells grown in culture display the differentiated properties appropriate to their origin: fibroblasts, a major cell type in connective tissue, continue to secrete collagen; cells derived from embryonic skeletal muscle fuse to form muscle fibers, which contract spontaneously in the culture dish; nerve cells extend axons that are electrically excitable and make synapses with other nerve cells; and epithelial cells form extensive sheets, with many of the properties of an intact epithelium (**Figure 1–38**). Because cultured cells are maintained in a controlled environment, they are accessible to study in ways that are often not possible *in vivo*. For example, cultured cells can be exposed to hormones or growth factors, and the effects that these signal molecules have on the shape or behavior of the cells can be easily explored.

In addition to studying human cells in culture, humans are also examined directly in clinics. Much of the research on human biology has been driven by medical interests, and the medical database on the human species is enormous. Although naturally occurring mutations in any given human gene are rare, the consequences of many mutations are well documented. This is because humans are unique among animals in that they report and record their own genetic defects: in no other species are billions of individuals so intensively examined, described, and investigated.

Nevertheless, the extent of our ignorance is still daunting. The mammalian body is enormously complex, being formed from thousands of

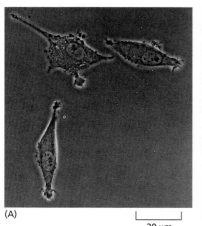 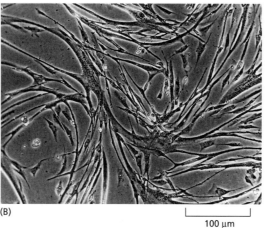 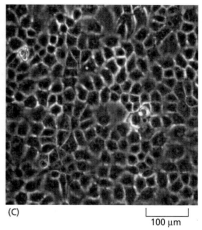

(A)

20 μm

(B)

100 μm

(C)

100 μm

billions of cells, and one might despair of ever understanding how the DNA in a fertilized mouse egg cell makes it generate a mouse rather than a fish, or how the DNA in a human egg cell directs the development of a human rather than a mouse. Yet the revelations of molecular biology have made the task seem eminently approachable. As much as anything, this new optimism has come from the realization that the genes of one type of animal have close counterparts in most other types of animals, apparently serving similar functions (**Figure 1–39**). We all have a common evolutionary origin, and under the surface it seems that we share the same molecular mechanisms. Flies, worms, fish, mice, and humans thus provide a key to understanding how animals in general are made and how their cells work.

Comparing Genome Sequences Reveals Life's Common Heritage

At a molecular level, evolutionary change has been remarkably slow. We can see in present-day organisms many features that have been preserved through more than 3 billion years of life on Earth—about one-fifth of the age of the universe. This evolutionary conservatism provides the foundation on which the study of molecular biology is built. To set the scene for the chapters that follow, therefore, we end this chapter by considering a little more closely the family relationships and basic similarities among all living things. This topic has been dramatically clarified in the past few years by technological advances that have allowed us to determine the complete genome sequences of thousands of organisms, including our own species (as discussed in more detail in Chapter 9).

The first thing we note when we look at an organism's genome is its overall size and how many genes it packs into that length of DNA. Prokaryotes carry very little superfluous genetic baggage and, nucleotide-

Figure 1–38 Cells in culture often display properties that reflect their origin. (A) Phase-contrast micrograph of fibroblasts in culture. (B) Micrograph of cultured myoblasts, some of which have fused to form multinucleate muscle cells that spontaneously contract in culture. (C) Cultured epithelial cells forming a cell sheet. Movie 1.7 shows a single heart muscle cell beating in culture. (A, courtesy of Daniel Zicha; B, courtesy of Rosalind Zalin; C, from K.B. Chua et al., *Proc. Natl Acad. Sci. USA* 104:11424–11429, 2007, with permission from the National Academy of Sciences.)

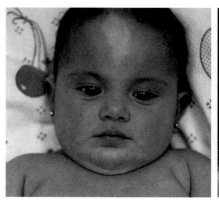 

Figure 1–39 Different species share similar genes. The human baby and the mouse shown here have similar white patches on their foreheads because they both have defects in the same gene (called *Kit*), which is required for the development and maintenance of some pigment cells. (Courtesy of R.A. Fleischman, from *Proc. Natl Acad. Sci. USA* 88:10885–10889, 1991. With permission from the National Academy of Sciences.)

Figure 1–40 Organisms vary enormously in the size of their genomes. Genome size is measured in nucleotide pairs of DNA per haploid genome, that is, per single copy of the genome. (The body cells of sexually reproducing organisms such as ourselves are generally diploid: they contain two copies of the genome, one inherited from the mother, the other from the father.) Closely related organisms can vary widely in the quantity of DNA in their genomes (as indicated by the length of the *green* bars), even though they contain similar numbers of functionally distinct genes. (Adapted from T.R. Gregory, 2008, Animal Genome Size Database: www.genomesize.com)

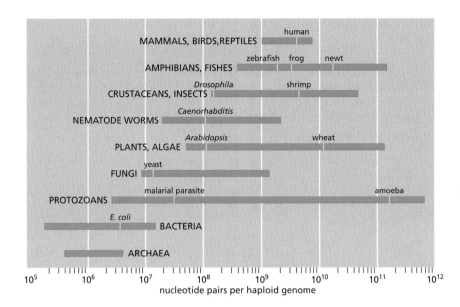

for-nucleotide, they squeeze a lot of information into their relatively small genomes. *E. coli*, for example, carries its genetic instructions in a single, circular, double-stranded molecule of DNA that contains 4.6 million nucleotide pairs and 4300 genes. The simplest known bacterium contains only about 500 genes, but most prokaryotes have genomes that contain at least 1 million nucleotide pairs and 1000–8000 genes. With these few thousand genes, prokaryotes are able to thrive in even the most hostile environments on Earth.

The compact genomes of typical bacteria are dwarfed by the genomes of typical eukaryotes. The human genome, for example, contains about 700 times more DNA than the *E. coli* genome, and the genome of an amoeba contains about 100 times more than ours (**Figure 1–40**). The rest of the model organisms we have described have genomes that fall somewhere in between *E. coli* and human in terms of size. *S. cerevisiae* contains about 2.5 times as much DNA as *E. coli*; *Drosophila* has about 10 times more DNA per cell than yeast; and mice have about 20 times more DNA per cell than the fruit fly (**Table 1–2**).

TABLE 1–2 SOME MODEL ORGANISMS AND THEIR GENOMES		
Organism	Genome size* (nucleotide pairs)	Approximate number of genes
Homo sapiens (human)	3200×10^6	30,000
Mus musculus (mouse)	2800×10^6	30,000
Drosophila melanogaster (fruit fly)	200×10^6	15,000
Arabidopsis thaliana (plant)	220×10^6	29,000
Caenorhabditis elegans (roundworm)	130×10^6	21,000
Saccharomyces cerevisiae (yeast)	13×10^6	6600
Escherichia coli (bacteria)	4.6×10^6	4300

*Genome size includes an estimate for the amount of highly repeated DNA sequence not in genome databases.

In terms of gene numbers, however, the differences are not so great. We have only about six times as many genes as *E. coli*. Moreover, many of our genes—and the proteins they encode—fall into closely related family groups, such as the family of hemoglobins, which has nine closely related members in humans. Thus the number of fundamentally different proteins in a human is not very many times more than in a bacterium, and the number of human genes that have identifiable counterparts in the bacterium is a significant fraction of the total.

This high degree of "family resemblance" is striking when we compare the genome sequences of different organisms. When genes from different organisms have very similar nucleotide sequences, it is highly probable that both descended from a common ancestral gene. Such genes (and their protein products) are said to be **homologous**. Now that we have the complete genome sequences of many different organisms from all three domains of life—archaea, bacteria, and eukaryotes—we can search systematically for homologies that span this enormous evolutionary divide. By taking stock of the common inheritance of all living things, scientists are attempting to trace life's origins back to the earliest ancestral cells.

Genomes Contain More Than Just Genes

Although our view of genome sequences tends to be "gene-centric," our genomes contain much more than just genes. The vast bulk of our DNA does not code for proteins or for functional RNA molecules. Instead, it includes a mixture of sequences that help regulate gene activity, plus sequences that seem to be dispensable. The large quantity of regulatory DNA contained in the genomes of eukaryotic multicellular organisms allows for enormous complexity and sophistication in the way different genes are brought into action at different times and places. Yet, in the end, the basic list of parts—the set of proteins that the cells can make, as specified by the DNA—is not much longer than the parts list of an automobile, and many of those parts are common not only to all animals, but also to the entire living world.

That DNA can program the growth, development, and reproduction of living cells and complex organisms is truly amazing. In the rest of this book, we will try to explain what is known about how cells work—by examining their component parts, how these parts work together, and how the genome of each cell directs the manufacture of the parts the cell needs to function and to reproduce.

ESSENTIAL CONCEPTS

- Cells are the fundamental units of life. All present-day cells are believed to have evolved from an ancestral cell that existed more than 3 billion years ago.

- All cells are enclosed by a plasma membrane, which separates the inside of the cell from its environment.

- All cells contain DNA as a store of genetic information and use it to guide the synthesis of RNA molecules and proteins.

- Cells in a multicellular organism, though they all contain the same DNA, can be very different. They turn on different sets of genes according to their developmental history and to signals they receive from their environment.

- Animal and plant cells are typically 5–20 μm in diameter and can be seen with a light microscope, which also reveals some of their internal components, including the larger organelles.

- The electron microscope reveals even the smallest organelles, but specimens require elaborate preparation and cannot be viewed while alive.

- Specific large molecules can be located in fixed or living cells with a fluorescence microscope.

- The simplest of present-day living cells are prokaryotes: although they contain DNA, they lack a nucleus and other organelles and probably resemble most closely the ancestral cell.

- Different species of prokaryotes are diverse in their chemical capabilities and inhabit an amazingly wide range of habitats. Two fundamental evolutionary subdivisions are recognized: bacteria and archaea.

- Eukaryotic cells possess a nucleus and other organelles not found in prokaryotes. They probably evolved in a series of stages, including the acquisition of mitochondria by engulfment of aerobic bacteria and (for plant cells) the acquisition of chloroplasts by engulfment of photosynthetic bacteria.

- The nucleus contains the genetic information of the eukaryotic organism, stored in DNA molecules.

- The cytoplasm includes all of the cell's contents outside the nucleus and contains a variety of membrane-enclosed organelles with specialized functions: mitochondria carry out the final oxidation of food molecules; in plant cells, chloroplasts perform photosynthesis; the endoplasmic reticulum and the Golgi apparatus synthesize complex molecules for export from the cell and for insertion in cell membranes; lysosomes digest large molecules.

- Outside the membrane-enclosed organelles in the cytoplasm is the cytosol, a very concentrated mixture of large and small molecules that carry out many essential biochemical processes.

- The cytoskeleton is composed of protein filaments that extend throughout the cytoplasm and are responsible for cell shape and movement and for the transport of organelles and other large molecular complexes from one location to another.

- Free-living, single-celled eukaryotic microorganisms are complex cells that can swim, mate, hunt, and devour other microorganisms.

- Animals, plants, and some fungi consist of diverse eukaryotic cell types, all derived from a single fertilized egg cell; the number of such cells cooperating to form a large multicellular organism such as a human runs into thousands of billions.

- Biologists have chosen a small number of model organisms to study closely, including the bacterium *E. coli*, brewer's yeast, a nematode worm, a fly, a small plant, a fish, a mouse, and humans themselves.

- The simplest known cell is a bacterium with about 500 genes, but most cells contain significantly more. The human genome has about 25,000 genes, which is only about twice as many as a fly and six times as many as *E. coli*.

QUESTIONS

QUESTION 1–8

By now you should be familiar with the following cellular components. Briefly define what they are and what function they provide for cells.

A. cytosol

B. cytoplasm

C. mitochondria

D. nucleus

E. chloroplasts

F. lysosomes

G. chromosomes

H. Golgi apparatus

I. peroxisomes

J. plasma membrane

K. endoplasmic reticulum

L. cytoskeleton

QUESTION 1–9

Which of the following statements are correct? Explain your answers.

A. The hereditary information of a cell is passed on by its proteins.

B. Bacterial DNA is found in the cytosol.

C. Plants are composed of prokaryotic cells.

D. All cells of the same organism have the same number of chromosomes (with the exception of egg and sperm cells).

E. The cytosol contains membrane-enclosed organelles, such as lysosomes.

F. The nucleus and mitochondria are surrounded by a double membrane.

G. Protozoans are complex organisms with a set of specialized cells that form tissues, such as flagella, mouthparts, stinging darts, and leglike appendages.

H. Lysosomes and peroxisomes are the sites of degradation of unwanted materials.

QUESTION 1–10

To get a feeling for the size of cells (and to practice the use of the metric system), consider the following: the human brain weighs about 1 kg and contains about 10^{12} cells. Calculate the average size of a brain cell (although we know that their sizes vary widely), assuming that each cell is entirely filled with water (1 cm^3 of water weighs 1 g). What would be the length of one side of this average-sized brain cell if it were a simple cube? If the cells were spread out as a thin layer that is only a single cell thick, how many pages of this book would this layer cover?

QUESTION 1–11

Identify the different organelles indicated with letters in the electron micrograph of a plant cell shown below. Estimate the length of the scale bar in the figure.

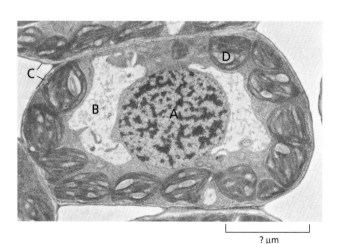

? µm

QUESTION 1–12

There are three major classes of filaments that make up the cytoskeleton. What are they, and what are the differences in their functions? Which cytoskeletal filaments would be most plentiful in a muscle cell or in an epidermal cell making up the outer layer of the skin? Explain your answers.

QUESTION 1–13

Natural selection is such a powerful force in evolution because cells with even a small proliferation advantage quickly outgrow their competitors. To illustrate this process, consider a cell culture that contains 1 million bacterial cells that double every 20 minutes. A single cell in this culture acquires a mutation that allows it to divide faster, with a generation time of only 15 minutes. Assuming that there is an unlimited food supply and no cell death, how long would it take before the progeny of the mutated cell became predominant in the culture? (Before you go through the calculation, make a guess: do you think it would take about a day, a week, a month, or a year?) How many cells of either type are present in the culture at this time? (The number of cells N in the culture at time t is described by the equation $N = N_0 \times 2^{t/G}$, where N_0 is the number of cells at zero time and G is the generation time.)

QUESTION 1–14

When bacteria are grown under adverse conditions, i.e., in the presence of a poison such as an antibiotic, most cells grow and proliferate slowly. But it is not uncommon that the growth rate of a bacterial culture kept in the presence of the poison is restored after a few days to that observed in its absence. Suggest why this may be the case.

QUESTION 1–15

Apply the principle of exponential growth of a culture as described in Question 1–13 to the cells in a multicellular organism, such as yourself. There are about 10^{13} cells in your body. Assume that one cell acquires a mutation that allows it to divide in an uncontrolled manner (i.e., it becomes a cancer cell). Some cancer cells can proliferate with a generation time of about 24 hours. If none of the cancer cells died, how long would it take before 10^{13} cells in your body would be cancer cells? (Use the equation $N = N_0 \times 2^{t/G}$, with t, the time, and G, the length of each generation. Hint: $10^{13} \approx 2^{43}$.)

QUESTION 1–16

Discuss the following statement: "The structure and function of a living cell are dictated by the laws of physics and chemistry."

QUESTION 1–17

What, if any, are the advantages in being multicellular?

QUESTION 1–18

Draw to scale the outline of two spherical cells, one a bacterium with a diameter of 1 μm, the other an animal cell with a diameter of 15 μm. Calculate the volume, surface area, and surface-to-volume ratio for each cell. How would the latter ratio change if you included the internal membranes of the cell in the calculation of surface area (assume internal membranes have 15 times the area of the plasma membrane)? (The volume of a sphere is given by $4\pi r^3/3$ and its surface by $4\pi r^2$, where r is its radius.) Discuss the following hypothesis: "Internal membranes allowed bigger cells to evolve."

QUESTION 1–19

What are the arguments that all living cells evolved from a common ancestor cell? Imagine the very early days of evolution of life on Earth. Would you assume that the primordial ancestor cell was the first and only cell to form?

QUESTION 1–20

In Figure 1–25, proteins are blue, nucleic acids are pink, lipids are yellow, and polysaccharides are green. Identify the major organelles and other important cellular structures shown in this slice through a eukaryotic cell.

QUESTION 1–21

Looking at some pond water under the microscope, you notice an unfamiliar rod-shaped cell about 200 μm long. Knowing that some exceptional bacteria can be as big as this or even bigger, you wonder whether your cell is a bacterium or a eukaryote. How will you decide? If it is not a eukaryote, how will you discover whether it is a bacterium or an archaeon?

CHAPTER **TWO**

2

Chemical Components of Cells

It is at first sight difficult to accept that living creatures are merely chemical systems. Their incredible diversity of form, their seemingly purposeful behavior, and their ability to grow and reproduce all seem to set them apart from the world of solids, liquids, and gases that chemistry normally describes. Indeed, until the nineteenth century, it was widely believed that animals contained a vital force—an "animus"—that was responsible for their distinctive properties.

We now know that there is nothing in living organisms that disobeys chemical or physical laws. However, the chemistry of life is indeed a special kind. First, it is based overwhelmingly on carbon compounds, the study of which is known as *organic chemistry*. Second, it depends almost exclusively on chemical reactions that take place in a watery, or *aqueous*, solution and in the relatively narrow range of temperatures experienced on Earth. Third, it is enormously complex: even the simplest cell is vastly more complicated in its chemistry than any other chemical system known. Fourth, it is dominated and coordinated by collections of enormous *polymeric molecules*—chains of chemical **subunits** linked end-to-end—whose unique properties enable cells and organisms to grow and reproduce and to do all the other things that are characteristic of life. Finally, the chemistry of life is tightly regulated: cells deploy a variety of mechanisms to make sure that all their chemical reactions occur at the proper place and time.

Because chemistry lies at the heart of all biology, in this chapter, we briefly survey the chemistry of the living cell. We will meet the molecules from which cells are made and examine their structures, shapes, and chemical properties. These molecules determine the size, structure, and functions

CHEMICAL BONDS

SMALL MOLECULES IN CELLS

MACROMOLECULES IN CELLS

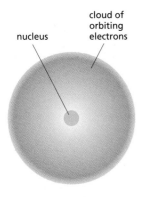

Figure 2–1 An atom consists of a nucleus surrounded by an electron cloud. The dense, positively charged nucleus contains most of the atom's mass. The much lighter and negatively charged electrons occupy space around the nucleus, as governed by the laws of quantum mechanics. The electrons are depicted as a continuous cloud, as there is no way of predicting exactly where an electron is at any given instant. The density of shading of the cloud is an indication of the probability that electrons will be found there. The diameter of the electron cloud ranges from about 0.1 nm (for hydrogen) to about 0.4 nm (for atoms of high atomic number). The nucleus is very much smaller: about 5×10^{-6} nm for carbon, for example.

of living cells. By understanding how they interact, we can begin to see how cells exploit the laws of chemistry and physics to survive, thrive, and reproduce.

CHEMICAL BONDS

Matter is made of combinations of *elements*—substances such as hydrogen or carbon that cannot be broken down or interconverted by chemical means. The smallest particle of an element that still retains its distinctive chemical properties is an *atom*. The characteristics of substances other than pure elements—including the materials from which living cells are made—depend on which atoms they contain and the way these atoms are linked together in groups to form *molecules*. To understand living organisms, therefore, it is crucial to know how the chemical bonds that hold atoms together in molecules are formed.

Cells Are Made of Relatively Few Types of Atoms

Each **atom** has at its center a dense, positively charged nucleus, which is surrounded at some distance by a cloud of negatively charged **electrons**, held there by electrostatic attraction to the nucleus (**Figure 2–1**). The nucleus consists of two kinds of subatomic particles: **protons**, which are positively charged, and *neutrons*, which are electrically neutral. The number of protons present in an atom's nucleus determines its *atomic number*. An atom of hydrogen has a nucleus composed of a single proton; so hydrogen, with an atomic number of 1, is the lightest element. An atom of carbon has six protons in its nucleus and an atomic number of 6 (**Figure 2–2**). The electric charge carried by each proton is exactly equal and opposite to the charge carried by a single electron. Because the whole atom is electrically neutral, the number of negatively charged electrons surrounding the nucleus is equal to the number of positively charged protons that the nucleus contains; thus the number of electrons in an atom also equals the atomic number. All atoms of a given element have the same atomic number, and we will see shortly that it is this number that dictates each atom's chemical behavior.

Neutrons have essentially the same mass as protons. They contribute to the structural stability of the nucleus—if there are too many or too few, the nucleus may disintegrate by radioactive decay—but they do not alter the chemical properties of the atom. Thus an element can exist in several physically distinguishable but chemically identical forms, called *isotopes*, each having a different number of neutrons but the same number of protons. Multiple isotopes of almost all the elements occur naturally,

Figure 2–2 The number of protons in an atom determines its atomic number. Schematic representations of an atom of carbon and an atom of hydrogen are shown. The nucleus of every atom except hydrogen consists of both positively charged protons and electrically neutral neutrons; the atomic weight equals the number of protons plus neutrons. The number of electrons in an atom is equal to the number of protons, so that the atom has no net charge. In contrast to Figure 2–1, the electrons are shown here as individual particles. The concentric *black* circles represent in a highly schematic form the "orbits" (that is, the different distributions) of the electrons. The neutrons, protons, and electrons are in reality minute in relation to the atom as a whole; their size is greatly exaggerated here.

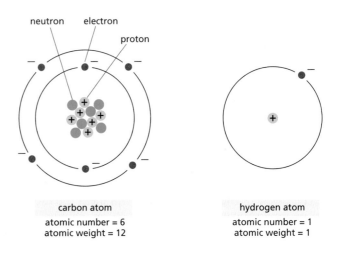

including some that are unstable—and thus radioactive. For example, while most carbon on Earth exists as the stable isotope carbon 12, with six protons and six neutrons, also present are small amounts of an unstable isotope, carbon 14, which has six protons and eight neutrons. Carbon 14 undergoes radioactive decay at a slow but steady rate, which allows archaeologists to estimate the age of organic material.

The **atomic weight** of an atom, or the **molecular weight** of a molecule, is its mass relative to that of a hydrogen atom. This is essentially equal to the number of protons plus neutrons that the atom or molecule contains, because the electrons are so light that they contribute almost nothing to the total mass. Thus the major isotope of carbon has an atomic weight of 12 and is written as ^{12}C. The unstable carbon isotope just mentioned has an atomic weight of 14 and is written as ^{14}C. The mass of an atom or a molecule is generally specified in *daltons*, one dalton being an atomic mass unit approximately equal to the mass of a hydrogen atom.

Atoms are so small that it is hard to imagine their size. An individual carbon atom is roughly 0.2 nm in diameter, so that it would take about 5 million of them, laid out in a straight line, to span a millimeter. One proton or neutron weighs approximately $1/(6 \times 10^{23})$ gram. As hydrogen has only one proton—thus an atomic weight of 1—1 gram of hydrogen contains 6×10^{23} atoms. For carbon—which has six protons and six neutrons, and an atomic weight of 12—12 grams contain 6×10^{23} atoms. This huge number, called **Avogadro's number**, allows us to relate everyday quantities of chemicals to numbers of individual atoms or molecules. If a substance has a molecular weight of M, M grams of the substance will contain 6×10^{23} molecules. This quantity is called one *mole* of the substance (**Figure 2–3**). The concept of mole is used widely in chemistry as a way to represent the number of molecules that are available to participate in chemical reactions.

There are about 90 naturally occurring elements, each differing from the others in the number of protons and electrons in its atoms. Living organisms, however, are made of only a small selection of these elements, four of which—carbon (C), hydrogen (H), nitrogen (N), and oxygen (O)—constitute 96% of an organism's weight. This composition differs markedly from that of the nonliving inorganic environment on Earth (**Figure 2–4**) and is evidence of a distinctive type of chemistry.

The Outermost Electrons Determine How Atoms Interact

To understand how atoms come together to form the molecules that make up living organisms, we have to pay special attention to the atoms' electrons. Protons and neutrons are welded tightly to one another in an atom's nucleus, and they change partners only under extreme conditions—during radioactive decay, for example, or in the interior of the sun or of a nuclear reactor. In living tissues, only the electrons of an atom undergo rearrangements. They form the accessible part of the atom and specify the rules of chemistry by which atoms combine to form molecules.

Electrons are in continuous motion around the nucleus, but motions on this submicroscopic scale obey different laws from those we are familiar with in everyday life. These laws dictate that electrons in an atom can exist only in certain discrete regions of movement—roughly speaking, in discrete orbits. Moreover, there is a strict limit to the number of electrons that can be accommodated in an orbit of a given type, a so-called *electron shell*. The electrons closest on average to the positive nucleus are attracted most strongly to it and occupy the inner, most tightly bound shell. This innermost shell can hold a maximum of two electrons. The second shell is farther away from the nucleus, and can hold up to eight

A mole is X grams of a substance, where X is the molecular weight of the substance. A mole will contain 6×10^{23} molecules of the substance.

1 mole of carbon weighs 12 g
1 mole of glucose weighs 180 g
1 mole of sodium chloride weighs 58 g

A one molar solution has a concentration of 1 mole of the substance in 1 liter of solution. A 1 M solution of glucose, for example, contains 180 g/l, and a one millimolar (1 mM) solution contains 180 mg/l.

The standard abbreviation for gram is g; the abbreviation for liter is L.

Figure 2–3 What's a mole? Some sample calculations of moles and molar solutions.

Figure 2–4 The distribution of elements in the Earth's crust differs radically from that in a living organism. The abundance of each element is expressed here as a percentage of the total number of atoms present in a biological or geological sample, including water. Thus, for example, more than 60% of the atoms in the human body are hydrogen atoms, and nearly 30% of the atoms in the Earth's crust are silicon atoms (Si). The relative abundance of elements is similar in all living things.

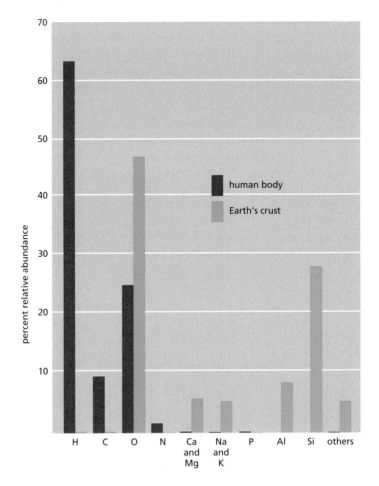

QUESTION 2–1

A cup of water, containing exactly 18 g, or 1 mole, of water, was emptied into the Aegean Sea 3000 years ago. What are the chances that the same quantity of water, scooped today from the Pacific Ocean, would include at least one of these ancient water molecules? Assume perfect mixing and an approximate volume for the world's oceans of 1.5 billion cubic kilometers (1.5×10^9 km^3).

electrons. The third shell can also hold up to eight electrons, which are even less tightly bound. The fourth and fifth shells can hold 18 electrons each. Atoms with more than four shells are very rare in biological molecules.

The arrangement of electrons in an atom is most stable when all the electrons are in the most tightly bound states that are possible for them—that is, when they occupy the innermost shells, closest to the nucleus. Therefore, with certain exceptions in the larger atoms, the electrons of an atom fill the shells in order—the first before the second, the second before the third, and so on. An atom whose outermost shell is entirely filled with electrons is especially stable and therefore chemically unreactive. Examples are helium with 2 electrons (atomic number 2), neon with 2 + 8 electrons (atomic number 10), and argon with 2 + 8 + 8 electrons (atomic number 18); these are all inert gases. Hydrogen, by contrast, has only one electron, which leaves its outermost shell half-filled, so it is highly reactive. The atoms found in living organisms all have outermost shells that are incompletely filled, and they are therefore able to react with one another to form molecules (**Figure 2–5**).

Because an incompletely filled electron shell is less stable than one that is completely filled, atoms with incomplete outer shells have a strong tendency to interact with other atoms so as to either gain or lose enough electrons to achieve a completed outermost shell. This electron exchange can be achieved either by transferring electrons from one atom to another or by sharing electrons between two atoms. These two strategies generate the two types of **chemical bonds** that bind atoms to one another: an *ionic bond* is formed when electrons are donated by one atom to another, whereas a *covalent bond* is formed when two atoms share a pair of electrons (**Figure 2–6**).

atomic number

element		I	II	III	IV
			electron shell		
1	Hydrogen (H)	●			
2	Helium (He)	●●			
6	Carbon (C)	●●	●●●●		
7	Nitrogen (N)	●●	●●●●●		
8	Oxygen (O)	●●	●●●●●●		
10	Neon (Ne)	●●	●●●●●●●●		
11	Sodium (Na)	●●	●●●●●●●●	●	
12	Magnesium (Mg)	●●	●●●●●●●●	●●	
15	Phosphorus (P)	●●	●●●●●●●●	●●●●●	
16	Sulfur (S)	●●	●●●●●●●●	●●●●●●	
17	Chlorine (Cl)	●●	●●●●●●●●	●●●●●●●	
18	Argon (Ar)	●●	●●●●●●●●	●●●●●●●●	
19	Potassium (K)	●●	●●●●●●●●	●●●●●●●●	●
20	Calcium (Ca)	●●	●●●●●●●●	●●●●●●●●	●●

Figure 2–5 An element's chemical reactivity depends on how its outermost electron shell is filled. All of the elements commonly found in living organisms have outermost shells that are not completely filled with electrons *(red)* and can thus participate in chemical reactions with other atoms. Inert gases *(yellow)*, in contrast, have completely filled outermost shells and are thus chemically unreactive.

An H atom, which needs only one more electron to fill its only shell, generally acquires it by sharing—forming one covalent bond with another atom. The other most common elements in living cells—C, N, and O, which have an incomplete second shell, and P and S, which have an incomplete third shell (see Figure 2–5)—generally share electrons and achieve a filled outer shell of eight electrons by forming several covalent bonds. The number of electrons an atom must acquire or lose (either by sharing or by transfer) to attain a filled outer shell determines the number of bonds the atom can make.

Because the state of the outer electron shell determines the chemical properties of an element, when the elements are listed in order of their atomic number we see a periodic recurrence of elements with similar properties: an element with, say, an incomplete second shell containing one electron will behave in much the same way as an element that has filled its second shell and has an incomplete third shell containing one electron. The metals, for example, have incomplete outer shells with just one or a few electrons, whereas, as we have just seen, the inert gases have full outer shells. This arrangement gives rise to the *periodic table* of the elements, outlined in **Figure 2–7**, which shows elements found in living organisms highlighted in color.

QUESTION 2–2

A carbon atom contains six protons and six neutrons.
A. What are its atomic number and atomic weight?
B. How many electrons does it have?
C. How many additional electrons must it add to fill its outermost shell? How does this affect carbon's chemical behavior?
D. Carbon with an atomic weight of 14 is radioactive. How does it differ in structure from nonradioactive carbon? How does this difference affect its chemical behavior?

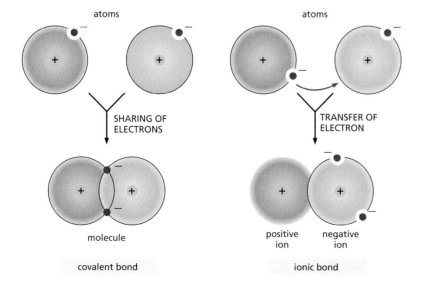

Figure 2–6 Atoms can attain a more stable arrangement of electrons in their outermost shell by interacting with one another. A covalent bond is formed when electrons are shared between atoms. An ionic bond is formed when electrons are transferred from one atom to the other. The two cases shown represent extremes; often, covalent bonds form with a partial transfer (unequal sharing of electrons), resulting in a polar covalent bond, as we discuss shortly.

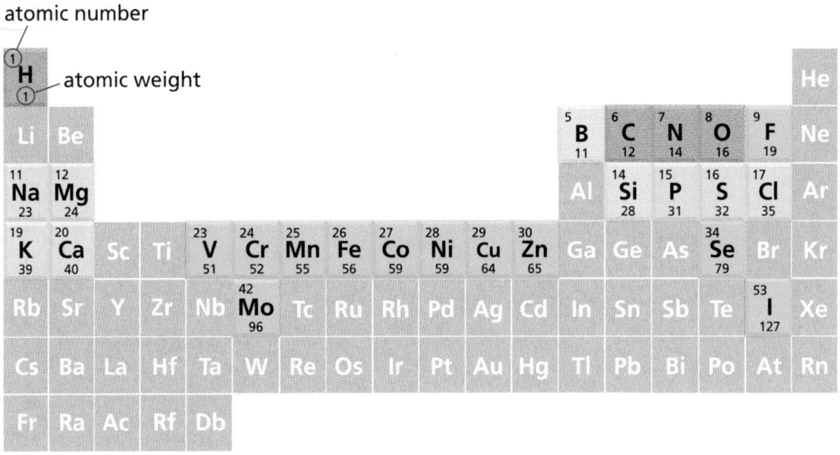

Figure 2–7 The chemistry of life is predominantly the chemistry of lighter elements. When ordered by their atomic number into a *periodic table*, elements fall into groups that show similar properties based on the number of electrons each element possesses in its outer shell. Atoms in the same vertical column must gain or lose the same number of electrons to attain a filled outer shell, and they thus behave similarly. Thus, both magnesium (Mg) and calcium (Ca) tend to give away the two electrons in their outer shells to form ionic bonds with atoms such as chlorine (Cl) that need extra electrons to complete their outer shells.

The four elements highlighted in *red* constitute 99% of the total number of atoms present in the human body and about 96% of our total weight. An additional seven elements, highlighted in *blue*, together represent about 0.9% of the total number of atoms. Other elements, shown in *green*, are required in trace amounts by humans. It remains unclear whether those elements shown in *yellow* are essential in humans or not.

The atomic weights shown here are those of the most common isotope of each element.

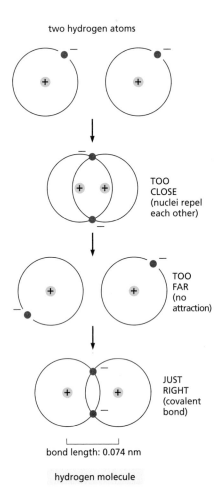

two hydrogen atoms

TOO CLOSE (nuclei repel each other)

TOO FAR (no attraction)

JUST RIGHT (covalent bond)

bond length: 0.074 nm

hydrogen molecule

Covalent Bonds Form by the Sharing of Electrons

All of the characteristics of a cell depend on the molecules it contains. A **molecule** is a cluster of atoms held together by **covalent bonds**, in which electrons are shared rather than transferred between atoms. The shared electrons complete the outer shells of the interacting atoms. In the simplest possible molecule—a molecule of hydrogen (H_2)—two H atoms, each with a single electron, share their electrons, thus filling their outermost shells. The shared electrons form a cloud of negative charge that is densest between the two positively charged nuclei. This electron density helps to hold the nuclei together by opposing the mutual repulsion between their positive charges that would otherwise force them apart. The attractive and repulsive forces are in balance when the nuclei are separated by a characteristic distance, called the *bond length* (**Figure 2–8**).

Whereas an H atom can form only a single covalent bond, the other common atoms that form covalent bonds in cells—O, N, S, and P, as well as the all-important C—can form more than one. The outermost shells of these atoms, as we have seen, can accommodate up to eight electrons, and they form covalent bonds with as many other atoms as necessary to reach this number. Oxygen, with six electrons in its outer shell, is most stable when it acquires two extra electrons by sharing with other atoms, and it therefore forms up to two covalent bonds. Nitrogen, with five outer electrons, forms a maximum of three covalent bonds, while carbon, with four outer electrons, forms up to four covalent bonds—thus sharing four pairs of electrons (see Figure 2–5).

When one atom forms covalent bonds with several others, these multiple bonds have definite orientations in space relative to one another, reflecting the orientations of the orbits of the shared electrons. Covalent bonds between multiple atoms are therefore characterized by specific bond angles, as well as by specific bond lengths and bond energies (**Figure 2–9**). The four covalent bonds that can form around a carbon

Figure 2–8 The hydrogen molecule is held together by a covalent bond. Each hydrogen atom in isolation has a single electron, which means that its first (and only) electron shell is incompletely filled. By coming together, the two atoms are able to share their electrons, so that each obtains a completely filled first shell, with the shared electrons adopting modified orbits around the two nuclei. The covalent bond between the two atoms has a definite length—0.074 nm, which is the distance between the two nuclei. If the atoms were closer together, the positive nuclei would repel each other; if they were farther apart, they would not be able to share electrons as effectively.

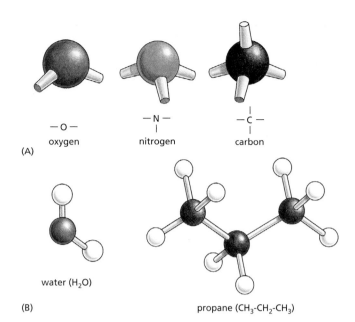

(A)

−O−
oxygen

−N−
  |
nitrogen

  |
−C−
  |
carbon

(B)

water (H₂O)

propane (CH₃-CH₂-CH₃)

Figure 2–9 Covalent bonds are characterized by particular geometries. (A) The spatial arrangement of the covalent bonds that can be formed by oxygen, nitrogen, and carbon. (B) Molecules formed from these atoms therefore have a precise three-dimensional structure defined by the bond angles and bond lengths for each covalent linkage. A water molecule, for example, forms a "V" shape with an angle close to 109°.

In these ball-and-stick models, the different colored balls represent different atoms, and the sticks represent the covalent bonds. The colors traditionally used to represent the different atoms— *black* for carbon, *white* for hydrogen, *blue* for nitrogen, and *red* for oxygen—were established by the chemist August Wilhelm Hofmann in 1865, when he used a set of colored croquet balls to build molecular models for a public lecture on "the combining power of atoms."

atom, for example, are arranged as if pointing to the four corners of a regular tetrahedron. The precise orientation of the covalent bonds around carbon produces the three-dimensional geometry of organic molecules.

There Are Different Types of Covalent Bonds

Most covalent bonds involve the sharing of two electrons, one donated by each participating atom; these are called *single bonds*. Some covalent bonds, however, involve the sharing of more than one pair of electrons. Four electrons can be shared, for example, two coming from each participating atom; such a bond is called a *double bond*. Double bonds are shorter and stronger than single bonds and have a characteristic effect on the three-dimensional geometry of molecules containing them. A single covalent bond between two atoms generally allows the rotation of one part of a molecule relative to the other around the bond axis. A double bond prevents such rotation, producing a more rigid and less flexible arrangement of atoms (**Figure 2–10**). This restriction has a major influence on the three-dimensional shape of many macromolecules. **Panel 2–1** (pp. 66–67) reviews the covalent bonds commonly encountered in biological molecules.

Some molecules contain atoms that share electrons in a way that produces bonds that are intermediate in character between single and double bonds. The highly stable benzene molecule, for example, is made up of a ring of six carbon atoms in which the bonding electrons are evenly distributed (although the arrangement is sometimes depicted as an alternating sequence of single and double bonds, as shown in Panel 2–1).

When the atoms joined by a single covalent bond belong to different elements, the two atoms usually attract the shared electrons to different degrees. Covalent bonds in which the electrons are shared unequally in this way are known as *polar covalent bonds*. A **polar** structure (in the electrical sense) is one in which the positive charge is concentrated toward one end of the molecule (the positive pole) and the negative charge is concentrated toward the other end (the negative pole). Oxygen and nitrogen atoms, for example, attract electrons relatively strongly, whereas an H atom attracts electrons relatively weakly (because of the relative differences in the positive charges of the nuclei of C, O, N, and H). Thus the

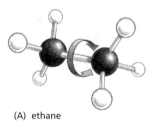

(A) ethane

(B) ethene

Figure 2–10 Carbon–carbon double bonds are shorter and more rigid than carbon–carbon single bonds. (A) The ethane molecule, with a single covalent bond between the two carbon atoms, shows the tetrahedral arrangement of the three single covalent bonds between each carbon atom and its three attached H atoms. The CH₃ groups, joined by a covalent C–C bond, can rotate relative to one another around the bond axis. (B) The double bond between the two carbon atoms in a molecule of ethene (ethylene) alters the bond geometry of the carbon atoms and brings all the atoms into the same plane; the double bond prevents the rotation of one CH₂ group relative to the other.

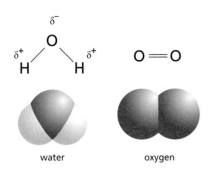

water   oxygen

Figure 2–11 In polar covalent bonds, the electrons are shared unequally. Comparison of electron distributions in the polar covalent bonds in a molecule of water (H_2O) and the nonpolar covalent bonds in a molecule of oxygen (O_2). In H_2O, electrons are more strongly attracted to the oxygen nucleus than to the H nucleus, as indicated by the distributions of the partial negative (δ^-) and partial positive (δ^+) charges.

covalent bond between O and H, O–H, or between N and H, N–H, is polar (**Figure 2–11**). An atom of C and an atom of H, by contrast, attract electrons more equally. Thus the bond between carbon and hydrogen, C–H, is relatively nonpolar.

Covalent Bonds Vary in Strength

We have already seen that the covalent bond between two atoms has a characteristic length that depends on the atoms involved. A further crucial property of any chemical bond is its strength. *Bond strength* is measured by the amount of energy that must be supplied to break the bond, usually expressed in units of either kilocalories per mole (kcal/mole) or kilojoules per mole (kJ/mole). A kilocalorie is the amount of energy needed to raise the temperature of 1 liter of water by 1°C. Thus, if 1 kilocalorie of energy must be supplied to break 6×10^{23} bonds of a specific type (that is, 1 mole of these bonds), then the strength of that bond is 1 kcal/mole. One kilocalorie is equal to about 4.2 kJ, which is the unit of energy universally employed by physical scientists and, increasingly, by cell biologists as well.

To get an idea of what bond strengths mean, it is helpful to compare them with the average energies of the impacts that molecules continually undergo owing to collisions with other molecules in their environment—their thermal, or heat, energy. Typical covalent bonds are stronger than these thermal energies by a factor of 100, so they are resistant to being pulled apart by thermal motions. In living organisms, they are normally broken only during specific chemical reactions that are carefully controlled by highly specialized protein catalysts, called *enzymes*.

When water is present, covalent bonds are much stronger than ionic bonds. In ionic bonds, electrons are transferred rather than shared, as we now discuss.

Ionic Bonds Form by the Gain and Loss of Electrons

Ionic bonds are usually formed between atoms that can attain a completely filled outer shell most easily by donating electrons to—or accepting electrons from—another atom, rather than by sharing them. For example, returning to Figure 2–5, we see that a sodium (Na) atom can achieve a filled outer shell by giving up the single electron in its third shell. By contrast, a chlorine (Cl) atom can complete its outer shell by gaining just one electron. Consequently, if a Na atom encounters a Cl atom, an electron can jump from the Na to the Cl, leaving both atoms with filled outer shells. The offspring of this marriage between sodium, a soft and intensely reactive metal, and chlorine, a toxic green gas, is table salt (NaCl).

When an electron jumps from Na to Cl, both atoms become electrically charged **ions**. The Na atom that lost an electron now has one less electron than it has protons in its nucleus; it therefore has a net single positive charge (Na^+). The Cl atom that gained an electron now has one more electron than it has protons and has a net single negative charge (Cl^-). Because of their opposite charges, the Na^+ and Cl^- ions are attracted

---

QUESTION 2–3

Discuss whether the following statement is correct: "An ionic bond can, in principle, be thought of as a very polar covalent bond. Polar covalent bonds, then, fall somewhere between ionic bonds at one end of the spectrum and nonpolar covalent bonds at the other end."

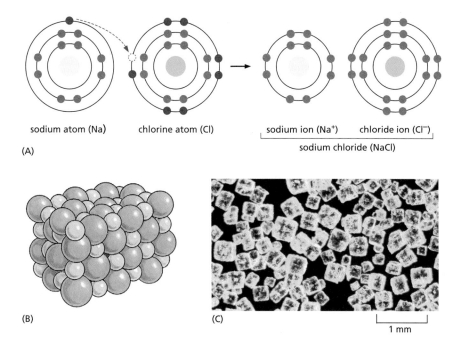

(A)

sodium atom (Na)   chlorine atom (Cl)   sodium ion (Na⁺)   chloride ion (Cl⁻)

sodium chloride (NaCl)

(B)   (C)

1 mm

Figure 2–12 Sodium chloride is held together by ionic bonds. (A) An atom of sodium (Na) reacts with an atom of chlorine (Cl). Electrons of each atom are shown in their different shells; electrons in the chemically reactive (incompletely filled) outermost shells are shown in *red*. The reaction takes place with transfer of a single electron from sodium to chlorine, forming two electrically charged atoms, or ions, each with complete sets of electrons in their outermost shells. The two ions have opposite charge and are held together by electrostatic attraction. (B) The product of the reaction between sodium and chlorine, crystalline sodium chloride, contains sodium and chloride ions packed closely together in a regular array in which the charges are exactly balanced. (C) Color photograph of crystals of sodium chloride.

to each other and are thereby held together by an ionic bond (**Figure 2–12A**). Ions held together solely by ionic bonds are generally called *salts* rather than molecules. A NaCl crystal contains astronomical numbers of Na⁺ and Cl⁻ packed together in a precise three-dimensional array with their opposite charges exactly balanced: a crystal only 1 mm across contains about 2×10^{19} ions of each type (**Figure 2–12B and C**).

Because of the favorable interaction between ions and water molecules (which are polar), many salts (including NaCl) are highly soluble in water. They dissociate into individual ions (such as Na⁺ and Cl⁻), each surrounded by a group of water molecules. Positive ions are called *cations*, and negative ions are called *anions*. Small inorganic ions such as Na⁺, Cl⁻, K⁺, and Ca²⁺ play important parts in many biological processes, including the electrical activity of nerve cells, as we discuss in Chapter 12.

Noncovalent Bonds Help Bring Molecules Together in Cells

In aqueous solution, ionic bonds are 10–100 times weaker than the covalent bonds that hold atoms together in molecules. But this weakness has its place: much of biology depends on specific but transient interactions between one molecule and another. These associations are mediated by **noncovalent bonds**. Although noncovalent bonds are individually quite weak, their energies can sum to create an effective force between two molecules.

The ionic bonds that hold together the Na⁺ and Cl⁻ ions in a salt crystal (see Figure 2–12) are a form of noncovalent bond called an *electrostatic attraction*. Electrostatic attractions are strongest when the atoms involved are fully charged, as are Na⁺ and Cl⁻. But a weaker electrostatic attraction also occurs between molecules that contain polar covalent bonds (see Figure 2–11). Polar covalent bonds are thus extremely important in biology because they allow molecules to interact through electrical forces. Any large molecule with many polar groups will have a pattern of partial positive and negative charges on its surface. When such a molecule encounters a second molecule with a complementary set of charges, the two will be attracted to each other by electrostatic attraction—even

---

QUESTION 2–4

What, if anything, is wrong with the following statement: "When NaCl is dissolved in water, the water molecules closest to the ions will tend to preferentially orient themselves so that their oxygen atoms face the sodium ions and face away from the chloride ions"? Explain your answer.

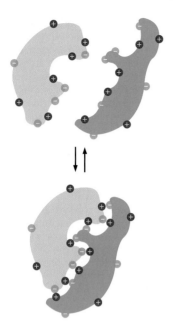

Figure 2–13 A large molecule, such as a protein, can bind to another protein through complementary charges on the surface of each molecule. In the aqueous environment of a cell, the many individual electrostatic attractions shown would help the two proteins stay bound to each other.

though water greatly reduces the attractiveness of these charges in most biological settings. When present in large numbers, however, weak non-covalent bonds on the surfaces of large molecules can promote strong and specific binding (**Figure 2–13**).

Hydrogen Bonds Are Important Noncovalent Bonds For Many Biological Molecules

Water accounts for about 70% of a cell's weight, and most intracellular reactions occur in an aqueous environment. Life on Earth is thought to have begun in the ocean. Thus the properties of water have put a permanent stamp on the chemistry of living things.

In each molecule of water (H_2O), the two H atoms are linked to the O atom by covalent bonds. The two H–O bonds are highly polar because the O is strongly attractive for electrons, whereas the H is only weakly attractive. Consequently, there is an unequal distribution of electrons in a water molecule, with a preponderance of positive charge on the two H atoms and negative charge on the O (see Figure 2–11). When a positively charged region of one water molecule (that is, one of its H atoms) comes close to a negatively charged region (that is, the O) of a second water molecule, the electrical attraction between them can establish a weak bond called a **hydrogen bond** (**Figure 2–14**). These bonds are much weaker than covalent bonds and are easily broken by random thermal motions. Thus each bond lasts only an exceedingly short time. But the combined effect of many weak bonds is far from trivial. Each water molecule can form hydrogen bonds through its two H atoms to two other water molecules, producing a network in which hydrogen bonds are being continually broken and formed. It is because of these interlocking hydrogen bonds that water at room temperature is a liquid—with a high boiling point and high surface tension—and not a gas. Without hydrogen bonds, life as we know it could not exist. The biologically significant properties of water are reviewed in **Panel 2–2** (pp. 68–69).

Hydrogen bonds are not limited to water. In general, a hydrogen bond can form whenever a positively charged H atom held in one molecule by a polar covalent linkage comes close to a negatively charged atom—typically an oxygen or a nitrogen—belonging to another molecule (see Figure 2–14). Hydrogen bonds can also occur between different parts of a single large molecule, where they often help the molecule fold into a particular shape. The length and strength of hydrogen bonds and of ionic bonds are compared to those of covalent bonds in **Table 2–1**.

Molecules, such as alcohols, that contain polar bonds and that can form hydrogen bonds mix well with water. As mentioned previously, molecules carrying positive or negative charges (ions) likewise dissolve readily in water. Such molecules are termed **hydrophilic**, meaning that they are

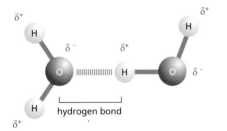

Figure 2–14 A hydrogen bond can form between two water molecules. These bonds are largely responsible for water's life-sustaining properties—including its ability to exist as a liquid at the temperatures inside the typical mammalian body.

TABLE 2–1 LENGTH AND STRENGTH OF SOME CHEMICAL BONDS			
Bond type	Length* (nm)	Strength (kcal/mole)	
		in vacuum	in water
Covalent	0.10	90 [377]**	90 [377]
Noncovalent: ionic bond	0.25	80 [335]	3 [12.6]
Noncovalent: hydrogen bond	0.17	4 [16.7]	1 [4.2]

*The bond lengths and strengths listed are approximate, because the exact values will depend on the atoms involved.
**Values in brackets are kJ/mole. 1 calorie = 4.184 joules.

"water-loving." A large proportion of the molecules in the aqueous environment of a cell fall into this category, including sugars, DNA, RNA, and a majority of proteins. **Hydrophobic** ("water-fearing") molecules, by contrast, are uncharged and form few or no hydrogen bonds, and they do not dissolve in water.

Hydrocarbons are important hydrophobic cell constituents (see Panel 2–1, pp. 66–67). In these molecules, the H atoms are covalently linked to C atoms by nonpolar bonds. Because the H atoms have almost no net positive charge, they cannot form effective hydrogen bonds to other molecules. This makes the hydrocarbon as a whole hydrophobic—a property that is exploited by cells, whose membranes are constructed largely from *lipid molecules* that have long hydrocarbon tails. Because lipids do not dissolve in water, they can form the thin membrane barriers that keep the aqueous interior of the cell separate from the surrounding aqueous environment, as we discuss later.

Some Polar Molecules Form Acids and Bases in Water

One of the simplest kinds of chemical reaction, and one that has profound significance in cells, takes place when a molecule possessing a highly polar covalent bond between a hydrogen and another atom dissolves in water. The hydrogen atom in such a bond has given up its electron almost entirely to the companion atom, so it exists as an almost naked positively charged hydrogen nucleus—in other words, a *proton* (H^+). When the polar molecule becomes surrounded by water molecules, the proton will be attracted to the partial negative charge on the oxygen atom of an adjacent water molecule (see Figure 2–11); this proton can dissociate from its original partner and associate instead with the oxygen atom of the water molecule, generating a **hydronium ion** (H_3O^+) (**Figure 2–15A**). The reverse reaction also takes place very readily, so one has to imagine an equilibrium state in which billions of protons are constantly flitting to and fro between one molecule and another in an aqueous solution.

Substances that release protons when they dissolve in water, thus forming H_3O^+, are termed **acids**. The higher the concentration of H_3O^+, the more acidic the solution. H_3O^+ is present even in pure water, at a concentration of 10^{-7} M, as a result of the movement of protons from one water molecule to another (**Figure 2–15B**). By tradition, the H_3O^+ concentration

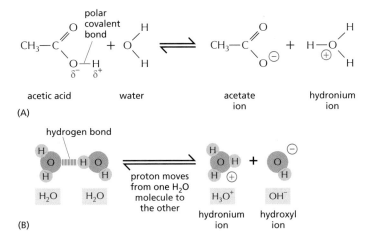

acetic acid   water   acetate ion   hydronium ion

(A)

(B)

Figure 2–15 Protons move continuously from one water molecule to another in aqueous solutions. (A) The reaction that takes place when a molecule of acetic acid dissolves in water. At pH 7, nearly all of the acetic acid molecules are present as acetate ions. (B) Water molecules are continually exchanging protons with each other to form hydronium and hydroxyl ions. These ions in turn rapidly recombine to form water molecules.

is usually referred to as the H^+ concentration, even though most protons in an aqueous solution are present as H_3O^+. To avoid the use of unwieldy numbers, the concentration of H^+ is expressed using a logarithmic scale called the **pH scale**, as illustrated in Panel 2–2. Pure water has a pH of 7.0 and is thus neutral—that is, neither acidic (pH < 7) nor basic (pH > 7).

Acids are characterized as being strong or weak, depending on how readily they give up their protons to water. Strong acids, such as hydrochloric acid (HCl), lose their protons easily. Acetic acid, on the other hand, is a weak acid because it holds on to its proton more tightly when dissolved in water. Many of the acids important in the cell—such as molecules containing a carboxyl (COOH) group—are weak acids (see Panel 2–2, pp. 68–69). Their tendency to give up a proton with some reluctance is a useful characteristic, as it renders the molecules sensitive to changes in pH in the cell—a property that can be exploited to regulate function.

Because protons can be passed readily to many types of molecules in cells, thus altering the molecules' character, the H^+ concentration inside a cell (the pH) must be closely controlled. Acids—especially weak acids—will give up their protons more readily if the H^+ concentration is low and will tend to accept them back if the concentration is high.

The opposite of an acid is a **base**, which includes any molecule that accepts a proton when dissolved in water. Just as the defining property of an acid is that it raises the concentration of H_3O^+ ions by donating a proton to a water molecule, so the defining property of a base is that it raises the concentration of hydroxyl (OH^-) ions by removing a proton from a water molecule. Thus sodium hydroxide (NaOH) is basic (the term *alkaline* is also used) because it dissociates in aqueous solution to form Na^+ ions and OH^- ions; because it does so readily, NaOH is called a strong base. Weak bases—which have a weak tendency to accept a proton from water—however, are actually more important in cells. Many biologically important weak bases contain an amino (NH_2) group, which can generate OH^- by taking a proton from water: $-NH_2 + H_2O \rightarrow -NH_3^+ + OH^-$ (see Panel 2–2, pp. 68–69).

Because an OH^- ion combines with a proton to form a water molecule, an increase in the OH^- concentration forces a decrease in the H^+ concentration, and vice versa. A pure solution of water thus contains an equal concentration (10^{-7} M) of both ions, rendering it neutral (pH 7). The interior of a cell is also kept close to neutral by the presence of **buffers**: mixtures of weak acids and bases that can adjust proton concentrations around pH 7 by releasing protons (acids) or taking them up (bases). This give-and-take keeps the pH of the cell relatively constant under a variety of conditions.

SMALL MOLECULES IN CELLS

Having looked at the ways atoms combine to form small molecules and how these molecules behave in an aqueous environment, we now examine the main classes of small molecules found in cells and their biological roles. Amazingly, we will see that a few basic categories of molecules, formed from a handful of different elements, give rise to all the extraordinary richness of form and behavior displayed by living things.

A Cell Is Formed from Carbon Compounds

If we disregard water, nearly all the molecules in a cell are based on carbon. Carbon is outstanding among all the elements in its ability to form large molecules; silicon—an element with the same number of electrons in its outer shell—is a poor second. Because a carbon atom is small and

---

QUESTION 2–5

A. Are there any H_3O^+ ions present in pure water at neutral pH (i.e., at pH = 7.0)? If so, how are they formed?

B. If they exist, what is the ratio of H_3O^+ ions to H_2O molecules at neutral pH? (Hint: the molecular weight of water is 18, and 1 liter of water weighs 1 kg.)

has four electrons and four vacancies in its outer shell, it can form four covalent bonds with other atoms (see Figure 2–9). Most importantly, one carbon atom can join to other carbon atoms through highly stable covalent C–C bonds to form chains and rings and hence generate large and complex molecules with no obvious upper limit to their size. The small and large carbon compounds made by cells are called **organic molecules**. By contrast, all other molecules, including water, are said to be **inorganic**.

Certain combinations of atoms, such as the methyl ($-CH_3$), hydroxyl ($-OH$), carboxyl ($-COOH$), carbonyl ($-C=O$), phosphoryl ($-PO_3^{2-}$), and amino ($-NH_2$) groups, occur repeatedly in organic molecules. Each such **chemical group** has distinct chemical and physical properties that influence the behavior of the molecule in which the group occurs, including whether the molecule tends to gain or lose protons and with which other molecules it will interact. Knowing these groups and their chemical properties greatly simplifies understanding the chemistry of life. The most common chemical groups and some of their properties are summarized in Panel 2–1 (pp. 67–68).

Cells Contain Four Major Families of Small Organic Molecules

The small organic molecules of the cell are carbon compounds with molecular weights in the range 100–1000 that contain up to 30 or so carbon atoms. They are usually found free in solution in the cytosol and have many different roles. Some are used as *monomer* subunits to construct the cell's giant polymeric *macromolecules*—its proteins, nucleic acids, and large polysaccharides. Others serve as energy sources, which are broken down and transformed into other small molecules in a maze of intracellular metabolic pathways. Many have more than one role in the cell—acting, for example, as both a potential subunit for a macromolecule and as an energy source. The small organic molecules are much less abundant than the organic macromolecules, accounting for only about one-tenth of the total mass of organic matter in a cell. As a rough guess, there may be a thousand different kinds of these small organic molecules in a typical animal cell.

All organic molecules are synthesized from—and are broken down into—the same set of simple compounds. Both their synthesis and their breakdown occur through sequences of simple chemical changes that are limited in variety and follow step-by-step rules. As a consequence, the compounds in a cell are chemically related, and most can be classified into a small number of distinct families. Broadly speaking, cells contain four major families of small organic molecules: the *sugars*, the *fatty acids*, the *amino acids*, and the *nucleotides* (**Figure 2–16**). Although many compounds present in cells do not fit into these categories, these four families of small organic molecules, together with the macromolecules made by linking them into long chains, account for a large fraction of a cell's mass (**Table 2–2**).

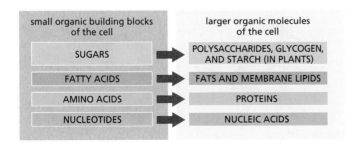

Figure 2–16 **Sugars, fatty acids, amino acids, and nucleotides are the four main families of small organic molecules in cells.** They form the monomeric building blocks, or subunits, for larger organic molecules, including most of the macromolecules and other molecular assemblies of the cell. Some, like the sugars and the fatty acids, are also energy sources.

TABLE 2–2 THE CHEMICAL COMPOSITION OF A BACTERIAL CELL

	Percent of total cell weight	Approximate number of types of each class of molecule
Water	70	1
Inorganic ions	1	20
Sugars and precursors	1	250
Amino acids and precursors	0.4	100
Nucleotides and precursors	0.4	100
Fatty acids and precursors	1	50
Other small molecules	0.2	300
Phospholipids	2	4*
Macromolecules (nucleic acids, proteins, and polysaccharides)	24	3000

*There are four classes of phospholipids, each of which exists in many varieties.

Sugars Are Both Energy Sources and Subunits of Polysaccharides

The simplest **sugars**—the *monosaccharides*—are compounds with the general formula $(CH_2O)_n$, where n is usually 3, 4, 5, or 6. Sugars, and the larger molecules made from them, are also called *carbohydrates* because of this simple formula. Glucose, for example, has the formula $C_6H_{12}O_6$ (**Figure 2–17**). The formula, however, does not fully define the molecule: the same set of carbons, hydrogens, and oxygens can be joined together by covalent bonds in a variety of ways, creating structures with different shapes. Thus glucose can be converted into a different sugar—mannose or galactose—simply by switching the orientations of specific –OH groups relative to the rest of the molecule (**Panel 2–3**, pp. 70–71). Each of these sugars, moreover, can exist in either of two forms, called the D-form and the L-form, which are mirror images of each other. Sets of molecules with the same chemical formula but different structures are called *isomers*, and mirror-image pairs of such molecules are called *optical isomers*. Isomers are widespread among organic molecules in general, and they play a

Figure 2–17 The structure of glucose, a monosaccharide, can be represented in several ways. (A) A structural formula in which the atoms are shown as chemical symbols, linked together by solid lines representing the covalent bonds. The thickened lines are used to indicate the plane of the sugar ring and to show that the –H and –OH groups are not in the same plane as the ring. (B) Another kind of structural formula that shows the three-dimensional structure of glucose in the so-called "chair configuration." (C) A ball-and-stick model in which the three-dimensional arrangement of the atoms in space is indicated. (D) A space-filling model, which, as well as depicting the three-dimensional arrangement of the atoms, also gives some idea of their relative sizes and of the surface contours of the molecule (Movie 2.1). The atoms in (C) and (D) are colored as in Figure 2–9: C, *black*; H, *white*; O, *red*. This is the conventional color coding for these atoms and will be used throughout this book.

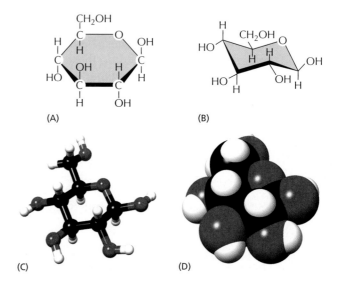

major part in generating the enormous variety of sugars. A more complete outline of sugar structures and chemistry is presented in Panel 2–3.

Monosaccharides can be linked by covalent bonds—called glycosidic bonds—to form larger carbohydrates. Two monosaccharides linked together make a disaccharide, such as sucrose, which is composed of a glucose and a fructose unit. Larger sugar polymers range from the *oligosaccharides* (trisaccharides, tetrasaccharides, and so on) up to giant *polysaccharides,* which can contain thousands of monosaccharide units. In most cases, the prefix *oligo-* is used to refer to molecules made of a small number of monomers, typically 2 to 10 in the case of oligosaccharides. Polymers, in contrast, can contain hundreds or thousands of subunits.

The way sugars are linked together illustrates some common features of biochemical bond formation. A bond is formed between an –OH group on one sugar and an –OH group on another by a **condensation reaction**, in which a molecule of water is expelled as the bond is formed. The subunits in other biological polymers, including nucleic acids and proteins, are also linked by condensation reactions in which water is expelled. The bonds created by all of these condensation reactions can be broken by the reverse process of **hydrolysis**, in which a molecule of water is consumed (**Figure 2–18**).

Because each monosaccharide has several free hydroxyl groups that can form a link to another monosaccharide (or to some other compound), sugar polymers can be branched, and the number of possible polysaccharide structures is extremely large. For this reason, it is much more difficult to determine the arrangement of sugars in a complex polysaccharide than to determine the nucleotide sequence of a DNA molecule or the amino acid sequence of a protein, in which each unit is joined to the next in exactly the same way.

The monosaccharide *glucose* has a central role as an energy source for cells. It is broken down to smaller molecules in a series of reactions, releasing energy that the cell can harness to do useful work, as we explain in Chapter 13. Cells use simple polysaccharides composed only of glucose units—principally *glycogen* in animals and *starch* in plants—as long-term stores of glucose, held in reserve for energy production.

Sugars do not function exclusively in the production and storage of energy. They are also used, for example, to make mechanical supports. The most abundant organic molecule on Earth—the *cellulose* that forms plant cell walls—is a polysaccharide of glucose. Another extraordinarily abundant organic substance, the *chitin* of insect exoskeletons and fungal cell walls, is also a polysaccharide—in this case, a linear polymer of a sugar derivative called *N*-acetylglucosamine (see Panel 2–3, pp. 70–71). Other polysaccharides, which tend to be slippery when wet, are the main components of slime, mucus, and gristle.

Smaller oligosaccharides can be covalently linked to proteins to form *glycoproteins,* or to lipids to form *glycolipids* (**Panel 2–4**, pp. 72–73), which are both found in cell membranes. The sugar side chains attached to glycoproteins and glycolipids in the plasma membrane are thought to help protect the cell surface and often help cells adhere to one another. Differences in the types of cell-surface sugars form the molecular basis for different human blood groups.

Fatty Acid Chains Are Components of Cell Membranes

A **fatty acid** molecule, such as *palmitic acid*, has two chemically distinct regions. One is a long hydrocarbon chain, which is hydrophobic and not very reactive chemically. The other is a carboxyl (–COOH) group,

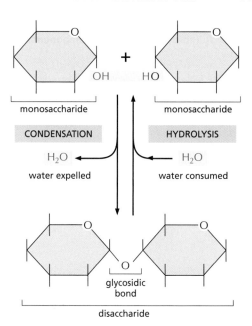

Figure 2–18 Two monosaccharides can be linked by a covalent glycosidic bond to form a disaccharide. This reaction belongs to a general category of reactions termed *condensation reactions*, in which two molecules join together as a result of the loss of a water molecule. The reverse reaction (in which water is added) is termed *hydrolysis.*

Figure 2–19 Fatty acids have both hydrophobic and hydrophilic components. The hydrophobic hydrocarbon chain is attached to a hydrophilic carboxylic acid group. Different fatty acids have different hydrocarbon tails. Palmitic acid is shown here. (A) Structural formula, showing the carboxylic acid head group in its ionized form, as it exists in water at pH 7. (B) Ball-and-stick model. (C) Space-filling model (Movie 2.2).

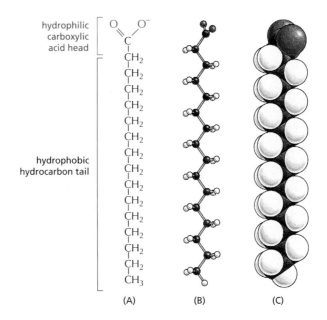

(A)   (B)   (C)

which behaves as an acid (carboxylic acid): in an aqueous solution, it is ionized (–COO⁻), extremely hydrophilic, and chemically reactive (**Figure 2–19**). Almost all the fatty acid molecules in a cell are covalently linked to other molecules by their carboxylic acid group (see Panel 2–4, pp. 72–73). Molecules—such as fatty acids—that possess both hydrophobic and hydrophilic regions are termed *amphipathic*.

The hydrocarbon tail of palmitic acid is *saturated*: it has no double bonds between its carbon atoms and contains the maximum possible number of hydrogens. Some other fatty acids, such as oleic acid, have *unsaturated* tails, with one or more double bonds along their length. The double bonds create kinks in the hydrocarbon tails, interfering with their ability to pack together, and it is the absence or presence of these double bonds that accounts for the difference between hard (saturated) and soft (polyunsaturated) margarine. Fatty acid tails are also found in cell membranes, where the tightness of their packing affects the fluidity of the membrane. The many different fatty acids found in cells differ only in the length of their hydrocarbon chains and in the number and position of the carbon–carbon double bonds (see Panel 2–4).

Fatty acids serve as a concentrated food reserve in cells: they can be broken down to produce about six times as much usable energy, weight for weight, as glucose. Fatty acids are stored in the cytoplasm of many cells in the form of fat droplets composed of *triacylglycerol* molecules—compounds made of three fatty acid chains covalently joined to a glycerol molecule (**Figure 2–20**, and see Panel 2–4). Triacylglycerols are the animal fats found in meat, butter, and cream, and the plant oils such as corn oil and olive oil. When a cell needs energy, the fatty acid chains can be released from triacylglycerols and broken down into two-carbon units. These two-carbon units are identical to those derived from the breakdown of glucose, and they enter the same energy-yielding reaction pathways, as described in Chapter 13.

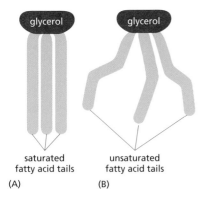

saturated fatty acid tails   unsaturated fatty acid tails
(A)   (B)

Figure 2–20 The properties of fats depend on the length and saturation of the fatty acid chains they carry. Fatty acids are stored in the cytoplasm of many cells in the form of droplets of *triacylglycerol* molecules made of three fatty acid chains joined to a glycerol molecule. (A) Saturated fats are found in meat and dairy products. (B) Plant oils, such as corn oil, contain unsaturated fatty acids, which may be monounsaturated (containing one double bond) or polyunsaturated (containing multiple double bonds); this is why plant oils are liquid at room temperature. Although fats are essential in the diet, saturated fats are not: they raise the concentration of cholesterol in the blood, which tends to clog the arteries, increasing the risk of heart attacks and strokes.

Fatty acids and their derivatives, including triacylglycerols, are examples of **lipids**. Lipids are loosely defined as molecules that are insoluble in water but soluble in fat and organic solvents such as benzene. They typically contain long hydrocarbon chains, as in the fatty acids, or multiple linked aromatic rings, as in the *steroids* (see Panel 2–4).

The most unique function of fatty acids is in the formation of the **lipid bilayer**, which is the basis for all cell membranes. These thin sheets,

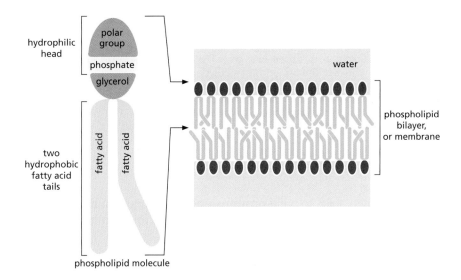

Figure 2–21 Phospholipids can aggregate to form cell membranes. Phospholipids are composed of two hydrophobic fatty acid tails joined to a hydrophilic head. In an aqueous environment, the hydrophobic tails pack together to exclude water, forming a lipid bilayer, with the hydrophilic heads of the phospholipid molecules on the outside, facing the aqueous environment, and the hydrophobic tails on the inside.

which enclose all cells and surround their internal organelles, are composed largely of *phospholipids* (**Figure 2–21**).

Like triacylglycerols, most phospholipids are constructed mainly from fatty acids and glycerol. In these phospholipids, however, the glycerol is joined to two fatty acid chains, rather than to three as in triacylglycerols. The remaining –OH group on the glycerol is linked to a hydrophilic phosphate group, which in turn is attached to a small hydrophilic compound such as choline (see Panel 2–4, pp. 72–73). With their two hydrophobic fatty acid tails and a hydrophilic, phosphate-containing head, phospholipids are strongly amphipathic. This characteristic amphipathic composition and shape gives them different physical and chemical properties from triacylglycerols, which are predominantly hydrophobic. In addition to phospholipids, cell membranes contain differing amounts of other lipids, including *glycolipids*, which contain one or more sugars instead of a phosphate group.

Thanks to their amphipathic nature, phospholipids readily form membranes in water. These lipids will spread over the surface of water to form a monolayer, with their hydrophobic tails facing the air and their hydrophilic heads in contact with the water. Two such molecular layers can readily combine tail-to-tail in water to form the phospholipid sandwich that is the lipid bilayer (see Chapter 11).

Amino Acids Are the Subunits of Proteins

Amino acids are small organic molecules with one defining property: they all possess a carboxylic acid group and an amino group, both linked to their α-carbon atom (**Figure 2–22**). Each amino acid also has a side chain attached to its α-carbon. The identity of this side chain is what distinguishes one amino acid from another.

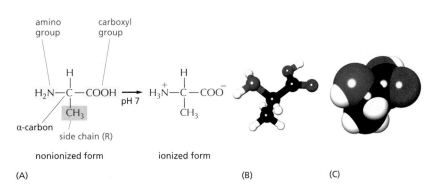

Figure 2–22 All amino acids have an amino group, a carboxyl group, and a side chain (R) attached to their α-carbon atom. In the cell, where the pH is close to 7, free amino acids exist in their ionized form; but, when they are incorporated into a polypeptide chain, the charges on their amino and carboxyl groups disappear. (A) The amino acid shown is alanine, one of the simplest amino acids, which has a methyl group (CH_3) as its side chain. (B) A ball-and-stick model and (C) a space-filling model of alanine. In (B) and (C), the N atom is *blue*.

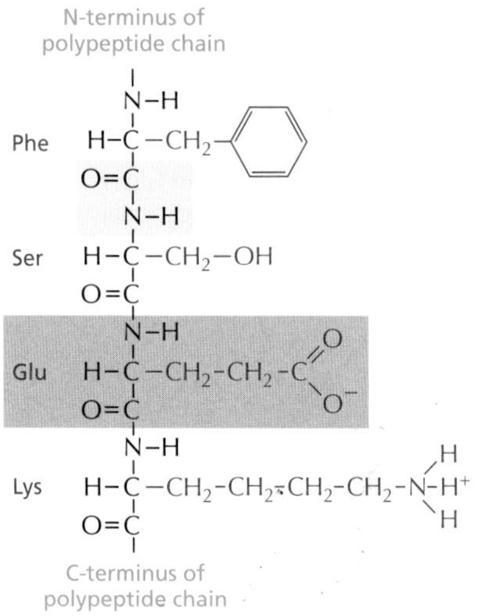

N-terminus of
polypeptide chain

C-terminus of
polypeptide chain

Figure 2–23 Amino acids in a protein are held together by peptide bonds. The four amino acids shown are linked together by three peptide bonds, one of which is highlighted in *yellow*. One of the amino acids, glutamic acid, is shaded in *gray*. The amino acid side chains are shown in *red*. The two ends of a polypeptide chain are chemically distinct. One end, the N-terminus, is capped by an amino group, and the other, the C-terminus, ends in a carboxyl group. The sequence of amino acids in a protein is abbreviated using either a three-letter or a one-letter code, and the sequence is always read from the N-terminus (see Panel 2–5, pp. 74–75). In the example given, the sequence is Phe-Ser-Glu-Lys (or FSEK).

Cells use amino acids to build **proteins**—polymers made of amino acids, which are joined head-to-tail in a long chain that folds up into a three-dimensional structure that is unique to each type of protein. The covalent bond between two adjacent amino acids in a protein chain is called a *peptide bond*; the chain of amino acids is also known as a *polypeptide*. Peptide bonds are formed by condensation reactions that link one amino acid to the next. Regardless of the specific amino acids from which it is made, the polypeptide always has an amino (NH_2) group at one end—its *N-terminus*—and a carboxyl (COOH) group at its other end—its *C-terminus* (**Figure 2–23**). This difference in the two ends gives a polypeptide a definite directionality—a structural (as opposed to electrical) polarity.

Twenty types of amino acids are commonly found in proteins, each with a different side chain attached to the α-carbon atom (**Panel 2–5**, pp. 74–75). The same 20 amino acids are found in all proteins, whether they hail from bacteria, plants, or animals. How this precise set of 20 amino acids came to be chosen is one of the mysteries surrounding the evolution of life; there is no obvious chemical reason why other amino acids could not have served just as well. But once the selection had been locked into place, it could not be changed, as too much chemistry had evolved to exploit it. Switching the types of amino acids used by cells would require a living creature to retool its entire metabolism to cope with the new building blocks.

Like sugars, all amino acids (except glycine) exist as optical isomers in D- and L-forms (see Panel 2–5). But only L-forms are ever found in proteins (although D-amino acids occur as part of bacterial cell walls and in some antibiotics, and D-serine is used as a signal molecule in the brain). The origin of this exclusive use of L-amino acids to make proteins is another evolutionary mystery.

The chemical versatility that the 20 standard amino acids provide is vitally important to the function of proteins. Five of the 20 amino acids—including lysine and glutamic acid, shown in Figure 2–23—have side chains that can form ions in solution and can therefore carry a charge. The others are uncharged. Some amino acids are polar and hydrophilic, and some are nonpolar and hydrophobic (see Panel 2–5). As we discuss in Chapter 4, the collective properties of the amino acid side chains underlie all the diverse and sophisticated functions of proteins.

Nucleotides Are the Subunits of DNA and RNA

DNA and RNA are built from subunits called **nucleotides**. *Nucleosides* are made of a nitrogen-containing ring compound linked to a five-carbon sugar, which can be either ribose or deoxyribose (**Panel 2–6**, pp. 76–77). Nucleotides are nucleosides that contain one or more phosphate groups attached to the sugar, and they come in two main forms: those containing ribose are known as *ribonucleotides*, and those containing deoxyribose are known as *deoxyribonucleotides*.

QUESTION 2–6

Why do you suppose only L-amino acids and not a random mixture of the L- and D-forms of each amino acid are used to make proteins?

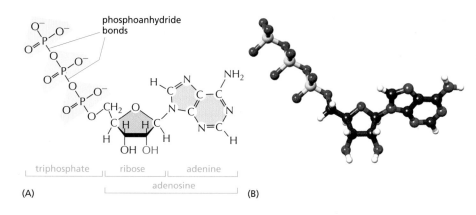

Figure 2–24 Adenosine triphosphate (ATP) is a crucially important energy carrier in cells. (A) Structural formula, in which the three phosphate groups are shaded in *yellow*. (B) Ball-and-stick model (Movie 2.3). In (B), the P atoms are *yellow*.

The nitrogen-containing rings of all these molecules are generally referred to as *bases* for historical reasons: under acidic conditions, they can each bind an H^+ (proton) and thereby increase the concentration of OH^- ions in aqueous solution. There is a strong family resemblance between the different nucleotide bases. *Cytosine* (C), *thymine* (T), and *uracil* (U) are called *pyrimidines*, because they all derive from a six-membered pyrimidine ring; *guanine* (G) and *adenine* (A) are *purines*, which bear a second, five-membered ring fused to the six-membered ring. Each nucleotide is named after the base it contains (see Panel 2–6, pp. 76–77).

Nucleotides can act as short-term carriers of chemical energy. Above all others, the ribonucleotide **adenosine triphosphate**, or **ATP** (**Figure 2–24**), participates in the transfer of energy in hundreds of metabolic reactions. ATP is formed through reactions that are driven by the energy released by the breakdown of foodstuffs. Its three phosphates are linked in series by two *phosphoanhydride bonds* (see Panel 2–6). Rupture of these phosphate bonds releases large amounts of useful energy. The terminal phosphate group in particular is frequently split off by hydrolysis (**Figure 2–25**). In many situations, transfer of this phosphate to other molecules releases energy that drives energy-requiring biosynthetic reactions. Other nucleotide derivatives serve as carriers for the transfer of other chemical groups. All of this is described in Chapter 3.

Nucleotides also have a fundamental role in the storage and retrieval of biological information. They serve as building blocks for the construction

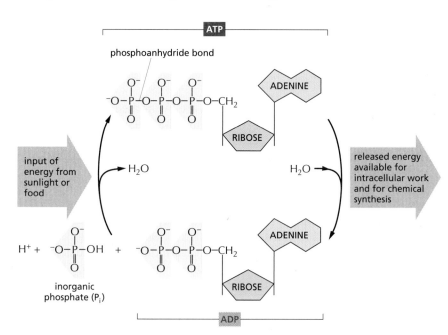

Figure 2–25 ATP is synthesized from ADP and inorganic phosphate, and it releases energy when it is hydrolyzed back to ADP and inorganic phosphate. The energy required for ATP synthesis is derived from either the energy-yielding oxidation of foodstuffs (in animal cells, fungi, and some bacteria) or the capture of light (in plant cells and some bacteria). The hydrolysis of ATP provides the energy to drive many processes inside cells. Together, the two reactions shown form the ATP cycle.

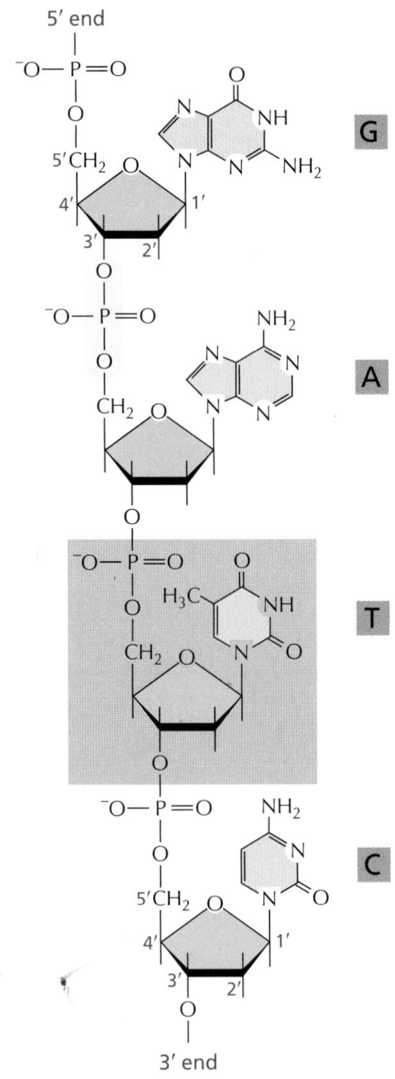

Figure 2–26 A short length of one chain of a deoxyribonucleic acid (DNA) molecule shows the covalent phosphodiester bonds linking four consecutive nucleotides. Because the bonds link specific carbon atoms in the sugar ring—known as the 5′ and 3′ atoms—one end of a polynucleotide chain, the 5′ end, has a free phosphate group and the other, the 3′ end, has a free hydroxyl group. One of the nucleotides, thymine (T), is shaded in *gray*, and one phosphodiester bond is highlighted in *yellow*. The linear sequence of nucleotides in a polynucleotide chain is commonly abbreviated by a one-letter code, and the sequence is always read from the 5′ end. In the example illustrated, the sequence is GATC.

of *nucleic acids*—long polymers in which nucleotide subunits are linked by the formation of covalent *phosphodiester bonds* between the phosphate group attached to the sugar of one nucleotide and a hydroxyl group on the sugar of the next nucleotide (**Figure 2–26**). Nucleic acid chains are synthesized from energy-rich nucleoside triphosphates by a condensation reaction that releases inorganic pyrophosphate during phosphodiester bond formation (see Panel 2–6, pp. 76–77).

There are two main types of nucleic acids, which differ in the type of sugar contained in their sugar–phosphate backbone. Those based on the sugar ribose are known as **ribonucleic acids**, or **RNA**, and contain the bases A, G, C, and U. Those based on deoxyribose (in which the hydroxyl at the 2′ position of the ribose carbon ring is replaced by a hydrogen) are known as **deoxyribonucleic acids**, or **DNA**, and contain the bases A, G, C, and T (T is chemically similar to the U in RNA; see Panel 2–6). RNA usually occurs in cells in the form of a single-stranded polynucleotide chain, but DNA is virtually always in the form of a double-stranded molecule: the DNA double helix is composed of two polynucleotide chains that run in opposite directions and are held together by hydrogen bonds between the bases of the two chains (**Panel 2–7**, pp. 78–79).

The linear sequence of nucleotides in a DNA or an RNA molecule encodes genetic information. The two nucleic acids, however, have different roles in the cell. DNA, with its more stable, hydrogen-bonded helices, acts as a long-term repository for hereditary information, while single-stranded RNA is usually a more transient carrier of molecular instructions. The ability of the bases in different nucleic acid molecules to recognize and pair with each other by hydrogen-bonding (called *base-pairing*)—G with C, and A with either T or U—underlies all of heredity and evolution, as explained in Chapter 5.

MACROMOLECULES IN CELLS

On the basis of weight, macromolecules are by far the most abundant of the organic molecules in a living cell (**Figure 2–27**). They are the principal building blocks from which a cell is constructed and also the components that confer the most distinctive properties on living things. Intermediate in size and complexity between small organic molecules and organelles, **macromolecules** are constructed simply by covalently linking small organic **monomers**, or *subunits*, into long chains, or **polymers** (**Figure 2–28** and **How We Know**, pp. 60–61). Yet they have many unexpected properties that could not have been predicted from their simple constituents. For example, it took a long time to determine that the nucleic acids DNA and RNA store and transmit hereditary information (see How We Know, pp. 174–176).

Proteins are especially versatile and perform thousands of distinct functions in cells. Many proteins act as enzymes that catalyze the chemical

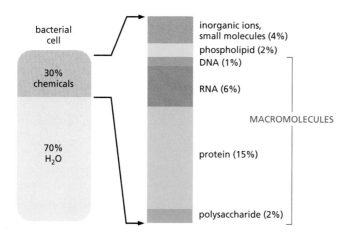

Figure 2–27 **Macromolecules are abundant in cells.** The approximate composition (by mass) of a bacterial cell is shown. The composition of an animal cell is similar.

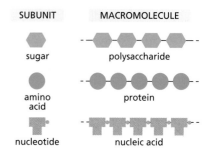

Figure 2–28 **Polysaccharides, proteins, and nucleic acids are made from monomeric subunits.** Each macromolecule is a polymer formed from small molecules (called monomers or subunits) that are linked together by covalent bonds.

reactions that take place in cells. For example, an enzyme in plants, called ribulose bisphosphate carboxylase, converts CO_2 to sugars, thereby creating most of the organic matter used by the rest of the living world. Other proteins are used to build structural components: tubulin, for example, self-assembles to make the cell's long, stiff microtubules (see Figure 1–27B), and histone proteins assemble into spool-like structures that help wrap up the cell's DNA in chromosomes. Yet other proteins, such as myosin, act as molecular motors to produce force and movement. We examine the molecular basis for many of these wide-ranging functions in later chapters. Here, we consider some of the general principles of macromolecular chemistry that make all of these activities possible.

Each Macromolecule Contains a Specific Sequence of Subunits

Although the chemical reactions for adding subunits to each polymer are different in detail for proteins, nucleic acids, and polysaccharides, they share important features. Each polymer grows by the addition of a monomer onto one end of the polymer chain via a condensation reaction, in which a molecule of water is lost with each subunit added (**Figure 2–29**). In all cases, the reactions are catalyzed by specific enzymes, which ensure that only the appropriate monomer is incorporated.

The stepwise polymerization of monomers into a long chain is a simple way to manufacture a large, complex molecule, because the subunits are added by the same reaction performed over and over again by the same set of enzymes. In a sense, the process resembles the repetitive operation of a machine in a factory—with some important differences. First, apart from some of the polysaccharides, most macromolecules are made from a set of monomers that are slightly different from one another; for example, proteins are constructed from 20 different amino acids (see Panel 2–5, pp. 74–75). Second, and most important, the polymer chain is not assembled at random from these subunits; instead the subunits are added in a particular order, or **sequence**.

The biological functions of proteins, nucleic acids, and many polysaccharides are absolutely dependent on the particular sequence of subunits in the linear chains. By varying the sequence of subunits, the cell can make an enormous diversity of the polymeric molecules. Thus, for a protein chain 200 amino acids long, there are 20^{200} possible combinations ($20 \times 20 \times 20 \times 20...$ multiplied 200 times), while for a DNA molecule

QUESTION 2–7

What is meant by "polarity" of a polypeptide chain and by "polarity" of a chemical bond? How do the meanings differ?

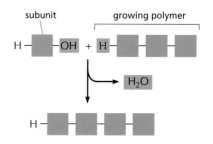

Figure 2–29 **Macromolecules are formed by adding subunits to one end.** In a condensation reaction, a molecule of water is lost with the addition of each monomer to one end of the growing chain. The reverse reaction—the breakdown of the polymer—occurs by the addition of water (hydrolysis). See also Figure 2–18.

HOW WE KNOW

WHAT ARE MACROMOLECULES?

The idea that proteins, polysaccharides, and nucleic acids are large molecules that are constructed from smaller subunits, linked one after another into long molecular chains, may seem fairly obvious today. But this was not always the case. In the early part of the twentieth century, few scientists believed in the existence of such biological polymers built from repeating units held together by covalent bonds. The notion that such "frighteningly large" macromolecules could be assembled from simple building blocks was considered "downright shocking" by chemists of the day. Instead, they thought that proteins and other seemingly large organic molecules were simply heterogeneous aggregates of small organic molecules held together by weak "association forces" (**Figure 2–30**).

The first hint that proteins and other organic polymers are large molecules came from observing their behavior in solution. At the time, scientists were working with various proteins and carbohydrates derived from foodstuffs and other organic materials—albumin from egg whites, casein from milk, collagen from gelatin, and cellulose from wood. Their chemical compositions seemed simple enough: like other organic molecules, they contained carbon, hydrogen, oxygen, and, in the case of proteins, nitrogen. But they behaved oddly in solution, showing, for example, an inability to pass through a fine filter.

Why these molecules misbehaved in solution was a puzzle. Were they really giant molecules, composed of an unusual number of covalently linked atoms? Or were they more like a colloidal suspension of particles—a big, sticky hodgepodge of small organic molecules that associate only loosely?

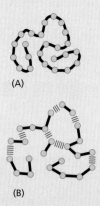

(A)

(B)

Figure 2–30 What might an organic macromolecule look like? Chemists in the early part of the twentieth century debated whether proteins, polysaccharides, and other apparently large organic molecules were (A) discrete particles made of an unusually large number of covalently linked atoms or (B) a loose aggregation of heterogeneous small organic molecules held together by weak forces.

One way to distinguish between the two possibilities was to determine the actual size of one of these molecules. If a protein such as albumin were made of molecules all identical in size, that would support the existence of true macromolecules. Conversely, if albumin were instead a miscellaneous conglomeration of small organic molecules, these should show a whole range of molecular sizes in solution.

Unfortunately, the techniques available to scientists in the early 1900s were not ideal for measuring the sizes of such large molecules. Some chemists estimated a protein's size by determining how much it would lower a solution's freezing point; others measured the osmotic pressure of protein solutions. These methods were susceptible to experimental error and gave variable results. Different techniques, for example, suggested that cellulose was anywhere from 6000 to 103,000 daltons in mass (where 1 dalton is approximately equal to the mass of a hydrogen atom). Such results helped to fuel the hypothesis that carbohydrates and proteins were loose aggregates of small molecules rather than true macromolecules.

Many scientists simply had trouble believing that molecules heavier than about 4000 daltons—the largest compound that had been synthesized by organic chemists—could exist at all. Take hemoglobin, the oxygen-carrying protein in red blood cells. Researchers tried to estimate its size by breaking it down into its chemical components. In addition to carbon, hydrogen, nitrogen, and oxygen, hemoglobin contains a small amount of iron. Working out the percentages, it appeared that hemoglobin had one atom of iron for every 712 atoms of carbon—and a minimum weight of 16,700 daltons. Could a molecule with hundreds of carbon atoms in one long chain remain intact in a cell and perform specific functions? Emil Fischer, the organic chemist who determined that the amino acids in proteins are linked by peptide bonds, thought that a polypeptide chain could grow no longer than about 30 or 40 amino acids. As for hemoglobin, with its purported 700 carbon atoms, the existence of molecular chains of such "truly fantastic lengths" was deemed "very improbable" by leading chemists.

Definitive resolution of the debate had to await the development of new techniques. Convincing evidence that proteins are macromolecules came from studies using the ultracentrifuge—a device that uses centrifugal force to separate molecules according to their size (see Panel 4–3, pp. 164–165). Theodor Svedberg, who designed the machine in 1925, performed the first studies. If a protein were really an aggregate of smaller molecules, he reasoned, it would appear as a smear of molecules of different sizes when sedimented in an ultracentrifuge. Using hemoglobin as his test protein, Svedberg found that the centrifuged sample revealed a single, sharp band with a molecular weight of 68,000 daltons. His results strongly supported the theory that proteins are true macromolecules (**Figure 2–31**).

Additional evidence continued to accumulate throughout the 1930s, as other researchers began to prepare crystals of pure protein that could be studied by X-ray diffraction. Only molecules with a uniform size and shape can form highly ordered crystals and diffract X-rays in such a way that their three-dimensional structure can be determined, as we discuss in Chapter 4. A heterogeneous suspension could not be studied in this way.

We now take it for granted that large macromolecules carry out many of the most important activities in living cells. But chemists once viewed the existence of such polymers with the same sort of skepticism that a zoologist might show on being told that "In Africa, there are elephants that are 100 meters long and 20 meters tall." It took decades for researchers to master the techniques required to convince everyone that molecules ten times larger than anything they had ever encountered were a cornerstone of biology. As we shall see throughout this book, such a labored pathway to discovery is not unusual, and progress in science is often driven by advances in technology.

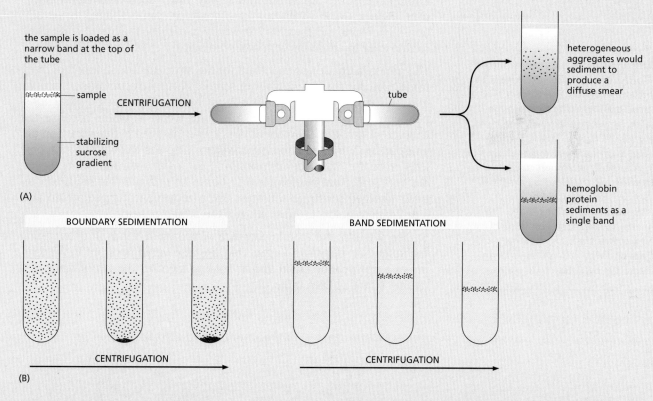

Figure 2–31 **The ultracentrifuge helped to settle the debate about the nature of macromolecules.** In the ultracentrifuge, centrifugal forces exceeding 500,000 times the force of gravity can be used to separate proteins or other large molecules. (A) In a modern ultracentrifuge, samples are loaded in a thin layer on top of a gradient of sucrose solution formed in a tube. The tube is placed in a metal rotor that is rotated at high speed. Molecules of different sizes sediment at different rates, and these molecules will therefore move as distinct bands in the sample tube. If hemoglobin were a loose aggregate of heterogeneous peptides, it would show a broad smear of sizes after centrifugation (*top* tube). Instead, it appears as a sharp band with a molecular weight of 68,000 daltons (*bottom* tube). Although the ultracentrifuge is now a standard, almost mundane, fixture in most biochemistry laboratories, its construction was a huge technological challenge. The centrifuge rotor must be capable of spinning centrifuge tubes at high speeds for many hours at constant temperature and with high stability; otherwise convection occurs in the sedimenting solution and ruins the experiment. In 1926, Svedberg won the Nobel Prize in Chemistry for his ultracentrifuge design and its application to chemistry. (B) In his actual experiment, Svedberg filled a special tube in the centrifuge with a homogeneous solution of hemoglobin; by shining light through the tube, he then carefully monitored the moving boundary between the sedimenting protein molecules and the clear aqueous solution left behind (so-called boundary sedimentation). The more recently developed method shown in (A) is a form of band sedimentation.

10,000 nucleotides long (small by DNA standards), with its four different nucleotides, there are $4^{10,000}$ different possibilities—an unimaginably large number. Thus the machinery of polymerization must be subject to a sensitive control that allows it to specify exactly which subunit should be added next to the growing polymer end. We discuss the mechanisms that specify the sequence of subunits in DNA, RNA, and protein molecules in Chapters 6 and 7.

Noncovalent Bonds Specify the Precise Shape of a Macromolecule

Most of the single covalent bonds that link together the subunits in a macromolecule allow rotation of the atoms they join; thus the polymer chain has great flexibility. In principle, this allows a single-chain macromolecule to adopt an almost unlimited number of shapes, or **conformations**, as the polymer chain writhes and rotates under the influence of random thermal energy. However, the shapes of most biological macromolecules are highly constrained because of weaker, noncovalent bonds that form between different parts of the molecule. In many cases, these weaker interactions ensure that the polymer chain preferentially adopts one particular conformation, determined by the linear sequence of monomers in the chain. Most protein molecules and many of the RNA molecules found in cells fold tightly into one highly preferred conformation in this way (**Figure 2–32**). These unique conformations—shaped by evolution—determine the chemistry and activity of these macromolecules and dictate their interactions with other biological molecules.

The noncovalent bonds important for the structure and function of macromolecules include two types described earlier: *electrostatic attractions* and *hydrogen bonds* (see Panel 2–7, pp. 78–79). Electrostatic attractions, although strong on their own, are quite weak in water because the charged or partially charged (polar) groups involved in the attraction are shielded by their interactions with water molecules and various inorganic ions present in the aqueous solution. Electrostatic attractions, however, are very important in biological systems. An enzyme that binds a positively charged substrate will often use a negatively charged amino acid side chain to guide its substrate into the proper position.

Earlier, we described the importance of hydrogen bonds in determining the unique properties of water. They are also very important in the folding of a polypeptide chain and in holding together the two strands of a double-stranded DNA molecule.

QUESTION 2–8

In principle, there are many different, chemically diverse ways in which small molecules can be linked to form polymers. For example, the small molecule ethene ($CH_2=CH_2$) is used commercially to make the plastic polyethylene (...–CH_2–CH_2–CH_2–CH_2–CH_2–...). The individual subunits of the three major classes of biological macromolecules, however, are all linked by similar reaction mechanisms, i.e., by condensation reactions that eliminate water. Can you think of any benefits that this chemistry offers and why it might have been selected in evolution?

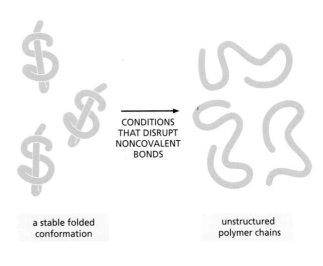

Figure 2–32 **Most proteins and many RNA molecules fold into a particularly stable three-dimensional shape, or conformation.** This shape is directed mostly by a multitude of weak, noncovalent intramolecular bonds. If the folded macromolecules are subjected to conditions that disrupt noncovalent bonds, the molecule becomes a flexible chain that loses both its conformation and its biological activity.

CONDITIONS THAT DISRUPT NONCOVALENT BONDS

a stable folded conformation

unstructured polymer chains

A third type of noncovalent interaction results from **van der Waals attractions**, which are a form of electrical attraction caused by fluctuating electric charges that arise whenever two atoms come within a very short distance of each other. Although van der Waals attractions are weaker than hydrogen bonds, in large numbers they play an important role in the attraction between macromolecules with complementary shapes. All of these noncovalent bonds are reviewed in Panel 2–7, pp. 78–79.

Another important noncovalent interaction is created by the three-dimensional structure of water, which forces together the hydrophobic portions of dissolved molecules in order to minimize their disruptive effect on the hydrogen-bonded network of water molecules (see Panel 2–7 and Panel 2–2, pp. 68–69). This expulsion from the aqueous solution generates what is sometimes thought of as a fourth kind of noncovalent bond, called a **hydrophobic interaction**. Such interactions hold together phospholipid molecules in cell membranes, for example, and they also play a crucial part in the folding of protein molecules into a compact globular shape.

Noncovalent Bonds Allow a Macromolecule to Bind Other Selected Molecules

As we discussed earlier, although noncovalent bonds are individually weak, they can add up to create a strong attraction between two molecules when these molecules fit together very closely, like a hand in a glove, so that many noncovalent bonds can occur between them (see Panel 2–7). This form of molecular interaction provides for great specificity in the binding of a macromolecule to other small and large molecules, because the multipoint contacts required for strong binding make it possible for a macromolecule to select just one of the many thousands of different molecules present inside a cell. Moreover, because the strength of the binding depends on the number of noncovalent bonds that are formed, associations of almost any strength are possible.

Binding of this type makes it possible for proteins to function as enzymes. It can also stabilize associations between any macromolecules, as long as their surfaces match closely (**Figure 2–33** and Movie 2.4). Noncovalent bonds thereby allow macromolecules to be used as building blocks for the formation of much larger structures. For example, proteins often bind

QUESTION 2–9

Why could covalent bonds not be used in place of noncovalent bonds to mediate most of the interactions of macromolecules?

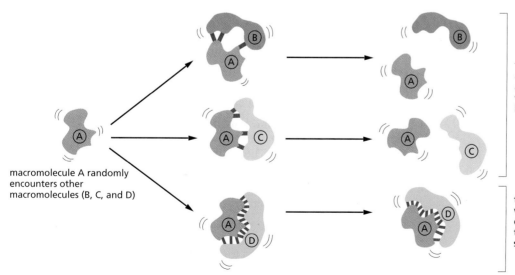

the surfaces of A and B, and A and C, are a poor match and are capable of forming only a few weak bonds; thermal motion rapidly breaks them apart

the surfaces of A and D match well and therefore can form enough weak bonds to withstand thermal jolting; they therefore stay bound to each other

macromolecule A randomly encounters other macromolecules (B, C, and D)

Figure 2–33 Noncovalent bonds mediate interactions between macromolecules. They can also mediate interactions between a macromolecule and small molecules (not shown).

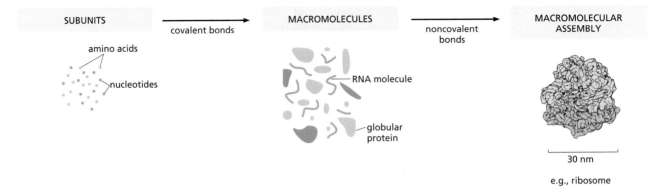

Figure 2–34 Both covalent bonds and noncovalent bonds are needed to form a macromolecular assembly such as a ribosome. Covalent bonds allow small organic molecules to join together to form macromolecules, which can assemble into large macromolecular complexes via noncovalent bonds. Ribosomes are large macromolecular machines that synthesize proteins inside cells. Each ribosome is composed of about 90 macromolecules (proteins and RNA molecules), and it is large enough to see in the electron microscope (see Figure 7–31). The subunits, macromolecules, and the ribosome here are shown roughly to scale.

together into multiprotein complexes that function as intricate machines with multiple moving parts, carrying out such complex tasks as DNA replication and protein synthesis (**Figure 2–34**). In fact, noncovalent bonds account for a great deal of the complex chemistry that makes life possible.

ESSENTIAL CONCEPTS

- Living cells obey the same chemical and physical laws as nonliving things. Like all other forms of matter, they are made of atoms, which are the smallest unit of a chemical element that retain the distinctive chemical properties of that element.

- Cells are made up of a limited number of elements, four of which—C, H, N, O—make up about 96% of a cell's mass.

- Each atom has a positively charged nucleus, which is surrounded by a cloud of negatively charged electrons. The chemical properties of an atom are determined by the number and arrangement of its electrons: it is most stable when its outer electron shell is completely filled.

- A covalent bond forms when a pair of outer-shell electrons is shared between two adjacent atoms; if two pairs of electrons are shared, a double bond is formed. Clusters of two or more atoms held together by covalent bonds are known as molecules.

- When an electron jumps from one atom to another, two ions of opposite charge are generated; these ions are held together by mutual attraction forming a noncovalent ionic bond.

- Living organisms contain a distinctive and restricted set of small carbon-based (organic) molecules, which are essentially the same for every living species. The main categories are sugars, fatty acids, amino acids, and nucleotides.

- Sugars are a primary source of chemical energy for cells and can also be joined together to form polysaccharides or shorter oligosaccharides.

- Fatty acids are an even richer energy source than sugars, but their most essential function is to form lipid molecules that assemble into cell membranes.

- The vast majority of the dry mass of a cell consists of macromolecules—mainly polysaccharides, proteins, and nucleic acids (DNA and RNA); these macromolecules are formed as polymers of sugars, amino acids, or nucleotides, respectively.

- The most diverse and versatile class of macromolecules are proteins, which are formed from 20 types of amino acids that are covalently linked by peptide bonds into long polypeptide chains.

- Nucleotides play a central part in energy-transfer reactions within cells; they are also joined together to form information-containing RNA and DNA molecules, each of which is composed of only four types of nucleotides.

- Protein, RNA, and DNA molecules are synthesized from subunits by repetitive condensation reactions, and it is the specific sequence of subunits that determines their unique functions.

- Four types of weak noncovalent bonds—hydrogen bonds, electrostatic attractions, van der Waals attractions, and hydrophobic interactions—enable macromolecules to bind specifically to other macromolecules or to selected small molecules.

- The same four types of noncovalent bonds between different regions of a polypeptide or RNA chain allow these chains to fold into unique shapes (conformations).

KEY TERMS

acid	inorganic molecule
amino acid	ion
atom	ionic bond
atomic weight	lipid
ATP	lipid bilayer
Avogadro's number	macromolecule
base	molecule
buffer	molecular weight
chemical bond	monomer
chemical group	noncovalent bond
condensation reaction	nucleotide
conformation	organic molecule
covalent bond	pH scale
DNA	polar
electron	polymer
electrostatic attraction	protein
fatty acid	proton
hydrogen bond	RNA
hydrolysis	sequence
hydronium ion	subunit
hydrophilic	sugar
hydrophobic	van der Waals attractions
hydrophobic interactions	

CARBON SKELETONS

Carbon has a unique role in the cell because of its ability to form strong covalent bonds with other carbon atoms. Thus carbon atoms can join to form:

chains

branched trees

rings

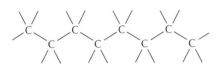

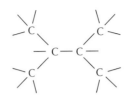

also written as

also written as

also written as

COVALENT BONDS

A covalent bond forms when two atoms come very close together and share one or more of their outer-shell electrons. Each atom forms a fixed number of covalent bonds in a defined spatial arrangement.

SINGLE BONDS: two electrons shared per bond

DOUBLE BONDS: four electrons shared per bond

The precise spatial arrangement of covalent bonds influences the three-dimensional structure and chemistry of molecules. In this review panel, we see how covalent bonds are used in a variety of biological molecules.

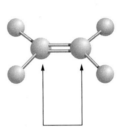

Atoms joined by two or more covalent bonds cannot rotate freely around the bond axis. This restriction has a major influence on the three-dimensional shape of many macromolecules.

C–H COMPOUNDS

Carbon and hydrogen together make stable compounds (or groups) called hydrocarbons. These are nonpolar, do not form hydrogen bonds, and are generally insoluble in water.

$$H-\underset{\underset{H}{|}}{\overset{\overset{H}{|}}{C}}-H$$

$$H-\underset{\underset{H}{|}}{\overset{\overset{H}{|}}{C}}-H$$

methane

methyl group

part of the hydrocarbon "tail" of a fatty acid molecule

ALTERNATING DOUBLE BONDS

A carbon chain can include double bonds. If these are on alternate carbon atoms, the bonding electrons move within the molecule, stabilizing the structure by a phenomenon called *resonance*.

the truth is somewhere between these two structures

Alternating double bonds in a ring can generate a very stable structure.

benzene

often written as

C–O COMPOUNDS

Many biological compounds contain a carbon covalently bonded to an oxygen. For example,

alcohol

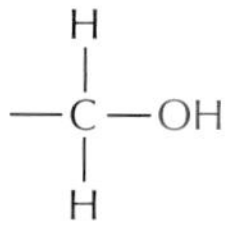

The –OH is called a hydroxyl group.

aldehyde

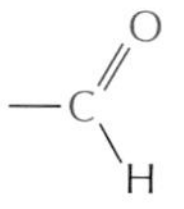

ketone

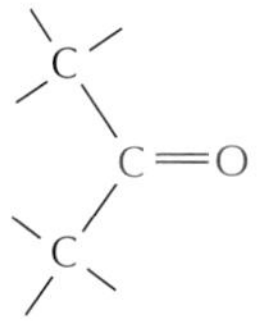

The C=O is called a carbonyl group.

carboxylic acid

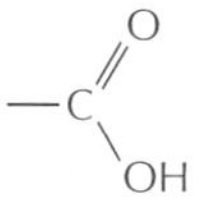

The –COOH is called a carboxyl group. In water, this loses an H$^+$ ion to become –COO$^-$.

esters

Esters are formed by combining an acid and an alcohol.

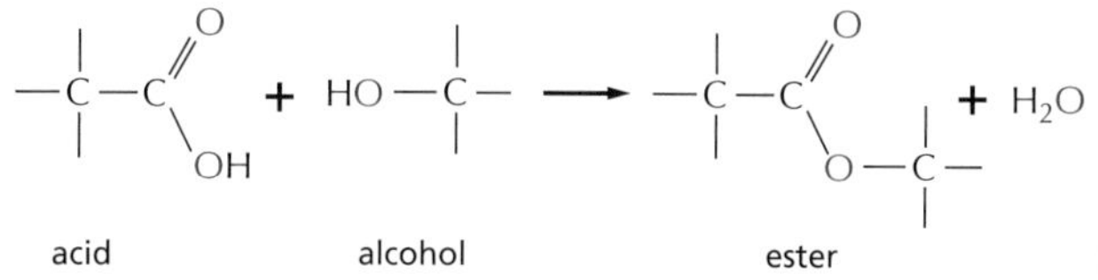

acid   alcohol   ester

C–N COMPOUNDS

Amines and amides are two important examples of compounds containing a carbon linked to a nitrogen.

Amines in water combine with an H$^+$ ion to become positively charged.

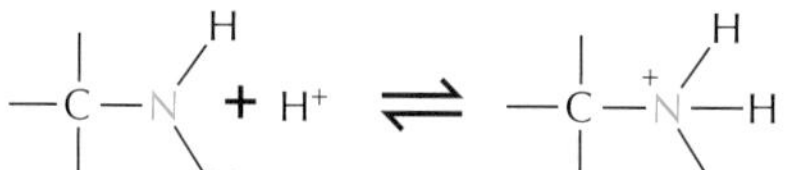

Amides are formed by combining an acid and an amine. Unlike amines, amides are uncharged in water. An example is the peptide bond that joins amino acids in a protein.

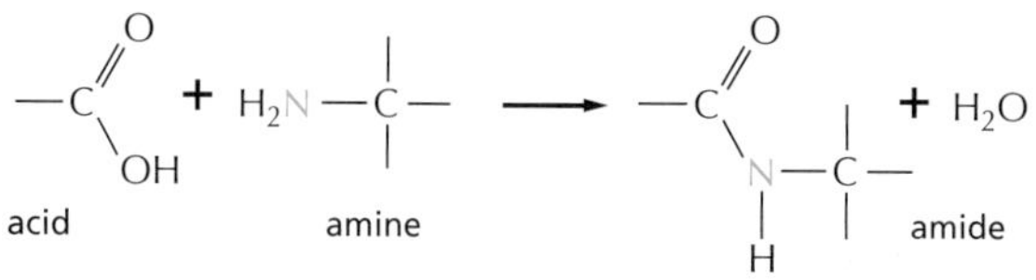

acid   amine   amide

Nitrogen also occurs in several ring compounds, including important constituents of nucleic acids: purines and pyrimidines.

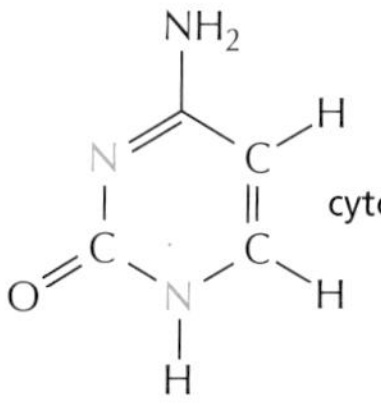

cytosine (a pyrimidine)

PHOSPHATES

Inorganic phosphate is a stable ion formed from phosphoric acid, H_3PO_4. It is also written as ⓟ$_i$.

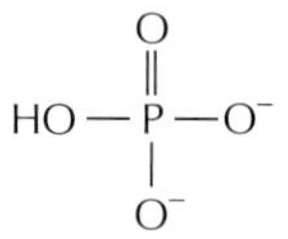

Phosphate esters can form between a phosphate and a free hydroxyl group. Phosphate groups are often covalently attached to proteins in this way.

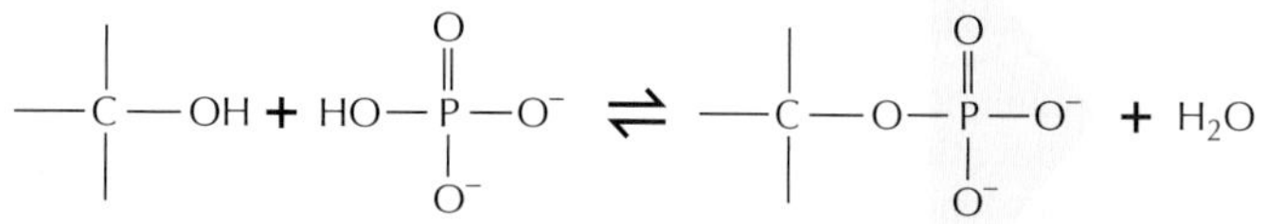

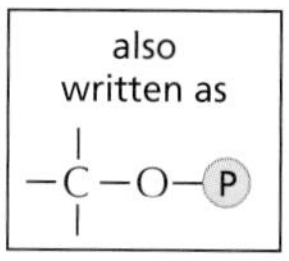

also written as

The combination of a phosphate and a carboxyl group, or two or more phosphate groups, gives an acid anhydride. Because compounds of this type release a large amount of energy when hydrolyzed in the cell, they are often said to contain a "high-energy" bond.

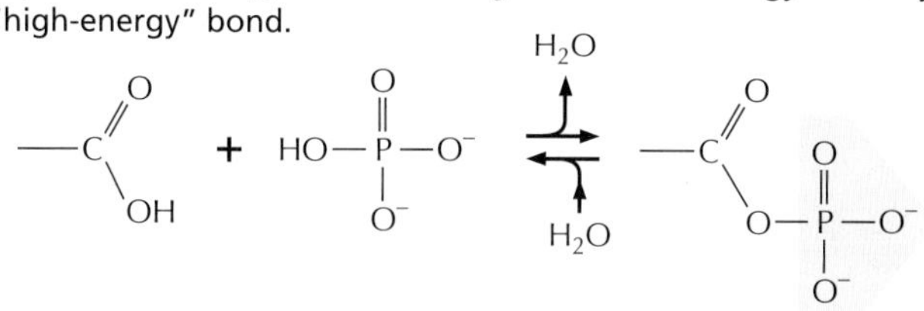

high-energy acyl phosphate bond (carboxylic–phosphoric acid anhydride) found in some metabolites

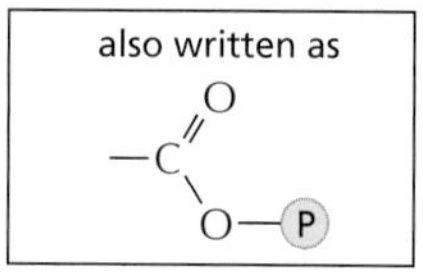

also written as

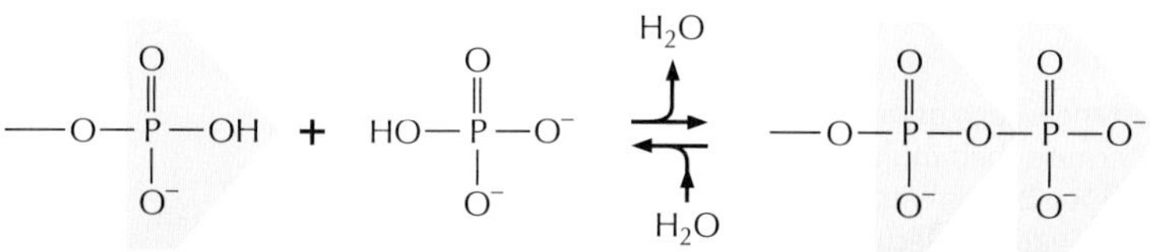

high-energy phosphoanhydride bond found in molecules such as ATP

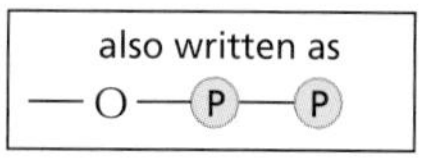

also written as

HYDROGEN BONDS

Because they are polarized, two adjacent H_2O molecules can form a noncovalent linkage known as a hydrogen bond. Hydrogen bonds have only about 1/20 the strength of a covalent bond.

Hydrogen bonds are strongest when the three atoms lie in a straight line.

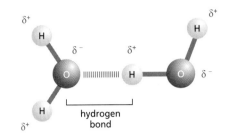

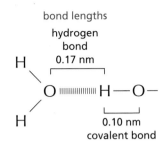

bond lengths

hydrogen bond
0.17 nm

0.10 nm
covalent bond

WATER

Two atoms connected by a covalent bond may exert different attractions for the electrons of the bond. In such cases, the bond is polar, with one end slightly negatively charged (δ^-) and the other slightly positively charged (δ^+).

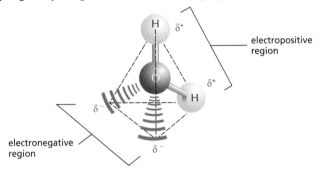

electropositive region

electronegative region

Although a water molecule has an overall neutral charge (having the same number of electrons and protons), the electrons are asymmetrically distributed, making the molecule polar. The oxygen nucleus draws electrons away from the hydrogen nuclei, leaving the hydrogen nuclei with a small net positive charge. The excess of electron density on the oxygen atom creates weakly negative regions at the other two corners of an imaginary tetrahedron. On these pages, we review the chemical properties of water and see how water influences the behavior of biological molecules.

WATER STRUCTURE

Molecules of water join together transiently in a hydrogen-bonded lattice.

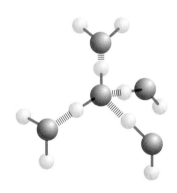

The cohesive nature of water is responsible for many of its unusual properties, such as high surface tension, high specific heat, and high heat of vaporization.

HYDROPHILIC MOLECULES

Substances that dissolve readily in water are termed hydrophilic. They include ions and polar molecules that attract water molecules through electrical charge effects. Water molecules surround each ion or polar molecule and carry it into solution.

Ionic substances such as sodium chloride dissolve because water molecules are attracted to the positive (Na^+) or negative (Cl^-) charge of each ion.

Polar substances such as urea dissolve because their molecules form hydrogen bonds with the surrounding water molecules.

HYDROPHOBIC MOLECULES

Substances that contain a preponderance of nonpolar bonds are usually insoluble in water and are termed hydrophobic. Water molecules are not attracted to such hydrophobic molecules and so have little tendency to surround them and bring them into solution.

Hydrocarbons, which contain many C–H bonds, are especially hydrophobic.

WATER AS A SOLVENT

Many substances, such as household sugar (sucrose), dissolve in water. That is, their molecules separate from each other, each becoming surrounded by water molecules.

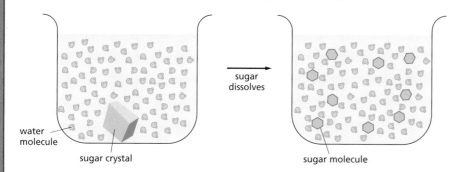

sugar dissolves

water molecule

sugar crystal

sugar molecule

When a substance dissolves in a liquid, the mixture is termed a solution. The dissolved substance (in this case sugar) is the solute, and the liquid that does the dissolving (in this case water) is the solvent. Water is an excellent solvent for hydrophilic substances because of its polar bonds.

ACIDS

Substances that release hydrogen ions (protons) into solution are called acids.

$$HCl \longrightarrow H^+ + Cl^-$$

hydrochloric acid (strong acid)   hydrogen ion   chloride ion

Many of the acids important in the cell are not completely dissociated, and they are therefore weak acids—for example, the carboxyl group (–COOH), which dissociates to give a hydrogen ion in solution.

$$-C{\overset{O}{\underset{OH}{}}} \rightleftharpoons H^+ + -C{\overset{O}{\underset{O^-}{}}}$$

(weak acid)

Note that this is a reversible reaction.

HYDROGEN ION EXCHANGE

Positively charged hydrogen ions (H$^+$) can spontaneously move from one water molecule to another, thereby creating two ionic species.

hydronium ion   hydroxyl ion

often written as: $H_2O \rightleftharpoons H^+ + OH^-$

hydrogen ion   hydroxyl ion

Because the process is rapidly reversible, hydrogen ions are continually shuttling between water molecules. Pure water contains equal concentrations of hydronium ions and hydroxyl ions (both 10^{-7} M).

pH

The acidity of a solution is defined by the concentration of hydronium ions (H$_3$O$^+$) it possesses, generally abbreviated as H$^+$. For convenience, we use the pH scale, where

$$pH = -\log_{10}[H^+]$$

For pure water

$$[H^+] = 10^{-7} \text{ moles/liter}$$

$$pH = 7.0$$

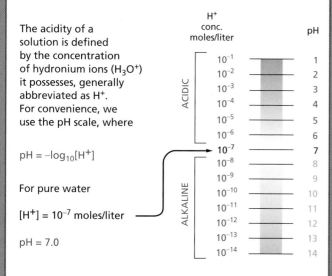

H$^+$ conc. moles/liter	pH
10^{-1}	1
10^{-2}	2
10^{-3}	3
10^{-4}	4
10^{-5}	5
10^{-6}	6
10^{-7}	7
10^{-8}	8
10^{-9}	9
10^{-10}	10
10^{-11}	11
10^{-12}	12
10^{-13}	13
10^{-14}	14

ACIDIC

ALKALINE

BASES

Substances that reduce the number of hydrogen ions in solution are called bases. Some bases, such as ammonia, combine directly with hydrogen ions.

$$NH_3 + H^+ \longrightarrow NH_4^+$$

ammonia   hydrogen ion   ammonium ion

Other bases, such as sodium hydroxide, reduce the number of H$^+$ ions indirectly, by making OH$^-$ ions that then combine directly with H$^+$ ions to make H$_2$O.

$$NaOH \longrightarrow Na^+ + OH^-$$

sodium hydroxide (strong base)   sodium ion   hydroxyl ion

Many bases found in cells are partially associated with H$^+$ ions and are termed weak bases. This is true of compounds that contain an amino group (–NH$_2$), which has a weak tendency to reversibly accept an H$^+$ ion from water, thereby increasing the concentration of free OH$^-$ ions.

$$-NH_2 + H^+ \rightleftharpoons -NH_3^+$$

MONOSACCHARIDES

Monosaccharides usually have the general formula $(CH_2O)_n$, where n can be 3, 4, 5, or 6, and have two or more hydroxyl groups. They either contain an aldehyde group ($-C{\scriptstyle\langle}^O_H$) and are called aldoses, or a ketone group ($\rangle C=O$) and are called ketoses.

3-carbon (TRIOSES)   5-carbon (PENTOSES)   6-carbon (HEXOSES)

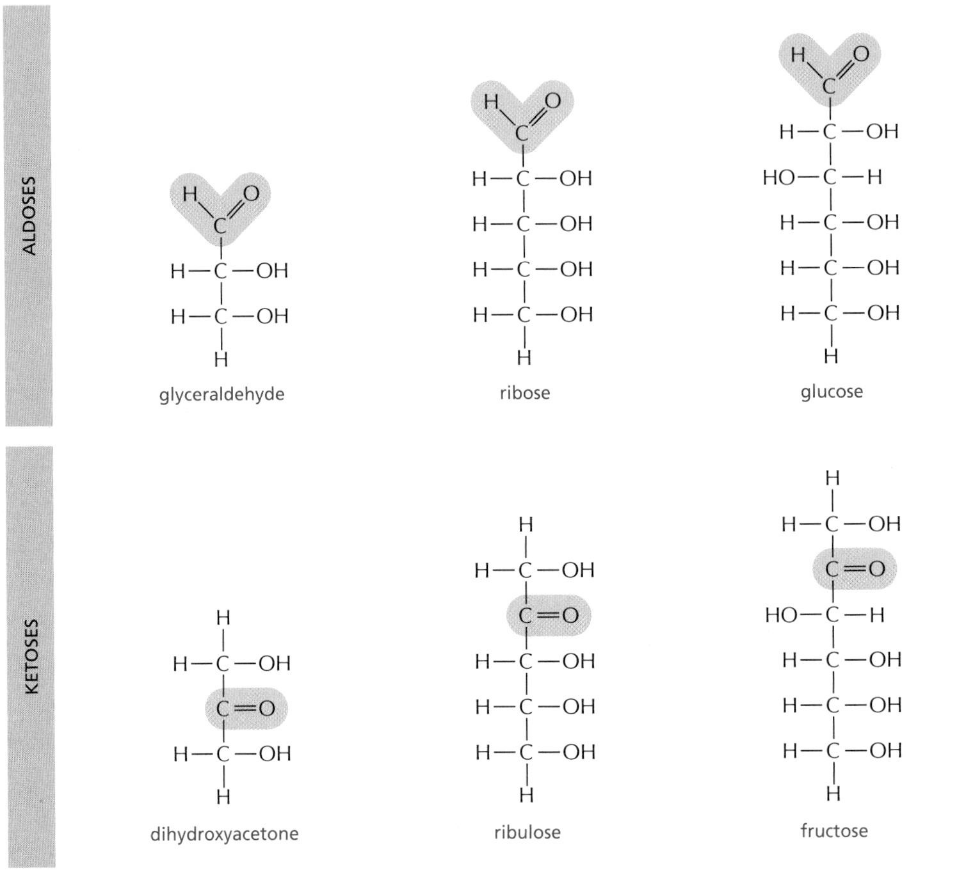

ALDOSES

glyceraldehyde   ribose   glucose

KETOSES

dihydroxyacetone   ribulose   fructose

RING FORMATION

In aqueous solution, the aldehyde or ketone group of a sugar molecule tends to react with a hydroxyl group of the same molecule, thereby closing the molecule into a ring.

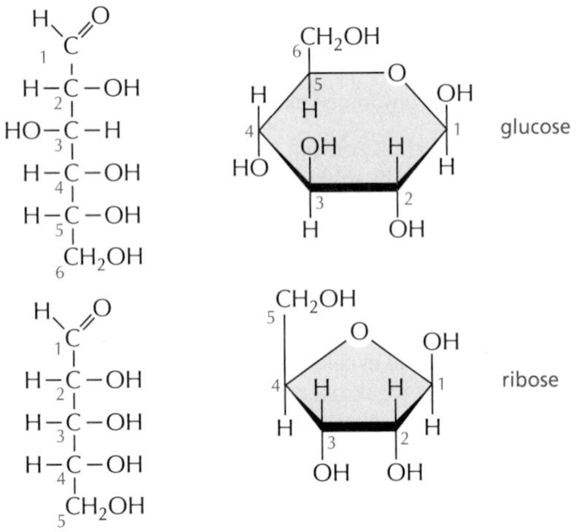

glucose

ribose

Note that each carbon atom has a number.

ISOMERS

Many monosaccharides differ only in the spatial arrangement of atoms—that is, they are isomers. For example, glucose, galactose, and mannose have the same formula ($C_6H_{12}O_6$) but differ in the arrangement of groups around one or two carbon atoms.

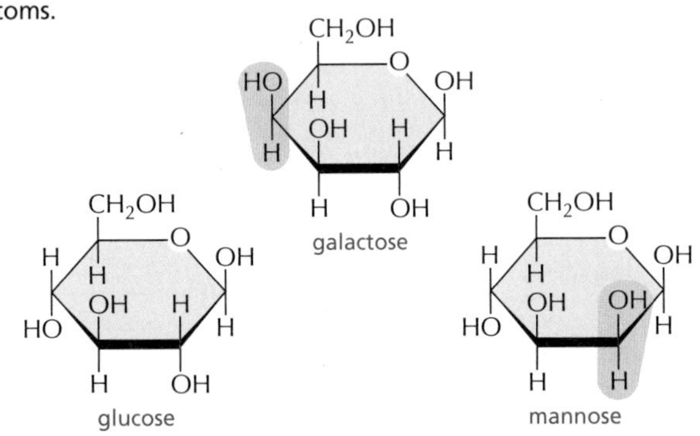

galactose

glucose

mannose

These small differences make only minor changes in the chemical properties of the sugars. But the differences are recognized by enzymes and other proteins and therefore can have major biological effects.

α AND β LINKS

The hydroxyl group on the carbon that carries the aldehyde or ketone can rapidly change from one position to the other. These two positions are called α and β.

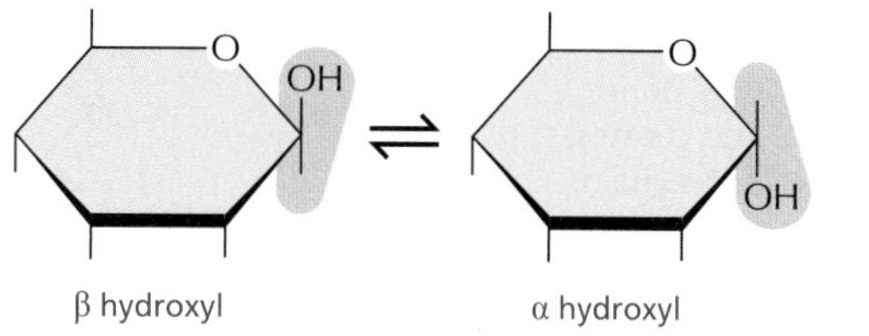

β hydroxyl   α hydroxyl

As soon as one sugar is linked to another, the α or β form is frozen.

SUGAR DERIVATIVES

The hydroxyl groups of a simple monosaccharide, such as glucose, can be replaced by other groups.

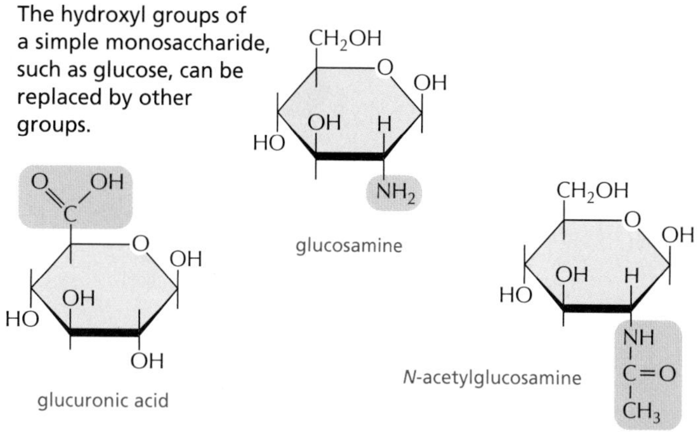

CH_2OH

NH_2

glucosamine

glucuronic acid

N-acetylglucosamine

DISACCHARIDES

The carbon that carries the aldehyde or the ketone can react with any hydroxyl group on a second sugar molecule to form a disaccharide. Three common disaccharides are

maltose (glucose + glucose)
lactose (galactose + glucose)
sucrose (glucose + fructose)

The reaction forming sucrose is shown here.

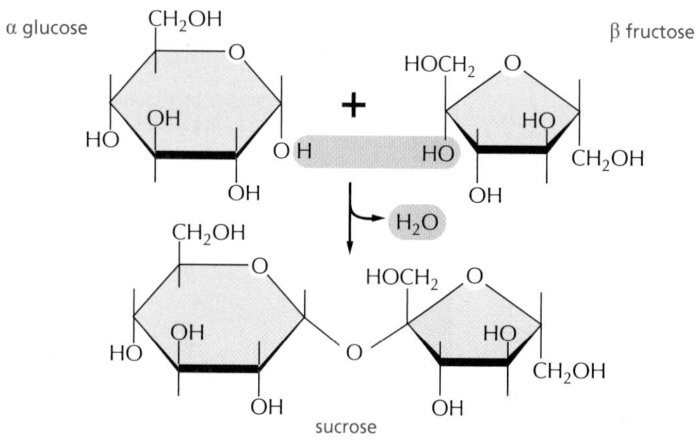

α glucose   β fructose

CH_2OH $HOCH_2$

+

H_2O

CH_2OH $HOCH_2$

sucrose

OLIGOSACCHARIDES AND POLYSACCHARIDES

Large linear and branched molecules can be made from simple repeating sugar units. Short chains are called oligosaccharides, and long chains are called polysaccharides. Glycogen, for example, is a polysaccharide made entirely of glucose units joined together.

branch points   glycogen

COMPLEX OLIGOSACCHARIDES

In many cases, a sugar sequence is nonrepetitive. Many different molecules are possible. Such complex oligosaccharides are usually linked to proteins or to lipids, as is this oligosaccharide, which is part of a cell-surface molecule that defines a particular blood group.

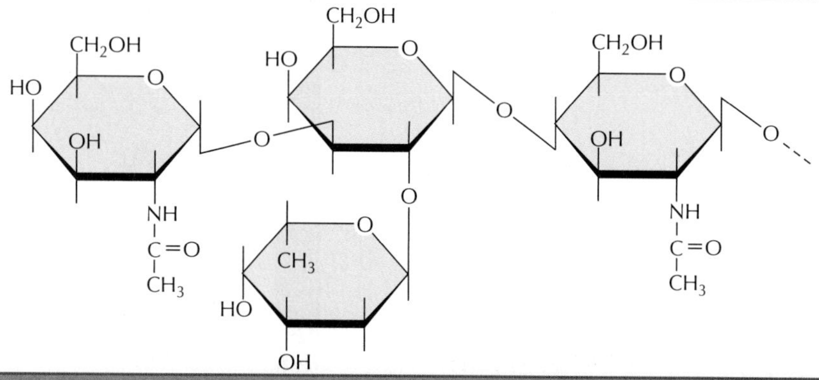

CH_2OH CH_2OH CH_2OH

CH_3

FATTY ACIDS

All fatty acids have carboxyl groups at one end and long hydrocarbon tails at the other.

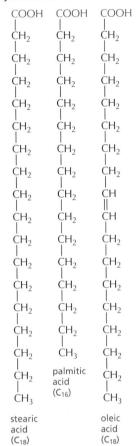

COOH   COOH   COOH

stearic acid (C₁₈) palmitic acid (C₁₆) oleic acid (C₁₈)

Hundreds of different kinds of fatty acids exist. Some have one or more double bonds in their hydrocarbon tail and are said to be unsaturated. Fatty acids with no double bonds are saturated.

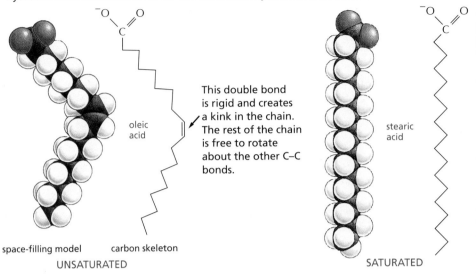

oleic acid

This double bond is rigid and creates a kink in the chain. The rest of the chain is free to rotate about the other C–C bonds.

space-filling model   carbon skeleton

UNSATURATED

stearic acid

SATURATED

TRIACYLGLYCEROLS

Fatty acids are stored in cells as an energy reserve (fats and oils) through an ester linkage to glycerol to form triacylglycerols.

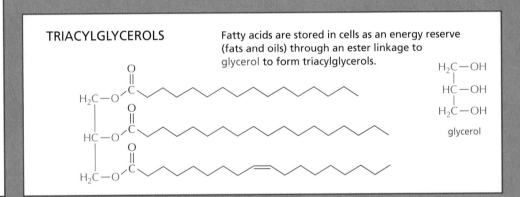

H_2C-OH
$HC-OH$
H_2C-OH

glycerol

CARBOXYL GROUP

If free, the carboxyl group of a fatty acid will be ionized.

But more often it is linked to other groups to form either esters

or amides.

PHOSPHOLIPIDS

Phospholipids are the major constituents of cell membranes.

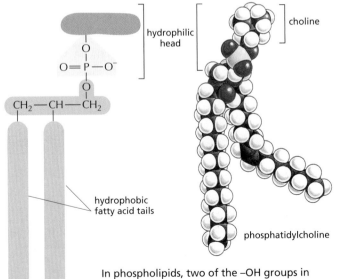

hydrophilic head

choline

$CH_2-CH-CH_2$

hydrophobic fatty acid tails

phosphatidylcholine

general structure of a phospholipid

In phospholipids, two of the –OH groups in glycerol are linked to fatty acids, while the third –OH group is linked to phosphoric acid. The phosphate is further linked to one of a variety of small polar groups, such as choline.

LIPID AGGREGATES

Fatty acids have a hydrophilic head and a hydrophobic tail.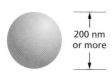

In water, they can form either a surface film or small, spherical micelles.

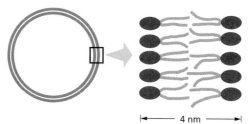

surface film

micelle

Their derivatives can form larger aggregates held together by hydrophobic forces:

Triacylglycerols form large spherical fat droplets in the cell cytoplasm.

200 nm or more

Phospholipids and glycolipids form self-sealing *lipid bilayers*, which are the basis for all cell membranes.

4 nm

OTHER LIPIDS

Lipids are defined as water-insoluble molecules that are soluble in organic solvents. Two other common types of lipids are steroids and polyisoprenoids. Both are made from isoprene units.

$$CH_3$$
$$C-CH=CH_2$$
$$CH_2 \quad \text{isoprene}$$

STEROIDS

Steroids have a common multiple-ring structure.

HO

cholesterol—found in many cell membranes

OH

O

testosterone—male sex hormone

GLYCOLIPIDS

Like phospholipids, these compounds are composed of a hydrophobic region, containing two long hydrocarbon tails, and a polar region, which contains one or more sugars and, unlike phospholipids, no phosphate.

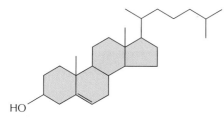

galactose

sugar

a simple glycolipid

POLYISOPRENOIDS

Long-chain polymers of isoprene

$$O^-$$
$$O=P-O^-$$
$$O$$

dolichol phosphate—used to carry activated sugars in the membrane-associated synthesis of glycoproteins and some polysaccharides

FAMILIES OF AMINO ACIDS

The common amino acids are grouped according to whether their side chains are

    acidic
    basic
    uncharged polar
    nonpolar

These 20 amino acids are given both three-letter and one-letter abbreviations.

Thus: alanine = Ala = A

BASIC SIDE CHAINS

lysine
(Lys, or K)

arginine
(Arg, or R)

histidine
(His, or H)

This group is very basic because its positive charge is stabilized by resonance (see Panel 2–1).

These nitrogens have a relatively weak affinity for an H^+ and are only partly positive at neutral pH.

THE AMINO ACID

The general formula of an amino acid is

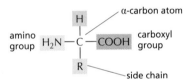

amino group — H_2N
carboxyl group — COOH
α-carbon atom
side chain — R

R is commonly one of 20 different side chains. At pH 7, both the amino and carboxyl groups are ionized.

$$\overset{\oplus}{H_3N}-\overset{\overset{H}{|}}{\underset{\underset{R}{|}}{C}}-COO^{\ominus}$$

OPTICAL ISOMERS

The α-carbon atom is asymmetric, allowing for two mirror-image (or stereo-) isomers, L and D.

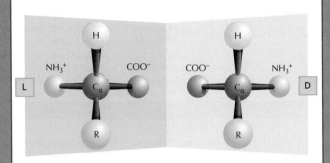

Proteins contain exclusively L-amino acids.

PEPTIDE BONDS

In proteins, amino acids are commonly joined together by an amide linkage, called a peptide bond.

The four atoms in each peptide bond *(red box)* form a rigid planar unit. There is no rotation around the C–N bond.

H_2O

Proteins are long polymers of amino acids linked by peptide bonds, and they are always written with the N-terminus toward the left. Peptides are shorter, usually fewer than 50 amino acids long. The sequence of this tripeptide is histidine-cysteine-valine.

amino terminus, or N-terminus

carboxyl terminus, or C-terminus

These two single bonds allow rotation, so that long chains of amino acids are very flexible.

ACIDIC SIDE CHAINS

aspartic acid
(Asp, or D)

glutamic acid
(Glu, or E)

UNCHARGED POLAR SIDE CHAINS

asparagine
(Asn, or N)

glutamine
(Gln, or Q)

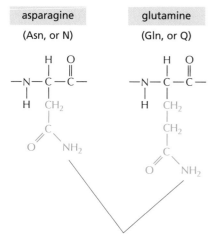

Although the amide N is not charged at neutral pH, it is polar.

serine
(Ser, or S)

threonine
(Thr, or T)

tyrosine
(Tyr, or Y)

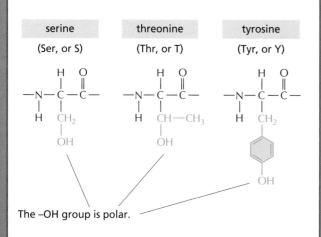

The –OH group is polar.

NONPOLAR SIDE CHAINS

alanine
(Ala, or A)

valine
(Val, or V)

leucine
(Leu, or L)

isoleucine
(Ile, or I)

proline
(Pro, or P)

(actually an imino acid)

phenylalanine
(Phe, or F)

methionine
(Met, or M)

tryptophan
(Trp, or W)

glycine
(Gly, or G)

cysteine
(Cys, or C)

Disulfide bonds can form between two cysteine side chains in proteins.

$$--CH_2—S—S—CH_2--$$

BASES

The bases are nitrogen-containing ring compounds, either pyrimidines or purines.

cytosine

uracil

thymine

adenine

guanine

PYRIMIDINE

PURINE

PHOSPHATES

The phosphates are normally joined to the C5 hydroxyl of the ribose or deoxyribose sugar (designated 5'). Mono-, di-, and triphosphates are common.

as in AMP

as in ADP

as in ATP

The phosphate makes a nucleotide negatively charged.

NUCLEOTIDES

A nucleotide consists of a nitrogen-containing base, a five-carbon sugar, and one or more phosphate groups.

BASE

PHOSPHATE

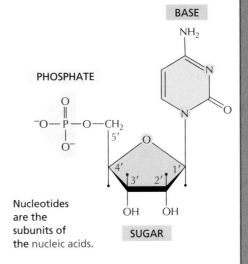

Nucleotides are the subunits of the nucleic acids.

SUGAR

BASE–SUGAR LINKAGE

N-glycosidic bond

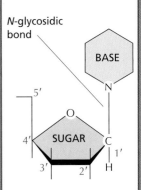

BASE

SUGAR

The base is linked to the same carbon (C1) used in sugar–sugar bonds.

SUGARS

PENTOSE

a five-carbon sugar

two kinds of pentoses are used

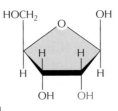

β-D-ribose
used in ribonucleic acid (RNA)

β-D-2-deoxyribose
used in deoxyribonucleic acid (DNA)

Each numbered carbon on the sugar of a nucleotide is followed by a prime mark; therefore, one speaks of the "5-prime carbon," etc.

NOMENCLATURE

The names can be confusing, but the abbreviations are clear.

BASE	NUCLEOSIDE	ABBR.
adenine	adenosine	A
guanine	guanosine	G
cytosine	cytidine	C
uracil	uridine	U
thymine	thymidine	T

Nucleotides are abbreviated by three capital letters. Some examples follow:

AMP = adenosine monophosphate
dAMP = deoxyadenosine monophosphate
UDP = uridine diphosphate
ATP = adenosine triphosphate

BASE + SUGAR = NUCLEO<u>S</u>IDE

BASE + SUGAR + PHOSPHATE = NUCLEO<u>T</u>IDE

NUCLEIC ACIDS

Nucleotides are joined together by phosphodiester bonds between 5' and 3' carbon atoms of the sugar ring, via a phosphate group, to form nucleic acids. The linear sequence of nucleotides in a nucleic acid chain is commonly abbreviated by a one-letter code, such as AGCTTACA, with the 5' end of the chain at the left.

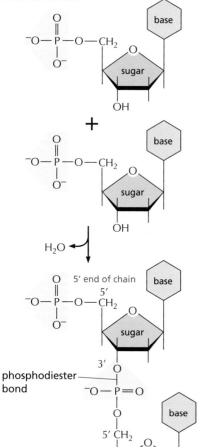

phosphodiester bond

example: DNA

NUCLEOTIDES HAVE MANY OTHER FUNCTIONS

1 They carry chemical energy in their easily hydrolyzed phosphoanhydride bonds.

phosphoanhydride bonds

example: ATP (or ATP)

2 They combine with other groups to form coenzymes.

example: coenzyme A (CoA)

3 They are used as small intracellular signaling molecules in the cell.

example: cyclic AMP

WEAK NONCOVALENT CHEMICAL BONDS

Organic molecules can interact with other molecules through three types of short-range attractive forces known as *noncovalent bonds*: van der Waals attractions, electrostatic attractions, and hydrogen bonds. The repulsion of hydrophobic groups from water is also important for these interactions and for the folding of biological macromolecules.

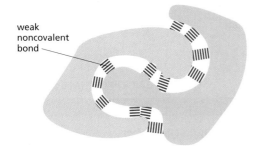

weak noncovalent bond

Weak noncovalent bonds have less than 1/20 the strength of a strong covalent bond. They are strong enough to provide tight binding only when many of them are formed simultaneously.

HYDROGEN BONDS

As already described for water (see Panel 2–2, pp. 68–69), hydrogen bonds form when a hydrogen atom is "sandwiched" between two electron-attracting atoms (usually oxygen or nitrogen).

Hydrogen bonds are strongest when the three atoms are in a straight line:

Examples in macromolecules:

Amino acids in a polypeptide chain can be hydrogen-bonded together in a folded protein.

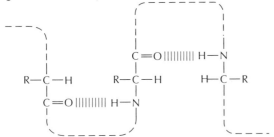

Two bases, G and C, are hydrogen-bonded in a DNA double helix.

VAN DER WAALS ATTRACTIONS

If two atoms are too close together they repel each other very strongly. For this reason, an atom can often be treated as a sphere with a fixed radius. The characteristic "size" for each atom is specified by a unique van der Waals radius. The contact distance between any two noncovalently bonded atoms is the sum of their van der Waals radii.

| 0.12 nm radius | 0.2 nm radius | 0.15 nm radius | 0.14 nm radius |

At very short distances, any two atoms show a weak bonding interaction due to their fluctuating electrical charges. The two atoms will be attracted to each other in this way until the distance between their nuclei is approximately equal to the sum of their van der Waals radii. Although they are individually very weak, such van der Waals attractions can become important when two macromolecular surfaces fit very close together, because many atoms are involved.

Note that when two atoms form a covalent bond, the centers of the two atoms (the two atomic nuclei) are much closer together than the sum of the two van der Waals radii. Thus,

| 0.4 nm two non-bonded carbon atoms | 0.15 nm two carbon atoms held by single covalent bond | 0.13 nm two carbon atoms held by double covalent bond |

HYDROGEN BONDS IN WATER

Any two atoms that can form hydrogen bonds to each other can alternatively form hydrogen bonds to water molecules. Because of this competition with water molecules, the hydrogen bonds formed in water between two peptide bonds, for example, are relatively weak.

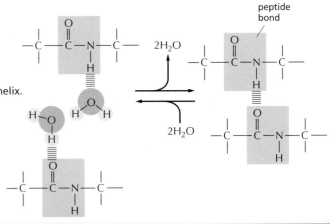

peptide bond

$2H_2O$

$2H_2O$

ELECTROSTATIC ATTRACTIONS

Attractive interactions occur both between fully charged groups (ionic bond) and between partially charged groups on polar molecules.

The force of attraction between the two partial charges, δ^+ and δ^-, falls off rapidly as the distance between the charges increases.

In the absence of water, ionic bonds are very strong. They are responsible for the strength of such minerals as marble and agate, and for crystal formation in common table salt, NaCl.

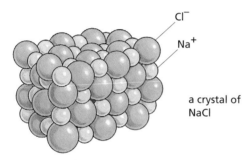

Cl⁻

Na⁺

a crystal of NaCl

ELECTROSTATIC ATTRACTIONS IN AQUEOUS SOLUTIONS

Charged groups are shielded by their interactions with water molecules. Electrostatic attractions are therefore quite weak in water.

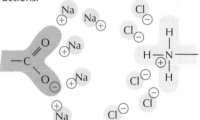

Inorganic ions in solution can also cluster around charged groups and further weaken these electrostatic attractions.

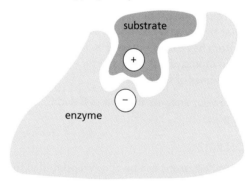

Despite being weakened by water and inorganic ions, electrostatic attractions are very important in biological systems. For example, an enzyme that binds a positively charged substrate will often have a negatively charged amino acid side chain at the appropriate place.

substrate

enzyme

HYDROPHOBIC INTERACTIONS

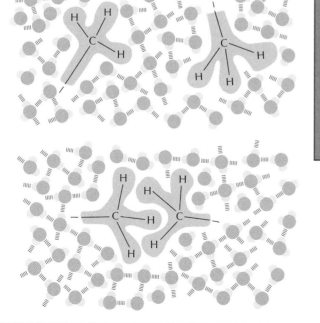

Water forces hydrophobic groups together in order to minimize their disruptive effects on the water network formed by the H bonds between water molecules. Hydrophobic groups held together in this way are sometimes said to be held together by "hydrophobic bonds," even though the attraction is actually caused by a repulsion from water.

QUESTIONS

QUESTION 2–10

Which of the following statements are correct? Explain your answers.

A. An atomic nucleus contains protons and neutrons.

B. An atom has more electrons than protons.

C. The nucleus is surrounded by a double membrane.

D. All atoms of the same element have the same number of neutrons.

E. The number of neutrons determines whether the nucleus of an atom is stable or radioactive.

F. Both fatty acids and polysaccharides can be important energy stores in the cell.

G. Hydrogen bonds are weak and can be broken by thermal energy, yet they contribute significantly to the specificity of interactions between macromolecules.

QUESTION 2–11

To gain a better feeling for atomic dimensions, assume that the page on which this question is printed is made entirely of the polysaccharide cellulose, whose molecules are described by the formula $(C_nH_{2n}O_n)$, where n can be a quite large number and is variable from one molecule to another. The atomic weights of carbon, hydrogen, and oxygen are 12, 1, and 16, respectively, and this page weighs 5 g.

A. How many carbon atoms are there in this page?

B. In cellulose, how many carbon atoms would be stacked on top of each other to span the thickness of this page (the size of the page is 21.2 cm × 27.6 cm, and it is 0.07 mm thick)?

C. Now consider the problem from a different angle. Assume that the page is composed only of carbon atoms. A carbon atom has a diameter of 2×10^{-10} m (0.2 nm); how many carbon atoms of 0.2 nm diameter would it take to span the thickness of the page?

D. Compare your answers from parts B and C and explain any differences.

QUESTION 2–12

A. How many electrons can be accommodated in the first, second, and third electron shells of an atom?

B. How many electrons would atoms of the elements listed below have to gain or lose to obtain a completely filled outer shell?

helium	gain __	lose __
oxygen	gain __	lose __
carbon	gain __	lose __
sodium	gain __	lose __
chlorine	gain __	lose __

C. What do the answers tell you about the reactivity of helium and the bonds that can form between sodium and chlorine?

QUESTION 2–13

The elements oxygen and sulfur have similar chemical properties because they both have six electrons in their outermost electron shells. Indeed, both elements form molecules with two hydrogen atoms, water (H_2O) and hydrogen sulfide (H_2S). Surprisingly, at room temperature, water is a liquid, yet H_2S is a gas, despite sulfur being much larger and heavier than oxygen. Explain why this might be the case.

QUESTION 2–14

Write the chemical formula for a condensation reaction of two amino acids to form a peptide bond. Write the formula for its hydrolysis.

QUESTION 2–15

Which of the following statements are correct? Explain your answers.

A. Proteins are so remarkably diverse because each is made from a unique mixture of amino acids that are linked in random order.

B. Lipid bilayers are macromolecules that are made up mostly of phospholipid subunits.

C. Nucleic acids contain sugar groups.

D. Many amino acids have hydrophobic side chains.

E. The hydrophobic tails of phospholipid molecules are repelled from water.

F. DNA contains the four different bases A, G, U, and C.

QUESTION 2–16

A. How many different molecules composed of (a) two, (b) three, and (c) four amino acids, linked together by peptide bonds, can be made from the set of 20 naturally occurring amino acids?

B. Assume you were given a mixture consisting of one molecule each of all possible sequences of a smallish protein of molecular weight 4800 daltons. If the average molecular weight of an amino acid is, say, 120 daltons, how much would the sample weigh? How big a container would you need to hold it?

C. What does this calculation tell you about the fraction of possible proteins that are currently in use by living organisms (the average molecular weight of proteins is about 30,000 daltons)?

QUESTION 2–17

This is a biology textbook. Explain why the chemical principles that are described in this chapter are important in the context of modern cell biology.

QUESTION 2–18

A. Describe the similarities and differences between van der Waals attractions and hydrogen bonds.

B. Which of the two bonds would form (a) between two hydrogens bound to carbon atoms, (b) between a nitrogen atom and a hydrogen bound to a carbon atom, and (c) between a nitrogen atom and a hydrogen bound to an oxygen atom?

QUESTION 2–19

What are the forces that determine the folding of a macromolecule into a unique shape?

QUESTION 2–20

Fatty acids are said to be "amphipathic." What is meant by this term, and how does an amphipathic molecule behave in water? Draw a diagram to illustrate your answer.

QUESTION 2–21

Are the formulas in **Figure Q2–21** correct or incorrect? Explain your answer in each case.

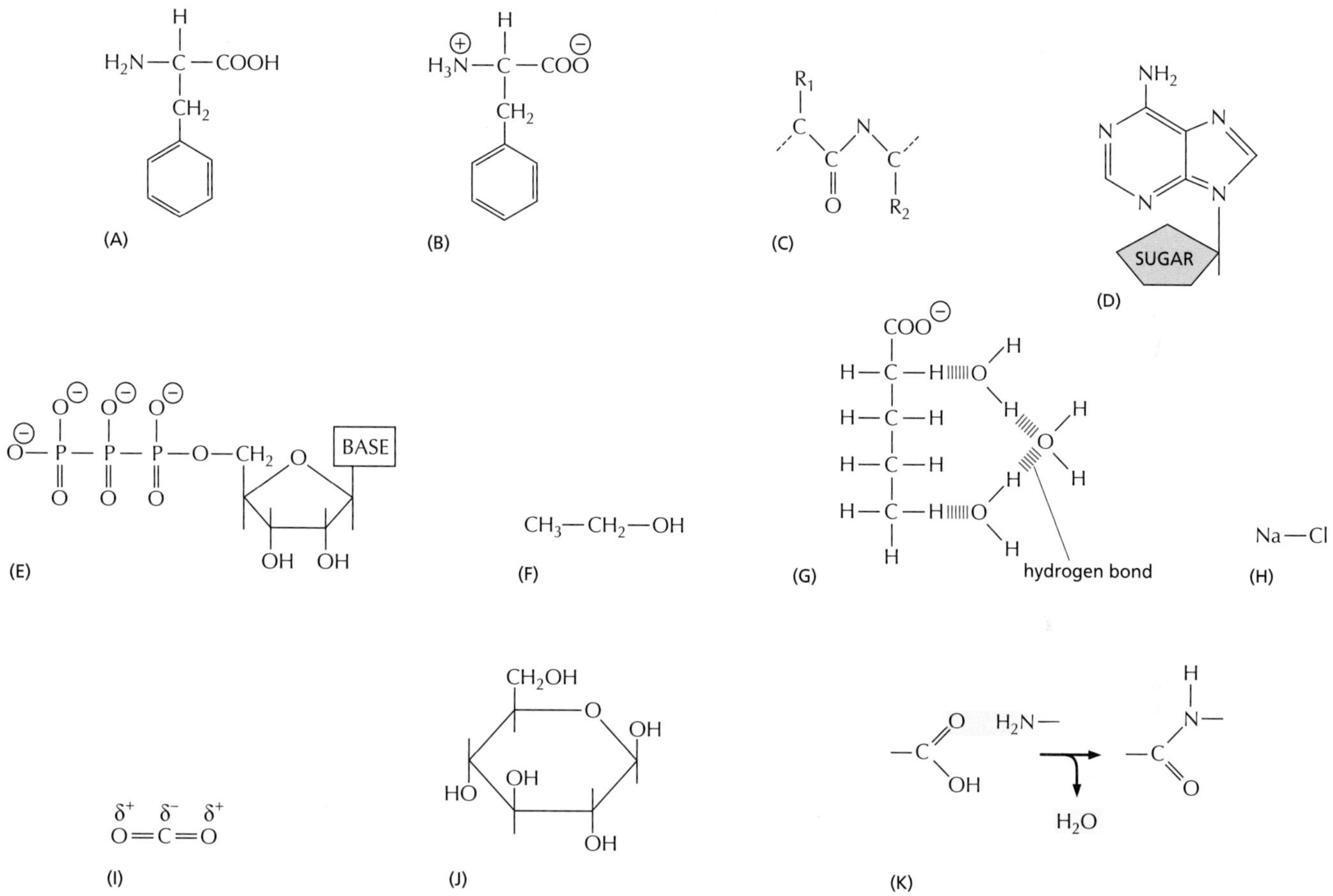

Figure Q2–21

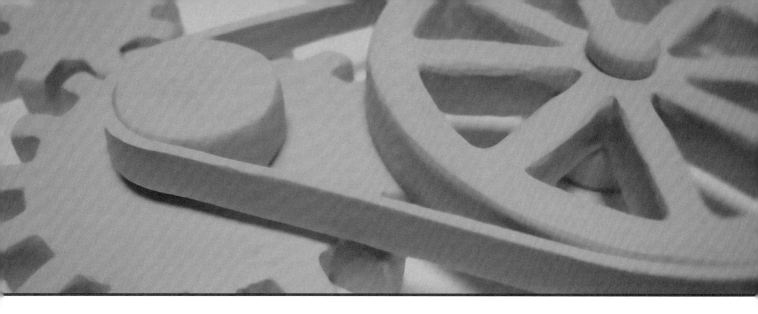

CHAPTER **THREE** 3

Energy, Catalysis, and Biosynthesis

One property above all makes living things seem almost miraculously different from nonliving matter: they create and maintain order in a universe that is tending always toward greater disorder. To accomplish this remarkable feat, the cells in a living organism must carry out a never-ending stream of chemical reactions that produce the molecules the organism requires to meet its metabolic needs. In some of these reactions, small organic molecules—amino acids, sugars, nucleotides, and lipids—are taken apart or modified to supply the many other small molecules that the cell requires. In other reactions, these small molecules are used to construct an enormously diverse range of larger molecules, including the proteins, nucleic acids, and other macromolecules that endow living systems with all of their most distinctive properties. Each cell can be viewed as a tiny chemical factory, performing many millions of these reactions every second.

To carry out the tremendous number of chemical reactions needed to sustain it, a living organism requires both a source of atoms in the form of food molecules and a source of energy. The atoms and the energy must both come, ultimately, from the nonliving environment. In this chapter, we discuss why cells require energy, and how they use energy and atoms from their environment to create the molecular order that makes life possible.

Most of the chemical reactions that cells perform would normally occur only at temperatures that are much higher than those inside a cell. Each reaction therefore requires a major boost in chemical reactivity to enable it to proceed rapidly within the cell. This boost is provided by specialized proteins called *enzymes*, each of which accelerates, or *catalyzes*, just one

THE USE OF ENERGY BY CELLS

FREE ENERGY AND CATALYSIS

ACTIVATED CARRIERS AND BIOSYNTHESIS

Figure 3–1 A series of enzyme-catalyzed reactions forms a metabolic pathway. Each enzyme catalyzes a chemical reaction involving a particular molecule. In this example, a set of enzymes acting in series converts molecule A to molecule F, forming a metabolic pathway.

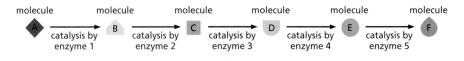

of the many possible kinds of reactions that a particular molecule might undergo. These enzyme-catalyzed reactions are usually connected in series, so that the product of one reaction becomes the starting material for the next (**Figure 3–1**). The long linear reaction pathways, or *metabolic pathways*, that result are in turn linked to one another, forming a complex web of interconnected reactions.

Rather than being an inconvenience, the necessity for *catalysis* is a benefit, as it allows the cell to precisely control its **metabolism**—the sum total of all the chemical reactions it needs to carry out to survive, grow, and reproduce. This control is central to the chemistry of life.

Two opposing streams of chemical reactions occur in cells, the *catabolic* pathways and the *anabolic* pathways. The catabolic pathways (**catabolism**) break down foodstuffs into smaller molecules, thereby generating both a useful form of energy for the cell and some of the small molecules that the cell needs as building blocks. The anabolic, or *biosynthetic*, pathways (**anabolism**) use the energy harnessed by catabolism to drive the synthesis of the many molecules that form the cell. Together, these two sets of reactions constitute the metabolism of the cell (**Figure 3–2**).

The details regarding the individual reactions that comprise cell metabolism are part of the subject matter of *biochemistry*, and they need not concern us here. But the general principles by which cells obtain energy from their environment and use it to create order are central to cell biology. We begin this chapter with a discussion of why a constant input of energy is needed to sustain living organisms. We then discuss how enzymes catalyze the reactions that produce biological order. Finally, we describe the molecules that carry the energy that makes life possible.

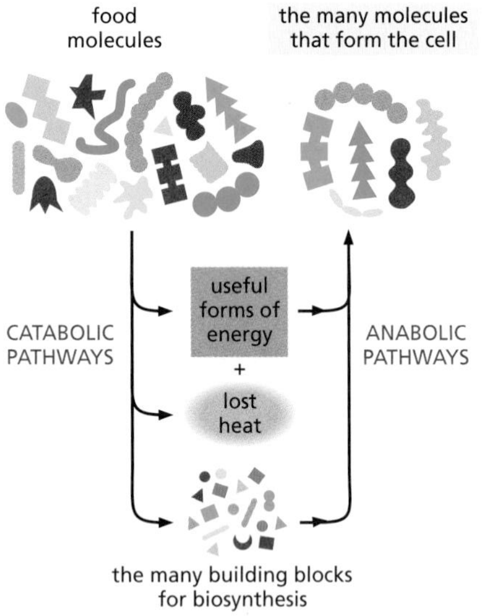

THE USE OF ENERGY BY CELLS

Nonliving things left to themselves eventually become disordered: buildings crumble and dead organisms decay. Living cells, by contrast, not only maintain, but actually generate order at every level, from the large-scale structure of a butterfly or a flower down to the organization of the molecules that make up these organisms (**Figure 3–3**). This property of life is made possible by elaborate molecular mechanisms that extract energy from the environment and convert it into the energy stored in chemical bonds. Biological structures are therefore able to maintain their form, even though the materials of which they are made are continually being broken down, replaced, and recycled. Your body has the same basic structure it had 10 years ago, even though you now contain atoms that, for the most part, were not in your body then.

Biological Order Is Made Possible by the Release of Heat Energy from Cells

The universal tendency of things to become disordered is expressed in a fundamental law of physics, the *second law of thermodynamics*. This law states that, in the universe or in any isolated system (a collection of matter that is completely isolated from the rest of the universe), the degree of disorder can only increase. The second law of thermodynamics has such profound implications for living things that it is worth restating in several ways.

Figure 3–2 Catabolic and anabolic pathways together constitute the cell's metabolism. Note that a major portion of the energy stored in the chemical bonds of food molecules is dissipated as heat. Thus, only some of this energy can be converted to the useful forms of energy needed to drive the synthesis of new molecules.

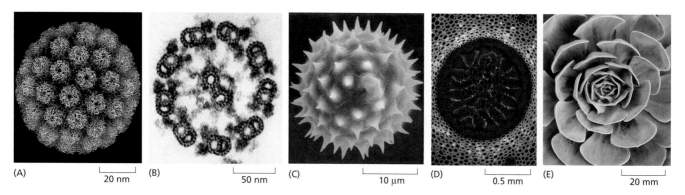

(A) ⊢—⊣ 20 nm   (B) ⊢—⊣ 50 nm   (C) ⊢—⊣ 10 μm   (D) ⊢—⊣ 0.5 mm   (E) ⊢—⊣ 20 mm

We can express the second law in terms of probability by stating that *systems will change spontaneously toward those arrangements that have the greatest probability.* Consider a box of 100 coins all lying heads up. A series of events that disturbs the box will tend to move the arrangement toward a mixture of 50 heads and 50 tails. The reason is simple: there are a huge number of possible arrangements of the individual coins that can achieve the 50–50 result, but only one possible arrangement that keeps them all oriented heads up. Because the 50–50 mixture accommodates a greater number of possibilities and places fewer constraints on the orientation of each individual coin, we say that it is more "disordered." For the same reason, one's living space will become increasingly disordered without an intentional effort to keep it organized. Movement toward disorder is a spontaneous process, requiring a periodic input of energy to reverse it **(Figure 3–4)**.

The measure of a system's disorder is called the **entropy** of the system, and the greater the disorder, the greater the entropy. Thus another way to express the second law of thermodynamics is to say that systems will change spontaneously toward arrangements with greater entropy. Living cells—by surviving, growing, and forming complex communities and even whole organisms—generate order and thus might appear to defy the second law of thermodynamics. This is not the case, however,

Figure 3–3 Biological structures are highly ordered. Well-defined, ornate, and beautiful spatial patterns can be found at every level of organization in living organisms. In order of increasing size: (A) protein molecules in the coat of a virus (a parasite that, although not technically alive, contains the same types of molecules as those found in living cells); (B) the regular array of microtubules seen in a cross section of a sperm tail; (C) surface contours of a pollen grain (a single cell); (D) cross section of a fern stem, showing the patterned arrangement of cells; and (E) flower with a spiral array of petals, each made of millions of cells. (A, courtesy of Robert Grant, Stéphane Crainic, and James M. Hogle; B, courtesy of Lewis Tilney; C, courtesy of Colin MacFarlane and Chris Jeffree; D, courtesy of Jim Haseloff.)

ORGANIZED EFFORT REQUIRING ENERGY INPUT

"SPONTANEOUS" REACTION

as time elapses

Figure 3–4 The spontaneous tendency toward disorder is an everyday experience. Reversing this natural tendency toward disorder requires an intentional effort and an input of energy. In fact, from the second law of thermodynamics, we can be certain that the human intervention required will release enough heat to the environment to more than compensate for the reestablishment of order in this room.

Figure 3–5 Living cells do not defy the second law of thermodynamics. In the diagram on the *left*, the molecules of both the cell and the rest of the universe (the environment) are depicted in a relatively disordered state. In the diagram on the *right*, the cell has taken in energy from food molecules and released heat by carrying out a reaction that orders the molecules that the cell contains. Because the heat increases the disorder in the environment around the cell—as depicted by the *longer, jagged red arrows*, which represent increased thermal motion, and the distorted molecules, which indicate enhanced molecular vibration and rotation—the second law of thermodynamics is satisfied, even as the cell grows and constructs larger molecules.

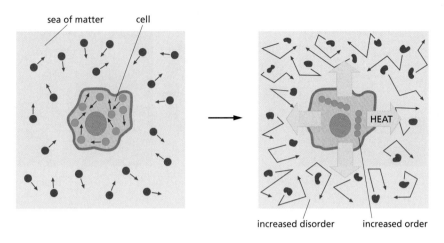

because a cell is not an isolated system. Rather, it takes in energy from its environment—in the form of food, inorganic molecules, or photons of light from the sun—and it then uses this energy to generate order within itself, forging new chemical bonds and building large macromolecules. In the course of performing the chemical reactions that generate order, some energy is lost in the form of heat. Heat is energy in its most disordered form—the random jostling of molecules (analogous to the random jostling of the coins in the box). Because the cell is not an isolated system, the heat energy that its reactions generate is quickly dispersed into the cell's surroundings. There, the heat increases the intensity of the thermal motions of nearby molecules, thereby increasing the entropy of the environment (**Figure 3–5**).

The amount of heat released by a cell must be great enough that the increased order generated inside the cell is more than compensated for by the increased disorder generated in the environment. Only in this case is the second law of thermodynamics satisfied, because the total entropy of the system—that of the cell plus its environment—increases as a result of the chemical reactions inside the cell.

Cells Can Convert Energy from One Form to Another

According to the *first law of thermodynamics*, energy cannot be created or destroyed—but it can be converted from one form to another (**Figure 3–6**). Cells take advantage of this law of thermodynamics, for example, when they convert the energy from sunlight into the energy in the chemical bonds of sugars and other small organic molecules during photosynthesis. Although chemical reactions that power such energy conversions can change how much energy is present in one form or another, the first law tells us that the total amount of energy in the universe must always be the same.

When an animal cell breaks down foodstuffs, some of the energy in the chemical bonds in the food molecules (chemical-bond energy) is converted into the thermal motion of molecules (heat energy). This conversion of chemical energy into heat energy causes the universe as a whole to become more disordered—as required by the second law of thermodynamics. But the cell cannot derive any benefit from the heat energy it produces unless the heat-generating reactions are directly linked to processes that maintain molecular order inside the cell. It is the tight coupling of heat production to an increase in order that distinguishes the metabolism of a cell from the wasteful burning of fuel in a fire. Later in this chapter, we illustrate how this coupling occurs. For the moment, it is

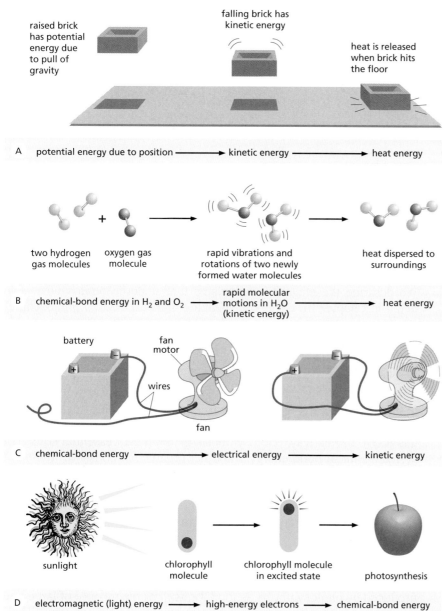

Figure 3–6 Different forms of energy are interconvertible, but the total amount of energy must be conserved. In (A), we can use the height and weight of the brick to predict exactly how much heat will be released when it hits the floor. In (B), the large amount of chemical-bond energy released when water (H_2O) is formed from H_2 and O_2 is initially converted to very rapid thermal motions in the two new H_2O molecules; however, collisions with other H_2O molecules almost instantaneously spread this kinetic energy evenly throughout the surroundings (heat transfer), making the new H_2O molecules indistinguishable from all the rest. (C) Cells can convert chemical-bond energy into kinetic energy to drive, for example, molecular motor proteins; however, this occurs without the intermediate conversion to electrical energy that a man-made appliance such as this fan requires. (D) Some cells can also harvest the energy from sunlight to form chemical bonds via photosynthesis.

sufficient to recognize that—by directly linking the "burning" of food molecules to the generation of biological order—cells are able to create and maintain an island of order in a universe tending toward chaos.

Photosynthetic Organisms Use Sunlight to Synthesize Organic Molecules

All animals live on energy stored in the chemical bonds of organic molecules, which they take in as food. These food molecules also provide the atoms that animals need to construct new living matter. Some animals obtain their food by eating other animals, others by eating plants. Plants, by contrast, obtain their energy directly from sunlight. Thus, the energy animals obtain by eating plants—or by eating animals that have eaten plants—ultimately comes from the sun (**Figure 3–7**).

Solar energy enters the living world through **photosynthesis**, a process that converts the electromagnetic energy in sunlight into chemical-bond energy in cells. Photosynthetic organisms—including plants, algae, and

Figure 3–7 With few exceptions, the radiant energy of sunlight sustains all life. Trapped by plants and some microorganisms through photosynthesis, light from the sun is the ultimate source of all energy for humans and other animals. (*Wheat Field Behind Saint-Paul Hospital with a Reaper* by Vincent van Gogh. Courtesy of Museum Folkwang, Essen.)

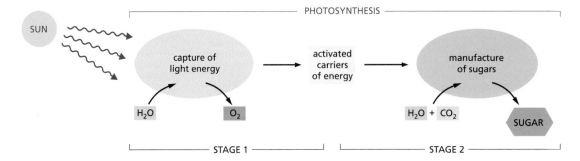

Figure 3–8 Photosynthesis takes place in two stages. The activated carriers generated in the first stage are two molecules that we will discuss shortly: ATP and NADPH.

some bacteria—use the energy they derive from sunlight to synthesize small chemical building blocks such as sugars, amino acids, nucleotides, and fatty acids. These small molecules in turn are converted into the macromolecules—the proteins, nucleic acids, polysaccharides, and lipids—that form the plant.

We describe the elegant mechanisms that underlie photosynthesis in detail in Chapter 14. Generally speaking, the reactions of photosynthesis take place in two stages. In the first stage, energy from sunlight is captured and transiently stored as chemical-bond energy in specialized molecules called *activated carriers*, which we discuss in more detail later in the chapter. All of the oxygen (O_2) in the air we breathe is generated by the splitting of water molecules during this first stage of photosynthesis.

In the second stage, the activated carriers are used to help drive a *carbon-fixation* process, in which sugars are manufactured from carbon dioxide gas (CO_2). In this way, photosynthesis generates an essential source of stored chemical-bond energy and other organic materials—for the plant itself and for any animals that eat it. The two stages of photosynthesis are summarized in **Figure 3–8**.

Cells Obtain Energy by the Oxidation of Organic Molecules

All animal and plant cells require the chemical energy stored in the chemical bonds of organic molecules—either the sugars that a plant has produced by photosynthesis as food for itself or the mixture of large and small molecules that an animal has eaten. To use this energy to live, grow, and reproduce, organisms must extract it in a usable form. In both plants and animals, energy is extracted from food molecules by a process of gradual *oxidation*, or controlled burning.

Earth's atmosphere is about 21% oxygen. In the presence of oxygen, the most energetically stable form of carbon is CO_2 and that of hydrogen is H_2O. A cell is therefore able to obtain energy from sugars or other organic molecules by allowing the carbon and hydrogen atoms in these molecules to combine with oxygen—that is, become *oxidized*—to produce CO_2 and H_2O, respectively—a process known as cellular **respiration**.

Photosynthesis and cellular respiration are complementary processes (**Figure 3–9**). This means that the transactions between plants and animals are not all one way. Plants, animals, and microorganisms have existed together on this planet for so long that they have become an essential part of each other's environments. The oxygen released by photosynthesis is consumed by nearly all organisms for the oxidative breakdown of organic molecules. And some of the CO_2 molecules that today are incorporated into organic molecules by photosynthesis in a green leaf were released yesterday into the atmosphere by the respiration of an animal, a fungus, or the plant itself, or by the burning of

QUESTION 3–1

Consider the equation
   light energy + CO_2 + H_2O →
      sugars + O_2 + heat energy
Would you expect this reaction to occur in a single step? Why must heat be generated in the reaction? Explain your answers.

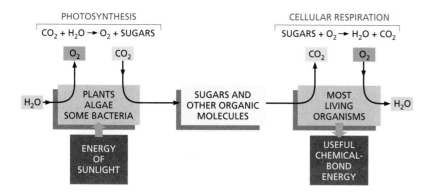

Figure 3–9 Photosynthesis and cellular respiration are complementary processes in the living world. The *left* side of the diagram shows how photosynthesis—carried out by plants and photosynthetic microorganisms—uses the energy of sunlight to produce sugars and other organic molecules from the carbon atoms in CO_2 in the atmosphere. In turn, these molecules serve as food for other organisms. The *right* side of the diagram shows how cellular respiration in most organisms—including plants and photosynthetic microorganisms—uses O_2 to oxidize food molecules, releasing the same carbon atoms in the form of CO_2 back to the atmosphere. In the process, the organisms obtain the useful chemical-bond energy that they need to survive. The first cells on Earth are thought to have been capable of neither photosynthesis nor cellular respiration (discussed in Chapter 14). However, photosynthesis must have preceded respiration on the Earth, because there is strong evidence that billions of years of photosynthesis were required to release enough O_2 to create an atmosphere that could support respiration.

fossil fuels. Carbon utilization therefore forms a huge cycle that involves the *biosphere* (all of the living organisms on Earth) as a whole, crossing boundaries between individual organisms (**Figure 3–10**).

Oxidation and Reduction Involve Electron Transfers

The cell does not oxidize organic molecules in one step, as occurs when organic material is burned in a fire. Through the use of enzyme catalysts, metabolism directs the molecules through a large number of reactions, few of which actually involve the direct addition of oxygen. Thus, before we consider some of these reactions, we should explain what is meant by oxidation.

The term **oxidation** literally means the addition of oxygen atoms to a molecule. More generally, though, oxidation is said to occur in any reaction in which electrons are transferred from one atom to another. Oxidation, in this sense, refers to the removal of electrons from an atom. The converse reaction, called **reduction**, involves the addition of electrons to an atom. Thus, Fe^{2+} is oxidized when it loses an electron to become Fe^{3+}, whereas a chlorine atom is reduced when it gains an electron to become Cl^-. Because the number of electrons is conserved in a chemical reaction (there is no net loss or gain), oxidation and reduction always occur simultaneously: that is, if one molecule gains an electron in a reaction (reduction), a second molecule must lose the electron (oxidation). When a sugar molecule is oxidized to CO_2 and H_2O, for example, the O_2 molecules involved in forming H_2O gain electrons and thus are said to have been reduced.

The terms oxidation and reduction apply even when there is only a partial shift of electrons between atoms linked by a covalent bond. When a carbon atom becomes covalently bonded to an atom with a strong affinity for electrons—oxygen, chlorine, or sulfur, for example—it gives up more

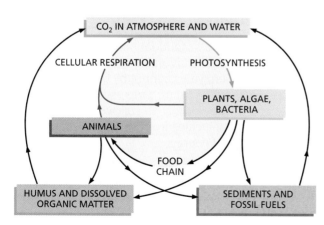

Figure 3–10 Carbon atoms cycle continuously through the biosphere. Individual carbon atoms are incorporated into organic molecules of the living world by the photosynthetic activity of plants, algae, and bacteria. They then pass to animals and microorganisms—as well as into organic material in soil and oceans—and are ultimately restored to the atmosphere in the form of CO_2 when organic molecules are oxidized by cells during respiration or burned by humans as fossil fuels.

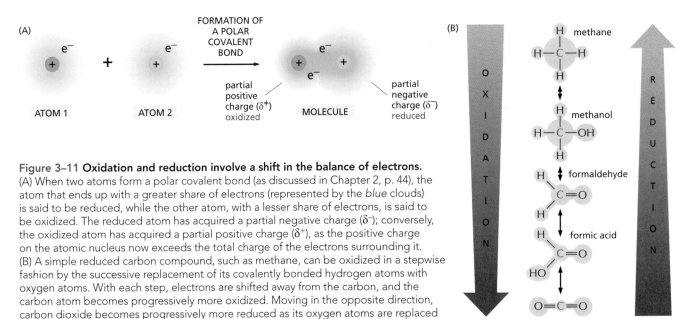

Figure 3–11 Oxidation and reduction involve a shift in the balance of electrons.
(A) When two atoms form a polar covalent bond (as discussed in Chapter 2, p. 44), the atom that ends up with a greater share of electrons (represented by the *blue* clouds) is said to be reduced, while the other atom, with a lesser share of electrons, is said to be oxidized. The reduced atom has acquired a partial negative charge (δ^-); conversely, the oxidized atom has acquired a partial positive charge (δ^+), as the positive charge on the atomic nucleus now exceeds the total charge of the electrons surrounding it. (B) A simple reduced carbon compound, such as methane, can be oxidized in a stepwise fashion by the successive replacement of its covalently bonded hydrogen atoms with oxygen atoms. With each step, electrons are shifted away from the carbon, and the carbon atom becomes progressively more oxidized. Moving in the opposite direction, carbon dioxide becomes progressively more reduced as its oxygen atoms are replaced by hydrogens to yield methane.

than its equal share of electrons and forms a *polar covalent bond*. The positive charge of the carbon nucleus now slightly exceeds the negative charge of its electrons, so that the carbon atom acquires a partial positive charge (δ^+) and is said to be oxidized. Conversely, the carbon atom in a C–H bond has somewhat more than its share of electrons; it acquires a partial negative charge (δ^-), and so is said to be reduced (**Figure 3–11A**).

When a molecule in a cell picks up an electron (e^-), it often picks up a proton (H^+) at the same time (protons being freely available in water). The net effect in this case is to add a hydrogen atom to the molecule:

$$A + e^- + H^+ \rightarrow AH$$

Even though a proton plus an electron is involved (instead of just an electron), such *hydrogenation* reactions are reductions, and the reverse, *dehydrogenation*, reactions are oxidations. An easy way to tell whether an organic molecule is being oxidized or reduced is to count its C–H bonds: reduction occurs when the number of C–H bonds increases, whereas oxidation occurs when the number of C–H bonds decreases (**Figure 3–11B**).

As we will see later in this chapter—and again in Chapter 13—cells use enzymes to catalyze the oxidation of organic molecules in small steps, through a sequence of reactions that allows energy to be harvested in useful forms.

FREE ENERGY AND CATALYSIS

Enzymes, like cells, obey the second law of thermodynamics. Although they can speed up energetically favorable reactions—those that produce disorder in the universe—enzymes cannot by themselves force energetically unfavorable reactions to occur. Cells, however, must do just that in order to grow and divide—or just to survive. They must build highly ordered and energy-rich molecules from small and simple ones—a process that requires an input of energy.

To understand how enzymes promote **catalysis**—the acceleration of the specific chemical reactions needed to sustain life—we first need to examine the energetics involved. In this section, we consider how the free energy of molecules contributes to their chemistry, and we see how free-energy changes—which reflect how much total disorder is generated in the universe by a reaction—influence whether and how the reaction will proceed. We then discuss how enzymes lower the activation energy needed to initiate reactions in the cell. And we describe how enzymes can exploit differences in the free-energy changes of different reactions to drive the energetically unfavorable reactions that produce biological order. Such enzyme-assisted catalysis is crucial for cells: without it, life could not exist.

Chemical Reactions Proceed in the Direction that Causes a Loss of Free Energy

Paper burns readily, releasing into the atmosphere water and carbon dioxide as gases, while simultaneously releasing energy as heat:

$$\text{paper} + O_2 \rightarrow \text{smoke} + \text{ashes} + \text{heat} + CO_2 + H_2O$$

This reaction occurs in only one direction: smoke and ashes never spontaneously gather carbon dioxide and water from the heated atmosphere and reconstitute themselves into paper. When paper burns, much of its chemical energy is dissipated as heat: it is not lost from the universe, since energy can never be created or destroyed; instead, it is irretrievably dispersed in the chaotic random thermal motions of molecules. At the same time, the atoms and molecules of the paper become dispersed and disordered. In the language of thermodynamics, there has been a release of *free energy*—that is, energy that can be harnessed to do work or drive chemical reactions. This release reflects a loss of orderliness in the way the energy and molecules had been stored in the paper. We will discuss free energy in more detail shortly, but the general principle can be summarized as follows: chemical reactions proceed only in the direction that leads to a loss of free energy. In other words, the spontaneous direction for any reaction is the direction that goes "downhill." A "downhill" reaction in this sense is said to be energetically favorable.

Enzymes Reduce the Energy Needed to Initiate Spontaneous Reactions

Although the most energetically favorable form of carbon under ordinary conditions is CO_2, and that of hydrogen is H_2O, a living organism will not disappear in a puff of smoke, and the book in your hands will not burst spontaneously into flames. This is because the molecules in both the living organism and the book are in a relatively stable state, and they cannot be changed to lower-energy states without an initial input of energy. In other words, a molecule requires a boost over an energy barrier before it can undergo a chemical reaction that moves it to a lower-energy (more stable) state (**Figure 3–12A**). This boost is known as the

QUESTION 3–2

In which of the following reactions does the *red* atom undergo an oxidation?

A. $Na \rightarrow Na^+$ (Na atom $\rightarrow$ Na$^+$ ion)
B. $Cl \rightarrow Cl^-$ (Cl atom $\rightarrow$ Cl$^-$ ion)
C. $CH_3CH_2OH \rightarrow CH_3CHO$
(ethanol $\rightarrow$ acetaldehyde)
D. $CH_3CHO \rightarrow CH_3COO^-$
(acetaldehyde $\rightarrow$ acetic acid)
E. $CH_2{=}CH_2 \rightarrow CH_3CH_3$
(ethene $\rightarrow$ ethane)

Figure 3–12 Even energetically favorable reactions require activation energy to get them started. (A) Compound Y (a reactant) is in a relatively stable state; thus energy is required to convert it to compound X (a product), even though X is at a lower overall energy level than Y. This conversion will not take place, therefore, unless compound Y can acquire enough *activation energy* (*energy a* minus *energy b*) from its surroundings to undergo the reaction that converts it into compound X. This energy may be provided by means of an unusually energetic collision with other molecules. For the reverse reaction, X → Y, the activation energy required will be much larger (*energy a* minus *energy c*); this reaction will therefore occur much more rarely. Activation energies are always positive. The total energy change for the energetically favorable reaction Y → X, is *energy c* minus *energy b*, a negative number, which corresponds to a loss of free energy. (B) Energy barriers for specific reactions can be lowered by catalysts, as indicated by the line marked *d*. Enzymes are particularly effective catalysts because they greatly reduce the activation energy for the reactions they catalyze.

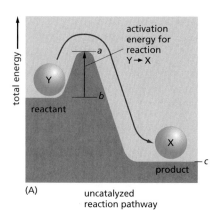

(A) uncatalyzed reaction pathway

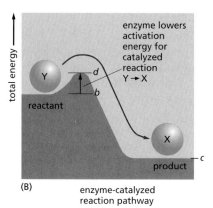

(B) enzyme-catalyzed reaction pathway

Figure 3–13 Lowering the activation energy greatly increases the probability that a reaction will occur. At any given instant, a population of identical substrate molecules will have a range of energies, distributed as shown on the graph. The varying energies come from collisions with surrounding molecules, which make the substrate molecules jiggle, vibrate, and spin. For a molecule to undergo a chemical reaction, the energy of the molecule must exceed the activation energy barrier for that reaction (*dashed lines*); for most biological reactions, this almost never happens without enzyme catalysis. Even with enzyme catalysis, only a small fraction of substrate molecules reach an energy state that is high enough for them to undergo a reaction (*red shaded* area).

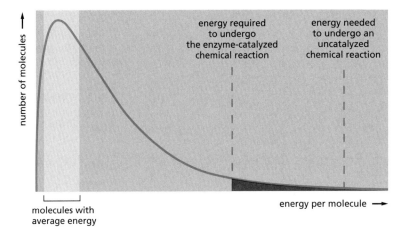

Figure 3–14 Enzymes catalyze reactions by lowering the activation energy barrier. (A) The dam represents the activation energy, which is lowered by enzyme catalysis. Each *green ball* represents a potential substrate molecule that is bouncing up and down in energy level owing to constant encounters with waves, an analogy for the thermal bombardment of substrate molecules by surrounding water molecules. When the barrier—the activation energy—is lowered significantly, the balls (substrate molecules) with sufficient energy can roll downhill, an energetically favorable movement. (B) The four walls of the box represent the activation energy barriers for four different chemical reactions that are all energetically favorable because the products are at lower energy levels than the substrates. In the left-hand box, none of these reactions occurs because even the largest waves are not large enough to surmount any of the energy barriers. In the right-hand box, enzyme catalysis lowers the activation energy for reaction number 1 only; now the jostling of the waves allows the substrate molecule to pass over this energy barrier, allowing reaction 1 to proceed (Movie 3.1). (C) A branching river with a set of barrier dams (*yellow* boxes) serves to illustrate how a series of enzyme-catalyzed reactions determines the exact reaction pathway followed by each molecule inside the cell by controlling specifically which reaction will be allowed at each junction.

activation energy. In the case of a burning book, the activation energy is provided by the heat of a lighted match. But cells can't raise their temperature to drive biological reactions. Inside cells, the push over the energy barrier is aided by specialized proteins called **enzymes**.

Each enzyme binds tightly to one or two molecules, called **substrates**, and holds them in a way that greatly reduces the activation energy needed to facilitate a specific chemical interaction between them (**Figure 3–12B**). A substance that can lower the activation energy of a reaction is termed a **catalyst**; catalysts increase the rate of chemical reactions because they allow a much larger proportion of the random collisions with surrounding molecules to kick the substrates over the energy barrier, as illustrated in **Figure 3–13** and **Figure 3–14A**. Enzymes are among the most effective catalysts known. They can speed up reactions by a factor of as much as 10^{14} (that is, trillions of times faster than the same reactions would proceed without an enzyme catalyst). Enzymes therefore allow reactions that would not otherwise occur to proceed rapidly at the normal temperature inside cells.

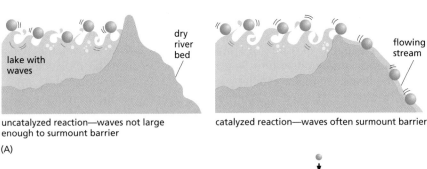

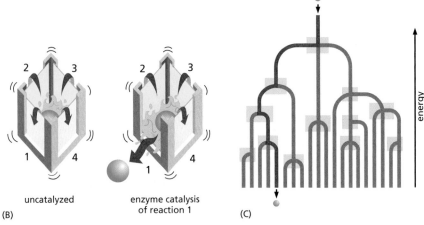

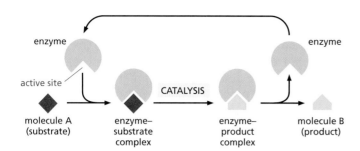

Figure 3–15 Enzymes convert substrates to products while remaining unchanged themselves. Each enzyme has an active site to which one or two substrate molecules bind, forming an enzyme–substrate complex. A reaction occurs at the active site, generating an enzyme–product complex. The product is then released, allowing the enzyme to bind additional substrate molecules and repeat the reaction. An enzyme thus serves as a catalyst, and it usually forms or breaks a single covalent bond in a substrate molecule.

Unlike the effects of temperature, enzymes are highly selective. Each enzyme usually speeds up only one particular reaction out of the several possible reactions that its substrate molecules could undergo. In this way, enzymes direct each of the many different molecules in a cell along specific reaction pathways (**Figure 3–14B and C**), thereby producing the compounds that the cell actually needs.

Like all catalysts, enzyme molecules themselves remain unchanged after participating in a reaction and therefore can function over and over again (**Figure 3–15**). In Chapter 4, we will discuss further how enzymes work, after we have looked in detail at the molecular structure of proteins.

The Free-Energy Change for a Reaction Determines Whether It Can Occur

According to the second law of thermodynamics, a chemical reaction can proceed only if it results in a net (overall) increase in the disorder of the universe (see Figure 3–5). Disorder increases when useful energy that could be harnessed to do work is dissipated as heat. The useful energy in a system is known as its **free energy**, or **G**. And because chemical reactions involve a transition from one molecular state to another, the term that is of most interest to chemists and cell biologists is the **free-energy change**, denoted **ΔG** ("Delta G").

Let's consider a collection of molecules. ΔG measures the amount of disorder created in the universe when a reaction involving these molecules takes place. *Energetically favorable* reactions, by definition, are those that create disorder by decreasing the free energy of the system to which they belong; in other words, they have a *negative* ΔG (**Figure 3–16**).

A reaction can occur spontaneously only if ΔG is negative. On a macroscopic scale, an energetically favorable reaction with a negative ΔG is the relaxation of a compressed spring into an expanded state, releasing its stored elastic energy as heat to its surroundings. On a microscopic scale, an energetically favorable reaction with a negative ΔG occurs when salt (NaCl) dissolves in water. Note that, just because a reaction can occur spontaneously, does not mean it will occur quickly. The decay of diamonds into graphite is a spontaneous process—but it takes millions of years.

Energetically unfavorable reactions, by contrast, create order in the universe; they have a *positive* ΔG. Such reactions—for example, the formation of a peptide bond between two amino acids—cannot occur spontaneously; they take place only when they are coupled to a second reaction with a negative ΔG large enough that the net ΔG of the entire process is negative (**Figure 3–17**). Life is possible because enzymes can create biological order by coupling energetically unfavorable reactions with energetically favorable ones. These critical concepts are summarized, with examples, in **Panel 3–1** (pp. 96–97).

The free energy of Y is greater than the free energy of X. Therefore ΔG is negative (< 0), and the disorder of the universe increases during the reaction Y→X.

this reaction can occur spontaneously

If the reaction X→Y occurred, ΔG would be positive (> 0), and the universe would become more ordered.

this reaction can occur only if it is coupled to a second, energetically favorable reaction

Figure 3–16 Energetically favorable reactions have a negative ΔG, whereas energetically unfavorable reactions have a positive ΔG.

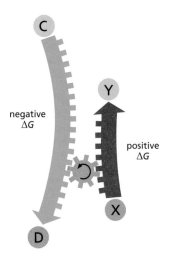

Figure 3–17 Reaction coupling can drive an energetically unfavorable reaction. The energetically unfavorable (ΔG > 0) reaction X → Y cannot occur unless it is coupled to an energetically favorable (ΔG < 0) reaction C → D, such that the net free-energy change for the coupled reactions is negative (less than 0).

ΔG Changes As a Reaction Proceeds Toward Equilibrium

It's easy to see how a tensed spring, when left to itself, will relax and release its stored energy to the environment as heat. But chemical reactions are a bit more complex—and harder to intuit. That's because whether a reaction will proceed depends not only on the energy stored in each individual molecule, but also on the concentrations of the molecules in the reaction mixture. Recalling our coin analogy, more coins in a jiggling box will flip from a head to a tail orientation when the box contains 90 heads and 10 tails, than when the box contains 10 heads and 90 tails.

The same is true for a chemical reaction. As the energetically favorable reaction Y → X proceeds, the concentration of the product X will increase and the concentration of the substrate Y will decrease. This change in relative concentrations of substrate and product will cause the ratio of Y to X to shrink, making the initially favorable ΔG less and less negative. Unless more Y is added, the reaction will slow and eventually stop.

Because ΔG changes as products accumulate and substrates are depleted, chemical reactions will generally proceed until they reach a state of **equilibrium**. At that point, the rates of the forward and reverse reactions are equal, and there is no further net change in the concentrations of substrate or product (**Figure 3–18**). For reactions at chemical equilibrium, ΔG = 0, so the reaction will not proceed forward or backward, and no work can be done.

Such a state of chemical inactivity would be incompatible with life. Living cells avoid reaching a state of complete chemical equilibrium because they are constantly exchanging materials with their environment: replenishing nutrients and eliminating waste products. Many of the individual reactions in the cell's complex metabolic network also exist in disequilibrium because the products of one reaction are continually being siphoned off to become the substrates in a subsequent reaction. Rarely do products and substrates reach concentrations at which the forward and reverse reaction rates are equal.

The Standard Free-Energy Change, ΔG°, Makes it Possible to Compare the Energetics of Different Reactions

Because ΔG depends on the concentrations of the molecules in the reaction mixture at any given time, it is not a particularly useful value for comparing the relative energies of different types of reactions. But such energetic assessments are necessary, for example, to predict whether an energetically favorable reaction is likely to have a ΔG negative enough to drive an energetically unfavorable reaction. To compare reactions in this way, we need to turn to the *standard free-energy change* of a reaction, **ΔG°**. The ΔG° is independent of concentration; it depends only on the intrinsic characters of the reacting molecules, based on their behavior under ideal conditions where the concentrations of all the reactants are set to the same fixed value of 1 mole/liter.

A large body of thermodynamic data has been collected from which ΔG° can be calculated for most metabolic reactions. Some common reactions are compared in terms of their ΔG° in Panel 3–1 (pp. 96–97).

The ΔG of a reaction can be calculated from ΔG° if the concentrations of the reactants and products are known. For the simple reaction Y → X, their relationship follows this equation:

$$\Delta G = \Delta G° + RT \ln \frac{[X]}{[Y]}$$

where ΔG is in kilocalories per mole, [Y] and [X] denote the concentrations

QUESTION 3–3

Consider the analogy of the jiggling box containing coins that was described on page 85. The reaction, the flipping of coins that either face heads up (H) or tails up (T), is described by the equation H ↔ T, where the rate of the forward reaction equals the rate of the reverse reaction.
A.   What are ΔG and ΔG° in this analogy?
B.   What corresponds to the temperature at which the reaction proceeds? What corresponds to the activation energy of the reaction? Assume you have an "enzyme," called jigglase, which catalyzes this reaction. What would the effect of jigglase be and what, mechanically, might jigglase do in this analogy?

FOR THE ENERGETICALLY FAVORABLE REACTION Y → X,

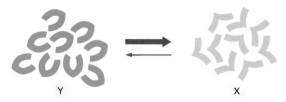

when X and Y are at equal concentrations, [Y] = [X], the formation of X is energetically favored. In other words, the ΔG of Y → X is negative and the ΔG of X → Y is positive. But because of thermal bombardments, there will always be some X converting to Y.

THUS, FOR EACH INDIVIDUAL MOLECULE,

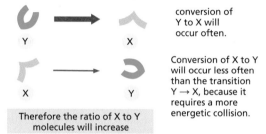

conversion of Y to X will occur often.

Conversion of X to Y will occur less often than the transition Y → X, because it requires a more energetic collision.

Therefore the ratio of X to Y molecules will increase

EVENTUALLY,   there will be a large enough excess of X over Y to just compensate for the slow rate of X → Y, such that the number of Y molecules being converted to X molecules each second is exactly equal to the number of X molecules being converted to Y molecules each second. At this point, the reaction will be at equilibrium.

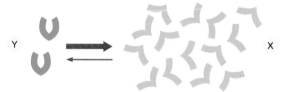

AT EQUILIBRIUM,   there is no net change in the ratio of Y to X, and the ΔG for both forward and backward reactions is zero.

Figure 3–18 Reactions will eventually reach a chemical equilibrium. At that point, the forward and the backward fluxes of reacting molecules are equal and opposite. The widths of the arrows indicate the relative rates at which an *individual molecule* converts.

of Y and X in moles/liter, ln is the natural logarithm, and RT is the product of the gas constant, R, and the absolute temperature, T. At 37°C, $RT = 0.616$. (A mole is 6×10^{23} molecules of a substance.)

From this equation, we can see that when the concentrations of reactants and products are equal, in other words, [X]/[Y] = 1, the value of ΔG equals the value of $\Delta G°$ (because ln 1 = 0). Thus when the reactants and products are present in equal concentrations, the direction of the reaction depends entirely on the intrinsic properties of the molecules.

The Equilibrium Constant Is Directly Proportional to $\Delta G°$

As mentioned earlier, all chemical reactions tend to proceed toward equilibrium. Knowing where that equilibrium lies for any given reaction will tell you which way the reaction will proceed—and how far it will go. For example, if a reaction is at equilibrium when the concentration of the product is ten times the concentration of the substrate, and we begin with a surplus of substrate and little or no product, the reaction will proceed forward for some time. For the simple reaction Y → X, that value—the ratio of substrate to product at equilibrium—is called the reaction's **equilibrium constant**, K. Expressed as an equation:

$$K = \frac{[X]}{[Y]}$$

where [X] is the concentration of the product and [Y] is the concentration of the substrate at equilibrium.

FREE ENERGY

This panel reviews the concept of free energy and offers examples showing how changes in free energy determine whether—and how—biological reactions occur.

The molecules of a living cell possess energy because of their vibrations, rotations, and movement through space, and because of the energy that is stored in the bonds between individual atoms.

The free energy, G (in kcal/mole), measures the energy of a molecule which could in principle be used to do useful work at constant temperature, as in a living cell. Energy can also be expressed in joules (1 cal = 4.184 joules).

REACTIONS CAUSE DISORDER

Think of a chemical reaction occurring in a cell that has a constant temperature and volume. This reaction can produce disorder in two ways.

1 Changes of bond energy of the reacting molecules can cause heat to be released, which disorders the environment around the cell.

2 The reaction can decrease the amount of order in the cell—for example, by breaking apart a long chain of molecules, or by disrupting an interaction that prevents bond rotations.

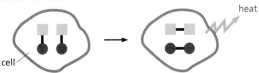

PREDICTING REACTIONS

To predict the outcome of a reaction (Will it proceed to the right or to the left? At what point will it stop?), we must measure its standard free-energy change ($\Delta G°$).
This quantity represents the gain or loss of free energy as one mole of reactant is converted to one mole of product under "standard conditions" (all molecules present at a concentration of 1 M and pH 7.0).

$\Delta G°$ for some reactions

glucose-1-P ⟶ glucose-6-P	–1.7 kcal/mole
sucrose ⟶ glucose + fructose	–5.5 kcal/mole
ATP ⟶ ADP + P_i	–7.3 kcal/mole
glucose + $6O_2$ ⟶ $6CO_2$ + $6H_2O$	–686 kcal/mole

driving force

ΔG ("DELTA G")

Changes in free energy occurring in a reaction are denoted by ΔG, where "Δ" indicates a difference. Thus, for the reaction

$$A + B \longrightarrow C + D$$

ΔG = free energy (C + D) minus free energy (A + B)

ΔG measures the amount of disorder caused by a reaction: the change in order inside the cell, plus the change in order of the surroundings caused by the heat released.

ΔG is useful because it measures how far away from equilibrium a reaction is. Thus the reaction

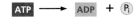

has a large negative ΔG because cells keep the reaction a long way from equilibrium by continually making fresh ATP. However, if the cell dies, then most of its ATP will be hydrolyzed, until equilibrium is reached; at equilibrium, the forward and backward reactions occur at equal rates and $\Delta G = 0$.

SPONTANEOUS REACTIONS

From the second law of thermodynamics, we know that the disorder of the universe can only increase. ΔG is *negative* if the disorder of the universe (reaction plus surroundings) *increases*.

In other words, a chemical reaction that occurs spontaneously must have a negative ΔG:

$$G_{products} - G_{reactants} = \Delta G < 0$$

EXAMPLE: The difference in free energy of 100 ml of 10 mM sucrose (common sugar) and 100 ml of 10 mM glucose plus 10 mM fructose is about –5.5 calories. Therefore, the hydrolysis reaction that produces two monosaccharides from a disaccharide (sucrose → glucose + fructose) can proceed spontaneously.

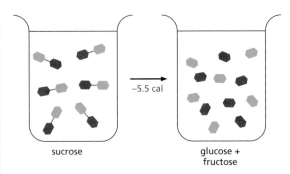

In contrast, the reverse reaction (glucose + fructose → sucrose), which has a ΔG of +5.5 calories, could not occur without an input of energy from a coupled reaction.

REACTION RATES

A spontaneous reaction is not necessarily an instantaneous reaction: a reaction with a negative free-energy change (ΔG) will not necessarily occur rapidly by itself. Consider, for example, the combustion of glucose in oxygen:

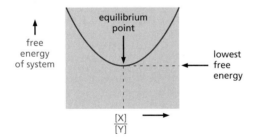

$$+ \ 6O_2 \longrightarrow 6CO_2 + 6H_2O$$

$$\Delta G^\circ = -686 \text{ kcal/mole}$$

Even this highly favorable reaction may not occur for centuries unless there are enzymes to speed up the process.
Enzymes are able to catalyze reactions and speed up their rate, but they cannot change the ΔG° of a reaction.

CHEMICAL EQUILIBRIA

A fixed relationship exists between the standard free-energy change of a reaction, ΔG°, and its equilibrium constant K. For example, the reversible reaction

$$Y \rightleftarrows X$$

will proceed until the ratio of concentrations [X]/[Y] is equal to K (note: square brackets [ ] indicate concentration). At this point, the free energy of the system will have its lowest value.

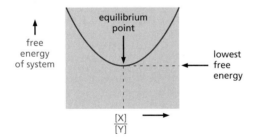

At 37°C, $\Delta G^\circ = -1.42 \log_{10} K$ (see text, p. 98)

$$K = 10^{-\Delta G^\circ/1.42}$$

For example, the reaction

glucose-1-P   glucose-6-P

has $\Delta G^\circ = -1.74$ kcal/mole. Therefore, its equilibrium constant

$$K = 10^{(1.74/1.42)} = 10^{(1.23)} = 17$$

So the reaction will reach steady state when
[glucose-6-P]/[glucose-1-P] = 17

COUPLED REACTIONS

Reactions can be "coupled" together if they share one or more intermediates. In this case, the overall free-energy change is simply the sum of the individual ΔG° values. A reaction that is unfavorable (has a positive ΔG°) can for this reason be driven by a second, highly favorable reaction.

SINGLE REACTION

$$\Delta G^\circ = +5.5 \text{ kcal/mole}$$

glucose   fructose   sucrose

NET RESULT: reaction will not occur

$$ATP \longrightarrow ADP + \text{P}   \Delta G^\circ = -7.3 \text{ kcal/mole}$$

NET RESULT: reaction is highly favorable

COUPLED REACTIONS

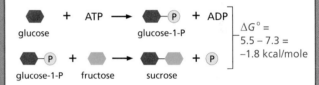

glucose   glucose-1-P   + ADP

$\Delta G^\circ =$
$5.5 - 7.3 =$
-1.8 kcal/mole

glucose-1-P   fructose   sucrose

NET RESULT: sucrose is made in a reaction driven by the hydrolysis of ATP

HIGH-ENERGY BONDS

One of the most common reactions in the cell is hydrolysis, in which a covalent bond is split by adding water.

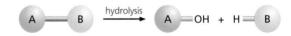

The ΔG° for this reaction is sometimes loosely termed the "bond energy." Compounds such as acetyl phosphate and ATP, which have a large negative ΔG° of hydrolysis in an aqueous solution, are said to have "high-energy" bonds.

		ΔG° (kcal/mole)
acetyl (P) ⟶	acetate + (Pᵢ)	−10.3
ATP ⟶	ADP + (Pᵢ)	−7.3
glucose-6-P ⟶	glucose + (Pᵢ)	−3.3

(Note that, for simplicity, H_2O is omitted from the above equations.)

But how do we know at what concentrations of substrate and product a reaction will reach equilibrium? It goes back to the intrinsic properties of the molecules involved, as expressed by $\Delta G°$. Let's see why.

At equilibrium, the rate of the forward reaction is exactly balanced by the rate of the reverse reaction. At that point, $\Delta G = 0$, and there is no net change of free energy to drive the reaction in either direction (see Panel 3–1, pp. 96–97).

Now, if we return to the equation presented on p. 94,

$$\Delta G = \Delta G° + RT \ln \frac{[X]}{[Y]}$$

we can see that, at equilibrium at 37°C, where $\Delta G = 0$ and the constant $RT = 0.616$, this equation becomes:

$$\Delta G° = -0.616 \ln \frac{[X]}{[Y]}$$

In other words, $\Delta G°$ is directly proportional to the equilibrium constant, K:

$$\Delta G° = -0.616 \ln K$$

If we convert this equation from natural log (ln) to the more commonly used base–10 logarithm (log), we get

$$\Delta G° = -1.42 \log K$$

This equation reveals how the equilibrium ratio of Y to X, expressed as the equilibrium constant K, depends on the intrinsic character of the molecules, as expressed in the value of $\Delta G°$ (**Table 3–1**). It tells us that for every 1.42 kcal/mole difference in free energy at 37°C, the equilibrium constant changes by a factor of 10. Thus, the more energetically favorable the reaction, the more product will accumulate if the reaction proceeds to equilibrium.

In Complex Reactions, the Equilibrium Constant Includes the Concentrations of All Reactants and Products

We have so far discussed the simplest of reactions, Y → X, in which a single substrate is converted into a single product. But inside cells, it is more common for two reactants to combine to form a single product: A + B ⇌ AB. How can we predict how this reaction will proceed?

The same principles apply, except that in this case the equilibrium constant K includes the concentrations of both of the reactants, in addition to the concentration of the product:

$$K = [AB]/[A][B]$$

As illustrated in **Figure 3–19**, the concentrations of both reactants are multiplied because the formation of product AB depends on the collision of A and B, and these encounters occur at a rate that is proportional to $[A] \times [B]$. As with single-substrate reactions, $\Delta G° = -1.42 \log K$ at 37°C.

The Equilibrium Constant Indicates the Strength of Molecular Interactions

The concept of free-energy change does not only apply to chemical reactions where covalent bonds are being broken and formed, but also to interactions where one molecule binds to another by means of noncovalent interactions (see Chapter 2, p. 63). Noncovalent interactions are immensely important to cells. They include the binding of substrates to enzymes, the binding of gene regulatory proteins to DNA, and the binding of one protein to another to make the many different structural and functional protein complexes that operate in a living cell.

TABLE 3–1 RELATIONSHIP BETWEEN THE STANDARD FREE-ENERGY CHANGE, $\Delta G°$, AND THE EQUILIBRIUM CONSTANT

Equilibrium Constant $\frac{[X]}{[Y]}$	Standard Free Energy ($\Delta G°$) of X minus Free Energy of Y in kcal/mole
10^5	–7.1
10^4	–5.7
10^3	–4.3
10^2	–2.8
10	–1.4
1	0
10^{-1}	1.4
10^{-2}	2.8
10^{-3}	4.3
10^{-4}	5.7
10^{-5}	7.1

Values of the equilibrium constant were calculated for the simple chemical reaction Y ↔ X, using the equation given in the text.

The $\Delta G°$ values given here are in kilocalories per mole at 37°C. As explained in the text, $\Delta G°$ represents the free-energy difference under standard conditions (where all components are present at a concentration of 1 mole/liter).

From this table, we see that, if there is a favorable free-energy change of –4.3 kcal/mole for the transition Y → X, there will be 1000 times more molecules of X than of Y at equilibrium.

$$A + B \xrightarrow{\text{association}} AB$$

$$\text{association rate} = \text{association rate constant} \times \text{concentration of A} \times \text{concentration of B}$$

$$\text{association rate} = k_{on}\,[A]\,[B]$$

$$AB \xrightarrow{\text{dissociation}} A + B$$

$$\text{dissociation rate} = \text{dissociation rate constant} \times \text{concentration of AB}$$

$$\text{dissociation rate} = k_{off}\,[AB]$$

AT EQUILIBRIUM:

association rate = dissociation rate

$$k_{on}\,[A]\,[B] = k_{off}\,[AB]$$

$$\frac{[AB]}{[A]\,[B]} = \frac{k_{on}}{k_{off}} = K = \textbf{equilibrium constant}$$

Figure 3–19 **The equilibrium constant (K) for the reaction A + B → AB depends on both the association and dissociation rate constants.** Molecules A and B must collide in order to interact, and the association rate is therefore proportional to the product of their individual concentrations [A] × [B]. As shown, the ratio of the rate constants k_{on} and k_{off} for the association and the dissociation reactions, respectively, is equal to the equilibrium constant, K, for the interaction. For two interacting components, K involves the concentrations of both substrates, in addition to that of the product. However, the relationship between K and $\Delta G°$ is the same as that shown in Table 3–1. The larger the value of K, the stronger is the binding between A and B.

Two molecules will bind to each other if the free-energy change for the interaction is negative; that is, the free energy of the resulting complex is lower than the sum of the free energies of the two partners when unbound. Because the equilibrium constant of a reaction is related directly to $\Delta G°$, K is commonly employed as a measure of the binding strength of a noncovalent interaction between two molecules. The binding strength is a very useful quantity to know because it also indicates how specific the interaction is between the two molecules.

Consider the reaction that was shown in Figure 3–19, where molecule A interacts with molecule B to form the complex AB. The reaction proceeds until it reaches equilibrium, at which point the number of association events precisely equals the number of dissociation events; at this point, the concentrations of reactants A and B, and of the complex AB, can be used to determine the equilibrium constant K.

K becomes larger as the *binding energy*—that is, the energy released in the binding interaction—increases. In other words, the larger K is, the greater is the drop in free energy between the dissociated and associated states, and the more tightly the two molecules will bind. Even a change of a few noncovalent bonds can have a striking effect on a binding interaction, as illustrated in **Figure 3–20**. In this example, eliminating a few hydrogen bonds from a binding interaction can be seen to cause a dramatic decrease in the amount of complex that exists at equilibrium.

Consider 1000 molecules of A and 1000 molecules of B in the cytosol of a eukaryotic cell. The concentration of both will be about 10^{-9} M.

If the equilibrium constant (K) for A + B ↔ AB is 10^{10} liters/mole, then at equilibrium there will be

270	270	730
A molecules	B molecules	AB complexes

If the equilibrium constant is a little weaker, say 10^8 liters/mole—a value that represents a loss of 2.8 kcal/mole of binding energy from the example above, or 2–3 fewer hydrogen bonds—then there will be

915	915	85
A molecules	B molecules	AB complexes

Figure 3–20 **Small changes in the number of weak bonds can have drastic effects on a binding interaction.** This example illustrates the dramatic effect of the presence or absence of a few weak noncovalent bonds in the interaction between two cytosolic proteins.

For Sequential Reactions, the Changes in Free Energy Are Additive

Now we return to our original concern: how can enzymes catalyze reactions that are energetically unfavorable? One way they do so is by directly coupling energetically unfavorable reactions with energetically favorable ones. Consider, for example, two sequential reactions,

$$X \rightarrow Y \text{ and } Y \rightarrow Z$$

where the $\Delta G°$ values are +5 and –13 kcal/mole, respectively. (Recall that a mole is 6×10^{23} molecules of a substance.) The unfavorable reaction, X → Y, will not occur spontaneously. However, it can be driven by the favorable reaction Y → Z, provided that the second reaction follows the first. That's because the overall free-energy change for the coupled reaction is equal to the sum of the free-energy changes for each individual reaction. In this case, the $\Delta G°$ for the coupled reaction will be –8 kcal/mole, making the overall pathway energetically favorable.

Cells can therefore cause the energetically unfavorable transition, X → Y, to occur if an enzyme catalyzing the X → Y reaction is supplemented by a second enzyme that catalyzes the energetically favorable reaction, Y → Z. In effect, the reaction Y → Z acts as a "siphon," pulling the conversion of all of molecule X to molecule Y, and then to molecule Z (**Figure 3–21**). For example, several of the reactions in the long pathway that converts sugars into CO_2 and H_2O are energetically unfavorable. The pathway nevertheless proceeds rapidly to completion, however, because the total $\Delta G°$ for the series of sequential reactions has a large negative value.

Forming a sequential pathway, however, is not the answer for all metabolic needs. Often the desired reaction is simply X → Y, without further conversion of Y to some other product. Fortunately, there are other, more general ways of using enzymes to couple reactions together, involving the production of activated carriers that can shuttle energy from one reaction site to another. We discuss these systems shortly. Before we do, let's pause to look at how enzymes find and recognize their substrates and how enzyme-catalyzed reactions proceed. After all, thermodynamic considerations merely establish whether chemical reactions can occur; enzymes actually make them happen.

Thermal Motion Allows Enzymes to Find Their Substrates

Enzymes and their substrates are both present in relatively small amounts in the cytosol of a cell, yet a typical enzyme can capture and process about a thousand substrate molecules every second. This means that an enzyme can release its product and bind a new substrate in a fraction of a millisecond. How do these molecules find each other so quickly in the crowded cytosol of the cell?

Rapid binding is possible because molecular motions are enormously fast. Because of heat energy, molecules are in constant motion and consequently will explore the cytosolic space very efficiently by wandering

Figure 3–21 An energetically unfavorable reaction can be driven by an energetically favorable follow-on reaction that acts as a chemical siphon. (A) At equilibrium, there are twice as many X molecules as Y molecules. (B) At equilibrium, there are 25 times more Z molecules than Y molecules. (C) If the reactions in (A) and (B) are coupled, nearly all of the X molecules will be converted to Z molecules, as shown. In terms of energetics, the $\Delta G°$ of the Y → Z reaction is so negative that, when coupled to the X → Y reaction, it lowers the ΔG of X → Y, because the ΔG of X → Y decreases as the ratio of Y to X declines. As shown in Figure 3–18, arrow widths reflect the relative rates at which an individual molecule converts; the arrow lengths are the same in both directions here, indicating that there is no net flux.

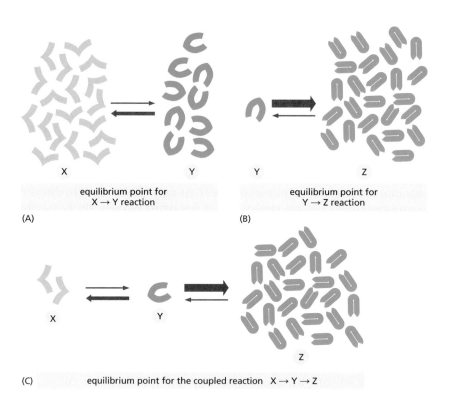

Figure 3–22 **A molecule traverses the cytosol by taking a random walk.** Molecules in solution move in a random fashion due to the continual buffeting they receive in collisions with other molecules. This movement allows small molecules to diffuse rapidly throughout the cell cytosol (**Movie 3.2**).

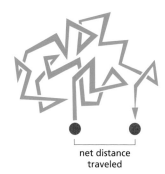

net distance
traveled

randomly through it—a process called **diffusion**. In this way, every molecule in the cytosol collides with a huge number of other molecules each second. As the molecules in a liquid collide and bounce off one another, an individual molecule moves first one way and then another, its path constituting a *random walk* (**Figure 3–22**).

Although the cytosol of a cell is densely packed with molecules of various shapes and sizes (**Figure 3–23**), experiments in which fluorescent dyes and other labeled molecules are injected into the cell cytosol show that small organic molecules diffuse through this aqueous gel nearly as rapidly as they do through water. A small organic molecule, such as a substrate, takes only about one-fifth of a second on average to diffuse a distance of 10 μm. Diffusion is therefore an efficient way for small molecules to move limited distances in the cell.

Because proteins diffuse through the cytosol much more slowly than do small molecules, the rate at which an enzyme will encounter its substrate depends on the concentration of the substrate. The most abundant substrates are present in the cell at a concentration of about 0.5 mM. Because pure water is 55 M, there is only about one such substrate molecule in the cell for every 10^5 water molecules. Nevertheless, the site on an enzyme that binds this substrate will be bombarded by about 500,000 random collisions with the substrate every second. For a substrate concentration tenfold lower (0.05 mM), the number of collisions drops to 50,000 per second, and so on.

The random encounters between an enzyme and its substrate often lead to the formation of an enzyme–substrate complex. This association is stabilized by the formation of multiple, weak bonds between the enzyme and substrate. These weak interactions—which can include hydrogen bonds, van der Waals attractions, and electrostatic attractions (discussed in Chapter 2)—persist until random thermal motion causes the molecules to dissociate again. When two colliding molecules have poorly matching surfaces, few noncovalent bonds are formed, and their total energy is negligible compared with that of thermal motion. In this case, the two molecules dissociate as rapidly as they come together (see Figure 2–33). This is what prevents incorrect and unwanted associations from forming between mismatched molecules, such as those between an enzyme and the wrong substrate. But when the enzyme and substrate are well matched, they form many weak interactions, which keep them held together long enough for a covalent bond in the substrate molecule to be formed or broken. Knowing the speed at which molecules collide and come apart, as well as how fast bonds can be formed and broken, makes the observed rate of enzymatic catalysis seem a little less amazing.

QUESTION 3–5

The enzyme carbonic anhydrase is one of the speediest enzymes known. It catalyzes the rapid conversion of CO_2 gas into the much more soluble bicarbonate ion (HCO_3^-). The reaction:

$$CO_2 + H_2O \leftrightarrow HCO_3^- + H^+$$

is very important for the efficient transport of CO_2 from tissue, where CO_2 is produced by respiration, to the lungs, where it is exhaled. Carbonic anhydrase accelerates the reaction 10^7-fold, hydrating 10^5 CO_2 molecules per second at its maximal speed. What do you suppose limits the speed of the enzyme? Sketch a diagram analogous to the one shown in Figure 3–13 and indicate which portion of your diagram has been designed to display the 10^7-fold acceleration.

Figure 3–23 **The cytosol is crowded with various molecules.** Only the macromolecules, which are drawn to scale, are shown. RNAs are *blue*, ribosomes are *green*, and proteins are *red*. Enzymes and other macromolecules diffuse relatively slowly in the cytosol, in part because they interact with so many other macromolecules. Small molecules, by contrast, can diffuse nearly as rapidly as they do in water. (Adapted from D.S. Goodsell, *Trends Biochem. Sci.* 16:203–206, 1991. With permission from Elsevier.)

100 nm

V_{max} and K_M Measure Enzyme Performance

To catalyze a reaction, an enzyme must first bind its substrate. The substrate then undergoes a reaction to form the product, which initially remains bound to the enzyme. Finally, the product is released and diffuses away, leaving the enzyme free to bind another substrate molecule and catalyze another reaction (see Figure 3–15). The rates of the different steps vary widely from one enzyme to another, and they can be measured by mixing purified enzymes and substrates together under carefully defined conditions in a test tube (see **How We Know**, pp. 104–106).

In such experiments, the substrate is introduced in increasing concentrations to a solution containing a fixed concentration of enzyme. At first, the concentration of the enzyme–substrate complex—and therefore the rate at which product is formed—rises in a linear fashion in direct proportion to substrate concentration. However, as more and more enzyme molecules become occupied by substrate, this rate increase tapers off, until at a very high concentration of substrate it reaches a maximum value, termed V_{max}. At this point, the active sites of all enzyme molecules in the sample are fully occupied by substrate, and the rate of product formation depends only on how rapidly the substrate molecule can undergo a reaction to form the product. For many enzymes, this **turnover number** is of the order of 1000 substrate molecules per second, although turnover numbers between 1 and 100,000 have been measured.

Because there is no clearly defined substrate concentration at which the enzyme can be deemed fully occupied, biochemists instead use a different parameter to gauge the concentration of substrate needed to make the enzyme work efficiently. This value is called the **Michaelis constant**, K_M, named after one of the biochemists who worked out the relationship. The K_M of an enzyme is defined as the concentration of substrate at which the enzyme works at half its maximum speed (**Figure 3–24**). In general, a small K_M indicates that a substrate binds very tightly to the enzyme, and a large K_M indicates weak binding.

Although an enzyme (or any catalyst) functions to lower the activation energy for a reaction such as $Y \rightarrow X$, it is important to note that the enzyme will also lower the activation energy for the reverse reaction $X \rightarrow Y$ to exactly the same degree. The forward and backward reactions will therefore be accelerated by the same factor by an enzyme, and the equilibrium point for the reaction—and thus its $\Delta G°$—remains unchanged (**Figure 3–25**).

QUESTION 3–6

In cells, an enzyme catalyzes the reaction AB → A + B. It was isolated, however, as an enzyme that carries out the opposite reaction A + B → AB. Explain the paradox.

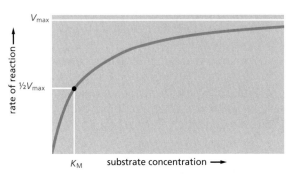

Figure 3–24 An enzyme's performance depends on how rapidly it can process its substrate. The rate of an enzyme reaction (*V*) increases as the substrate concentration increases, until a maximum value (V_{max}) is reached. At this point, all substrate-binding sites on the enzyme molecules are fully occupied, and the rate of the reaction is limited by the rate of the catalytic process on the enzyme surface. For most enzymes, the concentration of substrate at which the reaction rate is half-maximal (K_M) is a direct measure of how tightly the substrate is bound, with a large value of K_M (a large amount of substrate needed) corresponding to weak binding.

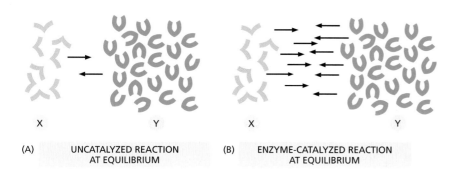

(A)   UNCATALYZED REACTION
AT EQUILIBRIUM

(B)   ENZYME-CATALYZED REACTION
AT EQUILIBRIUM

Figure 3–25 Enzymes cannot change the equilibrium point for reactions. Enzymes, like all catalysts, speed up the forward and reverse rates of a reaction by the same amount. Therefore, for both the (A) uncatalyzed and (B) catalyzed reactions shown here, the number of molecules undergoing the transition X → Y is equal to the number of molecules undergoing the transition Y → X when the ratio of Y molecules to X molecules is 3.5 to 1, as illustrated. In other words, both the catalyzed and uncatalyzed reactions will eventually reach the same equilibrium point, although the catalyzed reaction will reach equilibrium faster.

ACTIVATED CARRIERS AND BIOSYNTHESIS

The energy released by energetically favorable reactions such as the oxidation of food molecules must be stored temporarily before it can be used by cells to fuel energetically unfavorable reactions, such as the synthesis of all the other molecules needed by the cell. In most cases, the energy is stored as chemical-bond energy in a set of *activated carriers*, small organic molecules that contain one or more energy-rich covalent bonds. These molecules diffuse rapidly and carry their bond energy from the sites of energy generation to the sites where energy is used for **biosynthesis** or for other energy-requiring cell activities (**Figure 3–26**).

Activated carriers store energy in an easily exchangeable form, either as a readily transferable chemical group or as readily transferable ("high energy") electrons. They can serve a dual role as a source of both energy and chemical groups for biosynthetic reactions. The most important activated carriers are *ATP* and two molecules that are closely related to each other, *NADH* and *NADPH*. Cells use activated carriers like money to pay for the energetically unfavorable reactions that otherwise would not take place.

The Formation of an Activated Carrier Is Coupled to an Energetically Favorable Reaction

When a fuel molecule such as glucose is oxidized in a cell, enzyme-catalyzed reactions ensure that a large part of the free energy released is captured in a chemically useful form, rather than being released wastefully as heat. (Oxidizing sugar in a cell allows you to power metabolic reactions, whereas burning a chocolate bar in the street will get you nowhere, producing no metabolically useful energy.) In cells, energy capture is achieved by means of a **coupled reaction**, in which an energetically favorable reaction is used to drive an energetically unfavorable one that produces an activated carrier or some other useful molecule.

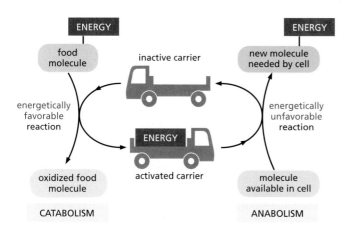

Figure 3–26 Activated carriers can store and transfer energy in a form that cells can use. By serving as intracellular energy shuttles, activated carriers perform their function as go-betweens that link the release of energy from the breakdown of food molecules (catabolism) to the energy-requiring biosynthesis of small and large organic molecules (anabolism).

MEASURING ENZYME PERFORMANCE

At first glance, it seems that a cell's metabolic pathways have been pretty well mapped out, with each reaction proceeding predictably to the next—substrate X is converted to product Y, which is passed along to enzyme Z. So why would anyone need to know exactly how tightly a particular enzyme clutches its substrate or whether it can process 100 or 1000 substrate molecules every second?

In reality, such elaborate metabolic maps merely suggest which pathways a cell might follow as it converts nutrients into small molecules, chemical energy, and the larger building blocks of life. Like a road map, they do not predict the density of traffic under a particular set of conditions: which pathways the cell will use when it is starving, when it is well fed, when oxygen is scarce, when it is stressed, or when it decides to divide. The study of an enzyme's *kinetics*—how fast it operates, how it handles its substrate, how its activity is controlled—makes it possible to predict how an individual catalyst will perform, and how it will interact with other enzymes in a network. Such knowledge leads to a deeper understanding of cell biology, and it opens the door to learning how to harness enzymes to perform desired reactions.

Speed

The first step to understanding how an enzyme performs involves determining the maximal velocity, V_{max}, for the reaction it catalyzes. This is accomplished by measuring, in a test tube, how rapidly the reaction proceeds in the presence of different concentrations of substrate (**Figure 3–27A**): the rate should increase as the amount of substrate rises until the reaction reaches its V_{max}. The velocity of the reaction is measured by monitoring either how quickly the substrate is consumed or how rapidly the product accumulates. In many cases, the appearance of product or the disappearance of substrate can be observed directly with a spectrophotometer. This instrument detects the presence of molecules that absorb light at a particular wavelength; NADH, for example, absorbs light at 340 nm, while its oxidized counterpart, NAD^+, does not. So, a reaction that generates NADH (by reducing NAD^+) can be monitored by following the formation of NADH at 340 nm in a spectrophotometer.

To determine the V_{max} of a reaction, you would set up a series of test tubes, where each tube contains a different concentration of substrate. For each tube, add the same amount of enzyme and then measure the velocity of the reaction—the number of micromoles of substrate consumed or product generated per minute. Because these numbers will tend to decrease over time, the rate used is the velocity measured early in the reaction. These initial velocity values (v) are then plotted against the substrate concentration, yielding a curve like the one shown in **Figure 3–27B**.

Looking at this plot, however, it is difficult to determine the exact value of V_{max}, as it is not clear where the reaction rate will reach its plateau. To get around this problem, the data are converted to their reciprocals

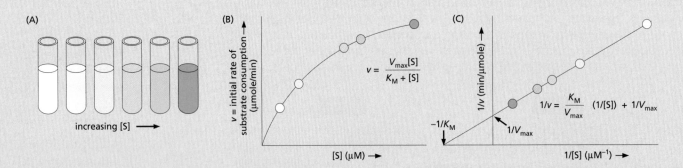

Figure 3–27 Measured reaction rates are plotted to determine V_{max} and K_M of an enzyme-catalyzed reaction. (A) A series of increasing substrate concentrations is prepared, a fixed amount of enzyme is added, and initial reaction rates (velocities) are determined. (B) The initial velocities (v) plotted against the substrate concentrations [S] give a curve described by the general equation $y = ax/(b + x)$. Substituting our kinetic terms, the equation becomes $v = V_{max}[S]/(K_M + [S])$, where V_{max} is the asymptote of the curve (the value of y at an infinite value of x), and K_M is equal to the substrate concentration where v is one-half V_{max}. This is called the *Michaelis–Menten equation*, named for the biochemists who provided evidence for this enzymatic relationship. (C) In a double-reciprocal plot, 1/v is plotted against 1/[S]. The equation describing this straight line is $1/v = (K_M/V_{max})(1/[S]) + 1/V_{max}$. When 1/[S] = 0, the y intercept (1/v) is $1/V_{max}$. When 1/v = 0, the x intercept (1/[S]) is $-1/K_M$. Plotting the data this way allows V_{max} and K_M to be calculated more precisely. By convention, lowercase letters are used for variables (hence v for velocity) and uppercase letters are used for constants (hence V_{max}).

and graphed in a "double-reciprocal plot," where the inverse of the velocity ($1/v$) appears on the y axis and the inverse of the substrate concentration ($1/[S]$) on the x axis (**Figure 3–27C**). This graph yields a straight line whose y intercept (the point where the line crosses the y axis) represents $1/V_{max}$ and whose x intercept corresponds to $-1/K_M$. These values are then converted to values for V_{max} and K_M.

Enzymologists use this technique to determine the kinetic parameters of many enzyme-catalyzed reactions (although these days computer programs automatically plot the data and spit out the sought-after values). Some reactions, however, happen too fast to be monitored in this way; the reaction is essentially complete—the substrate entirely consumed—within thousandths of a second. For these reactions, a special piece of equipment must be used to follow what happens during the first few milliseconds after enzyme and substrate meet (**Figure 3–28**).

Control

Substrates are not the only molecules that can influence how well or how quickly an enzyme works. In many cases, products, substrate lookalikes, inhibitors, and other small molecules can also increase or decrease enzyme activity. Such regulation allows cells to control when and how rapidly various reactions occur, a process we will consider in more detail in Chapter 4.

Determining how an inhibitor decreases an enzyme's activity can reveal how a metabolic pathway is regulated—and can suggest how those control points can be circumvented by carefully designed mutations in specific genes.

The effect of an inhibitor on an enzyme's activity is monitored in the same way that we measured the enzyme's kinetics. A curve is first generated showing the velocity of the uninhibited reaction between enzyme and substrate, as described previously. Additional curves are then produced for reactions in which the inhibitor molecule has been included in the mix.

Comparing these curves, with and without inhibitor, can also reveal how a particular inhibitor impedes enzyme activity. For example, some inhibitors bind to the same site on an enzyme as its substrate. These *competitive inhibitors* block enzyme activity by competing directly with the substrate for the enzyme's attention. They resemble the substrate enough to tie up the enzyme, but they differ enough in structure to avoid getting converted to product. This blockage can be overcome by adding enough substrate so that enzymes are more likely to encounter a substrate molecule than an inhibitor molecule. From the kinetic data, we can see that competitive inhibitors do not change the V_{max} of a reaction; in other words, add enough substrate and the enzyme will encounter mostly substrate molecules and will reach its maximum velocity (**Figure 3–29**).

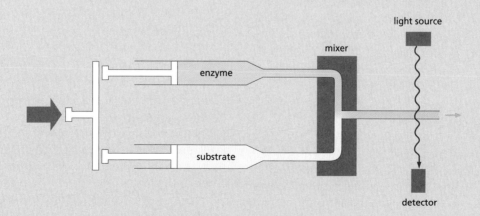

Figure 3–28 A stopped-flow apparatus is used to observe reactions during the first few milliseconds. In this piece of equipment, the enzyme and substrate are rapidly injected into a mixing chamber through two syringes. The enzyme and its substrate meet as they shoot through the mixing tube at flow rates that can easily reach 1000 cm/sec. They then enter another tube and zoom past a detector that monitors, say, the appearance of product. If the detector is located within a centimeter of where the enzyme and substrate meet, it is possible to observe reactions when they are only a few milliseconds old.

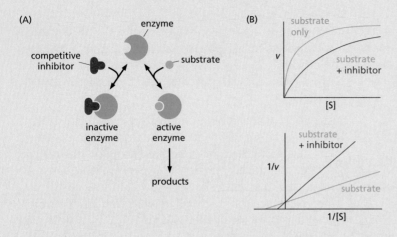

(A)

enzyme

competitive
inhibitor

substrate

inactive
enzyme

active
enzyme

products

(B)

substrate
only

substrate
+ inhibitor

v

[S]

substrate
+ inhibitor

$1/v$

substrate

$1/[S]$

Figure 3–29 A competitive inhibitor directly blocks substrate binding to an enzyme. (A) The active site of the enzyme can bind either the competitive inhibitor or the substrate, but not both together. (B) The upper plot shows that inhibition by a competitive inhibitor can be overcome by increasing the substrate concentration. The double-reciprocal plot below shows that the V_{max} of the reaction is not changed in the presence of the competitive inhibitor: the y intercept is identical for both the curves.

Competitive inhibitors can be used to treat patients who have been poisoned by ethylene glycol, an ingredient in commercially available antifreeze. Although ethylene glycol is itself not fatally toxic, a by-product of its metabolism—oxalic acid—can be lethal. To prevent oxalic acid from forming, the patient is given a large (though not quite intoxicating) dose of ethanol. Ethanol competes with the ethylene glycol for binding to alcohol dehydrogenase, the first enzyme in the pathway to oxalic acid formation. As a result, the ethylene glycol goes mostly unmetabolized and is safely eliminated from the body.

Other types of inhibitors may interact with sites on the enzyme distant from where the substrate binds. As we discuss in Chapter 4, many biosynthetic enzymes are regulated by feedback inhibition, whereby an enzyme early in a pathway will be shut down by a product generated later in the pathway. Because this type of inhibitor binds to a separate regulatory site on the enzyme, the substrate can still bind, but it might do so more slowly than it would in the absence of inhibitor. Such *noncompetitive inhibition* is not overcome by the addition of more substrate.

Design

With the kinetic data in hand, we can use computer modeling programs to predict how an enzyme will perform, and even how a cell will respond when exposed to different conditions—such as the addition of a particular sugar or amino acid to the culture medium, or the addition of a poison or a pollutant. Seeing how a cell manages its resources—which pathways it favors for dealing with particular biochemical challenges—can also suggest strategies for designing better catalysts for reactions of medical or commercial importance (e.g., for producing drugs or detoxifying industrial waste). Using such tactics, bacteria have even been genetically engineered to produce large amounts of indigo—the dye, originally extracted from plants, that makes your blue jeans blue.

Computer programs have been developed to facilitate the dissection of complex reaction pathways. They require information about the components in the pathway, including the K_M and V_{max} of the participating enzymes and the concentrations of enzymes, substrates, products, inhibitors, and other regulatory molecules. The program then predicts how molecules will flow through the pathway, which products will be generated, and where any bottlenecks might be. The process is not unlike balancing an algebraic equation, in which every atom of carbon, nitrogen, oxygen, and so on must be tallied. Such careful accounting makes it possible to rationally design ways to manipulate the pathway, such as re-routing it around a bottleneck, eliminating an important inhibitor, redirecting the reactions to favor the generation of predominantly one product, or extending the pathway to produce a novel molecule. Of course, such computer models must be validated in cells, which may not always behave as predicted.

Producing designer cells that spew out commercial products generally requires using genetic engineering techniques to introduce the gene or genes of choice into a cell, usually a bacterium, that can be manipulated and maintained in the laboratory. We discuss these methods at greater length in Chapter 10. Harnessing the power of cell biology for commercial purposes—even to produce something as simple as the amino acid tryptophan—is currently a multibillion-dollar industry. And, as more genome data come in, presenting us with more enzymes to exploit, it may not be long before vats of custom-made bacteria are churning out drugs and chemicals that represent the biological equivalent of pure gold.

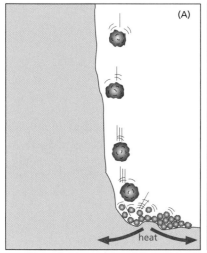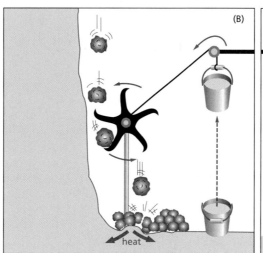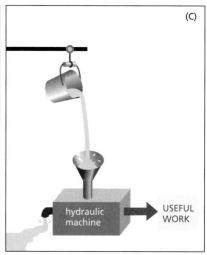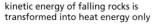

kinetic energy of falling rocks is transformed into heat energy only

part of the kinetic energy is used to lift a bucket of water, and a correspondingly smaller amount is transformed into heat

the potential energy stored in the raised bucket of water can be used to drive hydraulic machines that carry out a variety of useful tasks

Figure 3–30 A mechanical model illustrates the principle of coupled chemical reactions. The spontaneous reaction shown in (A) could serve as an analogy for the direct oxidation of glucose to CO_2 and H_2O, which produces only heat. In (B), the same reaction is coupled to a second reaction, which could serve as an analogy for the synthesis of activated carriers. The energy produced in (B) is in a more useful form than in (A) and can be used to drive a variety of otherwise energetically unfavorable reactions (C).

Such coupling requires enzymes, which are fundamental to all of the energy transactions in the cell.

The nature of a coupled reaction is illustrated by a mechanical analogy in **Figure 3–30**, in which an energetically favorable chemical reaction is represented by rocks falling from a cliff. The kinetic energy of falling rocks would normally be entirely wasted in the form of heat generated by friction when the rocks hit the ground (Figure 3–30A). By careful design, however, part of this energy could be used to drive a paddle wheel that lifts a bucket of water (Figure 3–30B). Because the rocks can now reach the ground only after moving the paddle wheel, we say that the energetically favorable reaction of rocks falling has been directly coupled to the energetically unfavorable reaction of lifting the bucket of water. Because part of the energy is used to do work in (B), the rocks hit the ground with less velocity than in (A), and correspondingly less energy is wasted as heat. The energy saved in the elevated bucket of water can then be used to do useful work (Figure 3–30C).

Analogous processes occur in cells, where enzymes play the role of the paddle wheel in Figure 3–30B. By mechanisms that we discuss in Chapter 13, enzymes couple an energetically favorable reaction, such as the oxidation of foodstuffs, to an energetically unfavorable reaction, such as the generation of activated carriers. As a result, the amount of heat released by the oxidation reaction is reduced by exactly the amount of energy that is stored in the energy-rich covalent bonds of the activated carrier. That saved energy can then be used to power a chemical reaction elsewhere in the cell.

ATP Is the Most Widely Used Activated Carrier

The most important and versatile of the activated carriers in cells is **ATP** (adenosine 5′-triphosphate). Just as the energy stored in the raised bucket of water in Figure 3–30B can be used to drive a wide variety of hydraulic machines, ATP serves as a convenient and versatile store, or currency, of energy that can be used to drive a variety of chemical reactions in cells.

QUESTION 3–7

Use Figure 3–30B to illustrate the following reaction driven by the hydrolysis of ATP:
X + ATP → Y + ADP + P$_i$
A. In this case, which molecule or molecules would be analogous to (i) rocks at top of cliff, (ii) broken debris at bottom of cliff, (iii) bucket at its highest point, and (iv) bucket on the ground?
B. What would be analogous to (i) the rocks hitting the ground in the absence of the paddle wheel in Figure 3–30A and (ii) the hydraulic machine in Figure 3–30C?

Figure 3–31 The interconversion of ATP and ADP occurs in a cycle. The two outermost phosphate groups in ATP are held to the rest of the molecule by high-energy phosphoanhydride bonds and are readily transferred to other organic molecules. Water can be added to ATP to form ADP and inorganic phosphate (P_i). Inside a cell, this hydrolysis of the terminal phosphate of ATP yields between 11 and 13 kcal/mole of usable energy. Although the $\Delta G°$ of this reaction is –7.3 kcal/mole, the ΔG is much more negative, because the ratio of ATP to the products ADP and P_i is so high inside the cell.

The large negative $\Delta G°$ of the reaction arises from a number of factors. Release of the terminal phosphate group removes an unfavorable repulsion between adjacent negative charges; in addition, the inorganic phosphate ion (P_i) released is stabilized by favorable hydrogen-bond formation with water. The formation of ATP from ADP and P_i reverses the hydrolysis reaction; because this condensation reaction is energetically unfavorable, it must be coupled to an energetically more favorable reaction to occur.

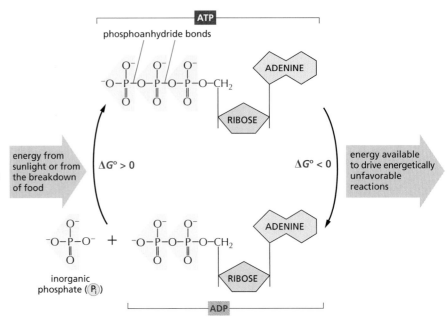

As shown in **Figure 3–31**, ATP is synthesized in an energetically unfavorable *phosphorylation* reaction, in which a phosphate group is added to **ADP** (adenosine 5′-diphosphate). When required, ATP gives up this energy packet in an energetically favorable hydrolysis to ADP and inorganic phosphate (P_i). The regenerated ADP is then available to be used for another round of the phosphorylation reaction that forms ATP, creating an ATP cycle in the cell.

The energetically favorable reaction of ATP hydrolysis is coupled to many otherwise unfavorable reactions through which other molecules are synthesized. We will encounter several of these reactions in this chapter, where we will see exactly how this is done. ATP hydrolysis is often coupled to the transfer of the terminal phosphate in ATP to another molecule, as illustrated in **Figure 3–32**. Any reaction that involves the transfer of a phosphate group to a molecule is termed a phosphorylation reaction. Phosphorylation reactions are examples of condensation reactions (see Figure 2–25), and they occur in many important cell processes: they activate substrates, mediate the exchange of chemical energy, and serve as key constituents of intracellular signaling pathways (discussed in Chapter 16).

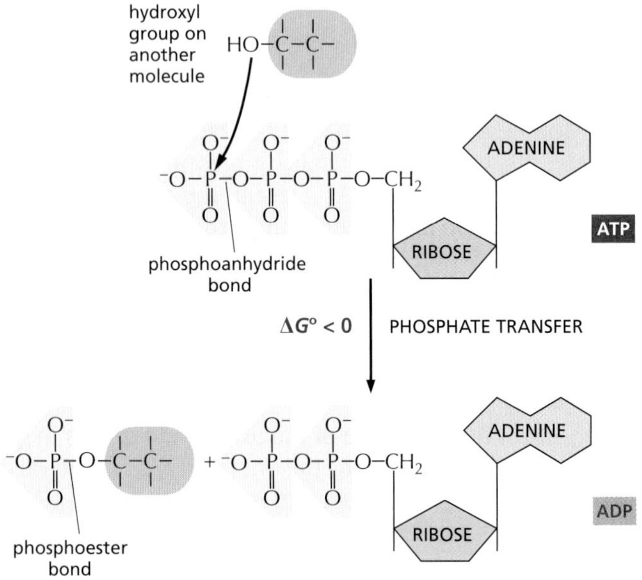

Figure 3–32 The terminal phosphate of ATP can be readily transferred to other molecules. Because an energy-rich phosphoanhydride bond in ATP is converted to a less energy-rich phosphoester bond in the phosphate-accepting molecule, this reaction is energetically favorable, having a large negative ΔG (see Panel 3–1, pp. 96–97). Phosphorylation reactions of this type are involved in the synthesis of phospholipids and in the initial steps of the breakdown of sugars, as well as in many other metabolic and intracellular signaling pathways.

ATP is the most abundant activated carrier in cells. It is used, for example, to supply energy for many of the pumps that actively transport substances into or out of the cell (discussed in Chapter 12); it also powers the molecular motors that enable muscle cells to contract and nerve cells to transport materials along their lengthy axons (discussed in Chapter 17). Why evolution selected this particular nucleotide over the others as the major carrier of energy, however, remains a mystery. The nucleotide GTP, although similar, has very different functions in the cell, as we discuss in later chapters.

Energy Stored in ATP Is Often Harnessed to Join Two Molecules Together

A common type of reaction that is needed for biosynthesis is one in which two molecules, A and B, are joined together by a covalent bond to produce A–B in the energetically unfavorable condensation reaction:

$$A–H + B–OH \rightarrow A–B + H_2O$$

ATP hydrolysis can be coupled indirectly to this reaction to make it go forward. In this case, energy from ATP hydrolysis is first used to convert B–OH to a higher-energy intermediate compound, which then reacts directly with A–H to give A–B. The simplest mechanism involves the transfer of a phosphate from ATP to B–OH to make $B–O–PO_3$, in which case the reaction pathway contains only two steps:

1. $B–OH + ATP \rightarrow B–O–PO_3 + ADP$
2. $A–H + B–O–PO_3 \rightarrow A–B + P_i$

Net result: $B–OH + ATP + A–H \rightarrow A–B + ADP + P_i$

The condensation reaction, which by itself is energetically unfavorable, has been forced to occur by being coupled to ATP hydrolysis in an enzyme-catalyzed reaction pathway (**Figure 3–33A**).

A biosynthetic reaction of exactly this type is employed to synthesize the amino acid glutamine, as illustrated in **Figure 3–33B**. We will see later in the chapter that very similar (but more complex) mechanisms are also used to produce nearly all of the large molecules of the cell.

NADH and NADPH Are Both Activated Carriers of Electrons

Other important activated carriers participate in oxidation–reduction reactions and are commonly part of coupled reactions in cells. These activated carriers are specialized to carry both high-energy electrons and hydrogen atoms. The most important of these *electron carriers* are **NADH** (nicotinamide adenine dinucleotide) and the closely related molecule **NADPH** (nicotinamide adenine dinucleotide phosphate). Both NADH and NADPH carry energy in the form of two high-energy electrons plus a proton (H^+), which together form a hydride ion (H^-). When these activated carriers pass their energy (in the form of a hydride ion) to a donor molecule, they become oxidized to form **NAD$^+$** and **NADP$^+$**, respectively.

Like ATP, NADPH is an activated carrier that participates in many important biosynthetic reactions that would otherwise be energetically unfavorable. NADPH is produced according to the general scheme shown in **Figure 3–34A**. During a special set of energy-yielding catabolic reactions, a hydride ion is removed from the substrate molecule and added to the nicotinamide ring of NADP$^+$ to form NADPH. This is a typical oxidation–reduction reaction; the substrate is oxidized and NADP$^+$ is reduced.

QUESTION 3–8

The phosphoanhydride bond that links two phosphate groups in ATP in a high-energy linkage has a $\Delta G°$ of –7.3 kcal/mole. Hydrolysis of this bond in a cell liberates from 11 to 13 kcal/mole of usable energy. How can this be? Why do you think a range of energies is given, rather than a precise number as for $\Delta G°$?

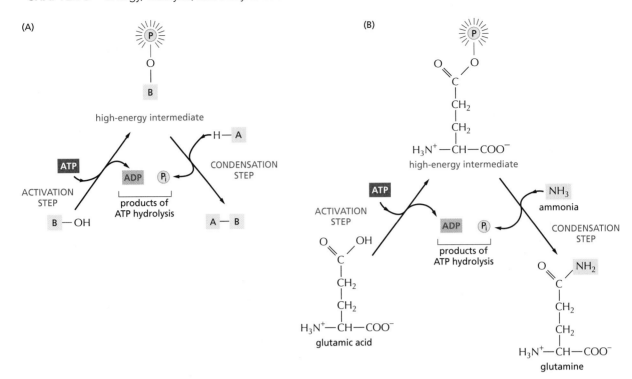

Figure 3–33 An energetically unfavorable biosynthetic reaction can be driven by ATP hydrolysis.
(A) Schematic illustration of the formation of A–B in the condensation reaction described in the text. (B) The biosynthesis of the amino acid glutamine from glutamic acid. Glutamic acid is first converted to a high-energy phosphorylated intermediate (corresponding to the compound B–O–PO$_3$ described in the text), which then reacts with ammonia (corresponding to A–H) to form glutamine. In this example, both steps occur on the surface of the same enzyme, glutamine synthetase (not shown). For clarity, the glutamic acid side chain is shown in its uncharged form. ATP hydrolysis can drive this energetically unfavorable reaction because it yields more energy ($\Delta G°$ of –7.3 kcal/mole) than the energy required for the synthesis of glutamine from glutamic acid plus NH$_3$ ($\Delta G°$ of +3.4 kcal/mole).

The hydride ion carried by NADPH is given up readily in a subsequent oxidation–reduction reaction, because the ring can achieve a more stable arrangement of electrons without it. In this subsequent reaction, which regenerates NADP$^+$, the NADPH becomes oxidized and the substrate becomes reduced—thus completing the NADPH cycle. NADPH is efficient at donating its hydride ion to other molecules for the same reason that ATP readily transfers a phosphate: in both cases, the transfer is accompanied by a large negative free-energy change. One example of the use of NADPH in biosynthesis is shown in **Figure 3–35**.

NADPH and NADH Have Different Roles in Cells

NADPH and NADH differ in a single phosphate group, which is located far from the region involved in electron transfer in NADPH (Figure 3–34B). Although this phosphate group has no effect on the electron-transfer properties of NADPH compared with NADH, it is nonetheless crucial for their distinctive roles, as it gives NADPH a slightly different shape from NADH. This subtle difference in conformation makes it possible for the two carriers to bind as substrates to different sets of enzymes and thereby deliver electrons (in the form of hydride ions) to different target molecules.

Why should there be this division of labor? The answer lies in the need to regulate two sets of electron-transfer reactions independently. NADPH operates chiefly with enzymes that catalyze anabolic reactions, supplying the high-energy electrons needed to synthesize energy-rich biological molecules. NADH, by contrast, has a special role as an intermediate in

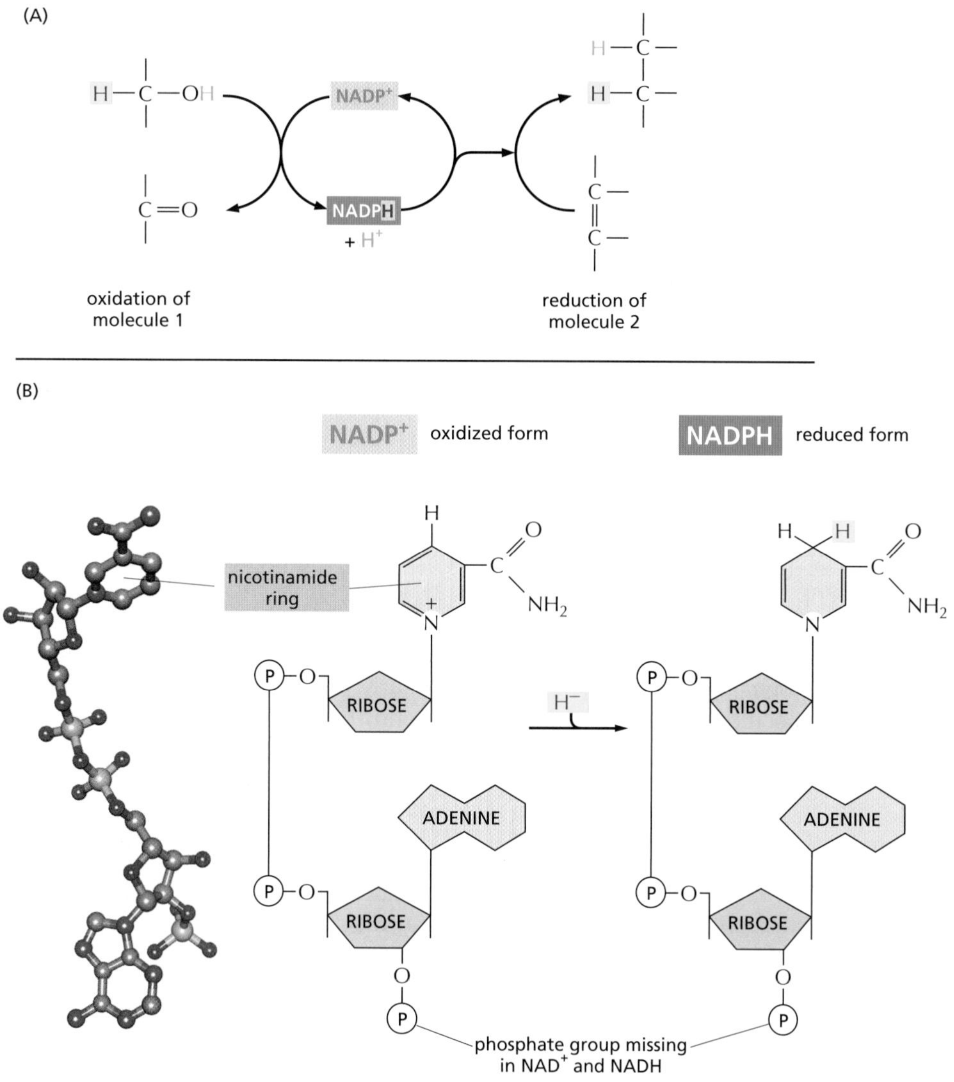

(A)

oxidation of
molecule 1

reduction of
molecule 2

(B)

NADP⁺ oxidized form   NADPH reduced form

nicotinamide
ring

RIBOSE

ADENINE

RIBOSE

P

H⁻

RIBOSE

ADENINE

RIBOSE

P

phosphate group missing
in NAD⁺ and NADH

Figure 3–34 NADPH is an activated carrier of electrons. (A) NADPH is produced in reactions of the general type shown on the left, in which two hydrogen atoms are removed from a substrate. The oxidized form of the carrier molecule, NADP⁺, receives one hydrogen atom plus an electron (a hydride ion), while the proton (H⁺) from the other H atom is released into solution. Because NADPH holds its hydride ion in a high-energy linkage, the ion can easily be transferred to other molecules, as shown on the right. (B) The structure of NADP⁺ and NADPH. On the left is a ball-and-stick model of NADP. The part of the NADP⁺ molecule known as the nicotinamide ring accepts two electrons, together with a proton (the equivalent of a hydride ion, H⁻), forming NADPH. NAD⁺ and NADH are identical in structure to NADP⁺ and NADPH, respectively, except that they lack the phosphate group, as indicated.

the catabolic system of reactions that generate ATP through the oxidation of food molecules, as we discuss in Chapter 13. The genesis of NADH from NAD⁺ and that of NADPH from NADP⁺ occurs by different pathways that are independently regulated, so that the cell can adjust the supply of electrons for these two contrasting purposes. Inside the cell, the ratio of NAD⁺ to NADH is kept high, whereas the ratio of NADP⁺ to NADPH is kept low. This arrangement provides plenty of NAD⁺ to act as an oxidizing agent and plenty of NADPH to act as a reducing agent—as required for their special roles in catabolism and anabolism, respectively.

Cells Make Use of Many Other Activated Carriers

In addition to ATP (which transfers a phosphate) and NADPH and NADH (which transfer electrons and hydrogen), cells make use of other activated carriers that pick up and carry a chemical group in an easily transferred, high-energy linkage. *FADH₂*, like NADH and NADPH, carries hydrogen and high-energy electrons (see Figure 13–13B). But other important reactions involve the transfers of acetyl, methyl, carboxyl, and glucose groups from activated carriers for the purpose of biosynthesis (**Table 3–2**). Coenzyme A, for example, can carry an acetyl group in a readily transferable linkage. This activated carrier, called **acetyl CoA** (acetyl coenzyme A), is shown in **Figure 3–36**. It is used, for example, to add sequentially two-carbon units in the biosynthesis of the hydrocarbon tails of fatty acids.

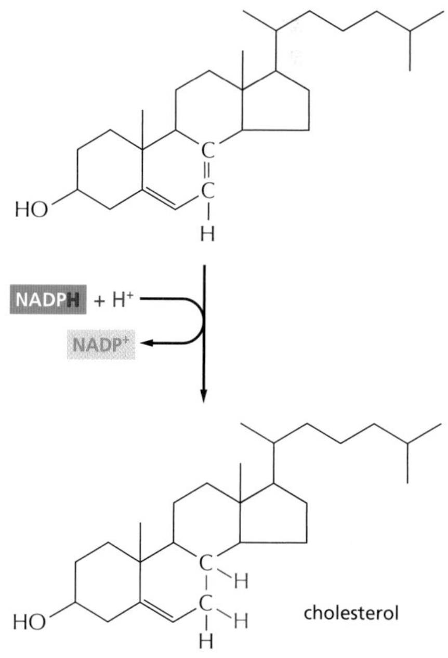

7-dehydrocholesterol

NADPH + H⁺

NADP⁺

cholesterol

Figure 3–35 NADPH participates in the final stage of one of the biosynthetic routes leading to cholesterol. As in many other biosynthetic reactions, the reduction of the C=C bond is achieved by the transfer of a hydride ion from the activated carrier NADPH, plus a proton (H⁺) from solution.

TABLE 3–2 SOME ACTIVATED CARRIERS WIDELY USED IN METABOLISM	
Activated Carrier	**Group Carried in High-Energy Linkage**
ATP	phosphate
NADH, NADPH, FADH$_2$	electrons and hydrogens
Acetyl CoA	acetyl group
Carboxylated biotin	carboxyl group
S-adenosylmethionine	methyl group
Uridine diphosphate glucose	glucose

In acetyl CoA and the other activated carriers in Table 3–2, the transferable group makes up only a small part of the molecule. The rest consists of a large organic portion that serves as a convenient "handle," facilitating the recognition of the carrier molecule by specific enzymes. As with acetyl CoA, this handle portion very often contains a nucleotide. This curious fact may be a relic from an early stage of cell evolution. It is thought that the main catalysts for early life forms on Earth were RNA molecules (or their close relatives) and that proteins were a later evolutionary addition. It is therefore tempting to speculate that many of the activated carriers that we find today originated in an earlier RNA world, where their nucleotide portions would have been useful for binding these carriers to RNA-based catalysts, or *ribozymes* (discussed in Chapter 7).

Activated carriers are usually generated in reactions coupled to ATP hydrolysis, as shown for biotin in **Figure 3–37**. Therefore, the energy that enables their groups to be used for biosynthesis ultimately comes from the catabolic reactions that generate ATP. Similar processes occur in the synthesis of the very large macromolecules—the nucleic acids, proteins, and polysaccharides—which we discuss next.

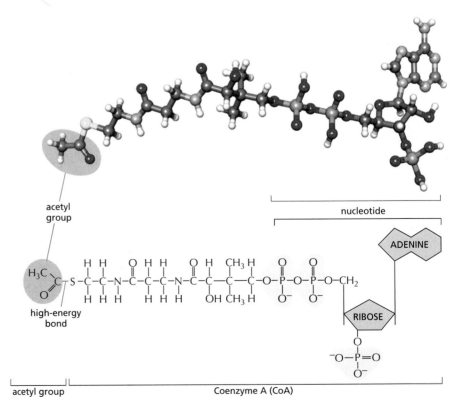

Figure 3–36 Acetyl coenzyme A (CoA) is another important activated carrier. A ball-and-stick model is shown above the structure of acetyl CoA. The sulfur atom (*yellow*) forms a thioester bond to acetate. Because the thioester bond is a high-energy linkage, it releases a large amount of free energy when it is hydrolyzed; thus the acetyl group carried by CoA can be readily transferred to other molecules.

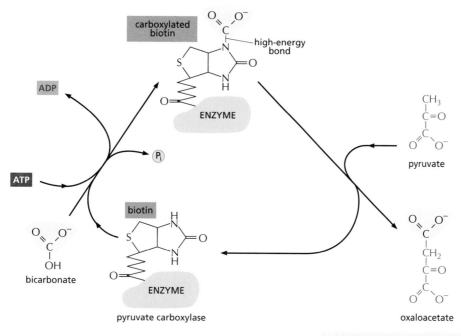

CARBOXYLATION OF BIOTIN

CARBOXYL GROUP TRANSFER

Figure 3–37 An activated carrier transfers a carboxyl group to a substrate. Biotin is a vitamin that is used by a number of enzymes, including *pyruvate carboxylase* shown here. Once it is carboxylated, biotin can transfer a carboxyl group to another molecule. Here, it transfers a carboxyl group to pyruvate, producing oxaloacetate, a molecule needed in the citric acid cycle (discussed in Chapter 13). Other enzymes use biotin to transfer carboxyl groups to other acceptor molecules. Note that the synthesis of carboxylated biotin requires energy derived from ATP hydrolysis—a general feature of many activated carriers.

The Synthesis of Biological Polymers Requires an Energy Input

The macromolecules of the cell constitute the vast majority of its dry mass—that is, the mass not due to water. These molecules are made from *subunits* (or monomers) that are linked together by bonds formed during an enzyme-catalyzed condensation reaction. The reverse reaction—the breakdown of polymers—occurs through enzyme-catalyzed hydrolysis reactions. These hydrolysis reactions are energetically favorable, whereas the corresponding biosynthetic reactions require an energy input and are more complex (**Figure 3–38**).

The nucleic acids (DNA and RNA), proteins, and polysaccharides are all polymers that are produced by the repeated addition of a subunit onto one end of a growing chain. The mode of synthesis of each of these macromolecules is outlined in **Figure 3–39**. As indicated, the condensation step in each case depends on energy provided by the hydrolysis of a nucleoside triphosphate. And yet, except for the nucleic acids, there are no phosphate groups left in the final product molecules. How, then, is the energy of ATP hydrolysis coupled to polymer synthesis?

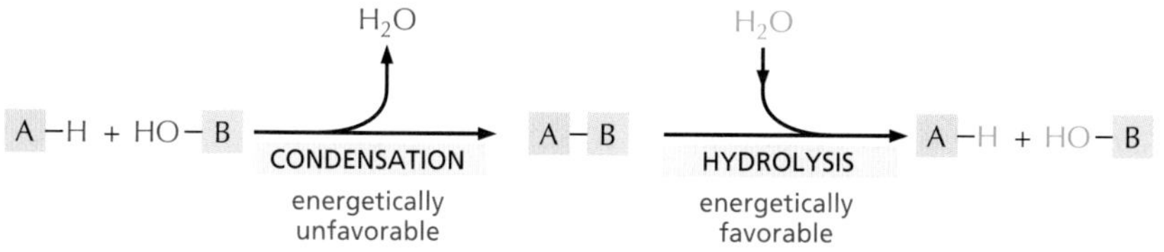

$$A-H + HO-B \xrightarrow[\substack{\text{CONDENSATION} \\ \text{energetically} \\ \text{unfavorable}}]{\text{H}_2\text{O} \uparrow} A-B \xrightarrow[\substack{\text{HYDROLYSIS} \\ \text{energetically} \\ \text{favorable}}]{\text{H}_2\text{O} \downarrow} A-H + HO-B$$

Figure 3–38 In cells, macromolecules are synthesized by condensation reactions and broken down by hydrolysis reactions. Condensation reactions are all energetically unfavorable, whereas hydrolysis reactions are all energetically favorable.

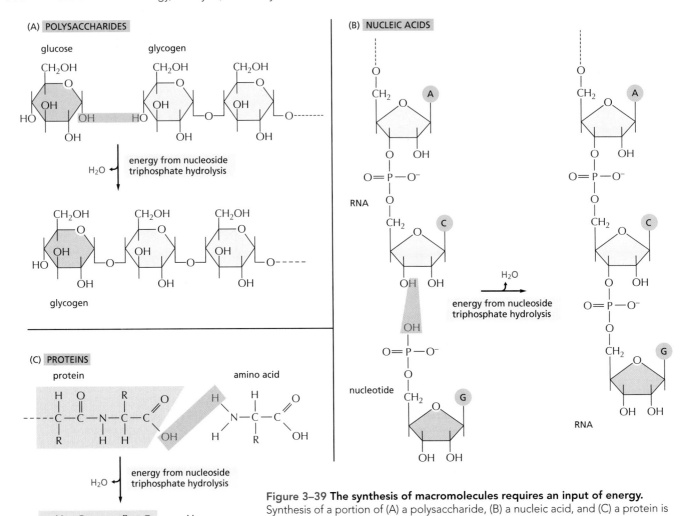

Figure 3–39 The synthesis of macromolecules requires an input of energy. Synthesis of a portion of (A) a polysaccharide, (B) a nucleic acid, and (C) a protein is shown here. In each case, synthesis involves a condensation reaction in which water is lost; the atoms involved are shaded in pink. Not shown is the consumption of high-energy nucleoside triphosphates that is required to activate each subunit prior to its addition. In contrast, the reverse reaction—the breakdown of all three types of polymers—occurs through the simple addition of water, or hydrolysis (not shown).

For each type of macromolecule, an enzyme-catalyzed pathway exists, which resembles that discussed previously for the synthesis of the amino acid glutamine (see Figure 3–33). The principle is exactly the same, in that the –OH group that will be removed in the condensation reaction is first activated by forming a high-energy linkage to a second molecule. The mechanisms used to link ATP hydrolysis to the synthesis of proteins and polysaccharides, however, are more complex than that used for glutamine synthesis. In the biosynthetic pathways leading to these macromolecules, a series of high-energy intermediates generates the final high-energy bond that is broken during the condensation step (as discussed in Chapter 7 for protein synthesis).

There are limits to what each activated carrier can do in driving biosynthesis. For example, the ΔG for the hydrolysis of ATP to ADP and inorganic phosphate (P_i) depends on the concentrations of all of the reactants, and under the usual conditions in a cell, is between –11 and –13 kcal/mole. In principle, this hydrolysis reaction can be used to drive an unfavorable reaction with a ΔG of, perhaps, +10 kcal/mole, provided that a suitable reaction path is available. For some biosynthetic reactions, however, even –13 kcal/mole may be insufficient. In these cases, the path of ATP

QUESTION 3–9

Which of the following reactions will occur only if coupled to a second, energetically favorable reaction?
A. glucose + O_2 → CO_2 + H_2O
B. CO_2 + H_2O → glucose + O_2
C. nucleoside triphosphates → DNA
D. nucleotide bases → nucleoside triphosphates
E. ADP + P_i → ATP

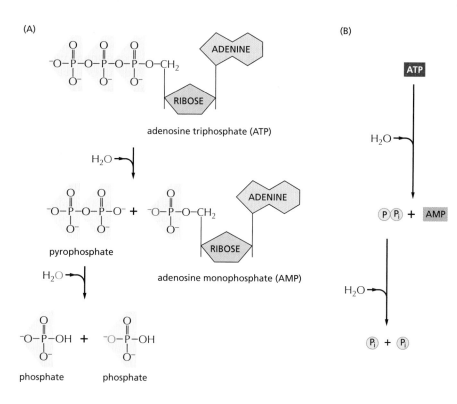

hydrolysis can be altered so that it initially produces AMP and pyrophosphate (PP$_i$), which is itself then hydrolyzed in solution in a subsequent step (**Figure 3–40**). The whole process makes available a total ΔG of about –26 kcal/mole. The biosynthetic reaction involved in the synthesis of nucleic acids (polynucleotides) is driven in this way (**Figure 3–41**).

ATP will make many appearances throughout the book as a molecule that powers reactions in the cell. And in Chapters 13 and 14, we discuss how the cell uses the energy from food to generate ATP. In the next chapter, we learn more about the proteins that make such reactions possible.

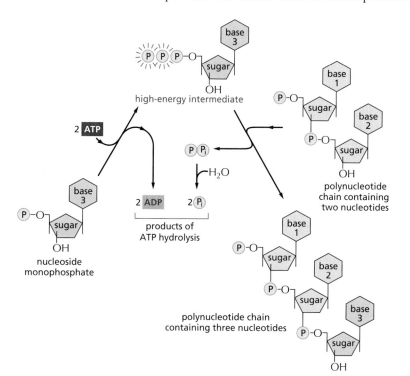

Figure 3–41 Synthesis of a polynucleotide, RNA or DNA, is a multistep process driven by ATP hydrolysis. In the first step, a nucleoside monophosphate is activated by the sequential transfer of the terminal phosphate groups from two ATP molecules. The high-energy intermediate formed—a nucleoside triphosphate—exists free in solution until it reacts with the growing end of an RNA or a DNA chain with release of pyrophosphate. Hydrolysis of the pyrophosphate to inorganic phosphate is highly favorable and helps to drive the overall reaction in the direction of polynucleotide synthesis.

ESSENTIAL CONCEPTS

- Living organisms are able to exist because of a continual input of energy. Part of this energy is used to carry out essential reactions that support cell metabolism, growth, movement, and reproduction; the remainder is lost in the form of heat.

- The ultimate source of energy for most living organisms is the sun. Plants, algae, and photosynthetic bacteria use solar energy to produce organic molecules from carbon dioxide. Animals obtain food by eating plants or by eating animals that feed on plants.

- Each of the many hundreds of chemical reactions that occur in a cell is specifically catalyzed by an enzyme. Large numbers of different enzymes work in sequence to form chains of reactions, called metabolic pathways, each performing a different function in the cell.

- Catabolic reactions release energy by breaking down organic molecules, including foods, through oxidative pathways. Anabolic reactions generate the many complex organic molecules needed by the cell, and they require an energy input. In animal cells, both the building blocks and the energy required for the anabolic reactions are obtained through catabolic reactions.

- Enzymes catalyze reactions by binding to particular substrate molecules in a way that lowers the activation energy required for making and breaking specific covalent bonds.

- The rate at which an enzyme catalyzes a reaction depends on how rapidly it finds its substrates and how quickly the product forms and then diffuses away. These rates vary widely from one enzyme to another.

- The only chemical reactions possible are those that increase the total amount of disorder in the universe. The free-energy change for a reaction, ΔG, measures this disorder, and it must be less than zero for a reaction to proceed spontaneously.

- The ΔG for a chemical reaction depends on the concentrations of the reacting molecules, and it may be calculated from these concentrations if the equilibrium constant (K) of the reaction (or the standard free-energy change, $\Delta G°$, for the reactants) is known.

- Equilibrium constants govern all of the associations (and dissociations) that occur between macromolecules and small molecules in the cell. The larger the binding energy between two molecules, the larger the equilibrium constant and the more likely that these molecules will be found bound to each other.

- By creating a reaction pathway that couples an energetically favorable reaction to an energetically unfavorable one, enzymes can make otherwise impossible chemical transformations occur.

- A small set of activated carriers, particularly ATP, NADH, and NADPH, plays a central part in these coupled reactions in cells. ATP carries high-energy phosphate groups, whereas NADH and NADPH carry high-energy electrons.

- Food molecules provide the carbon skeletons for the formation of macromolecules. The covalent bonds of these larger molecules are produced by condensation reactions that are coupled to energetically favorable bond changes in activated carriers such as ATP and NADPH.

KEY TERMS

acetyl CoA	free energy, G
activated carrier	free-energy change, ΔG
activation energy	hydrolysis
ADP, ATP	K_M
anabolism	metabolism
biosynthesis	Michaelis constant (K_M)
catabolism	NAD$^+$, NADH
catalysis	NADP$^+$, NADPH
catalyst	oxidation
condensation reaction	photosynthesis
coupled reaction	reduction
diffusion	respiration
entropy	standard free-energy change, $\Delta G°$
enzyme	substrate
equilibrium	turnover number
equilibrium constant, K	V_{max}

QUESTIONS

QUESTION 3–10

Which of the following statements are correct? Explain your answers.

A. Some enzyme-catalyzed reactions cease completely if their enzyme is absent.

B. High-energy electrons (such as those found in the activated carriers NADH and NADPH) move faster around the atomic nucleus.

C. Hydrolysis of ATP to AMP can provide about twice as much energy as hydrolysis of ATP to ADP.

D. A partially oxidized carbon atom has a somewhat smaller diameter than a more reduced one.

E. Some activated carrier molecules can transfer both energy and a chemical group to a second molecule.

F. The rule that oxidations release energy, whereas reductions require energy input, applies to all chemical reactions, not just those that occur in living cells.

G. Cold-blooded animals have an energetic disadvantage because they release less heat to the environment than warm-blooded animals do. This slows their ability to make ordered macromolecules.

H. Linking the reaction X → Y to a second, energetically favorable reaction Y → Z will shift the equilibrium constant of the first reaction.

QUESTION 3–11

Consider a transition of X → Y. Assume that the only difference between X and Y is the presence of three hydrogen bonds in Y that are absent in X. What is the ratio of X to Y when the reaction is in equilibrium? Approximate your answer by using Table 3–1 (p. 98), with 1 kcal/mole as the energy of each hydrogen bond. If Y instead has six hydrogen bonds that distinguish it from X, how would that change the ratio?

QUESTION 3–12

Protein A binds to protein B to form a complex, AB. At equilibrium in a cell the concentrations of A, B, and AB are all at 1 μM.

A. Referring to Figure 3–19, calculate the equilibrium constant for the reaction A + B ⇌ AB.

B. What would the equilibrium constant be if A, B, and AB were each present in equilibrium at the much lower concentrations of 1 nM each?

C. How many extra hydrogen bonds would be needed to hold A and B together at this lower concentration so that a similar proportion of the molecules are found in the AB complex? (Remember that each hydrogen bond contributes about 1 kcal/mole.)

QUESTION 3–13

Discuss the following statement: "Whether the ΔG for a reaction is larger, smaller, or the same as $\Delta G°$ depends on the concentration of the compounds that participate in the reaction."

QUESTION 3–14

A. How many ATP molecules could maximally be generated from one molecule of glucose, if the complete oxidation of 1 mole of glucose to CO_2 and H_2O yields 686 kcal of free energy and the useful chemical energy available in the high-energy phosphate bond of 1 mole of ATP is 12 kcal?

B. As we will see in Chapter 14 (Table 14–1), respiration produces 30 moles of ATP from 1 mole of glucose. Compare this number with your answer in part (A). What is the overall efficiency of ATP production from glucose?

C. If the cells of your body oxidize 1 mole of glucose, by how much would the temperature of your body (assume that your body consists of 75 kg of water) increase if the heat were not dissipated into the environment? [Recall that a kilocalorie (kcal) is defined as that amount of energy that heats 1 kg of water by 1°C.]

D. What would the consequences be if the cells of your body could convert the energy in food substances with only 20% efficiency? Would your body—as it is presently constructed—work just fine, overheat, or freeze?

E. A resting human hydrolyzes about 40 kg of ATP every 24 hours. The oxidation of how much glucose would produce this amount of energy? (Hint: Look up the structure of ATP in Figure 2–24 to calculate its molecular weight; the atomic weights of H, C, N, O, and P are 1, 12, 14, 16, and 31, respectively.)

QUESTION 3–15

A prominent scientist claims to have isolated mutant cells that can convert 1 molecule of glucose into 57 molecules of ATP. Should this discovery be celebrated, or do you suppose that something might be wrong with it? Explain your answer.

QUESTION 3–16

In a simple reaction $A \rightleftharpoons A^*$, a molecule is interconvertible between two forms that differ in standard free energy $G°$ by 4.3 kcal/mole, with A^* having the higher $G°$.

A. Use Table 3–1 (p. 98) to find how many more molecules will be in state A^* compared with state A at equilibrium.

B. If an enzyme lowered the activation energy of the reaction by 2.8 kcal/mole, how would the ratio of A to A^* change?

QUESTION 3–17

A reaction in a single-step biosynthetic pathway that converts a metabolite into a particularly vicious poison (metabolite $\rightleftharpoons$ poison) in a mushroom is energetically highly unfavorable. The reaction is normally driven by ATP hydrolysis. Assume that a mutation in the enzyme that catalyzes the reaction prevents it from utilizing ATP, but still allows it to catalyze the reaction.

A. Do you suppose it might be safe for you to eat a mushroom that bears this mutation? Base your answer on an estimation of how much less poison the mutant mushroom would produce, assuming the reaction is in equilibrium and most of the energy stored in ATP is used to drive the unfavorable reaction in nonmutant mushrooms.

B. Would your answer be different for another mutant mushroom whose enzyme couples the reaction to ATP hydrolysis but works 100 times more slowly?

QUESTION 3–18

Consider the effects of two enzymes, A and B. Enzyme A catalyzes the reaction

$$ATP + GDP \rightleftharpoons ADP + GTP$$

and enzyme B catalyzes the reaction

$$NADH + NADP^+ \rightleftharpoons NAD^+ + NADPH$$

Discuss whether the enzymes would be beneficial or detrimental to cells.

QUESTION 3–19

Discuss the following statement: "Enzymes and heat are alike in that both can speed up reactions that—although thermodynamically feasible—do not occur at an appreciable rate because they require a high activation energy. Diseases that seem to benefit from the careful application of heat—in the form of hot chicken soup, for example—are therefore likely to be due to the insufficient function of an enzyme."

QUESTION 3–20

The curve shown in Figure 3–24 is described by the Michaelis–Menten equation:

rate $(v) = V_{max} [S]/([S] + K_M)$

Can you convince yourself that the features qualitatively described in the text are accurately represented by this equation? In particular, how can the equation be simplified when the substrate concentration [S] is in one of the following ranges: (A) [S] is much smaller than the K_M, (B) [S] equals the K_M, and (C) [S] is much larger than the K_M?

QUESTION 3–21

The rate of a simple enzyme reaction is given by the standard Michaelis–Menten equation:

rate $= V_{max} [S]/([S] + K_M)$

If the V_{max} of an enzyme is 100 µmole/sec and the K_M is 1 mM, at what substrate concentration is the rate 50 µmole/sec? Plot a graph of rate versus substrate (S) concentration for [S] = 0 to 10 mM. Convert this to a plot of 1/rate versus 1/[S]. Why is the latter plot a straight line?

QUESTION 3–22

Select the correct options in the following and explain your choices. If [S] is much smaller than K_M, the active site of the enzyme is mostly occupied/unoccupied. If [S] is very much greater than K_M, the reaction rate is limited by the enzyme/substrate concentration.

QUESTION 3–23

A. The reaction rates of the reaction S → P catalyzed by enzyme E were determined under conditions such that only very little product was formed. The following data were measured:

Substrate concentration (μM)	Reaction rate (μmole/min)
0.08	0.15
0.12	0.21
0.54	0.7
1.23	1.1
1.82	1.3
2.72	1.5
4.94	1.7
10.00	1.8

Plot the above data as a graph. Use this graph to estimate the K_M and the V_{max} for this enzyme.

B. Recall from the How We Know essay (pp. 104–106) that to determine these values more precisely, a trick is generally used in which the Michaelis–Menten equation is transformed so that it is possible to plot the data as a straight line. A simple rearrangement yields

$$1/\text{rate} = (K_M/V_{max})\,(1/[S]) + 1/V_{max}$$

which is an equation of the form $y = ax + b$. Calculate 1/rate and 1/[S] for the data given in part (A) and then plot 1/rate versus 1/[S] as a new graph. Determine K_M and V_{max} from the intercept of the line with the axis, where $1/[S] = 0$, combined with the slope of the line. Do your results agree with the estimates made from the first graph of the raw data?

C. It is stated in part (A) that only very little product was formed under the reaction conditions. Why is this important?

D. Assume the enzyme is regulated such that upon phosphorylation its K_M increases by a factor of 3 without changing its V_{max}. Is this an activation or inhibition? Plot the data you would expect for the phosphorylated enzyme in both the graph for (A) and the graph for (B).

CHAPTER **FOUR**

4

Protein Structure and Function

When we look at a cell in a microscope or analyze its electrical or bio-chemical activity, we are, in essence, observing the handiwork of proteins. **Proteins** are the main building blocks from which cells are assembled, and they constitute most of the cell's dry mass. In addition to provid-ing the cell with shape and structure, proteins also execute nearly all its myriad functions. *Enzymes* promote intracellular chemical reactions by providing intricate molecular surfaces, contoured with particular bumps and crevices that can cradle or exclude specific molecules. Proteins embedded in the plasma membrane form the channels and pumps that control the passage of nutrients and other small molecules into and out of the cell. Other proteins carry messages from one cell to another, or act as signal integrators that relay information from the plasma mem-brane to the nucleus of individual cells. Some proteins act as motors that propel organelles through the cytoplasm, and others function as compo-nents of tiny molecular machines with precisely calibrated moving parts. Specialized proteins also act as antibodies, toxins, hormones, antifreeze molecules, elastic fibers, or luminescence generators. Before we can hope to understand how genes work, how muscles contract, how nerves conduct electricity, how embryos develop, or how our bodies function, we must understand proteins.

The multiplicity of functions carried out by proteins (**Panel 4–1**, p. 122) arises from the huge number of different shapes they adopt. We therefore begin our description of these remarkable macromolecules by discussing their three-dimensional structures and the properties that these struc-tures confer. We next look at how proteins work: how enzymes catalyze chemical reactions, how some proteins act as molecular switches, and how others generate orderly movement. We then examine how cells

THE SHAPE AND STRUCTURE
OF PROTEINS

HOW PROTEINS WORK

HOW PROTEINS ARE
CONTROLLED

HOW PROTEINS ARE STUDIED

ENZYMES

function: Catalyze covalent bond breakage or formation.

examples: Living cells contain thousands of different enzymes, each of which catalyzes (speeds up) one particular reaction. Examples include: *tryptophan synthetase*—makes the amino acid tryptophan; *pepsin*—degrades dietary proteins in the stomach; *ribulose bisphosphate carboxylase*—helps convert carbon dioxide into sugars in plants; *DNA polymerase*—copies DNA; *protein kinase*—adds a phosphate group to a protein molecule.

STRUCTURAL PROTEINS

function: Provide mechanical support to cells and tissues.

examples: Outside cells, *collagen* and *elastin* are common constituents of extracellular matrix and form fibers in tendons and ligaments. Inside cells, *tubulin* forms long, stiff microtubules, and *actin* forms filaments that underlie and support the plasma membrane; *keratin* forms fibers that reinforce epithelial cells and is the major protein in hair and horn.

TRANSPORT PROTEINS

function: Carry small molecules or ions.

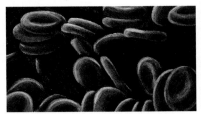

examples: In the bloodstream, *serum albumin* carries lipids, *hemoglobin* carries oxygen, and *transferrin* carries iron. Many proteins embedded in cell membranes transport ions or small molecules across the membrane. For example, the bacterial protein *bacteriorhodopsin* is a light-activated proton pump that transports H^+ ions out of the cell; *glucose carriers* shuttle glucose into and out of cells; and a *Ca^{2+} pump* clears Ca^{2+} from a muscle cell's cytosol after the ions have triggered a contraction.

MOTOR PROTEINS

function: Generate movement in cells and tissues.

examples: *Myosin* in skeletal muscle cells provides the motive force for humans to move; *kinesin* interacts with microtubules to move organelles around the cell; *dynein* enables eukaryotic cilia and flagella to beat.

STORAGE PROTEINS

function: Store amino acids or ions.

examples: Iron is stored in the liver by binding to the small protein *ferritin*; *ovalbumin* in egg white is used as a source of amino acids for the developing bird embryo; *casein* in milk is a source of amino acids for baby mammals.

SIGNAL PROTEINS

function: Carry extracellular signals from cell to cell.

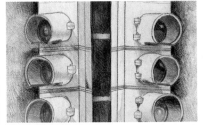

examples: Many of the hormones and growth factors that coordinate physiological functions in animals are proteins; *insulin*, for example, is a small protein that controls glucose levels in the blood; *netrin* attracts growing nerve cell axons to specific locations in the developing spinal cord; *nerve growth factor* (NGF) stimulates some types of nerve cells to grow axons; *epidermal growth factor* (EGF) stimulates the growth and division of epithelial cells.

RECEPTOR PROTEINS

function: Detect signals and transmit them to the cell's response machinery.

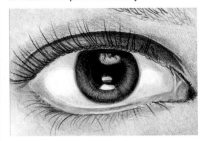

examples: *Rhodopsin* in the retina detects light; the *acetylcholine receptor* in the membrane of a muscle cell is activated by acetylcholine released from a nerve ending; the *insulin receptor* allows a cell to respond to the hormone insulin by taking up glucose; the *adrenergic receptor* on heart muscle increases the rate of the heartbeat when it binds to adrenaline.

GENE REGULATORY PROTEINS

function: Bind to DNA to switch genes on or off.

examples: The *lactose repressor* in bacteria silences the genes for the enzymes that degrade the sugar lactose; many different *homeodomain proteins* act as genetic switches to control development in multicellular organisms, including humans.

SPECIAL-PURPOSE PROTEINS

function: Highly variable.

examples: Organisms make many proteins with highly specialized properties. These molecules illustrate the amazing range of functions that proteins can perform. The *antifreeze proteins* of Arctic and Antarctic fishes protect their blood against freezing; *green fluorescent protein* from jellyfish emits a green light; *monellin*, a protein found in an African plant, has an intensely sweet taste; mussels and other marine organisms secrete *glue proteins* that attach them firmly to rocks, even when immersed in seawater.

control the activity and location of the proteins they contain. Finally, we present a brief description of the techniques that biologists use to work with proteins, including methods for purifying them—from tissues or cultured cells—and for determining their structures.

THE SHAPE AND STRUCTURE OF PROTEINS

From a chemical point of view, proteins are by far the most structurally complex and functionally sophisticated molecules known. This is perhaps not surprising, considering that the structure and activity of each protein has been developed and fine-tuned over billions of years of evolution. We start by considering how the position of each amino acid in the long string of amino acids that forms a protein determines its three-dimensional shape, which is stabilized by noncovalent interactions between different parts of the molecule. Understanding the structure of a protein at the atomic level allows us to see how the precise shape of the protein determines its function.

The Shape of a Protein Is Specified by Its Amino Acid Sequence

Proteins, as you may recall from Chapter 2, are assembled mainly from a set of 20 different amino acids, each with different chemical properties. A protein molecule is made from a long chain of these amino acids, held together by covalent **peptide bonds (Figure 4–1)**. Proteins are therefore referred to as **polypeptides**, and their amino acid chains are called **polypeptide chains**. In each type of protein, the amino acids are present in a unique order, called the **amino acid sequence**, which is exactly the same from one molecule of that protein to the next. One molecule of human insulin, for example, has the same amino acid sequence as every other molecule of human insulin. Many thousands of different proteins have been identified, each with its own distinct amino acid sequence.

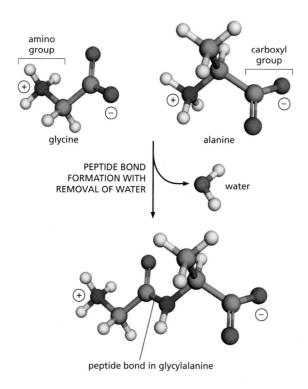

Figure 4–1 **Amino acids are linked together by peptide bonds.** A covalent peptide bond forms when the carbon atom of the carboxyl group of one amino acid (such as glycine) shares electrons with the nitrogen atom (*blue*) from the amino group of a second amino acid (such as alanine). Because a molecule of water is eliminated, peptide bond formation is classified as a condensation reaction (see Figure 2–29). In this diagram, carbon atoms are *gray*, nitrogen *blue*, oxygen *red*, and hydrogen *white*.

Figure 4–2 A protein is made of amino acids linked together into a polypeptide chain. The amino acids are linked by peptide bonds (see Figure 4–1) to form a polypeptide backbone of repeating structure (*gray boxes*), from which the side chain of each amino acid projects. The character and sequence of the chemically distinct side chains—for example, nonpolar (*green*), polar uncharged (*yellow*), and negative (*blue*) side chains—give each protein its distinct, individual properties. A small polypeptide of just four amino acids is shown here. Proteins are typically made up of chains of several hundred amino acids, whose sequence is always presented starting with the N-terminus reading from left to right.

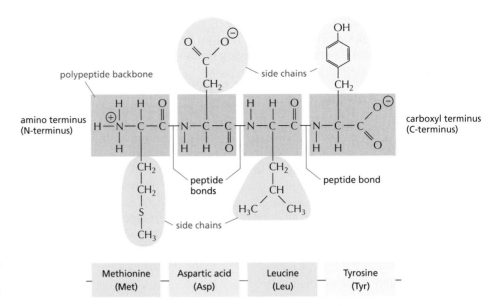

Each polypeptide chain consists of a backbone that is adorned with a variety of chemical side chains. This **polypeptide backbone** is formed from a repeating sequence of the core atoms (–N–C–C–) found in every amino acid (see Figure 4–1). Because the two ends of each amino acid are chemically different—one sports an amino group (NH_3^+, also written NH_2) and the other a carboxyl group (COO^-, also written COOH)—each polypeptide chain has a directionality: the end carrying the amino group is called the amino terminus, or **N-terminus**, and the end carrying the free carboxyl group is the carboxyl terminus, or **C-terminus**.

Projecting from the polypeptide backbone are the amino acid **side chains**—the part of the amino acid that is not involved in forming peptide bonds (**Figure 4–2**). The side chains give each amino acid its unique properties: some are nonpolar and hydrophobic ("water-fearing"), some are negatively or positively charged, some can be chemically reactive, and so on. The atomic formula for each of the 20 amino acids in proteins is presented in Panel 2–5 (pp. 74–75), and a brief list of the 20 common amino acids, with their abbreviations, is provided in **Figure 4–3**.

AMINO ACID			SIDE CHAIN		AMINO ACID			SIDE CHAIN
Aspartic acid	Asp	D	negatively charged		Alanine	Ala	A	nonpolar
Glutamic acid	Glu	E	negatively charged		Glycine	Gly	G	nonpolar
Arginine	Arg	R	positively charged		Valine	Val	V	nonpolar
Lysine	Lys	K	positively charged		Leucine	Leu	L	nonpolar
Histidine	His	H	positively charged		Isoleucine	Ile	I	nonpolar
Asparagine	Asn	N	uncharged polar		Proline	Pro	P	nonpolar
Glutamine	Gln	Q	uncharged polar		Phenylalanine	Phe	F	nonpolar
Serine	Ser	S	uncharged polar		Methionine	Met	M	nonpolar
Threonine	Thr	T	uncharged polar		Tryptophan	Trp	W	nonpolar
Tyrosine	Tyr	Y	uncharged polar		Cysteine	Cys	C	nonpolar

└────── POLAR AMINO ACIDS ──────┘   └────── NONPOLAR AMINO ACIDS ──┘

Figure 4–3 Twenty different amino acids are commonly found in proteins. Both three-letter and one-letter abbreviations are given, as well as the character of the side chain. There are equal numbers of polar (hydrophilic) and nonpolar (hydrophobic) side chains, and half of the polar side chains carry a positive or negative charge.

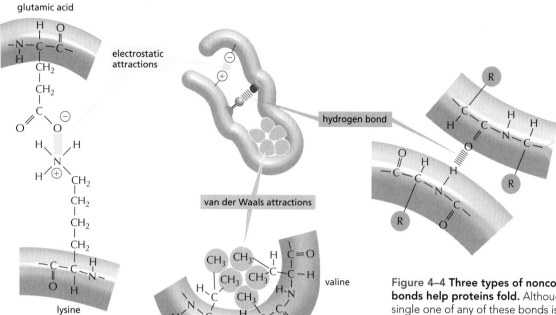

glutamic acid

electrostatic attractions

hydrogen bond

van der Waals attractions

valine

valine

alanine

lysine

R

Figure 4–4 Three types of noncovalent bonds help proteins fold. Although a single one of any of these bonds is quite weak, many of them together can create a strong bonding arrangement that stabilizes a particular three-dimensional structure, as in the small polypeptide shown in the *center*. R is often used as a general designation for an amino acid side chain. Protein folding is also aided by hydrophobic forces, as shown in Figure 4–5.

Long polypeptide chains are very flexible, as many of the peptide bonds that link the carbon atoms in the polypeptide backbone allow free rotation of the atoms they join. Thus, proteins can in principle fold in an enormous number of ways. The shape of each of these folded chains, however, is constrained by many sets of weak noncovalent bonds that form within proteins. These bonds involve atoms in the polypeptide backbone, as well as atoms in the amino acid side chains. The noncovalent bonds that help proteins fold up and maintain their shape include hydrogen bonds, electrostatic attractions, and van der Waals attractions, which are described in Chapter 2 (see Panel 2–7, pp. 78–79). Because a noncovalent bond is much weaker than a covalent bond, it takes many noncovalent bonds to hold two regions of a polypeptide chain tightly together. The stability of each folded shape is largely influenced by the combined strength of large numbers of noncovalent bonds (**Figure 4–4**).

A fourth weak force, *hydrophobic interaction*, also has a central role in determining the shape of a protein. In an aqueous environment, hydrophobic molecules, including the nonpolar side chains of particular amino acids, tend to be forced together to minimize their disruptive effect on the hydrogen-bonded network of the surrounding water molecules (see Panel 2–2, pp. 68–69). Therefore, an important factor governing the folding of any protein is the distribution of its polar and nonpolar amino acids. The nonpolar (hydrophobic) side chains—which belong to amino acids such as phenylalanine, leucine, valine, and tryptophan (see Figure 4–3)—tend to cluster in the interior of the folded protein (just as hydrophobic oil droplets coalesce to form one large drop). Tucked away inside the folded protein, hydrophobic side chains can avoid contact with the aqueous cytosol that surrounds them inside a cell. In contrast, polar side chains—such as those belonging to arginine, glutamine, and histidine—tend to arrange themselves near the outside of the folded protein, where they can form hydrogen bonds with water and with other polar molecules (**Figure 4–5**). When polar amino acids are buried within the protein, they are usually hydrogen-bonded to other polar amino acids or to the polypeptide backbone (**Figure 4–6**).

Figure 4–5 Hydrophobic forces help proteins fold into compact conformations. Polar amino acid side chains tend to be displayed on the outside of the folded protein, where they can interact with water; the nonpolar amino acid side chains are buried on the inside to form a highly packed hydrophobic core of atoms that are hidden from water.

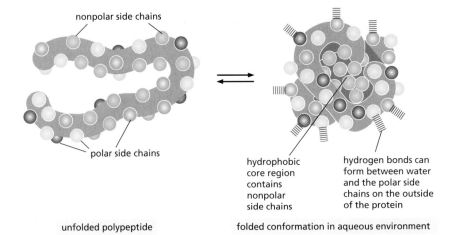

unfolded polypeptide

folded conformation in aqueous environment

Proteins Fold into a Conformation of Lowest Energy

Each type of protein has a particular three-dimensional structure, which is determined by the order of the amino acids in its polypeptide chain. The final folded structure, or **conformation**, adopted by any polypeptide chain is determined by energetic considerations: a protein generally folds into the shape in which its free energy (G) is minimized. The folding process is thus energetically favorable, as it releases heat and increases the disorder of the universe (see Panel 3–1, pp. 96–97).

Protein folding has been studied in the laboratory using highly purified proteins. A protein can be unfolded, or *denatured*, by treatment with solvents that disrupt the noncovalent interactions holding the folded chain together. This treatment converts the protein into a flexible polypeptide chain that has lost its natural shape. Under the right conditions, when the

Figure 4–6 Hydrogen bonds within a protein molecule help stabilize its folded shape. Large numbers of hydrogen bonds form between adjacent regions of the folded polypeptide chain. The structure shown is a portion of the enzyme lysozyme. Hydrogen bonds between backbone atoms are shown in *red*; those between the backbone and a side chain are shown in *yellow*; and those between atoms of two side chains are shown in *blue*. Note that the same amino acid side chain can make multiple hydrogen bonds (*red* arrow). The atoms are colored as in Figure 4–1, although the hydrogen atoms are not shown. (After C.K. Mathews, K.E. van Holde, and K.G. Ahern, Biochemistry, 3rd ed. San Francisco: Benjamin Cummings, 2000.)

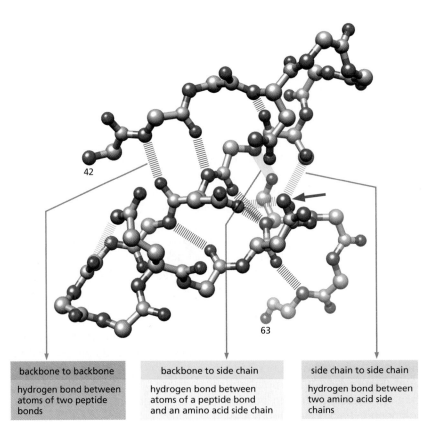

backbone to backbone	backbone to side chain	side chain to side chain
hydrogen bond between atoms of two peptide bonds	hydrogen bond between atoms of a peptide bond and an amino acid side chain	hydrogen bond between two amino acid side chains

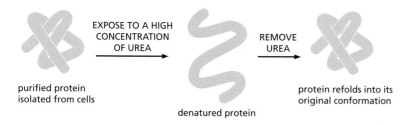

purified protein
isolated from cells

denatured protein

protein refolds into its
original conformation

EXPOSE TO A HIGH
CONCENTRATION
OF UREA

REMOVE
UREA

Figure 4–7 Denatured proteins can often recover their natural shapes. This type of experiment demonstrates that the conformation of a protein is determined solely by its amino acid sequence. Renaturation requires the correct conditions and works best for small proteins.

denaturing solvent is removed, the protein often refolds spontaneously into its original conformation—a process called *renaturation* (**Figure 4–7**). The fact that a denatured protein can, on its own, refold into the correct conformation indicates that all the information necessary to specify the three-dimensional shape of a protein is contained in its amino acid sequence.

Each protein normally folds into a single stable conformation. This conformation, however, often changes slightly when the protein interacts with other molecules in the cell. This change in shape is crucial to the function of the protein, as we discuss later.

When proteins fold incorrectly, they sometimes form aggregates that can damage cells and even whole tissues. Misfolded proteins are thought to contribute to a number of neurodegenerative disorders, such as Alzheimer's disease and Huntington's disease. Some infectious neurodegenerative diseases—including scrapie in sheep, bovine spongiform encephalopathy (BSE, or "mad cow" disease) in cattle, and Creutzfeldt–Jakob disease (CJD) in humans—are caused by misfolded proteins called prions. The misfolded prion form of a protein can convert the properly folded version of the protein in an infected brain into the abnormal conformation. This allows the misfolded prions, which tend to form aggregates, to spread rapidly from cell to cell, eventually causing the death of the affected animal or human (**Figure 4–8**). Prions are considered "infectious" because they can also spread from an affected individual to a normal individual via contaminated food, blood, or surgical instruments, for example.

Although a protein chain can fold into its correct conformation without outside help, protein folding in a living cell is generally assisted by special proteins called *chaperone proteins*. Some of these chaperones bind to partly folded chains and help them to fold along the most energetically favorable pathway (**Figure 4–9**). Others form "isolation chambers" in which single polypeptide chains can fold without the risk of forming aggregates in the crowded conditions of the cytoplasm (**Figure 4–10**). In either case, the final three-dimensional shape of the protein is still specified by its amino acid sequence; chaperones merely make the folding process more efficient and reliable.

Proteins Come in a Wide Variety of Complicated Shapes

Proteins are the most structurally diverse macromolecules in the cell. Although they range in size from about 30 amino acids to more than

QUESTION 4–1

Urea used in the experiment shown in Figure 4–7 is a molecule that disrupts the hydrogen-bonded network of water molecules. Why might high concentrations of urea unfold proteins? The structure of urea is shown here.

$$\begin{array}{c} O \\ \parallel \\ C \\ H_2N \quad\quad NH_2 \end{array}$$

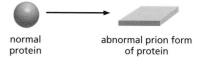

(A) normal protein can, on occasion, adopt an abnormal, misfolded prion form

normal
protein

abnormal prion form
of protein

(B) the prion form of the protein can bind to the normal form, inducing conversion to the abnormal conformation

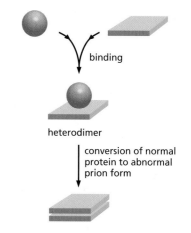

binding

heterodimer

conversion of normal protein to abnormal prion form

(C) abnormal prion proteins propagate and aggregate to form amyloid fibrils

amyloid fibril

Figure 4–8 Prion diseases are caused by proteins whose misfolding is infectious. (A) The protein undergoes a rare conformational change to give an abnormally folded prion form. (B) The abnormal form causes the conversion of normal proteins in the host's brain into a misfolded prion form. (C) The prions aggregate into amyloid fibrils, which disrupt brain cell function, causing a neurodegenerative disorder, such as "mad cow" disease (see also Figure 4–18).

Figure 4–9 Chaperone proteins can guide the folding of a newly synthesized polypeptide chain. The chaperones bind to newly synthesized or partially folded chains and helping them to fold along the most energetically favorable pathway. Association of these chaperones with the target protein requires an input of energy from ATP hydrolysis.

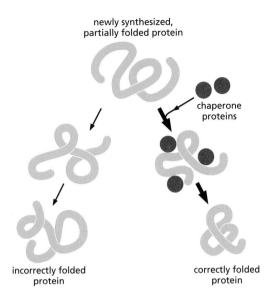

newly synthesized, partially folded protein

chaperone proteins

incorrectly folded protein

correctly folded protein

10,000, the vast majority are between 50 and 2000 amino acids long. Proteins can be globular or fibrous, and they can form filaments, sheets, rings, or spheres (**Figure 4–11**). We will encounter many of these structures later in this chapter and throughout the book.

To date, the structures of about 100,000 different proteins have been determined. We discuss how scientists unravel these structures later in the chapter. Most proteins have a three-dimensional conformation so intricate and irregular that their structure would require an entire chapter to describe in detail. But we can get some sense of the intricacies of polypeptide structure by looking at the conformation of a relatively small protein, such as the bacterial transport protein *HPr*.

This small protein is only 88 amino acids long, and it serves as a carrier protein that facilitates the transport of sugar into bacterial cells. In **Figure 4–12**, we present HPr's three-dimensional structure in four different ways, each of which emphasizes different features of the protein. The backbone model (Figure 4–12A) shows the overall organization of the polypeptide chain and provides a straightforward way to compare the structures of related proteins. The ribbon model (Figure 4–12B) shows the polypeptide backbone in a way that emphasizes its various folds, which we describe in detail shortly. The wire model (Figure 4–12C) includes the positions of all the amino acid side chains; this view is especially useful

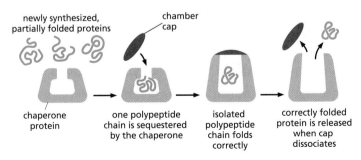

newly synthesized, partially folded proteins

chamber cap

chaperone protein

one polypeptide chain is sequestered by the chaperone

isolated polypeptide chain folds correctly

correctly folded protein is released when cap dissociates

Figure 4–10 Other chaperone proteins act as isolation chambers that help a polypeptide fold. In this case, the barrel of the chaperone provides an enclosed chamber in which a newly synthesized polypeptide chain can fold without the risk of aggregating with other polypeptides in the crowded conditions of the cytoplasm. This system also requires an input of energy from ATP hydrolysis, mainly for the association and subsequent dissociation of the cap that closes off the chamber.

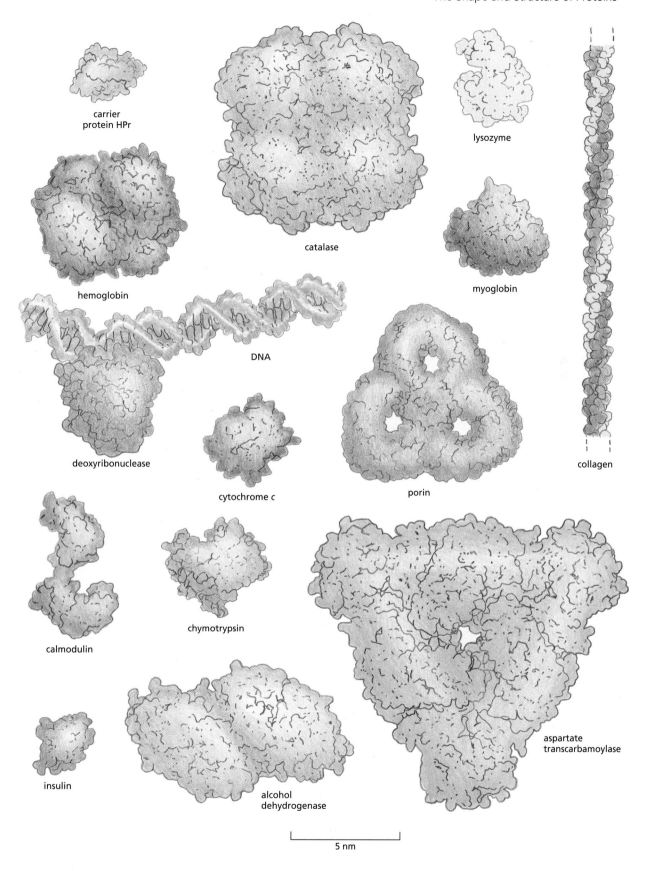

Figure 4–11 Proteins come in a variety of shapes and sizes. Each folded polypeptide is shown as a space-filling model, represented at the same scale. In the *top-left corner* is HPr, the small protein featured in detail in Figure 4–12. For comparison we also show a portion of a DNA molecule (*gray*) bound to the protein deoxyribonuclease. (After David S. Goodsell, *Our Molecular Nature*. New York: Springer-Verlag, 1996. With permission from Springer Science and Business Media.)

(A) backbone model

(B) ribbon model

(C) wire model

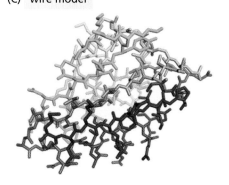

(D) space-filling model

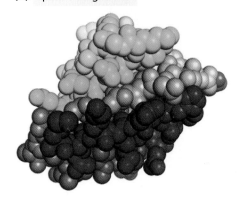

Figure 4–12 Protein conformation can be represented in a variety of ways. Shown here is the structure of the small bacterial transport protein HPr. The images are colored to make it easier to trace the path of the polypeptide chain. In these models, the region of polypeptide chain carrying the protein's N-terminus is *purple* and that near its C-terminus is *red.*

for predicting which amino acids might be involved in the protein's activity. Finally, the space-filling model (Figure 4–12D) provides a contour map of the protein surface, which reveals which amino acids are exposed on the surface and shows how the protein might look to a small molecule such as water or to another macromolecule in the cell.

The structures of larger proteins—or of multiprotein complexes—are even more complex. To visualize such detailed and complicated structures, scientists have developed various graphical and computer-based tools that generate a variety of images of a protein, only some of which are depicted in Figure 4–12. These images can be displayed on a computer screen and readily rotated and magnified to view all aspects of the structure (Movie 4.1).

When the three-dimensional structures of many different protein molecules are compared, it becomes clear that, although the overall conformation of each protein is unique, some regular folding patterns can be detected, as we discuss next.

The α Helix and the β Sheet Are Common Folding Patterns

More than 60 years ago, scientists studying hair and silk discovered two common folding patterns present in many different proteins. The first to be discovered, called the **α helix**, was found in the protein *α-keratin*, which is abundant in skin and its derivatives—such as hair, nails, and horns. Within a year of the discovery of the α helix, a second folded structure, called a **β sheet**, was found in the protein *fibroin*, the major constituent of silk. (Biologists often use Greek letters to name their discoveries, with the first example receiving the designation α, the second β, and so on.)

These two folding patterns are particularly common because they result from hydrogen bonds that form between the N–H and C=O groups in the polypeptide backbone (see Figure 4–6). Because the amino acid side chains are not involved in forming these hydrogen bonds, α helices and β sheets can be generated by many different amino acid sequences. In each case, the protein chain adopts a regular, repeating form. These structural features, and the shorthand cartoon symbols that are often used to represent them in models of protein structures, are presented in **Figure 4–13**.

Helices Form Readily in Biological Structures

The abundance of helices in proteins is, in a way, not surprising. A **helix** is a regular structure that resembles a spiral staircase. It is generated simply by placing many similar subunits next to one another, each in the same strictly repeated relationship to the one before. Because it is very rare for subunits to join up in a straight line, this arrangement will generally result in a helix (**Figure 4–14**). Depending on the twist of the staircase, a helix is said to be either right-handed or left-handed (Figure 4–14E). Handedness is not affected by turning the helix upside down, but it is reversed if the helix is reflected in a mirror.

An α helix is generated when a single polypeptide chain turns around itself to form a structurally rigid cylinder. A hydrogen bond is made between every fourth amino acid, linking the C=O of one peptide bond to

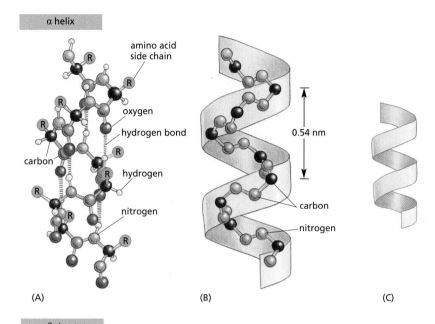

α helix

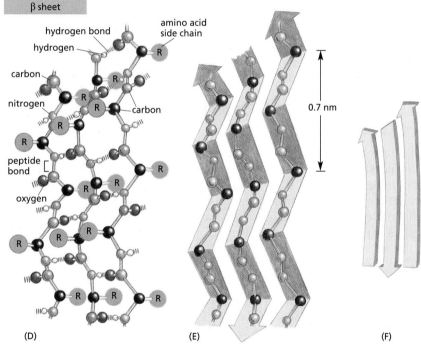

β sheet

Figure 4–13 Polypeptide chains often fold into one of two orderly repeating forms known as an α helix and a β sheet. (A–C) In an α helix, the N–H of every peptide bond is hydrogen-bonded to the C=O of a neighboring peptide bond located four amino acids away in the same chain. (D–F) In a β sheet, several segments (strands) of an individual polypeptide chain are held together by hydrogen-bonding between peptide bonds in adjacent strands. The amino acid side chains in each strand project alternately above and below the plane of the sheet. In the example shown, the adjacent chains run in opposite directions, forming an *antiparallel β sheet*. (A) and (D) show all of the atoms in the polypeptide backbone, but the amino acid side chains are denoted by R. (B) and (E) show only the carbon (*black* and *gray*) and nitrogen (*blue*) backbone atoms, while (C) and (F) display the cartoon symbols that are used to represent the α helix and the β sheet in ribbon models of proteins (see Figure 4–12B).

the N–H of another (see Figure 4–13A). This gives rise to a regular right-handed helix with a complete turn every 3.6 amino acids (Movie 4.2).

Short regions of α helix are especially abundant in proteins that are embedded in cell membranes, such as transport proteins and receptors. We will see in Chapter 11 that those portions of a transmembrane protein that cross the lipid bilayer usually form an α helix that is composed largely of amino acids with nonpolar side chains. The polypeptide backbone, which is hydrophilic, is hydrogen-bonded to itself in the α helix, and it is shielded from the hydrophobic lipid environment of the membrane by its protruding nonpolar side chains (**Figure 4–15**).

Sometimes two (or three) α helices will wrap around one another to form a particularly stable structure known as a **coiled-coil**. This structure forms when the α helices have most of their nonpolar (hydrophobic) side chains on one side, so that they can twist around each other with

QUESTION 4–2

Remembering that the amino acid side chains projecting from each polypeptide backbone in a β sheet point alternately above and below the plane of the sheet (see Figure 4–13D), consider the following protein sequence: Leu-Lys-Val-Asp-Ile-Ser-Leu-Arg-Leu-Lys-Ile-Arg-Phe-Glu. Do you find anything remarkable about the arrangement of the amino acids in this sequence when incorporated into a β sheet? Can you make any predictions as to how the β sheet might be arranged in a protein? (Hint: consult the properties of the amino acids listed in Figure 4–3.)

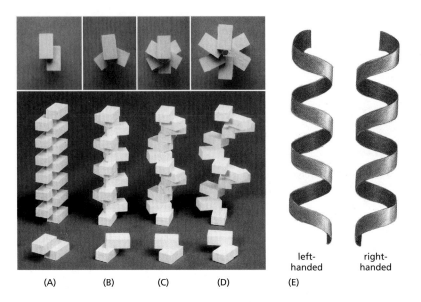

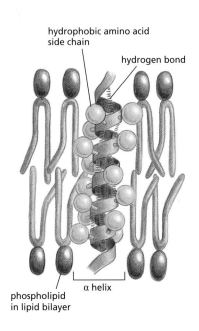

Figure 4–14 **The helix is a common, regular, biological structure.** A helix will form when a series of similar subunits bind to each other in a regular way. At the bottom, the interaction between two subunits is shown; behind them are the helices that result. These helices have two (A), three (B), or six (C and D) subunits per helical turn. At the top, the arrangement of subunits has been photographed from directly above the helix. Note that the helix in (D) has a wider path than that in (C), but the same number of subunits per turn. (E) A helix can be either right-handed or left-handed. As a reference, it is useful to remember that standard metal screws, which advance when turned clockwise, are right-handed. So to judge the handedness of a helix, imagine screwing it into a wall. Note that a helix preserves the same handedness when it is turned upside down.

these side chains facing inward—minimizing their contact with the aqueous cytosol (**Figure 4–16**). Long, rodlike coiled-coils form the structural framework for many elongated proteins. Examples include α-keratin, which forms the intracellular fibers that reinforce the outer layer of the skin, and myosin, the motor protein responsible for muscle contraction (discussed in Chapter 17).

β Sheets Form Rigid Structures at the Core of Many Proteins

A β sheet is made when hydrogen bonds form between segments of a polypeptide chain that lie side by side (see Figure 4–13D). When the neighboring segments run in the same orientation (say, from the N-terminus to the C-terminus), the structure is a *parallel β sheet*; when they run in opposite directions, the structure is an *antiparallel β sheet* (**Figure 4–17**). Both types of β sheet produce a very rigid, pleated structure, and they form the core of many proteins. Even the small bacterial protein HPr (see Figure 4–12) contains several β sheets.

β sheets have remarkable properties. They give silk fibers their extraordinary tensile strength. They also permit the formation of *amyloid fibers*—insoluble protein aggregates that include those associated with neurodegenerative disorders, such as Alzheimer's disease and prion diseases (see Figure 4–8). These structures, formed from abnormally folded proteins, are stabilized by β sheets that stack together tightly, with their amino acid side chains interdigitated like the teeth of a zipper (**Figure 4–18**). Although we tend to associate amyloid fibers with disease, many organisms take advantage of these stable structures to perform novel tasks. Infectious bacteria, for example, can use amyloid fibers to help form the biofilms that allow them to colonize host tissues. Other types of filamentous bacteria use amyloid fibers to extend filaments into the air, enabling the bacteria to disperse their spores far and wide.

Proteins Have Several Levels of Organization

A protein's structure does not end with α helices and β sheets; there are additional levels of organization. These levels are not independent but are built one upon the next to establish the three-dimensional structure of the entire protein. A protein's structure begins with its amino acid sequence, which is thus considered its **primary structure**. The next level of organization includes the α helices and β sheets that form within

Figure 4–15 **Many membrane-bound proteins cross the lipid bilayer as an α helix.** The hydrophobic side chains of the amino acids forming the α helix contact the hydrophobic hydrocarbon tails of the phospholipid molecules, while the hydrophilic parts of the polypeptide backbone form hydrogen bonds with one another in the interior of the helix. About 20 amino acids are required to span a membrane in this way. Note that, despite the appearance of a space along the interior of the helix in this schematic diagram, the helix is not a channel: no ions or small molecules can pass through it.

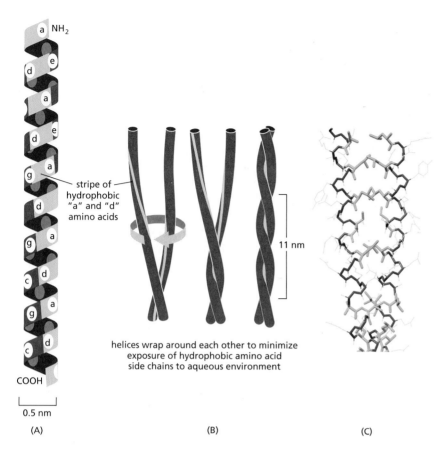

(A) (B) (C)

helices wrap around each other to minimize exposure of hydrophobic amino acid side chains to aqueous environment

stripe of hydrophobic "a" and "d" amino acids

11 nm

0.5 nm

Figure 4–16 Intertwined α helices can form a stiff coiled-coil. In (A), a single α helix is shown, with successive amino acid side chains labeled in a sevenfold repeating sequence "abcdefg." Amino acids "a" and "d" in such a sequence lie close together on the cylinder surface, forming a stripe (shaded in *green*) that winds slowly around the α helix. Proteins that form coiled-coils typically have nonpolar amino acids at positions "a" and "d." Consequently, as shown in (B), the two α helices can wrap around each other, with the nonpolar side chains of one α helix interacting with the nonpolar side chains of the other, while the more hydrophilic amino acid side chains (shaded in *red*) are left exposed to the aqueous environment. (C) A portion of the atomic structure of a coiled-coil made of two α helices, as determined by X-ray crystallography. In this structure, atoms that form the backbone of the helices are shown in *red*; the interacting, nonpolar side chains are *green*, and the remaining side chains are *gray*. Coiled-coils can also form from three α helices (Movie 4.3).

certain segments of the polypeptide chain; these folds are elements of the protein's **secondary structure**. The full, three-dimensional conformation formed by an entire polypeptide chain—including the α helices, β sheets, random coils, and any other loops and folds that form between the N- and C-termini—is sometimes referred to as the **tertiary structure**. Finally, if the protein molecule is formed as a complex of more than one polypeptide chain, then the complete structure is designated its **quaternary structure**.

Studies of the conformation, function, and evolution of proteins have also revealed the importance of a level of organization distinct from the four just described. This organizational unit is the **protein domain**, which is defined as any segment of a polypeptide chain that can fold independently into a compact, stable structure. A protein domain usually contains between 40 and 350 amino acids—folded into α helices and β sheets and other elements of secondary structure—and it is the modular unit from which many larger proteins are constructed (**Figure 4–19**). The different domains of a protein are often associated with different functions. For example, the bacterial *catabolite activator protein* (*CAP*), illustrated in Figure 4–19, has two domains: the small domain binds to DNA, while the large domain binds cyclic AMP, a small intracellular signaling molecule. When the large domain binds cyclic AMP, it causes a conformational change in the protein that enables the small domain to bind to a specific DNA sequence and thereby promote the expression of an adjacent gene. To provide a sense of the many different domain structures observed in proteins, ribbon models of three different domains are shown in **Figure 4–20**.

(A)

(B)

Figure 4–17 β sheets come in two varieties. (A) Antiparallel β sheet (see also Figure 4–13D). (B) Parallel β sheet. Both of these structures are common in proteins. By convention, the arrows point toward the C-terminus of the polypeptide chain (Movie 4.4).

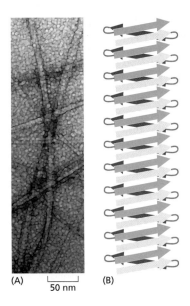

(A) |⎯ 50 nm ⎯| (B)

Figure 4–18 Stacking of β sheets allows some misfolded proteins to aggregate into amyloid fibers. (A) Electron micrograph shows an amyloid fiber formed from a segment of a yeast prion protein. (B) Schematic representation shows the stacking of β sheets that stabilize an individual amyloid fiber. (A, courtesy of David Eisenberg.)

Many Proteins Also Contain Unstructured Regions

Small protein molecules, such as the oxygen-carrying muscle protein myoglobin, contain only a single domain (see Figure 4–11). Larger proteins can contain as many as several dozen domains, which are usually connected by relatively unstructured lengths of polypeptide chain. Such regions of polypeptide chain lacking any definite structure, which continually bend and flex due to thermal buffeting, are abundant in cells. These **intrinsically disordered sequences** are often found as short stretches linking domains in otherwise highly ordered proteins. Other proteins, however, are almost entirely without secondary structure and exist as unfolded polypeptide chains in the cytosol.

Intrinsically disordered sequences remained undetected for many years. Their lack of folded structure makes them prime targets for the proteolytic enzymes that are released when cells are fractionated to isolate their molecular components (see Panel 4–3, pp. 164–165). Unstructured sequences also fail to form protein crystals and for this reason escape the attention of X-ray crystallographers (see How We Know, pp. 162–163). Indeed, the ubiquity of disordered sequences became appreciated only after bioinformatics methods were developed that could recognize them from their amino acid sequences. Present estimates suggest that a third of all eukaryotic proteins have long unstructured regions in their polypeptide chain (greater than 30 amino acids in length), while a substantial number of eukaryotic proteins are mostly disordered under normal conditions.

Unstructured sequences have a variety of important functions in cells. Being able to flex and bend, they can wrap around one or more target proteins like a scarf, binding with both high specificity and low affinity (**Figure 4–21**). By forming flexible tethers between the compact domains in a protein, they provide flexibility while increasing the frequency of encounters between the domains (Figure 4–21). They can help *scaffold proteins* bring together proteins in an intracellular signaling pathway, facilitating interactions (Figure 4–21). They also give proteins like elastin

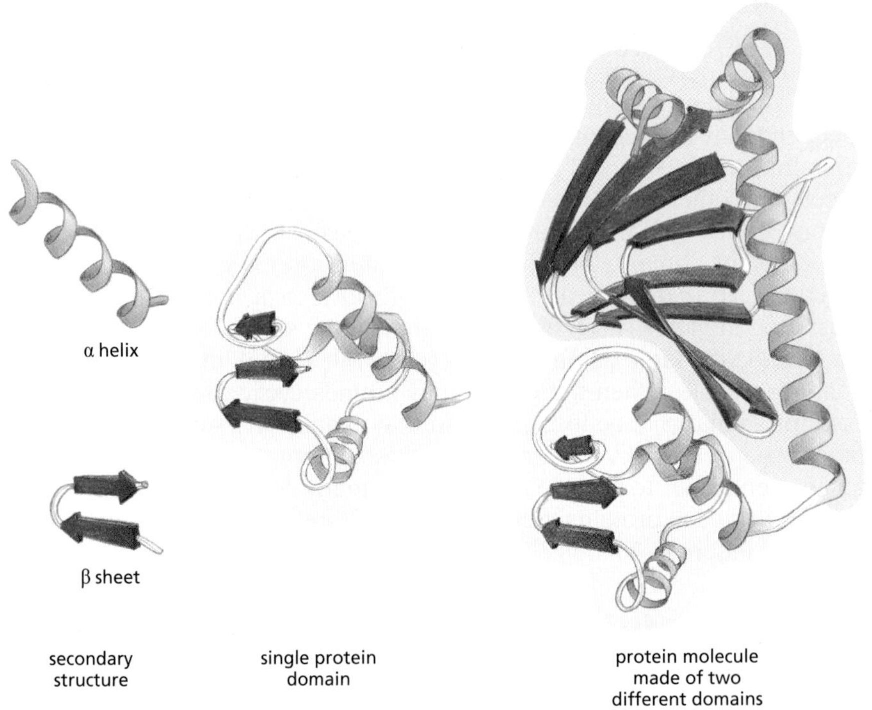

Figure 4–19 Many proteins are composed of separate functional domains. Elements of secondary structure such as α helices and β sheets pack together into stable, independently folding, globular elements called protein domains. A typical protein molecule is built from one or more domains, linked by a region of polypeptide chain that is often relatively unstructured. The ribbon diagram on the right represents the bacterial transcription regulator protein CAP, with one large domain (outlined in *blue*) and one small domain (outlined in *yellow*).

α helix

β sheet

secondary structure

single protein domain

protein molecule made of two different domains

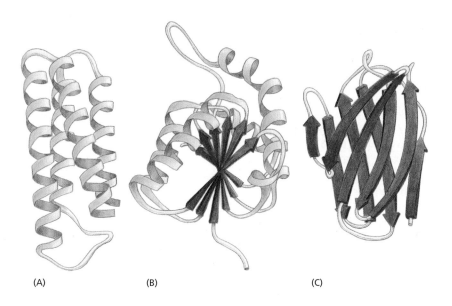

Figure 4–20 Ribbon models show three different protein domains. (A) Cytochrome b_{562} is a single-domain protein involved in electron transfer in *E. coli*. It is composed almost entirely of α helices. (B) The NAD-binding domain of the enzyme lactic dehydrogenase is composed of a mixture of α helices and β sheets. (C) An immunoglobulin domain of an antibody molecule is composed of a sandwich of two antiparallel β sheets. In these examples, the α helices are shown in *green*, while strands organized as β sheets are denoted by *red arrows*. The protruding loop regions (*yellow*) are often unstructured and can provide binding sites for other molecules. (Redrawn from originals courtesy of Jane Richardson.)

the ability to form rubberlike fibers, allowing our tendons and skin to recoil after being stretched. In addition to providing structural flexibility, unstructured sequences are also ideal substrates for the addition of chemical groups that control the way many proteins behave—a topic we discuss at length later in the chapter.

Few of the Many Possible Polypeptide Chains Will Be Useful

In theory, a vast number of different polypeptide chains could be made from 20 different amino acids. Because each amino acid is chemically distinct and could, in principle, occur at any position, a polypeptide chain four amino acids long has $20 \times 20 \times 20 \times 20 = 160,000$ different possible sequences. In other words, for a polypeptide that is *n* amino acids long, 20^n different chains are possible. For a typical protein length of 300 amino acids, more than 20^{300} (that's 10^{390}) different polypeptide chains could theoretically be made.

Of the unimaginably large collection of potential polypeptide sequences, only a miniscule fraction is actually present in cells. That's because many biological functions depend on proteins with stable, well-defined three-dimensional conformations. This requirement restricts the list of possible polypeptide sequences. Another constraint is that functional proteins

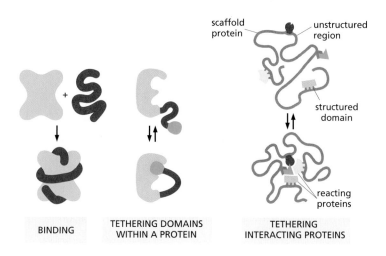

BINDING

TETHERING DOMAINS WITHIN A PROTEIN

TETHERING INTERACTING PROTEINS

scaffold protein

unstructured region

structured domain

reacting proteins

Figure 4–21 Unstructured regions of a polypeptide chain in proteins can peform many functions. A few of these functions are illustrated here.

must be "well-behaved" and not engage in unwanted associations with other proteins in the cell—forming insoluble protein aggregates, for example. Many potential proteins would therefore have been eliminated by natural selection through the long trial-and-error process that underlies evolution (discussed in Chapter 9).

Thanks to this rigorous process of selection, the amino acid sequences of many present-day proteins have evolved to guarantee that the polypeptide will adopt a stable conformation—one that bestows upon the protein the exact chemical properties that will enable it to perform a particular function. Such proteins are so precisely built that a change in even a few atoms in one amino acid can sometimes disrupt the structure of a protein and thereby eliminate its function. In fact, the structures of many proteins—and their constituent domains—are so stable and effective that they have been conserved throughout evolution among many diverse organisms. The three-dimensional structures of the DNA-binding domains from the yeast α2 protein and the *Drosophila* Engrailed protein, for example, are almost completely superimposable, even though these organisms are separated by more than a billion years of evolution. Other proteins, however, have changed their structure and function over evolutionary time, as we now discuss.

Proteins Can Be Classified into Families

Once a protein had evolved a stable conformation with useful properties, its structure could be modified over time to enable it to perform new functions. We know that this occurred quite often during evolution, because many present-day proteins can be grouped into **protein families**, in which each family member has an amino acid sequence and a three-dimensional conformation that closely resemble those of the other family members.

Consider, for example, the *serine proteases*, a family of protein-cleaving (proteolytic) enzymes that includes the digestive enzymes chymotrypsin, trypsin, and elastase, as well as several proteases involved in blood clotting. When any two of these enzymes are compared, portions of their amino acid sequences are found to be nearly the same. The similarity of their three-dimensional conformations is even more striking: most of the detailed twists and turns in their polypeptide chains, which are several hundred amino acids long, are virtually identical (**Figure 4–22**). The various serine proteases nevertheless have distinct enzymatic activities, each cleaving different proteins or the peptide bonds between different types of amino acids.

Figure 4–22 Serine proteases constitute a family of proteolytic enzymes. Backbone models of two serine proteases, elastase and chymotrypsin, are illustrated. Although only those amino acid sequences in the polypeptide chain shaded in *green* are the same in the two proteins, the two conformations are very similar nearly everywhere. Nonetheless, the two proteases prefer different substrates. The active site of each enzyme—where its substrates are bound and cleaved—is circled in *red*.

Serine proteases derive their name from the amino acid serine, which directly participates in the cleavage reaction. The two *black dots* on the *right side* of the chymotrypsin molecule mark the two ends created where the enzyme has cleaved its own backbone.

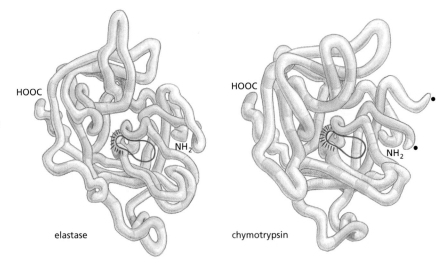

elastase

chymotrypsin

Large Protein Molecules Often Contain More Than One Polypeptide Chain

The same type of weak noncovalent bonds that enable a polypeptide chain to fold into a specific conformation also allow proteins to bind to each other to produce larger structures in the cell. Any region on a protein's surface that interacts with another molecule through sets of noncovalent bonds is termed a *binding site*. A protein can contain binding sites for a variety of molecules, large and small. If a binding site recognizes the surface of a second protein, the tight binding of two folded polypeptide chains at this site will create a larger protein, whose quaternary structure has a precisely defined geometry. Each polypeptide chain in such a protein is called a **subunit**, and each subunit may contain more than one domain.

In the simplest case, two identical, folded polypeptide chains form a symmetrical complex of two protein subunits (called a *dimer*) that is held together by interactions between two identical binding sites. The CAP protein in bacterial cells is a dimer (**Figure 4–23A**) formed from two identical copies of the protein subunit shown previously in Figure 4–19. Many other symmetrical protein complexes, formed from multiple copies of the same polypeptide chain, are commonly found in cells. The enzyme *neuraminidase*, for example, consists of a ring of four identical protein subunits (**Figure 4–23B**).

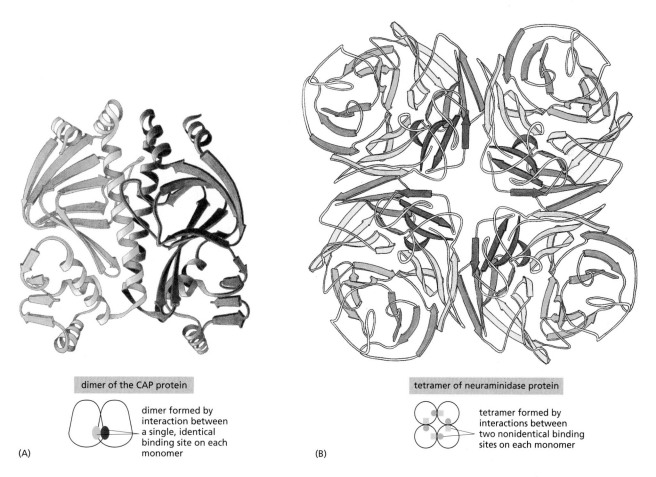

dimer of the CAP protein

dimer formed by interaction between a single, identical binding site on each monomer

(A)

tetramer of neuraminidase protein

tetramer formed by interactions between two nonidentical binding sites on each monomer

(B)

Figure 4–23 Many protein molecules contain multiple copies of the same protein subunit. (A) A symmetrical dimer. The CAP protein is a complex of two identical polypeptide chains (see also Figure 4–19). (B) A symmetrical homotetramer. The enzyme neuraminidase exists as a ring of four identical polypeptide chains. For both (A) and (B), a small schematic below the structure emphasizes how the repeated use of the same binding interaction forms the structure. In (A), the use of the same binding site on each monomer (represented by *brown* and *green ovals*) causes the formation of a symmetrical dimer. In (B), a pair of nonidentical binding sites (represented by *orange circles* and *blue squares*) causes the formation of a symmetrical tetramer.

Figure 4–24 Some proteins are formed as a symmetrical assembly of two different subunits. Hemoglobin, an oxygen-carrying protein abundant in red blood cells, contains two copies of α-globin (*green*) and two copies of β-globin (*blue*). Each of these four polypeptide chains contains a heme molecule (*red*), where oxygen (O$_2$) is bound. Thus, each molecule of hemoglobin in the blood carries four molecules of oxygen.

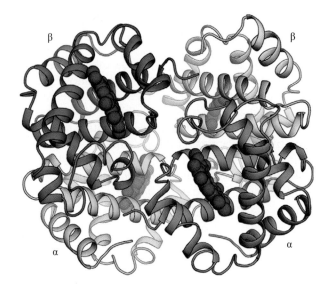

Other proteins contain two or more different polypeptide chains. *Hemoglobin*, the protein that carries oxygen in red blood cells, is a particularly well-studied example. The protein contains two identical α-globin subunits and two identical β-globin subunits, symmetrically arranged (**Figure 4–24**). Many proteins contain multiple subunits, and they can be very large (Movie 4.5).

Proteins Can Assemble into Filaments, Sheets, or Spheres

Proteins can form even larger assemblies than those discussed so far. Most simply, a chain of identical protein molecules can be formed if the binding site on one protein molecule is complementary to another region on the surface of another protein molecule of the same type. Because each protein molecule is bound to its neighbor in an identical way (see Figure 4–14), the molecules will often be arranged in a helix that can be extended indefinitely in either direction (**Figure 4–25**). This type of arrangement can produce an extended protein filament. An actin filament, for example, is a long, helical structure formed from many molecules of the protein actin (**Figure 4–26**). Actin is extremely abundant in eukaryotic cells, where it forms one of the major filament systems of the cytoskeleton (discussed in Chapter 17). Other sets of identical proteins associate to form tubes, as in the microtubules of the cytoskeleton (**Figure 4–27**), or cagelike spherical shells, as in the protein coats of virus particles (**Figure 4–28**).

Many large structures, such as viruses and ribosomes, are built from a mixture of one or more types of protein plus RNA or DNA molecules. These structures can be isolated in pure form and dissociated into their constituent macromolecules. It is often possible to mix the isolated components back together and watch them reassemble spontaneously into the original structure. This demonstrates that all the information needed for assembly of the complicated structure is contained in the macromolecules themselves. Experiments of this type show that much of the

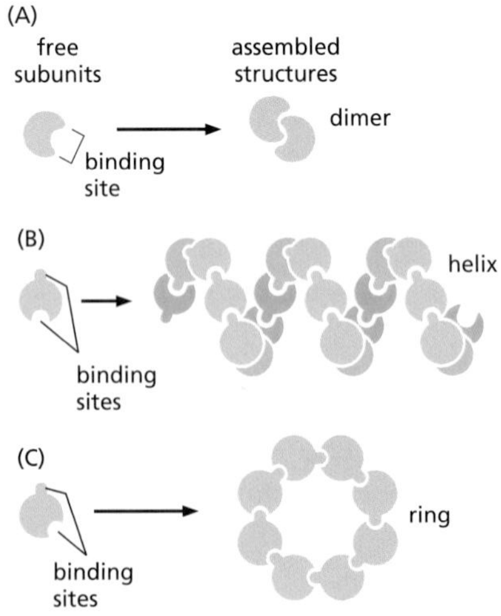

(A)

free subunits   assembled structures

binding site   dimer

(B)

helix

binding sites

(C)

ring

binding sites

Figure 4–25 Identical protein subunits can assemble into complex structures. (A) A protein with just one binding site can form a dimer with another identical protein. (B) Identical proteins with two different binding sites will often form a long, helical filament. (C) If the two binding sites are disposed appropriately in relation to each other, the protein subunits will form a closed ring instead of a helix (see also Figure 4–23B).

50 nm

Figure 4–26 An actin filament is composed of identical protein subunits. The helical array of actin molecules in a filament often contains thousands of molecules and extends for micrometers in the cell.

structure of a cell is self-organizing: if the required proteins are produced in the right amounts, the appropriate structures will form automatically.

Some Types of Proteins Have Elongated Fibrous Shapes

Most of the proteins we have discussed so far are **globular proteins**, in which the polypeptide chain folds up into a compact shape like a ball with an irregular surface. Enzymes, for example, tend to be globular proteins: even though many are large and complicated, with multiple subunits, most have a quaternary structure with an overall rounded shape (see Figure 4–11). In contrast, other proteins have roles in the cell that require them to span a large distance. These proteins generally have a relatively simple, elongated three-dimensional structure and are commonly referred to as **fibrous proteins**.

One large class of intracellular fibrous proteins resembles α-keratin, which we met earlier when we introduced the α-helix. Keratin filaments are extremely stable: long-lived structures such as hair, horns, and nails are composed mainly of this protein. An α-keratin molecule is a dimer of two identical subunits, with the long α helices of each subunit forming a coiled-coil (see Figure 4–16). These coiled-coil regions are capped at either end by globular domains containing binding sites that allow them to assemble into ropelike *intermediate filaments*—a component of the cytoskeleton that gives cells mechanical strength (discussed in Chapter 17).

Fibrous proteins are especially abundant outside the cell, where they form the gel-like *extracellular matrix* that helps bind cells together to form tissues. These proteins are secreted by the cells into their surroundings, where they often assemble into sheets or long fibrils. *Collagen* is the most abundant of these fibrous extracellular proteins in animal tissues. A collagen molecule consists of three long polypeptide chains, each containing the nonpolar amino acid glycine at every third position. This regular structure allows the chains to wind around one another to generate a long, regular, triple helix with glycine at its core (**Figure 4–29A**). Many such

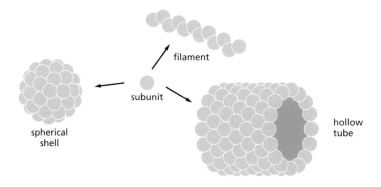

Figure 4–27 A single type of protein subunit can pack together to form a filament, a hollow tube, or a spherical shell. Actin subunits, for example, form actin filaments (see Figure 4–26), whereas tubulin subunits form hollow microtubules, and some virus proteins form a spherical shell (capsid) that encloses the viral genome (see Figure 4–28).

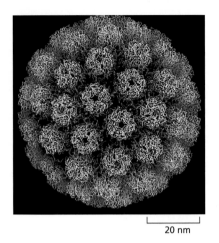

└─────┘
20 nm

Figure 4–28 Many viral capsids are more or less spherical protein assemblies. They are formed from many copies of a small set of protein subunits. The nucleic acid of the virus (DNA or RNA) is packaged inside. The structure of the simian virus SV40, shown here, was determined by X-ray crystallography and is known in atomic detail. (Courtesy of Robert Grant, Stephan Crainic, and James M. Hogle.)

collagen molecules bind to one another side-by-side and end-to-end to create long overlapping arrays called *collagen fibrils*, which are extremely strong and help hold tissues together, as described in Chapter 20.

In complete contrast to collagen is another fibrous protein in the extracellular matrix, *elastin*. Elastin molecules are formed from relatively loose and unstructured polypeptide chains that are covalently cross-linked into a rubberlike elastic meshwork. The resulting *elastic fibers* enable skin and other tissues, such as arteries and lungs, to stretch and recoil without tearing. As illustrated in **Figure 4–29B**, the elasticity is due to the ability of the individual protein molecules to uncoil reversibly whenever they are stretched.

Extracellular Proteins Are Often Stabilized by Covalent Cross-Linkages

Many protein molecules are either attached to the outside of a cell's plasma membrane or secreted as part of the extracellular matrix, which exposes them to extracellular conditions. To help maintain their structures, the polypeptide chains in such proteins are often stabilized by covalent cross-linkages. These linkages can either tie together two amino acids in the same polypeptide chain or join together many polypeptide chains in a large protein complex—as for the collagen fibrils and elastic fibers just described.

The most common covalent cross-links in proteins are sulfur–sulfur bonds. These **disulfide bonds** (also called *S–S bonds*) are formed before a protein is secreted by an enzyme in endoplasmic reticulum that links together two –SH groups from cysteine side chains that are adjacent in the folded protein (**Figure 4–30**). Disulfide bonds do not change a protein's conformation, but instead act as a sort of "atomic staple" to reinforce the protein's most favored conformation. For example, lysozyme—an

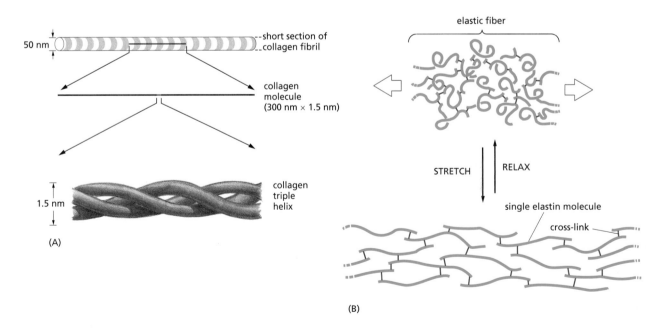

Figure 4–29 Collagen and elastin are abundant extracellular fibrous proteins. (A) A collagen molecule is a triple helix formed by three extended protein chains that wrap around one another. Many rodlike collagen molecules are cross-linked together in the extracellular space to form collagen fibrils (*top*), which have the tensile strength of steel. The striping on the collagen fibril is caused by the regular repeating arrangement of the collagen molecules within the fibril. (B) Elastin molecules are cross-linked together by covalent bonds (*red*) to form rubberlike, elastic fibers. Each elastin polypeptide chain uncoils into a more extended conformation when the fiber is stretched, and recoils spontaneously as soon as the stretching force is relaxed.

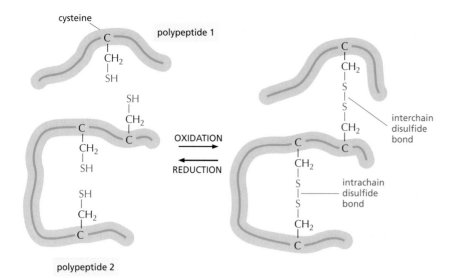

cysteine

polypeptide 1

OXIDATION

REDUCTION

interchain disulfide bond

intrachain disulfide bond

polypeptide 2

Figure 4–30 Disulfide bonds help stabilize a favored protein conformation. This diagram illustrates how covalent disulfide bonds form between adjacent cysteine side chains by the oxidation of their –SH groups. As indicated, these cross-links can join either two parts of the same polypeptide chain or two different polypeptide chains. Because the energy required to break one covalent bond is much larger than the energy required to break even a whole set of noncovalent bonds (see Table 2–1, p. 48), a disulfide bond can have a major stabilizing effect on a protein's folded structure (Movie 4.6).

enzyme in tears, saliva, and other secretions that can disrupt bacterial cell walls—retains its antibacterial activity for a long time because it is stabilized by such disulfide cross-links.

Disulfide bonds generally do not form in the cell cytosol, where a high concentration of reducing agents converts such bonds back to cysteine –SH groups. Apparently, proteins do not require this type of structural reinforcement in the relatively mild conditions in the cytosol.

HOW PROTEINS WORK

As we have just seen, proteins are made from an enormous variety of amino acid sequences and can fold into a unique shape. The surface topography of a protein's side chains endows each protein with a unique function, based on its chemical properties. The union of structure, chemistry, and function gives proteins the extraordinary ability to orchestrate the large number of dynamic processes that occur in cells.

Thus, for proteins, form and function are inextricably linked. But the fundamental question remains: How do proteins actually work? In this section, we will see that the activity of proteins depends on their ability to bind specifically to other molecules, allowing them to act as catalysts, structural supports, tiny motors, and so on. The examples we review here by no means exhaust the vast functional repertoire of proteins. However, the specialized functions of the proteins you will encounter elsewhere in this book are based on the same principles.

All Proteins Bind to Other Molecules

The biological properties of a protein molecule depend on its physical interaction with other molecules. Antibodies attach to viruses or bacteria as part of the body's defenses; the enzyme hexokinase binds glucose and ATP to catalyze a reaction between them; actin molecules bind to one another to assemble into long filaments; and so on. Indeed, all proteins stick, or bind, to other molecules in a specific manner. In some cases, this binding is very tight; in others, it is weak and short-lived. As we saw in Chapter 3, the affinity of an enzyme for its substrate is reflected in its K_M: the lower the K_M, the tighter the binding.

Regardless of its strength, the binding of a protein to other biological molecules always shows great *specificity*: each protein molecule can bind to just one or a few molecules out of the many thousands of different

QUESTION 4–4

Hair is composed largely of fibers of the protein keratin. Individual keratin fibers are covalently cross-linked to one another by many disulfide (S–S) bonds. If curly hair is treated with mild reducing agents that break a few of the cross-links, pulled straight, and then oxidized again, it remains straight. Draw a diagram that illustrates the three different stages of this chemical and mechanical process at the level of the keratin filaments, focusing on the disulfide bonds. What do you think would happen if hair were treated with strong reducing agents that break all the disulfide bonds?

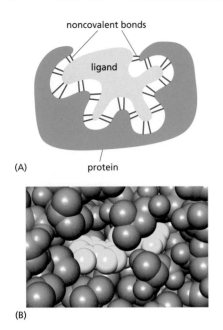

(A)

(B)

Figure 4–31 The binding of a protein to another molecule is highly selective. Many weak interactions are needed to enable a protein to bind tightly to a second molecule (a ligand). The ligand must therefore fit precisely into the protein's binding site, like a hand into a glove, so that a large number of noncovalent interactions can be formed between the protein and the ligand. (A) Schematic drawing shows the binding of a hypothetical protein and ligand; (B) space-filling model.

molecules it encounters. Any substance that is bound by a protein—whether it is an ion, a small organic molecule, or a macromolecule—is referred to as a **ligand** for that protein (from the Latin *ligare*, "to bind").

The ability of a protein to bind selectively and with high affinity to a ligand is due to the formation of a set of weak, noncovalent interactions—hydrogen bonds, electrostatic attractions, and van der Waals attractions—plus favorable hydrophobic forces (see Panel 2–7, pp. 78–79). Each individual noncovalent interaction is weak, so that effective binding requires many such bonds to be formed simultaneously. This is possible only if the surface contours of the ligand molecule fit very closely to the protein, matching it like a hand in a glove (**Figure 4–31**).

When molecules have poorly matching surfaces, few noncovalent interactions occur, and the two molecules dissociate as rapidly as they come together. This is what prevents incorrect and unwanted associations from forming between mismatched molecules. At the other extreme, when many noncovalent interactions are formed, the association can persist for a very long time. Strong binding between molecules occurs in cells whenever a biological function requires that the molecules remain tightly associated for a long time—for example, when a group of macromolecules come together to form a functional subcellular structure such as a ribosome.

The region of a protein that associates with a ligand, known as its **binding site**, usually consists of a cavity in the protein surface formed by a particular arrangement of amino acid side chains. These side chains can belong to amino acids that are widely separated on the linear polypeptide chain, but are brought together when the protein folds (**Figure 4–32**). Other regions on the surface often provide binding sites for different ligands that regulate the protein's activity, as we discuss later. Yet other parts of the protein may be required to attract or attach the protein to a particular location in the cell—for example, the hydrophobic α helix of a

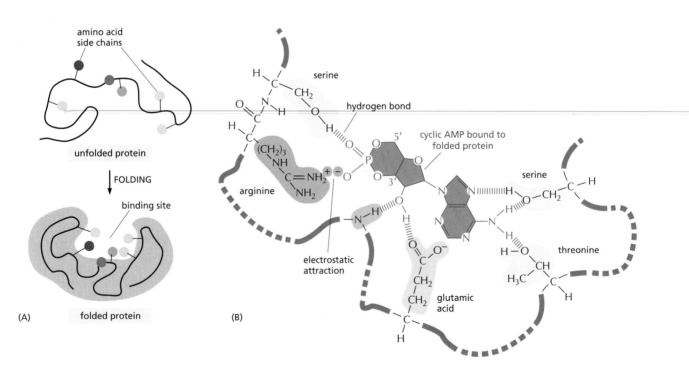

(A)

(B)

Figure 4–32 Binding sites allow proteins to interact with specific ligands. (A) The folding of the polypeptide chain typically creates a crevice or cavity on the folded protein's surface, where specific amino acid side chains are brought together in such a way that they can form a set of noncovalent bonds only with certain ligands. (B) Close-up view of an actual binding site showing the hydrogen bonds and an electrostatic interaction formed between a protein and its ligand (in this example, the bound ligand is cyclic AMP, shown in *dark brown*).

membrane-spanning protein allows it to be inserted into the lipid bilayer of a cell membrane (discussed in Chapter 11).

Although the atoms buried in the interior of a protein have no direct contact with the ligand, they provide an essential scaffold that gives the surface its contours and chemical properties. Even tiny changes to the amino acids in the interior of a protein can change the protein's three-dimensional shape and destroy its function.

There Are Billions of Different Antibodies, Each with a Different Binding Site

All proteins must bind to particular ligands to carry out their various functions. For antibodies, the universe of possible ligands is limitless. Each of us has the capacity to produce a huge variety of antibodies, among which there will be one that is capable of recognizing and binding tightly to almost any molecule imaginable.

Antibodies are immunoglobulin proteins produced by the immune system in response to foreign molecules, especially those on the surface of an invading microorganism. Each antibody binds to a particular target molecule extremely tightly, either inactivating the target directly or marking it for destruction. An antibody recognizes its target molecule—called an **antigen**—with remarkable specificity, and, because there are potentially billions of different antigens that a person might encounter, we have to be able to produce billions of different antibodies.

Antibodies are Y-shaped molecules with two identical antigen-binding sites, each of which is complementary to a small portion of the surface of the antigen molecule. A detailed examination of the antigen-binding sites of antibodies reveals that they are formed from several loops of polypeptide chain that protrude from the ends of a pair of closely juxtaposed protein domains (**Figure 4–33**). The amino acid sequence in these

Figure 4–33 An antibody is Y-shaped and has two identical antigen-binding sites, one on each arm of the Y.
(A) Schematic drawing of a typical antibody molecule. The protein is composed of four polypeptide chains (two identical heavy chains and two identical and smaller light chains), held together by disulfide bonds (*red*). Each chain is made up of several similar domains, here shaded either *blue* or *gray*. The antigen-binding site is formed where a heavy-chain variable domain (V_H) and a light-chain variable domain (V_L) come close together. These are the domains that differ most in their amino acid sequence in different antibodies—hence their name. (B) Ribbon drawing of a single light chain showing that the most variable parts of the polypeptide chain (*orange*) extend as loops at one end of the variable domain (V_L) to form half of one antigen-binding site of the antibody molecule shown in (A). Note that both the constant and variable domains are composed of a sandwich of two antiparallel β sheets (see also Figure 4–20C), connected by a disulfide bond (*red*).

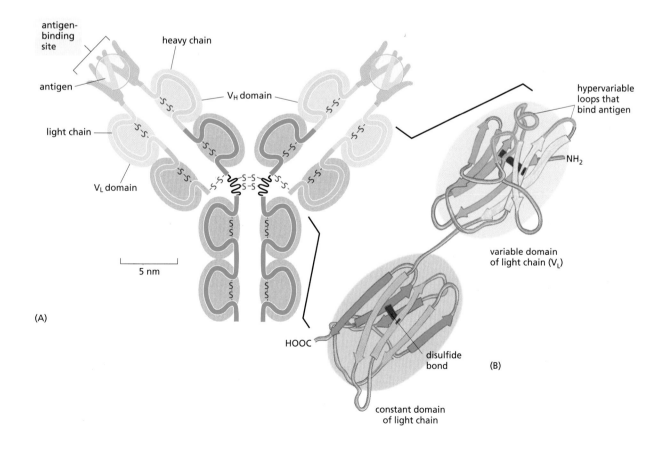

loops can vary greatly without altering the basic structure of the antibody. An enormous diversity of antigen-binding sites can be generated by changing only the length and amino acid sequence of the loops, which is how the wide variety of different antibodies is formed (Movie 4.7).

With their unique combination of specificity and diversity, antibodies are not only indispensable for fighting off infections, they are also invaluable in the laboratory, where they can be used to identify, purify, and study other molecules (**Panel 4–2**, pp. 146–147).

Enzymes Are Powerful and Highly Specific Catalysts

For many proteins, binding to another molecule is their main function. An actin molecule, for example, need only associate with other actin molecules to form a filament. There are proteins, however, for which ligand binding is simply a necessary first step in their function. This is the case for the large and very important class of proteins called **enzymes**. These remarkable molecules are responsible for nearly all of the chemical transformations that occur in cells. Enzymes bind to one or more ligands, called **substrates**, and convert them into chemically modified products, doing this over and over again with amazing rapidity. As we saw in Chapter 3, they speed up reactions, often by a factor of a million or more, without themselves being changed—that is, enzymes act as *catalysts* that permit cells to make or break covalent bonds at will. This catalysis of organized sets of chemical reactions by enzymes creates and maintains the cell, making life possible.

Enzymes can be grouped into functional classes based on the chemical reactions they catalyze (**Table 4–1**). Each type of enzyme is highly specific, catalyzing only a single type of reaction. Thus, *hexokinase* adds a phosphate group to D-glucose but not to its optical isomer L-glucose; the blood-clotting enzyme *thrombin* cuts one type of blood-clotting protein between a particular arginine and its adjacent glycine and nowhere

QUESTION 4–5

Use drawings to explain how an enzyme (such as hexokinase, mentioned in the text) can distinguish its normal substrate (here D-glucose) from the optical isomer L-glucose, which is not a substrate. (Hint: remembering that a carbon atom forms four single bonds that are tetrahedrally arranged and that the optical isomers are mirror images of each other around such a bond, draw the substrate as a simple tetrahedron with four different corners and then draw its mirror image. Using this drawing, indicate why only one optical isomer might bind to a schematic active site of an enzyme.)

TABLE 4–1 SOME COMMON FUNCTIONAL CLASSES OF ENZYMES	
Enzyme Class	**Biochemical Function**
Hydrolase	General term for enzymes that catalyze a hydrolytic cleavage reaction
Nuclease	Breaks down nucleic acids by hydrolyzing bonds between nucleotides
Protease	Breaks down proteins by hydrolyzing peptide bonds between amino acids
Ligase	Joins two molecules together; DNA ligase joins two DNA strands together end-to-end
Isomerase	Catalyzes the rearrangement of bonds within a single molecule
Polymerase	Catalyzes polymerization reactions such as the synthesis of DNA and RNA
Kinase	Catalyzes the addition of phosphate groups to molecules. Protein kinases are an important group of kinases that attach phosphate groups to proteins
Phosphatase	Catalyzes the hydrolytic removal of a phosphate group from a molecule
Oxido-reductase	General name for enzymes that catalyze reactions in which one molecule is oxidized while the other is reduced. Enzymes of this type are often called oxidases, reductases, or dehydrogenases
ATPase	Hydrolyzes ATP. Many proteins have an energy-harnessing ATPase activity as part of their function, including motor proteins such as myosin (discussed in Chapter 17) and membrane transport proteins such as the sodium pump (discussed in Chapter 12)

Enzyme names typically end in "-ase," with the exception of some enzymes, such as pepsin, trypsin, thrombin, lysozyme, and so on, which were discovered and named before the convention became generally accepted at the end of the nineteenth century. The name of an enzyme usually indicates the nature of the reaction catalyzed. For example, citrate synthase catalyzes the synthesis of citrate by a reaction between acetyl CoA and oxaloacetate.

else. As discussed in detail in Chapter 3, enzymes often work in tandem, with the product of one enzyme becoming the substrate for the next. The result is an elaborate network of *metabolic pathways* that provides the cell with energy and generates the many large and small molecules that the cell needs.

Lysozyme Illustrates How an Enzyme Works

To explain how enzymes catalyze chemical reactions, we will use the example of **lysozyme**—an enzyme that acts as a natural antibiotic in egg white, saliva, tears, and other secretions. Lysozyme severs the polysaccharide chains that form the cell walls of bacteria. Because the bacterial cell is under pressure due to intracellular osmotic forces, cutting even a small number of polysaccharide chains causes the cell wall to rupture and the bacterium to burst, or lyse. Lysozyme is a relatively small and stable protein, which can be isolated easily in large quantities. For these reasons it has been intensively studied, and it was the first enzyme whose structure was worked out in atomic detail by X-ray crystallography.

The reaction catalyzed by lysozyme is a hydrolysis: the enzyme adds a molecule of water to a single bond between two adjacent sugar groups in the polysaccharide chain, thereby causing the bond to break. The reaction is energetically favorable because the free energy of the severed polysaccharide chain is lower than the free energy of the intact chain. However, the pure polysaccharide can sit for years in water without being hydrolyzed to any detectable degree. This is because there is an energy barrier to such reactions, called the *activation energy* (discussed in Chapter 3, pp. 91–93). For a colliding water molecule to break a bond linking two sugars, the polysaccharide molecule has to be distorted into a particular shape—the **transition state**—in which the atoms around the bond have an altered geometry and electron distribution. To distort the polysaccharide in this way requires a large input of energy from random molecular collisions. In aqueous solution at room temperature, the energy of such collisions almost never exceeds the activation energy; therefore, hydrolysis occurs extremely slowly, if at all.

This is where the enzyme comes in. Like all enzymes, lysozyme has a binding site on its surface, termed an **active site**, that cradles the contours of its substrate molecule. Here, the catalysis of the chemical reaction occurs. Because its substrate is a polymer, lysozyme's active site is a long groove that holds six linked sugars in the polysaccharide chain at the same time. As soon as the enzyme–substrate complex forms, the enzyme cuts the polysaccharide by catalyzing the addition of a water molecule to one of its sugar–sugar bonds. The severed chain is then quickly released, freeing the enzyme for further cycles of cleavage (**Figure 4–34**).

The chemistry that underlies the binding of lysozyme to its substrate is the same as that for antibody binding to its antigen: the formation of

Figure 4–34 Lysozyme cleaves a polysaccharide chain. (A) Schematic view of the enzyme lysozyme (E), which catalyzes the cutting of a polysaccharide substrate molecule (S). The enzyme first binds to the polysaccharide to form an enzyme–substrate complex (ES), then it catalyzes the cleavage of a specific covalent bond in the backbone of the polysaccharide. The resulting enzyme–product complex (EP) rapidly dissociates, releasing the products (P) and leaving the enzyme free to act on another substrate molecule. (B) A space-filling model of lysozyme bound to a short length of polysaccharide chain prior to cleavage. (B, courtesy of Richard J. Feldmann.)

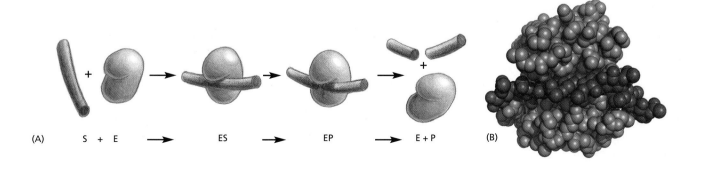

(A)   S + E → ES → EP → E + P   (B)

THE ANTIBODY MOLECULE

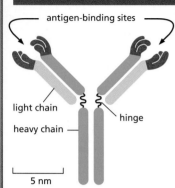

antigen-binding sites

light chain

hinge

heavy chain

5 nm

Antibodies are proteins that bind very tightly to their targets (antigens). They are produced in vertebrates as a defense against infection. Each antibody molecule is made of two identical light chains and two identical heavy chains, so the two antigen-binding sites are identical.

ANTIBODY SPECIFICITY

heavy chain

light chain

antigen

An individual human can make billions of different antibody molecules, each with a distinct antigen-binding site. Each antibody recognizes its antigen with great specificity.

ANTIBODIES DEFEND US AGAINST INFECTION

foreign molecules

viruses

bacteria

ANTIBODIES (Y) CROSS-LINK ANTIGENS INTO AGGREGATES

Antibody–antigen aggregates are ingested by phagocytic cells.

Special proteins in blood kill antibody-coated bacteria or viruses.

B CELLS PRODUCE ANTIBODIES

Antibodies are made by a class of white blood cells called B lymphocytes, or B cells. Each resting B cell carries a different membrane-bound antibody molecule on its surface that serves as a receptor for recognizing a specific antigen. When antigen binds to this receptor, the B cell is stimulated to divide and to secrete large amounts of the same antibody in a soluble form.

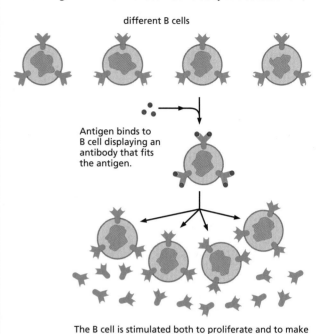

different B cells

Antigen binds to B cell displaying an antibody that fits the antigen.

The B cell is stimulated both to proliferate and to make and secrete more of same antibody.

RAISING ANTIBODIES IN ANIMALS

Antibodies can be made in the laboratory by injecting an animal (usually a mouse, rabbit, sheep, or goat) with antigen A.

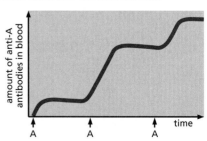

A

inject antigen A

take blood later

Repeated injections of the same antigen at intervals of several weeks stimulate specific B cells to secrete large amounts of anti-A antibodies into the bloodstream.

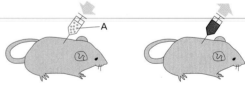

amount of anti-A antibodies in blood

time

A   A   A

Because many different B cells are stimulated by antigen A, the blood will contain a variety of anti-A antibodies, each of which binds A in a slightly different way.

USING ANTIBODIES TO PURIFY MOLECULES

IMMUNOPRECIPITATION

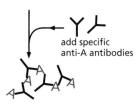

mixture of molecules

add specific
anti-A antibodies

collect aggregate of A molecules and
anti-A antibodies by centrifugation

IMMUNOAFFINITY
COLUMN
CHROMATOGRAPHY

bead coated with
anti-A antibodies

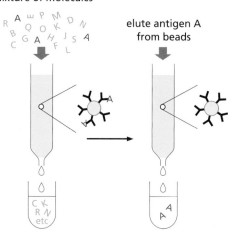

column packed
with these beads

mixture of molecules

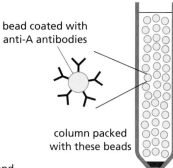

elute antigen A
from beads

discard flow-through   collect pure antigen A

MONOCLONAL ANTIBODIES

Large quantities of a single type of antibody
molecule can be obtained by fusing a B cell
(taken from an animal injected with antigen A)
with a tumor cell. The resulting hybrid cell
divides indefinitely and secretes anti-A
antibodies of a single (monoclonal) type.

B cell from animal
injected with antigen
A makes anti-A
antibody but does
not divide forever.

Tumor cells in
culture divide
indefinitely but
do not make
antibody.

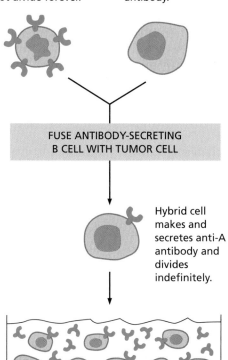

**FUSE ANTIBODY-SECRETING
B CELL WITH TUMOR CELL**

Hybrid cell
makes and
secretes anti-A
antibody and
divides
indefinitely.

USING ANTIBODIES AS MOLECULAR TAGS

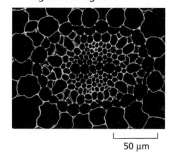

couple to fluorescent dye,
gold particle, or other
special tag

specific antibodies
against antigen A

labeled antibodies

MICROSCOPIC DETECTION

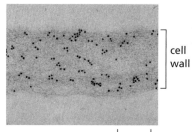

50 μm

Fluorescent antibody binds to
antigen A in tissue and is detected
in a fluorescence microscope. The
antigen here is pectin in the cell
walls of a slice of plant tissue.

200 nm

cell
wall

Gold-labeled antibody binds to
antigen A in tissue and is detected
in an electron microscope. The
antigen is pectin in the cell wall
of a single plant cell.

BIOCHEMICAL DETECTION

Antigen A is
separated from
other molecules
by electrophoresis.

Incubation with the
labeled antibodies
that bind to antigen A
allows the position of the
antigen to be determined.

Labeled second antibody
(*blue*) binds to first
antibody (*black*).

antigen

Note: In all cases, the sensitivity can
be greatly increased by using multiple
layers of antibodies. This "sandwich"
method enables smaller numbers of
antigen molecules to be detected.

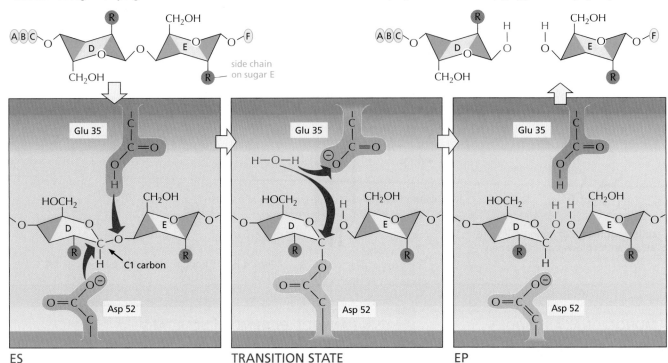

SUBSTRATE

This substrate is an oligosaccharide of six sugars, labeled A through F. Only sugars D and E are shown in detail.

PRODUCTS

The final products are an oligosaccharide of four sugars (left) and a disaccharide (right), produced by hydrolysis.

ES

In the enzyme–substrate complex (ES), the enzyme forces sugar D into a strained conformation. The Glu 35 in the enzyme is positioned to serve as an acid that attacks the adjacent sugar–sugar bond by donating a proton (H+) to sugar E; Asp 52 is poised to attack the C1 carbon atom of sugar D.

TRANSITION STATE

The Asp 52 has formed a covalent bond between the enzyme and the C1 carbon atom of sugar D. The Glu 35 then polarizes a water molecule (*red*), so that its oxygen can readily attack the C1 carbon atom of sugar D and displace Asp 52.

EP

The reaction of the water molecule (*red*) completes the hydrolysis and returns the enzyme to its initial state, forming the final enzyme–product complex (EP).

Figure 4–35 Enzymes bind to, and chemically alter, substrate molecules. In the active site of lysozyme, a covalent bond in a polysaccharide molecule is bent and then broken. The top row shows the free substrate and the free products. The three lower panels depict sequential events at the enzyme active site, during which a sugar–sugar covalent bond is broken. Note the change in the conformation of sugar D in the enzyme–substrate complex compared with the free substrate. This conformation favors the formation of the transition state shown in the middle panel, greatly lowering the activation energy required for the reaction. The reaction, and the structure of lysozyme bound to its product, are shown in Movie 4.8 and Movie 4.9. (Based on D.J. Vocadlo et al., *Nature* 412:835–838, 2001.)

multiple noncovalent bonds. However, lysozyme holds its polysaccharide substrate in such a way that one of the two sugars involved in the bond to be broken is distorted from its normal, most stable conformation. The bond to be broken is held close to two specific amino acids with acidic side chains—a glutamic acid and an aspartic acid—located within the active site of the enzyme. Conditions are thereby created in the microenvironment of the lysozyme active site that greatly reduce the activation energy necessary for the hydrolysis to take place (**Figure 4–35**). The overall chemical reaction, from the initial binding of the polysaccharide on the surface of the enzyme to the final release of the severed chains, occurs many millions of times faster than it would in the absence of enzyme.

Other enzymes use similar mechanisms to lower the activation energies and speed up the reactions they catalyze. In reactions involving two or more substrates, the active site also acts like a template or mold that brings the reactants together in the proper orientation for the reaction to occur (**Figure 4–36A**). As we saw for lysozyme, the active site of an enzyme contains precisely positioned chemical groups that speed up the reaction by altering the distribution of electrons in the substrates (**Figure 4–36B**). Binding to the enzyme also changes the shape of the substrate, bending bonds so as to drive the bound molecule toward a particular transition state (**Figure 4–36C**). Finally, like lysozyme, many enzymes participate intimately in the reaction by briefly forming a covalent bond between the substrate and an amino acid side chain in the active site. Subsequent steps in the reaction restore the side chain to its original state, so the enzyme remains unchanged after the reaction and can go on to catalyze many more reactions.

 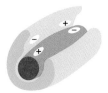 

(A) enzyme binds to two substrate molecules and orients them precisely to encourage a reaction to occur between them

(B) binding of substrate to enzyme rearranges electrons in the substrate, creating partial negative and positive charges that favor a reaction

(C) enzyme strains the bound substrate molecule, forcing it toward a transition state to favor a reaction

Figure 4–36 Enzymes can encourage a reaction in several ways. (A) Holding reacting substrates together in a precise alignment. (B) Rearranging the distribution of charge in a reaction intermediate. (C) Altering bond angles in the substrate to increase the rate of a particular reaction.

Many Drugs Inhibit Enzymes

Many of the drugs we take to treat or prevent illness work by blocking the activity of a particular enzyme. Cholesterol-lowering *statins* inhibit HMG-CoA reductase, an enzyme involved in the synthesis of cholesterol by the liver. *Methotrexate* kills some types of cancer cells by shutting down dihydrofolate reductase, an enzyme that produces a compound required for DNA synthesis during cell division. Because cancer cells have lost important intracellular control systems, some of them are unusually sensitive to treatments that interrupt chromosome replication, making them susceptible to methotrexate.

Pharmaceutical companies often develop drugs by first using automated methods to screen massive libraries of compounds to find chemicals that are able to inhibit the activity of an enzyme of interest. They can then chemically modify the most promising compounds to make them even more effective, enhancing their binding affinity and specificity for the target enzyme. As we discuss in Chapter 20, the anticancer drug Gleevec® was designed to specifically inhibit an enzyme whose aberrant behavior is required for the growth of a type of cancer called chronic myeloid leukemia. The drug binds tightly in the substrate-binding pocket of the enzyme, blocking its activity (see Figure 20–56).

Tightly Bound Small Molecules Add Extra Functions to Proteins

Although the order of amino acids in proteins gives these macromolecules their shape and functional versatility, sometimes the amino acids by themselves are not enough for a protein to do its job. Just as we use tools to enhance and extend the capabilities of our hands, so proteins often employ small, nonprotein molecules to perform functions that would be difficult or impossible using amino acids alone. Thus, the photoreceptor protein *rhodopsin*, which is the light-sensitive protein made by the rod cells in the retina, detects light by means of a small molecule, *retinal*, which is attached to the protein by a covalent bond to a lysine side chain (**Figure 4–37A**). Retinal changes its shape when it absorbs a photon of light, and this change is amplified by rhodopsin to trigger a cascade of reactions that eventually leads to an electrical signal being carried to the brain.

Another example of a protein that contains a nonprotein portion essential for its function is *hemoglobin* (see Figure 4–24). A molecule of hemoglobin carries four noncovalently bound *heme* groups, ring-shaped molecules each with a single central iron atom (**Figure 4–37B**). Heme gives hemoglobin (and blood) its red color. By binding reversibly to dissolved oxygen gas through its iron atom, heme enables hemoglobin to pick up oxygen in the lungs and release it in tissues that need it.

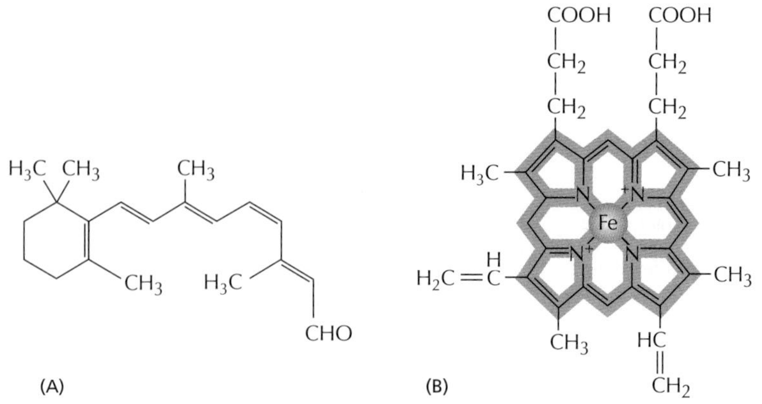

Figure 4–37 Retinal and heme are required for the function of certain proteins. (A) The structure of retinal, the light-sensitive molecule covalently attached to the rhodopsin protein in our eyes. (B) The structure of a heme group, shown with the carbon-containing heme ring colored *red* and the iron atom at its center in *orange*. A heme group is tightly, but noncovalently, bound to each of the four polypeptide chains in hemoglobin, the oxygen-carrying protein whose structure was shown in Figure 4–24.

When these small molecules are attached to their protein, they become an integral part of the protein molecule itself. We discuss in Chapter 11 how proteins can be anchored to cell membranes through covalently attached lipid molecules, and how proteins that are either secreted from the cell or bound to its surface can be modified by the covalent addition of sugars and oligosaccharides.

Enzymes, too, make use of nonprotein molecules: they frequently have a small molecule or metal atom associated with their active site that assists with their catalytic function. *Carboxypeptidase*, an enzyme that cuts polypeptide chains, carries a tightly bound zinc ion in its active site. During the cleavage of a peptide bond by carboxypeptidase, the zinc ion forms a transient bond with one of the substrate atoms, thereby assisting the hydrolysis reaction. In other enzymes, a small organic molecule serves a similar purpose. *Biotin*, for example, is found in enzymes that transfer a carboxyl group ($-COO^-$) from one molecule to another (see Figure 3–37). Biotin participates in these reactions by forming a transient covalent bond to the $-COO^-$ group to be transferred, thereby forming an activated carrier (see Table 3–2, p. 112). This small molecule is better suited for this function than any of the amino acids used to make proteins. Because biotin cannot be synthesized by humans, it must be provided in the diet; thus biotin is classified as a *vitamin*. Other vitamins are similarly needed to make small molecules that are essential components of our proteins; vitamin A, for example, is needed in the diet to make retinal, the light-sensitive part of rhodopsin just discussed.

HOW PROTEINS ARE CONTROLLED

So far, we have examined how proteins do their jobs: how binding to other proteins or small molecules allows them to perform their specific functions. But inside the cell, most proteins and enzymes do not work continuously, or at full speed. Instead, their activities are regulated in a coordinated fashion so the cell can maintain itself in an optimal state, producing only those molecules it requires to thrive under the current conditions. By coordinating when—and how vigorously—proteins function, the cell ensures that it does not deplete its energy reserves by accumulating molecules it does not need or waste its stockpiles of critical substrates. We now consider how cells control the activity of their enzymes and other proteins.

The regulation of protein activity occurs at many levels. At one level, the cell controls the amount of the protein it contains. It can do so by regulating the expression of the gene that encodes that protein (discussed in Chapter 8), and by regulating the rate at which the protein is degraded (discussed in Chapter 7). At another level, the cell controls enzymatic activities by confining sets of enzymes to particular subcellular compartments, often—but not always—enclosed by distinct membranes (discussed in Chapters 14 and 15). But the most rapid and general mechanism used to adjust the activity of a protein occurs at the level of the protein itself. Although proteins can be switched on or off in various ways, as we see next, all of these mechanisms cause the protein to alter its shape, and therefore its function.

The Catalytic Activities of Enzymes Are Often Regulated by Other Molecules

A living cell contains thousands of different enzymes, many of which are operating at the same time in the same small volume of the cytosol. By their catalytic action, enzymes generate a complex web of metabolic pathways, each composed of chains of chemical reactions in which the product of one enzyme becomes the substrate of the next. In this maze of pathways, there are many branch points where different enzymes compete for the same substrate. The system is so complex that elaborate controls are required to regulate when and how rapidly each reaction occurs.

A common type of control occurs when a molecule other than a substrate specifically binds to an enzyme at a special *regulatory site*, altering the rate at which the enzyme converts its substrate to product. In **feedback inhibition**, for example, an enzyme acting early in a reaction pathway is inhibited by a late product of that pathway. Thus, whenever large quantities of the final product begin to accumulate, the product binds to an earlier enzyme and slows down its catalytic action, limiting further entry of substrates into that reaction pathway (**Figure 4–38**). Where pathways branch or intersect, there are usually multiple points of control by different final products, each of which works to regulate its own synthesis (**Figure 4–39**). Feedback inhibition can work almost instantaneously and is rapidly reversed when product levels fall.

Feedback inhibition is a *negative regulation*: it prevents an enzyme from acting. Enzymes can also be subject to *positive regulation*, in which the enzyme's activity is stimulated by a regulatory molecule rather than being suppressed. Positive regulation occurs when a product in one branch of the metabolic maze stimulates the activity of an enzyme in another pathway.

Allosteric Enzymes Have Two or More Binding Sites That Influence One Another

One feature of feedback inhibition was initially puzzling to those who discovered it. Unlike what one expects to see for a competitive inhibitor (see Figure 3–29), the regulatory molecule often has a shape that is totally different from the shape of the enzyme's preferred substrate. Indeed, when this form of regulation was discovered in the 1960s, it was termed *allostery* (from the Greek *allo*, "other," and *stere*, "solid" or "shape"). As more was learned about feedback inhibition, researchers realized that many enzymes must have at least two different binding sites on their surface: the active site that recognizes the substrates and one or more sites that recognize regulatory molecules. And that these sites must somehow "communicate" to allow the catalytic events at the active site to be influenced by the binding of the regulatory molecule at its separate site.

The interaction between sites that are located in different regions on a protein molecule is now known to depend on *conformational changes* in the protein: binding of a ligand to one of the sites causes a shift in the protein's structure from one folded shape to a slightly different folded shape,

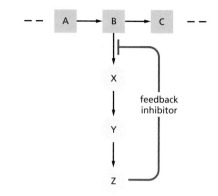

Figure 4–38 Feedback inhibition regulates the flow through biosynthetic pathways. B is the first metabolite in a pathway that gives the end product Z. Z inhibits the first enzyme that is specific to its own synthesis and thereby limits its own concentration in the cell. This form of negative regulation is called feedback inhibition.

Figure 4–39 Feedback inhibition at multiple points regulates connected metabolic pathways. The biosynthetic pathways for four different amino acids in bacteria are shown, starting from the amino acid aspartate. The *red* lines indicate points at which products feed back to inhibit enzymes and the blank boxes represent intermediates in each pathway. In this example, each amino acid controls the first enzyme specific to its own synthesis, thereby limiting its own concentrations and avoiding a wasteful buildup of intermediates. Some of the products also separately inhibit the initial set of reactions common to all the syntheses. Three different enzymes catalyze the initial reaction from aspartate to aspartyl phosphate, and each of these enzymes is inhibited by a different product.

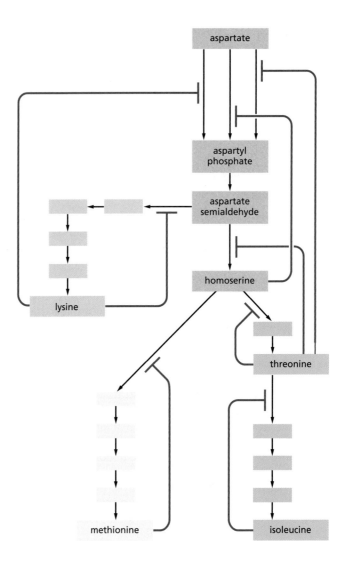

which alters the binding of a ligand to a second site. Many enzymes have two conformations that differ in activity, each stabilized by the binding of different ligands. During feedback inhibition, for example, the binding of an inhibitor at a regulatory site on the protein causes the protein to shift to a conformation in which its active site—located elsewhere in the protein—becomes less accommodating to the substrate molecule (**Figure 4–40**).

Many—if not most—protein molecules are **allosteric**: they can adopt two or more slightly different conformations, and their activity can be regulated by a shift from one to another. This is true not only for enzymes but also for many other proteins as well. The chemistry involved here is extremely simple in concept: because each protein conformation will have somewhat different contours on its surface, the protein's binding sites for ligands will be altered when the protein changes shape. Each ligand will stabilize the conformation that it binds to most strongly, and at high enough concentrations a ligand will tend to "switch" the population of proteins to the conformation that it favors (**Figure 4–41**).

Phosphorylation Can Control Protein Activity by Causing a Conformational Change

Enzymes are regulated solely by the binding of small molecules. Another method that eukaryotic cells use with great frequency to regulate protein

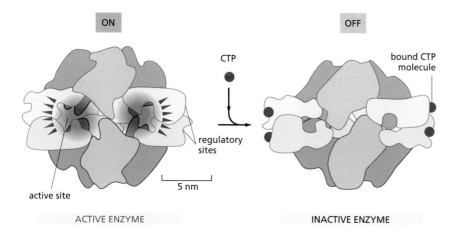

Figure 4–40 Feedback inhibition triggers a conformational change in an enzyme. The enzyme shown, aspartate transcarbamoylase from *E. coli*, was used in early studies of allosteric regulation. This large multisubunit enzyme (see Figure 4–11) catalyzes an important reaction that begins the synthesis of the pyrimidine ring of C, U, and T nucleotides (see Panel 2–6, p. 76–77). One of the final products of this pathway, cytosine triphosphate (CTP), binds to the enzyme to turn it off whenever CTP is plentiful. This diagram shows the conformational change that occurs when the enzyme is turned off by CTP binding to its four regulatory sites, which are distinct from the active site where the substrate binds. Note that the aspartate transcarbamoylase shown in Figure 4–11 is seen from the top. This figure depicts the enzyme as seen from the side.

activity involves attaching a phosphate group covalently to one or more of the protein's amino acid side chains. Because each phosphate group carries two negative charges, the enzyme-catalyzed addition of a phosphate group can cause a major conformational change in a protein by, for example, attracting a cluster of positively charged amino acid side chains from somewhere else in the same protein. This conformational change can, in turn, affect the binding of ligands elsewhere on the protein surface, thereby altering the protein's activity. Removal of the phosphate group by a second enzyme will return the protein to its original conformation and restore its initial activity.

This reversible **protein phosphorylation** controls the activity of many types of proteins in eukaryotic cells; indeed, it is used so extensively that more than one-third of the 10,000 or so proteins in a typical mammalian cell are phosphorylated at any one time. The addition and removal of phosphate groups from specific proteins often occur in response to signals that specify some change in a cell's state. For example, the complicated series of events that takes place as a eukaryotic cell divides is timed largely in this way (discussed in Chapter 18). And many of the intracellular signaling pathways activated by extracellular signals such as hormones depend on a network of protein phosphorylation events (discussed in Chapter 16).

Protein phosphorylation involves the enzyme-catalyzed transfer of the terminal phosphate group of ATP to the hydroxyl group on a serine, threonine, or tyrosine side chain of the protein. This reaction is catalyzed

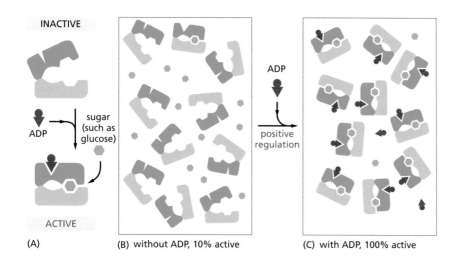

(A)   (B) without ADP, 10% active   (C) with ADP, 100% active

Figure 4–41 The equilibrium between two conformations of a protein is affected by the binding of a regulatory ligand. (A) Schematic diagram of a hypothetical, allosterically regulated enzyme for which a rise in the concentration of ADP molecules (*red* wedges) increases the rate at which the enzyme catalyzes the oxidation of sugar molecules (*blue* hexagons). (B) With no ADP present, only a small fraction of the enzyme molecules spontaneously adopt the active (closed) conformation; most are in the inactive (open) conformation. (C) Because ADP can bind to the protein only in its closed, active conformation, an increase in ADP concentration locks nearly all of the enzyme molecules in the active form. Such an enzyme could be used, for example, to sense when ADP is building up in the cell—which is usually a sign that ATP is decreasing. In this way, the increase in ADP would increase the oxidation of sugars to provide more energy for the synthesis of ATP from ADP—an example of positive regulation.

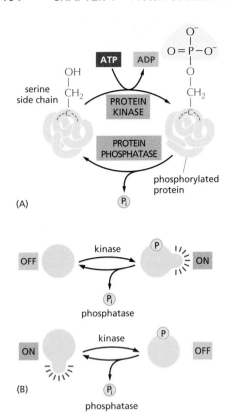

(A)

(B)

Figure 4–42 Protein phosphorylation is a very common mechanism for regulating protein activity. Many thousands of proteins in a typical eukaryotic cell are modified by the covalent addition of one or more phosphate groups. (A) The general reaction, shown here, entails transfer of a phosphate group from ATP to an amino acid side chain of the target protein by a protein kinase. Removal of the phosphate group is catalyzed by a second enzyme, a protein phosphatase. In this example, the phosphate is added to a serine side chain; in other cases, the phosphate is instead linked to the –OH group of a threonine or tyrosine side chain. (B) Phosphorylation can either increase or decrease the protein's activity, depending on the site of phosphorylation and the structure of the protein.

by a **protein kinase**. The reverse reaction—removal of the phosphate group, or *dephosphorylation*—is catalyzed by a **protein phosphatase** (**Figure 4–42A**). Phosphorylation can either stimulate protein activity or inhibit it, depending on the protein involved and the site of phosphorylation (**Figure 4–42B**). Cells contain hundreds of different protein kinases, each responsible for phosphorylating a different protein or set of proteins. Cells also contain a smaller set of different protein phosphatases; some of these are highly specific and remove phosphate groups from only one or a few proteins, whereas others act on a broad range of proteins. The state of phosphorylation of a protein at any moment in time, and thus its activity, will depend on the relative activities of the protein kinases and phosphatases that act on it.

For many proteins, a phosphate group is added to a particular side chain and then removed in a continuous cycle. Phosphorylation cycles of this kind allow proteins to switch rapidly from one state to another. The more rapidly the cycle is "turning," the faster the concentration of a phosphorylated protein can change in response to a sudden stimulus that increases its rate of phosphorylation. However, keeping the cycle turning costs energy, because one molecule of ATP is hydrolyzed with each turn of the cycle.

Covalent Modifications Also Control the Location and Interaction of Proteins

Phosphorylation can do more than control a protein's activity; it can create docking sites where other proteins can bind, thus promoting the assembly of proteins into larger complexes. For example, when extracellular signals stimulate a class of cell-surface, transmembrane proteins called *receptor tyrosine kinases*, they cause the receptor proteins to phosphorylate themselves on certain tyrosines. The phosphorylated tyrosines then serve as docking sites for the binding and activation of various intracellular signaling proteins, which pass along the message to the cell interior and change the behavior of the cell (see Figure 16–32).

Phosphorylation is not the only form of covalent modification that can affect a protein's activity or location. More than 100 types of covalent modifications can occur in the cell, each playing its own role in regulating protein function. Many proteins are modified by the addition of an acetyl group to a lysine side chain. And the addition of the fatty acid palmitate to a cysteine side chain drives a protein to associate with cell membranes. Attachment of ubiquitin, a 76-amino-acid polypeptide, can target a protein for degradation, as we discuss in Chapter 7. Each of these modifying groups is enzymatically added or removed depending on the needs of the cell.

A large number of proteins are modified on more than one amino acid side chain. The p53 protein, which plays a central part in controling how a cell responds to DNA damage and other stresses, can be modified at 20 sites (**Figure 4–43**). Because an enormous number of combinations of these 20 modifications is possible, the protein's behavior can in principle be altered in a huge number of ways.

The set of covalent modifications that a protein contains at any moment constitutes an important form of regulation. The attachment or removal of these modifying groups controls the behavior of a protein, changing its activity or stability, its binding partners, or its location inside the cell. In some cases, the modification alters the protein's conformation; in others, it serves as a docking site for other proteins to attach. This layer of control enables the cell to make optimal use of its proteins, and it allows the cell to respond rapidly to changes in its environment.

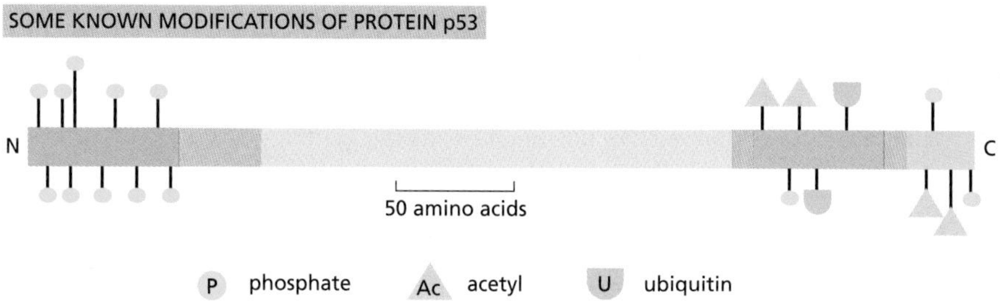

SOME KNOWN MODIFICATIONS OF PROTEIN p53

P phosphate   Ac acetyl   U ubiquitin

Figure 4–43 The modification of a protein at multiple sites can control the protein's behavior. This diagram shows some of the covalent modifications that control the activity and degradation of the protein p53, an important gene regulatory protein that regulates a cell's response to damage (discussed in Chapter 18). Not all of these modifications will be present at the same time. Colors along the body of the protein represent distinct protein domains, including one that binds to DNA (*green*) and one that activates gene transcription (*pink*). All of the modifications shown are located within relatively unstructured regions of the polypeptide chain.

GTP-Binding Proteins Are Also Regulated by the Cyclic Gain and Loss of a Phosphate Group

Eukaryotic cells have a second way to regulate protein activity by phosphate addition and removal. In this case, however, the phosphate is not enzymatically transferred from ATP to the protein. Instead, the phosphate is part of a guanine nucleotide—guanosine triphosphate (GTP)—that is bound tightly to various types of **GTP-binding proteins**. These proteins act as molecular switches: they are in their active conformation when GTP is bound, but they can hydrolyze this GTP to GDP, which releases a phosphate and flips the protein to an inactive conformation. As with protein phosphorylation, this process is reversible: the active conformation is regained by dissociation of the GDP, followed by the binding of a fresh molecule of GTP (**Figure 4–44**).

A large variety of such GTP-binding proteins function as molecular switches in cells. The dissociation of GDP and its replacement by GTP, which turns the switch on, is often stimulated in response to a signal received by the cell. The GTP-binding proteins in turn bind to other proteins to control their activities; their crucial role in intracellular signaling pathways is discussed in detail in Chapter 16.

ATP Hydrolysis Allows Motor Proteins to Produce Directed Movements in Cells

We have seen how conformational changes in proteins play a central part in enzyme regulation and cell signaling. But conformational changes also play another important role in the operation of the eukaryotic cell: they enable certain specialized proteins to drive directed movements of cells and their components. These **motor proteins** generate the forces responsible for muscle contraction and most other eukaryotic cell movements. They also power the intracellular movements of organelles and macromolecules. For example, they help move chromosomes to opposite ends of the cell during mitosis (discussed in Chapter 18), and they move organelles along cytoskeletal tracks (discussed in Chapter 17).

How are shape changes in proteins used to generate such orderly movements? If, for example, a protein is required to walk along a cytoskeletal fiber, it can move by undergoing a series of conformational changes. However, with nothing to drive these changes in an orderly sequence, the shape changes will be perfectly reversible. Thus the protein can only wander randomly back and forth (**Figure 4–45**).

QUESTION 4–7

Explain how phosphorylation and the binding of a nucleotide (such as ATP or GTP) can both be used to regulate protein activity. What do you suppose are the advantages of either form of regulation?

Figure 4–44 GTP-binding proteins function as molecular switches. A GTP-binding protein requires the presence of a tightly bound GTP molecule to be active (*switch ON*). The active protein can shut itself off by hydrolyzing its bound GTP to GDP and inorganic phosphate (P_i), which converts the protein to an inactive conformation (*switch OFF*). To reactivate the protein, the tightly bound GDP must dissociate, a slow step that can be greatly accelerated by specific signals; once the GDP dissociates, a molecule of GTP quickly replaces it, returning the protein to its active conformation.

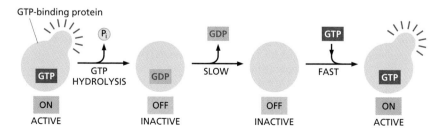

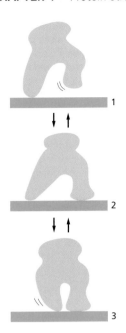

Figure 4–45 Changes in conformation can allow a protein to "walk" along a cytoskeletal filament. This protein's three different conformations allow it to wander randomly back and forth while bound to a filament. Without an input of energy to drive its movement in a single direction, the protein will only shuffle aimlessly, getting nowhere.

To make the conformational changes unidirectional—and force the entire cycle of movement to proceed in one direction—it is enough to make any one of the steps irreversible. For most proteins that are able to move in a single direction for long distances, this irreversibility is achieved by coupling one of the conformational changes to the hydrolysis of an ATP molecule bound to the protein—which is why motor proteins are also ATPases. A great deal of free energy is released when ATP is hydrolyzed, making it very unlikely that the protein will undergo a reverse shape change—as required for moving backward. (Such a reversal would require that the ATP hydrolysis be reversed, by adding a phosphate molecule to ADP to form ATP.) As a consequence, the protein moves steadily forward (**Figure 4–46**).

Many motor proteins generate directional movement by using the hydrolysis of a tightly bound ATP molecule to drive an orderly series of conformational changes. These movements can be rapid: the muscle motor protein *myosin* walks along actin filaments at about 6 μm/sec during muscle contraction (as discussed in Chapter 17).

Proteins Often Form Large Complexes That Function as Protein Machines

As one progresses from small, single-domain proteins to large proteins formed from many domains, the functions that the proteins can perform become more elaborate. The most complex tasks, however, are carried out by large protein assemblies formed from many protein molecules. Now that it is possible to reconstruct biological processes in cell-free systems in a test tube, it is clear that each central process in a cell—including DNA replication, gene transcription, protein synthesis, vesicle budding, and transmembrane signaling—is catalyzed by a highly coordinated, linked set of many proteins. In most such **protein machines**, the hydrolysis of bound nucleoside triphosphates (ATP or GTP) drives an ordered series of conformational changes in some of the individual protein subunits, enabling the ensemble of proteins to move coordinately. In this way, the appropriate enzymes can be positioned to carry out successive reactions in a series—as during the synthesis of proteins on a ribosome, for example (discussed in Chapter 7). Likewise, a large multiprotein complex moves rapidly along DNA to replicate the DNA double helix during cell division (discussed in Chapter 6). A simple mechanical analogy is illustrated in **Figure 4–47**.

Cells have evolved a large number of different protein machines suited to performing a variety of biological tasks. Cells employ protein machines for the same reason that humans have invented mechanical and electronic machines: for almost any job, manipulations that are spatially and temporally coordinated through linked processes are much more efficient than is the sequential use of individual tools.

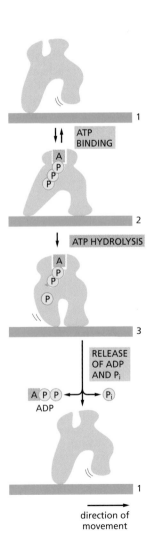

Figure 4–46 A schematic model of how a motor protein uses ATP hydrolysis to move in one direction along a cytoskeletal filament. An orderly transition among three conformations is driven by the hydrolysis of a bound ATP molecule and the release of the products: ADP and inorganic phosphate (Pᵢ). Because these transitions are coupled to the hydrolysis of ATP, the entire cycle is essentially irreversible. Through repeated cycles, the protein moves continuously to the right along the filament. The movement of a single molecule of myosin has been captured by atomic force microscopy.

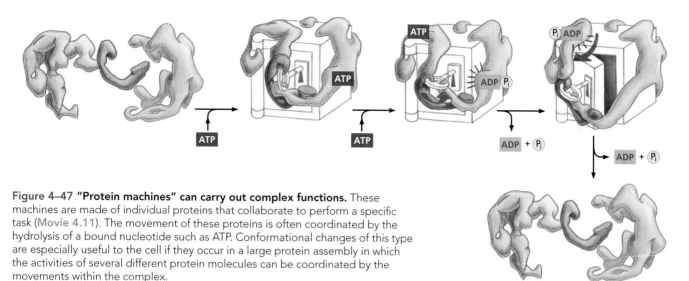

Figure 4–47 "Protein machines" can carry out complex functions. These machines are made of individual proteins that collaborate to perform a specific task (**Movie 4.11**). The movement of these proteins is often coordinated by the hydrolysis of a bound nucleotide such as ATP. Conformational changes of this type are especially useful to the cell if they occur in a large protein assembly in which the activities of several different protein molecules can be coordinated by the movements within the complex.

HOW PROTEINS ARE STUDIED

Understanding how a particular protein functions calls for detailed structural and biochemical analyses—both of which require large amounts of pure protein. But isolating a single type of protein from the thousands of other proteins present in a cell is a formidable task. For many years, proteins had to be purified directly from the source—the tissues in which they are most plentiful. That approach was inconvenient, entailing, for example, early-morning trips to the slaughterhouse. More important, the complexity of intact tissues and organs is a major disadvantage when trying to purify particular molecules, because a long series of chromatography steps is generally required. These procedures not only take weeks to perform, but they also yield only a few milligrams of pure protein.

Nowadays, proteins are more often isolated from cells that are grown in a laboratory (see, for example, Figure 1–38). Often these cells have been "tricked" into making large quantities of a given protein using the genetic engineering techniques that we describe in Chapter 10. Such engineered cells frequently allow large amounts of pure protein to be obtained in only a few days.

In this section, we outline how proteins are extracted and purified from cultured cells and other sources. We describe how these proteins are analyzed to determine their amino acid sequence and their three-dimensional structure. Finally, we discuss how technical advances are allowing proteins to be analyzed, cataloged, manipulated, and even designed from scratch.

Proteins Can be Purified from Cells or Tissues

Whether starting with a piece of liver, a dish of cultured cells, or a vat of bacterial, yeast, or animal cells that have been engineered to produce a protein of interest, the first step in any purification procedure is to break open the cells to release their contents. The resulting slurry is called a *cell homogenate* or *extract*. This physical disruption is followed by an initial fractionation procedure to separate out the class of molecules of interest—for example, all the soluble proteins in the cell (**Panel 4–3**, pp. 164–165).

With this collection of proteins in hand, the job is then to isolate the desired protein. The standard approach involves purifying the protein

QUESTION 4–8

Explain why the hypothetical enzymes in Figure 4–47 have a great advantage in opening the safe if they work together in a protein complex, as opposed to working individually in an unlinked, sequential manner.

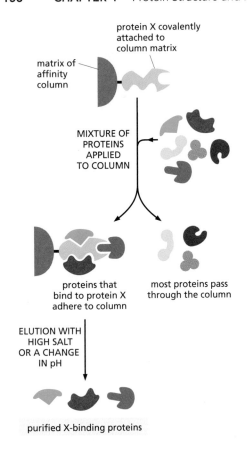

protein X covalently
attached to
column matrix

matrix of
affinity
column

MIXTURE OF
PROTEINS
APPLIED
TO COLUMN

proteins that
bind to protein X
adhere to column

most proteins pass
through the column

ELUTION WITH
HIGH SALT
OR A CHANGE
IN pH

purified X-binding proteins

Figure 4–48 Affinity chromatography can be used to isolate the binding partners of a protein of interest. The purified protein of interest (protein X) is covalently attached to the matrix of a chromatography column. An extract containing a mixture of proteins is then loaded onto the column. Those proteins that associate with protein X inside the cell will usually bind to it on the column. Proteins not bound to the column pass right through, and the proteins that are bound tightly to protein X can then be released by changing the pH or ionic composition of the washing solution.

through a series of **chromatography** steps, which use different materials to separate the individual components of a complex mixture into portions, or *fractions*, based on the properties of the protein—such as size, shape, or electrical charge. After each separation step, the fractions are examined to determine which ones contain the protein of interest. These fractions are then pooled and subjected to additional chromatography steps until the desired protein is obtained in pure form.

The most efficient forms of protein chromatography separate polypeptides on the basis of their ability to bind to a particular molecule—a process called *affinity chromatography* (**Panel 4–4**, p. 166). If large amounts of antibodies that recognize the protein are available, for example, they can be attached to the matrix of a chromatography column and used to help extract the protein from a mixture (see Panel 4–2, pp. 146–147).

Affinity chromatography can also be used to isolate proteins that interact physically with the protein being studied. In this case, a purified protein of interest is attached tightly to the column matrix; the proteins that bind to it will remain in the column and can then be removed by changing the composition of the washing solution (**Figure 4–48**).

Proteins can also be separated by **electrophoresis**. In this technique, a mixture of proteins is loaded onto a polymer gel and subjected to an electric field; the polypeptides will then migrate through the gel at different speeds depending on their size and net charge (**Panel 4–5**, p. 167). If too many proteins are present in the sample, or if the proteins are very similar in their migration rate, they can be resolved further using two-dimensional gel electrophoresis (see Panel 4–5). These electrophoretic approaches yield a number of bands or spots that can be visualized by staining; each band or spot contains a different protein. Chromatography and electrophoresis—both developed more than 50 years ago but greatly improved since—have been instrumental in building an understanding of what proteins look like and how they behave (**Table 4–2**). Both techniques are still frequently used in laboratories.

Once a protein has been obtained in pure form, it can be used in biochemical assays to study the details of its activity. It can also be subjected to techniques that reveal its amino acid sequence and precise three-dimensional structure.

Determining a Protein's Structure Begins with Determining Its Amino Acid Sequence

The task of determining the amino acid sequence of a protein can be accomplished in several ways. For many years, sequencing a protein was done by directly analyzing the amino acids in the purified protein. First, the protein was broken down into smaller pieces using a selective protease; the enzyme trypsin, for example, cleaves polypeptide chains on the carboxyl side of a lysine or an arginine. Then the identities of the amino acids in each fragment were determined chemically. The first protein sequenced in this way was the hormone *insulin*, in 1955.

TABLE 4–2 HISTORICAL LANDMARKS IN OUR UNDERSTANDING OF PROTEINS

1838	The name "protein" (from the Greek *proteios*, "primary") was suggested by Berzelius for the complex nitrogen-rich substance found in the cells of all animals and plants.
1819–1904	Most of the 20 common amino acids found in proteins were discovered.
1864	Hoppe-Seyler crystallized, and named, the protein hemoglobin.
1894	Fischer proposed a lock-and-key analogy for enzyme–substrate interactions.
1897	Buchner and Buchner showed that cell-free extracts of yeast can break down sucrose to form carbon dioxide and ethanol, thereby laying the foundations of enzymology.
1926	Sumner crystallized urease in pure form, demonstrating that proteins could possess the catalytic activity of enzymes; Svedberg developed the first analytical ultracentrifuge and used it to estimate the correct molecular weight of hemoglobin.
1933	Tiselius introduced electrophoresis for separating proteins in solution.
1934	Bernal and Crowfoot presented the first detailed X-ray diffraction patterns of a protein, obtained from crystals of the enzyme pepsin.
1942	Martin and Synge developed chromatography, a technique now widely used to separate proteins.
1951	Pauling and Corey proposed the structure of a helical conformation of a chain of amino acids—the α helix—and the structure of the β sheet, both of which were later found in many proteins.
1955	Sanger determined the order of amino acids in insulin, the first protein whose amino acid sequence was determined.
1956	Ingram produced the first protein fingerprints, showing that the difference between sickle-cell hemoglobin and normal hemoglobin is due to a change in a single amino acid (**Movie 4.12**).
1960	Kendrew described the first detailed three-dimensional structure of a protein (sperm whale myoglobin) to a resolution of 0.2 nm, and Perutz proposed a lower-resolution structure for hemoglobin.
1963	Monod, Jacob, and Changeux recognized that many enzymes are regulated through allosteric changes in their conformation.
1966	Phillips described the three-dimensional structure of lysozyme by X-ray crystallography, the first enzyme to be analyzed in atomic detail.
1973	Nomura reconstituted a functional bacterial ribosome from purified components.
1975	Henderson and Unwin determined the first three-dimensional structure of a transmembrane protein (bacteriorhodopsin), using a computer-based reconstruction from electron micrographs.
1976	Neher and Sakmann developed patch-clamp recording to measure the activity of single ion-channel proteins.
1984	Wüthrich used nuclear magnetic resonance (NMR) spectroscopy to solve the three-dimensional structure of a soluble sperm protein.
1988	Tanaka and Fenn separately developed methods for the analysis of proteins and other biological macromolecules.
1996–2013	Mann, Aebersold, Yates, and others developed efficient methods for using mass spectrometry to identify proteins in complex mixtures, exploiting the availability of complete genome sequences.

A much faster way to determine the amino acid sequence of proteins that have been isolated from organisms for which the full genome sequence is known is a method called *mass spectrometry*. This technique determines the exact mass of every peptide fragment in a purified protein, which then allows the protein to be identified from a database that contains a list of every protein thought to be encoded by the genome of the organism in question. Such lists are computed by taking the genome sequence of the organism and applying the genetic code (discussed in Chapter 7).

To perform **mass spectrometry**, the peptides derived from digestion with trypsin are blasted with a laser. This treatment heats the peptides, causing them to become electrically charged (ionized) and ejected in the form of a gas. Accelerated by a powerful electric field, the peptide ions then fly toward a detector; the time it takes them to arrive is related to their mass and their charge. (The larger the peptide is, the more slowly it moves; the

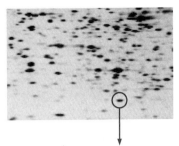

single protein spot excised from gel

N ▬▬▬▬▬▬▬▬▬▬▬▬▬▬▬▬▬ C

PEPTIDES PRODUCED
BY TRYPTIC DIGESTION
HAVE THEIR MASSES
MEASURED USING A
MASS SPECTROMETER

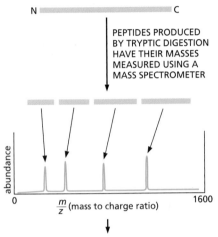

PROTEINS PREDICTED FROM GENOME
SEQUENCES ARE SEARCHED FOR MATCHES
WITH THEORETICAL MASSES CALCULATED
FOR ALL TRYPSIN-RELEASED PEPTIDES

↓

IDENTIFICATION OF PROTEIN
SUBSEQUENTLY ALLOWS ISOLATION
OF CORRESPONDING GENE

↓

THE GENE SEQUENCE ALLOWS LARGE
AMOUNTS OF THE PROTEIN TO BE OBTAINED
BY GENETIC ENGINEERING TECHNIQUES

Figure 4–49 Mass spectrometry can be used to identify proteins by determining the precise masses of peptides derived from them. As indicated, this in turn allows the proteins to be produced in the large amounts needed for determining their three-dimensional structure. In this example, the protein of interest is excised from a polyacrylamide gel after two-dimensional electrophoresis (see Panel 4–5, p. 167) and then digested with trypsin. The peptide fragments are loaded into the mass spectrometer, and their exact masses are measured. Genome sequence databases are then searched to find the protein encoded by the organism in question whose profile matches this peptide fingerprint. Mixtures of proteins can also be analyzed in this way. (Image courtesy of Patrick O'Farrell.)

more highly charged it is, the faster it moves.) The set of exact masses of the protein fragments produced by trypsin cleavage then serves as a "fingerprint" that identifies the protein—and its corresponding gene—from publicly accessible databases (**Figure 4–49**).

This approach can even be applied to complex mixtures of proteins, for example, starting with an extract containing all the proteins made by yeast cells grown under a particular set of conditions. To obtain the increased resolution required to distinguish individual proteins, such mixtures are frequently analyzed using *tandem mass spectrometry*. In this case, after the peptides pass through the first mass spectrometer, they are broken into even smaller fragments and analyzed by a second mass spectrometer.

Although all the information required for a polypeptide chain to fold is contained in its amino acid sequence, we have not yet learned how to reliably predict a protein's detailed three-dimensional conformation—the spatial arrangement of its atoms—from its sequence alone. At present, the only way to discover the precise folding pattern of any protein is by experiment, using either **X-ray crystallography** or **nuclear magnetic resonance (NMR)** spectroscopy (**How We Know**, pp. 162–163).

Genetic Engineering Techniques Permit the Large-Scale Production, Design, and Analysis of Almost Any Protein

Advances in genetic engineering techniques now permit the production of large quantities of almost any desired protein. In addition to making life much easier for biochemists interested in purifying specific proteins, this ability to churn out huge quantities of a protein has given rise to an entire biotechnology industry (**Figure 4–50**). Bacteria, yeast, and cultured mammalian cells are now used to mass produce a variety of therapeutic proteins, such as insulin, human growth hormone, and even the fertility-enhancing drugs used to boost egg production in women undergoing *in vitro* fertilization. Preparing these proteins previously required the collection and processing of vast amounts of tissue and other biological products—including, in the case of the fertility drugs, the urine of postmenopausal nuns.

The same sorts of genetic engineering techniques can also be employed to produce new proteins and enzymes that contain novel structures or perform unusual tasks: metabolizing toxic wastes, synthesizing life-saving drugs, or operating under conditions that would destroy most biological catalysts (see Chapter 3 How We Know, pp. 104–106). Most of these synthetic catalysts are nowhere near as effective as naturally occurring enzymes in terms of their ability to speed the rate of selected chemical reactions. But, as we continue to learn more about how proteins and enzymes exploit their unique conformations to carry out their biological functions, our ability to make novel proteins with useful functions can only improve.

Of course, to be able to study—or benefit from—the activity of an engineered protein in a living organism, the DNA encoding that protein must somehow be introduced into cells. Again, thanks to genetic engineering techniques, we are able to do just that. We discuss these methods in great detail in Chapter 10.

The Relatedness of Proteins Aids the Prediction of Protein Structure and Function

Biochemists have made enormous progress in understanding the structure and function of proteins over the past 150 years (see Table 4–2, p. 159). These advances are the fruits of decades of painstaking research on isolated proteins, performed by individual scientists working tirelessly on single proteins or protein families, one by one, sometimes for their entire careers. In the future, however, more and more of these investigations of protein conformation and activity will likely take place on a larger scale.

Improvements in our ability to rapidly sequence whole genomes, and the development of methods such as mass spectrometry, have fueled our ability to determine the amino acid sequences of enormous numbers of proteins. Millions of unique protein sequences from thousands of different species have thereby been deposited into publicly available databases, and the collection is expected to double in size every two years. Comparing the amino acid sequences of all of these proteins reveals that the majority belong to protein families that share specific "sequence patterns"—stretches of amino acids that fold into distinct structural domains. In some of these families, the proteins contain only a single structural domain. In others, the proteins include multiple domains arranged in novel combinations (**Figure 4–51**).

Although the number of multidomain families is growing rapidly, the discovery of novel single domains appears to be leveling off. This plateau suggests that the vast majority of proteins may fold up into a limited number of structural domains—perhaps as few as 10,000 to 20,000. For many single-domain families, the structure of at least one family member is known. And knowing the structure of one family member allows us to say something about the structure of its relatives. By this account, we have some structural information for almost three-quarters of the proteins archived in databases (Movie 4.13).

A future goal is to acquire the ability to look at a protein's amino acid sequence and be able to deduce its structure and gain insight into its function. We are coming closer to being able to predict protein structure based on sequence information, but there is still a long way to go. Predicting how a protein will function, alone, as part of a complex, or as part of a network in the cell, is much more challenging. But, the closer we get to addressing these questions, the closer we should be to understanding the fundamental basis of life.

Figure 4–50 Biotechnology companies produce mass quantities of useful proteins. Shown in this photograph are the fermenters used to grow the cells needed for such large-scale protein production. (Courtesy of Bioengineering AG, Switzerland.)

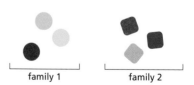

(A) single-domain protein families

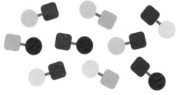

(B) a two-domain protein family

Figure 4–51 Most proteins belong to structurally related families. (A) More than two-thirds of all well-studied proteins contain a single structural domain. The members of these single-domain families can have different amino acid sequences but fold into a protein with a similar shape. (B) During evolution, structural domains have been combined in different ways to produce families of multidomain proteins. Almost all novelty in protein structure comes from the way these single domains are arranged. The number of multidomain families being added to the public databases is still rapidly increasing, unlike the number of novel single domains.

HOW WE KNOW

PROBING PROTEIN STRUCTURE

As you've no doubt already concluded in reading this chapter, for many proteins, their three-dimensional shape determines their function. So to learn more about how a protein works, it helps to know exactly what it looks like.

The problem is that most proteins are too small to be seen in any detail, even with a powerful electron microscope. To follow the path of an amino acid chain that is folded into a protein molecule, you need to be able to "see" its individual atoms. Scientists use two main methods to map the locations of atoms in a protein. The first involves the use of X-rays. Like light, X-rays are a form of electromagnetic radiation. But they have a wavelength that's much shorter: 0.1 nanometer (nm) as opposed to the 400–700 nm wavelength of visible light. That tiny wavelength—which is the approximate diameter of a hydrogen atom—allows scientists to probe the structure of very small objects at the atomic level.

A second method, called nuclear magnetic resonance (NMR) spectroscopy, takes advantage of the fact that—in many atoms—the nucleus is intrinsically magnetic. When exposed to a large magnet, these nuclei act like tiny bar magnets and align themselves with the magnetic field. If they are then excited with a blast of radio waves, the nuclei will wobble around their magnetic axes, and, as they relax back into the aligned position, they will give off a signal that can be used to reveal their relative positions in a protein.

Using these techniques, investigators have painstakingly pieced together many thousands of protein structures. With the help of computer graphics programs, they have been able to traverse the surfaces and climb inside these proteins, exploring the nooks where ATP likes to nestle, for example, or examining the loops and helices that proteins use to grab hold of a ligand or wrap around a segment of DNA. If the protein happens to belong to a virus or to a cancer cell, seeing its structure can provide clues to designing drugs that might thwart an infection or eliminate a tumor.

X-rays

To determine a protein's structure using X-ray crystallography, you first need to coax the protein into forming crystals: large, highly ordered arrays of the pure protein in which every molecule has the same conformation and is perfectly aligned with its neighbors. Growing high-quality protein crystals is still something of an art and is largely a matter of trial and error. Although robotic methods increase efficiency, it can still take years to find the right conditions—and some proteins resist crystallization altogether.

If you're lucky enough to get good crystals, you are ready for the X-ray analysis. When a narrow beam of X-rays is directed at a protein crystal, the atoms in the protein molecules scatter the incoming X-rays. These scattered waves either reinforce or cancel one another, producing a complex diffraction pattern that is collected by electronic detectors. The position and intensity of each spot in the diffraction pattern contains information about the position of the atoms in the protein crystal (Figure 4–52).

Because these patterns are so complex—even a small protein can generate 25,000 discrete spots—computers are used to interpret them and transform them by complex mathematical calculations into maps of the relative spatial positions of the atoms. By combining information obtained from such maps with the amino acid sequence of the protein, you can eventually generate an atomic model of the protein's structure. To determine whether the protein undergoes conformational changes in its structure when it binds a ligand that boosts its activity, you might subsequently try crystallizing it in the presence of its ligand. With crystals of sufficient quality, even small atomic movements can be detected by comparing the structures obtained in the presence and absence of stimulatory or inhibitory ligands.

Magnets

The trouble with X-ray crystallography is that you need crystals. And not all proteins like to form such orderly assemblies. Many have intrinsically disordered regions that wiggle around too much to stack neatly into a crystalline array. Others might not crystallize in the absence of the membranes in which they normally reside.

The other way to solve the structure of a protein does not require protein crystals. If the protein is small—say, 50,000 daltons or less—you can determine its structure by NMR spectroscopy. In this technique, a concentrated solution of pure protein is placed in a strong magnetic field and then bombarded with radio waves of different frequencies. Hydrogen nuclei, in particular, will generate an NMR signal that can be used to determine the distances between these atoms in different parts of the protein. This information is then used to build a model of how the hydrogens are arranged in space. Again, combined with the known amino acid sequence, an NMR spectrum can allow you to compute the three-dimensional structure of the protein (Figure 4–53). If the protein is larger than 50,000 daltons, you can try to break it up into its constituent functional domains and analyze each domain by NMR.

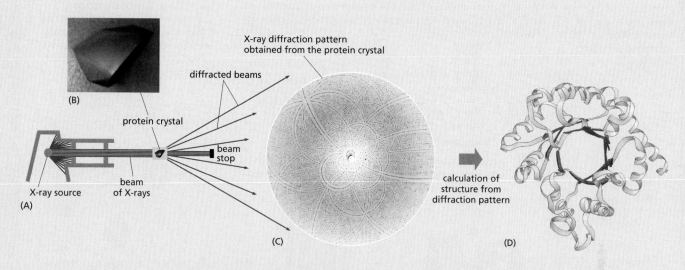

Figure 4–52 The structure of a protein can be determined by X-ray crystallography. Ribulose bisphosphate carboxylase is an enzyme that plays a central role in CO_2 fixation during photosynthesis. (A) X-ray diffraction apparatus; (B) photograph of crystal; (C) diffraction pattern; (D) three-dimensional structure determined from the pattern (α helices are shown in *green*, and β sheets in *red*). (B, courtesy of C. Branden; C, courtesy of J. Hajdu and I. Anderson; D, adapted from original provided by B. Furugren.)

Because determining the precise conformation of a protein is so time-consuming and costly—and the resulting insights so valuable—scientists routinely make their structures freely available by submitting the information to a publicly accessible database. Thanks to such databases, anyone interested in viewing the structure of, say, the ribosome—a complex macromolecular machine made of several RNAs and more than 50 proteins—can easily do so. In the future, improvements in X-ray crystallography and NMR spectroscopy should permit rapid analysis of many more proteins and protein machines. And once enough structures have been determined, it might become possible to generate algorithms for accurately predicting structure solely on the basis of a protein's amino acid sequence. After all, it is the sequence of the amino acids alone that determines how each protein folds up into its three-dimensional shape.

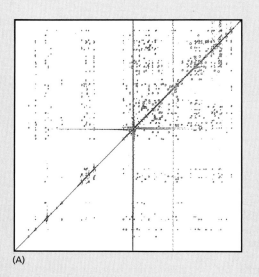

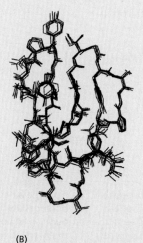

Figure 4–53 NMR spectroscopy can be used to determine the structure of small proteins or protein domains. (A) Two-dimensional NMR spectrum derived from the C-terminal domain of the enzyme cellulase, which breaks down cellulose. The spots represent interactions between neighboring hydrogen atoms. (B) The set of overlapping structures shown all satisfy the distance constraints equally well. (Courtesy of P. Kraulis.)

BREAKING CELLS AND TISSUES

The first step in the purification of most proteins is to disrupt tissues and cells in a controlled fashion.

Using gentle mechanical procedures, called homogenization, the plasma membranes of cells can be ruptured so that the cell contents are released. Four commonly used procedures are shown here.

The resulting thick soup (called a homogenate or an extract) contains large and small molecules from the cytosol, such as enzymes, ribosomes, and metabolites, as well as all of the membrane-enclosed organelles.

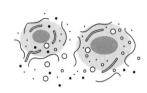

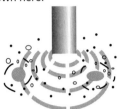

① Break cells with high-frequency sound (ultrasound).

② Use a mild detergent to make holes in the plasma membrane.

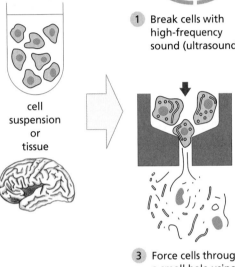

cell suspension or tissue

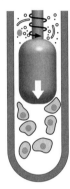

③ Force cells through a small hole using high pressure.

④ Shear cells between a close-fitting rotating plunger and the thick walls of a glass vessel.

When carefully conducted, homogenization leaves most of the membrane-enclosed organelles largely intact.

THE CENTRIFUGE

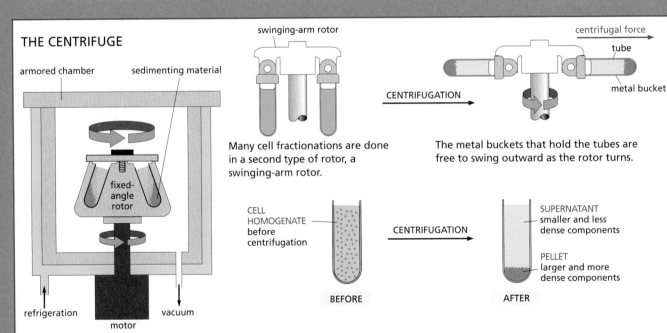

armored chamber

sedimenting material

swinging-arm rotor

centrifugal force

tube

CENTRIFUGATION

metal bucket

fixed-angle rotor

refrigeration

vacuum

motor

Many cell fractionations are done in a second type of rotor, a swinging-arm rotor.

The metal buckets that hold the tubes are free to swing outward as the rotor turns.

CELL HOMOGENATE before centrifugation

CENTRIFUGATION

SUPERNATANT smaller and less dense components

PELLET larger and more dense components

BEFORE

AFTER

Centrifugation is the most widely used procedure to separate a homogenate into different parts, or fractions. The homogenate is placed in test tubes and rotated at high speed in a centrifuge or ultracentrifuge. Present-day ultracentrifuges rotate at speeds up to 100,000 revolutions per minute and produce enormous forces, as high as 600,000 times gravity.

Such speeds require centrifuge chambers to be refrigerated and have the air evacuated so that friction does not heat up the homogenate. The centrifuge is surrounded by thick armor plating, because an unbalanced rotor can shatter with an explosive release of energy. A fixed-angle rotor can hold larger volumes than a swinging-arm rotor, but the pellet forms less evenly, as shown.

DIFFERENTIAL CENTRIFUGATION

Repeated centrifugation at progressively higher speeds will fractionate cell homogenates into their components.

Centrifugation separates cell components on the basis of size and density. The larger and denser components experience the greatest centrifugal force and move most rapidly. They sediment to form a pellet at the bottom of the tube, while smaller, less dense components remain in suspension above, a portion called the supernatant.

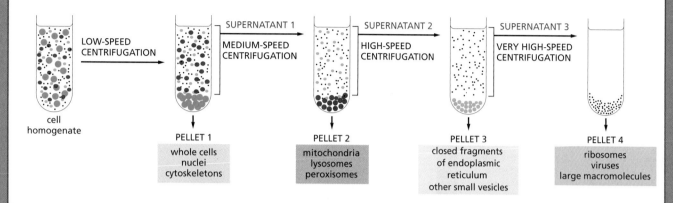

cell homogenate

LOW-SPEED CENTRIFUGATION

SUPERNATANT 1
MEDIUM-SPEED CENTRIFUGATION

SUPERNATANT 2
HIGH-SPEED CENTRIFUGATION

SUPERNATANT 3
VERY HIGH-SPEED CENTRIFUGATION

PELLET 1
whole cells
nuclei
cytoskeletons

PELLET 2
mitochondria
lysosomes
peroxisomes

PELLET 3
closed fragments
of endoplasmic
reticulum
other small vesicles

PELLET 4
ribosomes
viruses
large macromolecules

VELOCITY SEDIMENTATION

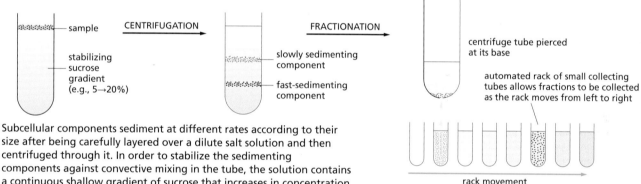

sample

CENTRIFUGATION

stabilizing sucrose gradient (e.g., 5→20%)

FRACTIONATION

slowly sedimenting component

fast-sedimenting component

centrifuge tube pierced at its base

automated rack of small collecting tubes allows fractions to be collected as the rack moves from left to right

rack movement

Subcellular components sediment at different rates according to their size after being carefully layered over a dilute salt solution and then centrifuged through it. In order to stabilize the sedimenting components against convective mixing in the tube, the solution contains a continuous shallow gradient of sucrose that increases in concentration toward the bottom of the tube. The gradient is typically 5→20% sucrose. When sedimented through such a dilute sucrose gradient, using a swinging-arm rotor, different cell components separate into distinct bands that can be collected individually.

After an appropriate centrifugation time, the bands may be collected, most simply by puncturing the plastic centrifuge tube and collecting drops from the bottom, as shown here.

EQUILIBRIUM SEDIMENTATION

The ultracentrifuge can also be used to separate cell components on the basis of their buoyant density, independently of their size or shape. The sample is usually either layered on top of, or dispersed within, a steep density gradient that contains a very high concentration of sucrose or cesium chloride. Each subcellular component will move up or down when centrifuged until it reaches a position where its density matches its surroundings and then will move no further. A series of distinct bands will eventually be produced, with those nearest the bottom of the tube containing the components of highest buoyant density. The method is also called density gradient centrifugation.

The sample is distributed throughout the sucrose density gradient.

At equilibrium, components have migrated to a region in the gradient that matches their own density.

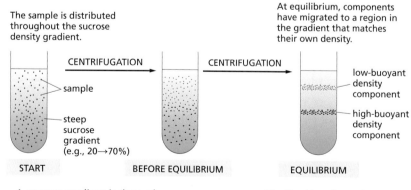

CENTRIFUGATION

sample

steep sucrose gradient (e.g., 20→70%)

CENTRIFUGATION

low-buoyant density component

high-buoyant density component

START

BEFORE EQUILIBRIUM

EQUILIBRIUM

A sucrose gradient is shown here, but denser gradients can be formed with cesium chloride that are particularly useful for separating nucleic acids (DNA and RNA).

The final bands can be collected from the base of the tube, as shown above for velocity sedimentation.

PROTEIN SEPARATION

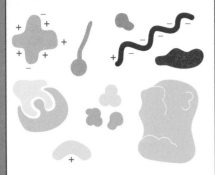

Proteins are very diverse. They differ in size, shape, charge, hydrophobicity, and their affinity for other molecules. All of these properties can be exploited to separate them from one another so that they can be studied individually.

THREE KINDS OF CHROMATOGRAPHY

Although the material used to form the matrix for column chromatography varies, it is usually packed in the column in the form of small beads. A typical protein purification strategy might employ in turn each of the three kinds of matrix described below, with a final protein purification of up to 10,000-fold.

Purity can easily be assessed by gel electrophoresis (Panel 4–5).

COLUMN CHROMATOGRAPHY

Proteins are often fractionated by column chromatography. A mixture of proteins in solution is applied to the top of a cylindrical column filled with a permeable solid matrix immersed in solvent. A large amount of solvent is then pumped through the column. Because different proteins are retarded to different extents by their interaction with the matrix, they can be collected separately as they flow out from the bottom. According to the choice of matrix, proteins can be separated according to their charge, hydrophobicity, size, or ability to bind to particular chemical groups (see below).

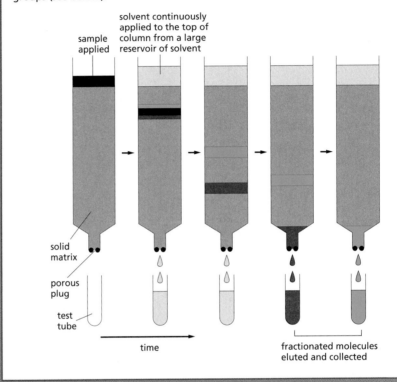

sample applied

solvent continuously applied to the top of column from a large reservoir of solvent

solid matrix

porous plug

test tube

time

fractionated molecules eluted and collected

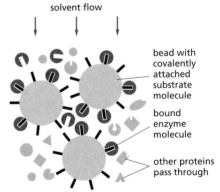

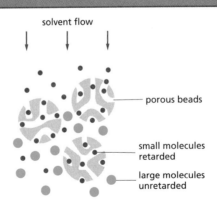

(A) ION-EXCHANGE CHROMATOGRAPHY

positively charged bead

bound negatively charged molecule

free positively charged molecule

Ion-exchange columns are packed with small beads carrying either positive or negative charges that retard proteins of the opposite charge. The association between a protein and the matrix depends on the pH and ionic strength of the solution passing down the column. These can be varied in a controlled way to achieve an effective separation.

(B) GEL-FILTRATION CHROMATOGRAPHY

porous beads

small molecules retarded

large molecules unretarded

Gel-filtration columns separate proteins according to their size. The matrix consists of tiny porous beads. Protein molecules that are small enough to enter the holes in the beads are delayed and travel more slowly through the column. Proteins that cannot enter the beads are washed out of the column first. Such columns also allow an estimate of protein size.

(C) AFFINITY CHROMATOGRAPHY

bead with covalently attached substrate molecule

bound enzyme molecule

other proteins pass through

Affinity columns contain a matrix covalently coupled to a molecule that interacts specifically with the protein of interest (e.g., an antibody, or an enzyme substrate). Proteins that bind specifically to such a column can subsequently be released by a pH change or by concentrated salt solutions, and they emerge highly purified (see also Figure 4–48).

GEL ELECTROPHORESIS

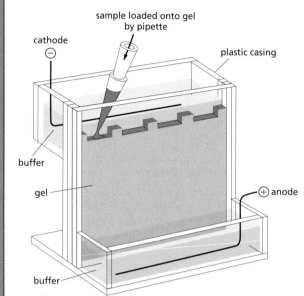

When an electric field is applied to a solution containing protein molecules, the molecules will migrate in a direction and at a speed that reflects their size and net charge. This forms the basis of the technique called electrophoresis.

The detergent sodium dodecyl sulfate (SDS) is used to solubilize proteins for SDS polyacrylamide-gel electrophoresis.

SDS

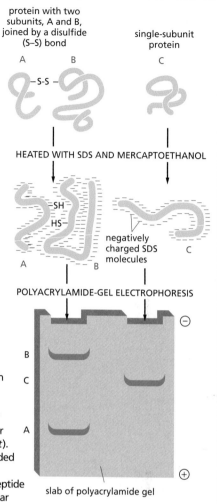

SDS polyacrylamide-gel electrophoresis (SDS-PAGE)

Individual polypeptide chains form a complex with negatively charged molecules of sodium dodecyl sulfate (SDS) and therefore migrate as negatively charged SDS–protein complexes through a slab of porous polyacrylamide gel. The apparatus used for this electrophoresis technique is shown above (*left*). A reducing agent (mercaptoethanol) is usually added to break any –S–S– linkages within or between proteins. Under these conditions, unfolded polypeptide chains migrate at a rate that reflects their molecular weight.

ISOELECTRIC FOCUSING

For any protein there is a characteristic pH, called the isoelectric point, at which the protein has no net charge and therefore will not move in an electric field. In isoelectric focusing, proteins are electrophoresed in a narrow tube of polyacrylamide gel in which a pH gradient is established by a mixture of special buffers. Each protein moves to a point in the pH gradient that corresponds to its isoelectric point and stays there.

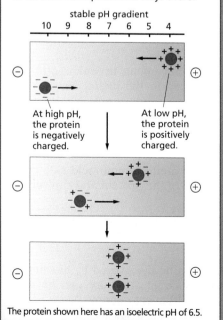

At high pH, the protein is negatively charged.

At low pH, the protein is positively charged.

The protein shown here has an isoelectric pH of 6.5.

TWO-DIMENSIONAL POLYACRYLAMIDE-GEL ELECTROPHORESIS

Complex mixtures of proteins cannot be resolved well on one-dimensional gels, but two-dimensional gel electrophoresis, combining two different separation methods, can be used to resolve more than 1000 proteins in a two-dimensional protein map. In the first step, native proteins are separated in a narrow gel on the basis of their intrinsic charge using isoelectric focusing (see *left*). In the second step, this gel is placed on top of a gel slab, and the proteins are subjected to SDS-PAGE (see *above*) in a direction perpendicular to that used in the first step. Each protein migrates to form a discrete spot.

All the proteins in an *E. coli* bacterial cell are separated in this two-dimensional gel, in which each spot corresponds to a different polypeptide chain. They are separated according to their isoelectric point from left to right and to their molecular weight from top to bottom. (Courtesy of Patrick O'Farrell.)

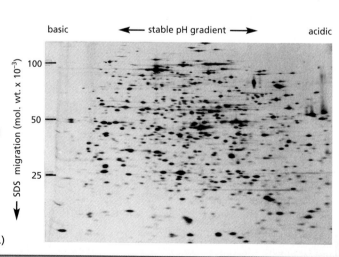

ESSENTIAL CONCEPTS

- Living cells contain an enormously diverse set of protein molecules, each made as a linear chain of amino acids linked together by covalent peptide bonds.

- Each type of protein has a unique amino acid sequence, which determines both its three-dimensional shape and its biological activity.

- The folded structure of a protein is stabilized by multiple noncovalent interactions between different parts of the polypeptide chain.

- Hydrogen bonds between neighboring regions of the polypeptide backbone often give rise to regular folding patterns, known as α helices and β sheets.

- The structure of many proteins can be subdivided into smaller globular regions of compact three-dimensional structure, known as protein domains.

- The biological function of a protein depends on the detailed chemical properties of its surface and how it binds to other molecules, called ligands.

- When a protein catalyzes the formation or breakage of a specific covalent bond in a ligand, the protein is called an enzyme and the ligand is called a substrate.

- At the active site of an enzyme, the amino acid side chains of the folded protein are precisely positioned so that they favor the formation of the high-energy transition states that the substrates must pass through to be converted to product.

- The three-dimensional structure of many proteins has evolved so that the binding of a small ligand can induce a significant change in protein shape.

- Most enzymes are allosteric proteins that can exist in two conformations that differ in catalytic activity, and the enzyme can be turned on or off by ligands that bind to a distinct regulatory site to stabilize either the active or the inactive conformation.

- The activities of most enzymes within the cell are strictly regulated. One of the most common forms of regulation is feedback inhibition, in which an enzyme early in a metabolic pathway is inhibited by the binding of one of the pathway's end products.

- Many thousands of proteins in a typical eukaryotic cell are regulated by cycles of phosphorylation and dephosphorylation.

- GTP-binding proteins also regulate protein function in eukaryotes; they act as molecular switches that are active when GTP is bound and inactive when GDP is bound; turning themselves off by hydrolyzing their bound GTP to GDP.

- Motor proteins produce directed movement in eukaryotic cells through conformational changes linked to the hydrolysis of ATP to ADP.

- Highly efficient protein machines are formed by assemblies of allosteric proteins in which the various conformational changes are coordinated to perform complex functions.

- Covalent modifications added to a protein's amino acid side chains can control the location and function of the protein and can serve as docking sites for other proteins.

- Starting from crude cell or tissue homogenates, individual proteins can be obtained in pure form by using a series of chromatography steps.

- The function of a purified protein can be discovered by biochemical analyses, and its exact three-dimensional structure can be determined by X-ray crystallography or NMR spectroscopy.

KEY TERMS

active site	mass spectrometry
allosteric	motor protein
α helix	N-terminus
amino acid sequence	nuclear magnetic resonance
antibody	(NMR) spectroscopy
antigen	peptide bond
β sheet	polypeptide, polypeptide chain
binding site	polypeptide backbone
C-terminus	primary structure
chromatography	protein
coiled-coil	protein domain
conformation	protein family
disulfide bond	protein kinase
electrophoresis	protein machine
enzyme	protein phosphatase
feedback inhibition	protein phosphorylation
fibrous protein	quaternary structure
globular protein	secondary structure
GTP-binding protein	side chain
helix	substrate
intrinsically disordered	subunit
sequence	tertiary structure
ligand	transition state
lysozyme	X-ray crystallography

QUESTIONS

QUESTION 4–9

Look at the models of the protein in Figure 4–12. Is the red α helix right- or left-handed? Are the three strands that form the large β sheet parallel or antiparallel? Starting at the N-terminus (the *purple* end), trace your finger along the peptide backbone. Are there any knots? Why, or why not?

QUESTION 4–10

Which of the following statements are correct? Explain your answers.

A. The active site of an enzyme usually occupies only a small fraction of the enzyme surface.

B. Catalysis by some enzymes involves the formation of a covalent bond between an amino acid side chain and a substrate molecule.

C. A β sheet can contain up to five strands, but no more.

D. The specificity of an antibody molecule is contained exclusively in loops on the surface of the folded light-chain domain.

E. The possible linear arrangements of amino acids are so vast that new proteins almost never evolve by alteration of old ones.

F. Allosteric enzymes have two or more binding sites.

G. Noncovalent bonds are too weak to influence the three-dimensional structure of macromolecules.

H. Affinity chromatography separates molecules according to their intrinsic charge.

I. Upon centrifugation of a cell homogenate, smaller organelles experience less friction and thereby sediment faster than larger ones.

QUESTION 4–11

What common feature of α helices and β sheets makes them universal building blocks for proteins?

QUESTION 4–12

Protein structure is determined solely by a protein's amino acid sequence. Should a genetically engineered protein in which the original order of all amino acids is reversed have the same structure as the original protein?

QUESTION 4–13

Consider the following protein sequence as an α helix: Leu-Lys-Arg-Ile-Val-Asp-Ile-Leu-Ser-Arg-Leu-Phe-Lys-Val. How many turns does this helix make? Do you find anything remarkable about the arrangement of the amino acids in this sequence when folded into an α helix? (Hint: consult the properties of the amino acids in Figure 4–3.)

QUESTION 4–14

Simple enzyme reactions often conform to the equation

$$E + S \rightleftharpoons ES \rightarrow EP \rightleftharpoons E + P$$

where E, S, and P are enzyme, substrate, and product, respectively.

A. What does ES represent in this equation?

B. Why is the first step shown with bidirectional arrows and the second step as a unidirectional arrow?

C. Why does E appear at both ends of the equation?

D. One often finds that high concentrations of P inhibit the enzyme. Suggest why this might occur.

E. If compound X resembles S and binds to the active site of the enzyme but cannot undergo the reaction catalyzed by it, what effects would you expect the addition of X to the reaction to have? Compare the effects of X and of the accumulation of P.

QUESTION 4–15

Which of the following amino acids would you expect to find more often near the center of a folded globular protein? Which ones would you expect to find more often exposed to the outside? Explain your answers. Ser, Ser-P (a Ser residue that is phosphorylated), Leu, Lys, Gln, His, Phe, Val, Ile, Met, Cys–S–S–Cys (two cysteines that are disulfide-bonded), and Glu. Where would you expect to find the most N-terminal amino acid and the most C-terminal amino acid?

QUESTION 4–16

Assume you want to make and study fragments of a protein. Would you expect that any fragment of the polypeptide chain would fold the same way as it would in the intact protein? Consider the protein shown in Figure 4–19. Which fragments do you suppose are most likely to fold correctly?

QUESTION 4–17

Neurofilament proteins assemble into long, intermediate filaments (discussed in Chapter 17), found in abundance running along the length of nerve cell axons. The C-terminal region of these proteins is an unstructured polypeptide, hundreds of amino acids long and heavily modified by the addition of phosphate groups. The term "polymer brush" has been applied to this part of the neurofilament. Can you suggest why?

QUESTION 4–18

An enzyme isolated from a mutant bacterium grown at 20°C works in a test tube at 20°C but not at 37°C (37°C is the temperature of the gut, where this bacterium normally lives). Furthermore, once the enzyme has been exposed to the higher temperature, it no longer works at the lower one. The same enzyme isolated from the normal bacterium works at both temperatures. Can you suggest what happens (at the molecular level) to the mutant enzyme as the temperature increases?

QUESTION 4–19

A motor protein moves along protein filaments in the cell. Why are the elements shown in the illustration not sufficient to mediate directed movement (**Figure Q4–19**)? With reference to Figure 4–46, modify the illustration shown here to include other elements that are required to create a unidirectional motor, and justify each modification you make to the illustration.

Figure Q4–19

QUESTION 4–20

Gel-filtration chromatography separates molecules according to their size (see Panel 4–4, p. 166). Smaller molecules diffuse faster in solution than larger ones, yet smaller molecules migrate more slowly through a gel-filtration column than larger ones. Explain this paradox. What should happen at very rapid flow rates?

QUESTION 4–21

As shown in Figure 4–16, both α helices and the coiled-coil structures that can form from them are helical structures, but do they have the same handedness in the figure? Explain why?

QUESTION 4–22

How is it possible for a change in a single amino acid in a protein of 1000 amino acids to destroy its function, even when that amino acid is far away from any ligand-binding site?

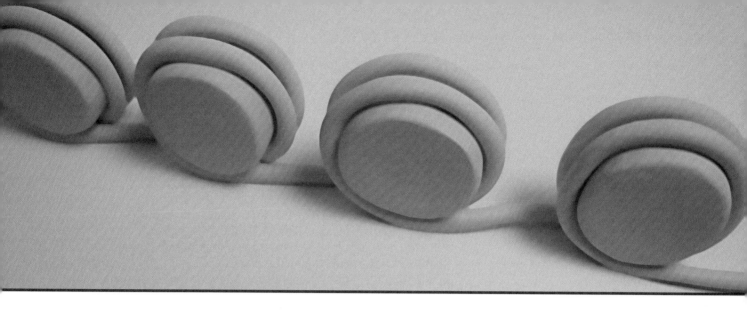

DNA and Chromosomes

Life depends on the ability of cells to store, retrieve, and translate the genetic instructions required to make and maintain a living organism. This hereditary information is passed on from a cell to its daughter cells at cell division, and from generation to generation in multicellular organisms through the reproductive cells—eggs and sperm. These instructions are stored within every living cell in its *genes*—the information-containing elements that determine the characteristics of a species as a whole and of the individuals within it.

At the beginning of the twentieth century, when genetics emerged as a science, scientists became intrigued by the chemical nature of genes. The information in genes is copied and transmitted from cell to daughter cells millions of times during the life of a multicellular organism, and it survives the process essentially unchanged. What kind of molecule could be capable of such accurate and almost unlimited replication, and also be able to direct the development of an organism and the daily life of a cell? What kind of instructions does the genetic information contain? How are these instructions physically organized so that the enormous amount of information required for the development and maintenance of even the simplest organism can be contained within the tiny space of a cell?

The answers to some of these questions began to emerge in the 1940s, when it was discovered from studies in simple fungi that genetic information consists primarily of instructions for making proteins. Proteins perform most of the cell's functions: they serve as building blocks for cell structures; they form the enzymes that catalyze the cell's chemical reactions; they regulate the activity of genes; and they enable cells to move and to communicate with one another. With hindsight, it is hard to imagine what other type of instructions the genetic information could have contained.

The other crucial advance made in the 1940s was the recognition that deoxyribonucleic acid (DNA) was the likely carrier of this genetic information. But the mechanism whereby the hereditary information is copied for transmission from one generation of cells to the next, and how proteins are specified by the instructions in DNA, remained completely mysterious until 1953, when the structure of DNA was determined by James Watson and Francis Crick. The structure immediately revealed how DNA might be copied, or replicated, and it provided the first clues about how a molecule of DNA might encode the instructions for making proteins. Today, the fact that DNA is the genetic material is so fundamental to our understanding of life that it is difficult to appreciate what an enormous intellectual gap this discovery filled.

In this chapter, we begin by describing the structure of DNA. We see how, despite its chemical simplicity, the structure and chemical properties of DNA make it ideally suited for carrying genetic information. The genes of every cell on Earth are made of DNA, and insights into the relationship between DNA and genes have come from experiments in a wide variety of organisms. We then consider how genes and other important segments of DNA are arranged in the single, long DNA molecule that forms the core of each chromosome in the cell. Finally, we discuss how eukaryotic cells fold these long DNA molecules into compact chromosomes inside the nucleus. This packing has to be done in an orderly fashion so that the chromosomes can be duplicated and apportioned correctly between the two daughter cells at each cell division. It must also allow the DNA to be accessed by the proteins that replicate and repair DNA, and regulate the activity of its many genes.

This is the first of five chapters that deal with basic genetic mechanisms—the ways in which the cell maintains and makes use of the genetic information carried in its DNA. In Chapter 6, we discuss the mechanisms by which the cell accurately replicates and repairs its DNA. In Chapter 7, we consider gene expression—how genes are used to produce RNA and protein molecules. In Chapter 8, we describe how a cell controls gene expression to ensure that each of the many thousands of proteins encoded in its DNA is manufactured at the proper time and place. In Chapter 9, we discuss how present-day genes evolved from distant ancestors, and, in Chapter 10, we consider some of the experimental techniques used to study both DNA and its role in fundamental cell processes.

An enormous amount has been learned about these subjects in the past 60 years. Much less obvious, but equally important, is that our knowledge is very incomplete; thus a great deal still remains to be discovered about how DNA provides the instructions to build living things.

THE STRUCTURE OF DNA

Well before biologists understood the structure of DNA, they had recognized that inherited traits and the genes that determine them were associated with the chromosomes. Chromosomes (named from the Greek *chroma*, "color," because of their staining properties) were discovered in the nineteenth century as threadlike structures in the nucleus of eukaryotic cells that become visible as the cells begin to divide (**Figure 5–1**). As biochemical analysis became possible, researchers learned that chromosomes contain both DNA and protein. But which of these components encoded the organism's genetic information was not clear.

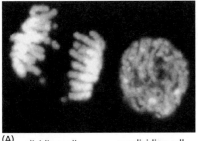

(A)   dividing cell   nondividing cell

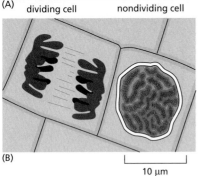

(B)

|— 10 μm —|

Figure 5–1 Chromosomes become visible as eukaryotic cells prepare to divide. (A) Two adjacent plant cells photographed in a fluorescence microscope. The DNA is labeled with a fluorescent dye (DAPI) that binds to it. The DNA is packaged into chromosomes, which become visible as distinct structures only when they condense in preparation for cell division, as shown on the *left*. The cell on the *right*, which is not dividing, contains the identical chromosomes, but they cannot be distinguished as individual entities because the DNA is in a much more extended conformation at this phase in the cell's life cycle. (B) Schematic diagram of the outlines of the two cells and their chromosomes. (A, courtesy of Peter Shaw.)

We now know that the DNA carries the hereditary information of the cell and that the protein components of chromosomes function largely to package and control the enormously long DNA molecules. But biologists in the 1940s had difficulty accepting DNA as the genetic material because of the apparent simplicity of its chemistry (see **How We Know**, pp. 174–176). DNA, after all, is simply a long polymer composed of only four types of nucleotide subunits, which are chemically very similar to one another.

Then, early in the 1950s, DNA was examined by X-ray diffraction analysis, a technique for determining the three-dimensional atomic structure of a molecule (see Figure 4–52). The early results indicated that DNA is composed of two strands wound into a helix. The observation that DNA is double-stranded was of crucial significance. It provided one of the major clues that led, in 1953, to a correct model for the structure of DNA. This structure immediately suggested how DNA could encode the instructions necessary for life, and how these instructions could be copied and passed along when cells divide. In this section, we examine the structure of DNA and explain in general terms how it is able to store hereditary information.

A DNA Molecule Consists of Two Complementary Chains of Nucleotides

A molecule of **deoxyribonucleic acid** (**DNA**) consists of two long polynucleotide chains. Each *chain*, or *strand*, is composed of four types of nucleotide subunits, and the two strands are held together by hydrogen bonds between the base portions of the nucleotides (**Figure 5–2**).

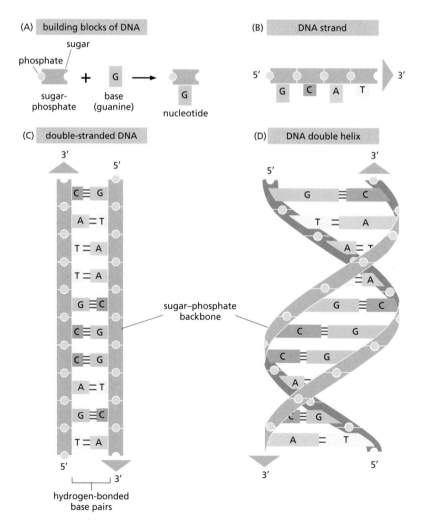

Figure 5–2 **DNA is made of four nucleotide building blocks.** (A) Each nucleotide is composed of a sugar–phosphate covalently linked to a base—guanine (G) in this figure. (B) The nucleotides are covalently linked together into polynucleotide chains, with a sugar–phosphate backbone from which the bases (A, C, G, and T) extend. (C) A DNA molecule is composed of two polynucleotide chains (DNA strands) held together by hydrogen bonds between the paired bases. The *arrows* on the DNA strands indicate the polarities of the two strands, which run antiparallel to each other in the DNA molecule. (D) Although the DNA is shown straightened out in (C), in reality, it is wound into a double helix, as shown here.

GENES ARE MADE OF DNA

By the 1920s, scientists generally agreed that genes reside on chromosomes, and they knew that chromosomes are composed of both DNA and proteins. But because DNA is so chemically simple, they naturally assumed that genes had to be made of proteins, which are much more chemically diverse than DNA molecules. Even when the experimental evidence suggested otherwise, this assumption proved hard to shake.

Messages from the dead

The case for DNA began to emerge in the late 1920s, when a British medical officer named Fred Griffith made an astonishing discovery. He was studying *Streptococcus pneumoniae* (pneumococcus), a bacterium that causes pneumonia. As antibiotics had not yet been discovered, infection with this organism was usually fatal. When

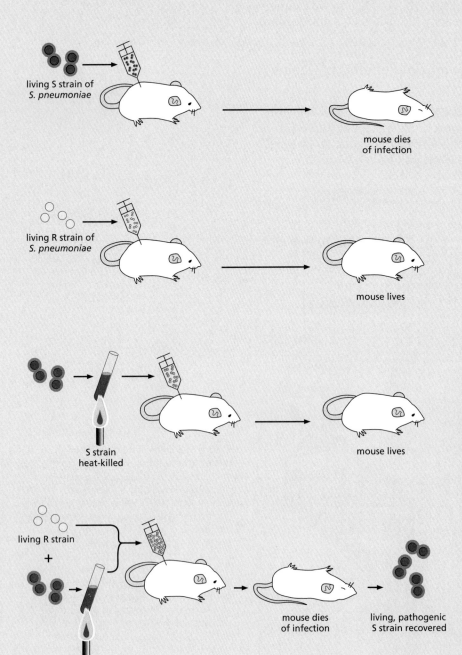

living S strain of
S. pneumoniae

mouse dies
of infection

living R strain of
S. pneumoniae

mouse lives

S strain
heat-killed

mouse lives

living R strain

+

S strain
heat-killed

mouse dies
of infection

living, pathogenic
S strain recovered

Figure 5–3 Griffith showed that heat-killed, infectious bacteria can transform harmless, living bacteria into pathogenic ones. The bacterium *Streptococcus pneumoniae* comes in two forms that differ from one another in their microscopic appearance and in their ability to cause disease. Cells of the pathogenic strain, which are lethal when injected into mice, are encased in a slimy, glistening polysaccharide capsule. When grown on a plate of nutrients in the laboratory, this disease-causing bacterium forms colonies that look dome-shaped and smooth; hence it is designated the S form. The harmless strain of the pneumococcus, on the other hand, lacks this protective coat; it forms colonies that appear flat and rough—hence, it is referred to as the R form. As illustrated, Griffith found that a substance present in the pathogenic S strain could permanently change, or transform, the nonlethal R strain into the deadly S strain.

grown in the laboratory, pneumococci come in two forms: a pathogenic form that causes a lethal infection when injected into animals, and a harmless form that is easily conquered by the animal's immune system and does not produce an infection.

In the course of his investigations, Griffith injected various preparations of these bacteria into mice. He showed that pathogenic pneumococci that had been killed by heating were no longer able to cause infection. The surprise came when Griffith injected both heat-killed pathogenic bacteria and live harmless bacteria into the same mouse. This combination proved lethal: not only did the animals die of pneumonia, but Griffith found that their blood was teeming with live bacteria of the pathogenic form (**Figure 5–3**). The heat-killed pneumococci had somehow converted the harmless bacteria into the lethal form. What's more, Griffith found that the change was permanent: he could grow these "transformed" bacteria in culture, and they remained pathogenic. But what was this mysterious material that turned harmless bacteria into killers? And how was this change passed on to progeny bacteria?

Transformation

Griffith's remarkable finding set the stage for the experiments that would provide the first strong evidence that genes are made of DNA. The American bacteriologist Oswald Avery, following up on Griffith's work, discovered that the harmless pneumococcus could be transformed into a pathogenic strain in a culture tube by exposing it to an extract prepared from the pathogenic strain. It would take another 15 years, however, for Avery and his colleagues Colin MacLeod and Maclyn McCarty to successfully purify the "transforming principle" from this soluble extract and to demonstrate that the active ingredient was DNA. Because the transforming principle caused a heritable change in the bacteria that received it, DNA must be the very stuff of which genes are made.

The 15-year delay was in part a reflection of the academic climate—and the widespread supposition that the genetic material was likely to be made of protein. Because of the potential ramifications of their work, the researchers wanted to be absolutely certain that the transforming principle was DNA before they announced their findings. As Avery noted in a letter to his brother, also a bacteriologist, "It's lots of fun to blow bubbles, but it's wiser to prick them yourself before someone else tries to." So the researchers subjected the transforming material to a battery of chemical tests (**Figure 5–4**). They found that it exhibited all the chemical properties

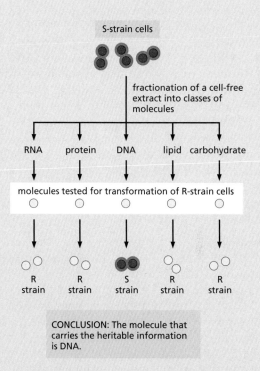

CONCLUSION: The molecule that carries the heritable information is DNA.

Figure 5–4 Avery, MacLeod, and McCarty demonstrated that DNA is the genetic material. The researchers prepared an extract from the disease-causing S strain of pneumococci and showed that the "transforming principle" that would permanently change the harmless R-strain pneumococci into the pathogenic S strain is DNA. This was the first evidence that DNA could serve as the genetic material.

characteristic of DNA; furthermore, they showed that enzymes that destroy proteins and RNA did not affect the ability of the extract to transform bacteria, while enzymes that destroy DNA inactivated it. And like Griffith before them, the investigators found that their purified preparation changed the bacteria permanently: DNA from the pathogenic species was taken up by the harmless species, and this change was faithfully passed on to subsequent generations of bacteria.

This landmark study offered rigorous proof that purified DNA can act as genetic material. But the resulting paper, published in 1944, drew remarkably little attention. Despite the meticulous care with which these experiments were performed, geneticists were not immediately convinced that DNA is the hereditary material. Many argued that the transformation might have been caused by some trace protein contaminant in the preparations. Or that the extract might contain a mutagen that alters the genetic material of the harmless bacteria—converting them to the pathogenic form—rather than containing the genetic material itself.

Virus cocktails

The debate was not settled definitively until 1952, when Alfred Hershey and Martha Chase fired up their laboratory blender and demonstrated, once and for all, that genes are made of DNA. The researchers were studying T2—a virus that infects and eventually destroys the bacterium *E. coli*. These bacteria-killing viruses behave like little molecular syringes: they inject their genetic material into the bacterial host cell, while the empty virus heads remain attached outside (**Figure 5–5A**). Once inside the bacterial cell, the viral genes direct the formation of new virus particles. In less than an hour, the infected cells explode, spewing thousands of new viruses into the medium. These then infect neighboring bacteria, and the process begins again.

The beauty of T2 is that these viruses contain only two kinds of molecules: DNA and protein. So the genetic material had to be one or the other. But which? The experiment was fairly straightforward. Because the viral DNA enters the bacterial cell, while the rest of the virus particle remains outside, the researchers decided to radioactively label the protein in one batch of virus and the DNA in another. Then, all they had to do was follow the radioactivity to see whether viral DNA or viral protein wound up inside the bacteria. To do this, Hershey and Chase incubated their radiolabeled viruses with *E. coli*; after allowing a few minutes for infection to take place, they poured the mix into a Waring blender and hit "puree." The blender's spinning blades sheared the empty virus heads from the surfaces of the bacterial cells. The researchers then centrifuged the sample to separate the heavier, infected bacteria, which formed a pellet at the bottom of the centrifuge tube, from the empty viral coats, which remained in suspension (**Figure 5–5B**).

As you have probably guessed, Hershey and Chase found that the radioactive DNA entered the bacterial cells, while the radioactive proteins remained outside with the empty virus heads. They found that the radioactive DNA was also incorporated into the next generation of virus particles.

This experiment demonstrated conclusively that viral DNA enters bacterial host cells, whereas viral protein does not. Thus, the genetic material in this virus had to be made of DNA. Together with the studies done by Avery, MacLeod, and McCarty, this evidence clinched the case for DNA as the agent of heredity.

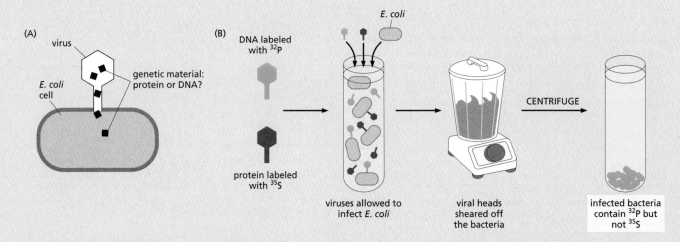

Figure 5–5 Hershey and Chase showed definitively that genes are made of DNA. (A) The researchers worked with T2 viruses, which are made entirely of protein and DNA. Each virus acts as a molecular syringe, injecting its genetic material into a bacterium; the empty viral capsule remains attached to the outside of the cell. (B) To determine whether the genetic material of the virus is protein or DNA, the researchers radioactively labeled the DNA in one batch of viruses with ^{32}P and the proteins in a second batch of viruses with ^{35}S. Because DNA lacks sulfur and the proteins lack phosphorus, these radioactive isotopes provided a handy way for the researchers to distinguish these two types of molecules. These labeled viruses were allowed to infect and replicate inside *E. coli*, and the mixture was then disrupted by brief pulsing in a Waring blender and separated to part the infected bacteria from the empty viral heads. When the researchers measured the radioactivity, they found that much of the ^{32}P-labeled DNA had entered the bacterial cells, while the vast majority of the ^{35}S-labeled proteins remained in solution with the spent viral particles.

As we saw in Chapter 2 (Panel 2–6, pp. 76–77), nucleotides are composed of a nitrogen-containing base and a five-carbon sugar, to which is attached one or more phosphate groups. For the nucleotides in DNA, the sugar is deoxyribose (hence the name deoxyribonucleic acid), and the base can be either *adenine* (A), *cytosine* (C), *guanine* (G), or *thymine* (T). The nucleotides are covalently linked together in a chain through the sugars and phosphates, which thus form a backbone of alternating sugar–phosphate–sugar–phosphate (see Figure 5–2B). Because it is only the base that differs in each of the four types of subunits, each polynucleotide chain in DNA can be thought of as a necklace: a sugar–phosphate backbone strung with four types of beads (the four bases A, C, G, and T). These same symbols (A, C, G, and T) are also commonly used to denote the four different nucleotides—that is, the bases with their attached sugar phosphates.

The way in which the nucleotide subunits are linked together gives a DNA strand a chemical polarity. If we imagine that each nucleotide has a knob (the phosphate) and a hole (see Figure 5–2A), each strand, formed by interlocking knobs with holes, will have all of its subunits lined up in the same orientation. Moreover, the two ends of the strand can be easily distinguished, as one will have a hole (the 3′ hydroxyl) and the other a knob (the 5′ phosphate). This polarity in a DNA strand is indicated by referring to one end as the *3′ end* and the other as the *5′ end*. This convention is based on the details of the chemical linkage between the nucleotide subunits.

The two polynucleotide chains in the DNA **double helix** are held together by hydrogen-bonding between the bases on the different strands. All the bases are therefore on the inside of the double helix, with the sugar–phosphate backbones on the outside (see Figure 5–2D). The bases do not pair at random, however: A always pairs with T, and G always pairs with C (**Figure 5–6**). In each case, a bulkier two-ring base (a purine, see Panel 2–6, pp. 76–77) is paired with a single-ring base (a pyrimidine). Each purine-pyrimidine pair is called a **base pair**, and this *complementary base-pairing* enables the base pairs to be packed in the energetically most favorable

Figure 5–6 The two strands of the DNA double helix are held together by hydrogen bonds between complementary base pairs. (A) The shapes and chemical structure of the bases allow hydrogen bonds to form efficiently only between A and T and between G and C, where atoms that are able to form hydrogen bonds (see Panel 2–2, pp. 68–69) can be brought close together without perturbing the double helix. Two hydrogen bonds form between A and T, whereas three form between G and C. The bases can pair in this way only if the two polynucleotide chains that contain them are antiparallel—that is, oriented in opposite directions. (B) A short section of the double helix viewed from its side. Four base pairs are shown. The nucleotides are linked together covalently by phosphodiester bonds through the 3′-hydroxyl (–OH) group of one sugar and the 5′-phosphate (–OPO₃) of the next (see Panel 2–6, pp. 76–77, to review how the carbon atoms in the sugar ring are numbered). This linkage gives each polynucleotide strand a chemical polarity; that is, its two ends are chemically different. The 3′ end carries an unlinked –OH group attached to the 3′ position on the sugar ring; the 5′ end carries a free phosphate group attached to the 5′ position on the sugar ring.

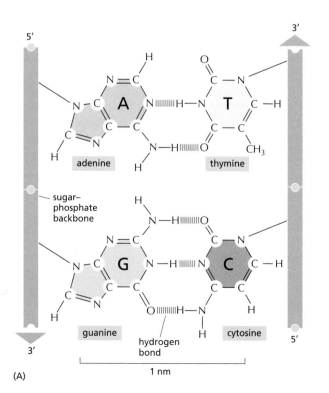

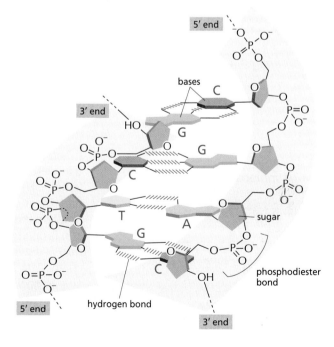

(A)

(B)

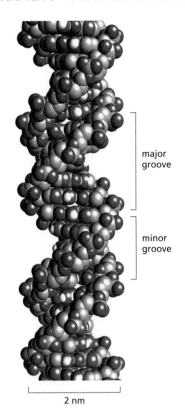

major groove

minor groove

2 nm

Figure 5–7 **A space-filling model shows the conformation of the DNA double helix.** The two DNA strands wind around each other to form a right-handed helix (see Figure 4–14) with 10 bases per turn. Shown here are 1.5 turns of the DNA double helix. The coiling of the two strands around each other creates two grooves in the double helix. The wider groove is called the major groove and the smaller one the minor groove. The colors of the atoms are: N, *blue*; O, *red*; P, *yellow*; and H, *white*.

arrangement in the interior of the double helix. In this arrangement, each base pair has a similar width, thus holding the sugar–phosphate backbones an equal distance apart along the DNA molecule. The members of each base pair can fit together within the double helix because the two strands of the helix run *antiparallel* to each other—that is, they are oriented with opposite polarities (see Figure 5–2C and D). The antiparallel sugar–phosphate strands then twist around each other to form a double helix containing 10 base pairs per helical turn (**Figure 5–7**). This twisting also contributes to the energetically favorable conformation of the DNA double helix.

A consequence of the base-pairing requirements is that each strand of a DNA double helix contains a sequence of nucleotides that is exactly **complementary** to the nucleotide sequence of its partner strand—an A always matches a T on the opposite strand, and a C always matches a G. This complementarity is of crucial importance when it comes to both copying and repairing the DNA, as we discuss in Chapter 6. An animated version of the DNA structure can be seen in Movie 5.1.

The Structure of DNA Provides a Mechanism for Heredity

The need for genes to encode information that must be copied and transmitted accurately when a cell divides raised two fundamental questions: how can the information for specifying an organism be carried in chemical form, and how can the information be accurately copied? The discovery of the structure of the DNA double helix was a landmark in biology because it immediately suggested the answers—and thereby resolved the problem of heredity at the molecular level. In this chapter, we outline the answer to the first question; in the next chapter, we address in detail the answer to the second.

Information is encoded in the order, or sequence, of the nucleotides along each DNA strand. Each base—A, C, T, or G—can be considered a letter in a four-letter alphabet that is used to spell out biological messages (**Figure 5–8**). Organisms differ from one another because their respective DNA molecules have different *nucleotide sequences* and, consequently, carry different biological messages. But how is the nucleotide alphabet used to make up messages, and what do they spell out?

It had already been established some time before the structure of DNA was determined that genes contain the instructions for producing proteins. DNA messages, therefore, must somehow be able to encode proteins. Consideration of the chemical character of proteins makes the problem easier to define. As discussed in Chapter 4, the function of a protein is determined by its three-dimensional structure, and this structure in turn is determined by the sequence of the amino acids in its polypeptide chain. The linear sequence of nucleotides in a gene must therefore be able to spell out the linear sequence of amino acids in a protein.

The exact correspondence between the 4-letter nucleotide alphabet of DNA and the 20-letter amino acid alphabet of proteins—the **genetic code**—is not obvious from the structure of the DNA molecule, and it took more than a decade after the discovery of the double helix to work it

QUESTION 5–1

Which of the following statements are correct? Explain your answers.
A. A DNA strand has a polarity because its two ends contain different bases.
B. G-C base pairs are more stable than A-T base pairs.

(A) molecular biology is...

(B)

(C) •━ •━•• ━ ━•• •

(D) 细胞生物学乐趣无穷

(E) TTCGAGCGACCTAACCTATAG

Figure 5–8 **Linear messages come in many forms.** The languages shown are (A) English, (B) a musical score, (C) Morse code, (D) Chinese, and (E) DNA.

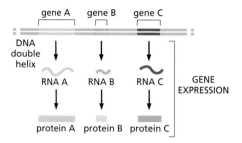

Figure 5–9 Most genes contain information to make proteins. As we discuss in Chapter 7, each protein-coding gene is used to produce RNA molecules, which then direct the production of the specific protein molecules.

out. In Chapter 7, we describe this code in detail when we discuss **gene expression**—the process by which the nucleotide sequence of a gene is *transcribed* into the nucleotide sequence of an RNA molecule, which, in most cases, is then *translated* into the amino acid sequence of a protein (**Figure 5–9**).

The amount of information in an organism's DNA is staggering: written out in the four-letter nucleotide alphabet, the nucleotide sequence of a very small protein-coding gene from humans occupies a quarter of a page of text, while the complete human DNA sequence would fill more than 1000 books the size of this one. Herein lies a problem that affects the architecture of all eukaryotic chromosomes: how can all this information be packed neatly into every cell nucleus? In the remainder of this chapter, we discuss the answer to this question.

THE STRUCTURE OF EUKARYOTIC CHROMOSOMES

Large amounts of DNA are required to encode all the information needed to make even a single-celled bacterium, and far more DNA is needed to encode the information to make a multicellular organism like you. Each human cell contains about 2 meters (m) of DNA; yet the cell nucleus is only 5–8 μm in diameter. Tucking all this material into such a small space is the equivalent of trying to fold 40 km (24 miles) of extremely fine thread into a tennis ball.

In eukaryotic cells, very long double-stranded DNA molecules are packaged into **chromosomes**. These DNA molecules not only fit readily inside the nucleus, but, after they are replicated, they can be easily apportioned between the two daughter cells at each cell division. The complex task of packaging DNA is accomplished by specialized proteins that bind to and fold the DNA, generating a series of coils and loops that provide increasingly higher levels of organization and prevent the DNA from becoming a tangled, unmanageable mess. Amazingly, the DNA is compacted in a way that allows it to remain accessible to all of the enzymes and other proteins that replicate it, repair it, and control the expression of its genes.

Bacteria typically carry their genes on a single, circular DNA molecule. This molecule is also associated with proteins that condense the DNA, but these proteins differ from the ones that package eukaryotic DNA. Although this prokaryotic DNA is called a bacterial "chromosome," it does not have the same structure as eukaryotic chromosomes, and less is known about how it is packaged. Our discussion of chromosome structure in this chapter will therefore focus entirely on eukaryotic chromosomes.

Eukaryotic DNA Is Packaged into Multiple Chromosomes

In eukaryotes, such as ourselves, the DNA in the nucleus is distributed among a set of different chromosomes. The DNA in a human nucleus, for example, contains approximately 3.2×10^9 nucleotides parceled out into 23 or 24 different types of chromosome (males, with their Y chromosome, have an extra type of chromosome that females do not have). Each chromosome consists of a single, enormously long, linear DNA molecule associated with proteins that fold and pack the fine thread of DNA into a more compact structure. The complex of DNA and protein is called *chromatin*. In addition to the proteins involved in packaging the DNA,

Figure 5–10 Each human chromosome can be "painted" a different color to allow its unambiguous identification. The chromosomes shown here were isolated from a cell undergoing nuclear division (mitosis) and are therefore in a highly compact (condensed) state. Chromosome painting is carried out by exposing the chromosomes to a collection of human DNA molecules that have been coupled to a combination of fluorescent dyes. For example, DNA molecules derived from Chromosome 1 are labeled with one specific dye combination, those from Chromosome 2 with another, and so on. Because the labeled DNA can form base pairs (hybridize) only to its chromosome of origin (discussed in Chapter 10), each chromosome is differently colored. For such experiments, the chromosomes are treated so that the individual strands of the double-helical DNA molecules partly separate to enable base-pairing with the labeled, single-stranded DNA, while keeping the chromosome structure relatively intact. (A) Micrograph shows the array of chromosomes as they originally spilled from the lysed cell. (B) The same chromosomes have been artificially lined up in order. In this so-called *karyotype*, the homologous chromosomes are numbered and arranged in pairs; the presence of a Y chromosome reveals that these chromosomes came from a male. (From E. Schröck et al., *Science* 273:494–497, 1996. With permission from the AAAS.)

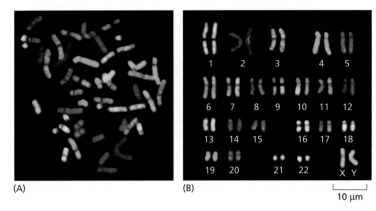

(A)   (B)

10 μm

chromosomes are also associated with many other proteins involved in DNA replication, DNA repair, and gene expression.

With the exception of the germ cells (sperm and eggs) and highly specialized cells that lack DNA entirely (such as mature red blood cells), human cells each contain two copies of each chromosome, one inherited from the mother and one from the father. The maternal and paternal chromosomes of a pair are called *homologous chromosomes (homologs)*. The only nonhomologous chromosome pairs are the sex chromosomes in males, where a *Y chromosome* is inherited from the father and an *X chromosome* from the mother. (Females inherit one X chromosome from each parent and have no Y chromosome.)

In addition to being different sizes, the different human chromosomes can be distinguished from one another by a variety of techniques. Each chromosome can be "painted" a different color using sets of chromosome-specific DNA molecules coupled to different fluorescent dyes (**Figure 5–10**). This involves a technique called *DNA hybridization*, which takes advantage of complementary base-pairing, as we will describe in detail in Chapter 10. A more traditional way of distinguishing one chromosome from another is to stain the chromosomes with dyes that bind to certain types of DNA sequences. These dyes mainly distinguish between DNA that is rich in A-T nucleotide pairs and DNA that is G-C rich, and they produce a predictable pattern of bands along each type of chromosome. The patterns that result allow each chromosome to be identified and numbered.

An ordered display of the full set of 46 human chromosomes is called the human **karyotype** (see Figure 5–10). If parts of a chromosome are lost, or switched between chromosomes, these changes can be detected. Cytogeneticists analyze karyotypes to detect chromosomal abnormalities that are associated with some inherited defects (**Figure 5–11**) and with certain types of cancer.

Chromosomes Contain Long Strings of Genes

The most important function of chromosomes is to carry the genes—the functional units of heredity (**Figure 5–12**). A **gene** is often defined as a

(A)   (B)

Figure 5–11 Abnormal chromosomes are associated with some inherited genetic defects. (A) A pair of Chromosomes 12 from a patient with inherited ataxia, a genetic disease of the brain characterized by progressive deterioration of motor skills. The patient has one normal Chromosome 12 (*left*) and one abnormally long Chromosome 12, which contains a piece of Chromosome 4 as identified by its banding pattern. (B) This interpretation was confirmed by chromosome painting, in which Chromosome 12 was painted *blue* and Chromosome 4 was painted *red*. (From E. Schröck et al., *Science* 273:494–497, 1996. With permission from the AAAS.)

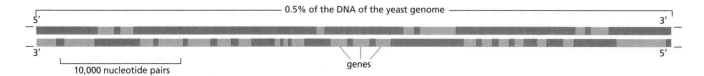

Figure 5–12 Genes are arranged along chromosomes. This figure shows a small region of the DNA double helix in one chromosome from the budding yeast *S. cerevisiae*. The *S. cerevisiae* genome contains about 12 million nucleotide pairs and 6600 genes—spread across 16 chromosomes. Note that, in each gene, only one of the two DNA strands actually encodes the information to make an RNA molecule, and this can be either strand, as indicated by the *light red* bars. However, a gene is generally denoted to contain both the "coding strand" and its complement, as in Figure 5–9. The high density of genes is characteristic of *S. cerevisiae*.

segment of DNA that contains the instructions for making a particular protein or RNA molecule. Most of the RNA molecules encoded by genes are subsequently used to produce a protein (see Figure 5–9). In some cases, however, the RNA molecule is the final product; like proteins, these RNA molecules have diverse functions in the cell, including structural, catalytic, and gene regulatory roles, as we discuss in later chapters.

Together, the total genetic information carried by all the chromosomes in a cell or organism constitutes its **genome**. Complete genome sequences have been determined for thousands of organisms, from *E. coli* to humans. As might be expected, some correlation exists between the complexity of an organism and the number of genes in its genome. For example, the total number of genes ranges from less than 500 for a simple bacterium to about 30,000 for humans. Bacteria and some single-celled eukaryotes, including *S. cerevisiae*, have especially compact genomes: the DNA molecules that make up their chromosomes are little more than strings of closely packed genes (see Figure 5–12). However, chromosomes from many eukaryotes—including humans—contain, in addition to genes and the specific nucleotide sequences required for normal gene expression, a large excess of interspersed DNA. This extra DNA is sometimes called "junk DNA," because the usefulness to the cell has not yet been demonstrated. Although the particular nucleotide sequence of most of this DNA might not be important, the DNA itself—acting as spacer material—may be crucial for the long-term evolution of the species and for the proper activity of the genes. In addition, comparisons of the genome sequences from many different species reveal that a portion of this extra DNA is highly conserved among related species, indicating that it serves an important function—although we don't yet know what that is.

In general, the more complex an organism, the larger is its genome. But this relationship does not always hold true. The human genome, for example, is 200 times larger than that of the yeast *S. cerevisiae*, but 30 times smaller than that of some plants and at least 60 times smaller than some species of amoeba (see Figure 1–40). Furthermore, how the DNA is apportioned over chromosomes also differs from one species to another. Humans have a total of 46 chromosomes (including both maternal and paternal sets), but a species of small deer has only 7, while some carp species have more than 100. Even closely related species with similar genome sizes can have very different numbers and sizes of chromosomes (**Figure 5–13**). Thus, although gene number is roughly correlated with species complexity, there is no simple relationship between gene number, chromosome number, and total genome size. The genomes and chromosomes of modern species have each been shaped by a unique history of seemingly random genetic events, acted on by specific selection pressures, as we discuss in Chapter 9.

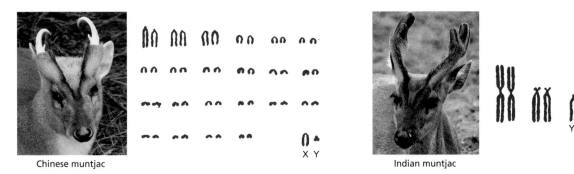

Chinese muntjac

Indian muntjac

Y_2 X Y_1

Figure 5–13 Two closely related species can have similar genome sizes but very different chromosome numbers. In the evolution of the Indian muntjac deer, chromosomes that were initially separate, and that remain separate in the Chinese species, fused without having a major effect on the number of genes—or the animal. (Courtesy of Deborah Carreno, Natural Wonders Photography.)

Specialized DNA Sequences Are Required for DNA Replication and Chromosome Segregation

To form a functional chromosome, a DNA molecule must do more than simply carry genes: it must be able to be replicated, and the replicated copies must be separated and partitioned equally and reliably into the two daughter cells at each cell division. These processes occur through an ordered series of events, known collectively as the **cell cycle**. This cycle of cell growth and division is briefly summarized in **Figure 5–14** and will be discussed in detail in Chapter 18. Only two broad stages of the cell cycle need concern us in this chapter: *interphase*, when chromosomes are duplicated, and *mitosis*, when they are distributed, or segregated, to the two daughter nuclei.

During interphase, the chromosomes are extended as long, thin, tangled threads of DNA in the nucleus and cannot be easily distinguished in the light microscope (see Figure 5–1). We refer to chromosomes in this extended state as *interphase chromosomes*. As we discuss in Chapter 6, specialized DNA sequences found in all eukaryotes ensure that DNA replication occurs efficiently during interphase. One type of nucleotide sequence acts as a **replication origin**, where replication of the DNA begins; eukaryotic chromosomes contain many replication origins to ensure that the long DNA molecules are replicated rapidly (**Figure 5–15**). Another DNA sequence forms the **telomeres** at each of the two ends of a chromosome. Telomeres contain repeated nucleotide sequences that are required for the ends of chromosomes to be replicated. They also cap the ends of the DNA molecule, preventing them from being mistaken by the cell as broken DNA in need of repair.

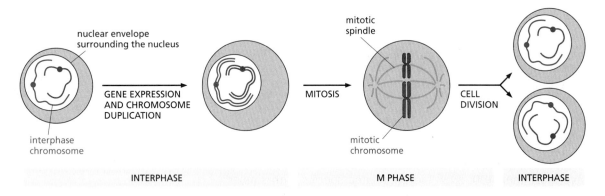

INTERPHASE   M PHASE   INTERPHASE

Figure 5–14 The duplication and segregation of chromosomes occurs through an ordered cell cycle in proliferating cells. During interphase, the cell expresses many of its genes, and—during part of this phase—it duplicates chromosomes. Once chromosome duplication is complete, the cell can enter *M phase*, during which nuclear division, or mitosis, occurs. In mitosis, the duplicated chromosomes condense, gene expression largely ceases, the nuclear envelope breaks down, and the mitotic spindle forms from microtubules and other proteins. The condensed chromosomes are then captured by the mitotic spindle, one complete set is pulled to each end of the cell, and a nuclear envelope forms around each chromosome set. In the final step of M phase, the cell divides to produce two daughter cells. Only two different chromosomes are shown here for simplicity.

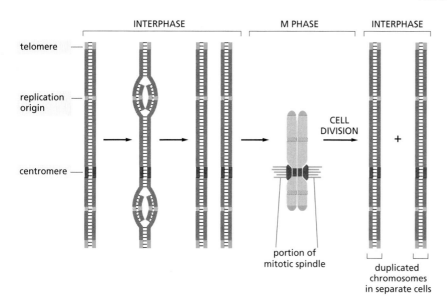

INTERPHASE   M PHASE   INTERPHASE

telomere

replication origin

centromere

CELL DIVISION

portion of mitotic spindle

duplicated chromosomes in separate cells

Figure 5–15 Three DNA sequence elements are needed to produce a eucaryotic chromosome that can be replicated and then segregated at mitosis. Each chromosome has multiple origins of replication, one centromere, and two telomeres. The sequence of events that a typical chromosome follows during the cell cycle is shown schematically. The DNA replicates in interphase, beginning at the origins of replication and proceeding bidirectionally from the origins across the chromosome. In M phase, the centromere attaches the duplicated chromosomes to the mitotic spindle so that one copy is distributed to each daughter cell when the cell divides. Prior to cell division, the centromere also helps to hold the compact, duplicated chromosomes together until they are ready to be pulled apart. Telomeres, which form special caps at the tips of each chromosome, aid in the replication of chromosome ends.

Eukaryotic chromosomes also contain a third type of specialized DNA sequence, called the **centromere**, that allows duplicated chromosomes to be separated during M phase (see Figure 5–15). During this stage of the cell cycle, the DNA coils up, adopting a more and more compact structure, ultimately forming highly compacted, or condensed, *mitotic chromosomes*. This is the state in which the duplicated chromosomes can be most easily visualized (**Figure 5–16** and see Figures 5–1 and 5–14). Once the chromosomes have condensed, the centromere attaches the mitotic spindle to each duplicated chromosome in a way that allows one copy of each chromosome to be segregated to each daughter cell (see Figure 5–15B). We describe the central role that centromeres play in cell division in Chapter 18.

Interphase Chromosomes Are Not Randomly Distributed Within the Nucleus

Inside the nucleus, the interphase chromosomes—although longer and finer than mitotic chromosomes—are nonetheless organized in various

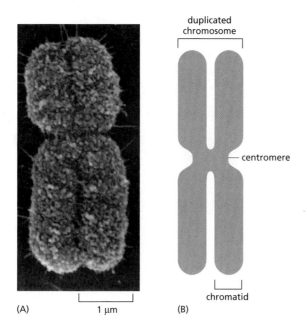

duplicated chromosome

centromere

chromatid

(A)   1 μm   (B)

Figure 5–16 A typical duplicated mitotic chromosome is highly compact. Because DNA is replicated during interphase, each duplicated mitotic chromosome contains two identical daughter DNA molecules (see Figure 5–15A). Each of these very long DNA molecules, with its associated proteins, is called a *chromatid*; once the two sister chromatids separate, they are considered individual chromosomes. (A) A scanning electron micrograph of a mitotic chromosome. The two chromatids are tightly joined together. The constricted region reveals the position of the centromere. (B) A cartoon representation of a mitotic chromosome. (A, courtesy of Terry D. Allen.)

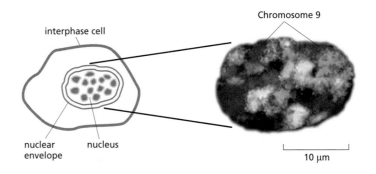

Figure 5–17 **Interphase chromosomes occupy their own distinct territories within the nucleus.** DNA probes coupled with different fluorescent markers were used to paint individual interphase chromosomes in a human cell. Viewed in a fluorescence microscope, each interphase chromosome is seen to occupy its own discrete territory within the nucleus, rather than being mixed with the other chromosomes like spaghetti in a bowl. Note that pairs of homologous chromosomes, such as the two copies of Chromosome 9 indicated, are not generally located in the same position. (From M.R. Speicher and N.P. Carter, *Nat. Rev. Genet.* 6:782–792, 2005. With permission from Macmillan Publishers Ltd.)

ways. First, each chromosome tends to occupy a particular region of the interphase nucleus, and so different chromosomes do not become extensively entangled with one another (**Figure 5–17**). In addition, some chromosomes are attached to particular sites on the *nuclear envelope—* the pair of concentric membranes that surround the nucleus—or to the underlying *nuclear lamina,* the protein meshwork that supports the envelope (discussed in Chapter 17).

The most obvious example of chromosome organization in the interphase nucleus is the **nucleolus** (**Figure 5–18**). The nucleolus is where the parts of the different chromosomes carrying genes that encode *ribosomal RNAs* cluster together. Here, ribosomal RNAs are synthesized and combine with proteins to form ribosomes, the cell's protein-synthesizing machines. As we discuss in Chapter 7, ribosomal RNAs play both structural and catalytic roles in the ribosome.

The DNA in Chromosomes Is Always Highly Condensed

As we have seen, all eukaryotic cells, whether in interphase or mitosis, package their DNA tightly into chromosomes. Human Chromosome 22, for example, contains about 48 million nucleotide pairs; stretched out end-to-end, its DNA would extend about 1.5 cm. Yet, during mitosis, Chromosome 22 measures only about 2 μm in length—that is, nearly 10,000 times more compact than the DNA would be if it were stretched to its full length. This remarkable feat of compression is performed by proteins that coil and fold the DNA into higher and higher levels of organization. The DNA of interphase chromosomes, although about 20 times less condensed than that of mitotic chromosomes (**Figure 5–19**), is still packed tightly.

Figure 5–18 **The nucleolus is the most prominent structure in the interphase nucleus.** Electron micrograph of a thin section through the nucleus of a human fibroblast. The nucleus is surrounded by the nuclear envelope. Inside the nucleus, the chromatin appears as a diffuse speckled mass, with regions that are especially dense, called heterochromatin (dark staining). Heterochromatin contains few genes and is located mainly around the periphery of the nucleus, immediately under the nuclear envelope. The large dark region is the nucleolus, which contains the genes for ribosomal RNAs; these genes are located on multiple chromosomes but are clustered together in the nucleolus. (Courtesy of E.G. Jordan and J. McGovern.)

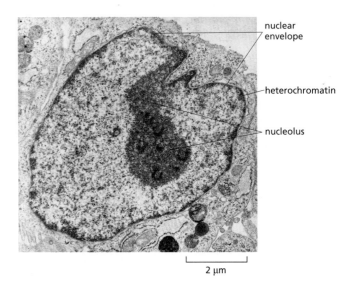

In the next sections, we introduce the specialized proteins that make this compression possible. Bear in mind, though, that chromosome structure is dynamic. Not only do chromosomes condense and decondense during the cell cycle, but chromosome packaging must be flexible enough to allow rapid, on-demand access to different regions of the interphase chromosome, unpacking enough to allow protein complexes access to specific, localized DNA sequences for replication, repair, or gene expression.

Nucleosomes Are the Basic Units of Eukaryotic Chromosome Structure

The proteins that bind to DNA to form eukaryotic chromosomes are traditionally divided into two general classes: the **histones** and the *nonhistone chromosomal proteins*. Histones are present in enormous quantities (more than 60 million molecules of several different types in each cell), and their total mass in chromosomes is about equal to that of the DNA itself. The complex of both classes of protein with nuclear DNA is called **chromatin**.

Histones are responsible for the first and most fundamental level of chromatin packing, the **nucleosome**, which was discovered in 1974. When interphase nuclei are broken open very gently and their contents examined with an electron microscope, much of the chromatin is in the form of *chromatin fibers* with a diameter of about 30 nm (**Figure 5–20A**). If this chromatin is subjected to treatments that cause it to unfold partially, it can then be seen in the electron microscope as a series of "beads on a string" (**Figure 5–20B**). The string is DNA, and each bead is a *nucleosome core particle*, which consists of DNA wound around a core of proteins formed from histones.

The structure of the nucleosome core particle was determined after first isolating nucleosomes by treating chromatin in its unfolded, "beads on a string" form with enzymes called nucleases, which break down DNA by cutting the phosphodiester bonds between nucleotides. After digestion for a short period, only the exposed DNA between the core particles—the *linker DNA*—is degraded, allowing the core particles to be isolated. An individual nucleosome core particle consists of a complex of eight histone proteins—two molecules each of histones H2A, H2B, H3, and H4—and a stretch of double-stranded DNA, 147 nucleotide pairs long, that winds around this *histone octamer* (**Figure 5–21**). The high-resolution structure of the nucleosome core particle was solved in 1997, revealing in atomic detail the disc-shaped histone octamer around which the DNA is tightly wrapped, making 1.7 turns in a left-handed coil (**Figure 5–22**).

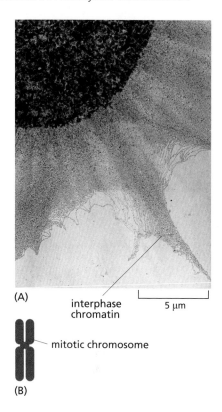

(A) interphase chromatin   5 μm

mitotic chromosome

(B)

Figure 5–19 DNA in interphase chromosomes is less compact than in mitotic chromosomes. (A) An electron micrograph showing an enormous tangle of chromatin (DNA with its associated proteins) spilling out of a lysed interphase nucleus. (B) Schematic drawing of a human mitotic chromosome drawn to the same scale. (Courtesy of Victoria Foe.)

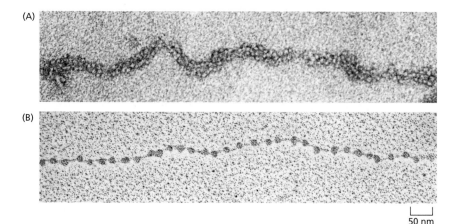

(A)

(B)

50 nm

Figure 5–20 Nucleosomes can be seen in the electron microscope. (A) Chromatin isolated directly from an interphase nucleus appears in the electron microscope as a chromatin fiber about 30-nm thick; a part of one such fiber is shown here. (B) This electron micrograph shows a length of a chromatin fiber that has been experimentally unpacked, or decondensed, after isolation to show the "beads-on-a-string" appearance of the nucleosomes. (A, courtesy of Barbara Hamkalo; B, courtesy of Victoria Foe.)

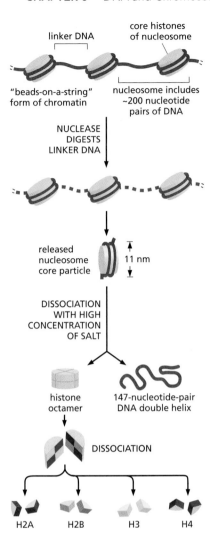

Figure 5–21 **Nucleosomes contain DNA wrapped around a protein core of eight histone molecules.** In a test tube, the nucleosome core particle can be released from chromatin by digestion of the linker DNA with a nuclease, which degrades the exposed DNA but not the DNA wound tightly around the nucleosome core. The DNA around each isolated nucleosome core particle can then be released and its length determined. With 147 nucleotide pairs in each fragment, the DNA wraps almost twice around each histone octamer.

The linker DNA between each nucleosome core particle can vary in length from a few nucleotide pairs up to about 80. (The term nucleosome technically refers to a nucleosome core particle plus one of its adjacent DNA linkers, as shown in Figure 5–21, but it is often used to refer to the nucleosome core particle itself.) The formation of nucleosomes converts a DNA molecule into a chromatin thread that is approximately one-third the length of the initial piece of DNA, and it provides the first level of DNA packing.

All four of the histones that make up the octamer are relatively small proteins, with a high proportion of positively charged amino acids (lysine and arginine). The positive charges help the histones bind tightly to the negatively charged sugar–phosphate backbone of DNA. These numerous electrostatic interactions explain in part why DNA of virtually any sequence can bind to a histone octamer. Each of the histones in the

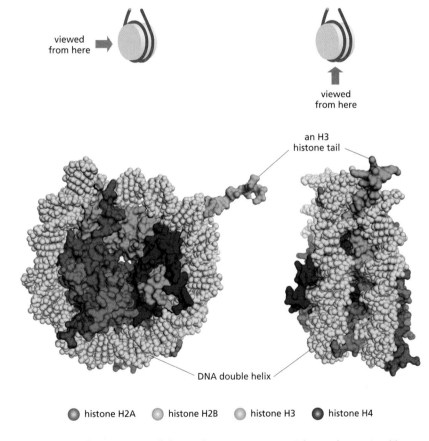

Figure 5–22 **The structure of the nucleosome core particle, as determined by X-ray diffraction analysis, reveals how DNA is tightly wrapped around a disc-shaped histone octamer.** Two views of a nucleosome core particle are shown here. The two strands of the DNA double helix are shown in *gray*. A portion of an H3 histone tail (*green*) can be seen extending from the nucleosome core particle, but the tails of the other histones have been truncated. (Reprinted by permission from K. Luger et al., *Nature* 389:251–260, 1997. With permission from Macmillan Publishers Ltd.)

octamer also has a long, unstructured N-terminal amino acid "tail" that extends out from the nucleosome core particle (see Figure 5–22). These histone tails are subject to several types of reversible, covalent chemical modifications that control many aspects of chromatin structure.

The histones that form the nucleosome core are among the most highly conserved of all known eukaryotic proteins: there are only two differences between the amino acid sequences of histone H4 from peas and cows, for example. This extreme evolutionary conservation reflects the vital role of histones in controlling eukaryotic chromosome structure.

Chromosome Packing Occurs on Multiple Levels

Although long strings of nucleosomes form on most chromosomal DNA, chromatin in the living cell rarely adopts the extended beads-on-a-string form seen in Figure 5–20B. Instead, the nucleosomes are further packed on top of one another to generate a more compact structure, such as the chromatin fiber shown in Figure 5–20A and Movie 5.2. This additional packing of nucleosomes into a chromatin fiber depends on a fifth histone called histone H1, which is thought to pull adjacent nucleosomes together into a regular repeating array. This "linker" histone changes the path the DNA takes as it exits the nucleosome core, allowing it to form a more condensed chromatin fiber (**Figure 5–23**).

We saw earlier that during mitosis chromatin becomes so highly condensed that individual chromosomes can be seen in the light microscope. How is a chromatin fiber folded to produce mitotic chromosomes? The answer is not yet known in detail, but it is known that the chromatin fiber is folded into a series of loops, and that these loops are further condensed to produce the interphase chromosome; finally, this compact string of loops is thought to undergo at least one more level of packing to form the mitotic chromosome (**Figure 5–24** and **Figure 5–25**).

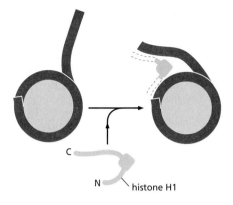

Figure 5–23 A linker histone helps to pull nucleosomes together and pack them into a more compact chromatin fiber. Histone H1 consists of a globular region plus a pair of long tails at its C-terminal and N-terminal ends. The globular region constrains an additional 20 base pairs of the DNA where it exits from the nucleosome core, an activity that is thought to be important for the formation of the chromatin fiber. The long C-terminal tail is required for H1 to bind to chromatin. The positions of the C-terminal and N-terminal tails in the nucleosome are not known.

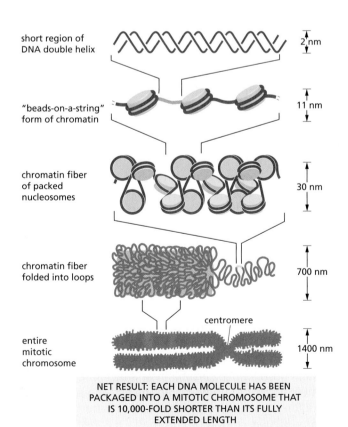

short region of
DNA double helix — 2 nm

"beads-on-a-string"
form of chromatin — 11 nm

chromatin fiber
of packed
nucleosomes — 30 nm

chromatin fiber
folded into loops — 700 nm

centromere

entire
mitotic
chromosome — 1400 nm

NET RESULT: EACH DNA MOLECULE HAS BEEN PACKAGED INTO A MITOTIC CHROMOSOME THAT IS 10,000-FOLD SHORTER THAN ITS FULLY EXTENDED LENGTH

QUESTION 5–2

Assuming that the histone octamer (shown in Figure 5–21) forms a cylinder 9 nm in diameter and 5 nm in height and that the human genome forms 32 million nucleosomes, what volume of the nucleus (6 μm in diameter) is occupied by histone octamers? (Volume of a cylinder is $\pi r^2 h$; volume of a sphere is $4/3\ \pi r^3$.) What fraction of the total volume of the nucleus do the histone octamers occupy? How does this compare with the volume of the nucleus occupied by human DNA?

Figure 5–24 DNA packing occurs on several levels in chromosomes. This schematic drawing shows some of the levels thought to give rise to the highly condensed mitotic chromosome. The actual structures are still uncertain.

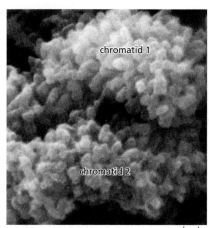

chromatid 1

chromatid 2

0.1 μm

Figure 5–25 The mitotic chromosome contains chromatin that is packed especially tightly. This scanning electron micrograph shows a region near one end of a typical mitotic chromosome. Each knoblike projection is believed to represent the tip of a separate loop of chromatin. The chromosome has duplicated, forming two sister chromatids that are still held close together (see Figure 5–16). The ends of the two chromatids can be distinguished on the right of the photo. (From M.P. Marsden and U.K. Laemmli, *Cell* 17:849–858, 1989. With permission from Elsevier.)

THE REGULATION OF CHROMOSOME STRUCTURE

So far, we have discussed how DNA is packed tightly into chromatin. We now turn to the question of how this packaging can be regulated to allow rapid access to the underlying DNA. The DNA in cells carries enormous amounts of coded information, and cells must be able to get to this information as needed.

In this section, we discuss how a cell can alter its chromatin structure to expose localized regions of DNA and allow access to specific proteins and protein complexes, particularly those involved in gene expression and in DNA replication and repair. We then discuss how chromatin structure is established and maintained—and how a cell can pass on some forms of this structure to its descendants. The regulation and inheritance of chromatin structure play crucial parts in the development of eukaryotic organisms.

Changes in Nucleosome Structure Allow Access to DNA

Eukaryotic cells have several ways to adjust the local structure of their chromatin rapidly. One way takes advantage of **chromatin-remodeling complexes**, protein machines that use the energy of ATP hydrolysis to change the position of the DNA wrapped around nucleosomes (**Figure 5–26A**). The complexes, which attach to both the histone octamer and the DNA wrapped around it, can locally alter the arrangement of nucleosomes on the DNA, making the DNA either more accessible (**Figure 5–26B**) or less accessible to other proteins in the cell. During mitosis, many of the chromatin-remodeling complexes are inactivated, which may help mitotic chromosomes maintain their tightly packed structure.

Another way of altering chromatin structure relies on the reversible chemical modification of the histones. The tails of all four of the core histones are particularly subject to these covalent modifications (**Figure 5–27A**). For example, acetyl, phosphate, or methyl groups can be added to and removed from the tails by enzymes that reside in the nucleus (**Figure 5–27B**). These and other modifications can have important consequences for the stability of the chromatin fiber. Acetylation of lysines, for instance, can reduce the affinity of the tails for adjacent nucleosomes, thereby loosening chromatin structure and allowing access to particular nuclear proteins.

Most importantly, however, these modifications can serve as docking sites on the histone tails for a variety of regulatory proteins. Different patterns of modifications attract different proteins to particular stretches of chromatin. Some of these proteins promote chromatin condensation, whereas others decondense chromatin and facilitate access to the DNA. Specific combinations of tail modifications and the proteins that bind to them have different meanings for the cell: one pattern, for example, indicates that a particular stretch of chromatin has been newly replicated;

QUESTION 5–3

Histone proteins are among the most highly conserved proteins in eukaryotes. Histone H4 proteins from a pea and a cow, for example, differ in only 2 of 102 amino acids. Comparison of the gene sequences shows many more differences, but only two change the amino acid sequence. These observations indicate that mutations that change amino acids must have been selected against during evolution. Why do you suppose that amino-acid-altering mutations in histone genes are deleterious?

(A)

ATP-dependent
chromatin-remodeling
complex

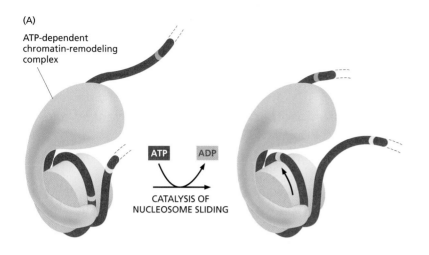

ATP → ADP

CATALYSIS OF
NUCLEOSOME SLIDING

Figure 5–26 Chromatin-remodeling complexes locally reposition the DNA wrapped around nucleosomes. (A) The complexes use energy derived from ATP hydrolysis to loosen the nucleosomal DNA and push it along the histone octamer, thereby exposing the DNA to other DNA-binding proteins. The *blue stripes* have been added to show how the nucleosome moves along the DNA. Many cycles of ATP hydrolysis are required to produce such a shift. (B) In the case shown, the repositioning of nucleosomes decondenses the chromatin in a particular chromosomal region; in other cases, it condenses the chromatin.

remodeling
complex

(B)

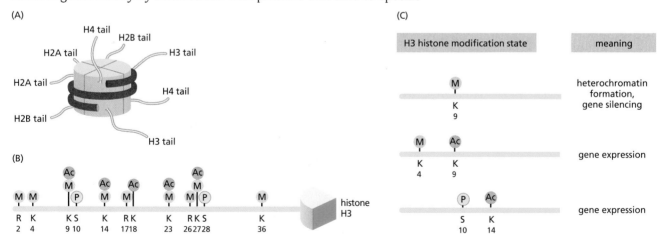

condensed chromatin   ATP → ADP   decondensed chromatin

REPEATED ROUNDS OF
NUCLEOSOME SLIDING

another indicates that the genes in that stretch of chromatin should be expressed; still others indicate that the nearby genes should be silenced (**Figure 5–27C**).

Like the chromatin-remodeling complexes, the enzymes that modify histone tails are tightly regulated. They are brought to particular chromatin regions mainly by interactions with proteins that bind to specific

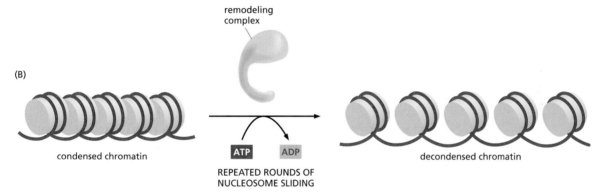

Figure 5–27 The pattern of modification of histone tails can dictate how a stretch of chromatin is treated by the cell. (A) Schematic drawing showing the positions of the histone tails that extend from each nucleosome. (B) Each histone can be modified by the covalent attachment of a number of different chemical groups, mainly to the tails. Histone H3, for example, can receive an acetyl group (Ac), a methyl group (M), or a phosphate group (P). The numbers denote the positions of the modified amino acids in the protein chain, with each amino acid designated by its one-letter code. Note that some positions, such as lysines (K) 9, 14, 23, and 27, can be modified in more than one way. Moreover, lysines can be modified with either one, two, or three methyl groups (not shown). Note that histone H3 contains 135 amino acids, most of which are in its globular portion (*green*), and that most modifications are on its N-terminal tail (*orange*). (C) Different combinations of histone tail modifications can confer a specific meaning on the stretch of chromatin on which they occur, as indicated. Only a few of these "meanings" are known.

sequences in DNA (we discuss these proteins in Chapter 8). The histone-modifying enzymes work in concert with the chromatin-remodeling complexes to condense or decondense stretches of chromatin, allowing local chromatin structure to change rapidly according to the needs of the cell.

Interphase Chromosomes Contain Both Condensed and More Extended Forms of Chromatin

The localized alteration of chromatin packing by remodeling complexes and histone modification has important effects on the large-scale structure of interphase chromosomes. Interphase chromatin is not uniformly packed. Instead, regions of the chromosome that contain genes that are being expressed are generally more extended, while those that contain silent genes are more condensed. Thus, the detailed structure of an interphase chromosome can differ from one cell type to the next, helping to determine which genes are expressed. Most cell types express about 20 to 30 % of the genes they contain.

The most highly condensed form of interphase chromatin is called **heterochromatin** (from the Greek *heteros*, "different," plus chromatin). It was first observed in the light microscope in the 1930s as discrete, strongly staining regions within the mass of chromatin. Heterochromatin typically makes up about 10% of an interphase chromosome, and in mammalian chromosomes, it is concentrated around the centromere region and in the telomeres at the ends of the chromosomes (see Figure 5–15).

The rest of the interphase chromatin is called **euchromatin** (from the Greek *eu*, "true" or "normal," plus chromatin). Although we use the term euchromatin to refer to chromatin that exists in a more decondensed state than heterochromatin, it is now clear that both euchromatin and heterochromatin are composed of mixtures of different chromatin structures (**Figure 5–28**).

Each type of chromatin structure is established and maintained by different sets of histone tail modifications that attract distinct sets of nonhistone proteins. The modifications that direct the formation of the most common type of heterochromatin, for example, include the methylation of lysine 9 in histone H3 (see Figure 5–27). Once it has been established, heterochromatin can spread because these histone tail modifications attract a set of heterochromatin-specific proteins, including histone-modifying enzymes, which then create the same histone tail modifications on adjacent nucleosomes. These modifications in turn recruit more of the heterochromatin-specific proteins, causing a wave of condensed chromatin to propagate along the chromosome. This heterochromatin will continue to spread until it encounters a barrier DNA sequence that stops the propagation (**Figure 5–29**). In this manner, extended regions of heterochromatin can be established along the DNA.

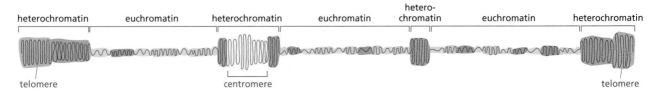

Figure 5–28 The structure of chromatin varies along a single interphase chromosome. As schematically indicated by different colors (and the path of the DNA molecule represented by the central *black line*), heterochromatin and euchromatin each represent a set of different chromatin structures with different degrees of condensation. Overall, heterochromatin is more condensed than euchromatin.

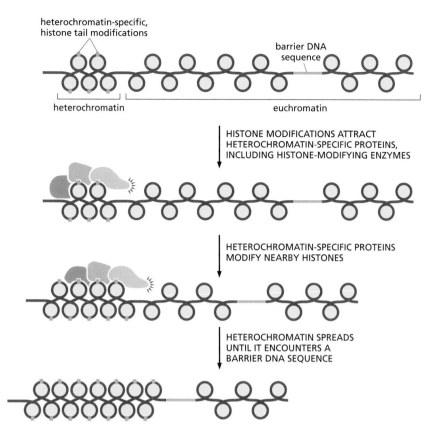

heterochromatin-specific,
histone tail modifications

barrier DNA
sequence

heterochromatin

euchromatin

HISTONE MODIFICATIONS ATTRACT
HETEROCHROMATIN-SPECIFIC PROTEINS,
INCLUDING HISTONE-MODIFYING ENZYMES

HETEROCHROMATIN-SPECIFIC PROTEINS
MODIFY NEARBY HISTONES

HETEROCHROMATIN SPREADS
UNTIL IT ENCOUNTERS A
BARRIER DNA SEQUENCE

Figure 5–29 Heterochromatin-specific modifications allow heterochromatin to form and to spread. These modifications attract heterochromatin-specific proteins that reproduce the same modifications on neighboring histones. In this manner, heterochromatin can spread until it encounters a barrier DNA sequence that blocks its propagation into regions of euchromatin.

Most DNA that is permanently folded into heterochromatin in the cell does not contain genes. Because heterochromatin is so compact, genes that accidentally become packaged into heterochromatin usually fail to be expressed. Such inappropriate packaging of genes in heterochromatin can cause disease: in humans, the gene that encodes β-globin—which forms part of the oxygen-carrying hemoglobin molecule—is situated next to a region of heterochromatin. If, because of an inherited DNA deletion, that heterochromatin spreads, the β-globin gene is poorly expressed and the person develops a severe form of anemia.

Perhaps the most striking example of the use of heterochromatin to keep genes shut down, or *silenced*, is found in the interphase X chromosomes of female mammals. In mammals, female cells contain two X chromosomes, whereas male cells contain one X and one Y. Because a double dose of X-chromosome products would be lethal, female mammals have evolved a mechanism for permanently inactivating one of the two X chromosomes in each cell. At random, one or other of the two X chromosomes in each cell becomes highly condensed into heterochromatin early in embryonic development. Thereafter, the condensed and inactive state of that X chromosome is inherited in all of the many descendants of those cells (**Figure 5–30**).

When a cell divides, it generally passes on its histone modifications, chromatin structure, and gene expression patterns to the two daughter cells. Such "cell memory" is critical for the establishment and maintenance of different cell types during the development of a complex multicellular organism. We discuss the mechanisms involved in cell memory in Chapter 8, where we consider the control of gene expression.

QUESTION 5–4

Mutations in a particular gene on the X chromosome result in color blindness in men. By contrast, most women carrying the mutation have proper color vision but see colored objects with reduced resolution, as though functional cone cells (the photoreceptor cells responsible for color vision) are spaced farther apart than normal in the retina. Can you give a plausible explanation for this observation? If a woman is color-blind, what could you say about her father? About her mother? Explain your answers.

Figure 5–30 One of the two X chromosomes is inactivated in the cells of mammalian females by heterochromatin formation. Each female cell contains two X chromosomes, one from the mother (X_m) and the other from the father (X_p). At an early stage in embryonic development, one of these two chromosomes becomes condensed into heterochromatin in each cell, apparently at random. At each cell division, the same X chromosome becomes condensed (and inactivated) in all the descendants of that original cell. Thus, all mammalian females end up as mixtures (mosaics) of cells bearing maternal or paternal inactivated X chromosomes. In most of their tissues and organs, about half the cells will be of one type, and the other half will be of the other.

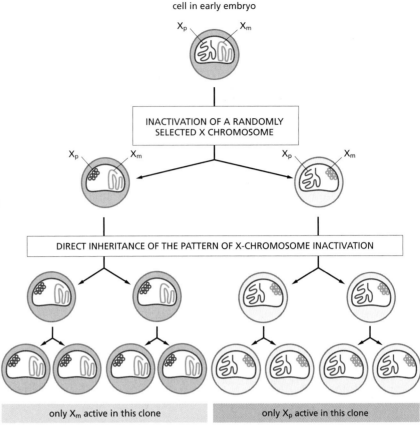

ESSENTIAL CONCEPTS

- Life depends on the stable storage and inheritance of genetic information.

- Genetic information is carried by very long DNA molecules and is encoded in the linear sequence of four nucleotides: A, T, G, and C.

- Each molecule of DNA is a double helix composed of a pair of antiparallel, complementary DNA strands, which are held together by hydrogen bonds between G-C and A-T base pairs.

- The genetic material of a eukaryotic cell is contained in a set of chromosomes, each formed from a single, enormously long DNA molecule that contains many genes.

- When a gene is expressed, part of its nucleotide sequence is transcribed into RNA molecules, many of which are translated into protein.

- The DNA that forms each eukaryotic chromosome contains, in addition to genes, many replication origins, one centromere, and two telomeres. These special DNA sequences ensure that, before cell division, each chromosome can be duplicated efficiently, and that the resulting daughter chromosomes are parceled out equally to the two daughter cells.

- In eukaryotic chromosomes, the DNA is tightly folded by binding to a set of histone and nonhistone proteins. This complex of DNA and protein is called chromatin.

- Histones pack the DNA into a repeating array of DNA–protein particles called nucleosomes, which further fold up into even more compact chromatin structures.

- A cell can regulate its chromatin structure—temporarily decondensing or condensing particular regions of its chromosomes—using chromatin-remodeling complexes and enzymes that covalently modify histone tails in various ways.

- The loosening of chromatin to a more decondensed state allows proteins involved in gene expression, DNA replication, and DNA repair to gain access to the necessary DNA sequences.

- Some forms of chromatin have a pattern of histone tail modification that causes the DNA to become so highly condensed that its genes cannot be expressed to produce RNA; such condensation occurs on all chromosomes during mitosis and in the heterochromatin of interphase chromosomes.

KEY TERMS

base pair	gene expression
cell cycle	genetic code
centromere	genome
chromatin	heterochromatin
chromatin-remodeling complex	histone
chromosome	karyotype
complementary	nucleolus
deoxyribonucleic acid (DNA)	nucleolus
double helix	replication origin
euchromatin	telomere gene
gene	

QUESTIONS

QUESTION 5–5

A. The nucleotide sequence of one DNA strand of a DNA double helix is

$$5'\text{-GGATTTTTGTCCACAATCA-}3'.$$

What is the sequence of the complementary strand?

B. In the DNA of certain bacterial cells, 13% of the nucleotides are adenine. What are the percentages of the other nucleotides?

C. How many possible nucleotide sequences are there for a stretch of DNA that is N nucleotides long, if it is (a) single-stranded or (b) double-stranded?

D. Suppose you had a method of cutting DNA at specific sequences of nucleotides. How many nucleotides long (on average) would such a sequence have to be in order to make just one cut in a bacterial genome of 3×10^6 nucleotide pairs? How would the answer differ for the genome of an animal cell that contains 3×10^9 nucleotide pairs?

QUESTION 5–6

An A-T base pair is stabilized by only two hydrogen bonds. Hydrogen-bonding schemes of very similar strengths can also be drawn between other base combinations that normally do not occur in DNA molecules, such as the A-C and the A-G pairs shown in Figure Q5–6.

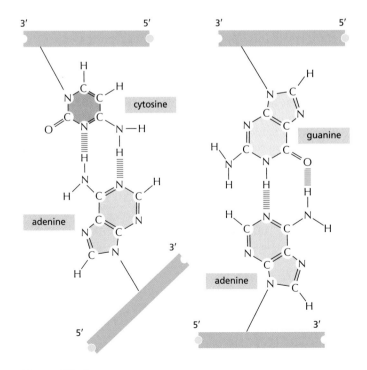

Figure Q5–6

What would happen if these pairs formed during DNA replication and the inappropriate bases were incorporated? Discuss why this does not often happen. (Hint: see Figure 5–6.)

QUESTION 5–7

A. A macromolecule isolated from an extraterrestrial source superficially resembles DNA, but closer analysis reveals that the bases have quite different structures (**Figure Q5–7**). Bases V, W, X, and Y have replaced bases A, T, G, and C. Look at these structures closely. Could these DNA-like molecules have been derived from a living organism that uses principles of genetic inheritance similar to those used by organisms on Earth?

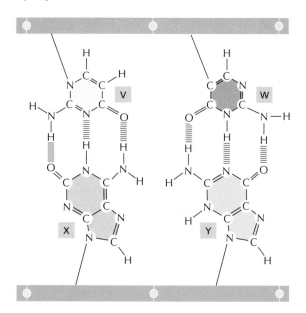

Figure Q5–7

B. Simply judged by their potential for hydrogen-bonding, could any of these extraterrestrial bases replace terrestrial A, T, G, or C in terrestrial DNA? Explain your answer.

QUESTION 5–8

The two strands of a DNA double helix can be separated by heating. If you raised the temperature of a solution containing the following three DNA molecules, in what order do you suppose they would "melt"? Explain your answer.

A. 5'-GCGGGCCAGCCCGAGTGGGTAGCCCAGG-3'

    3'-CGCCCGGTCGGGCTCACCCATCGGGTCC-5'

B. 5'-ATTATAAAATATTTAGATACTATATTTACAA-3'

    3'-TAATATTTTATAAATCTATGATATAAATGTT-5'

C. 5'-AGAGCTAGATCGAT-3'

    3'-TCTCGATCTAGCTA-5'

QUESTION 5–9

The total length of DNA in the human genome is about 1 m, and the diameter of the double helix is about 2 nm. Nucleotides in a DNA double helix are stacked (see

Figure 5–6B) at an interval of 0.34 nm. If the DNA were enlarged so that its diameter equaled that of an electrical extension cord (5 mm), how long would the extension cord be from one end to the other (assuming that it is completely stretched out)? How close would the bases be to each other? How long would a gene of 1000 nucleotide pairs be?

QUESTION 5–10

A compact disc (CD) stores about 4.8×10^9 bits of information in a 96 cm^2 area. This information is stored as a binary code—that is, every bit is either a 0 or a 1.

A. How many bits would it take to specify each nucleotide pair in a DNA sequence?

B. How many CDs would it take to store the information contained in the human genome?

QUESTION 5–11

Which of the following statements are correct? Explain your answers.

A. Each eukaryotic chromosome must contain the following DNA sequence elements: multiple origins of replication, two telomeres, and one centromere.

B. Nucleosome core particles are 30 nm in diameter.

QUESTION 5–12

Define the following terms and their relationships to one another:

A. Interphase chromosome

B. Mitotic chromosome

C. Chromatin

D. Heterochromatin

E. Histones

F. Nucleosome

QUESTION 5–13

Carefully consider the result shown in **Figure Q5–13**. Each of the two colonies shown on the *left* is a clump of approximately 100,000 yeast cells that has grown up from a single cell, which is now somewhere in the middle of the colony. The two yeast colonies are genetically different, as shown by the chromosomal maps on the right.

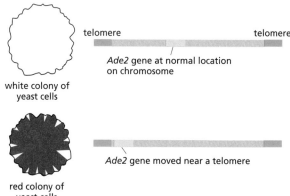

white colony of
yeast cells

telomere   telomere

Ade2 gene at normal location
on chromosome

red colony of
yeast cells
with white sectors

Ade2 gene moved near a telomere

Figure Q5–13

The yeast *Ade2* gene encodes one of the enzymes required for adenine biosynthesis, and the absence of the *Ade2* gene product leads to the accumulation of a red pigment. At its normal chromosome location, *Ade2* is expressed in all cells. When it is positioned near the telomere, which is highly condensed, *Ade2* is no longer expressed. How do you think the white sectors arise? What can you conclude about the propagation of the transcriptional state of the *Ade2* gene from mother to daughter cells?

QUESTION 5–14

The two electron micrographs in **Figure Q5–14** show nuclei of two different cell types. Can you tell from these pictures which of the two cells is transcribing more of its genes? Explain how you arrived at your answer. (Micrographs courtesy of Don W. Fawcett.)

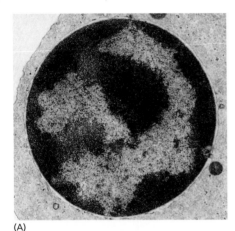

(A)

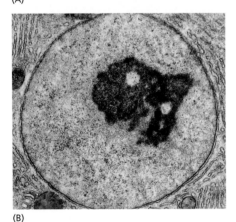

(B)

Figure Q5–14

QUESTION 5–15

DNA forms a right-handed helix. Pick out the right-handed helix from those shown in **Figure Q5–15**.

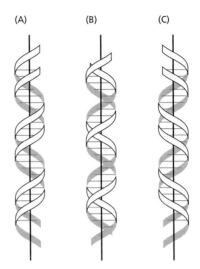

(A)   (B)   (C)

Figure Q5–15

QUESTION 5–16

A single nucleosome core particle is 11 nm in diameter and contains 147 bp of DNA (the DNA double helix measures 0.34 nm/bp). What packing ratio (ratio of DNA length to nucleosome diameter) has been achieved by wrapping DNA around the histone octamer? Assuming that there are an additional 54 bp of extended DNA in the linker between nucleosomes, how condensed is "beads-on-a-string" DNA relative to fully extended DNA? What fraction of the 10,000-fold condensation that occurs at mitosis does this first level of packing represent?

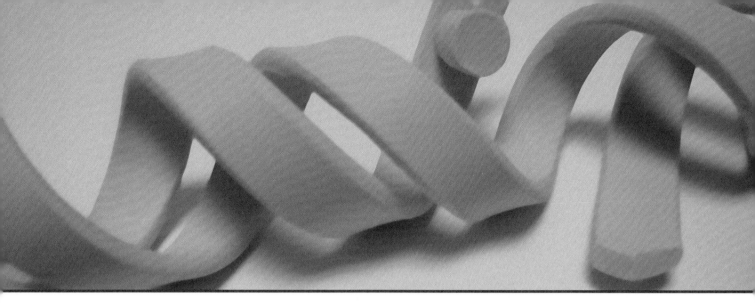

CHAPTER **SIX**

6

DNA Replication, Repair, and Recombination

The ability of a cell to survive and proliferate in a chaotic environment depends on the accurate duplication of the vast quantity of genetic information carried in its DNA. This duplication process, called *DNA replication*, must occur before a cell can divide to produce two genetically identical daughter cells. Maintaining order in a cell also requires the continual surveillance and repair of its genetic information, as DNA is subjected to unavoidable damage by chemicals and radiation in the environment and by reactive molecules that are generated inside the cell. In this chapter, we describe the protein machines that replicate and repair the cell's DNA. These machines catalyze some of the most rapid and accurate processes that take place within cells, and the strategies they have evolved to achieve this feat are marvels of elegance and efficiency.

Despite these systems for protecting a cell's DNA from copying errors and accidental damage, permanent changes—or *mutations*—sometimes do occur. Although most mutations do not affect the organism in any noticeable way, some have profound consequences. Occasionally, these changes can benefit the organism: for example, mutations can make bacteria resistant to antibiotics that are used to kill them. What is more, changes in DNA sequence can produce small variations that underlie the differences between individuals of the same species (**Figure 6–1**); when allowed to accumulate over millions of years, such changes provide the variety in genetic material that makes one species distinct from another, as we discuss in Chapter 9.

But, mutations are much more likely to be detrimental than beneficial: in humans, they are responsible for thousands of genetic diseases, including cancer. The survival of a cell or organism, therefore, depends on keeping

DNA REPLICATION

DNA REPAIR

Figure 6–1 Genetic information is passed from one generation to the next. Differences in DNA can produce the variations that underlie the differences between individuals of the same species—or, over time, the differences between one species and another. In this family photo, the children resemble one another and their parents more closely than they resemble other people because they inherit their genes from their parents. The cat shares many features with humans, but during the millions of years of evolution that have separated humans and cats, both have accumulated many changes in DNA that now make the two species different. The chicken is an even more distant relative.

changes in its DNA to a minimum. Without the protein machines that are continually monitoring and repairing damage to DNA, it is questionable whether life could exist at all.

DNA REPLICATION

At each cell division, a cell must copy its genome with extraordinary accuracy. In this section, we explore how the cell achieves this feat, while duplicating its DNA at rates as high as 1000 nucleotides per second.

Base-Pairing Enables DNA Replication

In the preceding chapter, we saw that each strand of a DNA double helix contains a sequence of nucleotides that is exactly complementary to the nucleotide sequence of its partner strand. Each strand can therefore serve as a **template**, or mold, for the synthesis of a new complementary strand. In other words, if we designate the two DNA strands as S and S′, strand S can serve as a template for making a new strand S′, while strand S′ can serve as a template for making a new strand S (**Figure 6–2**). Thus, the genetic information in DNA can be accurately copied by the beautifully simple process in which strand S separates from strand S′, and each separated strand then serves as a template for the production of a new complementary partner strand that is identical to its former partner.

The ability of each strand of a DNA molecule to act as a template for producing a complementary strand enables a cell to copy, or *replicate*, its genes before passing them on to its descendants. But the task is awe-inspiring, as it can involve copying billions of nucleotide pairs every time a cell divides. The copying must be carried out with incredible speed and accuracy: in about 8 hours, a dividing animal cell will copy the equivalent of 1000 books like this one and, on average, get no more than a few letters wrong. This impressive feat is performed by a cluster of proteins that together form a *replication machine*.

Figure 6–2 DNA acts as a template for its own duplication. Because the nucleotide A will successfully pair only with T, and G with C, each strand of a DNA double helix—labeled here as the S strand and its complementary S′ strand—can serve as a template to specify the sequence of nucleotides in its complementary strand. In this way, both strands of a DNA double helix can be copied precisely.

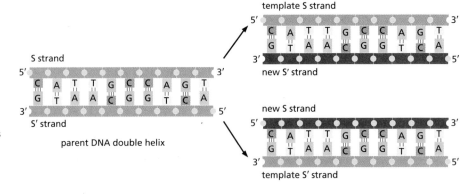

Figure 6–3 **In each round of DNA replication, each of the two strands of DNA is used as a template for the formation of a new, complementary strand.** DNA replication is "semiconservative" because each daughter DNA double helix is composed of one conserved strand and one newly synthesized strand.

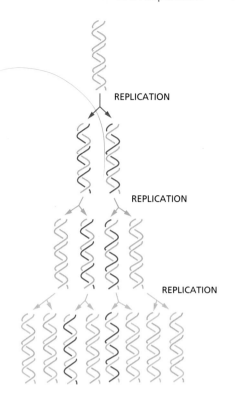

DNA replication produces two complete double helices from the original DNA molecule, with each new DNA helix being identical (except for rare copying errors) in nucleotide sequence to the original DNA double helix (see Figure 6–2). Because each parental strand serves as the template for one new strand, each of the daughter DNA double helices ends up with one of the original (old) strands plus one strand that is completely new; this style of replication is said to be *semiconservative* (**Figure 6–3**). In **How We Know**, pp. 200–202, we discuss the experiments that first demonstrated that DNA is replicated in this way.

DNA Synthesis Begins at Replication Origins

The DNA double helix is normally very stable: the two DNA strands are locked together firmly by the large numbers of hydrogen bonds between the bases on both strands (see Figure 5–2). As a result, only temperatures approaching those of boiling water provide enough thermal energy to separate the two strands. To be used as a template, however, the double helix must first be opened up and the two strands separated to expose unpaired bases. How does this occur at the temperatures found in living cells?

The process of DNA synthesis is begun by *initiator proteins* that bind to specific DNA sequences called **replication origins**. Here, the initiator proteins pry the two DNA strands apart, breaking the hydrogen bonds between the bases (**Figure 6–4**). Although the hydrogen bonds collectively make the DNA helix very stable, individually each hydrogen bond is weak (as discussed in Chapter 2). Separating a short length of DNA a few base pairs at a time therefore does not require a large energy input, and the initiator proteins can readily unzip the double helix at normal temperatures.

In simple cells such as bacteria or yeast, replication origins span approximately 100 nucleotide pairs. They are composed of DNA sequences that attract the initiator proteins and are especially easy to open. We saw in Chapter 5 that an A-T base pair is held together by fewer hydrogen bonds than is a G-C base pair. Therefore, DNA rich in A-T base pairs is relatively easy to pull apart, and A-T-rich stretches of DNA are typically found at replication origins.

A bacterial genome, which is typically contained in a circular DNA molecule of several million nucleotide pairs, has a single replication origin. The human genome, which is very much larger, has approximately 10,000 such origins—an average of 220 origins per chromosome. Beginning DNA replication at many places at once greatly shortens the time a cell needs to copy its entire genome.

Once an initiator protein binds to DNA at a replication origin and locally opens up the double helix, it attracts a group of proteins that carry out DNA replication. These proteins form a replication machine, in which each protein carries out a specific function.

Two Replication Forks Form at Each Replication Origin

DNA molecules in the process of being replicated contain Y-shaped junctions called **replication forks**. Two replication forks are formed at

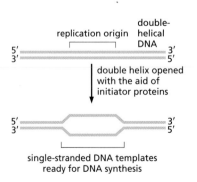

Figure 6–4 **A DNA double helix is opened at replication origins.** DNA sequences at replication origins are recognized by initiator proteins (not shown), which locally pry apart the two strands of the double helix. The exposed single strands can then serve as templates for copying the DNA.

THE NATURE OF REPLICATION

In 1953, James Watson and Francis Crick published their famous two-page paper describing a model for the structure of DNA (see Figure 5–2). In it, they proposed that complementary bases—adenine and thymine, guanine and cytosine—pair with one another along the center of the double helix, holding together the two strands of DNA. At the very end of this succinct scientific blockbuster, they comment, almost as an aside, "It has not escaped our notice that the specific pairing we have postulated immediately suggests a possible copying mechanism for the genetic material."

Indeed, one month after the classic paper appeared in print in the journal *Nature*, Watson and Crick published a second article, suggesting how DNA might be duplicated. In this paper, they proposed that the two strands of the double helix unwind, and that each strand serves as a template for the synthesis of a complementary daughter strand. In their model, dubbed semiconservative replication, each new DNA molecule consists of one strand derived from the original parent molecule and one newly synthesized strand (**Figure 6–5A**).

We now know that Watson and Crick's model for DNA replication was correct—but it was not universally accepted at first. Respected physicist-turned-geneticist Max Delbrück, for one, got hung up on what he termed "the untwiddling problem;" that is: how could the two strands of a double helix, twisted around each other

so many times all along their great length, possibly be unwound without making a big tangled mess? Watson and Crick's conception of the DNA helix opening up like a zipper seemed, to Delbrück, physically unlikely and simply "too inelegant to be efficient."

Instead, Delbrück proposed that DNA replication proceeds through a series of breaks and reunions, in which the DNA backbone is broken and the strands are copied in short segments—perhaps only 10 nucleotides at a time—before being rejoined. In this model, which was later dubbed dispersive, the resulting copies would be patchwork collections of old and new DNA, each strand containing a mixture of both (**Figure 6–5B**). No unwinding was necessary.

Yet a third camp promoted the idea that DNA replication might be *conservative*: that the parent helix would somehow remain entirely intact after copying, and the daughter molecule would contain two entirely new DNA strands (**Figure 6–5C**). To determine which of these models was correct, an experiment was needed—one that would reveal the composition of the newly synthesized DNA strands. That's where Matt Meselson and Frank Stahl came in.

As a graduate student working with Linus Pauling, Meselson was toying with a method for telling the difference between old and new proteins. After chatting with Delbrück about Watson and Crick's replication model, it

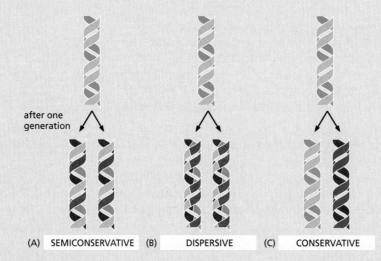

after one
generation

(A) SEMICONSERVATIVE   (B)   DISPERSIVE   (C)   CONSERVATIVE

Figure 6–5 Three models for DNA replication make different predictions. (A) In the semiconservative model, each parent strand serves as a template for the synthesis of a new daughter strand. The first round of replication would produce two hybrid molecules, each containing one strand from the original parent in addition to one newly synthesized strand. A subsequent round of replication would yield two hybrid molecules and two molecules that contain none of the original parent DNA (see Figure 6–3). (B) In the dispersive model, each generation of daughter DNA will contain a mixture of DNA from the parent strands and the newly synthesized DNA. (C) In the conservative model, the parent molecule remains intact after being copied. In this case, the first round of replication would yield the original parent double helix and an entirely new double helix. For each model, parent DNA molecules are shown in *orange*; newly replicated DNA is *red*. Note that only a very small segment of DNA is shown for each model.

occurred to Meselson that the approach he'd envisaged for exploring protein synthesis might also work for studying DNA. In the summer of 1954, Meselson met Stahl, who was then a graduate student in Rochester, NY, and they agreed to collaborate. It took a few years to get everything working, but the two eventually performed what has come to be known as "the most beautiful experiment in biology."

Their approach, in retrospect, was stunningly straightforward. They started by growing two batches of *E. coli* bacteria, one in a medium containing a heavy isotope of nitrogen, ^{15}N, the other in a medium containing the normal, lighter ^{14}N. The nitrogen in the nutrient medium gets incorporated into the nucleotide bases and, from there, makes its way into the DNA of the organism. After growing bacterial cultures for many generations in either the ^{15}N- or ^{14}N-containing medium, the researchers had two flasks of bacteria, one whose DNA was heavy, the other whose DNA was light. Meselson and Stahl then broke open the bacterial cells and loaded the DNA into tubes containing a high concentration of the salt cesium chloride. When these tubes are centrifuged at high speed, the cesium chloride forms a density gradient, and the DNA molecules float or sink within the solution until they reach the point at which their density equals that of the surrounding salt solution (see Panel 4–3, pp. 164–165). Using this method, called equilibrium density centrifugation, Meselson and Stahl found that they could distinguish between heavy (^{15}N-containing) DNA and light (^{14}N-containing) DNA by observing the positions of the DNA within the cesium chloride gradient. Because the heavy DNA was denser than the light DNA, it collected at a position nearer to the bottom of the centrifuge tube (Figure 6–6).

Once they had established this method for differentiating between light and heavy DNA, Meselson and Stahl set out to test the various hypotheses proposed for DNA replication. To do this, they took a flask of bacteria that had been grown in heavy nitrogen and transferred the bacteria into a medium containing the light isotope. At the start of the experiment, all the DNA would be heavy. But, as the bacteria divided, the newly synthesized DNA would be light. They could then monitor the accumulation of light DNA and see which model, if any, best fit the data. After one generation of growth, the researchers found that the parental, heavy DNA molecules—those made of two strands containing ^{15}N—had disappeared and were replaced by a new species of DNA that banded at a density halfway between those of ^{15}N-DNA and ^{14}N-DNA (Figure 6–7). These newly synthesized daughter helices, Meselson and Stahl reasoned, must be hybrids—containing both heavy and light isotopes.

Right away, this observation ruled out the conservative model of DNA replication, which predicted that

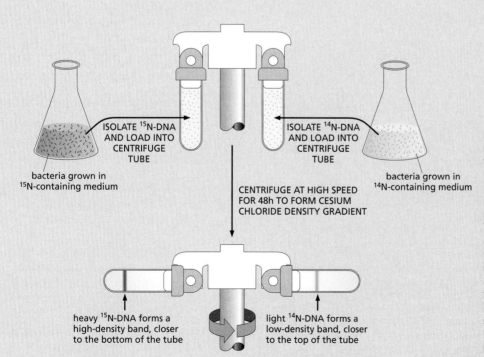

ISOLATE ^{15}N-DNA AND LOAD INTO CENTRIFUGE TUBE

ISOLATE ^{14}N-DNA AND LOAD INTO CENTRIFUGE TUBE

bacteria grown in ^{15}N-containing medium

CENTRIFUGE AT HIGH SPEED FOR 48h TO FORM CESIUM CHLORIDE DENSITY GRADIENT

bacteria grown in ^{14}N-containing medium

heavy ^{15}N-DNA forms a high-density band, closer to the bottom of the tube

light ^{14}N-DNA forms a low-density band, closer to the top of the tube

Figure 6–6 Centrifugation in a cesium chloride gradient allows the separation of heavy and light DNA. Bacteria are grown for several generations in a medium containing either ^{15}N (the heavy isotope) or ^{14}N (the light isotope) to label their DNA. The cells are then broken open, and the DNA is loaded into an ultracentrifuge tube containing a cesium chloride salt solution. These tubes are centrifuged at high speed for two days to allow the DNA to collect in a region where its density matches that of the salt surrounding it. The heavy and light DNA molecules collect in different positions in the tube.

CONDITION	RESULT	INTERPRETATION

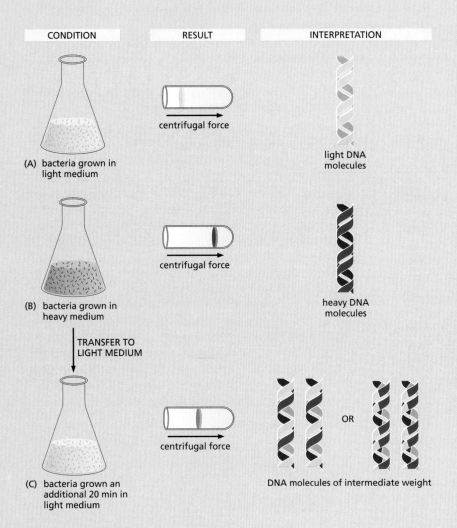

(A) bacteria grown in light medium

centrifugal force

light DNA molecules

(B) bacteria grown in heavy medium

TRANSFER TO LIGHT MEDIUM

centrifugal force

heavy DNA molecules

(C) bacteria grown an additional 20 min in light medium

centrifugal force

OR

DNA molecules of intermediate weight

Figure 6–7 The first part of the Meselson–Stahl experiment ruled out the conservative model of DNA replication. (A) Bacteria grown in light medium (containing ^{14}N) yield DNA that forms a band high up in the centrifuge tube, whereas bacteria grown in ^{15}N-containing heavy medium (B) produce DNA that migrates further down the tube. When bacteria grown in a heavy medium are transferred to a light medium and allowed to continue dividing, they produce a band whose position falls somewhere between that of the parent bands (C). These results rule out the conservative model of replication but do not distinguish between the semiconservative and dispersive models, both of which predict the formation of hybrid daughter DNA molecules.

The fact that the results came out looking so clean—with discrete bands forming at the expected positions for newly replicated hybrid DNA molecules—was a happy accident of the experimental protocol. The researchers used a hypodermic syringe to load their DNA samples into the ultracentrifuge tubes (see Figure 6–6). In the process, they unwittingly sheared the large bacterial chromosome into smaller fragments. Had the chromosomes remained whole, the researchers might have isolated DNA molecules that were only partially replicated, because many cells would have been caught in the middle of copying their DNA. Molecules in such an intermediate stage of replication would not have separated into such discrete bands. But because the researchers were instead working with smaller pieces of DNA, the likelihood that any given fragment had been fully replicated—and contained a complete parent and daughter strand—was high, thus yielding nice, clean results.

the parental DNA would remain entirely heavy, while the daughter DNA would be entirely light (see Figure 6–5C). The data matched with the semiconservative model, which predicted the formation of hybrid molecules containing one strand of heavy DNA and one strand of light (see Figure 6–5A). The results, however, were also consistent with the dispersive model, in which hybrid DNA strands would contain a mixture of heavy and light DNA (see Figure 6–5B).

To distinguish between the two models, Meselson and Stahl turned up the heat. When DNA is subjected to high temperature, the hydrogen bonds holding the two strands together break and the helix comes apart, leaving a collection of single-stranded DNAs. When the researchers heated their hybrid molecules before centrifuging, they discovered that one strand of the DNA was heavy, whereas the other was light. This observation supported only the semiconservative model; if the dispersive model were correct, the resulting strands, each containing a mottled assembly of heavy and light DNA, would have all banded together at an intermediate density.

According to historian Frederic Lawrence Holmes, the experiment was so elegant and the results so clean that Stahl—when being interviewed for a position at Yale University—was unable to fill the 50 minutes allotted for his talk. "I was finished in 25 minutes," said Stahl, "because that is all it takes to tell that experiment. It's so totally simple and contained." Stahl did not get the job at Yale, but the experiment convinced biologists that Watson and Crick had been correct. In fact, the results were accepted so widely and rapidly that the experiment was described in a textbook before Meselson and Stahl had even published the data.

each replication origin (**Figure 6–8**). At each fork, a replication machine moves along the DNA, opening up the two strands of the double helix and using each strand as a template to make a new daughter strand. The two forks move away from the origin in opposite directions, unzipping the DNA double helix and replicating the DNA as they go (**Figure 6–9**). DNA replication in bacterial and eukaryotic chromosomes is therefore termed *bidirectional*. The forks move very rapidly—at about 1000 nucleotide pairs per second in bacteria and 100 nucleotide pairs per second in humans. The slower rate of fork movement in humans (indeed, in all eukaryotes) may be due to the difficulties in replicating DNA through the more complex chromatin structure of eukaryotic chromosomes.

DNA Polymerase Synthesizes DNA Using a Parental Strand as Template

The movement of a replication fork is driven by the action of the replication machine, at the heart of which is an enzyme called **DNA polymerase**. This enzyme catalyzes the addition of nucleotides to the 3′ end of a growing DNA strand, using one of the original, parental DNA strands as a template. Base pairing between an incoming nucleotide and the template strand determines which of the four nucleotides (A, G, T, or C) will be selected. The final product is a new strand of DNA that is complementary in nucleotide sequence to the template (**Figure 6–10**).

The polymerization reaction involves the formation of a phosphodiester bond between the 3′ end of the growing DNA chain and the 5′-phosphate group of the incoming nucleotide, which enters the reaction as a *deoxyribonucleoside triphosphate*. The energy for polymerization is provided

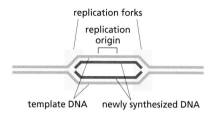

Figure 6–8 DNA synthesis occurs at Y-shaped junctions called replication forks. Two replication forks are formed at each replication origin.

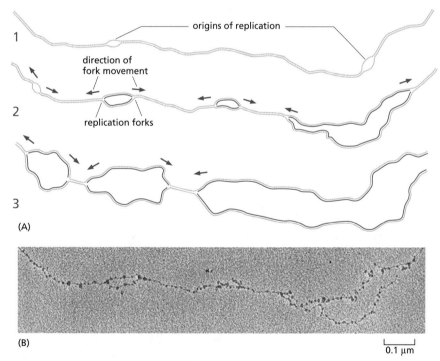

Figure 6–9 The two replication forks move away in opposite directions at each replication origin. (A) These drawings represent the same portion of a DNA molecule as it might appear at different times during replication. The *orange* lines represent the two parental DNA strands; the *red* lines represent the newly synthesized DNA strands. (B) An electron micrograph showing DNA replicating in an early fly embryo. The particles visible along the DNA are nucleosomes, structures made of DNA and the protein complexes around which the DNA is wrapped (discussed in Chapter 5). The chromosome in this micrograph is the one that was redrawn in sketch (2) above. (Electron micrograph courtesy of Victoria Foe.)

QUESTION 6–1

Look carefully at the micrograph and drawing 2 in Figure 6–9.
A.   Using the scale bar, estimate the lengths of the DNA strands between the replication forks. Numbering the replication forks sequentially from the left, how long will it take until forks 4 and 5, and forks 7 and 8, respectively, collide with each other? (Recall that the distance between the bases in DNA is 0.34 nm, and eukaryotic replication forks move at about 100 nucleotides per second.) For this question, disregard the nucleosomes seen in the micrograph and assume that the DNA is fully extended.
B.   The fly genome is about 1.8×10^8 nucleotide pairs in size. What fraction of the genome is shown in the micrograph?

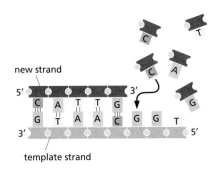

Figure 6–10 A new DNA strand is synthesized in the 5′–to–3′ direction. At each step, the appropriate incoming nucleotide is selected by forming base pairs with the next nucleotide in the template strand: A with T, T with A, C with G, and G with C. Each is added to the 3′ end of the growing new strand, as indicated.

by the incoming deoxyribonucleoside triphosphate itself: hydrolysis of one of its high-energy phosphate bonds fuels the reaction that links the nucleotide monomer to the chain, releasing pyrophosphate (**Figure 6–11**). Pyrophosphate is further hydrolyzed to inorganic phosphate (P_i), which makes the polymerization reaction effectively irreversible (see Figure 3–41).

DNA polymerase does not dissociate from the DNA each time it adds a new nucleotide to the growing strand; rather, it stays associated with the DNA and moves along the template strand stepwise for many cycles of the polymerization reaction (Movie 6.1). We will see later that a special protein keeps the polymerase attached to the DNA, as it repeatedly adds new nucleotides to the growing strand.

The Replication Fork Is Asymmetrical

The 5′-to-3′ direction of the DNA polymerization reaction poses a problem at the replication fork. As illustrated in Figure 5–2, the sugar–phosphate backbone of each strand of a DNA double helix has a unique chemical direction, or polarity, determined by the way each sugar residue is linked to the next, and the two strands in the double helix are antiparallel; that is, they run in opposite directions. As a consequence, at each replication fork, one new DNA strand is being made on a template that runs in one direction (3′ to 5′), whereas the other new strand is being made on a template that runs in the opposite direction (5′ to 3′) (**Figure 6–12**). The replication fork is therefore asymmetrical. Looking at Figure 6–9A, however, it appears that both of the new DNA strands are growing in the same direction; that is, the direction in which the replication fork is moving. That observation suggests that one strand is being synthesized in the 5′-to-3′ direction and the other in the 3′-to-5′ direction.

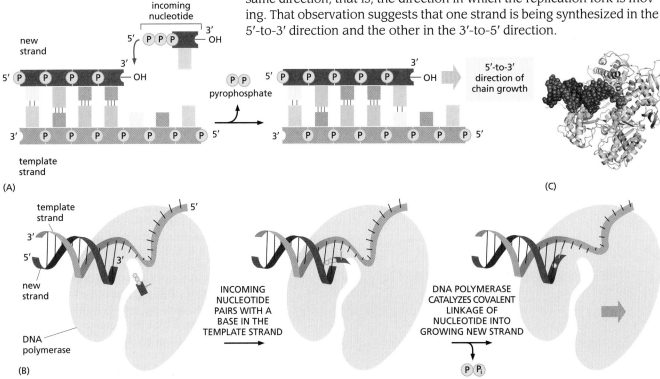

Figure 6–11 DNA polymerase adds a deoxyribonucleotide to the 3′ end of a growing DNA chain. (A) Nucleotides enter the reaction as deoxyribonucleoside triphosphates. This incoming nucleotide forms a base pair with its partner in the template strand. It is then linked to the free 3′ hydroxyl on the growing DNA strand. The new DNA strand is therefore synthesized in the 5′-to-3′ direction. Breakage of a high-energy phosphate bond in the incoming nucleoside triphosphate—accompanied by the release of pyrophosphate—provides the energy for the polymerization reaction. (B) The reaction is catalyzed by the enzyme DNA polymerase (*light green*). The polymerase guides the incoming nucleotide to the template strand and positions it such that its 5′ terminal phosphate will be able to react with the 3′-hydroxyl group on the newly synthesized strand. The *gray arrow* indicates the direction of polymerase movement. (C) Structure of DNA polymerase, as determined by X-ray crystallography, which shows the positioning of the DNA double helix. The template strand is the longer of the two DNA strands (Movie 6.1).

Does the cell have two types of DNA polymerase, one for each direction? The answer is no: all DNA polymerases add new subunits only to the 3′ end of a DNA strand (see Figure 6–11A). As a result, a new DNA chain can be synthesized only in a 5′-to-3′ direction. This can easily account for the synthesis of one of the two strands of DNA at the replication fork, but what happens on the other? This conundrum is solved by the use of a "backstitching" maneuver. The DNA strand that appears to grow in the incorrect 3′-to-5′ direction is actually made *discontinuously*, in successive, separate, small pieces—with the DNA polymerase moving backward with respect to the direction of replication-fork movement so that each new DNA fragment can be polymerized in the 5′-to-3′ direction.

The resulting small DNA pieces—called **Okazaki fragments** after the biochemists who discovered them—are later joined together to form a continuous new strand. The DNA strand that is made discontinuously in this way is called the **lagging strand**, because the backstitching imparts a slight delay to its synthesis; the other strand, which is synthesized continuously, is called the **leading strand** (**Figure 6–13**).

Although they differ in subtle details, the replication forks of all cells, prokaryotic and eukaryotic, have leading and lagging strands. This common feature arises from the fact that all DNA polymerases work only in the 5′-to-3′ direction—a restriction that provides cells with an important advantage, as we discuss next.

DNA Polymerase Is Self-correcting

DNA polymerase is so accurate that it makes only about one error in every 10^7 nucleotide pairs it copies. This error rate is much lower than can be explained simply by the accuracy of complementary base-pairing. Although A-T and C-G are by far the most stable base pairs, other, less stable base pairs—for example, G-T and C-A—can also be formed. Such incorrect base pairs are formed much less frequently than correct ones, but, if allowed to remain, they would result in an accumulation of mutations. This disaster is avoided because DNA polymerase has two special qualities that greatly increase the accuracy of DNA replication. First, the enzyme carefully monitors the base-pairing between each incoming nucleotide and the template strand. Only when the match is correct does DNA polymerase catalyze the nucleotide-addition reaction. Second,

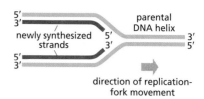

Figure 6–12 At a replication fork, the two newly synthesized DNA strands are of opposite polarities. This is because the two template strands are oriented in opposite directions.

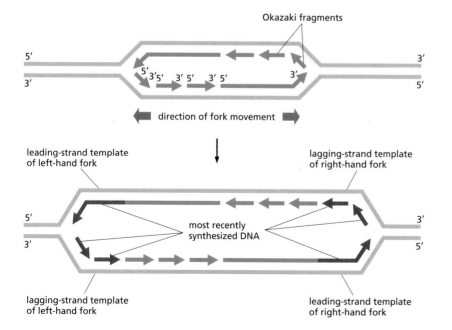

Figure 6–13 At each replication fork, the lagging DNA strand is synthesized in pieces. Because both of the new strands at a replication fork are synthesized in the 5′-to-3′ direction, the lagging strand of DNA must be made initially as a series of short DNA strands, which are later joined together. The upper diagram shows two replication forks moving in opposite directions; the lower diagram shows the same forks a short time later. To replicate the lagging strand, DNA polymerase uses a backstitching mechanism: it synthesizes short pieces of DNA (called Okazaki fragments) in the 5′-to-3′ direction and then moves back along the template strand (toward the fork) before synthesizing the next fragment.

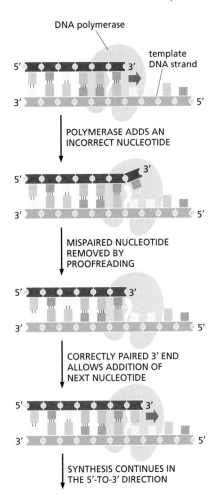

POLYMERASE ADDS AN INCORRECT NUCLEOTIDE

MISPAIRED NUCLEOTIDE REMOVED BY PROOFREADING

CORRECTLY PAIRED 3′ END ALLOWS ADDITION OF NEXT NUCLEOTIDE

SYNTHESIS CONTINUES IN THE 5′-TO-3′ DIRECTION

Figure 6–14 During DNA synthesis, DNA polymerase proofreads its own work. If an incorrect nucleotide is added to a growing strand, the DNA polymerase cleaves it from the strand and replaces it with the correct nucleotide before continuing.

Figure 6–15 DNA polymerase contains separate sites for DNA synthesis and proofreading. The diagrams are based on the structure of an *E. coli* DNA polymerase molecule, as determined by X-ray crystallography. DNA polymerase is shown with the replicating DNA molecule and the polymerase in the polymerizing mode (*left*) and in the proofreading mode (*right*). The catalytic sites for the polymerization activity (P) and error-correcting proofreading activity (E) are indicated. When the polymerase adds an incorrect nucleotide, the newly synthesized DNA strand (*red*) transiently unpairs from the template strand (*orange*), and its growing 3′ end moves into the error-correcting catalytic site (E) to be removed.

when DNA polymerase makes a rare mistake and adds the wrong nucleotide, it can correct the error through an activity called **proofreading**.

Proofreading takes place at the same time as DNA synthesis. Before the enzyme adds the next nucleotide to a growing DNA strand, it checks whether the previously added nucleotide is correctly base-paired to the template strand. If so, the polymerase adds the next nucleotide; if not, the polymerase clips off the mispaired nucleotide and tries again (**Figure 6–14**). This proofreading is carried out by a nuclease that cleaves the phosphodiester backbone. Polymerization and proofreading are tightly coordinated, and the two reactions are carried out by different catalytic domains in the same polymerase molecule (**Figure 6–15**).

This proofreading mechanism explains why DNA polymerases synthesize DNA only in the 5′-to-3′ direction, despite the need that this imposes for a cumbersome backstitching mechanism at the replication fork (see Figure 6–13). A hypothetical DNA polymerase that synthesized in the 3′-to-5′ direction (and would thereby circumvent the need for backstitching) would be unable to proofread: if it removed an incorrectly paired nucleotide, the polymerase would create a chemical dead end—a chain that could no longer be elongated. Thus, for a DNA polymerase to function as a self-correcting enzyme that removes its own polymerization errors as it moves along the DNA, it must proceed only in the 5′-to-3′ direction.

Short Lengths of RNA Act as Primers for DNA Synthesis

We have seen that the accuracy of DNA replication depends on the requirement of the DNA polymerase for a correctly base-paired 3′ end before it can add more nucleotides to a growing DNA strand. How then can the polymerase begin a completely new DNA strand? To get the process started, a different enzyme is needed—one that can begin a new polynucleotide strand simply by joining two nucleotides together without the need for a base-paired end. This enzyme does not, however, synthesize DNA. It makes a short length of a closely related type of nucleic acid—**RNA (ribonucleic acid)**—using the DNA strand as a template. This short length of RNA, about 10 nucleotides long, is base-paired to the template strand and provides a base-paired 3′ end as a starting point for DNA polymerase. It thus serves as a *primer* for DNA synthesis, and the enzyme that synthesizes the RNA primer is known as *primase*.

Primase is an example of an *RNA polymerase*, an enzyme that synthesizes RNA using DNA as a template. A strand of RNA is very similar chemically to a single strand of DNA except that it is made of ribonucleotide subunits, in which the sugar is ribose, not deoxyribose; RNA also differs from DNA in that it contains the base uracil (U) instead of thymine (T) (see Panel 2–6, pp. 76–77). However, because U can form a base pair with A, the RNA primer is synthesized on the DNA strand by complementary base-pairing in exactly the same way as is DNA (**Figure 6–16**).

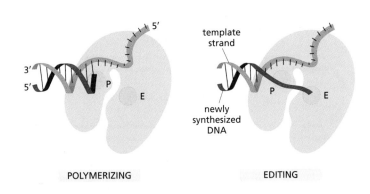

POLYMERIZING   **EDITING**

Figure 6–16 **RNA primers are synthesized by an RNA polymerase called primase, which uses a DNA strand as a template.** Like DNA polymerase, primase works in the 5′-to-3′ direction. Unlike DNA polymerase, however, primase can start a new polynucleotide chain by joining together two nucleoside triphosphates without the need for a base-paired 3′ end as a starting point. (In this case, ribonucleoside triphosphates, rather than deoxyribonucleoside triphosphates, provide the incoming nucleotides.)

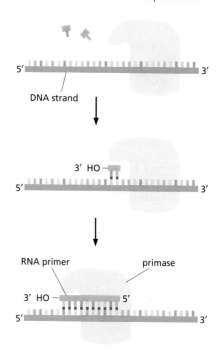

For the leading strand, an RNA primer is needed only to start replication at a replication origin; once a replication fork has been established, the DNA polymerase is continuously presented with a base-paired 3′ end as it tracks along the template strand. But on the lagging strand, where DNA synthesis is discontinuous, new primers are needed to keep polymerization going (see Figure 6–13). The movement of the replication fork continually exposes unpaired bases on the lagging strand template, and new RNA primers are laid down at intervals along the newly exposed, single-stranded stretch. DNA polymerase adds a deoxyribonucleotide to the 3′ end of each primer to start a new Okazaki fragment, and it will continue to elongate this fragment until it runs into the next RNA primer (**Figure 6–17**).

To produce a continuous new DNA strand from the many separate pieces of nucleic acid made on the lagging strand, three additional enzymes are needed. These act quickly to remove the RNA primer, replace it with DNA, and join the DNA fragments together. Thus, a nuclease degrades the RNA primer, a DNA polymerase called a *repair polymerase* then replaces this RNA with DNA (using the end of the adjacent Okazaki fragment as a primer), and the enzyme *DNA ligase* joins the 5′-phosphate end of one DNA fragment to the adjacent 3′-hydroxyl end of the next (**Figure 6–18**).

Primase can begin new polynucleotide chains, but this activity is possible because the enzyme does not proofread its work. As a result, primers frequently contain mistakes. But because primers are made of RNA instead of DNA, they stand out as "suspect copy" to be automatically removed and replaced by DNA. The repair DNA polymerases that make this DNA, like the replicative polymerases, proofread as they synthesize. In this way, the cell's replication machinery is able to begin new DNA chains and, at the same time, ensure that all of the DNA is copied faithfully.

Proteins at a Replication Fork Cooperate to Form a Replication Machine

DNA replication requires the cooperation of a large number of proteins that act in concert to open up the double helix and synthesize new DNA. These proteins form part of a remarkably complex replication machine. The first problem faced by the replication machine is accessing the

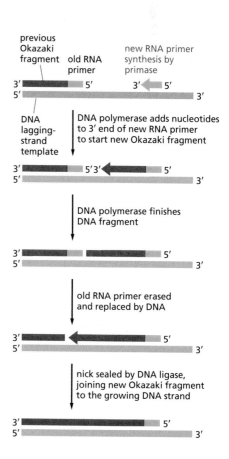

Figure 6–17 **Multiple enzymes are required to synthesize Okazaki fragments on the lagging DNA strand.** In eukaryotes, RNA primers are made at intervals of about 200 nucleotides on the lagging strand, and each RNA primer is approximately 10 nucleotides long. Primers are removed by nucleases that recognize an RNA strand in an RNA/DNA helix and degrade it; this leaves gaps that are filled in by a repair DNA polymerase that can proofread as it fills in the gaps. The completed fragments are finally joined together by an enzyme called DNA ligase, which catalyzes the formation of a phosphodiester bond between the 3′-OH end of one fragment and the 5′-phosphate end of the next, thus linking up the sugar–phosphate backbones. This nick-sealing reaction requires an input of energy in the form of ATP (not shown; see Figure 6–18).

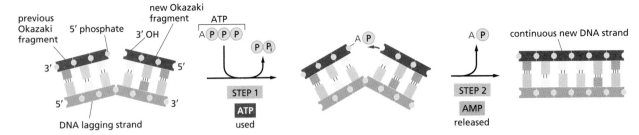

Figure 6–18 DNA ligase joins together Okazaki fragments on the lagging strand during DNA synthesis. The ligase enzyme uses a molecule of ATP to activate the 5′ end of one fragment (step 1) before forming a new bond with the 3′ end of the other fragment (step 2).

nucleotides that lie at the center of the helix. For DNA replication to occur, the double helix must be unzipped ahead of the replication fork so that the incoming nucleoside triphosphates can form base pairs with each template strand. Two types of replication proteins—*DNA helicases* and *single-strand DNA-binding proteins*—cooperate to carry out this task. The helicase sits at the very front of the replication machine where it uses the energy of ATP hydrolysis to propel itself forward, prying apart the double helix as it speeds along the DNA (**Figure 6–19A** and Movie 6.2). Single-strand DNA-binding proteins cling to the single-stranded DNA exposed by the helicase, transiently preventing the strands from re-forming base pairs and keeping them in an elongated form so that they can serve as efficient templates.

This localized unwinding of the DNA double helix itself presents a problem. As the helicase pries open the DNA within the replication fork, the

Figure 6–19 DNA synthesis is carried out by a group of proteins that act together as a replication machine.
(A) DNA polymerases are held on the leading and lagging strands by circular protein clamps that allow the polymerases to slide. On the lagging-strand template, the clamp detaches each time the polymerase completes an Okazaki fragment. A clamp loader (not shown) is required to attach a sliding clamp each time a new Okazaki fragment is begun. At the head of the fork, a DNA helicase unwinds the strands of the parental DNA double helix. Single-strand DNA-binding proteins keep the DNA strands apart to provide access for the primase and polymerase. For simplicity, this diagram shows the proteins working independently; in the cell, they are held together in a large replication machine, as shown in (B).

(B) This diagram shows a current view of how the replication proteins are arranged when a replication fork is moving. To generate this structure, the lagging strand shown in (A) has been folded to bring its DNA polymerase in contact with the leading-strand DNA polymerase. This folding process also brings the 3′ end of each completed Okazaki fragment close to the start site for the next Okazaki fragment. Because the lagging-strand DNA polymerase is bound to the rest of the replication proteins, it can be reused to synthesize successive Okazaki fragments; in this diagram, the lagging-strand DNA polymerase is about to let go of its completed Okazaki fragment and move to the RNA primer that is being synthesized by the nearby primase. To watch the replication complex in action, see Movies 6.4 and 6.5.

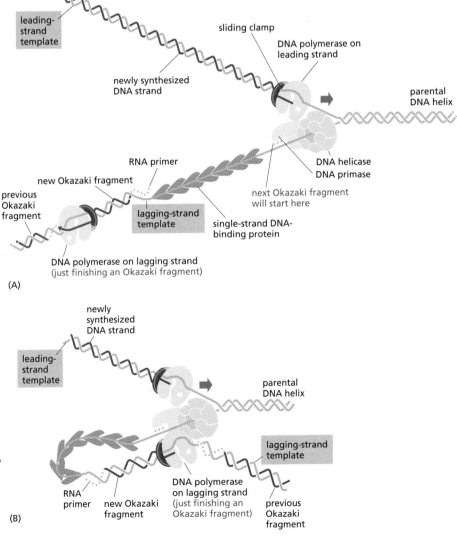

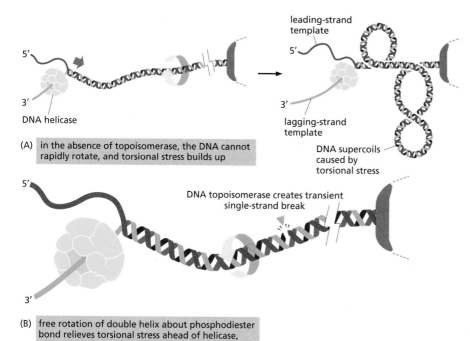

(A) in the absence of topoisomerase, the DNA cannot rapidly rotate, and torsional stress builds up

leading-strand template

lagging-strand template

DNA supercoils caused by torsional stress

DNA topoisomerase creates transient single-strand break

(B) free rotation of double helix about phosphodiester bond relieves torsional stress ahead of helicase, after which single-strand break is sealed

Figure 6–20 DNA topoisomerases relieve the tension that builds up in front of a replication fork. (A) As DNA helicase unwinds the DNA double helix, it generates a section of overwound DNA. Tension builds up because the chromosome is too large to rotate fast enough to relieve the buildup of torsional stress. The broken bars in the left-hand panel represent approximately 20 turns of DNA. (B) DNA topoisomerases relieve this stress by generating temporary nicks in the DNA.

DNA on the other side of the fork gets wound more tightly. This excess twisting in front of the replication fork creates tension in the DNA that—if allowed to build—makes unwinding the double helix increasingly difficult and impedes the forward movement of the replication machinery (**Figure 6–20A**). Cells use proteins called *DNA topoisomerases* to relieve this tension. These enzymes produce transient nicks in the DNA backbone, which temporarily release the tension; they then reseal the nick before falling off the DNA (**Figure 6–20B**).

An additional replication protein, called a *sliding clamp*, keeps DNA polymerase firmly attached to the template while it is synthesizing new strands of DNA. Left on their own, most DNA polymerase molecules will synthesize only a short string of nucleotides before falling off the DNA template strand. The sliding clamp forms a ring around the newly formed DNA double helix and, by tightly gripping the polymerase, allows the enzyme to move along the template strand without falling off as it synthesizes new DNA (see Figure 6–19A and Movie 6.3).

Assembly of the clamp around DNA requires the activity of another replication protein, the *clamp loader*, which hydrolyzes ATP each time it locks a sliding clamp around a newly formed DNA double helix. This loading needs to occur only once per replication cycle on the leading strand; on the lagging strand, however, the clamp is removed and then reattached each time a new Okazaki fragment is made.

Most of the proteins involved in DNA replication are held together in a large multienzyme complex that moves as a unit along the parental DNA double helix, enabling DNA to be synthesized on both strands in a coordinated manner. This complex can be likened to a miniature sewing machine composed of protein parts and powered by nucleoside triphosphate hydrolysis (Figure 6–19B and Movies 6.4 and 6.5).

Telomerase Replicates the Ends of Eukaryotic Chromosomes

Having discussed how DNA replication begins at origins and how movement of a replication fork proceeds, we now turn to the special problem

QUESTION 6–2

Discuss the following statement: "Primase is a sloppy enzyme that makes many mistakes. Eventually, the RNA primers it makes are disposed of and replaced with DNA synthesized by a polymerase with higher fidelity. This is wasteful. It would be more energy-efficient if a DNA polymerase made an accurate copy in the first place."

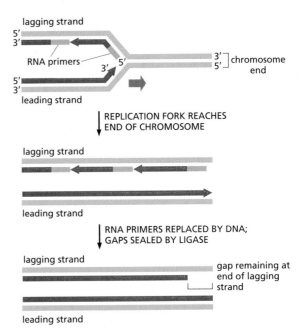

Figure 6–21 Without a special mechanism to replicate the ends of linear chromosomes, DNA would be lost during each round of cell division. DNA synthesis begins at origins of replication and continues until the replication machinery reaches the ends of the chromosome. The leading strand is reproduced in its entirety. But the ends of the lagging strand can't be completed, because once the final RNA primer has been removed there is no way to replace it with DNA. These gaps at the ends of the lagging strand must be filled in by a special mechanism to keep the chromosome ends from shrinking with each cell division.

QUESTION 6–3

A gene encoding one of the proteins involved in DNA replication has been inactivated by a mutation in a cell. In the absence of this protein, the cell attempts to replicate its DNA. What would happen during the DNA replication process if each of the following proteins were missing?

A. DNA polymerase
B. DNA ligase
C. Sliding clamp for DNA polymerase
D. Nuclease that removes RNA primers
E. DNA helicase
F. Primase

---

of replicating the very ends of chromosomes. As we discussed previously, because DNA replication proceeds only in the 5'-to-3' direction, the lagging strand of the replication fork has to be synthesized in the form of discontinuous DNA fragments, each of which is primed with an RNA primer laid down by a primase (see Figure 6–17). A serious problem arises, however, as the replication fork approaches the end of a chromosome: although the leading strand can be replicated all the way to the chromosome tip, the lagging strand cannot. When the final RNA primer on the lagging strand is removed, there is no way to replace it (**Figure 6–21**). Without a strategy to deal with this problem, the lagging strand would become shorter with each round of DNA replication; after repeated cell divisions, chromosomes would shrink—and eventually lose valuable genetic information.

Bacteria solve this "end-replication" problem by having circular DNA molecules as chromosomes. Eukaryotes solve it by having long, repetitive nucleotide sequences at the ends of their chromosomes which are incorporated into structures called **telomeres**. These telomeric DNA sequences attract an enzyme called **telomerase** to the chromosome ends. Using an RNA template that is part of the enzyme itself, telomerase extends the ends of the replicating lagging strand by adding multiple copies of the same short DNA sequence to the template strand. This extended template allows replication of the lagging strand to be completed by conventional DNA replication (**Figure 6–22**).

In addition to allowing replication of chromosome ends, telomeres form structures that mark the true ends of a chromosome. This allows the cell to distinguish unambiguously between the natural ends of chromosomes and the double-strand DNA breaks that sometimes occur accidentally in

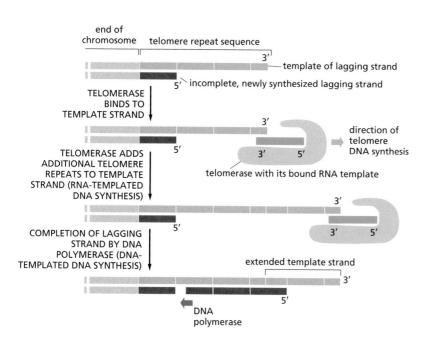

end of
chromosome   telomere repeat sequence

3'
template of lagging strand

5'   incomplete, newly synthesized lagging strand

TELOMERASE
BINDS TO
TEMPLATE STRAND

3'

direction of
telomere
DNA synthesis

TELOMERASE ADDS
ADDITIONAL TELOMERE
REPEATS TO TEMPLATE
STRAND (RNA-TEMPLATED
DNA SYNTHESIS)

5'   3'   5'

telomerase with its bound RNA template

3'

COMPLETION OF LAGGING
STRAND BY DNA
POLYMERASE (DNA-
TEMPLATED DNA SYNTHESIS)

5'   3'   5'

extended template strand

3'

5'

DNA
polymerase

Figure 6–22 Telomeres and telomerase prevent linear eukaryotic chromosomes from shortening with each cell division. For clarity, only the template DNA (*orange*) and newly synthesized DNA (*red*) of the lagging strand are shown (see bottom of Figure 6–21). To complete the replication of the lagging strand at the ends of a chromosome, the template strand is first extended beyond the DNA that is to be copied. To achieve this, the enzyme telomerase adds more repeats to the telomere repeat sequences at the 3' end of the template strand, which then allows the lagging strand to be completed by DNA polymerase, as shown. The telomerase enzyme carries a short piece of RNA (*blue*) with a sequence that is complementary to the DNA repeat sequence; this RNA acts as the template for telomere DNA synthesis. After the lagging-strand replication is complete, a short stretch of single-stranded DNA remains at the ends of the chromosome, as shown. To see telomerase in action, view Movie 6.6.

the middle of chromosomes. These breaks are dangerous and must be immediately repaired, as we see in the next section.

DNA REPAIR

The diversity of living organisms and their success in colonizing almost every part of the Earth's surface depend on genetic changes accumulated gradually over millions of years. Some of these changes allow organisms to adapt to changing conditions and to thrive in new habitats. However, in the short term, and from the perspective of an individual organism, genetic alterations can be detrimental. In a multicellular organism, such permanent changes in the DNA—called mutations—can upset the organism's extremely complex and finely tuned development and physiology.

To survive and reproduce, individuals must be genetically stable. This stability is achieved not only through the extremely accurate mechanism for replicating DNA that we have just discussed, but also through the work of a variety of protein machines that continually scan the genome for damage and fix it when it occurs. Although some changes arise from rare mistakes in the replication process, the majority of DNA damage is an unintended consequence of the vast number of chemical reactions that occur inside cells.

Most DNA damage is only temporary, because it is immediately corrected by processes collectively called **DNA repair**. The importance of these DNA repair processes is evident from the consequences of their malfunction. Humans with the genetic disease *xeroderma pigmentosum*, for example, cannot mend the damage done by ultraviolet (UV) radiation because they have inherited a defective gene for one of the proteins involved in this repair process. Such individuals develop severe skin lesions, including skin cancer, because of the accumulation of DNA damage in cells that are exposed to sunlight and the consequent mutations that arise in these cells.

In this section, we describe a few of the specialized mechanisms cells use to repair DNA damage. We then consider examples of what happens when these mechanisms fail—and discuss how the fidelity of DNA replication and repair are reflected in our genome.

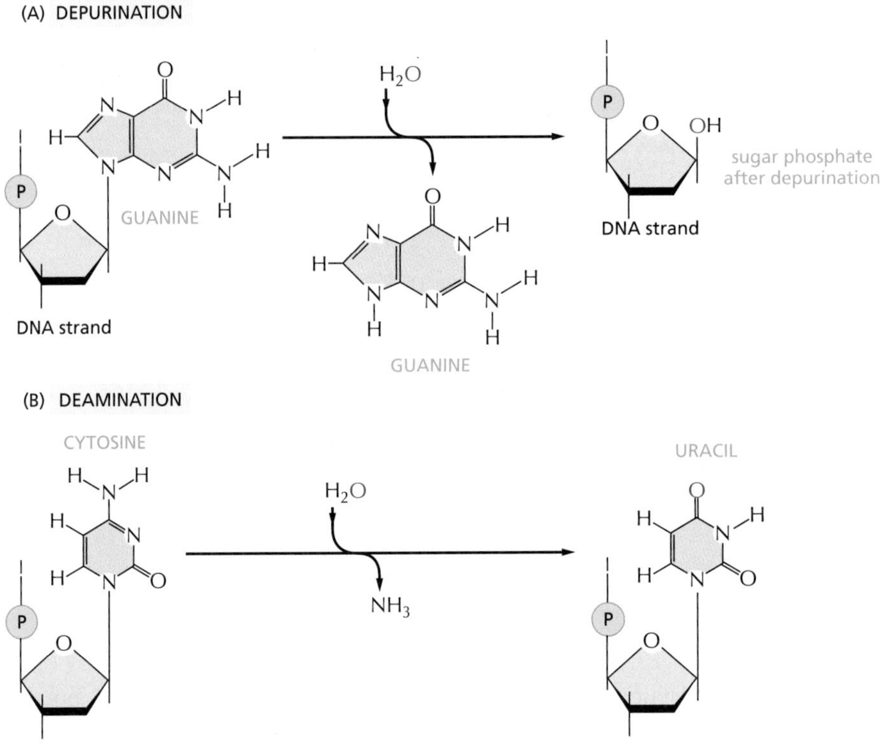

Figure 6–23 Depurination and deamination are the most frequent chemical reactions known to create serious DNA damage in cells. (A) Depurination can remove guanine (or adenine) from DNA. (B) The major type of deamination reaction converts cytosine to an altered DNA base, uracil; however, deamination can also occur on other bases as well. Both depurination and deamination take place on double-helical DNA, and neither break the phosphodiester backbone.

(A) DEPURINATION

(B) DEAMINATION

QUESTION 6–4

Discuss the following statement: "The DNA repair enzymes that fix deamination and depurination damage must preferentially recognize such damage on newly synthesized DNA strands."

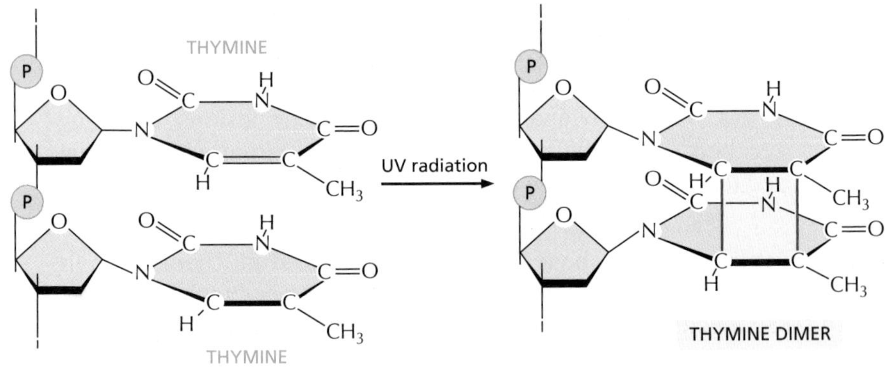

DNA Damage Occurs Continually in Cells

Just like any other molecule in the cell, DNA is continually undergoing thermal collisions with other molecules, often resulting in major chemical changes in the DNA. For example, during the time it takes to read this sentence, a total of about a trillion (10^{12}) purine bases (A and G) will be lost from DNA in the cells of your body by a spontaneous reaction called *depurination* (**Figure 6–23A**). Depurination does not break the DNA phosphodiester backbone but instead removes a purine base from a nucleotide, giving rise to lesions that resemble missing teeth (see Figure 6–25B). Another common reaction is the spontaneous loss of an amino group (*deamination*) from a cytosine in DNA to produce the base uracil (**Figure 6–23B**). Some chemically reactive by-products of cell metabolism also occasionally react with the bases in DNA, altering them in such a way that their base-pairing properties are changed.

The ultraviolet radiation in sunlight is also damaging to DNA; it promotes covalent linkage between two adjacent pyrimidine bases, forming, for example, the *thymine dimer* shown in **Figure 6–24**. It is the failure to repair thymine dimers that spells trouble for individuals with the disease xeroderma pigmentosum.

Figure 6–24 The ultraviolet radiation in sunlight can cause the formation of thymine dimers. Two adjacent thymine bases have become covalently attached to each other to form a thymine dimer. Skin cells that are exposed to sunlight are especially susceptible to this type of DNA damage.

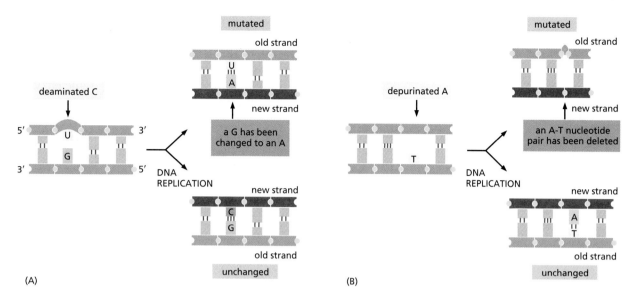

(A)   (B)

These are only a few of many chemical changes that can occur in our DNA. If left unrepaired, many of them would lead either to the substitution of one nucleotide pair for another as a result of incorrect base-pairing during replication (**Figure 6–25A**) or to deletion of one or more nucleotide pairs in the daughter DNA strand after DNA replication (**Figure 6–25B**). Some types of DNA damage (thymine dimers, for example) can stall the DNA replication machinery at the site of the damage.

In addition to this chemical damage, DNA can also be altered by replication itself. The replication machinery that copies the DNA can—quite rarely—incorporate an incorrect nucleotide that it fails to correct via proofreading (see Figure 6–14).

For each of these forms of DNA, cells possess a mechanism for repair, as we discuss next.

Cells Possess a Variety of Mechanisms for Repairing DNA

The thousands of random chemical changes that occur every day in the DNA of a human cell—through thermal collisions or exposure to reactive metabolic by-products, DNA-damaging chemicals, or radiation—are repaired by a variety of mechanisms, each catalyzed by a different set of enzymes. Nearly all these repair mechanisms depend on the double-helical structure of DNA, which provides two copies of the genetic information—one in each strand of the double helix. Thus, if the sequence in one strand is accidentally damaged, information is not lost irretrievably, because a backup version of the altered strand remains in the complementary sequence of nucleotides in the other strand. Most DNA damage creates structures that are never encountered in an undamaged DNA strand; thus the good strand is easily distinguished from the bad.

The basic pathway for repairing damage to DNA, illustrated schematically in **Figure 6–26**, involves three basic steps:

1. The damaged DNA is recognized and removed by one of a variety of mechanisms. These involve nucleases, which cleave the covalent bonds that join the damaged nucleotides to the rest of the DNA strand, leaving a small gap on one strand of the DNA double helix in the region.

2. A *repair DNA polymerase* binds to the 3′-hydroxyl end of the cut DNA strand. It then fills in the gap by making a complementary copy of the information stored in the undamaged strand. Although

Figure 6–25 Chemical modifications of nucleotides, if left unrepaired, produce mutations. (A) Deamination of cytosine, if uncorrected, results in the substitution of one base for another when the DNA is replicated. As shown in Figure 6–23B, deamination of cytosine produces uracil. Uracil differs from cytosine in its base-pairing properties and preferentially base-pairs with adenine. The DNA replication machinery therefore inserts an adenine when it encounters a uracil on the template strand. (B) Depurination, if uncorrected, can lead to the loss of a nucleotide pair. When the replication machinery encounters a missing purine on the template strand, it can skip to the next complete nucleotide, as shown, thus producing a daughter DNA molecule that is missing one nucleotide pair. In other cases (not shown), the replication machinery places an incorrect nucleotide across from the missing base, again resulting in a mutation.

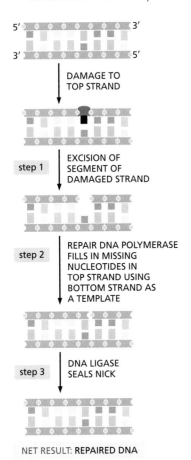

5′ ——————— 3′

3′ ——————— 5′

↓ DAMAGE TO
    TOP STRAND

step 1   EXCISION OF
             SEGMENT OF
             DAMAGED STRAND

step 2   REPAIR DNA POLYMERASE
             FILLS IN MISSING
             NUCLEOTIDES IN
             TOP STRAND USING
             BOTTOM STRAND AS
             A TEMPLATE

step 3   DNA LIGASE
             SEALS NICK

NET RESULT: **REPAIRED DNA**

Figure 6–26 The basic mechanism of DNA repair involves three steps. In step 1 (excision), the damage is cut out by one of a series of nucleases, each specialized for a type of DNA damage. In step 2 (resynthesis), the original DNA sequence is restored by a repair DNA polymerase, which fills in the gap created by the excision events. In step 3 (ligation), DNA ligase seals the nick left in the sugar–phosphate backbone of the repaired strand. Nick sealing, which requires energy from ATP hydrolysis, remakes the broken phosphodiester bond between the adjacent nucleotides (see Figure 6–18).

different from the DNA polymerase that replicates DNA, repair DNA polymerases synthesize DNA strands in the same way. For example, they elongate chains in the 5′-to-3′ direction and have the same type of proofreading activity to ensure that the template strand is copied accurately. In many cells, this is the same enzyme that fills in the gap left after the RNA primers are removed during the normal DNA replication process (see Figure 6–17).

3. When the repair DNA polymerase has filled in the gap, a break remains in the sugar–phosphate backbone of the repaired strand. This nick in the helix is sealed by DNA ligase, the same enzyme that joins the Okazaki fragments during replication of the lagging DNA strand.

Steps 2 and 3 are nearly the same for most types of DNA damage, including the rare errors that arise during DNA replication. However, step 1 uses a series of different enzymes, each specialized for removing different types of DNA damage. Humans produce hundreds of different proteins that function in DNA repair.

A DNA Mismatch Repair System Removes Replication Errors That Escape Proofreading

Although the high fidelity and proofreading abilities of the cell's replication machinery generally prevent replication errors from occurring, rare mistakes do happen. Fortunately, the cell has a backup system—called **mismatch repair**—which is dedicated to correcting these errors. The replication machine makes approximately one mistake per 10^7 nucleotides copied; DNA mismatch repair corrects 99% of these replication errors, increasing the overall accuracy to one mistake in 10^9 nucleotides copied. This level of accuracy is much, much higher than that generally encountered in our day-to-day lives (**Table 6–1**).

Whenever the replication machinery makes a copying mistake, it leaves behind a mispaired nucleotide (commonly called a *mismatch*). If left uncorrected, the mismatch will result in a permanent mutation in the next round of DNA replication (**Figure 6–27**). A complex of mismatch repair proteins recognizes such a DNA mismatch, removes a portion of the DNA strand containing the error, and then resynthesizes the missing DNA. This repair mechanism restores the correct sequence (**Figure 6–28**).

To be effective, the mismatch repair system must be able to recognize which of the DNA strands contains the error. Removing a segment from the strand of DNA that contains the correct sequence would only

TABLE 6–1 ERROR RATES	
US Postal Service on-time delivery of local first-class mail	13 late deliveries per 100 parcels
Airline luggage system	1 lost bag per 150
A professional typist typing at 120 words per minute	1 mistake per 250 characters
Driving a car in the United States	1 death per 10^4 people per year
DNA replication (without proofreading)	1 mistake per 10^5 nucleotides copied
DNA replication (with proofreading; without mismatch repair)	1 mistake per 10^7 nucleotides copied
DNA replication (with mismatch repair)	1 mistake per 10^9 nucleotides copied

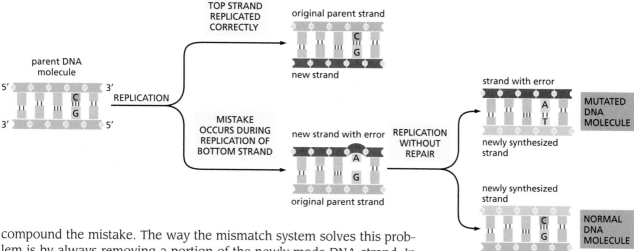

Figure 6–27 Errors made during DNA replication must be corrected to avoid mutations. If uncorrected, a mismatch will lead to a permanent mutation in one of the two DNA molecules produced by the next round of DNA replication.

compound the mistake. The way the mismatch system solves this problem is by always removing a portion of the newly made DNA strand. In bacteria, newly synthesized DNA lacks a type of chemical modification that is present on the preexisting parent DNA. Other cells use other strategies for distinguishing their parent DNA from a newly replicated strand.

Mismatch repair plays an important role in preventing cancer. An inherited predisposition to certain cancers (especially some types of colon cancer) is caused by mutations in genes that encode mismatch repair proteins. Humans inherit two copies of these genes (one from each parent), and individuals who inherit one damaged mismatch repair gene are unaffected until the undamaged copy of the same gene is randomly mutated in a somatic cell. This mutant cell—and all of its progeny—are then deficient in mismatch repair; they therefore accumulate mutations more rapidly than do normal cells. Because cancers arise from cells that have accumulated multiple mutations, a cell deficient in mismatch repair has a greatly enhanced chance of becoming cancerous. Thus, inheriting a damaged mismatch repair gene strongly predisposes an individual to cancer.

Double-Strand DNA Breaks Require a Different Strategy for Repair

The repair mechanisms we have discussed thus far rely on the genetic redundancy built into every DNA double helix. If nucleotides on one strand are damaged, they can be repaired using the information present in the complementary strand.

But what happens when both strands of the double helix are damaged at the same time? Radiation, mishaps at the replication fork, and various chemical assaults can all fracture the backbone of DNA, creating a

Figure 6–28 Mismatch repair eliminates replication errors and restores the original DNA sequence. When mistakes occur during DNA replication, the repair machinery must replace the incorrect nucleotide on the newly synthesized strand, using the original parent strand as its template. This mechanism eliminates the mutation.

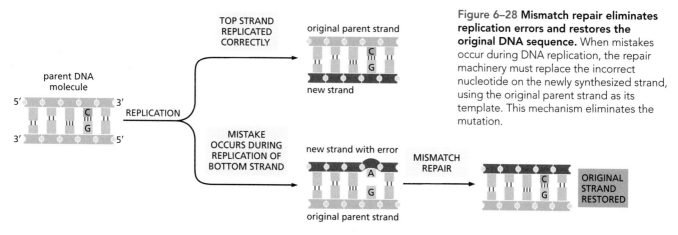

Figure 6–29 Cells can repair double-strand breaks in one of two ways. (A) In nonhomologous end joining, the break is first "cleaned" by a nuclease that chews back the broken ends to produce flush ends. The flush ends are then stitched together by a DNA ligase. Some nucleotides are lost in the repair process, as indicated by the black lines in the repaired DNA. (B) If a double-strand break occurs in one of two daughter DNA double helices after DNA replication has occurred, but before the daughter chromosomes have been separated, the undamaged double helix can be readily used as a template to repair the damaged double helix by homologous recombination. This is a more involved process than non-homologous end joining, but it accurately restores the original DNA sequence at the site of the break. The detailed mechanism is presented in Figure 6–30.

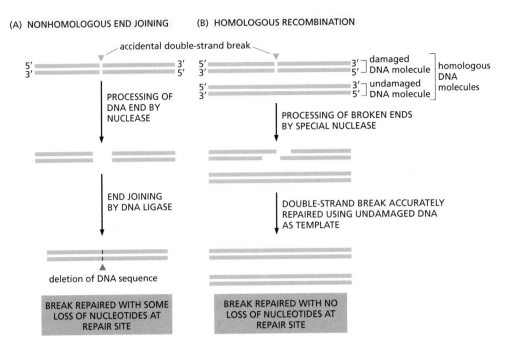

double-strand break. Such lesions are particularly dangerous, because they can lead to the fragmentation of chromosomes and the subsequent loss of genes.

This type of damage is especially difficult to repair. Each chromosome contains unique information; if a chromosome undergoes a double-strand break, and the broken pieces become separated, the cell has no spare copy it can use to reconstruct the information that is now missing.

To handle this potentially disastrous type of DNA damage, cells have evolved two basic strategies. The first involves rapidly sticking the broken ends back together, before the DNA fragments drift apart and get lost. This repair mechanism, called **nonhomologous end joining**, occurs in many cell types and is carried out by a specialized group of enzymes that "clean" the broken ends and rejoin them by DNA ligation. This "quick and dirty" mechanism rapidly repairs the damage, but it comes with a price: in "cleaning" the break to make it ready for ligation, nucleotides are often lost at the site of repair (**Figure 6–29A**).

In most cases, this emergency repair mechanism mends the damage without creating any additional problems. But if the imperfect repair disrupts the activity of a gene, the cell could suffer serious consequences. Thus, nonhomologous end joining can be a risky strategy for fixing broken chromosomes. So cells have an alternative, error-free strategy for repairing double-strand breaks, called homologous recombination (**Figure 6–29B**), as we discuss next.

Homologous Recombination Can Flawlessly Repair DNA Double-Strand Breaks

The problem with repairing a double-strand break, as we mentioned, is finding an intact template to guide the repair. However, if a double-strand break occurs in one double helix shortly after a stretch of DNA has been replicated, the undamaged double helix can readily serve as a template to guide the repair of the broken DNA: information on the undamaged strand of the intact double helix is used to repair the complementary broken strand in the other. Because the two DNA molecules are homologous—they have identical nucleotide sequences outside the broken region—this mechanism is known as **homologous recombination**. It results in a flawless repair of the double-strand break, with no loss of genetic information (see Figure 6–29B).

Homologous recombination most often occurs shortly after a cell's DNA has been replicated before cell division, when the duplicated helices are still physically close to each other (**Figure 6–30A**). To initiate the repair, a nuclease chews back the 5′ ends of the two broken strands at the break (**Figure 6–30B**). Then, with the help of specialized enzymes, one of the broken 3′ ends "invades" the unbroken homologous DNA duplex and searches for a complementary sequence through base-pairing (**Figure 6–30C**). Once an extensive, accurate match is found, the invading strand is elongated by a repair DNA polymerase, using the complementary strand as a template (**Figure 6–30D**). After the repair polymerase has passed the point where the break occurred, the newly repaired strand rejoins its original partner, forming base pairs that hold the two strands of the broken double helix together (**Figure 6–30E**). Repair is then completed by additional DNA synthesis at the 3′ ends of both strands of the broken double helix (**Figure 6–30F**), followed by DNA ligation (**Figure 6–30G**).

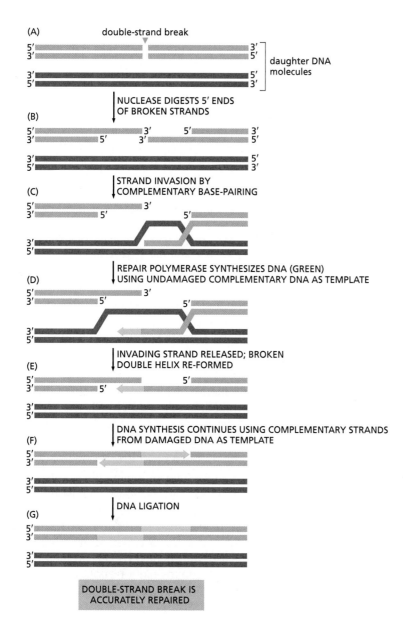

Figure 6–30 **Homologous recombination allows the flawless repair of DNA double-strand breaks.** This is the preferred method for repairing double-strand breaks that arise shortly after the DNA has been replicated but before the cell has divided. See text for details. (Adapted from M. McVey et al., *Proc. Natl. Acad. Sci. USA* 101:15694–15699, 2004. With permission from the National Academy of Sciences.)

The net result is two intact DNA helices, where the genetic information from one was used as a template to repair the other.

Homologous recombination can also be used to repair many other types of DNA damage, making it perhaps the most handy DNA repair mechanism available to the cell: all that is needed is an intact homologous chromosome to use as a partner—a situation that occurs transiently each time a chromosome is duplicated. The "all-purpose" nature of homologous recombinational repair probably explains why this mechanism, and the proteins that carry it out, have been conserved in virtually all cells on Earth.

Homologous recombination is versatile, and has a crucial role in the exchange of genetic information during the formation of the germ cells—sperm and eggs. This specialized process, called *meiosis*, enhances the generation of genetic diversity within a species during sexual reproduction. We will discuss it when we talk about sex in Chapter 19.

Failure to Repair DNA Damage Can Have Severe Consequences for a Cell or Organism

On occasion, the cell's DNA replication and repair processes fail and give rise to a mutation. This permanent change in the DNA sequence can have profound consequences. A mutation that affects just a single nucleotide pair can severely compromise an organism's fitness if the change occurs in a vital position in the DNA sequence. Because the structure and activity of each protein depend on its amino acid sequence, a protein with an altered sequence may function poorly or not at all. For example, humans use the protein hemoglobin to transport oxygen in the blood (see Figure 4–24). A permanent change in a single nucleotide in a hemoglobin gene can cause cells to make hemoglobin with an incorrect sequence of amino acids. One such mutation causes the disease *sickle-cell anemia*. The sickle-cell hemoglobin is less soluble than normal hemoglobin and forms fibrous intracellular precipitates, which produce the characteristic sickle shape of affected red blood cells (**Figure 6–31**). Because these cells are more fragile and frequently tear as they travel through the bloodstream, patients with this potentially life-threatening disease have fewer red blood cells than usual—that is, they are anemic. This anemia can cause weakness, dizziness, headaches, and breathlessness. Moreover, the abnormal red blood cells can aggregate and block small vessels, causing pain and organ failure. We know about sickle-cell hemoglobin because individuals with the mutation survive; the mutation even provides a benefit—an increased resistance to malaria. Over the course of evolution, many other mutations in the hemoglobin gene have arisen, but only those that do not completely destroy the protein remain in the population.

The example of sickle-cell anemia, which is an inherited disease, illustrates the importance of protecting reproductive cells (*germ cells*) against mutation. A mutation in a germ cell will be passed on to all the cells in the body of the multicellular organism that develops from it, including the germ cells responsible for the production of the next generation.

The many other cells in a multicellular organism (its *somatic cells*) must also be protected against mutation—in this case, against mutations that arise during the life of an individual. Nucleotide changes that occur in somatic cells can give rise to variant cells, some of which grow and divide in an uncontrolled fashion at the expense of the other cells in the organism. In the extreme case, an unchecked cell proliferation known as **cancer** results. Cancers are responsible for about 30% of the deaths that occur in Europe and North America, and they are caused largely by a gradual accumulation of random mutations in a somatic cell and its

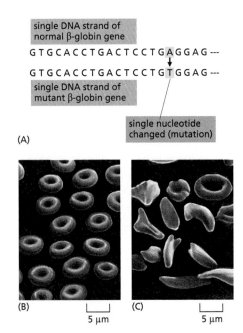

single DNA strand of normal β-globin gene

G T G C A C C T G A C T C C T G A G G A G ---

G T G C A C C T G A C T C C T G T G G A G ---

single DNA strand of mutant β-globin gene

single nucleotide changed (mutation)

(A)

(B)   (C)

5 μm   5 μm

Figure 6–31 A single nucleotide change causes the disease sickle-cell anemia. (A) β-globin is one of the two types of protein subunits that form hemoglobin (see Figure 4–24). A single nucleotide change (mutation) in the β-globin gene produces a β-globin subunit that differs from normal β-globin only by a change from glutamic acid to valine at the sixth amino acid position. (Only a small portion of the gene is shown here; the β-globin subunit contains a total of 146 amino acids.) Humans carry two copies of each gene (one inherited from each parent); a sickle-cell mutation in one of the two β-globin genes generally causes no harm to the individual, as it is compensated for by the normal gene. However, an individual who inherits two copies of the mutant β-globin gene will have sickle-cell anemia. Normal red blood cells are shown in (B), and those from an individual suffering from sickle-cell anemia in (C). Although sickle-cell anemia can be a life-threatening disease, the mutation responsible can also be beneficial. People with the disease, or those who carry one normal gene and one sickle-cell gene, are more resistant to malaria than unaffected individuals, because the parasite that causes malaria grows poorly in red blood cells that contain the sickle-cell form of hemoglobin.

Figure 6–32 Cancer incidence increases dramatically with age. The number of newly diagnosed cases of cancer of the colon in women in England and Wales in one year is plotted as a function of age at diagnosis. Colon cancer, like most human cancers, is caused by the accumulation of multiple mutations. Because cells are continually experiencing accidental changes to their DNA—which accumulate and are passed on to progeny cells when the mutated cells divide—the chance that a cell will become cancerous increases greatly with age. (Data from C. Muir et al., Cancer Incidence in Five Continents, Vol. V. Lyon: International Agency for Research on Cancer, 1987.)

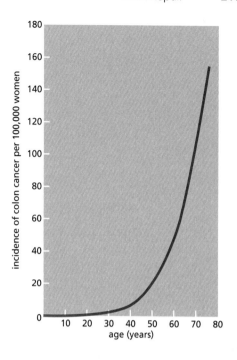

Figure 6–32 Cancer incidence increases dramatically with age. The number of newly diagnosed cases of cancer of the colon in women in England and Wales in one year is plotted as a function of age at diagnosis. Colon cancer, like most human cancers, is caused by the accumulation of multiple mutations. Because cells are continually experiencing accidental changes to their DNA—which accumulate and are passed on to progeny cells when the mutated cells divide—the chance that a cell will become cancerous increases greatly with age. (Data from C. Muir et al., Cancer Incidence in Five Continents, Vol. V. Lyon: International Agency for Research on Cancer, 1987.)

descendants (**Figure 6–32**). Increasing the mutation frequency even two- or threefold could cause a disastrous increase in the incidence of cancer by accelerating the rate at which such somatic cell variants arise.

Thus, the high fidelity with which DNA sequences are replicated and maintained is important both for reproductive cells, which transmit the genes to the next generation, and for somatic cells, which normally function as carefully regulated members of the complex community of cells in a multicellular organism. We should therefore not be surprised to find that all cells possess a very sophisticated set of mechanisms to reduce the number of mutations that occur in their DNA, devoting hundreds of genes to these repair processes.

A Record of the Fidelity of DNA Replication and Repair Is Preserved in Genome Sequences

Although the majority of mutations do neither harm nor good to an organism, those that have harmful consequences are usually eliminated from the population through natural selection; individuals carrying the altered DNA may die or experience decreased fertility, in which case these changes will be lost. By contrast, favorable changes will tend to persist and spread.

But even where no selection operates—at the many sites in the DNA where a change of nucleotide has no effect on the fitness of the organism—the genetic message has been faithfully preserved over tens of millions of years. Thus humans and chimpanzees, after about 5 million years of divergent evolution, still have DNA sequences that are at least 98% identical. Even humans and whales, after 10 or 20 times this amount of time, have chromosomes that are unmistakably similar in their DNA sequence, and many proteins have amino acid sequences that are almost identical (**Figure 6–33**). Thus our genome—and those of our relatives—contains a message from the distant past. Thanks to the faithfulness of DNA replication and repair, 100 million years of evolution have scarcely changed its essential content.

Figure 6–33 The sex-determination genes from humans and whales are unmistakably similar. Although their body plans are strikingly different, humans and whales are built from the same proteins. Despite the many millions of years that have passed since humans and whales diverged, the nucleotide sequences of many of their genes are closely similar. The DNA sequences of a part of the gene that determines maleness in humans and in whales are shown, one above the other; the positions where the two are identical are shaded in *green*.

ESSENTIAL CONCEPTS

- Before a cell divides, it must accurately replicate the vast quantity of genetic information carried in its DNA.

- Because the two strands of a DNA double helix are complementary, each strand can act as a template for the synthesis of the other. Thus DNA replication produces two identical, double-helical DNA molecules, enabling genetic information to be copied and passed on from a cell to its daughter cells and from a parent to its offspring.

- During replication, the two strands of a DNA double helix are pulled apart at a replication origin to form two Y-shaped replication forks. DNA polymerases at each fork produce a new complementary DNA strand on each parental strand.

- DNA polymerase replicates a DNA template with remarkable fidelity, making only about one error in every 10^7 nucleotides copied. This accuracy is made possible, in part, by a proofreading process in which the enzyme corrects its own mistakes as it moves along the DNA.

- Because DNA polymerase synthesizes new DNA in only one direction, only the leading strand at the replication fork can be synthesized in a continuous fashion. On the lagging strand, DNA is synthesized in a discontinuous backstitching process, producing short fragments of DNA that are later joined together by DNA ligase.

- DNA polymerase is incapable of starting a new DNA chain from scratch. Instead, DNA synthesis is primed by an RNA polymerase called primase, which makes short lengths of RNA primers that are then elongated by DNA polymerase. These primers are subsequently erased and replaced with DNA.

- DNA replication requires the cooperation of many proteins that form a multienzyme replication machine that copies both DNA strands as it moves along the double helix.

- In eukaryotes, a special enzyme called telomerase replicates the DNA at the ends of the chromosomes.

- The rare copying mistakes that escape proofreading are dealt with by mismatch repair proteins, which increase the accuracy of DNA replication to one mistake per 10^9 nucleotides copied.

- Damage to one of the two DNA strands, caused by unavoidable chemical reactions, is repaired by a variety of DNA repair enzymes that recognize damaged DNA and excise a short stretch of the damaged strand. The missing DNA is then resynthesized by a repair DNA polymerase, using the undamaged strand as a template.

- If both DNA strands are broken, the double-strand break can be rapidly repaired by nonhomologous end joining. Nucleotides are lost in the process, altering the DNA sequence at the repair site.

- Homologous recombination can flawlessly repair double-strand breaks using an undamaged homologous double helix as a template.

- Highly accurate DNA replication and DNA repair processes play a key role in protecting us from the uncontrolled growth of somatic cells known as cancer.

KEY TERMS

cancer	nonhomologous end joining
DNA ligase	Okazaki fragment
DNA polymerase	primase
DNA repair	proofreading
DNA replication	replication fork
homologous recombination	replication origin
lagging strand	RNA (ribonucleic acid)
leading strand	telomerase
mismatch repair	telomere
mutation	template

QUESTIONS

QUESTION 6–5

DNA mismatch repair enzymes preferentially repair bases on the newly synthesized DNA strand, using the old DNA strand as a template. If mismatches were simply repaired without regard for which strand served as template, would this reduce replication errors? Explain your answer.

QUESTION 6–6

Suppose a mutation affects an enzyme that is required to repair the damage to DNA caused by the loss of purine bases. The loss of a purine occurs about 5000 times in the DNA of each of your cells per day. As the average difference in DNA sequence between humans and chimpanzees is about 1%, how long will it take you to turn into an ape? What is wrong with this argument?

QUESTION 6–7

Which of the following statements are correct? Explain your answers.

A. A bacterial replication fork is asymmetrical because it contains two DNA polymerase molecules that are structurally distinct.

B. Okazaki fragments are removed by a nuclease that degrades RNA.

C. The error rate of DNA replication is reduced both by proofreading by DNA polymerase and by DNA mismatch repair.

D. In the absence of DNA repair, genes are unstable.

E. None of the aberrant bases formed by deamination occur naturally in DNA.

F. Cancer can result from the accumulation of mutations in somatic cells.

QUESTION 6–8

The speed of DNA replication at a replication fork is about 100 nucleotides per second in human cells. What is the minimum number of origins of replication that a human cell must have if it is to replicate its DNA once every 24 hours? Recall that a human cell contains two copies of the human genome, one inherited from the mother, the other from the father, each consisting of 3×10^9 nucleotide pairs.

QUESTION 6–9

Look carefully at Figure 6–11 and at the structures of the compounds shown in **Figure Q6–9**.

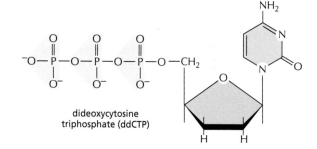

deoxycytosine
triphosphate (dCTP)

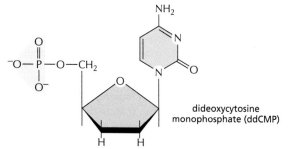

dideoxycytosine
triphosphate (ddCTP)

dideoxycytosine
monophosphate (ddCMP)

Figure Q6–9

A. What would you expect if ddCTP were added to a DNA replication reaction in large excess over the concentration of the available deoxycytosine triphosphate (dCTP), the normal deoxycytosine triphosphate?

B. What would happen if it were added at 10% of the concentration of the available dCTP?

C. What effects would you expect if ddCMP were added under the same conditions?

QUESTION 6–10

Figure Q6–10 shows a snapshot of a replication fork in which the RNA primer has just been added to the lagging strand. Using this diagram as a guide, sketch the path of the DNA as the next Okazaki fragment is synthesized. Indicate the sliding clamp and the single-strand DNA-binding protein as appropriate.

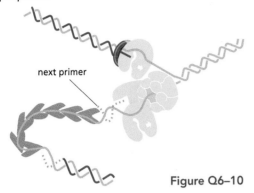

next primer

Figure Q6–10

QUESTION 6–11

Approximately how many high-energy bonds does DNA polymerase use to replicate a bacterial chromosome (ignoring helicase and other enzymes associated with the replication fork)? Compared with its own dry weight of 10^{-12} g, how much glucose does a single bacterium need to provide enough energy to copy its DNA once? The number of nucleotide pairs in the bacterial chromosome is 3×10^6. Oxidation of one glucose molecule yields about 30 high-energy phosphate bonds. The molecular weight of glucose is 180 g/mole. (Recall from Figure 2–3 that a mole consists of 6×10^{23} molecules.)

QUESTION 6–12

What, if anything, is wrong with the following statement: "DNA stability in both reproductive cells and somatic cells is essential for the survival of a species." Explain your answer.

QUESTION 6–13

A common type of chemical damage to DNA is produced by a spontaneous reaction termed *deamination*, in which a nucleotide base loses an amino group (NH_2). The amino group is replaced by a keto group (C=O), by the general reaction shown in **Figure Q6–13**. Write the structures of the bases A, G, C, T, and U and predict the products that will be produced by deamination. By looking at the products of this reaction—and remembering that, in the cell, these will need to be recognized and repaired—can you propose an explanation for why DNA cannot contain uracil?

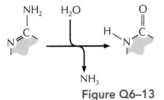

Figure Q6–13

QUESTION 6–14

A. Explain why telomeres and telomerase are needed for replication of eukaryotic chromosomes but not for replication of a circular bacterial chromosome. Draw a diagram to illustrate your explanation.

B. Would you still need telomeres and telomerase to complete eukaryotic chromosome replication if primase always laid down the RNA primer at the very 3′ end of the template for the lagging strand?

QUESTION 6–15

Describe the consequences that would arise if a eukaryotic chromosome

A. Contained only one origin of replication:
   (i) at the exact center of the chromosome
   (ii) at one end of the chromosome

B. Lacked one or both telomeres

C. Had no centromere

Assume that the chromosome is 150 million nucleotide pairs in length, a typical size for an animal chromosome, and that DNA replication in animal cells proceeds at about 100 nucleotides per second.

QUESTION 6–16

Because DNA polymerase proceeds only in the 5′-to-3′ direction, the enzyme is able to correct its own polymerization errors as it moves along the DNA (**Figure Q6–16**). A hypothetical DNA polymerase that synthesized in the 3′-to-5′ direction would be unable to proofread. Given what you know about nucleic acid chemistry and DNA synthesis, draw a sketch similar to Figure Q6–16 that shows what would happen if a DNA polymerase operating in the 3′-to-5′ direction were to remove an incorrect nucleotide from a growing DNA strand. Why would the edited strand be unable to be elongated?

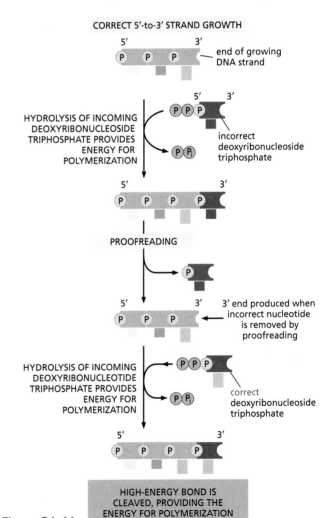

Figure Q6–16

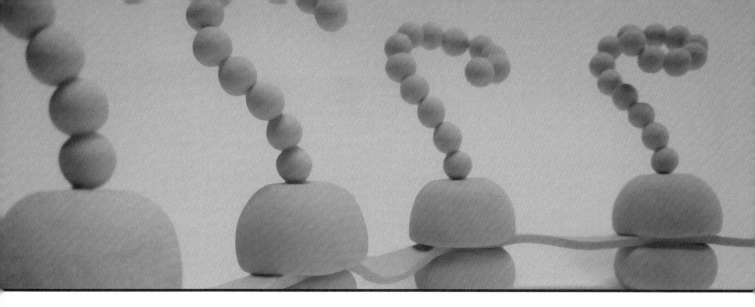

CHAPTER **SEVEN**

7

From DNA to Protein: How Cells Read the Genome

Once the double-helical structure of DNA (deoxyribonucleic acid) had been determined in the early 1950s, it became clear that the hereditary information in cells is encoded in the linear order—or *sequence*—of the four different nucleotide subunits that make up the DNA. We saw in Chapter 6 how this information can be passed on unchanged from a cell to its descendants through the process of DNA replication. But how does the cell decode and use the information? How do genetic instructions written in an alphabet of just four "letters" direct the formation of a bacterium, a fruit fly, or a human? We still have a lot to learn about how the information stored in an organism's genes produces even the simplest unicellular bacterium, let alone how it directs the development of complex multicellular organisms like ourselves. But the DNA code itself has been deciphered, and we have come a long way in understanding how cells read it.

Even before the DNA code was broken, it was known that the information contained in genes somehow directed the synthesis of proteins. Proteins are the principal constituents of cells and determine not only cell structure but also cell function. In previous chapters, we encountered some of the thousands of different kinds of proteins that cells can make. We saw in Chapter 4 that the properties and function of a protein molecule are determined by the sequence of the 20 different amino acid subunits in its polypeptide chain: each type of protein has its own unique amino acid sequence, which dictates how the chain will fold to form a molecule with a distinctive shape and chemistry. The genetic instructions carried by DNA must therefore specify the amino acid sequences of proteins. We will see in this chapter exactly how this is done.

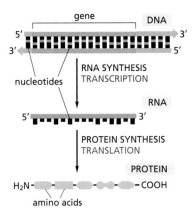

Figure 7–1 Genetic information directs the synthesis of proteins. The flow of genetic information from DNA to RNA (transcription) and from RNA to protein (translation) occurs in all living cells. It was Francis Crick who dubbed this flow of information "the central dogma." The segments of DNA that are transcribed into RNA are called genes.

DNA does not synthesize proteins itself, but it acts like a manager, delegating the various tasks to a team of workers. When a particular protein is needed by the cell, the nucleotide sequence of the appropriate segment of a DNA molecule is first copied into another type of nucleic acid—*RNA* (*ribonucleic acid*). That segment of DNA is called a **gene**, and the resulting RNA copies are then used to direct the synthesis of the protein. Many thousands of these conversions from DNA to protein occur every second in each cell in our body. The flow of genetic information in cells is therefore from DNA to RNA to protein (**Figure 7–1**). All cells, from bacteria to humans, express their genetic information in this way—a principle so fundamental that it has been termed the *central dogma* of molecular biology.

In this chapter, we explain the mechanisms by which cells copy DNA into RNA (a process called *transcription*) and then use the information in RNA to make protein (a process called *translation*). We also discuss a few of the key variations on this basic scheme. Principal among these is *RNA splicing*, a process in eukaryotic cells in which segments of an *RNA transcript* are removed—and the remaining segments stitched back together—before the RNA is translated into protein. In the final section, we consider how the present scheme of information storage, transcription, and translation might have arisen from much simpler systems in the earliest stages of cell evolution.

FROM DNA TO RNA

Transcription and translation are the means by which cells read out, or *express*, the instructions in their *genes*. Many identical RNA copies can be made from the same gene, and each RNA molecule can direct the synthesis of many identical protein molecules. This successive amplification enables cells to rapidly synthesize large amounts of protein whenever necessary. At the same time, each gene can be transcribed, and its RNA translated, at different rates, providing the cell with a way to make vast quantities of some proteins and tiny quantities of others (**Figure 7–2**). Moreover, as we discuss in Chapter 8, a cell can change (or regulate) the expression of each of its genes according to the needs of the moment. In this section, we discuss the production of RNA, the first step in *gene expression*.

QUESTION 7–1

Consider the expression "central dogma," which refers to the flow of genetic information from DNA to RNA to protein. Is the word "dogma" appropriate in this context?

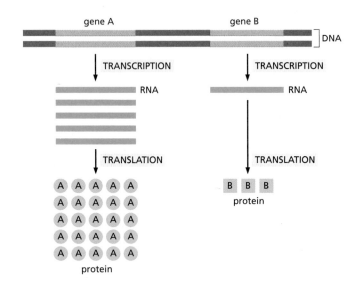

Figure 7–2 A cell can express different genes at different rates. In this and later figures, the untranscribed portions of the DNA are shown in *gray*.

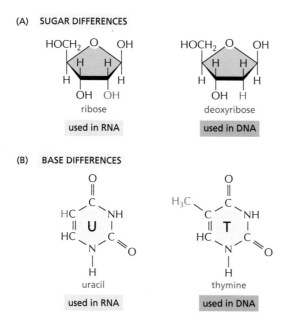

(A) SUGAR DIFFERENCES

ribose
used in RNA

deoxyribose
used in DNA

(B) BASE DIFFERENCES

uracil
used in RNA

thymine
used in DNA

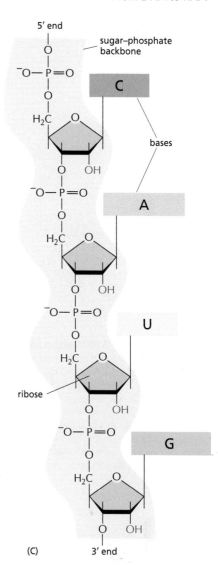

5′ end

sugar–phosphate backbone

bases

ribose

(C)   3′ end

Figure 7–3 The chemical structure of RNA differs slightly from that of DNA. (A) RNA contains the sugar ribose, which differs from deoxyribose, the sugar used in DNA, by the presence of an additional –OH group. (B) RNA contains the base uracil, which differs from thymine, the equivalent base in DNA, by the absence of a –CH₃ group. (C) A short length of RNA. The chemical linkage between nucleotides in RNA—a phosphodiester bond—is the same as that in DNA.

Portions of DNA Sequence Are Transcribed into RNA

The first step a cell takes in expressing one of its many thousands of genes is to copy the nucleotide sequence of that gene into RNA. The process is called **transcription** because the information, though copied into another chemical form, is still written in essentially the same language—the language of nucleotides. Like DNA, **RNA** is a linear polymer made of four different nucleotide subunits, linked together by phosphodiester bonds. It differs from DNA chemically in two respects: (1) the nucleotides in RNA are *ribonucleotides*—that is, they contain the sugar ribose (hence the name *ribo*nucleic acid) rather than deoxyribose; (2) although, like DNA, RNA contains the bases adenine (A), guanine (G), and cytosine (C), it contains uracil (U) instead of the thymine (T) found in DNA (**Figure 7–3**). Because U, like T, can base-pair by hydrogen-bonding with A (**Figure 7–4**), the complementary base-pairing properties described for DNA in Chapter 5 apply also to RNA.

Although their chemical differences are small, DNA and RNA differ quite dramatically in overall structure. Whereas DNA always occurs in cells as a double-stranded helix, RNA is single-stranded. This difference has important functional consequences. Because an RNA chain is single-stranded, it can fold up into a variety of shapes, just as a polypeptide chain folds up to form the final shape of a protein (**Figure 7–5**); double-stranded DNA cannot fold in this fashion. As we discuss later, the ability to fold into a complex three-dimensional shape allows RNA to carry out various functions in cells, in addition to conveying information between DNA and protein. Whereas DNA functions solely as an information store, some RNAs have structural, regulatory, or catalytic roles.

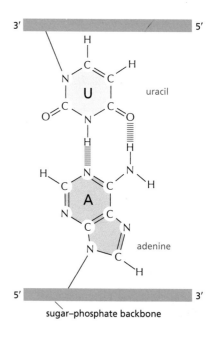

3′   5′

uracil

adenine

5′   3′

sugar–phosphate backbone

Figure 7–4 Uracil forms a base pair with adenine. The hydrogen bonds that hold the base pair together are shown in *red*. Uracil has the same base-pairing properties as thymine. Thus U-A base pairs in RNA closely resemble T-A base pairs in DNA (see Figure 5–6A).

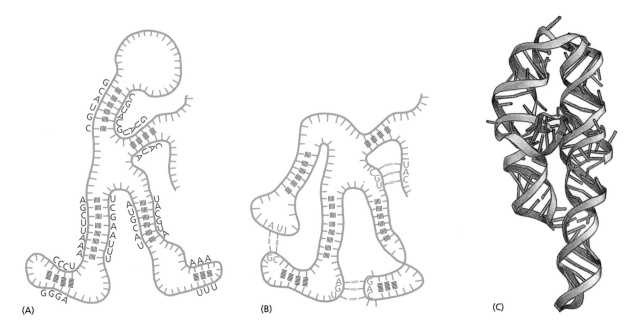

(A) (B) (C)

Figure 7–5 RNA molecules can form intramolecular base pairs and fold into specific structures. RNA is single-stranded, but it often contains short stretches of nucleotides that can base-pair with complementary sequences found elsewhere on the same molecule. These interactions—along with some "nonconventional base-pair interactions (e.g., A-G)—allow an RNA molecule to fold into a three-dimensional structure that is determined by its sequence of nucleotides. (A) A diagram of a hypothetical, folded RNA structure showing only conventional (G-C and A-U) base-pair interactions. (B) Incorporating nonconventional base-pair interactions (*green*) changes the structure of the hypothetical RNA shown in (A). (C) Structure of an actual RNA molecule that is involved in RNA splicing. This RNA contains a considerable amount of double-helical structure. The sugar–phosphate backbone is *blue* and the bases are *red*; the conventional base-pair interactions are indicated by red "rungs" that are continuous, and nonconventional base pairs are indicated by broken red rungs. For an additional view of RNA structure, see Movie 7.1.

Transcription Produces RNA That Is Complementary to One Strand of DNA

All the RNA in a cell is made by transcription, a process that has certain similarities to DNA replication (discussed in Chapter 6). Transcription begins with the opening and unwinding of a small portion of the DNA double helix to expose the bases on each DNA strand. One of the two strands of the DNA double helix then acts as a template for the synthesis of RNA. Ribonucleotides are added, one by one, to the growing RNA chain; as in DNA replication, the nucleotide sequence of the RNA chain is determined by complementary base-pairing with the DNA template. When a good match is made, the incoming ribonucleotide is covalently linked to the growing RNA chain by the enzyme *RNA polymerase*. The RNA chain produced by transcription—the **RNA transcript**—is therefore elongated one nucleotide at a time and has a nucleotide sequence exactly complementary to the strand of DNA used as the template (**Figure 7–6**).

Transcription differs from DNA replication in several crucial respects. Unlike a newly formed DNA strand, the RNA strand does not remain hydrogen-bonded to the DNA template strand. Instead, just behind the region where the ribonucleotides are being added, the RNA chain is displaced and the DNA helix re-forms. For this reason—and because only one strand of the DNA molecule is transcribed—RNA molecules are single-stranded. Further, because RNAs are copied from only a limited region of DNA, RNA molecules are much shorter than DNA molecules; DNA molecules in a human chromosome can be up to 250 million nucleotide pairs long, whereas most mature RNAs are no more than a few thousand nucleotides long, and many are much shorter than that.

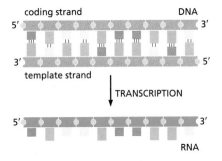

Figure 7–6 Transcription of a gene produces an RNA complementary to one strand of DNA. The transcribed strand of the gene, the *bottom* strand in this example, is called the template strand. The nontemplate strand of the gene (here, shown at the *top*) is sometimes called the *coding strand* because its sequence is equivalent to the RNA product, as shown. Which DNA strand serves as the template varies, depending on the gene, as we discuss later. By convention, an RNA molecule is always depicted with its 5′ end—the first part to be synthesized—to the left.

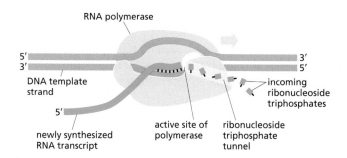

RNA polymerase

5′
3′

DNA template strand

5′

newly synthesized RNA transcript

active site of polymerase

ribonucleoside triphosphate tunnel

3′
5′

incoming ribonucleoside triphosphates

Figure 7–7 DNA is transcribed into RNA by the enzyme RNA polymerase. RNA polymerase (*pale blue*) moves stepwise along the DNA, unwinding the DNA helix in front of it. As it progresses, the polymerase adds ribonucleotides one by one to the RNA chain, using an exposed DNA strand as a template. The resulting RNA transcript is thus single-stranded and complementary to this template strand (see Figure 7–6). As the polymerase moves along the DNA template (in the 3′-to-5′ direction), it displaces the newly formed RNA, allowing the two strands of DNA behind the polymerase to rewind. A short region of hybrid DNA/RNA helix (approximately nine nucleotides in length) therefore forms only transiently, causing a "window" of DNA/RNA helix to move along the DNA with the polymerase (Movie 7.2).

Like the DNA polymerase that carries out DNA replication (discussed in Chapter 6), **RNA polymerases** catalyze the formation of the phosphodiester bonds that link the nucleotides together and form the sugar–phosphate backbone of the RNA chain (see Figure 7–3). The RNA polymerase moves stepwise along the DNA, unwinding the DNA helix just ahead to expose a new region of the template strand for complementary base-pairing. In this way, the growing RNA chain is extended by one nucleotide at a time in the 5′-to-3′ direction (**Figure 7–7**). The incoming ribonucleoside triphosphates (ATP, CTP, UTP, and GTP) provide the energy needed to drive the reaction forward (see Figure 6–11).

The almost immediate release of the RNA strand from the DNA as it is synthesized means that many RNA copies can be made from the same gene in a relatively short time; the synthesis of the next RNA is usually started before the first RNA has been completed (**Figure 7–8**). A medium-sized gene—say, 1500 nucleotide pairs—requires approximately 50 seconds for a molecule of RNA polymerase to transcribe it (Movie 7.2). At any given time, there could be dozens of polymerases speeding along this single stretch of DNA, hard on one another's heels, allowing more than 1000 transcripts to be synthesized in an hour. For most genes, however, the amount of transcription is much less than this.

Although RNA polymerase catalyzes essentially the same chemical reaction as DNA polymerase, there are some important differences between the two enzymes. First, and most obviously, RNA polymerase uses ribonucleoside for phosphates as substrates, so it catalyzes the linkage of ribonucleotides, not deoxyribonucleotides. Second, unlike the DNA polymerase involved in DNA replication, RNA polymerases can start an RNA chain without a primer. This difference likely evolved because transcription need not be as accurate as DNA replication; unlike DNA, RNA is not used as the permanent storage form of genetic information in cells, so mistakes in RNA transcripts have relatively minor consequences for a cell. RNA polymerases make about one mistake for every 10^4 nucleotides copied into RNA, whereas DNA polymerase makes only one mistake for every 10^7 nucleotides copied.

Cells Produce Various Types of RNA

The vast majority of genes carried in a cell's DNA specify the amino acid sequences of proteins. The RNA molecules encoded by these genes—which

QUESTION 7–2

In the electron micrograph in Figure 7–8, are the RNA polymerase molecules moving from right to left or from left to right? Why are the RNA transcripts so much shorter than the DNA segments (genes) that encode them?

Figure 7–8 Transcription can be visualized in the electron microscope. The micrograph shows many molecules of RNA polymerase simultaneously transcribing two adjacent ribosomal genes on a single DNA molecule. Molecules of RNA polymerase are barely visible as a series of tiny dots along the spine of the DNA molecule; each polymerase has an RNA transcript (a short, fine thread) radiating from it. The RNA molecules being transcribed from the two ribosomal genes—ribosomal RNAs (rRNAs)—are not translated into protein, but are instead used directly as components of ribosomes, macromolecular machines made of RNA and protein. The large particles that can be seen at the free, 5′ end of each rRNA transcript are believed to be ribosomal proteins that have assembled on the ends of the growing transcripts. (Courtesy of Ulrich Scheer.)

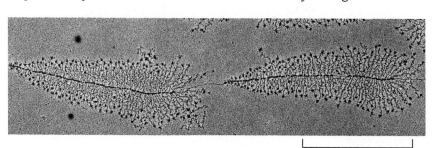

1 µm

ultimately direct the synthesis of proteins—are called **messenger RNAs (mRNAs)**. In eukaryotes, each mRNA typically carries information transcribed from just one gene, which codes for a single protein; in bacteria, a set of adjacent genes is often transcribed as a single mRNA, which therefore carries the information for several different proteins.

The final product of other genes, however, is the RNA itself. As we see later, these nonmessenger RNAs, like proteins, have various roles, serving as regulatory, structural, and catalytic components of cells. They play key parts, for example, in translating the genetic message into protein: *ribosomal RNAs* (*rRNAs*) form the structural and catalytic core of the ribosomes, which translate mRNAs into protein, and *transfer RNAs* (*tRNAs*) act as adaptors that select specific amino acids and hold them in place on a ribosome for their incorporation into protein. Other small RNAs, called *microRNAs* (*miRNAs*), serve as key regulators of eukaryotic gene expression, as we discuss in Chapter 8. The most common types of RNA are summarized in **Table 7–1**.

In the broadest sense, the term **gene expression** refers to the process by which the information encoded in a DNA sequence is translated into a product that has some effect on a cell or organism. In cases where the final product of the gene is a protein, gene expression includes both transcription and translation. When an RNA molecule is the gene's final product, however, gene expression does not require translation.

Signals in DNA Tell RNA Polymerase Where to Start and Finish Transcription

The initiation of transcription is an especially critical process because it is the main point at which the cell selects which proteins or RNAs are to be produced. To begin transcription, RNA polymerase must be able to recognize the start of a gene and bind firmly to the DNA at this site. The way in which RNA polymerases recognize the *transcription start site* of a gene differs somewhat between bacteria and eukaryotes. Because the situation in bacteria is simpler, we describe it first.

When an RNA polymerase collides randomly with a DNA molecule, the enzyme sticks weakly to the double helix and then slides rapidly along its length. RNA polymerase latches on tightly only after it has encountered a gene region called a **promoter**, which contains a specific sequence of nucleotides that lies immediately upstream of the starting point for RNA synthesis. Once bound tightly to this sequence, the RNA polymerase opens up the double helix immediately in front of the promoter to expose the nucleotides on each strand of a short stretch of DNA. One of the two exposed DNA strands then acts as a template for complementary base-pairing with incoming ribonucleoside triphosphates, two of which are

TABLE 7–1 TYPES OF RNA PRODUCED IN CELLS	
Type of RNA	**Function**
messenger RNAs (mRNAs)	code for proteins
ribosomal RNAs (rRNAs)	form the core of the ribosome's structure and catalyze protein synthesis
microRNAs (miRNAs)	regulate gene expression
transfer RNAs (tRNAs)	serve as adaptors between mRNA and amino acids during protein synthesis
other noncoding RNAs	used in RNA splicing, gene regulation, telomere maintenance, and many other processes

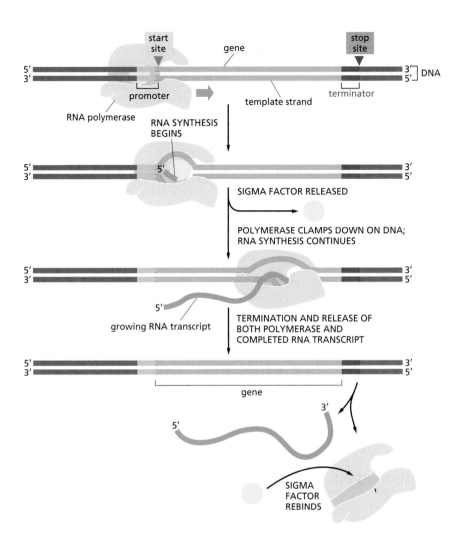

Figure 7–9 Signals in the nucleotide sequence of a gene tell bacterial RNA polymerase where to start and stop transcription. Bacterial RNA polymerase (*light blue*) contains a subunit called sigma factor (*yellow*) that recognizes the promoter of a gene (*green*). Once transcription has begun, sigma factor is released, and the polymerase moves forward and continues synthesizing the RNA. Chain elongation continues until the polymerase encounters a sequence in the gene called the terminator (*red*). There the enzyme halts and releases both the DNA template and the newly made RNA transcript. The polymerase then reassociates with a free sigma factor and searches for another promoter to begin the process again.

joined together by the polymerase to begin synthesis of the RNA chain. Chain elongation then continues until the enzyme encounters a second signal in the DNA, the *terminator* (or stop site), where the polymerase halts and releases both the DNA template and the newly made RNA transcript (**Figure 7–9**). This terminator sequence is contained within the gene and is transcribed into the 3′ end of the newly made RNA.

Because the polymerase must bind tightly before transcription can begin, a segment of DNA will be transcribed only if it is preceded by a promoter. This ensures that those portions of a DNA molecule that contain a gene will be transcribed into RNA. The nucleotide sequences of a typical promoter—and a typical terminator—are presented in **Figure 7–10**.

In bacteria, it is a subunit of RNA polymerase, the *sigma (σ) factor* (see Figure 7–9), that is primarily responsible for recognizing the promoter sequence on the DNA. But how can this factor "see" the promoter, given that the base-pairs in question are situated in the interior of the DNA double helix? It turns out that each base presents unique features to the outside of the double helix, allowing the sigma factor to find the promoter sequence without having to separate the entwined DNA strands.

The next problem an RNA polymerase faces is determining which of the two DNA strands to use as a template for transcription: each strand has a different nucleotide sequence and would produce a different RNA transcript. The secret lies in the structure of the promoter itself. Every promoter has a certain polarity: it contains two different nucleotide sequences upstream of the transcriptional start site that position the RNA polymerase, ensuring that it binds to the promoter in only one orientation

Figure 7–10 Bacterial promoters and terminators have specific nucleotide sequences that are recognized by RNA polymerase.
(A) The *green*-shaded regions represent the nucleotide sequences that specify a promoter. The numbers above the DNA indicate the positions of nucleotides counting from the first nucleotide transcribed, which is designated +1. The polarity of the promoter orients the polymerase and determines which DNA strand is transcribed. All bacterial promoters contain DNA sequences at –10 and –35 that closely resemble those shown here. (B) The *red*-shaded regions represent sequences in the gene that signal the RNA polymerase to terminate transcription. Note that the regions transcribed into RNA contain the terminator but not the promoter nucleotide sequences. By convention, the sequence of a gene is that of the non-template strand, as this strand has the same sequence as the transcribed RNA (with T substituting for U).

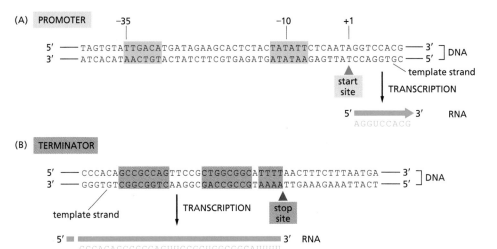

(see Figure 7–10A). Because the polymerase can only synthesize RNA in the 5′-to-3′ direction once the enzyme is bound it must use the DNA strand oriented in the 3′-to-5′ direction as its template.

This selection of a template strand does not mean that on a given chromosome, transcription always proceeds in the same direction. With respect to the chromosome as a whole, the direction of transcription varies from gene to gene. But because each gene typically has only one promoter, the orientation of its promoter determines in which direction that gene is transcribed and therefore which strand is the template strand (**Figure 7–11**).

Initiation of Eukaryotic Gene Transcription Is a Complex Process

Many of the principles we just outlined for bacterial transcription also apply to eukaryotes. However, transcription initiation in eukaryotes differs in several important ways from that in bacteria:

- The first difference lies in the RNA polymerases themselves. While bacteria contain a single type of RNA polymerase, eukaryotic cells have three—*RNA polymerase I*, *RNA polymerase II*, and *RNA polymerase III*. These polymerases are responsible for transcribing different types of genes. RNA polymerases I and III transcribe the genes encoding transfer RNA, ribosomal RNA, and various other RNAs that play structural and catalytic roles in the cell (**Table 7–2**). RNA polymerase II transcribes the vast majority of eukaryotic genes, including all those that encode proteins and miRNAs (**Movie 7.3**). Our subsequent discussion will therefore focus on RNA polymerase II.

- A second difference is that, whereas the bacterial RNA polymerase (along with its sigma subunit) is able to initiate transcription on its own, eukaryotic RNA polymerases require the assistance of a large set of accessory proteins. Principal among these are the *general transcription factors*, which must assemble at each promoter, along with the polymerase, before the polymerase can begin transcription.

Figure 7–11 On an individual chromosome, some genes are transcribed using one DNA strand as a template, and others are transcribed from the other DNA strand. RNA polymerase always moves in the 3′-to-5′ direction and the selection of the template strand is determined by the orientation of the promoter (*green* arrowheads) at the beginning of each gene. Thus the genes transcribed from left to right use the bottom DNA strand as the template (see Figure 7–10); those transcribed from right to left use the top strand as the template.

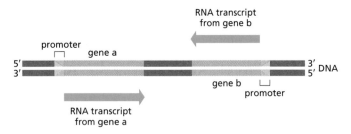

TABLE 7–2 THE THREE RNA POLYMERASES IN EUKARYOTIC CELLS	
Type of Polymerase	Genes Transcribed
RNA polymerase I	most rRNA genes
RNA polymerase II	all protein-coding genes, miRNA genes, plus genes for other noncoding RNAs (e.g., those in spliceosomes)
RNA polymerase III	tRNA genes 5S rRNA gene genes for many other small RNAs

- A third distinctive feature of transcription in eukaryotes is that the mechanisms that control its initiation are much more elaborate than those in prokaryotes—a point we discuss in detail in Chapter 8. In bacteria, genes tend to lie very close to one another in the DNA, with only very short lengths of nontranscribed DNA between them. But in plants and animals, including humans, individual genes are spread out along the DNA, with stretches of up to 100,000 nucleotide pairs between one gene and the next. This architecture allows a single gene to be controlled by a large variety of *regulatory DNA sequences* scattered along the DNA, and it enables eukaryotes to engage in more complex forms of transcriptional regulation than do bacteria.

- Last but not least, eukaryotic transcription initiation must take into account the packing of DNA into *nucleosomes* and more compact forms of chromatin structure, as we describe in Chapter 8.

We now turn to the general transcription factors and discuss how they help eukaryotic RNA polymerase II initiate transcription.

Eukaryotic RNA Polymerase Requires General Transcription Factors

The initial finding that, unlike bacterial RNA polymerase, purified eukaryotic RNA polymerase II could not initiate transcription on its own in a test tube led to the discovery and purification of the **general transcription factors**. These accessory proteins assemble on the promoter, where they position the RNA polymerase and pull apart the DNA double helix to expose the template strand, allowing the polymerase to begin transcription. Thus the general transcription factors have a similar role in eukaryotic transcription as sigma factor has in bacterial transcription.

Figure 7–12 shows how the general transcription factors assemble at a promoter used by RNA polymerase II. The assembly process typically begins with the binding of the general transcription factor TFIID to a short

QUESTION 7–3

Could the RNA polymerase used for transcription be used as the polymerase that makes the RNA primer required for DNA replication (discussed in Chapter 6)?

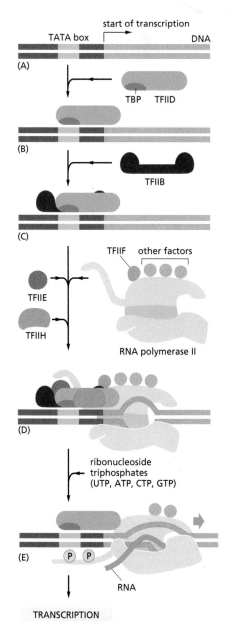

Figure 7–12 To begin transcription, eukaryotic RNA polymerase II requires a set of general transcription factors. These transcription factors are called TFIIB, TFIID, and so on. (A) Many eukaryotic promoters contain a DNA sequence called the TATA box. (B) The TATA box is recognized by a subunit of the general transcription factor TFIID, called the TATA-binding protein (TBP). For simplicity, the DNA distortion produced by the binding of the TBP (see Figure 7–13) is not shown. (C) The binding of TFIID enables the adjacent binding of TFIIB. (D) The rest of the general transcription factors, as well as the RNA polymerase itself, assemble at the promoter. (E) TFIIH then pries apart the double helix at the transcription start point, using the energy of ATP hydrolysis, which exposes the template strand of the gene (not shown). TFIIH also phosphorylates RNA polymerase II, releasing the polymerase from most of the general transcription factors, so it can begin transcription. The site of phosphorylation is a long polypeptide "tail" that extends from the polymerase.

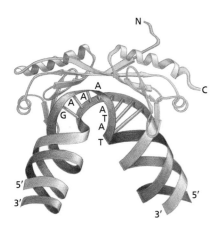

Figure 7–13 TATA-binding protein (TBP) binds to the TATA box (indicated by letters) and bends the DNA double helix. The unique distortion of DNA caused by TBP, which is a subunit of TFIID (see Figure 7–12), helps attract the other general transcription factors. TBP is a single polypeptide chain that is folded into two very similar domains (*blue* and *green*). The protein sits atop the DNA double helix like a saddle on a bucking horse (Movie 7.4). (Adapted from J.L. Kim et al., *Nature* 365:520–527, 1993. With permission from Macmillan Publishers Ltd.)

segment of DNA double helix composed primarily of T and A nucleotides; because of its composition, this part of the promoter is known as the *TATA box*. Upon binding to DNA, TFIID causes a dramatic local distortion in the DNA double helix (**Figure 7–13**), which helps to serve as a landmark for the subsequent assembly of other proteins at the promoter. The TATA box is a key component of many promoters used by RNA polymerase II, and it is typically located 25 nucleotides upstream from the transcription start site. Once TFIID has bound to the TATA box, the other factors assemble, along with RNA polymerase II, to form a complete *transcription initiation complex*. Although Figure 7–12 shows the general transcription factors piling onto the promoter in a certain order, the actual order of assembly probably differs from one promoter to the next.

After RNA polymerase II has been positioned on the promoter, it must be released from the complex of general transcription factors to begin its task of making an RNA molecule. A key step in liberating the RNA polymerase is the addition of phosphate groups to its "tail" (see Figure 7–12E). This liberation is initiated by the general transcription factor TFIIH, which contains a protein kinase as one of its subunits. Once transcription has begun, most of the general transcription factors dissociate from the DNA and then are available to initiate another round of transcription with a new RNA polymerase molecule. When RNA polymerase II finishes transcribing a gene, it too is released from the DNA; the phosphates on its tail are stripped off by protein phosphatases, and the polymerase is then ready to find a new promoter. Only the dephosphorylated form of RNA polymerase II can initiate RNA synthesis.

Eukaryotic mRNAs Are Processed in the Nucleus

Although the templating principle by which DNA is transcribed into RNA is the same in all organisms, the way in which the RNA transcripts are handled before they can be used by the cell to make protein differs greatly between bacteria and eukaryotes. Bacterial DNA lies directly exposed to the cytoplasm, which contains the *ribosomes* on which protein synthesis takes place. As an mRNA molecule in a bacterium starts to be synthesized, ribosomes immediately attach to the free 5′ end of the RNA transcript and begin translating it into protein.

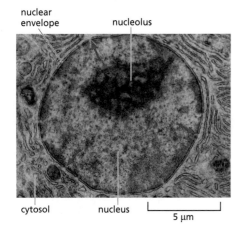

nuclear envelope

nucleolus

cytosol

nucleus

5 μm

Figure 7–14 Before they can be translated, mRNA molecules made in the nucleus must be exported to the cytosol via pores in the nuclear envelope (*red arrows*). Shown here is a section of a liver cell nucleus. The nucleolus is where ribosomal RNAs are synthesized and combined with proteins to form ribosomes, which are then exported to the cytoplasm. (From D.W. Fawcett, A Textbook of Histology, 11th ed. Philadelphia: Saunders, 1986. With permission from Elsevier.)

In eukaryotic cells, by contrast, DNA is enclosed within the *nucleus*. Transcription takes place in the nucleus, but protein synthesis takes place on ribosomes in the cytoplasm. So, before a eukaryotic mRNA can be translated into protein, it must be transported out of the nucleus through small pores in the nuclear envelope (**Figure 7–14**). Before it can be exported to the cytosol, however, a eukaryotic RNA must go through several **RNA processing** steps, which include *capping*, *splicing*, and *polyadenylation*, as we discuss shortly. These steps take place as the RNA is being synthesized. The enzymes responsible for RNA processing ride on the phosphorylated tail of eukaryotic RNA polymerase II as it synthesizes an RNA molecule (see Figure 7–12), and they process the transcript as it emerges from the polymerase (**Figure 7–15**).

Different types of RNA are processed in different ways before leaving the nucleus. Two processing steps, capping and polyadenylation, occur only on RNA transcripts destined to become mRNA molecules (called *precursor mRNAs*, or *pre-mRNAs*).

1. **RNA capping** modifies the 5' end of the RNA transcript, the end that is synthesized first. The RNA is capped by the addition of an atypical nucleotide—a guanine (G) nucleotide bearing a methyl group, which is attached to the 5' end of the RNA in an unusual way (**Figure 7–16**). This capping occurs after RNA polymerase II has produced about 25 nucleotides of RNA, long before it has completed transcribing the whole gene.

2. **Polyadenylation** provides a newly transcribed mRNA with a special structure at its 3' end. In contrast with bacteria, where the 3' end of an mRNA is simply the end of the chain synthesized by the RNA polymerase, the 3' end of a forming eukaryotic mRNA is first trimmed by an enzyme that cuts the RNA chain at a particular sequence of nucleotides. The transcript is then finished off by a second enzyme that adds a series of repeated adenine (A) nucleotides to the cut end. This *poly-A tail* is generally a few hundred nucleotides long (see Figure 7–16A).

These two modifications—capping and polyadenylation—increase the stability of a eukaryotic mRNA molecule, facilitate its export from the nucleus to the cytoplasm, and generally mark the RNA molecule as an mRNA. They are also used by the protein-synthesis machinery to make sure that both ends of the mRNA are present and that the message is therefore complete before protein synthesis begins.

In Eukaryotes, Protein-Coding Genes Are Interrupted by Noncoding Sequences Called Introns

Most eukaryotic pre-mRNAs have to undergo an additional processing step before they are functional mRNAs. This step involves a far more radical modification of the pre-mRNA transcript than capping or polyadenylation, and it is the consequence of a surprising feature of most eukaryotic genes. In bacteria, most proteins are encoded by an uninterrupted stretch of DNA sequence that is transcribed into an mRNA that, without any further processing, can be translated into protein. Most protein-coding eukaryotic genes, in contrast, have their coding sequences interrupted by long, noncoding, *intervening sequences* called **introns**. The scattered pieces of coding sequence—called *expressed sequences* or

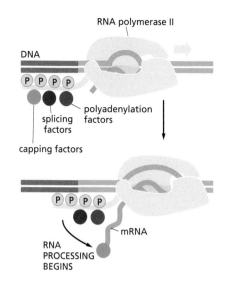

Figure 7–15 Phosphorylation of the tail of RNA polymerase II allows RNA-processing proteins to assemble there. Note that the phosphates shown here are in addition to the ones required for transcription initiation (see Figure 7–12). Capping, polyadenylation, and splicing are all modifications that occur during RNA processing in the nucleus.

Figure 7–16 Eukaryotic pre-mRNA molecules are modified by capping and polyadenylation. (A) A eukaryotic mRNA has a cap at the 5' end and a poly-A tail at the 3' end. Note that not all of the RNA transcript shown codes for protein. (B) The structure of the 5' cap. Many eukaryotic mRNA caps carry an additional modification: the 2'-hydroxyl group on the second ribose sugar in the mRNA is methylated (not shown).

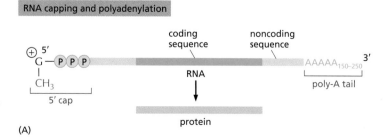

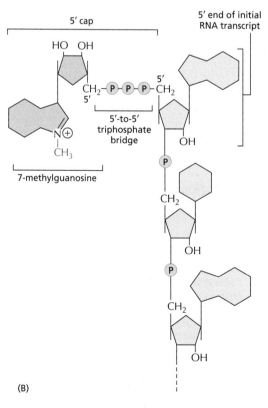

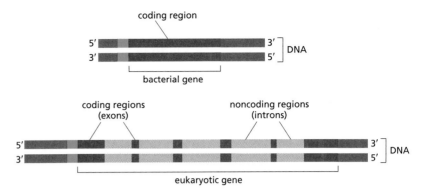

Figure 7–17 Eukaryotic and bacterial genes are organized differently. A bacterial gene consists of a single stretch of uninterrupted nucleotide sequence that encodes the amino acid sequence of a protein (or more than one protein). In contrast, the protein-coding sequences of most eukaryotic genes (*exons*) are interrupted by noncoding sequences (*introns*). Promoters for transcription are indicated in *green*.

exons—are usually shorter than the introns, and they often represent only a small fraction of the total length of the gene (**Figure 7–17**). Introns range in length from a single nucleotide to more than 10,000 nucleotides. Some protein-coding eukaryotic genes lack introns altogether, and some have only a few; but most have many (**Figure 7–18**). Note that the terms "exon" and "intron" apply to both the DNA and the corresponding RNA sequences.

Introns Are Removed From Pre-mRNAs by RNA Splicing

To produce an mRNA in a eukaryotic cell, the entire length of the gene, introns as well as exons, is transcribed into RNA. After capping, and as RNA polymerase II continues to transcribe the gene, the process of **RNA splicing** begins, in which the introns are removed from the newly synthesized RNA and the exons are stitched together. Each transcript ultimately receives a poly-A tail; in some cases, this happens after splicing, whereas in other cases, it occurs before the final splicing reactions have been completed. Once a transcript has been spliced and its 5′ and 3′ ends have been modified, the RNA is now a functional mRNA molecule that can leave the nucleus and be translated into protein.

How does the cell determine which parts of the RNA transcript to remove during splicing? Unlike the coding sequence of an exon, most of the nucleotide sequence of an intron is unimportant. Although there is little overall resemblance between the nucleotide sequences of different introns, each intron contains a few short nucleotide sequences that act as cues for its removal from the pre-mRNA. These special sequences are found at or near each end of the intron and are the same or very similar in all introns (**Figure 7–19**). Guided by these sequences, an elaborate splicing machine cuts out the intron in the form of a "lariat" structure (**Figure 7–20**), formed by the reaction of the "A" nucleotide highlighted in red in Figures 7–19 and 7–20.

Figure 7–18 Most protein-coding human genes are broken into multiple exons and introns. (A) The β-globin gene, which encodes one of the subunits of the oxygen-carrying protein hemoglobin, contains 3 exons. (B) The Factor VIII gene, which encodes a protein (Factor VIII) that functions in the blood-clotting pathway, contains 26 exons. Mutations in this large gene are responsible for the most prevalent form of the blood disorder hemophilia.

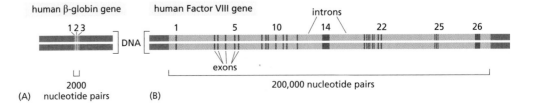

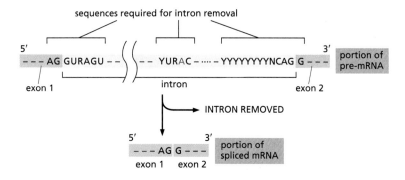

Figure 7–19 Special nucleotide sequences in a pre-mRNA transcript signal the beginning and the end of an intron. Only the nucleotide sequences shown are required to remove an intron; the other positions in an intron can be occupied by any nucleotide. The special sequences are recognized primarily by small nuclear ribonucleoproteins (snRNPs), which direct the cleavage of the RNA at the intron–exon borders and catalyze the covalent linkage of the exon sequences. Here, in addition to the standard symbols for nucleotides (A, C, G, U), R stands for either A or G; Y stands for either C or U; N stands for any nucleotide. The A shown in *red* forms the branch point of the lariat produced in the splicing reaction shown in Figure 7–20. The distances along the RNA between the three splicing sequences are highly variable; however, the distance between the branch point and the 5′ splice junction is typically much longer than that between the 3′ splice junction and the branch point (see Figure 7–20). The splicing sequences shown are from humans; similar sequences direct RNA splicing in other eukaryotes.

We will not describe the splicing machinery in detail, but it is worthwhile to note that, unlike the other steps of mRNA production we have discussed, RNA splicing is carried out largely by RNA molecules rather than proteins. These RNA molecules, called **small nuclear RNAs (snRNAs)**, are packaged with additional proteins to form *small nuclear ribonucleoproteins (snRNPs*, pronounced "snurps"). The snRNPs recognize splice-site sequences through complementary base-pairing between their RNA components and the sequences in the pre-mRNA, and they also participate intimately in the chemistry of splicing (**Figure 7–21**). Together, these snRNPs form the core of the **spliceosome**, the large assembly of RNA and protein molecules that carries out RNA splicing in the nucleus. To watch the spliceosome in action, see Movie 7.5.

The intron–exon type of gene arrangement in eukaryotes may, at first, seem wasteful. It does, however, have a number of important benefits. First, the transcripts of many eukaryotic genes can be spliced in different ways, each of which can produce a distinct protein. Such **alternative splicing** thereby allows many different proteins to be produced from the same gene (**Figure 7–22**). About 95% of human genes are thought to undergo alternative splicing. Thus RNA splicing enables eukaryotes to increase the already enormous coding potential of their genomes.

RNA splicing also provides another advantage to eukaryotes, one that is likely to have been profoundly important in the early evolutionary history of genes. As we discuss in detail in Chapter 9, the intron–exon structure of genes is thought to have sped up the emergence of new and useful proteins: novel proteins appear to have arisen by the mixing and matching of different exons of preexisting genes, much like the assembly of a new type of machine from a kit of preexisting functional components. Indeed, many proteins in present-day cells resemble patchworks composed from a common set of protein pieces, called protein *domains* (see Figure 4–51).

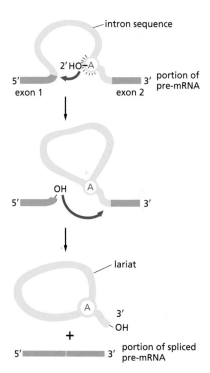

Figure 7–20 An intron in a pre-mRNA molecule forms a branched structure during RNA splicing. In the first step, the branch point adenine (*red* A) in the intron sequence attacks the 5′ splice site and cuts the sugar–phosphate backbone of the RNA at this point (this is the same A highlighted in *red* in Figure 7–19). In this process, the cut 5′ end of the intron becomes covalently linked to the 2′-OH group of the ribose of the A nucleotide to form a branched structure. The free 3′-OH end of the exon sequence then reacts with the start of the next exon sequence, joining the two exons together into a continuous coding sequence and releasing the intron in the form of a lariat structure, which is eventually degraded in the nucleus.

Figure 7–21 **Splicing is carried out by a collection of RNA–protein complexes called snRNPs.** There are five snRNPs, called U1, U2, U4, U5, and U6. As shown here, U1 and U2 bind to the 5′ splice site (U1) and the lariat branch point (U2) through complementary base-pairing. Additional snRNPs are attracted to the splice site, and interactions between their protein components drive the assembly of the complete spliceosome. Rearrangements in the base pairs that hold together the snRNPs and the RNA transcript then reorganize the spliceosome to form the active site that excises the intron, leaving the spliced mRNA behind (see also Figure 7–20).

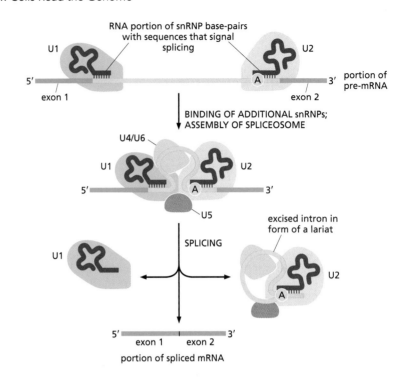

Mature Eukaryotic mRNAs Are Exported from the Nucleus

We have seen how eukaryotic pre-mRNA synthesis and processing take place in an orderly fashion within the cell nucleus. However, these events create a special problem for eukaryotic cells: of the total number of pre-mRNA transcripts that are synthesized, only a small fraction—the mature mRNAs—will be useful to the cell. The remaining RNA fragments—excised introns, broken RNAs, and aberrantly spliced transcripts—are not only useless, but they could be dangerous to the cell if allowed to leave the nucleus. How, then, does the cell distinguish between the relatively rare mature mRNA molecules it needs to export to the cytosol and the overwhelming amount of debris generated by RNA processing?

The answer is that the transport of mRNA from the nucleus to the cytosol, where mRNAs are translated into protein, is highly selective: only correctly processed mRNAs are exported. This selective transport is mediated by *nuclear pore complexes*, which connect the nucleoplasm with the cytosol and act as gates that control which macromolecules can enter or leave the nucleus (discussed in Chapter 15). To be "export ready," an mRNA molecule must be bound to an appropriate set of proteins, each of which recognizes different parts of a mature mRNA molecule. These proteins include poly-A–binding proteins, a cap-binding complex, and

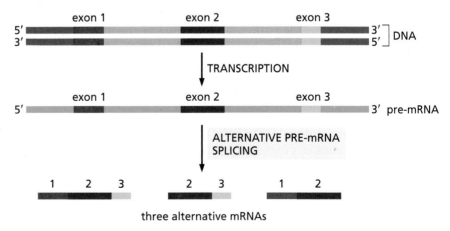

Figure 7–22 **Some pre-mRNAs undergo alternative RNA splicing to produce various mRNAs and proteins from the same gene.** Whereas all exons are present in a pre-mRNA, some exons can be excluded from the final mRNA molecule. In this example, three of four possible mRNAs are produced. The 5′ caps and poly-A tails on the mRNAs are not shown.

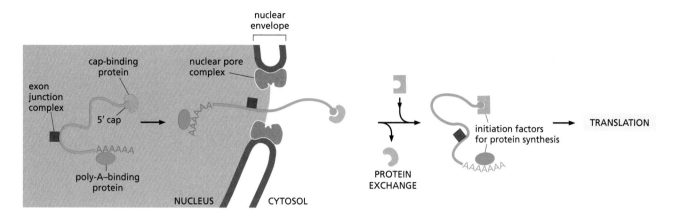

proteins that bind to mRNAs that have been appropriately spliced (**Figure 7–23**). The entire set of bound proteins, rather than any single protein, ultimately determines whether an mRNA molecule will leave the nucleus. The "waste RNAs" that remain behind in the nucleus are degraded there, and their nucleotide building blocks are reused for transcription.

mRNA Molecules Are Eventually Degraded in the Cytosol

Because a single mRNA molecule can be translated into protein many times (see Figure 7–2), the length of time that a mature mRNA molecule persists in the cell affects the amount of protein it produces. Each mRNA molecule is eventually degraded into nucleotides by ribonucleases (RNAses) present in the cytosol, but the lifetimes of mRNA molecules differ considerably—depending on the nucleotide sequence of the mRNA and the type of cell. In bacteria, most mRNAs are degraded rapidly, having a typical lifetime of about 3 minutes. The mRNAs in eukaryotic cells usually persist longer: some, such as those encoding β-globin, have lifetimes of more than 10 hours, whereas others have lifetimes of less than 30 minutes.

These different lifetimes are in part controlled by nucleotide sequences in the mRNA itself, most often in the portion of RNA called the *3′ untranslated region*, which lies between the 3′ end of the coding sequence and the poly-A tail. The different lifetimes of mRNAs help the cell control the amount of each protein that it synthesizes. In general, proteins made in large amounts, such as β-globin, are translated from mRNAs that have long lifetimes, whereas proteins made in smaller amounts, or whose levels must change rapidly in response to signals, are typically synthesized from short-lived mRNAs.

The Earliest Cells May Have Had Introns in Their Genes

The process of transcription is universal: all cells use RNA polymerase and complementary base-pairing to synthesize RNA from DNA. Indeed, bacterial and eukaryotic RNA polymerases are almost identical in overall structure and clearly evolved from a shared ancestral polymerase. It may therefore seem puzzling that the resulting RNA transcripts are handled so differently in eukaryotes and in prokaryotes (**Figure 7–24**). In particular, RNA splicing seems to mark a fundamental difference between those two types of cells. But how did this dramatic difference arise?

As we have seen, RNA splicing provides eukaryotes with the ability to produce a variety of proteins from a single gene. It also allows them to evolve new genes by mixing-and-matching exons from preexisting genes, as we discuss in Chapter 9. However, these advantages come with a cost: the cell has to maintain a larger genome and has to discard a

Figure 7–23 A specialized set of RNA-binding proteins signals that a mature mRNA is ready for export to the cytosol. As indicated on the left, the cap and poly-A tail of a mature mRNA molecule are "marked" by proteins that recognize these modifications. In addition, a group of proteins called the *exon junction complex* is deposited on the pre-mRNA after each successful splice has occurred. Once the mRNA is deemed "export ready," a nuclear transport receptor (discussed in Chapter 15) associates with the mRNA and guides it through the nuclear pore. In the cytosol, the mRNA can shed some of these proteins and bind new ones, which, along with poly-A–binding protein, act as initiation factors for protein synthesis, as we discuss later.

(A)   EUKARYOTES

(B)   PROKARYOTES

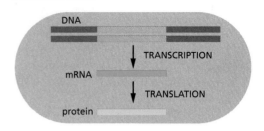

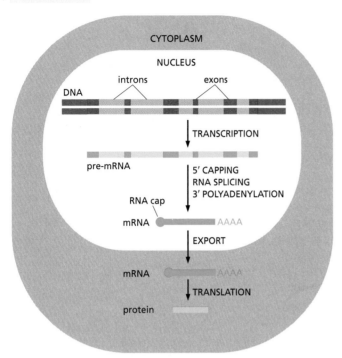

Figure 7–24 Prokaryotes and eukaryotes handle their RNA transcripts differently.
(A) In eukaryotic cells, the pre-mRNA molecule produced by transcription contains both intron and exon sequences. Its two ends are modified, and the introns are removed by RNA splicing. The resulting mRNA is then transported from the nucleus to the cytoplasm, where it is translated into protein. Although these steps are depicted as occurring in sequence, one at a time, in reality they occur simultaneously. For example, the RNA cap is usually added and splicing usually begins before transcription has been completed. Because of this overlap, transcripts of the entire gene (including all introns and exons) do not typically exist in the cell. (B) In prokaryotes, the production of mRNA molecules is simpler. The 5' end of an mRNA molecule is produced by the initiation of transcription by RNA polymerase, and the 3' end is produced by the termination of transcription. Because prokaryotic cells lack a nucleus, transcription and translation take place in a common compartment. Translation of a bacterial mRNA can therefore begin before its synthesis has been completed. In both eukaryotes and prokaryotes, the amount of a protein in a cell depends on the rates of each of these steps, as well as on the rates of degradation of the mRNA and protein molecules.

large fraction of the RNA it synthesizes without ever using it. According to one school of thought, early cells—the common ancestors of prokaryotes and eukaryotes—contained introns that were lost in prokaryotes during subsequent evolution. By shedding their introns and adopting a smaller, more streamlined genome, prokaryotes would have been able to reproduce more rapidly and efficiently. Consistent with this idea, simple eukaryotes that reproduce rapidly (some yeasts, for example) have relatively few introns, and these introns are usually much shorter than those found in higher eukaryotes.

On the other hand, some argue that introns were originally parasitic mobile genetic elements (discussed in Chapter 9) that happened to invade an early eukaryotic ancestor, colonizing its genome. These host cells then unwittingly replicated the "stowaway" nucleotide sequences along with their own DNA; modern eukaryotes simply never bothered to sweep away the genetic clutter left from that ancient infection. The issue, however, is far from settled; whether introns evolved early—and were lost in prokaryotes—or evolved later in eukaryotes is still a topic of scientific debate, and we return to it in Chapter 9.

FROM RNA TO PROTEIN

By the end of the 1950s, biologists had demonstrated that the information encoded in DNA is copied first into RNA and then into protein. The debate then shifted to the "coding problem": How is the information in a linear sequence of nucleotides in an RNA molecule translated into the linear sequence of a chemically quite different set of subunits—the amino acids in a protein? This fascinating question intrigued scientists at the time. Here was a cryptogram set up by nature that, after more than 3 billion years of evolution, could finally be solved by one of the products of evolution—human beings! Indeed, scientists have not only cracked the code but have revealed, in atomic detail, the precise workings of the machinery by which cells read this code.

codons																				
	AGA									UUA					AGC					
	AGG									UUG					AGU					
GCA	CGA					GGA				CUA			CCA	UCA	ACA			GUA		
GCC	CGC					GGC		AUA	CUC				CCC	UCC	ACC			GUC	UAA	
GCG	CGG	GAC	AAC	UGC	GAA	CAA	GGG	CAC	AUC	CUG	AAA		UUC	CCG	UCG	ACG		UAC	GUG	UAG
GCU	CGU	GAU	AAU	UGU	GAG	CAG	GGU	CAU	AUU	CUU	AAG	AUG	UUU	CCU	UCU	ACU	UGG	UAU	GUU	UGA

amino acids	Ala	Arg	Asp	Asn	Cys	Glu	Gln	Gly	His	Ile	Leu	Lys	Met	Phe	Pro	Ser	Thr	Trp	Tyr	Val	stop
	A	R	D	N	C	E	Q	G	H	I	L	K	M	F	P	S	T	W	Y	V	

Figure 7–25 The nucleotide sequence of an mRNA is translated into the amino acid sequence of a protein via the genetic code. All the three-nucleotide codons in mRNAs that specify a given amino acid are listed above that amino acid, which is given in both its three-letter and one-letter abbreviations (see Panel 2–5, pp. 74–75, for the full name of each amino acid and its structure). Like RNA molecules, codons are always written with the 5'-terminal nucleotide to the left. Note that most amino acids are represented by more than one codon, and there are some regularities in the set of codons that specify each amino acid. Codons for the same amino acid tend to contain the same nucleotides at the first and second positions and to vary at the third position. There are three codons that do not specify any amino acid but act as termination sites (*stop codons*), signaling the end of the protein-coding sequence in an mRNA. One codon—AUG—acts both as an initiation codon, signaling the start of a protein-coding message, and as the codon that specifies the amino acid methionine.

An mRNA Sequence Is Decoded in Sets of Three Nucleotides

Transcription as a means of information transfer is simple to understand: DNA and RNA are chemically and structurally similar, and DNA can act as a direct template for the synthesis of RNA through complementary base-pairing. As the term transcription signifies, it is as if a message written out by hand were being converted, say, into a typewritten text. The language itself and the form of the message do not change, and the symbols used are closely related.

In contrast, the conversion of the information in RNA into protein represents a **translation** of the information into another language that uses different symbols. Because there are only 4 different nucleotides in mRNA but 20 different types of amino acids in a protein, this translation cannot be accounted for by a direct one-to-one correspondence between a nucleotide in RNA and an amino acid in protein. The rules by which the nucleotide sequence of a gene, through an intermediary mRNA molecule, is translated into the amino acid sequence of a protein are known as the **genetic code**.

In 1961, it was discovered that the sequence of nucleotides in an mRNA molecule is read consecutively in groups of three. And because RNA is made of 4 different nucleotides, there are $4 \times 4 \times 4 = 64$ possible combinations of three nucleotides: AAA, AUA, AUG, and so on. However, only 20 different amino acids are commonly found in proteins. Either some nucleotide triplets are never used, or the code is redundant, with some amino acids being specified by more than one triplet. The second possibility turned out to be correct, as shown by the completely deciphered genetic code shown in **Figure 7–25**. Each group of three consecutive nucleotides in RNA is called a **codon**, and each codon specifies one amino acid. The strategy by which this code was cracked is described in **How We Know**, pp. 240–241.

The same genetic code is used in nearly all present-day organisms. Although a few slight differences have been found, these occur chiefly in the mRNA of mitochondria and of some fungi and protozoa. Mitochondria have their own DNA replication, transcription, and protein-synthesis machinery, which operates independently from the corresponding machinery in the rest of the cell (discussed in Chapter 14), and they have been able to accommodate minor changes to the otherwise universal genetic code. Even in fungi and protozoa, the similarities in the code far outweigh the differences.

In principle, an mRNA sequence can be translated in any one of three different **reading frames**, depending on where the decoding process begins (**Figure 7–26**). However, only one of the three possible reading frames

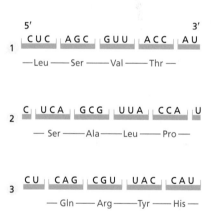

Figure 7–26 In principle, an mRNA molecule can be translated in three possible reading frames. In the process of translating a nucleotide sequence (*blue*) into an amino acid sequence (*red*), the sequence of nucleotides in an mRNA molecule is read from the 5' to the 3' end in sequential sets of three nucleotides. In principle, therefore, the same mRNA sequence can specify three completely different amino acid sequences, depending on where translation begins— that is, on the reading frame used. In reality, however, only one of these reading frames encodes the actual message and is therefore used in translation, as we discuss later.

CRACKING THE GENETIC CODE

By the beginning of the 1960s, the *central dogma* had been accepted as the pathway along which information flows from gene to protein. It was clear that genes encode proteins, that genes are made of DNA, and that mRNA serves as an intermediary, carrying the information from DNA to the ribosome, where the RNA is translated into protein.

Even the general format of the genetic code had been worked out: each of the 20 amino acids found in proteins is represented by a triplet codon in an mRNA molecule. But an even greater challenge remained: biologists, chemists, and even physicists set their sights on breaking the genetic code—attempting to figure out which amino acid each of the 64 possible nucleotide triplets designates. The most straightforward path to the solution would have been to compare the sequence of a segment of DNA or of mRNA with its corresponding polypeptide product. Techniques for sequencing nucleic acids, however, would not be devised for another 10 years.

So researchers decided that, to crack the genetic code, they would have to synthesize their own simple RNA molecules. If they could feed these RNA molecules to ribosomes—the machines that make proteins—and then analyze the resulting polypeptide product, they would be on their way to deciphering which triplets encode which amino acids.

Losing the cells

Before researchers could test their synthetic mRNAs, they needed to perfect a cell-free system for protein synthesis. This would allow them to translate their messages into polypeptides in a test tube. (Generally speaking, when working in the laboratory, the simpler the system, the easier it is to interpret the results.) To isolate the molecular machinery they needed for such a cell-free translation system, researchers broke open *E. coli* cells and loaded their contents into a centrifuge tube. Spinning these samples at high speed caused the membranes and other large chunks of cellular debris to be dragged to the bottom of the tube; the lighter cellular components required for protein synthesis—including mRNA, the tRNA adaptors, ribosomes, enzymes, and other small molecules—were left floating in the supernatant. Researchers found that simply adding radioactive amino acids to this cell "soup" would trigger the production of radiolabeled polypeptides. By centrifuging this supernatant again, at a higher speed, the researchers could force the ribosomes, and any newly synthesized peptides attached to them, to the bottom of the tube; the labeled polypeptides could then be detected by measuring the radioactivity in the sediment remaining in the tube after the top layer had been discarded.

The trouble with this particular system was that it produced proteins encoded by the cell's own mRNAs already present in the extract. But researchers wanted to use their own synthetic messages to direct protein synthesis. This problem was solved when Marshall Nirenberg discovered that he could destroy the cells' mRNA in the extract by adding a small amount of ribonuclease—an enzyme that degrades RNA—to the mix. Now all he needed to do was prepare large quantities of synthetic mRNA, add it to the cell-free system, and see what peptides came out.

Faking the message

Producing a synthetic polynucleotide with a defined sequence was not as simple as it sounds. Again, it would be years before chemists and bioengineers developed machines that could synthesize any given string of nucleic acids quickly and cheaply. Nirenberg decided to use polynucleotide phosphorylase, an enzyme that would join ribonucleotides together in the absence of a template. The sequence of the resulting RNA would then depend entirely on which nucleotides were presented to the enzyme. A mixture of nucleotides would be sewn into a random sequence; but a single type of nucleotide would yield a homogeneous polymer containing only that one nucleotide. Thus Nirenberg, working with his collaborator Heinrich Matthaei, first produced synthetic mRNAs made entirely of uracil—poly U.

Together, the researchers fed this poly U to their cell-free translation system. They then added a single type of radioactively labeled amino acid to the mix. After testing each amino acid—one at a time, in 20 different experiments—they determined that poly U directs the synthesis of a polypeptide containing only phenylalanine (**Figure 7–27**). With this electrifying result, the first word in the genetic code had been deciphered (see Figure 7–25).

Nirenberg and Matthaei then repeated the experiment with poly A and poly C and determined that AAA codes for lysine and CCC for proline. The meaning of poly G could not be ascertained by this method because this polynucleotide forms an odd triple-stranded helix that did not serve as a template in the cell-free system.

Feeding ribosomes with synthetic RNA seemed a fruitful technique. But with the single-nucleotide possibilities exhausted, researchers had nailed down only three codons; they had 61 still to go. The other codons, however, were harder to decipher, and a new synthetic approach was needed. In the 1950s, the organic chemist Gobind Khorana had been developing methods for preparing mixed polynucleotides of defined sequence—but his techniques worked only for DNA. When he

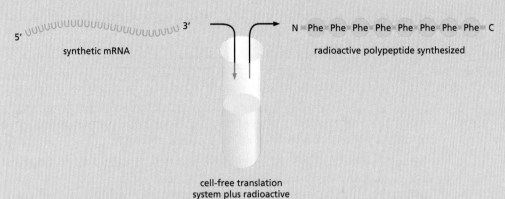

5' UUUUUUUUUUUUUUUUUUUUUUUUUUUU 3'

synthetic mRNA

cell-free translation
system plus radioactive
amino acids

N — Phe — Phe — Phe — Phe — Phe — Phe — Phe — Phe — C

radioactive polypeptide synthesized

Figure 7–27 UUU codes for phenylalanine. Synthetic mRNAs are fed into a cell-free translation system containing bacterial ribosomes, tRNAs, enzymes, and other small molecules. Radioactive amino acids are added to this mix and the resulting polypeptides analyzed. In this case, poly U is shown to encode a polypeptide containing only phenylalanine.

learned of Nirenberg's work with synthetic RNAs, Khorana directed his energies and skills to producing polyribonucleotides. He found that if he started out by making DNAs of a defined sequence, he could then use RNA polymerase to produce RNAs from those. In this way, Khorana prepared a collection of different RNAs of defined repeating sequence: he generated sequences of repeating dinucleotides (such as poly UC), trinucleotides (such as poly UUC), or tetranucleotides (such as poly UAUC).

These mixed polynucleotides, however, yielded results that were much more difficult to decode than the mononucleotide messages that Nirenberg had used. Take poly UG, for example. When this repeating dinucleotide is added to the translation system, researchers discovered that it codes for a polypeptide of alternating cysteines and valines. This RNA, of course, contains two different alternating codons: UGU and GUG. So researchers could say that UGU and GUG code for cysteine and valine, although they could not tell which went with which. Thus these mixed messages provided useful information, but they did not definitively reveal which codons specified which amino acids (**Figure 7–28**).

Trapping the triplets

These final ambiguities in the code were resolved when Nirenberg and a young medical graduate named Phil Leder discovered that RNA fragments that were only three nucleotides in length—the size of a single codon—could bind to a ribosome and attract the appropriate amino-acid-containing tRNA molecule to the protein-making machinery. These complexes—containing one ribosome, one mRNA codon, and one radiolabeled aminoacyl-tRNA—could then be captured on a piece of filter paper and the attached amino acid identified.

Their trial run with UUU—the first word—worked splendidly. Leder and Nirenberg primed the usual cell-free translation system with snippets of UUU. These

trinucleotides bound to the ribosomes, and Phe-tRNAs bound to the UUU. The new system was up and running, and the researchers had confirmed that UUU codes for phenylalanine.

All that remained was for researchers to produce all 64 possible codons—a task that was quickly accomplished in both Nirenberg's and Khorana's laboratories. Because these small trinucleotides were much simpler to synthesize chemically, and the triplet-trapping tests were easier to perform and analyze than the previous decoding experiments, the researchers were able to work out the complete genetic code within the next year.

MESSAGE	PEPTIDES PRODUCED	CODON ASSIGNMENTS	
poly UG	...Cys–Val–Cys–Val...	UGU GUG	Cys, Val*
poly AG	...Arg–Glu–Arg–Glu...	AGA GAG	Arg, Glu
poly UUC	...Phe–Phe–Phe... + ...Ser–Ser–Ser... + ...Leu–Leu–Leu...	UUC UCU CUU	Phe, Ser, Leu
poly UAUC	...Tyr–Leu–Ser–Ile...	UAU CUA UCU AUC	Tyr, Leu, Ser, Ile

* One codon specifies Cys, the other Val, but which is which? The same ambiguity exists for the other codon assignments shown here.

Figure 7–28 Using synthetic RNAs of mixed, repeating ribonucleotide sequences, scientists further narrowed the coding possibilities. Although these mixed messages produced mixed polypeptides, they did not permit the unambiguous assignment of a single codon to a specific amino acid. For example, the results of the poly-UG experiment cannot distinguish whether UGU or GUG encodes cysteine. As indicated, the same type of ambiguity confounded the interpretation of all the experiments using di-, tri-, and tetranucleotides.

in an mRNA specifies the correct protein. We discuss later how a special punctuation signal at the beginning of each mRNA molecule sets the correct reading frame.

tRNA Molecules Match Amino Acids to Codons in mRNA

The codons in an mRNA molecule do not directly recognize the amino acids they specify: the group of three nucleotides does not, for example, bind directly to the amino acid. Rather, the translation of mRNA into protein depends on adaptor molecules that can recognize and bind to a codon at one site on their surface and to an amino acid at another site. These adaptors consist of a set of small RNA molecules known as **transfer RNAs (tRNAs)**, each about 80 nucleotides in length.

We saw earlier that an RNA molecule generally folds into a three-dimensional structure by forming base pairs between different regions of the molecule. If the base-paired regions are sufficiently extensive, they will fold back on themselves to form a double-helical structure, like that of double-stranded DNA. The tRNA molecule provides a striking example of this. Four short segments of the folded tRNA are double-helical, producing a molecule that looks like a cloverleaf when drawn schematically (**Figure 7–29A**). For example, a 5′-GCUC-3′ sequence in one part of a polynucleotide chain can base-pair with a 5′-GAGC-3′ sequence in another region of the same molecule. The cloverleaf undergoes further folding to form a compact, L-shaped structure that is held together by additional hydrogen bonds between different regions of the molecule (**Figure 7–29B and C**).

Two regions of unpaired nucleotides situated at either end of the L-shaped tRNA molecule are crucial to the function of tRNAs in protein synthesis. One of these regions forms the **anticodon**, a set of three consecutive nucleotides that bind, through base-pairing, to the complementary codon

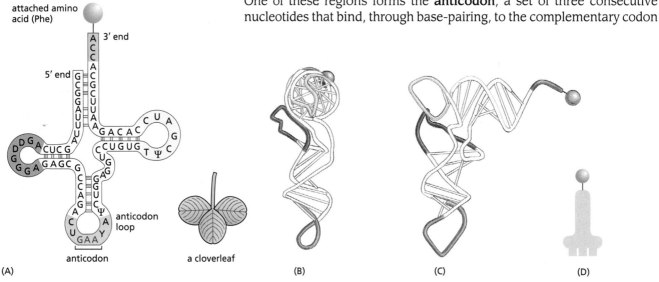

5′ GCGGAUUUAGCUCAGDDGGGAGAGCGCCAGACUGAAYAΨCUGGAGGUCCUGUGTΨCGAUCCACAGAAUUCGCACCA 3′
(E)   anticodon

Figure 7–29 tRNA molecules are molecular adaptors, linking amino acids to codons. In this series of diagrams, the same tRNA molecule—in this case, a tRNA specific for the amino acid phenylalanine (Phe)—is depicted in various ways. (A) The conventional "cloverleaf" structure shows the complementary base-pairing (*red lines*) that creates the double-helical regions of the molecule. The anticodon loop (*blue*) contains the sequence of three nucleotides (*red letters*) that base-pairs with a codon in mRNA. The amino acid matching the codon–anticodon pair is attached at the 3′ end of the tRNA. tRNAs contain some unusual bases, which are produced by chemical modification after the tRNA has been synthesized. The bases denoted Ψ (for pseudouridine) and D (for dihydrouridine) are derived from uracil. (B and C) Views of the actual L-shaped molecule, based on X-ray diffraction analysis. These two images are rotated 90° with respect to each other. (D) Schematic representation of tRNA, emphasizing the anticodon, that will be used in subsequent figures. (E) The linear nucleotide sequence of the tRNA molecule, color-coded to match A, B, and C.

in an mRNA molecule. The other is a short single-stranded region at the 3′ end of the molecule; this is the site where the amino acid that matches the codon is covalently attached to the tRNA.

We saw in the previous section that the genetic code is redundant; that is, several different codons can specify a single amino acid (see Figure 7–25). This redundancy implies either that there is more than one tRNA for many of the amino acids or that some tRNA molecules can base-pair with more than one codon. In fact, both situations occur. Some amino acids have more than one tRNA, and some tRNAs are constructed so that they require accurate base-pairing only at the first two positions of the codon and can tolerate a mismatch (or *wobble*) at the third position. This wobble base-pairing explains why so many of the alternative codons for an amino acid differ only in their third nucleotide (see Figure 7–25). Wobble base-pairings make it possible to fit the 20 amino acids to their 61 codons with as few as 31 kinds of tRNA molecules. The exact number of different kinds of tRNAs, however, differs from one species to the next. For example, humans have nearly 500 different tRNA genes, but only 48 anticodons are represented among them.

Specific Enzymes Couple tRNAs to the Correct Amino Acid

For a tRNA molecule to carry out its role as an adaptor, it must be linked—or charged—with the correct amino acid. How does each tRNA molecule recognize the one amino acid in 20 that is its right partner? Recognition and attachment of the correct amino acid depend on enzymes called **aminoacyl-tRNA synthetases**, which covalently couple each amino acid to its appropriate set of tRNA molecules. In most organisms, there is a different synthetase enzyme for each amino acid. That means that there are 20 synthetases in all: one attaches glycine to all tRNAs that recognize codons for glycine, another attaches phenylalanine to all tRNAs that recognize codons for phenylalanine, and so on. Each synthetase enzyme recognizes specific nucleotides in both the anticodon and the amino-acid-accepting arm of the correct tRNA (Movie 7.6). The synthetases are thus equal in importance to the tRNAs in the decoding process, because it is the combined action of the synthetases and tRNAs that allows each codon in the mRNA molecule to specify its proper amino acid (**Figure 7–30**).

Figure 7–30 The genetic code is translated by the cooperation of two adaptors: aminoacyl-tRNA synthetases and tRNAs. Each synthetase couples a particular amino acid to its corresponding tRNAs, a process called charging. The anticodon on the charged tRNA molecule then forms base pairs with the appropriate codon on the mRNA. An error in either the charging step or the binding of the charged tRNA to its codon will cause the wrong amino acid to be incorporated into a protein chain. In the sequence of events shown, the amino acid tryptophan (Trp) is selected by the codon UGG on the mRNA.

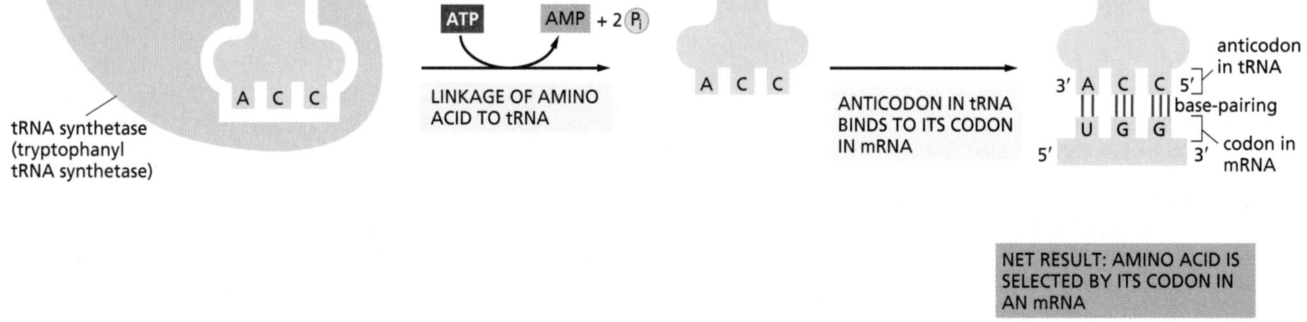

NET RESULT: AMINO ACID IS SELECTED BY ITS CODON IN AN mRNA

The synthetase-catalyzed reaction that attaches the amino acid to the 3′ end of the tRNA is one of many reactions in cells coupled to the energy-releasing hydrolysis of ATP (see Figure 3–33). The reaction produces a high-energy bond between the charged tRNA and the amino acid. The energy of this bond is later used to link the amino acid covalently to the growing polypeptide chain.

The mRNA Message Is Decoded by Ribosomes

The recognition of a codon by the anticodon on a tRNA molecule depends on the same type of complementary base-pairing used in DNA replication and transcription. However, accurate and rapid translation of mRNA into protein requires a molecular machine that can move along the mRNA, capture complementary tRNA molecules, hold the tRNAs in position, and then covalently link the amino acids that they carry to form a polypeptide chain. In both prokaryotes and eukaryotes, the machine that gets the job done is the **ribosome**—a large complex made from dozens of small proteins (the *ribosomal proteins*) and several crucial RNA molecules called **ribosomal RNAs** (**rRNAs**). A typical eukaryotic cell contains millions of ribosomes in its cytoplasm (**Figure 7–31**).

Eukaryotic and prokaryotic ribosomes are very similar in structure and function. Both are composed of one large subunit and one small subunit, which fit together to form a complete ribosome with a mass of several million daltons (**Figure 7–32**); for comparison, an average-sized protein has a mass of 30,000 daltons. The small ribosomal subunit matches the tRNAs to the codons of the mRNA, while the large subunit catalyzes the formation of the peptide bonds that covalently link the amino acids together into a polypeptide chain. These two subunits come together on an mRNA molecule near its 5′ end to start the synthesis of a protein. The mRNA is then pulled through the ribosome like a long piece of tape. As the mRNA inches forward in a 5′-to-3′ direction, the ribosome translates its nucleotide sequence into an amino acid sequence, one codon at a time, using the tRNAs as adaptors. Each amino acid is thereby added in the correct sequence to the end of the growing polypeptide chain (Movie 7.7). When synthesis of the protein is finished, the two subunits of the ribosome separate. Ribosomes operate with remarkable efficiency: a eukaryotic ribosome adds about 2 amino acids to a polypeptide chain each second; a bacterial ribosome operates even faster, adding about 20 amino acids per second.

> ### QUESTION 7–4
>
> In a clever experiment performed in 1962, a cysteine already attached to its tRNA was chemically converted to an alanine. These "hybrid" tRNA molecules were then added to a cell-free translation system from which the normal cysteine-tRNAs had been removed. When the resulting protein was analyzed, it was found that alanine had been inserted at every point in the polypeptide chain where cysteine was supposed to be. Discuss what this experiment tells you about the role of aminoacyl-tRNA synthetases during the normal translation of the genetic code.

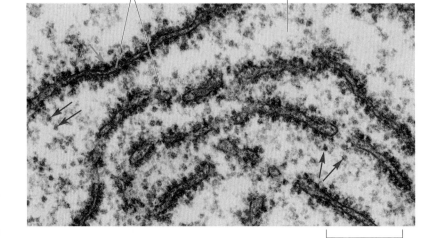

endoplasmic reticulum   cytosol

400 nm

Figure 7–31 Ribosomes are located in the cytoplasm of eukaryotic cells. This electron micrograph shows a thin section of a small region of cytoplasm. The ribosomes appear as small gray blobs. Some are free in the cytosol (*red arrows*); others are attached to membranes of the endoplasmic reticulum (*green arrows*). (Courtesy of George Palade.)

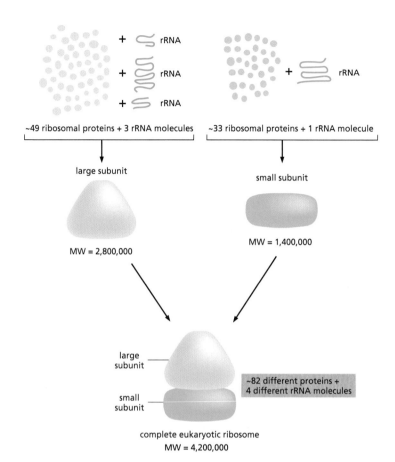

Figure 7–32 The eukaryotic ribosome is a large complex of four rRNAs and more than 80 small proteins. Prokaryotic ribosomes are very similar: both are formed from a large and small subunit, which only come together after the small subunit has bound an mRNA. Although ribosomal proteins greatly outnumber rRNAs, the RNAs account for most of the mass of the ribosome and give it its overall shape and structure.

How does the ribosome choreograph all the movements required for translation? In addition to a binding site for an mRNA molecule, each ribosome contains three binding sites for tRNA molecules, called the A site, the P site, and the E site (**Figure 7–33**). To add an amino acid to a growing peptide chain, the appropriate charged tRNA enters the A site by base-pairing with the complementary codon on the mRNA molecule. Its amino acid is then linked to the peptide chain held by the tRNA in the neighboring P site. Next, the large ribosomal subunit shifts forward, moving the spent tRNA to the E site before ejecting it (**Figure 7–34**). This cycle of reactions is repeated each time an amino acid is added to the polypeptide chain, with the new protein growing from its amino to its carboxyl end until a stop codon in the mRNA is encountered.

Figure 7–33 Each ribosome has a binding site for mRNA and three binding sites for tRNA. The tRNA sites are designated the A, P, and E sites (short for aminoacyl-tRNA, peptidyl-tRNA, and exit, respectively). (A) Three-dimensional structure of a bacterial ribosome, as determined by X-ray crystallography, with the small subunit in *dark green* and the large subunit in *light green*. Both the rRNAs and the ribosomal proteins are shown in *green*. tRNAs are shown bound in the E site (*red*), the P site (*orange*), and the A site (*yellow*). Although all three tRNA sites are shown occupied here, during the process of protein synthesis only two of these sites are occupied at any one time (see Figure 7–34). (B) Highly schematized representation of a ribosome (in the same orientation as A), which will be used in subsequent figures. Note that both the large and small subunits are involved in forming the A, P, and E sites, while only the small subunit forms the binding site for an mRNA. (B, adapted from M.M. Yusupov et al., *Science* 292:883–896, 2001, with permission from AAAS. Courtesy of Albion Baucom and Harry Noller.)

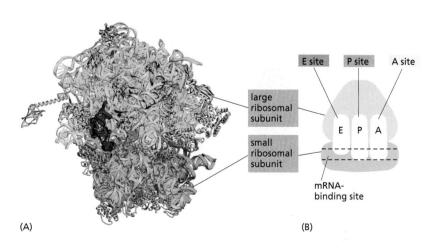

(A) (B)

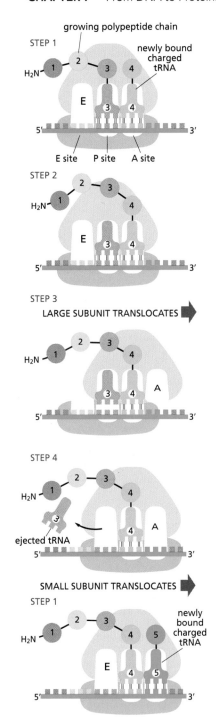

STEP 1

growing polypeptide chain

newly bound charged tRNA

E site   P site   A site

STEP 2

STEP 3

LARGE SUBUNIT TRANSLOCATES

STEP 4

ejected tRNA

SMALL SUBUNIT TRANSLOCATES

STEP 1

newly bound charged tRNA

Figure 7–34 Translation takes place in a four-step cycle. This cycle is repeated over and over during the synthesis of a protein. In *step 1*, a charged tRNA carrying the next amino acid to be added to the polypeptide chain binds to the vacant A site on the ribosome by forming base pairs with the mRNA codon that is exposed there. Because only the appropriate tRNA molecules can base-pair with each codon, this codon determines the specific amino acid added. The A and P sites are sufficiently close together that their two tRNA molecules are forced to form base pairs with codons that are contiguous, with no stray bases in between. This positioning of the tRNAs ensures that the correct reading frame will be preserved throughout the synthesis of the protein. In *step 2*, the carboxyl end of the polypeptide chain (amino acid 3 in step 1) is uncoupled from the tRNA at the P site and joined by a peptide bond to the free amino group of the amino acid linked to the tRNA at the A site. This reaction is catalyzed by an enzymatic site in the large subunit. In *step 3*, a shift of the large subunit relative to the small subunit moves the two tRNAs into the E and P sites of the large subunit. In *step 4*, the small subunit moves exactly three nucleotides along the mRNA molecule, bringing it back to its original position relative to the large subunit. This movement ejects the spent tRNA and resets the ribosome with an empty A site so that the next charged tRNA molecule can bind (Movie 7.8).

As indicated, the mRNA is translated in the 5'-to-3' direction, and the N-terminal end of a protein is made first, with each cycle adding one amino acid to the C-terminus of the polypeptide chain. To watch the translation cycle in atomic detail, see Movie 7.9.

The Ribosome Is a Ribozyme

The ribosome is one of the largest and most complex structures in the cell, composed of two-thirds RNA and one-third protein by weight. The determination of the entire three-dimensional structure of its large and small subunits in 2000 was a major triumph of modern biology. The structure confirmed earlier evidence that the rRNAs—not the proteins—are responsible for the ribosome's overall structure and its ability to choreograph and catalyze protein synthesis.

The rRNAs are folded into highly compact, precise three-dimensional structures that form the core of the ribosome (**Figure 7–35**). In marked contrast to the central positioning of the rRNAs, the ribosomal proteins are generally located on the surface, where they fill the gaps and crevices of the folded RNA. The main role of the ribosomal proteins seems to be

Figure 7–35 Ribosomal RNAs give the ribosome its overall shape. Shown here are the detailed structures of the two rRNAs that form the core of the large subunit of a bacterial ribosome—the 23S rRNA (*blue*) and the 5S rRNA (*purple*). One of the protein subunits of the ribosome (L1) is included as a reference point, as this protein forms a characteristic protrusion on the ribosome surface. Ribosomal components are commonly designated by their "S values," which refer to their rate of sedimentation in an ultracentrifuge. (Adapted from N. Ban et al., *Science* 289:905–920, 2000. With permission from AAAS.)

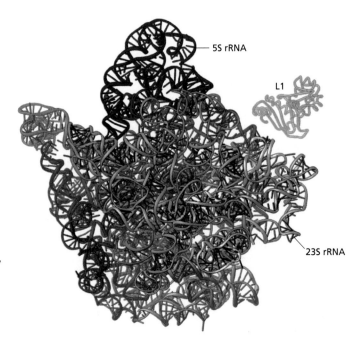

5S rRNA

L1

23S rRNA

to help fold and stabilize the RNA core, while permitting the changes in rRNA conformation that are necessary for this RNA to catalyze efficient protein synthesis.

Not only are the three tRNA-binding sites (the A, P, and E sites) on the ribosome formed primarily by the rRNAs, but the catalytic site for peptide bond formation is formed by the 23S rRNA of the large subunit; the nearest ribosomal protein is located too far away to make contact with the incoming charged tRNA or with the growing polypeptide chain. The catalytic site in this rRNA—a peptidyl transferase—is similar in many respects to that found in some protein enzymes: it is a highly structured pocket that precisely orients the two reactants—the elongating polypeptide and the charged tRNA—thereby greatly increasing the probability of a productive reaction.

RNA molecules that possess catalytic activity are called **ribozymes**. Later, in the final section of this chapter, we will consider other ribozymes and discuss what the existence of RNA-based catalysis might mean for the early evolution of life on Earth. Here we need only note that there is good reason to suspect that RNA rather than protein molecules served as the first catalysts for living cells. If so, the ribosome, with its catalytic RNA core, could be viewed as a relic of an earlier time in life's history, when cells were run almost entirely by ribozymes.

Specific Codons in mRNA Signal the Ribosome Where to Start and to Stop Protein Synthesis

In the test tube, ribosomes can be forced to translate any RNA molecule (see How We Know, pp. 240–241). In a cell, however, a specific start signal is required to initiate translation. The site at which protein synthesis begins on an mRNA is crucial, because it sets the reading frame for the whole length of the message. An error of one nucleotide either way at this stage will cause every subsequent codon in the mRNA to be misread, resulting in a nonfunctional protein with a garbled sequence of amino acids (see Figure 7–26). And the rate of initiation determines the rate at which the protein is synthesized from the mRNA.

The translation of an mRNA begins with the codon AUG, and a special charged tRNA is required to initiate translation. This **initiator tRNA** always carries the amino acid methionine (or a modified form of methionine, formyl-methionine, in bacteria). Thus newly made proteins all have methionine as the first amino acid at their N-terminal end, the end of a protein that is synthesized first. This methionine is usually removed later by a specific protease.

In eukaryotes, an initiator tRNA, charged with methionine, is first loaded into the P site of the small ribosomal subunit, along with additional proteins called **translation initiation factors** (Figure 7–36). The initiator tRNA is distinct from the tRNA that normally carries methionine. Of all the tRNAs in the cell, only a charged initiator tRNA molecule is capable of binding tightly to the P site in the absence of the large ribosomal subunit. Next, the small ribosomal subunit loaded with the initiator tRNA binds to

QUESTION 7–5

A sequence of nucleotides in a DNA strand—5′-TTAACGGCTTTTTTC-3′— was used as a template to synthesize an mRNA that was then translated into protein. Predict the C-terminal amino acid and the N-terminal amino acid of the resulting polypeptide. Assume that the mRNA is translated without the need for a start codon.

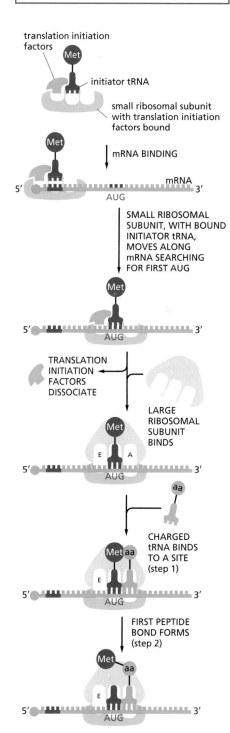

Figure 7–36 Initiation of protein synthesis in eukaryotes requires translation initiation factors and a special initiator tRNA. Although not shown here, efficient translation initiation also requires additional proteins that are bound at the 5′ cap and poly-A tail of the mRNA (see Figure 7–23). In this way, the translation apparatus can ascertain that both ends of the mRNA are intact before initiating translation. Following initiation, the protein is elongated by the reactions outlined in Figure 7–34.

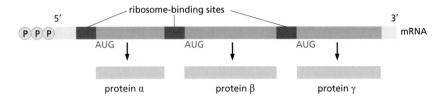

Figure 7–37 A single prokaryotic mRNA molecule can encode several different proteins. In prokaryotes, genes directing the different steps in a process are often organized into clusters (operons) that are transcribed together into a single mRNA. A prokaryotic mRNA does not have the same sort of 5′ cap as a eukaryotic mRNA, but instead has a triphosphate at its 5′ end. Prokaryotic ribosomes initiate translation at ribosome-binding sites (*dark blue*), which can be located in the interior of an mRNA molecule. This feature enables prokaryotes to synthesize different proteins from a single mRNA molecule, with each protein made by a different ribosome.

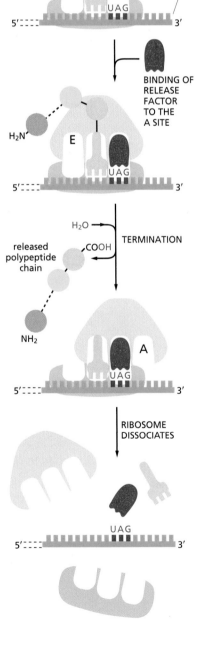

the 5′ end of an mRNA molecule, which is marked by the 5′ cap that is present on all eukaryotic mRNAs (see Figure 7–16). The small ribosomal subunit then moves forward (5′ to 3′) along the mRNA searching for the first AUG. When this AUG is encountered and recognized by the initiator tRNA, several initiation factors dissociate from the small ribosomal subunit to make way for the large ribosomal subunit to bind and complete ribosomal assembly. Because the initiator tRNA is bound to the P site, protein synthesis is ready to begin with the addition of the next charged tRNA to the A site (see Figure 7–34).

The mechanism for selecting a start codon is different in bacteria. Bacterial mRNAs have no 5′ caps to tell the ribosome where to begin searching for the start of translation. Instead, they contain specific ribosome-binding sequences, up to six nucleotides long, that are located a few nucleotides upstream of the AUGs at which translation is to begin. Unlike a eukaryotic ribosome, a prokaryotic ribosome can readily bind directly to a start codon that lies in the interior of an mRNA, as long as a ribosome-binding site precedes it by several nucleotides. Such ribosome-binding sequences are necessary in bacteria, as prokaryotic mRNAs are often *polycistronic*—that is, they encode several different proteins, each of which is translated from the same mRNA molecule (**Figure 7–37**). In contrast, a eukaryotic mRNA usually carries the information for a single protein.

The end of translation in both prokaryotes and eukaryotes is signaled by the presence of one of several codons, called *stop codons*, in the mRNA (see Figure 7–25). The stop codons—UAA, UAG, and UGA—are not recognized by a tRNA and do not specify an amino acid, but instead signal to the ribosome to stop translation. Proteins known as *release factors* bind to any stop codon that reaches the A site on the ribosome; this binding alters the activity of the peptidyl transferase in the ribosome, causing it to catalyze the addition of a water molecule instead of an amino acid to the peptidyl-tRNA (**Figure 7–38**). This reaction frees the carboxyl end of the polypeptide chain from its attachment to a tRNA molecule; because this is the only attachment that holds the growing polypeptide to the ribosome, the completed protein chain is immediately released. At this point, the ribosome also releases the mRNA and dissociates into its two separate subunits, which can then assemble on another mRNA molecule to begin a new round of protein synthesis.

Figure 7–38 Translation halts at a stop codon. In the final phase of protein synthesis, the binding of release factor to an A site bearing a stop codon terminates translation of an mRNA molecule. The completed polypeptide is released, and the ribosome dissociates into its two separate subunits. Note that only the 3′ end of the mRNA molecule is shown here.

We saw in Chapter 4 that many proteins can fold into their three-dimensional shape spontaneously, and some do so as they are spun out of the ribosome. Most proteins, however, require *chaperone proteins* to help them fold correctly in the cell. Chaperones can "steer" proteins along productive folding pathways and prevent them from aggregating inside the cell (see Figures 4–9 and 4–10). Newly synthesized proteins are typically met by their chaperones as they emerge from the ribosome.

Proteins Are Made on Polyribosomes

The synthesis of most protein molecules takes between 20 seconds and several minutes. But even during this short period, multiple ribosomes usually bind to each mRNA molecule being translated. If the mRNA is being translated efficiently, a new ribosome hops onto the 5′ end of the mRNA molecule almost as soon as the preceding ribosome has translated enough of the nucleotide sequence to move out of the way. The mRNA molecules being translated are therefore usually found in the form of *polyribosomes*, also known as *polysomes*. These large cytoplasmic assemblies are made up of many ribosomes spaced as close as 80 nucleotides apart along a single mRNA molecule (**Figure 7–39**). With multiple ribosomes working simultaneously on a single mRNA, many more protein molecules can be made in a given time than would be possible if each polypeptide had to be completed before the next could be started.

Polysomes operate in both bacteria and eukaryotes, but bacteria can speed up the rate of protein synthesis even further. Because bacterial mRNA does not need to be processed and is also physically accessible to ribosomes while it is being made, ribosomes will typically attach to the free end of a bacterial mRNA molecule and start translating it even before the transcription of that RNA is complete; these ribosomes follow closely behind the RNA polymerase as it moves along DNA.

Inhibitors of Prokaryotic Protein Synthesis Are Used as Antibiotics

The ability to translate mRNAs accurately into proteins is a fundamental feature of all life on Earth. Although the ribosome and other molecules that carry out this complex task are very similar among organisms, we

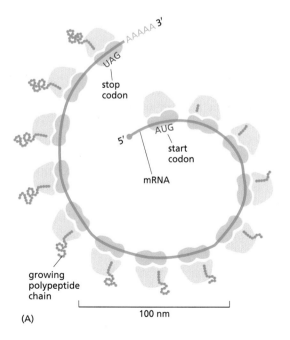

(A)

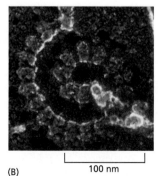

(B)

Figure 7–39 **Proteins are synthesized on polyribosomes.** (A) Schematic drawing showing how a series of ribosomes can simultaneously translate the same mRNA molecule (Movie 7.10). (B) Electron micrograph of a polyribosome in the cytosol of a eukaryotic cell. (B, courtesy of John Heuser.)

TABLE 7–3 ANTIBIOTICS THAT INHIBIT BACTERIAL PROTEIN OR RNA SYNTHESIS	
Antibiotic	Specific Effect
Tetracycline	blocks binding of aminoacyl-tRNA to A site of ribosome (step 1 in Figure 7–34)
Streptomycin	prevents the transition from initiation complex to chain elongation (see Figure 7–36); also causes miscoding
Chloramphenicol	blocks the peptidyl transferase reaction on ribosomes (step 2 in Figure 7–34)
Cycloheximide	blocks the translocation reaction on ribosomes (step 3 in Figure 7–34)
Rifamycin	blocks initiation of transcription by binding to RNA polymerase

have seen that there are some subtle differences in the way that bacteria and eukaryotes synthesize RNA and proteins. Through a quirk of evolution, these differences form the basis of one of the most important advances in modern medicine.

Many of our most effective antibiotics are compounds that act by inhibiting bacterial, but not eukaryotic, RNA and protein synthesis. Some of these drugs exploit the small structural and functional differences between bacterial and eukaryotic ribosomes, so that they interfere preferentially with bacterial protein synthesis. These compounds can thus be taken in doses high enough to kill bacteria without being toxic to humans. Because different antibiotics bind to different regions of the bacterial ribosome, these drugs often inhibit different steps in protein synthesis. A few of the antibiotics that inhibit bacterial RNA and protein synthesis are listed in **Table 7–3**.

Many common antibiotics were first isolated from fungi. Fungi and bacteria often occupy the same ecological niches; to gain a competitive edge, fungi have evolved, over time, potent toxins that kill bacteria but are harmless to themselves. Because fungi and humans are both eukaryotes, and are thus more closely related to each other than either is to bacteria (see Figure 1–28), we have been able to borrow these weapons to combat our own bacterial foes.

Controlled Protein Breakdown Helps Regulate the Amount of Each Protein in a Cell

After a protein is released from the ribosome, a cell can control its activity and longevity in various ways. The number of copies of a protein in a cell depends, like the human population, not only on how quickly new individuals are made but also on how long they survive. So controlling the breakdown of proteins into their constituent amino acids helps cells regulate the amount of each particular protein. Proteins vary enormously in their life-span. Structural proteins that become part of a relatively stable tissue such as bone or muscle may last for months or even years, whereas other proteins, such as metabolic enzymes and those that regulate cell growth and division (discussed in Chapter 18), last only for days, hours, or even seconds. How does the cell control these lifetimes?

Cells possess specialized pathways that enzymatically break proteins down into their constituent amino acids (a process termed *proteolysis*). The enzymes that degrade proteins, first to short peptides and finally to individual amino acids, are known collectively as **proteases**. Proteases

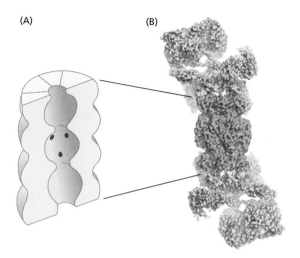

(A)   (B)

Figure 7–40 A proteasome degrades short-lived and misfolded proteins. The structures shown were determined by X-ray crystallography. (A) A cut-away view of the central cylinder of the proteasome, with the active sites of the proteases indicated by *red dots*. (B) The structure of the entire proteasome, in which access to the central cylinder (*yellow*) is regulated by a stopper (*blue*) at each end. (B, adapted from P.C.A da Fonseca et al., *Mol. Cell* 46:54–66, 2012.)

act by cutting (hydrolyzing) the peptide bonds between amino acids (see Panel 2–5, pp. 74–75). One function of proteolytic pathways is to rapidly degrade those proteins whose lifetimes must be kept short. Another is to recognize and remove proteins that are damaged or misfolded. Eliminating improperly folded proteins is critical for an organism, as misfolded proteins tend to aggregate, and protein aggregates can damage cells and even trigger cell death. Eventually, all proteins—even long-lived ones—accumulate damage and are degraded by proteolysis.

In eukaryotic cells, proteins are broken down by large protein machines called **proteasomes**, present in both the cytosol and the nucleus. A proteasome contains a central cylinder formed from proteases whose active sites face into an inner chamber. Each end of the cylinder is stoppered by a large protein complex formed from at least 10 types of protein subunits (**Figure 7–40**). These protein stoppers bind the proteins destined for degradation and then—using ATP hydrolysis to fuel this activity—unfold the doomed proteins and thread them into the inner chamber of the cylinder. Once the proteins are inside, proteases chop them into short peptides, which are then jettisoned from either end of the proteasome. Housing proteases inside these molecular destruction chambers makes sense, as it prevents the enzymes from running rampant in the cell.

How do proteasomes select which proteins in the cell should be degraded? In eukaryotes, proteasomes act primarily on proteins that have been marked for destruction by the covalent attachment of a small protein called *ubiquitin*. Specialized enzymes tag selected proteins with a short chain of ubiquitin molecules; these ubiquitylated proteins are then recognized, unfolded, and fed into proteasomes by proteins in the stopper (**Figure 7–41**).

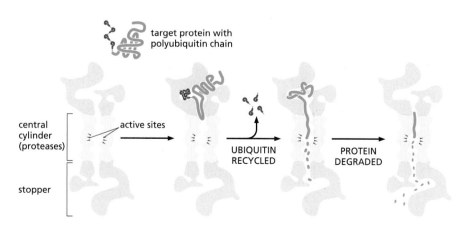

target protein with polyubiquitin chain

central cylinder (proteases)

active sites

UBIQUITIN RECYCLED

PROTEIN DEGRADED

stopper

Figure 7–41 Proteins marked by a polyubiquitin chain are degraded by the proteasome. Proteins in the stopper of a proteasome (*blue*) recognize target proteins marked by a specific type of polyubiquitin chain. The stopper then unfolds the target protein and threads it into the proteasome's central cylinder (*yellow*), which is lined with proteases that chop the protein to pieces.

Proteins that are meant to be short-lived often contain a short amino acid sequence that identifies the protein as one to be ubiquitylated and degraded in proteasomes. Damaged or misfolded proteins, as well as proteins containing oxidized or otherwise abnormal amino acids, are also recognized and degraded by this ubiquitin-dependent proteolytic system. The enzymes that add a polyubiquitin chain to such proteins recognize signals that become exposed on these proteins as a result of the misfolding or chemical damage—for example, amino acid sequences or conformational motifs that remain buried and inaccessible in the normal "healthy" protein.

There Are Many Steps Between DNA and Protein

We have seen that many types of chemical reactions are required to produce a protein from the information contained in a gene. The final concentration of a protein in a cell therefore depends on the rate at which each of the many steps is carried out (**Figure 7–42**). In addition, many proteins—once they leave the ribosome—require further attention before they are useful to the cell. Examples of such *post-translational modifications* include covalent modification (such as phosphorylation), the binding of small-molecule cofactors, or association with other protein subunits, which are often needed for a newly synthesized protein to become fully functional (**Figure 7–43**).

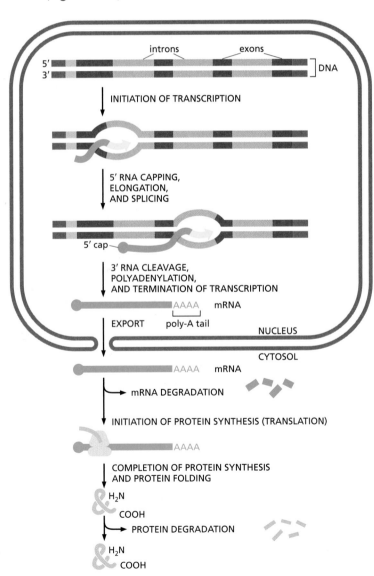

Figure 7–42 Protein production in a eukaryotic cell requires many steps. The final concentration of each protein depends on the rate of each step depicted. Even after an mRNA and its corresponding protein have been produced, their concentrations can be regulated by degradation. Although not shown here, the activity of the protein can also be regulated by other post-translational modifications or the binding of small molecules (see Figure 7–43).

We will see in the next chapter that cells have the ability to change the concentrations of most of their proteins according to their needs. In principle, all of the steps in Figure 7–42 can be regulated by the cell—and many of them, in fact, are. However, as we will see in the next chapter, the initiation of transcription is the most common point for a cell to regulate the expression of its genes.

Transcription and translation are universal processes that lie at the heart of life. However, when scientists came to consider how the flow of information from DNA to protein might have originated, they came to some unexpected conclusions.

RNA AND THE ORIGINS OF LIFE

The central dogma—that DNA makes RNA that makes protein—presented evolutionary biologists with a knotty puzzle: if nucleic acids are required to direct the synthesis of proteins, and proteins are required to synthesize nucleic acids, how could this system of interdependent components have arisen? One view is that an **RNA world** existed on Earth before cells containing DNA and proteins appeared. According to this hypothesis, RNA—which today serves largely as an intermediate between genes and proteins—both stored genetic information and catalyzed chemical reactions in primitive cells. Only later in evolutionary time did DNA take over as the genetic material and proteins become the major catalysts and structural components of cells (**Figure 7–44**). If this idea is correct, then the transition out of the RNA world was never completed; as we have seen, RNA still catalyzes several fundamental reactions in modern cells. These RNA catalysts—or ribozymes—including those that operate in the ribosome and in the RNA-splicing machinery, can thus be viewed as molecular fossils of an earlier world.

Life Requires Autocatalysis

The origin of life requires molecules that possess, if only to a small extent, one crucial property: the ability to catalyze reactions that lead—directly or indirectly—to the production of more molecules like themselves. Catalysts with this self-producing property, once they had arisen by chance, would divert raw materials from the production of other substances to make more of themselves. In this way, one can envisage the gradual development of an increasingly complex chemical system of organic monomers and polymers that function together to generate more molecules of the same types, fueled by a supply of simple raw materials in the primitive environment on Earth. Such an *autocatalytic* system would have many of the properties we think of as characteristic of living matter: the system would contain a far-from-random selection of interacting molecules; it would tend to reproduce itself; it would compete with other systems dependent on the same raw materials; and, if deprived of its raw materials or maintained at a temperature that upset the balance of reaction rates, it would decay toward chemical equilibrium and "die."

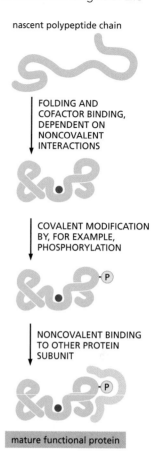

nascent polypeptide chain

FOLDING AND COFACTOR BINDING, DEPENDENT ON NONCOVALENT INTERACTIONS

COVALENT MODIFICATION BY, FOR EXAMPLE, PHOSPHORYLATION

NONCOVALENT BINDING TO OTHER PROTEIN SUBUNIT

mature functional protein

Figure 7–43 Many proteins require various modifications to become fully functional. To be useful to the cell, a completed polypeptide must fold correctly into its three-dimensional conformation and then bind any required cofactors (*red*) and protein partners—all via noncovalent bonding. Many proteins also require one or more covalent modifications to become active—or to be recruited to specific membranes or organelles (not shown). Although phosphorylation and glycosylation are the most common, more than 100 types of covalent modifications of proteins are known.

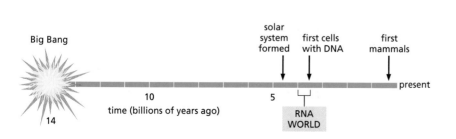

Figure 7–44 An RNA world may have existed before modern cells with DNA and proteins evolved.

But what molecules could have had such autocatalytic properties? In present-day living cells, the most versatile catalysts are proteins, which are able to adopt diverse three-dimensional forms that bristle with chemically reactive sites on their surface. However, there is no known way in which a protein can reproduce itself directly. RNA molecules, by contrast, could—at least, in principle—catalyze their own synthesis.

RNA Can Both Store Information and Catalyze Chemical Reactions

We have seen that complementary base-pairing enables one nucleic acid to act as a template for the formation of another. Thus a single strand of RNA or DNA can specify the sequence of a complementary polynucleotide, which, in turn, can specify the sequence of the original molecule, allowing the original nucleic acid to be replicated (**Figure 7–45**). Such complementary templating mechanisms lie at the heart of both DNA replication and transcription in modern-day cells.

But the efficient synthesis of polynucleotides by such complementary templating mechanisms also requires catalysts to promote the polymerization reaction: without catalysts, polymer formation is slow, error-prone, and inefficient. Today, nucleotide polymerization is catalyzed by protein enzymes—such as DNA and RNA polymerases. But how could this reaction be catalyzed before proteins with the appropriate catalytic ability existed? The beginnings of an answer were obtained in 1982, when it was discovered that RNA molecules themselves can act as catalysts. The unique potential of RNA molecules to act both as information carriers and as catalysts is thought to have enabled them to have a central role in the origin of life.

In present-day cells, RNA is synthesized as a single-stranded molecule, and we have seen that complementary base-pairing can occur between nucleotides in the same chain. This base-pairing, along with nonconventional hydrogen bonds, can cause each RNA molecule to fold up in a unique way that is determined by its nucleotide sequence (see Figure 7–5). Such associations produce complex three-dimensional shapes.

As we discuss in Chapter 4, protein enzymes are able to catalyze biochemical reactions because they have surfaces with unique contours and chemical properties. In the same way, RNA molecules, with their unique folded shapes, can serve as catalysts (**Figure 7–46**). RNAs do not have the same structural and functional diversity as do protein enzymes; they are, after all, built from only four different subunits. Nonetheless, ribozymes can catalyze many types of chemical reactions. Most of the ribozymes that have been studied were constructed in the laboratory and selected for their catalytic activity in a test tube (**Table 7–4**), as relatively few catalytic RNAs exist in present-day cells. But the processes in which catalytic RNAs still seem to have major roles include some of the most

Figure 7–45 An RNA molecule can in principle guide the formation of an exact copy of itself. In the first step, the original RNA molecule acts as a template to form an RNA molecule of complementary sequence. In the second step, this complementary RNA molecule itself acts as a template to form an RNA molecule of the original sequence. Since each template molecule can produce many copies of the complementary strand, these reactions can result in the amplification of the original sequence.

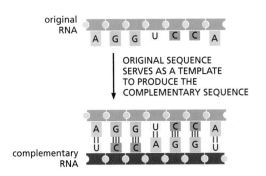
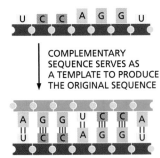

TABLE 7–4 BIOCHEMICAL REACTIONS THAT CAN BE CATALYZED BY RIBOZYMES

Activity	Ribozymes
Peptide bond formation in protein synthesis	ribosomal RNA
DNA ligation	*in vitro* selected RNA
RNA splicing	self-splicing RNAs, small nuclear RNAs
RNA polymerization	*in vitro* selected RNA
RNA phosphorylation	*in vitro* selected RNA
RNA aminoacylation	*in vitro* selected RNA
RNA alkylation	*in vitro* selected RNA
C–C bond rotation (isomerization)	*in vitro* selected RNA

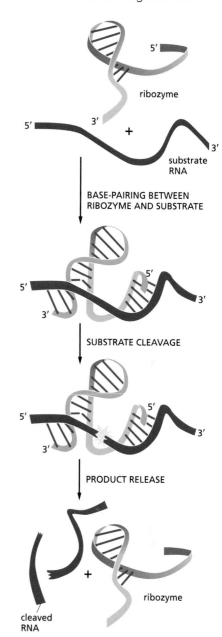

fundamental steps in the expression of genetic information—especially those steps where RNA molecules themselves are spliced or translated into protein.

RNA, therefore, has all the properties required of a molecule that could catalyze its own synthesis (**Figure 7–47**). Although self-replicating systems of RNA molecules have not been found in nature, scientists appear to be well on the way to constructing them in the laboratory. Although this demonstration would not prove that self-replicating RNA molecules were essential to the origin of life on Earth, it would establish that such a scenario is possible.

RNA Is Thought to Predate DNA in Evolution

The first cells on Earth would presumably have been much less complex and less efficient in reproducing themselves than even the simplest present-day cells. They would have consisted of little more than a simple membrane enclosing a set of self-replicating molecules and a few other components required to provide the materials and energy for this autocatalytic replication. If the evolutionary role for RNA proposed above is correct, these earliest cells would also have differed fundamentally from the cells we know today in having their hereditary information stored in RNA rather than DNA.

Evidence that RNA arose before DNA in evolution can be found in the chemical differences between them. Ribose (see Figure 7–3A), like

Figure 7–46 A ribozyme is an RNA molecule that possesses catalytic activity. The RNA molecule shown catalyzes the cleavage of a second RNA at a specific site. Similar ribozymes are found embedded in large RNA genomes—called viroids—that infect plants, where the cleavage reaction is one step in the replication of the viroid. (Adapted from T.R. Cech and O.C. Uhlenbeck, *Nature* 372:39–40, 1994. With permission from Macmillan Publishers Ltd.)

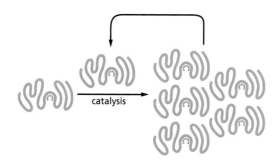

Figure 7–47 Could an RNA molecule catalyze its own synthesis? This hypothetical process would require that the RNA catalyze both steps shown in Figure 7–45. The *red* rays represent the active site of this ribozyme.

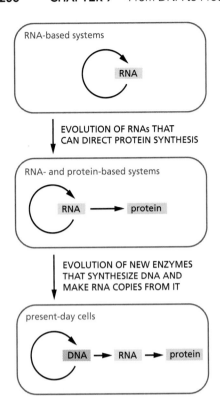

Figure 7–48 RNA may have preceded DNA and proteins in evolution. According to this hypothesis, RNA molecules provided genetic, structural, and catalytic functions in the earliest cells. DNA is now the repository of genetic information, and proteins carry out almost all catalysis in cells. RNA now functions mainly as a go-between in protein synthesis, while remaining a catalyst for a few crucial reactions (including protein synthesis).

QUESTION 7–6

Discuss the following: "During the evolution of life on Earth, RNA lost its glorious position as the first self-replicating catalyst. Its role now is as a mere messenger in the information flow from DNA to protein."

---

glucose and other simple carbohydrates, is readily formed from formaldehyde (HCHO), which is one of the principal products of experiments simulating conditions on the primitive Earth. The sugar deoxyribose is harder to make, and in present-day cells it is produced from ribose in a reaction catalyzed by a protein enzyme, suggesting that ribose predates deoxyribose in cells. Presumably, DNA appeared on the scene after RNA, and then proved more suited than RNA as a permanent repository of genetic information. In particular, the deoxyribose in its sugar–phosphate backbone makes chains of DNA chemically much more stable than chains of RNA, so that greater lengths of DNA can be maintained without breakage.

The other differences between RNA and DNA—the double-helical structure of DNA and the use of thymine rather than uracil—further enhance DNA stability by making the molecule easier to repair. We saw in Chapter 6 that a damaged nucleotide on one strand of the double helix can be repaired by using the other strand as a template. Furthermore, deamination, one of the most common unwanted chemical changes occurring in polynucleotides, is easier to detect and repair in DNA than in RNA (see Figure 6–23). This is because the product of the deamination of cytosine is, by chance, uracil, which already exists in RNA, so that such damage would be impossible for repair enzymes to detect in an RNA molecule. However, in DNA, which has thymine rather than uracil, any uracil produced by the accidental deamination of cytosine is easily detected and repaired.

Taken together, the evidence we have discussed supports the idea that RNA—with its ability to provide genetic, structural, and catalytic functions—preceded DNA in evolution. As cells more closely resembling present-day cells appeared, it is believed that many of the functions originally performed by RNA were taken over by DNA and proteins: DNA took over the primary genetic function, and proteins became the major catalysts, while RNA remained primarily as the intermediary connecting the two (**Figure 7–48**). With the advent of DNA, cells were able to become more complex, for they could then carry and transmit more genetic information than could be stably maintained by RNA alone. Because of the greater chemical complexity of proteins and the variety of chemical reactions they can catalyze, the shift (albeit incomplete) from RNA to proteins also provided a much richer source of structural components and enzymes. This enabled cells to evolve the great diversity of structure and function that we see in life today.

ESSENTIAL CONCEPTS

- The flow of genetic information in all living cells is DNA → RNA → protein. The conversion of the genetic instructions in DNA into RNAs and proteins is termed gene expression.

- To express the genetic information carried in DNA, the nucleotide sequence of a gene is first transcribed into RNA. Transcription is catalyzed by the enzyme RNA polymerase, which uses nucleotide sequences in the DNA molecule to determine which strand to use as a template, and where to start and stop transcribing.

- RNA differs in several respects from DNA. It contains the sugar ribose instead of deoxyribose and the base uracil (U) instead of thymine (T). RNAs in cells are synthesized as single-stranded molecules, which often fold up into complex three-dimensional shapes.

- Cells make several functional types of RNAs, including messenger RNAs (mRNAs), which carry the instructions for making proteins; ribosomal RNAs (rRNAs), which are the crucial components of ribosomes; and transfer RNAs (tRNAs), which act as adaptor molecules in protein synthesis.

- To begin transcription, RNA polymerase binds to specific DNA sites called promoters that lie immediately upstream of genes. To initiate transcription, eukaryotic RNA polymerases require the assembly of a complex of general transcription factors at the promoter, whereas bacterial RNA polymerase requires only an additional subunit, called sigma factor.

- Most protein-coding genes in eukaryotic cells are composed of a number of coding regions, called exons, interspersed with larger noncoding regions, called introns. When a eukaryotic gene is transcribed from DNA into RNA, both the exons and introns are copied.

- Introns are removed from the RNA transcripts in the nucleus by RNA splicing, a reaction catalyzed by small ribonucleoprotein complexes known as snRNPs. Splicing removes the introns from the RNA and joins together the exons—often in a variety of combinations, allowing multiple proteins to be produced from the same gene.

- Eukaryotic pre-mRNAs go through several additional RNA processing steps before they leave the nucleus as mRNAs, including 5′ RNA capping and 3′ polyadenylation. These reactions, along with splicing, take place as the pre-mRNA is being transcribed.

- Translation of the nucleotide sequence of an mRNA into a protein takes place in the cytoplasm on large ribonucleoprotein assemblies called ribosomes. As the mRNA moves through the ribosome, its message is translated into protein.

- The nucleotide sequence in mRNA is read in sets of three nucleotides called codons; each codon corresponds to one amino acid.

- The correspondence between amino acids and codons is specified by the genetic code. The possible combinations of the 4 different nucleotides in RNA give 64 different codons in the genetic code. Most amino acids are specified by more than one codon.

- tRNAs act as adaptor molecules in protein synthesis. Enzymes called aminoacyl-tRNA synthetases covalently link amino acids to their appropriate tRNAs. Each tRNA contains a sequence of three nucleotides, the anticodon, which recognizes a codon in an mRNA through complementary base-pairing.

- Protein synthesis begins when a ribosome assembles at an initiation codon (AUG) in an mRNA molecule, a process that depends on proteins called translation initiation factors. The completed protein chain is released from the ribosome when a stop codon (UAA, UAG, or UGA) in the mRNA is reached.

- The stepwise linking of amino acids into a polypeptide chain is catalyzed by an rRNA molecule in the large ribosomal subunit, which thus acts as a ribozyme.

- The concentration of a protein in a cell depends on the rate at which the mRNA and protein are synthesized and degraded. Protein degradation in the cytosol and nucleus occurs inside large protein complexes called proteasomes.

- From our knowledge of present-day organisms and the molecules they contain, it seems likely that life on Earth began with the evolution of RNA molecules that could catalyze their own replication.

- It has been proposed that RNA served as both the genome and the catalysts in the first cells, before DNA replaced RNA as a more stable molecule for storing genetic information, and proteins replaced RNAs as the major catalytic and structural components. RNA catalysts in modern cells are thought to provide a glimpse into an ancient, RNA-based world.

KEY TERMS

alternative splicing
aminoacyl-tRNA synthetase
anticodon
codon
exon
gene
gene expression
general transcription factors
genetic code
initiator tRNA
intron

messenger RNA (mRNA)
polyadenylation
promoter
protease
proteasome
reading frame
ribosomal RNA (rRNA)
ribosome
ribozyme
RNA
RNA capping

RNA polymerase
RNA processing
RNA splicing
RNA transcript
RNA world
small nuclear RNA (snRNA)
spliceosome
transcription
transfer RNA (tRNA)
translation
translation initiation factor

QUESTIONS

QUESTION 7–7

Which of the following statements are correct? Explain your answers.

A. An individual ribosome can make only one type of protein.

B. All mRNAs fold into particular three-dimensional structures that are required for their translation.

C. The large and small subunits of an individual ribosome always stay together and never exchange partners.

D. Ribosomes are cytoplasmic organelles that are encapsulated by a single membrane.

E. Because the two strands of DNA are complementary, the mRNA of a given gene can be synthesized using either strand as a template.

F. An mRNA may contain the sequence **ATTGACCCCGGTCAA.**

G. The amount of a protein present in a cell depends on its rate of synthesis, its catalytic activity, and its rate of degradation.

QUESTION 7–8

The Lacheinmal protein is a hypothetical protein that causes people to smile more often. It is inactive in many chronically unhappy people. The mRNA isolated from a number of different unhappy individuals in the same family was found to lack an internal stretch of 173 nucleotides that is present in the Lacheinmal mRNA isolated from happy members of the same family. The DNA sequences of the *Lacheinmal* genes from the happy and unhappy family members were determined and compared. They differed by a single nucleotide substitution, which lay in an intron. What can you say about the molecular basis of unhappiness in this family?

(Hints: [1] Can you hypothesize a molecular mechanism by which a single nucleotide substitution in a gene could cause the observed deletion in the mRNA? Note that the deletion is *internal* to the mRNA. [2] Assuming the 173-base-pair deletion removes coding sequences from the Lacheinmal mRNA, how would the Lacheinmal protein differ between the happy and unhappy people?)

QUESTION 7–9

Use the genetic code shown in Figure 7–25 to identify which of the following nucleotide sequences would code for the polypeptide sequence arginine-glycine-aspartate:

1. 5′-AGA-GGA-GAU-3′

2. 5′-ACA-CCC-ACU-3′

3. 5′-GGG-AAA-UUU-3′

4. 5′-CGG-GGU-GAC-3′

QUESTION 7–10

"The bonds that form between the anticodon of a tRNA molecule and the three nucleotides of a codon in mRNA are _____." Complete this sentence with each of the following options and explain why each of the resulting statements is correct or incorrect.

A. Covalent bonds formed by GTP hydrolysis

B. Hydrogen bonds that form when the tRNA is at the A site

C. Broken by the translocation of the ribosome along the mRNA

QUESTION 7–11

List the ordinary, dictionary definitions of the terms *replication, transcription,* and *translation.* By their side, list the special meaning each term has when applied to the living cell.

QUESTION 7–12

In an alien world, the genetic code is written in pairs of nucleotides. How many amino acids could such a code specify? In a different world, a triplet code is used, but the sequence of nucleotides is not important; it only matters which nucleotides are present. How many amino acids could this code specify? Would you expect to encounter any problems translating these codes?

QUESTION 7–13

One remarkable feature of the genetic code is that amino acids with similar chemical properties often have similar codons. Thus codons with U or C as the second nucleotide tend to specify hydrophobic amino acids. Can you suggest a possible explanation for this phenomenon in terms of the early evolution of the protein-synthesis machinery?

QUESTION 7–14

A mutation in DNA generates a UGA stop codon in the middle of the mRNA coding for a particular protein. A second mutation in the cell's DNA leads to a single nucleotide change in a tRNA that allows the correct translation of the protein; that is, the second mutation "suppresses" the defect caused by the first. The altered tRNA translates the UGA as tryptophan. What nucleotide change has probably occurred in the mutant tRNA molecule? What consequences would the presence of such a mutant tRNA have for the translation of the normal genes in this cell?

QUESTION 7–15

The charging of a tRNA with an amino acid can be represented by the following equation:

amino acid + tRNA + ATP $\rightarrow$ aminoacyl-tRNA + AMP + PP$_i$

where PP$_i$ is pyrophosphate (see Figure 3–40). In the aminoacyl-tRNA, the amino acid and tRNA are linked with a high-energy covalent bond; a large portion of the energy derived from the hydrolysis of ATP is thus stored in this bond and is available to drive peptide bond formation at the later stages of protein synthesis. The free-energy change of the charging reaction shown in the equation is close to zero and therefore would not be expected to favor attachment of the amino acid to tRNA. Can you suggest a further step that could drive the reaction to completion?

QUESTION 7–16

A. The average molecular weight of a protein in the cell is about 30,000 daltons. A few proteins, however, are much larger. The largest known polypeptide chain made by any cell is a protein called titin (made by mammalian muscle cells), and it has a molecular weight of 3,000,000 daltons. Estimate how long it will take a muscle cell to translate an mRNA coding for titin (assume the average molecular weight of an amino acid to be 120, and a translation rate of two amino acids per second for eukaryotic cells).

B. Protein synthesis is very accurate: for every 10,000 amino acids joined together, only one mistake is made. What is the fraction of average-sized protein molecules and of titin molecules that are synthesized without any errors? (Hint: the probability P of obtaining an error-free protein is given by $P = (1 - E)^n$, where E is the error frequency and n the number of amino acids.)

C. The molecular weight of all eukaryotic ribosomal proteins combined is about 2.5×10^6 daltons. Would it be advantageous to synthesize them as a single protein?

D. Transcription occurs at a rate of about 30 nucleotides per second. Is it possible to calculate the time required to synthesize a titin mRNA from the information given here?

QUESTION 7–17

Which of the following types of mutations would be predicted to harm an organism? Explain your answers.

A. Insertion of a single nucleotide near the end of the coding sequence.

B. Removal of a single nucleotide near the beginning of the coding sequence.

C. Deletion of three consecutive nucleotides in the middle of the coding sequence.

D. Deletion of four consecutive nucleotides in the middle of the coding sequence.

E. Substitution of one nucleotide for another in the middle of the coding sequence.

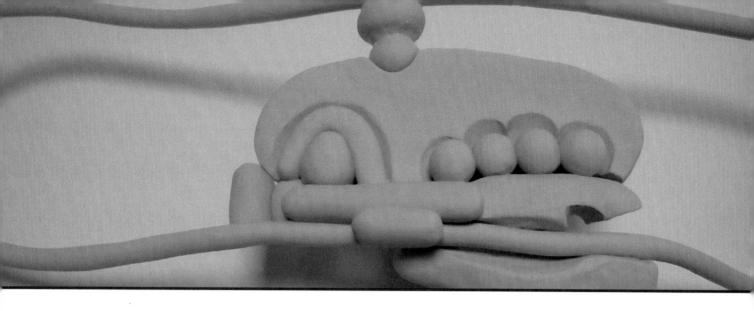

CHAPTER EIGHT

8

Control of Gene Expression

An organism's DNA encodes all of the RNA and protein molecules that are needed to make its cells. Yet a complete description of the DNA sequence of an organism—be it the few million nucleotides of a bacterium or the few billion nucleotides in each human cell—does not enable us to reconstruct that organism any more than a list of all the English words in a dictionary enables us to reconstruct a play by Shakespeare. We need to know how the elements in the DNA sequence or the words on a list work together to make the masterpiece.

For cells, the question involves *gene expression*. Even the simplest single-celled bacterium can use its genes selectively—for example, switching genes on and off to make the enzymes needed to digest whatever food sources are available. In multicellular plants and animals, however, gene expression is under much more elaborate control. Over the course of embryonic development, a fertilized egg cell gives rise to many cell types that differ dramatically in both structure and function. The differences between an information-processing nerve cell and an infection-fighting white blood cell, for example, are so extreme that it is difficult to imagine that the two cells contain the same DNA (**Figure 8–1**). For this reason, and because cells in an adult organism rarely lose their distinctive characteristics, biologists originally suspected that certain genes might be selectively lost when a cell becomes specialized. We now know, however, that nearly all the cells of a multicellular organism contain the same genome. Cell *differentiation* is instead achieved by changes in gene expression.

In mammals, hundreds of different cell types carry out a range of specialized functions that depend upon genes that are switched on in that

AN OVERVIEW OF GENE EXPRESSION

HOW TRANSCRIPTIONAL SWITCHES WORK

THE MOLECULAR MECHANISMS THAT CREATE SPECIALIZED CELL TYPES

POST-TRANSCRIPTIONAL CONTROLS

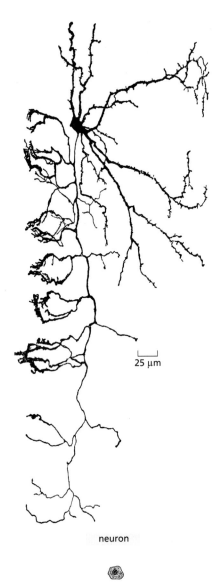

25 μm

neuron

liver cell

Figure 8–1 A neuron and a liver cell share the same genome. The long branches of this neuron from the retina enable it to receive electrical signals from many other neurons and carry them to many neighboring neurons. The liver cell, which is drawn to the same scale, is involved in many metabolic processes, including digestion and the detoxification of alcohol and other drugs. Both of these mammalian cells contain the same genome, but they express many different RNAs and proteins. (Neuron adapted from S. Ramón y Cajal, Histologie du Système Nerveux de l'Homme et de Vertébrés, 1909–1911. Paris: Maloine; reprinted, Madrid: C.S.I.C., 1972.)

cell type but not in most others: for example, the β cells of the pancreas make the protein hormone insulin, while the α cells of the pancreas make the hormone glucagon; the B lymphocytes of the immune system make antibodies, while developing red blood cells make the oxygen-transport protein hemoglobin. The differences between a neuron, a white blood cell, a pancreatic β cell, and a red blood cell depend upon the precise control of gene expression. A typical differentiated cell expresses only about half the genes in its total repertoire.

In this chapter, we discuss the main ways in which gene expression is regulated, with a focus on those genes that encode proteins as their final product. Although some of these control mechanisms apply to both eukaryotes and prokaryotes, eukaryotic cells—with their more complex chromosomal structure—have some ways of controlling gene expression that are not available to bacteria.

AN OVERVIEW OF GENE EXPRESSION

Gene expression is a complex process by which cells selectively direct the synthesis of the many thousands of proteins and RNAs encoded in their genome. But how do cells coordinate and control such an intricate process—and how does an individual cell specify which of its genes to express? This decision is an especially important problem for animals because, as they develop, their cells become highly specialized, ultimately producing an array of muscle, nerve, and blood cells, along with the hundreds of other cell types seen in the adult. Such cell **differentiation** arises because cells make and accumulate different sets of RNA and protein molecules: that is, they express different genes.

The Different Cell Types of a Multicellular Organism Contain the Same DNA

The evidence that cells have the ability to change which genes they express without altering the nucleotide sequence of their DNA comes from experiments in which the genome from a differentiated cell is made to direct the development of a complete organism. If the chromosomes of the differentiated cell were altered irreversibly during development, they would not be able to accomplish this feat.

Consider, for example, an experiment in which the nucleus is taken from a skin cell in an adult frog and injected into a frog egg from which the nucleus has been removed. In at least some cases, that doctored egg will develop into a normal tadpole (**Figure 8–2**). Thus, the transplanted skin-cell nucleus cannot have lost any critical DNA sequences. Nuclear transplantation experiments carried out with differentiated cells taken from adult mammals—including sheep, cows, pigs, goats, and mice—have shown similar results. And in plants, individual cells removed from a carrot, for example, can regenerate an entire adult carrot plant. These experiments all show that the DNA in specialized cell types of multicellular organisms still contains the entire set of instructions needed to form

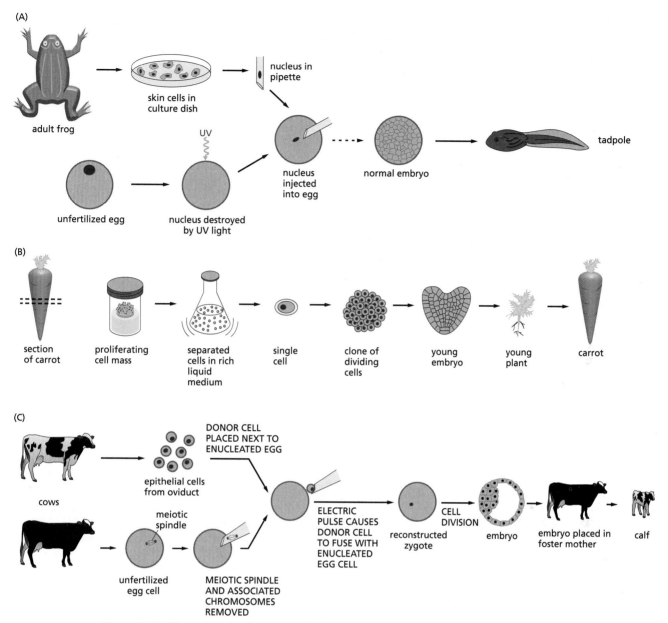

Figure 8–2 Differentiated cells contain all the genetic instructions necessary to direct the formation of a complete organism. (A) The nucleus of a skin cell from an adult frog transplanted into an egg whose nucleus has been destroyed can give rise to an entire tadpole. The broken arrow indicates that to give the transplanted genome time to adjust to an embryonic environment, a further transfer step is required in which one of the nuclei is taken from the early embryo that begins to develop and is put back into a second enucleated egg. (B) In many types of plants, differentiated cells retain the ability to "de-differentiate," so that a single cell can proliferate to form a clone of progeny cells that later give rise to an entire plant. (C) A nucleus removed from a differentiated cell from an adult cow can be introduced into an enucleated egg from a different cow to give rise to a calf. Different calves produced from the same differentiated cell donor are all clones of the donor and are therefore genetically identical. (A, modified from J.B. Gurdon, *Sci. Am.* 219:24–35, 1968, with permission from the Estate of Bunji Tagawa.)

a whole organism. The various cell types of an organism therefore differ not because they contain different genes, but because they express them differently.

Different Cell Types Produce Different Sets of Proteins

The extent of the differences in gene expression between different cell types may be roughly gauged by comparing the protein composition of cells in liver, heart, brain, and so on. In the past, such analysis was performed by two-dimensional gel electrophoresis (see Panel 4–5, p. 167). Nowadays, the total protein content of a cell can be rapidly analyzed by a method called mass spectrometry (see Figure 4–49). This technique is much more sensitive than electrophoresis and it enables the detection of even proteins that are produced in minor quantities.

Both techniques reveal that many proteins are common to all the cells of a multicellular organism. These *housekeeping* proteins include, for example, the structural proteins of chromosomes, RNA polymerases, DNA repair enzymes, ribosomal proteins, enzymes involved in glycolysis and other basic metabolic processes, and many of the proteins that form the cytoskeleton. In addition, each different cell type also produces specialized proteins that are responsible for the cell's distinctive properties. In mammals, for example, hemoglobin is made almost exclusively in developing red blood cells.

Gene expression can also be studied by cataloging a cell's RNAs, including the mRNAs that encode protein. The most comprehensive methods for such analyses involve determining the nucleotide sequence of every RNA molecule made by the cell, an approach that can also reveal their relative abundance. Estimates of the number of different mRNA sequences in human cells suggest that, at any one time, a typical differentiated human cell expresses perhaps 5000–15,000 protein-coding genes from a total of about 21,000. It is the expression of a different collection of genes in each cell type that causes the large variations seen in the size, shape, behavior, and function of differentiated cells.

A Cell Can Change the Expression of Its Genes in Response to External Signals

The specialized cells in a multicellular organism are capable of altering their patterns of gene expression in response to extracellular cues. For example, if a liver cell is exposed to the steroid hormone cortisol, the production of several proteins is dramatically increased. Released by the adrenal gland during periods of starvation, intense exercise, or prolonged stress, cortisol signals liver cells to boost the production of glucose from amino acids and other small molecules. The set of proteins whose production is induced by cortisol includes enzymes such as tyrosine aminotransferase, which helps convert tyrosine to glucose. When the hormone is no longer present, the production of these proteins returns to its resting level.

Other cell types respond to cortisol differently. In fat cells, for example, the production of tyrosine aminotransferase is reduced, while some other cell types do not respond to cortisol at all. The fact that different cell types often respond in different ways to the same extracellular signal contributes to the specialization that gives each cell type its distinctive character.

Gene Expression Can Be Regulated at Various Steps from DNA to RNA to Protein

If differences among the various cell types of an organism depend on the particular genes that the cells express, at what level is the control of gene expression exercised? As we saw in the last chapter, there are many steps in the pathway leading from DNA to protein, and all of them can in principle be regulated. Thus a cell can control the proteins it contains by (1) controlling when and how often a given gene is transcribed, (2) controlling how an RNA transcript is spliced or otherwise processed, (3) selecting which mRNAs are exported from the nucleus to the cytosol, (4) regulating how quickly certain mRNA molecules are degraded, (5) selecting which mRNAs are translated into protein by ribosomes, or

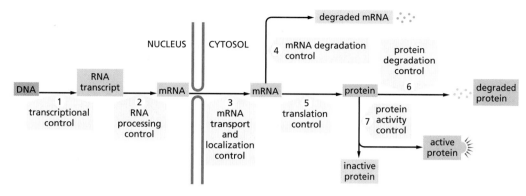

(6) regulating how rapidly specific proteins are destroyed after they have been made; in addition, the activity of individual proteins can be further regulated in a variety of ways. These steps are illustrated in **Figure 8–3**.

Gene expression can be regulated at each of these steps. For most genes, however, the control of transcription (step number 1 in Figure 8–3) is paramount. This makes sense because only transcriptional control can ensure that no unnecessary intermediates are synthesized. So it is the regulation of transcription—and the DNA and protein components that determine which genes a cell transcribes into RNA—that we address first.

HOW TRANSCRIPTIONAL SWITCHES WORK

Until 50 years ago, the idea that genes could be switched on and off was revolutionary. This concept was a major advance, and it came originally from studies of how *E. coli* bacteria adapt to changes in the composition of their growth medium. Many of the same principles apply to eukaryotic cells. However, the enormous complexity of gene regulation in higher organisms, combined with the packaging of their DNA into chromatin, creates special challenges and some novel opportunities for control—as we will see. We begin with a discussion of the *transcription regulators,* proteins that bind to DNA and control gene transcription.

Transcription Regulators Bind to Regulatory DNA Sequences

Control of transcription is usually exerted at the step at which the process is initiated. In Chapter 7, we saw that the **promoter** region of a gene binds the enzyme *RNA polymerase* and correctly orients the enzyme to begin its task of making an RNA copy of the gene. The promoters of both bacterial and eukaryotic genes include a *transcription initiation site*, where RNA synthesis begins, plus a sequence of approximately 50 nucleotide pairs that extends upstream from the initiation site (if one likens the direction of transcription to the flow of a river). This upstream region contains sites that are required for the RNA polymerase to recognize the *promoter*, although they do not bind to RNA polymerase directly. Instead, these sequences contain recognition sites for proteins that associate with the active polymerase—sigma factor in bacteria (see Figure 7–9) or the general transcription factors in eukaryotes (see Figure 7–12).

In addition to the promoter, nearly all genes, whether bacterial or eukaryotic, have **regulatory DNA sequences** that are used to switch the gene on or off. Some regulatory DNA sequences are as short as 10 nucleotide pairs and act as simple switches that respond to a single signal; such simple regulatory switches predominate in bacteria. Other regulatory DNA sequences, especially those in eukaryotes, are very long (sometimes spanning more than 10,000 nucleotide pairs) and act as molecular

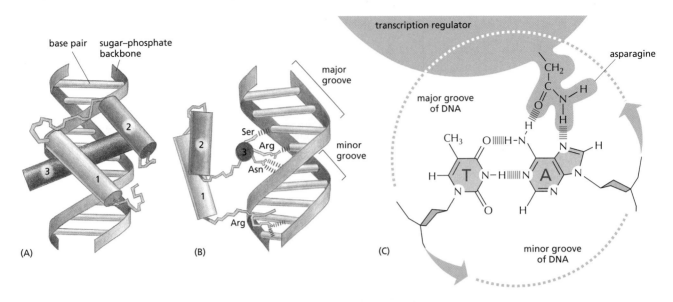

Figure 8–4 A transcription regulator interacts with the major groove of a DNA double helix. (A) This regulator recognizes DNA via three α helices, shown as numbered cylinders, which allow the protein to fit into the major groove and form tight associations with the base pairs in a short stretch of DNA. This particular structural motif, called a *homeodomain*, is found in many eukaryotic DNA-binding proteins (Movie 8.1). (B) Most of the contacts with the DNA bases are made by helix 3 (*red*), which is shown here end-on. The protein interacts with the edges of the nucleotides without disrupting the hydrogen bonds that hold the base pairs together. (C) An asparagine residue from helix 3 forms two hydrogen bonds with the adenine in an A-T base pair. The view is end-on looking down the DNA double helix, and the protein contacts the base pair from the major groove side. For simplicity, only one amino acid–base contact is shown; in reality, transcription regulators form hydrogen bonds (as shown here), ionic bonds, and hydrophobic interactions with individual bases in the major groove. Typically, the protein–DNA interface would consist of 10–20 such contacts, each involving a different amino acid and each contributing to the overall strength of the protein–DNA interaction.

microprocessors, integrating information from a variety of signals into a command that dictates how often transcription of the gene is initiated.

Regulatory DNA sequences do not work by themselves. To have any effect, these sequences must be recognized by proteins called **transcription regulators**. It is the binding of a transcription regulator to a regulatory DNA sequence that acts as the switch to control transcription. The simplest bacterium produces several hundred different transcription regulators, each of which recognizes a different DNA sequence and thereby regulates a distinct set of genes. Humans make many more—several thousand—indicating the importance and complexity of this form of gene regulation in the development and function of a complex organism.

Proteins that recognize a specific nucleotide sequence do so because the surface of the protein fits tightly against the surface features of the DNA double helix in that region. Because these surface features will vary depending on the nucleotide sequence, different DNA-binding proteins will recognize different nucleotide sequences. In most cases, the protein inserts into the major groove of the DNA helix and makes a series of intimate molecular contacts with the nucleotide pairs within the groove (**Figure 8–4**). Although each individual contact is weak, the 10 to 20 contacts that are typically formed at the protein–DNA interface combine to ensure that the interaction is both highly specific and very strong; indeed, protein–DNA interactions are among the tightest and most specific molecular interactions known in biology.

Many transcription regulators bind to the DNA helix as dimers (**Figure 8–5**). Such dimerization roughly doubles the area of contact with the DNA, thereby greatly increasing the strength and specificity of the protein–DNA interaction.

Transcriptional Switches Allow Cells to Respond to Changes in Their Environment

The simplest and best understood examples of gene regulation occur in bacteria and in the viruses that infect them. The genome of the bacterium *E. coli* consists of a single circular DNA molecule of about 4.6×10^6 nucleotide pairs. This DNA encodes approximately 4300 proteins, although only a fraction of these are made at any one time. Bacteria regulate the expression of many of their genes according to the food sources that are available in the environment. For example, in *E. coli*, five genes code for enzymes that manufacture the amino acid tryptophan. These genes are arranged in a cluster on the chromosome and are transcribed from a single promoter as one long mRNA molecule; such coordinately transcribed clusters are called *operons* (**Figure 8–6**). Although operons are common in bacteria, they are rare in eukaryotes, where genes are transcribed and regulated individually (see Figure 7–2).

When tryptophan concentrations are low, the operon is transcribed; the resulting mRNA is translated to produce a full set of biosynthetic enzymes, which work in tandem to synthesize tryptophan. When tryptophan is abundant, however—for example, when the bacterium is in the gut of a mammal that has just eaten a protein-rich meal—the amino acid is imported into the cell and shuts down production of the enzymes, which are no longer needed.

We now understand in considerable detail how this repression of the tryptophan operon comes about. Within the operon's promoter is a short DNA sequence, called the operator (see Figure 8–6), that is recognized by a transcription regulator. When this regulator binds to the *operator*, it blocks access of RNA polymerase to the promoter, preventing transcription of the operon and production of the tryptophan-producing enzymes. The transcription regulator is known as the *tryptophan repressor*, and it is controlled in an ingenious way: the repressor can bind to DNA only if it has also bound several molecules of tryptophan (**Figure 8–7**).

The tryptophan repressor is an allosteric protein (see Figure 4–41): the binding of tryptophan causes a subtle change in its three-dimensional structure so that the protein can bind to the operator sequence. When the concentration of free tryptophan in the bacterium drops, the repressor no longer binds to DNA, and the tryptophan operon is transcribed. The repressor is thus a simple device that switches production of a set of biosynthetic enzymes on and off according to the availability of the end product of the pathway that the enzymes catalyze.

The tryptophan repressor protein itself is always present in the cell. The gene that encodes it is continuously transcribed at a low level, so that a small amount of the repressor protein is always being made. Thus the bacterium can respond very rapidly to a rise in tryptophan concentration.

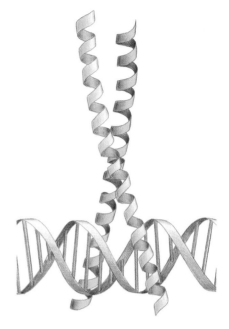

Figure 8–5 Many transcription regulators bind to DNA as dimers. This transcription regulator contains a *leucine zipper* motif, which is formed by two α helices, each contributed by a different protein subunit. Leucine zipper proteins thus bind to DNA as dimers, gripping the double helix like a clothespin on a clothesline (Movie 8.2).

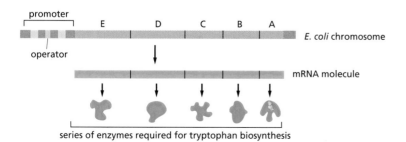

series of enzymes required for tryptophan biosynthesis

Figure 8–6 A cluster of bacterial genes can be transcribed from a single promoter. Each of these five genes encodes a different enzyme; all of the enzymes are needed to synthesize the amino acid tryptophan. The genes are transcribed as a single mRNA molecule, a feature that allows their expression to be coordinated. Clusters of genes transcribed as a single mRNA molecule are common in bacteria. Each of these clusters is called an *operon* because its expression is controlled by a regulatory DNA sequence called the *operator* (*green*), situated within the promoter. The *yellow* blocks in the promoter represent DNA sequences that bind RNA polymerase.

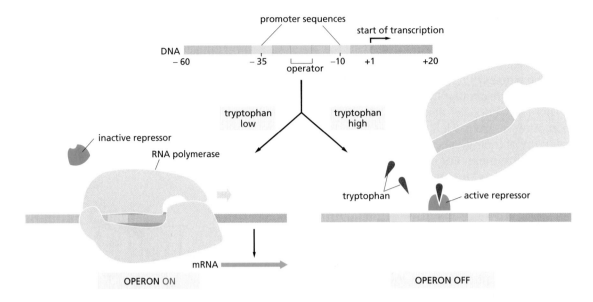

Figure 8–7 Genes can be switched off by repressor proteins. If the concentration of tryptophan inside a bacterium is low (*left*), RNA polymerase (*blue*) binds to the promoter and transcribes the five genes of the tryptophan operon. However, if the concentration of tryptophan is high (*right*), the repressor protein (*dark green*) becomes active and binds to the operator (*light green*), where it blocks the binding of RNA polymerase to the promoter. Whenever the concentration of intracellular tryptophan drops, the repressor falls off the DNA, allowing the polymerase to again transcribe the operon. The promoter contains two key blocks of DNA sequence information, the –35 and –10 regions, highlighted in *yellow*, which are recognized by RNA polymerase (see Figure 7–10). The complete operon is shown in Figure 8–6.

Repressors Turn Genes Off and Activators Turn Them On

The tryptophan repressor, as its name suggests, is a **transcriptional repressor** protein: in its active form, it switches genes off, or *represses* them. Some bacterial transcription regulators do the opposite: they switch genes on, or *activate* them. These **transcriptional activator** proteins work on promoters that—in contrast to the promoter for the tryptophan operon—are only marginally able to bind and position RNA polymerase on their own. However, these poorly functioning promoters can be made fully functional by activator proteins that bind nearby and contact the RNA polymerase to help it initiate transcription (**Figure 8–8**).

Like the tryptophan repressor, activator proteins often have to interact with a second molecule to be able to bind DNA. For example, the bacterial activator protein *CAP* has to bind cyclic AMP (cAMP) before it can bind to DNA (see Figure 4–19). Genes activated by CAP are switched on in response to an increase in intracellular cAMP concentration, which rises when glucose, the bacterium's preferred carbon source, is no longer available; as a result, CAP drives the production of enzymes that allow the bacterium to digest other sugars.

An Activator and a Repressor Control the *Lac* Operon

In many instances, the activity of a single promoter is controlled by two different transcription regulators. The *Lac operon* in *E. coli*, for example,

Figure 8–8 Genes can be switched on by activator proteins. An activator protein binds to a regulatory sequence on the DNA and then interacts with the RNA polymerase to help it initiate transcription. Without the activator, the promoter fails to initiate transcription efficiently. In bacteria, the binding of the activator to DNA is often controlled by the interaction of a metabolite or other small molecule (*red* triangle) with the activator protein. The *Lac* operon works in this manner, as we discuss shortly.

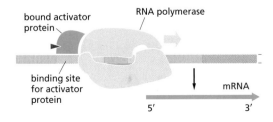

is controlled by both the *Lac repressor* and the CAP activator that we just discussed. The *Lac* operon encodes proteins required to import and digest the disaccharide lactose. In the absence of glucose, the bacterium makes cAMP, which activates CAP to switch on genes that allow the cell to utilize alternative sources of carbon—including lactose. It would be wasteful, however, for CAP to induce expression of the *Lac* operon if lactose itself were not present. Thus the Lac repressor shuts off the operon in the absence of lactose. This arrangement enables the control region of the *Lac* operon to integrate two different signals, so that the operon is highly expressed only when two conditions are met: glucose must be absent and lactose must be present (**Figure 8–9**). This genetic circuit thus behaves much like a switch that carries out a logic operation in a computer. When lactose is present AND glucose is absent, the cell executes the appropriate program—in this case, transcription of the genes that permit the uptake and utilization of lactose.

The elegant logic of the *Lac* operon first attracted the attention of biologists more than 50 years ago. The molecular basis of the switch in *E. coli* was uncovered by a combination of genetics and biochemistry, providing the first insight into how transcription is controlled. In a eukaryotic cell, similar transcription regulatory devices are combined to generate increasingly complex circuits, including those that enable a fertilized egg to form the tissues and organs of a multicellular organism.

QUESTION 8–1

Bacterial cells can take up the amino acid tryptophan (Trp) from their surroundings, or if there is an insufficient external supply they can synthesize tryptophan from other small molecules. The Trp repressor is a transcription regulator that shuts off the transcription of genes that code for the enzymes required for the synthesis of tryptophan (see Figure 8–7).
A. What would happen to the regulation of the tryptophan operon in cells that express a mutant form of the tryptophan repressor that (1) cannot bind to DNA, (2) cannot bind tryptophan, or (3) binds to DNA even in the absence of tryptophan?
B. What would happen in scenarios (1), (2), and (3) if the cells, in addition, produced normal tryptophan repressor protein from a second, normal gene?

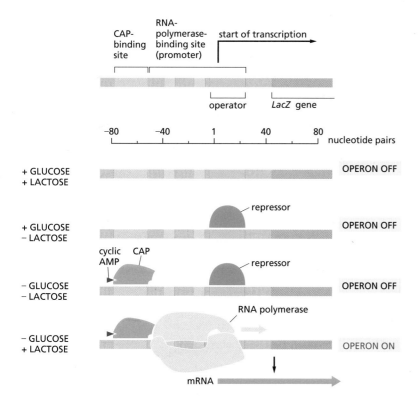

Figure 8–9 The *Lac* operon is controlled by two transcription regulators, the Lac repressor and CAP. When lactose is absent, the Lac repressor binds to the *Lac* operator and shuts off expression of the operon. Addition of lactose increases the intracellular concentration of a related compound, allolactose; allolactose binds to the Lac repressor, causing it to undergo a conformational change that releases its grip on the operator DNA (not shown). When glucose is absent, cyclic AMP (*red* triangle) is produced by the cell, and CAP binds to DNA. *LacZ*, the first gene of the operon, encodes the enzyme β-galactosidase, which breaks down lactose to galactose and glucose.

QUESTION 8–2

Explain how DNA-binding proteins can make sequence-specific contacts to a double-stranded DNA molecule without breaking the hydrogen bonds that hold the bases together. Indicate how, through such contacts, a protein can distinguish a T-A from a C-G pair. Indicate the parts of the nucleotide base pairs that could form noncovalent interactions—hydrogen bonds, electrostatic attractions, or hydrophobic interactions (see Panel 2–7, pp. 78–79)—with a DNA-binding protein. The structures of all the base pairs in DNA are given in Figure 5–6.

Eukaryotic Transcription Regulators Control Gene Expression from a Distance

Eukaryotes, too, use transcription regulators—both activators and repressors—to regulate the expression of their genes. The DNA sites to which eukaryotic gene activators bind are termed *enhancers*, because their presence dramatically enhances the rate of transcription. It was surprising to biologists when, in 1979, it was discovered that these activator proteins could enhance transcription even when they are bound thousands of nucleotide pairs away from a gene's promoter. They also work when bound either upstream or downstream from the gene. These observations raised several questions. How do enhancer sequences and the proteins bound to them function over such long distances? How do they communicate with the promoter?

Many models for this "action at a distance" have been proposed, but the simplest of these seems to apply in most cases. The DNA between the enhancer and the promoter loops out to allow eukaryotic activator proteins to influence directly events that take place at the promoter (**Figure 8–10**). The DNA thus acts as a tether, allowing a protein that is bound to an enhancer—even one that is thousands of nucleotide pairs away—to interact with the proteins in the vicinity of the promoter—including RNA polymerase and the general transcription factors (see Figure 7–12). Often, additional proteins serve to link the distantly bound transcription regulators to these proteins at the promoter; the most important of these regulators is a large complex of proteins known as *Mediator* (see Figure 8–10). One of the ways in which these proteins function is by aiding the assembly of the general transcription factors and RNA polymerase to form a large *transcription complex* at the promoter. Eukaryotic repressor proteins do the opposite: they decrease transcription by preventing the assembly of the same protein complex.

In addition to promoting—or repressing—the assembly of a transcription initiation complex directly, eukaryotic transcription regulators have an additional mechanism of action: they attract proteins that modify chromatin structure and thereby affect the accessibility of the promoter to the general transcription factors and RNA polymerase, as we discuss next.

Figure 8–10 In eukaryotes, gene activation can occur at a distance. An activator protein bound to a distant enhancer attracts RNA polymerase and general transcription factors to the promoter. Looping of the intervening DNA permits contact between the activator and the transcription initiation complex bound to the promoter. In the case shown here, a large protein complex called Mediator serves as a go-between. The broken stretch of DNA signifies that the length of DNA between the enhancer and the start of transcription varies, sometimes reaching tens of thousands of nucleotide pairs in length. The TATA box is a DNA recognition sequence for the first general transcription factor that binds to the promoter (see Figure 7–12).

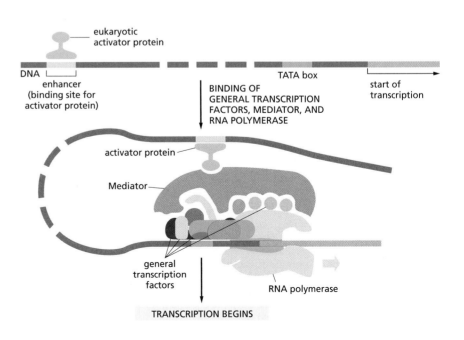

Eukaryotic Transcription Regulators Help Initiate Transcription by Recruiting Chromatin-Modifying Proteins

Initiation of transcription in eukaryotic cells must also take into account the packaging of DNA into chromosomes. As discussed in Chapter 5, eukaryotic DNA is packed into nucleosomes, which, in turn, are folded into higher-order structures. How do transcription regulators, general transcription factors, and RNA polymerase gain access to such DNA? Nucleosomes can inhibit the initiation of transcription if they are positioned over a promoter, because they physically block the assembly of the general transcription factors or RNA polymerase on the promoter. Such chromatin packaging may have evolved in part to prevent leaky gene expression by blocking the initiation of transcription in the absence of the proper activator proteins.

In eukaryotic cells, activator and repressor proteins exploit chromatin structure to help turn genes on and off. As we saw in Chapter 5, chromatin structure can be altered by *chromatin-remodeling complexes* and by enzymes that covalently modify the histone proteins that form the core of the nucleosome (see Figures 5–26 and 5–27). Many gene activators take advantage of these mechanisms by recruiting such chromatin-modifying proteins to promoters. For example, the recruitment of *histone acetyltransferases* promotes the attachment of acetyl groups to selected lysines in the tail of histone proteins. This modification alters chromatin structure, allowing greater accessibility to the underlying DNA; moreover, the acetyl groups themselves attract proteins that promote transcription, including some of the general transcription factors (**Figure 8–11**).

Likewise, gene repressor proteins can modify chromatin in ways that reduce the efficiency of transcription initiation. For example, many repressors attract *histone deacetylases*—enzymes that remove the acetyl groups from histone tails, thereby reversing the positive effects that acetylation has on transcription initiation. Although some eukaryotic repressor proteins work on a gene-by-gene basis, others can orchestrate the formation of large swathes of transcriptionally inactive chromatin containing many

QUESTION 8–3

Some transcription regulators bind to DNA and cause the double helix to bend at a sharp angle. Such "bending proteins" can stimulate the initiation of transcription without contacting either the RNA polymerase, any of the general transcription factors, or any other transcription regulators. Can you devise a plausible explanation for how these proteins might work to modulate transcription? Draw a diagram that illustrates your explanation.

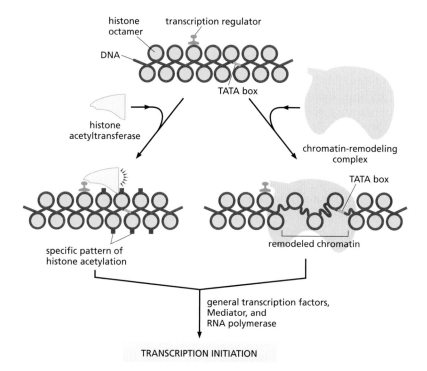

Figure 8–11 Eukaryotic transcriptional activators can recruit chromatin-modifying proteins to help initiate gene transcription. On the right, chromatin-remodeling complexes render the DNA packaged in chromatin more accessible to other proteins in the cell, including those required for transcription initiation; notice, for example, the increased exposure of the TATA box. On the left, the recruitment of histone-modifying enzymes such as histone acetyltransferases adds acetyl groups to specific histones, which can then serve as binding sites for proteins that stimulate transcription initiation (not shown).

genes. As discussed in Chapter 5, these transcription-resistant regions of DNA include the heterochromatin found in interphase chromosomes and the inactive X chromosome in the cells of female mammals.

THE MOLECULAR MECHANISMS THAT CREATE SPECIALIZED CELL TYPES

All cells must be able to switch genes on and off in response to signals in their environment. But the cells of multicellular organisms have evolved this capacity to an extreme degree and in highly specialized ways to form organized arrays of differentiated cell types. In particular, once a cell in a multicellular organism becomes committed to differentiate into a specific cell type, the choice of fate is generally maintained through subsequent cell divisions. This means that the changes in gene expression, which are often triggered by a transient signal, must be remembered by the cell. This phenomenon of *cell memory* is a prerequisite for the creation of organized tissues and for the maintenance of stably differentiated cell types. In contrast, the simplest changes in gene expression in both eukaryotes and bacteria are often only transient; the tryptophan repressor, for example, switches off the tryptophan operon in bacteria only in the presence of tryptophan; as soon as the amino acid is removed from the medium, the genes switch back on, and the descendants of the cell will have no memory that their ancestors had been exposed to tryptophan.

In this section, we discuss some of the special features of transcriptional regulation that are found in multicellular organisms. Our focus will be on how these mechanisms create and maintain the specialized cell types that give a worm, a fly, or a human its distinctive characteristics.

Eukaryotic Genes Are Controlled by Combinations of Transcription Regulators

Because eukaryotic transcription regulators can control transcription initiation when bound to DNA many base pairs away from the promoter, the nucleotide sequences that control the expression of a gene can be spread over long stretches of DNA. In animals and plants, it is not unusual to find the regulatory DNA sequences of a gene dotted over tens of thousands of nucleotide pairs, although much of the intervening DNA serves as "spacer" sequence and is not directly recognized by the transcription regulators.

So far in this chapter, we have treated transcription regulators as though each functions individually to turn a gene on or off. While this idea holds true for many simple bacterial activators and repressors, most eukaryotic transcription regulators work as part of a "committee" of regulatory proteins, all of which are necessary to express the gene in the right place, in the right cell type, in response to the right conditions, at the right time, and in the required amount.

The term **combinatorial control** refers to the way that groups of transcription regulators work together to determine the expression of a single gene. We saw a simple example of such regulation by multiple regulators when we discussed the bacterial *Lac* operon (see Figure 8–9). In eukaryotes, the regulatory inputs have been amplified, and a typical gene is controlled by dozens of transcription regulators. These help assemble chromatin-remodeling complexes, histone-modifying enzymes, RNA polymerase, and general transcription factors via the multiprotein Mediator complex (**Figure 8–12**). In many cases, both repressors and activators will be present in the same complex; how the cell integrates the effects of all of these proteins to determine the final level of gene

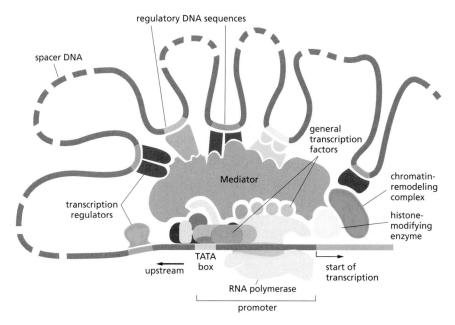

spacer DNA

regulatory DNA sequences

general transcription factors

Mediator

transcription regulators

chromatin-remodeling complex

histone-modifying enzyme

TATA box

upstream

RNA polymerase

start of transcription

promoter

Figure 8–12 Transcription regulators work together as a "committee" to control the expression of a eukaryotic gene. Whereas the general transcription factors that assemble at the promoter are the same for all genes transcribed by RNA polymerase (see Figure 7–12), the transcription regulators and the locations of their DNA binding sites relative to the promoters are different for different genes. These regulators, along with chromatin-modifying proteins, are assembled at the promoter by the Mediator. The effects of multiple transcription regulators combine to determine the final rate of transcription initiation.

expression is only now beginning to be understood. An example of such a complex regulatory system—one that participates in the development of a fruit fly from a fertilized egg—is described in **How We Know**, pp. 274–275.

The Expression of Different Genes Can Be Coordinated by a Single Protein

In addition to being able to switch individual genes on and off, all cells—whether prokaryote or eukaryote—need to coordinate the expression of different genes. When a eukaryotic cell receives a signal to divide, for example, a number of hitherto unexpressed genes are turned on together to set in motion the events that lead eventually to cell division (discussed in Chapter 18). As discussed earlier, one way in which bacteria coordinate the expression of a set of genes is by having them clustered together in an operon under the control of a single promoter (see Figure 8–6). Such clustering is not seen in eukaryotic cells, where each gene is transcribed and regulated individually. So how do these cells coordinate gene expression? In particular, given that a eukaryotic cell uses a committee of transcription regulators to control each of its genes, how can it rapidly and decisively switch whole groups of genes on or off?

The answer is that even though control of gene expression is combinatorial, the effect of a single transcription regulator can still be decisive in switching any particular gene on or off, simply by completing the combination needed to activate or repress that gene. This is like dialing in the final number of a combination lock: the lock will spring open if the other numbers have been previously entered. Just as the same number can complete the combination for different locks, the same protein can complete the combination for several different genes. As long as different genes contain regulatory DNA sequences that are recognized by the same transcription regulator, they can be switched on or off together, as a coordinated unit.

An example of such coordinated regulation in humans is seen with the *cortisol receptor protein*. In order to bind to regulatory sites in DNA, this

GENE REGULATION—THE STORY OF *EVE*

The ability to regulate gene expression is crucial to the proper development of a multicellular organism from a fertilized egg to a fertile adult. Beginning at the earliest moments in development, a succession of transcriptional programs guides the differential expression of genes that allows an animal to form a proper body plan—helping to distinguish its back from its belly, and its head from its tail. These programs ultimately direct the correct placement of a wing or a leg, a mouth or an anus, a neuron or a sex cell.

A central challenge in development, then, is to understand how an organism generates these patterns of gene expression, which are laid down within hours of fertilization. Among the most important genes involved in these early stages of development are those that encode transcription regulators. By interacting with different regulatory DNA sequences, these proteins instruct every cell in the embryo to switch on the genes that are appropriate for that cell at each time point during development. How can a protein binding to a piece of DNA help direct the development of a complex multicellular organism? To see how we can address that large question, we review the story of *Eve*.

Seeing *Eve*

Even-skipped—*Eve*, for short—is a gene whose expression plays an important part in the development of the *Drosophila* embryo. If this gene is inactivated by mutation, many parts of the embryo fail to form and the fly larva dies early in development. But *Eve* is not expressed uniformly throughout the embryo. Instead, the Eve protein is produced in a striking series of seven neat stripes, each of which occupies a very precise position along the length of the embryo. These seven stripes correspond to seven of the fourteen segments that define the body plan of the fly—three for the head, three for the thorax, and eight for the abdomen.

This pattern never varies: *Eve* can be found in the very same places in every *Drosophila* embryo (see Figure 8–13B). How can the expression of a gene be regulated with such spatial precision—such that one cell will produce a protein while a neighboring cell does not? To find out, researchers took a trip upstream.

Dissecting the DNA

As we have seen in this chapter, regulatory DNA sequences control which cells in an organism will express a particular gene, and at what point during development that gene will be turned on. In eukaryotes, these regulatory sequences are frequently located upstream of the gene itself. One way to locate a regulatory DNA sequence—and study how it operates—is to remove a piece of DNA from the region upstream of a gene of interest and insert that DNA upstream of a **reporter gene**—one that encodes a protein with an activity that is easy to monitor experimentally. If the piece of DNA contains a regulatory sequence, it will drive the expression of the reporter gene. When this patchwork piece of DNA is subsequently introduced into a cell or organism, the reporter gene will be expressed in the same cells and tissues that normally express the gene from which the regulatory sequence was derived (see Figure 10–31).

By excising various segments of the DNA sequences upstream of *Eve*, and coupling them to a reporter gene, researchers found that the expression of the gene is controlled by a series of seven regulatory modules—each of which specifies a single stripe of *Eve* expression. In this way, researchers identified, for example, a single segment of regulatory DNA that specifies stripe 2. They could excise this regulatory segment, link it to a reporter gene, and introduce the resulting DNA segment into the fly. When they examined embryos that carried this engineered DNA, they found that the reporter gene is expressed in the precise position of stripe 2 (**Figure 8–13**). Similar experiments revealed the existence of six other regulatory modules, one for each of the other Eve stripes.

The next question is: How does each of these seven regulatory segments direct the formation of a single stripe in a specific position? The answer, researchers found, is that each segment contains a unique combination of regulatory sequences that bind different combinations of transcription regulators. These regulators, like Eve itself, are distributed in unique patterns within the embryo—some toward the head, some toward the rear, some in the middle.

The regulatory segment that defines stripe 2, for example, contains regulatory DNA sequences for four transcription regulators: two that activate *Eve* transcription and two that repress it (**Figure 8–14**). In the narrow band of tissue that constitutes stripe 2, it just so happens the repressor proteins are not present—so the *Eve* gene is expressed; in the bands of tissue on either side of the stripe, the repressors keep *Eve* quiet. And so a stripe is formed.

The regulatory segments controlling the other stripes are thought to function along similar lines; each regulatory segment reads "positional information" provided

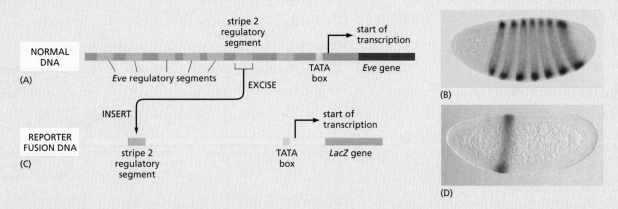

(A) NORMAL DNA

Eve regulatory segments

stripe 2 regulatory segment

EXCISE

TATA box

start of transcription

Eve gene

(B)

INSERT

(C) REPORTER FUSION DNA

stripe 2 regulatory segment

TATA box

start of transcription

LacZ gene

(D)

Figure 8–13 An experimental approach that involves the use of a reporter gene reveals the modular construction of the *Eve* gene regulatory region. (A) Expression of the *Eve* gene is controlled by a series of regulatory segments (*orange*) that direct the production of Eve protein in stripes along the embryo. (B) Embryos stained with antibodies to the Eve protein show the seven characteristic stripes of Eve expression. (C) In the laboratory, the regulatory segment that directs the formation of stripe 2 can be excised from the DNA shown in part A and inserted upstream of the *E. coli LacZ* gene, which encodes the enzyme β-galactosidase (see Figure 8–9). (D) When the engineered DNA containing the stripe 2 regulatory segment is introduced into the genome of a fly, the resulting embryo expresses β-galactosidase precisely in the position of the second *Eve* stripe. Enzyme activity is assayed by the addition of X-gal, a modified sugar that when cleaved by β-galactosidase generates an insoluble blue product. (B and D, courtesy of Stephen Small and Michael Levine.)

by some unique combination of transcription regulators in the embryo and expresses *Eve* on the basis of this information. The entire regulatory region is strung out over 20,000 nucleotide pairs of DNA and, altogether, binds more than 20 transcription regulators. This large regulatory region is built from a series of smaller regulatory segments, each of which consists of a unique arrangement of regulatory DNA sequences recognized by specific transcription regulators. In this way, the

Eve gene can respond to an enormous combination of inputs.

The Eve protein is itself a transcription regulator, and it—in combination with many other regulatory proteins—controls key events in the development of the fly. This complex organization of a discrete number of regulatory elements begins to explain how the development of an entire organism can be orchestrated by repeated applications of a few basic principles.

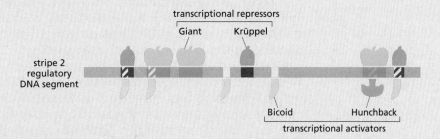

transcriptional repressors

Giant   Krüppel

stripe 2 regulatory DNA segment

Bicoid   Hunchback

transcriptional activators

Figure 8–14 The regulatory segment that specifies Eve stripe 2 contains binding sites for four different transcription regulators. All four regulators are responsible for the proper expression of *Eve* in stripe 2. Flies that are deficient in the two activators, called Bicoid and Hunchback, fail to form stripe 2 efficiently; in flies deficient in either of the two repressors, called Giant and Krüppel, stripe 2 expands and covers an abnormally broad region of the embryo. As indicated in the diagram, in some cases the binding sites for the transcription regulators overlap, and the proteins compete for binding to the DNA. For example, the binding of Bicoid and Krüppel to the site at the far right is thought to be mutually exclusive. The regulatory segment is 480 base pairs in length.

Figure 8–15 **A single transcription regulator can coordinate the expression of many different genes.**

Figure 8–15 **A single transcription regulator can coordinate the expression of many different genes.** The action of the cortisol receptor is illustrated. On the left is a series of genes, each of which has a different gene activator protein bound to its respective regulatory DNA sequences. However, these bound proteins are not sufficient on their own to activate transcription efficiently. On the right is shown the effect of adding an additional transcription regulator—the cortisol–receptor complex—that can bind to the same regulatory DNA sequence in each gene. The activated cortisol receptor completes the combination of transcription regulators required for efficient initiation of transcription, and the genes are now switched on as a set.

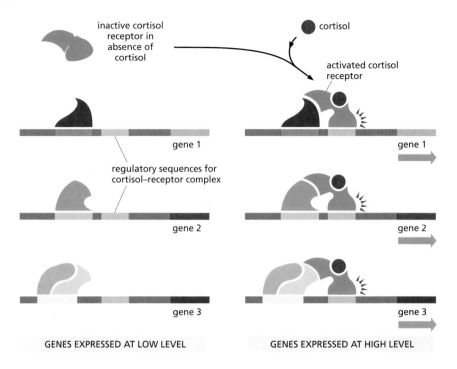

transcription regulator must first form a complex with a molecule of cortisol (see Table 16–1, p. 529). In response to cortisol, liver cells increase the expression of many genes, one of which encodes the enzyme tyrosine aminotransferase, as discussed earlier. All these genes are regulated by the binding of the cortisol–receptor complex to a regulatory sequence in the DNA of each gene. When the cortisol concentration decreases again, the expression of all of these genes drops to its normal level. In this way, a single transcription regulator can coordinate the expression of many different genes (**Figure 8–15**).

Combinatorial Control Can Also Generate Different Cell Types

The ability to switch many different genes on or off using a limited number of transcription regulators is not only useful in the day-to-day regulation of cell function. It is also one of the means by which eukaryotic cells diversify into particular types of cells during embryonic development. A striking example is the development of muscle cells. A mammalian skeletal muscle cell is distinguished from other cells by the production of a large number of characteristic proteins, such as the muscle-specific forms of actin and myosin that make up the contractile apparatus (discussed in Chapter 17), as well as the receptor proteins and ion channel proteins in the plasma membrane that make the muscle cell sensitive to nerve stimulation. The genes encoding these muscle-specific proteins are all switched on coordinately as the muscle cell differentiates. Studies of developing muscle cells in culture have identified a small number of key transcription regulators, expressed only in potential muscle cells, that coordinate muscle-specific gene expression and are thus crucial for muscle-cell differentiation. This set of regulators activates the transcription of the genes that code for muscle-specific proteins by binding to specific DNA sequences present in their regulatory regions.

Some transcription regulators can even convert one specialized cell type to another. For example, when the gene encoding the transcription regulator MyoD is artificially introduced into fibroblasts cultured from skin

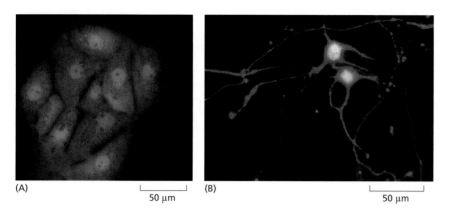

Figure 8–16 A small number of transcription regulators can convert one differentiated cell type directly into another. In this experiment, liver cells grown in culture (A) were converted into neuronal cells (B) via the artificial introduction of three nerve-specific transcription regulators. The cells are labeled with a fluorescent dye. (From S. Marro et al., *Cell Stem Cell* 9:374–378, 2011. With permission from Elsevier.)

(A)   50 μm   (B)   50 μm

connective tissue, the fibroblasts form musclelike cells. It appears that the fibroblasts, which are derived from the same broad class of embryonic cells as muscle cells, have already accumulated many of the other necessary transcription regulators required for the combinatorial control of the muscle-specific genes, and that addition of MyoD completes the unique combination required to direct the cells to become muscle.

This type of reprogramming can produce even more dramatic effects. For example, a set of nerve-specific transcription regulators, when artificially expressed in cultured liver cells, can convert them into functional neurons (**Figure 8–16**). Such dramatic results suggest that it may someday be possible to produce in the laboratory any cell type for which the correct combination of transcription regulators can be identified. How these transcription regulators can then lead to the generation of different cell types is illustrated schematically in **Figure 8–17**.

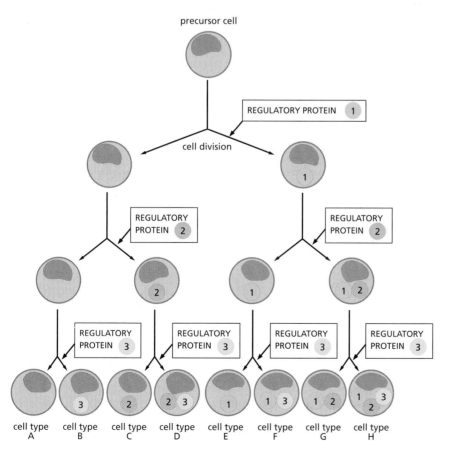

Figure 8–17 Combinations of a few transcription regulators can generate many cell types during development. In this simple scheme, a "decision" to make a new transcription regulator (shown as a *numbered* circle) is made after each cell division. Repetition of this simple rule can generate eight cell types (A through H), using only three transcription regulators. Each of these hypothetical cell types would then express many different genes, as dictated by the combination of transcription regulators that each cell type produces.

Figure 8–18 A combination of transcription regulators can induce a differentiated cell to de-differentiate into a pluripotent cell. The artificial expression of a set of four genes, each of which encodes a transcription regulator, can reprogram a fibroblast into a pluripotent cell with ES cell-like properties. Like ES cells, such *iPS cells* can proliferate indefinitely in culture and can be stimulated by appropriate extracellular signal molecules to differentiate into almost any cell type in the body.

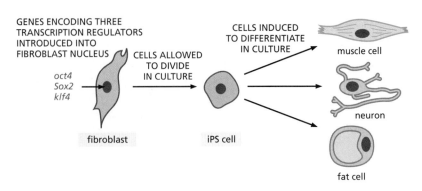

Specialized Cell Types Can Be Experimentally Reprogrammed to Become Pluripotent Stem Cells

We have seen that, in some cases, one type of differentiated cell can be experimentally converted into another type by the artificial expression of specific transcription regulators (see Figure 8–16). Even more surprising, transcription regulators can coax various differentiated cells to *de-differentiate* into **pluripotent stem cells** that are capable of giving rise to all the specialized cell types in the body, much like the embryonic stem (ES) cells discussed in Chapter 20 (see pp. 708–711).

Using a defined set of transcription regulators, cultured mouse fibroblasts have been reprogrammed to become **induced pluripotent stem (iPS) cells**—cells that look and behave like the pluripotent ES cells that are derived from embryos (**Figure 8–18**). The approach was quickly adapted to produce iPS cells from a variety of specialized cell types, including cells taken from humans. Such human iPS cells can then be directed to generate a population of differentiated cells for use in the study or treatment of disease, as we discuss in Chapter 20.

The Formation of an Entire Organ Can Be Triggered by a Single Transcription Regulator

We have seen that a small number of transcription regulators can control the expression of whole sets of genes and can even convert one cell type into another. But an even more stunning example of the power of transcriptional control comes from studies of eye development in *Drosophila*. In this case, a single "master" transcription regulator called Ey could be used to trigger the formation of not just a single cell type but a whole organ. In the laboratory, the *Ey* gene can be artificially expressed in fruit fly embryos in cells that would normally give rise to a leg. When these modified embryos develop into adult flies, some have an eye in the middle of a leg (**Figure 8–19**).

How the Ey protein coordinates the specification of each type of cell found in the eye—and directs their proper organization in three-dimensional space—is an actively studied topic in developmental biology. In essence, however, Ey functions like any other transcription regulator, controlling the expression of multiple genes by binding to DNA sequences in their regulatory regions. Some of the genes controlled by Ey encode additional transcription regulators that, in turn, control the expression of other genes. In this way, the action of a single transcription regulator can produce a cascade of regulators that, working in combination, lead to the formation of an organized group of many different types of cells. One can begin to imagine how, by repeated applications of this principle, a complex organism self-assembles, piece by piece.

eye structure on leg

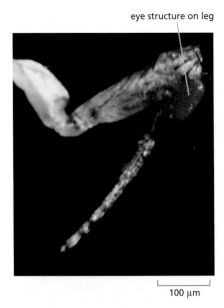

100 μm

Figure 8–19 Artificially induced expression of the *Drosophila Ey* gene in the precursor cells of the leg triggers the misplaced development of an eye on a fly's leg. (Courtesy of Walter Gehring.)

Epigenetic Mechanisms Allow Differentiated Cells to Maintain Their Identity

Once a cell has become differentiated into a particular cell type, it will generally remain differentiated, and all its progeny cells will remain that same cell type. Some highly specialized cells, including skeletal muscle cells and neurons, never divide again once they have differentiated—that is, they are *terminally differentiated* (as discussed in Chapter 18). But many other differentiated cells—such as fibroblasts, smooth muscle cells, and liver cells—will divide many times in the life of an individual. When they do, these specialized cell types give rise only to cells like themselves: smooth muscle cells do not give rise to liver cells, nor liver cells to fibroblasts.

For a proliferating cell to maintain its identity—a property called **cell memory**—the patterns of gene expression responsible for that identity must be remembered and passed on to its daughter cells through all subsequent cell divisions. Thus, in the model illustrated in Figure 8–17, the production of each transcription regulator, once begun, has to be continued in the daughter cells of each cell division. How is such perpetuation accomplished?

Cells have several ways of ensuring that their daughters "remember" what kind of cells they are. One of the simplest and most important is through a **positive feedback loop**, where a master transcription regulator activates transcription of its own gene, in addition to that of other cell-type–specific genes. Each time a cell divides the regulator is distributed to both daughter cells, where it continues to stimulate the positive feedback loop. The continued stimulation ensures that the regulator will continue to be produced in subsequent cell generations. The Ey protein discussed earlier functions in such a positive feedback loop. Positive feedback is crucial for establishing the "self-sustaining" circuits of gene expression that allow a cell to commit to a particular fate—and then to transmit that information to its progeny (**Figure 8–20**).

Although positive feedback loops are probably the most prevalent way of ensuring that daughter cells remember what kind of cells they are meant to be, there are other ways of reinforcing cell identity. One involves the methylation of DNA. In vertebrate cells, **DNA methylation** occurs on certain cytosine bases (**Figure 8–21**). This covalent modification generally

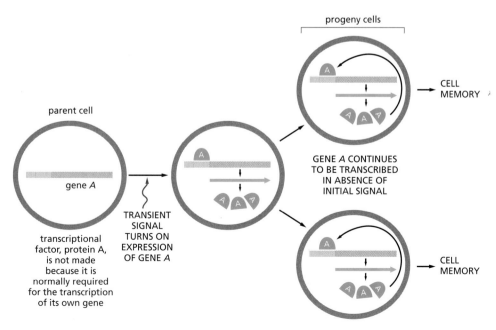

Figure 8–20 A positive feedback loop can create cell memory. Protein A is a master transcription regulator that activates the transcription of its own gene—as well as other cell-type-specific genes (not shown). All of the descendants of the original cell will therefore "remember" that the progenitor cell had experienced a transient signal that initiated the production of protein A.

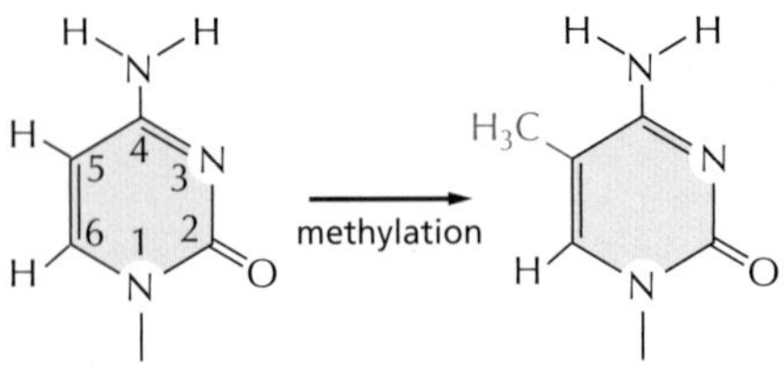

cytosine

5-methylcytosine

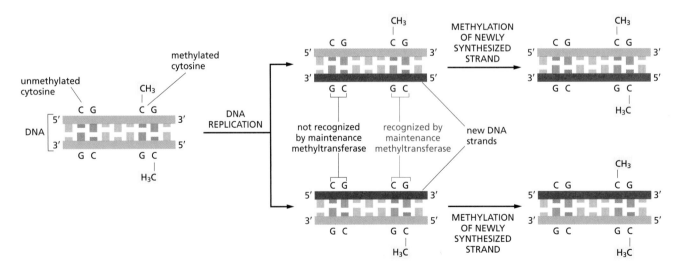

Figure 8–21 Formation of 5-methylcytosine occurs by methylation of a cytosine base in the DNA double helix. In vertebrates, this modification is confined to selected cytosine (C) nucleotides that fall next to a guanine (G) in the sequence CG.

turns off genes by attracting proteins that bind to methylated cytosines and block gene transcription. DNA methylation patterns are passed on to progeny cells by the action of an enzyme that copies the methylation pattern on the parent DNA strand to the daughter DNA strand as it is synthesized (**Figure 8–22**).

Another mechanism for inheriting gene expression patterns involves the modification of histones. When a cell replicates its DNA, each daughter double helix receives half of its parent's histone proteins, which contain the covalent modifications of the parent chromosome. Enzymes responsible for these modifications may bind to the parental histones and confer the same modifications to the new histones nearby. This cycle of modification reestablishes the pattern of chromatin structure found in the parent chromosome (**Figure 8–23**).

Because all of these cell-memory mechanisms transmit patterns of gene expression from parent to daughter cell without altering the actual nucleotide sequence of the DNA, they are considered to be forms of **epigenetic inheritance**. Such epigenetic changes play an important part in controlling patterns of gene expression, allowing transient signals from the environment to be permanently recorded by our cells—a fact that has important implications for understanding how cells operate and how they malfunction in disease.

POST-TRANSCRIPTIONAL CONTROLS

We have seen that transcription regulators control gene expression by promoting or hindering the transcription of specific genes. The vast majority of genes in all organisms are regulated in this way. But many additional points of control can come into play later in the pathway from DNA to protein, giving cells a further opportunity to regulate the amount or activity of the gene products that they make (see Figure 8–3). These **post-transcriptional controls**, which operate after transcription has begun, play a crucial part in regulating the expression of almost all genes.

We have already encountered a few examples of such post-transcriptional control. We have seen how alternative RNA splicing allows different

Figure 8–22 DNA methylation patterns can be faithfully inherited when a cell divides. An enzyme called a maintenance methyltransferase guarantees that once a pattern of DNA methylation has been established, it is inherited by newly made DNA. Immediately after DNA replication, each daughter double helix will contain one methylated DNA strand—inherited from the parent double helix—and one unmethylated, newly synthesized strand. The maintenance methyltransferase interacts with these hybrid double helices and methylates only those CG sequences that are base-paired with a CG sequence that is already methylated.

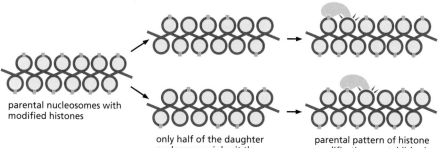

parental nucleosomes with modified histones

only half of the daughter nucleosomes inherit the modified parental histones

parental pattern of histone modification reestablished by enzymes that recognize the same modifications they catalyze

Figure 8–23 Histone modifications may be inherited by daughter chromosomes. When a chromosome is replicated, its resident histones are distributed more or less randomly to each of the two daughter DNA double helices. Thus, each daughter chromosome will inherit about half of its parent's collection of modified histones. The remaining stretches of DNA receive newly synthesized, not-yet-modified histones. If the enzymes responsible for each type of modification bind to the specific modification they create, they can catalyze the spread of this modification on the new histones. This cycle of modification and recognition can restore the parental histone modification pattern and, ultimately, allow the inheritance of the parental chromatin structure. This mechanism may apply to some but not all types of histone modifications.

forms of a protein, encoded by the same gene, to be made in different tissues (Figure 7–22). And we have discussed how various post-translational modifications of a protein can regulate its concentration and activity (see Figure 4–43). In the remainder of this chapter, we consider several other examples—some only recently discovered—of the many ways in which cells can manipulate the expression of a gene after transcription has commenced.

Each mRNA Controls Its Own Degradation and Translation

The more time an mRNA persists in the cell before it is degraded, the more protein it will produce. In bacteria, most mRNAs last only a few minutes before being destroyed. This instability allows a bacterium to adapt quickly to environmental changes. Eukaryotic mRNAs are generally more stable. The mRNA that encodes β-globin, for example, has a half-life of more than 10 hours. Most eukaryotic mRNAs, however, have half-lives of less than 30 minutes, and the most short-lived are those that encode proteins whose concentrations need to change rapidly based on the cell's needs, such as transcription regulators. Whether bacterial or eukaryotic, an mRNA's lifetime is dictated by specific nucleotide sequences within the untranslated regions that lie both upstream and downstream of the protein-coding sequence. These sequences often harbor binding sites for proteins that are involved in RNA degradation.

In addition to the nucleotide sequences that regulate its half-life, each mRNA possesses sequences that help control how often or how efficiently it will be translated into protein. These sequences control translation initiation. Although the details differ between eukaryotes and bacteria, the general strategy is similar for both.

Bacterial mRNAs contain a short ribosome-binding sequence located a few nucleotide pairs upstream of the AUG codon where translation begins (see Figure 7–37). This binding sequence forms base pairs with the RNA in the small ribosomal subunit, correctly positioning the initiating AUG codon within the ribosome. Because this interaction is needed for efficient translation initiation, it provides an ideal target for translational control. By blocking—or exposing—the ribosome-binding sequence, the bacterium can either inhibit—or promote—the translation of an mRNA (**Figure 8–24**).

Eukaryotic mRNAs possess a 5′ cap that helps guide the ribosome to the first AUG, the codon where translation will start (see Figure 7–36). Eukaryotic repressor proteins can inhibit translation initiation by binding to specific nucleotide sequences in the 5′ untranslated region of the mRNA, thereby preventing the ribosome from finding the first AUG—a mechanism similar to that in bacteria. When conditions change, the cell can inactivate the repressor to initiate translation of the mRNA.

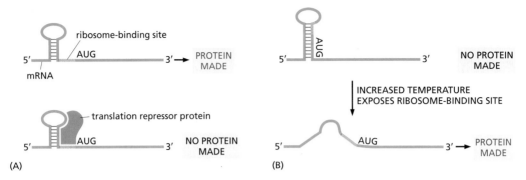

(A)

(B)

Figure 8–24 A bacterial gene's expression can be controlled by regulating translation of its mRNA. (A) Sequence-specific RNA-binding proteins can repress the translation of specific mRNAs by keeping the ribosome from binding to the ribosome-binding sequence (*orange*) in the mRNA. Some ribosomal proteins exploit this mechanism to inhibit the translation of their own mRNA. In this way, "extra" ribosomal proteins—those not incorporated into ribosomes—serve as a signal to halt their synthesis. (B) An mRNA from the pathogen *Listeria monocytogenes* contains a "thermosensor" RNA sequence that controls the translation of a set of mRNAs produced by virulence genes. At the warmer temperature that the bacterium encounters inside its human host, the thermosensor sequence denatures, exposing the ribosome-binding sequence, so the virulence proteins are made.

Regulatory RNAs Control the Expression of Thousands of Genes

As we saw in Chapter 7, RNAs perform many critical tasks in cells. In addition to the mRNAs, which code for proteins, *noncoding RNAs* have various functions. It has long been known that some have key structural and catalytic roles, particularly in protein synthesis by ribosomes (see pp. 246–247). But a recent series of surprising discoveries has revealed several new classes of noncoding RNAs and shown that these RNAs are far more prevalent than previously suspected.

What, then, are all these newly discovered noncoding RNAs doing? Many have unanticipated but important roles in regulating gene expression and are therefore referred to as **regulatory RNAs**. There are at least three major types of regulatory RNAs—*microRNAs*, *small interfering RNAs*, and *long noncoding RNAs*. We discuss each one in turn.

MicroRNAs Direct the Destruction of Target mRNAs

MicroRNAs, or **miRNAs**, are tiny RNA molecules that control gene expression by base-pairing with specific mRNAs and reducing both their stability and their translation into protein. In humans, miRNAs are thought to regulate the expression of at least one-third of all protein-coding genes.

Like other noncoding RNAs, such as tRNA and rRNA, a precursor miRNA transcript undergoes a special type of processing to yield the mature, functional miRNA molecule, which is only about 22 nucleotides in length. This small but mature miRNA is packaged with specialized proteins to form an *RNA-induced silencing complex* (*RISC*), which patrols the cytoplasm in search of mRNAs that are complementary to the bound miRNA molecule (**Figure 8–25**). Once a target mRNA forms base pairs with an miRNA, it is either destroyed immediately by a nuclease present within the RISC or its translation is blocked. In the latter case, the bound mRNA molecule is delivered to a region of the cytoplasm where other nucleases eventually degrade it. Destruction of the mRNA releases the RISC and allows it to seek out additional mRNA targets. Thus, a single miRNA—as part of a RISC—can eliminate one mRNA molecule after another, thereby efficiently blocking production of the protein that the mRNAs encode.

Two features of miRNAs make them especially useful regulators of gene expression. First, a single miRNA can inhibit the transcription of a whole set of different mRNAs so long as all the mRNAs carry a common sequence, usually located in either their 5′ or 3′ untranslated regions. In humans, some individual miRNAs influence the transcription of hundreds of different mRNAs in this manner. Second, a gene that encodes an miRNA occupies relatively little space in the genome compared with one that encodes a transcription regulator. Indeed, their very small size is one reason that miRNAs were discovered only recently. There are thought

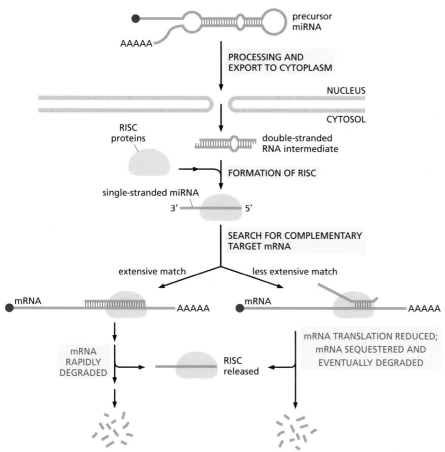

Figure 8–25 An miRNA targets a complementary mRNA molecule for destruction. Each precursor miRNA transcript is processed to form a double-stranded intermediate, which is further processed to form a mature, single-stranded miRNA. This miRNA assembles with a set of proteins into a complex called RISC, which then searches for mRNAs that have a nucleotide sequence complementary to its bound miRNA. Depending on how extensive the region of complementarity is, the target mRNA is either rapidly degraded by a nuclease within the RISC or transferred to an area of the cytoplasm where other cellular nucleases destroy it.

to be roughly 500 different miRNAs encoded by the human genome. Although we are only beginning to understand the full impact of these miRNAs, it is clear that they play a critical part in regulating gene expression and thereby influence many cell functions.

Small Interfering RNAs Are Produced From Double-Stranded, Foreign RNAs to Protect Cells From Infections

Some of the same components that process and package miRNAs also play another crucial part in the life of a cell: they serve as a powerful cell defense mechanism. In this case, the system is used to eliminate "foreign" RNA molecules—in particular, the double-stranded RNAs produced by many viruses and transposable genetic elements (discussed in Chapter 9). The process is called **RNA interference (RNAi)**.

In the first step of RNAi, the double-stranded, foreign RNAs are cut into short fragments (approximately 22 nucleotide pairs in length) by a protein called Dicer—the same protein used to generate the double-stranded RNA intermediate in miRNA production (see Figure 8–25). The resulting double-stranded RNA fragments, called **small interfering RNAs (siRNAs)**, are then taken up by the same RISCs that carry miRNAs. The RISC discards one strand of the siRNA duplex and uses the remaining single-stranded RNA to seek and destroy complementary foreign RNA molecules (**Figure 8–26**). In this way, the infected cell turns the foreign RNA back on itself.

RNAi operates in a wide variety of organisms, including single-celled fungi, plants, and worms, indicating that it is an evolutionarily ancient defense mechanism. In some organisms, including plants, the RNAi defense response can spread from tissue to tissue, allowing the entire organism to become resistant to a virus after only a few of its cells

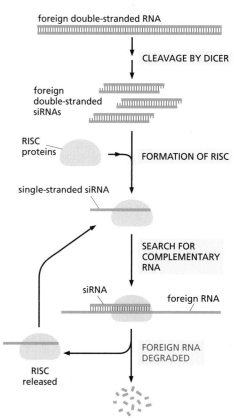

Figure 8–26 siRNAs are produced from double-stranded, foreign RNAs in the process of RNA interference. Double-stranded RNAs from a virus or transposable genetic element are first cleaved by a nuclease called Dicer. The resulting double-stranded fragments are incorporated into RISCs, which discard one strand of the foreign RNA duplex and use the other strand to locate and destroy foreign RNAs with a complementary sequence.

have been infected. In this sense, RNAi resembles certain aspects of the adaptive immune responses of vertebrates; in both cases, an invading pathogen elicits the production of molecules—either siRNAs or antibodies—that are custom-made to inactivate the specific invader and thereby protect the host.

Thousands of Long Noncoding RNAs May Also Regulate Mammalian Gene Activity

At the other end of the size spectrum are the **long noncoding RNAs**, a class of RNA molecules that are more than 200 nucleotides in length. There are thought to be upwards of 8000 of these RNAs encoded in the human and mouse genomes. Yet, with few exceptions, their roles in the biology of the organism are not entirely clear.

One of the best understood of the long noncoding RNAs is *Xist*. This enormous RNA molecule, some 17,000 nucleotides long, is a key player in X inactivation—the process by which one of the two X chromosomes in the cells of female mammals is permanently silenced (see Figure 5–30). Early in development, Xist is produced by only one of the X chromosomes in each female nucleus. The transcript then "sticks around," coating the chromosome and presumably attracting the enzymes and chromatin-remodeling complexes that promote the formation of highly condensed heterochromatin. Other long noncoding RNAs may promote the silencing of specific genes in a similar manner.

Some long noncoding RNAs arise from protein-coding regions of the genome, but are transcribed from the "wrong" DNA strand. Some of these *antisense* transcripts are known to bind to the mRNAs produced from that DNA segment, regulating their translation and stability—in some cases by producing siRNAs (see Figure 8–26).

Regardless of how the various long noncoding RNAs operate—or what exactly they do—the discovery of this large class of RNAs reinforces the idea that a eukaryotic genome is densely packed with information that provides not only an inventory of the molecules and structures every cell must make, but a set of instructions for how and when to assemble these parts to guide the growth and development of a complete organism.

ESSENTIAL CONCEPTS

- A typical eukaryotic cell expresses only a fraction of its genes, and the distinct types of cells in multicellular organisms arise because different sets of genes are expressed as cells differentiate.

- In principle, gene expression can be controlled at any of the steps between a gene and its ultimate functional product. For the majority of genes, however, the initiation of transcription is the most important point of control.

- The transcription of individual genes is switched on and off in cells by transcription regulator proteins, which bind to short stretches of DNA called regulatory DNA sequences.

- In bacteria, transcription regulators usually bind to regulatory DNA sequences close to where RNA polymerase binds. This binding can either activate or repress transcription of the gene. In eukaryotes, regulatory DNA sequences are often separated from the promoter by many thousands of nucleotide pairs.

- Eukaryotic transcription regulators act in two main ways: (1) they can directly affect the assembly process that requires RNA polymerase and the general transcription factors at the promoter, and (2) they can locally modify the chromatin structure of promoter regions.

- In eukaryotes, the expression of a gene is generally controlled by a combination of different transcription regulator proteins.

- In multicellular plants and animals, the production of different transcription regulators in different cell types ensures the expression of only those genes appropriate to the particular type of cell.

- One differentiated cell type can be converted to another by artificially expressing an appropriate set of transcription regulators. A differentiated cell can also be reprogrammed into a stem cell by artificially expressing a particular set of such regulators.

- Cells in multicellular organisms have mechanisms that enable their progeny to "remember" what type of cell they should be. A prominent mechanism for propagating cell memory relies on transcription regulators that perpetuate transcription of their own gene—a form of positive feedback.

- A master transcription regulator, if expressed in the appropriate precursor cell, can trigger the formation of a specialized cell type or even an entire organ.

- The pattern of DNA methylation can be transmitted from one cell generation to the next, producing a form of epigenetic inheritance that helps a cell remember the state of gene expression in its parent cell. There is also evidence for a form of epigenetic inheritance based on transmitted chromatin structures.

- Cells can regulate gene expression by controlling events that occur after transcription has begun. Many of these post-transcriptional mechanisms rely on RNA molecules that can influence their own stability or translation.

- MicroRNAs (miRNAs) control gene expression by base-pairing with specific mRNAs and inhibiting their stability and translation.

- Cells have a defense mechanism for destroying "foreign" double-stranded RNAs, many of which are produced by viruses. It makes use of small interfering RNAs (siRNAs) that are produced from the foreign RNAs in a process called RNA interference (RNAi).

- Scientists can take advantage of RNAi to inactivate specific genes of interest.

- The recent discovery of thousands of long noncoding RNAs in mammals has opened a new window to the roles of RNAs in gene regulation.

KEY TERMS

combinatorial control	promoter
differentiation	regulatory DNA sequence
DNA methylation	regulatory RNA
epigenetic inheritance	reporter gene
gene expression	RNA interference (RNAi)
long noncoding RNA	small interfering RNA (siRNA)
microRNA (miRNA)	transcription regulator
positive feedback loop	transcriptional activator
post-transcriptional control	transcriptional repressor

QUESTIONS

QUESTION 8–4

A virus that grows in bacteria (bacterial viruses are called bacteriophages) can replicate in one of two ways. In the prophage state, the viral DNA is inserted into the bacterial chromosome and is copied along with the bacterial genome each time the cell divides. In the lytic state, the viral DNA is released from the bacterial chromosome and replicates many times in the cell. This viral DNA then produces viral coat proteins that together with the replicated viral DNA form many new virus particles that burst out of the bacterial cell. These two forms of growth are controlled by two transcription regulators, called c1 ("c one") and Cro, that are encoded by the virus. In the prophage state, cI is expressed; in the lytic state, Cro is expressed. In addition to regulating the expression of other genes, c1 represses the *Cro* gene, and Cro represses the *c1* gene (**Figure Q8–4**). When bacteria containing a phage in the prophage state are briefly irradiated with UV light, c1 protein is degraded.

A. What will happen next?

B. Will the change in (A) be reversed when the UV light is switched off?

C. Why might this response to UV light have evolved?

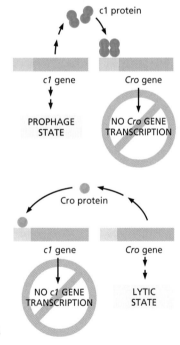

Figure Q8–4

QUESTION 8–5

Which of the following statements are correct? Explain your answers.

A. In bacteria, but not in eukaryotes, many mRNAs contain the coding region for more than one gene.

B. Most DNA-binding proteins bind to the major groove of the DNA double helix.

C. Of the major control points in gene expression (transcription, RNA processing, RNA transport, translation, and control of a protein's activity), transcription initiation is one of the most common.

QUESTION 8–6

Your task in the laboratory of Professor Quasimodo is to determine how far an enhancer (a binding site for an activator protein) could be moved from the promoter of the *straightspine* gene and still activate transcription. You systematically vary the number of nucleotide pairs between these two sites and then determine the amount of transcription by measuring the production of Straightspine mRNA. At first glance, your data look confusing (**Figure Q8–6**). What would you have expected for the results of this experiment? Can you save your reputation and explain these results to Professor Quasimodo?

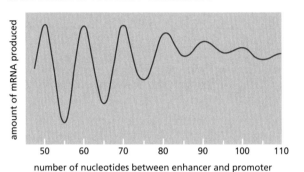

Figure Q8–6

QUESTION 8–7

The λ repressor binds as a dimer to critical sites on the λ genome to repress the virus's lytic genes. This is necessary to maintain the prophage (integrated) state. Each molecule of the repressor consists of an N-terminal DNA-binding domain and a C-terminal dimerization domain (**Figure Q8–7**). Upon induction (for example, by irradiation with UV light), the genes for lytic growth are expressed, λ progeny are produced, and the bacterial cell is lysed (see Question 8–4). Induction is initiated by cleavage of the λ repressor at a site between the DNA-binding domain and the dimerization domain, which causes the repressor to dissociate from the DNA. In the absence of bound repressor, RNA polymerase binds and initiates lytic growth. Given that the number (concentration) of DNA-binding domains is unchanged by cleavage of the repressor, why do you suppose its cleavage results in its dissociation from the DNA?

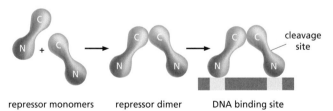

repressor monomers   repressor dimer   DNA binding site

Figure Q8–7

QUESTION 8–8

The genes that encode the enzymes for arginine biosynthesis are located at several positions around the genome of *E. coli*, and they are regulated coordinately by a transcription regulator encoded by the *ArgR* gene.

The activity of the ArgR protein is modulated by arginine. Upon binding arginine, ArgR alters its conformation, dramatically changing its affinity for the DNA sequences in the promoters of the genes for the arginine biosynthetic enzymes. Given that ArgR is a repressor protein, would you expect that ArgR would bind more tightly or less tightly to the DNA sequences when arginine is abundant? If ArgR functioned instead as an activator protein, would you expect the binding of arginine to increase or to decrease its affinity for its regulatory DNA sequences? Explain your answers.

QUESTION 8–9

When enhancers were initially found to influence transcription many thousands of nucleotide pairs from the promoters they control, two principal models were invoked to explain this action at a distance. In the "DNA looping" model, direct interactions between proteins bound at enhancers and promoters were proposed to stimulate transcription initiation. In the "scanning" or "entry-site" model, RNA polymerase (or another component of the transcription machinery) was proposed to bind at the enhancer and then scan along the DNA until it reached the promoter. These two models were tested using an enhancer on one piece of DNA and a β-globin gene and promoter on a separate piece of DNA (Figure Q8–9). The β-globin gene was not expressed from the mixture of pieces. However, when the two segments of DNA were joined via a linker (made of a protein that binds to a small molecule called biotin), the β-globin gene was expressed.

   Does this experiment distinguish between the DNA looping model and the scanning model? Explain your answer.

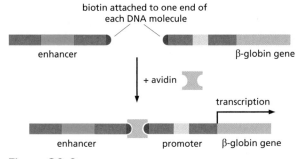

Figure Q8–9

QUESTION 8–10

Differentiated cells of an organism contain the same genes. (Among the few exceptions to this rule are the cells of the mammalian immune system, in which the formation of specialized cells is based on limited rearrangements of the genome.) Describe an experiment that substantiates the first sentence of this question, and explain why it does.

QUESTION 8–11

Figure 8–17 shows a simple scheme by which three transcription regulators are used during development to create eight different cell types. How many cell types could you create, using the same rules, with four different transcription regulators? As described in the text, MyoD is a transcription regulator that by itself is sufficient to induce muscle-specific gene expression in fibroblasts. How does this observation fit the scheme in Figure 8–17?

QUESTION 8–12

Imagine the two situations shown in Figure Q8–12. In cell I, a transient signal induces the synthesis of protein A, which is a transcriptional activator that turns on many genes including its own. In cell II, a transient signal induces the synthesis of protein R, which is a transcriptional repressor that turns off many genes including its own. In which, if either, of these situations will the descendants of the original cell "remember" that the progenitor cell had experienced the transient signal? Explain your reasoning.

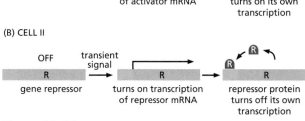

Figure Q8–12

QUESTION 8–13

Discuss the following argument: "If the expression of every gene depends on a set of transcription regulators, then the expression of these regulators must also depend on the expression of other regulators, and their expression must depend on the expression of still other regulators, and so on. Cells would therefore need an infinite number of genes, most of which would code for transcription regulators." How does the cell get by without having to achieve the impossible?

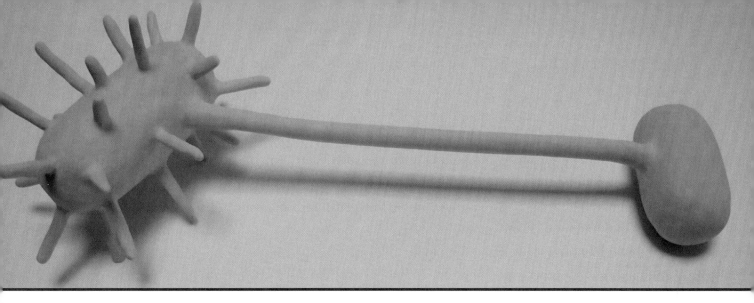

CHAPTER NINE

9

How Genes and Genomes Evolve

For a given individual, the nucleotide sequence of the genome in virtually every one of its cells is the same. But compare the DNA of two individuals—even parent and child—and that is no longer the case: the genomes of individuals within a species contain slightly different information. And between members of different species, the deviations are even more extensive.

Such differences in DNA sequence are responsible for the diversity of life on Earth, from the subtle variations in hair color, eye color, and skin color that characterize members our own species (**Figure 9–1**) to the dramatic differences in phenotype that distinguish a fish from a fungus or a robin from a rose. But if all life emerged from a common ancestor—a single-celled organism that existed some 3.5 billion years ago—where did these genetic improvisations come from? How did they arise, why were they preserved, and how do they contribute to the breathtaking biological diversity that surrounds us?

Improvements in the methods used to sequence and analyze whole genomes—from pufferfish and the plague bacterium to people from around the world—are now allowing us to address some of these questions. In Chapter 10, we describe these revolutionary technologies, which continue to transform the modern era of genomics. In this chapter, we present some of the fruits of these technological innovations. Our ability to compare the genomes of a wide-ranging collection of organisms has provided striking confirmation of Darwin's explanations for the diversity of life on Earth—revealing how processes of mutation and natural selection have been sculpting DNA sequences for billions of years, giving rise to the spectacular menagerie of present-day life-forms that crowd every corner of the planet.

GENERATING GENETIC VARIATION

RECONSTRUCTING LIFE'S FAMILY TREE

TRANSPOSONS AND VIRUSES

EXAMINING THE HUMAN GENOME

Figure 9–1 Small differences in DNA sequence account for differences in appearance between one individual and the next. A group of English schoolchildren displays a sampling of the characteristics that define the unity and diversity of our own species. (Courtesy of Fiona Pragoff, Wellcome Images.)

In this chapter, we discuss how genes and genomes change over time. We examine the molecular mechanisms that generate genetic diversity, and we consider how the information in present-day genomes can be deciphered to yield a historical record of the evolutionary processes that have shaped these DNA sequences. We take a brief look at mobile genetic elements and consider how these elements, along with modern-day viruses, can carry genetic information from place to place and from organism to organism. Finally, we end the chapter by taking a closer look at the human genome to see what our own DNA sequences tell us about who we are and where we come from.

GENERATING GENETIC VARIATION

Evolution is more a tinkerer than an inventor: it uses as its raw materials the DNA sequences that each organism inherits from its ancestors. There is no natural mechanism for making long stretches of entirely novel nucleotide sequences. In this sense, no gene or genome is ever entirely new. Instead, the astonishing diversity in form and function in the living world is all the result of variations on preexisting themes. As genetic variations pile up over millions of generations, they can produce radical change.

Several basic types of genetic change are especially crucial in evolution (**Figure 9–2**):

- *Mutation within a gene*: An existing gene can be modified by a mutation that changes a single nucleotide or deletes or duplicates one or more nucleotides. These mutations can alter the splicing of a gene's transcript or change the stability, activity, location, or interactions of its encoded protein or RNA product.

- *Mutation within regulatory DNA*: When and where a gene is expressed can be affected by a mutation in the stretches of DNA sequence that regulate the gene's activity (described in Chapter 8). For example, humans and fish have a surprisingly large number of genes in common, but changes in the regulation of those shared genes underlie many of the most dramatic differences between those species.

- *Gene duplication*: An existing gene, a larger segment of DNA, or even a whole genome can be duplicated, creating a set of closely related genes within a single cell. As this cell and its progeny divide, the original DNA sequence and its duplicate can acquire additional mutations and thereby assume new functions and patterns of expression.

- *Exon shuffling*: Two or more existing genes can be broken and rejoined to make a hybrid gene containing DNA segments that originally belonged to separate genes. In eukaryotes, the breaking and rejoining often occurs within the long intron sequences, which do not encode protein. Because intron sequences are removed by RNA splicing, the breaking and joining do not have to be precise to result in a functional gene.

- *Mobile genetic elements*: Specialized DNA sequences that can move from one chromosomal location to another can alter the activity or regulation of a gene; they can also promote gene duplication, exon shuffling, and other genome rearrangements.

- *Horizontal gene transfer*: A piece of DNA can be transferred from the genome of one cell to that of another—even to that of another species. This process, which is rare among eukaryotes but common among bacteria, differs from the usual "vertical" transfer of genetic information from parent to progeny.

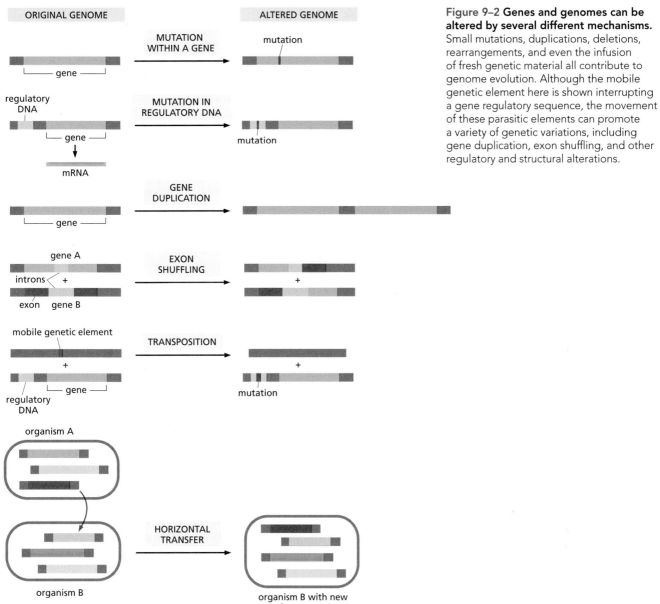

Figure 9–2 Genes and genomes can be altered by several different mechanisms. Small mutations, duplications, deletions, rearrangements, and even the infusion of fresh genetic material all contribute to genome evolution. Although the mobile genetic element here is shown interrupting a gene regulatory sequence, the movement of these parasitic elements can promote a variety of genetic variations, including gene duplication, exon shuffling, and other regulatory and structural alterations.

Each of these forms of genetic variation—from the simple mutations that occur within a gene to the more extensive duplications, deletions, rearrangements, and additions that occur within a genome—has played an important part in the evolution of modern organisms. And they still play that part today, as organisms continue to evolve. In this section, we discuss these basic mechanisms of genetic change, and we consider their consequences for genome evolution. But first, we pause to consider the contribution of sex—the mechanism that many organisms use to pass genetic information on to future generations.

In Sexually Reproducing Organisms, Only Changes to the Germ Line Are Passed On To Progeny

For bacteria and unicellular organisms that reproduce mainly asexually, the inheritance of genetic information is fairly straightforward. Each individual duplicates its genome and donates one copy to each daughter cell when the individual divides in two. The family tree of such unicellular organisms is simply a branching diagram of cell divisions that directly links each individual to its progeny and to its ancestors.

QUESTION 9–1

In this chapter, it is argued that genetic variability is beneficial for a species because it enhances that species' ability to adapt to changing conditions. Why, then, do you think that cells go to such great lengths to ensure the fidelity of DNA replication?

Figure 9–3 Germ-line cells and somatic cells have fundamentally different functions. In sexually reproducing organisms, genetic information is propagated into the next generation exclusively by germ-line cells (*red*). This cell lineage includes the specialized reproductive cells—the germ cells (eggs and sperm, *red* half circles)—which contain only half the number of chromosomes than do the other cells in the body (full circles). When two germ cells come together during fertilization, they form a fertilized egg or zygote (*purple*), which once again contains a full set of chromosomes (discussed in Chapter 19). The zygote gives rise to both germ-line cells and to somatic cells (*blue*). Somatic cells form the body of the organism but do not contribute their DNA to the next generation.

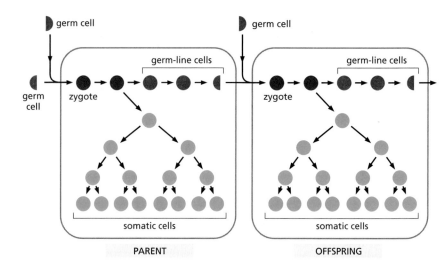

For a multicellular organism that reproduces sexually, however, the family connections are considerably more complex. Although individual cells within that organism divide, only the specialized reproductive cells—the **germ cells**—carry a copy of its genome to the next generation of organisms (discussed in Chapter 19). All the other cells of the body—the **somatic cells**—are doomed to die without leaving evolutionary descendants of their own (**Figure 9–3**). In a sense, somatic cells exist only to help the germ cells survive and propagate.

A mutation that occurs in a somatic cell—although it might have unfortunate consequences for the individual in which it occurs (causing cancer, for example)—will not be transmitted to the organism's offspring. For a mutation to be passed on to the next generation, it must alter the **germ line**—the cell lineage that gives rise to the germ cells (**Figure 9–4**). Thus, when we track the genetic changes that accumulate during the evolution of sexually reproducing organisms, we are looking at events that took place in a germ-line cell. It is through a series of germ-line cell divisions that sexually reproducing organisms trace their descent back to their ancestors and, ultimately, back to the ancestors of us all—the first cells that existed, at the origin of life more than 3.5 billion years ago.

In addition to perpetuating a species, sex also introduces its own form of genetic change: when germ cells from a male and female unite during fertilization, they generate offspring that are genetically distinct from either parent. We discuss this form of genetic diversification in detail in Chapter 19. In the meantime, aside from this mating-based genome

Figure 9–4 Mutations in germ-line cells and somatic cells have different consequences. A mutation that occurs in a germ-line cell (A) can be passed on to the cell's progeny and, ultimately, to the progeny of the organism (*green*). By contrast, a mutation that arises in a somatic cell (B) affects only the progeny of that cell (*orange*) and will not be passed on to the organism's progeny. As we discuss in Chapter 20, somatic mutations are responsible for most human cancers (see pp. 714–717).

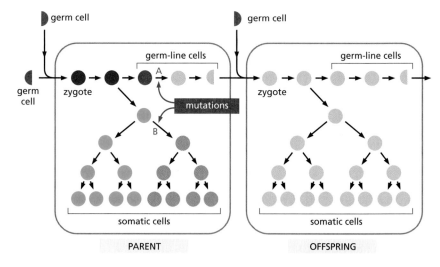

reshuffling, which influences how mutations are inherited in organisms that reproduce sexually, most of the mechanisms that generate genetic change are the same for all living things, as we now discuss.

Point Mutations Are Caused by Failures of the Normal Mechanisms for Copying and Repairing DNA

Despite the elaborate mechanisms that exist to faithfully copy and repair DNA sequences, each nucleotide pair in an organism's genome runs a small risk of changing each time a cell divides. Changes that affect a single nucleotide pair are called **point mutations**. These typically arise from rare errors in DNA replication or repair (discussed in Chapter 6).

The point mutation rate has been determined directly in experiments with bacteria such as *E. coli*. Under laboratory conditions, *E. coli* divides about once every 20–25 minutes; in less than a day, a single *E. coli* can produce more descendants than there are humans on Earth—enough to provide a good chance for almost any conceivable point mutation to occur. A culture containing 10^9 *E. coli* cells thus harbors millions of mutant cells whose genomes differ subtly from the ancestor cell. Some of these mutations may confer a selective advantage on individual cells: resistance to a poison, for example, or the ability to survive when deprived of a standard nutrient. By exposing the culture to a selective condition—adding an antibiotic or removing an essential nutrient, for example—one can find these needles in the haystack; that is, the cells that have undergone a specific mutation enabling them to survive in conditions where the original cells cannot (**Figure 9–5**). Such experiments have revealed that the overall point mutation frequency in *E. coli* is about 3 changes per 10^{10} nucleotide pairs each cell generation. The mutation rate in humans, as determined by comparing the DNA sequences of children and their parents (and estimating how many times the parental germ cells divided), is

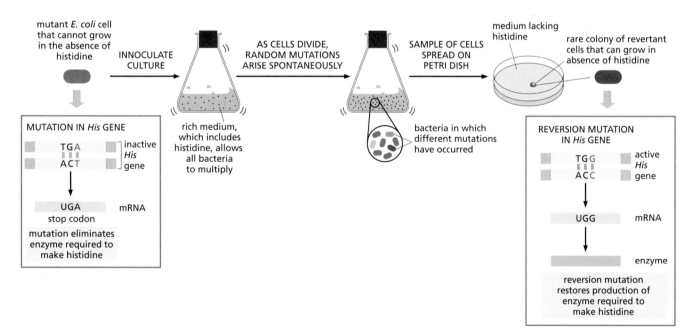

Figure 9–5 Mutation rates can be measured in the laboratory. In this experiment, an *E. coli* strain that carries a deleterious point mutation in the *His* gene—which is needed to manufacture the amino acid histidine—is used. The mutation converts a G-C nucleotide pair to an A-T, resulting in a premature stop signal in the mRNA produced from the mutant gene (*left* box). As long as histidine is supplied in the growth medium, this strain can grow and divide normally. If a large number of mutant cells (say 10^{10}) is spread on an agar plate that lacks histidine, the great majority will die. The rare survivors will contain a "reversion" mutation (in which the A-T is changed back to a G-C). This reversion corrects the original defect and now allows the bacterium to make the enzyme it needs to survive in the absence of histidine. Such mutations happen by chance and only rarely, but the ability to work with very large numbers of *E. coli* cells makes it possible to detect this change and to accurately measure its frequency.

about one-third that of *E. coli*—which suggests that the mechanisms that evolved to maintain genome integrity operate with an efficiency that does not differ significantly in even distantly related species.

Point mutations can destroy a gene's activity or—very rarely—improve it (as shown in Figure 9–5). More often, however, they do neither of these things. At many sites in the genome, a point mutation has absolutely no effect on the organism's appearance, viability, or ability to reproduce. Such *neutral mutations* often fall in regions of the gene where the DNA sequence is unimportant, including most of an intron's sequence. In cases where they occur within an exon, neutral mutations can change the third position of a codon such that the amino acid it specifies is unchanged—or is so similar that the protein's function is unaffected.

Point Mutations Can Change the Regulation of a Gene

Mutations in the coding sequences of genes are fairly easy to spot because they change the amino acid sequence of the encoded protein in predictable ways. But mutations in regulatory DNA are more difficult to recognize, because they don't affect protein sequence and can be located some distance from the coding sequence of the gene.

Despite these difficulties, many examples have been discovered where point mutations in regulatory DNA have a profound effect on the protein's production and thereby on the organism. For example, a small number of people are resistant to malaria because of a point mutation that affects the expression of a cell-surface receptor to which the malaria parasite *Plasmodium vivax* binds. The mutation prevents the receptor from being produced in red blood cells, rendering the individuals who carry this mutation immune to malarial infection.

Point mutations in regulatory DNA also have a role in our ability to digest lactose, the main sugar in milk. Our earliest ancestors were lactose intolerant, because the enzyme that breaks down lactose—called lactase—was made only during infancy. Adults, who were no longer exposed to breast milk, did not need the enzyme. When humans began to get milk from domestic animals some 10,000 years ago, variant genes—produced by random mutation—enabled those who carried the variation to continue to express lactase as adults. We now know that people who retain the ability to digest milk as adults contain a point mutation in the regulatory DNA of the lactase gene, allowing it to be efficiently transcribed throughout life. In a sense, these milk-drinking adults are "mutants" with respect to their ability to digest lactose. It is remarkable how quickly this trait spread through the human population, especially in societies that depended heavily on milk for nutrition (**Figure 9–6**).

These evolutionary changes in the regulatory sequence of the lactase gene occurred relatively recently (10,000 years ago), well after humans became a distinct species. However, much more ancient changes in regulatory sequences have occurred in other genes, and some of these are thought to underlie many of the profound differences among species (**Figure 9–7**).

DNA Duplications Give Rise to Families of Related Genes

Point mutations can influence the activity of an existing gene, but how do new genes with new functions come into being? Gene duplication is perhaps the most important mechanism for generating new genes from old ones. Once a gene has been duplicated, each of the two copies is free to accumulate mutations that might allow it to perform a slightly different function—as long as the original activity of the gene is not lost. This specialization of duplicated genes occurs gradually, as mutations

percentage of population
that is
lactose tolerant

- 100%
- 90–99%
- 80–89%
- 70–79%
- 60–69%
- 50–59%
- 40–49%
- 30–39%
- 20–29%
- 10–19%
- 0–9%
- no data

Native Americans

Indigenous Australians

Figure 9–6 The ability of adult humans to digest milk followed the domestication of cattle. Approximately 10,000 years ago, humans in northern Europe and central Africa began to raise cattle. The subsequent availability of cow's milk—particularly during periods of starvation—gave a selective advantage to those humans able to digest lactose as adults. Two independent point mutations that allow the expression of lactase in adults arose in human populations—one in northern Europe and another in central Africa. These mutations have since spread through different regions of the world. For example, the migration of Northern Europeans to North America and Australia explains why most people living on these continents can digest lactose as adults; the native populations of North America and Australia, however, remain lactose intolerant.

accumulate in the descendants of the original cell in which gene duplication occurred. By repeated rounds of this process of **gene duplication and divergence** over many millions of years, one gene can give rise to a whole family of genes, each with a specialized function, within a single genome. Analysis of genome sequences reveals many examples of such **gene families**: in *Bacillus subtilis*, for example, nearly half of the genes have one or more obvious relatives elsewhere in the genome. And in vertebrates, the globin family of genes, which encode oxygen-carrying proteins, clearly arose from a single primordial gene, as we see shortly. But how does gene duplication occur in the first place?

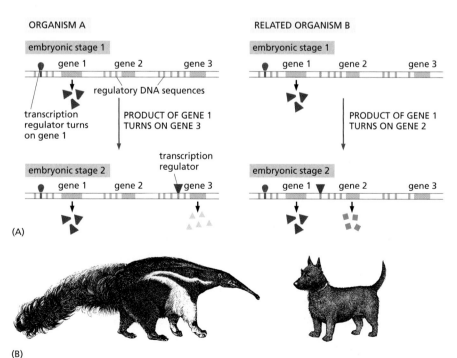

(A)

(B)

Figure 9–7 Changes in regulatory DNA sequences can have dramatic consequences for the development of an organism. (A) In this hypothetical example, the genomes of organisms A and B contain the same three genes (1, 2, and 3) and encode the same two transcription regulators (*red oval, brown triangle*). However, the regulatory DNA controlling expression of genes 2 and 3 is different in the two organisms. Although both express the same gene—gene 1—during embryonic stage 1, the differences in their regulatory DNA cause them to express different genes in stage 2. (B) In principle, a collection of such regulatory changes can have profound effects on an organism's developmental program—and, ultimately, on the appearance of the adult.

Figure 9–8 Gene duplication can be caused by crossovers between short, repeated DNA sequences in adjacent homologous chromosomes. The two chromosomes shown here undergo homologous recombination at short repeated sequences (*red*), that bracket a gene (*orange*). These repeated sequences can be remnants of mobile genetic elements, which are present in many copies in the human genome, as we discuss shortly. When crossing-over occurs unequally, as shown, one chromosome will get two copies of the gene, while the other will get none. The type of homologous recombination that produces gene duplications is called *unequal crossing-over* because the resulting products are unequal in size. If this process occurs in the germ line, some progeny will inherit the long chromosome, while others will inherit the short one.

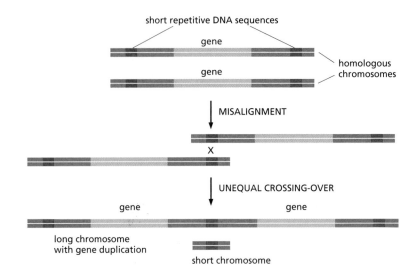

Many gene duplications are believed to be generated by *homologous recombination*. As discussed in Chapter 6, homologous recombination provides an important mechanism for mending a broken double helix; it allows an intact chromosome to be used as a template to repair a damaged sequence on its homolog. Homologous recombination normally takes place only after two long stretches of nearly identical DNA become paired, so that the information in the intact piece of DNA can be used to "restore" the sequence in the broken DNA. On rare occasions, however, a recombination event can occur between a pair of shorter DNA sequences—identical or very similar—that fall on either side of a gene. If these short sequences are not aligned properly during recombination, a lopsided exchange of genetic information can occur. Such *unequal crossovers* can generate one chromosome that has an extra copy of the gene and another with no copy (**Figure 9–8**). Once a gene has been duplicated in this way, subsequent unequal crossovers can readily add extra copies to the duplicated set by the same mechanism. As a result, entire sets of closely related genes, arranged in series, are commonly found in genomes.

The Evolution of the Globin Gene Family Shows How Gene Duplication and Divergence Can Produce New Proteins

The evolutionary history of the globin gene family provides a striking example of how gene duplication and divergence has generated new proteins. The unmistakable similarities in amino acid sequence and structure among the present-day globin proteins indicate that all the globin genes must derive from a single ancestral gene.

The simplest globin protein has a polypeptide chain of about 150 amino acids, which is found in many marine worms, insects, and primitive fish. Like our hemoglobin, this protein transports oxygen molecules throughout the animal's body. The oxygen-carrying protein in the blood of adult mammals and most other vertebrates, however, is more complex; it is composed of four globin chains of two distinct types—α globin and β globin (**Figure 9–9**). The four oxygen-binding sites in the $\alpha_2\beta_2$ molecule interact, allowing an allosteric change in the molecule as it binds and releases oxygen. This structural shift enables the four-chain hemoglobin molecule to efficiently take up and release four oxygen molecules in an all-or-none fashion, a feat not possible for the single-chain version. This efficiency is particularly important for large multicellular animals, which

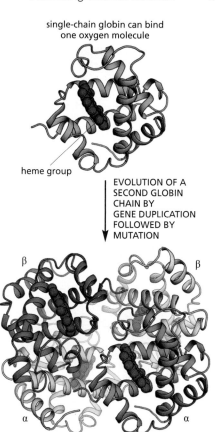

single-chain globin can bind
one oxygen molecule

heme group

EVOLUTION OF A
SECOND GLOBIN
CHAIN BY
GENE DUPLICATION
FOLLOWED BY
MUTATION

β β

α α

four-chain hemoglobin can bind
four oxygen molecules in a
cooperative way

Figure 9–9 An ancestral globin gene encoding a single-chain globin molecule is thought to have given rise to the pair of genes that produce four-chain hemoglobin proteins of modern humans and other mammals. The mammalian hemoglobin molecule is a complex of two α- and two β-globin chains. Each chain has a bound heme group (*red*) that is responsible for binding oxygen.

cannot rely on the simple diffusion of oxygen through the body to oxygenate their tissues adequately.

The α- and β-globin genes are the result of gene duplications that occurred early in vertebrate evolution. Genome analyses suggest that one of our ancient ancestors had a single globin gene. But about 500 million years ago, gene duplications followed by mutation are thought to have given rise to two slightly different globin genes, one encoding α globin, the other encoding β globin. Still later, as the different mammals began diverging from their common ancestor, the β-globin gene underwent its own duplication and divergence to give rise to a second β-like globin gene that is expressed specifically in the fetus (**Figure 9–10**). The resulting fetal hemoglobin molecule has a higher affinity for oxygen compared with adult hemoglobin, a property that helps transfer oxygen from mother to fetus.

Subsequent rounds of duplication in both the α- and β-globin genes gave rise to additional members of these families. Each of these duplicated genes has been modified by point mutations that affect the properties of the final hemoglobin molecule, and by changes in regulatory DNA that determine when—and how strongly—each gene is expressed. As a result, each globin differs slightly in its ability to bind and release oxygen and in the stage of development during which it is expressed.

In addition to these specialized globin genes, there are several duplicated DNA sequences in the α- and β-globin gene clusters that are not functional genes. They are similar in DNA sequence to the functional globin genes, but they have been disabled by the accumulation of many mutations that inactivate them. The existence of such *pseudogenes* makes it clear that, as might be expected, not every DNA duplication leads to a new functional gene. Most gene duplication events are unsuccessful in that one copy is gradually inactivated by mutation. Although we have focused here on the evolution of the globin genes, similar rounds of gene duplication and divergence have clearly taken place in many other gene families present in the human genome.

Figure 9–10 Repeated rounds of duplication and mutation are thought to have generated the globin gene family in humans. About 500 million years ago, an ancestral globin gene duplicated and gave rise to the β-globin gene family (including the five genes shown) and the related α-globin gene family. In most vertebrates, a molecule of hemoglobin (see Figure 9–9) is formed from two chains of α globin and two chains of β globin—which can be any one of the five subtypes of the β family listed here.

The evolutionary scheme shown was worked out by comparing globin genes from many different organisms. The nucleotide sequences of the γG and γA genes—which produce the β-globin-like chains that form fetal hemoglobin—are much more similar to each other than either of them is to the adult β gene. And the δ-globin gene that arose during primate evolution encodes a minor β-globin form that's only made in adult primates. In humans, the β-globin genes are located in a cluster on Chromosome 11. A subsequent chromosome breakage event, which occurred about 300 million years ago, is believed to have separated the α- and β-globin genes; the α-globin genes now reside on human Chromosome 16 (not shown).

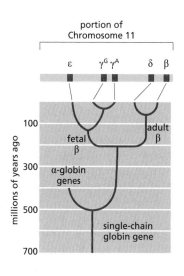

portion of
Chromosome 11

ε   γG γA   δ β

millions of years ago

100 fetal β / adult β

300 α-globin genes

500 single-chain globin gene

700

Figure 9–11 **Different species of the frog *Xenopus* have different DNA contents.** *X. tropicalis* (*above*) has an ordinary diploid genome with two sets of chromosomes in every somatic cell; the tetraploid *X. laevis* (*below*) has a duplicated genome containing twice as much DNA per cell. (Courtesy of Enrique Amaya.)

Whole-Genome Duplications Have Shaped the Evolutionary History of Many Species

Almost every gene in the genomes of vertebrates exists in multiple versions, suggesting that, rather than single genes being duplicated in a piecemeal fashion, the whole vertebrate genome was long ago duplicated in one fell swoop. Early in vertebrate evolution, it appears that the entire genome actually underwent duplication twice in succession, giving rise to four copies of every gene. In some groups of vertebrates, such as the salmon and carp families (including the zebrafish; see Figure 1–37), there may have been yet another duplication, creating an eightfold multiplicity of genes.

The precise history of whole-genome duplications in vertebrate evolution is difficult to chart because many other changes have occurred since these ancient evolutionary events. In some organisms, however, full genome duplications are especially obvious, as they have occurred relatively recently—evolutionarily speaking. The frog genus *Xenopus*, for example, comprises a set of closely similar species related to one another by repeated duplications or triplications of the whole genome (**Figure 9–11**). Such large-scale duplications can happen if cell division fails to occur following a round of genome replication in the germ line of a particular individual. Once an accidental doubling of the genome occurs in a germ-line cell, it will be faithfully passed on to germ-line progeny cells in that individual and, ultimately, to any offspring these cells might produce.

Novel Genes Can Be Created by Exon Shuffling

As we discussed in Chapter 4, many proteins are composed of a set of smaller functional *domains*. In eukaryotes, each of these protein domains is usually encoded by a separate exon, which is surrounded by long stretches of noncoding introns (see Figures 7–17 and 7–18). This organization of eukaryotic genes can facilitate the evolution of new proteins by allowing exons from one gene to be added to another—a process called **exon shuffling**.

This duplication and movement of exons is promoted by the same type of recombination that gives rise to gene duplications (see Figure 9–8). In this case, recombination occurs within the introns that surround the exons. If the introns in question are from two different genes, this recombination can generate a hybrid gene that includes complete exons from both. The presumed results of such exon shuffling are seen in many present-day proteins, which contain a patchwork of many different protein domains (**Figure 9–12**).

It has been proposed that all the proteins encoded by the human genome (approximately 21,000) arose from the duplication and shuffling of a few thousand distinct exons, each encoding a protein domain of approximately 30–50 amino acids. This remarkable idea suggests that the great

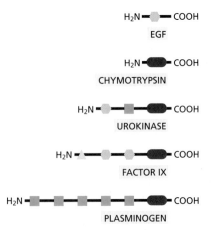

Figure 9–12 **Exon shuffling during evolution can generate proteins with new combinations of protein domains.** Each type of colored symbol represents a different protein domain. These different domains are thought to have been joined together by exon shuffling during evolution to create the modern-day human proteins shown here.

diversity of protein structures is generated from a quite small universal "list of parts," pieced together in different combinations.

The Evolution of Genomes Has Been Profoundly Influenced by the Movement of Mobile Genetic Elements

Mobile genetic elements—DNA sequences that can move from one chromosomal location to another—are an important source of genomic change and have profoundly affected the structure of modern genomes. These parasitic DNA sequences can colonize a genome and then spread within it. In the process, they often disrupt the function or alter the regulation of existing genes; sometimes they even create novel genes through fusions between mobile sequences and segments of existing genes.

The insertion of a mobile genetic element into the coding sequence of a gene or into its regulatory region can cause the "spontaneous" mutations that are observed in many of today's organisms. Mobile genetic elements can severely disrupt a gene's activity if they land directly within its coding sequence. Such an *insertion mutation* destroys the gene's capacity to encode a useful protein—as is the case for a number of mutations that cause hemophilia in humans, for example.

The activity of mobile genetic elements can also change the way existing genes are regulated. An insertion of an element into a regulatory DNA region, for instance, will often have a striking effect on where and when genes are expressed (**Figure 9–13**). Many mobile genetic elements carry DNA sequences that are recognized by specific transcription regulators; if these elements insert themselves near a gene, that gene can be brought under the control of these transcription regulators, thereby changing the gene's expression pattern. Thus, mobile genetic elements can be a major source of developmental changes: They are thought to have been particularly important in the evolution of the body plans of multicellular plants and animals.

Finally, mobile genetic elements provide opportunities for genome rearrangements by serving as targets of homologous recombination (see Figure 9–8). For example, the duplications that gave rise to the β-globin gene cluster are thought to have occurred by crossovers between the abundant mobile genetic elements sprinkled throughout the human genome. Later in the chapter, we describe these elements in more detail and discuss the mechanisms that have allowed them to establish a stronghold within our genome.

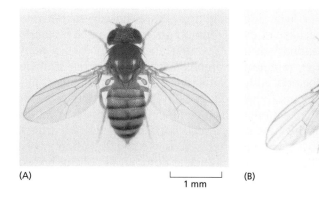

(A)   1 mm   (B)

Figure 9–13 Mutation due to a mobile genetic element can induce dramatic alterations in the body plan of an organism. (A) A normal fruit fly (*Drosophila melanogaster*). (B) A mutant fly in which the antennae have been transformed into legs because of a mutation in a regulatory DNA sequence that causes genes for leg formation to be activated in the positions normally reserved for antennae. Although this particular change is not advantageous to the fly, it illustrates how the movement of a transposable element can produce a major change in the appearance of an organism. (A, courtesy of E.B. Lewis; B, courtesy of Matthew Scott.)

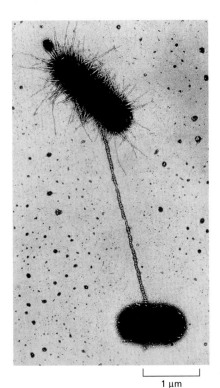

1 µm

Figure 9–14 Bacterial cells can exchange DNA through conjugation. Conjugation begins when a donor cell (*top*) attaches to a recipient cell (*bottom*) by a fine appendage, called a sex pilus. DNA from the donor cell then moves through the pilus into the recipient cell. In this electron micrograph, the sex pilus has been labeled along its length by viruses that specifically bind to it and make the structure more visible. Conjugation is one of several ways in which bacteria carry out horizontal gene transfer. (Courtesy of Charles C. Brinton Jr. and Judith Carnahan.)

Genes Can Be Exchanged Between Organisms by Horizontal Gene Transfer

So far we have considered genetic changes that take place within the genome of an individual organism. However, genes and other portions of genomes can also be exchanged between individuals of different species. This mechanism of **horizontal gene transfer** is rare among eukaryotes but common among bacteria, which can exchange DNA by the process of conjugation (**Figure 9–14** and Movie 9.1).

E. coli, for example, has acquired about one-fifth of its genome from other bacterial species within the past 100 million years. And such genetic exchanges are currently responsible for the rise of new and potentially dangerous strains of drug-resistant bacteria. Genes that confer resistance to antibiotics are readily transferred from species to species, providing the recipient bacterium with an enormous selective advantage in evading the antimicrobial compounds that constitute modern medicine's frontline attack against bacterial infection. As a result, many antibiotics are no longer effective against the common bacterial infections for which they were originally used; as an example, most strains of *Neisseria gonorrhoeae*, the bacterium that causes gonorrhea, are now resistant to penicillin, which is therefore no longer the primary drug used to treat this disease.

RECONSTRUCTING LIFE'S FAMILY TREE

We have seen how genomes can change over evolutionary time. The nucleotide sequences of present-day genomes provide a record of those changes that conferred biological success. By comparing the genomes of a variety of living organisms, we can thus begin to decipher our evolutionary history, seeing how our ancestors veered off in adventurous new directions that led us to where we are today.

The most astonishing revelation of such genome comparisons has been that **homologous genes**—those that are similar in nucleotide sequence because of their common ancestry—can be recognized across vast evolutionary distances. Unmistakable homologs of many human genes are easy to detect in organisms such as worms, fruit flies, yeasts, and even bacteria. Although the lineage that led to the evolution of vertebrates is thought to have diverged from the one that led to nematode worms and insects more than 600 million years ago, when we compare the genomes of the nematode *Caenorhabditis elegans*, the fruit fly *Drosophila melanogaster*, and *Homo sapiens*, we find that about 50% of the genes in each of these species have clear homologs in one or both of the other two species. In other words, clearly recognizable versions of at least half of all human genes were already present in the common ancestor of worms, flies, and humans.

By tracing such relationships among genes, we can begin to define the evolutionary relationships among different species, placing each bacterium, animal, plant, or fungus in a single vast family tree of life. In this

QUESTION 9–2

Why do you suppose that horizontal gene transfer is more prevalent in single-celled organisms than in multicellular organisms?

section, we discuss how these relationships are determined and what they tell us about our genetic heritage.

Genetic Changes That Provide a Selective Advantage Are Likely to Be Preserved

Evolution is commonly thought of as progressive, but at the molecular level the process is random. Consider the fate of a point mutation that occurs in a germ-line cell. On rare occasions, the mutation might cause a change for the better. But most often it will either have no consequence or cause serious damage. Mutations of the first type will tend to be perpetuated, because the organism that inherits them will have an increased likelihood of reproducing itself. Mutations that are *selectively neutral* may or may not be passed on. And mutations that are deleterious will be lost. Through endless repetition of such cycles of error and trial—of mutation and natural selection—organisms gradually evolve. Their genomes change and they develop new ways to exploit the environment—to outcompete others and to reproduce successfully.

Clearly, some parts of the genome can accumulate mutations more easily than others in the course of evolution. A segment of DNA that does not code for protein or RNA and has no significant regulatory role is free to change at a rate limited only by the frequency of random mutation. In contrast, deleterious alterations in a gene that codes for an essential protein or RNA molecule cannot be accommodated so easily: when mutations occur, the faulty organism will almost always be eliminated or fail to reproduce. Genes of this latter sort are therefore *highly conserved*; that is, the proteins they encode are very similar from organism to organism. Throughout the 3.5 billion years or more of evolutionary history, the most highly conserved genes remain perfectly recognizable in all living species. They encode crucial proteins such as DNA and RNA polymerases, and they are the ones we turn to when we wish to trace family relationships among the most distantly related organisms in the tree of life.

Closely Related Organisms Have Genomes That Are Similar in Organization As Well As Sequence

For species that are closely related, it is often most informative to focus on selectively neutral mutations. Because they accumulate steadily at a rate that is unconstrained by selection pressures, these mutations provide a metric for gauging how much modern species have diverged from their common ancestor. Such comparisons of nucleotide changes allow the construction of a **phylogenetic tree**, a diagram that depicts the evolutionary relationships among a group of organisms. **Figure 9–15** presents a phylogenetic tree that lays out the relationships among higher primates.

> **QUESTION 9–3**
>
> Highly conserved genes such as those for ribosomal RNA are present as clearly recognizable relatives in all organisms on Earth; thus, they have evolved very slowly over time. Were such genes "born" perfect?

Figure 9–15 Phylogenetic trees display the relationships among modern life-forms. In this family tree of higher primates, humans fall closer to chimpanzees than to gorillas or orangutans, as there are fewer differences between human and chimp DNA sequences than there are between those of humans and gorillas, or of humans and orangutans. As indicated, the genome sequences of each of these four species are estimated to differ from the sequence of the last common ancestor of higher primates by about 1.5%. Because changes occur independently in each lineage, the divergence between any two species will be twice as much as the amount of change that takes place between each of the species and their last common ancestor. For example, although humans and orangutans differ from their common ancestor by about 1.5% in terms of nucleotide sequence, they typically differ from one another by slightly more than 3%; human and chimp genomes differ by about 1.2%. Although this phylogenetic tree is based solely on nucleotide sequences, the estimated dates of divergence, shown on the *right* side of the graph, derive from data obtained from the fossil record. (Modified from F.C. Chen and W.H. Li, *Am. J. Hum. Genet.* 68:444–456, 2001. With permission from Elsevier.)

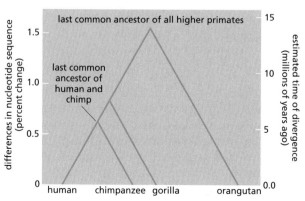

Figure 9–16 Ancestral gene sequences can be reconstructed by comparing closely related present-day species. Shown here, in five contiguous segments of DNA, are nucleotide sequences from the protein-coding region of the leptin gene from humans and chimpanzees. Leptin is a hormone that regulates food intake and energy utilization. As indicated by the codons boxed in *green*, only 5 out of a total 441 nucleotides differ between the chimp and human sequences. Only one of these changes (marked with an asterisk) results in a change in the amino acid sequence. The nucleotide sequence of the last common ancestor was probably the same as the human and chimp sequences where they agree; in the few places where they disagree, the gorilla sequence (*red*) can be used as a "tiebreaker." This strategy is based on the relationship shown in Figure 9–15: differences between humans and chimpanzees reflect relatively recent events in evolutionary history, and the gorilla sequence reveals the most likely precursor sequence. For convenience, only the first 300 nucleotides of the leptin-coding sequences are shown. The last 141 nucleotides are identical between humans and chimpanzees.

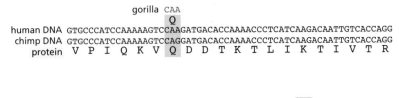

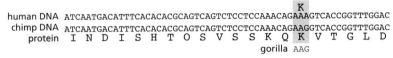

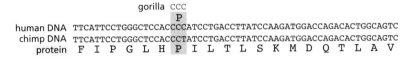

It is clear from this figure that chimpanzees are our closest living relative among the higher primates. Not only do chimpanzees seem to have essentially the same set of genes as we do, but their genes are arranged in nearly the same way. The only substantial exception is human Chromosome 2, which arose from a fusion of two chromosomes that remain separate in the chimpanzee, gorilla, and orangutan. Humans and chimpanzees are so closely related that it is possible to use DNA sequence comparisons to reconstruct the sequence of genes that must have been present in the now-extinct, common ancestor of the two species (**Figure 9–16**).

Even the rearrangement of genomes by recombination, which we described earlier, has produced only minor differences between the human and chimp genomes. For example, both the chimp and human genomes contain a million copies of a type of mobile genetic element called an *Alu* sequence. More than 99% of these elements are in corresponding positions in both genomes, indicating that most of the *Alu* sequences in our genome were in place before humans and chimpanzees diverged.

Functionally Important Genome Regions Show Up As Islands of Conserved DNA Sequence

As we delve back further into our evolutionary history and compare our genomes with those of more distant relatives, the picture begins to change. The lineages of humans and mice, for example, diverged about 75 million years ago. These genomes are about the same size, contain practically the same genes, and are both riddled with mobile genetic elements. However, the mobile genetic elements found in mouse and human DNA, although similar in sequence, are distributed differently, as they have had more time to proliferate and move around the two genomes since these species diverged (**Figure 9–17**).

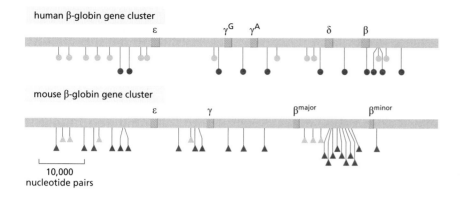

Figure 9–17 The positions of mobile genetic elements in the human and mouse genomes reflect the long evolutionary time separating the two species. This stretch of human Chromosome 11 (introduced in Figure 9–10) contains five functional β-globin-like genes (*orange*); the comparable region from the mouse genome contains only four. The positions of two types of mobile genetic element—*Alu* sequences (*green*) and *L1* sequences (*red*)—are shown in each genome. Although the mobile genetic elements in human (*circles*) and mouse (*triangles*) are not identical, they are closely related. The absence of these elements within the globin genes can be attributed to natural selection, which most likely eliminated any insertion that compromised gene function. (The mobile genetic element that falls inside the human β-globin gene (*far right*) is actually located within an intron.) (Courtesy of Ross Hardison and Webb Miller.)

In addition to the movement of mobile genetic elements, the large-scale organization of the human and mouse genomes has been scrambled by many episodes of chromosome breakage and recombination in the past 75 million years: it is estimated that about 180 such "break-and-join" events have dramatically altered chromosome structure. For example, in humans most centromeres lie near the middle of the chromosome, whereas those of mouse are located at the chromosome ends.

In spite of this significant degree of genetic shuffling, one can nevertheless still recognize many blocks of **conserved synteny**, regions where corresponding genes are strung together in the same order in both species. These genes were neighbors in the ancestral species and, despite all the chromosomal upheavals, they remain neighbors in the two present-day species. More than 90% of the mouse and human genomes can be partitioned into such corresponding regions of conserved synteny. Within these regions, we can align the DNA of mouse with that of humans so that we can compare the nucleotide sequences in detail. Such genome-wide sequence comparisons reveal that, in the roughly 75 million years since humans and mice diverged from their common ancestor, about 50% of the nucleotides have changed. Against this background of dissimilarity, however, one can now begin to see very clearly the regions where changes are not tolerated, so that the human and mouse sequences have remained nearly the same (**Figure 9–18**). Here, the sequences have been conserved by **purifying selection**—that is, by the elimination of individuals carrying mutations that interfere with important functions.

The power of *comparative genomics* can be increased by stacking our genome up against the genomes of additional animals, including the rat, chicken, and dog. Such comparisons take advantage of the results of the "natural experiment" that has lasted for hundreds of millions of years, and they highlight some of the most important regions of these genomes. These comparisons reveal that roughly 4.5% of the human genome consists of DNA sequences that are highly conserved in many other mammals (**Figure 9–19**). Surprisingly, only about one-third of these sequences code for proteins. Some of the conserved noncoding sequences correspond

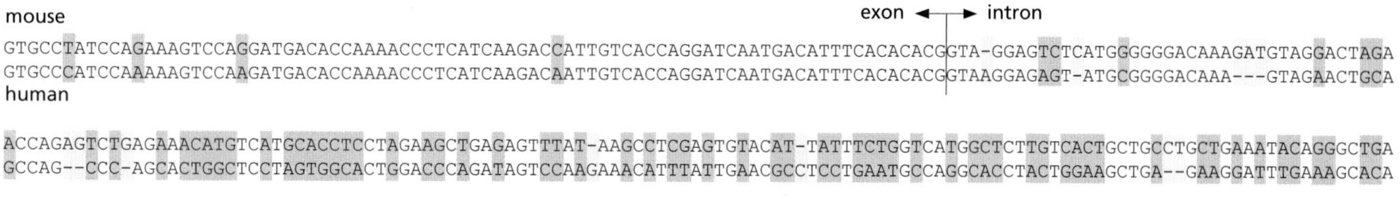

mouse   exon ◄—► intron

```
GTGCCTATCCAGAAAGTCCAGGATGACACCAAAACCCTCATCAAGACCATTGTCACCAGGATCAATGACATTTCACACACGGTA-GGAGTCTCATGGGGGGACAAAGATGTAGGACTAGA
GTGCCCATCCAAAAAGTCCAAGATGACACCAAAACCCTCATCAAGACAATTGTCACCAGGATCAATGACATTTCACACACGGTAAGGAGAGT-ATGCGGGGACAAA---GTAGAACTGCA
```
human

```
ACCAGAGTCTGAGAAACATGTCATGCACCTCCTAGAAGCTGAGAGTTTAT-AAGCCTCGAGTGTACAT-TATTTCTGGTCATGGCTCTTGTCACTGCTGCCTGCTGAAATACAGGGCTGA
GCCAG--CCC-AGCACTGGCTCCTAGTGGCACTGGACCCAGATAGTCCAAGAAACATTTATTGAACGCCTCCTGAATGCCAGGCACCTACTGGAAGCTGA--GAAGGATTTGAAAGCACA
```

Figure 9–18 Accumulated mutations have resulted in considerable divergence in the nucleotide sequences of the human and the mouse genomes. Shown here in two contiguous segments of DNA are portions of the human and mouse leptin gene sequences. Positions where the sequences differ by a single nucleotide substitution are boxed in *green*, and positions where they differ by the addition or deletion of nucleotides are boxed in *yellow*. Note that the coding sequence of the exon is much more conserved than the adjacent intron sequence.

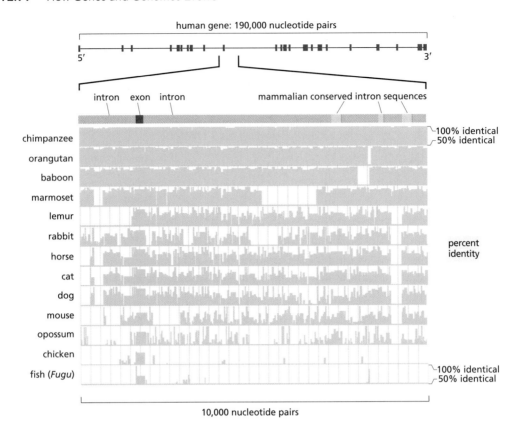

Figure 9–19 Comparison of nucleotide sequences from many different vertebrates reveals regions of high conservation. The nucleotide sequence examined in this diagram is a small segment of the human gene for a plasma membrane transporter protein. Exons in the complete gene (*top*) and in the expanded region of the gene are indicated in *red*. Three blocks of intron sequence that are conserved in mammals are shown in *blue*. In the lower part of the figure, the expanded human DNA sequence is aligned with the corresponding sequences of different vertebrates; the percent identity with the human sequences for successive stretches of 100 nucleotide pairs is plotted in *green*, with only identities above 50% shown. Note that the sequence of the exon is highly conserved in all the species, including chicken and fish, but the three intron sequences that are conserved in mammals are not conserved in chickens or fish. The functions of most conserved intron sequences in the human genome (including these three) are not known. (Courtesy of Eric D. Green.)

to regulatory DNA, whereas others are transcribed to produce RNA molecules that are not translated into protein but serve regulatory functions (discussed in Chapter 8). The functions of the majority of these conserved noncoding sequences, however, remain unknown. The unexpected discovery of these mysterious conserved DNA sequences suggests that we understand much less about the cell biology of mammals than we had previously imagined. With the plummeting cost and accelerating speed of whole-genome sequencing, we can expect many more surprises that will lead to an increased understanding in the years ahead.

Genome Comparisons Show That Vertebrate Genomes Gain and Lose DNA Rapidly

Going back even further in evolution, we can compare our genome with those of more distantly related vertebrates. The lineages of fish and mammals diverged about 400 million years ago. This is long enough for random sequence changes and differing selection pressures to have obliterated almost every trace of similarity in nucleotide sequence—except where purifying selection has operated to prevent change. Regions of the genome conserved between humans and fishes thus stand out even more strikingly than those conserved between different mammals. In fishes, one can still recognize most of the same genes as in humans and even many of the same segments of regulatory DNA. On the other hand, the extent of duplication of any given gene is often different, resulting in different numbers of members of gene families in the two species.

But even more striking is the finding that although all vertebrate genomes contain roughly the same number of genes, their overall size varies considerably. Whereas human, dog, and mouse are all in the same size range (around 3×10^9 nucleotide pairs), the chicken genome is only one-third this size. An extreme example of genome compression is the pufferfish *Fugu rubripes* (**Figure 9–20**), whose tiny genome is one-tenth the size of mammalian genomes, largely because of the small size of

Figure 9–20 The pufferfish, *Fugu rubripes*, has a remarkably compact genome. At 400 million nucleotide pairs, the *Fugu* genome is only one-quarter the size of the zebrafish genome, even though the two species have nearly the same genes. (From a woodcut by Hiroshige, courtesy of Arts and Designs of Japan.)

its introns. *Fugu* introns, as well as other noncoding segments in the animal's genome, lack the repetitive DNA that makes up a large portion of most mammalian genomes. Nonetheless, the positions of most *Fugu* introns are perfectly conserved when compared with their positions in mammalian genomes (**Figure 9–21**). Clearly, the intron structure of most vertebrate genes was already in place in the common ancestor of fish and mammals.

What factors could be responsible for the size differences among modern vertebrate genomes? Detailed comparisons of many genomes have led to the unexpected finding that small blocks of sequence are being lost from and added to genomes at a surprisingly rapid rate. It seems likely, for example, that the *Fugu* genome is so tiny because it lost DNA sequences faster than it gained them. Over long periods, this imbalance apparently cleared out those DNA sequences whose loss could be tolerated. This "cleansing" process has been enormously helpful to biologists: by "trimming the fat" from the *Fugu* genome, evolution has provided a conveniently slimmed-down version of a vertebrate genome in which the only DNA sequences that remain are those that are very likely to have important functions.

Sequence Conservation Allows Us to Trace Even the Most Distant Evolutionary Relationships

As we go back further still to the genomes of our even more distant relatives—beyond apes, mice, fish, flies, worms, plants, and yeasts, all the way to bacteria—we find fewer and fewer resemblances to our own genome. Yet even across this enormous evolutionary divide, purifying selection has maintained a few hundred fundamentally important genes. By comparing the sequences of these genes in different organisms and seeing how far they have diverged, we can attempt to construct a phylogenetic tree that goes all the way back to the ultimate ancestors—the cells at the very origins of life, from which we all derive.

To construct such a tree, biologists have focused on one particular gene that is conserved in all living species: the gene that codes for the ribosomal RNA (rRNA) of the small ribosomal subunit (see Figure 7–32). Because the process of translation is fundamental to all living cells, this

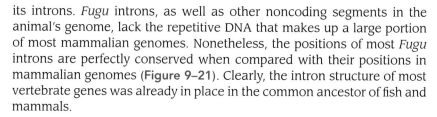

human gene

Fugu gene

| 0.0 | 100.0 | 180.0 |

thousands of nucleotide pairs

Figure 9–21 The positions of introns and exons are conserved between *Fugu* and humans. Comparison of the nucleotide sequences of the genes that encode the huntingtin protein in human and in *Fugu*. Both genes (*red*) contain 67 short exons, which align in 1:1 correspondence with one another; the corresponding exons are connected by the curved black lines. The human gene is 7.5 times larger than the *Fugu* gene (180,000 versus 24,000 nucleotide pairs), due entirely to the larger introns in the human sequence. The larger size of the human introns is due in part to mobile genetic elements, whose positions are represented by the *blue* vertical lines. These elements are absent in *Fugu*. In humans, mutation of this gene causes Huntington's disease, an inherited neurodegenerative disorder of the brain. (Adapted from S. Baxendale et al., *Nat. Genet.* 10:67–76, 1995. With permission from Macmillan Publishers Ltd.)

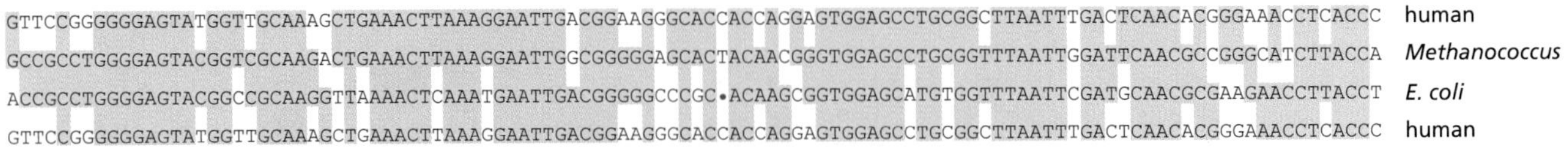

GTTCCGGGGGGAGTATGGTTGCAAAGCTGAAACTTAAAGGAATTGACGGAAGGGCACCACCAGGAGTGGAGCCTGCGGCTTAATTTGACTCAACACGGGAAACCTCACCC human
GCCGCCTGGGGAGTACGGTCGCAAGACTGAAACTTAAAGGAATTGGCGGGGGAGCACTACAACGGGTGGAGCCTGCGGTTTAATTGGATTCAACGCCGGGCATCTTACCA Methanococcus
ACCGCCTGGGGAGTACGGCCGCAAGGTTAAAACTCAAATGAATTGACGGGGGCCCGC•ACAAGCGGTGGAGCATGTGGTTTAATTCGATGCAACGCGAAGAACCTTACCT E. coli
GTTCCGGGGGGAGTATGGTTGCAAAGCTGAAACTTAAAGGAATTGACGGAAGGGCACCACCAGGAGTGGAGCCTGCGGCTTAATTTGACTCAACACGGGAAACCTCACCC human

Figure 9–22 Some genetic information has been conserved since the beginnings of life. A part of the gene for the small subunit rRNA (see Figure 7–32) is shown. Corresponding segments of nucleotide sequence from this gene in three distantly related species (*Methanococcus jannaschii*, an archaeon; *Escherichia coli*, a bacterium; and *Homo sapiens*, a eukaryote) are aligned in parallel. Sites where the nucleotides are identical between species are indicated by *green* shading; the human sequence is repeated at the bottom of the alignment so that all three two-way comparisons can be seen. The red dot halfway along the *E. coli* sequence denotes a site where a nucleotide has been either deleted from the bacterial lineage in the course of evolution or inserted in the other two lineages. Note that the three sequences have all diverged from one another to a roughly similar extent, while still retaining unmistakable similarities.

component of the ribosome has been highly conserved since early in the history of life on Earth (**Figure 9–22**).

By applying the same principles used to construct the primate family tree (see Figure 9–15), the small subunit rRNA nucleotide sequences have been used to create a single, all-encompassing tree of life. Although many aspects of this phylogenetic tree were anticipated by classical taxonomy (which is based on the outward appearance of organisms), there were also many surprises. Perhaps the most important was the realization that some of the organisms that were traditionally classed as "bacteria" are as widely divergent in their evolutionary origins as is any prokaryote from any eukaryote. As discussed in Chapter 1, it is now apparent that the prokaryotes comprise two distinct groups—the *bacteria* and the *archaea*—that diverged early in the history of life on Earth. The living world therefore has three major divisions or *domains*: bacteria, archaea, and eukaryotes (**Figure 9–23**).

Although we humans have been classifying the visible world since antiquity, we now realize that most of life's genetic diversity lies in the world of microscopic organisms. These microbes have tended to go unnoticed, unless they cause disease or rot the timbers of our houses. Yet they make up most of the total mass of living matter on our planet. Many of these organisms cannot be grown under laboratory conditions. Thus it is only through the analysis of DNA sequences, obtained from around the globe, that we are beginning to obtain a more detailed understanding of all life on Earth—knowledge that is less distorted by our biased perspective as large animals living on dry land.

ARCHAEA
BACTERIA
EUKARYOTES
Sulfolobus
Aeropyrum
Haloferax
Methanothermobacter
Methanococcus
human
maize
yeast
Paramecium
Dictyostelium
Euglena
Trypanosoma
Giardia
Trichomonas
cyanobacteria
Bacillus
E. coli
Thermotoga
Aquifex
common ancestor cell
1 change/10 nucleotides

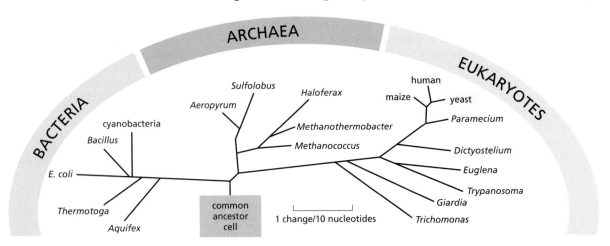

Figure 9–23 The tree of life has three major divisions. Each branch on the tree is labeled with the name of a representative member of that group, and the length of each branch corresponds to the degree of difference in the DNA sequences that encode their small subunit rRNAs (see Figure 9–22). Note that all the organisms we can see with the unaided eye—animals, plants, and some fungi (highlighted in *yellow*)—represent only a small subset of the diversity of life.

TRANSPOSONS AND VIRUSES

The tree of life depicted in Figure 9–23 includes representatives from life's most distant branches, from the cyanobacteria that release oxygen into the atmosphere to the animals, like us, that use that oxygen to boost their metabolism. What the diagram does not encompass, however, are the parasitic genetic elements that operate on the outskirts of life. Although these elements are built from the same nucleic acids contained in all life-forms and can multiply and move from place to place, they do not cross the threshold of actually being alive. Yet because of their prevalence and behavior, these diminutive genetic parasites have major implications for the evolution of species and for human health.

Mobile genetic elements, known informally as jumping genes, are found in virtually all cells. Their DNA sequences make up almost half of the human genome. Although they can insert themselves into virtually any DNA sequence, most mobile genetic elements lack the ability to leave the cell in which they reside. This is not the case for their relatives, the *viruses*. Not much more than strings of genes wrapped in a protective coat, viruses can escape from one cell and infect another.

In this section, we briefly discuss mobile genetic elements as well as viruses. We review their structure and outline how they operate—and we consider the effects they have on gene expression, genome evolution, and the transmission of disease.

Mobile Genetic Elements Encode the Components They Need for Movement

Mobile genetic elements, also called **transposons**, are typically classified according to the mechanism by which they move or *transpose*. In bacteria, the most common mobile genetic elements are the *DNA-only transposons*. The name is derived from the fact that the element moves from one place to another as a piece of DNA, as opposed to being converted into an RNA intermediate—which is the case for another type of mobile element we discuss below. Bacteria contain many different DNA-only transposons. Some move to the target site using a simple cut-and-paste mechanism, whereby the element is simply excised from the genome and inserted into a different site; other DNA-only transposons replicate their DNA before inserting into the new chromosomal site, leaving the original copy intact at its previous location (**Figure 9–24**).

Each mobile genetic element typically encodes a specialized enzyme, called a *transposase*, that mediates its movement. These enzymes recognize and act on unique DNA sequences that are present on each mobile genetic element. Many mobile genetic elements also carry additional genes: some

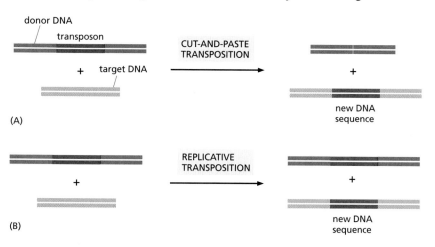

Figure 9–24 The most common mobile genetic elements in bacteria, DNA-only transposons, move by two types of mechanism. (A) In cut-and-paste transposition, the element is cut out of the donor DNA and inserted into the target DNA, leaving behind a broken donor DNA molecule, which is subsequently repaired. (B) In replicative transposition, the mobile genetic element is copied by DNA replication. The donor molecule remains unchanged, and the target molecule receives a copy of the mobile genetic element. In general, a particular type of transposon moves by only one of these mechanisms. However, the two mechanisms have many enzymatic similarities, and a few transposons can move by either mechanism. The donor and target DNAs can be part of the same DNA molecule or reside on different DNA molecules.

Figure 9–25 Transposons contain the components they need for transposition. Shown here are three types of bacterial DNA-only transposons. Each carries a gene that encodes a transposase (*blue*)—the enzyme that catalyzes the element's movement—as well as DNA sequences (*red*) that are recognized by that transposase.

Some transposons carry additional genes (*yellow*) that encode enzymes that inactivate antibiotics such as ampicillin (*AmpR*) and tetracycline (*TetR*). The spread of these transposons is a serious problem in medicine, as it has allowed many disease-causing bacteria to become resistant to antibiotics developed during the twentieth century.

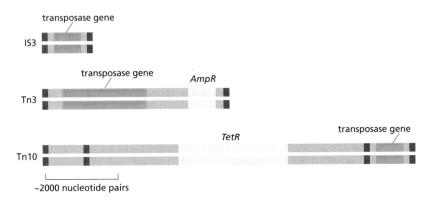

QUESTION 9–4

Many transposons move within a genome by replicative mechanisms (such as those shown in Figure 9–24B). They therefore increase in copy number each time they transpose. Although individual transposition events are rare, many transposons are found in multiple copies in genomes. What do you suppose keeps the transposons from completely overrunning their hosts' genomes?

mobile genetic elements, for example, carry antibiotic-resistance genes, which have contributed greatly to the widespread dissemination of antibiotic resistance in bacterial populations (**Figure 9–25**).

In addition to relocating themselves, mobile genetic elements occasionally rearrange the DNA sequences of the genome in which they are embedded. For example, if two mobile genetic elements that are recognized by the same transposase integrate into neighboring regions of the same chromosome, the DNA between them can be accidentally excised and inserted into a different gene or chromosome (**Figure 9–26**). In eukaryotic genomes, such accidental transposition provides a pathway for generating novel genes, both by altering gene expression and by duplicating existing genes.

The Human Genome Contains Two Major Families of Transposable Sequences

The sequencing of human genomes has revealed many surprises, as we describe in detail in the next section. But one of the most stunning was the finding that a large part of our DNA is not entirely our own. Nearly half of the human genome is made up of mobile genetic elements, which number in the millions. Some of these elements have moved from place to place within the human genome using the cut-and-paste mechanism discussed earlier (see Figure 9–24A). However, most have moved not as DNA, but via an RNA intermediate. These **retrotransposons** appear to be unique to eukaryotes.

Figure 9–26 Mobile genetic elements can move exons from one gene to another. When two mobile genetic elements of the same type (*red*) happen to insert near each other in a chromosome, the transposition mechanism occasionally recognizes the ends of two different elements (instead of the two ends of the same element). As a result, the chromosomal DNA that lies between the mobile genetic elements gets excised and moved to a new site. Such inadvertent transposition of chromosomal DNA can either generate novel genes, as shown, or alter gene regulation (not shown).

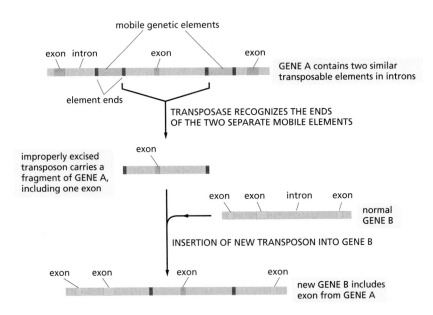

One abundant human retrotransposon, the **L1 element** (sometimes referred to as *LINE-1*, a long interspersed nuclear element), is transcribed into RNA by a host cell's RNA polymerase. A double-stranded DNA copy of this RNA is then made using an enzyme called **reverse transcriptase**, an unusual DNA polymerase that can use RNA as a template. The reverse transcriptase is encoded by the *L1* element itself. The DNA copy of the element is then free to reintegrate into another site in the genome (**Figure 9–27**).

L1 elements constitute about 15% of the human genome. Although most copies have been immobilized by the accumulation of deleterious mutations, a few still retain the ability to transpose. Their movement can sometimes precipitate disease: for example, about 40 years ago, movement of an *L1* element into the gene that encodes Factor VIII—a protein essential for proper blood clotting—caused hemophilia in an individual with no family history of the disease.

Another type of retrotransposon, the **Alu sequence**, is present in about 1 million copies, making up about 10% of our genome. *Alu* elements do not encode their own reverse transcriptase and thus depend on enzymes already present in the cell to help them move.

Comparisons of the sequence and locations of the *L1* and *Alu* elements in different mammals suggest that these sequences have proliferated in primates relatively recently in evolutionary history (see Figure 9–17). Given that the placement of mobile genetic elements can have profound effects on gene expression, it is humbling to contemplate how many of our uniquely human qualities we might owe to these prolific genetic parasites.

Viruses Can Move Between Cells and Organisms

Viruses are also mobile, but unlike the transposons we have discussed so far, they can actually escape from cells and move to other cells and organisms. Viruses were first categorized as disease-causing agents that, by virtue of their tiny size, passed through ultrafine filters that can hold back even the smallest bacterial cell. We now know that viruses are essentially genomes enclosed by a protective protein coat, and that they must enter a cell and coopt its molecular machinery to express their genes, make their proteins, and reproduce. Although the first viruses that were discovered attack mammalian cells, it is now recognized that many types of viruses exist, and virtually all organisms—including plants, animals, and bacteria—can serve as viral hosts.

Viral reproduction is often lethal to the host cells; in many cases, the infected cell breaks open (lyses), releasing progeny viruses, which can then infect neighboring cells. Many of the symptoms of viral infections reflect this lytic effect of the virus. The cold sores formed by herpes simplex virus and the blisters caused by the chickenpox virus, for example, reflect the localized killing of human skin cells.

Most viruses that cause human disease have genomes made of either double-stranded DNA or single-stranded RNA (**Table 9–1**). However, viral genomes composed of single-stranded DNA and of double-stranded RNA are also known. The simplest viruses found in nature have a small genome, composed of as few as three genes, enclosed by a protein coat built from many copies of a single polypeptide chain. More complex viruses have larger genomes of up to several hundred genes, surrounded by an elaborate shell composed of many different proteins (**Figure 9–28**). The amount of genetic material that can be packaged inside a viral protein shell is limited. Because these shells are too small to encode the

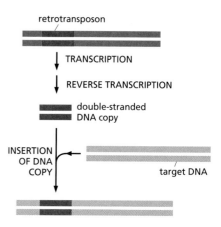

Figure 9–27 Retrotransposons move via an RNA intermediate. These transposable elements are first transcribed into an RNA intermediate. Next, a double-stranded DNA copy of this RNA is synthesized by the enzyme reverse transcriptase. This DNA copy is then inserted into the target location, which can be on either the same or a different DNA molecule. The donor retrotransposon remains at its original location, so each time it transposes, it duplicates itself. These mobile genetic elements are called retrotransposons because at one stage in their transposition their genetic information flows backward, from RNA to DNA.

QUESTION 9–5

Discuss the following statement: "Viruses exist in the twilight zone of life: outside cells they are simply dead assemblies of molecules; inside cells, however, they are alive."

Figure 9–28 Viruses come in different shapes and sizes. These electron micrographs of virus particles are all shown at the same scale. (A) T4 bacteriophage, a large DNA-containing virus that infects *E. coli* cells. The DNA is stored in the viral head and is injected into the bacterium through the cylindrical tail. (B) Potato virus X, a tubelike plant virus that contains an RNA genome. (C) Adenovirus, a DNA-containing animal virus that can infect human cells. (D) Influenza virus, a large RNA-containing animal virus whose protein coat is further enclosed in a lipid-bilayer-based envelope. The spikes protruding from the envelope are viral coat proteins embedded in the lipid bilayer. (A, courtesy of James R. Paulson; B, courtesy of Graham Hills; C, courtesy of Mei Lie Wong; D, courtesy of R.C. Williams and H.W. Fisher.)

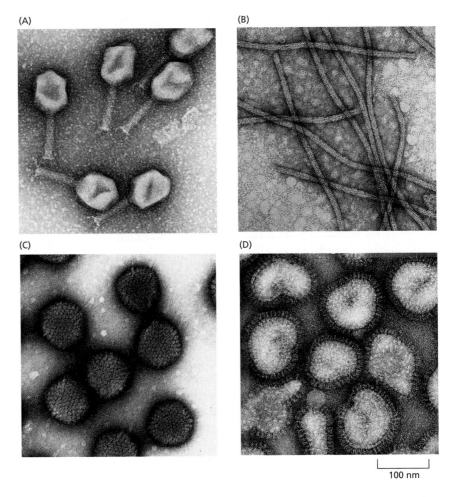

(A) (B) (C) (D)

100 nm

many enzymes and other proteins that are required to replicate even the simplest virus, viruses must hijack their host's biochemical machinery to reproduce themselves (**Figure 9–29**). The viral genome will typically encode both viral coat proteins and proteins that help them to coopt the host enzymes needed to replicate their genetic material.

Retroviruses Reverse the Normal Flow of Genetic Information

Although there are many similarities between bacterial and eukaryotic viruses, one important class of viruses—the **retroviruses**—is found only in eukaryotic cells. In many respects, retroviruses resemble the retrotransposons we just discussed. A key feature of the life cycle of both is a step in which DNA is synthesized using RNA as a template—hence the prefix *retro*, which refers to the reversal of the usual flow of DNA information to RNA. Retroviruses are thought to have derived from a retrotransposon that long ago acquired additional genes encoding the coat proteins and other proteins required to make a virus particle. The RNA stage of its replicative cycle could then be packaged into a viral particle that could leave the cell. The complete life cycle of a retrovirus is shown in **Figure 9–30**.

Like retrotransposons, retroviruses use the enzyme reverse transcriptase to convert RNA into DNA. The enzyme is encoded by the retroviral genome, and a few molecules of the enzyme are packaged along with the RNA genome in each virus particle. When the single-stranded RNA genome of the retrovirus enters a cell, the reverse transcriptase brought in with it makes a complementary DNA strand to form a DNA/RNA hybrid double helix. The RNA strand is removed, and the reverse transcriptase

TABLE 9–1 VIRUSES THAT CAUSE HUMAN DISEASE

Virus	Genome Type	DISEASE
Herpes simplex virus	double-stranded DNA	recurrent cold sores
Epstein–Barr virus (EBV)	double-stranded DNA	infectious mononucleosis
Varicella-zoster virus	double-stranded DNA	chickenpox and shingles
Smallpox virus	double-stranded DNA	smallpox
Hepatitis B virus	part single-, part double-stranded DNA	serum hepatitis
Human immunodeficiency virus (HIV)	single-stranded RNA	acquired immune deficiency syndrome (AIDS)
Influenza virus type A	single-stranded RNA	respiratory disease (flu)
Poliovirus	single-stranded RNA	poliomyelitis
Rhinovirus	single-stranded RNA	common cold
Hepatitis A virus	single-stranded RNA	infectious hepatitis
Hepatitis C virus	single-stranded RNA	non-A, non-B type hepatitis
Yellow fever virus	single-stranded RNA	yellow fever
Rabies virus	single-stranded RNA	rabies encephalitis
Mumps virus	single-stranded RNA	mumps
Measles virus	single-stranded RNA	measles

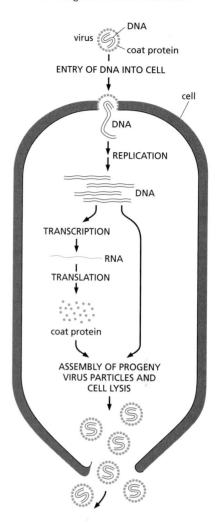

Figure 9–29 Viruses commandeer the host cell's molecular machinery to reproduce. The hypothetical simple virus illustrated here consists of a small double-stranded DNA molecule that encodes just a single type of viral coat protein. To reproduce, the viral genome must first enter a host cell, where it is replicated to produce multiple copies, which are transcribed and translated to produce the viral coat protein. The viral genomes can then assemble spontaneously with the coat protein to form new virus particles, which escape from the cell by lysing it.

(which can use either DNA or RNA as a template) now synthesizes a complementary DNA strand to produce a DNA double helix. This DNA is then inserted, or integrated, into a randomly selected site in the host genome by a virally encoded *integrase* enzyme. In this integrated state, the virus is *latent*: each time the host cell divides, it passes on a copy of the integrated viral genome, which is known as a *provirus*, to its progeny cells.

The next step in the replication of a retrovirus—which can take place long after its integration into the host genome—is the copying of the integrated viral DNA into RNA by a host-cell RNA polymerase, which produces large numbers of single-stranded RNAs identical to the original infecting genome. These viral RNAs are then translated by the host-cell ribosomes to produce the viral shell proteins, the envelope proteins, and reverse transcriptase—all of which are assembled with the RNA genome into new virus particles.

The human immunodeficiency virus (HIV), which is the cause of AIDS, is a retrovirus. As with other retroviruses, the HIV genome can persist in a latent state as a provirus embedded in the chromosomes of an infected cell. This ability to hide in host cells complicates attempts to treat the infection with antiviral drugs. But because the HIV reverse transcriptase is not used by cells for any purpose of their own, it is one of the prime targets of drugs currently used to treat AIDS.

EXAMINING THE HUMAN GENOME

The human genome contains an enormous amount of information about who we are and where we came from (**Figure 9–31**). Its 3.2×10^9 nucleotide pairs, spread out over 23 sets of chromosomes—22 autosomes and

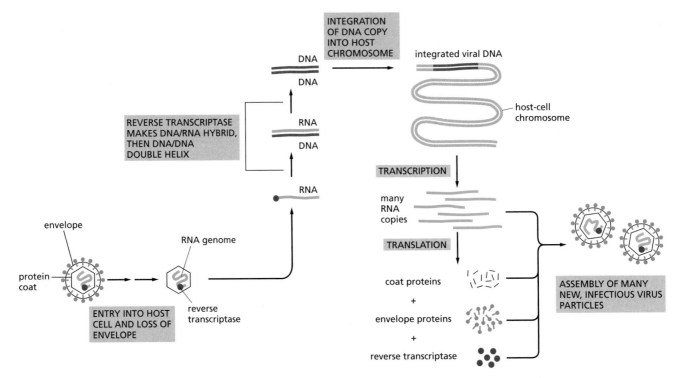

Figure 9–30 The life cycle of a retrovirus includes reverse transcription and integration of the viral genome into the host cell's DNA. The retrovirus genome consists of an RNA molecule (*blue*) that is typically between 7000 and 12,000 nucleotides in size. It is packaged inside a protein coat, which is surrounded by a lipid-based envelope that contains virus-encoded envelope proteins (*green*). The enzyme reverse transcriptase (*red* circle), encoded by the viral genome and packaged with its RNA, first makes a single-stranded DNA copy of the viral RNA molecule and then a second DNA strand, generating a double-stranded DNA copy of the RNA genome. This DNA double helix is then integrated into a host chromosome, a step required for the synthesis of new viral RNA molecules by a host-cell RNA polymerase.

a pair of sex chromosomes (X and Y)—provide the instructions needed to build a human being. Yet, 25 years ago, biologists actively debated the value of determining the *human genome sequence*—the complete list of nucleotides contained in our DNA.

The task was not simple. An international consortium of investigators labored tirelessly for the better part of a decade—and spent nearly $3 billion—to give us our first glimpse of this genetic blueprint. But the effort turned out to be well worth the cost, as the data continue to shape our thinking about how our genome functions and how it has evolved.

The first human genome sequence was just the beginning. Spectacular improvements in sequencing technologies, coupled with powerful new tools for handling massive amounts of data, are taking genomics to a whole new level. The cost of DNA sequencing has dropped about 100,000-fold since the human genome project was launched in 1990, such that a whole human genome can now be sequenced in a few days for about $1000. Investigators around the world are collaborating to collect and compare the nucleotide sequences of thousands of human genomes. This resulting deluge of data promises to tell us what makes us human, and what makes each of us unique.

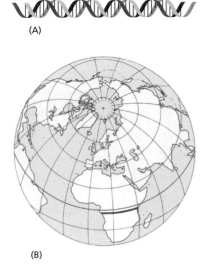

Figure 9–31 The 3 billion nucleotide pairs of the human genome contain a vast amount of information, including clues about our origins. If each nucleotide pair is drawn to span 1 mm, as shown in (A), the human genome would extend 3200 km (approximately 2000 miles)—far enough to stretch across central Africa, where humans first arose (*red* line in B). At this scale, there would be, on average, a protein-coding gene every 150 m. An average gene would extend for 30 m, but the coding sequences (exons) in this gene would add up to only just over a meter; the rest would be introns.

Although it will take decades to analyze the rapidly accumulating genome data, the recent findings have already influenced the content of every chapter in this book. In this section, we describe some of the most striking features of the human genome—many of which were entirely unexpected. We review what genome comparisons can tell us about how we evolved, and we discuss some of the mysteries that still remain.

The Nucleotide Sequences of Human Genomes Show How Our Genes Are Arranged

When the DNA sequence of human Chromosome 22, one of the smallest human chromosomes, was completed in 1999, it became possible for the first time to see exactly how genes are arranged along an entire vertebrate chromosome (**Figure 9–32**). The subsequent publication of the whole human genome sequence—a first draft in 2001 and a finished draft in 2004—provided a more panoramic view of the complete genetic landscape, including how many genes we have, what those genes look like, and how they are distributed across the genome (**Table 9–2**).

The first striking feature of the human genome is how little of it—less than 2%—codes for proteins (**Figure 9–33**). In addition, almost half of our DNA is made up of mobile genetic elements that have colonized our genome over evolutionary time. Because these elements have accumulated mutations, most can no longer move; rather, they are relics from an earlier evolutionary era when mobile genetic elements ran rampant through our genome.

It was a surprise to discover how few protein-coding genes our genome actually contains. Earlier estimates had been in the neighborhood of 100,000 (see **How We Know**, pp. 316–317). Although the exact count is still being refined, current estimates place the number of human

> ### QUESTION 9–6
>
> Mobile genetic elements, such as the *Alu* sequences, are found in many copies in human DNA. In what ways could the presence of an *Alu* sequence affect a nearby gene?

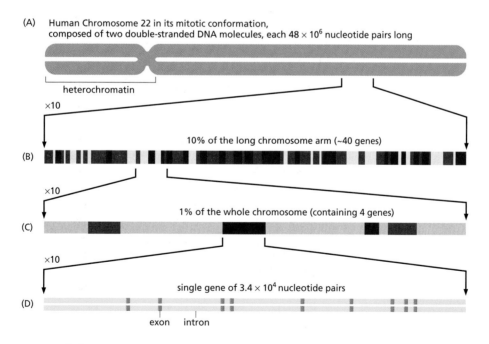

Figure 9–32 The sequence of Chromosome 22 shows how human chromosomes are organized. (A) Chromosome 22, one of the smallest human chromosomes, contains 48×10^6 nucleotide pairs and makes up approximately 1.5% of the entire human genome. Most of the left arm of Chromosome 22 consists of short repeated sequences of DNA that are packaged in a particularly compact form of chromatin (heterochromatin), as discussed in Chapter 5. (B) A tenfold expansion of a portion of Chromosome 22 shows about 40 genes. Those in *dark brown* are known genes, and those in *red* are predicted genes. (C) An expanded portion of (B) shows the entire length of several genes. (D) The intron–exon arrangement of a typical gene is shown after a further tenfold expansion. Each exon (*orange*) codes for a portion of the protein, while the DNA sequence of the introns (*yellow*) is relatively unimportant. (Adapted from The International Human Genome Sequencing Consortium, *Nature* 409:860–921, 2001. With permission from Macmillan Publishers Ltd.)

TABLE 9–2 SOME VITAL STATISTICS FOR THE HUMAN GENOME

DNA length	3.2×10^9 nucleotide pairs*
Number of protein-coding genes	approximately 21,000
Number of non-protein-coding genes**	approximately 9000
Largest gene	2.4×10^6 nucleotide pairs
Mean gene size	27,000 nucleotide pairs
Smallest number of exons per gene	1
Largest number of exons per gene	178
Mean number of exons per gene	10.4
Largest exon size	17,106 nucleotide pairs
Mean exon size	145 nucleotide pairs
Number of pseudogenes***	approximately 11,000
Percentage of DNA sequence in exons (protein-coding sequences)	1.5%
Percentage of DNA conserved with other mammals that does not encode protein****	3.5%
Percentage of DNA in high-copy repetitive elements	approximately 50%

*The sequence of 2.85 billion nucleotide pairs is known precisely (error rate of only about one in 100,000 nucleotides). The remaining DNA consists primarily of short, highly repeated sequences that are tandemly repeated, with repeat numbers differing from one individual to the next.

**These include genes that encode structural, catalytic, and regulatory RNAs.

***A pseudogene is a DNA sequence that closely resembles that of a functional gene but contains numerous mutations that prevent its proper expression. Most pseudogenes arise from the duplication of a functional gene, followed by the accumulation of damaging mutations in one copy.

****This includes DNA encoding 5′ and 3′ UTRs (untranslated regions of mRNAs), regulatory DNA, and conserved regions of unknown function.

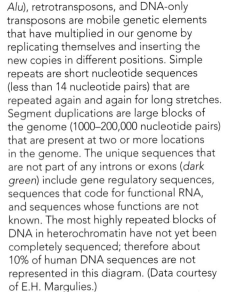

Figure 9–33 The bulk of the human genome is made of repetitive nucleotide sequences and other noncoding DNA. The LINEs (which include *L1*), SINEs (short interspersed nuclear element, which include *Alu*), retrotransposons, and DNA-only transposons are mobile genetic elements that have multiplied in our genome by replicating themselves and inserting the new copies in different positions. Simple repeats are short nucleotide sequences (less than 14 nucleotide pairs) that are repeated again and again for long stretches. Segment duplications are large blocks of the genome (1000–200,000 nucleotide pairs) that are present at two or more locations in the genome. The unique sequences that are not part of any introns or exons (*dark green*) include gene regulatory sequences, sequences that code for functional RNA, and sequences whose functions are not known. The most highly repeated blocks of DNA in heterochromatin have not yet been completely sequenced; therefore about 10% of human DNA sequences are not represented in this diagram. (Data courtesy of E.H. Margulies.)

protein-coding genes at about 21,000. Perhaps another 9000 genes encode functional RNAs that are not translated into proteins. The estimate of 30,000 total genes brings us much closer to the gene numbers for simpler multicellular animals—for example, 13,000 for *Drosophila*, 21,000 for *C. elegans*, and 28,000 for the small weed *Arabidopsis* (see Table 1–2).

The number of protein-coding genes we have may be unexpectedly small, but their relative size is unusually large. Only about 1300 nucleotide pairs are needed to encode an average-sized human protein of about 430 amino acids. Yet the average length of a human gene is 27,000 nucleotide

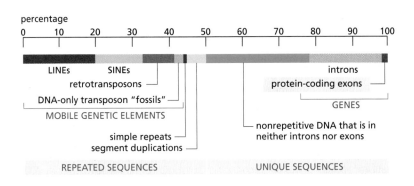

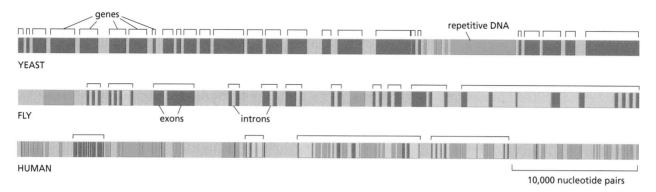

pairs. Most of this DNA is in noncoding introns. In addition to the voluminous introns (see Figure 9–32D), each gene is associated with regulatory DNA sequences that ensure that the gene is expressed at the proper level, time, and place. In humans, these regulatory DNA sequences are typically interspersed along tens of thousands of nucleotide pairs, much of which seems to be "spacer" DNA. Indeed, compared to many other eukaryotic genomes, the human genome is much less densely packed (**Figure 9–34**).

Although exons and their associated gene regulatory sequences comprise less than 2% of the human genome, comparative studies indicate that about 5% of the human genome is highly conserved when compared with other mammalian genomes (see Figure 9–19). An additional 4% of the genome shows reduced variation in the human population, as determined by comparing the DNA sequence of thousands of individuals. Taken together, this conservation suggests that about 9% of the human genome contains sequences that are likely to be functionally important—but we do not yet know the function of much of this DNA.

Accelerated Changes in Conserved Genome Sequences Help Reveal What Makes Us Human

When the chimpanzee genome sequence became available in 2005, scientists began searching for DNA sequence changes that might account for the striking differences between us and them (**Figure 9–35**). With about 3 billion nucleotide pairs to compare between the two species, the task is daunting. But the search is made much easier by confining the comparison to those sequences that are highly conserved across multiple mammalian species (see Figure 9–19). These conserved sequences represent parts of the genome that are most likely to be functionally important—and are thus areas of particular interest when we search for genetic changes that make humans different from our mammalian cousins.

Although these sequences are conserved, they are not identical: when the version from one mammal is compared with that of another, they are typically found to have drifted apart by a small amount, which corresponds to the time elapsed since the species diverged during evolution. In a small proportion of cases, however, the sequences show signs of a sudden evolutionary spurt. For example, some DNA sequences that have been highly conserved in most mammalian species are found to have changed exceptionally fast during the last six million years of human evolution. Such *human accelerated regions* are thought to reflect functions that have been especially important in making us the unique animal that we are.

One study identified about 50 such sites—one-quarter of which were located near genes associated with brain development. The sequence

Figure 9–34 Genes are sparsely distributed in the human genome. Compared to these other eukaryotic genomes, the human genome is less gene-dense. Shown here are DNA segments about 50,000 nucleotide pairs in length from yeast, *Drosophila*, and human. The human segment contains only 4 genes, compared to 26 in the yeast and 11 in the fly. Exons are shown in *orange*, introns in *yellow*, repetitive elements in *blue*, and "spacer" DNA in *gray*. The genes of yeast and flies are generally more compact, with fewer introns, than the genes of humans.

Figure 9–35 DNA sequences that have changed rapidly in the past six million years may account for the differences between chimps and humans. Many of these changes may have affected the way human brains develop. Shown here is anthropologist Jane Goodall with one of her chimpanzee subjects. (Courtesy of the Jane Goodall Institute of Canada.)

COUNTING GENES

How many genes does it take to make a human? It seems a natural thing to wonder. If 6000 genes can produce a yeast and 13,000 a fly, how many are needed to make a human being—a creature curious and clever enough to study its own genome? Until researchers completed the first draft of the human genome sequence, the most frequently cited estimate was 100,000. But where did that figure come from? And how was the revised estimate of only 21,000 protein-coding genes derived?

Walter Gilbert, a physicist-turned-biologist who won a Nobel Prize for developing techniques for sequencing DNA, was one of the first to throw out a ballpark estimate of the number of human genes. In the mid-1980s, Gilbert suggested that humans could have 100,000 genes, an estimate based on the average size of the few human genes known at the time (about 3×10^4 nucleotide pairs) and the size of our genome (about 3×10^9 nucleotide pairs). This back-of-the-envelope calculation yielded a number with such a pleasing roundness that it wound up being quoted widely in articles and textbooks.

The calculation provides an estimate of the number of genes a human could have in principle, but it does not address the question of how many genes we actually have. As it turns out, that question is not so easy to answer, even with the complete human genome sequence in hand. The problem is, how does one identify a gene? Consider protein-coding genes, which comprise only 1.5% of the human genome. Looking at a given piece of raw DNA sequence—an apparently random string of As, Ts, Gs, and Cs—how can one tell which parts represent protein-coding segments? Being able to accurately and reliably distinguish the rare coding sequences from the more plentiful noncoding sequences in a genome is necessary before one can hope to locate and count its genes.

Signals and chunks

As always, the situation is simplest in bacteria and simple eukaryotes such as yeasts. In these genomes, genes that encode proteins are identified by searching through the entire DNA sequence looking for **open reading frames (ORFs)**. These are long sequences—say, 100 codons or more—that lack stop codons. A random sequence of nucleotides will by chance encode a stop codon about once every 20 codons (as there are three stop codons in the set of 64 possible codons—see Figure 7–25). So finding an ORF—a continuous nucleotide sequence that encodes more than 100 amino acids—is the first step in identifying a good candidate for a protein-coding gene. Today, computer programs are used to search for such ORFs, which begin with an initiation codon, usually ATG, and end with a termination codon, TAA, TAG, or TGA (**Figure 9–36**).

In animals and plants, the process of identifying ORFs is complicated by the presence of large intron sequences, which interrupt the protein-coding portions of genes. As we have seen, these introns are generally much larger than the exons, which might represent only a few percent of the gene. In human DNA, exons sometimes contain as few as 50 codons (150 nucleotide pairs), while introns may exceed 10,000 nucleotide pairs in length. Fifty codons is too short to generate a statistically significant

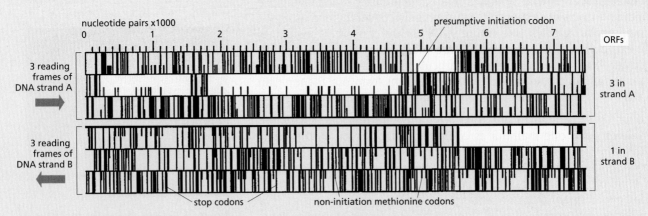

Figure 9–36 Computer programs are used to identify protein-coding genes. In this example, a DNA sequence of 7500 nucleotide pairs from the pathogenic yeast *Candida albicans* was fed into a computer, which then calculated the proteins that could, in theory, be produced from each of its six possible reading frames—three on each of the two strands (see Figure 7–26). The output shows the location of start and stop codons for each reading frame. The reading frames are laid out in horizontal columns. Stop, or termination, codons (TGA, TAA, and TAG) are represented by tall, vertical black lines, and methionine codons (ATG) are represented by shorter black lines. Four open-reading frames, or ORFs (shaded *yellow*), can be clearly identified by the statistically significant absence of stop codons. For each ORF, the presumptive initiation codon (ATG) is indicated in *red*. The additional ATG codons in the ORFs code for methionine in the protein.

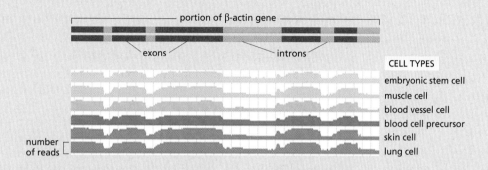

Figure 9–37 RNA sequencing can be used to identify protein-coding genes. Presented here is a set of data corresponding to RNAs produced from a segment of the gene for β-actin, which is depicted schematically at the top. Millions of RNA "sequence reads," each approximately 200 nucleotides long, were collected from a variety of cell types (*right*) and matched to DNA sequences within the β-actin gene. The height of each trace is proportional to how often each sequence appears in a read. Exon sequences are present at high levels, reflecting their presence in mature β-actin mRNAs. Intron sequences are present at low levels, most likely reflecting their presence in pre-mRNA molecules that have not yet been spliced or spliced introns that have not yet been degraded.

"ORF signal," as it is not all that unusual for 50 random codons to lack a stop signal. Moreover, introns are so long that they are likely to contain by chance quite a bit of "ORF noise," numerous stretches of sequence lacking stop signals. Finding the true ORFs in this sea of information in which the noise often outweighs the signal can be difficult. To make the task more manageable, computers are used to search for other distinctive features that mark the presence of a protein-coding gene. These include the splicing sequences that signal an intron–exon boundary (see Figure 7–19), gene regulatory sequences, or conservation with coding sequences from other organisms.

In 1992, researchers used a computer program to predict protein-coding regions in a preliminary human sequence. They found two genes in a 58,000-nucleotide-pair segment of Chromosome 4, and five genes in a 106,000-nucleotide-pair segment of Chromosome 19. That works out to an average of 1 gene every 23,000 nucleotide pairs. Extrapolating from that density to the whole genome would give humans nearly 130,000 genes. It turned out, however, that the chromosomes the researchers analyzed had been chosen for sequencing precisely because they appeared to be gene-rich. When the estimate was adjusted to take into account the gene-poor regions of the human genome—guessing that half of the human genome had maybe one-tenth of that gene-rich density—the estimated number dropped to 71,000.

Matching RNAs

Of course, these estimates are based on what we think genes look like; to get around this bias, we must employ more direct, experiment-based methods for locating genes. Because genes are transcribed into RNA, the preferred strategy for finding genes involves isolating all of the RNAs produced by a particular cell type and determining their nucleotide sequence—a technique called RNA Seq. These sequences are then mapped back to the genome to locate their genes. For protein-coding genes, exon segments are more highly represented among the sequenced transcripts, as intron sequences tend to be spliced out and destroyed. Because different cell types express different genes, and splice their RNA transcripts differently, a variety of cell types are used in the analysis **(Figure 9–37)**.

RNA Seq also offers a few additional benefits. First, the relative abundance of each sequence can be used to assess how highly its gene is expressed. Furthermore, the approach also locates genes that do not code for proteins, but instead encode functional or regulatory RNAs. Many noncoding RNAs were first identified through RNA Seq.

Human gene countdown

Based on a combination of all of these computational and experimental techniques, current estimates of the number of human genes are now converging around 30,000. It could be many years, however, before we have the final answer to how many genes it takes to make a human. In the end, having an exact count will not be nearly as important as understanding the functions of each gene and how they interact to build the living organism.

exhibiting the most rapid change (18 changes between human and chimp, compared with only two changes between chimp and chicken) was examined further and found to encode a short, non-protein-coding RNA that is produced in the human cerebral cortex at a critical time during brain development. Although the function of this RNA is not yet known, this exciting finding is stimulating further studies that might help shed light on features of the human brain that distinguish us from chimps.

Similar studies have identified genes that may have played a role in even more recent human evolution. In 2010, investigators completed their analysis of the first Neanderthal genome. Our closest evolutionary relative, Neanderthals lived side by side with the ancestors of modern humans in Europe and Western Asia. By comparing the Neanderthal genome sequence—obtained from DNA that was extracted from a fossilized bone fragment found in a cave in Croatia—with those of five people from different parts of the world, the researchers identified a handful of genomic regions that have undergone a sudden spurt of changes in modern humans. These regions include genes involved in metabolism, brain development, and the shape of the skeleton, particularly the rib cage and head—all features thought to differ between modern humans and our extinct cousins.

Remarkably, these studies also revealed that some modern humans—those that hail from Europe and Asia—share from 1 to 4 percent of their genomes with Neanderthals. This genetic overlap suggests that our ancestors may have mated with Neanderthals—before outcompeting or actively exterminating them—on the way out of Africa, a relationship that left a permanent mark in the human genome.

Genome Variation Contributes to Our Individuality—But How?

With the possible exception of some identical twins, no two people have exactly the same genome sequence. When the same region of the genome from two different humans is compared, the nucleotide sequences typically differ by about 0.1%. That might seem an insignificant degree of variation, but considering the size of the human genome, it amounts to some 3 million genetic differences per genome between one person and the next. Detailed analyses of human genetic variation suggest that the bulk of this variation was already present early in our evolution, perhaps 100,000 years ago, when the human population was still small. This means that a great deal of the genetic variation in present-day humans was inherited from our early human ancestors.

Most of the genetic variation in the human genome takes the form of single base changes called **single-nucleotide polymorphisms** (**SNPs**, pronounced snips). These polymorphisms are simply points in the genome that differ in nucleotide sequence between one portion of the population and another—positions where more than 1% of the population has a G-C nucleotide pair, for example, while another has an A-T (**Figure 9–38**). Two human genomes chosen at random from the world's population will differ by approximately 2.5×10^6 SNPs that are scattered throughout the genome.

Another important source of variation inherited from our ancestors involves the duplication and deletion of large segments of DNA. When the genome of any person is compared with a standard reference genome, one observes roughly 100 instances in which a relatively long stretch of DNA has been gained or lost. Some of these **copy-number variations** (**CNVs**) are very common, whereas others are present in only a small minority of people. From an initial sampling, nearly half of

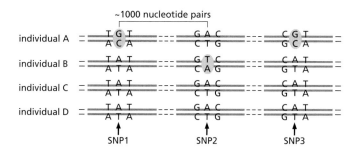

these segments contain known genes and can affect one's susceptibility to certain diseases. In retrospect, this type of structural variation is not surprising, given the extensive history of DNA addition and DNA loss in vertebrate genomes discussed earlier. Exactly how it contributes to our individuality, however, remains to be determined.

In addition to the SNPs and the CNVs that we inherited from our ancestors, humans also possess repetitive nucleotide sequences that are particularly prone to new mutations. CA repeats, for example, are ubiquitous in the human genome. Nucleotide sequences containing large numbers of CA repeats are often replicated inaccurately (imagine trying to copy a word that is nothing more than a string of CACACACAC…); hence, the precise length of such repeats can vary widely between individuals and can increase from one generation to the next. Because they show such exceptional variability, and because this variability has arisen so recently in human history, CA repeats, and others like them, make ideal markers for distinguishing the DNA of individual humans. For this reason, differences in the numbers of *short tandem repeats* at different positions in the genome are used to identify individuals by *DNA fingerprinting* in crime investigations, paternity suits, and other forensic applications (see Figure 10–18).

Most of the variations in the human genome sequence are genetically silent, as they fall within noncritical regions of the genome. Such variations have no effect on how we look or how our cells function. This means that only a small subset of the variation we observe in our DNA is responsible for the heritable differences from one human to the next. It remains a major challenge to identify those genetic variations that are functionally important—a problem we return to in Chapter 19.

Differences in Gene Regulation May Help Explain How Animals With Similar Genomes Can Be So Different

The finding that humans, chimps, and mice contain essentially the same protein-coding genes has raised a fundamental question: What makes these creatures so different from one another?

To a large extent, the instructions needed to produce a multicellular animal from a fertilized egg are provided by the regulatory DNA associated with each gene. These noncoding DNA sequences contain, scattered within them, dozens of separate regulatory elements, including short DNA segments that serve as binding sites for specific transcription regulators (discussed in Chapter 8). Regulatory DNA ultimately dictates each organism's developmental program—the rules its cells follow as they proliferate, assess their positions in the embryo, and specialize by switching on and off specific genes at the right time and place. The evolution of species is likely to have more to do with innovations in gene regulatory sequences than in the proteins or functional RNAs those genes encode.

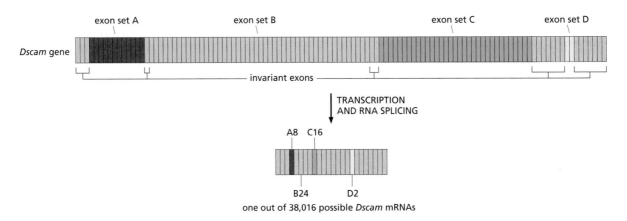

one out of 38,016 possible *Dscam* mRNAs

Figure 9–39 Alternative splicing of RNA transcripts can produce many distinct proteins. The *Drosophila* Dscam proteins are receptors that help nerve cells make their appropriate connections. The final mRNA transcript contains 24 exons, four of which (denoted A, B, C, and D) are present in the *Dscam* gene as arrays of alternative exons. Each mature mRNA contains 1 of 12 alternatives for exon A (*red*), 1 of 48 alternatives for exon B (*green*), 1 of 33 alternatives for exon C (*blue*), 1 of 2 alternatives for exon D (*yellow*), and all of the 19 invariant exons (*gray*). If all possible splicing combinations were used, 38,016 different proteins could in principle be produced from the *Dscam* gene. Only one of the many possible splicing patterns and the mature mRNA it produces is shown. (Adapted from D.L. Black, *Cell* 103:367–370, 2000. With permission from Elsevier.)

Although we have made great strides in recognizing many of these regulatory sequences amidst the excess of noncritical "spacer" DNA, we still do not know how to "read" these sequences so that we can predict exactly how they operate in cells to control development. For example, the same short stretch of regulatory DNA may be recognized by several different transcription regulators, so simply knowing its nucleotide sequence will not reveal which transcription regulator—or regulators—might bind to the sequence in a particular cell at a particular time or place. In addition, gene expression is controlled by complex combinations of proteins (see Figure 8–12), which further complicates our attempts to decipher when in development and in which type of cell any given gene will be expressed.

Even if we could predict when a particular protein-coding gene would be expressed, we would not necessarily be able to predict what protein that gene would produce. Recent studies suggest that more than 90% of human genes undergo alternative RNA splicing, which allows cells to produce a range of related but distinct proteins from a single gene (see Figure 7–22). RNA splicing is often regulated, so that one form of a protein is produced in one type of cell, while other forms are produced preferentially in other cell types. In one extreme example, from *Drosophila*, a single gene can produce thousands of different protein variants through alternative RNA splicing (**Figure 9–39**). Thus an organism can produce far more proteins than it has genes. We do not yet know enough about alternative splicing to predict exactly which human genes are subject to this process—and when, where, and how during development such regulation occurs. Nonetheless, it seems likely that these differences in alternative RNA splicing could help explain how animals with very similar protein-coding genes develop so differently.

Another part of the explanation may involve regulatory RNAs, such as the microRNAs and long noncoding RNAs discussed in Chapter 8. Thus for example, microRNAs have diverse roles in controlling gene expression, especially during development. They regulate as many as one-third of all human genes, for example, yet few of them have been studied in any detail—and new ones are still being found. And even less is known about the long noncoding RNAs.

The information that guides the countless decisions made by developing cells as they divide and specialize is all contained within the genome sequence of an organism. But we are only just beginning to learn the grammar and rules by which this genetic information orchestrates development. Deciphering this code—which has been shaped by evolution and refined by individual variation—is one of the great challenges facing the next generation of cell biologists.

ESSENTIAL CONCEPTS

- By comparing the DNA and protein sequences of contemporary organisms, we are beginning to reconstruct how genomes have evolved in the billions of years that have elapsed since the appearance of the first cells.

- Genetic variation—the raw material for evolutionary change—arises through a variety of mechanisms that alter the nucleotide sequence of genomes. These changes in sequence range from simple point mutations to larger-scale deletions, duplications, and rearrangements.

- Genetic changes that give an organism a selective advantage are the most likely to be perpetuated. Changes that compromise an organism's fitness or ability to reproduce are eliminated through natural selection.

- Gene duplication is one of the most important sources of genetic diversity. Once duplicated, the two genes can accumulate different mutations and thereby diversify to perform different roles.

- Repeated rounds of gene duplication and divergence during evolution have produced many large gene families.

- The evolution of new proteins is thought to have been greatly facilitated by the swapping of exons between genes to create hybrid proteins with new functions.

- The human genome contains 3.2×10^9 nucleotide pairs distributed among 23 pairs of chromosomes—22 autosomes and a pair of sex chromosomes. Less than a tenth of this DNA is transcribed to produce protein-coding or otherwise functional RNAs.

- Individual humans differ from one another by an average of 1 nucleotide pair in every 1000; this and other genetic variation underlies most of our individuality and provides the basis for identifying individuals by DNA analysis.

- Nearly half of the human genome consists of mobile genetic elements that can move from one site to another within a genome. Two classes of these elements have multiplied to especially high copy numbers.

- Viruses are genes packaged in protective coats that can move from cell to cell and organism to organism, but they require host cells to reproduce themselves.

- Some viruses have RNA instead of DNA as their genetic material. Retroviruses copy their RNA genomes into DNA before integrating into the host-cell genome.

- Comparing genome sequences of different species provides a powerful way to identify conserved, functionally important DNA sequences.

- Related species, such as human and mouse, have many genes in common; evolutionary changes in the regulatory DNA sequences that affect how these genes are expressed are especially important in determining the differences between species.

KEY TERMS

Alu sequence	germ line	purifying selection
conserved synteny	homologous gene	retrotransposon
copy-number variation	horizontal gene transfer	retrovirus
divergence	*L1* element	reverse transcriptase
exon shuffling	mobile genetic element	single-nucleotide polymorphism (SNP)
gene duplication and divergence	open reading frame (ORF)	somatic cell
gene family	phylogenetic tree	transposon
germ cell	point mutation	virus

QUESTIONS

QUESTION 9–7

Discuss the following statement: "Mobile genetic elements are parasites. They are harmful to the host organism and therefore place it at an evolutionary disadvantage."

QUESTION 9–8

Human Chromosome 22 (48×10^6 nucleotide pairs in length) has about 700 protein-coding genes, which average 19,000 nucleotide pairs in length and contain an average of 5.4 exons, each of which averages 266 nucleotide pairs. What fraction of the average protein-coding gene is converted into mRNA? What fraction of the chromosome do these genes occupy?

QUESTION 9–9

(True/False) The majority of human DNA is unimportant junk. Explain your answer.

QUESTION 9–10

Mobile genetic elements make up nearly half of the human genome and are inserted more or less randomly throughout it. However, in some spots these elements are rare, as illustrated for a cluster of genes called HoxD, which lies on Chromosome 2 (**Figure Q9–10**). This cluster is about 100 kb in length and contains nine genes whose differential expression along the length of the developing embryo helps establish the basic body plan for humans (and for other animals). Why do you suppose that mobile genetic elements are so rare in this cluster? In Figure Q9–10, lines that project *upward* indicate exons of known genes. Lines that project *downward* indicate mobile genetic elements; they are so numerous they merge into nearly a solid block outside the HoxD cluster. For comparison, an equivalent region of Chromosome 22 is shown.

Chromosome 22

Chromosome 2

100 kb

HoxD cluster

Figure Q9–10

QUESTION 9–11

An early graphical method for comparing nucleotide sequences—the so-called diagon plot—still yields one of the best visual comparisons of sequence relatedness. An example is illustrated in **Figure Q9–11**, in which the human β-globin gene is compared with the human cDNA for β globin (which contains only the coding portion of the gene; Figure Q9–11A) and to the mouse β-globin gene (Figure Q9–11B). Diagon plots are generated by comparing blocks of sequence, in this case blocks of 11 nucleotides at a time. If 9 or more of the nucleotides match, a dot is placed on the diagram at the coordinates corresponding to the blocks being compared. A comparison of all possible blocks generates diagrams such as the ones shown in Figure Q9–11, in which sequence similarities show up as diagonal lines.

A. From the comparison of the human β-globin gene with the human β-globin cDNA (Figure Q9–11A), can you deduce the positions of exons and introns in the β-globin gene?

B. Are the exons of the human β-globin gene (indicated by shading in Figure Q9–11B) similar to those of the mouse β-globin gene? Identify and explain any key differences.

C. Is there any sequence similarity between the human and mouse β-globin genes that lies outside the exons? If so, identify its location and offer an explanation for its preservation during evolution.

D. Did the mouse or human gene undergo a change of intron length during their evolutionary divergence? How can you tell?

QUESTION 9–12

Your advisor, a brilliant bioinformatician, has high regard for your intellect and industry. She suggests that you write a computer program that will identify the exons of protein-coding genes directly from the sequence of the human genome. In preparation for that task, you decide to write down a list of the features that might distinguish protein-coding sequences from intronic DNA and from other sequences in the genome. What features would you list? (You may wish to review basic aspects of gene expression in Chapter 7.)

QUESTION 9–13

You are interested in finding out the function of a particular gene in the mouse genome. You have determined the nucleotide sequence of the gene, defined the portion that

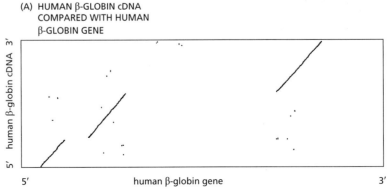

(A) HUMAN β-GLOBIN cDNA COMPARED WITH HUMAN β-GLOBIN GENE

human β-globin cDNA 5′ → 3′

5′ human β-globin gene 3′

(B) MOUSE β-GLOBIN GENE COMPARED WITH HUMAN β-GLOBIN GENE

mouse β-globin gene 5′ → 3′

5′ human β-globin gene 3′

Figure Q9–11

codes for its protein product, and searched the relevant database for similar sequences; however, neither the gene nor the encoded protein resembles anything previously described. What types of additional information about the gene and the encoded protein would you like to know in order to narrow down its function, and why? Focus on the information you would want, rather than on the techniques you might use to get that information.

QUESTION 9–14

Why do you expect to encounter a stop codon about every 20 codons or so in a random sequence of DNA?

QUESTION 9–15

The genetic code (see Figure 7–25) relates the nucleotide sequence of mRNA to the amino acid sequence of encoded proteins. Ever since the code was deciphered, some have claimed it must be a frozen accident—that is, the system randomly fell into place in some ancestral organism and was then perpetuated unchanged throughout evolution; others have argued that the code has been shaped by natural selection.

A striking feature of the genetic code is its inherent resistance to the effects of mutation. For example, a change in the third position of a codon often specifies the same amino acid or one with similar chemical properties. But is the natural code more resistant to mutation than other possible versions? The answer is an emphatic "Yes," as illustrated in Figure Q9–15. Only one in a million computer-generated "random" codes is more error-resistant than the natural genetic code.

Does the resistance to mutation of the actual genetic code argue in favor of its origin as a frozen accident or as a result of natural selection? Explain your reasoning.

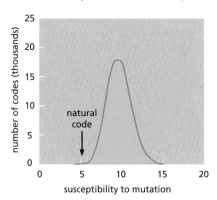

Figure Q9–15

QUESTION 9–16

Which of the processes listed below contribute significantly to the evolution of new protein-coding genes?

A. Duplication of genes to create extra copies that can acquire new functions.

B. Formation of new genes *de novo* from noncoding DNA in the genome.

C. Horizontal transfer of DNA between cells of different species.

D. Mutation of existing genes to create new functions.

E. Shuffling of protein domains by gene rearrangement.

QUESTION 9–17

Some genes evolve more rapidly than others. But how can this be demonstrated? One approach is to compare several genes from the same two species, as shown for rat and human in the table above. Two measures of rates of nucleotide substitution are indicated in the table. Nonsynonymous changes refer to single-nucleotide changes in the DNA sequence that alter the encoded amino acid (ATC → TTC, which gives isoleucine → phenylalanine, for example). Synonymous changes refer to those that do not alter the encoded amino acid (ATC → ATT, which gives isoleucine → isoleucine, for example). (As is apparent in the genetic code, Figure 7–25, there are many cases where several codons correspond to the same amino acid.)

Gene	Amino Acids	Rates of Change	
		Nonsynonymous	Synonymous
Histone H3	135	0.0	4.5
Hemoglobin α	141	0.6	4.4
Interferon γ	136	3.1	5.5

Rates were determined by comparing rat and human sequences and are expressed as nucleotide changes per site per 10^9 years. The average rate of nonsynonymous changes for several dozen rat and human genes is about 0.8.

A. Why are there such large differences between the synonymous and nonsynonymous rates of nucleotide substitution?

B. Considering that the rates of synonymous changes are about the same for all three genes, how is it possible for the histone H3 gene to resist so effectively those nucleotide changes that alter its amino acid sequence?

C. In principle, a protein might be highly conserved because its gene exists in a "privileged" site in the genome that is subject to very low mutation rates. What feature of the data in the table argues against this possibility for the histone H3 protein?

QUESTION 9–18

Plant hemoglobins were found initially in legumes, where they function in root nodules to lower the oxygen concentration, allowing the resident bacteria to fix nitrogen. These hemoglobins impart a characteristic pink color to the root nodules. The discovery of hemoglobin in plants was initially surprising because scientists regarded hemoglobin as a distinctive feature of animal blood. It was hypothesized that the plant hemoglobin gene was acquired by horizontal transfer from an animal. Many more hemoglobin genes have now been sequenced from a variety of organisms, and a phylogenetic tree of hemoglobins is shown in Figure Q9–18.

A. Does the evidence in the tree support or refute the hypothesis that the plant hemoglobins arose by horizontal gene transfer?

B. Supposing that the plant hemoglobin genes were originally derived by horizontal transfer (from a parasitic nematode, for example), what would you expect the phylogenetic tree to look like?

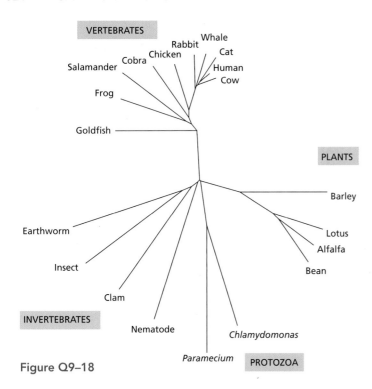

Figure Q9–18

QUESTION 9–19

The accuracy of DNA replication in the human germ-cell line is such that on average only about 0.6 out of the 6 billion nucleotides is altered at each cell division. Because most of our DNA is not subject to any precise constraint on its sequence, most of these changes are selectively neutral. Any two modern humans picked at random will show about 1 difference of nucleotide sequence in every 1000 nucleotides. Suppose we are all descended from a single pair of ancestors—Adam and Eve—who were genetically identical and homozygous (each chromosome was identical to its homolog). Assuming that all germ-line mutations that arise are preserved in descendants, how many cell generations must have elapsed since the days of Adam and Eve for 1 difference per 1000 nucleotides to have accumulated in modern humans? Assuming that each human generation corresponds on average to 200 cell-division cycles in the germ-cell lineage and allowing 30 years per human generation, how many years ago would this ancestral couple have lived?

QUESTION 9–20

Reverse transcriptases do not proofread as they synthesize DNA using an RNA template. What do you think the consequences of this are for the treatment of AIDS?

CHAPTER **TEN**

10

Modern Recombinant DNA Technology

Since the turn of the century, biologists have amassed an unprecedented wealth of information on the genes that direct the development and behavior of living things. Thanks to advances in our ability to rapidly determine the nucleotide sequence of entire genomes, we now have access to the complete molecular blueprints for thousands of different organisms, from the platypus to the plague bacterium, and for thousands of different people from all over the world.

This information explosion would not have been possible without the technological revolution that enabled us to manipulate DNA molecules. In the early 1970s, it became possible, for the first time, to isolate a selected piece of DNA from the many millions of nucleotide pairs in a typical chromosome—and to replicate, sequence, and modify this DNA. These modified DNA molecules can then be introduced into another organism's genome, where they become a functional and heritable part of that organism's genetic instructions.

These technical breakthroughs—dubbed **recombinant DNA technology**, or *genetic engineering*—have had a dramatic impact on all aspects of cell biology. They have advanced our understanding of the organization and evolutionary history of complex eukaryotic genomes (as discussed in Chapter 9) and have led to the discovery of whole new classes of genes, RNAs, and proteins. They continue to generate new ways of determining the functions of genes and proteins in living organisms, and they provide an important set of tools for unraveling the mechanisms—still poorly understood—by which a complex organism can develop from a single fertilized egg.

Recombinant DNA technology has also had a profound influence on our understanding and treatment of disease: it is used, for example, to detect

MANIPULATING AND ANALYZING DNA MOLECULES

DNA CLONING IN BACTERIA

DNA CLONING BY PCR

EXPLORING AND EXPLOITING GENE FUNCTION

the mutations in human genes that are responsible for inherited disorders or that predispose us to a variety of common diseases, including cancer; it is used to produce an increasing number of pharmaceuticals, such as insulin for diabetics and blood-clotting proteins for hemophiliacs. But recombinant DNA technology also has applications outside the clinic. It allows, for example, forensic science to identify or acquit suspects in a crime. Even our laundry detergents contain heat-stable, stain-removing proteases, courtesy of DNA technology. Of all the discoveries described in this book, those that led to the development of recombinant DNA technology have the greatest impact on our everyday lives.

In this chapter, we present a brief overview of how we learned to manipulate DNA, identify genes, and produce many copies of any given nucleotide sequence in the laboratory. We discuss several approaches to exploring gene function, including new ways to monitor gene expression and to inactivate or modify genes in cells, animals, and plants. These methods—which are continuously being improved and made ever-more powerful—are not only revolutionizing the way we do science, they are transforming our understanding of cell biology and human disease. Indeed, they are responsible for a substantial portion of the information we present in this book.

MANIPULATING AND ANALYZING DNA MOLECULES

Humans have been experimenting with DNA, albeit without realizing it, for millennia. The roses in our gardens, the corn on our plate, and the dogs in our yards are all the product of selective breeding that has taken place over many, many generations (**Figure 10–1**). But it wasn't until the development of recombinant DNA techniques in the 1970s that we could begin to engineer organisms with desired properties by directly tinkering with their genes.

Isolating and manipulating individual genes is not a trivial matter. Unlike a protein, a gene does not exist as a discrete entity in cells; it is a small part of a much larger DNA molecule. Even bacterial genomes, which are much less complex than the chromosomes of eukaryotes, are enormously long. The *E. coli* genome, for example, contains 4.6 million nucleotide pairs.

How, then, can a single gene be separated from a eukaryotic genome—which is considerably larger—so that it can be handled in the laboratory? The solution to this problem emerged, in large part, with the discovery of a class of bacterial enzymes known as *restriction nucleases*. These

Figure 10–1 By breeding plants and animals, humans have been unwittingly experimenting with DNA for millennia. (A) The oldest known depiction of a rose in Western art, from the palace of Knossos in Crete, around 2000 BC. Modern roses are the result of centuries of breeding between such wild roses. (B) Dogs have been bred to exhibit a wide variety of characteristics, including different head shapes, coat colors, and of course size. All dogs, regardless of breed, belong to a single species that was domesticated from the gray wolf some 10,000 to 15,000 years ago. (B, from A.L. Shearin & E.A. Ostrander, *PLoS Biol.* 8:e1000310, 2010.)

(A)   (B)

enzymes cut double-stranded DNA at particular sequences. They can therefore be used to produce a reproducible set of specific DNA fragments from any genome. In this section, we describe how these enzymes work and how the DNA fragments they produce can be separated and visualized. We then discuss how these fragments can be probed to identify the ones that contain the DNA sequence of interest.

Restriction Nucleases Cut DNA Molecules at Specific Sites

Like many of the tools of recombinant DNA technology, restriction nucleases were discovered by researchers trying to understand an intriguing biological phenomenon. It had been observed that certain bacteria always degraded "foreign" DNA that was introduced into them experimentally. A search for the mechanism responsible revealed a novel class of bacterial nucleases that cleave DNA at specific nucleotide sequences. The bacteria's own DNA is protected from cleavage by chemical modification of these specific sequences. Because these enzymes function to restrict the transfer of DNA between strains of bacteria, they were called **restriction nucleases**. The pursuit of this seemingly arcane biological puzzle set off the development of technologies that have forever changed the way cell and molecular biologists study living things.

Different bacterial species produce different restriction nucleases, each cutting at a different, specific nucleotide sequence (**Figure 10–2**). Because these target sequences are short—generally four to eight nucleotide pairs—many sites of cleavage will occur, purely by chance, in any long DNA molecule. The reason restriction nucleases are so useful in the laboratory is that each enzyme will cut a particular DNA molecule, at the same sites. Thus for a given sample of DNA, a particular restriction nuclease will reliably generate the same set of DNA fragments.

The size of the resulting fragments depends on the target sequences of the restriction nucleases. As shown in Figure 10–2, the enzyme HaeIII cuts at a sequence of four nucleotide pairs; a sequence this long would be expected to occur purely by chance approximately once every 256 nucleotide pairs (1 in 4^4). In comparison, a restriction nuclease with a target sequence that is eight nucleotides long would be expected to cleave DNA on average once every 65,536 nucleotide pairs (1 in 4^8). This difference in sequence selectivity makes it possible to cleave a long DNA molecule into the fragment sizes that are most suitable for a given application.

Gel Electrophoresis Separates DNA Fragments of Different Sizes

After a large DNA molecule is cleaved into smaller pieces with a restriction nuclease, the DNA fragments can be separated from one another on

Figure 10–2 Restriction nucleases cleave DNA at specific nucleotide sequences. Target sequences are often palindromic (that is, the nucleotide sequence is symmetrical around a central point). Here, both strands of the DNA double helix are cut at specific points within the target sequence (*orange*). Some enzymes, such as HaeIII, cut straight across the double helix and leave two blunt-ended DNA molecules; with others, such as EcoRI and HindIII, the cuts on each strand are staggered. These staggered cuts generate "sticky ends"—short, single-stranded overhangs that help the cut DNA molecules join back together through complementary base-pairing. This rejoining of DNA molecules becomes important for DNA cloning, as we discuss later. Restriction nucleases are usually obtained from bacteria, and their names reflect their origins: for example, the enzyme EcoRI comes from *Escherichia coli*.

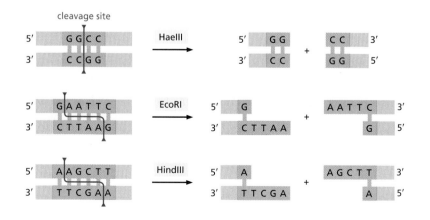

Figure 10–3 **DNA molecules can be separated by size using gel electrophoresis.** (A) Schematic illustration compares the results of cutting the same DNA molecule (in this case, the genome of a virus that infects parasitic wasps) with two different restriction nucleases, EcoRI (*middle*) and HindIII (*right*). The fragments are then separated by gel electrophoresis. Because larger fragments migrate more slowly than smaller ones, the lowermost bands on the gel contain the smallest DNA fragments. The sizes of the fragments can be estimated by comparing them to a set of DNA fragments of known sizes (*left*). (B) Photograph of an actual gel shows the positions of DNA bands that have been labeled with a fluorescent dye. (B, from U. Albrecht et al., *J. Gen. Virol.* 75:3353–3363, 1994.)

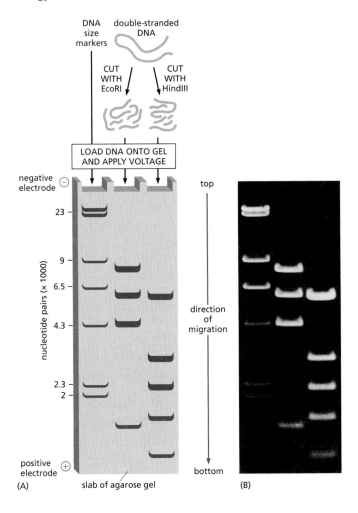

the basis of their length by gel electrophoresis—the same method used to separate mixtures of proteins (see Panel 4–5, p. 167). A mixture of DNA fragments is loaded at one end of a slab of agarose or polyacrylamide gel, which contains a microscopic network of pores. When a voltage is applied across the gel, the negatively charged DNA fragments migrate toward the positive electrode; larger fragments will migrate more slowly because their progress is impeded to a greater extent by the gel matrix. Over several hours, the DNA fragments become spread out across the gel according to size, forming a ladder of discrete bands, each composed of a collection of DNA molecules of identical length (**Figure 10–3**). To isolate a desired DNA fragment, the small section of the gel that contains the band is excised with a scalpel or a razor blade, and the DNA is then extracted.

QUESTION 10–2

Which products result when the double-stranded DNA molecule *below* is digested with (A) EcoRI, (B) HaeIII, (C) HindIII, or (D) all three of these enzymes together? (See Figure 10–2 for the target sequences of these enzymes.)

5'-AAGAATTGCGGAATTCGGGCCTTAAGCGCCGCGTCGAGGCCTTAAA-3'
3'-TTCTTAACGCCTTAAGCCCGGAATTCGCGGCGCAGCTCCGGAATTT-5'

Bands of DNA in a Gel Can Be Visualized Using Fluorescent Dyes or Radioisotopes

The separated DNA bands on an agarose or polyacrylamide gel are not, by themselves, visible. To see these bands, the DNA must be labeled or stained in some way. One sensitive method involves exposing the gel to a dye that fluoresces under ultraviolet (UV) light when it is bound to DNA. When the gel is placed on a UV light box, the individual bands glow bright orange—or bright white when the gel is photographed in black and white (see Figure 10–3B).

An even more sensitive detection method involves incorporating a radioisotope into the DNA molecules before they are separated by electrophoresis; ^{32}P is often used, as it can be incorporated into the phosphates of DNA. Because the β particles emitted from ^{32}P can activate the radiation-sensitive particles in photographic film, a sheet of film placed flat on top of the agarose gel will, when developed, show the position of all the DNA bands.

Exposing a gel to a fluorescent dye that binds to DNA—or starting with DNA that has been pre-labeled with ^{32}P—will allow every band on the gel to be seen. But it does not reveal which of those bands contains a DNA sequence of interest. To do that, a probe is designed to bind specifically to the desired nucleotide sequence by complementary base-pairing, as we see next.

Hybridization Provides a Sensitive Way to Detect Specific Nucleotide Sequences

Under normal conditions, the two strands of a DNA double helix are held together by hydrogen bonds between the complementary base pairs (see Figure 5–6). But these relatively weak, noncovalent bonds can be fairly easily broken. Such *DNA denaturation* will release the two strands from each other, but does not break the covalent bonds that link together the nucleotides within each strand. Perhaps the simplest way to achieve this separation involves heating the DNA to around 90°C. When the conditions are reversed—by slowly lowering the temperature—the complementary strands will readily come back together to re-form a double helix. This **hybridization**, or *DNA renaturation*, is driven by the re-formation of the hydrogen bonds between complementary base pairs (**Figure 10–4**).

This fundamental capacity of a single-stranded nucleic acid molecule, either DNA or RNA, to form a double helix with a single-stranded molecule of a complementary sequence provides a very powerful and sensitive technique for detecting specific nucleotide sequences in both DNA and RNA. Today, one simply designs a short, single-stranded *DNA probe* that is complementary to the nucleotide sequence of interest. Because the nucleotide sequences of so many genomes are known—and are stored in publicly accessible databases—designing such a probe is straightforward. The desired probe can then be synthesized in the laboratory—usually by a

Figure 10–4 A molecule of DNA can undergo denaturation and renaturation (hybridization). For two single-stranded molecules to hybridize, they must have complementary nucleotide sequences that allow base-pairing. In this example, the *red* and *orange* strands are complementary to each other, and the *blue* and *green* strands are complementary to each other. Although denaturation by heating is shown, DNA can also be renatured after being denatured by alkali treatment. The 1961 discovery that single strands of DNA could readily re-form a double helix in this way was a big surprise to scientists.

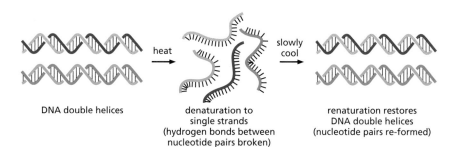

DNA double helices → heat → denaturation to single strands (hydrogen bonds between nucleotide pairs broken) → slowly cool → renaturation restores DNA double helices (nucleotide pairs re-formed)

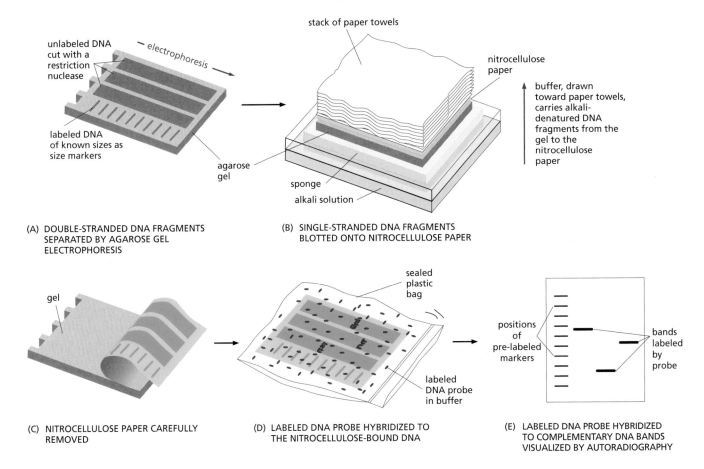

(A) DOUBLE-STRANDED DNA FRAGMENTS SEPARATED BY AGAROSE GEL ELECTROPHORESIS

(B) SINGLE-STRANDED DNA FRAGMENTS BLOTTED ONTO NITROCELLULOSE PAPER

(C) NITROCELLULOSE PAPER CAREFULLY REMOVED

(D) LABELED DNA PROBE HYBRIDIZED TO THE NITROCELLULOSE-BOUND DNA

(E) LABELED DNA PROBE HYBRIDIZED TO COMPLEMENTARY DNA BANDS VISUALIZED BY AUTORADIOGRAPHY

Figure 10–5 Gel-transfer hybridization, or Southern blotting, is used to detect specific DNA fragments. (A) The mixture of double-stranded DNA fragments generated by restriction nuclease treatment of DNA is separated according to length by gel electrophoresis. (B) A sheet of nitrocellulose paper is laid over the gel, and the separated DNA fragments are denatured with alkali and transferred to the sheet by blotting. In this process, a stack of absorbent paper towels is used to suck buffer up through the gel, transferring the single-stranded DNA fragments from the gel to the nitrocellulose paper. (C) The nitrocellulose sheet is carefully peeled off the gel. (D) The sheet containing the bound single-stranded DNA fragments is exposed to a radioactive, single-stranded DNA probe specific for the DNA sequence of interest under conditions that favor hybridization. (E) The sheet is washed thoroughly, so that only probe molecules that have hybridized to the DNA on the paper remain attached. After autoradiography, the DNA that has hybridized to the labeled probe will show up as a band on the autoradiograph. An adaptation of this technique, used to detect specific RNA sequences, is called *Northern blotting*. In this case, RNA molecules are electrophoresed through the gel, and the probe is usually a single-stranded DNA molecule. The same procedures can be carried out with non-radioactive probes using an appropriate method of detection.

commercial organization or a centralized academic facility. Such probes carry a fluorescent or radioactive label to facilitate detection of the nucleotide sequence to which they bind.

Once a suitable probe has been obtained, it can be used in a variety of situations to search for nucleic acids with a complementary sequence—for example, finding a sequence of interest among DNA fragments that have been separated on an agarose gel. In this case, the fragments are first transferred to a special sheet of paper, which is then exposed to the labeled probe. This common technique, called *Southern blotting*, was named after the scientist who invented it (**Figure 10–5**).

DNA probes are widely used in cell biology. Later in the chapter, we discuss how they can be used to determine in which tissues and at what stages of development a gene is transcribed. But first, we consider how hybridization facilitates the process of DNA cloning.

DNA CLONING IN BACTERIA

The term **DNA cloning** refers to the production of many identical copies of a DNA sequence. It is this amplification that makes it possible to separate a defined segment of DNA—often a gene of interest—from the rest of a cell's genome. DNA cloning is one of the most important feats of recombinant DNA technology, as it is the starting point for understanding the function of any stretch of DNA within the genome.

In this section, we describe the classical approach to DNA cloning, in which one copies all of the DNA from a cell or tissue and then finds and isolates the specific DNA of interest. Later, we discuss how the development of the *polymerase chain reaction* (PCR) has facilitated a more direct approach to cloning, allowing one to copy, in a test tube, only the DNA fragment of interest.

DNA Cloning Begins with Genome Fragmentation and Production of Recombinant DNAs

Whole genomes, even small ones, are too large and unwieldy to be handled easily in the laboratory. Thus the first step in cloning any gene is to break the genome into smaller, more manageable pieces. These fragments can then be joined together, or recombined, to produce the DNA molecules that will be amplified. Our ability to generate such **recombinant DNA molecules** is made possible by the use of molecular tools that are provided by cells themselves.

As we discussed earlier, bacterial restriction nucleases can be used to cut long DNA molecules into conveniently sized fragments (see Figure 10–2). These fragments can then be joined to one another—or to any piece of DNA—using **DNA ligase**, an enzyme that reseals the nicks that arise in the DNA backbone during DNA replication and DNA repair in cells (see Figure 6–18). DNA ligase allows investigators to join together any two pieces of DNA in a test tube, producing recombinant DNA molecules that are not found in nature (**Figure 10–6**).

The production of recombinant DNA molecules in this way is a key step in the classical approach to DNA cloning. It allows the DNA fragments generated by treatment with a restriction nuclease to be inserted into another, special DNA molecule that serves as a carrier, or *vector*, which can be copied—and thereby amplified—inside a cell, as we discuss next.

Recombinant DNA Can Be Inserted Into Plasmid Vectors

The vectors typically used for gene cloning are relatively small, circular DNA molecules called **plasmids**. (**Figure 10–7**). Each plasmid contains a replication origin, which enables it to replicate in a bacterial cell independently of the bacterial chromosome. It also has cleavage sites for common restriction nucleases, so that the plasmid can be conveniently opened and a foreign DNA fragment inserted.

The plasmids used for cloning are basically streamlined versions of plasmids that occur naturally in many bacteria. Bacterial plasmids were first recognized by physicians and scientists because they often carry

Figure 10–6 **DNA ligase can join together any two DNA fragments *in vitro* to produce recombinant DNA molecules.** ATP provides the energy necessary for the ligase to reseal the sugar–phosphate backbone of DNA. (A) DNA ligase can readily join two DNA fragments produced by the same restriction nuclease, in this case EcoRI. Note that the staggered ends produced by this enzyme enable the ends of the two fragments to base-pair correctly with each other, greatly facilitating their rejoining. (B) DNA ligase can also be used to join DNA fragments produced by different restriction nucleases—for example, EcoRI and HaeIII. In this case, before the fragments undergo ligation, DNA polymerase plus a mixture of deoxyribonucleoside triphosphates (dNTPs) are used to fill in the staggered cut produced by EcoRI. Each DNA fragment shown in the figure is oriented so that its 5′ ends are the left end of the upper strand and the right end of the lower strand, as indicated.

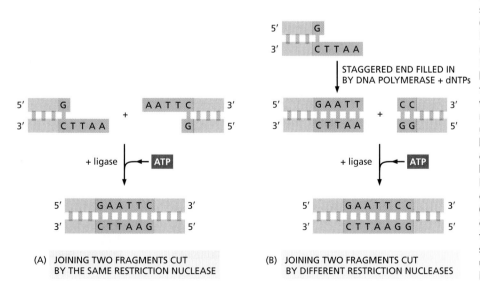

(A) JOINING TWO FRAGMENTS CUT BY THE SAME RESTRICTION NUCLEASE

(B) JOINING TWO FRAGMENTS CUT BY DIFFERENT RESTRICTION NUCLEASES

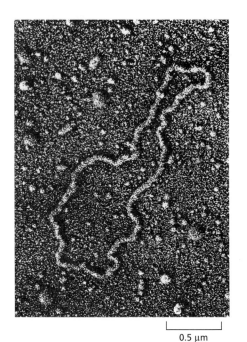

0.5 μm

Figure 10–7 Bacterial plasmids are commonly used as cloning vectors. This circular, double-stranded DNA molecule was the first plasmid for DNA cloning; it contains about nine thousand nucleotide pairs. The staining procedure used to make the DNA visible in this electron micrograph causes the DNA to appear much thicker than it actually is. (Courtesy of Stanley N. Cohen, Stanford University.)

genes that render their microbial host resistant to one or more antibiotics. Indeed, historically potent antibiotics—penicillin, for example—are no longer effective against many of today's bacterial infections because plasmids that confer resistance to the antibiotic have spread among bacterial species by horizontal gene transfer (see Figure 9–14).

To insert a piece of DNA into a plasmid vector, the purified plasmid DNA is opened up by a restriction nuclease that cleaves it at a single site, and the DNA fragment to be cloned is then spliced into that site using DNA ligase (**Figure 10–8**). This recombinant DNA molecule is now ready to be introduced into a bacterium, where it will be copied and amplified, as we see next.

Recombinant DNA Can Be Copied Inside Bacterial Cells

To introduce recombinant DNA into a bacterial cell, investigators take advantage of the fact that some bacteria naturally take up DNA molecules present in their surroundings. The mechanism that controls this uptake is called **transformation**, because early observations suggested it could "transform" one bacterial strain into another. Indeed, the first proof that genes are made of DNA came from an experiment in which DNA purified from a pathogenic strain of pneumococcus was used to transform a harmless bacterium into a deadly one (see How We Know, pp. 174–176).

In a natural bacterial population, a source of DNA for transformation is provided by bacteria that have died and released their contents, including DNA, into the environment. In a test tube, however, bacteria such as *E. coli* can be coaxed to take up recombinant DNA that has been created in the laboratory. These bacteria are then suspended in a nutrient-rich broth and allowed to proliferate.

Each time the bacterial population doubles—every 30 minutes or so—the number of copies of the recombinant DNA molecule also doubles. Thus, in 24 hours, the engineered cells will produce hundreds of millions of copies of the plasmid, along with the DNA fragment it contains. The bacteria can then be split open (lysed) and the plasmid DNA purified from

Figure 10–8 A DNA fragment is inserted into a bacterial plasmid by using the enzyme DNA ligase. The plasmid is first cut open at a single site with a restriction nuclease (in this case, one that produces staggered ends). It is then mixed with the DNA fragment to be cloned, which has been cut with the same restriction nuclease. DNA ligase and ATP are also added to the mix. The staggered ends base-pair, and the nicks in the DNA backbone are sealed by the DNA ligase to produce a complete recombinant DNA molecule. In the accompanying micrographs, we have colored the DNA fragment *red* to make it easier to see. (Micrographs courtesy of Huntington Potter and David Dressler.)

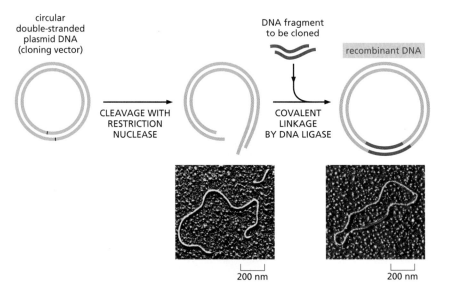

circular double-stranded plasmid DNA (cloning vector)

DNA fragment to be cloned

recombinant DNA

CLEAVAGE WITH RESTRICTION NUCLEASE

COVALENT LINKAGE BY DNA LIGASE

200 nm   200 nm

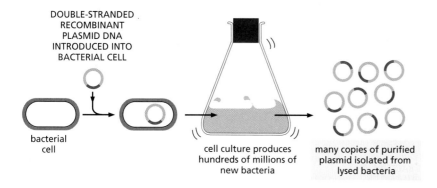

DOUBLE-STRANDED
RECOMBINANT
PLASMID DNA
INTRODUCED INTO
BACTERIAL CELL

bacterial
cell

cell culture produces
hundreds of millions of
new bacteria

many copies of purified
plasmid isolated from
lysed bacteria

Figure 10–9 A DNA fragment can be replicated inside a bacterial cell. To clone a particular fragment of DNA, it is first inserted into a plasmid vector, as shown in Figure 10–8. The resulting recombinant plasmid DNA is then introduced into a bacterium, where it is replicated many millions of times as the bacterium multiplies. For simplicity, the genome of the bacterial cell is not shown.

the rest of the cell contents, including the large bacterial chromosome (**Figure 10–9**).

The DNA fragment can be readily recovered by cutting it out of the plasmid DNA with the same restriction nuclease that was used to insert it, and then separating it from the plasmid DNA by gel electrophoresis (see Figure 10–3). Together, these steps allow the amplification and purification of any segment of DNA from the genome of any organism.

Genes Can Be Isolated from a DNA Library

Thus far, we have described the amplification of a single DNA fragment. In reality, when a genome is cut by a restriction nuclease, millions of different DNA fragments are generated. How can the single fragment that contains the DNA of interest be isolated from this collection? The solution involves introducing all of the fragments into bacteria and then selecting those bacterial cells that have amplified the desired DNA molecule.

The entire collection of DNA fragments can be ligated into plasmid vectors, using conditions that favor the insertion of a single DNA fragment into each plasmid molecule. These recombinant plasmids are then introduced into *E. coli* at a concentration that ensures that no more than one plasmid molecule is taken up by each bacterium. The collection of cloned DNA fragments in this bacterial culture is known as a **DNA library**. Because the DNA fragments were derived directly from the chromosomal DNA of the organism of interest, the resulting collection—called a **genomic library**—should represent the entire genome of that organism (**Figure 10–10**).

To find a particular gene within this library, one can use a labeled DNA probe designed to bind specifically to part of the gene's DNA sequence. Using such a probe, the rare bacterial clones in the DNA library that contain the gene—or a portion of it—can be identified by hybridization (**Figure 10–11**).

But before a gene has been cloned, how can one design a probe to detect it? In the early days of cloning, investigators wishing to study a protein-coding gene would first determine at least part of the protein's amino acid sequence. By applying the genetic code in reverse, they could use this amino acid sequence to deduce the corresponding gene sequence, which allowed them to generate an appropriate DNA probe.

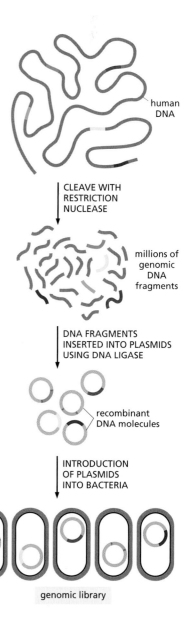

human
DNA

CLEAVE WITH
RESTRICTION
NUCLEASE

millions of
genomic
DNA
fragments

DNA FRAGMENTS
INSERTED INTO PLASMIDS
USING DNA LIGASE

recombinant
DNA molecules

INTRODUCTION
OF PLASMIDS
INTO BACTERIA

genomic library

Figure 10–10 Human genomic libraries containing DNA fragments representing the whole human genome can be constructed using restriction nucleases and DNA ligase. Such a genomic library consists of a set of bacteria, each carrying a different small fragment of human DNA. For simplicity, only the *colored* DNA fragments are shown in the library; in reality, all of the different *gray* fragments will also be represented.

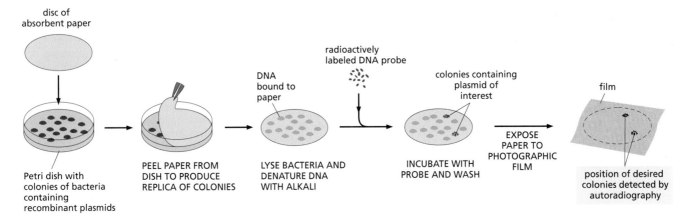

Figure 10–11 A bacterial colony carrying a particular DNA clone can be identified by hybridization. A replica of the arrangement of the bacterial colonies (clones) on the Petri dish is made by pressing a piece of absorbent paper against the surface of the dish. This replica is treated with alkali (to lyse the cells and dissociate the plasmid DNA into single strands), and the paper is then hybridized to a highly radioactive DNA probe. Those bacterial colonies that have bound the probe are identified by autoradiography. Living bacterial cells containing the plasmid can then be isolated from the original Petri dish.

Many genes were originally identified and cloned using variations on this basic approach. Now that the complete genome sequences of many organisms, including humans, are known, however, cloning genes is very much easier, faster, and cheaper. The sequence of any gene in an organism can be looked up in an electronic database, making it a simple matter to design a probe that can be synthesized to order. As we discuss shortly, gene cloning today is typically done directly on the original DNA sample, bypassing the use of a DNA library entirely.

cDNA Libraries Represent the mRNAs Produced by Particular Cells

For many applications—for example, when attempting to clone a protein-coding gene, it is advantageous to obtain the gene in a form that contains only the coding sequence; that is, a form that lacks the intron DNA. For some genes, the complete genomic clone—including introns and exons—is too large and unwieldy to handle conveniently in the laboratory (see, for example, Figure 7–18B). What's more, the bacterial or yeast cells typically used to amplify cloned DNA are unable to remove introns from mammalian RNA transcripts. So if the goal is to use a cloned mammalian gene to produce a large amount of the protein it encodes, it is essential to use only the coding sequence of the gene. Fortunately, it is relatively simple to isolate a gene free of all its introns, by using a different type of DNA library, called a **cDNA library**.

A cDNA library is similar to a genomic library in that it also contains numerous clones containing many different DNA sequences. But it differs in one important respect. The DNA that goes into a cDNA library is not genomic DNA; it is DNA copied from the mRNAs present in a particular type of cell. To prepare a **cDNA** library, all of the mRNAs are extracted, and double-stranded DNA copies of these mRNAs are produced by the enzymes *reverse transcriptase* and DNA polymerase (**Figure 10–12**). These **complementary DNA**—or **cDNA**—molecules are then introduced into bacteria and amplified, as described for genomic DNA fragments (see Figure 10–10). The gene of interest—in this case, without its introns—can then be isolated by using a probe that hybridizes to the DNA sequence (see Figure 10–11). We discuss later how such cDNAs can be used to produce purified proteins on a commercial scale.

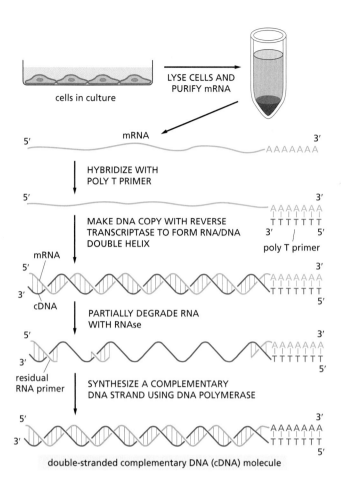

Figure 10–12 Complementary DNA (cDNA) is prepared from mRNA. Total mRNA is extracted from a selected type of cell, and double-stranded, complementary DNA (cDNA) is produced using reverse transcriptase (see Figure 9–30) and DNA polymerase. For simplicity, the copying of just one of these mRNAs into cDNA is illustrated here. Note that an RNA fragment that remains hybridized to the first cDNA strand after partial RNAse digestion serves as the primer needed for DNA polymerase to begin synthesis of the complementary DNA strand.

There are several important differences between genomic DNA clones and cDNA clones, as illustrated in **Figure 10–13**. Genomic clones represent a random sample of all of the DNA sequences found in an organism's genome and, with very rare exceptions, will contain the same sequences regardless of the cell type from which the DNA came. Also, genomic clones from eukaryotes contain large amounts of noncoding DNA, repetitive DNA sequences, introns, regulatory DNA, and spacer DNA; sequences that code for proteins will make up only a few percent of the library (see Figure 9–33). By contrast, cDNA clones contain predominantly protein-coding sequences, and only those for genes that have been transcribed into mRNA in the cells from which the cDNA was made. As different types of cells produce distinct sets of mRNA molecules, each yields a different cDNA library. Furthermore, patterns of gene expression change during development, so cells at different stages in their development will also yield different cDNA libraries.

As we discuss later, cDNAs are used to assess which genes are expressed in specific cells, at particular times in development, or under a particular set of conditions. In contrast, genomic clones—which include introns and exons, as well as regulatory DNA sequences—provide the starting material for determining the complete nucleotide sequence of an organism's genome.

DNA CLONING BY PCR

Genomic and cDNA libraries were once the only route to gene cloning, and they are still used for cloning very large genes and for sequencing whole genomes. However, a powerful and versatile method for amplifying DNA, known as the **polymerase chain reaction** (**PCR**), provides a

QUESTION 10–3

Discuss the following statement: "From the nucleotide sequence of a cDNA clone, the complete amino acid sequence of a protein can be deduced by applying the genetic code. Thus, protein biochemistry has become superfluous because there is nothing more that can be learned by studying the protein."

Figure 10–13 Genomic DNA clones and cDNA clones derived from the same region of the genome are different.
In this example, gene A is infrequently transcribed, whereas gene B is frequently transcribed, and both genes contain introns (*orange*). In the genomic DNA library, both introns and nontranscribed DNA (*gray*) are included in the clones, and most clones will contain either no coding sequence or only part of the coding sequence of a gene (*red*); the DNA sequences that regulate the expression of each gene are also included (not indicated). In the cDNA clones, the intron sequences have been removed by RNA splicing during the formation of the mRNA (*blue*), and a continuous coding sequence is therefore present in each clone. Because gene B is transcribed more frequently than gene A in the cells from which the cDNA library was made, it will be represented much more often than A in the cDNA library. In contrast, genes A and B should be represented equally in the genomic library.

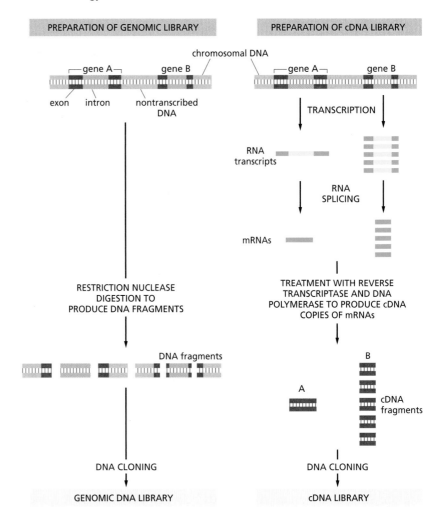

more rapid and straightforward approach to DNA cloning, particularly in organisms whose complete genome sequence is known. Today, most genes are cloned via PCR.

Invented in the 1980s, PCR revolutionized the way that DNA and RNA are analyzed. The technique can amplify any nucleotide sequence rapidly and selectively. Unlike the traditional approach of cloning using vectors—which relies on bacteria to make copies of the desired DNA sequences—PCR is performed entirely in a test tube. Eliminating the need for bacteria makes PCR convenient and incredibly quick—billions of copies of a nucleotide sequence can be generated in a matter of hours. At the same time, PCR is remarkably sensitive: the method can be used to detect the trace amounts of DNA in a drop of blood left at a crime scene or in a few copies of a viral genome in a patient's blood sample. Because of its sensitivity, speed, and ease of use, PCR has many applications in addition to DNA cloning, including forensics and diagnostics.

In this section, we provide a brief overview of how PCR works and how it is used for a range of purposes that require the amplification of specific DNA sequences.

PCR Uses a DNA Polymerase to Amplify Selected DNA Sequences in a Test Tube

The success of PCR depends on the exquisite selectivity of DNA hybridization, along with the ability of DNA polymerase to copy a DNA template reliably, through repeated rounds of replication *in vitro*. The enzyme works by adding nucleotides to the 3′ end of a growing strand of DNA (see Figure 6–11). To initiate the reaction, the polymerase requires a primer—a short nucleotide sequence that provides a 3′ end from which synthesis can begin. The beauty of PCR is that the primers that are added to the reaction mixture not only serve as starting points, they also direct the polymerase to the specific DNA sequence to be amplified. These primers, like the DNA probes used to identify specific nucleotide sequences as discussed earlier, are designed by the experimenter based on the DNA sequence of interest and then synthesized chemically. Thus, PCR can only be used to clone a DNA segment for which the sequence is known in advance. With the large and growing number of genome sequences available in public databases, this requirement is rarely a drawback.

Multiple Cycles of Amplification *In Vitro* Generate Billions of Copies of the Desired Nucleotide Sequence

PCR is an iterative process in which the cycle of amplification is repeated dozens of times. At the start of each cycle, the two strands of the double-stranded DNA template are separated and a unique primer is annealed to each. DNA polymerase is then allowed to replicate each strand independently (**Figure 10–14**). In subsequent cycles, all the newly synthesized DNA molecules produced by the polymerase serve as templates for the next round of replication (**Figure 10–15**). Through this iterative amplification process, many copies of the original sequence can be made—billions after about 20 to 30 cycles.

PCR is now the method of choice for cloning relatively short DNA fragments (say, under 10,000 nucleotide pairs). Each cycle takes only about five minutes, and automation of the whole procedure enables cell-free cloning of a DNA fragment in a few hours, compared with the several days required for cloning in bacteria. The original template for PCR can be either DNA or RNA, so this method can be used to obtain either a full genomic clone (complete with introns and exons) or a cDNA copy of an mRNA (**Figure 10–16**). A major benefit of PCR is that genes can be cloned directly from any piece of DNA or RNA without the time and effort needed to first construct a DNA library.

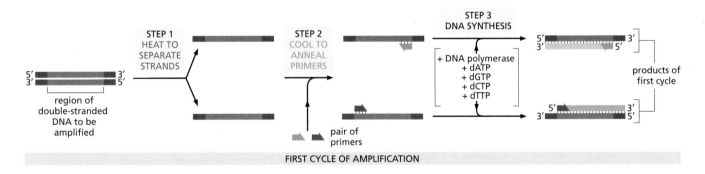

FIRST CYCLE OF AMPLIFICATION

Figure 10–14 A pair of PCR primers directs the amplification of a desired segment of DNA in a test tube. Each cycle of PCR includes three steps: (1) The double-stranded DNA is heated briefly to separate the two strands. (2) The DNA is exposed to a large excess of a pair of specific primers—designed to bracket the region of DNA to be amplified—and the sample is cooled to allow the primers to hybridize to complementary sequences in the two DNA strands. (3) This mixture is incubated with DNA polymerase and the four deoxyribonucleoside triphosphates so that DNA can be synthesized, starting from the two primers. The cycle can then be repeated by reheating the sample to separate the newly synthesized DNA strands (see Figure 10–15).
 The technique depends on the use of a special DNA polymerase isolated from a thermophilic bacterium; this polymerase is stable at much higher temperatures than eukaryotic DNA polymerases, so it is not denatured by the heat treatment shown in step 1. The enzyme therefore does not have to be added again after each cycle.

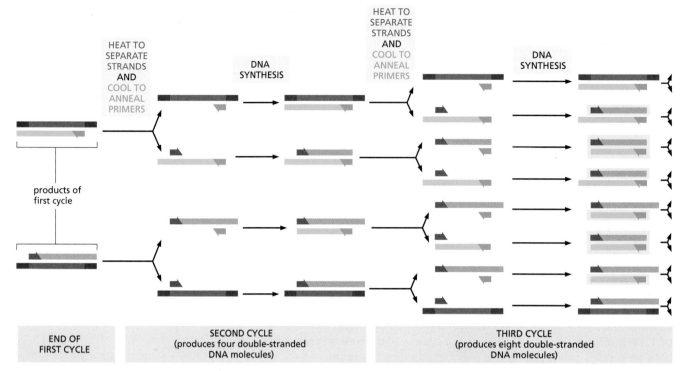

HEAT TO SEPARATE STRANDS AND COOL TO ANNEAL PRIMERS

DNA SYNTHESIS

HEAT TO SEPARATE STRANDS AND COOL TO ANNEAL PRIMERS

DNA SYNTHESIS

products of first cycle

| END OF FIRST CYCLE | SECOND CYCLE (produces four double-stranded DNA molecules) | THIRD CYCLE (produces eight double-stranded DNA molecules) |

Figure 10–15 PCR uses repeated rounds of strand separation, hybridization, and synthesis to amplify DNA. As the procedure outlined in Figure 10–14 is repeated, all the newly synthesized fragments serve as templates in their turn. Because the polymerase and the primers remain in the sample after the first cycle, PCR involves simply heating and then cooling the same sample, in the same test tube, again and again. Each cycle doubles the amount of DNA synthesized in the previous cycle, so that within a few cycles, the predominant DNA is identical to the sequence bracketed by and including the two primers in the original template. In the example illustrated here, three cycles of reaction produce 16 DNA chains, 8 of which (boxed in *yellow*) correspond exactly to one or the other strand of the original bracketed sequence. After four more cycles, 240 of the 256 DNA chains will correspond exactly to the original sequence, and after several more cycles, essentially all of the DNA strands will be this length. The whole procedure is shown in Movie 10.1.

PCR is Also Used for Diagnostic and Forensic Applications

In addition to its use in gene cloning, PCR is frequently employed to amplify DNA for other, more practical purposes. Because of its extraordinary sensitivity, PCR can be used to detect invading microorganisms at very early stages of infection. In this case, short sequences complementary to a segment of the infectious agent's genome are used as primers, and following many cycles of amplification, even a few copies of an invading bacterial or viral genome in a patient sample can be detected (**Figure 10–17**). For many infections, PCR has replaced the use of antibodies against microbial molecules to detect the presence of pathogens. PCR can also be used to track epidemics, detect bioterrorist attacks, and test food products for the presence of potentially harmful microbes. It is also used to verify the authenticity of a food source—for example, whether a sample of beef actually came from a cow.

Finally, PCR is now widely used in forensic medicine. The method's extreme sensitivity allows forensic investigators to isolate DNA from minute traces of human blood or other tissue to obtain a *DNA fingerprint* of the person who left the sample behind. With the possible exception of identical twins, the genome of each human differs in DNA sequence from that of every other person on Earth. Using primer pairs targeted at genome sequences that are known to be highly variable in the human population, PCR makes it possible to generate a distinctive DNA fingerprint for any individual (**Figure 10–18**). Such forensic analyses can be used not only to point the finger at those who have done wrong, but—equally important—to help exonerate those who have been wrongfully convicted.

QUESTION 10–4

A.   If the PCR shown in Figure 10–15 is carried through an additional two rounds of amplification, how many of the DNA fragments labeled in gray, green, or red or outlined in yellow are produced? If many additional cycles are carried out, which fragments will predominate?

B.   Assume you start with one double-stranded DNA molecule and amplify a 500-nucleotide-pair sequence contained within it. Approximately how many cycles of PCR amplification will you need to produce 100 ng of this DNA? 100 ng is an amount that can be easily detected after staining with a fluorescent dye. (Hint: for this calculation, you need to know that each nucleotide has an average molecular mass of 330 g/mole.)

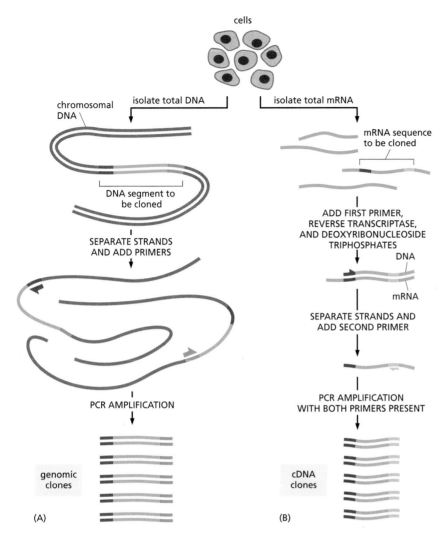

Figure 10–16 PCR can be used to obtain either genomic or cDNA clones. (A) To use PCR to clone a segment of chromosomal DNA, total DNA is first purified from cells. PCR primers that flank the stretch of DNA to be cloned are added, and many cycles of PCR are completed (see Figure 10–15). Because only the DNA between (and including) the primers is amplified, PCR provides a way to obtain selectively any short stretch of chromosomal DNA in an effectively pure form. (B) To use PCR to obtain a cDNA clone of a gene, total mRNA is first purified from cells. The first primer is added to the population of mRNAs, and reverse transcriptase is used to make a DNA strand complementary to the specific RNA sequence of interest. The second primer is then added, and the DNA molecule is amplified through many cycles of PCR.

EXPLORING AND EXPLOITING GENE FUNCTION

The procedures we have described thus far enable biologists to obtain large amounts of DNA in a form that is easy to work with in the laboratory. Whether present as fragments stored in a DNA library in bacteria or as a collection of PCR products nestled in the bottom of a test tube, this DNA also provides the raw material for experiments designed to unravel how individual genes—and the RNA molecules and proteins they encode—function in cells and organisms.

This is where creativity comes in. There are as many ways to study gene function as there are scientists interested in studying it. The techniques

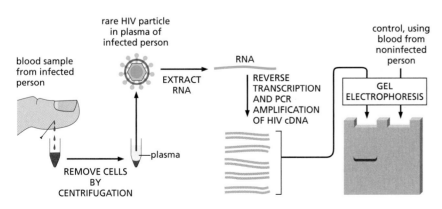

Figure 10–17 PCR can be used to detect the presence of a viral genome in a sample of blood. Because of its ability to amplify enormously the signal from every single molecule of nucleic acid, PCR is an extraordinarily sensitive method for detecting trace amounts of virus in a sample of blood or tissue without the need to purify the virus. For HIV, the virus that causes AIDS, the genome is a single-stranded molecule of RNA, as illustrated here. In addition to HIV, many other viruses that infect humans are now detected in this way.

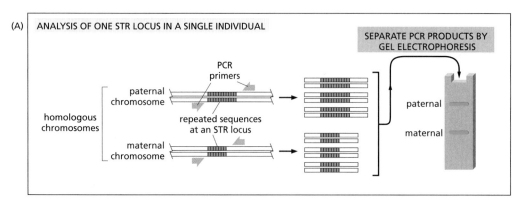

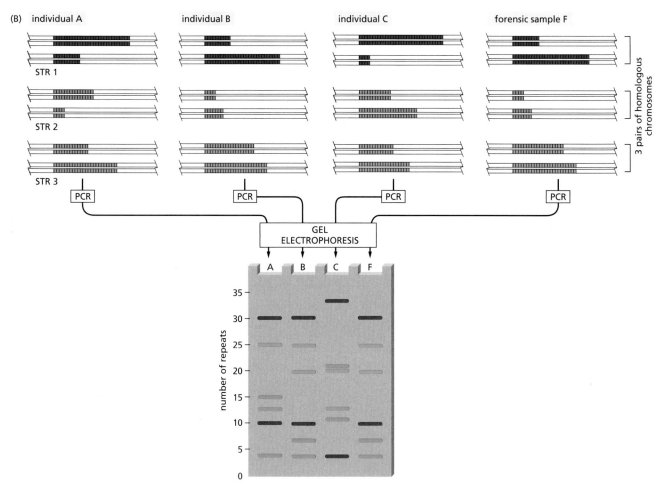

Figure 10–18 PCR is used in forensic science to distinguish one individual from another. The DNA sequences analyzed are short tandem repeats (STRs) composed of sequences such as CACACA… or GTGTGT…. STRs are found in various positions (loci) in the human genome. The number of repeats in each STR locus is highly variable in the population, ranging from 4 to 40 in different individuals. Because of the variability in these sequences, individuals will usually inherit a different number of repeats at each STR locus from their mother and from their father; two unrelated individuals, therefore, rarely contain the same pair of sequences at a given STR locus. (A) PCR using primers that recognize unique sequences on either side of one particular STR locus produces a pair of bands of amplified DNA from each individual, one band representing the maternal STR variant and the other representing the paternal STR variant. The length of the amplified DNA, and thus its position after gel electrophoresis, will depend on the exact number of repeats at the locus. (B) In the schematic example shown here, the same three STR loci are analyzed in samples from three suspects (individuals A, B, and C), producing six bands for each individual. Although different people can have several bands in common, the overall pattern is quite distinctive for each person. The band pattern can therefore serve as a *DNA fingerprint* to identify an individual nearly uniquely. The fourth lane (F) contains the products of the same PCR amplifications carried out on a hypothetical forensic DNA sample, which could have been obtained from a single hair or a tiny spot of blood left at a crime scene.

The more loci that are examined, the more confident one can be about the results. When examining the variability at 5–10 different STR loci, the odds that two random individuals would share the same fingerprint by chance are approximately one in 10 billion. In the case shown here, individuals A and C can be eliminated from inquiries, while B is a clear suspect. A similar approach is now used routinely in paternity testing.

an investigator chooses often depend on his or her background and training: a geneticist might, for example, engineer mutant organisms in which the activity of the gene has been disrupted, whereas a biochemist might take the same gene and produce large amounts of its protein to determine its three-dimensional structure.

In this section, we present a few of the methods that investigators currently use to study the function of a gene—all of which depend on recombinant DNA technology. Because a gene's activity is specified by its nucleotide sequence, we begin by outlining the techniques used to determine—and begin to interpret—the nucleotide sequence of a stretch of DNA. We then explore a variety of approaches for investigating when and where a gene is expressed. We describe how disrupting the activity of a gene in a cell, tissue, or whole plant or animal can provide insights into what that gene normally does. Finally, we explain how recombinant DNA technology can be harnessed to produce large amounts of any protein. Together, the methods we discuss have revolutionized all aspects of cell biology.

Whole Genomes Can Be Sequenced Rapidly

In the late 1970s, researchers developed several schemes for determining, simply and quickly, the nucleotide sequence of any purified DNA fragment. The one that became the most widely used is called **dideoxy sequencing** or **Sanger sequencing** (after the scientist who invented it). The technique uses DNA polymerase, along with special chain-terminating nucleotides called dideoxyribonucleoside triphosphates (**Figure 10–19**), to make partial copies of the DNA fragment to be sequenced. It ultimately produces a collection of different DNA copies that terminate at every position in the original DNA sequence.

Until recently, these DNA copies, which differ in length by a single nucleotide, would then be separated by gel electrophoresis, and the nucleotide sequence of the original DNA would be determined manually from the order of labeled DNA fragments in the gel (**Figure 10–20**). These days, however, Sanger sequencing is fully automated: robotic devices mix the reagents—including the four different chain-terminating dideoxynucleotides, each tagged with a different-colored fluorescent dye—and load the reaction samples onto long, thin capillary gels, which have replaced the flat gel slabs used since the 1970s. A detector then records the color of each band in the gel, and a computer translates the information into a nucleotide sequence (**Figure 10–21**). How such sequence information is then analyzed to assemble a complete genome sequence—for example, the first draft of the human genome—is described in **How We Know**, pp. 344–345.

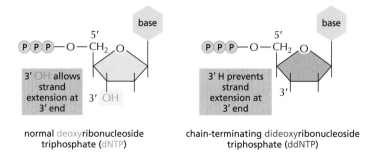

normal deoxyribonucleoside triphosphate (dNTP)   chain-terminating dideoxyribonucleoside triphosphate (ddNTP)

Figure 10–19 The dideoxy, or Sanger, method of sequencing DNA relies on chain-terminating dideoxynucleoside triphosphates (ddNTPs). These ddNTPs are derivatives of the normal deoxyribonucleoside triphosphates that lack the 3′ hydroxyl group. When incorporated into a growing DNA strand, they block further elongation of that strand.

Figure 10–20 The Sanger method produces four sets of labeled DNA molecules. To determine the complete sequence of a single-stranded fragment of DNA (gray), the DNA is first hybridized with a short DNA primer (orange) that is labeled with a fluorescent dye or radioisotope. DNA polymerase and an excess of all four normal deoxyribonucleoside triphosphates (blue A, C, G, and T) are added to the primed DNA, which is then divided into four reaction tubes. Each of these tubes receives a small amount of a single chain-terminating dideoxyribonucleoside triphosphate (red A, C, G, or T). Because the chain-terminating ddNTPs will be incorporated only occasionally, each reaction produces a set of DNA copies that terminate at different points in the sequence. The products of these four reactions are separated by electrophoresis in four parallel lanes of a polyacrylamide gel (labeled here A, T, C, and G). In each lane, the bands represent fragments that have terminated at a given nucleotide (e.g., A in the leftmost lane) but at different positions in the DNA. By reading off the bands in order, starting at the bottom of the gel and reading across all lanes, the DNA sequence of the newly synthesized strand can be determined. The sequence, which is given in the green arrow to the right of the gel, is complementary to the sequence of the original gray single-stranded DNA, as shown on the bottom.

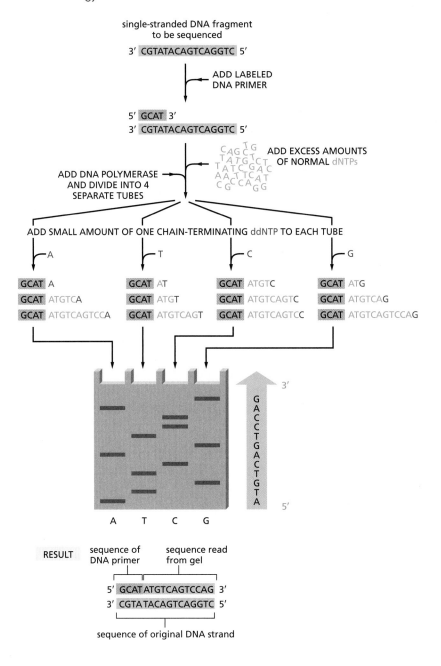

Figure 10–21 Fully automated machines can set up and run Sanger sequencing reactions. (A) The automated method uses an excess amount of normal dNTPs plus a mixture of four different chain-terminating ddNTPs, each of which is labeled with a fluorescent tag of a different color. The reaction products are loaded onto a long, thin capillary gel and separated by electrophoresis. A camera reads the color of each band on the gel and feeds the data to a computer that assembles the sequence (not shown). (B) A tiny part of the data from such an automated sequencing run. Each colored peak represents a nucleotide in the DNA sequence.

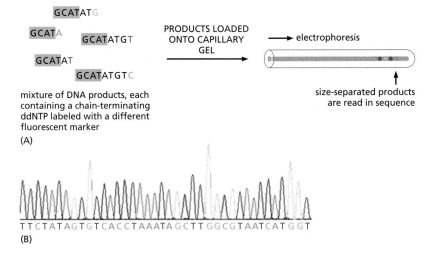

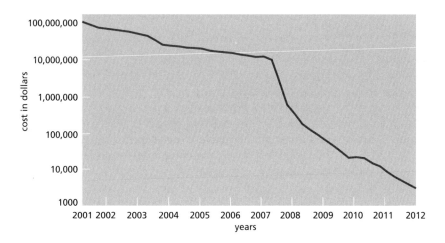

Figure 10–22 The cost of DNA sequencing has dropped precipitously since the advent of next-generation sequencing technologies. Shown here are the costs of sequencing a human genome which was $100 million in 2001 and not much more than a thousand dollars by the end of 2012. (Data from the National Human Genome Research Initiative.)

Next-Generation Sequencing Techniques Make Genome Sequencing Faster and Cheaper

The Sanger method has made it possible to sequence the genomes of humans and of many other organisms including most of those discussed in this book. But newer methods, developed since 2005, have made genome sequencing even more rapid—and very much cheaper. With these so-called *second-generation sequencing methods*, the cost of sequencing DNA has plummeted (**Figure 10–22**). At the same time, the number of genomes that have been sequenced has skyrocketed. These rapid methods allow multiple genomes to be sequenced in parallel in a matter of weeks, enabling investigators to examine thousands of human genomes, catalog the variation in nucleotide sequences from people around the world, and uncover the mutations that increase the risk of various diseases—from cancer to autism—as we discuss in Chapter 19.

Although each method differs in detail, most rely on PCR amplification of a random collection of DNA fragments attached to a solid support, such as a glass slide or a microwell plate. For each fragment, the amplification generates a "cluster" that contains about 1000 copies of an individual DNA fragment. These clusters—tens of millions of which can fit on a single slide or plate—are then sequenced at the same time (**Figure 10–23**).

Even more remarkable are the newest, *third-generation sequencing methods*, which permit the sequencing of just a single molecule of DNA. In one of these techniques, for example, each DNA molecule is slowly pulled through a very tiny channel, like thread through the eye of a needle. Because each of the four nucleotides has a different, characteristic shape, the way a nucleotide obstructs the pore as it passes through reveals

Figure 10–23 Second-generation sequencing methods rely on massively parallel sequencing reactions carried out on clusters of PCR-amplified DNA. Each spot on a slide or plate contains about a thousand copies of a single DNA fragment. In the first step, the plate is incubated with DNA polymerase and a special set of four nucleoside triphosphates (NTPs) that terminate DNA synthesis in a reversible manner, each of which carries a fluorescent marker of a different color; no normal dNTPs are present. A camera then images and records the fluorescence at each position on the plate. In the second step, the DNA is chemically treated to remove the fluorescent markers and chemical blockers from each nucleoside; strand synthesis then continues after a new batch of fluorescent NTPs is added. These steps are repeated until the sequence is complete. The snapshots of each round of synthesis are compiled by computer to yield the sequence of the cluster of fragments located at each of the potentially millions of positions on the plate.

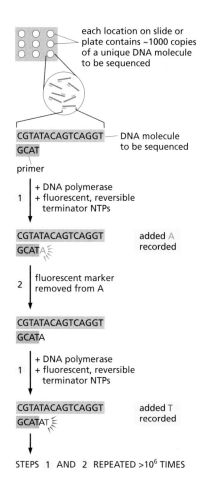

SEQUENCING THE HUMAN GENOME

When DNA sequencing techniques became fully automated, determining the order of the nucleotides in a piece of DNA went from being an elaborate Ph.D. thesis project to a routine laboratory chore. Feed DNA into the sequencing machine, add the necessary reagents, and out comes the sought-after result: the order of As, Ts, Gs, and Cs. Nothing could be simpler.

So why was sequencing the human genome such a formidable task? Largely because of its size. The DNA sequencing methods employed at the time were limited by the physical size of the gel used to separate the labeled fragments (see Figure 10–20). At most, only a few hundred nucleotides could be read from a single gel. How, then, do you handle a genome that contains billions of nucleotide pairs?

The solution is to break the genome into fragments and sequence these smaller pieces. The main challenge then comes in piecing the short fragments together in the correct order to yield a comprehensive sequence of a whole chromosome, and ultimately a whole genome. There are two main strategies for accomplishing this genomic breakage and reassembly: the shotgun method and the clone-by-clone approach.

Shotgun sequencing

The most straightforward approach to sequencing a genome is to break it into random fragments, separate and sequence each of the single-stranded fragments, and then use a powerful computer to order these pieces using sequence overlaps to guide the assembly (**Figure 10–24**). This approach is called the shotgun sequencing strategy. As an analogy, imagine shredding several

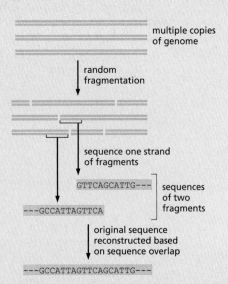

Figure 10–24 Shotgun sequencing is the method of choice for small genomes. The genome is first broken into much smaller, overlapping fragments. Each fragment is then sequenced, and the genome is assembled based on overlapping sequences.

copies of *Essential Cell Biology* (*ECB*), mixing up the pieces, and then trying to put one whole copy of the book back together again by matching up the words or phrases or sentences that appear on each piece. (Several copies would be needed to generate enough overlap for reassembly.) It could be done, but it would be much easier if the book were, say, only two pages long.

For this reason, a straight-out shotgun approach is the strategy of choice only for sequencing small genomes. The method proved its worth in 1995, when it was used to sequence the genome of the infectious bacterium *Haemophilus influenzae*, the first organism to have its complete genome sequence determined. The trouble with shotgun sequencing is that the reassembly process can be derailed by repetitive nucleotide sequences. Although rare in bacteria, these sequences make up a large fraction of vertebrate genomes (see Figure 9–33). Highly repetitive DNA segments make it difficult to piece DNA sequences back together accurately (**Figure 10–25**). Returning to the *ECB* analogy, this chapter alone contains more than a few instances of the phrase "the human genome." Imagine that one slip of paper from the shredded *ECB*s contains the information: "So why was sequencing the human genome" (which appears at the start of this section); another contains the information: "the human genome sequence consortium combined shotgun sequencing with a clone-by-clone approach" (which appears below). You might be tempted to join these two segments together based on the overlapping phrase "the human genome." But you would wind up with the nonsensical statement: "So why was sequencing the human genome sequence consortium combined shotgun sequencing with a clone-by-clone approach." You would also lose the several paragraphs of important text that originally appeared between these two instances of "the human genome."

And that's just in this section. The phrase "the human genome" appears in many chapters of this book. Such repetition compounds the problem of placing each fragment in its correct context. To circumvent these assembly problems, researchers in the human genome sequence consortium combined shotgun sequencing with a clone-by-clone approach.

Clone-by-clone

In this approach, researchers started by preparing a genomic DNA library. They broke the human genome into overlapping fragments, 100–200 kilobase pairs in size. They then plugged these segments into bacterial artificial chromosomes (BACs) and inserted them into *E. coli*. (BACs are similar to the bacterial plasmids discussed earlier, except they can carry much larger pieces of DNA.) As the bacteria divided, they copied the BACs,

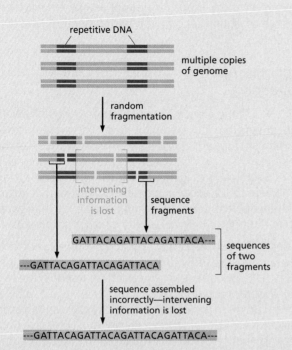

Figure 10–25 Repetitive DNA sequences in a genome make it difficult to accurately assemble its fragments. In this example, the DNA contains two segments of repetitive DNA, each made of many copies of the sequence GATTACA. When the resulting sequences are examined, two fragments from different parts of the DNA appear to overlap. Assembling these sequences incorrectly would result in a loss of the information (in brackets) that lies between the original repeats.

nucleotide sequence of each BAC separately using the shotgun method. They could then assemble the whole genome sequence by stitching together the sequences of thousands of individual BACs that span the length of the genome.

The beauty of this approach was that it was relatively easy to accurately determine where the BAC fragments belong in the genome. This mapping step reduces the likelihood that regions containing repetitive sequences will be assembled incorrectly, and it virtually eliminates the possibility that sequences from different chromosomes will be mistakenly joined together. Returning to the textbook analogy, the BAC-based approach is akin to first separating your copies of *ECB* into individual pages and then shredding each page into its own separate pile. It should be much easier to put the book back together when one pile of fragments contains words from page 1, a second pile from page 2, and so on. And there's virtually no chance of mistakenly sticking a sentence from page 40 into the middle of a paragraph on page 412.

All together now

The clone-by-clone approach produced the first draft of the human genome sequence in 2000 and the completed sequence in 2004. As the set of instructions that specify all of the RNA and protein molecules needed to build a human being, this string of genetic bits holds the secrets to human development and physiology. But the sequence was also of great value to researchers interested in comparative genomics or in the physiology of other organisms: it eased the assembly of nucleotide sequences from other mammalian genomes—mice, rats, dogs, and other primates. It also made it much easier to determine the nucleotide sequences of the genomes of individual humans by providing a framework on which the new sequences could be simply superimposed.

The first human sequence was the only mammalian genome completed in this methodical way. But the human genome project was an unqualified success in that it provided the techniques, confidence, and momentum that drove the development of the next generation of DNA sequencing methods, which are now rapidly transforming all areas of biology.

thus producing a collection of overlapping cloned fragments (see Figure 10–10).

The researchers then determined where each of these DNA fragments fit into the existing map of the human genome. To do this, different restriction nucleases were used to cut each clone to generate a unique restriction-site "signature." The locations of the restriction sites in each fragment allowed researchers to map each BAC clone onto a restriction map of a whole human genome that had been generated previously using the same set of restriction nucleases (**Figure 10–26**).

Knowing the relative positions of the cloned fragments, the researchers then selected some 30,000 BACs, sheared each into smaller fragments, and determined the

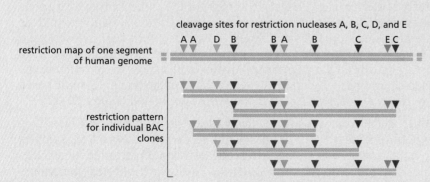

Figure 10–26 Individual BAC clones are positioned on the physical map of the human genome sequence on the basis of their restriction site "signatures." Clones are digested with five different restriction nucleases, and the sites at which the different enzymes cut each clone are recorded. The distinctive pattern of restriction sites allows investigators to order the fragments and place them on a restriction map of a human genome that had been previously generated using the same nucleases.

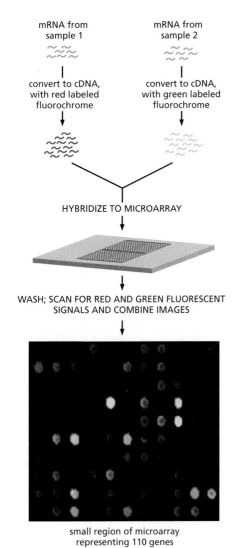

Figure 10–27 DNA microarrays are used to analyze the production of thousands of different mRNAs in a single experiment. In this example, mRNA is collected from two different cell samples—for example, cells treated with a hormone and untreated cells of the same type—to allow for a direct comparison of the specific genes expressed under both conditions. The mRNAs are converted to cDNAs that are labeled with a red fluorescent dye for one sample, and a green fluorescent dye for the other. The labeled samples are mixed and then allowed to hybridize to the microarray. After incubation, the array is washed and the fluorescence scanned. Only a small proportion of the microarray, representing 110 genes, is shown. *Red* spots indicate that the gene in sample 1 is expressed at a higher level than the corresponding gene in sample 2, and *green* spots indicate the opposite. *Yellow* spots reveal genes that are expressed at about equal levels in both cell samples. The intensity of the fluorescence provides an estimate of how much RNA is present from a gene. *Dark* spots indicate little or no expression of the gene whose fragment is located at that position in the array.

its identity—information that is then used to compile the sequence of the DNA molecule. Such methods require no amplification or chemical labeling, and thereby reduce the cost and time of sequencing even further, making it possible to obtain a complete human genome sequence for under $1000 in hours.

Comparative Genome Analyses Can Identify Genes and Predict Their Function

Strings of nucleotides, at first glance, reveal nothing about how that genetic information directs the development of a living organism—or even what type of organism it might encode. One way to learn something about the function of a particular nucleotide sequence is to compare it with the multitude of sequences available in public databases. Using a computer program to search for sequence similarity, one can determine whether a nucleotide sequence contains a gene and what that gene is likely to do—based on the gene's known activity in other organisms.

Comparative analyses have revealed that the coding regions of genes from a wide variety of organisms show a large degree of sequence conservation (see Figure 9–19). The sequences of noncoding regions, however, tend to diverge over evolutionary time (see Figure 9–18). Thus, a search for sequence similarity can often indicate from which organism a particular piece of DNA was derived, and which species are most closely related. Such information is particularly useful when the origin of a DNA sample is unknown—because it was extracted, for example, from a sample of soil or seawater or the blood of a patient with an undiagnosed infection.

But knowing where a nucleotide sequence comes from—or even what activity it might have—is only the first step toward determining what role it has in the development or physiology of the organism. The knowledge that a particular DNA sequence encodes a transcription regulator, for example, does not reveal when and where that protein is produced, or which genes it might regulate. To learn that, investigators must head back to the laboratory.

Analysis of mRNAs By Microarray or RNA-Seq Provides a Snapshot of Gene Expression

As we discussed in Chapter 8, a cell expresses only a subset of the thousands of genes available in its genome. This subset differs from one cell type to another. One way to determine which genes are being expressed in a population of cells or in a tissue is to analyze which mRNAs are being produced.

The first tool that allowed investigators to analyze simultaneously the thousands of different RNAs produced by cells or tissues was the **DNA microarray**. Developed in the 1990s, DNA microarrays are glass microscope slides that contain hundreds of thousands of DNA fragments, each of which serves as a probe for the mRNA produced by a specific gene. Such microarrays allow investigators to monitor the expression of every gene in an entire genome in a single experiment. To do the analysis, mRNAs are extracted from cells or tissues and converted to cDNAs (see Figure 10–12). The cDNAs are fluorescently labeled and allowed to hybridize to the fragments on the microarray. An automated fluorescence microscope then determines which mRNAs were present in the original sample based on the array positions to which the cDNAs are bound (**Figure 10–27**).

Although microarrays are relatively inexpensive and easy to use, they suffer from one obvious drawback: the sequences of the mRNA samples to be analyzed must be known in advance and represented by a corresponding probe on the array. With the development of next-generation sequencing technologies, investigators increasingly use a more direct approach for cataloging the RNAs produced by a cell. The RNAs are converted to cDNAs, which are then sequenced using second-generation sequencing methods. The approach, called **RNA-Seq**, provides a more quantitative analysis of the *transcriptome*—the complete collection of RNAs produced by a cell under a certain set of conditions. It also determines the number of times a particular sequence appears in a sample and detects rare mRNAs, RNA transcripts that are alternatively spliced, mRNAs that harbor sequence variations, and noncoding RNAs. For these reasons, RNA-Seq is replacing microarrays as the method of choice for analyzing the transcriptome.

In Situ Hybridization Can Reveal When and Where a Gene Is Expressed

Although microarrays and RNA-Seq provide a list of genes that are being expressed by a cell or tissue, they do not reveal exactly where in the cell or tissue those mRNAs are produced. To see where a particular RNA is made, investigators use a technique called *in situ* **hybridization** (from the Latin *in situ*, "in place"), which allows a specific nucleic acid sequence— either DNA or RNA—to be visualized in its normal location.

In situ hybridization uses single-stranded DNA or RNA probes, labeled with either fluorescent dyes or radioactive isotopes, to detect complementary nucleic acid sequences within a tissue, a cell (**Figure 10–28**), or even an isolated chromosome (**Figure 10–29**). The latter application is used in the clinic to determine, for example, whether fetuses carry abnormal chromosomes.

In situ hybridization is frequently used to study the expression patterns of a particular gene or collection of genes in an adult or developing tissue. In one particularly ambitious project, neuroscientists are using the method to assemble a three-dimensional map of all the genes expressed in both the mouse and human brain (**Figure 10–30**). Knowing where and when a gene is expressed can provide important clues about its function.

Reporter Genes Allow Specific Proteins to be Tracked in Living Cells

For a gene that encodes a protein, the location of the protein within the cell, tissue, or organism yields clues to the gene's function. Traditionally, the most effective way to visualize a protein within a cell or tissue involved using a labeled antibody. That approach requires the generation of an antibody that specifically recognizes the protein of interest—a process that can be time-consuming and has no guarantee of success.

An alternative approach is to use the regulatory DNA sequences of the protein-coding gene to drive the expression of some type of

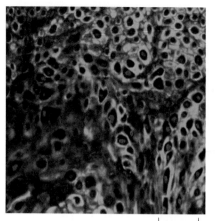

50 μm

Figure 10–28 *In situ* hybridization can be used to detect the presence of a virus in cells. In this micrograph, the nuclei of cultured epithelial cells infected with the human papillomavirus (HPV) are stained *pink* by a fluorescent probe that recognizes a viral DNA sequence. The cytoplasm of all cells is stained *green*. (Courtesy of Hogne Røed Nilsen.)

Figure 10–29 *In situ* hybridization can be used to locate genes on isolated chromosomes. Here, six different DNA probes have been used to mark the locations of their respective nucleotide sequences on human Chromosome 5 isolated from a mitotic cell in metaphase (see Figure 5–16 and Panel 18–1, pp. 622–623). The DNA probes have been labeled with different chemical groups and are detected using fluorescent antibodies specific for those groups. Both the maternal and paternal copies of Chromosome 5 are shown, aligned side by side. Each probe produces two dots on each chromosome because chromosomes undergoing mitosis have already replicated their DNA; therefore, each chromosome contains two identical DNA helices. The technique employed here is nicknamed FISH, for *fluorescence* in situ *hybridization*. (Courtesy of David C. Ward.)

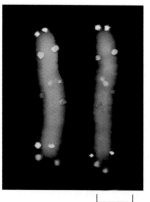

2 μm

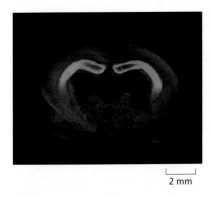

Figure 10–30 *In situ* hybridization has been used to generate an atlas of gene expression in the mouse brain. This computer-generated image shows the expression of genes specific to an area of the brain associated with learning and memory. Similar maps of expression patterns of all known genes in the mouse brain are compiled in the brain atlas project, which is available for free online. (From M. Hawrylycz et al., *PLoS Comput. Biol.* 7:e1001065, 2011.)

reporter gene, one that encodes a protein that can be easily monitored by its fluorescence or enzymatic activity. A recombinant gene of this type usually mimics the expression of the gene of interest, producing the reporter protein when, where, and in the same amounts as the normal protein would be made (**Figure 10–31A**). The same approach can be used to study the regulatory DNA sequences that control the gene's expression (**Figure 10–31B**).

One of the most popular reporter proteins used today is **green fluorescent protein** (**GFP**), the molecule that gives luminescent jellyfish their greenish glow. In many cases, the gene that encodes GFP is simply attached to one end of the gene of interest. The resulting *GFP fusion protein* often behaves in the same way as the normal protein produced by the gene of interest, and its location can be monitored by fluorescence microscopy (**Figure 10–32**). GFP fusion has become a standard strategy for tracking not only the location but also the movement of specific proteins in living cells. In addition, the use of multiple GFP variants that fluoresce at different wavelengths can provide insights into how different cells interact in a living tissue (**Figure 10–33**).

The Study of Mutants Can Help Reveal the Function of a Gene

Although it may seem counterintuitive, one of the best ways to determine a gene's function is to see what happens to an organism when the gene is inactivated by a mutation. Before the advent of gene cloning, geneticists

Figure 10–31 Reporter genes can be used to determine the pattern of a gene's expression. (A) Suppose the goal is to find out which cell types (A–F) express protein X, but it is difficult to detect the protein directly—with antibodies, for example. Using recombinant DNA techniques, the coding sequence for protein X can be replaced with the coding sequence for reporter protein Y, which can be easily monitored visually; two commonly used reporter proteins are the enzyme β-galactosidase (see Figure 8–13C) and green fluorescent protein (GFP, see Figure 10–32). The expression of the reporter protein Y will now be controlled by the regulatory sequences (here labeled 1, 2, and 3) that control the expression of the normal protein X. (B) To determine which regulatory sequences normally control expression of gene X in particular cell types, reporters with various combinations of the regulatory regions associated with gene X can be constructed. These recombinant DNA molecules are then tested for expression after their introduction into the different cell types.

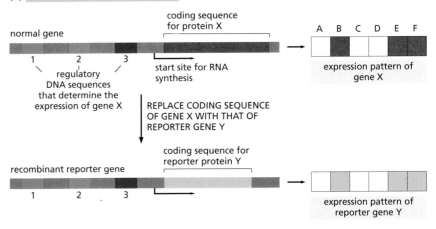

(A) **CONSTRUCTING A REPORTER GENE**

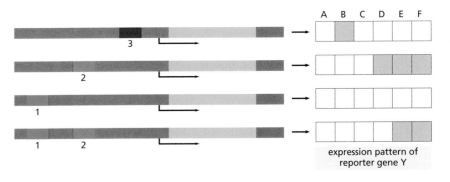

(B) **USING A REPORTER GENE TO STUDY GENE X REGULATORY SEQUENCES**

CONCLUSIONS —regulatory sequence 3 turns on gene X in cell B
—regulatory sequence 2 turns on gene X in cells D, E, and F
—regulatory sequence 1 turns off gene X in cell D

Figure 10–32 Green fluorescent protein (GFP) can be used to identify specific cells in a living animal. For this experiment, carried out in the fruit fly, recombinant DNA techniques were used to join the gene encoding GFP to the regulatory DNA sequences that direct the production of a particular *Drosophila* protein. Both the GFP and the normal fly protein are made only in a specialized set of neurons. This image of a live fly embryo was captured by a fluorescence microscope and shows approximately 20 neurons, each with long projections (axons and dendrites) that communicate with other (nonfluorescent) cells. These neurons, located just under the embryo's surface, allow the organism to sense its immediate environment. (From W.B. Grueber et al., *Curr. Biol.* 13:618–626, 2003. With permission from Elsevier.)

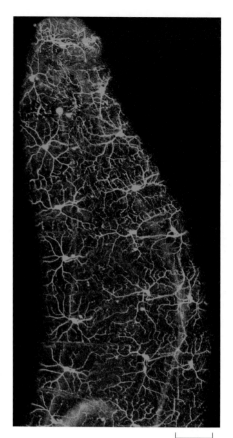

200 μm

studied the mutant organisms that arise spontaneously in a population. The mutants of most interest were often selected because of their unusual *phenotype*—fruit flies with white eyes or curly wings, for example. The gene responsible for the mutant phenotype could then be studied by breeding experiments, as Gregor Mendel did with peas in the nineteenth century (discussed in Chapter 19).

Although mutant organisms can arise spontaneously, they do so infrequently. The process can be accelerated by treating organisms with either radiation or chemical mutagens, which randomly disrupt gene activity. Such random mutagenesis generates large numbers of mutant organisms, each of which can then be studied individually. This "classical genetic approach," which we discuss in detail in Chapter 19, is most applicable to organisms that reproduce rapidly and can be analyzed genetically in the laboratory—such as bacteria, yeasts, nematode worms, and fruit flies—although it has also been used in zebrafish and mice.

RNA Interference (RNAi) Inhibits the Activity of Specific Genes

Recombinant DNA technology has made possible a more targeted genetic approach to studying gene function. Instead of beginning with a randomly generated mutant and then identifying the responsible gene, a gene of known sequence can be inactivated deliberately and the effects on the cell or organism's phenotype can be observed. Because this strategy is essentially the reverse of that used in classical genetics—which goes from mutants to genes—it is often referred to as *reverse genetics*.

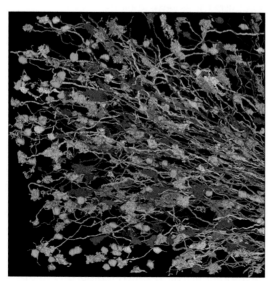

30 μm

Figure 10–33 GFPs that fluoresce at different wavelengths help reveal the connections that individual neurons make within the brain. This image shows differently colored neurons in one region of a mouse brain. The neurons randomly express different combinations of differently colored GFPs, making it possible to distinguish and trace many individual neurons within a population. The stunning appearance of these labeled neurons have earned these animals the colorful nickname "brainbow mice." (From J. Livet et al., *Nature* 450:56–62, 2007. With permission from Macmillan Publishers Ltd.)

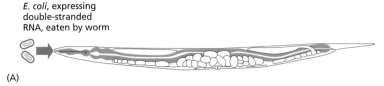

(A)

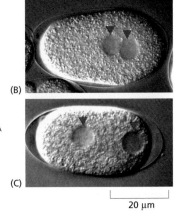

(B)

(C)

20 μm

Figure 10–34 Gene function can be tested by RNA interference.
(A) Double-stranded RNA (dsRNA) can be introduced into *C. elegans* by
(1) feeding the worms *E. coli* that express the dsRNA or (2) injecting the dsRNA
directly into the animal's gut. (B) In a wild-type worm embryo, the egg and
sperm pronuclei (*red* arrowheads) come together in the posterior half of the
embryo shortly after fertilization. (C) In an embryo in which a particular gene
has been silenced by RNAi, the pronuclei fail to migrate. This experiment
revealed an important but previously unknown function of this gene in
embryonic development. (B and C, from P. Gönczy et al., *Nature* 408:331–336,
2000. With permission from Macmillan Publishers Ltd.)

One of the fastest and easiest ways to silence genes in cells and organisms is via **RNA interference** (**RNAi**). Discovered in 1998, RNAi exploits a natural mechanism used in a wide variety of plants and animals to protect themselves against certain viruses and the proliferation of mobile genetic elements (discussed in Chapter 9). The technique involves introducing into a cell or organism double-stranded RNA molecules with a nucleotide sequence that matches the gene to be inactivated. The double-stranded RNA is cleaved and processed by special RNAi machinery to produce shorter, double-stranded fragments called small interfering RNAs (siRNAs). These siRNAs are unwound to form single-stranded RNA fragments that hybridize with the target gene's mRNAs and direct their degradation (see Figure 8–26). In some organisms, the same fragments can direct the production of more siRNAs allowing continued inactivation of the target mRNAs.

RNAi is frequently used to inactivate genes in cultured mammalian cell lines, *Drosophila*, and the nematode *C. elegans*. Introducing double-stranded RNAs into *C. elegans* is particularly easy: the worm can be fed with *E. coli* that have been genetically engineered to produce the double-stranded RNAs that trigger RNAi (**Figure 10–34**). These RNAs get converted into siRNAs, which get distributed throughout the animal's body to inhibit expression of the target gene in various tissues. For the many organisms whose genomes have been completely sequenced, RNAi can, in principle, be used to explore the function of any gene, and large collections of DNA vectors that produce these double-stranded RNAs are available for several species.

A Known Gene Can Be Deleted or Replaced With an Altered Version

Despite its usefulness, RNAi has some limitations. Non-target genes are sometimes inhibited along with the gene of interest, and certain cell types are resistant to RNAi entirely. Even for cell types in which the mechanism functions effectively, gene inactivation by RNAi is often temporary, earning the description "gene knockdown."

Fortunately, there are other, more specific and effective means of eliminating gene activity in cells and organisms. Using recombinant DNA techniques, the coding sequence of a cloned gene can be mutated *in vitro* to change the functional properties of its protein product. Alternatively, the coding region can be left intact and the regulatory region of the gene changed, so that the amount of protein made will be altered or the gene will be expressed in a different type of cell or at a different time during development. By re-introducing this altered gene back into the organism from which it originally came, one can produce a mutant organism that can be studied to determine the gene's function. Often the altered gene is inserted into the genome of reproductive cells so that it can be stably inherited by subsequent generations. Organisms whose genomes have been altered in this way are known as **transgenic organisms**, or *genetically modified organisms* (*GMOs*); the introduced gene is called a *transgene*.

To study the function of a gene that has been altered *in vitro*, ideally one would like to generate an organism in which the normal gene has been replaced by the altered one. In this way, the function of the mutant protein can be analyzed in the absence of the normal protein. A common way of doing this in mice makes use of cultured mouse embryonic stem (ES) cells (discussed in Chapter 20). These cells are first subjected to targeted gene replacement before being transplanted into a developing embryo to produce a mutant mouse, as illustrated in **Figure 10–35**.

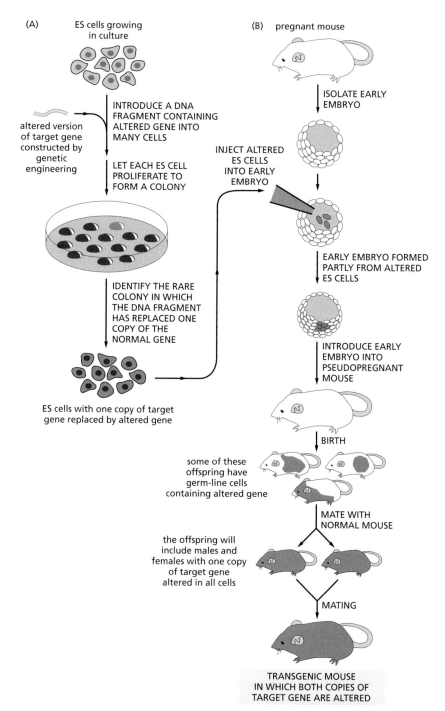

Figure 10–35 Targeted gene replacement in mice utilizes embryonic stem (ES) cells. (A) First, an altered version of the gene is introduced into cultured ES cells. In a few rare ES cells, the altered gene will replace the corresponding normal gene through homologous recombination. Although the procedure is often laborious, these rare cells can be identified and cultured to produce many descendants, each of which carries an altered gene in place of one of its two normal corresponding genes. (B) Next, the altered ES cells are injected into a very early mouse embryo; the cells are incorporated into the growing embryo, which then develops into a mouse that contains some somatic cells (colored *orange*) that carry the altered gene. Some of these mice may also have germ-line cells that contain the altered gene; when bred with a normal mouse, some of the progeny of these mice will contain the altered gene in all of their cells. Such a mouse is called a "knock-in" mouse. If two such mice are bred, one can obtain progeny that contain two copies of the altered gene—one on each chromosome—in all of their cells.

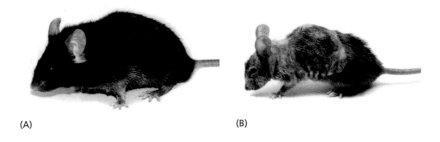

(A)   (B)

Figure 10–36 Transgenic mice with a mutant DNA helicase show premature aging. The helicase, encoded by the *Xpd* gene, is involved in both transcription and DNA repair. Compared with a wild-type mouse (A), a transgenic mouse that expresses a defective version of *Xpd* (B) exhibits many of the symptoms of premature aging, including osteoporosis, emaciation, early graying, infertility, and reduced life-span. The mutation in *Xpd* used here impairs the activity of the helicase and mimics a human mutation that causes trichothiodystrophy, a disorder characterized by brittle hair, skeletal abnormalities, and a greatly reduced life expectancy. These results support the hypothesis that an accumulation of DNA damage contributes to the aging process in both humans and mice. (From J. de Boer et al., *Science* 296:1276–1279, 2002. With permission from the AAAS.)

Using a similar strategy, the activity of both copies of a gene can also be eliminated entirely, creating a "**gene knockout.**" To do this, one can either introduce an inactive, mutant version of the gene into cultured ES cells or delete the gene altogether. The ability to use ES cells to produce such "knockout mice" revolutionized the study of gene function, and the technique is now being used to systematically determine the function of every mouse gene (**Figure 10–36**). A variation of this technique is used to produce *conditional knockout mice*, in which a known gene can be disrupted more selectively—only in a particular cell type or at a certain time in development. Such conditional knockouts are useful for studying genes with a critical function during development, because mice missing these crucial genes often die before birth.

Mutant Organisms Provide Useful Models of Human Disease

Technically speaking, transgenic approaches could be used to alter genes in the human germ line. For ethical reasons, such manipulations are unlawful. But transgenic technologies are widely used to generate animal models of human diseases in which mutant genes play a major part.

With the explosion of DNA sequencing technologies, investigators can rapidly search the genomes of patients for mutations that cause or greatly increase the risk of their disease (discussed in Chapter 19). These mutations can then be introduced into animals, such as mice, that can be studied in the laboratory. The resulting transgenic animals, which often mimic some of the phenotypic abnormalities associated with the condition in patients, can be used to explore the cellular and molecular basis of the disease and to screen for drugs that could potentially be used therapeutically in humans.

An encouraging example is provided by *fragile X syndrome*, a neuropsychiatric disorder associated with intellectual impairment, neurological abnormalities, and often autism. The disease is caused by a mutation in the *fragile X mental retardation gene* (*FMR1*), which encodes a protein that inhibits the translation of mRNAs into proteins at synapses—the junctions where nerve cells communicate with one another (see Figure 12–38). Transgenic mice in which the *FMR1* gene has been disabled show many of the same neurological and behavioral abnormalities seen in patients with the disorder, and drugs that return synaptic protein synthesis to near-normal levels also reverse many of the problems seen in these mutant mice. Preliminary studies suggest that at least one of these drugs may benefit patients with the disease.

Transgenic Plants Are Important for Both Cell Biology and Agriculture

Although we tend to think of recombinant DNA research in terms of animal biology, these techniques have also had a profound impact on the study of plants. In fact, certain features of plants make them especially amenable to recombinant DNA methods.

When a piece of plant tissue is cultured in a sterile medium containing nutrients and appropriate growth regulators, some of the cells are stimulated to proliferate indefinitely in a disorganized manner, producing a mass of relatively undifferentiated cells called a *callus*. If the nutrients and growth regulators are carefully manipulated, one can induce the formation of a shoot within the callus, and in many species a whole new plant can be regenerated from such shoots. In a number of plants—including tobacco, petunia, carrot, potato, and *Arabidopsis*—a single cell from such a callus can be grown into a small clump of cells from which a whole plant can be regenerated (see Figure 8–2B). Just as mutant mice can be derived by the genetic manipulation of embryonic stem cells in culture, so transgenic plants can be created from plant cells transfected with DNA in culture (**Figure 10–37**).

The ability to produce transgenic plants has greatly accelerated progress in many areas of plant cell biology. It has played an important part, for example, in isolating receptors for growth regulators and in analyzing the mechanisms of morphogenesis and of gene expression in plants. These techniques have also opened up many new possibilities in agriculture that could benefit both the farmer and the consumer. They have made it possible, for example, to modify the ratio of lipid, starch, and protein in seeds, to impart pest and virus resistance to plants, and to create modified plants that tolerate extreme habitats such as salt marshes or water-stressed soil. One variety of rice has been genetically engineered to produce β-carotene, the precursor of vitamin A. If it replaced conventional rice, this "golden rice"—so called because of its faint yellow color—could help to alleviate severe vitamin A deficiency, which causes blindness in hundreds of thousands of children in the developing world each year.

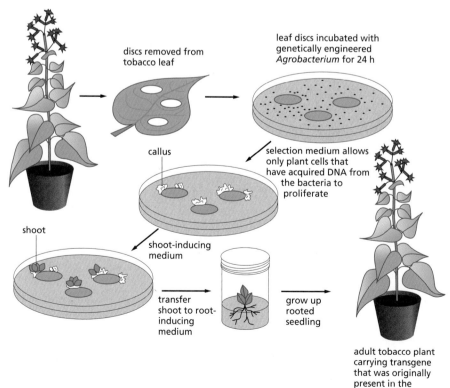

Figure 10–37 **Transgenic plants can be made using recombinant DNA techniques optimized for plants.** A disc is cut out of a leaf and incubated in a culture of *Agrobacterium* that carries a recombinant plasmid with both a selectable marker and a desired genetically engineered gene. The wounded plant cells at the edge of the disc release substances that attract the bacteria, which inject their DNA into the plant cells. Only those plant cells that take up the appropriate DNA and express the selectable marker gene survive and proliferate and form a callus. The manipulation of growth factors supplied to the callus induces it to form shoots, which subsequently root and grow into adult plants carrying the engineered gene.

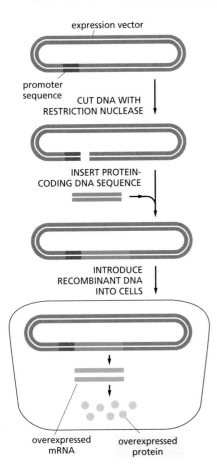

expression vector

promoter sequence

CUT DNA WITH RESTRICTION NUCLEASE

INSERT PROTEIN-CODING DNA SEQUENCE

INTRODUCE RECOMBINANT DNA INTO CELLS

overexpressed mRNA

overexpressed protein

Figure 10–38 Large amounts of a protein can be produced from a protein-coding DNA sequence inserted into an expression vector and introduced into cells. Here, a plasmid vector has been engineered to contain a highly active promoter, which causes unusually large amounts of mRNA to be produced from the inserted protein-coding gene. Depending on the characteristics of the cloning vector, the plasmid is introduced into bacterial, yeast, insect, or mammalian cells, where the inserted gene is efficiently transcribed and translated into protein.

Even Rare Proteins Can Be Made in Large Amounts Using Cloned DNA

One of the most important contributions of DNA cloning and genetic engineering to cell biology is that they make it possible to produce any protein, including the rare ones, in nearly unlimited amounts. Such high-level production is usually accomplished by using specially designed vectors known as *expression vectors*. These vectors include transcription and translation signals that direct an inserted gene to be expressed at very high levels. Different expression vectors are designed for use in bacterial, yeast, insect, or mammalian cells, each containing the appropriate regulatory sequences for transcription and translation in these cells (**Figure 10–38**). The expression vector is replicated at each round of cell division, so that the transfected cells in the culture are able to synthesize very large amounts of the protein of interest—often comprising 1–10% of the total cell protein. It is usually a simple matter to purify this protein away from the other proteins made by the host cell.

This technology is now used to make large amounts of many medically useful proteins, including hormones (such as insulin), growth factors, and viral coat proteins for use in vaccines. Expression vectors also allow scientists to produce many proteins of biological interest in large enough amounts for detailed structural and functional studies that were once impossible—especially for proteins that are normally present in very small amounts, such as some receptors and transcription regulators. Recombinant DNA techniques thus allow scientists to move with ease from protein to gene, and vice versa, so that the functions of both can be explored on multiple fronts (**Figure 10–39**).

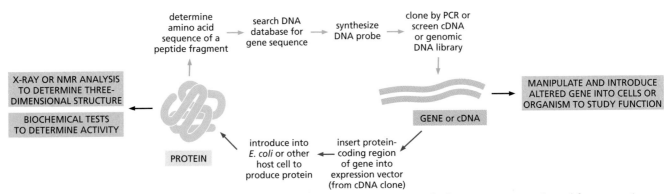

determine amino acid sequence of a peptide fragment

search DNA database for gene sequence

synthesize DNA probe

clone by PCR or screen cDNA or genomic DNA library

X-RAY OR NMR ANALYSIS TO DETERMINE THREE-DIMENSIONAL STRUCTURE

BIOCHEMICAL TESTS TO DETERMINE ACTIVITY

MANIPULATE AND INTRODUCE ALTERED GENE INTO CELLS OR ORGANISM TO STUDY FUNCTION

GENE or cDNA

PROTEIN

introduce into *E. coli* or other host cell to produce protein

insert protein-coding region of gene into expression vector (from cDNA clone)

Figure 10–39 Recombinant DNA techniques make it possible to move experimentally from gene to protein and from protein to gene. A small quantity of a purified protein or peptide fragment is used to obtain a partial amino acid sequence, which is used to search a DNA database for the corresponding nucleotide sequence. This sequence is used to synthesize a DNA probe, which can be used either to pick out the corresponding gene from a DNA library by DNA hybridization (see Figure 10–11) or to clone the gene by PCR from a sequenced genome (see Figure 10–16). Once the gene has been isolated and sequenced, its protein-coding sequence can be inserted into an expression vector to produce large quantities of the protein (see Figure 10–38), which can then be studied biochemically or structurally. In addition to producing protein, the gene or DNA can also be manipulated and introduced into cells or organisms to study its function. (NMR stands for nuclear magnetic resonance; see How We Know, pp. 162–163.)

ESSENTIAL CONCEPTS

- Recombinant DNA technology has revolutionized the study of cells, making it possible to pick out any gene at will from the thousands of genes in a cell and to determine its nucleotide sequence.

- A crucial element in this technology is the ability to cut a large DNA molecule into a specific and reproducible set of DNA fragments using restriction nucleases, each of which cuts the DNA double helix only at a particular nucleotide sequence.

- DNA fragments can be separated from one another on the basis of size by gel electrophoresis.

- Nucleic acid hybridization can detect any given DNA or RNA sequence in a mixture of nucleic acid fragments. This technique depends on highly specific base-pairing between a labeled, single-stranded DNA or RNA probe and another nucleic acid with a complementary sequence.

- DNA cloning techniques enable any DNA sequence to be selected from millions of other sequences and produced in unlimited amounts in pure form.

- DNA fragments can be joined together *in vitro* by using DNA ligase to form recombinant DNA molecules that are not found in nature.

- DNA fragments can be maintained and amplified by inserting them into a larger DNA molecule capable of replication, such as a plasmid. This recombinant DNA molecule is then introduced into a rapidly dividing host cell, usually a bacterium, so that the DNA is replicated at each cell division.

- A collection of cloned fragments of chromosomal DNA representing the complete genome of an organism is known as a genomic library. The library is often maintained as millions of clones of bacteria, each different clone carrying a different fragment of the organism's genome.

- cDNA libraries contain cloned DNA copies of the total mRNA of a particular type of cell or tissue. Unlike genomic DNA clones, cDNA clones contain predominantly protein-coding sequences; they lack introns, regulatory DNA sequences, and promoters. Thus they are useful when the cloned gene is needed to make a protein.

- The polymerase chain reaction (PCR) is a powerful form of DNA amplification that is carried out *in vitro* using a purified DNA polymerase. PCR requires prior knowledge of the sequence to be amplified, because two synthetic oligonucleotide primers must be synthesized that bracket the portion of DNA to be replicated.

- Historically, genes were cloned using hybridization techniques to identify the bacteria carrying the desired sequence in a DNA library. Today, a gene is usually cloned using PCR to specifically amplify it from a sample of DNA or mRNA.

- DNA sequencing techniques have become increasingly fast and cheap, so that the entire genomes of thousands of different organisms are now known, including thousands of individual humans.

- Using recombinant DNA techniques, a protein can be joined to a molecular tag, such as green fluorescent protein (GFP), which allows its movement to be tracked inside a cell and, in some cases, inside a living organism.

- *In situ* nucleic acid hybridization can be used to detect the precise location of genes on chromosomes and of RNAs in cells and tissues.

- DNA microarrays and RNA-Seq can be used to monitor the expression of tens of thousands of genes at once.

- Cloned genes can be altered *in vitro* and stably inserted into the genome of a cell or an organism to study their function. Such mutants are called transgenic organisms.

- The expression of particular genes can be inhibited in cells or organisms by the technique of RNA interference (RNAi), which prevents an mRNA from being translated into protein.

- Bacteria, yeasts, and mammalian cells can be engineered to synthesize large quantities of any protein whose gene has been cloned, making it possible to study proteins that are otherwise rare or difficult to isolate.

KEY TERMS

cDNA	hybridization
cDNA library	*in situ* hybridization
dideoxy (Sanger) DNA	plasmid
sequencing	polymerase chain reaction (PCR)
DNA cloning	recombinant DNA
DNA library	recombinant DNA technology
DNA ligase	reporter gene
DNA microarray	restriction nuclease
gene knockout	RNA interference (RNAi)
gene replacement	RNA-Seq
genomic DNA library	transformation
green fluorescent protein (GFP)	transgenic organism

QUESTIONS

QUESTION 10–5

What are the consequences for a DNA sequencing reaction if the ratio of dideoxyribonucleoside triphosphates to deoxyribonucleoside triphosphates is increased? What happens if this ratio is decreased?

QUESTION 10–6

Almost all the cells in an individual animal contain identical genomes. In an experiment, a tissue composed of several different cell types is fixed and subjected to *in situ* hybridization with a DNA probe to a particular gene. To your surprise, the hybridization signal is much stronger in some cells than in others. How might you explain this result?

QUESTION 10–7

After decades of work, Dr. Ricky M. isolated a small amount of attractase—an enzyme that produces a powerful human pheromone—from hair samples of Hollywood celebrities. To take advantage of attractase for his personal use, he obtained a complete genomic clone of the attractase gene, connected it to a strong bacterial promoter on an expression plasmid, and introduced the plasmid into *E. coli* cells. He was devastated to find that no attractase was produced in the cells. What is a likely explanation for his failure?

QUESTION 10–8

Which of the following statements are correct? Explain your answers.

A. Restriction nucleases cut DNA at specific sites that are always located between genes.

B. DNA migrates toward the positive electrode during electrophoresis.

C. Clones isolated from cDNA libraries contain promoter sequences.

D. PCR utilizes a heat-stable DNA polymerase because for each amplification step, double-stranded DNA must be heat-denatured.

E. Digestion of genomic DNA with AluI, a restriction enzyme that recognizes a four-nucleotide sequence, produces fragments that are all exactly 256 nucleotides in length.

F. To make a cDNA library, both a DNA polymerase and a reverse transcriptase must be used.

G. DNA fingerprinting by PCR relies on the fact that different individuals have different numbers of repeats in STR regions in their genome.

H. It is possible for a coding region of a gene to be present in a genomic library prepared from a particular tissue but to be absent from a cDNA library prepared from the same tissue.

QUESTION 10–9

A. What is the sequence of the DNA that was used in the sequencing reaction shown in **Figure Q10–9**? The four lanes show the products of sequencing reactions that contained ddG (lane 1), ddA (lane 2), ddT (lane 3), and ddC (lane 4). The numbers to the right of the autoradiograph represent the positions of marker DNA fragments of 50 and 116 nucleotides.

B. This DNA was derived from the middle of a cDNA clone of a mammalian protein. Using the genetic code table (see Figure 7–25), can you determine the amino acid sequence of this portion of the protein?

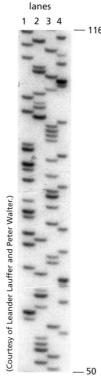

lanes

1 2 3 4 — 116

(Courtesy of Leander Lauffer and Peter Walter.)

— 50

Figure Q10–9

QUESTION 10–10

A. How many different DNA fragments would you expect to obtain if you cleaved human genomic DNA with HaeIII? (Recall that there are 3×10^9 nucleotide pairs per haploid genome.) How many fragments would you expect with EcoRI?

B. Human genomic libraries used for DNA sequencing are often made from fragments obtained by cleaving human DNA with HaeIII in such a way that the DNA is only partially digested; that is, not all the possible HaeIII sites have been cleaved. What is a possible reason for doing this?

QUESTION 10–11

A molecule of double-stranded DNA was cleaved with restriction nucleases, and the resulting products were separated by gel electrophoresis (**Figure Q10–11**). DNA fragments of known sizes were electrophoresed on the same gel for use as size markers (*left* lane). The size of the

DNA markers is given in kilobase pairs (kb), where 1 kb = 1000 nucleotide pairs. Using the size markers as a guide, estimate the length of each restriction fragment obtained. From this information, construct a map of the original DNA molecule indicating the relative positions of all the restriction enzyme cleavage sites.

QUESTION 10–12

You have isolated a small amount of a rare protein. You cleaved the protein into fragments using proteases, separated some of the fragments by chromatography, and determined their amino acid sequence. Unfortunately, as is often the case when only small amounts of protein are available, you obtained only three short stretches of amino acid sequence from the protein:

1. Trp-Met-His-His-Lys

2. Leu-Ser-Arg-Leu-Arg

3. Tyr-Phe-Gly-Met-Gln

A. Using the genetic code (see Figure 7–25), design a collection of DNA probes specific for each peptide that could be used to detect the gene in a cDNA library by hybridization. Which of the three collections of oligonucleotide probes would it be preferable to use first? Explain your answer. (Hint: the genetic code is redundant, so each peptide has multiple potential coding sequences.)

B. You have also been able to determine that the Gln of your peptide #3 is the C-terminal (i.e., the final) amino acid of your protein. How would you go about designing oligonucleotide primers that could be used to amplify a portion of the gene from a cDNA library using PCR?

C. Suppose the PCR amplification in (B) yields a DNA that is precisely 300 nucleotides long. Upon determining the nucleotide sequence of this DNA, you find the sequence CTATCACGCCTTAGG approximately in its middle. What would you conclude from these observations?

QUESTION 10–13

Assume that a DNA sequencing reaction is carried out as shown in Figure 10–20, except that the four different dideoxyribonucleoside triphosphates are modified so that each contains a covalently attached dye of a different color (which does not interfere with its incorporation into the DNA chain). What would the products be if you added a mixture of all four of these labeled dideoxyribonucleoside triphosphates along with the four unlabeled deoxyribonucleoside triphosphates into a single sequencing reaction? What would the results look like if you electrophoresed these products in a single lane of a gel?

QUESTION 10–14

Genomic DNA clones are often used to "walk" along a chromosome. In this approach, one cloned DNA is used to isolate other clones that contain overlapping DNA sequences (**Figure Q10–14**). Using this method, it is possible to build up a long stretch of DNA and thus identify new genes in near proximity to a previously cloned gene.

A. Would it be faster to use cDNA clones in this method, because they do not contain any intron sequences?

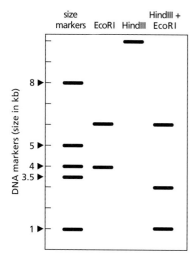

size
markers   EcoRI   HindIII   HindIII +
                              EcoRI

DNA markers (size in kb)

8 ▶

5 ▶

4 ▶
3.5 ▶

1 ▶

Figure Q10–11

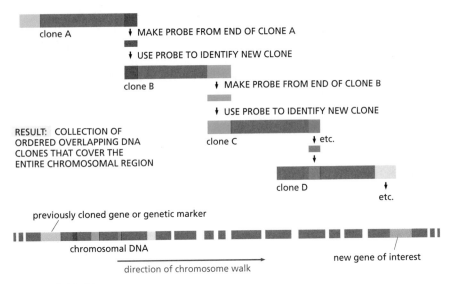

clone A

↓ MAKE PROBE FROM END OF CLONE A

↓ USE PROBE TO IDENTIFY NEW CLONE

clone B

↓ MAKE PROBE FROM END OF CLONE B

↓ USE PROBE TO IDENTIFY NEW CLONE

RESULT: COLLECTION OF
ORDERED OVERLAPPING DNA
CLONES THAT COVER THE
ENTIRE CHROMOSOMAL REGION

clone C   ↓ etc.

↓

clone D   ↓

etc.

previously cloned gene or genetic marker

chromosomal DNA

new gene of interest

→ direction of chromosome walk

Figure Q10–14

Figure Q10–16

(Courtesy of John Bedbrook and DNA Plant Technology Corporation.)

B. What would happen if you encountered a repetitive DNA sequence, like the *L1* transposon (see Figure 9–17), which is found in many copies and in many different places in the genome?

QUESTION 10–15

There has been a colossal snafu in the maternity ward of your local hospital. Four sets of male twins, born within an hour of each other, were inadvertently shuffled in the excitement occasioned by that unlikely event. You have been called in to set things straight. As a first step, you would like to match each baby with his twin. (Many newborns look alike so you don't want to rely on appearance alone.) To that end you analyze a small blood sample from each infant using a hybridization probe that detects short tandem repeats (STRs) located in widely scattered regions of the genome. The results are shown in Figure Q10–15.

A. Which infants are twins? Which are identical twins?

B. How could you match a pair of twins to the correct parents?

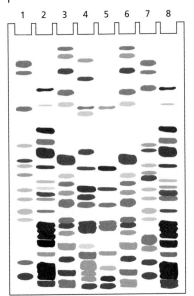

Figure Q10–15

QUESTION 10–16

One of the first organisms that was genetically modified using recombinant DNA technology was a bacterium that normally lives on the surface of strawberry plants. This bacterium makes a protein, called ice-protein, that causes the efficient formation of ice crystals around it when the temperature drops to just below freezing. Thus, strawberries harboring this bacterium are particularly susceptible to frost damage because their cells are destroyed by the ice crystals. Consequently, strawberry farmers have a considerable interest in preventing ice crystallization.

A genetically engineered version of this bacterium was constructed in which the ice-protein gene was knocked out. The mutant bacteria were then introduced in large numbers into strawberry fields, where they displaced the normal bacteria by competition for their ecological niche. This approach has been successful: strawberries bearing the mutant bacteria show a much reduced susceptibility to frost damage.

At the time they were first carried out, the initial open-field trials triggered an intense debate because they represented the first release into the environment of an organism that had been genetically engineered using recombinant DNA technology. Indeed, all preliminary experiments were carried out with extreme caution and in strict containment (Figure Q10–16).

Do you think that bacteria lacking the ice-protein could be isolated without the use of modern DNA technology? Is it likely that such mutations have already occurred in nature? Would the use of a mutant bacterial strain isolated from nature be of lesser concern? Should we be concerned about the risks posed by the application of recombinant DNA techniques in agriculture and medicine? Explain your answers.

CHAPTER **ELEVEN**

11

Membrane Structure

A living cell is a self-reproducing system of molecules held inside a container. That container is the **plasma membrane**—a protein-studded, fatty film so thin that it cannot be seen directly in the light microscope. Every cell on Earth uses such a membrane to separate and protect its chemical components from the outside environment. Without membranes, there would be no cells, and thus no life.

The structure of the plasma membrane is simple: it consists of a two-ply sheet of lipid molecules about 5 nm—or 50 atoms—thick, into which proteins have been inserted. Its properties, however, are unlike those of any sheet of material we are familiar with in the everyday world. Although it serves as a barrier to prevent the contents of the cell from escaping and mixing with the surrounding medium (**Figure 11–1**), the plasma membrane does much more than that. If a cell is to survive and grow, nutrients must pass inward across the plasma membrane, and waste products must pass out. To facilitate this exchange, the membrane is penetrated by highly selective channels and transporters—proteins that allow specific,

THE LIPID BILAYER

MEMBRANE PROTEINS

Figure 11–1 Cell membranes act as selective barriers. The plasma membrane separates a cell from its surroundings, enabling the molecular composition of a cell to differ from that of its environment. (A) In some bacteria, the plasma membrane is the only membrane. (B) Eukaryotic cells also have internal membranes that enclose individual organelles. All cell membranes prevent molecules on one side from freely mixing with those on the other, as schematically indicated by the colored dots.

(A) BACTERIAL CELL

(B) EUKARYOTIC CELL

Figure 11–2 The plasma membrane is involved in cell communication, import and export of molecules, and cell growth and motility. (1) Receptor proteins in the plasma membrane enable the cell to receive signals from the environment; (2) transport proteins in the membrane enable the import and export of small molecules; (3) the flexibility of the membrane and its capacity for expansion allow the cell to grow, change shape, and move.

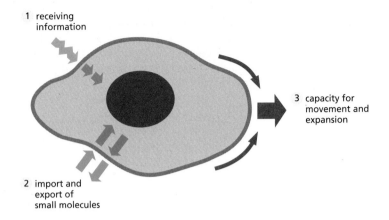

1 receiving
information

2 import and
export of
small molecules

3 capacity for
movement and
expansion

small molecules and ions to be imported and exported. Other proteins in the membrane act as sensors, or receptors, that enable the cell to receive information about changes in its environment and respond to them in appropriate ways. The mechanical properties of the plasma membrane are equally remarkable. When a cell grows or changes shape, so does its membrane: it enlarges in area by adding new membrane without ever losing its continuity, and it can deform without tearing (**Figure 11–2**). If the membrane is pierced, it neither collapses like a balloon nor remains torn; instead, it quickly reseals.

As shown in Figure 11–1, the simplest bacteria have only a single membrane—the plasma membrane—whereas eukaryotic cells also contain internal membranes that enclose intracellular compartments. The internal membranes form various organelles, including the endoplasmic reticulum, Golgi apparatus, and mitochondria (**Figure 11–3**). Although these internal membranes are constructed on the same principles as the plasma membrane, there are subtle differences in their composition, especially in their resident proteins.

Regardless of their location, all cell membranes are composed of lipids and proteins and share a common general structure (**Figure 11–4**). The lipids are arranged in two closely apposed sheets, forming a *lipid bilayer* (see Figure 11–4B and C). This lipid bilayer serves as a permeability barrier to most water-soluble molecules. The proteins carry out the other functions of the membrane and give different membranes their individual characteristics.

In this chapter, we consider the structure of biological membranes and the organization of their two main constituents: lipids and proteins. Although we focus mainly on the plasma membrane, most of the concepts we discuss also apply to internal membranes. The functions of cell membranes, including their role in cell communication, the transport of small molecules, and energy generation, are considered in later chapters.

THE LIPID BILAYER

Because cells are filled with—and surrounded by—water, the structure of cell membranes is determined by the way membrane lipids behave in a watery (aqueous) environment. In this section, we take a closer look at the **lipid bilayer**, which constitutes the fundamental structure of all cell membranes. We consider how lipid bilayers form, how they are maintained, and how their properties establish the general properties of all cell membranes.

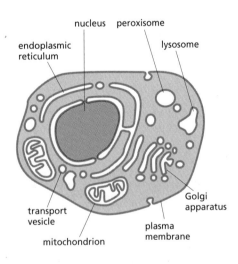

endoplasmic
reticulum

nucleus   peroxisome

lysosome

transport
vesicle

mitochondrion

plasma
membrane

Golgi
apparatus

Figure 11–3 Internal membranes form many different compartments in a eukaryotic cell. Some of the main membrane-enclosed organelles in a typical animal cell are shown here. Note that the nucleus and mitochondria are each enclosed by two membranes.

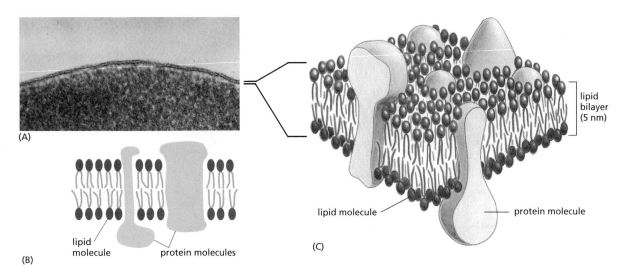

(A)

(B)

(C)

lipid
bilayer
(5 nm)

lipid molecule

protein molecule

lipid
molecule

protein molecules

Figure 11–4 A cell membrane can be viewed in a number of ways. (A) An electron micrograph of a plasma membrane of a human red blood cell seen in cross section. (B and C) Schematic drawings showing two-dimensional and three-dimensional views of a cell membrane. (A, courtesy of Daniel S. Friend.)

Membrane Lipids Form Bilayers in Water

The lipids in cell membranes combine two very different properties in a single molecule: each lipid has a hydrophilic ("water-loving") head and a hydrophobic ("water-fearing") tail. The most abundant lipids in cell membranes are the **phospholipids**, which have a phosphate-containing, hydrophilic head linked to a pair of hydrophobic tails (**Figure 11–5**). **Phosphatidylcholine**, for example, has the small molecule choline attached to a phosphate group as its hydrophilic head (**Figure 11–6**).

Molecules with both hydrophilic and hydrophobic parts are termed **amphipathic**, a property shared by other types of membrane lipids, including the cholesterol, which is found in animal cell membranes and the glycolipids, which have sugars as part of their hydrophilic head (**Figure 11–7**). Having both hydrophilic and hydrophobic parts plays a crucial part in driving these lipid molecules to assemble into bilayers in an aqueous environment.

As discussed in Chapter 2 (see Panel 2–2, pp. 68–69), hydrophilic molecules dissolve readily in water because they contain either charged groups or uncharged polar groups that can form either electrostatic attractions or hydrogen bonds with water molecules (**Figure 11–8**). Hydrophobic molecules, by contrast, are insoluble in water because all—or almost all—of their atoms are uncharged and nonpolar; they therefore cannot form favorable interactions with water molecules. Instead, they force adjacent water molecules to reorganize into a cagelike structure around them (**Figure 11–9**). Because this cagelike structure is more highly ordered than the rest of the water, its formation requires free energy. This energy cost is minimized when the hydrophobic molecules cluster together, limiting their contacts with the surrounding water molecules. Thus purely hydrophobic molecules, like the fats found in animal fat cells and the oils found in plant seeds (**Figure 11–10A**), coalesce into a single large drop when dispersed in water.

Amphipathic molecules, such as phospholipids (**Figure 11–10B**), are subject to two conflicting forces: the hydrophilic head is attracted to water, while the hydrophobic tails shun water and seek to aggregate with other hydrophobic molecules. This conflict is beautifully resolved by the

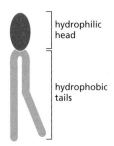

hydrophilic
head

hydrophobic
tails

Figure 11–5 A typical membrane lipid molecule has a hydrophilic head and two hydrophobic tails.

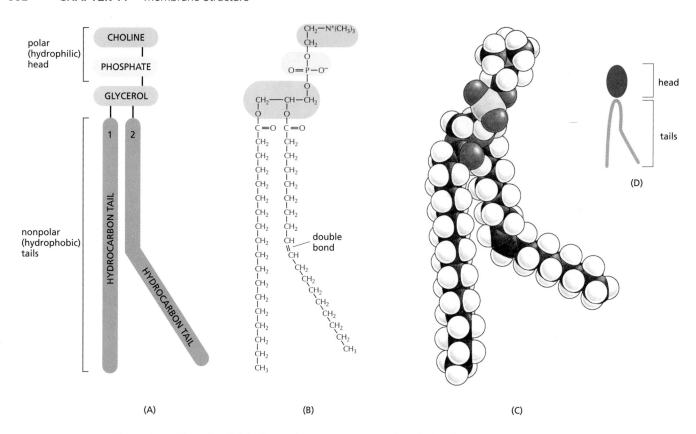

Figure 11–6 Phosphatidylcholine is the most common phospholipid in cell membranes. It is represented schematically in (A), as a chemical formula in (B), as a space-filling model in (C), and as a symbol in (D). This particular phospholipid is built from five parts: the hydrophilic head, which consists of *choline* linked to a *phosphate group*; two *hydrocarbon chains*, which form the hydrophobic tails; and a molecule of glycerol, which links the head to the tails. Each of the hydrophobic tails is a *fatty acid*—a hydrocarbon chain with a –COOH group at one end—which has been attached to glycerol via this group. A kink in one of the hydrocarbon chains occurs where there is a double bond between two carbon atoms. (The "phosphatidyl" part of the name of a phospholipid refers to the phosphate–glycerol–fatty acid portion of the molecule.)

Figure 11–7 Different types of membrane lipids are all amphipathic. Each of the three types shown here has a hydrophilic head and one or two hydrophobic tails. The hydrophilic head (shaded *blue* and *yellow*) is serine phosphate in phosphatidylserine, an –OH group in cholesterol, and a sugar (galactose) plus an –OH group in galactocerebroside. See also Panel 2–4, pp. 72–73.

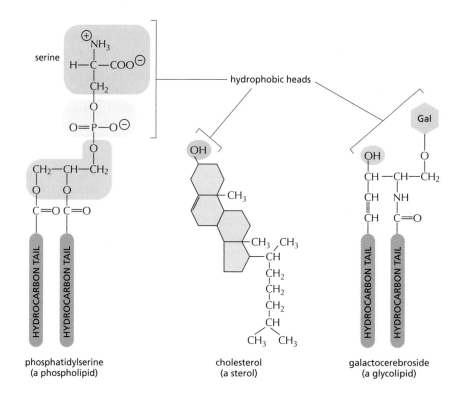

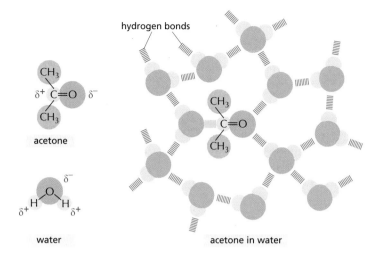

Figure 11–8 A hydrophilic molecule attracts water molecules. Both acetone and water are polar molecules: thus acetone readily dissolves in water. Polar atoms are shown in *red* and *blue*, with δ^- indicating a partial negative charge, and δ^+ indicating a partial positive charge. Hydrogen bonds (*red*) and an electrostatic attraction (*yellow*) form between acetone and the surrounding water molecules. Nonpolar groups are shown in *gray*.

formation of a lipid bilayer—an arrangement that satisfies all parties and is energetically most favorable. The hydrophilic heads face water on both surfaces of the bilayer; but the hydrophobic tails are all shielded from the water, as they lie next to one another in the interior, like the filling in a sandwich (**Figure 11–11**).

The same forces that drive the amphipathic molecules to form a bilayer help to make the bilayer self-sealing. Any tear in the sheet will create a free edge that is exposed to water. Because this situation is energetically unfavorable, the molecules of the bilayer will spontaneously rearrange to eliminate the free edge. If the tear is small, this spontaneous re-arrangement will exclude the water molecules and lead to repair of the bilayer, restoring a single continuous sheet. If the tear is large, the sheet may begin to fold in on itself and break up into separate closed vesicles. In either case, the overriding principle is that free edges are quickly eliminated.

The prohibition on free edges has a profound consequence: the only way a finite amphipathic sheet can avoid having free edges is to bend and seal,

QUESTION 11–1

Water molecules are said "to reorganize into a cagelike structure" around hydrophobic compounds (e.g., see Figure 11–9). This seems paradoxical because water molecules do not interact with the hydrophobic compound. So how could they "know" about its presence and change their behavior to interact differently with one another? Discuss this argument and, in doing so, develop a clear concept of what is meant by a "cagelike" structure. How does it compare to ice? Why would this cagelike structure be energetically unfavorable?

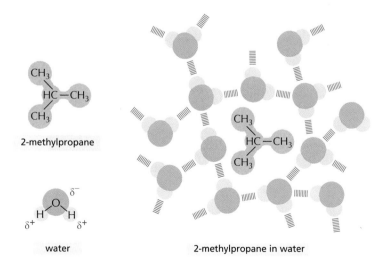

Figure 11–9 A hydrophobic molecule tends to avoid water. Because the 2-methylpropane molecule is entirely hydrophobic, it cannot form favorable interactions with water. This causes the adjacent water molecules to reorganize into a cagelike structure around it, in order to maximize their hydrogen bonds with each other.

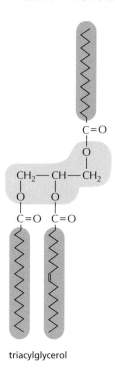

triacylglycerol

Figure 11–10 Fat molecules are hydrophobic, unlike phospholipids. Triacylglycerols, which are the main constituents of animal fats and plant oils, are entirely hydrophobic. Here, the third hydrophobic tail of the triacylglycerol molecule in (A) is drawn facing upward for comparison with the phospholipid (see Figure 11–6), although normally it is depicted facing down (see Panel 2–4, pp. 72–73).

forming a boundary around a closed space (**Figure 11–12**). Therefore, amphipathic molecules such as phospholipids necessarily assemble into self-sealing containers that define closed compartments. This remarkable behavior, fundamental to the creation of a living cell, is simply a result of the property that each molecule is hydrophilic at one end and hydrophobic at the other.

The Lipid Bilayer Is a Flexible Two-dimensional Fluid

The aqueous environment inside and outside a cell prevents membrane lipids from escaping from the bilayer, but nothing stops these molecules from moving about and changing places with one another within the plane of the bilayer. The membrane therefore behaves as a two-dimensional fluid, a fact that is crucial for membrane function and integrity (Movie 11.1).

The lipid bilayer is also flexible—that is, it is able to bend. Like fluidity, flexibility is important for membrane function, and it sets a lower limit of about 25 nm to the size of vesicle that cell membranes can form.

The fluidity of lipid bilayers can be studied using synthetic lipid bilayers, which are easily produced by the spontaneous aggregation of amphipathic lipid molecules in water. Pure phospholipids, for example, will form closed spherical vesicles, called liposomes, when added to water; they vary in size from about 25 nm to 1 mm in diameter (**Figure 11–13**).

Such simple synthetic bilayers allow the movements of the lipid molecules to be measured. These measurements reveal that some types of movement are rare, while others are frequent and rapid. Thus, in synthetic lipid bilayers, phospholipid molecules very rarely tumble from one half of the bilayer, or monolayer, to the other. Without proteins to facilitate the process, it is estimated that this event, called "flip-flop," occurs less than once a month for any individual lipid molecule under conditions

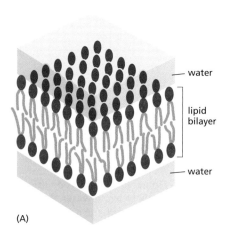

(A)

water

lipid bilayer

water

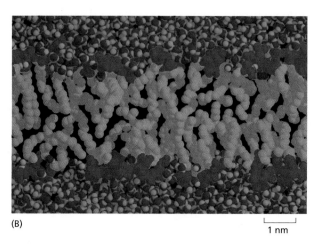

(B)

1 nm

Figure 11–11 Amphipathic phospholipids form a bilayer in water. (A) Schematic drawing of a phospholipid bilayer in water. (B) Computer simulation showing the phospholipid molecules (*red* heads and *orange* tails) and the surrounding water molecules (*blue*) in a cross section of a lipid bilayer. (B, adapted from *Science* 262:223–228, 1993, with permission from the AAAS; courtesy of R. Venable and R. Pastor.)

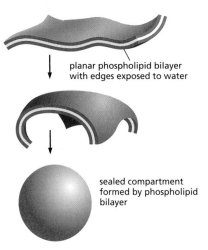

planar phospholipid bilayer
with edges exposed to water

sealed compartment
formed by phospholipid
bilayer

Figure 11–12 Phospholipid bilayers spontaneously close in on themselves to form sealed compartments. The closed structure is stable because it avoids the exposure of the hydrophobic hydrocarbon tails to water, which would be energetically unfavorable.

similar to those in a cell. On the other hand, as the result of random thermal motions, lipid molecules continuously exchange places with their neighbors in the same monolayer. This exchange leads to rapid lateral diffusion of lipid molecules within the plane of each monolayer, so that, for example, a lipid in an artificial bilayer may diffuse a length equal to that of an entire bacterial cell (~2 µm) in about one second.

Similar studies show that individual lipid molecules not only flex their hydrocarbon tails, but they also rotate rapidly about their long axis—some reaching speeds of 500 revolutions per second. Studies of whole cells—and isolated cell membranes—indicate that lipid molecules in cell membranes undergo the same movements as they do in synthetic bilayers. The movements of membrane phospholipid molecules are summarized in **Figure 11–14**.

The Fluidity of a Lipid Bilayer Depends on Its Composition

The fluidity of a cell membrane—the ease with which its lipid molecules move within the plane of the bilayer—is important for membrane function and has to be maintained within certain limits. Just how fluid a lipid bilayer is at a given temperature depends on its phospholipid composition and, in particular, on the nature of the hydrocarbon tails: the closer and more regular the packing of the tails, the more viscous and less fluid the bilayer will be. Two major properties of hydrocarbon tails affect how tightly they pack in the bilayer: their length and the number of double bonds they contain.

A shorter chain length reduces the tendency of the hydrocarbon tails to interact with one another and therefore increases the fluidity of the bilayer. The hydrocarbon tails of membrane phospholipids vary in length between 14 and 24 carbon atoms, with 18–20 atoms being most usual. Most phospholipids contain one hydrocarbon tail that has one or more double bonds between adjacent carbon atoms, and a second tail with single bonds only (see Figure 11–6). The chain that harbors a double bond does not contain the maximum number of hydrogen atoms that could, in principle, be attached to its carbon backbone; it is thus said to be **unsaturated** with respect to hydrogen. The hydrocarbon tail with no double bonds has a full complement of hydrogen atoms and is said to be **saturated**. Each double bond in an unsaturated tail creates a small kink in the tail (see Figure 11–6), which makes it more difficult for the tails to pack against one another. For this reason, lipid bilayers that contain a large proportion of unsaturated hydrocarbon tails are more fluid than those with lower proportions.

In bacterial and yeast cells, which have to adapt to varying temperatures, both the lengths and the unsaturation of the hydrocarbon tails in the bilayer are constantly adjusted to maintain the membrane at a relatively constant fluidity: at higher temperatures, for example, the cell makes

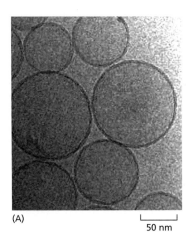

(A) | 50 nm

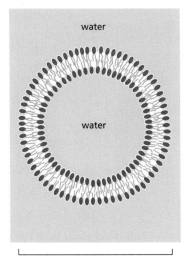

water

water

| 25 nm

(B)

Figure 11–13 Pure phospholipids can form closed, spherical liposomes. (A) An electron micrograph of phospholipid vesicles (liposomes) showing the bilayer structure of the membrane. (B) A drawing of a small, spherical liposome seen in cross section. (A, courtesy of Jean Lepault.)

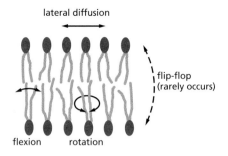

Figure 11–14 Membrane phospholipids are motile. The drawing shows the types of movement that phospholipid molecules undergo in a lipid bilayer. Because of these movements, the bilayer behaves as a two-dimensional fluid, in which the individual lipid molecules are able to move in their own monolayer. Note that lipid molecules do not move spontaneously from one monolayer to the other.

QUESTION 11–2

Five students in your class always sit together in the front row. This could be because (A) they really like each other or (B) nobody else in your class wants to sit next to them. Which explanation holds for the assembly of a lipid bilayer? Explain. Suppose, instead, that the other explanation held for lipid molecules. How would the properties of the lipid bilayer be different?

---

membrane lipids with tails that are longer and that contain fewer double bonds. A similar trick is used in the manufacture of margarine from vegetable oils. The fats produced by plants are generally unsaturated and therefore liquid at room temperature, unlike animal fats such as butter or lard, which are generally saturated and therefore solid at room temperature. Margarine is made of hydrogenated vegetable oils; their double bonds have been removed by the addition of hydrogen, so that they are more solid and butterlike at room temperature.

In animal cells, membrane fluidity is modulated by the inclusion of the sterol **cholesterol**. This molecule is present in especially large amounts in the plasma membrane, where it constitutes approximately 20% of the lipids in the membrane by weight. Because cholesterol molecules are short and rigid, they fill the spaces between neighboring phospholipid molecules left by the kinks in their unsaturated hydrocarbon tails (**Figure 11–15**). In this way, cholesterol tends to stiffen the bilayer, making it less flexible, as well as less permeable. The chemical properties of membrane lipids—and how they affect membrane fluidity—are reviewed in Movie 11.2.

For all cells, membrane fluidity is important for many reasons. It enables many membrane proteins to diffuse rapidly in the plane of the bilayer and to interact with one another, as is crucial, for example, in cell signaling (discussed in Chapter 16). It permits membrane lipids and proteins to diffuse from sites where they are inserted into the bilayer after their synthesis to other regions of the cell. It ensures that membrane molecules are distributed evenly between daughter cells when a cell divides. And, under appropriate conditions, it allows membranes to fuse with one another and mix their molecules (discussed in Chapter 15). If biological membranes were not fluid, it is hard to imagine how cells could live, grow, and reproduce.

Membrane Assembly Begins in the ER

In eukaryotic cells, new phospholipids are manufactured by enzymes bound to the cytosolic surface of the *endoplasmic reticulum* (*ER*; see Figure 11–3). Using free fatty acids as substrates (see Panel 2–4, pp. 72–73), the enzymes deposit the newly made phospholipids exclusively in the cytosolic half of the bilayer.

Despite this preferential treatment, cell membranes manage to grow evenly. So how do new phospholipids make it to the opposite monolayer?

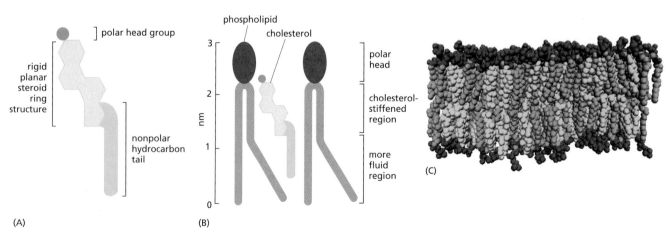

Figure 11–15 Cholesterol tends to stiffen cell membranes. (A) The shape of a cholesterol molecule. (B) How cholesterol fits into the gaps between phospholipid molecules in a lipid bilayer. (C) Space-filling model of the bilayer, with cholesterol molecules in *green*. The chemical formula of cholesterol is shown in Figure 11–7. (C, from H.L. Scott, *Curr. Opin. Struct. Biol.* 12: 499, 2002.)

As we saw in Figure 11–14, the transfer of lipids from one monolayer to the other rarely occur spontaneously. Instead, they are catalyzed by enzymes called *scramblases*, which remove randomly selected phospholipids from one half of the lipid bilayer and insert them in the other. As a result of this scrambling, newly made phospholipids are redistributed equally between each monolayer of the ER membrane (**Figure 11–16A**).

Some of this newly assembled membrane will remain in the ER; the rest will be used to supply fresh membrane to other compartments in the cell. Bits of membrane are continually pinching off the ER to form small, spherical vesicles that then fuse with other membranes, such as those of the Golgi apparatus. Additional vesicles bubble from the Golgi to become incorporated into the plasma membrane. We discuss this dynamic process of membrane transport in detail in Chapter 15.

Certain Phospholipids Are Confined to One Side of the Membrane

Most cell membranes are asymmetrical: the two halves of the bilayer often include strikingly different sets of phospholipids. But if membranes emerge from the ER with an evenly scrambled set of phospholipids, where does this asymmetry arise? It begins in the Golgi apparatus. The Golgi membrane contains another family of phospholipid-handling enzyme, called *flippases*. These enzymes remove specific phospholipids from the side of the bilayer facing the exterior space and flip them into the monolayer that faces the cytosol (**Figure 11–16B**).

The action of these flippases—and similar enzymes in the plasma membrane—initiates and maintains the asymmetric arrangement of phospholipids that is characteristic of the membranes of animal cells. This asymmetry is preserved as membranes bud from one organelle and fuse with another—or with the plasma membrane. This means that all

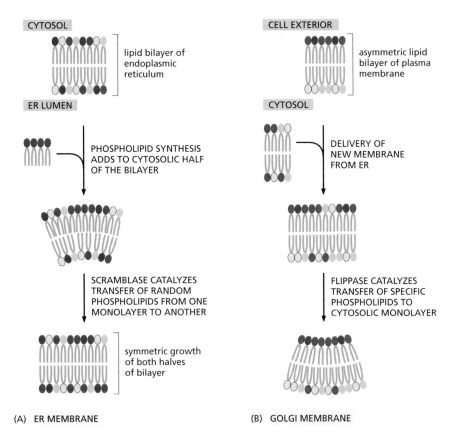

(A) ER MEMBRANE

(B) GOLGI MEMBRANE

Figure 11–16 Newly synthesized phospholipids are added to the cytosolic side of the ER membrane and then redistributed by enzymes that transfer them from one half of the lipid bilayer to the other. (A) Biosynthetic enzymes bound to the cytosolic monolayer of the ER membrane (not shown) produce new phospholipids from free fatty acids and insert them into the cytosolic monlayer. Enzymes called scramblases then randomly transfer phospholipid molecules from one monolayer to the other, allowing the membrane to grow as a bilayer. (B) When membranes leave the ER and are incorporated in the Golgi, they encounter enzymes called flippases, which selectively remove phosphatidylserine (*light green*) and phosphatidylethanolamine (*yellow*) from the noncytosolic monolayer and flip them to the cytosolic side. This transfer leaves phosphatidylcholine (*red*) and sphingomyelin (*brown*) concentrated in the noncytosolic monolayer. The resulting curvature of the membrane may actually help drive subsequent vesicle budding.

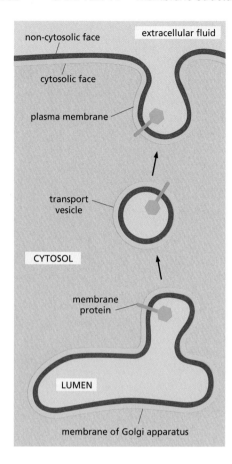

non-cytosolic face

extracellular fluid

cytosolic face

plasma membrane

transport vesicle

CYTOSOL

membrane protein

LUMEN

membrane of Golgi apparatus

Figure 11–17 Membranes retain their orientation during transfer between cell compartments. Membranes are transported by a process of vesicle budding and fusing. Here, a vesicle is shown budding from the Golgi apparatus and fusing with the plasma membrane. Note that the orientations of both the membrane lipids and proteins are preserved during the process: the original cytosolic surface of the lipid bilayer (*green*) remains facing the cytosol, and the noncytosolic surface (*red*) continues to face away from the cytosol, toward the lumen of the Golgi or transport vesicle—or toward the extracellular compartment. Similarly, the glycoprotein shown here remains in the same orientation, with its attached sugar facing the noncytosolic side.

cell membranes have distinct "inside" and "outside" faces: the cytosolic monolayer always faces the cytosol, while the noncytosolic monolayer is exposed to either the cell exterior—in the case of the plasma membrane—or to the interior space (*lumen*) of an organelle. This conservation of orientation applies not only to the phospholipids that make up the membrane, but to any proteins that might be inserted in the membrane **(Figure 11–17)**. For membrane proteins, this positioning is very important, as their orientation within the lipid bilayer is often crucial for their function (see Figure 11–19).

Among lipids, those that show the most dramatically lopsided distribution in cell membranes are the glycolipids, which are located mainly in the plasma membrane, and only in the noncytosolic half of the bilayer **(Figure 11–18)**. Their sugar groups face the cell exterior, where they form part of a continuous coat of carbohydrate that surrounds and protects animal cells. Glycolipid molecules acquire their sugar groups in the Golgi apparatus, where the enzymes that engineer this chemical modification are confined. These enzymes are oriented such that sugars are added only to lipid molecules in the noncytosolic half of the bilayer. Once a glycolipid molecule has been created in this way, it remains trapped in this monolayer, as there are no flippases that transfer glycolipids to the cytosolic side. Thus, when a glycolipid molecule is finally delivered to the plasma membrane, it displays its sugars to the exterior of the cell.

Other lipid molecules show different types of asymmetric distributions, which relate to their specific functions. For example, the inositol phospholipids—a minor component of the plasma membrane—have a special

QUESTION 11–3

It seems paradoxical that a lipid bilayer can be fluid yet asymmetrical. Explain.

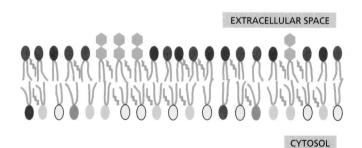

EXTRACELLULAR SPACE

CYTOSOL

Figure 11–18 Phospholipids and glycolipids are distributed asymmetrically in the lipid bilayer of a eukaryotic plasma membrane. Phosphatidylcholine (*red*) and sphingomyelin (*brown*) are concentrated in the noncytosolic monolayer, whereas phosphatidylserine (*light green*), and phosphatidylethanolamine (*yellow*) are found mainly on the cytosolic side. In addition to these phospholipids, phosphatidylinositols (*dark green*), a minor constituent of the plasma membrane, are shown in the cytosolic monolayer, where they participate in cell signaling. Glycolipids are drawn with hexagonal *blue* head groups to represent sugars; these are found exclusively in the noncytosolic monolayer of the membrane. Within the bilayer, cholesterol (*green*) is distributed almost equally in both monolayers.

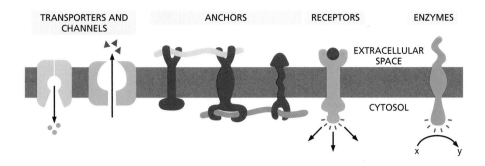

Figure 11–19 **Plasma membrane proteins have a variety of functions.**

role in relaying signals from the cell surface to the cell interior (discussed in Chapter 16); thus they are concentrated in the cytosolic half of the lipid bilayer.

MEMBRANE PROTEINS

Although the lipid bilayer provides the basic structure of all cell membranes and serves as a permeability barrier to the hydrophilic molecules on either side of it, most membrane functions are carried out by **membrane proteins**. In animals, proteins constitute about 50% of the mass of most plasma membranes, the remainder being lipid plus the relatively small amounts of carbohydrate found on some of the lipids (glycolipids) and many of the proteins (glycoproteins). Because lipid molecules are much smaller than proteins, however, a cell membrane typically contains about 50 times more lipid molecules than protein molecules (see Figure 11–4C).

Membrane proteins serve many functions. Some transport particular nutrients, metabolites, and ions across the lipid bilayer. Others anchor the membrane to macromolecules on either side. Still others function as receptors that detect chemical signals in the cell's environment and relay them into the cell interior, or work as enzymes to catalyze specific reactions at the membrane (**Figure 11–19** and **Table 11–1**). Each type of cell membrane contains a different set of proteins, reflecting the specialized functions of the particular membrane. In this section, we discuss the structure of membrane proteins and how they associate with the lipid bilayer.

TABLE 11–1 SOME EXAMPLES OF PLASMA MEMBRANE PROTEINS AND THEIR FUNCTIONS		
Functional Class	Protein Example	Specific Function
Transporters	Na$^+$ pump	actively pumps Na$^+$ out of cells and K$^+$ in (discussed in Chapter 12)
Ion channels	K$^+$ leak channel	allows K$^+$ ions to leave cells, thereby having a major influence on cell excitability (discussed in Chapter 12)
Anchors	integrins	link intracellular actin filaments to extracellular matrix proteins (discussed in Chapter 20)
Receptors	platelet-derived growth factor (PDGF) receptor	binds extracellular PDGF and, as a consequence, generates intracellular signals that cause the cell to grow and divide (discussed in Chapters 16 and 18)
Enzymes	adenylyl cyclase	catalyzes the production of the small intracellular signaling molecule cyclic AMP in response to extracellular signals (discussed in Chapter 16)

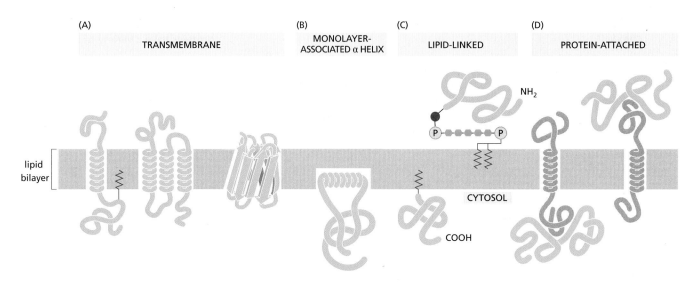

Figure 11–20 Membrane proteins can associate with the lipid bilayer in different ways. (A) Transmembrane proteins extend across the bilayer as a single α helix, as multiple α helices, or as a rolled-up β sheet (called a β barrel). (B) Some membrane proteins are anchored to the cytosolic half of the lipid bilayer by an amphipathic α helix. (C) Others are linked to either side of the bilayer solely by a covalently attached lipid molecule (*red* zigzag lines). (D) Many proteins are attached to the membrane only by relatively weak, noncovalent interactions with other membrane proteins. All except (D) are *integral membrane proteins.*

Membrane Proteins Associate with the Lipid Bilayer in Different Ways

Proteins can be associated with the lipid bilayer of a cell membrane in any one of the ways illustrated in **Figure 11–20**.

1. Many membrane proteins extend through the bilayer, with part of their mass on either side (Figure 11–20A). Like their lipid neighbors, these *transmembrane proteins* are amphipathic, having both hydrophobic and hydrophilic regions. Their hydrophobic regions lie in the interior of the bilayer, nestled against the hydrophobic tails of the lipid molecules. Their hydrophilic regions are exposed to the aqueous environment on either side of the membrane.

2. Other membrane proteins are located almost entirely in the cytosol and are associated with the cytosolic half of the lipid bilayer by an amphipathic α helix exposed on the surface of the protein (Figure 11–20B).

3. Some proteins lie entirely outside the bilayer, on one side or the other, attached to the membrane only by one or more covalently attached lipid groups (Figure 11–20C).

4. Yet other proteins are bound indirectly to one or the other face of the membrane, held in place only by their interactions with other membrane proteins (Figure 11–20D).

Proteins that are directly attached to the lipid bilayer—whether they are transmembrane, associated with the lipid monolayer, or lipid-linked—can be removed only by disrupting the bilayer with detergents, as discussed shortly. Such proteins are known as *integral membrane proteins.* The remaining membrane proteins are known as *peripheral membrane proteins;* they can be released from the membrane by more gentle extraction procedures that interfere with protein–protein interactions but leave the lipid bilayer intact.

A Polypeptide Chain Usually Crosses the Lipid Bilayer as an α Helix

All membrane proteins have a unique orientation in the lipid bilayer, which is essential for their function. For a transmembrane receptor protein, for example, the part of the protein that receives a signal from the environment must be on the outside of the cell, whereas the part that passes along the signal must be in the cytosol (see Figure 11–19). This orientation is a consequence of the way in which membrane proteins are synthesized (discussed in Chapter 15). The portions of a transmembrane protein located on either side of the lipid bilayer are connected by specialized membrane-spanning segments of the polypeptide chain (see Figure 11–20A). These segments, which run through the hydrophobic environment of the interior of the lipid bilayer, are composed largely of amino acids with hydrophobic side chains. Because these side chains cannot form favorable interactions with water molecules, they prefer to interact with the hydrophobic tails of the lipid molecules, where no water is present.

In contrast to the hydrophobic side chains, however, the peptide bonds that join the successive amino acids in a protein are normally polar, making the polypeptide backbone hydrophilic (**Figure 11–21**). Because water is absent from the interior of the bilayer, atoms forming the backbone are driven to form hydrogen bonds with one another. Hydrogen-bonding is maximized if the polypeptide chain forms a regular α helix, and so the great majority of the membrane-spanning segments of polypeptide chains traverse the bilayer as α helices (see Figure 4–13). In these membrane-spanning α helices, the hydrophobic side chains are exposed on the outside of the helix, where they contact the hydrophobic lipid tails, while atoms in the polypeptide backbone form hydrogen bonds with one another on the inside of the helix (**Figure 11–22**).

In many transmembrane proteins, the polypeptide chain crosses the membrane only once (see Figure 11–20A). Many of these *single-pass* transmembrane proteins are receptors for extracellular signals. Other transmembrane proteins function as channels, forming aqueous pores across the lipid bilayer to allow small, water-soluble molecules to cross the membrane. Such channels cannot be formed by proteins with a single transmembrane α helix. Instead, they usually consist of a series of α helices that cross the bilayer a number of times (see Figure 11–20A). In many of these *multipass* transmembrane proteins, one or more of the membrane-spanning regions are amphipathic—formed from α helices that contain both hydrophobic and hydrophilic amino acid side chains. These amino acids tend to be arranged so that the hydrophobic side chains fall on one side of the helix, while the hydrophilic side chains are concentrated on the other side. In the hydrophobic environment of the lipid bilayer, α helices of this sort pack side by side in a ring, with the hydrophobic side chains exposed to the lipids of the membrane and the hydrophilic side chains forming the lining of a hydrophilic pore through the lipid bilayer (**Figure 11–23**). How such channels function in the selective transport of small, water-soluble molecules, especially inorganic ions, is discussed in Chapter 12.

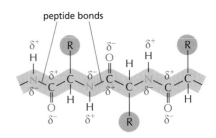

Figure 11–21 The backbone of a polypeptide chain is hydrophilic. The atoms on either side of a peptide bond (*red line*) are polar and carry partial positive or negative charges (δ^+ or δ^-). These charges allow these atoms to hydrogen-bond with one another when the polypeptide folds into an α helix that spans the lipid bilayer (see Figure 11–22).

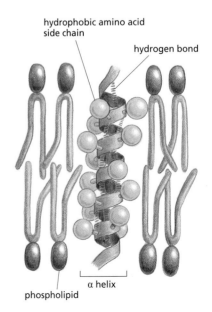

Figure 11–22 A transmembrane polypeptide chain usually crosses the lipid bilayer as an α helix. In this segment of a transmembrane protein, the hydrophobic side chains (*light green*) of the amino acids forming the α helix contact the hydrophobic hydrocarbon tails of the phospholipid molecules, while the hydrophilic parts of the polypeptide backbone form hydrogen bonds with one another in the interior of the helix. An α helix containing about 20 amino acids is required to completely traverse a cell membrane.

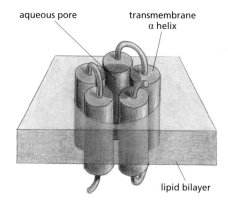

aqueous pore

transmembrane
α helix

lipid bilayer

**Figure 11–23 A transmembrane hydrophilic pore can be formed
by multiple amphipathic α helices.** In this example, five such
transmembrane α helices form a water-filled channel across the lipid
bilayer. The hydrophobic amino acid side chains (*green*) on one side of
each helix contact the hydrophobic lipid tails, while the hydrophilic side
chains (*red*) on the opposite side of the helices form a water-filled pore.

QUESTION 11–4

Explain why the polypeptide chain
of most transmembrane proteins
crosses the lipid bilayer as an α helix
or a β barrel.

Although the α helix is by far the most common form in which a
polypeptide chain crosses a lipid bilayer, the polypeptide chain of some
transmembrane proteins crosses the lipid bilayer as a β sheet that is
rolled into a cylinder, forming a keglike structure called a β barrel (see
Figure 11–20A). As expected, the amino acid side chains that face the
inside of the barrel, and therefore line the aqueous channel, are mostly
hydrophilic, while those on the outside of the barrel, which contact the
hydrophobic core of the lipid bilayer, are exclusively hydrophobic. The
most striking example of a β-barrel structure is found in the *porin* proteins, which form large, water-filled pores in mitochondrial and bacterial
outer membranes (**Figure 11–24**). Mitochondria and some bacteria are
surrounded by a double membrane, and porins allow the passage of
small nutrients, metabolites, and inorganic ions across their outer membranes, while preventing unwanted larger molecules from crossing.

Membrane Proteins Can Be Solubilized in Detergents

To understand a protein fully, one needs to know its structure in detail.
For membrane proteins, this presents special problems. Most biochemical procedures are designed for studying molecules in aqueous solution.
Membrane proteins, however, are built to operate in an environment that
is partly aqueous and partly fatty, and taking them out of this environment and purifying them while preserving their essential structure is no
easy task.

Before an individual protein can be studied in detail, it must be separated from all the other cell proteins. For most membrane proteins, the
first step in this separation process involves solubilizing the membrane
with agents that destroy the lipid bilayer by disrupting hydrophobic
associations. The most widely used disruptive agents are **detergents**
(Movie 11.3). These small, amphipathic, lipidlike molecules differ from
membrane phospholipids in that they have only a single hydrophobic
tail (**Figure 11–25**). Because they have one tail, detergent molecules are
shaped like cones; in water, they thus tend to aggregate into small clusters called *micelles*, rather than forming a bilayer as do the phospholipids,
which—with their two tails—are more cylindrical in shape.

When mixed in great excess with membranes, the hydrophobic ends of
detergent molecules interact with the membrane-spanning hydrophobic
regions of the transmembrane proteins, as well as with the hydrophobic

**Figure 11–24 Porin proteins form water-
filled channels in the outer membrane
of a bacterium.** The protein illustrated is
from *E. coli*, and it consists of a 16-stranded
β sheet curved around on itself to form a
transmembrane water-filled channel. The
three-dimensional structure was determined
by X-ray crystallography. Although not
shown in the drawing, three porin proteins
associate to form a trimer, which has three
separate channels.

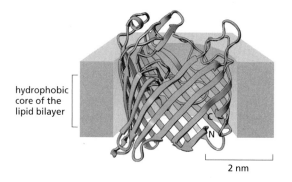

hydrophobic
core of the
lipid bilayer

2 nm

Figure 11–25 SDS and Triton X-100 are two commonly used detergents. Sodium dodecyl sulfate (SDS) is a strong ionic detergent—that is, it has an ionized (charged) group at its hydrophilic end. Triton X-100 is a mild nonionic detergent—that is, it has a nonionized but polar structure at its hydrophilic end. The hydrophobic portion of each detergent is shown in *blue*, and the hydrophilic portion in *red*. The bracketed portion of Triton X-100 is repeated about eight times. Strong ionic detergents like SDS not only displace lipid molecules from proteins but also unfold the proteins (see Panel 4–5, p. 167).

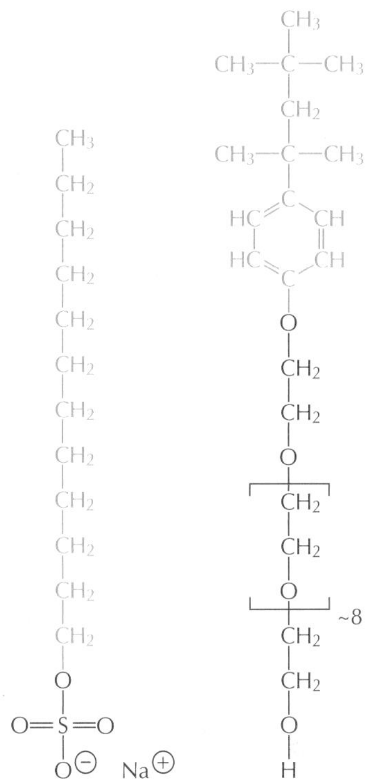

tails of the phospholipid molecules, thereby disrupting the lipid bilayer and separating the proteins from most of the phospholipids. Because the other end of the detergent molecule is hydrophilic, these interactions bring the membrane proteins into solution as protein–detergent complexes; at the same time, the detergent solubilizes the phospholipids (**Figure 11–26**). The protein–detergent complexes can then be separated from one another and from the lipid–detergent complexes for further analysis.

We Know the Complete Structure of Relatively Few Membrane Proteins

For many years, much of what we knew about the structure of membrane proteins was learned by indirect means. The standard method for determining a protein's three-dimensional structure directly is X-ray crystallography (see Figure 4–52), but this requires ordered crystalline arrays of the molecule. Because membrane proteins have to be purified in detergent micelles that are often heterogeneous in size, they are harder to crystallize than the soluble proteins that inhabit the cell cytosol or extracellular fluids. Nevertheless, with recent advances in protein preparation and X-ray crystallography, the structures of an increasing number of membrane proteins have now been determined to high resolution.

One example is bacteriorhodopsin, the structure of which first revealed exactly how α helices cross the lipid bilayer. **Bacteriorhodopsin** is a small protein (about 250 amino acids) found in large amounts in the plasma membrane of an archaean, called *Halobacterium halobium*, that lives in salt marshes. Bacteriorhodopsin acts as a membrane transport protein that pumps H+ (protons) out of the cell. Pumping requires energy, and bacteriorhodopsin gets its energy directly from sunlight. Each bacteriorhodopsin molecule contains a single light-absorbing nonprotein

sodium dodecyl sulfate (SDS)   Triton X-100

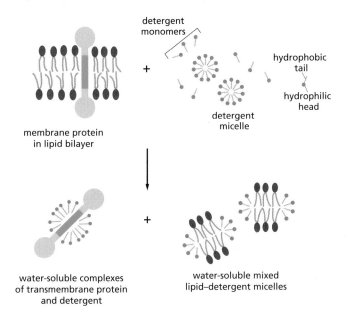

membrane protein in lipid bilayer

detergent monomers

detergent micelle

hydrophobic tail

hydrophilic head

water-soluble complexes of transmembrane protein and detergent

water-soluble mixed lipid–detergent micelles

QUESTION 11–5

For the two detergents shown in Figure 11–25, explain why the red portions of the molecules are hydrophilic and the blue portions hydrophobic. Draw a short stretch of a polypeptide chain made up of three amino acids with hydrophobic side chains (see Panel 2–5, pp. 74–75) and apply a similar color scheme.

Figure 11–26 Membrane proteins can be solubilized by a mild detergent such as Triton X-100. The detergent molecules (*gold*) are shown as both monomers and micelles, the form in which detergent molecules tend to aggregate in water. The detergent disrupts the lipid bilayer and brings the proteins into solution as protein–detergent complexes. As illustrated, the phospholipids in the membrane are also solubilized by the detergents, forming lipid–detergent micelles.

Figure 11–27 Bacteriorhodopsin acts as a proton pump. The polypeptide chain crosses the lipid bilayer as seven α helices. The location of the retinal (*purple*) and the probable pathway taken by protons during the light-activated pumping cycle (*red* arrows) are highlighted. Strategically placed polar amino acid side chains, shown in *red, yellow,* and *blue,* guide the movement of the proton across the bilayer, allowing the proton to avoid contact with the lipid environment. The proton-transfer steps are shown in Movie 11.4. Retinal is also used to detect light in our own eyes, where it is attached to a protein with a structure very similar to bacteriorhodopsin. (Adapted from H. Luecke et al., *Science* 286:255–260, 1999. With permission from the AAAS.)

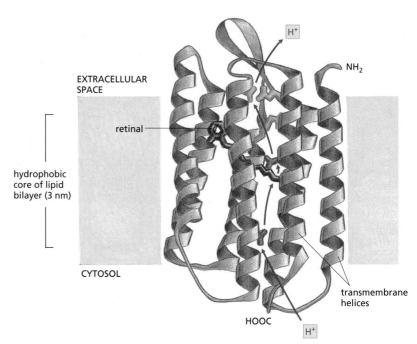

molecule, called *retinal,* that gives the protein—and the bacterium—a deep purple color. This small hydrophobic molecule is covalently attached to one of bacteriorhodopsin's seven transmembrane α helices (**Figure 11–27**). When retinal absorbs a photon of light, it changes shape, and in doing so, it causes the protein embedded in the lipid bilayer to undergo a series of small conformational changes. These changes result in the transfer of one H^+ from the retinal to the outside of the bacterium (see Figure 11–27). The retinal is then regenerated by taking up a H^+ from the cytosol, returning the protein to its original conformation so that it can repeat the cycle. The overall outcome is the movement of one H^+ from inside to outside the cell.

In the presence of sunlight, thousands of bacteriorhodopsin molecules pump H^+ out of the cell, generating a concentration gradient of H^+ across the plasma membrane. The cell uses this proton gradient to store energy and convert it into ATP, as we discuss in detail in Chapter 14. Bacteriorhodopsin is a pump protein, a class of transmembrane protein that actively moves small organic molecules and inorganic ions into and out of cells (see Figure 11–19). We will meet other pump proteins in Chapter 12.

The Plasma Membrane Is Reinforced by the Underlying Cell Cortex

A cell membrane by itself is extremely thin and fragile. It would require nearly 10,000 cell membranes laid on top of one another to achieve the thickness of this paper. Most cell membranes are therefore strengthened and supported by a framework of proteins, attached to the membrane via transmembrane proteins. For plants, yeasts, and bacteria, the cell's shape and mechanical properties are conferred by a rigid cell wall—a meshwork of proteins, sugars, and other macromolecules that encases the plasma membrane. By contrast, the plasma membrane of animal cells is stabilized by a meshwork of fibrous proteins, called the cell cortex, that is attached to the underside of the membrane.

The cortex of human red blood cells is a relatively simple and regular structure and has been especially well studied. These cells are small and have a distinctive flattened shape (**Figure 11–28**). The main component of their cortex is the dimeric protein spectrin, a long, thin, flexible rod

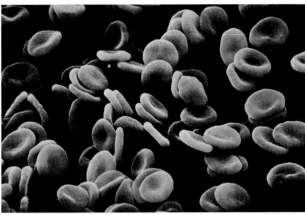

Figure 11–28 Human red blood cells have a characteristic flattened biconcave shape, as seen in this scanning electron micrograph. These cells lack a nucleus and other intracellular organelles. (Courtesy of Bernadette Chailley.)

5 μm

QUESTION 11–6

Look carefully at the transmembrane proteins shown in Figure 11–29. What can you say about their mobility in the membrane?

---

about 100 nm in length. It forms a meshwork that provides support for the plasma membrane and maintains the cell's biconcave shape. The spectrin meshwork is connected to the membrane through intracellular attachment proteins that link the spectrin to specific transmembrane proteins (**Figure 11–29** and Movie 11.5). The importance of this meshwork is seen in mice and humans that have genetic abnormalities in spectrin structure. These individuals are anemic: they have fewer red blood cells than normal. The red cells they do have are spherical instead of flattened and are abnormally fragile.

Proteins similar to spectrin and to its associated attachment proteins are present in the cortex of most animal cells. But the cortex in these cells is especially rich in actin and the motor protein *myosin*, and it is much more complex than that of red blood cells. While red blood cells need their cortex mainly to provide mechanical strength as they are pumped through blood vessels, other cells also need their cortex to allow them to selectively take up materials from their environment, to change their shape actively, and to move, as we discuss in Chapter 17. In addition, cells use their cortex to restrain the diffusion of proteins within the plasma membrane, as we see next.

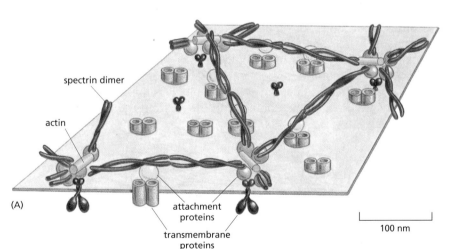

(A)

spectrin dimer

actin

attachment proteins

transmembrane proteins

100 nm

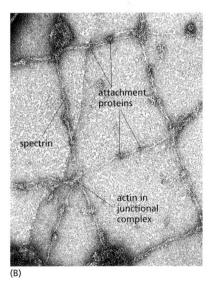

attachment proteins

spectrin

actin in junctional complex

(B)

Figure 11–29 A spectrin meshwork forms the cell cortex in human red blood cells. (A) Spectrin dimers are linked end-to-end to form longer tetramers. The spectrin tetramers, together with a smaller number of actin molecules, are linked together into a mesh. This network is attached to the plasma membrane by the binding of at least two types of attachment proteins (shown here in *yellow* and *blue*) to two kinds of transmembrane proteins (shown here in *green* and *brown*). (B) Electron micrograph showing the spectrin meshwork on the cytoplasmic side of a red blood cell membrane. The meshwork has been stretched out to show the details of its structure; in the normal cell, the meshwork shown would be much more crowded and would occupy only about one-tenth of this area. (B, courtesy of T. Byers and D. Branton, *Proc. Natl. Acad. Sci. USA* 82:6153–6157, 1985. With permission from the National Academy of Sciences.)

Figure 11–30 **Formation of mouse–human hybrid cells shows that some plasma membrane proteins can move laterally in the lipid bilayer.** When the mouse and human cells are first fused, their proteins are confined to their own halves of the newly formed hybrid-cell plasma membrane. Within a short time, however, they completely intermix. To monitor the movement of a selected sampling of these proteins, the cells are labeled with antibodies that bind to either human or mouse proteins; the antibodies are coupled to two different fluorescent tags—rhodamine (*red*) or fluorescein (*green*)—so they can be distinguished in a fluorescence microscope (see Panel 4–2, pp. 146–147). (Based on observations of L.D. Frye and M. Edidin, *J. Cell Sci.* 7:319–335, 1970. With permission from The Company of Biologists Ltd.)

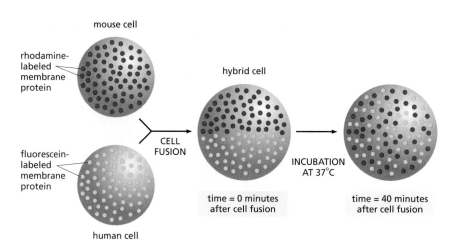

A Cell Can Restrict the Movement of Its Membrane Proteins

Because a membrane is a two-dimensional fluid, many of its proteins, like its lipids, can move freely within the plane of the lipid bilayer. This lateral diffusion was initially demonstrated by experimentally fusing a mouse cell with a human cell to form a double-sized hybrid cell and then monitoring the distribution of certain mouse and human plasma membrane proteins. At first, the mouse and human proteins are confined to their own halves of the newly formed hybrid cell, but within half an hour or so the two sets of proteins become evenly mixed over the entire cell surface (**Figure 11–30**). We describe some other techniques for studying the movement of membrane proteins in **How We Know**, pp. 378–379.

The picture of a cell membrane as a sea of lipid in which all proteins float freely is too simple, however. Cells have ways of confining particular proteins to localized areas within the bilayer membrane, thereby creating functionally specialized regions, or **membrane domains**, on the cell or organelle surface.

As illustrated in **Figure 11–31**, plasma membrane proteins can be tethered to structures outside the cell—for example, to molecules in the extracellular matrix or on an adjacent cell (discussed in Chapter 20)—or to relatively immobile structures inside the cell, especially to the cell cortex (see Figure 11–29). Additionally, cells can create barriers that restrict particular membrane components to one membrane domain. In epithelial cells that line the gut, for example, it is important that transport proteins involved in the uptake of nutrients from the gut be confined to

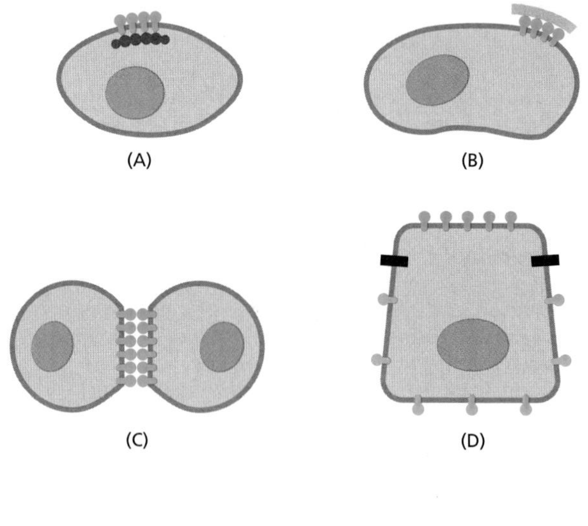

Figure 11–31 **The lateral mobility of plasma membrane proteins can be restricted in several ways.** Proteins can be tethered to the cell cortex inside the cell (A), to extracellular matrix molecules outside the cell (B), or to proteins on the surface of another cell (C). Diffusion barriers (shown as *black* bars) can restrict proteins to a particular membrane domain (D).

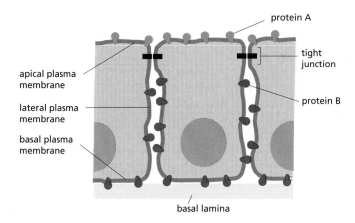

Figure 11–32 Membrane proteins are restricted to particular domains of the plasma membrane of epithelial cells in the gut. Protein A (in the apical membrane) and protein B (in the basal and lateral membranes) can diffuse laterally in their own membrane domains but are prevented from entering the other domain by a specialized cell junction called a tight junction. The basal lamina is a mat of extracellular matrix that supports all epithelial sheets (discussed in Chapter 20).

the *apical* surface of the cells (the surface that faces the gut contents) and that other transport proteins involved in the export of solutes out of the epithelial cell into the tissues and bloodstream be confined to the *basal* and *lateral* surfaces (see Figure 12–17). This asymmetric distribution of membrane proteins is maintained by a barrier formed along the line where the cell is sealed to adjacent epithelial cells by a so-called *tight junction* (**Figure 11–32**). At this site, specialized junctional proteins form a continuous belt around the cell where the cell contacts its neighbors, creating a seal between adjacent plasma membranes (see Figure 20–23). Membrane proteins cannot diffuse past the junction.

The Cell Surface Is Coated with Carbohydrate

We saw earlier that some of the lipids in the outer layer of the plasma membrane have sugars covalently attached to them. The same is true for most of the proteins in the plasma membrane. The great majority of these proteins have short chains of sugars, called oligosaccharides, linked to them; they are called *glycoproteins*. Other membrane proteins, the *proteoglycans*, contain one or more long polysaccharide chains. All of the carbohydrate on the glycoproteins, proteoglycans, and glycolipids is located on the outside of the plasma membrane, where it forms a sugar coating called the *carbohydrate layer* or **glycocalyx** (**Figure 11–33**).

This layer of carbohydrate helps protect the cell surface from mechanical damage. As the oligosaccharides and polysaccharides adsorb water, they also give the cell a slimy surface, which helps motile cells such as white blood cells squeeze through narrow spaces and prevents blood cells from sticking to one another or to the walls of blood vessels.

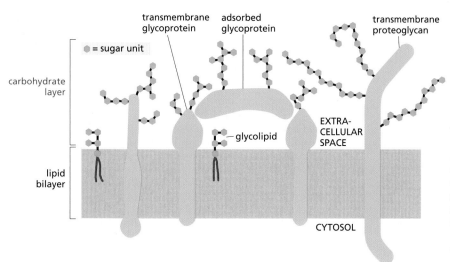

Figure 11–33 Eukaryotic cells are coated with sugars. The carbohydrate layer is made of the oligosaccharide side chains attached to membrane glycolipids and glycoproteins, and of the polysaccharide chains on membrane proteoglycans. As shown, glycoproteins that have been secreted by the cell and then adsorbed back onto its surface can also contribute. Note that all the carbohydrate is on the external (noncytosolic) surface of the plasma membrane.

HOW WE KNOW

MEASURING MEMBRANE FLOW

An essential feature of the lipid bilayer is its fluidity, which is crucial for cell membrane integrity and function. This property allows many membrane-embedded proteins to move laterally in the plane of the bilayer, so that they can engage in the various protein–protein interactions on which cells depend. The fluid nature of cell membranes is so central to their proper function that it may seem surprising that this property was not recognized until the early 1970s.

Given its importance for membrane structure and function, how do we measure and study the fluidity of cell membranes? The most common methods are visual: simply label some of the molecules native to the membrane and then watch them move. Such an approach first demonstrated the lateral movement of membrane proteins that had been tagged with labeled antibodies (see Figure 11–30). This experiment seemed to suggest that membrane proteins diffuse freely, without restriction, in an open sea of lipids. We now know that this image is not entirely accurate. To probe membrane fluidity more thoroughly, researchers had to invent more precise methods for tracking the movement of proteins within a membrane such as the plasma membrane of a living cell.

The FRAP attack

One such technique, called fluorescence recovery after photobleaching (FRAP), involves uniformly labeling the components of the cell membrane—its lipids or, more often, its proteins—with some sort of fluorescent marker. Labeling membrane proteins can be accomplished by incubating living cells with a fluorescent antibody or by covalently attaching a fluorescent protein such as green fluorescent protein (GFP) to a membrane protein of interest using recombinant DNA techniques (discussed in Chapter 10).

Once a protein has been labeled, a small patch of membrane is irradiated with an intense pulse of light from a sharply focused laser beam. This treatment irreversibly "bleaches" the fluorescence from the labeled proteins in that small patch of membrane, typically an area about 1 µm square. The fluorescence of this irradiated membrane is monitored in a fluorescence microscope, and the amount of time it takes for the neighboring, unbleached fluorescent proteins to migrate into the bleached region of the membrane is measured (**Figure 11–34**). The rate of this "fluorescence recovery" is a direct measure of the rate at which the protein molecules can diffuse within the membrane (Movie 11.6). Such experiments have revealed that, generally speaking, a cell membrane is about as viscous as olive oil.

One-by-one

One drawback to the FRAP approach is that the technique monitors the movement of fairly large populations of proteins—hundreds or thousands—across a relatively

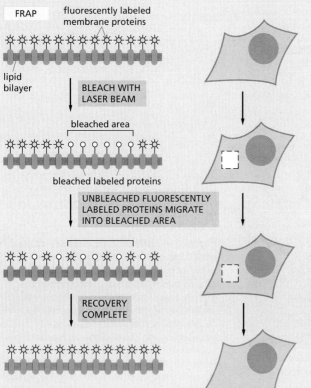

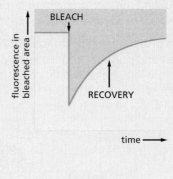

Figure 11–34 **Photobleaching techniques can be used to measure the rate of lateral diffusion of a membrane protein.** A specific protein of interest can be labeled with a fluorescent antibody (as shown here) or can be produced—using genetic engineering techniques—as a fusion protein tagged with green fluorescent protein (GFP), which is intrinsically fluorescent. In the FRAP technique, fluorescent molecules are bleached in a small area of membrane using a laser beam. The fluorescence intensity recovers as the bleached molecules diffuse away and unbleached, fluorescent molecules diffuse in (shown here in side and top views). The diffusion coefficient is calculated from a graph of the rate of fluorescence recovery: the greater the diffusion coefficient of the membrane protein, the faster the recovery.

large area of the membrane. With this technique it is impossible to track the motion of individual molecules. If the labeled proteins fail to migrate into the bleached zone over the course of a FRAP study, for example, is it because they are immobile, essentially anchored in one place in the membrane? Or, alternatively, are they restricted to movement within a very small region—fenced in by cytoskeletal proteins—and thus only appear motionless?

To get around this problem, researchers have developed methods for labeling and observing the movement of individual molecules or small clusters of molecules. One such technique, dubbed single-particle tracking (SPT) microscopy, relies on tagging protein molecules with antibody-coated gold nanoparticles. The gold particles look like tiny black dots when seen with a light microscope, and their movement, and thus the movement of individually tagged protein molecules, can be followed using video microscopy.

From the studies carried out to date, it appears that membrane proteins can display a variety of patterns of movement, from random diffusion to complete immobility (**Figure 11–35**). Some proteins rapidly switch between these different kinds of motion.

Freed from cells

In many cases, researchers wish to study the behavior of a particular type of membrane protein in a synthetic lipid bilayer, in the absence of other proteins that might restrain its movement or alter its activity. For such studies, membrane proteins can be isolated from cells and the protein of interest purified and reconstituted in artificial phospholipid vesicles (**Figure 11–36**). The lipids

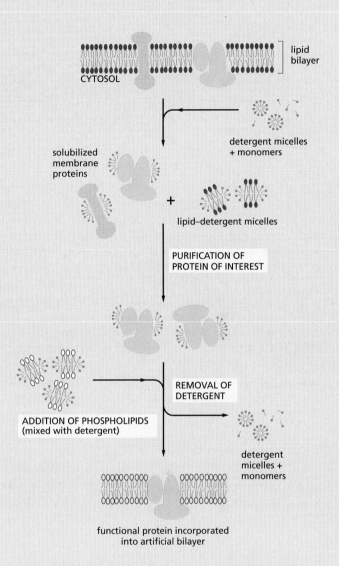

Figure 11–36 Mild detergents can be used to solubilize and reconstitute functional membrane proteins.

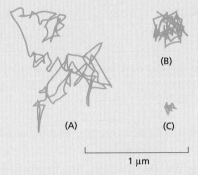

Figure 11–35 Proteins show different patterns of diffusion. Single-particle tracking studies reveal some of the pathways that real proteins follow on the surface of a living cell. Shown here are some trajectories representative of different kinds of proteins in the plasma membrane. (A) Tracks made by a protein that is free to diffuse randomly in the lipid bilayer. (B) Tracks made by a protein that is corralled within a small membrane domain by other proteins. (C) Tracks made by a protein that is tethered to the cytoskeleton and hence is essentially immobile. The movement of the proteins is monitored over a period of seconds.

allow the purified protein to maintain its proper structure and function, so that its activity and behavior can be analyzed in detail.

It is apparent from such studies that membrane proteins diffuse more freely and rapidly in artificial lipid bilayers than in cell membranes. The fact that most proteins show reduced mobility in a cell membrane makes sense, as these membranes are crowded with many types of proteins and contain a greater variety of lipids than an artificial lipid bilayer. Furthermore, many membrane proteins in a cell are tethered to proteins in the extracellular matrix, or anchored to the cell cortex just under the plasma membrane, or both (as illustrated in Figure 11–31).

Taken together, such studies have revolutionized our understanding of membrane proteins and of the architecture and organization of cell membranes.

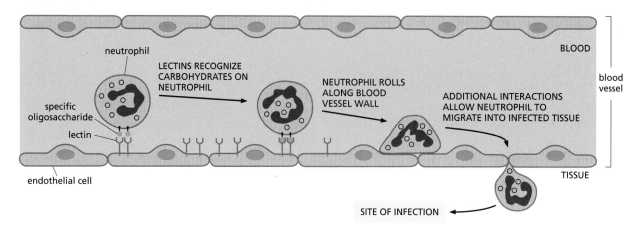

Figure 11–37 **The recognition of the cell-surface carbohydrate on neutrophils is the first stage of their migration out of the blood at sites of infection.** Specialized transmembrane proteins (called lectins) are made by the endothelial cells lining the blood vessel in response to chemical signals emanating from a site of infection. These proteins recognize particular sugar groups carried by glycolipids and glycoproteins on the surface of neutrophils (a type of white blood cell) circulating in the blood. The neutrophils consequently stick to the endothelial cells that line the blood vessel wall. This association is not very strong, but it leads to another, much stronger protein–protein interaction (not shown) that helps the neutrophil slip between the endothelial cells, so it can migrate out of the bloodstream and into the tissue at the site of infection (Movie 11.7).

Cell-surface carbohydrates do more than just protect and lubricate the cell, however. They have an important role in cell–cell recognition and adhesion. Just as many proteins will recognize a particular site on another protein, proteins called *lectins* are specialized to bind to particular oligosaccharide side chains. The oligosaccharide side chains of glycoproteins and glycolipids, although short (typically fewer than 15 sugar units), are enormously diverse. Unlike proteins, in which the amino acids are all joined together in a linear chain by identical peptide bonds, sugars can be joined together in many different arrangements, often forming elaborate branched structures (see Panel 2–3, pp. 70–71). Using a variety of covalent linkages, even three different sugars can form hundreds of different trisaccharides.

The carbohydrate layer on the surface of cells in a multicellular organism serves as a kind of distinctive clothing, like a police officer's uniform. It is characteristic of each cell type and is recognized by other cell types that interact with it. Specific oligosaccharides in the carbohydrate layer are involved, for example, in the recognition of an egg by a sperm (discussed in Chapter 19). Similarly, in the early stages of a bacterial infection, the carbohydrate on the surface of white blood cells called neutrophils is recognized by a lectin on the cells lining the blood vessels at the site of infection; this recognition causes the neutrophils to adhere to the blood vessel wall and then migrate from the bloodstream into the infected tissue, where they help destroy the invading bacteria (Figure 11–37).

ESSENTIAL CONCEPTS

- Cell membranes enable cells to create barriers that confine particular molecules to specific compartments. They consist of a continuous double layer—a bilayer—of lipid molecules in which proteins are embedded.

- The lipid bilayer provides the basic structure and barrier function of all cell membranes.

- Membrane lipid molecules are amphipathic, having both hydrophobic and hydrophilic regions. This property promotes their spontaneous assembly into bilayers when placed in water, forming closed compartments that reseal if torn.

- There are three major classes of membrane lipid molecules: phospholipids, sterols, and glycolipids.

- The lipid bilayer is fluid, and individual lipid molecules are able to diffuse within their own monolayer; they do not, however, spontaneously flip from one monolayer to the other.

- The two lipid monolayers of a cell membrane have different lipid compositions, reflecting the different functions of the two faces of the membrane.

- Cells that live at different temperatures maintain their membrane fluidity by modifying the lipid composition of their membranes.

- Membrane proteins are responsible for most of the functions of cell membranes, including the transport of small, water-soluble molecules across the lipid bilayer.

- Transmembrane proteins extend across the lipid bilayer, usually as one or more α helices but sometimes as a β sheet rolled into the form of a barrel.

- Other membrane proteins do not extend across the lipid bilayer but are attached to one or the other side of the membrane, either by noncovalent association with other membrane proteins, by covalent attachment of lipids, or by association of an exposed amphipathic α helix with a single lipid monolayer.

- Most cell membranes are supported by an attached framework of proteins. An especially important example is the meshwork of fibrous proteins that forms the cell cortex underneath the plasma membrane.

- Although many membrane proteins can diffuse rapidly in the plane of the membrane, cells have ways of confining proteins to specific membrane domains. They can also immobilize particular membrane proteins by attaching them to intracellular or extracellular macromolecules.

- Many of the proteins and some of the lipids exposed on the surface of cells have attached sugar chains, which form a carbohydrate layer that helps protect and lubricate the cell surface, while also being involved in specific cell–cell recognition.

KEY TERMS

amphipathic	membrane protein
bacteriorhodopsin	phosphatidylcholine
cholesterol	phospholipid
detergent	plasma membrane
glycocalyx	saturated
lipid bilayer	unsaturated
membrane domain	

QUESTIONS

QUESTION 11–7

Describe the different methods that cells use to restrict proteins to specific regions of the plasma membrane. Is a membrane with many of its proteins restricted still fluid?

QUESTION 11–8

Which of the following statements are correct? Explain your answers.

A. Lipids in a lipid bilayer spin rapidly around their long axis.

B. Lipids in a lipid bilayer rapidly exchange positions with one another in their own monolayer.

C. Lipids in a lipid bilayer do not flip-flop readily from one lipid monolayer to the other.

D. Hydrogen bonds that form between lipid head groups and water molecules are continually broken and re-formed.

E. Glycolipids move between different membrane-enclosed compartments during their synthesis but remain restricted to one side of the lipid bilayer.

F. Margarine contains more saturated lipids than the vegetable oil from which it is made.

G. Some membrane proteins are enzymes.

H. The sugar layer that surrounds all cells makes cells more slippery.

QUESTION 11–9

What is meant by the term "two-dimensional fluid"?

QUESTION 11–10

The structure of a lipid bilayer is determined by the particular properties of its lipid molecules. What would happen if

A. Phospholipids had only one hydrocarbon tail instead of two?

B. The hydrocarbon tails were shorter than normal, say, about 10 carbon atoms long?

C. All of the hydrocarbon tails were saturated?

D. All of the hydrocarbon tails were unsaturated?

E. The bilayer contained a mixture of two kinds of phospholipid molecules, one with two saturated hydrocarbon tails and the other with two unsaturated hydrocarbon tails?

F. Each phospholipid molecule were covalently linked through the end carbon atom of one of its hydrocarbon tails to a phospholipid tail in the opposite monolayer?

QUESTION 11–11

What are the differences between a phospholipid molecule and a detergent molecule? How would the structure of a phospholipid molecule need to change to make it a detergent?

QUESTION 11–12

A. Membrane lipid molecules exchange places with their lipid neighbors every 10^{-7} second. A lipid molecule diffuses from one end of a 2-μm-long bacterial cell to the other in about 1 second. Are these two numbers in agreement (assume that the diameter of a lipid head group is about 0.5 nm)? If not, can you think of a reason for the difference?

B. To get an appreciation for the great speed of molecular diffusion, assume that a lipid head group is about the size of a ping-pong ball (4 cm in diameter) and that the floor of your living room (6 m × 6 m) is covered wall-to-wall with these balls. If two neighboring balls exchanged positions once every 10^{-7} second, what would their speed be in kilometers per hour? How long would it take for a ball to move from one side of the room to the opposite side?

QUESTION 11–13

Why does a red blood cell plasma membrane need transmembrane proteins?

QUESTION 11–14

Consider a transmembrane protein that forms a hydrophilic pore across the plasma membrane of a eukaryotic cell, allowing Na^+ to enter the cell when it is activated upon binding a specific ligand on its extracellular side. It is made of five similar transmembrane subunits, each containing a membrane-spanning α helix with hydrophilic amino acid side chains on one surface of the helix and hydrophobic amino acid side chains on the opposite surface. Considering the function of the protein as a channel for Na^+ ions to enter the cell, propose a possible arrangement of the five membrane-spanning α helices in the membrane.

QUESTION 11–15

In the membrane of a human red blood cell, the ratio of the mass of protein (average molecular weight 50,000) to phospholipid (molecular weight 800) to cholesterol (molecular weight 386) is about 2:1:1. How many lipid molecules are there for every protein molecule?

QUESTION 11–16

Draw a schematic diagram that shows a close-up view of two plasma membranes as they come together during cell fusion, as shown in Figure 11–30. Show membrane proteins in both cells that were labeled from the outside by the binding of differently colored fluorescent antibody molecules. Indicate in your drawing the fates of these color tags as the cells fuse. Will they remain on the outside of the hybrid cell after cell fusion and still be there after the mixing of membrane proteins that occurs during the incubation at 37°C? How would the experimental outcome be different if the incubation were done at 0°C?

QUESTION 11–17

Compare the hydrophobic forces that hold a membrane protein in the lipid bilayer with those that help proteins fold into a unique three-dimensional structure.

QUESTION 11–18

Predict which one of the following organisms will have the highest percentage of unsaturated phospholipids in its membranes. Explain your answer.

A. Antarctic fish

B. Desert snake

C. Human being

D. Polar bear

E. Thermophilic bacterium that lives in hot springs at 100°C.

QUESTION 11–19

Which of the three 20-amino-acid sequences listed below in the single-letter amino acid code is the most likely candidate to form a transmembrane region (α helix) of a transmembrane protein? Explain your answer.

A. I T L I Y F G N M S S V T Q T I L L I S

B. L L L I F F G V M A L V I V V I L L I A

C. L L K K F F R D M A A V H E T I L E E S

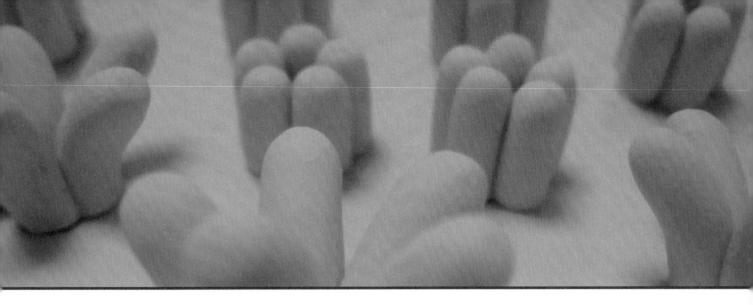

CHAPTER TWELVE

12

Transport Across Cell Membranes

To survive and grow, cells must be able to exchange molecules with their environment. They must import nutrients such as sugars and amino acids and eliminate metabolic waste products. They must also regulate the concentrations of a variety of inorganic ions in their cytosol and organelles. A few molecules, such as CO_2 and O_2, can simply diffuse across the lipid bilayer of the plasma membrane. But the vast majority cannot. Instead, their transfer depends on specialized **membrane transport proteins** that span the lipid bilayer, providing private passageways across the membrane for select substances (**Figure 12–1**).

In this chapter, we consider how cell membranes control the traffic of inorganic ions and small, water-soluble molecules into and out of the cell and its membrane-enclosed organelles. Cells can also selectively transfer macromolecules such as proteins across their membranes, but this transport requires more elaborate machinery and is discussed in Chapter 15.

We begin by outlining some of the general principles that guide the passage of ions and small molecules through cell membranes. We then examine, in turn, the two main classes of membrane proteins that mediate this transfer: transporters and channels. *Transporters* shift small organic molecules or inorganic ions from one side of the membrane to the other by changing shape. *Channels*, in contrast, form tiny hydrophilic pores across the membrane through which such substances can pass by diffusion. Most channels only permit passage of inorganic ions and are therefore called *ion channels*. Because these ions are electrically charged, their movements can create a powerful electric force—or voltage—across the membrane. In the final part of the chapter, we discuss how these voltage differences enable nerve cells to communicate—and ultimately to shape our behavior.

PRINCIPLES OF TRANSMEMBRANE TRANSPORT

TRANSPORTERS AND THEIR FUNCTIONS

ION CHANNELS AND THE MEMBRANE POTENTIAL

ION CHANNELS AND NERVE CELL SIGNALING

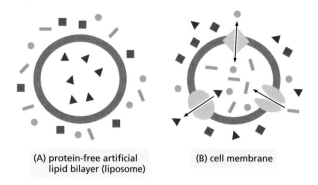

(A) protein-free artificial lipid bilayer (liposome)

(B) cell membrane

Figure 12–1 Cell membranes contain specialized membrane transport proteins that facilitate the passage of selected small water-soluble molecules. (A) Protein-free, artificial lipid bilayers such as liposomes (see Figure 11–13) are impermeable to most water-soluble molecules. (B) Cell membranes, by contrast, contain transport proteins, each of which transfers a particular type of molecule. This selective transport can include the active pumping of specific molecules either out of (*purple triangles*) or into (*green bars*) the cell. The combined action of different transport proteins allows a specific set of solutes to build up inside a membrane-enclosed compartment, such as the cytosol or an organelle.

PRINCIPLES OF TRANSMEMBRANE TRANSPORT

As we saw in Chapter 11, the hydrophobic interior of the lipid bilayer creates a barrier to the passage of most hydrophilic molecules, including all ions. These molecules are as reluctant to enter a fatty environment as hydrophobic molecules are reluctant to enter water. But cells and organelles must also allow the passage of many hydrophilic, water-soluble molecules, such as inorganic ions, sugars, amino acids, nucleotides, and other cell metabolites. These molecules cross lipid bilayers far too slowly by *simple diffusion*, so their passage across cell membranes must be accelerated by specialized membrane transport proteins—a process called *facilitated transport*. In this section, we review the basic principles of such facilitated transmembrane transport and introduce the various types of membrane transport proteins that mediate this movement. We also discuss why the transport of inorganic ions, in particular, is of such fundamental importance for all cells.

Lipid Bilayers Are Impermeable to Ions and Most Uncharged Polar Molecules

Given enough time, virtually any molecule will diffuse across a lipid bilayer. The rate at which it diffuses, however, varies enormously depending on the size of the molecule and its solubility properties. In general, the smaller the molecule and the more hydrophobic, or nonpolar, it is, the more rapidly it will diffuse across the membrane.

Of course, many of the molecules that are of interest to cells are polar and water-soluble. These *solutes*—substances that, in this case, are dissolved in water—are unable to cross the lipid bilayer without the aid of membrane transport proteins. The relative ease with which a variety of solutes can cross cell membranes is shown in **Figure 12–2**.

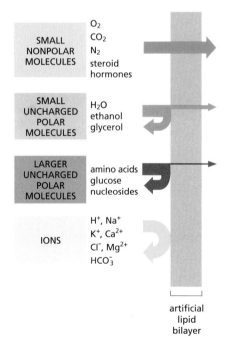

SMALL NONPOLAR MOLECULES — O_2, CO_2, N_2, steroid hormones

SMALL UNCHARGED POLAR MOLECULES — H_2O, ethanol, glycerol

LARGER UNCHARGED POLAR MOLECULES — amino acids, glucose, nucleosides

IONS — H^+, Na^+, K^+, Ca^{2+}, Cl^-, Mg^{2+}, HCO_3^-

artificial lipid bilayer

Figure 12–2 The rate at which a molecule crosses a protein-free artificial lipid bilayer by simple diffusion depends on its size and solubility. The smaller the molecule and, more importantly, the fewer its favorable interactions with water (that is, the less polar it is), the more rapidly the molecule diffuses across the bilayer. Note that many of the organic molecules that a cell uses as nutrients (shaded in *red*) are too large and polar to pass through an artificial lipid bilayer that does not contain the appropriate membrane-transport proteins.

1. *Small nonpolar molecules*, such as molecular oxygen (O_2, molecular mass 32 daltons) and carbon dioxide (CO_2, 44 daltons), dissolve readily in lipid bilayers and therefore rapidly diffuse across them; indeed, cells depend on this permeability to gases for the cell respiration processes discussed in Chapter 14.

2. *Uncharged polar molecules* (molecules with an uneven distribution of electric charge) also diffuse readily across a bilayer if they are small enough. Water (H_2O, 18 daltons) and ethanol (46 daltons), for example, cross at a measureable rate, while glycerol (92 daltons) crosses less rapidly. Larger uncharged polar molecules such as glucose (180 daltons) cross hardly at all.

3. In contrast, lipid bilayers are highly impermeable to all *charged molecules,* including all inorganic ions, no matter how small. These molecules' charges and their strong electrical attraction to water

molecules inhibit their entry into the inner, hydrocarbon phase of the bilayer. Thus synthetic lipid bilayers are a billion (10^9) times more permeable to water than they are to even small ions such as Na^+ or K^+.

The Ion Concentrations Inside a Cell Are Very Different from Those Outside

Because cell membranes are impermeable to inorganic ions, living cells are able to maintain internal ion concentrations that are very different from the concentrations of ions in the media that surrounds them. These differences in ion concentration are crucial for a cell's survival and function. Among the most important inorganic ions for cells are Na^+, K^+, Ca^{2+}, Cl^-, and H^+ (protons). The movement of these ions across cell membranes plays an essential part in many biological processes, but is perhaps most striking in the production of ATP by all cells, and in communication by nerve cells (to be discussed later).

Na^+ is the most plentiful positively charged ion (cation) outside the cell, whereas K^+ is the most abundant inside (**Table 12–1**). For a cell to avoid being torn apart by electrical forces, the quantity of positive charge inside the cell must be balanced by an almost exactly equal quantity of negative charge, and the same is true for the charge in the surrounding fluid. The high concentration of Na^+ outside the cell is electrically balanced chiefly by extracellular Cl^-, whereas the high concentration of K^+ inside is balanced by a variety of negatively charged organic and inorganic ions (anions) including nucleic acids, proteins, and many cell metabolites (see Table 12–1).

Differences in the Concentration of Inorganic Ions Across a Cell Membrane Create a Membrane Potential

Although the electrical charges inside and outside the cell are generally kept in balance, tiny excesses of positive or negative charge, concentrated in the neighborhood of the plasma membrane, do occur. Such electrical imbalances generate a voltage difference across the membrane called the **membrane potential**.

TABLE 12–1 A COMPARISON OF ION CONCENTRATIONS INSIDE AND OUTSIDE A TYPICAL MAMMALIAN CELL		
Component	Intracellular Concentration (mM)	Extracellular Concentration (mM)
Cations		
Na^+	5–15	145
K^+	140	5
Mg^{2+}	0.5*	1–2
Ca^{2+}	10^{-4}*	1–2
H^+	7×10^{-5} ($10^{-7.2}$ M or pH 7.2)	4×10^{-5} ($10^{-7.4}$ M or pH 7.4)
Anions**		
Cl^-	5–15	110

*The concentrations of Mg^{2+} and Ca^{2+} given are for the free ions. There is a total of about 20 mM Mg^{2+} and 1–2 mM Ca^{2+} in cells, but these ions are mostly bound to proteins and other organic molecules and, for Ca^{2+}, stored within various organelles.
**In addition to Cl^-, a cell contains many other anions not listed in this table. In fact, most cell constituents are negatively charged (HCO_3^-, PO_4^{3-}, proteins, nucleic acids, metabolites carrying phosphate and carboxyl groups, etc.).

When a cell is "unstimulated," the exchange of anions and cations across the membrane will be precisely balanced. In such steady-state conditions, the voltage difference across the cell membrane—called the *resting membrane potential*—holds steady. But it is not zero. In animal cells, for example, the resting membrane potential can be anywhere between –20 and –200 millivolts (mV), depending on the organism and cell type. The value is expressed as a negative number because the interior of the cell is more negatively charged than the exterior. This membrane potential allows cells to power the transport of certain metabolites and provides those cells that are excitable with a means to communicate with their neighbors.

It is the activity of membrane transport proteins embedded in the bilayer that enables cells to establish and maintain their membrane potential, as we discuss next.

Cells Contain Two Classes of Membrane Transport Proteins: Transporters and Channels

Membrane transport proteins occur in many forms and are present in all cell membranes. Each provides a private portal across the membrane for a particular small, water-soluble molecule—an ion, sugar, or amino acid, for example. Most of these proteins allow passage of only select members of a particular molecular class: some permit transit of Na^+ but not K^+, others K^+ but not Na^+, and so on. Each type of cell membrane has its own characteristic set of transport proteins, which determines exactly which solutes can pass into and out of the cell or an organelle.

As discussed in Chapter 11, most membrane transport proteins have polypeptide chains that traverse the lipid bilayer multiple times—that is, they are multipass transmembrane proteins (see Figure 11–23). By crisscrossing back and forth across the bilayer, the polypeptide chain forms a continuous protein-lined pathway that allows selected small, hydrophilic molecules to cross the membrane without coming into direct contact with the hydrophobic interior of the lipid bilayer.

There are two main classes of membrane transport proteins: transporters and channels. These proteins differ in the way they discriminate between solutes, transporting some but not others (**Figure 12–3**). *Channels* discriminate mainly on the basis of size and electric charge: when the channel is open, any ion or molecule that is small enough and carries the appropriate charge can pass through. A *transporter*, on the other hand, transfers only those molecules or ions that fit into specific binding sites on the protein. Transporters bind their solutes with great specificity, in the same way an enzyme binds its substrate, and it is this requirement for specific binding that gives transporters their selectivity.

Solutes Cross Membranes by Either Passive or Active Transport

Transporters and channels allow small hydrophilic molecules to cross the cell membrane, but what controls whether these solutes move into the

Figure 12–3 Inorganic ions and small, polar organic molecules can cross a cell membrane through either a transporter or a channel. (A) A transporter undergoes a series of conformational changes to transfer small solutes across the lipid bilayer. (B) A channel, when open, forms a pore across the bilayer through which specific inorganic ions or, in some cases, polar organic molecules can diffuse. As would be expected, channels transfer solutes at a much greater rate than transporters.

Ion channels can exist in either an open or a closed conformation, and they transport only in the open conformation, which is shown here. Channel opening and closing is usually controlled by an external stimulus or by conditions within the cell.

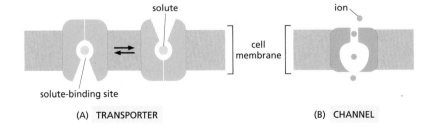

(A) TRANSPORTER   (B) CHANNEL

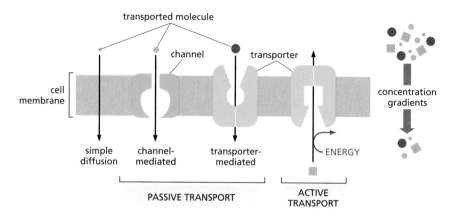

Some small nonpolar molecules such as CO_2 (see Figure 12–2) can move passively down their concentration gradient across the lipid bilayer by simple diffusion, without the help of a transport protein. Most solutes, however, require the assistance of a channel or transporter. Passive transport, which allows molecules to move down their concentration gradients, occurs spontaneously; whereas active transport against a concentration gradient requires an input of energy. Only transporters can carry out active transport.

cell or organelle—or out of it? In many cases, the direction of transport depends only on the relative concentrations of the solute on either side of the membrane. Molecules will spontaneously flow "downhill" from a region of high concentration to a region of low concentration, provided a pathway exists. Such movements are called passive, because they need no additional driving force. If, for example, a solute is present at a higher concentration outside the cell than inside, and an appropriate channel or transporter is present in the plasma membrane, the solute will move into the cell by **passive transport**, without expenditure of energy by the transport protein. This is because even though the solute moves in both directions across the membrane, more solute will move in than out until the two concentrations equilibrate. All channels and many transporters act as conduits for such passive transport.

To move a solute against its concentration gradient, a membrane transport protein must do work: it has to drive the flow "uphill" by coupling it to some other process that provides an input of energy (as discussed in Chapter 3 for enzyme-catalyzed reactions). The movement of a solute against its concentration gradient in this way is termed **active transport**, and it is carried out by special types of transporters called *pumps*, which harness an energy source to power the transport process (**Figure 12–4**). As discussed later, this energy can come from ATP hydrolysis, a transmembrane ion gradient, or sunlight.

Both the Concentration Gradient and Membrane Potential Influence the Passive Transport of Charged Solutes

For an uncharged molecule, the direction of passive transport is determined solely by its concentration gradient, as we have implied above. But for electrically charged molecules, whether inorganic ions or small organic molecules, an additional force comes into play. As mentioned earlier, most cell membranes have a voltage across them—a difference in charge referred to as a membrane potential. The membrane potential exerts a force on any molecule that carries an electric charge. The cytosolic side of the plasma membrane is usually at a negative potential relative to the extracellular side, so the membrane potential tends to pull positively charged solutes into the cell and drive negatively charged ones out.

At the same time, a charged solute will also tend to move down its concentration gradient. The net force driving a charged solute across a cell membrane is therefore a composite of two forces, one due to the concentration gradient and the other due to the membrane potential. This net driving force, called the solute's **electrochemical gradient**, determines

Figure 12–5 An electrochemical gradient has two components. The net driving force (the electrochemical gradient) tending to move a charged solute (ion) across a cell membrane is the sum of a force from the concentration gradient of the solute and a force from the membrane potential. The membrane potential is represented here by the + and – signs on opposite sides of the membrane. The width of the *green* arrow represents the magnitude of the electrochemical gradient for a positively charged solute in two different situations. In (A), the concentration gradient and membrane potential work together to increase the driving force for movement of the solute. In (B), the membrane potential acts against the concentration gradient, decreasing the electrochemical driving force.

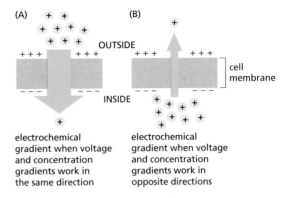

electrochemical gradient when voltage and concentration gradients work in the same direction

electrochemical gradient when voltage and concentration gradients work in opposite directions

the direction that each solute will flow across the membrane by passive transport. For some ions, the voltage and concentration gradients work in the same direction, creating a relatively steep electrochemical gradient (**Figure 12–5A**). This is the case for Na⁺, which is positively charged and at a higher concentration outside cells than inside (see Table 12–1). Na⁺ therefore tends to enter cells if given an opportunity. If, however, the voltage and concentration gradients have opposing effects, the resulting electrochemical gradient can be small (**Figure 12–5B**). This is the case for K⁺, which is present at a much higher concentration inside cells than outside. Because of its small electrochemical gradient across the resting plasma membrane, there is little net movement of K⁺ across the membrane even when K⁺ channels are open.

Water Moves Passively Across Cell Membranes Down Its Concentration Gradient—a Process Called Osmosis

Cells are mostly water (generally about 70% by weight), and so the movement of water across cell membranes is crucially important for living things. Because water molecules are small and uncharged, they can diffuse directly across the lipid bilayer—although slowly (see Figure 12–2). However, some cells also contain specialized channel proteins called *aquaporins* in their plasma membrane, which greatly facilitate this flow (**Figure 12–6** and Movie 12.1).

But which way does water tend to flow? As we saw in Table 12–1, cells contain a high concentration of solutes, including many charged molecules and ions. Thus the total concentration of solute particles inside the cell—also called its *osmolarity*—generally exceeds solute concentration outside the cell. The resulting osmotic gradient tends to "pull" water into the cell. This movement of water down its concentration gradient—from an area of low solute concentration (high water concentration) to an area of high solute concentration (low water concentration)—is called **osmosis**.

Osmosis, if it occurs without constraint, can make a cell swell. Different cells cope with this osmotic challenge in different ways. Most animal

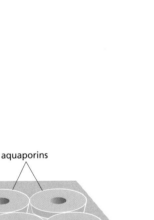

plasma membrane   aquaporins

(A)

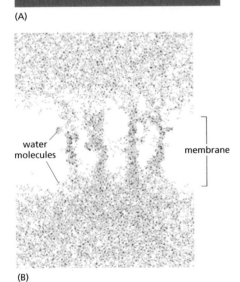

water molecules

membrane

(B)

Figure 12–6 Water molecules diffuse rapidly through aquaporin channels in the plasma membrane of some cells. (A) Shaped like an hourglass, each aquaporin channel forms a pore across the bilayer, allowing the selective passage of water molecules. Shown here is an aquaporin tetramer, the biologically active form of the protein. (B) In this snapshot, taken from a real-time, molecular dynamics simulation, four columns of water molecules can be seen passing though the pores of an aquaporin tetramer (not shown). The space where the membrane would be located is indicated. (B, adapted from B. de Groot and H. Grubmüller, *Science* 294:2353–2357, 2001.)

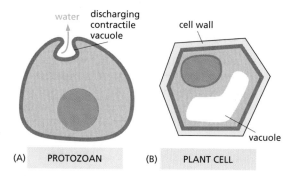

(A)   PROTOZOAN   (B)   PLANT CELL

Figure 12–7 Cells use different tactics to avoid osmotic swelling. (A) A fresh water amoeba avoids swelling by periodically ejecting the water that moves into the cell and accumulates in contractile vacuoles. The contractile vacuole first accumulates solutes, which cause water to follow by osmosis; it then pumps most of the solutes back into the cytosol before emptying its contents at the cell surface. (B) The plant cell's tough cell wall prevents swelling.

cells have a gel-like cytoplasm (see Figure 1–25) that resists osmotic swelling. Some fresh water protozoans, such as amoebae, eliminate excess water using contractile vacuoles that periodically discharge their contents to the exterior (**Figure 12–7A**). Plant cells are prevented from swelling by their tough cell walls and so can tolerate a large osmotic difference across their plasma membrane (**Figure 12–7B**); indeed, plant cells make use of osmotic swelling pressure, or *turgor pressure*, to keep their cell walls tense, so that the stems of the plant are rigid and its leaves are extended. If turgor pressure is lost, plants wilt.

TRANSPORTERS AND THEIR FUNCTIONS

Transporters are responsible for the movement of most small, water-soluble, organic molecules and some inorganic ions across cell membranes. Each transporter is highly selective, often transferring just one type of molecule. To guide and propel the complex traffic of solutes into and out of the cell, and between the cytosol and the different membrane-enclosed organelles, each cell membrane contains a characteristic set of different transporters appropriate to that particular membrane. For example, the plasma membrane contains transporters that import nutrients such as sugars, amino acids, and nucleotides; the lysosome membrane contains an H^+ transporter that imports H^+ to acidify the lysosome interior and other transporters that move digestion products out of the lysosome into the cytosol; the inner membrane of mitochondria contains transporters for importing the pyruvate that mitochondria use as fuel for generating ATP, as well as transporters for exporting ATP once it is synthesized (**Figure 12–8**).

In this section, we describe the general principles that govern the function of transporters, and we present a more detailed view of the molecular mechanisms that drive the movement of a few key solutes.

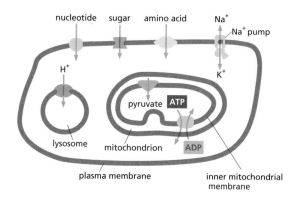

Figure 12–8 Each cell membrane has its own characteristic set of transporters. Only a few of these are indicated here.

Passive Transporters Move a Solute Along Its Electrochemical Gradient

An important example of a transporter that mediates passive transport is the *glucose transporter* in the plasma membrane of many mammalian cell types. The protein, which consists of a polypeptide chain that crosses the membrane at least 12 times, can adopt several conformations—and it switches reversibly and randomly between them. In one conformation, the transporter exposes binding sites for glucose to the exterior of the cell; in another, it exposes the sites to the cell interior.

Because glucose is uncharged, the chemical component of its electrochemical gradient is zero. Thus the direction in which it is transported is determined by its concentration gradient alone. When glucose is plentiful outside cells, as it is after a meal, the sugar binds to the transporter's externally displayed binding sites; when the protein switches conformation—spontaneously and at random—it carries the bound sugar inward and releases it into the cytosol, where the glucose concentration is low (**Figure 12–9**). Conversely, when blood glucose levels are low as they are when you are hungry—the hormone glucagon stimulates liver cells to produce large amounts of glucose by the breakdown of glycogen. As a result, the glucose concentration is higher inside liver cells than outside. This glucose binds to the internally displayed binding sites on the transporter. When the protein switches conformation in the opposite direction, the glucose is transported out of the cells, where it is made available for others to import. The net flow of glucose can thus go either way, according to the direction of the glucose concentration gradient across the plasma membrane: inward if glucose is more concentrated outside the cell than inside, and outward if the opposite is true.

Although passive transporters of this type play no part in determining the direction of transport, they are highly selective. For example, the binding sites in the glucose transporter bind only D-glucose and not its mirror image, L-glucose, which the cell cannot use for glycolysis.

Pumps Actively Transport a Solute Against Its Electrochemical Gradient

Cells cannot rely solely on passive transport. An active transport of solutes against their electrochemical gradient is essential to maintain the appropriate intracellular ionic composition of cells and to import solutes that are at a lower concentration outside the cell than inside. For these purposes, cells depend on transmembrane **pumps**, which can carry out

> ### QUESTION 12–1
>
> A simple enzyme reaction can be described by the equation
> E + S ↔ ES ↔ E + P, where E is the enzyme, S the substrate, P the product, and ES the enzyme–substrate complex.
> A. Write a corresponding equation describing the workings of a transporter (T) that mediates the transport of a solute (S) down its concentration gradient.
> B. What does this equation tell you about the function of a transporter?
> C. Why would this equation be an inappropriate description of channel function?

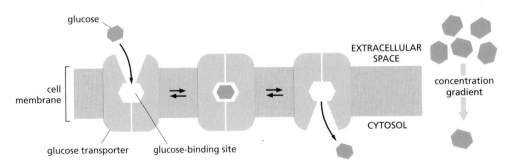

Figure 12–9 Conformational changes in a transporter mediate the passive transport of a solute such as glucose. The transporter is shown in three conformational states: in the outward–open state (*left*), the binding sites for solute are exposed on the outside; in the inward–open state (*right*), the sites are exposed on the inside of the bilayer, and in the occluded state (*center*), the sites are not accessible from either side. The transition between the states occurs randomly, is completely reversible, and—most importantly for the function of the transporter shown—does not depend on whether the solute-binding site is occupied. Therefore, if the solute concentration is higher on the outside of the bilayer, more solute will bind to the transporter in the outward–open conformation than in the inward–open conformation, and there will be a net transport of glucose down its concentration gradient.

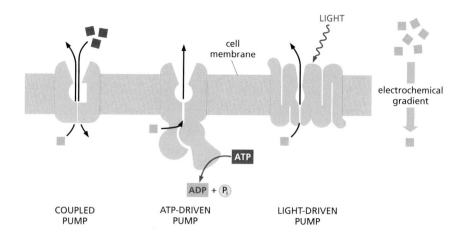

Figure 12–10 Pumps carry out active transport in three main ways. The actively transported generic molecule is shown in *yellow*, and the energy source is shown in *red*.

active transport in three main ways (**Figure 12–10**): (i) *ATP-driven pumps* hydrolyze ATP to drive uphill transport. (ii) *Coupled pumps* link the uphill transport of one solute across a membrane to the downhill transport of another. (iii) *Light-driven pumps*, which are found mainly in bacterial cells, use energy derived from sunlight to drive uphill transport, as discussed in Chapter 11 for bacteriorhodopsin (see Figure 11–27).

The different forms of active transport are often linked. Thus, in the plasma membrane of an animal cell, an ATP-driven Na^+ pump transports Na^+ out of the cell against its electrochemical gradient; this Na^+ can then flow back into the cell, down its electrochemical gradient. As the ion flows back in through various Na^+-coupled pumps, the influx of Na^+ provides the energy for the active transport of many other substances into the cell against their electrochemical gradients. If the Na^+ pump ceased operating, the Na^+ gradient would soon run down, and transport through Na^+-coupled pumps would come to a halt. For this reason, the ATP-driven Na^+ pump has a central role in the active transport of small molecules across the plasma membrane of animal cells. Plant cells, fungi, and many bacteria, use ATP-driven H^+ pumps in an analogous way: in pumping H^+ out of the cell, these proteins create an electrochemical gradient of H^+ across the plasma membrane that is subsequently harnessed for solute transport, as we discuss later.

The Na^+ Pump in Animal Cells Uses Energy Supplied by ATP to Expel Na^+ and Bring in K^+

The ATP-driven **Na^+ pump** plays such a central part in the energy economy of animal cells, that it typically accounts for 30% or more of their total ATP consumption. This pump uses the energy derived from ATP hydrolysis to transport Na^+ out of the cell as it carries K^+ in. The pump is therefore also known as the *Na^+-K^+ ATPase* or the *Na^+-K^+ pump*.

The energy from ATP hydrolysis induces a series of protein conformational changes that drive the Na^+/K^+ ion exchange. As part of the process, the phosphate group removed from ATP gets transferred to the pump itself (**Figure 12–11**).

The ion transport (Na^+ out, K^+ in) involves a reaction cycle, in which each step depends on the one before. If any of the individual steps is prevented from occurring, the entire cycle halts. The toxin, ouabain, for example, inhibits the pump by preventing the binding of extracellular K^+, arresting the cycle. The process is very efficient: the whole cycle takes only 10 milliseconds. Furthermore, the tight coupling between steps in the pumping cycle ensures that the pump operates only when the appropriate ions are available to be transported, thereby avoiding useless ATP hydrolysis.

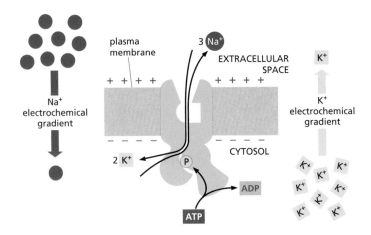

The Na⁺ Pump Generates a Steep Concentration Gradient of Na⁺ Across the Plasma Membrane

The Na⁺ pump functions like a bilge pump in a leaky ship, ceaselessly expeling the Na⁺ that is constantly entering the cell through other transporters and ion channels in the plasma membrane. In this way, the pump keeps the Na⁺ concentration in the cytosol about 10–30 times lower than in the extracellular fluid and the K⁺ concentration about 10–30 times higher (see Table 12–1, p. 385).

The steep concentration gradient of Na⁺ across the plasma membrane acts together with the membrane potential to create a large Na⁺ electrochemical gradient, which tends to pull Na⁺ back into the cell (see Figure 12–5A). This high concentration of Na⁺ outside the cell, on the uphill side of its electrochemical gradient, is like a large volume of water behind a high dam: it represents a very large store of energy (**Figure 12–12**). Even if one artificially halts the operation of the Na⁺ pump with ouabain, this stored energy is sufficient to sustain for many minutes the various pumps in the plasma membrane that are driven by the downhill flow of Na⁺, which we discuss shortly.

Ca²⁺ Pumps Keep the Cytosolic Ca²⁺ Concentration Low

Ca²⁺, like Na⁺, is also kept at a low concentration in the cytosol compared with its concentration in the extracellular fluid, but it is much less plentiful than Na⁺, both inside and outside cells (see Table 12–1). The movement of Ca²⁺ across cell membranes is nonetheless crucial, because Ca²⁺ can bind tightly to a variety of proteins in the cell, altering their activities. An influx of Ca²⁺ into the cytosol through Ca²⁺ channels, for example, is used by different cells as an intracellular signal to trigger various cell processes, such as muscle contraction (discussed in Chapter 17), fertilization (discussed in Chapters 16 and 19), and nerve cell communication, discussed later.

Figure 12–12 The high concentration of Na⁺ outside the cell is like water behind a high dam. The water in the dam has potential energy, which can be used to drive energy-requiring processes. In the same way, an ion gradient across a membrane can be used to drive active processes in a cell, including the active transport of other molecules across the plasma membrane. Shown here is the Table Rock Dam in Branson, Missouri, USA. (Courtesy of K. Trimble.)

The lower the background concentration of free Ca²⁺ in the cytosol, the more sensitive the cell is to an increase in cytosolic Ca²⁺. Thus eukaryotic cells in general maintain a very low concentration of free Ca²⁺ in their cytosol (about 10^{-4} mM) in the face of a very much higher extracellular Ca²⁺ concentration (typically 1–2 mM). This huge concentration difference is achieved mainly by means of ATP-driven **Ca²⁺ pumps** in both the plasma membrane and the endoplasmic reticulum membrane, which actively pump Ca²⁺ out of the cytosol.

Ca²⁺ pumps are ATPases that work in much the same way as the Na⁺ pump depicted in Figure 12–11. The main difference is that Ca²⁺ pumps

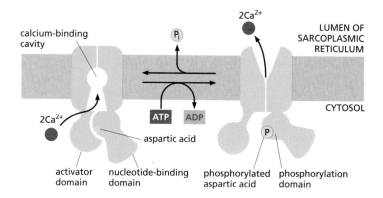

Figure 12–13 The Ca²⁺ pump in the sarcoplasmic reticulum was the first ATP-driven ion pump to have its three-dimensional structure determined by X-ray crystallography. When a muscle cell is stimulated, Ca²⁺ floods into the cytosol from the sarcoplasmic reticulum—a specialized form of endoplasmic reticulum. The influx of Ca²⁺ stimulates the cell to contract; to recover from the contraction, Ca²⁺ must be pumped back into the sarcoplasmic reticulum by this Ca²⁺ pump.

The Ca²⁺ pump uses ATP to phosphorylate itself, inducing a series of conformational changes that—when the pump is open to the lumen of the sarcoplasmic reticulum—eliminate the Ca²⁺-binding sites, ejecting the two Ca²⁺ ions into the organelle.

return to their original conformation without a requirement for binding and transporting a second ion (**Figure 12–13**). The Na^+ and Ca^{2+} pumps have similar amino acid sequences and structures, indicating that they share a common evolutionary origin.

Coupled Pumps Exploit Solute Gradients to Mediate Active Transport

A gradient of any solute across a membrane, like the electrochemical Na^+ gradient generated by the Na^+ pump, can be used to drive the active transport of a second molecule. The downhill movement of the first solute down its gradient provides the energy to power the uphill transport of the second. The active transporters that work in this way are called **coupled pumps** (see Figure 12–10). They can couple the movement of one inorganic ion to that of another, the movement of an inorganic ion to that of a small organic molecule, or the movement of one small organic molecule to that of another. If the pump moves both solutes in the same direction across the membrane, it is called a *symport*. If it moves them in opposite directions, it is called an *antiport*. A transporter that ferries only one type of solute across the membrane (and is therefore not a coupled transporter) is called a *uniport* (**Figure 12–14**). The passive glucose transporter described earlier (see Figure 12–9) is an example of a uniport.

The Electrochemical Na^+ Gradient Drives Coupled Pumps in the Plasma Membrane of Animal Cells

Symports that make use of the inward flow of Na^+ down its steep electrochemical gradient have an especially important role in driving the import of other solutes into animal cells. The epithelial cells that line the gut, for example, pump glucose from the gut lumen across the gut epithelium and, ultimately, into the blood. If these cells had only the passive glucose uniport just mentioned, they would release glucose into the gut

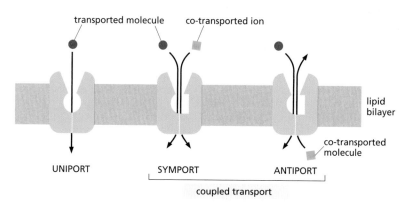

Figure 12–14 Transporters can function as uniports, symports, or antiports. Transporters that carry a single solute across the membrane are called uniports. Transporters that move multiple solutes are called coupled transporters. In coupled transport, the solutes can be transferred either in the same direction, by symports, or in the opposite direction, by antiports (Movie 12.3). Uniports, symports, and antiports can be used for either passive or active transport. Some coupled transporters, for example, act as pumps, coupling the uphill transport of one solute to the downhill transport of another.

after fasting as freely as they take it up from the gut after a feast (see Figure 12–9). But these epithelial cells also possess a *glucose–Na⁺ symport*, which they can use to take up glucose from the gut lumen, even when the concentration of glucose is higher in the cell's cytosol than it is in the gut lumen. Because the electrochemical gradient for Na⁺ is steep, when Na⁺ moves into the cell down its gradient, glucose is, in a sense, "dragged" into the cell with it. Because the binding of Na⁺ and glucose is cooperative—the binding of one enhances the binding of the other—if one of the two solutes is missing, the other fails to bind; therefore both molecules must be present for coupled transport to occur (**Figure 12–15**).

If the gut epithelial cells had only this symport, however, they could never release glucose for use by the other cells of the body. These cells, therefore, have two types of glucose transporters located at opposite ends of the cell. In the apical domain of the plasma membrane, which faces the gut lumen, they have the glucose–Na⁺ symports. These take up glucose actively, creating a high glucose concentration in the cytosol. In the basal and lateral domains of the plasma membrane, the cells have the passive glucose uniports, which release the glucose down its concentration gradient for use by other tissues (**Figure 12–16**). As shown in the figure, the two types of glucose transporters are kept segregated in their proper domains of the plasma membrane by a diffusion barrier formed by a tight junction around the apex of the cell. This prevents mixing of membrane components between the two domains, as discussed in Chapter 11 (see Figure 11–32).

Cells in the lining of the gut and in many other organs, including the kidney, contain a variety of active symports in their plasma membrane that are similarly driven by the electrochemical gradient of Na⁺; each of these coupled pumps specifically imports a small group of related sugars or amino acids into the cell. But Na⁺-driven pumps that operate as antiports are also important for cells. For example, the *Na⁺–H⁺ exchanger* in the

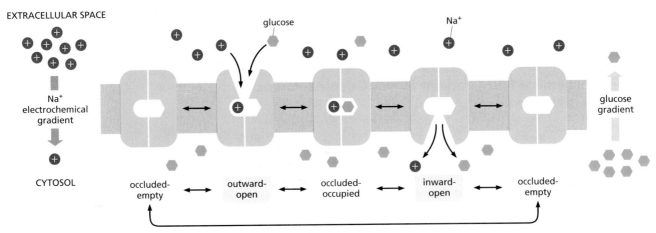

Figure 12–15 A glucose–Na⁺ symport protein uses the electrochemical Na⁺ gradient to drive the active import of glucose. The pump oscillates randomly between alternate states. In one state ("outward-open") the protein is open to the extracellular space; in another state ("inward-open") it is open to the cytosol. Although Na⁺ and glucose can each bind to the pump in either of these "open" states, the pump can transition between them only through an "occluded" state. For their symport, the occluded state can only be reached when both glucose and Na⁺ are bound ("occluded-occupied") or when neither is bound ("occluded-empty"). Because the Na⁺ concentration is high in the extracellular space, the Na⁺ binding site is readily occupied in the outward-open state, and the transporter will have to wait for a rare glucose molecule to bind. When that happens, the pump flips to the occluded-occupied state, trapping both solutes.

Because conformational transitions are reversible, one of two things can happen: the transporter could flip back to the outward-open state. In this case, the solutes would dissociate, and nothing would be gained. Alternatively, it could flip into the inward-open state, exposing the solute binding sites to the cytosol where the Na⁺ concentration is very low. Thus sodium readily dissociates and then is pumped back out of the cell by the Na⁺ pump (shown in Figure 12–11) to maintain the steep Na⁺ gradient. The transporter is now trapped with a partially occupied binding site until the glucose molecule also dissociates. At this point, with no solute bound, it can transition into the "occluded-empty" state and from there back to the outward-open state to repeat the transport cycle.

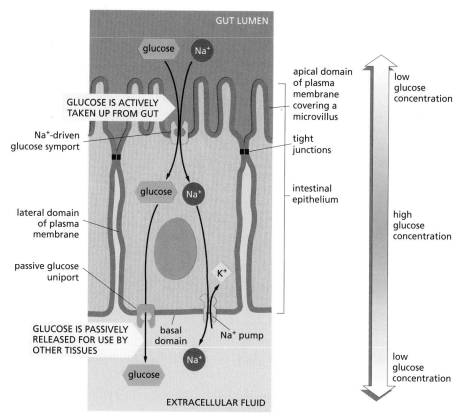

Figure 12–16 Two types of glucose transporters enable gut epithelial cells to transfer glucose across the epithelial lining of the gut. In addition, to keep the concentration of Na^+ in the cytosol low—and the Na^+ electrochemical gradient steep—Na^+ that enters the cell via the Na^+-driven glucose symport is pumped out by Na^+ pumps in the basal and lateral plasma membranes, as indicated. The diet provides ample Na^+ in the gut lumen to drive the Na^+-coupled glucose symport. The process is shown in Movie 12.4.

plasma membranes of many animal cells uses the downhill influx of Na^+ to pump H^+ out of the cell; it is one of the main devices that animal cells use to control the pH in their cytosol—preventing the cell interior from becoming too acidic.

Electrochemical H⁺ Gradients Drive Coupled Pumps in Plants, Fungi, and Bacteria

Plant cells, bacteria, and fungi (including yeasts) do not have Na^+ pumps in their plasma membrane. Instead of an electrochemical Na^+ gradient, they rely mainly on an electrochemical gradient of H^+ to import solutes into the cell. The gradient is created by **H^+ pumps** in the plasma membrane that pump H^+ out of the cell, thus setting up an electrochemical proton gradient across this membrane and creating an acid pH in the medium surrounding the cell. The import of many sugars and amino acids into bacterial cells is then mediated by H^+ symports, which use the electrochemical H^+ gradient in much the same way that animal cells use the electrochemical Na^+ gradient to import these nutrients.

In some photosynthetic bacteria, the H^+ gradient is created by the activity of light-driven H^+ pumps such as bacteriorhodopsin (see Figure 11–27). In other bacteria, fungi, and plants, the H^+ gradient is generated by H^+ pumps in the plasma membrane that use the energy of ATP hydrolysis to pump H^+ out of the cell; these H^+ pumps resemble the Na^+ pumps and Ca^{2+} pumps in animal cells discussed earlier.

A different type of ATP-dependent H^+ pump is found in the membranes of some intracellular organelles, such as the lysosomes of animal cells and the central vacuole of plant and fungal cells. These pumps—which resemble the turbine-like enzyme that synthesizes ATP in mitochondria and chloroplasts (discussed in Chapter 14)—actively transport H^+ out of the cytosol into the organelle, thereby helping to keep the pH of the

QUESTION 12–2

A rise in the intracellular Ca^{2+} concentration causes muscle cells to contract. In addition to an ATP-driven Ca^{2+} pump, muscle cells that contract quickly and regularly, such as those of the heart, have an additional type of Ca^{2+} pump—an antiport that exchanges Ca^{2+} for extracellular Na^+ across the plasma membrane. The majority of the Ca^{2+} ions that have entered the cell during contraction are rapidly pumped back out of the cell by this antiport, thus allowing the cell to relax. Ouabain and digitalis are used for treating patients with heart disease because they make heart muscle cells contract more strongly. Both drugs function by partially inhibiting the Na^+ pump in the plasma membrane of these cells. Can you propose an explanation for the effects of the drugs in the patients? What will happen if too much of either drug is taken?

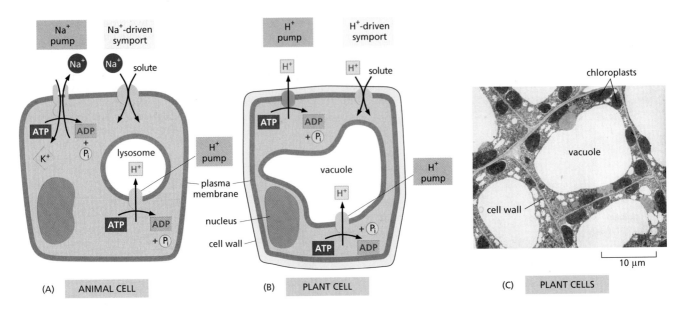

(A)   ANIMAL CELL

(B)   PLANT CELL

(C)   PLANT CELLS

Figure 12–17 Animal and plant cells use a variety of transmembrane pumps to drive the active transport of solutes.
(A) In animal cells, an electrochemical Na^+ gradient across the plasma membrane generated by the Na^+ pump, is used by symports to import various solutes. (B) In plant cells, an electrochemical gradient of H^+, set up by an H^+ pump, is often used for this purpose; a similar strategy is used by bacteria and fungi (not shown). The lysosomes in animal cells and the vacuoles in plant and fungal cells contain a similar H^+ pump in their membrane that pumps in H^+, helping to keep the internal environment of these organelles acidic. (C) An electron micrograph shows the vacuole in plant cells in a young tobacco leaf. (C, courtesy of J. Burgess.)

cytosol neutral and the pH of the interior of the organelle acidic. The acid environment in many organelles is crucial to their function, as we discuss in Chapter 15.

Some of the transmembrane pumps considered in this chapter are shown in **Figure 12–17** and are listed in **Table 12–2**.

ION CHANNELS AND THE MEMBRANE POTENTIAL

In principle, the simplest way to allow a small water-soluble molecule to cross from one side of a membrane to the other is to create a hydrophilic channel through which the molecule can pass. Channel proteins, or **channels**, perform this function in cell membranes, forming transmembrane pores that allow the passive movement of small water-soluble molecules into or out of the cell or organelle.

TABLE 12–2 SOME EXAMPLES OF TRANSMEMBRANE PUMPS			
Transporter	Location	Energy Source	Function
Na^+-driven glucose pump (glucose-Na^+ symport)	apical plasma membrane of kidney and intestinal cells	Na^+ gradient	active import of glucose
Na^+-H^+ exchanger	plasma membrane of animal cells	Na^+ gradient	active export of H^+ ions, pH regulation
Na^+ pump (Na^+-K^+ ATPase)	plasma membrane of most animal cells	ATP hydrolysis	active export of Na^+ and import of K^+
Ca^{2+} pump (Ca^{2+} ATPase)	plasma membrane of eukaryotic cells	ATP hydrolysis	active export of Ca^{2+}
Ca^{2+} pump (Ca^{2+} ATPase)	sarcoplasmic reticulum membrane of muscle cells and endoplasmic reticulum of most animal cells	ATP hydrolysis	active import of Ca^{2+} into sarcoplasmic reticulum
H^+ pump (H^+ ATPase)	plasma membrane of plant cells, fungi, and some bacteria	ATP hydrolysis	active export of H^+
H^+ pump (H^+ ATPase)	membranes of lysosomes in animal cells and of vacuoles in plant and fungal cells	ATP hydrolysis	active export of H^+ from cytosol into vacuole
Bacteriorhodopsin	plasma membrane of some bacteria	light	active export of H^+

A few channels form relatively large, aqueous pores: examples are the proteins that form *gap junctions* between two adjacent cells (see Figure 20–29) and the *porins* that form pores in the outer membrane of mitochondria and some bacteria (see Figure 11–24). But such large, permissive channels would lead to disastrous leaks if they directly connected the cytosol of a cell to the extracellular space. Thus most of the channels in the plasma membrane form narrow, highly selective pores. The *aquaporins* discussed earlier, for example, facilitate the flow of water across the plasma membrane of some prokaryotic and eukaryotic cells. These pores are structured in such a way that they allow the passive diffusion of uncharged water molecules, while prohibiting the movement of ions, including even the smallest ion, H^+.

The bulk of a cell's channels facilitate the passage of select inorganic ions. It is these *ion channels* we discuss in this section.

Ion Channels Are Ion-selective and Gated

Two important properties distinguish **ion channels** from simple holes in the membrane. First, they show *ion selectivity*, permitting some inorganic ions to pass but not others. Ion selectivity depends on the diameter and shape of the ion channel and on the distribution of the charged amino acids that line it. Each ion in aqueous solution is surrounded by a small shell of water molecules, most of which have to be shed for the ions to pass, in single file, through the selectivity filter in the narrowest part of the channel (**Figure 12–18**). An ion channel is narrow enough in places to force ions into contact with the channel wall so that only those ions of appropriate size and charge are able to pass (Movie 12.5).

The second important distinction between simple holes and ion channels is that ion channels are not continuously open. Ion transport would be of no value to the cell if the many thousands of ion channels in a cell membrane were open all the time and there were no means of controlling the flow of ions through them. Instead, ion channels open only briefly and then close again (**Figure 12–19**). As we discuss later, most ion channels are *gated*: a specific stimulus triggers them to switch between a closed and an open state by a change in their conformation.

Unlike a transporter, an open ion channel does not need to undergo conformational changes with each ion it passes, and so it has a large advantage over a transporter with respect to its maximum rate of

> **QUESTION 12–3**
>
> A transmembrane protein has the following properties: it has two binding sites, one for solute A and one for solute B. The protein can undergo a conformational change to switch between two states: either both binding sites are exposed exclusively on one side of the membrane or both binding sites are exposed exclusively on the other side of the membrane. The protein can switch between the two conformational states only if both binding sites are occupied or if both binding sites are empty, but cannot switch if only one binding site is occupied.
>
> A. What kind of protein do these properties define?
>
> B. Do you need to specify any additional properties to turn this protein into a symport that couples the movement of solute A up its concentration gradient to the movement of solute B down its electrochemical gradient?
>
> C. Write a set of rules that defines an antiport.

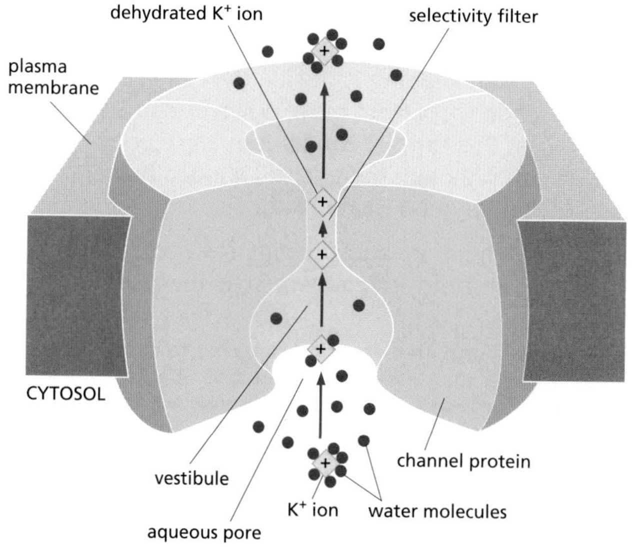

Figure 12–18 An ion channel has a selectivity filter that controls which inorganic ions it will allow to cross the membrane. Shown here is a portion of a bacterial K^+ channel. One of the four protein subunits has been omitted from the drawing to expose the interior structure of the pore (*blue*). From the cytosolic side, the pore opens into a vestibule that sits in the middle of the membrane. K^+ ions in the vestibule are still partially cloaked in their associated water molecules. The narrow selectivity filter, which connects the vestibule with the outside of the cell, is lined with polar groups (not shown) that form transient binding sites for the K^+ ions once the ions have shed their water shell. To observe this selectivity in action, see Movie 12.5.) (Adapted from D.A. Doyle et al., *Science* 280:69–77, 1998. With permission from the AAAS.)

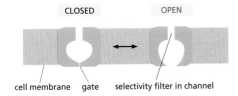

CLOSED   OPEN

cell membrane   gate   selectivity filter in channel

Figure 12–19 A typical ion channel fluctuates between closed and open conformations. The channel shown here in cross section forms a hydrophilic pore across the lipid bilayer only in the "open" conformation. As illustrated in Figure 12–18, the pore narrows to atomic dimensions in the selectivity filter, where the ion selectivity of the channel is largely determined.

transport. More than a million ions can pass through an open channel each second, which is 1000 times greater than the fastest rate of transfer known for any transporter. On the other hand, channels cannot couple the ion flow to an energy source to carry out active transport: most simply make the membrane transiently permeable to selected inorganic ions, mainly Na^+, K^+, Ca^{2+}, or Cl^-.

Thanks to active transport by pumps, the concentrations of most ions are far from equilibrium across a cell membrane. When an ion channel opens, therefore, ions usually flow through it, moving rapidly down their electrochemical gradients. This rapid shift of ions changes the membrane potential, as we discuss next.

Membrane Potential Is Governed by the Permeability of a Membrane to Specific Ions

Changes in membrane potential are the basis of electrical signaling in many types of cells, whether they are the nerve or muscle cells in animals, or the touch-sensitive cells of a carnivorous plant (**Figure 12–20**). Such electrical changes are mediated by alterations in the permeability of membranes to ions. In an animal cell that is in an unstimulated, or "resting," state, the negative charges on the organic molecules inside the cell are largely balanced by K^+, the predominant intracellular ion (see Table 12–1). K^+ is actively imported into the cell by the Na^+ pump, which generates a K^+ gradient across the plasma membrane. The plasma membrane, however, also contains a set of K^+ channels known as **K^+ leak channels**. These channels randomly flicker between open and closed states no matter what the conditions are inside or outside the cell; when they are open, they allow K^+ to move freely. In a resting cell, these are the main ion channels open in the plasma membrane, rendering the membrane much more permeable to K^+ than to other ions.

When the channels are open, K^+ has a tendency to flow out of the cell down its steep concentration gradient. This transfer of K^+ across the plasma membrane leaves behind unbalanced negative charges on the other side, creating a voltage difference, or membrane potential (**Figure 12–21**). Because this charge imbalance will oppose any further movement of K^+ out of the cell, an equilibrium condition is established in which the membrane potential keeping K^+ inside the cell is just strong enough to counteract the tendency of K^+ to move down its concentration gradient and out of the cell. In this state of equilibrium, the electrochemical gradient for K^+ is zero, even though there is still a much higher concentration of K^+ inside the cell than out (**Figure 12–22**).

The membrane potential in such steady-state conditions—in which the flow of positive and negative ions across the plasma membrane is

Figure 12–20 A Venus flytrap uses electrical signaling to capture its prey. The leaves snap shut in less than half a second when an insect moves on them. The response is triggered by touching any two of the three trigger hairs in succession in the center of each leaf. This mechanical stimulation opens ion channels in the plasma membrane and thereby sets off an electrical signal, which, by an unknown mechanism, leads to a rapid change in turgor pressure that closes the leaf. (Courtesy of Gabor Izso, Getty Images.)

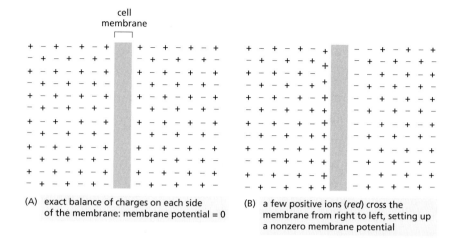

(A) exact balance of charges on each side of the membrane: membrane potential = 0

(B) a few positive ions (*red*) cross the membrane from right to left, setting up a nonzero membrane potential

Figure 12–21 The distribution of ions on either side of a cell membrane gives rise to its membrane potential. The membrane potential results from a thin (<1 nm) layer of ions close to the membrane, held in place by their electrical attraction to oppositely charged ions on the other side of the membrane. (A) When there is an exact balance of charges on either side of the membrane, there is no membrane potential. (B) When ions of one type cross the membrane, they set up a charge difference between the two sides of the membrane that creates a membrane potential. The number of ions that must move across the membrane to set up a membrane potential is a tiny fraction of all those present on either side. In the case of the plasma membrane in animal cells, for example, 6000 K^+ ions crossing 1 μm^2 of membrane are enough to shift the membrane potential by about 100 mV; the number of K^+ ions in 1 μm^3 of cytosol is 70,000 times larger than this.

precisely balanced, so that no further difference in charge accumulates across the membrane—is called the **resting membrane potential**. A simple formula called the **Nernst equation** expresses this equilibrium quantitatively and makes it possible to calculate the theoretical resting membrane potential if the ion concentrations on either side of the membrane are known (**Figure 12–23**). In animal cells, the resting membrane potential—which varies between –20 to –200 mV—is chiefly a reflection of the electrochemical K^+ gradient across the plasma membrane, because, at rest, the plasma membrane is chiefly permeable to K^+, and K^+ is the main positive ion inside the cell.

When a cell is stimulated, other ion channels in the plasma membrane open, changing the membrane's permeability to those ions. Whether the

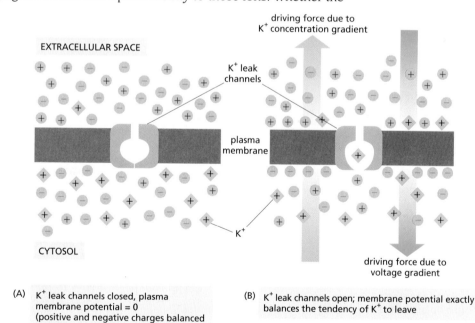

(A) K^+ leak channels closed, plasma membrane potential = 0 (positive and negative charges balanced exactly)

(B) K^+ leak channels open; membrane potential exactly balances the tendency of K^+ to leave

Figure 12–22 The K^+ concentration gradient and K^+ leak channels play major parts in generating the resting membrane potential across the plasma membrane in animal cells. (A) A hypothetical situation in which the K^+ leak channels are closed, and the membrane potential is zero. (B) As soon as the channels open, K^+ will tend to leave the cell, moving down its concentration gradient. Assuming the membrane contains no open channels permeable to other ions, K^+ will cross the membrane, but negative ions will be unable to follow. The resulting charge imbalance gives rise to a membrane potential that tends to drive K^+ back into the cell. At equilibrium, the effect of the K^+ concentration gradient is exactly balanced by the effect of the membrane potential, and there is no net movement of K^+ across the membrane.
  The Na^+ pump also contributes to the resting potential—both by helping to establish the K^+ gradient and by pumping 3 Na^+ ions out of the cell for every 2 K^+ ions it pumps in (see Figure 12–11), thereby helping to keep the inside of the cell more negative than the outside (not shown here).

The force tending to drive an ion across a membrane is made up of two components: one due to the electrical membrane potential and one due to the concentration gradient of the ion. At equilibrium, the two forces are balanced and satisfy a simple mathematical relationship given by the

Nernst equation

$$V = 62 \log_{10}(C_o/C_i)$$

where V is the membrane potential in millivolts, and C_o and C_i are the outside and inside concentrations of the ion, respectively. This form of the equation assumes that the ion carries a single positive charge and that the temperature is 37°C.

Figure 12–23 The Nernst equation can be used to calculate the resting potential of a membrane. The relevant ion concentrations are those on either side of the membrane. From this equation, we see that each tenfold change in the ion concentration ratio (C_o/C_i) alters the membrane potential by 62 millivolts.

ions enter or leave the cell depends on the direction of their electrochemical gradients. Thus the membrane potential at any time depends on both the state of the membrane's ion channels and the ion concentrations on either side of the plasma membrane. Bulk changes in ion concentrations cannot occur quickly enough to drive the rapid changes in membrane potential that are associated with electrical signaling. Instead, it is the rapid opening and closing of ion channels, which occurs within milliseconds, that matters most for this type of cell signaling.

Ion Channels Randomly Snap Between Open and Closed States

Measuring changes in electrical current is the main method used to study ion movements and ion channels in living cells. Amazingly, electrical recording techniques can detect and measure the electric current flowing through a single channel molecule. The procedure developed for doing this is known as **patch-clamp recording**, and it provides a direct and surprising picture of how individual ion channels behave.

In patch-clamp recording, a fine glass tube is used as a *microelectrode* to isolate and make electrical contact with a small area of the membrane at the surface of the cell (**Figure 12–24**). The technique makes it

Figure 12–24 Patch-clamp recording is used to monitor ion channel activity. First, a microelectrode is made by heating a glass tube and pulling it to create an extremely fine tip with a diameter of no more than a few micrometers; the tube is then filled with an aqueous conducting solution, and the tip is pressed against the cell surface. (A) With gentle suction, a tight electrical seal is formed where the cell membrane contacts the mouth of the microelectrode. Because of the extremely tight seal, current can enter or leave the microelectrode only by passing through the ion channel or channels in the patch of membrane covering its tip. (B) To expose the cytosolic face of the membrane, the patch of membrane held in the microelectrode can be torn from the cell. The advantage of the detached patch is that it is easy to alter the composition of the solution on either side of the membrane to test the effect of various solutes on channel behavior. (C) A micrograph showing an isolated nerve cell held in a suction pipette (the tip of which is shown on the left), while a microelectrode is being used for patch-clamp recording. (D) The circuitry for patch-clamp recording. At the open end of the microelectrode, a metal wire is inserted. Current that enters the microelectrode through ion channels in the small patch of membrane covering its tip passes via the wire, through measuring instruments, back into the bath of medium surrounding the cell or the detached patch. (C, from T.D. Lamb, H.R. Mathews, and V. Torre, *J. Physiol.* 37:315–349, 1986. With permission from Blackwell Publishing.)

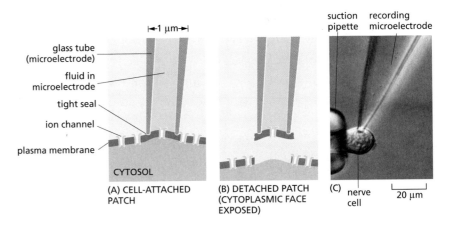

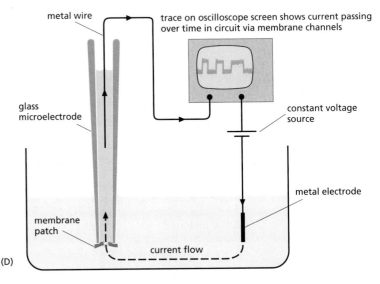

possible to record the activity of ion channels in all sorts of cell types—particularly in large nerve and muscle cells, which are famous for their electrical activities. By varying the concentrations of ions on either side of the patch, one can test which ions will go through the channels in the patch. With the appropriate electronic circuitry, the voltage across the membrane patch—that is, the membrane potential—can also be set and held "clamped" at any chosen value (hence the term "patch-clamp"). The ability to expose the membrane to different voltages makes it possible to see how changes in membrane potential affect the opening and closing of the ion channels in the membrane.

With a sufficiently small area of membrane in the patch, sometimes only a single ion channel will be present. Modern electrical instruments are sensitive enough to reveal the ion flow through a single channel, detected as a minute electric current (of the order of 10^{-12} ampere or 1 picoampere). Monitoring individual ion channels in this way revealed something surprising about the way they behave: even when conditions are held constant, the currents abruptly appear and disappear, as though an on/off switch were being jiggled randomly (**Figure 12–25**). This behavior indicates that the channel has moving parts and is snapping back and forth between open and closed conformations (see Figure 12–19) as the channel is knocked from one conformation to the other by the random thermal movements of the molecules in its environment. Patch-clamp recording was the first technique that could monitor such conformational changes, and the picture it paints—of a jerky piece of machinery subjected to constant external buffeting—is now known to apply also to other proteins with moving parts.

The activity of each ion channel is very much "all-or-none": when an ion channel is open, it is fully open; when it is closed, it is fully closed. That raises a fundamental question: If ion channels randomly snap between open and closed conformations even when conditions on each side of the membrane are held constant, how can their state be regulated by conditions inside or outside the cell? The answer is that, when the appropriate conditions change, the random behavior continues but with a greatly changed bias: if the altered conditions tend to open the channel, for example, the channel will now spend a much greater proportion of its time in the open conformation, although it will not remain open continuously (see Figure 12–25).

Different Types of Stimuli Influence the Opening and Closing of Ion Channels

There are more than a hundred types of ion channels, and even simple organisms can possess many different types. The nematode worm *C. elegans*, for example, has genes that encode 68 different but related K^+ channels alone. Ion channels differ from one another primarily with

Figure 12–25 The behavior of a single ion channel can be observed using the patch-clamp technique. The voltage (the membrane potential) across the isolated patch of membrane is held constant during the recording. In this example, the neurotransmitter acetylcholine is present, and the membrane patch from a muscle cell contains a single channel protein that is responsive to acetylcholine (discussed later, see Figure 12–41). As seen, this ion channel opens to allow passage of positive ions when acetylcholine binds to the exterior face of the channel. But even when acetylcholine is bound to the channel, as is the case during the three channel openings shown here, the channel does not remain open all the time. Instead, it flickers between open and closed states. Note that how long the channel remains open is variable. If acetylcholine were not present, the channel would only rarely open. (Courtesy of David Colquhoun.)

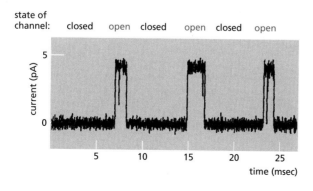

Figure 12–26 Different types of gated ion channels respond to different types of stimuli. Depending on the type of channel, the probability of gate opening is controlled by (A) a change in the voltage difference across the membrane, (B) the binding of a chemical ligand to the extracellular face of a channel, (C) ligand binding to the intracellular face of a channel, or (D) mechanical stress.

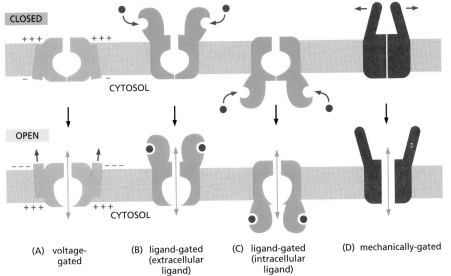

(A) voltage-gated

(B) ligand-gated (extracellular ligand)

(C) ligand-gated (intracellular ligand)

(D) mechanically-gated

respect to their *ion selectivity*—the type of ions they allow to pass—and their *gating*—the conditions that influence their opening and closing. For a **voltage-gated channel**, the probability of being open is controlled by the membrane potential (**Figure 12–26A**). For a **ligand-gated channel**, opening is controlled by the binding of some molecule (the ligand) to the channel (**Figure 12–26B and C**). For a **mechanically-gated channel**, opening is controlled by a mechanical force applied to the channel (**Figure 12–26D**).

The *auditory hair* cells in the ear are an important example of cells that depend on mechanically-gated channels. Sound vibrations pull the channels open, causing ions to flow into the hair cells; this ion flow sets up an electrical signal that is transmitted from the hair cell to the auditory nerve, which conveys the signal to the brain (**Figure 12–27**).

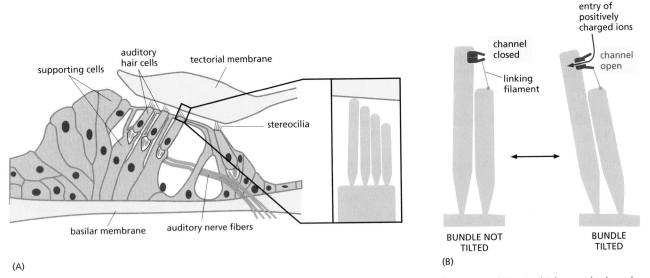

Figure 12–27 Mechanically-gated ion channels allow us to hear. (A) A section through the organ of Corti, which runs the length of the cochlea, the auditory portion of the inner ear. Each auditory hair cell has a tuft of spiky extensions called stereocilia projecting from its upper surface. The hair cells are embedded in an epithelial sheet of supporting cells, which is sandwiched between the *basilar membrane* below and the *tectorial membrane* above. (These are not lipid bilayer membranes but sheets of extracellular matrix.) (B) Sound vibrations cause the basilar membrane to vibrate up and down, causing the stereocilia to tilt. Each stereocilium in the staggered array of stereocilia on each hair cell is attached to the next shorter stereocilium by a fine filament. The tilting stretches the filaments, which pull open mechanically-gated ion channels in the stereocilium plasma membrane, allowing positively charged ions to enter from the surrounding fluid (Movie 12.6). The influx of ions activates the hair cells, which stimulate underlying nerve endings of the auditory nerve fibers that relay the auditory signal to the brain.

The hair-cell mechanism is astonishingly sensitive: the faintest sounds we can hear have been estimated to stretch the filaments by an average of about 0.04 nm, which is less than the diameter of a hydrogen ion (Movie 12.7).

(A) time 0 sec (B) 1 sec (C) 3 sec (D) 5 sec

Voltage-gated Ion Channels Respond to the Membrane Potential

Voltage-gated ion channels play a major role in propagating electrical signals along all nerve cell processes, such as those that relay signals from our brain to our toe muscles. But voltage-gated ion channels are present in many other cell types, too, including muscle cells, egg cells, protozoans, and even plant cells, where they enable electrical signals to travel from one part of the plant to another, as in the leaf-closing response of a *Mimosa pudica* plant (**Figure 12–28**).

Voltage-gated ion channels have domains called *voltage sensors* that are extremely sensitive to changes in the membrane potential: changes above a certain threshold value exert sufficient electrical force on these domains to encourage the channel to switch from its closed to its open conformation. As discussed earlier, a change in the membrane potential does not affect how wide the channel is open, but instead alters the probability that it will open (see Figure 12–25). Thus, in a large patch of membrane containing many molecules of the channel protein, one might find that on average 10% of them are open at any instant when the membrane is at one potential, whereas 90% are open after this potential changes.

When one type of voltage-gated ion channel opens, the membrane potential of the cell can change. This in turn can activate or inactivate other voltage-gated ion channels. This control circuit, from ion channels → membrane potential → ion channels, is fundamental to all electrical signaling in cells. To see how such a circuit can be used for electrical signaling, we now turn to nerve cells: they—more than any other cell type—have made a profession of electrical signaling, and they employ ion channels in very sophisticated ways.

ION CHANNELS AND NERVE CELL SIGNALING

The fundamental task of a nerve cell, or **neuron**, is to receive, integrate, and transmit signals. Neurons carry signals inward from sense organs, such as eyes and ears, to the *central nervous system*—the brain and spinal cord. In the central nervous system, neurons signal from one to another through networks of enormous complexity, allowing the brain and spinal cord to analyze, interpret, and respond to the signals coming in from the sense organs.

Every neuron consists of a *cell body*, which contains the nucleus and has a number of long, thin extensions radiating outward from it. Usually, a neuron has one long extension called an **axon**, which conducts electrical signals away from the cell body toward distant target cells; it also usually has several shorter, branching extensions called **dendrites**, which radiate from the cell body like antennae and provide an enlarged surface area to receive signals from the axons of other neurons (**Figure 12–29**). The

Figure 12–28 Both mechanically-gated and voltage-gated ion channels underlie the leaf-closing response in the touch-sensitive plant, *Mimosa pudica*. (A) Resting leaf. (B and C) Successive responses to touch. A few seconds after the leaf is touched, the leaflets snap shut. The response involves the opening of mechanically-gated ion channels in touch-sensitive sensory cells, which then pass a signal to cells containing voltage-gated ion channels, generating an electric impulse. When the impulse reaches specialized hinge cells at the base of each leaflet, a rapid loss of water by these cells occurs, causing the leaflets to fold into a closed conformation suddenly and progressively down the leaf stalk.

QUESTION 12–4

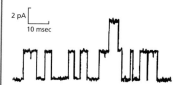

2 pA
10 msec

Figure Q12–4 (above) shows a recording from a patch-clamp experiment in which the electrical current passing across a patch of membrane is measured as a function of time. The membrane patch was plucked from the plasma membrane of a muscle cell by the technique shown in Figure 12–24 and contains molecules of the acetylcholine receptor, which is a ligand-gated cation channel that is opened by the binding of acetylcholine to the extracellular face of the channel. To obtain a recording, acetylcholine was added to the solution inside the microelectrode. (A) Describe what you can learn about the channels from this recording. (B) How would the recording differ if acetylcholine were (i) omitted or (ii) added to the solution outside the microelectrode only?

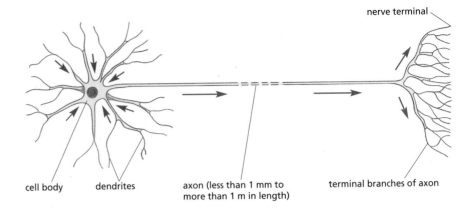

Figure 12–29 **A typical neuron has a cell body, a single axon, and multiple dendrites.** The axon conducts electrical signals away from the cell body toward its target cells, while the multiple dendrites receive signals from the axons of other neurons. The *red* arrows indicate the direction in which signals travel.

cell body   dendrites   axon (less than 1 mm to more than 1 m in length)   terminal branches of axon

nerve terminal

QUESTION 12–5

Using the Nernst equation and the ion concentrations given in Table 12–1 (p. 385), calculate the equilibrium membrane potential of K^+ and Na^+—that is, the membrane potential where there would be no net movement of the ion across the plasma membrane (assume that the concentration of intracellular Na^+ is 10 mM). What membrane potential would you predict in a resting animal cell? Explain your answer. What would happen if a large number of Na^+ channels suddenly opened, making the membrane much more permeable to Na^+ than to K^+? (Note that because few ions need to move across the membrane to change the charge distribution across the membrane drastically, you can safely assume that the ion concentrations on either side of the membrane do not change significantly.) What would you predict would happen next if the Na^+ channels closed again?

---

axon commonly divides at its far end into many branches, each of which ends in a **nerve terminal**, so that the neuron's message can be passed simultaneously to many target cells—muscle or gland cells or other neurons. Likewise, the branching of the dendrites can be extensive, in some cases, sufficient to receive as many as 100,000 inputs on a single neuron.

No matter what the meaning of the signal a neuron carries—whether it is visual information from the eye, a motor command to a muscle, or one step in a complex network of neural processing in the brain—the form of the signal is always the same: it consists of changes in the electrical potential across the neuron's plasma membrane.

Action Potentials Allow Rapid Long-Distance Communication Along Axons

A neuron is stimulated by a signal—typically from another neuron—delivered to a localized site on its surface. This signal initiates a change in the membrane potential at that site. To transmit the signal onward, this local change in membrane potential has to spread from this point, which is usually on a dendrite or the cell body, to the axon terminals, which relay the signal to the next cells in the pathway—forming a *neural circuit*. The distances required can be substantial: a signal that leaves a motor neuron in your spinal cord may have to travel a meter or more before it reaches a muscle in your foot.

The local change in membrane potential generated by a signal can spread passively along an axon or a dendrite to adjacent regions of the plasma membrane. Such a passively spread signal, however, rapidly becomes weaker with increasing distance from the source. Over short distances, this weakening is unimportant. But for long-distance communication, such *passive spread* is inadequate.

Neurons solve this long-distance communication problem by employing an active signaling mechanism. Here a local electrical stimulus of sufficient strength triggers an explosion of electrical activity in the plasma membrane that propagates rapidly along the membrane of the axon, continuously renewing itself all along the way. This traveling wave of electrical excitation, known as an **action potential**, or a *nerve impulse*, can carry a message, without the signal weakening, all the way from one end of a neuron to the other, at speeds of up to 100 meters per second.

The early research that established this mechanism of electrical signaling along axons was done on the giant axon of the squid (**Figure 12–30**). This axon has such a large diameter that it is possible to record its electrical activity from an electrode inserted directly into it (**How We Know**, pp. 406–407). From such studies, it was deduced how action potentials are the direct consequence of the properties of voltage-gated ion channels in the axonal plasma membrane, as we now explain.

Action Potentials Are Mediated by Voltage-gated Cation Channels

When a neuron is stimulated, the membrane potential of the plasma membrane shifts to a less negative value (that is, toward zero). If this **depolarization** is sufficiently large, it will cause **voltage-gated Na^+ channels** in the membrane to open transiently at the site. As these channels flicker open, they allow a small amount of Na^+ to enter the cell down its steep electrochemical gradient. The influx of positive charge depolarizes the membrane further (that is, it makes the membrane potential even less negative), thereby opening additional voltage-gated Na^+ channels and causing still further depolarization. This process continues in an explosive, self-amplifying fashion until, within about a millisecond, the membrane potential in the local region of the neuron's plasma membrane has shifted from its resting value of about –60 mV to about +40 mV (**Figure 12–31**).

The voltage of +40 mV is close to the membrane potential at which the electrochemical driving force for movement of Na^+ across the membrane is zero—that is, at which the effects of the membrane potential and the concentration gradient for Na^+ are equal and opposite, so that Na^+ has no further tendency to enter or leave the cell. If the channels continued to respond to the altered membrane potential, the cell would get stuck with most of its voltage-gated Na^+ channels open.

The cell is saved from this fate because the Na^+ channels have an automatic inactivating mechanism—a kind of "timer" that causes them to rapidly adopt (within a millisecond or so) a special inactivated conformation, in which the channel is closed, even though the membrane is still depolarized. The Na^+ channels remain in this *inactivated state* until the membrane potential has returned to its initial negative value. A

Figure 12–30 The squid *Loligo* has a nervous system that is adept at responding rapidly to threats in the animal's environment. Among the nerve cells that make up this escape system is one that possesses a "giant axon," with a very large diameter. Long before patch clamping allowed recordings from single ion channels in small cells (see Figure 12–24), the squid giant axon was routinely used to record and study action potentials.

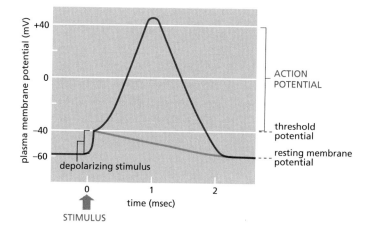

Figure 12–31 An action potential is triggered by a depolarization of a neuron's plasma membrane. The resting membrane potential in this neuron is –60 mV, and a stimulus that depolarizes the plasma membrane to about –40 mV (the threshold potential) is sufficient to open voltage-gated Na^+ channels in the membrane and thereby trigger an action potential. The membrane rapidly depolarizes further, and the membrane potential (*red* curve) swings past zero, reaching +40 mV before it returns to its resting negative value as the action potential terminates. The *green* curve shows how the membrane potential would simply have relaxed back to the resting value after the initial depolarizing stimulus if there had been no amplification by voltage-gated ion channels in the plasma membrane.

SQUID REVEAL SECRETS OF MEMBRANE EXCITABILITY

Each spring, *Loligo pealei* migrate to the shallow waters off Cape Cod on the eastern coast of the United States. There they spawn, launching the next generation of squid. But more than just meeting and breeding, these animals provide neuroscientists summering at the Marine Biological Laboratory in Woods Hole, Massachusetts, with a golden opportunity to study the mechanism of electrical signaling along nerve axons.

Like most animals, squid survive by catching prey and escaping predators. Fast reflexes and an ability to accelerate rapidly and make sudden changes in swimming direction help them avoid danger while chasing down a decent meal. Squid derive their speed and agility from a specialized biological jet propulsion system: they draw water into their mantle cavity and then contract their muscular body wall to expel the collected water rapidly through a tubular siphon, thus propelling themselves through the water.

Controlling such quick and coordinated muscle contraction requires a nervous system that can convey signals with great speed down the length of the animal's body. Indeed, *Loligo pealei* possesses some of the largest nerve cell axons found in nature. Squid giant axons can reach 10 cm in length and are over 100 times the diameter of a mammalian axon—about the width of a pencil lead. Generally speaking, the larger the diameter of an axon, the more rapidly signals can travel along its length.

In the 1930s, scientists first started to take advantage of the squid giant axon for studying the electrophysiology of the nerve cell. Because of its relatively large size, an investigator can isolate an individual axon and insert an electrode into it to measure the axon's membrane potential and monitor its electrical activity. This experimental system allowed researchers to address a variety of questions, including which ions are important for establishing the resting membrane potential and for initiating and propagating an action potential, and how changes in the membrane potential control ion permeability.

Setup for action

Because the squid axon is so long and wide, an electrode made from a glass capillary tube containing a conducting solution can be thrust down the axis of the isolated axon so that its tip lies deep in the cytoplasm (**Figure 12–32A**). This setup allowed investigators to measure the voltage difference between the inside and the outside of the axon—that is, the membrane potential—as an action potential sweeps past the tip of the electrode (**Figure 12–32B**). The action potential itself was triggered by applying a brief electrical stimulus to one end of the axon. It didn't matter which end was stimulated, as the action potential could travel in either direction; it also didn't matter how big the stimulus was, as long as it exceeded a certain threshold (see Figure 12–31), indicating that an action potential is all or nothing.

Once researchers could reliably generate and measure an action potential, they could use the preparation to answer other questions about membrane excitability. For example, which ions are critical for an action potential? The three most plentiful ions, both inside and outside an axon, are Na^+, K^+, and Cl^-. Do they have

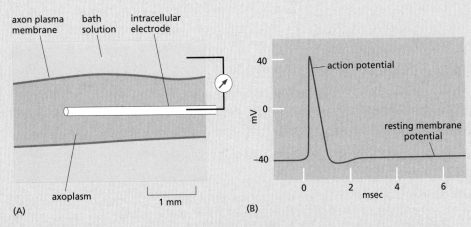

Figure 12–32 Scientists can study nerve cell excitability using an isolated axon from squid. An electrode can be inserted into the cytoplasm (axoplasm) of a squid giant axon (A) to measure the resting membrane potential and monitor action potentials induced when the axon is electrically stimulated (B).

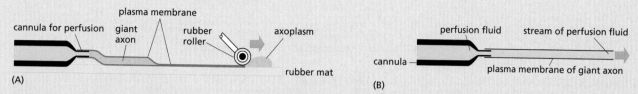

Figure 12–33 The cytoplasm in a squid axon can be removed and replaced with an artificial solution of pure ions. (A) The axon cytoplasm is extruded using a rubber roller. (B) A perfusion fluid containing the desired concentration of ions is pumped gently through the emptied-out axon.

equal importance when it comes to the action potential? Because the squid axon is so large and robust, it was possible to extrude the cytoplasm from the axon like toothpaste from a tube (**Figure 12–33A**). The emptied-out axon could then be reinflated by perfusing it with a pure solution of Na^+, K^+, or Cl^-, (**Figure 12–33B**). Thus, the ions inside the axon and in the bath solution (see Figure 12–32) could be varied independently. In this way, the researchers could show that the axon would generate a normal action potential if and only if the concentrations of Na^+ and K^+ approximated the natural concentrations found inside and outside the cell. Thus, they concluded that the cell components crucial to the action potential are the plasma membrane, Na^+ and K^+ ions, and the energy provided by the concentration gradients of these ions across the membrane; all other components, including other sources of metabolic energy, were presumably removed by the perfusion.

Channel traffic

Once Na^+ and K^+ had been singled out as critical for an action potential, the question then became: What does each of these ions contribute to the action potential? How permeable is the membrane to each, and how does the membrane permeability change as an action potential sweeps by? Again, the squid giant axon provided some answers. The concentrations of Na^+ and K^+ inside and outside the axon could be altered, and the effects of these changes on the membrane potential could be measured directly. From such studies, it was determined that, at rest, the membrane potential of an axon is close to the equilibrium potential for K^+: when the external concentration of K^+ was varied, the resting potential of the axon changed roughly in accordance with the Nernst equation (see Figure 12–23). They concluded that at rest, the membrane is chiefly permeable to K^+; we now know that K^+ leak channels provide the main pathway these ions can take through the resting plasma membrane.

The situation for Na^+ is very different. When the external concentration of Na^+ was varied, there was no effect on the resting potential of the axon. However, the height of

the peak of the action potential varied with the concentration of Na^+ outside the axon (**Figure 12–34**). During the action potential, therefore, the membrane appeared to be chiefly permeable to Na^+, presumably as the result of the opening of Na^+ channels. In the aftermath of the action potential, the Na^+ permeability decreased and the membrane potential reverted to a negative value, which depended on the external concentration of K^+. As the membrane lost its permeability to Na^+, it became even more permeable to K^+ than before, presumably because additional K^+ channels opened, accelerating the resetting of the membrane potential to the resting state, and readying the membrane for the next action potential.

These studies on the squid giant axon made an enormous contribution to our understanding of nerve cell excitability, and the researchers who made these discoveries in the 1940s and 1950s—Alan Hodgkin and Andrew Huxley—received a Nobel Prize in 1963. However, it was years before the various ion channel proteins that they had hypothesized to exist would be biochemically identified. We now know the three-dimensional structures of many of these channel proteins, allowing us to marvel at the fundamental beauty of these molecular machines.

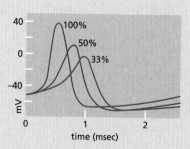

Figure 12–34 The shape of the action potential depends on the concentration of Na^+ outside the squid axon. Shown here are action potentials recorded when the external medium contains 100%, 50%, or 33% of the normal extracellular concentration of Na^+.

Figure 12–35 A voltage-gated Na⁺ channel can flip from one conformation to another, depending on the membrane potential. When the membrane is at rest and highly polarized, positively charged amino acids in its voltage sensors (*red bars*) are oriented by the membrane potential in a way that keeps the channel in its closed conformation. When the membrane is depolarized, the voltage sensors shift, changing the channel's conformation so the channel has a high probability of opening. But in the depolarized membrane, the inactivated conformation is even more stable than the open conformation, and so, after a brief period spent in the open conformation, the channel becomes temporarily inactivated and cannot open. The *red* arrows indicate the sequence that follows a sudden depolarization, and the *black* arrow indicates the return to the original conformation after the membrane has repolarized.

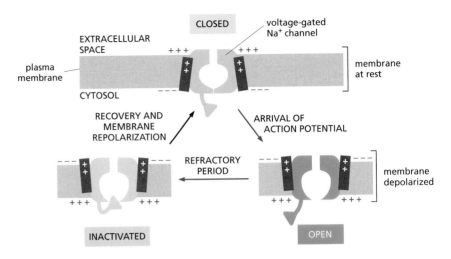

QUESTION 12–6

Explain as precisely as you can but in no more than 100 words the ionic basis of an action potential and how it is passed along an axon.

schematic illustration of these three distinct states of the voltage-gated Na⁺ channel—*closed, open,* and *inactivated*—is shown in **Figure 12–35**. How they contribute to the rise and fall of an action potential is shown in **Figure 12–36**.

During an action potential, Na⁺ channels do not act alone. The depolarized axonal membrane is helped to return to its resting potential by the opening of *voltage-gated K⁺ channels.* These also open in response to depolarization, but not as promptly as the Na⁺ channels, and they stay open as long as the membrane remains depolarized. As the local depolarization reaches its peak, K⁺ ions (carrying positive charge) therefore start to flow out of the cell through these newly opened K⁺ channels down their electrochemical gradient, temporarily unhindered by the negative membrane potential that normally restrains them in the resting cell. The rapid outflow of K⁺ through the voltage-gated K⁺ channels brings the membrane back to its resting state much more quickly than could be achieved by K⁺ outflow through the K⁺ leak channels alone.

Once it begins, the self-amplifying depolarization of a small patch of plasma membrane quickly spreads outward: Na⁺ flowing in through open Na⁺ channels begins to depolarize the neighboring region of the membrane, which then goes through the same self-amplifying cycle. In this way, an action potential spreads outward as a traveling wave from the initial site of depolarization, eventually reaching the axon terminals (**Figure 12–37**).

Faced with the consequences of the Na⁺ and K⁺ fluxes caused by a passing action potential, Na⁺ pumps in the axon plasma membrane labor

Figure 12–36 Voltage-gated Na⁺ channels change their conformation during an action potential. In this example, the action potential is triggered by a brief pulse of electric current (A), which partially depolarizes the membrane, as shown in the plot of membrane potential versus time in (B). (B) The course of the action potential (*red* curve), which reflects the opening and subsequent inactivation of voltage-gated Na⁺ channels, whose state is shown in (C). Even if restimulated, the plasma membrane cannot produce a second action potential until the Na⁺ channels have returned from the inactivated to the closed conformation (see Figure 12–35). Until then, the membrane is resistant, or refractory, to stimulation.

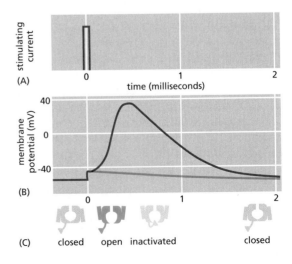

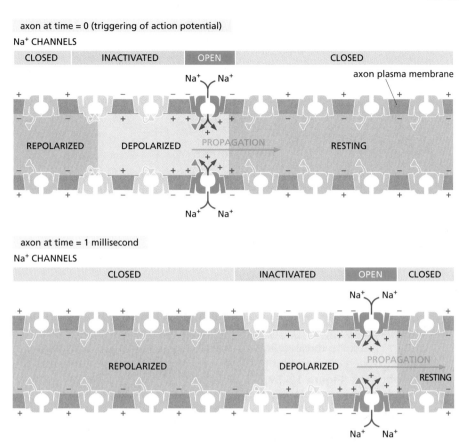

axon at time = 0 (triggering of action potential)

Na⁺ CHANNELS

Figure 12–37 An action potential propagates along the length of an axon. The changes in the Na^+ channels and the consequent flow of Na^+ across the membrane (*red* arrows) alters the membrane potential and gives rise to the traveling action potential, as shown here and in Movie 12.8. The region of the axon with a depolarized membrane is shaded in *blue*. Note that an action potential can only travel forward from the site of depolarization. This is because Na^+-channel inactivation in the aftermath of an action potential prevents the advancing front of depolarization from spreading backward (see also Figure 12–36).

continuously to restore the ion gradients of the resting cell. It is remarkable that the human brain consumes 20% of the total energy generated from the metabolism of food, mostly to power this pump.

Voltage-gated Ca^{2+} Channels in Nerve Terminals Convert an Electrical Signal into a Chemical Signal

When an action potential reaches the nerve terminals at the end of an axon, the signal must somehow be relayed to the *target cells* that the terminals contact—usually neurons or muscle cells. The signal is transmitted to the target cells at specialized junctions known as **synapses**. At most synapses, the plasma membranes of the cells transmitting and receiving the message—the *presynaptic* and the *postsynaptic* cells, respectively—are separated from each other by a narrow *synaptic cleft* (typically 20 nm across), which the electrical signal cannot cross. To transmit the message across this gap, the electrical signal is converted into a chemical signal, in the form of a small, secreted signal molecule known as a **neurotransmitter**. Neurotransmitters are initially stored in the nerve terminals within membrane-enclosed **synaptic vesicles** (**Figure 12–38**).

When an action potential reaches the nerve terminal, some of the synaptic vesicles fuse with the plasma membrane, releasing their neurotransmitters into the synaptic cleft. This link between the arrival of an action potential and the secretion of neurotransmitter involves the activation of yet another type of voltage-gated cation channel. The depolarization of the nerve-terminal plasma membrane caused by the arrival of the action potential transiently opens *voltage-gated Ca^{2+} channels*, which are concentrated in the plasma membrane of the presynaptic nerve terminal. Because the Ca^{2+} concentration outside the terminal is more than 1000 times greater than the free Ca^{2+} concentration in its cytosol (see Table 12–1), Ca^{2+} rushes into the nerve terminal through the open channels.

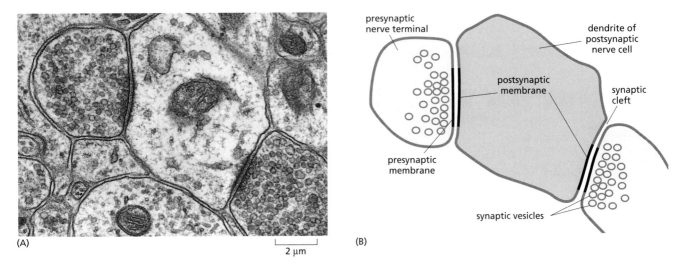

(A)

2 μm

(B)

Figure 12–38 Neurons connect to their target cells at synapses. An electron micrograph (A) and drawing (B) of a cross section of two nerve terminals (*yellow*) forming synapses on a single nerve cell dendrite (*blue*) in the mammalian brain. Neurotransmitters carry the signal across the synaptic cleft that separates the presynaptic and postsynaptic cell. The neurotransmitter in the presynaptic terminal is contained within synaptic vesicles, which release it into the synaptic cleft. Note that both the presynaptic and postsynaptic membranes are thickened and highly specialized at the synapse. (A, courtesy of Cedric Raine.)

The resulting increase in Ca^{2+} concentration in the cytosol of the terminal immediately triggers the membrane fusion that releases the neurotransmitter. Thanks to these voltage-gated Ca^{2+} channels, the electrical signal has now been converted into a chemical signal that is secreted into the synaptic cleft (**Figure 12–39**).

Transmitter-gated Ion Channels in the Postsynaptic Membrane Convert the Chemical Signal Back into an Electrical Signal

The released neurotransmitter rapidly diffuses across the synaptic cleft and binds to *neurotransmitter receptors* concentrated in the postsynaptic plasma membrane of the target cell. The binding of neurotransmitter to its receptors causes a change in the membrane potential of the target cell, which—if large enough—triggers the cell to fire an action potential. The neurotransmitter is then quickly removed from the synaptic cleft—either by enzymes that destroy it, by pumping it back into the nerve terminals that released it, or by uptake into neighboring non-neuronal cells. This rapid removal of the neurotransmitter limits the duration and spread of the signal and ensures that, when the presynaptic cell falls quiet, the postsynaptic cell will do the same.

Figure 12–39 An electrical signal is converted into a secreted chemical signal at a nerve terminal. When an action potential reaches a nerve terminal, it opens voltage-gated Ca^{2+} channels in the plasma membrane, allowing Ca^{2+} to flow into the terminal. The increased Ca^{2+} in the nerve terminal stimulates the synaptic vesicles to fuse with the plasma membrane, releasing their neurotransmitter into the synaptic cleft—a process called exocytosis (discussed in Chapter 15).

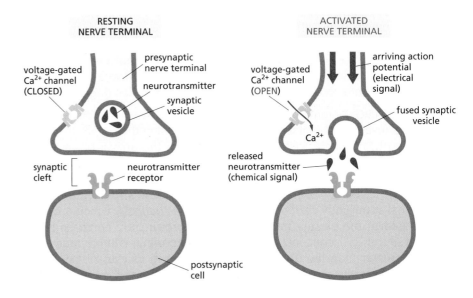

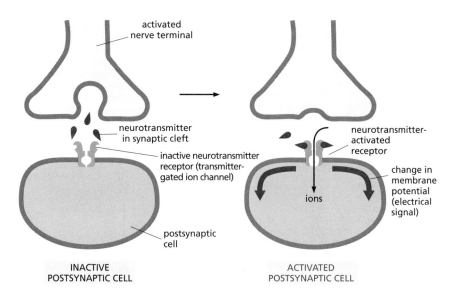

Figure 12–40 A chemical signal is converted into an electrical signal by postsynaptic transmitter-gated ion channels at a synapse. The released neurotransmitter binds to and opens the transmitter-gated ion channels in the plasma membrane of the postsynaptic cell. The resulting ion flows alter the membrane potential of the postsynaptic cell, thereby converting the chemical signal back into an electrical one (Movie 12.9).

Neurotransmitter receptors can be of various types; some mediate relatively slow effects in the target cell, whereas others trigger more rapid responses. Rapid responses—on a time scale of milliseconds—depend on receptors that are **transmitter-gated ion channels** (also called ion-channel-coupled receptors). These constitute a subclass of ligand-gated ion channels (see Figure 12–26B), and their function is to convert the chemical signal carried by a neurotransmitter back into an electrical signal. The channels open transiently in response to the binding of the neurotransmitter, thus changing the ion permeability of the postsynaptic membrane. This in turn causes a change in the membrane potential (**Figure 12–40**). If the change is big enough, it will depolarize the postsynaptic membrane and trigger an action potential in the postsynaptic cell.

A well-studied example of a transmitter-gated ion channel is found at the *neuromuscular junction*—the specialized synapse formed between a motor neuron and a skeletal muscle cell. In vertebrates, the neurotransmitter here is *acetylcholine*, and the transmitter-gated ion channel is an *acetylcholine receptor* (**Figure 12–41**). But not all neurotransmitters excite the postsynaptic cell, as we consider next.

Neurotransmitters Can Be Excitatory or Inhibitory

Neurotransmitters can either excite or inhibit a postsynaptic cell, and it is the character of the receptor that recognizes the neurotransmitter that determines how the postsynaptic cell will respond. The chief receptors for excitatory neurotransmitters, such as *acetylcholine* and *glutamate*, are ligand-gated cation channels. When a neurotransmitter binds, these channels open to allow an influx of Na^+, which depolarizes the plasma membrane and thus tends to activate the postsynaptic cell, encouraging it to fire an action potential. By contrast, the main receptors for inhibitory neurotransmitters, such as *γ-aminobutyric acid (GABA)* and *glycine*, are ligand-gated Cl^- channels. When neurotransmitters bind, these channels open, increasing the membrane permeability to Cl^-; this change in permeability inhibits the postsynaptic cell by making its plasma membrane harder to depolarize.

Toxins that bind to one of these excitatory or inhibitory neurotransmitter receptors can have dramatic effects on humans. *Curare*, for example, causes muscle paralysis by blocking excitatory acetylcholine receptors at the neuromuscular junction. This drug was used by South American

QUESTION 12–7

In the disease myasthenia gravis, the human body makes—by mistake—antibodies to its own acetylcholine receptor molecules. These antibodies bind to and inactivate acetylcholine receptors on the plasma membrane of muscle cells. The disease leads to a devastating progressive weakening of the people affected. Early on, they may have difficulty opening their eyelids, for example, and, in an animal model of the disease, rabbits have difficulty holding their ears up. As the disease progresses, most muscles weaken, and people with myasthenia gravis have difficulty speaking and swallowing. Eventually, impaired breathing can cause death. Explain which step of muscle function is affected.

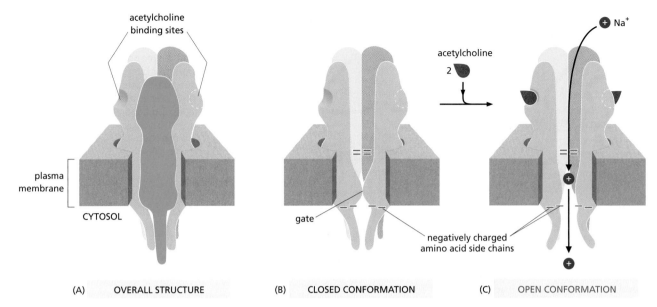

    (B) CLOSED CONFORMATION   (C) OPEN CONFORMATION

Figure 12–41 The acetylcholine receptor in the plasma membrane of vertebrate skeletal muscle cells opens when it binds the neurotransmitter acetylcholine. (A) This transmitter-gated ion channel is composed of five transmembrane protein subunits, two of which (*green*) are identical. The subunits combine to form a transmitter-gated aqueous pore across the lipid bilayer. The pore is lined by five transmembrane α helices, one contributed by each subunit. There are two acetylcholine-binding sites, one formed by parts of a *green* and *blue* subunit, the other by parts of a *green* and *orange* subunit, as shown. (B) The closed conformation. The *blue* subunit has been removed here and in (C) to show the interior of the pore. Negatively charged amino acid side chains at either end of the pore (indicated here by *red* minus signs) ensure that only positively charged ions, mainly Na^+ and K^+, can pass. But when acetylcholine is not bound and the channel is in its closed conformation, the pore is occluded (blocked) by hydrophobic amino acid side chains in the region called the gate. (C) The open conformation. When acetylcholine, released by a motor neuron, binds to both binding sites, the channel undergoes a conformational change; the hydrophobic side chains move apart and the gate opens, allowing Na^+ to flow across the membrane down its electrochemical gradient, depolarizing the membrane. Even with acetylcholine bound, the channel flickers randomly between the open and closed states (see Figure 12–25); without acetylcholine bound, it rarely opens.

Indians to make poison arrows and is still used by surgeons to relax muscles during an operation. By contrast, *strychnine*—a common ingredient in rat poisons—causes muscle spasms, convulsions, and death by blocking inhibitory glycine receptors on neurons in the brain and spinal cord.

The locations and functions of the ion channels discussed in this chapter are summarized in **Table 12–3**.

TABLE 12–3 SOME EXAMPLES OF ION CHANNELS		
Ion Channel	**Typical Location**	**Function**
K^+ leak channel	plasma membrane of most animal cells	maintenance of resting membrane potential
Voltage-gated Na^+ channel	plasma membrane of nerve cell axon	generation of action potentials
Voltage-gated K^+ channel	plasma membrane of nerve cell axon	return of membrane to resting potential after initiation of an action potential
Voltage-gated Ca^{2+} channel	plasma membrane of nerve terminal	stimulation of neurotransmitter release
Acetylcholine receptor (acetylcholine-gated cation channel)	plasma membrane of muscle cell (at neuromuscular junction)	excitatory synaptic signaling
Glutamate receptors (glutamate-gated cation channels)	plasma membrane of many neurons (at synapses)	excitatory synaptic signaling
GABA receptor (GABA-gated Cl^- channel)	plasma membrane of many neurons (at synapses)	inhibitory synaptic signaling
Glycine receptor (glycine-gated Cl^- channel	plasma membrane of many neurons (at synapses)	inhibitory synaptic signaling
Mechanically-activated cation channel	auditory hair cell in inner ear	detection of sound vibrations

Most Psychoactive Drugs Affect Synaptic Signaling by Binding to Neurotransmitter Receptors

Many drugs used in the treatment of insomnia, anxiety, depression, and schizophrenia act by binding to transmitter-gated ion channels in the brain. Sedatives and tranquilizers such as barbiturates, Valium, Ambien, and Restoril, for example, bind to GABA-gated Cl^- channels. Their binding makes the channels easier to open by GABA, rendering the neuron more sensitive to GABA's inhibitory action. By contrast, the antidepressant Prozac blocks the Na^+-driven symport responsible for the reuptake of the excitatory neurotransmitter *serotonin*, increasing the amount of serotonin available in the synapses that use it. This drug has changed the lives of many people who suffer from depression—although why boosting serotonin can elevate mood is still unknown.

The number of distinct types of neurotransmitter receptors is very large, although they fall into a small number of families. There are, for example, many subtypes of acetylcholine, glutamate, GABA, glycine, and serotonin receptors; they are usually located on different neurons and often differ only subtly in their electrophysiological properties. With such a large variety of receptors, it may be possible to design a new generation of psychoactive drugs that will act more selectively on specific sets of neurons to mitigate the mental illnesses that devastate so many people's lives. One percent of the human population, for example, have schizophrenia, another 1% have bipolar disorder, about 1% have an autistic disorder, and many more suffer from anxiety or depressive disorders. Mutations in genes that affect synaptic function can greatly increase the risk of the most serious of these disorders. The fact that these disorders are so prevalent suggests that the complexity of synaptic signaling may make the brain especially vulnerable to genetic abnormalities. But complexity also provides some distinct advantages, as we discuss next.

> ### QUESTION 12–8
>
> When an inhibitory neurotransmitter such as GABA opens Cl^- channels in the plasma membrane of a postsynaptic neuron, why does this make it harder for an excitatory neurotransmitter to excite the neuron?

The Complexity of Synaptic Signaling Enables Us to Think, Act, Learn, and Remember

For a process so critical for animal survival, the mechanism that governs synaptic signaling seems unnecessarily cumbersome, as well as error-prone. For a signal to pass from one neuron to the next, the nerve terminal of the presynaptic cell must convert an electrical signal into a secreted chemical. This chemical signal must then diffuse across the synaptic cleft so that a postsynaptic cell can convert it back into an electric one. Why would evolution have favored such an apparently inefficient and vulnerable way to pass a signal between cells? It would seem more efficient and robust to have a direct electrical connection between them—or to do away with the synapse altogether and use a single continuous cell.

The value of synapses that rely on secreted chemical signals becomes clear when we consider how they function in the context of the nervous system—a huge network of neurons, interconnected by many branching circuits, performing complex computations, storing memories, and generating plans for action. To carry out these functions, neurons have to do more than merely generate and relay signals: they must also combine them, interpret them, and record them. Chemical synapses make these activities possible. A motor neuron in the spinal cord, for example, receives inputs from hundreds or thousands of other neurons that make synapses on it (**Figure 12–42**). Some of these signals tend to stimulate the neuron, while others inhibit it. The motor neuron has to combine all of the information it receives and react, either by stimulating a muscle to contract or by remaining quiet.

Figure 12–42 Thousands of synapses form on the cell body and dendrites of a motor neuron in the spinal cord. (A) Many thousands of nerve terminals synapse on this neuron, delivering signals from other parts of the animal to control the firing of action potentials along the neuron's axon. (B) A rat nerve cell in culture. Its cell body and dendrites (*green*) are stained with a fluorescent antibody that recognizes a cytoskeletal protein. Thousands of axon terminals (*red*) from other nerve cells (not visible) make synapses on the cell's surface; they are stained with a fluorescent antibody that recognizes a protein in synaptic vesicles, which are located in the terminals (see Figure 12–38). (B, courtesy of Olaf Mundigl and Pietro de Camilli.)

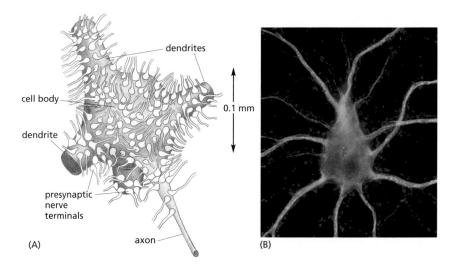

This task of computing an appropriate output from a babble of inputs is achieved by a complicated interplay between different types of ion channels in the neuron's plasma membrane. Each of the hundreds of types of neurons in your brain has its own characteristic set of receptors and ion channels that enables the cell to respond in a particular way to a certain set of inputs and thus to perform its specialized task.

In addition to integrating a variety of chemical inputs, a synapse can also adjust the magnitude of its response—reacting more vigorously (or less vigorously) to an incoming action potential—based on how heavily that synapse has been used in the past. This ability to adapt, called **synaptic plasticity**, is triggered by the entry of Ca^{2+} through special cation channels in the postsynaptic plasma membrane, which can lead to functional alterations on either side of the synapse—in the amount of neurotransmitter released from the axon terminal, the way the postsynaptic cell responds to the transmitter, or both. These synaptic changes can last hours, days, weeks, or longer, and they are thought to play an important part in learning and memory.

Synapses are thus critical components of the machinery that enables us to act, think, feel, speak, learn, and remember. Given that they operate in neuronal circuits that are so dauntingly complex, will it ever be possible to deeply understand the circuits that drive complex human behaviors? Although cracking this problem in humans is still far in the future, we now have increasingly powerful ways to study the neural circuits—and molecules—that underlie behavior in experimental animals. One of the most promising techniques makes use of a type of light-gated ion channel borrowed from unicellular algae, as we now discuss.

Optogenetics Uses Light-gated Ion Channels to Transiently Activate or Inactivate Neurons in Living Animals

Photosynthetic green algae use light-gated channels to sense and navigate toward sunlight. In response to blue light, one of these channels—called *channelrhodopsin*—allows Na^+ to flow into the cell. This depolarizes the plasma membrane and, ultimately, modulates the beating of the flagella the cell uses to swim. Although these channels are peculiar to unicellular green algae, they continue to function properly even when they are artificially transferred into other cell types, thereby rendering those cells responsive to light.

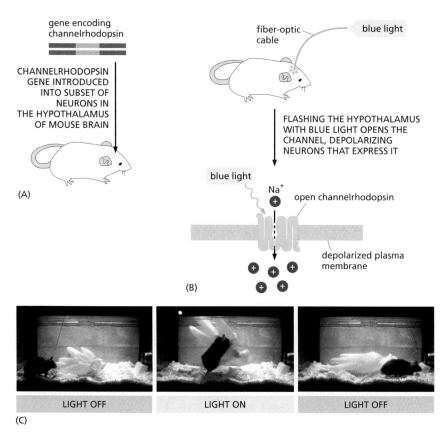

(A)

(B)

(C)

gene encoding
channelrhodopsin

CHANNELRHODOPSIN
GENE INTRODUCED
INTO SUBSET OF
NEURONS IN
THE HYPOTHALAMUS
OF MOUSE BRAIN

fiber-optic
cable

blue light

FLASHING THE HYPOTHALAMUS
WITH BLUE LIGHT OPENS THE
CHANNEL, DEPOLARIZING
NEURONS THAT EXPRESS IT

blue light

Na$^+$

open channelrhodopsin

depolarized plasma
membrane

LIGHT OFF   LIGHT ON   LIGHT OFF

Figure 12–43 Light-gated ion channels can control the activity of specific neurons in a living animal. (A) In this experiment, the gene encoding channelrhodopsin was introduced into a subset of neurons in the mouse hypothalamus. (B) When the neurons are exposed to blue light using a tiny fiber-optic cable implanted into the animal's brain, channelrhodopsin opens, depolarizing and stimulating the channel-containing neurons. (C) When the light is switched on, the mouse immediately becomes aggressive; when the light is switched off, its behavior immediately returns to normal. (C, from D. Lin et al., *Nature* 470:221–226, 2011. With permission from Macmillan Publishers Ltd.)

Because nerve cells are activated by a depolarizing influx of Na$^+$ (see Figure 12–36), channelrhodopsin can be used to manipulate the activity of neurons and neural circuits. It has even been used to control the behavior of living animals. In one particularly stunning experiment, the channel-rhodopsin gene was introduced into a select subpopulation of neurons in the mouse hypothalamus—a brain region involved in many functions, including aggression. When the channels were subsequently illuminated by a thin, optic fiber implanted in the animal's brain, the mouse launched an attack on any object in its path—including other mice or, in one comical instance, an inflated rubber glove. When the light was switched off, the neurons fell silent, and the mouse's behavior immediately returned to normal (**Figure 12–43** and Movie 12.10).

Because the approach uses light to control neurons into which chan-nelrhodopsin—or any other light-gated channel—has been introduced by genetic engineering techniques (discussed in Chapter 10), the method has been dubbed **optogenetics**. This new tool is revolutionizing neuro-biology, allowing investigators to dissect the neural circuits that govern even the most complex behaviors in a variety of experimental animals, from fruit flies to monkeys. But its implications extend beyond the labora-tory. As genetic studies continue to identify genes associated with various human neurological and psychiatric disorders, the ability to exploit light-gated ion channels to study where and how these genes function in model organisms promises to greatly advance our understanding of the molecular and cellular basis of human behavior.

ESSENTIAL CONCEPTS

- The lipid bilayer of cell membranes is highly permeable to small nonpolar molecules such as oxygen and carbon dioxide and, to a lesser extent, to very small polar molecules such as water. It is highly impermeable to most large water-soluble molecules and to all ions.

- Transfer of nutrients, metabolites, and inorganic ions across cell membranes depends on membrane transport proteins.

- Cell membranes contain a variety of transport proteins that function either as transporters or channels, each responsible for the transfer of a particular type of solute.

- Channel proteins form pores across the lipid bilayer through which solutes can passively diffuse.

- Both transporters and channels can mediate passive transport, in which an uncharged solute moves spontaneously down its concentration gradient.

- For the passive transport of a charged solute, its electrochemical gradient determines its direction of movement, rather than its concentration alone.

- Transporters can act as pumps to mediate active transport, in which solutes are moved uphill against their concentration or electrochemical gradients; this process requires energy that is provided by ATP hydrolysis, a downhill flow of Na^+ or H^+ ions, or sunlight.

- Transporters transfer specific solutes across a membrane by undergoing conformational changes that expose the solute-binding site first on one side of the membrane and then on the other.

- The Na^+ pump in the plasma membrane of animal cells is an ATPase; it actively transports Na^+ out of the cell and K^+ in, maintaining a steep Na^+ gradient across the plasma membrane that is used to drive other active transport processes and to convey electrical signals.

- Ion channels allow inorganic ions of appropriate size and charge to cross the membrane. Most are gated and open transiently in response to a specific stimulus.

- Even when activated by a specific stimulus, ion channels do not remain continuously open: they flicker randomly between open and closed conformations. An activating stimulus increases the proportion of time that the channel spends in the open state.

- The membrane potential is determined by the unequal distribution of charged ions on the two sides of a cell membrane; it is altered when these ions flow through open ion channels in the membrane.

- In most animal cells, the negative value of the resting membrane potential across the plasma membrane depends mainly on the K^+ gradient and the operation of K^+-selective leak channels; at this resting potential, the driving force for the movement of K^+ across the membrane is almost zero.

- Neurons propagate electrical impulses in the form of action potentials, which can travel long distances along an axon without weakening. Action potentials are mediated by voltage-gated Na^+ channels that open in response to depolarization of the plasma membrane.

- Voltage-gated Ca^{2+} channels in a nerve terminal couple the arrival of an action potential to neurotransmitter release at a synapse. Transmitter-gated ion channels convert this chemical signal back into an electrical one in the postsynaptic target cell.

- Excitatory neurotransmitters open transmitter-gated cation channels that allow the influx of Na^+, which depolarizes the postsynaptic cell's plasma membrane and encourages the cell to fire an action potential. Inhibitory neurotransmitters open transmitter-gated Cl^- channels in the postsynaptic cell plasma membrane, making it harder for the membrane to depolarize and fire an action potential.

- Complex sets of nerve cells in the human brain exploit all of the above mechanisms to make human behaviors possible.

KEY TERMS

action potential	Nernst equation
active transport	nerve terminal
antiport	neuron
axon	neurotransmitter
Ca^{2+} pump (or Ca^{2+}-ATPase)	optogenetics
channel	osmosis
coupled pumps	passive transport
dendrite	patch-clamp recording
depolarization	pump
electrochemical gradient	resting membrane potential
H^+ pump (or H^+ ATPase)	symport
ion channel	synapse
K^+ leak channels	synaptic plasticity
ligand-gated channel	synaptic vesicle
mechanically-gated channel	transmitter-gated ion channel
membrane potential	voltage-gated channel
membrane transport protein	voltage-gated Na^+ channel
Na^+ pump (or Na^+-K^+ ATPase)	

QUESTIONS

QUESTION 12–9

The diagram in Figure 12–9 shows a passive transporter that mediates the transfer of a solute down its concentration gradient across the membrane. How would you need to change the diagram to convert the transporter into a pump that moves the solute up its concentration gradient by hydrolyzing ATP? Explain the need for each of the steps in your new illustration.

QUESTION 12–10

Which of the following statements are correct? Explain your answers.

A. The plasma membrane is highly impermeable to all charged molecules.

B. Channels have specific binding pockets for the solute molecules they allow to pass.

C. Transporters allow solutes to cross a membrane at much faster rates than do channels.

D. Certain H^+ pumps are fueled by light energy.

E. The plasma membrane of many animal cells contains open K^+ channels, yet the K^+ concentration in the cytosol is much higher than outside the cell.

F. A symport would function as an antiport if its orientation in the membrane were reversed (i.e., if the portion of the molecule normally exposed to the cytosol faced the outside of the cell instead).

G. The membrane potential of an axon temporarily becomes more negative when an action potential excites it.

QUESTION 12–11

List the following compounds in order of increasing lipid bilayer permeability: RNA, Ca^{2+}, glucose, ethanol, N_2, water.

QUESTION 12–12

Name at least one similarity and at least one difference between the following (it may help to review the definitions of the terms using the Glossary):

A. Symport and antiport

B. Active transport and passive transport

C. Membrane potential and electrochemical gradient

D. Pump and transporter

E. Axon and telephone wire

F. Solute and ion

QUESTION 12–13

Discuss the following statement: "The differences between a channel and a transporter are like the differences between a bridge and a ferry."

QUESTION 12–14

The neurotransmitter acetylcholine is made in the cytosol and then transported into synaptic vesicles, where its concentration is more than 100-fold higher than in the cytosol. When synaptic vesicles are isolated from neurons, they can take up additional acetylcholine added to the solution in which they are suspended, but only when ATP is present. Na^+ ions are not required for the uptake, but, curiously, raising the pH of the solution in which the synaptic vesicles are suspended increases the rate of uptake.

Furthermore, transport is inhibited when drugs are added that make the membrane permeable to H^+ ions. Suggest a mechanism that is consistent with all of these observations.

QUESTION 12–15

The resting membrane potential of a typical animal cell is about –70 mV, and the thickness of a lipid bilayer is about 4.5 nm. What is the strength of the electric field across the membrane in V/cm? What do you suppose would happen if you applied this field strength to two metal electrodes separated by a 1-cm air gap?

QUESTION 12–16

Phospholipid bilayers form sealed spherical vesicles in water (discussed in Chapter 11). Assume you have constructed lipid vesicles that contain Na^+ pumps as the sole membrane protein, and assume for the sake of simplicity that each pump transports one Na^+ one way and one K^+ the other way in each pumping cycle. All the Na^+ pumps have the portion of the molecule that normally faces the cytosol oriented toward the outside of the vesicles. With the help of Figure 12–11, determine what would happen if:

A. Your vesicles were suspended in a solution containing both Na^+ and K^+ ions and had a solution with the same ionic composition inside them.

B. You add ATP to the suspension described in (A).

C. You add ATP, but the solution—outside as well as inside the vesicles—contains only Na^+ ions and no K^+ ions.

D. The concentrations of Na^+ and K^+ were as in (A), but half of the pump molecules embedded in the membrane of each vesicle were oriented the other way around so that the normally cytosolic portions of these molecules faced the inside of the vesicles. You then add ATP to the suspension.

E. You add ATP to the suspension described in (A), but in addition to Na^+ pumps, the membrane of your vesicles also contains K^+ leak channels.

QUESTION 12–17

Name the three ways in which an ion channel can be gated.

QUESTION 12–18

One thousand Ca^{2+} channels open in the plasma membrane of a cell that is 1000 μm^3 in size and has a cytosolic Ca^{2+} concentration of 100 nM. For how long would the channels need to stay open in order for the cytosolic Ca^{2+} concentration to rise to 5 μM? There is virtually unlimited Ca^{2+} available in the outside medium (the extracellular Ca^{2+} concentration in which most animal cells live is a few millimolar), and each channel passes 10^6 Ca^{2+} ions per second.

QUESTION 12–19

Amino acids are taken up by animal cells using a symport in the plasma membrane. What is the most likely ion whose electrochemical gradient drives the import? Is ATP consumed in the process? If so, how?

QUESTION 12–20

We will see in Chapter 15 that endosomes, which are membrane-enclosed intracellular organelles, need an acidic lumen in order to function. Acidification is achieved by an H^+ pump in the endosomal membrane, which also contains Cl^- channels. If the channels do not function properly (e.g., because of a mutation in the genes encoding the channel proteins), acidification is also impaired.

A. Can you explain how Cl^- channels might help acidification?

B. According to your explanation, would the Cl^- channels be absolutely required to lower the pH inside the endosome?

QUESTION 12–21

Some bacterial cells can grow on either ethanol (CH_3CH_2OH) or acetate (CH_3COO^-) as their only carbon source. Dr. Schwips measured the rate at which the two compounds traverse the bacterial plasma membrane but, due to excessive inhalation of one of the compounds (which one?), failed to label his data accurately.

A. Plot the data from the table below.

Concentration of Carbon Source (mM)	Rate of Transport ($\mu mol/min$)	
	Compound A	Compound B
0.1	2.0	18
0.3	6.0	46
1.0	20	100
3.0	60	150
10.0	200	182

B. Determine from your graph whether the data describing compound A correspond to the uptake of ethanol or acetate.

C. Determine the rates of transport for compounds A and B at 0.5 mM and 100 mM. (This part of the question requires that you be familiar with the principles of enzyme kinetics discussed in Chapter 3.)

Explain your answers.

QUESTION 12–22

Acetylcholine-gated cation channels do not discriminate between Na^+, K^+, and Ca^{2+} ions, allowing all to pass through them freely. So why is it that when acetylcholine binds to this protein in the plasma membrane of muscle cells, the channel opens and there is a large net influx of primarily Na^+ ions?

QUESTION 12–23

The ion channels that are regulated by binding of neurotransmitters, such as acetylcholine, glutamate, GABA, or glycine, have a similar overall structure. Yet, each class of these channels consists of a very diverse set of subtypes with different transmitter affinities, different channel conductances, and different rates of opening and closing. Do you suppose that such extreme diversity is a good or a bad thing from the standpoint of the pharmaceutical industry?

CHAPTER **THIRTEEN**

13

How Cells Obtain Energy From Food

As we discussed in Chapter 3, cells require a constant supply of energy to generate and maintain the biological order that allows them to grow, divide, and carry out their day-to-day activities. This energy comes from the chemical-bond energy in food molecules, which thereby serve as fuel for cells.

Perhaps the most important fuel molecules are the sugars. Plants make their own sugars from CO_2 by photosynthesis. Animals obtain sugars—and other organic molecules that can be chemically transformed into sugars—by eating plants and other organisms. Nevertheless, the process whereby all these sugars are broken down to generate energy is very similar in both animals and plants. In both cases, the organism's cells harvest useful energy from the chemical-bond energy locked in sugars as the sugar molecule is broken down and oxidized to carbon dioxide (CO_2) and water (H_2O)—a process called **cell respiration**. The energy released during these reactions is captured in the form of "high-energy" chemical bonds—covalent bonds that release large amounts of energy when hydrolyzed—in *activated carriers* such as ATP and NADH. These carriers in turn serve as portable sources of the chemical groups and electrons needed for biosynthesis (discussed in Chapter 3).

In this chapter, we trace the major steps in the breakdown of sugars and show how ATP, NADH, and other activated carriers are produced along the way. We concentrate on the breakdown of glucose because it generates most of the energy produced in the majority of animal cells. A very similar pathway operates in plants, fungi, and many bacteria. Other molecules, such as fatty acids and proteins, can also serve as energy sources if they are funneled through appropriate enzymatic pathways. We will

THE BREAKDOWN AND UTILIZATION OF SUGARS AND FATS

REGULATION OF METABOLISM

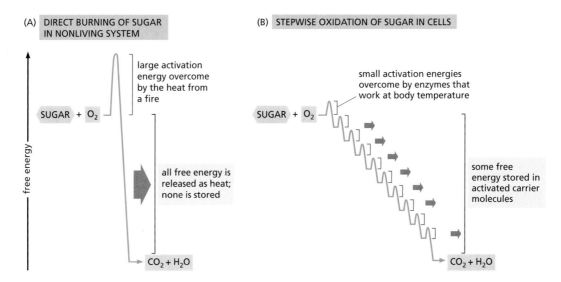

(A) DIRECT BURNING OF SUGAR IN NONLIVING SYSTEM

(B) STEPWISE OXIDATION OF SUGAR IN CELLS

large activation energy overcome by the heat from a fire

SUGAR + O_2

all free energy is released as heat; none is stored

$CO_2 + H_2O$

free energy

small activation energies overcome by enzymes that work at body temperature

SUGAR + O_2

some free energy stored in activated carrier molecules

$CO_2 + H_2O$

Figure 13–1 The controlled, stepwise oxidation of sugar in cells captures useful energy, unlike the simple burning of the same fuel molecule. (A) The direct burning of sugar in nonliving systems generates more energy than can be stored by any carrier molecule. This energy is thus released as heat. (B) In a cell, enzymes catalyze the breakdown of sugars via a series of small steps, in which a portion of the free energy released is captured by the formation of activated carriers—most often ATP and NADH. Each step is catalyzed by an enzyme that lowers the activation energy barrier that must be surmounted by the random collision of molecules at the temperature of cells (body temperature), so as to allow the reaction to occur. The total free energy released by the oxidative breakdown of glucose—686 kcal/mole (2880 kJ/mole)—is exactly the same in (A) and (B).

see how cells use many of the molecules generated from the breakdown of sugars and fats as starting points to make other organic molecules.

Finally, we examine how cells regulate their metabolism and how they store food molecules for their future metabolic needs. We will save our discussion of the elaborate mechanism cells use to produce the bulk of their ATP for Chapter 14.

THE BREAKDOWN AND UTILIZATION OF SUGARS AND FATS

If a fuel molecule such as glucose were oxidized to CO_2 and H_2O in a single step—by, for example, the direct application of fire—it would release an amount of energy many times larger than any carrier molecule could capture (**Figure 13–1A**). Instead, cells use enzymes to carry out the oxidation of sugars in a tightly controlled series of reactions. Thanks to the action of enzymes—which operate at temperatures typical of living things—cells degrade each glucose molecule step by step, paying out energy in small packets to activated carriers by means of coupled reactions (**Figure 13–1B**). In this way, much of the energy released by the breakdown of glucose is saved in the high-energy bonds of ATP and other activated carriers, which can then be made available to do useful work for the cell.

Animal cells make ATP in two ways. First, certain energetically favorable, enzyme-catalyzed reactions involved in the breakdown of foods are directly coupled to the energetically unfavorable reaction **ADP** + P_i → **ATP**. Thus the oxidation of food molecules can provide energy for the immediate production of ATP. Most ATP synthesis, however, requires an intermediary. In this second pathway to making ATP, the energy from other activated carriers is used to drive ATP production. This process, called *oxidative phosphorylation*, takes place on the inner mitochondrial membrane (**Figure 13–2**), and it is described in detail in Chapter 14. In this chapter, we focus on the first sequence of reactions by which food molecules are oxidized—both in the cytosol and in the mitochondrial matrix (see Figure 13–2). These reactions produce both ATP and the additional

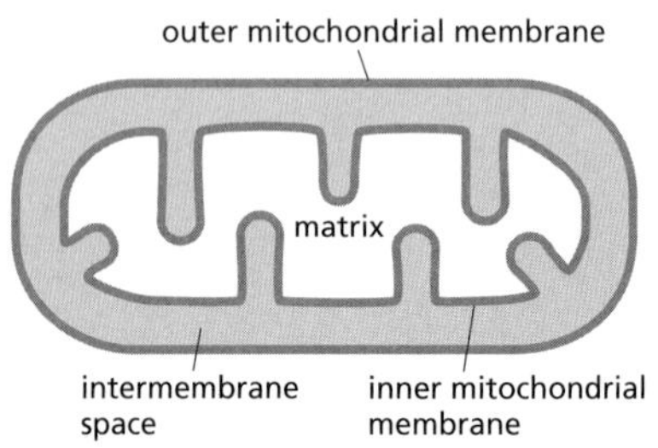

outer mitochondrial membrane

matrix

intermembrane space

inner mitochondrial membrane

Figure 13–2 A mitochondrion has two membranes and a large internal space called the matrix. Most of the energy from food molecules is harvested in mitochondria—both in the matrix and in the inner mitochondrial membrane.

activated carriers that will subsequently help drive the production of much larger amounts of ATP by oxidative phosphorylation.

Food Molecules Are Broken Down in Three Stages

The proteins, fats, and polysaccharides that make up most of the food we eat must be broken down into smaller molecules before our cells can use them—either as a source of energy or as building blocks for making other organic molecules. This breakdown process—in which enzymes degrade complex organic molecules into simpler ones—is called **catabolism**. The process takes place in three stages, as illustrated in **Figure 13–3**.

Figure 13–3 The breakdown of food molecules occurs in three stages.
(A) Stage 1 mostly occurs outside cells in the mouth and the gut—although intracellular lysosomes can also digest large organic molecules. Stage 2 occurs mainly in the cytosol, except for the final step of conversion of pyruvate to acetyl groups on acetyl CoA, which occurs in the mitochondrial matrix. Stage 3 begins with the citric acid cycle in the mitochondrial matrix and concludes with oxidative phosphorylation on the mitochondrial inner membrane. The NADH generated in stage 2—during glycolysis and the conversion of pyruvate to acetyl CoA—adds to the NADH produced by the citric acid cycle to drive the production of ATP by oxidative phosphorylation.
(B) The net products of the complete oxidation of food include ATP, NADH, CO_2, and H_2O. The ATP and NADH provide the energy and electrons needed for biosynthesis; the CO_2 and H_2O are waste products.

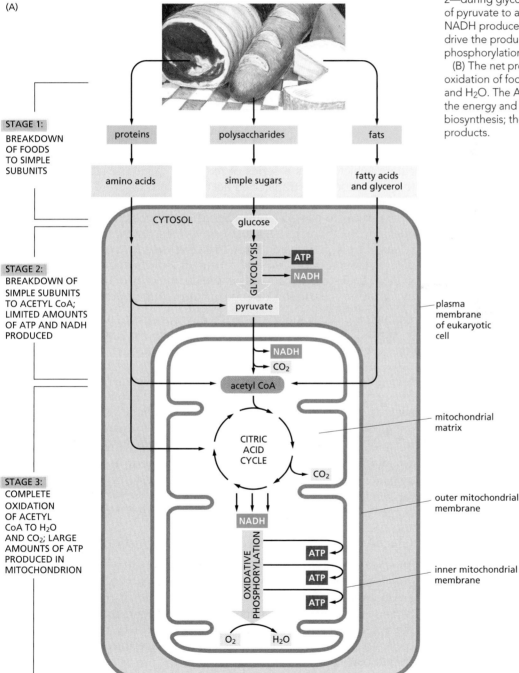

(A)

STAGE 1:
BREAKDOWN OF FOODS TO SIMPLE SUBUNITS

STAGE 2:
BREAKDOWN OF SIMPLE SUBUNITS TO ACETYL CoA; LIMITED AMOUNTS OF ATP AND NADH PRODUCED

STAGE 3:
COMPLETE OXIDATION OF ACETYL CoA TO H_2O AND CO_2; LARGE AMOUNTS OF ATP PRODUCED IN MITOCHONDRION

proteins   polysaccharides   fats

amino acids   simple sugars   fatty acids and glycerol

CYTOSOL   glucose

GLYCOLYSIS → ATP
→ NADH

pyruvate

plasma membrane of eukaryotic cell

NADH
CO_2

acetyl CoA

mitochondrial matrix

CITRIC ACID CYCLE CO_2

outer mitochondrial membrane

NADH

OXIDATIVE PHOSPHORYLATION   ATP
ATP
ATP

inner mitochondrial membrane

O_2 H_2O

(B)   NET RESULT: FOOD + O_2 → ATP + NADH + CO_2 + H_2O

In *stage 1* of catabolism, enzymes convert the large polymeric molecules in food into simpler monomeric subunits: proteins into amino acids, polysaccharides into sugars, and fats into fatty acids and glycerol. This stage—also called *digestion*—occurs either outside cells (in the intestine) or in specialized organelles within cells called lysosomes (discussed in Chapter 15). After digestion, the small organic molecules derived from food enter the cytosol of a cell, where their gradual oxidative breakdown begins.

In *stage 2* of catabolism, a chain of reactions called *glycolysis* splits each molecule of *glucose* into two smaller molecules of *pyruvate*. Sugars other than glucose can also be used, after first being converted into one of the intermediates in this sugar-splitting pathway. Glycolysis takes place in the cytosol and, in addition to producing pyruvate, it generates two types of activated carriers: ATP and NADH. The pyruvate is transported from the cytosol into the mitochondrion's large, internal compartment called the *matrix*. There, a giant enzyme complex converts each pyruvate molecule into CO_2 plus *acetyl CoA*, another of the activated carriers discussed in Chapter 3 (see Figure 3–36). In the same compartment, large amounts of acetyl CoA are also produced by the stepwise oxidative breakdown of fatty acids derived from fats (see Figure 13–3).

Stage 3 of catabolism takes place entirely in mitochondria. The acetyl group in acetyl CoA is transferred to an oxaloacetate molecule to form citrate, which enters a series of reactions called the *citric acid cycle*. In these reactions, the transferred acetyl group is oxidized to CO_2 with the production of large amounts of NADH. Finally, the high-energy electrons from NADH are passed along a series of enzymes within the mitochondrial inner membrane called an *electron-transport chain*, where the energy released by their transfer is used to drive oxidative phosphorylation—a process that produces ATP and consumes molecular oxygen (O_2 gas). It is in these final steps of catabolism that the majority of the energy released by oxidation is harnessed to produce most of the cell's ATP.

Through the production of ATP, the energy derived from the breakdown of sugars and fats is redistributed into packets of chemical energy in a form convenient for use in the cell. In total, nearly half of the energy that could, in theory, be derived from the breakdown of glucose or fatty acids to H_2O and CO_2 is captured and used to drive the energetically unfavorable reaction $ADP + P_i \rightarrow ATP$. By contrast, a modern combustion engine, such as a car engine, can convert no more than 20% of the available energy in its fuel into useful work. In both cases, the remaining energy is released as heat, which in animals helps to keep the body warm.

Roughly 10^9 molecules of ATP are in solution in a typical cell at any instant. In many cells, all of this ATP is turned over (that is, consumed and replaced) every 1–2 minutes. An average person at rest will hydrolyze his or her weight in ATP molecules every 24 hours.

Glycolysis Extracts Energy from the Splitting of Sugar

The central process in stage 2 of catabolism is the oxidative breakdown of **glucose** in the sequence of reactions known as **glycolysis**. Glycolysis produces ATP without the involvement of oxygen. It occurs in the cytosol of most cells, including many anaerobic microorganisms that thrive in the absence of oxygen. Glycolysis probably evolved early in the history of life on Earth, before photosynthetic organisms introduced oxygen into the atmosphere.

The term "glycolysis" comes from the Greek *glykys*, "sweet," and *lysis*, "splitting." It is an appropriate name, as glycolysis splits a molecule of glucose, which has six carbon atoms, to form two molecules of pyruvate, each of which contains three carbon atoms. The series of chemical

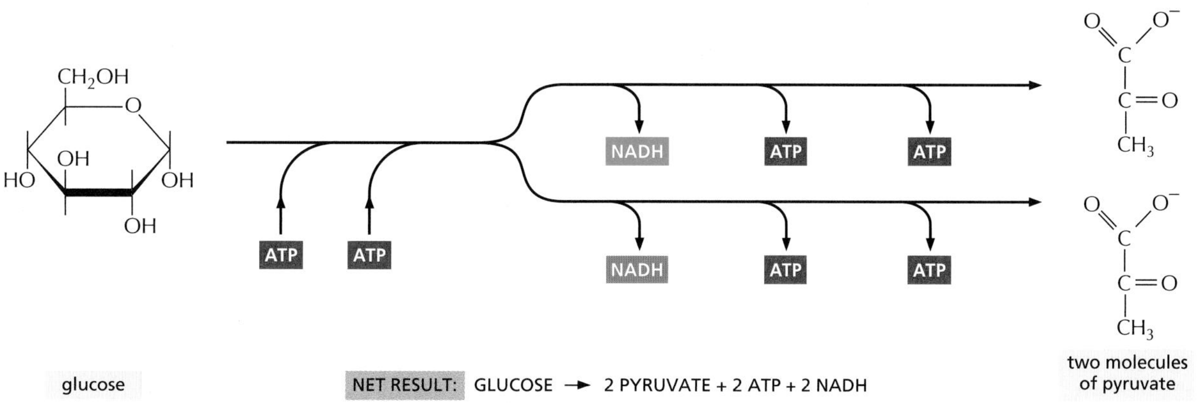

glucose

NET RESULT: GLUCOSE ⟶ 2 PYRUVATE + 2 ATP + 2 NADH

two molecules
of pyruvate

rearrangements that ultimately generate pyruvate release energy because the electrons in a molecule of pyruvate are, overall, at a lower energy state than those in a molecule of glucose. Nevertheless, for each molecule of glucose that enters glycolysis, two molecules of ATP are initially consumed to provide the energy needed to prepare the sugar to be split. This investment of energy is more than recouped in the later steps of glycolysis, when four molecules of ATP are produced. Energy is also captured in this "payoff phase" in the form of NADH. Thus, at the end of glycolysis, there is a net gain of two molecules of ATP and two molecules of NADH for each glucose molecule broken down (**Figure 13–4**).

Figure 13–4 Glycolysis splits a molecule of glucose to form two molecules of pyruvate. The process requires an input of energy, in the form of ATP, at the start. This energy investment is later recouped by the production of two NADHs and four ATPs.

Glycolysis Produces Both ATP and NADH

Piecing together the complete glycolytic pathway in the 1930s was a major triumph of biochemistry, as the pathway consists of a sequence of 10 separate reactions, each producing a different sugar intermediate and each catalyzed by a different enzyme. Like most enzymes, those that catalyze glycolysis all have names ending in -*ase*—like isomerase and dehydrogenase—which specify the type of reaction they catalyze (**Table 13–1**). The reactions of the glycolytic pathway are presented in outline in **Figure 13–5** and in detail in **Panel 13–1** (pp. 428–429).

TABLE 13–1 SOME TYPES OF ENZYMES INVOLVED IN GLYCOLYSIS

Enzyme type	General function	Role in glycolysis
Kinase	catalyzes the addition of a phosphate group to molecules	a kinase transfers a phosphate group from ATP to a substrate in steps 1 and 3; other kinases transfer a phosphate to ADP to form ATP in steps 7 and 10
Isomerase	catalyzes the rearrangement of bonds within a single molecule	isomerases in steps 2 and 5 prepare molecules for the chemical alterations to come
Dehydrogenase	catalyzes the oxidation of a molecule by removing a hydrogen atom plus an electron (a hydride ion, H^-)	the enzyme glyceraldehyde 3-phosphate dehydrogenase generates NADH in step 6
Mutase	catalyzes the shifting of a chemical group from one position to another within a molecule	the movement of a phosphate by phosphoglycerate mutase in step 8 helps prepare the substrate to transfer this group to ADP to make ATP in step 10

Figure 13–5 The stepwise breakdown of sugars begins with glycolysis. Each of the 10 steps of glycolysis is catalyzed by a different enzyme. Note that step 4 cleaves a six-carbon sugar into two three-carbon sugars, so that the number of molecules at every stage after this doubles. Note also that one of the products of step 4 needs to be modified (isomerized) in step 5 before it can proceed to step 6 (see Panel 13–1). As indicated, step 6 begins the energy-generation phase of glycolysis, which results in the net synthesis of ATP and NADH (see also Figure 13–4). Glycolysis is also sometimes referred to as the Embden–Meyerhof pathway, named for the chemists who first described it. All the steps of glycolysis are reviewed in Movie 13.1.

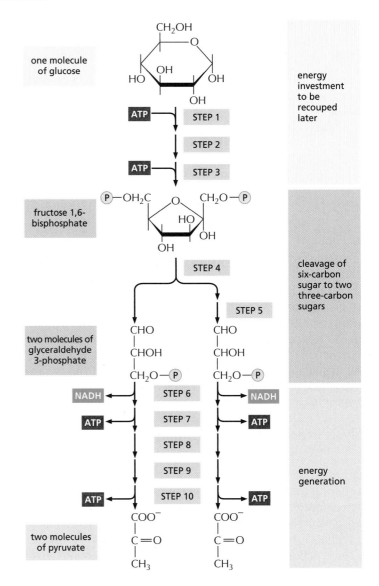

Much of the energy released by the breakdown of glucose is used to drive the synthesis of ATP molecules from ADP and P_i. This form of ATP synthesis, which takes place in steps 7 and 10 in glycolysis, is known as *substrate-level phosphorylation* because it occurs by the transfer of a phosphate group directly from a substrate molecule—one of the sugar intermediates—to ADP. By contrast, most phosphorylations in cells occur by the transfer of phosphate from ATP to a substrate molecule.

The remainder of the energy released during glycolysis is stored in the electrons in the **NADH** molecule produced in step 6 by an oxidation reaction. As discussed in Chapter 3, oxidation does not always involve oxygen; it occurs in any reaction in which electrons are lost from one atom and transferred to another. So, although no molecular oxygen is involved in glycolysis, oxidation does occur: in step 6, a hydrogen atom plus an electron is removed from the sugar intermediate, glyceraldehyde 3-phosphate, and transferred to **NAD+**, producing NADH (see Panel 13 1, p. 428).

Over the course of glycolysis, two molecules of NADH are formed per molecule of glucose. In aerobic organisms, these NADH molecules donate their electrons to the electron-transport chain in the inner mitochondrial membrane, as described in detail in Chapter 14. Such electron transfers release energy as the electrons fall from a state of higher energy to a lower one. The electrons that are passed along the electron-transport chain are ultimately passed on to O_2, forming water.

In giving up its electrons, NADH is converted back into NAD^+, which is then available to be used again for glycolysis. In the absence of oxygen, NAD^+ can be regenerated by an alternate type of energy-yielding reaction called a fermentation, as we discuss next.

Fermentations Can Produce ATP in the Absence of Oxygen

For most animal and plant cells, glycolysis is only a prelude to the third and final stage of the breakdown of food molecules, in which large amounts of ATP are generated in mitochondria by oxidative phosphorylation, a process that requires the consumption of oxygen. However, for many anaerobic microorganisms, which can grow and divide in the absence of oxygen, glycolysis is the principal source of ATP. The same is true for certain animal cells, such as skeletal muscle cells, which can continue to function at low levels of oxygen.

In these anaerobic conditions, the pyruvate and NADH made by glycolysis remain in the cytosol. The pyruvate is converted into products that are excreted from the cell: lactate in muscle cells, for example, or ethanol and CO_2 in the yeast cells used in brewing and breadmaking. The NADH gives up its electrons in the cytosol, and is converted back to the NAD^+ required to maintain the reactions of glycolysis (**Figure 13–6**). Such energy-yielding pathways that break down sugar in the absence of oxygen are called **fermentations**. Scientific studies of the commercially important fermentations carried out by yeasts laid the foundations for early biochemistry.

> ### QUESTION 13–1
>
> At first glance, the final steps in fermentation appear to be unnecessary: the generation of lactate or ethanol does not produce any additional energy for the cell. Explain why cells growing in the absence of oxygen could not simply discard pyruvate as a waste product. Which products derived from glucose would accumulate in cells unable to generate either lactate or ethanol by fermentation?

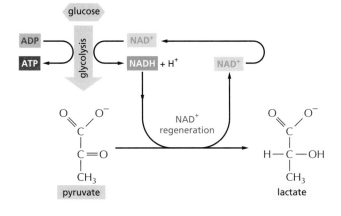

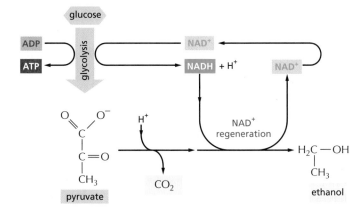

Figure 13–6 Pyruvate is broken down in the absence of oxygen by fermentation. (A) When inadequate oxygen is present, for example, in a muscle cell undergoing vigorous contraction, the pyruvate produced by glycolysis is converted to lactate in the cytosol. This reaction restores the NAD^+ consumed in step 6 of glycolysis, but the whole pathway yields much less energy overall than if the pyruvate were oxidized in mitochondria. (B) In microorganisms that can grow anaerobically, pyruvate is converted into carbon dioxide and ethanol. Again, this pathway regenerates NAD^+ from NADH, as required to enable glycolysis to continue. Both (A) and (B) are examples of fermentations. Note that in both cases, for each molecule of glucose that enters glycolysis, two molecules of pyruvate are generated (only a single pyruvate is shown here). Fermentation of these two pyruvates subsequently yields two molecules of lactate—or two molecules of CO_2 and ethanol—plus two molecules of NAD^+.

Figure 13–7 A pair of coupled reactions drives the energetically unfavorable formation of ATP in steps 6 and 7 of glycolysis. In this diagram, energetically favorable reactions are represented by *blue arrows*; energetically costly reactions by *red arrows*. In step 6, the energy released by the energetically favorable oxidation of a C–H bond in glyceraldehyde 3-phosphate (*blue arrow*) is large enough to drive two energetically costly reactions: the formation of both NADH and a high-energy phosphate bond in 1,3-bisphosphoglycerate (*red arrows*). The subsequent energetically favorable hydrolysis of that high-energy phosphate bond in step 7 then drives the formation of ATP.

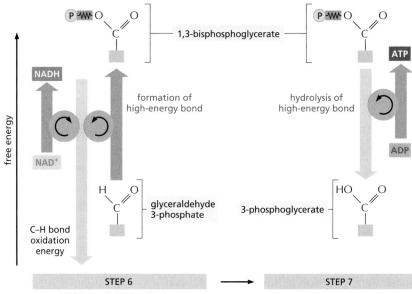

TOTAL ENERGY CHANGE for step 6 followed by step 7 is a favorable –3 kcal/mole

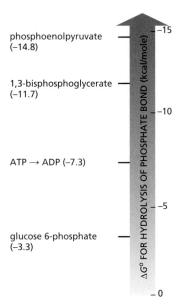

Figure 13–8 Differences in the energies of different phosphate bonds allow the formation of ATP by substrate-level phosphorylation. Examples of molecules containing different types of phosphate bonds are shown, along with the free-energy change for hydrolysis of those bonds in kcal/mole (1 kcal = 4.184 kJ). The transfer of a phosphate group from one molecule to another is energetically favorable if the standard free-energy change (ΔG°) for hydrolysis of the phosphate bond is more negative for the donor molecule than for the acceptor. (The hydrolysis reactions can be thought of as the transfer of the phosphate group to water.) Thus, a phosphate group is readily transferred from 1,3-bisphosphoglycerate to ADP to form ATP. Transfer reactions involving the phosphate groups in these molecules are detailed in Panel 13–1 (pp. 428–429).

Many bacteria and archaea can also generate ATP in the absence of oxygen by *anaerobic respiration*, a process that uses a molecule other than oxygen as a final electron acceptor. Anaerobic respiration differs from fermentation in that it involves an electron-transport chain embedded in a membrane—in this case, the plasma membrane of the microbe.

Glycolytic Enzymes Couple Oxidation to Energy Storage in Activated Carriers

The "paddle wheel" analogy in Chapter 3 explained how cells harvest useful energy from the oxidation of organic molecules by coupling an energetically unfavorable reaction to an energetically favorable one (see Figure 3–30). Here, we take a closer look at a key pair of glycolytic reactions that demonstrate how enzymes—the paddle wheel in our analogy—allow coupled reactions to facilitate the transfer of chemical energy to ATP and NADH.

The reactions in question—steps 6 and 7 in Panel 13–1—convert the three-carbon sugar intermediate glyceraldehyde 3-phosphate (an aldehyde) into 3-phosphoglycerate (a carboxylic acid). This conversion, which entails the oxidation of an aldehyde group to a carboxylic acid group, occurs in two steps. The overall reaction releases enough free energy to transfer two electrons from the aldehyde to NAD$^+$ to form NADH and to transfer a phosphate group to a molecule of ADP to form ATP. It also releases enough heat to the environment to make the overall reaction energetically favorable: the ΔG° for step 6 followed by step 7 is –3.0 kcal/mole (**Figure 13–7**).

The energy contained in any phosphate bond can be determined by measuring the standard free-energy change (ΔG°) when that bond is broken by hydrolysis. Molecules that contain phosphate bonds that have more energy than those found in ATP—including the high-energy 1,3-bisphosphoglycerate generated in step 6 of glycolysis—readily transfer their phosphate group to ADP to form ATP. **Figure 13–8** compares the high-energy phosphoanhydride bond in ATP with a few of the other phosphate bonds that are generated during glycolysis. As explained in Panel 13–1, we describe these bonds as "high energy" only in that their hydrolysis is particularly energetically favorable.

The reaction in step 6 is the only one in glycolysis that creates a high-energy phosphate linkage directly from inorganic phosphate—an example of the substrate-level phosphorylation mentioned earlier. How this high-energy linkage is generated in step 6—and then consumed in step 7 to produce ATP—is detailed in **Figure 13–9**.

(A) STEPS 6 AND 7 OF GYCOLYSIS

glyceraldehyde 3-phosphate

A short-lived covalent bond is formed between glyceraldehyde 3-phosphate and the –SH group of a cysteine side chain of the enzyme glyceraldehyde 3-phosphate dehydrogenase. The enzyme also binds noncovalently to NAD+.

Glyceraldehyde 3-phosphate is oxidized as the enzyme removes a hydrogen atom (*yellow*) and transfers it, along with an electron, to NAD+, forming NADH (see Figure 3.34). Part of the energy released by the oxidation of the aldehyde is thus stored in NADH, and part is stored in the high-energy thioester bond that links glyceraldehyde 3-phosphate to the enzyme.

high-energy thioester bond

high-energy phosphate bond

inorganic phosphate

A molecule of inorganic phosphate displaces the high-energy thioester bond to create 1,3-bisphospho-glycerate, which contains a high-energy phosphate bond.

1,3-bisphosphoglycerate

The high-energy phosphate group is transferred to ADP to form ATP.

3-phosphoglycerate

(B) SUMMARY OF STEPS 6 AND 7

aldehyde → carboxylic acid

The oxidation of an aldehyde to a carboxylic acid releases energy, much of which is captured in the activated carriers ATP and NADH.

QUESTION 13–2

Arsenate (AsO_4^{3-}) is chemically very similar to phosphate (PO_4^{3-}) and is used as an alternative substrate by many phosphate-requiring enzymes. In contrast to phosphate, however, an anhydride bond between arsenate and carbon is very quickly hydrolyzed nonenzymatically in water. Knowing this, suggest why arsenate is a compound of choice for murderers but not for cells. Formulate your explanation in the context of Figure 13–7.

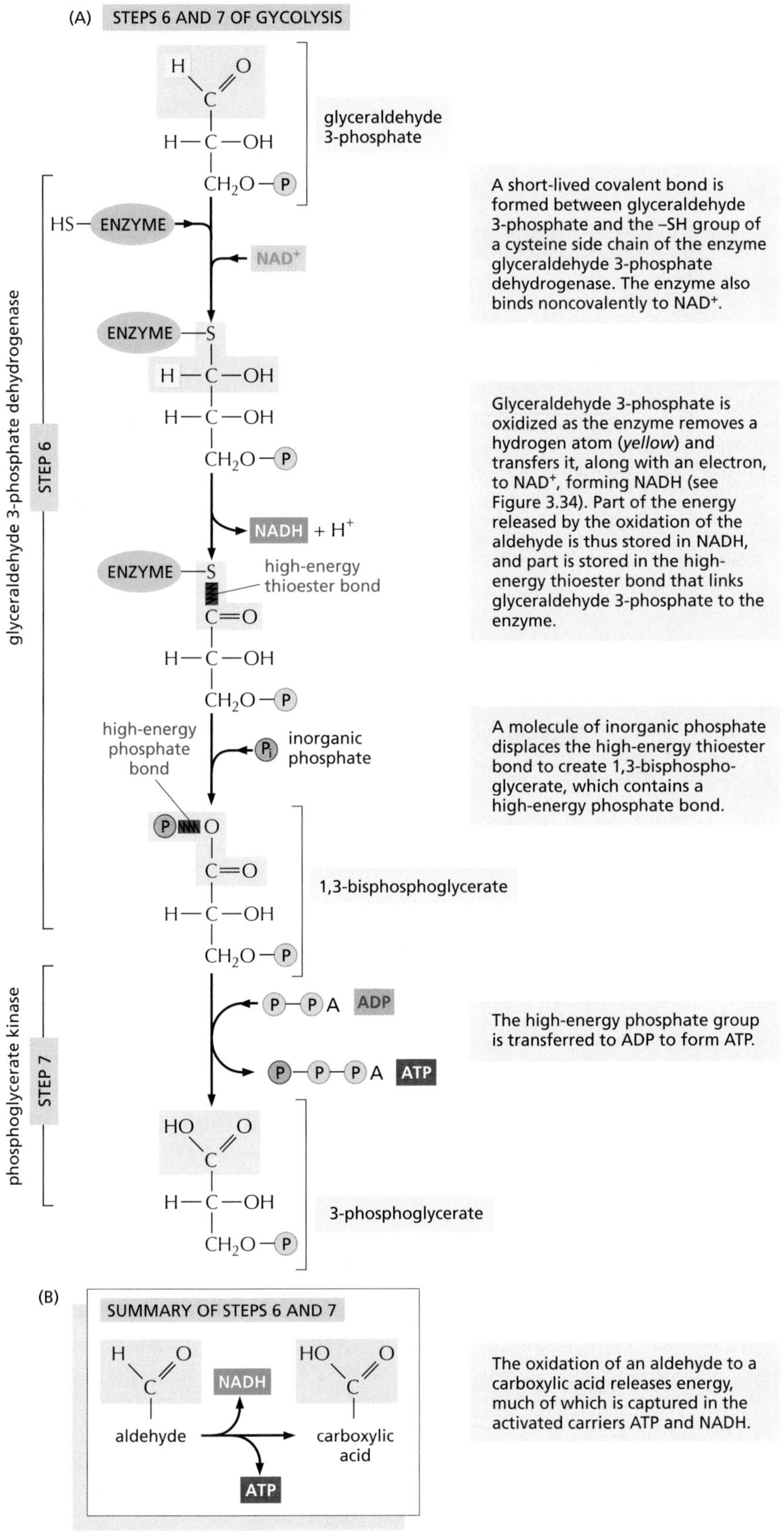

Figure 13–9 The oxidation of glyceraldehyde 3-phosphate is coupled to the formation of ATP and NADH in steps 6 and 7 of glycolysis. (A) In step 6, the enzyme glyceraldehyde 3-phosphate dehydrogenase couples the energetically favorable oxidation of an aldehyde to the energetically unfavorable formation of a high-energy phosphate bond. At the same time, it enables energy to be stored in NADH. The formation of the high-energy phosphate bond is driven by the oxidation reaction, and the enzyme thereby acts like the "paddle wheel" coupler in Figure 3–30B. In step 7, the newly formed high-energy phosphate bond in 1,3-bisphosphoglycerate is transferred to ADP, forming a molecule of ATP and leaving a free carboxylic acid group on the oxidized sugar. The part of the molecule that undergoes a change is shaded in *blue*; the rest of the molecule remains unchanged throughout all these reactions. (B) Summary of the overall chemical change produced by the reactions of steps 6 and 7.

For each step, the part of the molecule that undergoes a change is shadowed in blue, and the name of the enzyme that catalyzes the reaction is in a yellow box. To watch a video of the reactions of glycolysis, see Movie 13.1.

Step 1   Glucose is phosphorylated by ATP to form a sugar phosphate. The negative charge of the phosphate prevents passage of the sugar phosphate through the plasma membrane, trapping glucose inside the cell.

glucose   +   ATP   →   glucose 6-phosphate   +   ADP   +   H$^+$

hexokinase

Step 2   A readily reversible rearrangement of the chemical structure (isomerization) moves the carbonyl oxygen from carbon 1 to carbon 2, forming a ketose from an aldose sugar. (See Panel 2–3, pp. 70–71.)

glucose 6-phosphate (ring form / open-chain form)   ⇌   phosphoglucose isomerase   ⇌   fructose 6-phosphate (open-chain form / ring form)

Step 3   The new hydroxyl group on carbon 1 is phosphorylated by ATP, in preparation for the formation of two three-carbon sugar phosphates. The entry of sugars into glycolysis is controlled at this step, through regulation of the enzyme *phosphofructokinase*.

fructose 6-phosphate   +   ATP   →   phosphofructokinase   →   fructose 1,6-bisphosphate   +   ADP   +   H$^+$

Step 4   The six-carbon sugar is cleaved to produce two three-carbon molecules. Only the glyceraldehyde 3-phosphate can proceed immediately through glycolysis.

fructose 1,6-bisphosphate (ring form / open-chain form)   ⇌   aldolase   ⇌   dihydroxyacetone phosphate   +   glyceraldehyde 3-phosphate

Step 5   The other product of step 4, dihydroxyacetone phosphate, is isomerized to form glyceraldehyde 3-phosphate.

dihydroxyacetone phosphate   ⇌   triose phosphate isomerase   ⇌   glyceraldehyde 3-phosphate

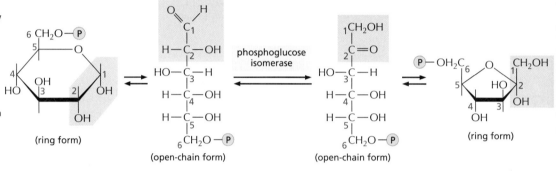

Step 6 The two molecules of glyceraldehyde 3-phosphate are oxidized. The energy-generation phase of glycolysis begins, as NADH and a new high-energy anhydride linkage to phosphate are formed (see Figure 13–5).

glyceraldehyde 3-phosphate dehydrogenase

glyceraldehyde 3-phosphate

1,3-bisphosphoglycerate

Step 7 The transfer to ADP of the high-energy phosphate group that was generated in step 6 forms ATP.

phosphoglycerate kinase

1,3-bisphosphoglycerate

3-phosphoglycerate

Step 8 The remaining phosphate ester linkage in 3-phosphoglycerate, which has a relatively low free energy of hydrolysis, is moved from carbon 3 to carbon 2 to form 2-phosphoglycerate.

phosphoglycerate mutase

3-phosphoglycerate

2-phosphoglycerate

Step 9 The removal of water from 2-phosphoglycerate creates a high-energy enol phosphate linkage.

enolase

2-phosphoglycerate

phosphoenolpyruvate

Step 10 The transfer to ADP of the high-energy phosphate group that was generated in step 9 forms ATP, completing glycolysis.

pyruvate kinase

phosphoenolpyruvate

pyruvate

NET RESULT OF GLYCOLYSIS

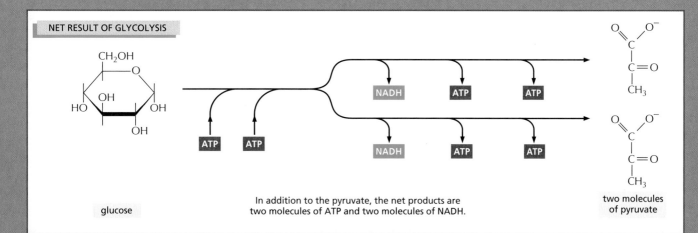

glucose

In addition to the pyruvate, the net products are two molecules of ATP and two molecules of NADH.

two molecules of pyruvate

(A)

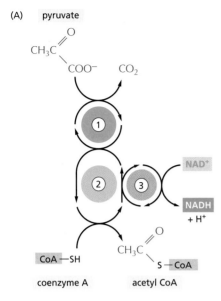

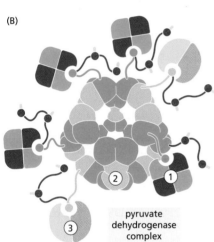

Figure 13–10 Pyruvate is converted into acetyl CoA and CO_2 by the pyruvate dehydrogenase complex in the mitochondrial matrix. (A) The pyruvate dehydrogenase complex, which contains multiple copies of three different enzymes—*pyruvate dehydrogenase* (1), *dihydrolipoyl transacetylase* (2), and *dihydrolipoyl dehydrogenase* (3)—converts pyruvate to acetyl CoA; NADH and CO_2 are also produced in this reaction. Pyruvate and its products are shown in *red lettering*. (B) In this large multienzyme complex, reaction intermediates are passed directly from one enzyme to another via flexible tethers. Only one-tenth of the subunits labeled 1 and 3, attached to the core formed by subunit 2, are shown here. To get a sense of scale, the pyruvate dehydrogenase complex is larger than a ribosome.

Several Organic Molecules Are Converted to Acetyl CoA in the Mitochondrial Matrix

In aerobic metabolism in eukaryotic cells, the **pyruvate** produced by glycolysis is actively pumped into the mitochondrial matrix (see Figure 13–3). There, it is rapidly decarboxylated by a giant complex of three enzymes, called the *pyruvate dehydrogenase complex*. The products of pyruvate decarboxylation are CO_2 (a waste product), NADH, and **acetyl CoA** (**Figure 13–10**).

In addition to sugar, which is broken down during glycolysis, **fat** is a major source of energy for most nonphotosynthetic organisms, including humans. Like the pyruvate derived from glycolysis, the fatty acids derived from fat are also converted into acetyl CoA in the mitochondrial matrix (see Figure 13–3). Fatty acids are first activated by covalent linkage to CoA and are then broken down completely by a cycle of reactions that trims two carbons at a time from their carboxyl end, generating one molecule of acetyl CoA for each turn of the cycle. Two activated carriers—NADH and another high-energy electron carrier, $FADH_2$—are also produced in this process (**Figure 13–11**).

In addition to pyruvate and fatty acids, some amino acids are transported from the cytosol into the mitochondrial matrix, where they are also converted into acetyl CoA or one of the other intermediates of the citric acid cycle (see Figure 13–3). Thus, in the eukaryotic cell, the mitochondrion is the center toward which all energy-yielding catabolic processes lead, whether they begin with sugars, fats, or proteins. In aerobic bacteria—which have no mitochondria—glycolysis and acetyl CoA production, as well as the citric acid cycle, take place in the cytosol.

Catabolism does not end with the production of acetyl CoA. In the process of converting food molecules to acetyl CoA, only a small part of their stored energy is extracted and converted into ATP, NADH, or $FADH_2$. Most of that energy is still locked up in acetyl CoA. The next stage in cell respiration is the citric acid cycle, in which the acetyl group in acetyl CoA is oxidized to CO_2 and H_2O in the mitochondrial matrix, as we now discuss.

The Citric Acid Cycle Generates NADH by Oxidizing Acetyl Groups to CO_2

The **citric acid cycle** accounts for about two-thirds of the total oxidation of carbon compounds in most cells, and its major end products are CO_2 and high-energy electrons in the form of NADH. The CO_2 is released as a waste product, while the high-energy electrons from NADH are passed to the electron-transport chain in the inner mitochondrial membrane. At the end of the chain, these electrons combine with O_2 to produce H_2O.

The citric acid cycle, which takes place in the mitochondrial matrix, does not itself use O_2. However, it requires O_2 to proceed because the

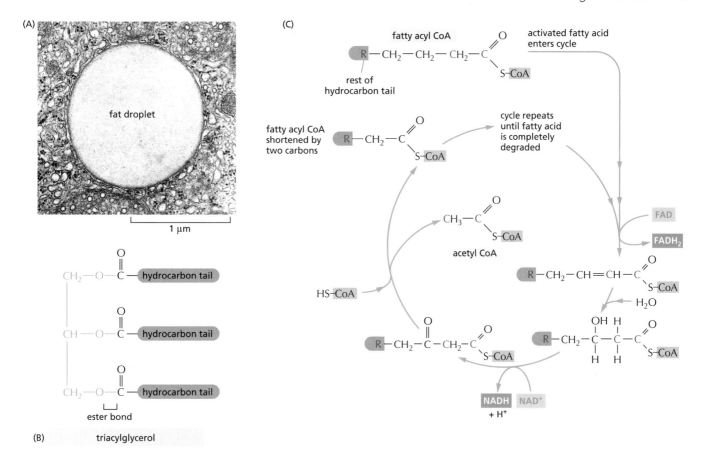

Figure 13–11 Fatty acids derived from fats are also converted to acetyl CoA in the mitochondrial matrix.
(A) Fats are insoluble in water and spontaneously form large lipid droplets in specialized fat cells called adipocytes. This electron micrograph shows a lipid droplet in the cytoplasm of an adipocyte. (B) Fats are stored in the form of triacylglycerol. The glycerol portion, to which three fatty acid chains (shaded in *red*) are linked through ester bonds, is shown in *blue*. Enzymes called lipases can cleave the ester bonds that link the fatty acid chains to glycerol when fatty acids are needed for energy. (C) Fatty acids are first coupled to coenzyme A in a reaction requiring ATP (not shown). The activated fatty acid chains (fatty acyl CoA) are then oxidized in a cycle containing four enzymes. Each turn of the cycle shortens a fatty acyl CoA molecule by two carbons (*red*) and generates one molecule of acetyl CoA and one molecule each of NADH and $FADH_2$. (A, courtesy of Daniel S. Friend.)

electron-transport chain—which uses O_2 as its final acceptor—allows NADH to get rid of its electrons and thus regenerate the NAD^+ needed to keep the cycle going. Although living organisms have inhabited Earth for the past 3.5 billion years, the planet is thought to have developed an atmosphere containing O_2 gas only some 1 to 2 billion years ago (see Figure 14–45). Many of the energy-generating reactions of the citric acid cycle—also called the *tricarboxylic acid cycle* or the *Krebs cycle*—are therefore likely to be of relatively recent origin.

The citric acid cycle catalyzes the complete oxidation of the carbon atoms of the acetyl groups in acetyl CoA, converting them into CO_2. The acetyl group is not oxidized directly, however. Instead, it is transferred from acetyl CoA to a larger four-carbon molecule, oxaloacetate, to form the six-carbon tricarboxylic acid, citric acid, for which the subsequent cycle of reactions is named. The citric acid molecule (also called citrate) is then progressively oxidized, and the energy of this oxidation is harnessed to produce activated carriers in much the same manner as we described for glycolysis. The chain of eight reactions forms a cycle, because the oxaloacetate that began the process is regenerated at the end (**Figure 13–12**). The citric acid cycle is presented in detail in **Panel 13–2** (pp. 434–435), and the experiments that first revealed the cyclic nature of this series of oxidative reactions are described in **How We Know**, pp. 436–437.

QUESTION 13–3

Many catabolic and anabolic reactions are based on reactions that are similar but work in opposite directions, such as the hydrolysis and condensation reactions described in Figure 3–38. This is true for fatty acid breakdown and fatty acid synthesis. From what you know about the mechanism of fatty acid breakdown outlined in Figure 13–11, would you expect the fatty acids found in cells to most commonly have an even or an odd number of carbon atoms?

Figure 13–12 The citric acid cycle catalyzes the complete oxidation of acetyl groups derived from food. The cycle begins with the reaction of acetyl CoA (derived from pyruvate as shown in Figure 13–10) with oxaloacetate to produce citrate (citric acid). The number of carbon atoms in each intermediate is shaded in *yellow*. (See also Panel 13–2, pp. 434–435.) The steps of the citric acid cycle are reviewed in Movie 13.2.

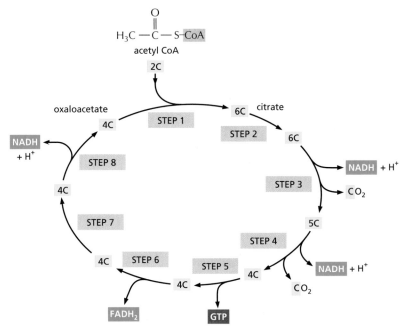

NET RESULT: ONE TURN OF THE CYCLE PRODUCES THREE NADH, ONE GTP, AND ONE FADH$_2$, AND RELEASES TWO MOLECULES OF CO$_2$

Thus far, we have discussed only one of the three types of activated carriers that are produced by the citric acid cycle—NADH. In addition to three molecules of NADH, each turn of the cycle also produces one molecule of **FADH$_2$ (reduced flavin adenine dinucleotide)** from FAD and one molecule of the ribonucleoside triphosphate **GTP (guanosine triphosphate)** from **GDP** (see Figure 13–12). The structures of these two activated carriers are illustrated in **Figure 13–13**. GTP is a close relative of ATP, and the transfer of its terminal phosphate group to ADP produces one ATP molecule in each cycle. Like NADH, FADH$_2$ is a carrier of high-energy electrons and hydrogen. As we discuss shortly, the energy stored in the readily transferred high-energy electrons of NADH and FADH$_2$ is

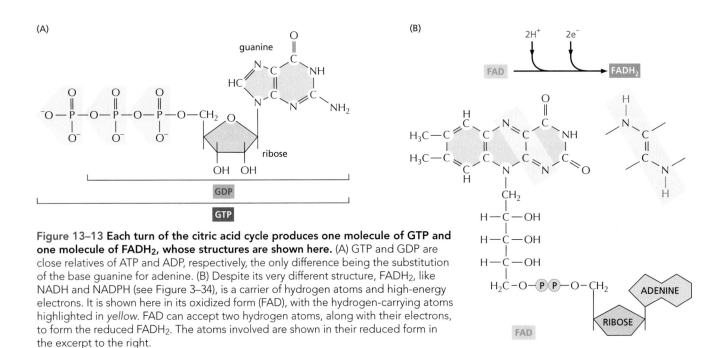

Figure 13–13 Each turn of the citric acid cycle produces one molecule of GTP and one molecule of FADH$_2$, whose structures are shown here. (A) GTP and GDP are close relatives of ATP and ADP, respectively, the only difference being the substitution of the base guanine for adenine. (B) Despite its very different structure, FADH$_2$, like NADH and NADPH (see Figure 3–34), is a carrier of hydrogen atoms and high-energy electrons. It is shown here in its oxidized form (FAD), with the hydrogen-carrying atoms highlighted in *yellow*. FAD can accept two hydrogen atoms, along with their electrons, to form the reduced FADH$_2$. The atoms involved are shown in their reduced form in the excerpt to the right.

subsequently used to produce ATP through oxidative phosphorylation on the inner mitochondrial membrane, the only step in the oxidative catabolism of foodstuffs that directly requires O_2 from the atmosphere.

A common misconception about the citric acid cycle is that the atmospheric O_2 required for the process to proceed is converted into the CO_2 that is released as a waste product. In fact, the oxygen atoms required to make CO_2 from the acetyl groups entering the citric acid cycle are supplied not by O_2 but by water. As illustrated in Panel 13–2, three molecules of water are split in each cycle, and the oxygen atoms of some of them are ultimately used to make CO_2. As we see shortly, the O_2 that we breathe is actually reduced to water by the electron-transport chain; it is not incorporated directly into the CO_2 we exhale.

Many Biosynthetic Pathways Begin with Glycolysis or the Citric Acid Cycle

Catabolic reactions, such as those of glycolysis and the citric acid cycle, produce both energy for the cell and the building blocks from which many other organic molecules are made. So far, we have emphasized energy production rather than the provision of starting materials for biosynthesis. But many of the intermediates formed in glycolysis and the citric acid cycle are siphoned off by such **anabolic pathways**, in which they are converted by series of enzyme-catalyzed reactions into amino acids, nucleotides, lipids, and other small organic molecules that the cell needs. Oxaloacetate and α-ketoglutarate from the citric acid cycle, for example, are transferred from the mitochondrial matrix back to the cytosol, where they serve as precursors for the production of many essential molecules, such as the amino acids aspartate and glutamate, respectively. An idea of the complexity of this process can be gathered from **Figure 13–14,** which illustrates some of the branches leading from the central catabolic reactions to biosyntheses.

QUESTION 13–4

Looking at the chemistry detailed in Panel 13–2 (pp. 434–435), why do you suppose it is useful to link the acetyl group first to another, larger carbon skeleton, oxaloacetate, before completely oxidizing both carbons to CO_2?

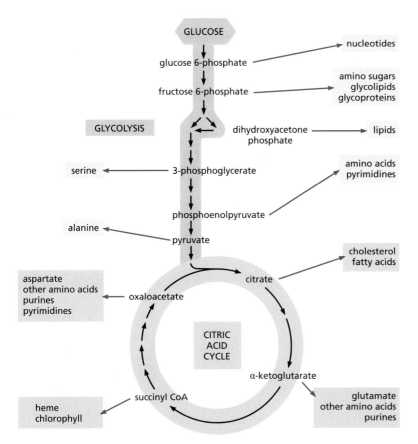

Figure 13–14 **Glycolysis and the citric acid cycle provide the precursors needed for cells to synthesize many important organic molecules.** The amino acids, nucleotides, lipids, sugars, and other molecules—shown here as products—in turn serve as the precursors for many of the cell's macromolecules. Each *black* arrow in this diagram denotes a single enzyme-catalyzed reaction; the *red* arrows generally represent pathways with many steps that are required to produce the indicated products.

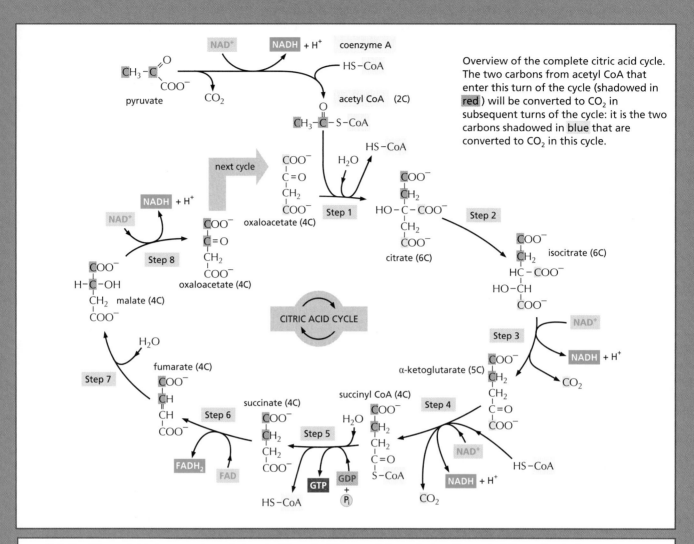

Overview of the complete citric acid cycle. The two carbons from acetyl CoA that enter this turn of the cycle (shadowed in red) will be converted to CO_2 in subsequent turns of the cycle: it is the two carbons shadowed in blue that are converted to CO_2 in this cycle.

Details of these eight steps are shown below. In this part of the panel, for each step, the part of the molecule that undergoes a change is shadowed in blue, and the name of the enzyme that catalyzes the reaction is in a yellow box. To watch a video of the reactions of the citric acid cycle, see Movie 13.2.

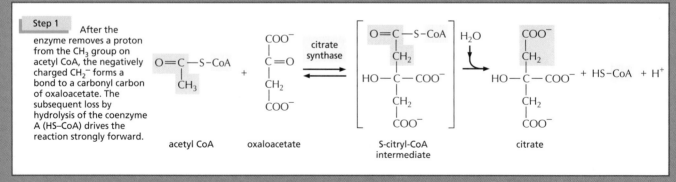

Step 1 After the enzyme removes a proton from the CH_3 group on acetyl CoA, the negatively charged CH_2^- forms a bond to a carbonyl carbon of oxaloacetate. The subsequent loss by hydrolysis of the coenzyme A (HS–CoA) drives the reaction strongly forward.

acetyl CoA   oxaloacetate   S-citryl-CoA intermediate   citrate

Step 2 An isomerization reaction, in which water is first removed and then added back, moves the hydroxyl group from one carbon atom to its neighbor.

citrate   cis-aconitate intermediate   isocitrate

Step 3 In the first of four oxidation steps in the cycle, the carbon carrying the hydroxyl group is converted to a carbonyl group. The immediate product is unstable, losing CO_2 while still bound to the enzyme.

isocitrate → (isocitrate dehydrogenase; NAD^+ → $NADH + H^+$) → oxalosuccinate intermediate → (H^+; CO_2) → α-ketoglutarate

Step 4 The *α-ketoglutarate dehydrogenase complex* closely resembles the large enzyme complex that converts pyruvate to acetyl CoA, the *pyruvate dehydrogenase* complex in Figure 13–10. It likewise catalyzes an oxidation that produces NADH, CO_2, and a high-energy thioester bond to coenzyme A (CoA).

α-ketoglutarate + HS–CoA → (α-ketoglutarate dehydrogenase complex; NAD^+ → $NADH + H^+$; CO_2) → succinyl-CoA

Step 5 A phosphate molecule from solution displaces the CoA, forming a high-energy phosphate linkage to succinate. This phosphate is then passed to GDP to form GTP. (In bacteria and plants, ATP is formed instead.)

succinyl-CoA → (succinyl-CoA synthetase; H_2O; P_i; GDP → GTP) → succinate + HS–CoA

Step 6 In the third oxidation step in the cycle, FAD accepts two hydrogen atoms from succinate.

succinate → (succinate dehydrogenase; FAD → $FADH_2$) → fumarate

Step 7 The addition of water to fumarate places a hydroxyl group next to a carbonyl carbon.

fumarate → (fumarase; H_2O) → malate

Step 8 In the last of four oxidation steps in the cycle, the carbon carrying the hydroxyl group is converted to a carbonyl group, regenerating the oxaloacetate needed for step 1.

malate → (malate dehydrogenase; NAD^+ → $NADH + H^+$) → oxaloacetate

UNRAVELING THE CITRIC ACID CYCLE

"I have often been asked how the work on the citric acid cycle arose and developed," stated biochemist Hans Krebs in a lecture and review article in which he described his Nobel Prize-winning discovery of the cycle of reactions that lies at the center of cell metabolism. Did the concept stem from a sudden inspiration, a revelatory vision? "It was nothing of the kind," answered Krebs. Instead, his realization that these reactions occur in a cycle—rather than a set of linear pathways, as in glycolysis—arose from a "very slow evolutionary process" that occurred over a five-year period, during which Krebs coupled insight and reasoning to careful experimentation to discover one of the central pathways that underlies energy metabolism.

Minced tissues, curious catalysis

By the early 1930s, Krebs and other investigators had discovered that a select set of small organic molecules are oxidized extraordinarily rapidly in various types of tissue preparations—slices of kidney or liver, or suspensions of minced pigeon muscle. Because these reactions were seen to depend on the presence of oxygen, the researchers surmised that this set of molecules might include intermediates that are important in *cell respiration*—the consumption of O_2 and production of CO_2 that occurs when tissues break down foodstuffs.

Using the minced-tissue preparations, Krebs and others made the following observations. First, in the presence of oxygen, certain organic acids—citrate, succinate, fumarate, and malate—were readily oxidized to CO_2. These reactions depended on a continuous supply of oxygen.

Second, the oxidation of these acids occurred in two linear, sequential pathways:

citrate → α-ketoglutarate → succinate

and

succinate → fumarate → malate → oxaloacetate

Third, the addition of small amounts of several of these compounds to the minced-muscle suspensions stimulated an unusually large uptake of O_2—far greater than that needed to oxidize only the added molecules. To explain this surprising observation, Albert Szent-Györgyi (the Nobel laureate who worked out the second pathway above) suggested that a single molecule of each compound must somehow act catalytically to stimulate the oxidation of many molecules of some endogenous substance in the muscle.

At this point, most of the reactions central to the citric acid cycle were known. What was not yet clear—and caused great confusion, even to future Nobel laureates—was how these apparently linear reactions could

drive such a catalytic consumption of oxygen, where each molecule of metabolite fuels the oxidation of many more molecules. To simplify the discussion of how Krebs ultimately solved this puzzle—by linking these linear reactions together into a circle—we will now refer to the molecules involved by a sequence of letters, A through H (**Figure 13–15**).

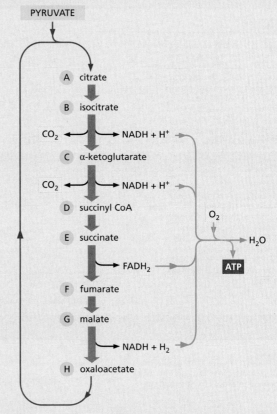

Figure 13–15 In this simplified representation of the citric acid cycle, O_2 is consumed and CO_2 is liberated as the molecular intermediates become oxidized. Krebs and others did not initially realize that these oxidation reactions occur in a cycle, as shown here.

A poison suggests a cycle

Many of the clues that Krebs used to work out the citric acid cycle came from experiments using malonate—a poisonous compound that specifically inhibits the enzyme succinate dehydrogenase, which converts E to F. Malonate closely resembles succinate (E) in its structure (**Figure 13–16**), and it serves as a competitive inhibitor of

```
COO⁻          COO⁻
|             |
CH₂           CH₂
|             |
COO⁻          CH₂
              |
malonate      COO⁻

              succinate
```

Figure 13–16 The structure of malonate closely resembles that of succinate.

the enzyme. Because the addition of malonate poisons cell respiration in tissues, Krebs concluded that succinate dehydrogenase (and the entire pathway linked to it) must play a critical role in the respiration process.

Krebs then discovered that when A, B, or C was added to malonate-poisoned tissue suspensions, E accumulated (**Figure 13–17A**). This observation reinforced the importance of succinate dehydrogenase for successful cell respiration. However, he found that E also accumulated when F, G, or H was added to malonate-poisoned muscle (**Figure 13–17B**). The latter result suggested that an additional set of reactions must exist that can convert F, G, and H molecules into E, since E was previously shown to be a precursor for F, G, and H, rather than a product of their reactions.

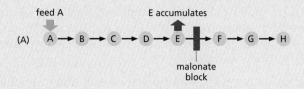

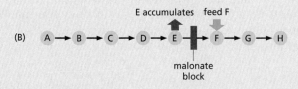

Figure 13–17 Poisoning muscle preparations with malonate provided clues to the cyclic nature of these oxidative reactions. (A) Adding A (or B or C—not shown) to malonate-poisoned muscle causes an accumulation of E. (B) Addition of F (or G or H—not shown) to a malonate-poisoned preparation also causes an accumulation of E, suggesting that enzymatic reactions can convert these molecules into E. The discovery that citrate (A) can be formed from oxaloacetate (H) and pyruvate allowed Krebs to join these two reaction pathways into a complete circle.

At about this time, Krebs also determined that when muscle suspensions were incubated with pyruvate and oxaloacetate, citrate formed: pyruvate + H → A.

This observation led Krebs to postulate that when oxygen is present, pyruvate and H condense to form A, converting the previously delineated string of linear reactions into a cyclic sequence (see Figure 13–15).

Explaining the mysterious stimulatory effects

The cycle of reactions that Krebs proposed clearly explained how the addition of small amounts of any of the intermediates A through H could cause the large increase in the uptake of O_2 that had been observed. Pyruvate is abundant in minced tissues, being readily produced by glycolysis (see Figure 13–4), using glucose derived from stored glycogen. Its oxidation requires a functioning citric acid cycle, in which each turn of the cycle results in the oxidation of one molecule of pyruvate. If the intermediates A through H are in small enough supply, the rate at which the entire cycle turns will be restricted. Adding a supply of any one of these intermediates will then have a dramatic effect on the rate at which the entire cycle operates. Thus, it is easy to see how a large number of pyruvate molecules can be oxidized, and a great deal of oxygen consumed, for every molecule of a citric acid cycle intermediate that is added (**Figure 13–18**).

Krebs went on to demonstrate that all of the individual enzymatic reactions in his postulated cycle took place in tissue preparations. Furthermore, they occured at rates high enough to account for the rate of pyruvate and oxygen consumption in these tissues. Krebs therefore concluded that this series of reactions is the major, if not the sole, pathway for the oxidation of pyruvate—at least in muscle. By fitting together pieces of information like a jigsaw puzzle, he arrived at a coherent picture of the intricate metabolic processes that underlie the oxidation—and took home a share of the 1953 Nobel Prize in Physiology or Medicine.

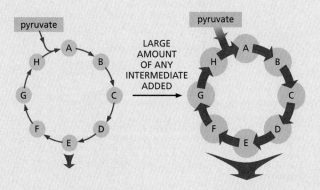

Figure 13–18 Replenishing the supply of any single intermediate has a dramatic effect on the rate at which the entire citric acid cycle operates. When the concentrations of intermediates are limiting, the cycle turns slowly and little pyruvate is used. O_2 uptake is low because only small amounts of NADH and $FADH_2$ are produced to feed oxidative phosphorylation (see Figure 13–19). But when a large amount of any one intermediate is added, the cycle turns rapidly; more of all the intermediates is made, and O_2 uptake is high.

Electron Transport Drives the Synthesis of the Majority of the ATP in Most Cells

We now return briefly to the final stage in the oxidation of food molecules: **oxidative phosphorylation**. It is in this stage that the chemical energy captured by the activated carriers produced during glycolysis and the citric acid cycle is used to generate ATP. During oxidative phosphorylation, NADH and $FADH_2$ transfer their high-energy electrons to the **electron-transport chain**—a series of electron carriers embedded in the inner mitochondrial membrane in eukaryotic cells (and in the plasma membrane of aerobic bacteria). As the electrons pass through the series of electron acceptor and donor molecules that form the chain, they fall to successively lower energy states. At specific sites in the chain, the energy released is used to drive H^+ (protons) across the inner membrane, from the mitochondrial matrix to the intermembrane space (see Figure 13–2). This movement generates a proton gradient across the inner membrane, which serves as a source of energy (like a battery) that can be tapped to drive a variety of energy-requiring reactions (discussed in Chapter 12). The most prominent of these reactions is the phosphorylation of ADP to generate ATP on the matrix side of the inner membrane (**Figure 13–19**).

At the end of the transport chain, the electrons are added to molecules of O_2 that have diffused into the mitochondrion, and the resulting reduced oxygen molecules immediately combine with protons (H^+) from the surrounding solution to produce water (see Figure 13–19). The electrons have now reached their lowest energy level, with all the available energy extracted from the food molecule being oxidized. In total, the complete oxidation of a molecule of glucose to H_2O and CO_2 can produce about 30 molecules of ATP. In contrast, only two molecules of ATP are produced per molecule of glucose by glycolysis alone.

Oxidative phosphorylation occurs in both eukaryotic cells and in aerobic bacteria. It represents a remarkable evolutionary achievement, and the ability to extract energy from food with such great efficiency has shaped the entire character of life on Earth. In the next chapter, we describe the mechanisms behind this game-changing molecular process and discuss how it likely arose.

QUESTION 13–5

What, if anything, is wrong with the following statement: "The oxygen consumed during the oxidation of glucose in animal cells is returned as part of CO_2 to the atmosphere." How could you support your answer experimentally?

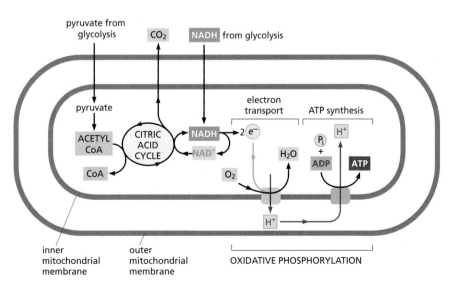

Figure 13–19 **Oxidative phosphorylation completes the catabolism of food molecules and generates the bulk of the ATP made by the cell.** Electron-bearing activated carriers produced by the citric acid cycle and glycolysis donate their high-energy electrons to an electron-transport chain in the inner mitochondrial membrane (or in the plasma membrane of aerobic bacteria). This electron transfer pumps protons across the inner membrane (*red arrows*). The resulting proton gradient is then used to drive the synthesis of ATP through the process of oxidative phosphorylation.

REGULATION OF METABOLISM

A cell is an intricate chemical machine, and our discussion of metabolism—with a focus on glycolysis and the citric acid cycle—has considered only a tiny fraction of the many enzymatic reactions that can take place in a cell at any time (**Figure 13–20**). For all these pathways to work together smoothly, as is required to allow the cell to survive and to respond to its environment, the choice of which pathway each metabolite will follow must be carefully regulated at every branch point.

Many sets of reactions need to be coordinated and controlled. For example, to maintain order within their cells, all organisms need to replenish their ATP pools continuously through the oxidation of sugars or fats. Yet animals have only periodic access to food, and plants need to survive without sunlight overnight, when they are unable to produce sugar through photosynthesis. Animals and plants have evolved several ways to cope with this problem. One is to synthesize food reserves in times of plenty that can be later consumed when other energy sources are scarce. Thus, depending on conditions, a cell must decide whether to route key metabolites into anabolic or catabolic pathways—in other words, whether to use them to build other molecules or burn them to provide

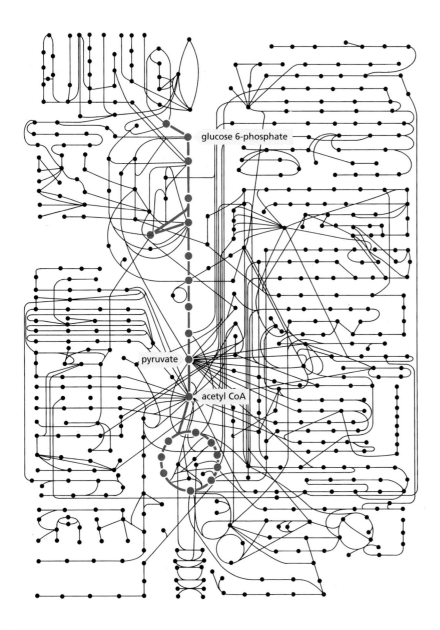

glucose 6-phosphate

pyruvate

acetyl CoA

> ### QUESTION 13–6
>
> A cyclic reaction pathway requires that the starting material be regenerated and available at the end of each cycle. If compounds of the citric acid cycle are siphoned off as building blocks to make other organic molecules via a variety of metabolic reactions, why does the citric acid cycle not quickly grind to a halt?

Figure 13–20 Glycolysis and the citric acid cycle constitute a small fraction of the reactions that occur in a cell. In this diagram, the filled circles represent molecules in various metabolic pathways, and the lines that connect them represent the enzymatic reactions that convert one metabolite to another. The reactions of glycolysis and the citric acid cycle are shown in *red*. Many other reactions either lead into these two central catabolic pathways—delivering small organic molecules to be oxidized for energy—or lead outward to the anabolic pathways that supply carbon compounds for biosynthesis.

immediate energy. In this section, we discuss how a cell regulates its intricate web of interconnected metabolic pathways to best serve both its immediate and long-term needs.

Catabolic and Anabolic Reactions Are Organized and Regulated

All the reactions shown in Figure 13–20 occur in a cell that is less than 0.1 mm in diameter, and each step requires a different enzyme. To add to the complexity, the same substrate is often a part of many different pathways. Pyruvate, for example, is a substrate for half a dozen or more different enzymes, each of which modifies it chemically in a different way. We have already seen that the pyruvate dehydrogenase complex converts pyruvate to acetyl CoA, and that, during fermentation, lactate dehydrogenase converts it to lactate. A third enzyme converts pyruvate to oxaloacetate, a fourth to the amino acid alanine, and so on. All these pathways compete for pyruvate molecules, and similar competitions for thousands of other small molecules go on at the same time.

To balance the activities of these interrelated reactions—and to allow organisms to adapt swiftly to changes in food availability or energy expenditure—an elaborate network of *control mechanisms* regulates and coordinates the activity of the enzymes that catalyze the myriad metabolic reactions that go on in a cell. As we discuss in Chapter 4, the activity of enzymes can be controlled by covalent modification—such as the addition or removal of a phosphate group (see Figure 4–41)—and by the binding of small regulatory molecules, often a metabolite (see pp. 150–151). Such regulation can either enhance the activity of the enzyme or inhibit it. As we see next, both types of regulation—positive and negative—control the activity of key enzymes involved in the breakdown and synthesis of glucose.

Feedback Regulation Allows Cells to Switch from Glucose Breakdown to Glucose Synthesis

Animals need an ample supply of glucose. Active muscles need glucose to power their contraction, and brain cells depend almost completely on glucose for energy. During periods of fasting or intense physical exercise, the body's glucose reserves get used up faster than they can be replenished from food. One way to increase available glucose is to synthesize it from pyruvate by a process called **gluconeogenesis**.

Gluconeogenesis is, in many ways, a reversal of glycolysis: it builds glucose from pyruvate, whereas glycolysis does the opposite. Indeed, gluconeogenesis makes use of many of the same enzymes as glycolysis; it simply runs them in reverse. For example, the isomerase that converts glucose 6-phosphate to fructose 6-phosphate in step 2 of glycolysis (see Panel 13–1, pp. 428–429) will readily catalyze the reverse reaction. There are, however, three steps in glycolysis that so strongly favor the direction of glucose breakdown that they are effectively irreversible. To get around these one-way steps, gluconeogenesis uses a special set of enzymes to catalyze a set of bypass reactions. In step 3 of glycolysis, for example, the enzyme phosphofructokinase catalyzes the phosphorylation of fructose 6-phosphate to produce the intermediate fructose 1,6-bisphosphate. In gluconeogenesis, the enzyme fructose 1,6-bisphosphatase removes a phosphate from this intermediate to produce fructose 6-phosphate (**Figure 13–21**).

How does a cell decide whether to synthesize glucose or to degrade it? Part of the decision centers on the reactions shown in Figure 13–21. The activity of the enzyme phosphofructokinase is allosterically regulated by

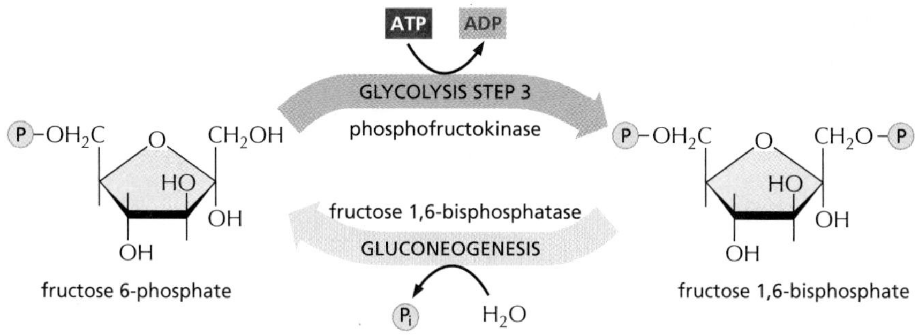

Figure 13–21 Gluconeogenesis uses specific enzymes to bypass those steps in glycolysis that are essentially irreversible. The enzyme phosphofructokinase catalyzes the phosphorylation of fructose 6-phosphate to form fructose 1, 6-bisphosphate in step 3 of glycolysis. This reaction is so energetically favorable that the enzyme will not work in reverse. To produce fructose 6-phosphate in gluconeogenesis, the enzyme fructose 1,6-bisphosphatase removes the phosphate from fructose 1,6-bisphosphate. Coordinated feedback regulation of these two enzymes helps control the flow of metabolites toward glucose synthesis or glucose breakdown.

the binding of a variety of metabolites, which provide both positive and negative *feedback regulation*. The enzyme is activated by byproducts of ATP hydrolysis—including ADP, AMP, and inorganic phosphate—and it is inhibited by ATP. Thus, when ATP is depleted and its metabolic byproducts accumulate, phosphofructokinase is turned on and glycolysis proceeds to generate ATP; when ATP is abundant, the enzyme is turned off and glycolysis shuts down. The enzyme that catalyzes the reverse reaction, fructose 1, 6–bisphosphatase (see Figure 13–21), is regulated by the same molecules but in the opposite direction. Thus this enzyme is activated when phosphofructokinase is turned off, allowing gluconeogenesis to proceed. Many such coordinated regulatory mechanisms enable a cell to respond rapidly to changing conditions and to adjust its metabolism accordingly.

Some of the biosynthetic bypass reactions required for gluconeogenesis are energetically costly. Production of a single molecule of glucose by gluconeogenesis consumes four molecules of ATP and two molecules of GTP. Thus a cell must tightly regulate the balance between glycolysis and gluconeogenesis. If both processes were to proceed simultaneously, they would shuttle metabolites back and forth in a futile cycle that would consume large amounts of energy and generate heat for no purpose.

Cells Store Food Molecules in Special Reservoirs to Prepare for Periods of Need

As we have seen, gluconeogenesis is a costly process, requiring substantial amounts of energy from the hydrolysis of ATP and GTP. During periods when food is scarce, this expensive way of producing glucose is suppressed if alternatives are available. Thus fasting cells can mobilize glucose that has been stored in the form of **glycogen**, a branched polymer of glucose (**Figure 13–22A** and see Panel 2–3, pp. 70–71). This large polysaccharide is stored as small granules in the cytoplasm of many animal cells, but mainly in liver and muscle cells (**Figure 13–22B**). The synthesis and degradation of glycogen occur by separate metabolic pathways, which can be rapidly and coordinately regulated according to need. When more ATP is needed than can be generated from food molecules taken in from the bloodstream, cells break down glycogen in a reaction that is catalyzed by the enzyme *glycogen phosphorylase*. This enzyme produces *glucose 1-phosphate*, which is then converted to the glucose 6-phosphate that feeds into the glycolytic pathway (**Figure 13–22C**).

The glycogen synthetic and degradative pathways are coordinated by feedback regulation. Enzymes in each pathway are allosterically regulated by glucose 6-phosphate, but in opposite directions: *glycogen synthetase* in the synthetic pathway is activated by glucose 6-phosphate, whereas glycogen phosphorylase, which breaks down glycogen (see Figure 13–22C), is inhibited by glucose 6-phosphate, as well as by ATP. This regulation

Figure 13–22 Animal cells store glucose in the form of glycogen to provide energy in times of need. (A) The structure of glycogen (starch in plants is a very similar branched polymer of glucose but has many fewer branch points). (B) An electron micrograph showing glycogen granules in the cytoplasm of a liver cell; each granule contains both glycogen and the enzymes required for glycogen synthesis and breakdown. (C) The enzyme glycogen phosphorylase breaks down glycogen when cells need more glucose. (B, courtesy of Robert Fletterick and Daniel S. Friend.)

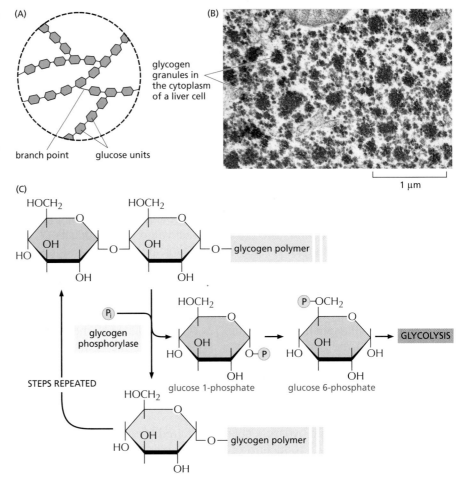

Figure 13–23 Fats are stored in the form of fat droplets in animal cells. The fat droplets (stained *red*) shown here are in the cytoplasm of developing adipocytes. (Courtesy of Peter Tontonoz and Ronald M. Evans.)

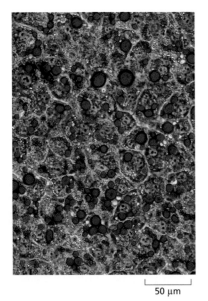

helps to prevent glycogen breakdown when ATP is plentiful and to favor glycogen synthesis when glucose 6-phosphate concentration is high. The balance between glycogen synthesis and breakdown is further regulated by intracellular signaling pathways that are controlled by the hormones insulin, adrenaline, and glucagon (see Table 16–1, p. 529 and Figure 16–25, p. 546).

Quantitatively, fat is a far more important storage material than glycogen, in part because the oxidation of a gram of fat releases about twice as much energy as the oxidation of a gram of glycogen. Moreover, glycogen binds a great deal of water, producing a sixfold difference in the actual mass of glycogen required to store the same amount of energy as fat. An average adult human stores enough glycogen for only about a day of normal activity, but enough fat to last nearly a month. If our main fuel reserves had to be carried as glycogen instead of fat, body weight would need to be increased by an average of about 60 pounds (nearly 30 kilograms).

Most of our fat is stored as droplets of water-insoluble triacylglycerols in specialized fat cells called *adipocytes* (**Figure 13–23** and see Figure 13–11 A and B). In response to hormonal signals, fatty acids can be released from these depots into the bloodstream for other cells to use as required. Such a need arises after a period of not eating. Even a normal overnight fast results in the mobilization of fat: in the morning, most of the acetyl CoA that enters the citric acid cycle is derived from fatty acids rather than from glucose. After a meal, however, most of the acetyl CoA entering the citric acid cycle comes from glucose derived from food, and any excess

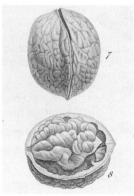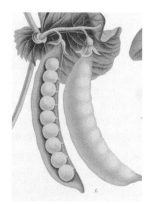

Figure 13–24 Some plant seeds serve as important foods for humans. Corn, nuts, and peas all contain rich stores of starch and fats, which provide the plant embryo in the seed with energy and building blocks for biosynthesis. (Courtesy of the John Innes Foundation.)

glucose is used to make glycogen or fat. (Although animal cells can readily convert sugars to fats, they cannot convert fatty acids to sugars.)

The food reserves in both animals and plants form a vital part of the human diet. Plants convert some of the sugars they make through photosynthesis during daylight into fats and into **starch**, a branched polymer of glucose very similar to animal glycogen. The fats in plants are triacylglycerols, as they are in animals, and they differ only in the types of fatty acids that predominate (see Figures 2–19 and 2–20).

The embryo inside a plant seed must live on stored food reserves for a long time, until the seed germinates to produce a plant with leaves that can harvest the energy in sunlight. The embryo uses these food stores as sources of energy and of small molecules to build cell walls and to synthesize many other biological molecules as it develops. For this reason, plant seeds often contain especially large amounts of fats and starch—which make them a major food source for animals, including ourselves (**Figure 13–24**). Germinating seeds convert the stored fat and starch into glucose as needed.

In plant cells, fats and starch are both stored in chloroplasts—specialized organelles that carry out photosynthesis (**Figure 13–25**). These energy-rich molecules serve as food reservoirs that are mobilized by the cell to produce ATP in mitochondria during periods of darkness. In the next chapter, we take a closer look at chloroplasts and mitochondria, and review the elaborate mechanisms by which they harvest energy from sunlight and from food.

QUESTION 13–7

After looking at the structures of sugars and fatty acids (discussed in Chapter 2), give an intuitive explanation as to why oxidation of a sugar yields only about half as much energy as the oxidation of an equivalent dry weight of a fatty acid.

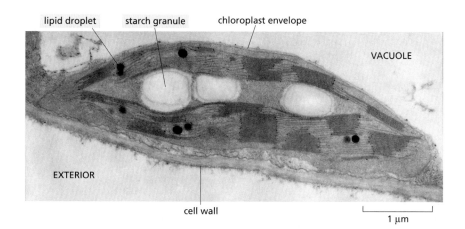

lipid droplet   starch granule   chloroplast envelope

VACUOLE

EXTERIOR

cell wall

1 μm

Figure 13–25 Plant cells store both starch and fats in their chloroplasts. An electron micrograph of a single chloroplast in a plant cell shows the starch granules and lipid droplets (fats) that have been synthesized in the organelle. (Courtesy of K. Plaskitt.)

ESSENTIAL CONCEPTS

- Food molecules are broken down in successive steps, in which energy is captured in the form of activated carriers such as ATP and NADH.

- In plants and animals, these catabolic reactions occur in different cell compartments: glycolysis in the cytosol, the citric acid cycle in the mitochondrial matrix, and oxidative phosphorylation on the inner mitochondrial membrane.

- During glycolysis, the six-carbon sugar glucose is split to form two molecules of the three-carbon sugar pyruvate, producing small amounts of ATP and NADH.

- In the presence of oxygen, eukaryotic cells convert pyruvate into acetyl CoA plus CO_2 in the mitochondrial matrix. The citric acid cycle then converts the acetyl group in acetyl CoA to CO_2 and H_2O, capturing much of the energy released as high-energy electrons in the activated carriers NADH and $FADH_2$.

- Fatty acids produced from the digestion of fats are also imported into mitochondria and converted to acetyl CoA molecules, which are then further oxidized through the citric acid cycle.

- In the mitochondrial matrix, NADH and $FADH_2$ pass their high-energy electrons to an electron-transport chain in the inner mitochondrial membrane, where a series of electron transfers is used to drive the formation of ATP. Most of the energy captured during the breakdown of food molecules is harvested during this process of oxidative phosphorylation (described in detail in Chapter 14).

- Many intermediates of glycolysis and the citric acid cycle are starting points for the anabolic pathways that lead to the synthesis of proteins, nucleic acids, and the many other organic molecules of the cell.

- The thousands of different reactions carried out simultaneously by a cell are regulated and coordinated by positive and negative feedback, enabling the cell to adapt to changing conditions; for example, such feedback allows a cell to switch from glucose breakdown to glucose synthesis when food is scarce.

- Cells store food molecules in special reserves. Glucose subunits are stored as glycogen in animal cells and as starch in plant cells; both animal and plant cells store fatty acids as fats. The food reserves stored by plants are major sources of food for animals, including humans.

KEY TERMS

acetyl CoA	GDP, GTP
ADP, ATP	gluconeogenesis
anabolic pathways	glucose
catabolism	glycogen
cell respiration	glycolysis
citric acid cycle	NAD^+, NADH
electron-transport chain	oxidative phosphorylation
FAD, $FADH_2$	pyruvate
fat	starch
fermentation	

QUESTIONS

QUESTION 13–8

The oxidation of sugar molecules by the cell takes place according to the general reaction $C_6H_{12}O_6$ (glucose) + $6O_2 \rightarrow 6CO_2 + 6H_2O$ + energy. Which of the following statements are correct? Explain your answers.

A. All of the energy produced is in the form of heat.

B. None of the produced energy is in the form of heat.

C. The energy is produced by a process that involves the oxidation of carbon atoms.

D. The reaction supplies the cell with essential water.

E. In cells, the reaction takes place in more than one step.

F. Many steps in the oxidation of sugar molecules involve reaction with oxygen gas.

G. Some organisms carry out the reverse reaction.

H. Some cells that grow in the absence of O_2 produce CO_2.

QUESTION 13–9

An exceedingly sensitive instrument (yet to be devised) shows that one of the carbon atoms in Charles Darwin's last breath is resident in your bloodstream, where it forms part of a hemoglobin molecule. Suggest how this carbon atom might have traveled from Darwin to you, and list some of the molecules it could have entered en route.

QUESTION 13–10

Yeast cells can grow both in the presence of O_2 (aerobically) and in its absence (anaerobically). Under which of the two conditions could you expect the cells to grow better? Explain your answer.

QUESTION 13–11

During movement, muscle cells require large amounts of ATP to fuel their contractile apparatus. These cells contain high levels of creatine phosphate (**Figure Q13–11**), which has a standard free-energy change ($\Delta G°$) for hydrolysis of its phosphate bond of –10.3 kcal/mole. Why is this a useful compound to store energy? Justify your answer with the information shown in Figure 13–8.

creatine phosphate   **Figure Q13–11**

QUESTION 13–12

Identical pathways that make up the complicated sequence of reactions of glycolysis, shown in Panel 13–1 (pp. 428–429), are found in most living cells, from bacteria to humans. One could envision, however, countless alternative chemical reaction mechanisms that would allow the oxidation of sugar molecules and that could, in principle, have evolved to take the place of glycolysis. Discuss this fact in the context of evolution.

QUESTION 13–13

An animal cell, roughly cubical in shape with side length of 10 µm, uses 10^9 ATP molecules every minute. Assume that the cell replaces this ATP by the oxidation of glucose according to the overall reaction $6O_2 + C_6H_{12}O_6 \rightarrow 6CO_2 + 6H_2O$ and that complete oxidation of each glucose molecule produces 30 ATP molecules. How much oxygen does the cell consume every minute? How long will it take before the cell has used up an amount of oxygen gas equal to its own volume? (Recall that one mole of a gas has a volume of 22.4 liter.)

QUESTION 13–14

Under the conditions existing in the cell, the free energies of the first few reactions in glycolysis (in Panel 13–1, pp. 428–429) are:

step 1 ΔG = –8.0 kcal/mole

step 2 ΔG = –0.6 kcal/mole

step 3 ΔG = –5.3 kcal/mole

step 4 ΔG = –0.3 kcal/mole

Are these reactions energetically favorable? Using these values, draw to scale an energy diagram (A) for the overall reaction and (B) for the pathway composed of the four individual reactions.

QUESTION 13–15

The chemistry of most metabolic reactions was deciphered by synthesizing metabolites containing atoms that are different isotopes from those occurring naturally. The products of reactions starting with isotopically labeled metabolites can be analyzed to determine precisely which atoms in the products are derived from which atoms in the starting material. The methods of detection exploit, for example, the fact that different isotopes have different masses that can be distinguished using biophysical techniques such as mass spectrometry. Moreover, some isotopes are radioactive and can therefore be readily recognized with electronic counters or photographic film that becomes exposed by radiation.

A. Assume that pyruvate containing radioactive ^{14}C in its carboxyl group is added to a cell extract that can support oxidative phosphorylation. Which of the molecules produced should contain the vast majority of the ^{14}C that was added?

B. Assume that oxaloacetate containing radioactive ^{14}C in its keto group (refer to Panel 13–2, pp. 434–435) is added to the extract. Where should the ^{14}C atom be located after precisely one turn of the cycle?

QUESTION 13–16

In cells that can grow both aerobically and anaerobically, fermentation is inhibited in the presence of O_2. Suggest a reason for this observation.

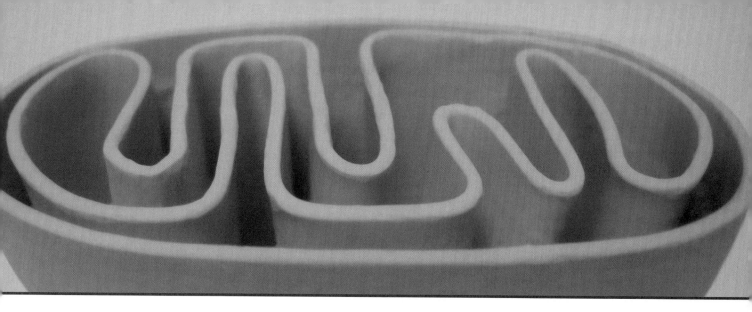

CHAPTER **FOURTEEN**

14

Energy Generation in Mitochondria and Chloroplasts

The fundamental need to generate energy efficiently has had a profound influence on the history of life on Earth. Much of the structure, function, and evolution of cells and organisms can be related to their need for energy. With no oxygen in the atmosphere, it is thought that the earliest cells may have produced ATP by breaking down organic molecules that had been generated by geochemical processes. Such fermentation reactions, discussed in Chapter 13, occur in the cytosol of present-day cells, where they use the energy derived from the partial oxidation of energy-rich food molecules to form ATP.

But early in the history of life, a much more efficient mechanism for generating energy and synthesizing ATP appeared—one based on the transport of electrons along membranes. Billions of years later, this mechanism is so central to the survival of life on Earth that we devote this entire chapter to it. As we will see, membrane-based electron-transport mechanisms are used by cells to extract energy from a wide variety of sources. These mechanisms are central to both the conversion of light energy into chemical-bond energy in photosynthesis and to the generation of large amounts of ATP from food molecules during **cell respiration**. Although membrane-based electron transport first appeared in bacteria more than 3 billion years ago, the descendants of these pioneering cells now crowd every corner and crevice of our planet's land and oceans with a wild menagerie of living forms. Perhaps even more remarkably, remnants of these bacteria survive within every living eukaryotic cell in the form of chloroplasts and mitochondria.

In this chapter, we consider the molecular mechanisms by which electron transport enables cells to generate the energy they need to survive.

MITOCHONDRIA AND
OXIDATIVE PHOSPHORYLATION

MOLECULAR MECHANISMS OF
ELECTRON TRANSPORT AND
PROTON PUMPING

CHLOROPLASTS AND
PHOTOSYNTHESIS

THE EVOLUTION OF ENERGY-
GENERATING SYSTEMS

Figure 14–1 Membrane-based mechanisms use the energy provided by food or sunlight to generate ATP. In oxidative phosphorylation, which occurs in mitochondria, an electron-transport system uses energy derived from the oxidation of food to generate a proton (H⁺) gradient across a membrane. In photosynthesis, which occurs in chloroplasts, an electron-transport system uses energy derived from the sun to generate a proton gradient across a membrane. In both cases, this proton gradient is then used to drive ATP synthesis.

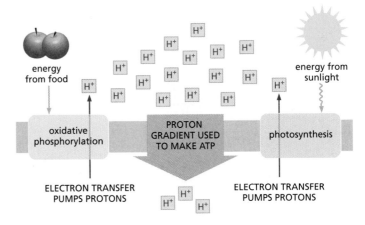

We describe how such systems operate in both mitochondria and chloroplasts, and we review the chemical principles that allow the transfer of electrons to release large amounts of energy. Finally, we trace the evolutionary pathways that gave rise to these mechanisms.

But first, we take a brief look at the general principles central to the generation of energy in all living things: the use of a membrane to harness the energy of moving electrons.

Cells Obtain Most of Their Energy by a Membrane-based Mechanism

The main chemical energy currency in cells is ATP (see Figure 3–32). Small amounts of ATP are generated during glycolysis in the cytosol of all cells (discussed in Chapter 13). But for the majority of cells, most of their ATP is produced by *oxidative phosphorylation*. The generation of ATP by oxidative phosphorylation differs from the way ATP is produced during glycolysis, in that it requires a membrane. In eukaryotic cells, oxidative phosphorylation takes place in mitochondria, and it depends on an electron-transport process that drives the transport of protons (H⁺) across the inner mitochondrial membrane. A related membrane-based process produces ATP during photosynthesis in plants, algae, and photosynthetic bacteria (**Figure 14–1**).

This membrane-based process for making ATP consists of two linked stages: one sets up an electrochemical proton gradient, the other uses that gradient to generate ATP. Both stages are carried out by special protein complexes in the membrane.

1. In Stage 1, high-energy electrons derived from the oxidation of food molecules (discussed in Chapter 13), from sunlight, or from other sources (discussed later) are transferred along a series of electron carriers—called an **electron-transport chain**—embedded in the membrane. These electron transfers release energy that is used to pump protons, derived from the water that is ubiquitous in cells, across the membrane and thus generate an electrochemical proton gradient (**Figure 14–2A**). An ion gradient across a membrane is a form of stored energy that can be harnessed to do useful work when the ions are allowed to flow back across the membrane down their electrochemical gradient (discussed in Chapter 12).

2. In Stage 2 of oxidative phosphorylation, protons flow back down their electrochemical gradient through a protein complex called *ATP synthase*, which catalyzes the energy-requiring synthesis of ATP from ADP and inorganic phosphate (P$_i$). This ubiquitous enzyme functions like a turbine, permitting the proton gradient to drive the production of ATP (**Figure 14–2B**).

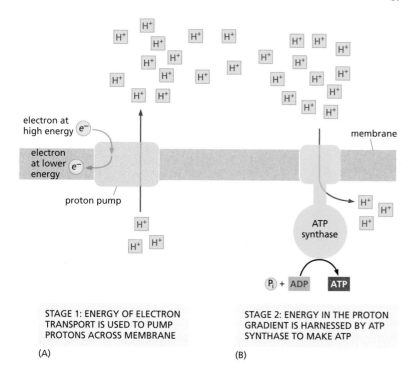

Figure 14–2 Membrane-based systems use the energy stored in an electrochemical proton gradient to synthesize ATP. The process occurs in two stages. (A) In the first stage, a proton pump harnesses the energy of electron transfer (details not shown here) to pump protons (H⁺) derived from water, creating a proton gradient across the membrane. The *blue arrow* shows the direction of electron movement. These high-energy electrons can come from organic or inorganic molecules, or they can be produced by the action of light on special molecules such as chlorophyll. (B) The proton gradient produced in (A) serves as a versatile energy store. It is used to drive a variety of energy-requiring reactions in mitochondria, chloroplasts, and prokaryotes—including the synthesis of ATP by an ATP synthase.

STAGE 1: ENERGY OF ELECTRON TRANSPORT IS USED TO PUMP PROTONS ACROSS MEMBRANE

(A)

STAGE 2: ENERGY IN THE PROTON GRADIENT IS HARNESSED BY ATP SYNTHASE TO MAKE ATP

(B)

When it was first proposed in 1961, this mechanism for generating energy was called the *chemiosmotic hypothesis*, because it linked the chemical bond-forming reactions that synthesize ATP ("chemi-") with the membrane transport processes that pump protons ("osmotic," from the Greek *osmos*, "to push"). Thanks to this chemiosmotic mechanism, now known as **chemiosmotic coupling**, cells can harness the energy of electron transfers in much the same way that the energy stored in a battery can be harnessed to do useful work (**Figure 14–3**).

Chemiosmotic Coupling is an Ancient Process, Preserved in Present-Day Cells

The membrane-based, chemiosmotic mechanism for making ATP arose very early in life's history. The exact same type of ATP-generating processes occur in the plasma membrane of modern bacteria and archaea. Apparently, the mechanism was so successful that its essential features have been retained in the long evolutionary journey from early prokaryotes to present-day cells.

This remarkable resemblance can be attributed in part to the fact that the organelles that produce ATP in eukaryotic cells—the chloroplasts and mitochondria—evolved from bacteria that were engulfed by ancestral cells more than a billion years ago (see Figures 1–18 and 1–20). As evidence of their bacterial ancestry, both chloroplasts and mitochondria reproduce in a manner similar to that of most prokaryotes (**Figure 14–4**).

Figure 14–3 Batteries can use the energy of electron transfer to perform work. (A) If a battery's terminals are directly connected to each other, the energy released by electron transfer is all converted into heat. (B) If the battery is connected to a pump, much of the energy released by electron transfers can be harnessed to do work instead (in this case, to pump water). Cells can similarly harness the energy of electron transfer to do work—for example, pumping H⁺ (see Figure 14–2A).

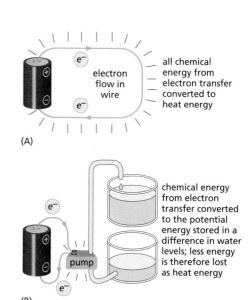

(A) electron flow in wire — all chemical energy from electron transfer converted to heat energy

(B) pump — chemical energy from electron transfer converted to the potential energy stored in a difference in water levels; less energy is therefore lost as heat energy

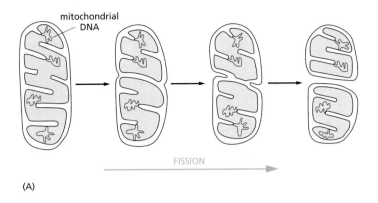

FISSION

(A)

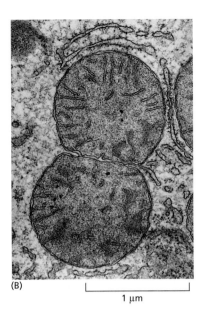

(B)

1 μm

Figure 14–4 A mitochondrion can divide like a bacterium. (A) It undergoes a fission process that is conceptually similar to bacterial division. (B) An electron micrograph of a dividing mitochondrion in a liver cell. (B, courtesy of Daniel S. Friend.)

They also harbor bacterial-like biosynthetic machinery for making RNA and proteins, and they retain their own genomes (**Figure 14–5**). Many chloroplast genes are strikingly similar to those of cyanobacteria—the photosynthetic bacteria from which chloroplasts are thought to have been derived.

Although mitochondria and chloroplasts still contain DNA, the bacteria that gave rise to these organelles gave up many of the genes required for independent living as they developed the symbiotic relationships that led to the evolution of eukaryotic animal and plant cells. These jettisoned genes were not lost, however; many moved to the cell nucleus, where they continue to direct the production of proteins that mitochondria and chloroplasts import to carry out their specialized functions—including the generation of ATP, a process we discuss in detail throughout the remainder of the chapter.

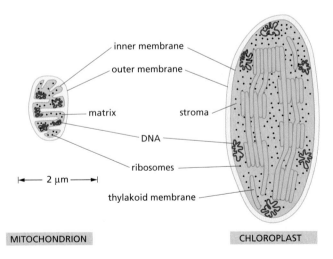

MITOCHONDRION

CHLOROPLAST

Figure 14–5 Mitochondria and chloroplasts share many of the features of their bacterial ancestors. Both organelles contain their own DNA-based genome and the machinery to copy this DNA and to make RNA and protein. The inner compartments of these organelles—the mitochondrial matrix and the chloroplast stroma—contain the DNA (*red*) and a special set of ribosomes. Membranes in both organelles—the mitochondrial inner membrane and the chloroplast thylakoid membrane—contain the protein complexes involved in ATP production.

MITOCHONDRIA AND OXIDATIVE PHOSPHORYLATION

Mitochondria are present in nearly all eukaryotic cells, where they produce the bulk of the cell's ATP. Without mitochondria, eukaryotes would have to rely on the relatively inefficient process of glycolysis for all of their ATP production. When glucose is converted to pyruvate by glycolysis in the cytosol, the net result is that only two molecules of ATP are produced per glucose molecule, which is less than 10% of the total free energy potentially available from oxidizing the sugar. By contrast, about 30 molecules of ATP are produced when mitochondria are recruited to complete the oxidation of glucose that begins in glycolysis. Had ancestral cells not established the relationship with the bacteria that gave rise to modern mitochondria, it seems unlikely that complex multicellular organisms could have evolved.

The importance of mitochondria is further highlighted by the dire consequences of mitochondrial dysfunction. For example, patients with an inherited disorder called *myoclonic epilepsy and ragged red fiber disease (MERRF)* are deficient in multiple proteins required for electron transport. As a result, they typically experience muscle weakness, heart problems, epilepsy, and often dementia. Muscle and nerve cells are especially sensitive to mitochondrial defects, because they need so much ATP to function normally.

In this section, we review the structure and function of mitochondria. We outline how this organelle makes use of an electron-transport chain, embedded in its inner membrane, to generate the proton gradient needed to drive the synthesis of ATP. And we consider the overall efficiency with which this membrane-based system converts the energy stored in food molecules into the energy contained in the phosphate bonds of ATP.

Mitochondria Can Change Their Shape, Location, and Number to Suit a Cell's Needs

Isolated mitochondria are generally similar in size and shape to their bacterial ancestors. Although they are no longer capable of living independently, mitochondria are remarkably adaptable and can adjust their location, shape, and number to suit the needs of the cell. In some cells, mitochondria remain fixed in one location, where they supply ATP directly to a site of unusually high energy consumption. In a heart muscle cell, for example, mitochondria are located close to the contractile apparatus, whereas in a sperm they are wrapped tightly around the motile flagellum (**Figure 14–6**). In other cells, mitochondria fuse to form elongated, dynamic tubular networks, which are diffusely distributed through

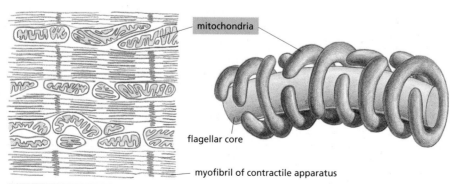

mitochondria

flagellar core

myofibril of contractile apparatus

(A) CARDIAC MUSCLE CELL

(B) SPERM TAIL

Figure 14–6 Some mitochondria are located near sites of high ATP utilization. (A) In a cardiac muscle cell, mitochondria are located close to the contractile apparatus, in which ATP hydrolysis provides the energy for contraction. (B) In a sperm, mitochondria are located in the tail, wrapped around a portion of the motile flagellum that requires ATP for its movement.

Figure 14–7 Mitochondria often fuse to form elongated tubular networks, which can extend throughout the cytoplasm. (A) Mitochondria (*red*) are fluorescently labeled in this cultured mouse fibroblast. (B) In a yeast cell, the mitochondria (*red*) form a continuous network, tucked against the plasma membrane. (A, courtesy of Michael W. Davidson, Carl Zeiss Microscopy Online Campus; B, from J. Nunnari et al., *Mol. Biol. Cell.* 8:1233–1242, 1997. With permission by The American Society for Cell Biology.)

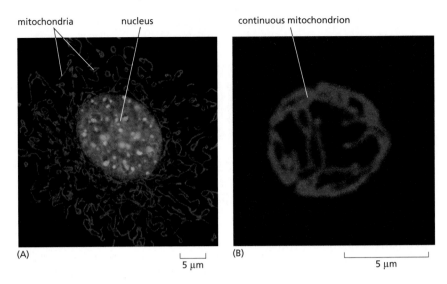

mitochondria   nucleus   continuous mitochondrion

(A)   (B)

5 μm   5 μm

the cytoplasm (**Figure 14–7**). These networks are dynamic, continually breaking apart by fission (see Figure 14–4) and fusing again.

Mitochondria are present in large numbers—1000 to 2000 in a liver cell, for example. But their numbers vary depending on the cell type and can change with the energy needs of the cell. In skeletal muscle cells, for example, mitochondria can divide until their numbers increase five- to tenfold if the muscle has been repeatedly stimulated to contract.

Regardless of their varied appearance, location, and number, however, all mitochondria have the same basic internal structure—a design that supports the efficient production of ATP, as we see next.

A Mitochondrion Contains an Outer Membrane, an Inner Membrane, and Two Internal Compartments

An individual mitochondrion is bounded by two highly specialized membranes—one surrounding the other. These membranes, called the outer and inner mitochondrial membranes, create two mitochondrial compartments: a large internal space called the **matrix** and a much narrower *intermembrane space* (**Figure 14–8**). When purified mitochondria are gently fractionated into separate components and their contents analyzed (see Panel 4–3, pp. 164–165), each of the membranes, and the spaces they enclose, are found to contain a unique collection of proteins.

The *outer membrane* contains many molecules of a transport protein called *porin*, which, forms wide aqueous channels through the lipid bilayer (described in Chapter 11). As a result, the outer membrane is like a sieve that is permeable to all molecules of 5000 daltons or less, including small proteins. This makes the intermembrane space chemically equivalent to the cytosol with respect to the small molecules and inorganic ions it contains. In contrast, the *inner membrane*, like other membranes in the cell, is impermeable to the passage of ions and most small molecules, except where a path is provided by specific membrane transport proteins. The mitochondrial matrix therefore contains only molecules that are selectively transported into the matrix across the inner membrane, and so its contents are highly specialized.

The inner mitochondrial membrane is the site of oxidative phosphorylation, and it contains the proteins of the electron-transport chain, the proton pumps, and the ATP synthase required for ATP production. It also contains a variety of transport proteins that allow the entry of selected small molecules—such as pyruvate and fatty acids that will be oxidized by the mitochondrion—into the matrix.

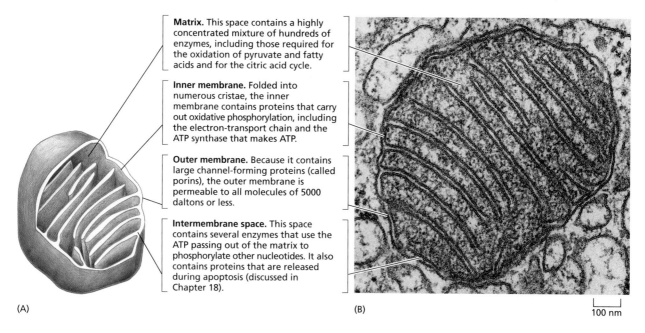

Matrix. This space contains a highly concentrated mixture of hundreds of enzymes, including those required for the oxidation of pyruvate and fatty acids and for the citric acid cycle.

Inner membrane. Folded into numerous cristae, the inner membrane contains proteins that carry out oxidative phosphorylation, including the electron-transport chain and the ATP synthase that makes ATP.

Outer membrane. Because it contains large channel-forming proteins (called porins), the outer membrane is permeable to all molecules of 5000 daltons or less.

Intermembrane space. This space contains several enzymes that use the ATP passing out of the matrix to phosphorylate other nucleotides. It also contains proteins that are released during apoptosis (discussed in Chapter 18).

(A)   (B)

100 nm

The inner membrane is highly convoluted, forming a series of infoldings—known as *cristae*—that project into the matrix space (see Figure 14–8 and Movie 14.1). These folds greatly increase the surface area of the membrane. In a liver cell, for example, the inner membranes of all the mitochondria make up about one-third of the total membranes of the cell. And the number of cristae in a mitochondrion of a cardiac muscle cell is three times greater than that in a mitochondrion from a liver cell.

Figure 14–8 A mitochondrion is organized into four separate compartments. (A) A schematic drawing and (B) an electron micrograph of a mitochondrion. Each compartment contains a unique set of proteins, enabling it to perform its distinct functions. In liver mitochondria, an estimated 67% of the total mitochondrial protein is located in the matrix, 21% in the inner membrane, 6% in the outer membrane, and 6% in the intermembrane space. (B, courtesy of Daniel S. Friend.)

The Citric Acid Cycle Generates the High-Energy Electrons Required for ATP Production

The generation of ATP is powered by the flow of electrons that are derived from the burning of carbohydrates, fats, and other foodstuffs during glycolysis and the citric acid cycle (discussed in Chapter 13). These high-energy electrons are provided by activated carriers generated during these two stages of catabolism, with the majority being churned out by the citric acid cycle that operates in the mitochondrial matrix.

The citric acid cycle gets the fuel it needs to produce these activated carriers from food-derived molecules that make their way into mitochondria from the cytosol. Both the pyruvate produced by glycolysis, which takes place in the cytosol, and the fatty acids derived from the breakdown of fats (see Figure 13–3) can enter the mitochondrial intermembrane space through the porins in the outer mitochondrial membrane. These fuel molecules are then transported across the inner mitochondrial membrane into the matrix, where they are converted into the crucial metabolic intermediate, acetyl CoA (**Figure 14–9**). The acetyl groups in acetyl CoA are

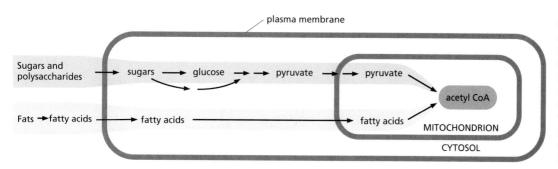

plasma membrane

| Sugars and polysaccharides | sugars → glucose → → pyruvate → → pyruvate |
| Fats → fatty acids → fatty acids → fatty acids |

acetyl CoA

MITOCHONDRION

CYTOSOL

Figure 14–9 In eukaryotic cells, acetyl CoA is produced in the mitochondria from molecules derived from sugars and fats. Most of the cell's oxidation reactions occur in these organelles, and most of its ATP is made here.

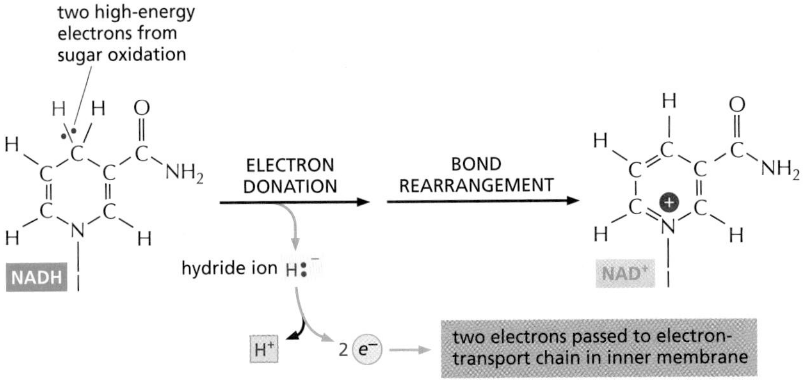

Figure 14–10 NADH donates its high-energy electrons to the electron-transport chain. In this drawing, the electrons being transferred are shown as two *red dots* on a *red* hydrogen atom. A hydride ion (a hydrogen atom with an extra electron) is removed from NADH and is converted into a proton and two electrons. Only the part of NADH that carries these high-energy electrons is shown; for the complete structure and the conversion of NAD$^+$ back to NADH, see the structure of the closely related NADPH in Figure 3–34. Electrons are also carried in a similar way by FADH$_2$, whose structure is shown in Figure 13–13B.

then oxidized to CO$_2$ via the citric acid cycle (see Figure 13–12). Some of the energy derived from this oxidation is saved in the form of high-energy electrons, held by the activated carriers NADH and FADH$_2$. These activated carriers can then donate their high-energy electrons to the electron-transport chain in the inner mitochondrial membrane (**Figure 14–10**).

The Movement of Electrons is Coupled to the Pumping of Protons

The chemiosmotic generation of energy begins when the activated carriers NADH and FADH$_2$ donate their high-energy electrons to the electron-transport chain in the inner mitochondrial membrane, becoming oxidized to NAD$^+$ and FAD in the process (see Figure 14–10). The electrons are quickly passed along the chain to molecular oxygen (O$_2$) to form water (H$_2$O). The stepwise movement of these high-energy electrons through the components of the electron-transport chain releases energy that can then be used to pump protons across the inner membrane (**Figure 14–11**). The resulting proton gradient, in turn, is used to drive the synthesis of ATP. The full sequence of reactions is shown in **Figure 14–12**. The inner mitochondrial membrane thus serves as a device that converts the energy contained in the high-energy electrons of NADH (and FADH$_2$) into the phosphate bond of ATP molecules (**Figure 14–13**). This chemiosmotic mechanism for ATP synthesis is called **oxidative phosphorylation**, because it involves both the consumption of O$_2$ and the addition of a phosphate group to ADP to form ATP.

The source of the high-energy electrons that power the proton pumping differs widely between different organisms and different processes. In cell respiration—which takes place in both mitochondria and aerobic bacteria—the high-energy electrons are ultimately derived from sugars

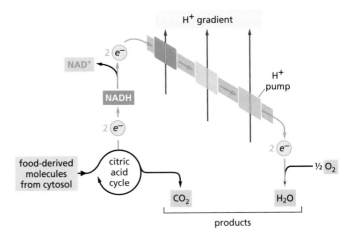

Figure 14–11 As electrons are transferred from activated carriers to oxygen, protons are pumped across the inner mitochondrial membrane. This is stage 1 of chemiosmotic coupling (see Figure 14–2). The path of electron flow is indicated by *blue arrows*.

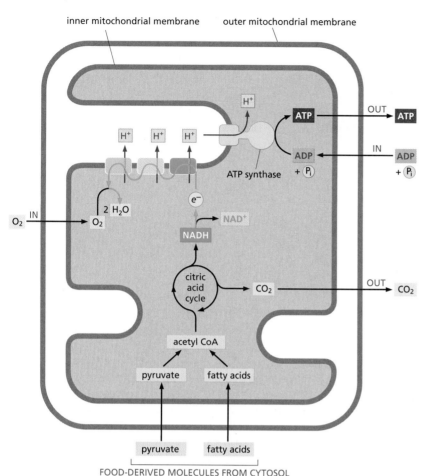

inner mitochondrial membrane   outer mitochondrial membrane

Figure 14–12 Activated carriers generated during the citric acid cycle power the production of ATP. Pyruvate and fatty acids enter the mitochondrial matrix (*bottom*), where they are converted to acetyl CoA. The acetyl CoA is then metabolized by the citric acid cycle, which produces NADH (and $FADH_2$, not shown). During oxidative phosphorylation, high-energy electrons donated by NADH (and $FADH_2$) are then passed along the electron-transport chain in the inner membrane to oxygen (O_2); this electron transport generates a proton gradient across the inner membrane, which is used to drive the production of ATP by ATP synthase. The exact ratios of "reactants" and "products" are not indicated in this diagram: for example, we will see shortly that it requires four electrons from four NADH molecules to convert O_2 to two H_2O molecules.

or fats. In photosynthesis, the high-energy electrons come from the organic green pigment *chlorophyll*, which captures energy from sunlight. And many single-celled organisms (archaea and bacteria) use inorganic substances such as hydrogen, iron, and sulfur as the source of the high-energy electrons that they need to make ATP (see, for example, Figure 1–12).

Regardless of the electron source, the vast majority of living organisms use a chemiosmotic mechanism to generate ATP. In the following sections, we describe in detail how this process occurs.

Protons Are Pumped Across the Inner Mitochondrial Membrane by Proteins in the Electron-Transport Chain

The electron-transport chain—or *respiratory chain*—that carries out oxidative phosphorylation is present in many copies in the inner mitochondrial membrane. Each chain contains over 40 proteins, grouped into three large **respiratory enzyme complexes**. These complexes each contain multiple individual proteins, including transmembrane proteins that anchor the complex firmly in the inner mitochondrial membrane.

The three respiratory enzyme complexes, in the order in which they receive electrons, are: (1) *NADH dehydrogenase complex*, (2) *cytochrome* c *reductase complex*, and (3) *cytochrome* c *oxidase complex* (**Figure 14–14**). Each complex contains metal ions and other chemical groups that act as stepping stones to facilitate the passage of electrons. The movement of electrons through these respiratory complexes is accompanied by the pumping of protons from the mitochondrial matrix to the intermembrane space. Thus each complex can be thought of as a proton pump.

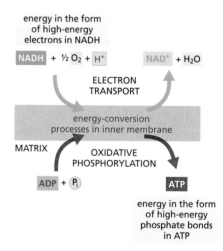

Figure 14–13 Mitochondria catalyze a major conversion of energy. In oxidative phosphorylation, the energy released by the oxidation of NADH to NAD^+ is harnessed—through energy-conversion processes in the inner mitochondrial membrane—to drive the energy-requiring phosphorylation of ADP to form ATP. The net equation for this process, in which four electrons pass from NADH to oxygen, is $2NADH + O_2 + 2H^+ \rightarrow 2NAD^+ + 2H_2O$.

Figure 14–14 High-energy electrons are transferred through three respiratory enzyme complexes in the inner mitochondrial membrane. The relative size and shape of each complex are indicated, although the numerous individual protein components that form each complex are not. During the transfer of high-energy electrons from NADH to oxygen (*blue lines*), protons derived from water are pumped across the membrane from the matrix into the intermembrane space by each of the complexes (Movie 14.2). Ubiquinone (Q) and cytochrome c (c) serve as mobile carriers that ferry electrons from one complex to the next.

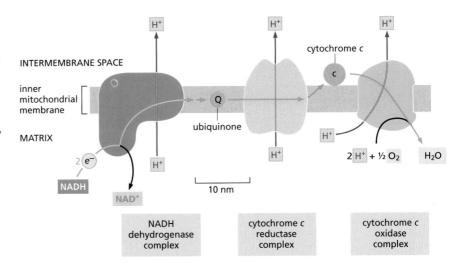

The first respiratory complex in the chain, NADH dehydrogenase, accepts electrons from NADH. These electrons are extracted from NADH in the form of a hydride ion (H⁻), which is then converted into a proton and two high-energy electrons. That reaction, $H^- \rightarrow H^+ + 2e^-$ (see Figure 14–10), is catalyzed by the NADH dehydrogenase complex. The electrons are then passed along the chain to each of the other enzyme complexes in turn, using mobile electron carriers to ferry electrons between complexes (see Figure 14–14). This transfer of electrons is energetically favorable: the electrons are passed from electron carriers with weaker electron affinity to those with stronger electron affinity, until they combine with a molecule of O_2 to form water. This final reaction is the only oxygen-requiring step in cell respiration, and it consumes nearly all of the oxygen that we breathe.

Proton Pumping Produces a Steep Electrochemical Proton Gradient Across the Inner Mitochondrial Membrane

Without a mechanism for harnessing the energy released by the energetically favorable transfer of electrons from NADH to O_2, this energy would simply be liberated as heat. Cells are able to recover much of this energy because the three respiratory enzyme complexes in the electron-transport chain use it to pump protons across the inner mitochondrial membrane, from the matrix into the intermembrane space (see Figure 14–14). Later, we will outline the molecular mechanisms involved. For now, we focus on the consequences of this nifty maneuver. First, the pumping of protons generates a H⁺ gradient—or pH gradient—across the inner membrane. As a result, the pH in the matrix (around 7.9) is about 0.7 unit higher than it is in the intermembrane space (which is 7.2, the same pH as the cytosol). Second, proton pumping generates a voltage gradient—or membrane potential—across the inner membrane; as H⁺ flows outward, the matrix side of the membrane becomes negative and the side facing the intermembrane space becomes positive.

As discussed in Chapter 12, the force that drives the passive flow of an ion across a membrane is proportional to the ion's *electrochemical gradient*. The strength of that electrochemical gradient depends both on the voltage across the membrane, as measured by the membrane potential, and on the ion's concentration gradient (see Figure 12–5). Because protons are positively charged, they will more readily cross a membrane if there is an excess of negative charge on the other side. In the case of the inner mitochondrial membrane, the pH gradient and membrane potential work

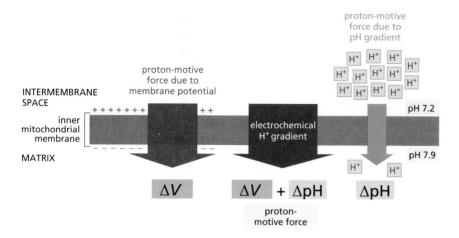

Figure 14–15 The electrochemical H⁺ gradient across the inner mitochondrial membrane includes a large force due to the membrane potential (ΔV) and a smaller force due to the H⁺ concentration gradient—that is, the pH gradient (ΔpH). Both forces combine to generate the proton-motive force, which pulls H⁺ back into the mitochondrial matrix. The exact, mathematical relationship between these forces is expressed by the Nernst equation (see Figure 12–23).

together to create a steep electrochemical proton gradient that makes it energetically very favorable for H⁺ to flow back into the mitochondrial matrix. The membrane potential contributes significantly to this *proton-motive force*, which pulls H⁺ back across the membrane; the greater the membrane potential, the more energy is stored in the proton gradient (**Figure 14–15**).

ATP Synthase Uses the Energy Stored in the Electrochemical Proton Gradient to Produce ATP

If protons in the intermembrane space were allowed simply to flow back into the mitochondrial matrix, the energy stored in the electrochemical proton gradient would be lost as heat. Such a seemingly wasteful process allows hibernating bears to stay warm, as we discuss further in How We Know (pp. 462–463). In most cells, however, the electrochemical proton gradient across the inner mitochondrial membrane is used to drive the synthesis of ATP from ADP and Pᵢ (see Figure 2–25). The device that makes this possible is **ATP synthase**, a large, multisubunit protein embedded in the inner mitochondrial membrane.

ATP synthase is of ancient origin; the same enzyme generates ATP in the mitochondria of animal cells, the chloroplasts of plants and algae, and the plasma membrane of bacteria. The part of the protein that catalyzes the phosphorylation of ADP is shaped like a lollipop head and projects into the mitochondrial matrix; it is attached by a central stalk to a transmembrane H⁺ carrier (**Figure 14–16**). The passage of protons through the carrier causes the carrier and its stalk to spin rapidly, like a tiny motor. As the stalk rotates, it rubs against proteins in the stationary head, altering their conformation and prompting them to produce ATP. In this way, a mechanical deformation gets converted into the chemical-bond energy of ATP (Movie 14.3). This fine-tuned sequence of interactions allows ATP synthase to produce more than 100 molecules of ATP per second—3 molecules of ATP per revolution.

ATP synthase can also operate in reverse—using the energy of ATP hydrolysis to pump protons "uphill," against their electrochemical gradient across the membrane (**Figure 14–17**). In this mode, ATP synthase functions like the H⁺ pumps described in Chapter 12. Whether ATP synthase primarily makes ATP—or consumes it to pump protons—depends on the magnitude of the electrochemical proton gradient across the membrane in which the enzyme is embedded. In many bacteria that can grow either aerobically or anaerobically, the direction in which the ATP

> **QUESTION 14–3**
>
> When the drug dinitrophenol (DNP) is added to mitochondria, the inner membrane becomes permeable to protons (H⁺). In contrast, when the drug nigericin is added to mitochondria, the inner membrane becomes permeable to K⁺. (A) How does the electrochemical proton gradient change in response to DNP? (B) How does it change in response to nigericin?

Figure 14–16 ATP synthase acts like a motor to convert the energy of protons flowing down their electrochemical gradient to chemical-bond energy in ATP.
(A) The multisubunit protein is composed of a stationary head, called the F_1 ATPase, and a rotating portion called F_0. Both F_1 and F_0 are formed from multiple subunits. Driven by the electrochemical proton gradient, the F_0 part of the protein—which consists of the transmembrane H^+ carrier (*blue*) plus a central stalk (*purple*)—spins rapidly within the stationary head of the F_1 ATPase (*green*), causing it to generate ATP from ADP and P_i. The stationary head is secured to the inner membrane by an elongated protein "arm" called the peripheral stalk (*orange*). The F_1 ATPase is so named because it can carry out the reverse reaction—the hydrolysis of ATP to ADP and P_i—when detached from the F_0 portion of the complex.

(B) The three-dimensional structure of ATP synthase, as determined by X-ray crystallography. The peripheral stalk is fixed to the membrane with the help of the subunit denoted by the *pink* oval, which is the only part of the complex still lacking structural details. At its other end, this stalk is tied to the F_1 ATPase head via the small *red* subunit. (B, courtesy of K. Davies.)

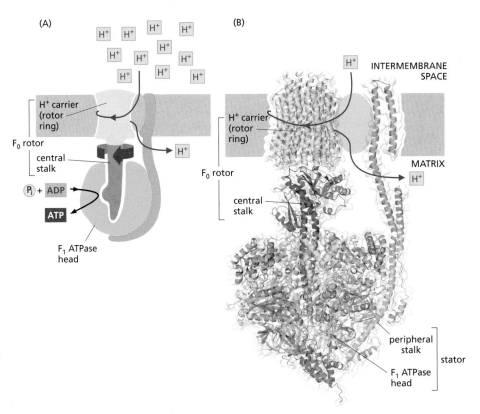

synthase works is routinely reversed when the bacterium runs out of O_2. Under these conditions, the ATP synthase uses some of the ATP generated inside the cell by glycolysis to pump protons out of the cell, creating the proton gradient that the bacterial cell needs to import its essential nutrients by coupled transport. A similar mechanism is used to drive the transport of small molecules in and out of the mitochondrial matrix, as we discuss next.

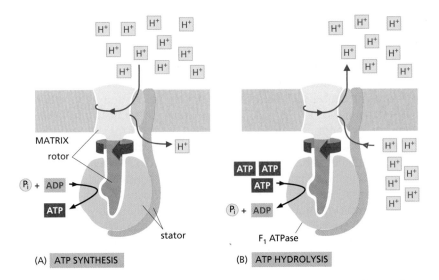

Figure 14–17 ATP synthase is a reversible coupling device. It can either synthesize ATP by harnessing the electrochemical H^+ gradient (A) or pump protons against this gradient by hydrolyzing ATP (B). The direction of operation at any given instant depends on the net free-energy change (ΔG, discussed in Chapter 3) for the coupled processes of H^+ translocation across the membrane and the synthesis of ATP from ADP and P_i. For example, if the electrochemical proton gradient falls below a certain level, the ΔG for H^+ transport into the matrix will no longer be large enough to drive ATP production; instead, ATP will be hydrolyzed by the ATP synthase to rebuild the proton gradient. A tribute to the activity of ATP synthase is shown in Movie 14.4.

Coupled Transport Across the Inner Mitochondrial Membrane Is Also Driven by the Electrochemical Proton Gradient

The synthesis of ATP is not the only process driven by the electrochemical proton gradient in mitochondria. Many small, charged molecules, such as pyruvate, ADP, and inorganic phosphate (P_i), are imported into the mitochondrial matrix from the cytosol, while others, such as ATP, must be transported in the opposite direction. Carrier proteins that bind these molecules can couple their transport to the energetically favorable flow of H^+ into the matrix (see the "coupled transporters" in Figure 12–14). Pyruvate and P_i, for example, are each co-transported inward along with protons, as the protons move down their electrochemical gradient into the matrix.

Other transporters take advantage of the membrane potential generated by the electrochemical proton gradient, which makes the matrix side of the inner mitochondrial membrane more negatively charged than the side that faces the intermembrane space. An antiport carrier protein exploits this voltage gradient to export ATP from the mitochondrial matrix and to bring ADP in. This exchange allows the ATP synthesized in the mitochondrion to be exported rapidly (**Figure 14–18**).

In eukaryotic cells, therefore, the electrochemical proton gradient is used to drive both the formation of ATP and the transport of selected metabolites across the inner mitochondrial membrane. In bacteria, the proton gradient across the plasma membrane is similarly used to drive ATP synthesis and metabolite transport. But it also serves as an important source of directly usable energy: in motile bacteria, for instance, the flow of protons into the cell drives the rapid rotation of the bacterial flagellum, which propels the bacterium along (Movie 14.5).

The Rapid Conversion of ADP to ATP in Mitochondria Maintains a High ATP/ADP Ratio in Cells

As a result of the nucleotide exchange shown in Figure 14–18, ADP molecules—produced by hydrolysis of ATP in the cytosol—are rapidly drawn back into mitochondria for recharging, while the bulk of the ATP molecules produced in mitochondria are exported into the cytosol, where they are most needed. (A small amount of ATP is used within mitochondria

QUESTION 14–4

The remarkable properties that allow ATP synthase to run in either direction allow the interconversion of energy stored in the H^+ gradient and energy stored in ATP to proceed in either direction. (A) If ATP synthase making ATP can be likened to a water-driven turbine producing electricity, what would be an appropriate analogy when it works in the opposite direction? (B) Under what conditions would one expect the ATP synthase to stall, running neither forward nor backward? (C) What determines the direction in which the ATP synthase operates?

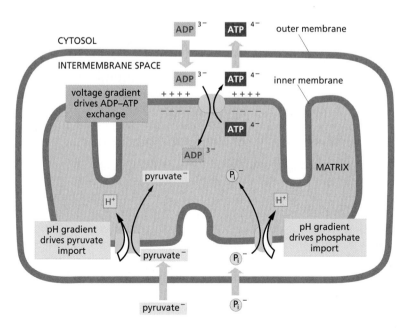

Figure 14–18 The electrochemical proton gradient across the inner mitochondrial membrane is used to drive some coupled transport processes. The charge on each of the transported molecules is indicated for comparison with the membrane potential, which is negative inside, as shown. Pyruvate and inorganic phosphate (P_i) are moved into the matrix along with protons, as the protons move down their electrochemical gradient. Both are negatively charged, so their movement is opposed by the negative membrane potential; however, the H^+ concentration gradient—the pH gradient—is harnessed in a way that nevertheless drives their inward transport. ADP is pumped into the matrix and ATP is pumped out by an antiport process that uses the voltage gradient across the membrane to drive this exchange. The outer mitochondrial membrane is freely permeable to all of these compounds due to the presence of porins in the membrane (not shown). The active transport of molecules across membranes by carrier proteins and the generation of a membrane potential are discussed in Chapter 12.

themselves to power DNA replication, protein synthesis, and other energy-consuming reactions that occur there.) With all of this back-and-forth, a typical ATP molecule in a human cell will shuttle out of a mitochondrion and back in (as ADP) more than once every minute.

As discussed in Chapter 3, most biosynthetic enzymes drive energetically unfavorable reactions by coupling them to the energetically favorable hydrolysis of ATP (see Figure 3–33A). The pool of ATP in a cell is thus used to drive a huge variety of cell processes in much the same way that a battery is used to drive an electric engine. To be useful in this way, the concentration of ATP in the cytosol must be kept about 10 times higher than that of ADP. If the activity of mitochondria were halted, ATP levels would fall dramatically and the cell's battery would run down. Eventually, energetically unfavorable reactions could no longer take place and the cell would die. The poison cyanide, which blocks electron transport in the inner mitochondrial membrane, causes cell death in exactly this way.

Cell Respiration Is Amazingly Efficient

The oxidation of sugars to produce ATP may seem unnecessarily complex. Surely the process could be accomplished more directly—perhaps by eliminating the citric acid cycle or some of the steps in the respiratory chain. Such simplification would certainly make the chemistry easier for students to learn—but it would be bad news for the cell. As discussed in Chapter 13, the oxidative pathways that allow cells to extract energy from food efficiently and in a usable form involve many intermediates, each differing only slightly from its predecessor. In this way, the huge amounts of energy locked up in food molecules can be parceled out into small packets that can be captured in activated carriers, such as NADH and $FADH_2$ (see Figure 13–1).

Much of the energy carried by NADH and $FADH_2$ is ultimately converted into the bond energy of ATP. How much ATP each of these activated carriers can produce depends on several factors, including where its electrons enter the respiratory chain. The NADH molecules produced in the mitochondrial matrix during the citric acid cycle pass their high-energy electrons to the NADH dehydrogenase complex—the first complex in the chain. As the electrons pass from one enzyme complex to the next, they promote the pumping of protons across the inner mitochondrial membrane at each step along the way. In this way, each NADH molecule provides enough net energy to generate about 2.5 molecules of ATP (see Question 14–5 and its answer).

$FADH_2$ molecules, on the other hand, bypass the NADH dehydrogenase complex and pass their electrons to the membrane-embedded mobile carrier ubiquinone (see Figure 14–14). Because these electrons enter further down the respiratory chain than those donated by NADH, they promote the pumping of fewer protons: each molecule of $FADH_2$ thus produces only 1.5 molecules of ATP. **Table 14–1** provides a full accounting of the ATP produced by the complete oxidation of glucose.

Although the biological oxidation of glucose to CO_2 and H_2O consists of many interdependent steps, the overall process is remarkably efficient. Almost 50% of the total energy that could be released by burning sugars or fats is captured and stored in the phosphate bonds of ATP during cell respiration. That might not seem impressive, but it is considerably better than most nonbiological energy-conversion devices. Electric motors and gasoline engines operate at about 10–20% efficiency. If cells operated at this efficiency, an organism would have to eat voraciously just to maintain itself. Moreover, because the wasted energy is liberated as heat, large organisms (including ourselves) would need far better mechanisms

TABLE 14–1 PRODUCT YIELDS FROM GLUCOSE OXIDATION

Process	Direct product	Final ATP yield per molecule of glucose
Glycolysis	2 NADH (cytosolic)	3*
	2 ATP	2
Pyruvate oxidation to acetyl CoA (two per glucose)	2 NADH (mitochondrial matrix)	5
Complete acetyl CoA oxidation (two per glucose)	6 NADH (mitochondrial matrix)	15
	2 FADH$_2$	3
	2 GTP	2
	TOTAL	30

*NADH produced in the cytosol yields fewer ATP molecules than NADH produced in the mitochondrial matrix because the mitochondrial inner membrane is impermeable to NADH. Transporting NADH into the mitochondrial matrix—where it encounters NADH dehydrogenase—thus requires energy.

for cooling themselves. It is hard to imagine how animals could have evolved without the elaborate yet economical mechanisms that allow cells to extract a maximum amount of energy from food.

MOLECULAR MECHANISMS OF ELECTRON TRANSPORT AND PROTON PUMPING

For many years, biochemists struggled to understand why electron-transport chains had to be embedded in membranes to function in ATP production. The puzzle was essentially solved in the 1960s, when it was discovered that transmembrane proton gradients drive the process. The concept of chemiosmotic coupling was so novel, however, that it was not widely accepted until many years later, when additional experiments with artificial energy-generating systems put the power of proton gradients to the test (see **How We Know**, pp. 462–463).

Although investigators are still unraveling some of the details of chemiosmotic coupling at the atomic level, the fundamentals are now clear. In this section, we examine the basic principles that drive the movement of electrons, and we explain in molecular detail how electron transport can generate a proton gradient. Because very similar mechanisms are used by mitochondria, chloroplasts, and prokaryotes, these principles apply to nearly all living things.

Protons Are Readily Moved by the Transfer of Electrons

Although protons resemble other positive ions such as Na$^+$ and K$^+$ in the way they move across membranes, in some respects they are unique. Hydrogen atoms are by far the most abundant atom in living organisms: they are plentiful not only in all carbon-containing biological molecules but also in the water molecules that surround them. The protons in water are highly mobile: by rapidly dissociating from one water molecule and associating with its neighbor, they can rapidly flit through a hydrogen-bonded network of water molecules (see Figure 2–15B). Thus water, which is everywhere in cells, serves as a ready reservoir for donating and accepting protons.

QUESTION 14–5

Calculate the number of usable ATP molecules produced per pair of electrons transferred from NADH to oxygen, if (i) five protons are pumped across the inner mitochondrial membrane for each electron passed through the three respiratory enzyme complexes, (ii) three protons must pass through the ATP synthase for each ATP molecule that it produces from ADP and inorganic phosphate inside the mitochondrion, and (iii) one proton is used to produce the voltage gradient needed to transport each ATP molecule out of the mitochondrion to the cytosol where it is used.

HOW WE KNOW

HOW CHEMIOSMOTIC COUPLING DRIVES ATP SYNTHESIS

In 1861, Louis Pasteur discovered that yeast cells grow and divide more vigorously when air is present—the first demonstration that aerobic metabolism is more efficient than anaerobic metabolism. His observations make sense now that we know that oxidative phosphorylation is a much more efficient means of generating ATP than is glycolysis, producing about 30 molecules of ATP for each molecule of glucose oxidized, compared with the 2 ATPs generated by glycolysis alone. But it took another hundred years for researchers to determine that it is the process of chemiosmotic coupling—using proton pumping to power ATP synthesis—that allows cells to generate energy with such efficiency.

Imaginary intermediates

In the 1950s, many researchers believed that the oxidative phosphorylation that takes place in mitochondria generates ATP via a mechanism similar to that used in glycolysis. During glycolysis, ATP is produced when a molecule of ADP receives a phosphate group directly from a "high-energy" intermediate. Such substrate-level phosphorylation occurs in steps 7 and 10 of glycolysis, where the high-energy phosphate groups from 1,3-bisphosphoglycerate and phosphoenolpyruvate, respectively, are transferred to ADP to form ATP (see Panel 13–1, pp. 428–429). It was assumed that the electron-transport chain in mitochondria would similarly generate some phosphorylated intermediate that could then donate its phosphate group directly to ADP. This model inspired a long and frustrating search for this mysterious high-energy intermediate. Investigators occasionally claimed to have discovered the missing intermediate, but the compounds turned out to be either unrelated to electron transport or, as one researcher put it in a review of the history of bioenergetics, "products of high-energy imagination."

Harnessing the force

It wasn't until 1961 that Peter Mitchell suggested that the "high-energy intermediate" his colleagues were seeking was, in fact, the electrochemical proton gradient generated by the electron-transport system. His proposal, dubbed the chemiosmotic hypothesis, stated that the energy of an electrochemical proton gradient formed during the transfer of electrons through the electron-transport chain could be tapped to drive ATP synthesis.

Several lines of evidence offered support for Mitchell's proposed mechanism. First, mitochondria do generate an electrochemical proton gradient across their inner membrane. But what does this gradient—also called the proton-motive force—actually do? If the gradient is required to drive ATP synthesis, as the chemiosmotic hypothesis posits, then either disrupting the inner membrane or eliminating the proton gradient across it should inhibit ATP production. In fact, researchers found both these predictions to be true. Physical disruption of the inner mitochondrial membrane halts ATP synthesis in that organelle. Similarly, dissipation of the proton gradient by a chemical "uncoupling" agent such as 2,4-dinitrophenol (DNP) also inhibits mitochondrial ATP production. Such gradient-busting chemicals carry H^+ across the inner mitochondrial membrane, forming a shuttle system for the movement of H^+ that bypasses the ATP synthase (**Figure 14–19**). In this way, compounds such as DNP uncouple electron transport from ATP synthesis. As a result of this short-circuiting, the proton-motive force is dissipated completely, and the organelle can no longer make ATP.

Such uncoupling occurs naturally in some specialized fat cells. In these cells, called *brown fat cells*, most of the energy from the oxidation of fat is dissipated as heat rather than converted into ATP. The inner membranes

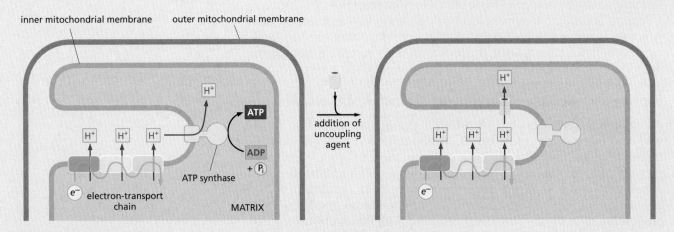

Figure 14–19 Uncoupling agents are H^+ carriers that can insert into the inner mitochondrial membrane. They render the membrane permeable to protons, allowing H^+ to flow into the mitochondrial matrix without passing through ATP synthase. This short circuit effectively uncouples electron transport from ATP synthesis.

of the large mitochondria in these cells contain a carrier protein that allows protons to move down their electrochemical gradient, circumventing ATP synthase. As a result, the cells oxidize their fat stores at a rapid rate and produce more heat than ATP. Tissues containing brown fat serve as biological heating pads, helping to revive hibernating animals and to protect sensitive areas of newborn human babies (such as the backs of their necks) from the cold.

Artificial ATP generation

If disrupting the electrochemical proton gradient across the mitochondrial inner membrane terminates ATP synthesis, then, conversely, generating an artificial proton gradient should stimulate ATP synthesis. Again, this is exactly what happens. When a proton gradient is imposed artificially by lowering the pH on the outside of the mitochondrial inner membrane, out pours ATP.

How does the electrochemical proton gradient drive ATP production? This is where the ATP synthase comes in. In 1974, Efraim Racker and Walther Stoeckenius demonstrated that they could reconstitute a complete artificial ATP-generating system by combining an ATP synthase isolated from the mitochondria of cow heart muscle with a proton pump purified from the purple membrane of the prokaryote *Halobacterium halobium*. As discussed

in Chapter 11, the plasma membrane of this archaean is packed with bacteriorhodopsin, a protein that pumps H^+ out of the cell in response to sunlight (see Figure 11–27).

When bacteriorhodopsin was reconstituted into artificial lipid vesicles (liposomes), Racker and Stoeckenius showed that, in the presence of light, the bacterial protein pumps H^+ into the vesicles, generating a proton gradient. (The orientation of the protein is reversed in these membranes, so that protons are transported into the vesicles; in the bacterium, protons are pumped out.) When the bovine ATP synthase was then incorporated into these vesicles, much to the amazement of many biochemists, the system could catalyze the synthesis of ATP from ADP and inorganic phosphate in response to light. This ATP formation showed an absolute dependence on an intact proton gradient, as either eliminating bacteriorhodopsin from the system or adding uncoupling agents such as DNP abolished ATP synthesis (**Figure 14–20**).

This remarkable experiment demonstrated without a doubt that a proton gradient could stimulate ATP synthase to make ATP. Thus, although biochemists had initially hoped to discover a high-energy intermediate involved in oxidative phosphorylation, the experimental evidence eventually convinced them that their search was in vain and that the chemiosmotic hypothesis was correct. Mitchell was awarded a Nobel Prize in 1978.

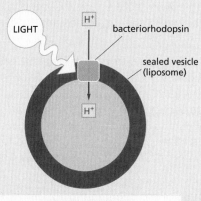

(A)

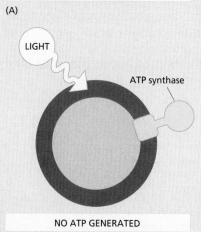

(C)

Figure 14–20 Experiments in which bacteriorhodopsin and bovine mitochondrial ATP synthase were introduced into liposomes provided direct evidence that proton gradients can power ATP production. (A) When bacteriorhodopsin is added to artificial lipid vesicles (liposomes), the protein generates a proton gradient in response to light. (B) In artificial vesicles containing both bacteriorhodopsin and an ATP synthase, a light-generated proton gradient drives the formation of ATP from ADP and P_i. (C) Artificial vesicles containing only ATP synthase do not on their own produce ATP in response to light. (D) In vesicles containing both bacteriorhodopsin and ATP synthase, uncoupling agents that abolish the proton gradient eliminate light-induced ATP synthesis.

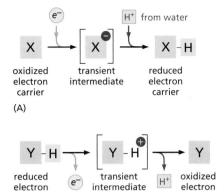

(A)

(B)

Figure 14–21 Electron transfers can cause the movement of entire hydrogen atoms, because protons are readily accepted from or donated to water. In these examples, an oxidized electron carrier molecule, X, picks up an electron plus a proton when it is reduced (A), and a reduced electron carrier molecule, Y, loses an electron plus a proton when it is oxidized (B).

These protons often accompany the electrons that are transferred during oxidation and reduction. When a molecule is reduced by acquiring an electron (e^-), the electron brings with it a negative charge; in many cases, this charge is immediately neutralized by the addition of a proton from water, so that the net effect of the reduction is to transfer an entire hydrogen atom, $H^+ + e^-$ (**Figure 14–21A**). Similarly, when a molecule is oxidized, it often loses an electron from one of its hydrogen atoms: in most instances, the electron is transferred to an electron carrier, and the proton is passed on to water (**Figure 14–21B**). Therefore, in a membrane in which electrons are being passed along an electron-transport chain, it is a relatively simple matter, in principle, to move protons from one side of the membrane to the other. All that is required is that the electron carrier be oriented in the membrane in such a way that it accepts an electron—along with a proton from water—on one side of the membrane, and then releases that proton on the other side of the membrane when the electron is passed on to the next electron carrier molecule in the chain (**Figure 14–22**).

The Redox Potential Is a Measure of Electron Affinities

The proteins of the respiratory chain guide the electrons so that they move sequentially from one enzyme complex to another—with no short circuits that skip a complex. Each electron transfer is an oxidation–reduction reaction: as described in Chapter 3, the molecule or atom donating the electron becomes oxidized, while the receiving molecule or atom becomes reduced (see pp. 89–90). Electrons will pass spontaneously from molecules that have a relatively low affinity for their outer-shell electrons, and thus lose them easily, to molecules that have a higher affinity for electrons. For example, NADH has a low electron affinity, so that its electrons are readily passed to the NADH dehydrogenase complex (see Figure 14–14). The batteries we use to power our electronic gadgets are based on similar electron transfers between chemical substances with different electron affinities.

In biochemical reactions, any electrons removed from one molecule are always passed to another, so that whenever one molecule is oxidized, another is reduced. Like any other chemical reaction, the tendency of such oxidation–reduction reactions, or **redox reactions**, to proceed spontaneously depends on the free-energy change (ΔG) for the electron transfer, which in turn depends on the relative affinities of the two molecules for electrons. (The role of free energy in chemical reactions is discussed in Chapter 3, pp. 90–100.)

Because electron transfers provide most of the energy in living things, it is worth taking time to understand them. Molecules that donate protons are known as acids; those that accept protons are called bases (see Panel 2–2, pp. 68–69). These molecules exist in conjugate acid–base pairs, in which the acid is readily converted into the base by the loss of a proton. For example, acetic acid (CH_3COOH) is converted into its conjugate base (CH_3COO^-) in the reaction

$$CH_3COOH \rightleftharpoons CH_3COO^- + H^+$$

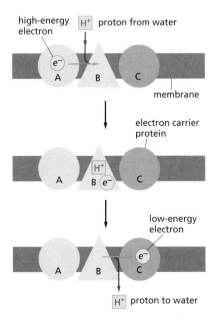

Figure 14–22 The orientation of a membrane-embedded electron carrier allows electron transfer to drive proton pumping. As an electron passes along an electron-transport chain, it can bind and release a proton at each step. In this schematic diagram, the electron carrier, protein B, picks up a proton (H^+) from one side of the membrane when it accepts an electron (e^-) from protein A; protein B releases the proton to the other side of the membrane when it donates its electron to the electron carrier, protein C.

In the same way, pairs of compounds such as NADH and NAD$^+$ are called **redox pairs**, because NADH is converted to NAD$^+$ by the loss of electrons in the reaction

$$NADH \rightleftharpoons NAD^+ + H^+ + 2e^-$$

NADH is a strong electron donor. Its electrons can be said to be held at high-energy because the ΔG for passing them to many other molecules is favorable. Conversely, it is difficult to produce the high-energy electrons in NADH, so its partner, NAD$^+$, is of necessity a weak electron acceptor.

The tendency for a redox pair such as NADH/NAD$^+$ to donate or accept electrons can be determined experimentally by measuring its **redox potential (Panel 14–1**, p. 466). Electrons will move spontaneously from a redox pair with a low redox potential (or low affinity for electrons), such as NADH/NAD$^+$, to a redox pair with a high redox potential (or high affinity for electrons), such as O$_2$/H$_2$O. Thus, NADH is an excellent molecule to donate electrons to the respiratory chain, while O$_2$ is well suited to act as an electron "sink" at the end of the pathway. As explained in Panel 14–1, the difference in redox potential, $\Delta E_0'$, is a direct measure of the standard free-energy change ($\Delta G°$) for the transfer of an electron from one molecule to another. In fact, $\Delta E_0'$ is equal to $\Delta G°$ times a negative number that is a constant.

Electron Transfers Release Large Amounts of Energy

The amount of energy that can be released by an electron transfer can be determined by comparing the redox potentials of the molecules involved. Again, let's look at the transfer of electrons from NADH and to O$_2$. As shown in Panel 14–1, a 1:1 mixture of NADH and NAD$^+$ has a redox potential of –320 mV, indicating that NADH has a weak affinity for electrons—and a strong tendency to donate them; a 1:1 mixture of H$_2$O and ½O$_2$ has a redox potential of +820 mV, indicating that O$_2$ has a strong affinity for electrons—and a strong tendency to accept them. The difference in redox potential between these two pairs is 1.14 volts (1140 mV), which means that the transfer of each electron from NADH to O$_2$ under these standard conditions is enormously favorable: the $\Delta G°$ for that electron transfer is –26.2 kcal/mole per electron—or –52.4 kcal/mole for the two electrons that are donated from each NADH molecule (see Panel 14–1). If we compare this free-energy change with that needed for the formation of the phosphoanhydride bonds in ATP in cells (about 13 kcal/mole), we see that enough energy is released by the oxidization of one NADH molecule to synthesize a couple of molecules of ATP.

Living systems could have evolved enzymes that would allow NADH to donate electrons directly to O$_2$ to make water. But because of the huge free-energy drop, this reaction would proceed with almost explosive force and nearly all of the energy would be released as heat. Instead, as we have seen, the transfer of electrons from NADH to O$_2$ is made in many small steps along the electron-transport chain, enabling nearly half of the released energy to be stored in the proton gradient across the inner membrane rather than getting lost to the environment as heat.

Metals Tightly Bound to Proteins Form Versatile Electron Carriers

Each of the three respiratory enzyme complexes includes metal atoms that are tightly bound to the proteins. Once an electron has been donated to a respiratory complex, it moves within the complex by skipping from one embedded metal ion to another with a greater affinity for electrons.

HOW REDOX POTENTIALS ARE MEASURED

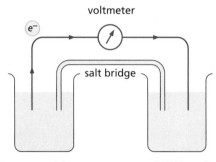

Areduced and Aoxidized
in equimolar amounts

1 M H+ and
1 atmosphere H2 gas

One beaker (*left*) contains substance A with an equimolar mixture of the reduced (Areduced) and oxidized (Aoxidized) members of its redox pair. The other beaker contains the hydrogen reference standard ($2H^+ + 2e^- \rightleftharpoons H_2$), whose redox potential is arbitrarily assigned as zero by international agreement. (A salt bridge formed from a concentrated KCl solution allows K^+ and Cl^- to move between the beakers; as required to neutralize the charges when electrons flow between the beakers.) The metal wire (*dark blue*) provides a resistance-free path for electrons, and a voltmeter then measures the redox potential of substance A. If electrons flow from Areduced to H^+, as indicated here, the redox pair formed by substance A is said to have a negative redox potential. If they instead flow from H_2 to Aoxidized, the redox pair is said to have a positive redox potential.

THE STANDARD REDOX POTENTIAL, E_0'

The standard redox potential for a redox pair, defined as E_0, is measured for a standard state where all of the reactants are at a concentration of 1 M, including H^+. Since biological reactions occur at pH 7, biologists instead define the standard state as Areduced = Aoxidized and $H^+ = 10^{-7}$ M. This standard redox potential is designated by the symbol E_0', in place of E_0.

examples of redox reactions	standard redox potential E_0'
$NADH \rightleftharpoons NAD^+ + H^+ + 2e^-$	–320 mV
reduced ubiquinone $\rightleftharpoons$ oxidized ubiquinone $+ 2H^+ + 2e^-$	+30 mV
reduced cytochrome *c* $\rightleftharpoons$ oxidized cytochrome *c* $+ e^-$	+230 mV
$H_2O \rightleftharpoons \frac{1}{2}O_2 + 2H^+ + 2e^-$	+820 mV

CALCULATION OF $\Delta G°$ FROM REDOX POTENTIALS

To determine the energy change for an electron transfer, the $\Delta G°$ of the reaction (kcal/mole) is calculated as follows:

$\Delta G° = -n(0.023)\,\Delta E_0'$, where n is the number of electrons transferred across a redox potential change of $\Delta E_0'$ millivolts (mV), and

$\Delta E_0' = E_0'(\text{acceptor}) - E_0'(\text{donor})$

EXAMPLE:

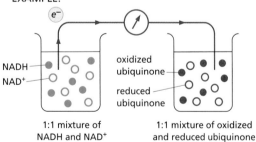

NADH
NAD+

oxidized ubiquinone

reduced ubiquinone

1:1 mixture of
NADH and NAD+

1:1 mixture of oxidized
and reduced ubiquinone

For the transfer of one electron from NADH to ubiquinone:

$$\Delta E_0' = +30 - (-320) = +350 \text{ mV}$$

$\Delta G° = -n(0.023)\Delta E_0' = -1(0.023)(350) = -8.0$ kcal/mole

The same calculation reveals that the transfer of one electron from ubiquinone to oxygen has an even more favorable $\Delta G°$ of –18.2 kcal/mole. The $\Delta G°$ value for the transfer of one electron from NADH to oxygen is the sum of these two values, –26.2 kcal/mole.

EFFECT OF CONCENTRATION CHANGES

As explained in Chapter 3 (see p. 94), the actual free-energy change for a reaction, ΔG, depends on the concentration of the reactants and generally will be different from the standard free-energy change, $\Delta G°$. The standard redox potentials are for a 1:1 mixture of the redox pair. For example, the standard redox potential of –320 mV is for a 1:1 mixture of NADH and NAD+. But when there is an excess of NADH over NAD+, electron transfer from NADH to an electron acceptor becomes more favorable. This is reflected by a more negative redox potential and a more negative ΔG for electron transfer.

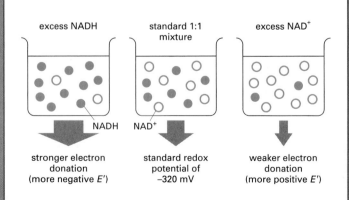

excess NADH

standard 1:1
mixture

excess NAD+

NADH   NAD+

stronger electron
donation
(more negative *E'*)

standard redox
potential of
–320 mV

weaker electron
donation
(more positive *E'*)

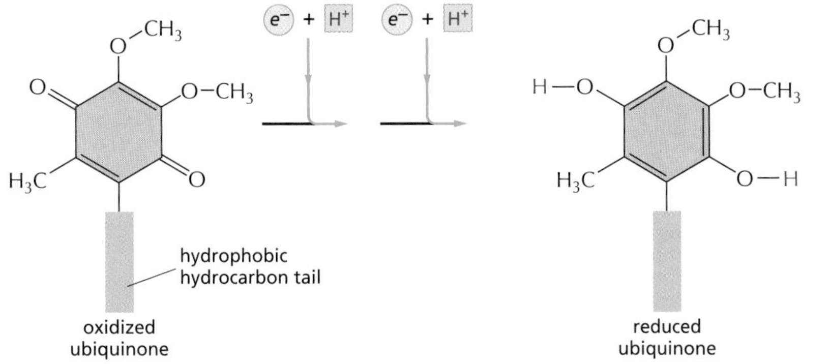

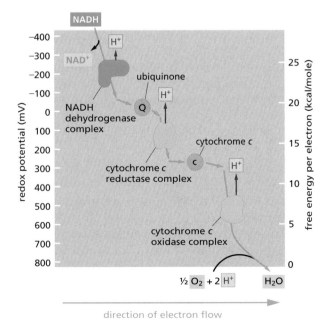

Figure 14–23 Quinones carry electrons within the lipid bilayer. The quinone in the mitochondrial electron-transport chain is called ubiquinone. It picks up one H^+ from the aqueous environment for every electron it accepts, and it can carry two electrons as part of its hydrogen atoms (*red*). When this reduced ubiquinone donates its electrons to the next carrier in the chain, the protons are released. Its long, hydrophobic hydrocarbon tail confines ubiquinone to the inner mitochondrial membrane.

When passing from one respiratory complex to the next, in contrast, the electrons are ferried by electron carriers that diffuse freely within the lipid bilayer. These mobile molecules pick up electrons from one complex and deliver them to the next in line. In the mitochondrial respiratory chain, for example, a small, hydrophobic molecule called ubiquinone picks up electrons from the NADH dehydrogenase complex and delivers them to the cytochrome *c* reductase complex (see Figure 14–14). A related **quinone** functions similarly during electron transport in photosynthesis. Ubiquinone can accept or donate either one or two electrons, and it picks up one H^+ from water with each electron that it carries (**Figure 14–23**). Its redox potential of +30 mV places ubiquinone between the NADH dehydrogenase complex and the cytochrome *c* reductase complex in terms of its tendency to gain or lose electrons—which explains why ubiquinone receives electrons from the former and donates them to the latter (**Figure 14–24**). Ubiquinone also serves as the entry point for electrons donated by the $FADH_2$ that is generated during the citric acid cycle and from fatty acid oxidation (see Figures 13–11 and 13–12).

The redox potentials of different metal complexes influence where they are used along the electron-transport chain. **Iron–sulfur centers** have relatively low affinities for electrons and thus are prominent in the electron carriers that operate in the early part of the chain. An iron–sulfur center in the NADH dehydrogenase complex, for example, passes electrons to ubiquinone. Later in the pathway, iron atoms held in the heme groups bound to cytochrome proteins are commonly used as electron carriers (**Figure 14–25**). These heme groups give **cytochromes**, such as

---

QUESTION 14–6

At many steps in the electron-transport chain, Fe ions are used as part of heme or FeS clusters to bind the electrons in transit. Why do these functional groups that carry out the chemistry of electron transfer need to be bound to proteins? Provide several different reasons why this is necessary.

Figure 14–24 Redox potential increases along the mitochondrial electron-transport chain. The big increases in redox potential occur across each of the three respiratory enzyme complexes, as required for each of them to pump protons. To convert free-energy values to kJ/mole, recall that 1 kilocalorie is equal to about 4.2 kilojoules.

Figure 14–25 The iron in a heme group can serve as an electron acceptor.
(A) Ribbon structure shows the position of the heme group (*red*) associated with cytochrome c (*green*). (B) The porphyrin ring of the heme group (*light red*) is attached covalently to side chains in the protein. The heme groups of different cytochromes have different electron affinities because they differ slightly in structure and are held in different local environments within each protein.

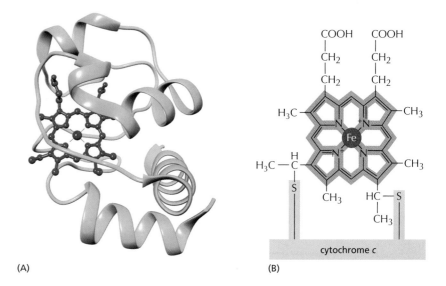

(A)   (B)

the cytochrome *c* reductase and cytochrome *c* oxidase complexes, their color ("cytochrome" from the Greek *chroma*, "color"). Like other electron carriers, the cytochrome proteins increase in redox potential the further down the mitochondrial electron-transport chain they are located. For example, *cytochrome* c, a small protein that accepts electrons from the cytochrome *c* reductase complex and transfers them to the cytochrome *c* oxidase complex, has a redox potential of +230 mV—a value about midway between those of the cytochromes with which it interacts (see Figure 14–24).

Cytochrome c Oxidase Catalyzes the Reduction of Molecular Oxygen

Cytochrome *c* oxidase, the final electron carrier in the respiratory chain, has the highest redox potential of all. This protein complex removes electrons from cytochrome *c*, thereby oxidizing it—hence the name "cytochrome *c* oxidase." These electrons are then handed off to O_2 to produce H_2O. In total, four electrons donated by cytochrome *c* and four protons from the aqueous environment are added to each O_2 molecule in the reaction $4e^- + 4H^+ + O_2 \rightarrow 2H_2O$.

In addition to the protons that combine with O_2, four other protons are pumped across the membrane during the transfer of four electrons from cytochrome *c* to O_2. This transfer of electrons drives allosteric changes in the conformation of the protein that move protons out of the mitochondrial matrix. A special oxygen-binding site within this protein complex—which contains both a heme group plus a copper atom—serves as the final repository for all of the electrons donated by NADH at the start of the electron-transport chain (**Figure 14–26**). It is here that nearly all the oxygen we breathe is consumed.

Oxygen is useful as an electron sink because of its very high affinity for electrons. However, once O_2 picks up one electron, it forms the superoxide radical O_2^-; this radical is dangerously reactive and will avidly take up another three electrons wherever it can find them, a tendency that can cause serious damage to nearby DNA, proteins, and lipid membranes. The active site of cytochrome *c* oxidase holds on tightly to an oxygen molecule until it receives all four of the electrons needed to convert it to two molecules of H_2O. This retention helps prevent superoxide radicals from attacking macromolecules throughout the cell—damage that has been postulated to contribute to human aging.

QUESTION 14–7

Two different diffusible electron carriers, ubiquinone and cytochrome c, shuttle electrons between the three protein complexes of the electron-transport chain. Could the same diffusible carrier, in principle, be used for both steps? Explain your answer.

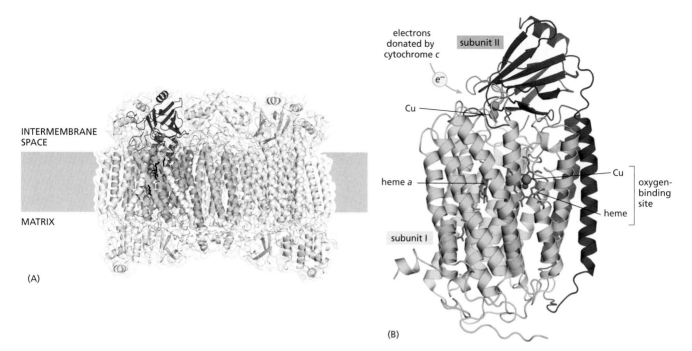

Figure 14–26 Cytochrome *c* oxidase is a finely tuned protein machine. The protein is a dimer formed from a monomer with 13 different protein subunits. (A) The entire protein is shown positioned in the inner mitochondrial membrane. The three colored subunits that form the functional core of the complex are encoded by the mitochondrial genome; the remaining subunits are encoded by the nuclear genome. (B) As electrons pass through this protein on the way to its bound O_2 molecule, they cause the protein to pump protons across the membrane. As indicated, a heme and copper atom (Cu) form the site where a tightly bound O_2 molecule is reduced to H_2O.

The evolution of cytochrome *c* oxidase was crucial to the formation of cells that could use O_2 as an electron acceptor, and this protein complex is therefore essential for all aerobic life. Poisons such as cyanide are extremely toxic because they bind tightly to cytochrome *c* oxidase complexes, thereby halting electron transport and the production of ATP.

CHLOROPLASTS AND PHOTOSYNTHESIS

Virtually all the organic material in present-day cells is produced by **photosynthesis**—the series of light-driven reactions that creates organic molecules from atmospheric carbon dioxide (CO_2). Plants, algae, and photosynthetic bacteria such as cyanobacteria use electrons from water and the energy of sunlight to convert atmospheric CO_2 into organic compounds. In the course of these reactions, water molecules are split, releasing vast quantities of O_2 gas into the atmosphere. This oxygen in turn supports oxidative phosphorylation—not only in animals but also in plants and aerobic bacteria. Thus the activity of early photosynthetic bacteria, which filled the atmosphere with oxygen, enabled the evolution of the myriad life-forms that use aerobic metabolism to make their ATP (**Figure 14–27**).

In plants, photosynthesis is carried out in a specialized intracellular organelle—the **chloroplast**, which contains light-capturing pigments such as the green pigment *chlorophyll*. For most plants, the leaves are the major sites of photosynthesis. Photosynthesis occurs only during the daylight hours, producing ATP and NADPH. These activated carriers can then be used, at any time of day, to convert CO_2 into sugar inside the chloroplast—a process called *carbon fixation*.

Given the chloroplast's central role in photosynthesis, we begin this section by describing the structure of this highly specialized organelle. We then provide an overview of photosynthesis, followed by a detailed accounting of the mechanism by which chloroplasts harvest energy from sunlight to produce huge amounts of ATP and NADPH. We next describe how plants use these two activated carriers to synthesize the sugars and other food molecules that sustain them—and the many organisms that eat plants.

Figure 14–27 Microorganisms that carry out oxygen-producing photosynthesis changed Earth's atmosphere. (A) Living stromatolites from a lagoon in Western Australia. These structures are formed in specialized environments by large colonies of oxygen-producing photosynthetic cyanobacteria, which form mats that trap sand or minerals in thin layers. (B) Cross section of a modern stromatolite, showing its stratification. A similar structure is seen in fossilized stromatolites (not shown). These ancient accretions, some more than 3.5 billion years old, contain the remnants of the photosynthetic bacteria whose O₂-liberating activities transformed the Earth's atmosphere. (A, courtesy of Cambridge Carbonates Ltd.; B, courtesy of Roger Perkins, Virtual Fossil Museum.)

(A)

(B)

Chloroplasts Resemble Mitochondria but Have an Extra Compartment—the Thylakoid

Chloroplasts are larger than mitochondria, but both are organized along structurally similar principles. Chloroplasts have a highly permeable outer membrane and a much less permeable inner membrane, in which various membrane transport proteins are embedded. Together, these two membranes—and the narrow, intermembrane space that separates them—form the chloroplast envelope. The inner membrane surrounds a large space called the **stroma**, which is analogous to the mitochondrial matrix and contains many metabolic enzymes (see Figure 14–5).

There is, however, one important difference between the organization of mitochondria and that of chloroplasts. The inner membrane of the chloroplast does not contain the photosynthetic machinery. Instead, the light-capturing systems, electron-transport chain, and ATP synthase that produce ATP during photosynthesis are all contained in the *thylakoid membrane*. This third membrane is folded to form a set of flattened, disclike sacs, called the **thylakoids**, which are arranged in stacks called *grana* (**Figure 14–28**). The space inside each thylakoid is thought to be connected with that of other thylakoids, creating a third internal compartment, the *thylakoid space*, which is separate from the stroma.

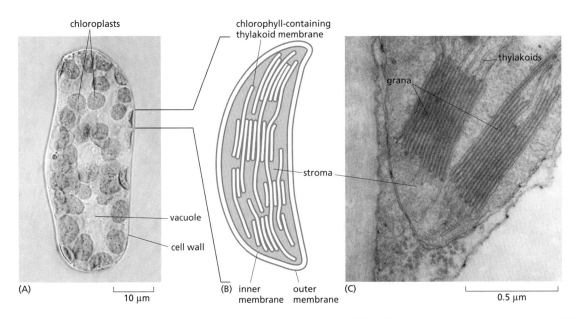

Figure 14–28 Chloroplasts, like mitochondria, are composed of a set of specialized membranes and compartments. (A) Light micrograph shows chloroplasts (*green*) in the cell of a flowering plant. (B) Drawing of a single chloroplast shows the organelle's three sets of membranes, including the thylakoid membrane, which contains the light-capturing and ATP-generating systems. (C) A high-magnification view of an electron micrograph shows the thylakoids arranged in stacks called *grana*; a single thylakoid stack is called a *granum*. (A, courtesy of Preeti Dahiya; C, courtesy of K. Plaskitt.)

Photosynthesis Generates—Then Consumes—ATP and NADPH

The chemistry carried out by photosynthesis can be summarized in one simple equation:

$$\text{light energy} + CO_2 + H_2O \rightarrow \text{sugars} + O_2 + \text{heat energy}$$

On its surface, the equation accurately represents the process by which light energy drives the production of sugars from CO_2. But this superficial accounting leaves out two of the most important players in photosynthesis: the activated carriers ATP and NADPH. In the first stage of photosynthesis, the energy from sunlight is used to produce ATP and NADPH; in the second stage, these activated carriers are consumed to fuel the synthesis of sugars.

1. *Stage 1* of photosynthesis is, in large part, equivalent to the oxidative phosphorylation that takes place on the mitochondrial inner membrane. In this stage, an electron-transport chain in the thylakoid membrane harnesses the energy of electron transport to pump protons into the thylakoid space; the resulting proton gradient then drives the synthesis of ATP by ATP synthase. What makes photosynthesis different is that the high-energy electrons donated to the *photosynthetic electron-transport chain* come from a molecule of **chlorophyll** that has absorbed energy from sunlight. Thus the energy-producing reactions of stage 1 are sometimes called the **light reactions** (**Figure 14–29**). The other major difference between photosynthesis and oxidative phosphorylation is where the high-energy electrons ultimately wind up: those that make their way down the photosynthetic electron-transport chain in chloroplasts are donated not to O_2, but to $NADP^+$, to produce NADPH.

2. In *Stage 2* of photosynthesis, the ATP and the NADPH produced by the photosynthetic electron-transfer reactions of stage 1 are used to drive the manufacture of sugars from CO_2 (see Figure 14–29). These *carbon-fixation reactions* can occur in the absence of sunlight and are thus also called the **dark reactions**. They begin in the chloroplast stroma, where they generate a three-carbon sugar called *glyceraldehyde 3-phosphate*. This simple sugar is exported to the cytosol, where it is used to produce sucrose and a large number of other organic molecules in the leaves of the plant.

Although the formation of ATP and NADPH during stage 1, and the conversion of CO_2 to carbohydrate during stage 2, are mediated by two separate sets of reactions, they are linked by elaborate feedback mechanisms that

QUESTION 14–8

Chloroplasts have a third internal compartment, the thylakoid space, bounded by the thylakoid membrane. This membrane contains the photosystems, reaction centers, electron-transport chain, and ATP synthase. In contrast, mitochondria use their inner membrane for electron transport and ATP synthesis. In both organelles, protons are pumped out of the largest internal compartment (the matrix in mitochondria and the stroma in chloroplasts). The thylakoid space is completely sealed off from the rest of the cell. Why does this arrangement allow a larger H^+ gradient in chloroplasts than can be achieved for mitochondria?

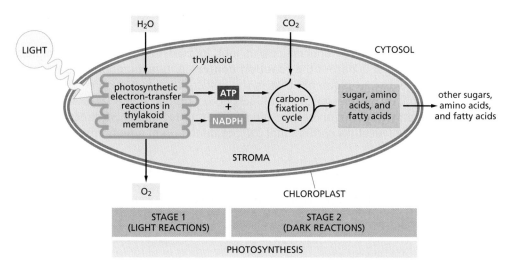

Figure 14–29 Both stages of photosynthesis depend on the chloroplast. In stage 1, a series of photosynthetic electron-transfer reactions produce ATP and NADPH; in the process, electrons are extracted from water and oxygen is released as a by-product, as we discuss shortly. In stage 2, carbon dioxide is assimilated (fixed) to produce sugars and a variety of other organic molecules. Stage 1 occurs in the thylakoid membrane, whereas stage 2 begins in the chloroplast stroma (as shown) and continues in the cytosol.

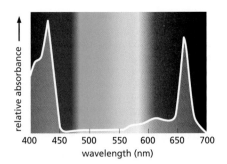

Figure 14–30 Chlorophylls absorb light of blue and red wavelengths. As shown in this absorption spectrum, one form of chlorophyll preferentially absorbs light around wavelengths of 430 nm (*blue*) and 660 nm (*red*). Green light, in contrast, is absorbed poorly by this pigment. Other chlorophylls can absorb light of slightly different wavelengths.

allow a plant to manufacture sugars only when it is appropriate to do so. Several of the enzymes required for carbon fixation, for example, are inactivated in the dark and reactivated by light-stimulated electron transport.

Chlorophyll Molecules Absorb the Energy of Sunlight

Visible light is a form of electromagnetic radiation composed of many wavelengths, ranging from violet (wavelength 400 nm) to deep red (700 nm). Most chlorophylls best absorb light in the blue and red wavelengths (**Figure 14–30**). Because these pigments absorb green light poorly, plants look green to us: the green light is reflected back to our eyes.

Chlorophyll's ability to harness energy derived from sunlight stems from its unique structure. The electrons in a chlorophyll molecule are distributed in a decentralized cloud around the molecule's light-absorbing porphyrin ring (**Figure 14–31**). When light of an appropriate wavelength hits a molecule of chlorophyll, it excites electrons in this diffuse network, perturbing the way the electrons are distributed. This perturbed high-energy state is unstable, and an excited chlorophyll molecule will seek to get rid of this excess energy so it can return to its more stable, unexcited state.

A molecule of chlorophyll, on its own in solution, would simply release its absorbed energy in the form of light or heat—accomplishing nothing useful. However, chlorophyll molecules in a chloroplast are able to convert light energy into a form of energy useful to the cell because they are associated with a special set of photosynthetic proteins in the thylakoid membrane, as we see next.

Excited Chlorophyll Molecules Funnel Energy into a Reaction Center

In the thylakoid membrane of plants and the plasma membrane of photosynthetic bacteria, chlorophyll molecules are held in large multiprotein complexes called **photosystems**. Each photosystem consists of a set of *antenna complexes*, which capture light energy, and a *reaction center*, which converts that light energy into chemical energy.

In each **antenna complex**, hundreds of chlorophyll molecules are arranged so that the light energy captured by one chlorophyll molecule can be transferred to a neighboring chlorophyll molecule in the network. In this way, energy jumps randomly from one chlorophyll molecule to the next—either within the same antenna or in an adjacent antenna. At some point, this wandering energy will encounter a chlorophyll dimer called the *special pair*, which holds its electrons at a lower energy than do the other chlorophyll molecules. Thus when energy is accepted by this special pair, it becomes effectively trapped there.

The chlorophyll special pair is not located in an antenna complex. Instead, it is part of the **reaction center**—a transmembrane complex of proteins and pigments that is thought to have first evolved more than 3 billion years ago in primitive photosynthetic bacteria (Movie 14.6). Within the reaction center, the special pair is positioned directly next to a set of electron carriers that are poised to accept a high-energy electron

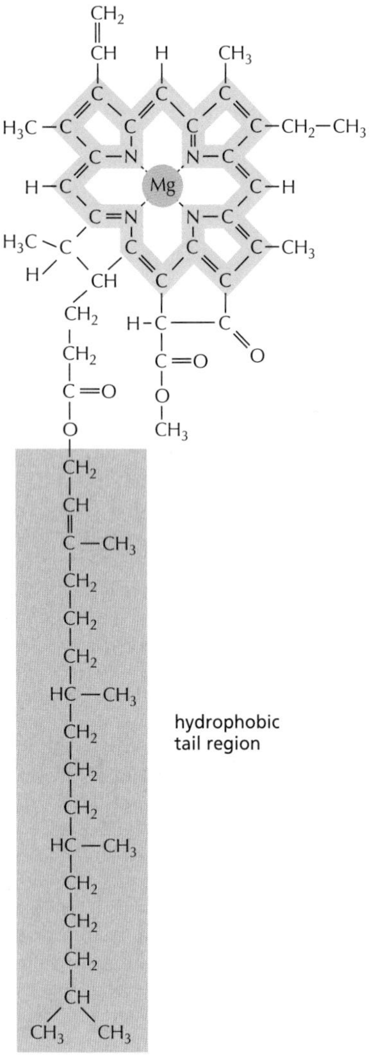

Figure 14–31 Chlorophyll's structure allows it to absorb energy from light. Each chlorophyll molecule contains a porphyrin ring with a magnesium atom (*pink*) at its center. This porphyrin ring is structurally similar to the one that binds iron in heme (see Figure 14–25). Light is absorbed by electrons within the bond network shown in *blue*, while the long, hydrophobic tail (*gray*) helps hold the chlorophyll in the thylakoid membrane.

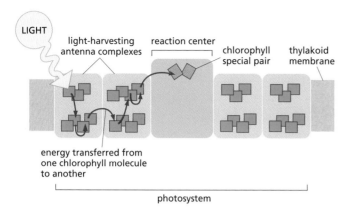

Figure 14–32 A photosystem consists of a reaction center surrounded by chlorophyll-containing antenna complexes. Once light energy has been captured by a chlorophyll molecule in an antenna complex, it will pass randomly from one chlorophyll molecule to another (*red lines*), until it gets trapped by a chlorophyll dimer called the *special pair*, located in the reaction center. The chlorophyll special pair holds its electrons at a lower energy than the antenna chlorophylls, so the energy transferred to it from the antenna gets trapped there. Note that in the antenna complex only energy moves from one chlorophyll molecule to another, not electrons.

from the excited chlorophyll special pair (**Figure 14–32**). This electron transfer lies at the heart of photosynthesis, because it converts the light energy that came into the special pair into chemical energy in the form of a transferable electron. As soon as the high-energy electron is handed off, the chlorophyll special pair becomes positively charged, and the electron carrier that accepts the electron becomes negatively charged. The rapid movement of this electron along a set of electron carriers in the reaction center then creates a *charge separation* that sets in motion the flow of electrons from the reaction center to an electron-transport chain (**Figure 14–33**).

A Pair of Photosystems Cooperate to Generate Both ATP and NADPH

Photosynthesis is ultimately a biosynthetic process, and to build organic molecules from CO_2, a plant cell requires a huge input of energy, in the form of ATP, and a very large amount of reducing power, in the form of the activated carrier NADPH (see Figure 3–34). To generate both ATP and NADPH, plant cells—and free-living photosynthetic organisms such as cyanobacteria—use a pair of photosystems that are similar in structure, but that do different things with the high-energy electrons that leave their reaction center chlorophylls.

When the first photosystem (which, paradoxically, is called photosystem II for historical reasons) absorbs light energy, its reaction center passes electrons to a mobile electron carrier called *plastoquinone*, which is part of the photosynthetic electron-transport chain. This carrier transfers the high-energy electrons to a proton pump, which—like the proton pumps in the mitochondrial inner membrane—uses the movement of electrons to generate an electrochemical proton gradient. The electrochemical proton gradient then drives the production of ATP by an ATP synthase located in the thylakoid membrane (**Figure 14–34**).

At the same time, a second nearby photosystem—called photosystem I—has been also busy capturing the energy from sunlight. The reaction center of this photosystem passes its high-energy electrons to a different mobile electron carrier, which brings them to an enzyme that uses them to reduce $NADP^+$ to NADPH (**Figure 14–35**). The combined action of these

Figure 14–33 In a reaction center, a high-energy electron is transferred from the special pair to a carrier that becomes part of an electron-transport chain. Not shown is a set of intermediary carriers embedded in the reaction center that provide the path from the special pair to this carrier (*orange*). As illustrated, the transfer of the high-energy electron from the excited chlorophyll special pair leaves behind a positive charge that creates a charge-separated state, thereby converting light energy to chemical energy. Once the electron in the special pair has been replaced (an event we will discuss in detail shortly), the carrier diffuses away from the reaction center, transferring the high-energy electron to the transport chain.

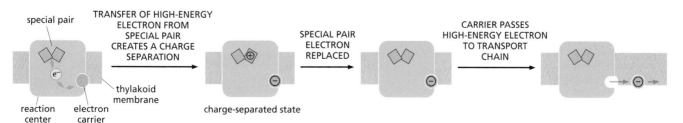

Figure 14–34 Photosystem II feeds electrons to a photosynthetic proton pump, leading to ATP synthesis by ATP synthase. When light energy is captured by photosystem II, a high-energy electron is transferred to a mobile electron carrier called plastoquinone (Q), which closely resembles the ubiquinone of mitochondria. This carrier transfers its electrons to a proton pump called the cytochrome b_6-f complex, which resembles the cytochrome c reductase complex of mitochondria and is the sole site of active proton pumping in the chloroplast electron-transport chain. As in mitochondria, an ATP synthase embedded in the membrane then uses the energy of the electrochemical proton gradient to produce ATP.

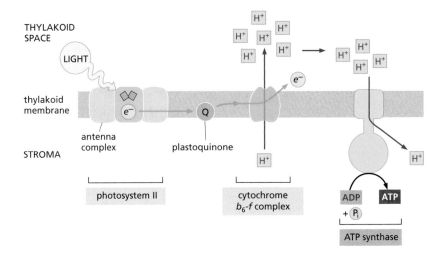

two photosystems thus produces both the ATP (photosystem II) and the NADPH (photosystem I) that will be used in stage 2 of photosynthesis (see Figure 14–29).

Oxygen Is Generated by a Water-Splitting Complex Associated with Photosystem II

The scheme that we have thus far described for photosynthesis has ignored a major chemical conundrum. When a mobile electron carrier removes an electron from a reaction center (whether in photosystem I or photosystem II), it leaves behind a positively charged chlorophyll special pair (see Figure 14–33). To reset the system and allow photosynthesis to proceed, this missing electron must be replaced.

For photosystem II, the missing electron is replaced by a special protein complex that removes the electrons from water. This *water-splitting enzyme* contains a cluster of manganese atoms that holds onto two water molecules from which electrons are extracted one at a time. Once four electrons have been removed from these two water molecules—and used to replace the electrons lost by four excited chlorophyll special pairs—O_2 is released (**Figure 14–36**).

This "waiting for four electrons" maneuver ensures that no partly oxidized water molecules are released as dangerous highly reactive chemicals. The same trick is used by the cytochrome c oxidase that catalyzes the reverse reaction—the transfer of electrons to O_2 to produce water—during oxidative phosphorylation (see Figure 14–26).

It is astounding to realize that essentially all of the oxygen in the Earth's atmosphere has been produced by the water-splitting enzyme of photosystem II.

Figure 14–35 Photosystem I transfers high-energy electrons to an enzyme that produces NADPH. When light energy is captured by photosystem I, a high-energy electron is passed to a mobile electron carrier called ferredoxin (Fd), a small protein that contains an iron–sulfur center. Ferredoxin carries its electrons to ferredoxin-NADP reductase (FNR), the final protein in the electron-transport chain that generates NADPH.

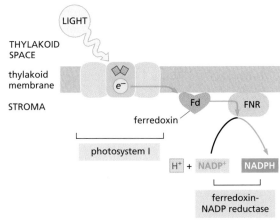

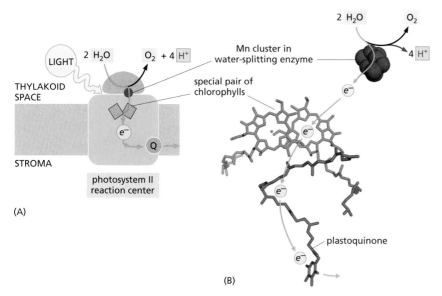

(A)

(B)

Figure 14–36 The reaction center of photosystem II includes an enzyme that catalyzes the extraction of electrons from water. (A) Schematic diagram shows the flow of electrons through the reaction center of photosystem II. When light energy excites the chlorophyll special pair, an electron is passed to the mobile electron carrier plastoquinone (Q). An electron is then returned to the special pair by a water-splitting enzyme that extracts electrons from water. The Mn cluster that participates in the electron extraction is shown as a *red spot*. Once four electrons have been withdrawn from two water molecules, O_2 is released into the atmosphere. (B) The structure and position of some of the electron carriers involved.

The Special Pair in Photosystem I Receives its Electrons from Photosystem II

We have seen that photosystem II receives electrons from water. But where does photosystem I get the electrons it needs to reset its special pair? It gets them from photosystem II: the chlorophyll special pair in photosystem I serves as the final electron acceptor for the electron-transport chain that carries electrons from photosystem II. The overall flow of electrons is shown in **Figure 14–37**. Electrons removed from water by photosystem II are passed, through a proton pump (the cytochrome b_6-f complex), to a mobile electron carrier called plastocyanin. Plastocyanin then carries these electrons to photosystem I, to replace the electrons lost by its excited chlorophyll special pair. When light is again absorbed by this photosystem, this electron will be boosted to the very high-energy level needed to reduce $NADP^+$ to NADPH.

Having these two photosystems operating in tandem effectively couples their two electron-energizing steps. This extra boost of energy—provided by the light harvested by both photosystems—allows an electron to be moved from water, which normally holds onto its electrons very tightly (redox potential = +820 mV), to NADPH, which normally holds onto its electrons loosely (redox potential = −320 mV). There is even enough energy left over to enable the electron-transport chain that links the two photosystems to pump H^+ across the thylakoid membrane, so that ATP

Figure 14–37 The movement of electrons along the photosynthetic electron-transport chain powers the production of both ATP and NADPH. Electrons are supplied to photosystem II by a water-splitting complex that extracts four electrons from two molecules of water, producing O_2 as a by-product. After their energy is raised by the absorption of light, these electrons power the pumping of protons by the cytochrome b_6-f complex. Electrons that pass through this complex are then donated to a copper-containing protein, the mobile electron carrier plastocyanin (pC), which ferries them to the reaction center of photosystem I. After an additional energy boost from light, these electrons are used to generate NADPH. An overview of these reactions is shown in Movie 14.7.

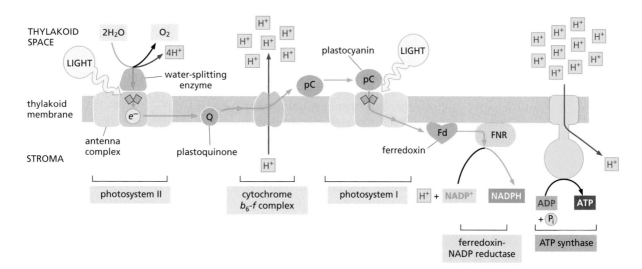

Figure 14–38 The combined actions of photosystems I and II boost electrons to the energy level needed to produce both ATP and NADPH. The redox potential for each molecule is indicated by its position on the vertical axis. Electron transfers are shown with non-wavy *blue arrows*. Photosystem II passes electrons from its excited chlorophyll special pair to an electron-transport chain in the thylakoid membrane that leads to photosystem I (see Figure 14–37). The net electron flow through the two photosystems linked in series is from water to NADP$^+$, to form NADPH.

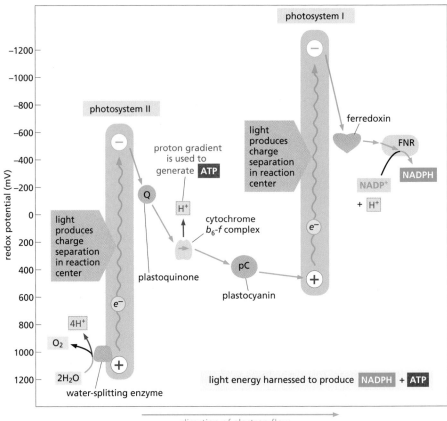

synthase can harness some of the light-derived energy for ATP production (**Figure 14–38**).

Carbon Fixation Uses ATP and NADPH to Convert CO$_2$ into Sugars

The light reactions of photosynthesis generate ATP and NADPH in the chloroplast stroma, as we have just seen. But the inner membrane of the chloroplast is impermeable to both of these compounds, which means that they cannot be exported directly to the cytosol. To provide energy and reducing power for the rest of the cell, the ATP and NADPH are instead used within the chloroplast stroma to produce sugars, which can be exported by specific carrier proteins in the chloroplast inner membrane. This production of sugar from CO$_2$ and water, which occurs during the dark reactions (stage 2) of photosynthesis, is called **carbon fixation**.

In the central reaction of photosynthetic carbon fixation, CO$_2$ from the atmosphere is attached to a five-carbon sugar derivative, ribulose 1,5-bisphosphate, to yield two molecules of the three-carbon compound *3-phosphoglycerate*. This carbon-fixing reaction, which was discovered in 1948, is catalyzed in the chloroplast stroma by a large enzyme called ribulose bisphosphate carboxylase or *Rubisco* (**Figure 14–39**). Rubisco works much more slowly than most other enzymes: it processes about three molecules of substrate per second—compared with 1000 molecules per second for a typical enzyme. To compensate for this sluggish behavior, plants maintain a surplus of Rubisco to ensure the efficient production of sugars. The enzyme often represents more than 50% of the total chloroplast protein, and it is widely claimed to be the most abundant protein on Earth.

Although the production of carbohydrates from CO$_2$ and H$_2$O is energetically unfavorable, the fixation of CO$_2$ catalyzed by Rubisco is an

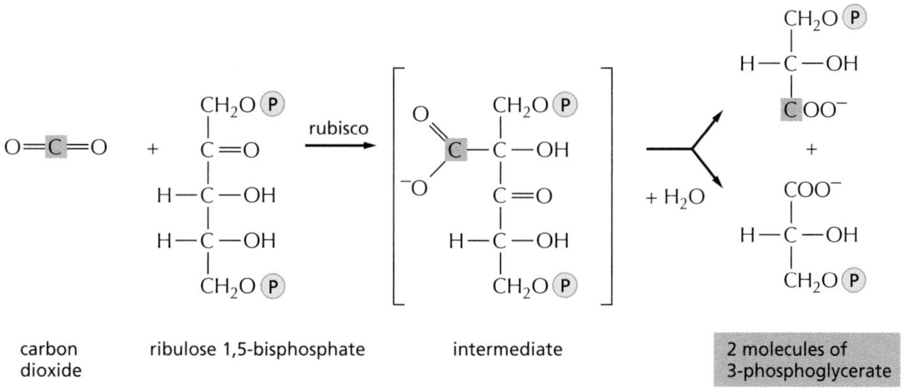

Figure 14–39 Carbon fixation involves the formation of a covalent bond that attaches carbon dioxide to ribulose 1,5-bisphosphate. The reaction is catalyzed in the chloroplast stroma by the abundant enzyme ribulose bisphosphate carboxylase, or Rubisco. As shown, the product is two molecules of 3-phosphoglycerate.

energetically favorable reaction. Carbon fixation is energetically favorable because a continuous supply of the energy-rich ribulose 1,5-bisphosphate is fed into it. As this compound is consumed—by the addition of CO_2 (see Figure 14–39)—it must be replenished. The energy and reducing power needed to regenerate ribulose 1,5-bisphosphate come from the ATP and NADPH produced by the photosynthetic light reactions.

The elaborate series of reactions in which CO_2 combines with ribulose 1,5-bisphosphate to produce a simple sugar—a portion of which is used to regenerate ribulose 1,5-bisphosphate—forms a cycle, called the *carbon-fixation cycle*, or the Calvin cycle (**Figure 14–40**). For every three

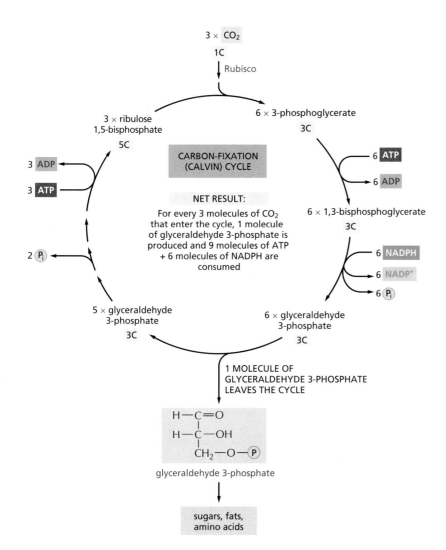

Figure 14–40 **The carbon-fixation cycle consumes ATP and NADPH to form glyceraldehyde 3-phosphate from CO_2 and H_2O.** In the first stage of the cycle, CO_2 is added to ribulose 1,5-bisphosphate (as shown in Figure 14–39). In the second stage, ATP and NADPH are consumed to produce glyceraldehyde 3-phosphate. In the final stage, some of the glyceraldehyde 3-phosphate produced is used to regenerate ribulose 1,5-bisphosphate; the rest is transported out of the chloroplast stroma into the cytosol. The number of carbon atoms in each type of molecule is indicated in *yellow*. There are many intermediates between glyceraldehyde 3-phosphate and ribulose 5-phosphate, but they have been omitted here for clarity. The entry of water into the cycle is also not shown.

QUESTION 14–10

A. How do cells in plant roots survive, since they contain no chloroplasts and are not exposed to light?

B. Unlike mitochondria, chloroplasts do not have a transporter that allows them to export ATP to the cytosol. How, then, do plant cells obtain the ATP that they need to carry out energy-requiring metabolic reactions in the cytosol?

Figure 14–41 Chloroplasts often contain large stores of carbohydrates and fatty acids. A thin section of a single chloroplast shows the chloroplast envelope, starch granules, and fat droplets that have accumulated in the stroma as a result of the biosynthetic processes that occur there.

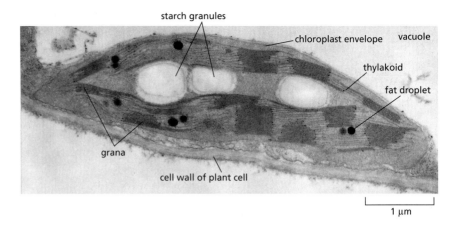

molecules of CO_2 that enter the cycle, one molecule of glyceraldehyde 3-phosphate is produced, and nine molecules of ATP and six molecules of NADPH are consumed. *Glyceraldehyde 3-phosphate*, the three-carbon sugar that is the final product of the cycle, then provides the starting material for the synthesis of many other sugars and other organic molecules.

Sugars Generated by Carbon Fixation Can Be Stored As Starch or Consumed to Produce ATP

The glyceraldehyde 3-phosphate generated by carbon fixation in the chloroplast stroma can be used in a number of ways, depending on the needs of the plant. During periods of excess photosynthetic activity, much of it is retained in the chloroplast stroma and converted to *starch*. Like glycogen in animal cells, starch is a large polymer of glucose that serves as a carbohydrate reserve, and it is stored as large granules in the chloroplast stroma. Starch forms an important part of the diet of all animals that eat plants. Other glyceraldehyde 3-phosphate molecules are converted to fat in the stroma. This material, which accumulates as fat droplets, likewise serves as an energy reserve (**Figure 14–41**).

At night, this stored starch and fat can be broken down to sugars and fatty acids, which are exported to the cytosol to help support the metabolic needs of the plant. Some of the exported sugar enters the glycolytic pathway (see Figure 13–5), where it is converted to pyruvate. That pyruvate, along with the fatty acids, can enter the plant cell mitochondria and be fed into the citric acid cycle, ultimately leading to the production of ATP by oxidative phosphorylation (**Figure 14–42**). Plants use this ATP in the same way that animal cells and other nonphotosynthetic organisms do to power a variety of metabolic reactions.

Figure 14–42 In plants, the chloroplasts and mitochondria collaborate to supply cells with metabolites and ATP. The chloroplast's inner membrane is impermeable to the ATP and NADPH that are produced in the stroma during the light reactions of photosynthesis. These molecules are therefore funneled into the carbon-fixation cycle, where they are used to make sugars. The resulting sugars and their metabolites are either stored within the chloroplast—in the form of starch or fat—or exported to the rest of the plant cell. There, they can enter the energy-generating pathway that ends in ATP synthesis in the mitochondria. Mitochondrial membranes are permeable to ATP, as indicated. Note that the O_2 released to the atmosphere by photosynthesis in chloroplasts is used for oxidative phosphorylation in mitochondria; similarly, the CO_2 released by the citric acid cycle in mitochondria is used for carbon fixation in chloroplasts.

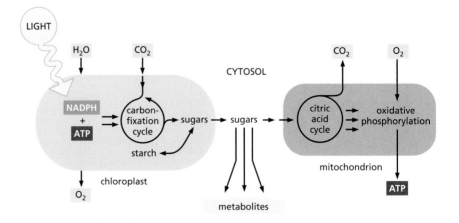

The glyceraldehyde 3-phosphate exported from chloroplasts into the cytosol can also be converted into many other metabolites, including the disaccharide *sucrose*. Sucrose is the major form in which sugar is transported between the cells of a plant: just as glucose is transported in the blood of animals, so sucrose is exported from the leaves via the vascular bundle to provide carbohydrate to the rest of the plant.

THE EVOLUTION OF ENERGY-GENERATING SYSTEMS

The ability to sequence the genomes of microorganisms that are difficult, if not impossible, to grow in culture has made it possible to identify a huge variety of previously mysterious life-forms. Some of these unicellular organisms thrive in the most inhospitable habitats on the planet, including sulfurous hot springs and hydrothermal vents that lie deep on the ocean floor. In these remarkable microbes, we are finding clues to life's history. Like fingerprints left at the scene of a crime, the proteins and small molecules these organisms produce provide evidence that allows us to trace the history of ancient biological events, including those that gave rise to the ATP-generating systems present in the mitochondria and chloroplasts of modern eukaryotic cells. We therefore end this chapter with a brief review of what has been learned about the origins of present-day energy-harvesting systems, which have played such a critical part in fueling the evolution of life on Earth.

Oxidative Phosphorylation Evolved in Stages

As we mentioned earlier, the first living cells on Earth—both prokaryotes and primitive eukaryotes—may have consumed geochemically produced organic molecules and generated ATP by fermentation. Because oxygen was not yet present in the atmosphere, such anaerobic fermentation reactions would have dumped organic acids—such as lactic or formic acids, for example—into the environment (see Figure 13–6A).

Perhaps such acids lowered the pH of the environment, favoring the survival of cells that evolved transmembrane proteins that could pump H⁺ out of the cytosol, thereby preventing the cell from becoming too acidic (stage 1 in **Figure 14–43**). One of these pumps may have used the energy available from ATP hydrolysis to eject H⁺ from the cell; such a proton pump could have been the ancestor of present-day ATP synthases.

As the Earth's supply of geochemically produced nutrients began to dwindle, organisms that could find a way to pump H⁺ without consuming ATP would have been at an advantage: they could save the small amounts of ATP they derived from the fermentation of increasingly scarce foodstuffs to fuel other important activities. This need to conserve resources might have led to the evolution of electron-transport proteins that allowed cells to use the movement of electrons between molecules of different redox potentials as a source of energy for pumping H⁺ across the plasma membrane (stage 2 in Figure 14–43). Some of these cells might have used the nonfermentable organic acids that neighboring cells had excreted as waste to provide the electrons needed to feed this electron-transport system. Some present-day bacteria grow on formic acid, for example, using the small amount of redox energy derived from the transfer of electrons from formic acid to fumarate to pump H⁺.

Eventually, some bacteria would have developed H⁺-pumping electron-transport systems that were so efficient that they could harvest more redox energy than they needed to maintain their internal pH. Such cells would probably have generated large electrochemical proton gradients,

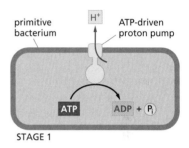

STAGE 1

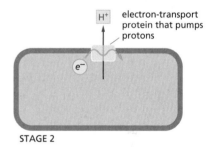

STAGE 2

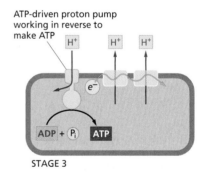

STAGE 3

Figure 14–43 Oxidative phosphorylation might have evolved in stages. The first stage could have involved the evolution of an ATPase that pumped protons out of the cell using the energy of ATP hydrolysis. Stage 2 could have involved the evolution of a different proton pump, driven by an electron-transport chain. Stage 3 would then have linked these two systems together to generate an ATP synthase that uses the protons pumped by the electron-transport chain to synthesize ATP. A bacterium with this final system would have had a selective advantage over bacteria with neither of the systems or only one.

which they could then use to produce ATP. Protons could leak back into the cell through the ATP-driven H^+ pumps, essentially running them in reverse so that they synthesized ATP (stage 3 in Figure 14–43). Because such cells would require much less of the dwindling supply of fermentable nutrients, they would have proliferated at the expense of their neighbors.

Photosynthetic Bacteria Made Even Fewer Demands on Their Environment

The major evolutionary breakthrough in energy metabolism, however, was almost certainly the formation of photochemical reaction centers that could use the energy of sunlight to produce molecules such as NADH. It is thought that this development occurred early in the process of evolution—more than 3 billion years ago, in the ancestors of green sulfur bacteria. Present-day green sulfur bacteria use light energy to transfer hydrogen atoms (as an electron plus a proton) from H_2S to NADPH, thereby creating the strong reducing power required for carbon fixation (**Figure 14–44**).

The next step is thought to have involved the evolution of organisms capable of using water instead of H_2S as the electron source for photosynthesis. This entailed the evolution of a water-splitting enzyme and the addition of a second photosystem, acting in conjunction with the first, to bridge the enormous gap in redox potential between H_2O and NADPH (see Figure 14–38). The biological consequences of this evolutionary step were far-reaching. For the first time, there were organisms that made only minimal chemical demands on their environment. These cells—including the first cyanobacteria (see Figure 14–27)—could spread and evolve in ways denied to the earlier photosynthetic bacteria, which needed H_2S, organic acids, or other sources as a source of electrons. Consequently, large amounts of fermentable organic materials—produced by these cells and their ancestors—began to accumulate. Moreover, O_2 entered the atmosphere in large amounts (**Figure 14–45**).

The availability of O_2 made possible the development of bacteria that relied on aerobic metabolism to make their ATP. As explained previously, these organisms could harness the large amount of energy released when they break down carbohydrates and other reduced organic molecules all the way to CO_2 and H_2O.

As organic materials accumulated as a by-product of photosynthesis, some photosynthetic bacteria—including the ancestors of the bacterium

Figure 14–44 Photosynthesis in green sulfur bacteria uses hydrogen sulfide (H_2S) as an electron donor rather than water. Electrons are easier to extract from H_2S than from H_2O, because H_2S has a much higher redox potential (compare with Figure 14–38). Therefore, only one photosystem is needed to produce NADPH, and elemental sulfur is formed as a by-product instead of O_2. The photosystem in green sulfur bacteria resembles photosystem I in plants and cyanobacteria, in that they all use a series of iron–sulfur centers as the electron carriers that eventually donate their high-energy electrons to ferredoxin (Fd). A bacterium of this type is *Chlorobium tepidum*, which can thrive at high temperatures and low light intensities in hot springs.

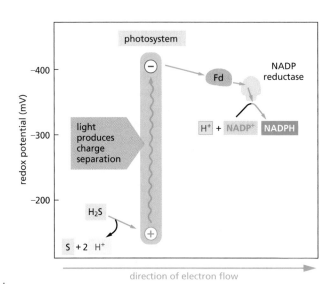

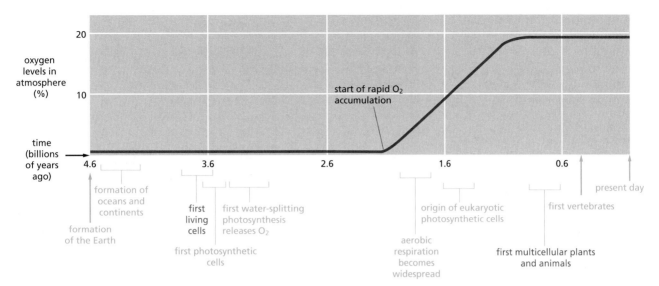

Figure 14–45 Oxygen entered Earth's atmosphere billions of years ago.
With the evolution of photosynthesis in prokaryotes more than 3 billion years ago, organisms would have no longer depended on preformed organic chemicals. They could have then made their own organic molecules from CO_2. The delay of more than a billion years between the appearance of bacteria that split water and released O_2 during photosynthesis and the accumulation of high levels of O_2 in the atmosphere is thought to have been due to the initial reaction of the O_2 with abundant ferrous iron (Fe^{2+}) dissolved in the early oceans. Only when the iron was used up would large amounts of O_2 have started to accumulate in the atmosphere. In response to the rising amount of O_2 in the atmosphere, nonphotosynthetic, aerobic organisms appeared, and the concentration of O_2 in the atmosphere eventually leveled out.

E. coli—lost their ability to survive on light energy alone and came to rely entirely on cell respiration. Mitochondria probably arose when a pre-eukaryotic cell engulfed such an aerobic bacterium (see Figure 1–18). Plants arose somewhat later, when a descendant of this early aerobic eukaryote captured a photosynthetic bacterium, which became the precursor of chloroplasts (see Figure 1–20). Once eukaryotes had acquired the bacterial symbionts that became mitochondria and chloroplasts, they could then embark on the spectacular pathway of evolution that eventually led to complex multicellular organisms.

The Lifestyle of *Methanococcus* Suggests That Chemiosmotic Coupling Is an Ancient Process

The conditions today that most resemble those under which cells are thought to have lived 3.5–3.8 billion years ago may be those near deep-ocean hydrothermal vents. These vents represent places where the Earth's molten mantle is breaking through the overlying crust, expanding the width of the ocean floor. Indeed, the modern organisms that appear to be most closely related to the hypothetical cells from which all life evolved live at 75°C to 95°C, temperatures approaching that of boiling water. This ability to thrive at such extreme temperatures suggests that life's common ancestor—the cell that gave rise to bacteria, archaea, and eukaryotes—lived under very hot, anaerobic conditions.

One of the archaea that live in this environment today is *Methanococcus jannaschii*. Originally isolated from a hydrothermal vent more than a mile beneath the ocean surface, the organism grows in the complete absence of light and gaseous oxygen, using as nutrients the inorganic gases—hydrogen (H_2), CO_2, and nitrogen (N_2)—that bubble up from the vent (**Figure 14–46**). Its mode of existence gives us a hint of how early cells might have used electron transport to derive energy and to extract carbon molecules from inorganic materials that were freely available on the hot early Earth.

Methanococcus relies on N_2 gas as its source of nitrogen for making organic molecules such as amino acids. The organism reduces N_2 to ammonia (NH_3) by the addition of hydrogen, a process called **nitrogen fixation**. Nitrogen fixation requires a large amount of energy, as does the carbon-fixation process that converts CO_2 and H_2O into sugars. Much of the energy that *Methanococcus* requires for both processes is derived from the transfer of electrons from H_2 to CO_2, with the release of large amounts of methane (CH_4) as a waste product (thus producing natural gas

Figure 14–46 *Methanococcus* represents life-forms that might have existed early in Earth's history. (A) This deep-sea archaean lives in hydrothermal vents, such as the one shown, where temperatures reach near that of boiling water. (B) Scanning electron micrograph shows individual *Methanococcus* cells. These organisms use the hydrogen gas (H_2) that bubbles from deep-sea vents as the source of reducing power to generate energy via chemiosmotic coupling. (A, courtesy of the National Oceanic and Atmospheric Administration's Pacific Marine Environmental Laboratory Vents Program; B, courtesy of Chan B. Park.)

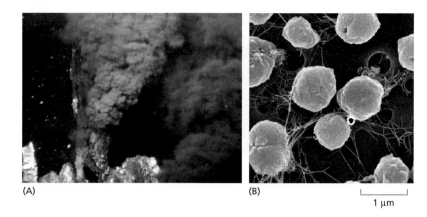

(A)

(B)

⊢——————⊣
1 µm

and giving the organism its name). Part of this electron transfer occurs in the plasma membrane and results in the pumping of protons (H^+) across it. The resulting electrochemical proton gradient drives an ATP synthase in the same membrane to make ATP.

The fact that such chemiosmotic coupling exists in an organism like *Methanococcus* suggests that the storage of energy in a proton gradient derived from electron transport is an extremely ancient process. Thus, chemiosmotic coupling may have fueled the evolution of nearly all life-forms on Earth.

ESSENTIAL CONCEPTS

- Mitochondria, chloroplasts, and many prokaryotes generate energy by a membrane-based mechanism known as chemiosmotic coupling, which involves using an electrochemical proton gradient to drive the synthesis of ATP.

- Mitochondria produce most of an animal cell's ATP, using energy derived from oxidation of sugars and fatty acids.

- Mitochondria have an inner and an outer membrane. The inner membrane encloses the mitochondrial matrix, where the citric acid cycle produces large amounts of NADH and $FADH_2$ from the oxidation of acetyl CoA.

- In the inner mitochondrial membrane, high-energy electrons donated by NADH and $FADH_2$ pass along an electron-transport chain and eventually combine with molecular oxygen (O_2) to form water.

- Much of the energy released by electron transfers along the electron-transport chain is harnessed to pump protons (H^+) out of the matrix, creating an electrochemical proton gradient. The proton pumping is carried out by three large respiratory enzyme complexes embedded in the inner membrane.

- The electrochemical proton gradient across the inner mitochondrial membrane is harnessed to make ATP when protons move back into the matrix through an ATP synthase located in the inner membrane.

- The electrochemical proton gradient also drives the active transport of selected metabolites into and out of the mitochondrial matrix.

- In photosynthesis in chloroplasts and photosynthetic bacteria, the energy of sunlight is captured by chlorophyll molecules embedded in large protein complexes known as photosystems; in plants, these photosystems are located in the thylakoid membranes of chloroplasts in leaf cells.

- Electron-transport chains associated with photosystems transfer high-energy electrons from water to $NADP^+$ to form NADPH, which produces O_2 as a by-product.

- The photosynthetic electron-transport chains in chloroplasts also generate a proton gradient across the thylakoid membrane, which is used by an ATP synthase embedded in the membrane to generate ATP.

- The ATP and the NADPH made by photosynthesis are used within the chloroplast stroma to drive the carbon-fixation cycle, which produces carbohydrate from CO_2 and water.

- Carbohydrate is exported from the stroma to the cell cytosol, where it provides the starting material for the synthesis of other organic molecules.

- Both mitochondria and chloroplasts are thought to have evolved from bacteria that were endocytosed by other cells. Each retains its own genome and divides by processes that resemble a bacterial cell division.

- Chemiosmotic coupling mechanisms are of ancient origin. Modern microorganisms that live in environments similar to those thought to have been present on the early Earth also use chemiosmotic coupling to produce ATP.

KEY TERMS

antenna complex	mitochondrion
ATP synthase	nitrogen fixation
carbon fixation	oxidative phosphorylation
cell respiration	photosynthesis
chemiosmotic coupling	photosystem
chlorophyll	quinone
chloroplast	reaction center
cytochrome	redox pair
cytochrome c oxidase	redox potential
dark reactions	redox reaction
electron-transport chain	respiratory enzyme complex
iron–sulfur center	stroma
light reactions	thylakoids
matrix	

QUESTIONS

QUESTION 14–11

Which of the following statements are correct? Explain your answers.

A. After an electron has been removed by light, the affinity for electrons of the positively charged chlorophyll in the reaction center of the first photosystem (photosystem II) is even greater than the electron affinity of O_2.

B. Photosynthesis is the light-driven transfer of an electron from chlorophyll to a second molecule that normally has a much lower affinity for electrons.

C. Because it requires the removal of four electrons to release one O_2 molecule from two H_2O molecules, the water-splitting enzyme in photosystem II has to keep the reaction intermediates tightly bound so as to prevent partly reduced, and therefore hazardous, superoxide radicals from escaping.

QUESTION 14–12

Which of the following statements are correct? Explain your answers.

A. Many, but not all, electron-transfer reactions involve metal ions.

B. The electron-transport chain generates an electrical potential across the membrane because it moves electrons from the intermembrane space into the matrix.

C. The electrochemical proton gradient consists of two components: a pH difference and an electrical potential.

D. Ubiquinone and cytochrome c are both diffusible electron carriers.

E. Plants have chloroplasts and therefore can live without mitochondria.

F. Both chlorophyll and heme contain an extensive system of double bonds that allows them to absorb visible light.

G. The role of chlorophyll in photosynthesis is equivalent to that of heme in mitochondrial electron transport.

H. Most of the dry weight of a tree comes from the minerals that are taken up by the roots.

QUESTION 14–13

A single proton moving down its electrochemical gradient into the mitochondrial matrix space liberates 4.6 kcal/mole of free energy (ΔG). How many protons have to flow across the inner mitochondrial membrane to synthesize one molecule of ATP if the ΔG for ATP synthesis under intracellular conditions is between 11 and 13 kcal/mole? (ΔG is discussed in Chapter 3, pp. 90–100.) Why is a range given for this latter value, and not a precise number? Under which conditions would the lower value apply?

QUESTION 14–14

In the following statement, choose the correct one of the alternatives in italics and justify your answer. "If no O_2 is available, all components of the mitochondrial electron-transport chain will accumulate in their *reduced/oxidized* form. If O_2 is suddenly added again, the electron carriers in cytochrome *c* oxidase will become *reduced/oxidized before/after* those in NADH dehydrogenase."

QUESTION 14–15

Assume that the conversion of oxidized ubiquinone to reduced ubiquinone by NADH dehydrogenase occurs on the matrix side of the inner mitochondrial membrane and that its oxidation by cytochrome *c* reductase occurs on the intermembrane space side of the membrane (see Figures 14–14 and 14–23). What are the consequences of this arrangement for the generation of the H^+ gradient across the membrane?

QUESTION 14–16

If a voltage is applied to two platinum wires (electrodes) immersed in water, then water molecules become split into H_2 and O_2 gas. At the negative electrode, electrons are donated and H_2 gas is released; at the positive electrode, electrons are accepted and O_2 gas is produced. When photosynthetic bacteria and plant cells split water, they produce O_2, but no H_2. Why?

QUESTION 14–17

In an insightful experiment performed in the 1960s, chloroplasts were first soaked in an acidic solution at pH 4, so that the stroma and thylakoid space became acidified (Figure Q14–17). They were then transferred to a basic solution (pH 8). This quickly increased the pH of the stroma to 8, while the thylakoid space temporarily remained at pH 4. A burst of ATP synthesis was observed, and the pH difference between the thylakoid and the stroma then disappeared.

A. Explain why these conditions lead to ATP synthesis.

B. Is light needed for the experiment to work?

C. What would happen if the solutions were switched so that the first incubation is in the pH 8 solution and the second one in the pH 4 solution?

Figure Q14–17

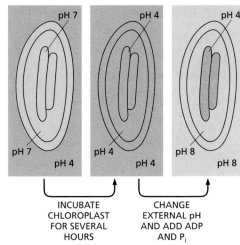

INCUBATE CHLOROPLAST FOR SEVERAL HOURS

CHANGE EXTERNAL pH AND ADD ADP AND P_i

D. Does the experiment support or question the chemiosmotic model?

Explain your answers.

QUESTION 14–18

As your first experiment in the laboratory, your adviser asks you to reconstitute purified bacteriorhodopsin, a light-driven H^+ pump from the plasma membrane of photosynthetic bacteria, and purified ATP synthase from ox-heart mitochondria together into the same membrane vesicles—as shown in **Figure Q14–18**. You are then asked to add ADP and P_i to the external medium and shine light into the suspension of vesicles.

A. What do you observe?

B. What do you observe if not all the detergent is removed and the vesicle membrane therefore remains leaky to ions?

C. You tell a friend over dinner about your new experiments, and he questions the validity of an approach that utilizes components from so widely divergent, unrelated organisms: "Why would anybody want to mix vanilla pudding with brake fluid?" Defend your approach against his critique.

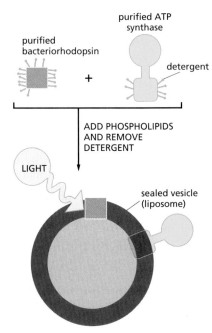

purified bacteriorhodopsin

purified ATP synthase

detergent

+

ADD PHOSPHOLIPIDS AND REMOVE DETERGENT

LIGHT

sealed vesicle (liposome)

Figure Q14–18

QUESTION 14–19

FADH$_2$ is produced in the citric acid cycle by a membrane-embedded enzyme complex, called succinate dehydrogenase, that contains bound FAD and carries out the reactions

$$succinate + FAD \rightarrow fumarate + FADH_2$$

and

$$FADH_2 \rightarrow FAD + 2H^+ + 2e^-$$

The redox potential of FADH$_2$, however, is only –220 mV. Referring to Panel 14–1 (p. 466) and Figure 14–24, suggest a plausible mechanism by which its electrons could be fed into the electron-transport chain. Draw a diagram to illustrate your proposed mechanism.

QUESTION 14–20

Some bacteria have become specialized to live in an environment of high pH (pH ~10). Do you suppose that these bacteria use a proton gradient across their plasma membrane to produce their ATP? (Hint: all cells must maintain their cytoplasm at a pH close to neutrality.)

QUESTION 14–21

Figure Q14–21 summarizes the circuitry used by mitochondria and chloroplasts to interconvert different forms of energy. Is it accurate to say

A. that the products of chloroplasts are the substrates for mitochondria?

B. that the activation of electrons by the photosystems enables chloroplasts to drive electron transfer from H$_2$O to carbohydrate, which is the opposite direction of electron transfer in the mitochondrion?

C. that the citric acid cycle is the reverse of the normal carbon-fixation cycle?

QUESTION 14–22

A manuscript has been submitted for publication to a prestigious scientific journal. In the paper, the authors describe an experiment in which they have succeeded in trapping an individual ATP synthase molecule and then mechanically rotating its head by applying a force to it. The authors show that upon rotating the head of the ATP synthase, ATP is produced, in the absence of an H$^+$ gradient. What might this mean about the mechanism whereby ATP synthase functions? Should this manuscript be considered for publication in one of the best journals?

QUESTION 14–23

You mix the following components in a solution. Assuming that the electrons must follow the path specified in Figure 14–14, in which experiments would you expect a net transfer of electrons to cytochrome c? Discuss why electron transfer does not occur in the other experiments.

A. reduced ubiquinone and oxidized cytochrome c

B. oxidized ubiquinone and oxidized cytochrome c

C. reduced ubiquinone and reduced cytochrome c

D. oxidized ubiquinone and reduced cytochrome c

E. reduced ubiquinone, oxidized cytochrome c, and cytochrome c reductase complex

F. oxidized ubiquinone, oxidized cytochrome c, and cytochrome c reductase complex

G. reduced ubiquinone, reduced cytochrome c, and cytochrome c reductase complex

H. oxidized ubiquinone, reduced cytochrome c, and cytochrome c reductase complex

(A) MITOCHONDRION

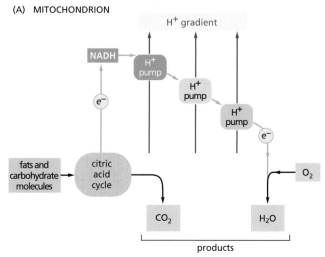

(B) CHLOROPLAST

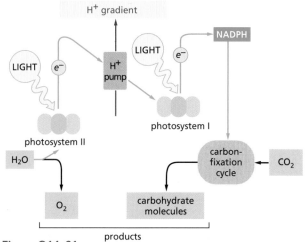

Figure Q14–21

CHAPTER **FIFTEEN**

15

Intracellular Compartments and Protein Transport

At any one time, a typical eukaryotic cell carries out thousands of different chemical reactions, many of which are mutually incompatible. One series of reactions makes glucose, for example, while another breaks it down; some enzymes synthesize peptide bonds, whereas others hydrolyze them, and so on. Indeed, if the cells of an organ such as the liver are broken apart and their contents mixed together in a test tube, chemical chaos results, and the cells' enzymes and other proteins are quickly degraded by their own proteolytic enzymes. For a cell to operate effectively, the different intracellular processes that occur simultaneously must somehow be segregated.

Cells have evolved several strategies for isolating and organizing their chemical reactions. One strategy used by both prokaryotic and eukaryotic cells is to aggregate the different enzymes required to catalyze a particular sequence of reactions into large, multicomponent complexes. Such complexes are used, for example, in the synthesis of DNA, RNA, and proteins. A second strategy, which is most highly developed in eukaryotic cells, is to confine different metabolic processes—and the proteins required to perform them—within different membrane-enclosed compartments. As discussed in Chapters 11 and 12, cell membranes provide selectively permeable barriers through which the transport of most molecules can be controlled. In this chapter, we consider this strategy of membrane-dependent compartmentalization.

In the first section, we describe the principal membrane-enclosed compartments, or *membrane-enclosed organelles*, of eukaryotic cells and briefly consider their main functions. In the second section, we discuss how the protein composition of the different compartments is set up and

MEMBRANE-ENCLOSED
ORGANELLES

PROTEIN SORTING

VESICULAR TRANSPORT

SECRETORY PATHWAYS

ENDOCYTIC PATHWAYS

Figure 15–1 In eukaryotic cells, internal membranes create enclosed compartments that segregate different metabolic processes. Examples of many of the major membrane-enclosed organelles can be identified in this electron micrograph of part of a liver cell, seen in cross section. The small black granules between the compartments are aggregates of glycogen and the enzymes that control its synthesis and breakdown. (Courtesy of Daniel S. Friend.)

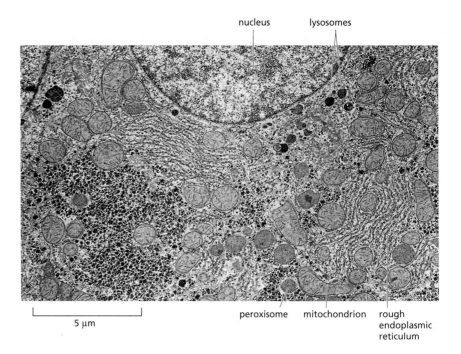

nucleus   lysosomes

5 µm

peroxisome   mitochondrion   rough endoplasmic reticulum

maintained. Each compartment contains a unique set of proteins that have to be transferred selectively from the cytosol, where they are made, to the compartment where they are used. This transfer process, called *protein sorting*, depends on signals built into the amino acid sequence of the proteins. In the third section, we describe how certain membrane-enclosed compartments in a eukaryotic cell communicate with one another by forming small, membrane-enclosed sacs, or *vesicles*. These vesicles pinch off from one compartment, move through the cytosol, and fuse with another compartment in a process called *vesicular transport*. In the last two sections, we discuss how this constant vesicular traffic also provides the main routes for releasing proteins from the cell by the process of *exocytosis* and for importing them by the process of *endocytosis*.

MEMBRANE-ENCLOSED ORGANELLES

Whereas a prokaryotic cell usually consists of a single compartment enclosed by the plasma membrane, eukaryotic cells are elaborately subdivided by internal membranes. When a cross section through a plant or an animal cell is examined in the electron microscope, numerous small, membrane-enclosed sacs, tubes, spheres, and irregularly shaped structures can be seen, often arranged without much apparent order (**Figure 15–1**). These structures are all distinct, membrane-enclosed organelles, or parts of such organelles, each of which contains a unique set of large and small molecules and carries out a specialized function. In this section, we review these functions and discuss how different membrane-enclosed organelles may have evolved.

Eukaryotic Cells Contain a Basic Set of Membrane-enclosed Organelles

The major **membrane-enclosed organelles** of an animal cell are illustrated in **Figure 15–2**, and their functions are summarized in **Table 15–1**. These organelles are surrounded by the *cytosol*, which is enclosed by the plasma membrane. The *nucleus* is generally the most prominent organelle in eukaryotic cells. It is surrounded by a double membrane, known as the *nuclear envelope*, and communicates with the cytosol via *nuclear pores* that perforate the envelope. The outer nuclear membrane

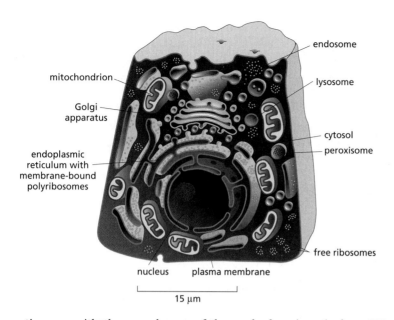

The nucleus, endoplasmic reticulum (ER), Golgi apparatus, lysosomes, endosomes, mitochondria, and peroxisomes are distinct compartments separated from the cytosol (*gray*) by at least one selectively permeable membrane. Ribosomes are shown bound to the cytosolic surface of portions of the ER, called the rough ER; the ER that lacks ribosomes is called smooth ER. Additional ribosomes can be found free in the cytosol.

is continuous with the membrane of the *endoplasmic reticulum* (*ER*), a system of interconnected sacs and tubes of membrane that often extends throughout most of the cell. The ER is the major site of synthesis of new membranes in the cell. Large areas of the ER have ribosomes attached to the cytosolic surface and are designated *rough endoplasmic reticulum* (*rough ER*). The ribosomes are actively synthesizing proteins that are delivered into the ER membrane or into the ER interior, a space called the *lumen*. The *smooth endoplasmic reticulum* (*smooth ER*) lacks ribosomes. It is scanty in most cells but is highly developed for performing particular functions in others: for example, it is the site of steroid hormone synthesis in some endocrine cells of the adrenal gland and the site where a variety of organic molecules, including alcohol, are detoxified in liver cells. In many eukaryotic cells, the smooth ER also sequesters Ca^{2+} from the cytosol; the release and reuptake of Ca^{2+} from the ER is involved in the rapid response to many extracellular signals, as discussed in Chapters 12 and 16.

TABLE 15–1 THE MAIN FUNCTIONS OF MEMBRANE-ENCLOSED COMPARTMENTS OF A EUKARYOTIC CELL	
Compartment	**Main Function**
Cytosol	contains many metabolic pathways (Chapters 3 and 13); protein synthesis (Chapter 7); the cytoskeleton (Chapter 17)
Nucleus	contains main genome (Chapter 5); DNA and RNA synthesis (Chapters 6 and 7)
Endoplasmic reticulum (ER)	synthesis of most lipids (Chapter 11); synthesis of proteins for distribution to many organelles and to the plasma membrane (this chapter)
Golgi apparatus	modification, sorting, and packaging of proteins and lipids for either secretion or delivery to another organelle (this chapter)
Lysosomes	intracellular degradation (this chapter)
Endosomes	sorting of endocytosed material (this chapter)
Mitochondria	ATP synthesis by oxidative phosphorylation (Chapter 14)
Chloroplasts (in plant cells)	ATP synthesis and carbon fixation by photosynthesis (Chapter 14)
Peroxisomes	oxidation of toxic molecules

The *Golgi apparatus*, which is usually situated near the nucleus, receives proteins and lipids from the ER, modifies them, and then dispatches them to other destinations in the cell. Small sacs of digestive enzymes called *lysosomes* degrade worn-out organelles, as well as macromolecules and particles taken into the cell by endocytosis. On their way to lysosomes, endocytosed materials must first pass through a series of compartments called *endosomes*, which sort the ingested molecules and recycle some of them back to the plasma membrane. *Peroxisomes* are small organelles that contain enzymes used in a variety of oxidative reactions that break down lipids and destroy toxic molecules. *Mitochondria* and (in plant cells) *chloroplasts* are each surrounded by a double membrane and are the sites of oxidative phosphorylation and photosynthesis, respectively (discussed in Chapter 14); both contain internal membranes that are highly specialized for the production of ATP.

Many of the membrane-enclosed organelles, including the ER, Golgi apparatus, mitochondria, and chloroplasts, are held in their relative locations in the cell by attachment to the cytoskeleton, especially to microtubules. Cytoskeletal filaments provide tracks for moving the organelles around and for directing the traffic of vesicles between one organelle and another. These movements are driven by motor proteins that use the energy of ATP hydrolysis to propel the organelles and vesicles along the filaments, as discussed in Chapter 17.

On average, the membrane-enclosed organelles together occupy nearly half the volume of a eukaryotic cell (**Table 15–2**), and the total amount of membrane associated with them is enormous. In a typical mammalian cell, for example, the area of the endoplasmic reticulum membrane is 20–30 times greater than that of the plasma membrane. In terms of its area and mass, the plasma membrane is only a minor membrane in most eukaryotic cells.

Much can be learned about the composition and function of an organelle once it has been isolated from other cell structures. For the most part, organelles are far too small to be isolated by hand, but it is possible to separate one type of organelle from another by differential centrifugation (described in Panel 4–3, pp. 164–165). Once a purified sample of one type of organelle has been obtained, the organelle's proteins can be identified. In many cases, the organelle itself can be incubated in a test tube under conditions that allow its functions to be studied. Isolated mitochondria, for example, can produce ATP from the oxidation of pyruvate to CO_2 and water, provided they are adequately supplied with ADP, inorganic phosphate, and O_2.

TABLE 15–2 THE RELATIVE VOLUMES AND NUMBERS OF THE MAJOR MEMBRANE-ENCLOSED ORGANELLES IN A LIVER CELL (HEPATOCYTE)		
Intracellular Compartment	Percentage of Total Cell Volume	Approximate Number per Cell
Cytosol	54	1
Mitochondria	22	1700
Endoplasmic reticulum	12	1
Nucleus	6	1
Golgi apparatus	3	1
Peroxisomes	1	400
Lysosomes	1	300
Endosomes	1	200

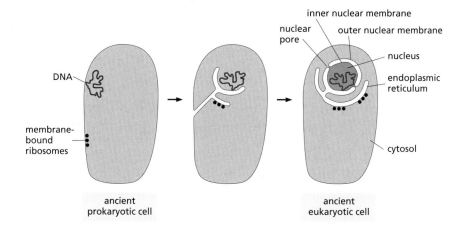

In bacteria, the single DNA molecule is typically attached to the plasma membrane. It is possible that in a very ancient prokaryotic cell, the plasma membrane, with its attached DNA, could have invaginated and, in subsequent generations, formed a two-layered envelope of membrane completely surrounding the DNA. This envelope is presumed to have eventually pinched off completely from the plasma membrane, ultimately producing a nuclear compartment penetrated by channels called nuclear pores, which enable communication with the cytosol. Other portions of the invaginated membrane may have formed the ER, which would explain why the space between the inner and outer nuclear membranes is continuous with the ER lumen.

Membrane-enclosed Organelles Evolved in Different Ways

In trying to understand the relationships between the different compartments of a modern eukaryotic cell, it is helpful to consider how they evolved. The compartments probably evolved in stages. The precursors of the first eukaryotic cells are thought to have been simple microorganisms, resembling bacteria, which had a plasma membrane but no internal membranes. The plasma membrane in such cells would have provided all membrane-dependent functions, including ATP synthesis and lipid synthesis, as does the plasma membrane in most modern bacteria. Bacteria can get by with this arrangement because of their small size, which gives them a high surface-to-volume ratio: their plasma membrane area is thus sufficient to sustain all the vital functions for which membranes are required. Present-day eukaryotic cells, by contrast, have volumes 1000 to 10,000 times greater than that of a typical bacterium such as *E. coli*. Such a large cell has a small surface-to-volume ratio and presumably could not survive with a plasma membrane as its only membrane. Thus, the increase in size typical of eukaryotic cells probably could not have occurred without the development of internal membranes.

Membrane-enclosed organelles are thought to have arisen in evolution in at least two ways. The nuclear membranes and the membranes of the ER, Golgi apparatus, endosomes, and lysosomes most likely originated by invagination of the plasma membrane, as illustrated for the nuclear and ER membranes in **Figure 15–3**. The ER, Golgi apparatus, peroxisomes, endosomes, and lysosomes are all part of what is collectively called the **endomembrane system**. As we discuss later, the interiors of these organelles communicate extensively with one another and with the outside of the cell by means of small vesicles that bud off from one of these organelles and fuse with another. Consistent with this proposed evolutionary origin, the interiors of these organelles are treated by the cell in many ways as "extracellular," as we will see. The hypothetical scheme shown in Figure 15–3 also explains why the nucleus is surrounded by two membranes.

Mitochondria and chloroplasts are thought to have originated in a different way. They differ from all other organelles in that they possess their own small genomes and can make some of their own proteins, as discussed in Chapter 14. The similarity of their genomes to those of bacteria and the close resemblance of some of their proteins to bacterial proteins strongly suggest that both these organelles evolved from bacteria that were engulfed by primitive pre-eukaryotic cells with which they initially lived in symbiosis (**Figure 15–4**). As might be expected from their origins, mitochondria and chloroplasts remain isolated from the extensive vesicular traffic that connects the interiors of most of the other membrane-enclosed organelles to one another and to the outside of the cell.

QUESTION 15–1

As shown in the drawings in Figure 15–3, the lipid bilayer of the inner and outer nuclear membranes forms a continuous sheet, joined around the nuclear pores. As membranes are two-dimensional fluids, this would imply that membrane proteins can diffuse freely between the two nuclear membranes. Yet each of these two nuclear membranes has a different protein composition, reflecting different functions. How could you reconcile this apparent contradiction?

Figure 15–4 Mitochondria are thought to have originated when an aerobic prokaryote was engulfed by a larger pre-eukaryotic cell. Chloroplasts are thought to have originated later in a similar way, when a eukaryotic cell with mitochondria engulfed a photosynthetic prokaryote. This theory would explain why these organelles have two membranes, possess their own genomes, and do not participate in the vesicular traffic that connects the compartments of the endomembrane system.

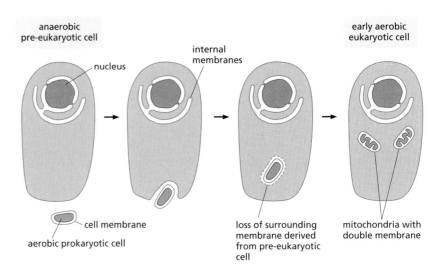

PROTEIN SORTING

Before a eukaryotic cell divides, it must duplicate its membrane-enclosed organelles. As cells grow, membrane-enclosed organelles enlarge by incorporation of new molecules; the organelles then divide and, during cell division, are distributed between the two daughter cells. Organelle growth requires a supply of new lipids to make more membrane and a supply of the appropriate proteins—both membrane proteins and the soluble proteins that will occupy the interior of the organelle. Even in cells that are not dividing, proteins are being produced continually. These newly synthesized proteins must be accurately delivered to their appropriate organelle—some for eventual secretion from the cell and some to replace organelle proteins that have been degraded. Directing newly made proteins to their correct organelle is therefore necessary for any cell to grow and divide, or just to function properly.

For some organelles, including mitochondria, chloroplasts, peroxisomes, and the interior of the nucleus, proteins are delivered directly from the cytosol. For others, including the Golgi apparatus, lysosomes, endosomes, and the inner nuclear membrane, proteins and lipids are delivered indirectly via the ER, which is itself a major site of lipid and protein synthesis. Proteins enter the ER directly from the cytosol: some are retained there, but most are transported by vesicles to the Golgi apparatus and then onward to the plasma membrane or other organelles. Peroxisomes acquire some of their membrane proteins from the ER, but the bulk of their enzymes enter directly from the cytosol.

In this section, we discuss the mechanisms by which proteins directly enter membrane-enclosed organelles from the cytosol. Proteins made in the cytosol are dispatched to different locations in the cell according to specific address labels contained in their amino acid sequence. Once at the correct address, the protein enters either the membrane or the interior lumen of its designated organelle.

Proteins Are Transported into Organelles by Three Mechanisms

The synthesis of virtually all proteins in the cell begins on ribosomes in the cytosol. The exceptions are the few mitochondrial and chloroplast proteins that are synthesized on ribosomes inside these organelles; most mitochondrial and chloroplast proteins, however, are made in the cytosol and subsequently imported. The fate of any protein molecule synthesized in the cytosol depends on its amino acid sequence, which can contain a *sorting signal* that directs the protein to the organelle in which it is required. Proteins that lack such signals remain as permanent residents in the cytosol; those that possess a sorting signal move from the cytosol to the appropriate organelle. Different sorting signals direct proteins into the nucleus, mitochondria, chloroplasts (in plants), peroxisomes, and the ER.

When a membrane-enclosed organelle imports a water-soluble protein to its interior—either from the cytosol or from another organelle—it faces a problem: how can it transport the protein across its membrane (or membranes), which are normally impermeable to hydrophilic macromolecules? This task is accomplished in different ways by different organelles.

1. Proteins moving from the cytosol into the nucleus are transported through the nuclear pores, which penetrate both the inner and outer nuclear membranes. The pores function as selective gates that actively transport specific macromolecules but also allow free diffusion of smaller molecules (mechanism 1 in **Figure 15–5**).

2. Proteins moving from the cytosol into the ER, mitochondria, or chloroplasts are transported across the organelle membrane by *protein translocators* located in the membrane. Unlike transport through nuclear pores, the transported protein must usually unfold in order to snake across the membrane through the translocator (mechanism 2 in Figure 15–5). Bacteria have similar protein translocators in their plasma membrane, which they use to export proteins from the cytosol to the cell exterior.

3. Proteins moving onward from the ER—and from one compartment of the endomembrane system to another—are transported by a mechanism that is fundamentally different. These proteins are ferried by *transport vesicles*, which pinch off from the membrane of one compartment and then fuse with the membrane of a second

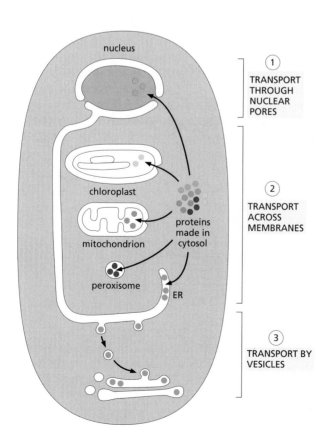

Figure 15–5 **Membrane-enclosed organelles import proteins by one of three mechanisms.** All of these processes require energy. The protein remains folded during transport in mechanisms 1 and 3 but usually has to be unfolded during mechanism 2.

TABLE 15–3 SOME TYPICAL SIGNAL SEQUENCES

Function of Signal	Example of Signal Sequence
Import into ER	+H3N-Met-Met-Ser-Phe-Val-Ser-Leu-Leu-Leu-Val-Gly-Ile-Leu-Phe-Trp-Ala-Thr-Glu-Ala-Glu-Gln-Leu-Thr-Lys-Cys-Glu-Val-Phe-Gln-
Retention in lumen of ER	-Lys-Asp-Glu-Leu-COO⁻
Import into mitochondria	+H3N-Met-Leu-Ser-Leu-Arg-Gln-Ser-Ile-Arg-Phe-Phe-Lys-Pro-Ala-Thr-Arg-Thr-Leu-Cys-Ser-Ser-Arg-Tyr-Leu-Leu-
Import into nucleus	-Pro-Pro-Lys-Lys-Lys-Arg-Lys-Val-
Export from nucleus	-Met-Glu-Glu-Leu-Ser-Gln-Ala-Leu-Ala-Ser-Ser-Phe-
Import into peroxisomes	-Ser-Lys-Leu-

Positively charged amino acids are shown in *red* and negatively charged amino acids in *blue*. Important hydrophobic amino acids are shown in *green*.
+H3N indicates the N-terminus of a protein; COO⁻ indicates the C-terminus.

compartment (mechanism 3 in Figure 15–5). In this process, transport vesicles deliver soluble cargo proteins, as well as the proteins and lipids that are part of the vesicle membrane.

Signal Sequences Direct Proteins to the Correct Compartment

The typical sorting signal on a protein is a continuous stretch of amino acid sequence, typically 15–60 amino acids long. This **signal sequence** is often (but not always) removed from the finished protein once it has been sorted. Some of the signal sequences used to specify different destinations in the cell are shown in **Table 15–3**.

Signal sequences are both necessary and sufficient to direct a protein to a particular destination. This has been shown by experiments in which the sequence is either deleted or transferred from one protein to another by genetic engineering techniques (discussed in Chapter 10). Deleting a signal sequence from an ER protein, for example, converts it into a cytosolic protein, while placing an ER signal sequence at the beginning of a cytosolic protein redirects the protein to the ER (**Figure 15–6**). The signal sequences specifying the same destination can vary greatly even though they have the same function; physical properties such as hydrophobicity

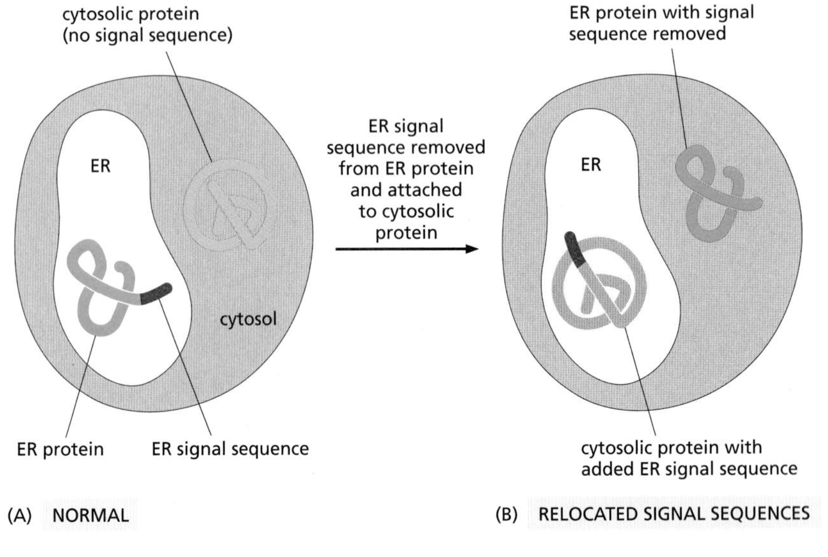

Figure 15–6 Signal sequences direct proteins to the correct destination. (A) Proteins destined for the ER possess an N-terminal signal sequence that directs them to that organelle, whereas those destined to remain in the cytosol lack any such signal sequence. (B) Recombinant DNA techniques can be used to change the destination of the two proteins: if the signal sequence is removed from an ER protein and attached to a cytosolic protein, both proteins are reassigned to the expected, inappropriate location.

(A) NORMAL

(B) RELOCATED SIGNAL SEQUENCES

or the placement of charged amino acids often appear to be more important for the function of these signals than the exact amino acid sequence.

Proteins Enter the Nucleus Through Nuclear Pores

The **nuclear envelope**, which encloses the nuclear DNA and defines the nuclear compartment, is formed from two concentric membranes. The *inner nuclear membrane* contains some proteins that act as binding sites for the chromosomes (discussed in Chapter 5) and others that provide anchorage for the *nuclear lamina*, a finely woven meshwork of protein filaments that lines the inner face of this membrane and provides structural support for the nuclear envelope (discussed in Chapter 17). The composition of the *outer nuclear membrane* closely resembles the membrane of the ER, with which it is continuous (**Figure 15–7**).

The nuclear envelope in all eukaryotic cells is perforated by **nuclear pores** that form the gates through which molecules enter or leave the nucleus. A nuclear pore is a large, elaborate structure composed of a complex of about 30 different proteins (**Figure 15–8**). Many of the proteins that line the nuclear pore contain extensive, unstructured regions in which the polypeptide chains are largely disordered. These disordered segments form a soft, tangled meshwork—like a kelp forest—that fills the center of the channel, preventing the passage of large molecules but allowing small, water-soluble molecules to pass freely and nonselectively between the nucleus and the cytosol.

Selected larger molecules and macromolecular complexes also need to pass through nuclear pores. RNA molecules, which are synthesized in the nucleus, and ribosomal subunits, which are assembled there, must be exported to the cytosol (discussed in Chapter 7). And newly made proteins that are destined for the nucleus must be imported from the cytosol (Movie 15.1). To gain entry to a pore, these large molecules and

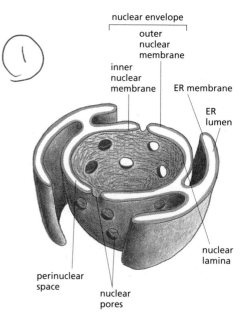

Figure 15–7 The outer nuclear membrane is continuous with the ER membrane. The double membrane of the nuclear envelope is penetrated by nuclear pores. The ribosomes that are normally bound to the cytosolic surface of the ER membrane and outer nuclear membrane are not shown.

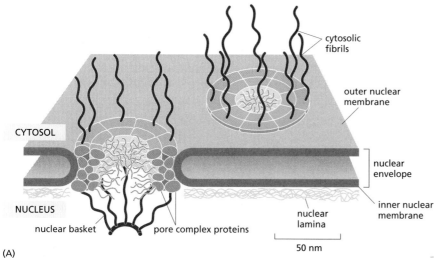

(A)

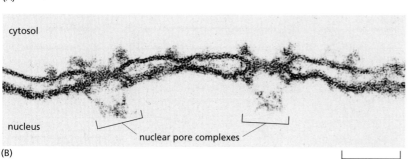

(B)

Figure 15–8 The nuclear pore complex forms a gate through which select ... macromolecules and larger complexes enter or exit the nucleus. (A) Drawing of a small region of the nuclear envelope showing two pores. Protein fibrils protrude from both sides of the pore complex; on the nuclear side, they converge to form a basketlike structure. The spacing between the fibrils is wide enough that the fibrils do not obstruct access to the pores. (B) Electron micrograph of a region of nuclear envelope showing a side view of two nuclear pores (brackets). (C) Electron micrograph showing a face-on view of nuclear pore protein complexes; the membranes have been extracted with detergent. (B, courtesy of Werner W. Franke; C, courtesy of Ron Milligan.)

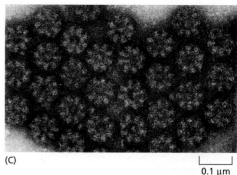

(C)

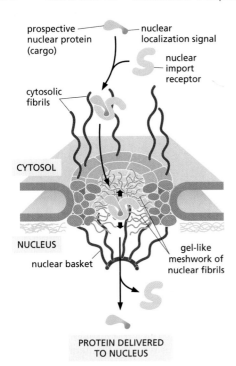

prospective nuclear protein (cargo)

nuclear localization signal

nuclear import receptor

cytosolic fibrils

CYTOSOL

NUCLEUS

nuclear basket

gel-like meshwork of nuclear fibrils

PROTEIN DELIVERED TO NUCLEUS

Figure 15–9 Prospective nuclear proteins are imported from the cytosol through nuclear pores. The proteins contain a nuclear localization signal that is recognized by nuclear import receptors, which interact with the cytosolic fibrils that extend from the rim of the pore. As indicated by the short *black* arrows, after being captured, the receptors move randomly with their cargo through the gel-like meshwork of nuclear fibrils, until nuclear entry triggers cargo release. After cargo delivery, the receptors return to the cytosol via nuclear pores for reuse. Similar types of transport receptors, operating in the reverse direction, export mRNAs from the nucleus (see Figure 7–23). These sets of import and export receptors have a similar basic structure.

macromolecular complexes must display an appropriate sorting signal. The signal sequence that directs a protein from the cytosol into the nucleus, called a *nuclear localization signal*, typically consists of one or two short sequences containing several positively charged lysines or arginines (see Table 15–3).

The nuclear localization signal on proteins destined for the nucleus is recognized by cytosolic proteins called *nuclear import receptors*. These receptors help direct a newly synthesized protein to a nuclear pore by interacting with the tentacle-like fibrils that extend from the rim of the pore into the cytosol (**Figure 15–9**). Once there, the nuclear import receptor penetrates the pore by grabbing onto short, repeated amino acid sequences within the tangle of nuclear pore proteins that fill the center of the pore. When the nuclear pore is empty, these repeated sequences bind to one another, forming a loosely packed gel. Nuclear import receptors interrupt these interactions, and they open a local passageway through the meshwork. The import receptors simply bump along from one repeat sequence to the next, until they enter the nucleus and deliver their cargo. The empty receptor then returns to the cytosol via the nuclear pore for reuse (see Figure 15–9).

Like any process that creates order, the import of nuclear proteins requires energy. In this case, the energy is provided by the hydrolysis of GTP, mediated by a monomeric GTPase named Ran. This GTP hydrolysis drives nuclear transport in the appropriate direction, as shown in **Figure 15–10**. Nuclear pore proteins operate this molecular gate at an amazing speed, rapidly pumping macromolecules in both directions through each pore.

Figure 15–10 Energy supplied by GTP hydrolysis drives nuclear transport. A nuclear import receptor picks up a prospective nuclear protein in the cytosol and enters the nucleus. There it encounters a small monomeric GTPase called Ran, which carries a molecule of GTP. This Ran-GTP binds to the import receptor, causing it to release the nuclear protein. Having discharged its cargo in the nucleus, the receptor—still carrying Ran-GTP—is transported back through the pore to the cytosol. There, an accessory protein (not shown) triggers Ran to hydrolyze its bound GTP. Ran-GDP falls off the import receptor, which is then free to bind another protein destined for the nucleus. A similar cycle operates to export mRNAs and ribosomal subunits from the nucleus into the cytosol, using nuclear export receptors that recognize nuclear export signals (see Table 15–3).

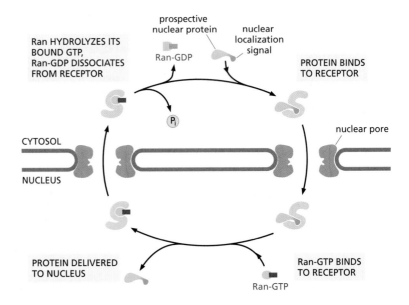

Ran HYDROLYZES ITS BOUND GTP, Ran-GDP DISSOCIATES FROM RECEPTOR

prospective nuclear protein

nuclear localization signal

Ran-GDP

PROTEIN BINDS TO RECEPTOR

P_i

CYTOSOL

NUCLEUS

nuclear pore

PROTEIN DELIVERED TO NUCLEUS

Ran-GTP BINDS TO RECEPTOR

Ran-GTP

Nuclear pores transport proteins in their fully folded conformation and ribosomal components as assembled particles. This feature distinguishes the nuclear transport mechanism from the mechanisms that transport proteins into most other organelles. Proteins have to unfold to cross the membranes of mitochondria and chloroplasts, as we discuss next.

Proteins Unfold to Enter Mitochondria and Chloroplasts

Both mitochondria and chloroplasts are surrounded by inner and outer membranes, and both organelles specialize in the synthesis of ATP. Chloroplasts also contain a third membrane system, the thylakoid membrane (discussed in Chapter 14). Although both organelles contain their own genomes and make some of their own proteins, most mitochondrial and chloroplast proteins are encoded by genes in the nucleus and are imported from the cytosol. These proteins usually have a signal sequence at their N-terminus that allows them to enter their specific organelle. Proteins destined for either organelle are translocated simultaneously across both the inner and outer membranes at specialized sites where the two membranes contact each other. Each protein is unfolded as it is transported, and its signal sequence is removed after translocation is complete (**Figure 15–11**).

Chaperone proteins (discussed in Chapter 4) inside the organelles help to pull the protein across the two membranes and to fold it once it is inside. Subsequent transport to a particular site within the organelle, such as the inner or outer membrane or the thylakoid membrane in chloroplasts, usually requires further sorting signals in the protein, which are often only exposed after the first signal sequence has been removed. The insertion of transmembrane proteins into the inner membrane, for example, is guided by signal sequences in the protein that start and stop the transfer process across the membrane, as we describe later for the insertion of transmembrane proteins in the ER membrane.

The growth and maintenance of mitochondria and chloroplasts require not only the import of new proteins but also the incorporation of new lipids into the organelle membranes. Most of their membrane phospholipids are thought to be imported from the ER, which is the main site of lipid synthesis in the cell. Phospholipids are transported to these organelles

QUESTION 15–2

Why do eukaryotic cells require a nucleus as a separate compartment when prokaryotic cells can manage perfectly well without?

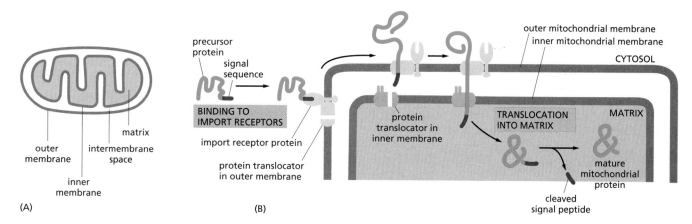

Figure 15–11 Mitochondrial precursor proteins are unfolded during import. (A) A mitochondrion has an outer and inner membrane, both of which must be crossed for a mitochondrial precursor protein to enter the organelle. (B) To initiate transport, the mitochondrial signal sequence on a mitochondrial precursor protein is recognized by a receptor in the outer mitochondrial membrane. This receptor is associated with a protein translocator. The complex of receptor, precursor protein, and translocator diffuses laterally in the outer membrane until it encounters a second translocator in the inner membrane. The two translocators then transport the protein across both the outer and inner membranes, unfolding the protein in the process (Movie 15.2). The signal sequence is finally cleaved off by a signal peptidase in the mitochondrial matrix. Proteins are imported into chloroplasts by a similar mechanism. The chaperone proteins that help pull the protein across the membranes and help it to refold are not shown.

by lipid-carrying proteins that extract a phospholipid molecule from one membrane and deliver it into another. Such transport may occur at specific junctions where mitochondrial and ER membranes are held in close proximity. Thanks to these lipid-carrying proteins, the different cell membranes are able to maintain different lipid compositions.

Proteins Enter Peroxisomes from Both the Cytosol and the Endoplasmic Reticulum

Peroxisomes generally contain one or more enzymes that produce hydrogen peroxide, hence their name. These organelles are present in all eukaryotic cells, where they break down a variety of molecules, including toxins, alcohol, and fatty acids. They also synthesize certain phospholipids, including those that are abundant in the myelin sheath that insulates nerve cell axons.

Peroxisomes acquire the bulk of their proteins via selective transport from the cytosol. A short sequence of only three amino acids serves as an import signal for many peroxisomal proteins. This sequence is recognized by receptor proteins in the cytosol, at least one of which escorts its cargo protein all the way into the peroxisome before returning to the cytosol. Like the membranes of mitochondria and chloroplasts, the peroxisomal membrane contains a protein translocator that aids in the transport. Unlike the mechanism that operates in mitochondria and chloroplasts, however, proteins do not need to unfold to enter the peroxisome—and the transport mechanism is still mysterious.

Although most peroxisomal proteins—including those embedded in the peroxisomal membrane—come from the cytosol, a few membrane proteins arrive via vesicles that bud from the ER membrane. The vesicles either fuse with preexisting peroxisomes or import peroxisomal proteins from the cytosol to grow into mature peroxisomes.

The most severe peroxisomal disease, called Zellweger syndrome, is caused by mutations that block peroxisomal protein import. Individuals with this disorder are born with severe abnormalities in their brain, liver, and kidneys. Most do not survive past the first six months of life—a grim reminder of the crucial importance of these underappreciated organelles for proper cell function and for the health of the organism.

Proteins Enter the Endoplasmic Reticulum While Being Synthesized

The endoplasmic reticulum is the most extensive membrane system in a eukaryotic cell (**Figure 15–12A**). Unlike the organelles discussed so far, it serves as an entry point for proteins destined for other organelles, as well as for the ER itself. Proteins destined for the Golgi apparatus, endosomes, and lysosomes, as well as proteins destined for the cell surface, all first enter the ER from the cytosol. Once inside the ER lumen, or embedded in the ER membrane, individual proteins will not re-enter the cytosol during their onward journey. They will instead be ferried by transport vesicles from organelle to organelle within the endomembrane system, or to the plasma membrane.

Two kinds of proteins are transferred from the cytosol to the ER: (1) water-soluble proteins are completely translocated across the ER membrane and are released into the ER lumen; (2) prospective transmembrane proteins are only partly translocated across the ER membrane and become embedded in it. The water-soluble proteins are destined either for secretion (by release at the cell surface) or for the lumen of an organelle of the endomembrane system. The transmembrane proteins are destined

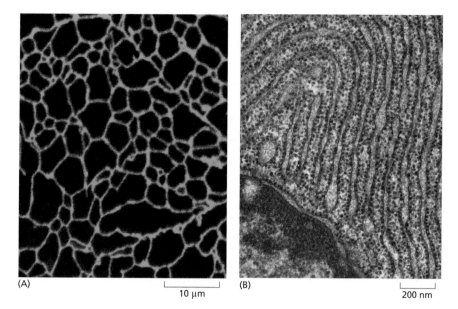

(A)

10 µm

(B)

200 nm

Figure 15–12 The endoplasmic reticulum is the most extensive membrane network in eukaryotic cells. (A) Fluorescence micrograph of a living plant cell showing the ER as a complex network of tubes. The cell shown here has been genetically engineered so that it contains a fluorescent protein in the ER lumen. Only part of the ER network in the cell is shown. (B) An electron micrograph showing the rough ER in a cell from a dog's pancreas, which makes and secretes large amounts of digestive enzymes. The cytosol is filled with closely packed sheets of ER, studded with ribosomes. A portion of the nucleus and its nuclear envelope can be seen at the bottom left; note that the outer nuclear membrane, which is continuous with the ER, is also studded with ribosomes. For a dynamic view of the ER network, watch Movie 15.3. (A, courtesy of Petra Boevink and Chris Hawes; B, courtesy of Lelio Orci.)

to reside in the membrane of one of these organelles or in the plasma membrane. All of these proteins are initially directed to the ER by an *ER signal sequence*, a segment of eight or more hydrophobic amino acids (see Table 15–3, p. 494), which is also involved in the process of translocation across the membrane.

Unlike the proteins that enter the nucleus, mitochondria, chloroplasts, or peroxisomes, most of the proteins that enter the ER begin to be threaded across the ER membrane before the polypeptide chain has been completely synthesized. This requires that the ribosome synthesizing the protein be attached to the ER membrane. These membrane-bound ribosomes coat the surface of the ER, creating regions termed **rough endoplasmic reticulum** because of its characteristic beaded appearance when viewed in an electron microscope (**Figure 15–12B**).

There are, therefore, two separate populations of ribosomes in the cytosol. *Membrane-bound ribosomes* are attached to the cytosolic side of the ER membrane (and outer nuclear membrane) and are making proteins that are being translocated into the ER. *Free ribosomes* are unattached to any membrane and are making all of the other proteins encoded by the nuclear DNA. Membrane-bound ribosomes and free ribosomes are structurally and functionally identical; they differ only in the proteins they are making at any given time. When a ribosome happens to be making a protein with an ER signal sequence, the signal sequence directs the ribosome to the ER membrane. Because proteins with an ER signal sequence are translocated as they are being made, no additional energy is required for their transport; the elongation of each polypeptide provides the thrust needed to push the growing chain through the ER membrane.

As an mRNA molecule is translated, many ribosomes bind to it, forming a *polyribosome* (discussed in Chapter 7). In the case of an mRNA molecule encoding a protein with an ER signal sequence, the polyribosome becomes riveted to the ER membrane by the growing polypeptide chains, which have become inserted into the ER membrane (**Figure 15–13**).

Soluble Proteins Made on the ER Are Released into the ER Lumen

Two protein components help guide ER signal sequences to the ER membrane: (1) a *signal-recognition particle (SRP)*, present in the cytosol, binds

Figure 15–13 A common pool of ribosomes is used to synthesize all the proteins encoded by the nuclear genome. Ribosomes that are translating proteins with no ER signal sequence remain free in the cytosol. Ribosomes that are translating proteins containing an ER signal sequence (*red*) on the growing polypeptide chain will be directed to the ER membrane. Many ribosomes bind to each mRNA molecule, forming a polyribosome. At the end of each round of protein synthesis, the ribosomal subunits are released and rejoin the common pool in the cytosol. As we see shortly, how the ribosome and signal sequence bind to the ER and translocation channel is more complicated than illustrated here.

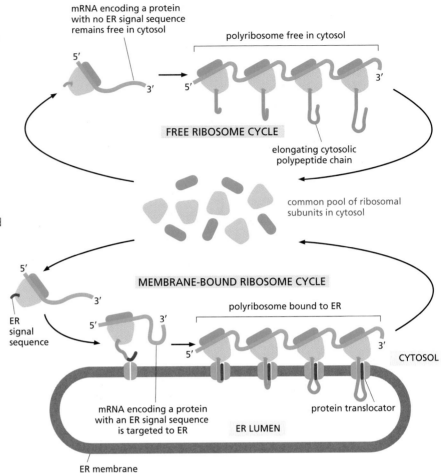

to both the ribosome and the ER signal sequence when it emerges from the ribosome, and (2) an *SRP receptor*, embedded in the ER membrane, recognizes the SRP. Binding of an SRP to a ribosome that displays an ER signal sequence slows protein synthesis by that ribosome until the SRP engages with an SRP receptor on the ER. Once bound, the SRP is released, the receptor passes the ribosome to a protein translocator in the ER membrane, and protein synthesis recommences. The polypeptide is then threaded across the ER membrane through a *channel* in the translocator (**Figure 15–14**). Thus the SRP and SRP receptor function as molecular matchmakers, uniting ribosomes that are synthesizing proteins with an ER signal sequence and available translocation channels in the ER membrane.

In addition to directing proteins to the ER, the signal sequence—which for soluble proteins is almost always at the N-terminus, the end synthesized first—functions to open the channel in the protein translocator. This sequence remains bound to the channel, while the rest of the polypeptide chain is threaded through the membrane as a large loop. It is removed by a transmembrane signal peptidase, which has an active site facing the lumenal side of the ER membrane. The cleaved signal sequence is then released from the translocation channel into the lipid bilayer and rapidly degraded.

Once the C-terminus of a soluble protein has passed through the translocation channel, the protein will be released into the ER lumen (**Figure 15–15**).

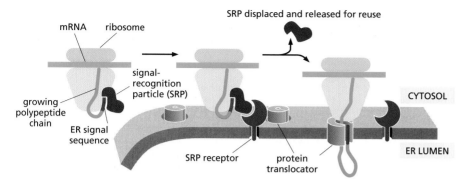

Figure 15–14 **An ER signal sequence and an SRP direct a ribosome to the ER membrane.** The SRP binds to both the exposed ER signal sequence and the ribosome, thereby slowing protein synthesis by the ribosome. The SRP–ribosome complex then binds to an SRP receptor in the ER membrane. The SRP is released, passing the ribosome from the SRP receptor to a protein translocator in the ER membrane. Protein synthesis resumes, and the translocator starts to transfer the growing polypeptide across the lipid bilayer.

Start and Stop Signals Determine the Arrangement of a Transmembrane Protein in the Lipid Bilayer

Not all proteins made by ER-bound ribosomes are released into the ER lumen. Some remain embedded in the ER membrane as transmembrane proteins. The translocation process for such proteins is more complicated than it is for soluble proteins, as some parts of the polypeptide chain must be translocated completely across the lipid bilayer, whereas other parts remain fixed in the membrane.

In the simplest case, that of a transmembrane protein with a single membrane-spanning segment, the N-terminal signal sequence initiates translocation—as it does for a soluble protein. But the transfer process is halted by an additional sequence of hydrophobic amino acids, a *stop-transfer sequence*, further along the polypeptide chain. At this point, the translocation channel releases the growing polypeptide chain sideways into the lipid bilayer. The N-terminal signal sequence is cleaved off, whereas the stop-transfer sequence remains in the bilayer, where it forms an α-helical membrane-spanning segment that anchors the protein in the membrane. As a result, the protein ends up as a single-pass transmembrane protein inserted in the membrane with a defined orientation—the N-terminus on the lumenal side of the lipid bilayer and the C-terminus on the cytosolic side (**Figure 15–16**). Once inserted into the membrane, a transmembrane protein does not change its orientation, which is retained throughout any subsequent vesicle budding and fusion events.

In some transmembrane proteins, an internal, rather than an N-terminal, signal sequence is used to start the protein transfer; this internal signal sequence, called a *start-transfer sequence*, is never removed from

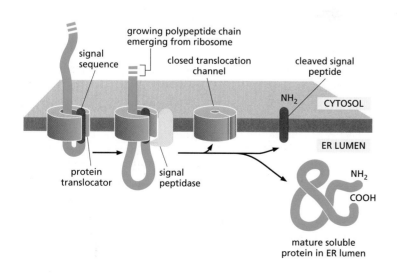

Figure 15–15 **A soluble protein crosses the ER membrane and enters the lumen.** The protein translocator binds the signal sequence and threads the rest of the polypeptide across the lipid bilayer as a loop. At some point during the translocation process, the signal peptide is cleaved from the growing protein by a signal peptidase. This cleaved signal sequence is ejected into the bilayer, where it is degraded. Once protein synthesis is complete, the translocated polypeptide is released as a soluble protein into the ER lumen, and the pore of the translocation channel closes. The membrane-bound ribosome is omitted from this and the following two figures for clarity.

Figure 15–16 A single-pass transmembrane protein is retained in the lipid bilayer. An N-terminal ER signal sequence (*red*) initiates transfer as in Figure 15–15. In addition, the protein also contains a second hydrophobic sequence, which acts as a stop-transfer sequence (*orange*). When this sequence enters the translocation channel, the channel discharges the growing polypeptide chain sideways into the lipid bilayer. The N-terminal signal sequence is cleaved off, leaving the transmembrane protein anchored in the membrane (Movie 15.4). Protein synthesis on the cytosolic side then continues to completion.

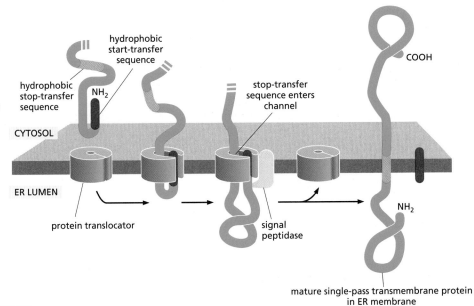

mature single-pass transmembrane protein in ER membrane

QUESTION 15–4

A. Predict the membrane orientation of a protein that is synthesized with an uncleaved, internal signal sequence (shown as the *red* start-transfer sequence in Figure 15–17) but does not contain a stop-transfer sequence.
B. Similarly, predict the membrane orientation of a protein that is synthesized with an N-terminal cleaved signal sequence followed by a stop-transfer sequence, followed by a start-transfer sequence.
C. What arrangement of signal sequences would enable the insertion of a multipass protein with ⎯⎯⎯ of transmembrane

the polypeptide. This arrangement occurs in some transmembrane proteins in which the polypeptide chain passes back and forth across the lipid bilayer. In these cases, hydrophobic signal sequences are thought to work in pairs: an internal start-transfer sequence serves to initiate translocation, which continues until a stop-transfer sequence is reached; the two hydrophobic sequences are then released into the bilayer, where they remain as membrane-spanning α helices (**Figure 15–17**). In complex multipass proteins, in which many hydrophobic α helices span the bilayer, additional pairs of start- and stop-transfer sequences come into play: one sequence reinitiates translocation further down the polypeptide chain, and the other stops translocation and causes polypeptide release, and so on for subsequent starts and stops. Thus, multipass membrane proteins are stitched into the lipid bilayer as they are being synthesized, by a mechanism resembling the workings of a sewing machine.

Having considered how proteins enter the ER lumen or become embedded in the ER membrane, we now discuss how they are carried onward by vesicular transport.

Figure 15–17 A double-pass transmembrane protein has an internal ER signal sequence. This internal sequence (*red*) not only acts as a start-transfer signal, it also helps to anchor the final protein in the membrane. Like the N-terminal ER signal sequence, the internal signal sequence is recognized by an SRP, which brings the ribosome to the ER membrane (not shown). When a stop-transfer sequence (*orange*) enters the translocation channel, the channel discharges both sequences into the lipid bilayer. Neither the start-transfer nor the stop-transfer sequence is cleaved off, and the entire polypeptide chain remains anchored in the membrane as a double-pass transmembrane protein. Proteins that span the membrane more times contain further pairs of start- and stop-transfer sequences, and the same process is repeated for each pair.

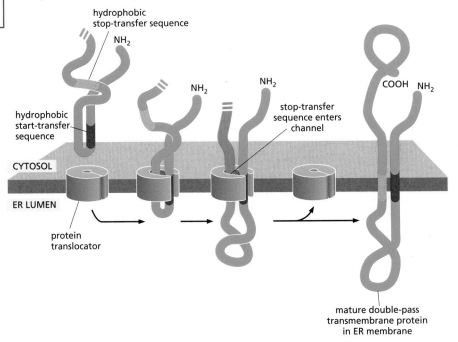

mature double-pass transmembrane protein in ER membrane

VESICULAR TRANSPORT

Entry into the ER lumen or membrane is usually only the first step on a pathway to another destination. That destination, initially at least, is generally the Golgi apparatus; there, proteins and lipids are modified and sorted for shipment to other sites. Transport from the ER to the Golgi apparatus—and from the Golgi apparatus to other compartments of the endomembrane system—is carried out by the continual budding and fusion of transport vesicles. This **vesicular transport** extends outward from the ER to the plasma membrane, and inward from the plasma membrane to lysosomes, and thus provides routes of communication between the interior of the cell and its surroundings. As proteins and lipids are transported outward along these pathways, many of them undergo various types of chemical modification, such as the addition of carbohydrate side chains.

In this section, we discuss how vesicles shuttle proteins and membranes between intracellular compartments, allowing cells to eat, drink, and secrete. We also consider how these transport vesicles are directed to their proper destination, be it an organelle of the endomembrane system or the plasma membrane.

Transport Vesicles Carry Soluble Proteins and Membrane Between Compartments

Vesicular transport between membrane-enclosed compartments of the endomembrane system is highly organized. A major outward *secretory pathway* starts with the synthesis of proteins on the ER membrane and their entry into the ER, and it leads through the Golgi apparatus to the cell surface; at the Golgi apparatus, a side branch leads off through endosomes to lysosomes. A major inward *endocytic pathway,* which is responsible for the ingestion and degradation of extracellular molecules, moves materials from the plasma membrane, through endosomes, to lysosomes (**Figure 15–18**).

To function optimally, each transport vesicle that buds off from a compartment must take with it only the proteins appropriate to its destination and must fuse only with the appropriate target membrane. A vesicle carrying cargo from the Golgi apparatus to the plasma membrane, for example, must exclude proteins that are to stay in the Golgi apparatus,

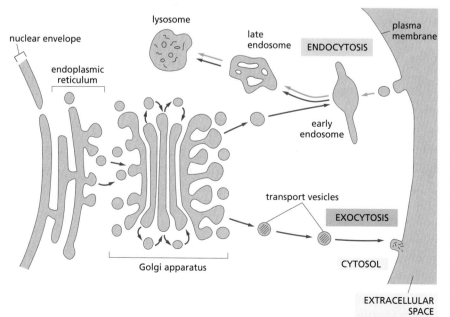

Figure 15–18 Transport vesicles bud from one membrane and fuse with another, carrying membrane components and soluble proteins between compartments of the endomembrane system and the plasma membrane. The membrane of each compartment or vesicle maintains its orientation, so the cytosolic side always faces the cytosol and the noncytosolic side faces the lumen of the compartment or the outside of the cell (see Figure 11–18). The extracellular space and each of the membrane-enclosed compartments (shaded *gray*) communicate with one another by means of transport vesicles, as shown. In the outward secretory pathway (*red* arrows), protein molecules are transported from the ER, through the Golgi apparatus, to the plasma membrane or (via early and late endosomes) to lysosomes. In the inward endocytic pathway (*green* arrows), extracellular molecules are ingested (endocytosed) in vesicles derived from the plasma membrane and are delivered to early endosomes and, usually, on to lysosomes via late endosomes.

and it must fuse only with the plasma membrane and not with any other organelle. While participating in this constant flow of membrane components, each organelle must maintain its own distinct identity, that is, its own distinctive protein and lipid composition. All of these recognition events depend on proteins displayed on the surface of the transport vesicle. As we will see, different types of transport vesicles shuttle between the various organelles, each carrying a distinct set of molecules.

Vesicle Budding Is Driven by the Assembly of a Protein Coat

Vesicles that bud from membranes usually have a distinctive protein coat on their cytosolic surface and are therefore called **coated vesicles**. After budding from its parent organelle, the vesicle sheds its coat, allowing its membrane to interact directly with the membrane to which it will fuse. Cells produce several kinds of coated vesicles, each with a distinctive protein coat. The coat serves at least two functions: it helps shape the membrane into a bud and captures molecules for onward transport.

The best-studied vesicles are those that have an outer coat made of the protein **clathrin**. These *clathrin-coated vesicles* bud from both the Golgi apparatus on the outward secretory pathway and from the plasma membrane on the inward endocytic pathway. At the plasma membrane, for example, each vesicle starts off as a *clathrin-coated pit*. Clathrin molecules assemble into a basketlike network on the cytosolic surface of the membrane, and it is this assembly process that starts shaping the membrane into a vesicle (**Figure 15–19**). A small GTP-binding protein called

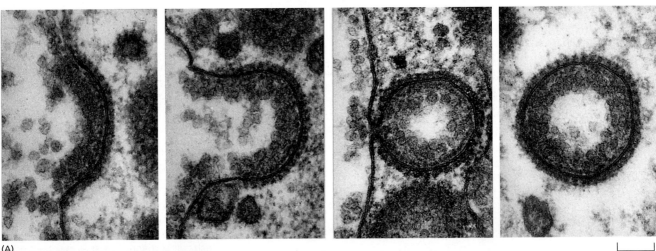

(A)

0.1 μm

Figure 15–19 Clathrin molecules form basketlike cages that help shape membranes into vesicles. (A) Electron micrographs showing the sequence of events in the formation of a clathrin-coated vesicle from a clathrin-coated pit. The clathrin-coated pits and vesicles shown here are unusually large and are being formed at the plasma membrane of a hen oocyte. They are involved in taking up particles made of lipid and protein into the oocyte to form yolk. (B) Electron micrograph showing numerous clathrin-coated pits and vesicles budding from the inner surface of the plasma membrane of cultured skin cells. (A, courtesy of M.M. Perry and A.B. Gilbert, *J. Cell Sci.* 39:257–272, 1979. With permission from The Company of Biologists Ltd; B, from J. Heuser, *J. Cell Biol.* 84:560–583, 1980. With permission from Rockefeller University Press.)

(B)

0.2 μm

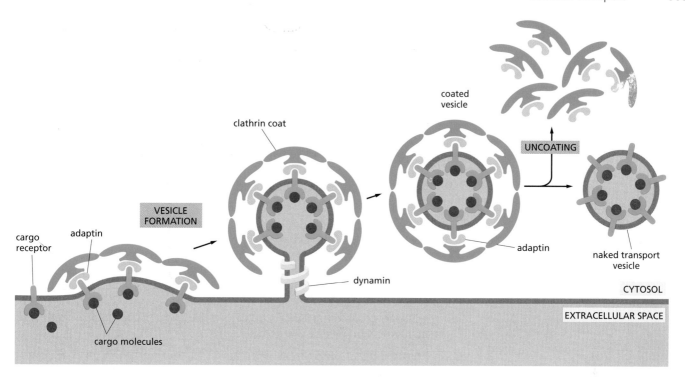

Figure 15–20 **Clathrin-coated vesicles transport selected cargo molecules.** Here, as in Figure 15–19, the vesicles are shown budding from the plasma membrane. Cargo receptors, with their bound cargo molecules, are captured by adaptins, which also bind clathrin molecules to the cytosolic surface of the budding vesicle (Movie 15.5). Dynamin proteins assemble around the neck of budding vesicles; once assembled, the dynamin molecules—which are monomeric GTPases (discussed in Chapter 16)—hydrolyze their bound GTP and, with the help of other proteins recruited to the neck (not shown), pinch off the vesicle. After budding is complete, the coat proteins are removed, and the naked vesicle can fuse with its target membrane. Functionally similar coat proteins are found in other types of coated vesicles.

dynamin assembles as a ring around the neck of each deeply invaginated coated pit. Together with other proteins recruited to the neck of the vesicle, the dynamin causes the ring to constrict, thereby pinching off the vesicle from its parent membrane. Other kinds of transport vesicles, with different coat proteins, are also involved in vesicular transport. They form in a similar way and carry their own characteristic sets of molecules between the endoplasmic reticulum, the Golgi apparatus, and the plasma membrane. But how does a transport vesicle select its particular cargo? The mechanism is best understood for clathrin-coated vesicles.

Clathrin itself plays no part in choosing specific molecules for transport. This is the function of a second class of coat proteins called *adaptins,* which both secure the clathrin coat to the vesicle membrane and help select cargo molecules for transport. Molecules for onward transport carry specific *transport signals* that are recognized by *cargo receptors* in the Golgi or plasma membrane. Adaptins help capture specific cargo molecules by trapping the cargo receptors that bind them. In this way, a selected set of cargo molecules, bound to their specific receptors, is incorporated into the lumen of each newly formed clathrin-coated vesicle (**Figure 15–20**). There are different types of adaptins: the adaptins that bind cargo receptors in the plasma membrane, for example, are not the same as those that bind cargo receptors in the Golgi apparatus, reflecting the differences in the cargo molecules to be transported from each of these sources.

Another class of coated vesicles, called *COP-coated vesicles* (COP being shorthand for "coat protein"), is involved in transporting molecules between the ER and the Golgi apparatus and from one part of the Golgi apparatus to another (**Table 15–4**).

Vesicle Docking Depends on Tethers and SNAREs

After a transport vesicle buds from a membrane, it must find its way to its correct destination to deliver its contents. Often, the vesicle is actively transported by motor proteins that move along cytoskeletal fibers, as discussed in Chapter 17.

TABLE 15–4 SOME TYPES OF COATED VESICLES			
Type of Coated Vesicle	Coat Proteins	Origin	Destination
Clathrin-coated	clathrin + adaptin 1	Golgi apparatus	lysosome (via endosomes)
Clathrin-coated	clathrin + adaptin 2	plasma membrane	endosomes
COP-coated	COP proteins	ER Golgi cisterna Golgi apparatus	Golgi apparatus Golgi cisterna ER

Once a transport vesicle has reached its target, it must recognize and dock with its specific organelle. Only then can the vesicle membrane fuse with the target membrane and unload the vesicle's cargo. The impressive specificity of vesicular transport suggests that each type of transport vesicle in the cell displays molecular markers on its surface that identify the vesicle according to its origin and cargo. These markers must be recognized by complementary receptors on the appropriate target membrane, including the plasma membrane.

The identification process depends on a diverse family of monomeric GTPases called **Rab proteins**. Specific Rab proteins on the surface of each type of vesicle are recognized by corresponding *tethering proteins* on the cytosolic surface of the target membrane. Each organelle and each type of transport vesicle carries a unique combination of Rab proteins, which serve as molecular markers for each membrane type. The coding system of matching Rab and tethering proteins helps to ensure that transport vesicles fuse only with the correct membrane.

Additional recognition is provided by a family of transmembrane proteins called **SNAREs**. Once the tethering protein has captured a vesicle by grabbing hold of its Rab protein, SNAREs on the vesicle (called v-SNAREs) interact with complementary SNAREs on the target membrane (called t-SNAREs), firmly docking the vesicle in place (**Figure 15–21**).

The same SNAREs involved in docking also play a central role in catalyzing the membrane fusion required for a transport vesicle to deliver its cargo. Fusion not only delivers the soluble contents of the vesicle into the interior of the target organelle, but it also adds the vesicle membrane to the membrane of the organelle (see Figure 15–21). After vesicle docking,

---

QUESTION 15–5

The budding of clathrin-coated vesicles from eukaryotic plasma membrane fragments can be observed when adaptins, clathrin, and dynamin-GTP are added to the membrane preparation. What would you observe if you omitted (A) adaptins, (B) clathrin, or (C) dynamin? (D) What would you observe if the plasma membrane fragments were from a prokaryotic cell?

Figure 15–21 Rab proteins, tethering proteins, and SNAREs help direct transport vesicles to their target membranes. A filamentous tethering protein on a membrane binds to a Rab protein on the surface of a vesicle. This interaction allows the vesicle to dock on its particular target membrane. A v-SNARE on the vesicle then binds to a complementary t-SNARE on the target membrane. Whereas Rab and tethering proteins provide the initial recognition between a vesicle and its target membrane, complementary SNARE proteins ensure that transport vesicles dock at their appropriate target membranes. These SNARE proteins also catalyze the final fusion of the two membranes (see Figure 15–22).

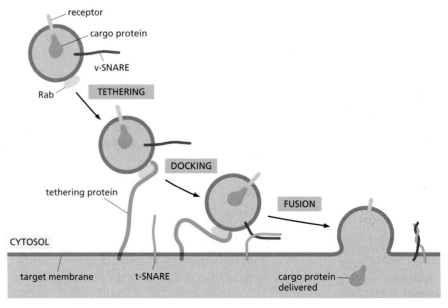

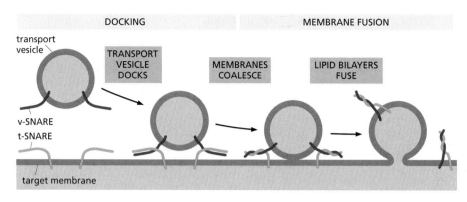

DOCKING

MEMBRANE FUSION

transport vesicle

TRANSPORT VESICLE DOCKS

MEMBRANES COALESCE

LIPID BILAYERS FUSE

v-SNARE
t-SNARE

target membrane

Figure 15–22 Following vesicle docking, SNARE proteins can catalyze the fusion of the vesicle and target membranes. Once appropriately triggered, the tight pairing of v-SNAREs and t-SNAREs draws the two lipid bilayers into close apposition. The force of the SNAREs winding together squeezes out any water molecules that remain trapped between the two membranes, allowing their lipids to flow together to form a continuous bilayer. In a cell, other proteins recruited to the fusion site help to complete the fusion process. After fusion, the SNAREs are pried apart so that they can be used again.

the fusion of a vesicle with its target membrane sometimes requires a special stimulatory signal. Whereas docking requires only that the two membranes come close enough for the SNAREs protruding from the two lipid bilayers to interact, fusion requires a much closer approach: the two bilayers must come within 1.5 nm of each other so that their lipids can intermix. For this close approach, water must be displaced from the hydrophilic surfaces of the membranes—a process that is energetically highly unfavorable and thus prevents membranes from fusing randomly. All membrane fusions in cells must therefore be catalyzed by specialized proteins that assemble to form a fusion complex, which provides the means to cross this energy barrier. The SNARE proteins themselves catalyze the fusion process: once fusion is triggered, the v-SNAREs and t-SNAREs wrap around each other, thereby acting like a winch that pulls the two lipid bilayers into close proximity (**Figure 15–22**).

SECRETORY PATHWAYS

Vesicular traffic is not confined to the interior of the cell. It extends to and from the plasma membrane. Newly made proteins, lipids, and carbohydrates are delivered from the ER, via the Golgi apparatus, to the cell surface by transport vesicles that fuse with the plasma membrane in the process of *exocytosis* (see Figure 15–18). Each molecule that travels along this route passes through a fixed sequence of membrane-enclosed compartments and is often chemically modified en route.

In this section, we follow the outward path of proteins as they travel from the ER, where they are made and modified, through the Golgi apparatus, where they are further modified and sorted, to the plasma membrane. As a protein passes from one compartment to another, it is monitored to check that it has folded properly and assembled with its appropriate partners, so that only correctly built proteins make it to the cell surface. Incorrect assemblies, which are often in the majority, are degraded inside the cell. Quality, it seems, is more important than economy when it comes to the production and transport of proteins via this pathway.

Most Proteins Are Covalently Modified in the ER

Most proteins that enter the ER are chemically modified there. *Disulfide bonds* are formed by the oxidation of pairs of cysteine side chains (see Figure 4–30), a reaction catalyzed by an enzyme that resides in the ER lumen. The disulfide bonds help to stabilize the structure of proteins that will encounter degradative enzymes and changes in pH outside the cell—either after they are secreted or after they are incorporated into the plasma membrane. Disulfide bonds do not form in the cytosol because the environment there is reducing.

Many of the proteins that enter the ER lumen or ER membrane are converted to glycoproteins in the ER by the covalent attachment of short branched oligosaccharide side chains composed of multiple sugars. This process of *glycosylation* is carried out by glycosylating enzymes present in the ER but not in the cytosol. Very few proteins in the cytosol are glycosylated, and those that are have only a single sugar attached to them. The oligosaccharides on proteins can serve various functions. They can protect a protein from degradation, hold it in the ER until it is properly folded, or help guide it to the appropriate organelle by serving as a transport signal for packaging the protein into appropriate transport vesicles. When displayed on the cell surface, oligosaccharides form part of the cell's outer carbohydrate layer or *glycocalyx* (see Figure 11–33) and can function in the recognition of one cell by another.

In the ER, individual sugars are not added one-by-one to the protein to create the oligosaccharide side chain. Instead, a preformed, branched oligosaccharide containing a total of 14 sugars is attached *en bloc* to all proteins that carry the appropriate site for glycosylation. The oligosaccharide is originally attached to a specialized lipid, called *dolichol*, in the ER membrane; it is then transferred to the amino (NH_2) group of an asparagine side chain on the protein, immediately after a target asparagine emerges in the ER lumen during protein translocation (**Figure 15–23**). The addition takes place in a single enzymatic step that is catalyzed by a membrane-bound enzyme (an oligosaccharyl transferase) that has its active site exposed on the lumenal side of the ER membrane—which explains why cytosolic proteins are not glycosylated in this way. A simple sequence of three amino acids, of which the asparagine is one, defines which asparagines in a protein receive the oligosaccharide. Oligosaccharide side chains linked to an asparagine NH_2 group in a protein are said to be *N-linked* and are by far the most common type of linkage found on glycoproteins.

The addition of the 14-sugar oligosaccharide in the ER is only the first step in a series of further modifications before the mature glycoprotein reaches the cell surface. Despite their initial similarity, the *N*-linked

---

QUESTION 15–6

Why might it be advantageous to add a preassembled block of 14 sugar residues to a protein in the ER, rather than building the sugar chains step-by-step on the surface of the protein by the sequential addition of sugars by individual enzymes?

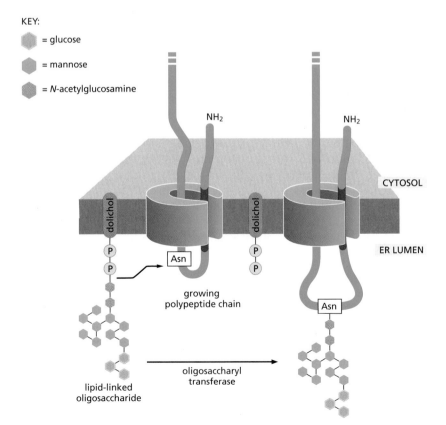

Figure 15–23 Many proteins are glycosylated on asparagines in the ER. When an appropriate asparagine enters the ER lumen, it is glycosylated by addition of a branched oligosaccharide side chain. Each oligosaccharide chain is transferred as an intact unit to the asparagine from a lipid called dolichol, catalyzed by the enzyme oligosaccharyl transferase. Asparagines that are glycosylated are always present in the tripeptide sequences asparagine-X-serine or asparagine-X-threonine, where X can be almost any amino acid.

oligosaccharides on mature glycoproteins are remarkably diverse. All of the diversity results from extensive modification of the original precursor structure shown in Figure 15–23. This oligosaccharide processing begins in the ER and continues in the Golgi apparatus.

Exit from the ER Is Controlled to Ensure Protein Quality

Some proteins made in the ER are destined to function there. They are retained in the ER (and are returned to the ER whenever they escape to the Golgi apparatus) by a C-terminal sequence of four amino acids called an *ER retention signal* (see Table 15–3, p. 494). This retention signal is recognized by a membrane-bound receptor protein in the ER and Golgi apparatus. Most proteins that enter the ER, however, are destined for other locations; they are packaged into transport vesicles that bud from the ER and fuse with the Golgi apparatus.

Exit from the ER is highly selective. Proteins that fail to fold correctly, and dimeric or multimeric proteins that do not assemble properly, are actively retained in the ER by binding to *chaperone proteins* that reside there. The chaperones hold these proteins in the ER until proper folding or assembly occurs. Chaperones prevent misfolded proteins from aggregating, which helps steer proteins along a path toward proper folding (**Figure 15–24** and see Figures 4–9 and 4–10); if proper folding and assembly still fail, the proteins are exported to the cytosol, where they are degraded. Antibody molecules, for example, are composed of four polypeptide chains (see Figure 4–33) that assemble into the complete antibody molecule in the ER. Partially assembled antibodies are retained in the ER until all four polypeptide chains have assembled; any antibody molecule that fails to assemble properly is degraded. In this way, the ER controls the quality of the proteins that it exports to the Golgi apparatus.

Sometimes, however, this quality control mechanism can be detrimental to the organism. For example, the predominant mutation that causes the common genetic disease *cystic fibrosis*, which leads to severe lung damage, produces a plasma-membrane transport protein that is slightly misfolded; even though the mutant protein could function normally as a chloride channel if it reached the plasma membrane, it is retained in the ER, with dire consequences. Thus this devastating disease comes about not because the mutation inactivates an important protein but because the active protein is discarded by the cells before it is given an opportunity to function.

The Size of the ER Is Controlled by the Demand for Protein

Although chaperones help proteins in the ER fold properly and retain those that do not, this quality control system can become overwhelmed. When that happens, misfolded proteins accumulate in the ER. If the buildup is large enough, it triggers a complex program called the **unfolded protein response** (**UPR**). This program prompts the cell to produce more

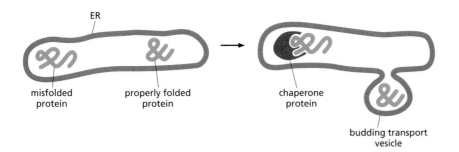

Figure 15–24 Chaperones prevent misfolded or partially assembled proteins from leaving the ER. Misfolded proteins bind to chaperone proteins in the ER lumen and are thus retained there, whereas normally folded proteins are transported in transport vesicles to the Golgi apparatus. If the misfolded proteins fail to refold normally, they are transported back into the cytosol, where they are degraded (not shown).

Figure 15–25 Accumulation of misfolded proteins in the ER lumen triggers an unfolded protein response (UPR). The misfolded proteins are recognized by several types of transmembrane sensor proteins in the ER membrane, each of which activates a different part of the UPR. Some sensors stimulate the production of transcription regulators that activate genes encoding chaperones or other proteins of the ER quality control system. Another sensor also inhibits protein synthesis, reducing the flow of proteins through the ER.

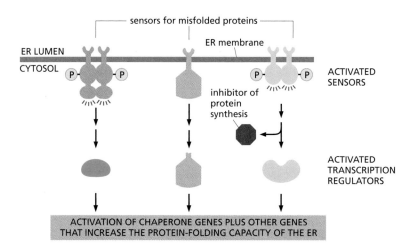

ER, including more chaperones and other proteins concerned with quality control (**Figure 15–25**).

The UPR allows a cell to adjust the size of its ER according to the load of proteins entering the secretory pathway. In some cases, however, even an expanded ER cannot cope, and the UPR directs the cell to self-destruct by undergoing apoptosis. Such a situation may occur in adult-onset diabetes, where tissues gradually become resistant to the effects of insulin. To compensate for this resistance, the insulin-secreting cells in the pancreas produce more and more insulin. Eventually, their ER reaches a maximum capacity, at which point the UPR can trigger cell death. As more insulin-secreting cells are eliminated, the demand on the surviving cells increases, making it more likely that they will die as well, further exacerbating the disease.

Proteins Are Further Modified and Sorted in the Golgi Apparatus

The **Golgi apparatus** is usually located near the cell nucleus, and in animal cells it is often close to the centrosome, a small cytoskeletal structure near the cell center (see Figure 17–12). The Golgi apparatus consists of a collection of flattened, membrane-enclosed sacs called cisternae, which are piled like stacks of pita bread. Each stack contains 3–20 cisternae (**Figure 15–26**). The number of Golgi stacks per cell varies greatly depending on the cell type: some cells contain one large stack, while others contain hundreds of very small ones.

Each Golgi stack has two distinct faces: an entry, or *cis*, face and an exit, or *trans,* face. The *cis* face is adjacent to the ER, while the *trans* face points toward the plasma membrane. The outermost cisterna at each face is connected to a network of interconnected membranous tubes and vesicles (see Figure 15–26A). Soluble proteins and membrane enter the *cis Golgi network* via transport vesicles derived from the ER. The proteins travel through the cisternae in sequence by means of transport vesicles that bud from one cisterna and fuse with the next. Proteins exit from the *trans Golgi network* in transport vesicles destined for either the cell surface or another organelle of the endomembrane system (see Figure 15–18).

Both the *cis* and *trans* Golgi networks are thought to be important for protein sorting: proteins entering the *cis* Golgi network can either move onward through the Golgi stack or, if they contain an ER retention signal, be returned to the ER; proteins exiting from the *trans* Golgi network are sorted according to whether they are destined for lysosomes (via endosomes) or for the cell surface. We discuss some examples of sorting by

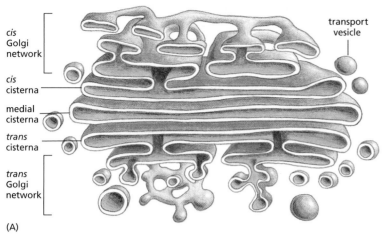

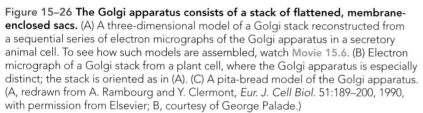

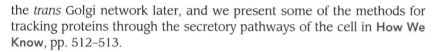

(A)

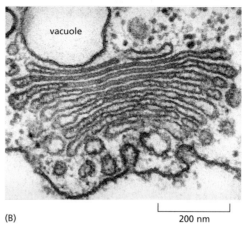

(B)

200 nm

(C)

Figure 15–26 The Golgi apparatus consists of a stack of flattened, membrane-enclosed sacs. (A) A three-dimensional model of a Golgi stack reconstructed from a sequential series of electron micrographs of the Golgi apparatus in a secretory animal cell. To see how such models are assembled, watch Movie 15.6. (B) Electron micrograph of a Golgi stack from a plant cell, where the Golgi apparatus is especially distinct; the stack is oriented as in (A). (C) A pita-bread model of the Golgi apparatus. (A, redrawn from A. Rambourg and Y. Clermont, *Eur. J. Cell Biol.* 51:189–200, 1990, with permission from Elsevier; B, courtesy of George Palade.)

the *trans* Golgi network later, and we present some of the methods for tracking proteins through the secretory pathways of the cell in **How We Know**, pp. 512–513.

Many of the oligosaccharide chains that are added to proteins in the ER (see Figure 15–23) undergo further modifications in the Golgi apparatus. On some proteins, for example, more complex oligosaccharide chains are created by a highly ordered process in which sugars are added and removed by a series of enzymes that act in a rigidly determined sequence as the protein passes through the Golgi stack. As would be expected, the enzymes that act early in the chain of processing events are located in cisternae close to the *cis* face, while enzymes that act late are located in cisternae near the *trans* face.

Secretory Proteins Are Released from the Cell by Exocytosis

In all eukaryotic cells, a steady stream of vesicles buds from the *trans* Golgi network and fuses with the plasma membrane in the process of **exocytosis**. This *constitutive exocytosis pathway* supplies the plasma membrane with newly made lipids and proteins (Movie 15.7), enabling the plasma membrane to expand prior to cell division and refreshing old lipids and proteins in nonproliferating cells. The constitutive pathway also carries soluble proteins to the cell surface to be released to the outside, a process called **secretion**. Some of these proteins remain attached to the cell surface; some are incorporated into the extracellular matrix; still others diffuse into the extracellular fluid to nourish or signal other cells. Entry into the constitutive pathway does not require a particular signal sequence like those that direct proteins to endosomes or back to the ER.

In addition to the constitutive exocytosis pathway, which operates continually in all eukaryotic cells, there is a *regulated exocytosis pathway*, which operates only in cells that are specialized for secretion. Each specialized *secretory cell* produces large quantities of a particular product—such as a hormone, mucus, or digestive enzymes—which is stored in

TRACKING PROTEIN AND VESICLE TRANSPORT

Over the years, biologists have taken advantage of a variety of techniques to untangle the pathways and mechanisms by which proteins are sorted and transported into and out of the cell and its resident organelles. Biochemical, genetic, molecular biological, and microscopic techniques all provide ways to monitor how proteins shuttle from one cell compartment to

another. Some can even track the migration of proteins and transport vesicles in real time in living cells.

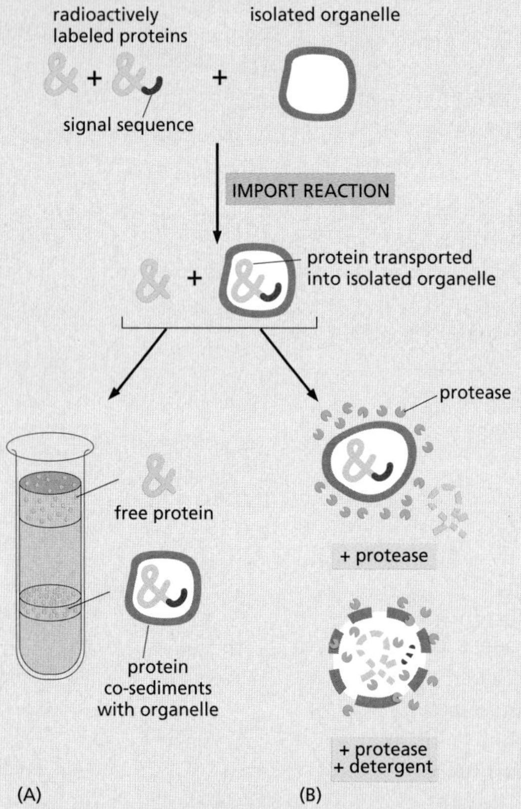

In a tube

A protein bearing a signal sequence can be introduced to a preparation of isolated organelles in a test tube. This mixture can then be tested to see whether the protein is taken up by the organelle. The protein is usually produced *in vitro* by cell-free translation of a purified mRNA encoding the polypeptide; in the process, radioactive amino acids can be used to label the protein so that it will be easy to isolate and to follow. The labeled protein is incubated with a selected organelle and its translocation is monitored by one of several methods (**Figure 15–27**).

Ask a yeast

Movement of proteins between different cell compartments via transport vesicles has been studied extensively using genetic techniques. Studies of mutant yeast cells that are defective for secretion at high temperatures have identified numerous genes involved in carrying proteins from the ER to the cell surface. Many of these mutant genes encode temperature-sensitive proteins (discussed in Chapter 19). These mutant proteins may function normally at 25°C, but, when the yeast cells are shifted to 35°C, the proteins are inactivated. As a result, when researchers raise the temperature, the various proteins destined for secretion instead accumulate inappropriately in the ER, Golgi apparatus, or transport vesicles—depending on the particular mutation (**Figure 15–28**).

At the movies

The most commonly used method for tracking a protein as it moves throughout the cell involves tagging the polypeptide with a fluorescent protein, such as green fluorescent protein (GFP). Using the genetic engineering techniques discussed in Chapter 10, this small protein can be fused to other cell proteins. Fortunately, for many proteins studied, the addition of GFP to one or other end does not perturb the protein's normal function or transport. The movement of a GFP-tagged protein can then be monitored in a living cell with a fluorescence microscope. In 2008, the Nobel Prize in Chemistry was awarded to Martin Chalfie and Roger Tsien for the development and refinement of this technology.

Such GFP fusion proteins are widely used to study the location and movement of proteins in cells (**Figure**

Figure 15–27 Several methods can be used to determine whether a labeled protein bearing a particular signal sequence is transported into a preparation of isolated organelles. (A) The labeled protein with or without a signal sequence is incubated with the organelles, and the preparation is centrifuged. Only those labeled proteins that contained a signal sequence will be transported and therefore will co-fractionate with the organelle. (B) The labeled proteins are incubated with the organelle, and a protease is added to the preparation. A transported protein will be selectively protected from digestion by the organelle membrane; adding a detergent that disrupts the organelle membrane will eliminate that protection, and the transported protein will also be degraded.

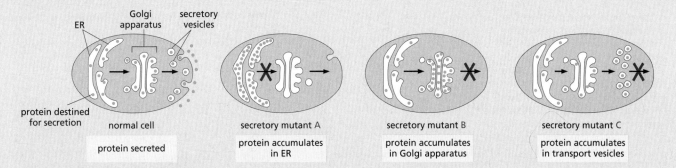

Figure 15–28 Temperature-sensitive mutants have been used to dissect the protein secretory pathway in yeast. Mutations in genes involved at different stages of the transport process, as indicated by the *red* X, result in the accumulation of proteins in the ER, the Golgi apparatus, or transport vesicles.

15–29). GFP fused to a protein that shuttles in and out of the nucleus, for example, can be used to study nuclear transport events. GFP fused to a plasma membrane protein can be used to measure the kinetics of its movement through the secretory pathway. Movies demonstrating the power and beauty of this technique are included on the DVD that accompanies this book (Movie 15.1, Movie 15.7, Movie 15.8, and Movie 15.11).

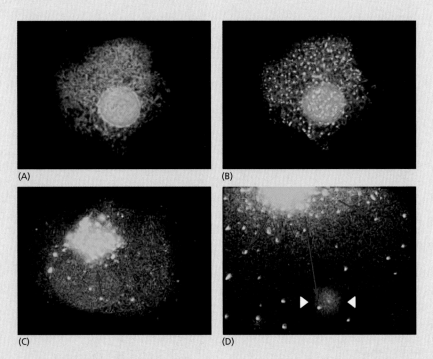

Figure 15–29 Tagging a protein with GFP allows the resulting fusion protein to be tracked throughout the cell. In this experiment, GFP is fused to a viral coat protein and expressed in cultured animal cells. In an infected cell, the viral protein moves through the secretory pathway from the ER to the cell surface, where the virus particles are assembled. *Red arrows* indicate the direction of protein movement. The viral coat protein used in this experiment contains a mutation that allows export from the ER only at a low temperature. (A) At high temperatures, the fusion protein labels the ER. (B) As the temperature is lowered, the GFP fusion protein rapidly accumulates at ER exit sites. (C) The fusion protein then moves to the Golgi apparatus. (D) Finally, the fusion protein is delivered to the plasma membrane, shown here in a more close-up view. The halo between the two *white arrowheads* marks the spot where a single vesicle has fused, allowing the fusion protein to incorporate into the plasma membrane. These images are stills taken from Movie 15.7. (A–D, courtesy of Jennifer Lippincott-Schwartz.)

Figure 15–30 In secretory cells, the regulated and constitutive pathways of exocytosis diverge in the *trans* Golgi network. Many soluble proteins are continually secreted from the cell by the constitutive secretory pathway, which operates in all eukaryotic cells (Movie 15.8). This pathway also continually supplies the plasma membrane with newly synthesized lipids and proteins. Specialized secretory cells have, in addition, a regulated exocytosis pathway by which selected proteins in the *trans* Golgi network are diverted into secretory vesicles, where the proteins are concentrated and stored until an extracellular signal stimulates their secretion. It is unclear how these special aggregates of secretory proteins (*red*) are segregated into secretory vesicles. Secretory vesicles have unique proteins in their membranes; perhaps some of these proteins act as receptors for secretory protein aggregates in the *trans* Golgi network.

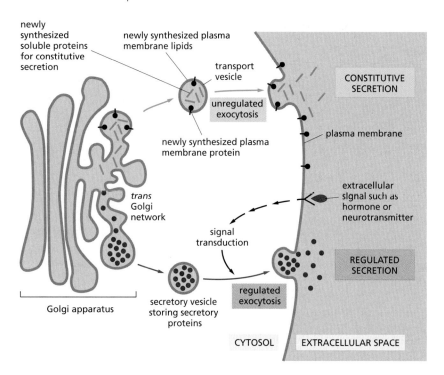

secretory vesicles for later release. These vesicles, which are part of the endomembrane system, bud off from the *trans* Golgi network and accumulate near the plasma membrane. There they wait for the extracellular signal that will stimulate them to fuse with the plasma membrane and release their contents to the cell exterior by exocytosis (**Figure 15–30**). An increase in blood glucose, for example, signals insulin-producing endocrine cells in the pancreas to secrete the hormone (**Figure 15–31**).

Proteins destined for regulated secretion are sorted and packaged in the *trans* Golgi network. Proteins that travel by this pathway have special surface properties that cause them to aggregate with one another under the ionic conditions (acidic pH and high Ca^{2+}) that prevail in the *trans* Golgi network. The aggregated proteins are packaged into secretory vesicles, which pinch off from the network and await a signal instructing them to fuse with the plasma membrane. Proteins secreted by the constitutive pathway, on the other hand, do not aggregate and are therefore carried automatically to the plasma membrane by the transport vesicles of the constitutive pathway. Selective aggregation has another function: it allows secretory proteins to be packaged into secretory vesicles at concentrations much higher than the concentration of the unaggregated protein in the Golgi lumen. This increase in concentration can reach

Figure 15–31 Secretory vesicles store insulin in a pancreatic β cell. The electron micrograph shows the release of insulin into the extracellular space in response to an increase in glucose levels in the blood. The insulin in each secretory vesicle is stored in a highly concentrated, aggregated form. After secretion, the insulin aggregates dissolve rapidly in the blood. (Courtesy of Lelio Orci, from L. Orci, J.D. Vassali, and A. Perrelet, *Sci. Am.* 259:85–94, 1988. With permission from Scientific American.)

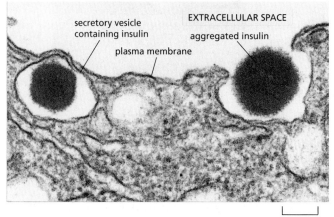

0.2 μm

200-fold, enabling secretory cells to release large amounts of the protein promptly when triggered to do so (see Figure 15–30).

When a secretory vesicle or transport vesicle fuses with the plasma membrane and discharges its contents by exocytosis, its membrane becomes part of the plasma membrane. Although this should greatly increase the surface area of the plasma membrane, it does so only transiently because membrane components are removed from other regions of the surface by endocytosis almost as fast as they are added by exocytosis. This removal returns both the lipids and the proteins of the vesicle membrane to the Golgi network, where they can be used again.

ENDOCYTIC PATHWAYS

Eukaryotic cells are continually taking up fluid, as well as large and small molecules, by the process of **endocytosis**. Specialized cells are also able to internalize large particles and even other cells. The material to be ingested is progressively enclosed by a small portion of the plasma membrane, which first buds inward and then pinches off to form an intracellular *endocytic vesicle*. The ingested materials, including the membrane components, are delivered to *endosomes*, from which they can be recycled to the plasma membrane or sent to lysosomes for digestion. The metabolites generated by digestion are transferred directly out of the lysosome into the cytosol, where they can be used by the cell.

Two main types of endocytosis are distinguished on the basis of the size of the endocytic vesicles formed. *Pinocytosis* ("cellular drinking") involves the ingestion of fluid and molecules via small pinocytic vesicles (<150 nm in diameter). *Phagocytosis* ("cellular eating") involves the ingestion of large particles, such as microorganisms and cell debris, via large vesicles called *phagosomes* (generally >250 nm in diameter). Whereas all eukaryotic cells are continually ingesting fluid and molecules by pinocytosis, large particles are ingested mainly by specialized *phagocytic cells*.

In this final section, we trace the endocytic pathway from the plasma membrane to lysosomes. We start by considering the uptake of large particles by phagocytosis.

Specialized Phagocytic Cells Ingest Large Particles

The most dramatic form of endocytosis, **phagocytosis**, was first observed more than a hundred years ago. In protozoa, phagocytosis is a form of feeding: these unicellular eukaryotes ingest large particles such as bacteria by taking them up into phagosomes (**Movie 15.9**). The phagosomes then fuse with lysosomes, where the food particles are digested. Few cells in multicellular organisms are able to ingest large particles efficiently. In the animal gut, for example, large particles of food have to be broken down to individual molecules by extracellular enzymes before they can be taken up by the absorptive cells lining the gut.

Nevertheless, phagocytosis is important in most animals for purposes other than nutrition. **Phagocytic cells**—including *macrophages*, which are widely distributed in tissues, and other white blood cells, such as *neutrophils*—defend us against infection by ingesting invading microorganisms. To be taken up by macrophages or neutrophils, particles must first bind to the phagocytic cell surface and activate one of a variety of surface receptors. Some of these receptors recognize antibodies, the proteins that help protect us against infection by binding to the surface of microorganisms. Binding of antibody-coated bacteria to these receptors induces the phagocytic cell to extend sheetlike projections of the plasma membrane, called *pseudopods*, that engulf the bacterium

Figure 15–32 Specialized phagocytic cells can ingest other cells. (A) Electron micrograph of a phagocytic white blood cell (a neutrophil) ingesting a bacterium, which is in the process of dividing. (B) Scanning electron micrograph showing a macrophage engulfing a pair of red blood cells. The *red* arrows point to the edges of the pseudopods that the phagocytic cells are extending like collars to envelop their prey. (A, courtesy of Dorothy F. Bainton; B, courtesy of Jean Paul Revel.)

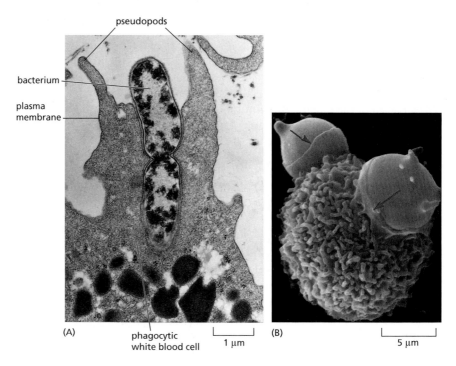

pseudopods

bacterium

plasma membrane

(A)

phagocytic white blood cell

1 μm

(B)

5 μm

(**Figure 15–32A**) and fuse at their tips to form a phagosome. The phagosome then fuses with a lysosome, and the microbe is destroyed. Some pathogenic bacteria have evolved tricks for subverting the system: for example, *Mycobacterium tuberculosis*, the agent responsible for tuberculosis, can inhibit the membrane fusion that unites the phagosome with a lysosome. Instead of being destroyed, the engulfed organism survives and multiplies within the macrophage. Although the mechanism is not completely understood, identifying the proteins involved will provide therapeutic targets for drugs that could restore the macrophages' ability to eliminate the infection.

Phagocytic cells also play an important part in scavenging dead and damaged cells and cell debris. Macrophages, for example, ingest more than 10^{11} of your worn-out red blood cells each day (**Figure 15–32B**).

Fluid and Macromolecules Are Taken Up by Pinocytosis

Eukaryotic cells continually ingest bits of their plasma membrane, along with small amounts of extracellular fluid, in the process of **pinocytosis.** The rate at which plasma membrane is internalized in **pinocytic vesicles** varies from cell type to cell type, but it is usually surprisingly large. A macrophage, for example, swallows 25% of its own volume of fluid each hour. This means that it removes 3% of its plasma membrane each minute, or 100% in about half an hour. Pinocytosis occurs more slowly in fibroblasts, but more rapidly in some phagocytic amoebae. Because a cell's total surface area and volume remain unchanged during this process, as much membrane is being added to the cell surface by exocytosis as is being removed by endocytosis (see Figure 15–18). It is not known how eukaryotic cells maintain this remarkable balance.

Pinocytosis is carried out mainly by the clathrin-coated pits and vesicles that we discussed earlier (see Figures 15–19 and 15–20). After they pinch off from the plasma membrane, clathrin-coated vesicles rapidly shed their coat and fuse with an endosome. Extracellular fluid is trapped in the coated pit as it invaginates to form a coated vesicle, and so substances dissolved in the extracellular fluid are internalized and delivered to endosomes. This fluid intake by clathrin-coated and other types of pinocytic vesicles is generally balanced by fluid loss during exocytosis.

Receptor-mediated Endocytosis Provides a Specific Route into Animal Cells

Pinocytosis, as just described, is indiscriminate. The endocytic vesicles simply trap any molecules that happen to be present in the extracellular fluid and carry them into the cell. In most animal cells, however, pinocytosis via clathrin-coated vesicles also provides an efficient pathway for taking up specific macromolecules from the extracellular fluid. These macromolecules bind to complementary receptors on the cell surface and enter the cell as receptor–macromolecule complexes in clathrin-coated vesicles. This process, called **receptor-mediated endocytosis**, provides a selective concentrating mechanism that increases the efficiency of internalization of particular macromolecules more than 1000-fold compared with ordinary pinocytosis, so that even minor components of the extracellular fluid can be taken up in large amounts without taking in a correspondingly large volume of extracellular fluid. An important example of receptor-mediated endocytosis is the ability of animal cells to take up the cholesterol they need to make new membrane.

Cholesterol is a lipid that is extremely insoluble in water (see Figure 11–7). It is transported in the bloodstream bound to protein in the form of particles called low-density lipoproteins, or LDL. Cholesterol-containing LDLs, which are secreted by the liver, bind to receptors located on cell surfaces causing the receptor–LDL complexes to be ingested by receptor-mediated endocytosis and delivered to endosomes. The interior of endosomes is more acidic than the surrounding cytosol or the extracellular fluid, and in this acidic environment the LDL dissociates from its receptor: the receptors are returned in transport vesicles to the plasma membrane for reuse, while the LDL is delivered to lysosomes. In the lysosomes, the LDL is broken down by hydrolytic enzymes. The cholesterol is released and escapes into the cytosol, where it is available for new membrane synthesis (**Figure 15–33**).

This pathway for cholesterol uptake is disrupted in individuals who inherit a defective gene encoding the LDL receptor protein. In some cases, the receptors are missing; in others, they are present but nonfunctional. In either case, because the cells are deficient in taking up LDL, cholesterol accumulates in the blood and predisposes the individuals to develop atherosclerosis. Unless they take drugs (statins) to reduce their

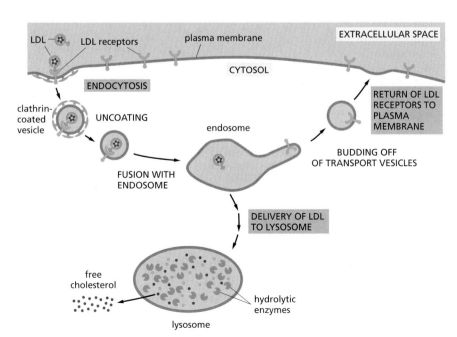

Figure 15–33 **LDL enters cells via receptor-mediated endocytosis.** LDL binds to LDL receptors on the cell surface and is internalized in clathrin-coated vesicles. The vesicles lose their coat and then fuse with endosomes. In the acidic environment of the endosome, LDL dissociates from its receptors. The LDL ends up in lysosomes, where it is degraded to release free cholesterol (*red dots*), but the LDL receptors are returned to the plasma membrane via transport vesicles to be used again (Movie 15.10). For simplicity, only one LDL receptor is shown entering the cell and returning to the plasma membrane. Whether it is occupied or not, an LDL receptor typically makes one round trip into the cell and back every 10 minutes, making a total of several hundred trips in its 20-hour life-span.

blood cholesterol, they will likely die at an early age of heart attacks, which result from cholesterol clogging the coronary arteries that supply the heart muscle.

Receptor-mediated endocytosis is also used to take up many other essential metabolites, such as vitamin B_{12} and iron, that cells cannot take up by the processes of transmembrane transport discussed in Chapter 12. Vitamin B_{12} and iron are both required, for example, for the synthesis of hemoglobin, which is the major protein in red blood cells; they enter immature red blood cells as part of a complex with their respective receptor proteins. Many cell-surface receptors that bind extracellular signal molecules are also ingested by this pathway: some are recycled to the plasma membrane for reuse, whereas others are degraded in lysosomes. Unfortunately, receptor-mediated endocytosis can also be exploited by viruses: the influenza virus, which causes the flu, gains entry into cells in this way.

Endocytosed Macromolecules Are Sorted in Endosomes

Because most extracellular material taken up by pinocytosis is rapidly delivered to **endosomes**, it is possible to visualize the endosomal compartment by incubating living cells in fluid containing an electron-dense marker that will show up when viewed in an electron microscope. When examined in this way, the endosomal compartment reveals itself to be a complex set of connected membrane tubes and larger vesicles. Two sets of endosomes can be distinguished in such loading experiments: the marker molecules appear first in *early endosomes*, just beneath the plasma membrane; 5–15 minutes later, they show up in *late endosomes,* closer to the nucleus (see Figure 15–18). Early endosomes mature gradually into late endosomes as they fuse with each other or with a preexisting late endosome (Movie 15.11). The interior of the endosome compartment is kept acidic (pH 5–6) by an ATP-driven H^+ (proton) pump in the endosomal membrane that pumps H^+ into the endosome lumen from the cytosol.

The endosomal compartment acts as the main sorting station in the inward endocytic pathway, just as the *trans* Golgi network serves this function in the outward secretory pathway. The acidic environment of the endosome plays a crucial part in the sorting process by causing many (but not all) receptors to release their bound cargo. The routes taken by receptors once they have entered an endosome differ according to the type of receptor: (1) most are returned to the same plasma membrane domain from which they came, as is the case for the LDL receptor discussed earlier; (2) some travel to lysosomes, where they are degraded; and (3) some proceed to a different domain of the plasma membrane, thereby transferring their bound cargo molecules across the cell from one extracellular space to another, a process called *transcytosis* (**Figure 15–34**).

QUESTION 15–8

Iron (Fe) is an essential trace metal that is needed by all cells. It is required, for example, for synthesis of the heme groups and iron-sulfur centers that are part of the active site of many proteins involved in electron-transfer reactions; it is also required in hemoglobin, the main protein in red blood cells. Iron is taken up by cells by receptor-mediated endocytosis. The iron-uptake system has two components: a soluble protein called transferrin, which circulates in the bloodstream; and a transferrin receptor—a transmembrane protein that, like the LDL receptor in Figure 15–33, is continually endocytosed and recycled to the plasma membrane. Fe ions bind to transferrin at neutral pH but not at acidic pH. Transferrin binds to the transferrin receptor at neutral pH only when it has an Fe ion bound, but it binds to the receptor at acidic pH even in the absence of bound iron. From these properties, describe how iron is taken up, and discuss the advantages of this elaborate scheme.

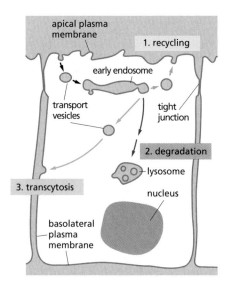

Figure 15–34 The fate of receptor proteins following their endocytosis depends on the type of receptor. Three pathways from the endosomal compartment in an epithelial cell are shown. Receptors that are not specifically retrieved from early endosomes follow the pathway from the endosomal compartment to lysosomes, where they are degraded. Retrieved receptors are returned either to the same plasma membrane domain from which they came (*recycling*) or to a different domain of the plasma membrane (*transcytosis*). Tight junctions separate the apical and basolateral plasma membranes preventing their resident receptor proteins from diffusing from one domain to another. If the ligand that is endocytosed with its receptor stays bound to the receptor in the acidic environment of the endosome, it will follow the same pathway as the receptor; otherwise it will be delivered to lysosomes for degradation.

Cargo proteins that remain bound to their receptors share the fate of their receptors. Cargo that dissociates from receptors in the endosome is doomed to destruction in lysosomes, along with most of the contents of the endosome lumen. Late endosomes contain some lysosomal enzymes, so digestion of cargo proteins and other macromolecules begins in the endosome and continues as the endosome gradually matures into a lysosome: once it has digested most of its ingested contents, the endosome takes on the dense, rounded appearance characteristic of a mature, "classical" lysosome.

Lysosomes Are the Principal Sites of Intracellular Digestion

Many extracellular particles and molecules ingested by cells end up in **lysosomes**, which are membranous sacs of hydrolytic enzymes that carry out the controlled intracellular digestion of both extracellular materials and worn-out organelles. They contain about 40 types of hydrolytic enzymes, including those that degrade proteins, nucleic acids, oligosaccharides, and lipids. All of these enzymes are optimally active in the acidic conditions (pH ~5) maintained within lysosomes. The membrane of the lysosome normally keeps these destructive enzymes out of the cytosol (whose pH is about 7.2), but the enzymes' acid dependence protects the contents of the cytosol against damage even if some of them should escape.

Like all other intracellular organelles, the lysosome not only contains a unique collection of enzymes but also has a unique surrounding membrane. The lysosomal membrane contains transporters that allow the final products of the digestion of macromolecules, such as amino acids, sugars, and nucleotides, to be transferred to the cytosol; from there, they can be either excreted or utilized by the cell. The membrane also contains an ATP-driven H$^+$ pump, which, like the ATPase in the endosome membrane, pumps H$^+$ into the lysosome, thereby maintaining its contents at an acidic pH (**Figure 15–35**). Most of the lysosomal membrane proteins are unusually highly glycosylated; the sugars, which cover much of the protein surfaces facing the lumen, protect the proteins from digestion by the lysosomal proteases.

The specialized digestive enzymes and membrane proteins of the lysosome are synthesized in the ER and transported through the Golgi apparatus to the *trans* Golgi network. While in the ER and the *cis* Golgi network, the enzymes are tagged with a specific phosphorylated sugar group (mannose 6-phosphate), so that when they arrive in the *trans* Golgi network they can be recognized by an appropriate receptor, the mannose 6-phosphate receptor. This tagging permits the lysosomal enzymes to be sorted and packaged into transport vesicles, which bud off and deliver their contents to lysosomes via endosomes (see Figure 15–18).

Depending on their source, materials follow different paths to lysosomes. We have seen that extracellular particles are taken up into phagosomes, which fuse with lysosomes, and that extracellular fluid and macromolecules are taken up into smaller endocytic vesicles, which deliver their contents to lysosomes via endosomes.

Cells have an additional pathway that supplies materials to lysosomes; this pathway, called **autophagy**, is used to degrade obsolete parts of the cell—the cell literally eats itself. In electron micrographs of liver cells, for example, one often sees lysosomes digesting mitochondria, as well as other organelles. The process begins with the enclosure of the organelle by a double membrane, creating an *autophagosome*, which then fuses with a lysosome (**Figure 15–36**). It is still debated where these membrane fragments originate, or how specific cell components are marked

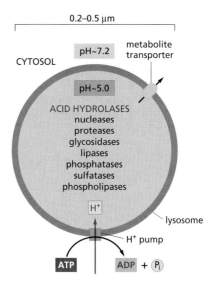

Figure 15–35 A lysosome contains a large variety of hydrolytic enzymes, which are only active under acidic conditions. The lumen of the lysosome is maintained at an acidic pH by an ATP-driven H$^+$ pump in the membrane that hydrolyzes ATP to pump H$^+$ into the lumen.

Figure 15–36 Materials destined for degradation in lysosomes follow different pathways to the lysosome. Each pathway leads to the intracellular digestion of materials derived from a different source. Early endosomes, phagosomes, and autophagosomes can fuse with either lysosomes or late endosomes, both of which contain acid-dependent hydrolytic enzymes.

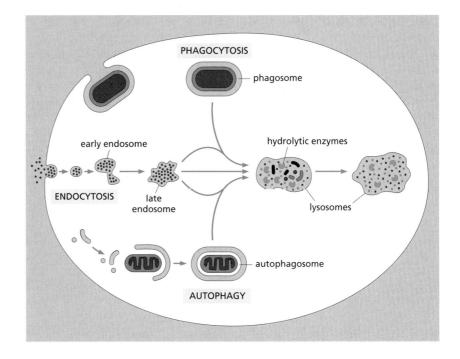

for such destruction, but autophagy of organelles and cytosolic proteins increases when eukaryotic cells are starved or when they remodel themselves extensively during development. The amino acids generated by this cannibalistic form of digestion can then be recycled to allow continued protein synthesis.

ESSENTIAL CONCEPTS

- Eukaryotic cells contain many membrane-enclosed organelles, including a nucleus, an endoplasmic reticulum (ER), a Golgi apparatus, lysosomes, endosomes, mitochondria, chloroplasts (in plant cells), and peroxisomes. The ER, Golgi apparatus, peroxisomes, endosomes, and lysosomes are all part of the *endomembrane system*.

- Most organelle proteins are made in the cytosol and transported into the organelle where they function. Sorting signals in the amino acid sequence guide the proteins to the correct organelle; proteins that function in the cytosol have no such signals and remain where they are made.

- Nuclear proteins contain nuclear localization signals that help direct their active transport from the cytosol into the nucleus through nuclear pores, which penetrate the double-membrane nuclear envelope. The proteins are transported in their fully folded conformation.

- Most mitochondrial and chloroplast proteins are made in the cytosol and are then transported into the organelles by protein translocators in their membranes. The proteins are unfolded during the transport process.

- The ER makes most of the cell's lipids and many of its proteins. The proteins are made by ribosomes that are directed to the ER by a signal-recognition particle (SRP) in the cytosol that recognizes an ER signal sequence on the growing polypeptide chain. The ribosome–SRP complex binds to a receptor on the ER membrane, which passes the ribosome to a protein translocator that threads the growing polypeptide across the ER membrane through a translocation channel.

- Water-soluble proteins destined for secretion or for the lumen of an organelle of the endomembrane system pass completely into the ER lumen, while transmembrane proteins destined for either the membrane of these organelles or for the plasma membrane remain anchored in the lipid bilayer by one or more membrane-spanning α helices.

- In the ER lumen, proteins fold up, assemble with their protein partners, form disulfide bonds, and become decorated with oligosaccharide chains.

- Exit from the ER is an important quality-control step; proteins that either fail to fold properly or fail to assemble with their normal partners are retained in the ER by chaperone proteins, which prevent their aggregation and help them fold; proteins that still fail to fold or assemble are transported to the cytosol, where they are degraded.

- Excessive accumulation of misfolded proteins triggers an unfolded protein response that expands the ER, increases its capacity to fold new proteins properly, and reduces protein synthesis.

- Protein transport from the ER to the Golgi apparatus and from the Golgi apparatus to other destinations is mediated by transport vesicles that continually bud off from one membrane and fuse with another, a process called vesicular transport.

- Budding transport vesicles have distinctive coat proteins on their cytosolic surface; the assembly of the coat helps drive both the budding process and the incorporation of cargo receptors, with their bound cargo molecules, into the forming vesicle.

- Coated vesicles rapidly lose their protein coat, enabling them to dock and then fuse with a particular target membrane; docking and fusion are mediated by proteins on the surface of the vesicle and target membrane, including Rab and SNARE proteins.

- The Golgi apparatus receives newly made proteins from the ER; it modifies their oligosaccharides, sorts the proteins, and dispatches them from the *trans* Golgi network to the plasma membrane, lysosomes (via endosomes), or secretory vesicles.

- In all eukaryotic cells, transport vesicles continually bud from the *trans* Golgi network and fuse with the plasma membrane; this process of constitutive exocytosis delivers proteins to the cell surface for secretion and incorporates lipids and proteins into the plasma membrane.

- Specialized secretory cells also have a regulated exocytosis pathway, in which molecules concentrated and stored in secretory vesicles are released from the cell by exocytosis when the cell is signaled to secrete.

- Cells ingest fluid, molecules, and sometimes even particles by endocytosis, in which regions of plasma membrane invaginate and pinch off to form endocytic vesicles.

- Much of the material that is endocytosed is delivered to endosomes, which mature into lysosomes, in which the material is degraded by hydrolytic enzymes; most of the components of the endocytic vesicle membrane, however, are recycled in transport vesicles back to the plasma membrane for reuse.

KEY TERMS	
autophagy	peroxisome
chaperone protein	phagocytic cell
clathrin	phagocytosis
coated vesicle	pinocytosis
endocytosis	Rab protein
endomembrane system	receptor-mediated endocytosis
endoplasmic reticulum (ER)	rough endoplasmic reticulum
endosome	secretion
exocytosis	secretory vesicle
Golgi apparatus	signal sequence
lysosome	SNARE
membrane-enclosed organelle	transport vesicle
nuclear envelope	unfolded protein response (UPR)
nuclear pore	vesicular transport

QUESTIONS

QUESTION 15–9

Which of the following statements are correct? Explain your answers.

A. Ribosomes are cytoplasmic structures that, during protein synthesis, become linked by an mRNA molecule to form polyribosomes.

B. The amino acid sequence Leu-His-Arg-Leu-Asp-Ala-Gln-Ser-Lys-Leu-Ser-Ser is a signal sequence that directs proteins to the ER.

C. All transport vesicles in the cell must have a v-SNARE protein in their membrane.

D. Transport vesicles deliver proteins and lipids to the cell surface.

E. If the delivery of prospective lysosomal proteins from the *trans* Golgi network to the late endosomes were blocked, lysosomal proteins would be secreted by the constitutive secretion pathways shown in Figure 15–30.

F. Lysosomes digest only substances that have been taken up by cells by endocytosis.

G. *N*-linked sugar chains are found on glycoproteins that face the cell surface, as well as on glycoproteins that face the lumen of the ER, *trans* Golgi network, and mitochondria.

QUESTION 15–10

Some proteins shuttle back and forth between the nucleus and the cytosol. They need a nuclear export signal to get out of the nucleus. How do you suppose they get into the nucleus?

QUESTION 15–11

Influenza viruses are surrounded by a membrane that contains a fusion protein, which is activated by acidic pH. Upon activation, the protein causes the viral membrane to fuse with cell membranes. An old folk remedy against flu recommends that one should spend a night in a horse's stable. Odd as it may sound, there is a rational explanation for this advice. Air in stables contains ammonia (NH_3) generated by bacteria in the horse's urine. Sketch a diagram showing the pathway (in detail) by which flu virus enters cells, and speculate how NH_3 may protect cells from virus infection. (Hint: NH_3 can neutralize acidic solutions by the reaction $NH_3 + H^+ \rightarrow NH_4^+$.)

QUESTION 15–12

Consider the v-SNAREs that direct transport vesicles from the *trans* Golgi network to the plasma membrane. They, like all other v-SNAREs, are membrane proteins that are integrated into the membrane of the ER during their biosynthesis and are then carried by transport vesicles to their destination. Thus, transport vesicles budding from the ER contain at least two kinds of v-SNAREs—those that target the vesicles to the *cis* Golgi cisternae, and those that are in transit to the *trans* Golgi network to be packaged in different transport vesicles destined for the plasma membrane. (A) Why might this be a problem? (B) Suggest possible ways in which the cell might solve it.

QUESTION 15–13

A particular type of *Drosophila* mutant becomes paralyzed when the temperature is raised. The mutation affects the structure of dynamin, causing it to be inactivated at the higher temperature. Indeed, the function of dynamin was discovered by analyzing the defect in these mutant fruit flies. The complete paralysis at the elevated temperature suggests that synaptic transmission between nerve and muscle cells (discussed in Chapter 12) is blocked. Suggest why signal transmission at a synapse might require dynamin. On the basis of your hypothesis, what would you expect to see in electron micrographs of synapses of flies that were exposed to the elevated temperature?

QUESTION 15–14

Edit each of the following statements, if required, to make them true: "Because nuclear localization sequences are not cleaved off by proteases following protein import into the nucleus, they can be reused to import nuclear proteins after mitosis, when cytosolic and nuclear proteins have become intermixed. This is in contrast to ER signal sequences, which are cleaved off by a signal peptidase once they reach the lumen of the ER. ER signal sequences cannot therefore be reused to import ER proteins after mitosis, when cytosolic and ER proteins have become intermixed; these ER proteins must therefore be degraded and resynthesized."

QUESTION 15–15

Consider a protein that contains an ER signal sequence at its N-terminus and a nuclear localization sequence in its middle. What do you think the fate of this protein would be? Explain your answer.

QUESTION 15–16

Compare and contrast protein import into the ER and into the nucleus. List at least two major differences in the mechanisms, and speculate why the ER mechanism might not work for nuclear import and vice versa.

QUESTION 15–17

During mitosis, the nuclear envelope breaks down and intranuclear proteins completely intermix with cytosolic proteins. Is this consistent with the evolutionary scheme proposed in Figure 15–3?

QUESTION 15–18

A protein that inhibits certain proteolytic enzymes (proteases) is normally secreted into the bloodstream by liver cells. This inhibitor protein, antitrypsin, is absent from the bloodstream of patients who carry a mutation that results in a single amino acid change in the protein. Antitrypsin deficiency causes a variety of severe problems, particularly in lung tissue, because of the uncontrolled activity of proteases. Surprisingly, when the mutant antitrypsin is synthesized in the laboratory, it is as active as the normal antitrypsin at inhibiting proteases. Why, then, does the mutation cause the disease? Think of more than one possibility, and suggest ways in which you could distinguish between them.

QUESTION 15–19

Dr. Outonalimb's claim to fame is her discovery of forgettin, a protein predominantly made by the pineal gland in human teenagers. The protein causes selective short-term unresponsiveness and memory loss when the auditory system receives statements like "Please take out the garbage!" Her hypothesis is that forgettin has a hydrophobic ER signal sequence at its C-terminus that is recognized by an SRP and causes it to be translocated across the ER membrane by the mechanism shown in Figure 15–14. She predicts that the protein is secreted from pineal cells into the bloodstream, from where it exerts its devastating systemic effects. You are a member of the committee deciding whether she should receive a grant for further work on her hypothesis. Critique her proposal, and remember that grant reviews should be polite and constructive.

QUESTION 15–20

Taking the evolutionary scheme in Figure 15–3 one step further, suggest how the Golgi apparatus could have evolved. Sketch a simple diagram to illustrate your ideas. For the Golgi apparatus to be functional, what else would be needed?

QUESTION 15–21

If membrane proteins are integrated into the ER membrane by means of the ER protein translocator (which is itself composed of membrane proteins), how do the first protein translocation channels become incorporated into the ER membrane?

QUESTION 15–22

The sketch in **Figure Q15–22** is a schematic drawing of the electron micrograph shown in the third panel of Figure 15–19A. Name the structures that are labeled in the sketch.

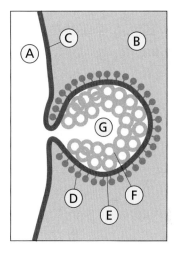

Q15–22

QUESTION 15–23

What would happen to proteins bound for the nucleus if there were insufficient energy to transport them?

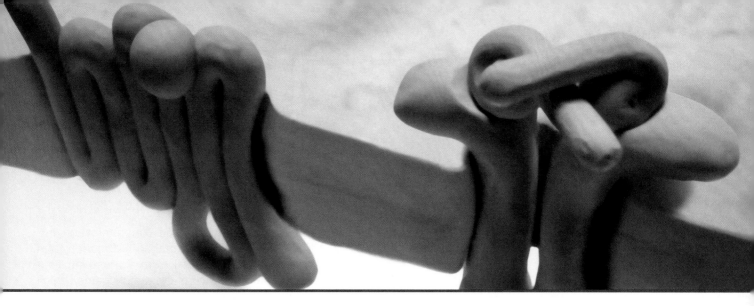

CHAPTER **SIXTEEN**

16

Cell Signaling

Individual cells, like multicellular organisms, need to sense and respond to their environment. A free-living cell—even a humble bacterium—must be able to track down nutrients, tell the difference between light and dark, and avoid poisons and predators. And if such a cell is to have any kind of "social life," it must be able to communicate with other cells. When a yeast cell is ready to mate, for example, it secretes a small protein called a mating factor. Yeast cells of the opposite "sex" detect this chemical mating call and respond by halting their progress through the cell-division cycle and reaching out toward the cell that emitted the signal (**Figure 16–1**).

In a multicellular organism, things are much more complicated. Cells must interpret the multitude of signals they receive from other cells to help coordinate their behaviors. During animal development, for example, cells in the embryo exchange signals to determine which specialized role each cell will adopt, what position it will occupy in the animal, and whether it will survive, divide, or die. Later in life, a large variety of signals coordinates the animal's growth and its day-to-day physiology and behavior. In plants as well, cells are in constant communication with one another. These cell–cell interactions allow the plant to coordinate what happens in its roots, stems, and leaves.

In this chapter, we examine some of the most important mechanisms by which cells send signals and interpret the signals they receive. First, we present an overview of the general principles of cell signaling. We then consider two of the main systems animal cells use to receive and interpret signals, followed by a brief discussion of cell signaling mechanisms in plants. Finally, we consider how extensive and intricate signaling networks interact to control complex behaviors.

GENERAL PRINCIPLES OF CELL SIGNALING

G-PROTEIN-COUPLED RECEPTORS

ENZYME-COUPLED RECEPTORS

Figure 16–1 Yeast cells respond to mating factor. Budding yeast (*Saccharomyces cerevisiae*) cells are normally spherical (A), but when they are exposed to an appropriate mating factor produced by neighboring yeast cells (B), they extend a protrusion toward the source of the factor. (Courtesy of Michael Snyder.)

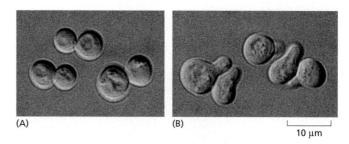

(A) (B) 10 μm

GENERAL PRINCIPLES OF CELL SIGNALING

Information can come in a variety of forms, and communication frequently involves converting the signals that carry that information from one form to another. When you receive a call from a friend on your mobile phone, for instance, the phone converts the radio signals, which travel through the air, into sound waves, which you hear. This process of conversion is called **signal transduction** (**Figure 16–2**).

The signals that pass between cells are simpler than the sorts of messages that humans ordinarily exchange. In a typical communication between cells, the *signaling cell* produces a particular type of *extracellular signal molecule* that is detected by the *target cell*. As in human conversation, most animal cells both send and receive signals, and they can therefore act as both signaling cells and target cells.

Target cells possess proteins called *receptors* that recognize and respond specifically to the signal molecule. Signal transduction begins when the receptor on a target cell receives an incoming extracellular signal and converts it to the *intracellular signaling molecules* that alter cell behavior. Most of this chapter is concerned with signal reception and transduction—the events that cell biologists have in mind when they refer to **cell signaling**. First, however, we look briefly at the different types of extracellular signals that cells send to one another.

Signals Can Act over a Long or Short Range

Cells in multicellular organisms use hundreds of kinds of *extracellular signal molecules* to communicate with one another. The signal molecules can be proteins, peptides, amino acids, nucleotides, steroids, fatty acid derivatives, or even dissolved gases—but they all rely on only a handful of basic styles of communication for getting the message across.

In multicellular organisms, the most "public" style of cell-to-cell communication involves broadcasting the signal throughout the whole body by secreting it into an animal's bloodstream or a plant's sap. Extracellular signal molecules used in this way are called **hormones**, and, in animals, the cells that produce hormones are called *endocrine* cells (**Figure 16–3A**). Part of the pancreas, for example, is an endocrine gland that produces several hormones—including insulin, which regulates glucose uptake in cells all over the body.

Figure 16–2 Signal transduction is the process whereby one type of signal is converted to another. (A) When a mobile telephone receives a radio signal, it converts it into a sound signal; when transmitting a signal, it does the reverse. (B) A target cell converts an extracellular signal molecule (molecule A) into an intracellular signaling molecule (molecule B).

sound OUT

radio signal IN

(A)

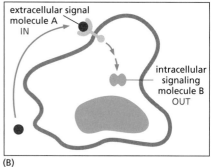

extracellular signal molecule A IN

intracellular signaling molecule B OUT

(B)

Somewhat less public is the process known as *paracrine signaling*. In this case, rather than entering the bloodstream, the signal molecules diffuse locally through the extracellular fluid, remaining in the neighborhood of the cell that secretes them. Thus, they act as **local mediators** on nearby cells (**Figure 16–3B**). Many of the signal molecules that regulate inflammation at the site of an infection or that control cell proliferation in a healing wound function in this way. In some cases, cells can respond to the local mediators that they themselves produce, a form of paracrine communication called *autocrine signaling*; cancer cells sometimes promote their own survival and proliferation in this way.

Neuronal signaling is a third form of cell communication. Like endocrine cells, nerve cells (neurons) can deliver messages over long distances. In the case of neuronal signaling, however, a message is not broadcast widely but is instead delivered quickly and specifically to individual target cells through private lines. As described in Chapter 12, the axon of a neuron terminates at specialized junctions (*synapses*) on target cells that can lie far from the neuronal cell body (**Figure 16–3C**). The axons that extend from the spinal cord to the big toe in an adult human, for example, can be more than a meter in length. When activated by signals from the environment or from other nerve cells, a neuron sends electrical impulses racing along its axon at speeds of up to 100 m/sec. On reaching the axon terminal, these electrical signals are converted into a chemical form: each electrical impulse stimulates the nerve terminal to release a pulse of an extracellular signal molecule called a **neurotransmitter**. The neurotransmitter then diffuses across the narrow (<100 nm) gap that separates the membrane of the axon terminal from that of the target cell, reaching its destination in less than 1 msec.

A fourth style of signal-mediated cell-to-cell communication—the most intimate and short-range of all—does not require the release of a secreted molecule. Instead, the cells make direct physical contact through signal

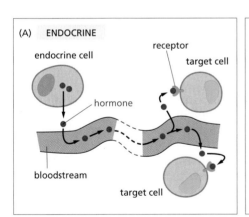
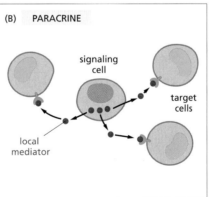
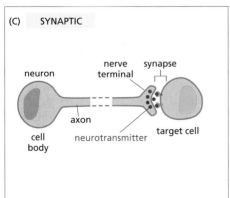
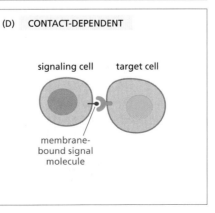

Figure 16–3 **Animal cells use extracellular signal molecules to communicate with one another in various ways.** (A) Hormones produced in endocrine glands are secreted into the bloodstream and are distributed widely throughout the body. (B) Paracrine signals are released by cells into the extracellular fluid in their neighborhood and act locally. (C) Neuronal signals are transmitted electrically along a nerve cell axon. When this electrical signal reaches the nerve terminal, it causes the release of neurotransmitters onto adjacent target cells. (D) In contact-dependent signaling, a cell-surface-bound signal molecule binds to a receptor protein on an adjacent cell. Many of the same types of signal molecules are used for endocrine, paracrine, and neuronal signaling. The crucial differences lie in the speed and selectivity with which the signals are delivered to their targets.

Figure 16–4 Contact-dependent signaling controls nerve-cell production in the fruit fly *Drosophila*. The fly nervous system originates in the embryo from a sheet of epithelial cells. Isolated cells in this sheet begin to specialize as neurons, while their neighbors remain non-neuronal and maintain the structure of the epithelial sheet. The signals that control this process are transmitted via direct cell–cell contacts: each future neuron delivers an inhibitory signal to the cells next to it, deterring them from specializing as neurons too—a process called lateral inhibition. Both the signal molecule (in this case, Delta) and the receptor molecule (called Notch) are transmembrane proteins.

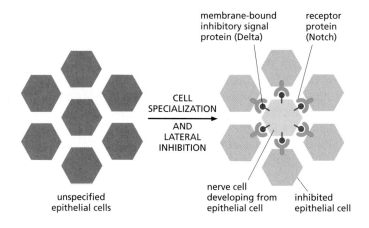

molecules lodged in the plasma membrane of the signaling cell and receptor proteins embedded in the plasma membrane of the target cell (**Figure 16–3D**). During embryonic development, for example, such *contact-dependent signaling* allows adjacent cells that are initially similar to become specialized to form different cell types (**Figure 16–4**).

To contrast these different signaling styles, imagine trying to advertise a potentially stimulating lecture—or a concert or football game. An endocrine signal would be akin to broadcasting the information over a radio station. A localized paracrine signal would be the equivalent of posting a flyer on selected notice boards in your neighborhood. Neuronal signals—long-distance but personal—would be similar to a phone call, a text message, or an e-mail, and contact-dependent signaling would be like a good old-fashioned, face-to-face conversation. In autocrine signaling, you might write a note to remind yourself to attend.

Table 16–1 lists some examples of hormones, local mediators, neurotransmitters, and contact-dependent signal molecules. The action of several of these is discussed in more detail later in the chapter.

Each Cell Responds to a Limited Set of Extracellular Signals, Depending on Its History and Its Current State

A typical cell in a multicellular organism is exposed to hundreds of different signal molecules in its environment. These may be free in the extracellular fluid, embedded in the extracellular matrix in which most cells reside, or bound to the surface of neighboring cells. Each cell must respond very selectively to this mixture of signals, disregarding some and reacting to others, according to the cell's specialized function.

Whether a cell responds to a signal molecule depends first of all on whether it possesses a **receptor** for that signal. Each receptor is usually activated by only one type of signal. Without the appropriate receptor, a cell will be deaf to the signal and will not respond to it. By producing only a limited set of receptors out of the thousands that are possible, a cell restricts the types of signals that can affect it.

Of course, even this restricted set of extracellular signal molecules could change the behavior of a target cell in a large variety of ways. They could alter the cell's shape, movement, metabolism, or gene expression, or some combination of these. As we will see, the signal from a cell-surface receptor is generally conveyed into the target cell interior via a set of intracellular signaling molecules. These molecules act in sequence and ultimately alter the activity of *effector proteins*, those that have some direct effect on the behavior of the target cell. This intracellular relay

QUESTION 16–1

To remain a local stimulus, paracrine signal molecules must be prevented from straying too far from their points of origin. Suggest different ways by which this could be accomplished. Explain your answers.

TABLE 16–1 SOME EXAMPLES OF SIGNAL MOLECULES

Signal Molecule	Site of Origin	Chemical Nature	Some Actions
Hormones			
Adrenaline (epinephrine)	adrenal gland	derivative of the amino acid tyrosine	increases blood pressure, heart rate, and metabolism
Cortisol	adrenal gland	steroid (derivative of cholesterol)	affects metabolism of proteins, carbohydrates, and lipids in most tissues
Estradiol	ovary	steroid (derivative of cholesterol)	induces and maintains secondary female sexual characteristics
Insulin	β cells of pancreas	protein	stimulates glucose uptake, protein synthesis, and lipid synthesis in various cell types
Testosterone	testis	steroid (derivative of cholesterol)	induces and maintains secondary male sexual characteristics
Thyroid hormone (thyroxine)	thyroid gland	derivative of the amino acid tyrosine	stimulates metabolism in many cell types
Local Mediators			
Epidermal growth factor (EGF)	various cells	protein	stimulates epidermal and many other cell types to proliferate
Platelet-derived growth factor (PDGF)	various cells, including blood platelets	protein	stimulates many cell types to proliferate
Nerve growth factor (NGF)	various innervated tissues	protein	promotes survival of certain classes of neurons; promotes their survival and growth of their axons
Histamine	mast cells	derivative of the amino acid histidine	causes blood vessels to dilate and become leaky, helping to cause inflammation
Nitric oxide (NO)	nerve cells; endothelial cells lining blood vessels	dissolved gas	causes smooth muscle cells to relax; regulates nerve-cell activity
Neurotransmitters			
Acetylcholine	nerve terminals	derivative of choline	excitatory neurotransmitter at many nerve–muscle synapses and in central nervous system
γ-Aminobutyric acid (GABA)	nerve terminals	derivative of the amino acid glutamic acid	inhibitory neurotransmitter in central nervous system
Contact-dependent Signal Molecules			
Delta	prospective neurons; various other developing cell types	transmembrane protein	inhibits neighboring cells from becoming specialized in same way as the signaling cell

system and the intracellular effector proteins on which it acts vary from one type of specialized cell to another, so that different types of cells respond to the same signal in different ways. For example, when a heart pacemaker cell is exposed to the neurotransmitter *acetylcholine*, its rate of firing decreases. When a salivary gland is exposed to the same signal, it secretes components of saliva, even though the receptors are the same on both cell types. In skeletal muscle, acetylcholine binds to a different receptor protein, causing the cell to contract (**Figure 16–5**). Thus, the extracellular signal molecule alone is not the message: the information conveyed by the signal depends on how the target cell receives and interprets the signal.

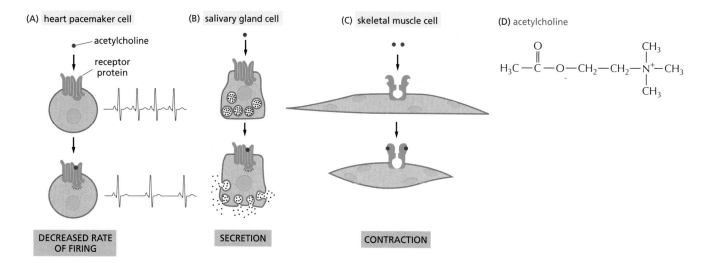

(A) heart pacemaker cell

acetylcholine

receptor protein

DECREASED RATE OF FIRING

(B) salivary gland cell

SECRETION

(C) skeletal muscle cell

CONTRACTION

(D) acetylcholine

$$H_3C-\overset{\overset{\textstyle O}{\|}}{C}-O-CH_2-CH_2-\overset{\overset{\textstyle CH_3}{|}}{\underset{\underset{\textstyle CH_3}{|}}{N^+}}-CH_3$$

Figure 16–5 The same signal molecule can induce different responses in different target cells. Different cell types are configured to respond to the neurotransmitter acetylcholine in different ways. Acetylcholine binds to similar receptor proteins on heart pacemaker cells (A) and salivary gland cells (B), but it evokes different responses in each cell type. Skeletal muscle cells (C) produce a different type of receptor protein for the same signal. (D) For such a versatile molecule, acetylcholine has a fairly simple chemical structure.

A typical cell possesses many sorts of receptors—each present in tens to hundreds of thousands of copies. Such variety makes the cell simultaneously sensitive to many different extracellular signals and allows a relatively small number of signal molecules, used in different combinations, to exert subtle and complex control over cell behavior. A combination of signals can evoke a response that is different from the sum of the effects that each signal would trigger on its own. As we discuss later, this "tailoring" of a cell's response occurs, in part, because the intracellular relay systems activated by the different signals interact. Thus the presence of one signal will often modify the effects of another. One combination of signals might enable a cell to survive; another might drive it to differentiate in some specialized way; and another might cause it to divide. In the absence of any signals, most animal cells are programmed to kill themselves (**Figure 16–6**).

Figure 16–6 An animal cell depends on multiple extracellular signals. Every cell type displays a set of receptor proteins that enables it to respond to a specific set of extracellular signal molecules produced by other cells. These signal molecules work in combinations to regulate the behavior of the cell. As shown here, cells may require multiple signals (*blue* arrows) to survive, additional signals (*red* arrows) to grow and divide, and still other signals (*green* arrows) to differentiate. If deprived of survival signals, most cells undergo a form of cell suicide known as apoptosis (discussed in Chapter 18).

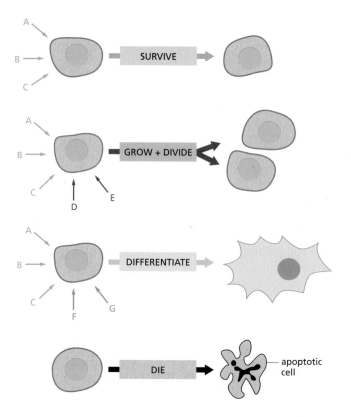

SURVIVE

GROW + DIVIDE

DIFFERENTIATE

DIE

apoptotic cell

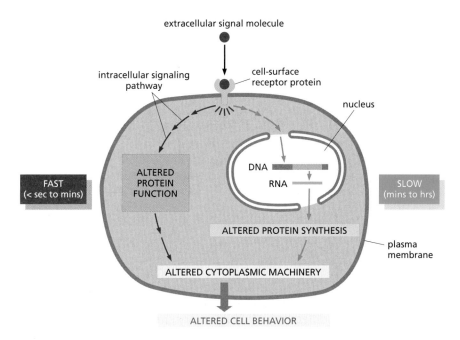

extracellular signal molecule

intracellular signaling pathway

cell-surface receptor protein

nucleus

DNA

RNA

FAST (< sec to mins)

ALTERED PROTEIN FUNCTION

SLOW (mins to hrs)

ALTERED PROTEIN SYNTHESIS

plasma membrane

ALTERED CYTOPLASMIC MACHINERY

ALTERED CELL BEHAVIOR

Figure 16–7 Extracellular signals can act slowly or rapidly. Certain types of cell responses—such as cell differentiation or increased cell growth and division (see Figure 16–6)—involve changes in gene expression and the synthesis of new proteins; they therefore occur relatively slowly. Other responses—such as changes in cell movement, secretion, or metabolism—need not involve changes in gene expression and therefore occur more quickly (see Figure 16–5).

A Cell's Response to a Signal Can Be Fast or Slow

The length of time a cell takes to respond to an extracellular signal can vary greatly, depending on what needs to happen once the message has been received. Some extracellular signals act swiftly: acetylcholine can stimulate a skeletal muscle cell to contract within milliseconds and a salivary gland cell to secrete within a minute or so. Such rapid responses are possible because, in each case, the signal affects the activity of proteins that are already present inside the target cell, awaiting their marching orders.

Other responses take more time. Cell growth and cell division, when triggered by the appropriate signal molecules, can take many hours to execute. This is because the response to these extracellular signals requires changes in gene expression and the production of new proteins (**Figure 16–7**). We will encounter additional examples of both fast and slow responses—and the signal molecules that stimulate them—later in the chapter.

Some Hormones Cross the Plasma Membrane and Bind to Intracellular Receptors

Extracellular signal molecules generally fall into two classes. The first and largest class consists of molecules that are too large or too hydrophilic to cross the plasma membrane of the target cell. They rely on receptors on the surface of the target cell to relay their message across the membrane (**Figure 16–8A**). The second, and smaller, class of signals

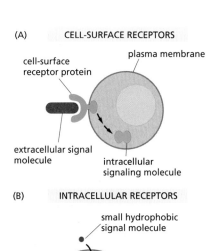

(A)   CELL-SURFACE RECEPTORS

plasma membrane

cell-surface receptor protein

extracellular signal molecule

intracellular signaling molecule

(B)   INTRACELLULAR RECEPTORS

small hydrophobic signal molecule

nucleus

intracellular receptor

Figure 16–8 Extracellular signal molecules bind either to cell-surface receptors or to intracellular enzymes or receptors. (A) Most extracellular signal molecules are large and hydrophilic and are therefore unable to cross the plasma membrane directly; instead, they bind to cell-surface receptors, which in turn generate one or more intracellular signaling molecules in the target cell. (B) Some small, hydrophobic, extracellular signal molecules, by contrast, pass through the target cell's plasma membrane and either activate intracellular enzymes directly or bind to intracellular receptors—in the cytosol or in the nucleus (as shown here)—that then regulate gene transcription or other functions.

consists of molecules that are small enough or hydrophobic enough to pass through the plasma membrane and into the cytosol. Once inside, these signal molecules usually activate intracellular enzymes or bind to intracellular receptor proteins that regulate gene expression (**Figure 16–8B**).

One important category of signal molecules that rely on intracellular receptor proteins is the family of **steroid hormones**—including *cortisol*, *estradiol*, and *testosterone*—and the *thyroid hormones* such as *thyroxine* (**Figure 16–9**). All of these hydrophobic molecules pass through the plasma membrane of the target cell and bind to receptor proteins located in either the cytosol or the nucleus. Both the cytosolic and nuclear receptors are referred to as **nuclear receptors**, because, when activated by hormone binding, they act as transcription regulators in the nucleus (discussed in Chapter 8). In unstimulated cells, nuclear receptors are typically present in an inactive form. When a hormone binds, the receptor undergoes a large conformational change that activates the protein, allowing it to promote or inhibit the transcription of specific target genes (**Figure 16–10**). Each hormone binds to a different nuclear receptor, and each receptor acts at a different set of regulatory sites in DNA (discussed in Chapter 8). Moreover, a given hormone usually regulates different sets of genes in different cell types, thereby evoking different physiological responses in different target cells.

Nuclear receptors and the hormones that activate them have essential roles in human physiology (see Table 16–1, p. 529). Loss of these signaling systems can have dramatic consequences, as illustrated by what happens in individuals who lack the receptor for the male sex hormone testosterone. Testosterone in humans shapes the formation of the external genitalia and influences brain development in the fetus; at puberty, the hormone triggers the development of male secondary sexual characteristics. Some very rare individuals are genetically male—that is, they have both an X and a Y chromosome—but lack the testosterone receptor as a result of a mutation in the corresponding gene; thus, they make testosterone, but their cells cannot respond to it. As a result, these individuals develop as females, which is the path that sexual and brain development would take if no male or female hormones were produced. Such a sex reversal demonstrates the crucial role of the testosterone receptor in sexual development, and it also shows that the receptor is required not just in one cell type to mediate one effect of testosterone, but in many cell types to help produce the whole range of features that distinguish men from women.

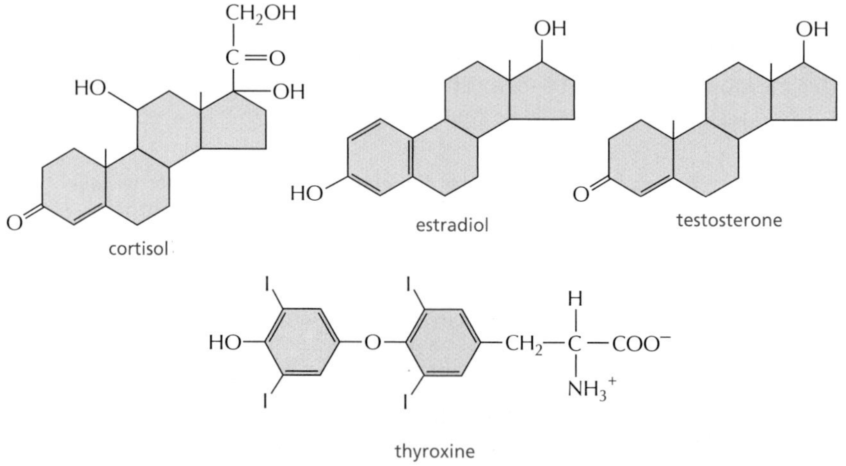

Figure 16–9 Some small hydrophobic hormones bind to intracellular receptors that act as transcription regulators. Although these signal molecules differ in their chemical structures and functions, they all act by binding to intracellular receptor proteins that act as transcription regulators. Their receptors are not identical, but they are evolutionarily related, belonging to the *nuclear receptor superfamily*. The sites of origin and functions of these hormones are given in Table 16–1 (p. 529).

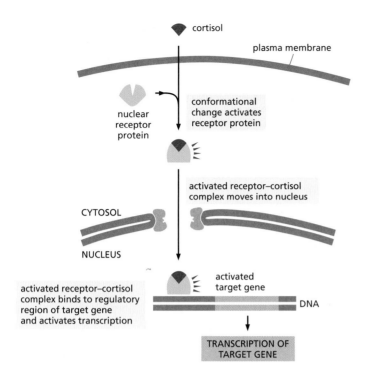

Figure 16–10 The steroid hormone cortisol acts by activating a transcription regulator. Cortisol is one of the hormones produced by the adrenal glands in response to stress. It crosses the plasma membrane and binds to its receptor protein, which is located in the cytosol. The receptor–hormone complex is then transported into the nucleus via the nuclear pores. Cortisol binding activates the receptor protein, which is then able to bind to specific regulatory sequences in DNA and activate (or repress, not shown) the transcription of specific target genes. Whereas the receptors for cortisol and some other steroid hormones are located in the cytosol, those for other steroid hormones and for thyroid hormones are already bound to DNA in the nucleus even in the absence of hormone.

Some Dissolved Gases Cross the Plasma Membrane and Activate Intracellular Enzymes Directly

Steroid hormones and thyroid hormones are not the only extracellular signal molecules that can pass through the plasma membrane. Some dissolved gases can diffuse across the membrane to the cell interior and directly regulate the activity of specific intracellular proteins. This direct approach allows such signals to alter a target cell within a few seconds or minutes. The gas **nitric oxide** (**NO**) acts in this way. NO is synthesized from the amino acid arginine and diffuses readily from its site of synthesis into neighboring cells. The gas acts only locally because it is quickly converted to nitrates and nitrites (with a half-life of about 5–10 seconds) by reacting with oxygen and water outside cells.

Endothelial cells—the flattened cells that line every blood vessel—release NO in response to neurotransmitters secreted by nearby nerve endings. This NO signal causes smooth muscle cells in the adjacent vessel wall to relax, allowing the vessel to dilate, so that blood flows through it more freely (**Figure 16–11**). The effect of NO on blood vessels accounts for the action of nitroglycerine, which has been used for almost 100 years to treat patients with angina—pain caused by inadequate blood flow to the heart muscle. In the body, nitroglycerine is converted to NO, which rapidly relaxes blood vessels, thereby reducing the workload on the heart and decreasing the muscle's need for oxygen-rich blood. Many nerve cells also use NO to signal neighboring cells: NO released by nerve terminals in the penis, for instance, acts as a local mediator to trigger the blood-vessel dilation responsible for penile erection.

Inside many target cells, NO binds to and activates the enzyme *guanylyl cyclase*, stimulating the formation of *cyclic GMP* from the nucleotide GTP (see Figure 16–11B). Cyclic GMP is itself a small intracellular signaling molecule that forms the next link in the NO signaling chain that leads to the cell's ultimate response. The impotence drug Viagra enhances penile erection by blocking the enzyme that degrades cyclic GMP, prolonging the NO signal. Cyclic GMP is very similar in its structure and mechanism

QUESTION 16–2

Consider the structure of cholesterol, a small hydrophobic molecule with a sterol backbone similar to that of three of the hormones shown in Figure 16–9, but possessing fewer polar groups such as –OH, =O, and –COO⁻. If cholesterol were not normally found in cell membranes, could it be used effectively as a hormone if an appropriate intracellular receptor evolved?

(A)

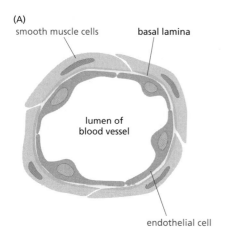

smooth muscle cells   basal lamina

lumen of
blood vessel

endothelial cell

Figure 16–11 Nitric oxide (NO) triggers smooth muscle relaxation in a blood-vessel wall. (A) Simplified drawing shows a cross section of a blood vessel with endothelial cells lining its lumen and smooth muscle cells surrounding the outside of the vessel. (B) The neurotransmitter acetylcholine causes the blood vessel to dilate by binding to receptors on the surface of the endothelial cells, stimulating the cells to make and release NO. The NO then diffuses out of the endothelial cells and into adjacent smooth muscle cells, where it regulates the activity of specific proteins, causing the muscle cells to relax. One key target protein that can be activated by NO in smooth muscle cells is guanylyl cyclase, which catalyzes the production of cyclic GMP from GTP. Note that NO gas is highly toxic when inhaled and should not be confused with nitrous oxide (N_2O), also known as laughing gas.

(B)

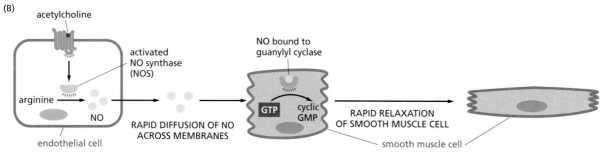

acetylcholine

activated
NO synthase
(NOS)

arginine →

NO

endothelial cell

RAPID DIFFUSION OF NO
ACROSS MEMBRANES

NO bound to
guanylyl cyclase

GTP → cyclic GMP

RAPID RELAXATION
OF SMOOTH MUSCLE CELL

smooth muscle cell

of action to *cyclic AMP*, a much more commonly used intracellular signaling molecule, which we discuss later.

Cell-Surface Receptors Relay Extracellular Signals via Intracellular Signaling Pathways

In contrast to NO and the steroid and thyroid hormones, the vast majority of signal molecules are too large or hydrophilic to cross the plasma membrane of the target cell. These proteins, peptides, and small hydrophilic molecules bind to cell-surface receptor proteins that span the plasma membrane (see Figure 16–8A). Transmembrane receptors detect a signal on the outside and relay the message, in a new form, across the membrane into the interior of the cell.

The receptor protein performs the primary step in signal transduction: it recognizes the extracellular signal and generates new intracellular signals in response (see Figure 16–2B). The resulting intracellular signaling process usually works like a molecular relay race, in which the message is passed "downstream" from one **intracellular signaling molecule** to another, each activating or generating the next signaling molecule in the pathway, until a metabolic enzyme is kicked into action, the cytoskeleton is tweaked into a new configuration, or a gene is switched on or off. This final outcome is called the response of the cell (**Figure 16–12**).

The components of these **intracellular signaling pathways** perform one or more crucial functions (**Figure 16–13**):

1. They can simply *relay* the signal onward and thereby help spread it through the cell.

2. They can *amplify* the signal received, making it stronger, so that a few extracellular signal molecules are enough to evoke a large intracellular response.

3. They can detect signals from more than one intracellular signaling pathway and *integrate* them before relaying a signal onward.

QUESTION 16–3

In principle, how might an intracellular signaling protein amplify a signal as it relays it onward?

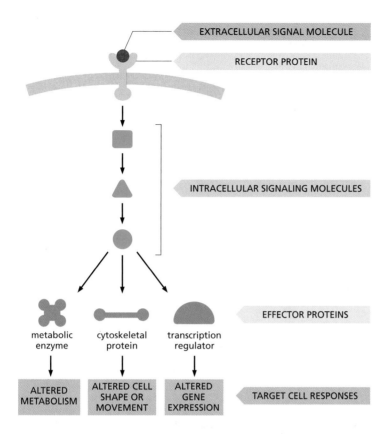

EXTRACELLULAR SIGNAL MOLECULE

RECEPTOR PROTEIN

INTRACELLULAR SIGNALING MOLECULES

EFFECTOR PROTEINS

metabolic enzyme

cytoskeletal protein

transcription regulator

ALTERED METABOLISM

ALTERED CELL SHAPE OR MOVEMENT

ALTERED GENE EXPRESSION

TARGET CELL RESPONSES

Figure 16–12 Many extracellular signals act via cell-surface receptors to change the behavior of the target cell. The receptor protein activates one or more intracellular signaling pathways, each mediated by a series of intracellular signaling molecules, which can be proteins or small messenger molecules; only one pathway is shown. Signaling molecules eventually interact with specific *effector proteins*, altering them to change the behavior of the cell in various ways.

4. They can *distribute* the signal to more than one effector protein, creating branches in the information flow diagram and evoking a complex response.

The steps in a signaling pathway are generally subject to modulation by *feedback regulation*. In positive feedback, a component that lies downstream in the pathway acts on an earlier component in the same pathway to enhance the response to the initial signal; in negative feedback, a downstream component acts to inhibit an earlier component in the pathway to diminish the response to the initial signal (**Figure 16–14**). Such feedback regulation is very common in biological systems and can lead to complex responses: positive feedback can generate all-or-none, switchlike responses, for example, whereas negative feedback can generate responses that oscillate on and off.

Some Intracellular Signaling Proteins Act as Molecular Switches

Many of the key intracellular signaling proteins behave as **molecular switches**: receipt of a signal causes them to toggle from an inactive to an active state. Once activated, these proteins can stimulate—or in other cases suppress—other proteins in the signaling pathway. They then persist in an active state until some other process switches them off again.

The importance of the switching-off process is often underappreciated: imagine the consequences if a signaling pathway that boosts your heart rate were to remain active indefinitely. If a signaling pathway is to recover after transmitting a signal and make itself ready to transmit another, every activated protein in the pathway must be reset to its original, unstimulated state. Thus, for every activation step along the pathway, there has to be an inactivation mechanism. The two are equally important for a signaling pathway to be useful.

Figure 16–13 Intracellular signaling proteins can relay, amplify, integrate, and distribute the incoming signal. In this example, a receptor protein located on the cell surface transduces an extracellular signal into an intracellular signal, which initiates one or more intracellular signaling pathways that relay the signal into the cell interior. Each pathway includes intracellular signaling proteins that can function in one of the various ways shown; some, for example, integrate signals from other intracellular signaling pathways. Many of the steps in the process can be modulated by other molecules or events in the cell (not shown). Note that some proteins in the pathway may be held in close proximity by a scaffold protein, which allows them to be activated at a specific location in the cell and with greater speed, efficiency, and selectivity. We discuss the production and function of small intracellular messenger molecules later in the chapter.

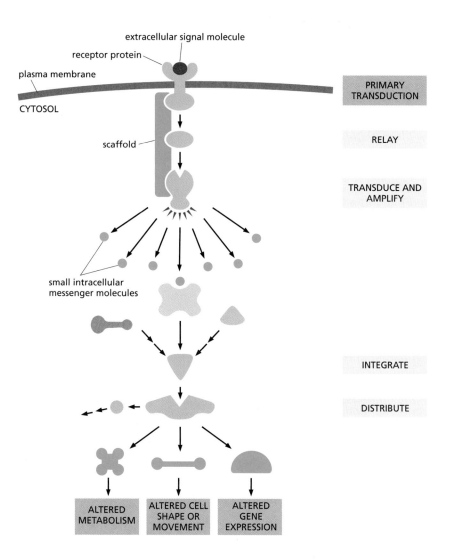

Figure 16–14 Feedback regulation within an intracellular signaling pathway can adjust the response to an extracellular signal. In these simple examples, a downstream protein in two signaling pathways, protein Y, acts to (A) increase via positive feedback or (B) decrease via negative feedback the activity of the protein that activated it.

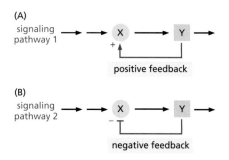

Proteins that act as molecular switches fall mostly into one of two classes. The first—and by far the largest—class consists of proteins that are activated or inactivated by phosphorylation, a chemical modification discussed in Chapter 4 (see Figure 4–42). For these molecules, the switch is thrown in one direction by a **protein kinase**, which covalently attaches a phosphate group onto the switch protein, and in the other direction by a **protein phosphatase**, which takes the phosphate off again (**Figure 16–15A**). The activity of any protein that is regulated by phosphorylation depends—moment by moment—on the balance between the activities of the protein kinases that phosphorylate it and the protein phosphatases that dephosphorylate it.

Many of the switch proteins controlled by phosphorylation are themselves protein kinases, and these are often organized into *phosphorylation cascades*: one protein kinase, activated by phosphorylation, phosphorylates the next protein kinase in the sequence, and so on, transmitting the signal onward and, in the process, amplifying, distributing, and regulating it. Two main types of protein kinases operate in intracellular signaling pathways: the most common are **serine/threonine kinases**, which—as the name implies—phosphorylate proteins on serines or threonines; others are **tyrosine kinases**, which phosphorylate proteins on tyrosines.

The other class of switch proteins involved in intracellular signaling pathways are **GTP-binding proteins**. These toggle between an active and an inactive state depending on whether they have GTP or GDP bound to

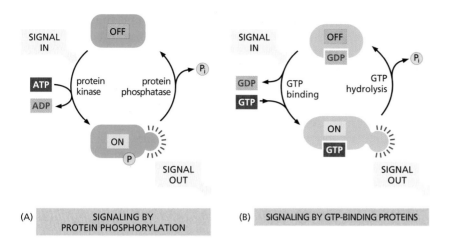

(A) **SIGNALING BY PROTEIN PHOSPHORYLATION**

(B) **SIGNALING BY GTP-BINDING PROTEINS**

Figure 16–15 Many intracellular signaling proteins act as molecular switches. These proteins can be activated—or in some cases inhibited—by the addition or removal of a phosphate group. (A) In one class of switch protein, the phosphate is added covalently by a protein kinase, which transfers the terminal phosphate group from ATP to the signaling protein; the phosphate is then removed by a protein phosphatase. (B) In the other class of switch protein, a GTP-binding protein is activated when it exchanges its bound GDP for GTP (which, in a sense, adds a phosphate to the protein); the protein then switches itself off by hydrolyzing its bound GTP to GDP.

them, respectively (**Figure 16–15B**). Once activated by GTP binding, these proteins have intrinsic GTP-hydrolyzing (*GTPase*) activity, and they shut themselves off by hydrolyzing their bound GTP to GDP.

Two main types of GTP-binding proteins participate in intracellular signaling pathways. Large, *trimeric GTP-binding proteins* (also called *G proteins*) relay messages from *G-protein-coupled receptors*; we discuss this major class of GTP-binding proteins in detail shortly. Other cell-surface receptors rely on small, *monomeric GTPases* to help relay their signals. These monomeric GTP-binding proteins are aided by two sets of regulatory proteins. *Guanine nucleotide exchange factors* (*GEFs*) activate the switch proteins by promoting the exchange of GDP for GTP, and *GTPase-activating proteins* (*GAPs*) turn them off by promoting GTP hydrolysis (**Figure 16–16**).

Cell-Surface Receptors Fall into Three Main Classes

All cell-surface receptor proteins bind to an extracellular signal molecule and transduce its message into one or more intracellular signaling molecules that alter the cell's behavior. Most of these receptors belong to one of three large classes, which differ in the transduction mechanism they use.

1. *Ion-channel-coupled receptors* change the permeability of the plasma membrane to selected ions, thereby altering the membrane potential and, if the conditions are right, producing an electrical current (**Figure 16–17A**).

2. *G-protein-coupled receptors* activate membrane-bound, trimeric GTP-binding proteins (G proteins), which then activate (or inhibit) an enzyme or an ion channel in the plasma membrane, initiating an intracellular signaling cascade (**Figure 16–17B**).

3. *Enzyme-coupled receptors* either act as enzymes or associate with enzymes inside the cell (**Figure 16–17C**); when stimulated, the enzymes can activate a wide variety of intracellular signaling pathways.

The number of different types of receptors in each of these three classes is even greater than the number of extracellular signals that act on them. This is because for many extracellular signal molecules there is more than one type of receptor, and these may belong to different receptor classes. The neurotransmitter acetylcholine, for example, acts on skeletal muscle cells via an ion-channel-coupled receptor, whereas in heart cells it acts through a G-protein-coupled receptor. These two types of

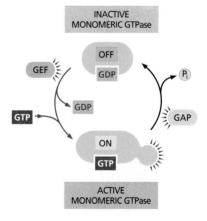

Figure 16–16 The activity of monomeric GTP-binding proteins is controlled by two types of regulatory proteins. Guanine nucleotide exchange factors (GEFs) promote the exchange of GDP for GTP, thereby switching the GTP-binding protein on. GTPase-activating proteins (GAPs) stimulate the hydrolysis of GTP to GDP, thereby switching the GTP-binding protein off.

receptors generate different intracellular signals and thus enable the two types of cells to react to acetylcholine in different ways, increasing contraction in skeletal muscle and decreasing the rate of contractions in heart (see Figure 16–5A and C).

This plethora of cell-surface receptors also provides targets for many foreign substances that interfere with our physiology, from heroin and nicotine to tranquilizers and chili peppers. These substances either block or overstimulate the receptor's natural activity. Many drugs and poisons act in this way (**Table 16–2**), and a large part of the pharmaceutical industry is devoted to producing drugs that will exert a precisely defined effect by binding to a specific type of cell-surface receptor.

Ion-channel–coupled Receptors Convert Chemical Signals into Electrical Ones

Of all the types of cell-surface receptors, **ion-channel-coupled receptors** (also known as transmitter-gated ion channels) function in the simplest and most direct way. As we discuss in detail in Chapter 12, these receptors are responsible for the rapid transmission of signals across synapses in the nervous system. They transduce a chemical signal, in the form of a pulse of secreted neurotransmitter molecules delivered to the outside of the target cell, directly into an electrical signal, in the

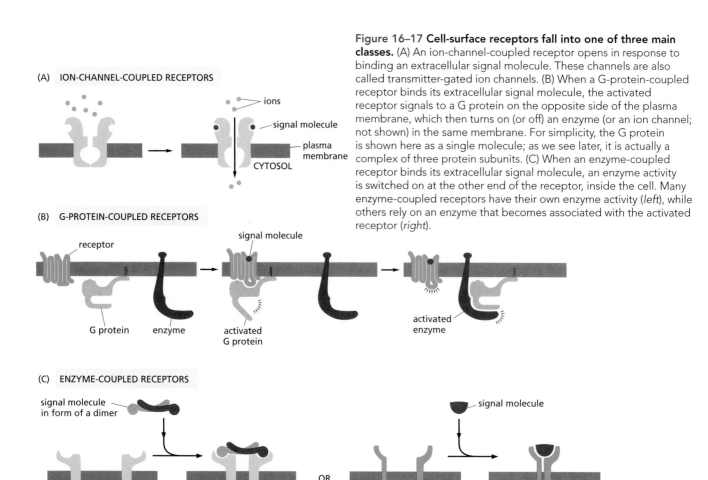

Figure 16–17 **Cell-surface receptors fall into one of three main classes.** (A) An ion-channel-coupled receptor opens in response to binding an extracellular signal molecule. These channels are also called transmitter-gated ion channels. (B) When a G-protein-coupled receptor binds its extracellular signal molecule, the activated receptor signals to a G protein on the opposite side of the plasma membrane, which then turns on (or off) an enzyme (or an ion channel; not shown) in the same membrane. For simplicity, the G protein is shown here as a single molecule; as we see later, it is actually a complex of three protein subunits. (C) When an enzyme-coupled receptor binds its extracellular signal molecule, an enzyme activity is switched on at the other end of the receptor, inside the cell. Many enzyme-coupled receptors have their own enzyme activity (*left*), while others rely on an enzyme that becomes associated with the activated receptor (*right*).

TABLE 16–2 SOME FOREIGN SUBSTANCES THAT ACT ON CELL-SURFACE RECEPTORS

Substance	Normal Signal	Receptor Action	Effect
Barbiturates and benzodiazepines (Valium and Ambien)	γ-aminobutyric acid (GABA)	stimulate GABA-activated ion-channel-coupled receptors	relief of anxiety; sedation
Nicotine	acetylcholine	stimulates acetylcholine-activated ion-channel-coupled receptors	constriction of blood vessels; elevation of blood pressure
Morphine and heroin	endorphins and enkephalins	stimulate G-protein-coupled opiate receptors	analgesia (relief of pain); euphoria
Curare	acetylcholine	blocks acetylcholine-activated ion-channel-coupled receptors	blockage of neuromuscular transmission, resulting in paralysis
Strychnine	glycine	blocks glycine-activated ion-channel-coupled receptors	blockage of inhibitory synapses in spinal cord and brain, resulting in seizures and muscle spasm
Capsaicin	heat	stimulates temperature-sensitive ion-channel-coupled receptors	induces painful, burning sensation; prolonged exposure paradoxically leads to analgesia
Menthol	cold	stimulates temperature-sensitive ion-channel-coupled receptors	in moderate amounts, induces a cool sensation; in higher doses, can cause burning pain

form of a change in voltage across the target cell's plasma membrane (see Figure 12–40). When the neurotransmitter binds, this type of receptor alters its conformation so as to open an ion channel in the plasma membrane, allowing the flow of specific types of ions, such as Na^+, K^+, or Ca^{2+} (see Figure 16–17A and Movie 16.1). Driven by their electrochemical gradients, the ions rush into or out of the cell, creating a change in the membrane potential within milliseconds. This change in potential may trigger a nerve impulse or make it easier (or harder) for other neurotransmitters to do so. As we discuss later, the opening of Ca^{2+} channels has additional important effects, as changes in the Ca^{2+} concentration in the target cell cytosol can profoundly alter the activities of many Ca^{2+}-responsive proteins.

Whereas ion-channel-coupled receptors are especially important in nerve cells and other electrically excitable cells such as muscle cells, G-protein-coupled receptors and enzyme-coupled receptors are important for practically every cell type in the body. Most of the remainder of this chapter deals with these two receptor families and with the signal transduction processes that they use.

G-PROTEIN-COUPLED RECEPTORS

G-protein-coupled receptors (**GPCRs**) form the largest family of cell-surface receptors. There are more than 700 GPCRs in humans, and mice have about 1000 involved in the sense of smell alone. These receptors mediate responses to an enormous diversity of extracellular signal molecules, including hormones, local mediators, and neurotransmitters. The signal molecules are as varied in structure as they are in function: they can be proteins, small peptides, or derivatives of amino acids or fatty acids, and for each one of them there is a different receptor or set of receptors. Because GPCRs are involved in such a large variety of cell processes, they are an attractive target for the development of drugs to treat many disorders. About one-third of all drugs used today work through GPCRs.

Despite the diversity of the signal molecules that bind to them, all GPCRs that have been analyzed have a similar structure: each is made of a single

QUESTION 16–4

The signaling mechanisms used by a steroid-hormone-type nuclear receptor and by an ion-channel-coupled receptor are relatively simple as they have few components. Can they lead to an amplification of the initial signal, and, if so, how?

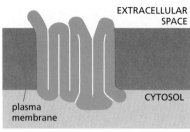

EXTRACTED_LABELS_ABOVE

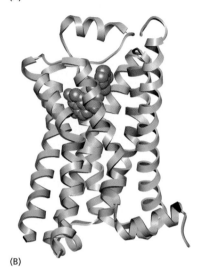

(B)

Figure 16–18 All GPCRs possess a similar structure. The polypeptide chain traverses the membrane as seven α helices. The cytoplasmic portions of the receptor bind to a G protein inside the cell. (A) For receptors that recognize small signal molecules, such as adrenaline or acetylcholine, the ligand usually binds deep within the plane of the membrane to a pocket that is formed by amino acids from several transmembrane segments. (B) Shown here is the structure of a GPCR that binds to adrenaline (*red*). Stimulation of this receptor by adrenaline makes the heart beat faster. Receptors that recognize signal molecules that are proteins usually have a large extracellular domain that, together with some of the transmembrane segments, binds the protein ligand (not shown).

polypeptide chain that threads back and forth across the lipid bilayer seven times (**Figure 16–18**). This superfamily of *seven-pass transmembrane receptor proteins* includes rhodopsin (the light-activated photoreceptor protein in the vertebrate eye), the olfactory (smell) receptors in the vertebrate nose, and the receptors that participate in the mating rituals of single-celled yeasts (see Figure 16–1). Evolutionarily speaking, GPCRs are ancient: even prokaryotes possess structurally similar membrane proteins—such as the bacteriorhodopsin that functions as a light-driven H+ pump (see Figure 11–27). Although they resemble eukaryotic GPCRs, these prokaryotic proteins do not act through G proteins, but are coupled to other signal transduction systems.

We begin this section with a discussion of how G proteins are activated by GPCRs. We then consider how activated G proteins stimulate ion channels and how they regulate membrane-bound enzymes that control the concentrations of small intracellular messenger molecules, including cyclic AMP and Ca2+—which in turn control the activity of important intracellular signaling proteins. We end with a discussion of how light-activated GPCRs in photoreceptors in our eyes enable us to see.

Stimulation of GPCRs Activates G-Protein Subunits

When an extracellular signal molecule binds to a GPCR, the receptor protein undergoes a conformational change that enables it to activate a **G protein** located on the other side of the plasma membrane. To explain how this activation leads to the transmission of a signal, we must first consider how G proteins are constructed and how they function.

There are several varieties of G proteins. Each is specific for a particular set of receptors and for a particular set of target enzymes or ion channels in the plasma membrane. All of these G proteins, however, have a similar general structure and operate in a similar way. They are composed of three protein subunits—α, β, and γ—two of which are tethered to the plasma membrane by short lipid tails. In the unstimulated state, the α subunit has GDP bound to it, and the G protein is idle (**Figure 16–19A**). When an extracellular signal molecule binds to its receptor, the altered receptor activates a G protein by causing the α subunit to decrease its affinity for GDP, which is then exchanged for a molecule of GTP. In some cases, this activation breaks up the G-protein subunits, so that the activated α subunit, clutching its GTP, detaches from the βγ complex, which is also activated (**Figure 16–19B**). The two activated parts of the G protein—the α subunit and the βγ complex—can then each interact directly with target proteins in the plasma membrane, which in turn may relay the signal to other destinations in the cell. The longer these target proteins remain bound to an α or a βγ subunit, the more prolonged the relayed signal will be.

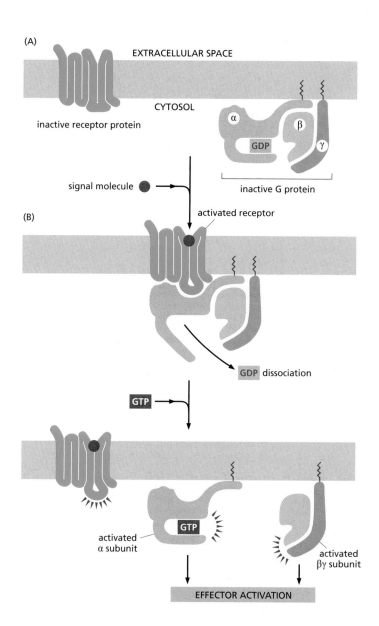

(A)

EXTRACELLULAR SPACE

CYTOSOL

inactive receptor protein

α

β

GDP

γ

signal molecule

inactive G protein

(B)

activated receptor

GDP dissociation

GTP

activated
α subunit

GTP

activated
βγ subunit

EFFECTOR ACTIVATION

Figure 16–19 An activated GPCR activates G proteins by encouraging the α subunit to expel its GDP and pick up GTP. (A) In the unstimulated state, the receptor and the G protein are both inactive. Although they are shown here as separate entities in the plasma membrane, in some cases at least, they are associated in a preformed complex. (B) Binding of an extracellular signal molecule to the receptor changes the conformation of the receptor, which in turn alters the conformation of the bound G protein. The alteration of the α subunit of the G protein allows it to exchange its GDP for GTP. This exchange triggers an additional conformational change that activates both the α subunit and a βγ complex, which dissociate to interact with their preferred target proteins in the plasma membrane (Movie 16.2). The receptor stays active as long as the external signal molecule is bound to it, and it can therefore catalyze the activation of many molecules of G protein. Note that both the α and γ subunits of the G protein have covalently attached lipid molecules (*red*) that help anchor the subunits to the plasma membrane.

The amount of time that the α and βγ subunits remain "switched on"—and hence available to relay signals—also determines how long a response lasts. This timing is controlled by the behavior of the α subunit. The α subunit has an intrinsic GTPase activity, and it eventually hydrolyzes its bound GTP to GDP, returning the whole G protein to its original, inactive conformation (**Figure 16–20**). GTP hydrolysis and inactivation usually occur within seconds after the G protein has been activated. The inactive G protein is now ready to be reactivated by another activated receptor.

Some Bacterial Toxins Cause Disease by Altering the Activity of G Proteins

G proteins demonstrate a general principle of cell signaling mentioned earlier: the mechanisms that shut a signal off are as important as the mechanisms that turn it on (see Figure 16–15B). The shut-off mechanisms also offer as many opportunities for control—and as many dangers for mishap. Consider cholera, for example. The disease is caused by a bacterium that multiplies in the human intestine, where it produces a protein called *cholera toxin*. This protein enters the cells that line the intestine and modifies the α subunit of a G protein called G_S—so named because

Figure 16–20 The G-protein α subunit switches itself off by hydrolyzing its bound GTP to GDP. When an activated α subunit interacts with its target protein, it activates that target protein (or in some cases inactivates it; not shown) for as long as the two remain in contact. Normally the α subunit hydrolyzes its bound GTP to GDP within seconds. This loss of GTP inactivates the α subunit, which dissociates from its target protein and—if the α subunit had separated from the βγ complex (as shown)—reassociates with a βγ complex to re-form an inactive G protein. The G protein is now ready to couple to another activated receptor, as in Figure 16–19B. Both the activated α subunit and the activated βγ complex can interact with target proteins in the plasma membrane. See also Movie 16.2.

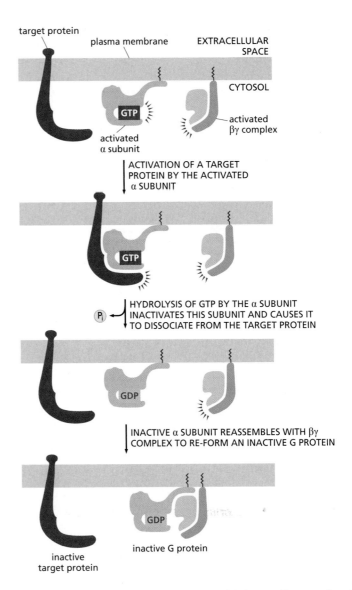

ACTIVATION OF A TARGET PROTEIN BY THE ACTIVATED α SUBUNIT

HYDROLYSIS OF GTP BY THE α SUBUNIT INACTIVATES THIS SUBUNIT AND CAUSES IT TO DISSOCIATE FROM THE TARGET PROTEIN

INACTIVE α SUBUNIT REASSEMBLES WITH βγ COMPLEX TO RE-FORM AN INACTIVE G PROTEIN

QUESTION 16–5

GPCRs activate G proteins by reducing the strength of GDP binding to the G protein. This results in rapid dissociation of bound GDP, which is then replaced by GTP, because GTP is present in the cytosol in much higher concentrations than GDP. What consequences would result from a mutation in the α subunit of a G protein that caused its affinity for GDP to be reduced without significantly changing its affinity for GTP? Compare the effects of this mutation with the effects of cholera toxin.

---

it *stimulates* the enzyme adenylyl cyclase, which we discuss shortly. The modification prevents G_s from hydrolyzing its bound GTP, thus locking the G protein in the active state, in which it continuously stimulates adenylyl cyclase. In intestinal cells, this stimulation causes a prolonged and excessive outflow of Cl⁻ and water into the gut, resulting in catastrophic diarrhea and dehydration. The condition often leads to death unless urgent steps are taken to replace the lost water and ions.

A similar situation occurs in whooping cough (pertussis), a common respiratory infection against which infants are now routinely vaccinated. In this case, the disease-causing bacterium colonizes the lung, where it produces a protein called *pertussis toxin*. This protein alters the α subunit of a different type of G protein, called G_i, because it *inhibits* adenylyl cyclase. In this case, however, modification by the toxin disables the G protein by locking it into its inactive GDP-bound state. Inhibiting G_i, like activating G_s, results in the prolonged and inappropriate activation of adenylyl cyclase, which, in this case, stimulates coughing. Both the diarrhea-producing effects of cholera toxin and the cough-provoking effects of pertussis toxin help the disease-causing bacteria move from host to host.

Some G Proteins Directly Regulate Ion Channels

The target proteins recognized by G-protein subunits are either enzymes or ion channels in the plasma membrane. There are about 20 different types of mammalian G proteins, each activated by a particular set of cell-surface receptors and dedicated to activating a particular set of target proteins. Consequently, the binding of an extracellular signal molecule to a GPCR leads to changes in the activities of a specific subset of the possible target proteins in the plasma membrane, leading to a response that is appropriate for that signal and that type of cell.

We look first at an example of direct G-protein regulation of ion channels. The heartbeat in animals is controlled by two sets of nerves: one speeds the heart up, the other slows it down. The nerves that signal a slow-down in heartbeat do so by releasing acetylcholine (see Figure 16–5A), which binds to a GPCR on the surface of the heart pacemaker cells. This GPCR activates the G protein, G_i. In this case, the $\beta\gamma$ complex binds to the intracellular face of a K^+ channel in the plasma membrane of the pacemaker cell, forcing the ion channel into an open conformation (**Figure 16–21A and B**). This channel opening slows the heart rate by increasing the plasma membrane's permeability to K^+, making it more difficult to electrically activate, as explained in Chapter 12. The original signal is terminated—and the K^+ channel recloses—when the α subunit inactivates itself by hydrolyzing its bound GTP, returning the G protein to its inactive state (**Figure 16–21C**).

Many G Proteins Activate Membrane-bound Enzymes that Produce Small Messenger Molecules

When G proteins interact with ion channels, they cause an immediate change in the state and behavior of the cell. Their interactions with enzymes, in contrast, have consequences that are less rapid and more complex, as they lead to the production of additional intracellular signaling molecules. The two most frequent target enzymes for G proteins are *adenylyl cyclase*, which produces the **small intracellular signaling molecule** *cyclic AMP*, and *phospholipase C*, which generates the small

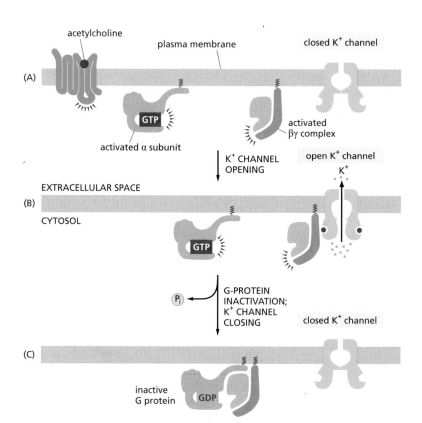

Figure 16–21 **A G_i protein directly couples receptor activation to the opening of K^+ channels in the plasma membrane of heart pacemaker cells.** (A) Binding of the neurotransmitter acetylcholine to its GPCR on the heart cells results in the activation of the G protein, G_i. (B) The activated $\beta\gamma$ complex directly opens a K^+ channel in the plasma membrane, increasing its permeability to K^+ and thereby making the membrane harder to activate and slowing the heart rate. (C) Inactivation of the α subunit by hydrolysis of its bound GTP returns the G protein to its inactive state, allowing the K^+ channel to close.

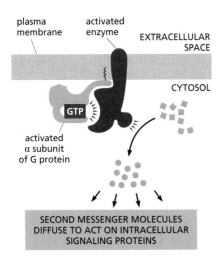

plasma membrane

activated enzyme

EXTRACELLULAR SPACE

CYTOSOL

GTP

activated α subunit of G protein

SECOND MESSENGER MOLECULES DIFFUSE TO ACT ON INTRACELLULAR SIGNALING PROTEINS

Figure 16–22 Enzymes activated by G proteins increase the concentrations of small intracellular signaling molecules. Because each activated enzyme generates many molecules of these second messengers, the signal is greatly amplified at this step in the pathway (see Figure 16–31). The signal is relayed onward by the second messenger molecules, which bind to specific signaling proteins in the cell and influence their activity.

intracellular signaling molecules *inositol trisphosphate* and *diacylglycerol*. Inositol trisphosphate, in turn, promotes the accumulation of cytosolic Ca^{2+}—yet another small intracellular signaling molecule.

Adenylyl cyclase and phospholipase C are activated by different types of G proteins, allowing cells to couple the production of these small intracellular signaling molecules to different extracellular signals. Although the coupling may be either stimulatory or inhibitory—as we saw in our discussion of the actions of cholera toxin and pertussis toxin—we concentrate here on G proteins that stimulate enzyme activity.

The small intracellular signaling molecules generated by these enzymes are often called *small messengers*, or *second messengers*—the "first messengers" being the extracellular signals that activated the enzymes in the first place. Once activated, the enzymes generate large quantities of small messengers, which rapidly diffuse away from their source, thereby amplifying and spreading the intracellular signal (**Figure 16–22**).

Different small messenger molecules produce different responses. We first examine the consequences of an increase in the cytosolic concentration of cyclic AMP. This will take us along one of the main types of signaling pathways that lead from the activation of GPCRs. We then discuss the actions of three other small messenger molecules—inositol trisphosphate, diacylglycerol, and Ca^{2+}—which will lead us along a different signaling route.

The Cyclic AMP Signaling Pathway Can Activate Enzymes and Turn On Genes

Many extracellular signals acting via GPCRs affect the activity of the enzyme **adenylyl cyclase** and thus alter the intracellular concentration of the small messenger molecule **cyclic AMP**. Most commonly, the activated G-protein α subunit switches on the adenylyl cyclase, causing a dramatic and sudden increase in the synthesis of cyclic AMP from ATP (which is always present in the cell). Because it stimulates the cyclase, this G protein is called G_S. To help terminate the signal, a second enzyme, called *cyclic AMP phosphodiesterase*, rapidly converts cyclic AMP to ordinary AMP (**Figure 16–23**). One way that caffeine acts as a stimulant is by inhibiting this phosphodiesterase in the nervous system, blocking cyclic AMP degradation and thereby keeping the concentration of this small messenger high.

Cyclic AMP phosphodiesterase is continuously active inside the cell. Because it eliminates cyclic AMP so quickly, the cytosolic concentration of this small messenger can change rapidly in response to extracellular

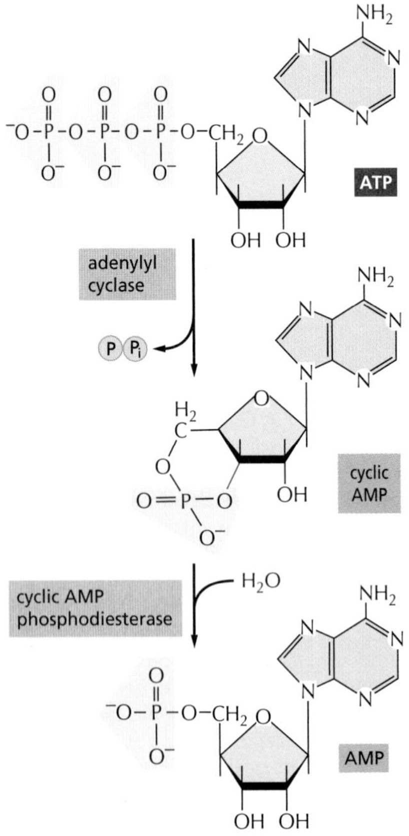

ATP

adenylyl cyclase

P P_i

cyclic AMP

cyclic AMP phosphodiesterase

H_2O

AMP

Figure 16–23 Cyclic AMP is synthesized by adenylyl cyclase and degraded by cyclic AMP phosphodiesterase. Cyclic AMP (abbreviated cAMP) is formed from ATP by a cyclization reaction that removes two phosphate groups from ATP and joins the "free" end of the remaining phosphate group to the sugar part of the AMP molecule (*red* bond). The degradation reaction breaks this new bond, forming AMP.

time 0 sec

time 20 sec

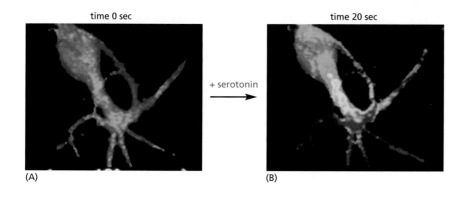

+ serotonin

(A)

(B)

Figure 16–24 Cyclic AMP concentration rises rapidly in response to an extracellular signal. A nerve cell in culture responds to the binding of the neurotransmitter serotonin to a GPCR by synthesizing cyclic AMP. The concentration of intracellular cyclic AMP was monitored by injecting into the cell a fluorescent protein whose fluorescence changes when it binds cyclic AMP. *Blue* indicates a low level of cyclic AMP, *yellow* an intermediate level, and *red* a high level. (A) In the resting cell, the cyclic AMP concentration is about 5×10^{-8} M. (B) Twenty seconds after adding serotonin to the culture medium, the intracellular concentration of cyclic AMP has risen more than twentyfold (to > 10^{-6} M) in the parts of the cell where the serotonin receptors are concentrated. (Courtesy of Roger Tsien.)

signals, rising or falling tenfold in a matter of seconds (**Figure 16–24**). Cyclic AMP is water-soluble, so it can, in some cases, carry the signal throughout the cell, traveling from the site on the membrane where it is synthesized to interact with proteins located in the cytosol, in the nucleus, or on other organelles.

Cyclic AMP exerts most of its effects by activating the enzyme **cyclic-AMP-dependent protein kinase** (**PKA**). This enzyme is normally held inactive in a complex with a regulatory protein. The binding of cyclic AMP to the regulatory protein forces a conformational change that releases the inhibition and unleashes the active kinase. Activated PKA then catalyzes the phosphorylation of particular serines or threonines on specific intracellular proteins, thus altering the activity of these target proteins. In different cell types, different sets of proteins are available to be phosphorylated, which largely explains why the effects of cyclic AMP vary with the type of target cell.

Many kinds of cell responses are mediated by cyclic AMP; a few are listed in **Table 16–3**. As the table shows, different target cells respond very differently to extracellular signals that change intracellular cyclic AMP concentrations. When we are frightened or excited, for example, the adrenal gland releases the hormone *adrenaline*, which circulates in the bloodstream and binds to a class of GPCRs called adrenergic receptors (see Figure 16–18B), which are present on many types of cells. The consequences vary from one cell type to another, but all the cell responses help prepare the body for sudden action. In skeletal muscle, for instance, adrenaline increases intracellular cyclic AMP, causing the breakdown of glycogen—the polymerized storage form of glucose. It does so by activating PKA, which leads to both the activation of an enzyme that promotes glycogen breakdown (**Figure 16–25**) and the inhibition of an enzyme that drives glycogen synthesis. By stimulating glycogen breakdown and

TABLE 16–3 SOME CELL RESPONSES MEDIATED BY CYCLIC AMP		
Extracellular Signal Molecule*	**Target Tissue**	**Major Response**
Adrenaline	heart	increase in heart rate and force of contraction
Adrenaline	skeletal muscle	glycogen breakdown
Adrenaline, glucagon	fat	fat breakdown
Adrenocorticotropic hormone (ACTH)	adrenal gland	cortisol secretion

*Although all of the signal molecules listed here are hormones, some responses to local mediators and to neurotransmitters are also mediated by cyclic AMP.

Figure 16–25 Adrenaline stimulates glycogen breakdown in skeletal muscle cells. The hormone activates a GPCR, which turns on a G protein (G_s) that activates adenylyl cyclase to boost the production of cyclic AMP. The increase in cyclic AMP activates PKA, which phosphorylates and activates an enzyme called phosphorylase kinase. This kinase activates glycogen phosphorylase, the enzyme that breaks down glycogen. Because these reactions do not involve changes in gene transcription or new protein synthesis, they occur rapidly.

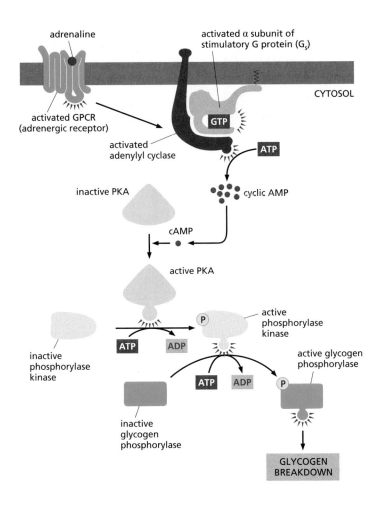

inhibiting its synthesis, the increase in cyclic AMP maximizes the amount of glucose available as fuel for anticipated muscular activity. Adrenaline also acts on fat cells, stimulating the breakdown of fat to fatty acids. These fatty acids can then be exported to fuel ATP production in other cells.

inhibiting its synthesis, the increase in cyclic AMP maximizes the amount of glucose available as fuel for anticipated muscular activity. Adrenaline also acts on fat cells, stimulating the breakdown of fat to fatty acids. These fatty acids can then be exported to fuel ATP production in other cells.

In some cases, the effects of increasing cyclic AMP are rapid; in skeletal muscle, for example, glycogen breakdown occurs within seconds of adrenaline binding to its receptor (see Figure 16–25). In other cases, cyclic AMP responses involve changes in gene expression that take minutes or hours to develop. In these slow responses, PKA typically phosphorylates transcription regulators, which then activate the transcription of selected genes. Thus an increase in cyclic AMP in certain neurons in the brain controls the production of proteins involved in some forms of learning. **Figure 16–26** illustrates a typical cyclic-AMP-mediated pathway from the plasma membrane to the nucleus.

We now turn to the other enzyme-mediated signaling pathway that leads from GPCRs—the pathway that begins with the activation of the membrane-bound enzyme *phospholipase C* and leads to an increase in the small messengers *diacylglycerol*, *inositol trisphosphate*, and Ca^{2+}.

The Inositol Phospholipid Pathway Triggers a Rise in Intracellular Ca^{2+}

Some GPCRs exert their effects via a G protein called G_q, which activates the membrane-bound enzyme **phospholipase C** instead of adenylyl cyclase. Examples of signal molecules that act through phospholipase C are given in **Table 16–4**.

QUESTION 16–6

Explain why cyclic AMP must be broken down rapidly in a cell to allow rapid signaling.

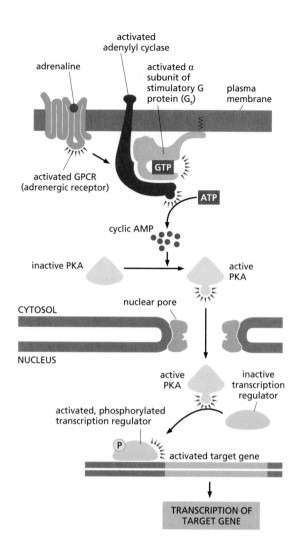

Figure 16–26 A rise in intracellular cyclic AMP can activate gene transcription. Binding of a signal molecule to its GPCR can lead to the activation of adenylyl cyclase and a rise in the concentration of cytosolic cyclic AMP. The increase in cyclic AMP activates PKA, which then moves into the nucleus and phosphorylates specific transcription regulators. Once phosphorylated, these proteins stimulate the transcription of a whole set of target genes (Movie 16.3). This type of signaling pathway controls many processes in cells, ranging from hormone synthesis in endocrine cells to the production of proteins involved in long-term memory in the brain. Activated PKA can also phosphorylate and thereby regulate other proteins and enzymes in the cytosol (as shown in Figure 16–25).

Once activated, phospholipase C propagates the signal by cleaving a lipid molecule that is a component of the plasma membrane. The molecule is an **inositol phospholipid** (a phospholipid with the sugar inositol attached to its head) that is present in small quantities in the cytosolic leaflet of the membrane lipid bilayer (see Figure 11–18). Because of the involvement of this phospholipid, the signaling pathway that begins with the activation of phospholipase C is often referred to as the *inositol phospholipid pathway*. It operates in almost all eukaryotic cells and can regulate a host of different effector proteins.

The action of phospholipase C generates two small messenger molecules: **inositol 1,4,5-trisphosphate (IP₃)** and **diacylglycerol (DAG)**. Both molecules play a crucial part in relaying the signal (**Figure 16–27**).

TABLE 16–4 SOME CELL RESPONSES MEDIATED BY PHOSPHOLIPASE C ACTIVATION

Signal Molecule	Target Tissue	Major Response
Vasopressin (a peptide hormone)	liver	glycogen breakdown
Acetylcholine	pancreas	secretion of amylase (a digestive enzyme)
Acetylcholine	smooth muscle	contraction
Thrombin (a proteolytic enzyme)	blood platelets	aggregation

Figure 16–27 Phospholipase C activates two signaling pathways. Two small messenger molecules are produced when a membrane inositol phospholipid is hydrolyzed by activated phospholipase C. Inositol 1,4,5-trisphosphate (IP₃) diffuses through the cytosol and triggers the release of Ca^{2+} from the ER by binding to and opening special Ca^{2+} channels in the ER membrane. The large electrochemical gradient for Ca^{2+} across this membrane causes Ca^{2+} to rush out of the ER and into the cytosol. Diacylglycerol remains in the plasma membrane and, together with Ca^{2+}, helps activate the enzyme protein kinase C (PKC), which is recruited from the cytosol to the cytosolic face of the plasma membrane. PKC then phosphorylates its own set of intracellular proteins, further propagating the signal. At the start of the pathway, both the α subunit and the βγ subunit of the G protein G_q are involved in activating phospholipase C.

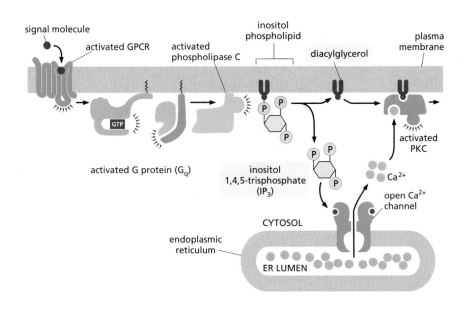

IP_3 is a water-soluble sugar phosphate that is released into the cytosol; there it binds to and opens Ca^{2+} channels that are embedded in the endoplasmic reticulum (ER) membrane. Ca^{2+} stored inside the ER rushes out into the cytosol through these open channels, causing a sharp rise in the cytosolic concentration of free Ca^{2+}, which is normally kept very low. This Ca^{2+} in turn signals to other proteins, as discussed below.

Diacylglycerol is a lipid that remains embedded in the plasma membrane after it is produced by phospholipase C; there, it helps recruit and activate a protein kinase, which translocates from the cytosol to the plasma membrane. This enzyme is called **protein kinase C** (**PKC**) because it also needs to bind Ca^{2+} to become active (see Figure 16–27). Once activated, PKC phosphorylates a set of intracellular proteins that varies depending on the cell type. PKC operates on the same principle as PKA, although the proteins it phosphorylates are different.

A Ca^{2+} Signal Triggers Many Biological Processes

Ca^{2+} has such an important and widespread role as an intracellular messenger that we will digress to consider its functions more generally. A surge in the cytosolic concentration of free Ca^{2+} is triggered by many kinds of cell stimuli, not only those that act through GPCRs. When a sperm fertilizes an egg cell, for example, Ca^{2+} channels open, and the resulting rise in cytosolic Ca^{2+} triggers the egg to start development (**Figure 16–28**); for muscle cells, a signal from a nerve triggers a rise in cytosolic Ca^{2+} that initiates muscle contraction; and in many secretory cells, including nerve cells, Ca^{2+} triggers secretion. Ca^{2+} stimulates all these responses by binding to and influencing the activity of various Ca^{2+}-responsive proteins.

The concentration of free Ca^{2+} in the cytosol of an unstimulated cell is extremely low (10^{-7} M) compared with its concentration in the extracellular fluid (about 10^{-3} M) and in the ER. These differences are maintained by membrane-embedded Ca^{2+} pumps that actively remove Ca^{2+} from the cytosol—sending it either into the ER or across the plasma membrane and out of the cell. As a result, a steep electrochemical gradient of Ca^{2+} exists across both the ER membrane and the plasma membrane (discussed in Chapter 12). When a signal transiently opens Ca^{2+} channels in either of these membranes, Ca^{2+} rushes down its electrochemical gradient into the cytosol, where it triggers changes in Ca^{2+}-responsive proteins in the

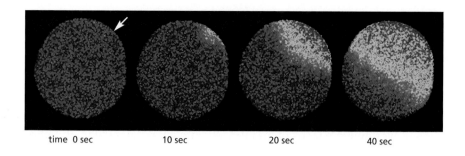

time 0 sec   10 sec   20 sec   40 sec

Figure 16–28 Fertilization of an egg by a sperm triggers an increase in cytosolic Ca^{2+} in the egg. This starfish egg was injected with a Ca^{2+}-sensitive fluorescent dye before it was fertilized. When a sperm enters the egg, a wave of cytosolic Ca^{2+} (*red*)—released from the ER—sweeps across the egg from the site of sperm entry (*arrow*). This Ca^{2+} wave provokes a change in the egg surface, preventing entry of other sperm, and it also initiates embryonic development. To catch this Ca^{2+} wave, go to Movie 16.4. (Courtesy of Stephen A. Stricker.)

cytosol. The same Ca^{2+} pumps that normally operate to keep cytosolic Ca^{2+} concentrations low also help to terminate the Ca^{2+} signal.

The effects of Ca^{2+} in the cytosol are largely indirect, in that they are mediated through the interaction of Ca^{2+} with various kinds of Ca^{2+}-responsive proteins. The most widespread and common of these is **calmodulin**, which is present in the cytosol of all eukaryotic cells that have been examined, including those of plants, fungi, and protozoa. When Ca^{2+} binds to calmodulin, the protein undergoes a conformational change that enables it to interact with a wide range of target proteins in the cell, altering their activities (**Figure 16–29**). One particularly important class of targets for calmodulin is the **Ca^{2+}/calmodulin-dependent protein kinases (CaM-kinases)**. When these kinases are activated by binding to calmodulin complexed with Ca^{2+}, they influence other processes in the cell by phosphorylating selected proteins. In the mammalian brain, for example, a neuron-specific CaM-kinase is abundant at synapses, where it is thought to play an important part in some forms of learning and memory. This CaM-kinase is activated by the pulses of Ca^{2+} signals that occur during neural activity, and mutant mice that lack the kinase show a marked inability to remember where things are.

GPCR-Triggered Intracellular Signaling Cascades Can Achieve Astonishing Speed, Sensitivity, and Adaptability

The steps in the *signaling cascades* associated with GPCRs take a long time to describe, but they often take only seconds to execute. Consider how quickly a thrill can make your heart race (when adrenaline stimulates the GPCRs in your cardiac pacemaker cells), or how fast the smell of food can make your mouth water (through the GPCRs for odors in your

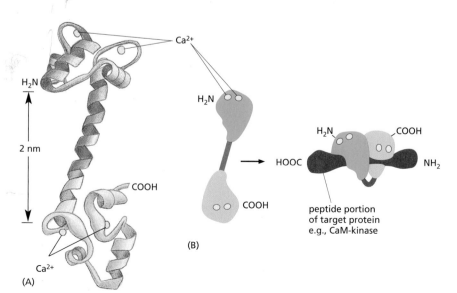

Figure 16–29 Calcium binding changes the shape of the calmodulin protein. (A) Calmodulin has a dumbbell shape, with two globular ends connected by a long α helix. Each end has two Ca^{2+}-binding domains. (B) Simplified representation of the structure, showing the conformational changes in Ca^{2+}/calmodulin that occur when it binds to an isolated segment of a target protein. In this conformation, the α helix jackknifes to surround the target (Movie 16.5). (A, from Y.S. Babu et al., *Nature* 315:37–40, 1985. With permission from Macmillan Publishers Ltd; B, from W.E. Meador, A.R. Means, and F.A. Quiocho, *Science* 257:1251–1255, 1992, and M. Ikura et al., *Science* 256:632–638, 1992. With permission from the AAAS.)

nose and the GPCRs for acetylcholine in salivary cells, which stimulate secretion). Among the fastest of all responses mediated by a GPCR, however, is the response of the eye to light: it takes only 20 msec for the most quickly responding photoreceptor cells of the retina (the cone photoreceptors, which are responsible for color vision in bright light) to produce their electrical response to a sudden flash of light.

This exceptional speed is achieved in spite of the necessity to relay the signal over multiple steps of an intracellular signaling cascade. But photoreceptors also provide a beautiful illustration of the positive advantages of intracellular signaling cascades: in particular, such cascades allow spectacular amplification of the incoming signal and also allow cells to adapt so as to be able to detect signals of widely varying intensity. The quantitative details have been most thoroughly analyzed for the rod photoreceptor cells in the eye, which are responsible for noncolor vision in dim light (**Figure 16–30**). In this photoreceptor cell, light is sensed by rhodopsin, a G-protein-coupled light receptor. Light-activated rhodopsin activates a G protein called transducin. The activated α subunit of transducin then activates an intracellular signaling cascade that causes cation channels to close in the plasma membrane of the photoreceptor cell. This produces a change in the voltage across the cell membrane, which alters neurotransmitter release and ultimately leads to a nerve impulse being sent to the brain.

The signal is repeatedly amplified as it is relayed along this intracellular signaling pathway (**Figure 16–31**). When lighting conditions are dim, as on a moonless night, the amplification is enormous: as few as a dozen photons absorbed in the entire retina will cause a perceptible signal to be delivered to the brain. In bright sunlight, when photons flood through each photoreceptor cell at a rate of billions per second, the signaling cascade undergoes a form of *adaptation*, stepping down the amplification more than 10,000-fold, so that the photoreceptor cells are not overwhelmed and can still register increases and decreases in the strong light. The adaptation depends on negative feedback: an intense response in the photoreceptor cell decreases the cytosolic Ca^{2+} concentration, inhibiting the enzymes responsible for signal amplification.

Adaptation frequently occurs in intracellular signaling pathways that respond to extracellular signal molecules, allowing cells to respond to fluctuations in the concentration of such molecules regardless of whether they are present in small or large amounts. By taking advantage of positive and negative feedback mechanisms (see Figure 16–14), adaptation thus allows a cell to respond both to messages that are whispered and to those that are shouted.

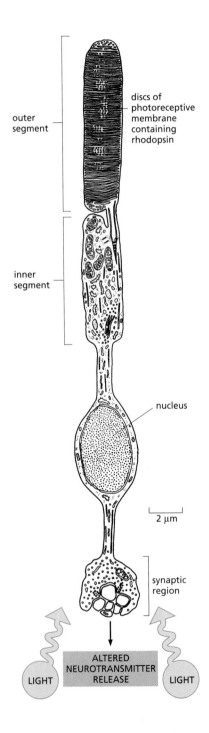

outer segment

discs of photoreceptive membrane containing rhodopsin

inner segment

nucleus

2 μm

synaptic region

LIGHT   ALTERED NEUROTRANSMITTER RELEASE   LIGHT

Figure 16–30 A rod photoreceptor cell from the retina is exquisitely sensitive to light. Drawing of a rod photoreceptor. The light-absorbing rhodopsin proteins are embedded in many pancake-shaped vesicles (*discs*) of membrane inside the outer segment of the cell. Neurotransmitter is released from the opposite end of the cell to control firing of the retinal nerve cells that pass on the signal to the nerve cells in the retina that connect to the brain. When the rod cell is stimulated by light, a signal is relayed from the rhodopsin molecules in the discs, through the cytosol of the outer segment, to ion channels that allow positive ions to flow through the plasma membrane of the outer segment. These cation channels close in response to the cytosolic signal, producing a change in the membrane potential of the rod cell. By mechanisms similar to those that control neurotransmitter release in ordinary nerve cells, the change in membrane potential alters the rate of neurotransmitter release from the synaptic region of the cell. (Adapted from T.L. Lentz, Cell Fine Structure. Philadelphia: Saunders, 1971. With permission from Elsevier.)

Figure 16–31 **The light-induced signaling cascade in rod photoreceptor cells greatly amplifies the light signal.** When rod photoreceptors are adapted for dim light, signal amplification is enormous. The intracellular signaling pathway from the G protein transducin uses components that differ from the ones in previous figures. The cascade functions as follows. In the absence of a light signal, the small messenger molecule cyclic GMP is continuously produced by an enzyme in the cytosol of the photoreceptor cell. The cyclic GMP then binds to cation channels in the photoreceptor cell plasma membrane, keeping them open. Activation of rhodopsin by light results in the activation of transducin α subunits. These turn on an enzyme called cyclic GMP phosphodiesterase, which breaks down cyclic GMP to GMP (much as cyclic AMP phosphodiesterase breaks down cyclic AMP; see Figure 16–23). The sharp fall in the cytosolic concentration of cyclic GMP causes the bound cyclic GMP to dissociate from the cation channels, which therefore close. Closing these channels decreases the influx of Na⁺, thereby altering the voltage gradient (membrane potential) across the plasma membrane and, ultimately, the rate of neurotransmitter release, as described in Chapter 12. The red arrows indicate the steps at which amplification occurs, with the thickness of the arrow roughly indicating the magnitude of the amplification.

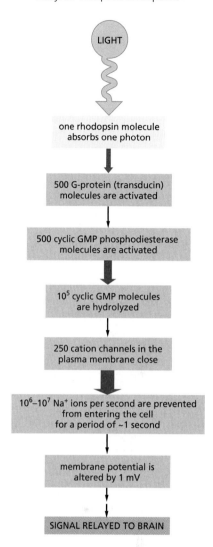

Taste and smell also depend on GPCRs. It seems likely that this mechanism of signal reception, invented early in the evolution of the eukaryotes, has its origins in the basic and universal need of cells to sense and respond to their environment. Of course, GPCRs are not the only receptors that activate intracellular signaling cascades. We now turn to another major class of cell-surface receptors—enzyme-coupled receptors—which play a key part in controlling cell numbers, cell differentiation, and cell movement in multicellular animals, especially during development.

ENZYME-COUPLED RECEPTORS

Like GPCRs, **enzyme-coupled receptors** are transmembrane proteins that display their ligand-binding domains on the outer surface of the plasma membrane (see Figure 16–17C). Instead of associating with a G protein, however, the cytoplasmic domain of the receptor either acts as an enzyme itself or forms a complex with another protein that acts as an enzyme. Enzyme-coupled receptors were discovered through their role in responses to extracellular signal proteins ("growth factors") that regulate the growth, proliferation, differentiation, and survival of cells in animal tissues (see Table 16–1, p. 529, for examples). Most of these signal proteins function as local mediators and can act at very low concentrations (about 10^{-9} to 10^{-11} M). Responses to them are typically slow (on the order of hours), and their effects may require many intracellular transduction steps that usually lead to a change in gene expression.

Enzyme-coupled receptors, however, can also mediate direct, rapid reconfigurations of the cytoskeleton, changing the cell's shape and the way that it moves. The extracellular signals that induce such changes are often not diffusible signal proteins, but proteins attached to the surfaces over which a cell is crawling.

The largest class of enzyme-coupled receptors consists of receptors with a cytoplasmic domain that functions as a tyrosine protein kinase, which phosphorylates particular tyrosines on specific intracellular signaling proteins. These receptors, called **receptor tyrosine kinases (RTKs)**, will be our main focus in this section.

We begin with a discussion of how RTKs are activated in response to extracellular signals. We then consider how activated RTKs transmit the

QUESTION 16–8

One important feature of any intracellular signaling pathway is its ability to be turned off. Consider the pathway shown in Figure 16–31. Where would off switches be required? Which ones do you suppose would be the most important?

signal along two major intracellular signaling pathways that terminate at various effector proteins in the target cell. Finally, we describe how some enzyme-coupled receptors bypass such intracellular signaling cascades and use a more direct mechanism to regulate gene transcription.

Abnormal cell growth, proliferation, differentiation, survival, and migration are fundamental features of a cancer cell, and abnormalities in signaling via RTKs and other enzyme-coupled receptors have a major role in the development of most cancers.

Activated RTKs Recruit a Complex of Intracellular Signaling Proteins

To do its job as a signal transducer, an enzyme-coupled receptor has to switch on the enzyme activity of its intracellular domain (or of an associated enzyme) when an external signal molecule binds to its extracellular domain. Unlike the seven-pass transmembrane GPCRs, enzyme-coupled receptor proteins usually have only one transmembrane segment, which spans the lipid bilayer as a single α helix. Because a single α helix is poorly suited to transmit a conformational change across the bilayer, enzyme-coupled receptors have a different strategy for transducing the extracellular signal. In many cases, the binding of an extracellular signal molecule causes two receptor molecules to come together in the plasma membrane, forming a dimer. This pairing brings the two intracellular tails of the receptors together, activating their kinase domains so that each receptor tail phosphorylates the other. In the case of RTKs, the phosphorylations occur on specific tyrosines.

This tyrosine phosphorylation then triggers the assembly of a transient but elaborate intracellular signaling complex on the cytosolic tails of the receptor. The newly phosphorylated tyrosines serve as docking sites for a whole zoo of intracellular signaling proteins—perhaps as many as 10 or 20 different molecules (**Figure 16–32**). Some of these proteins become phosphorylated and activated on binding to the receptor, and they then propagate the signal; others function solely as scaffolds, which couple the receptor to other signaling proteins, thereby helping to build the active signaling complex (see Figure 16–13). All of these docked

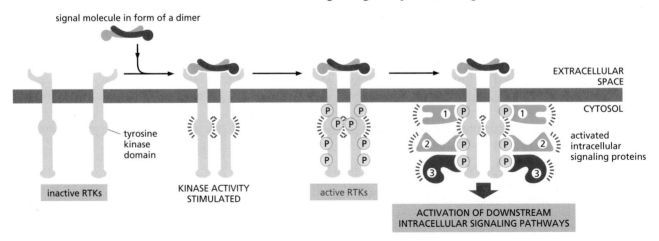

Figure 16–32 Activation of an RTK stimulates the assembly of an intracellular signaling complex. Typically, the binding of a signal molecule to the extracellular domain of an RTK causes two receptor molecules to associate into a dimer. The signal molecule shown here is itself a dimer and thus can physically cross-link two receptor molecules; other signal molecules induce a conformational change in the RTKs, causing the receptors to dimerize (not shown). In either case, dimer formation brings the kinase domains of each cytosolic receptor tail into contact with the other; this activates the kinases to phosphorylate the adjacent tail on several tyrosines. Each phosphorylated tyrosine serves as a specific docking site for a different intracellular signaling protein, which then helps relay the signal to the cell's interior; these proteins contain a specialized interaction domain—in this case, a module called an SH2 domain—that recognizes and binds to specific phosphorylated tyrosines on the cytosolic tail of an activated RTK or on another intracellular signaling protein.

intracellular signaling proteins possess a specialized *interaction domain*, which recognizes specific phosphorylated tyrosines on the receptor tails. Other interaction domains allow intracellular signaling proteins to recognize phosphorylated lipids that are produced on the cytosolic side of the plasma membrane in response to certain signals, as we discuss later.

While they last, the signaling protein complexes assembled on the cytosolic tails of the RTKs can transmit a signal along several routes simultaneously to many destinations in the cell, thus activating and coordinating the numerous biochemical changes that are required to trigger a complex response, such as cell proliferation or differentiation. To help terminate the response, the tyrosine phosphorylations are reversed by protein tyrosine phosphatases, which remove the phosphates that were added to the tyrosines of both the RTKs and other intracellular signaling proteins in response to the extracellular signal. In some cases, activated RTKs (as well as some GPCRs) are inactivated in a more brutal way: they are dragged into the interior of the cell by endocytosis and then destroyed by digestion in lysosomes.

Different RTKs recruit different collections of intracellular signaling proteins, producing different effects; however, certain components are used by most RTKs. These include, for example, a phospholipase C that functions in the same way as the phospholipase C activated by GPCRs to trigger the inositol phospholipid signaling pathway discussed earlier (see Figure 16–27). Another intracellular signaling protein that is activated by almost all RTKs is a small GTP-binding protein called Ras, as we discuss next.

Most RTKs Activate the Monomeric GTPase Ras

As we have seen, activated RTKs recruit and activate many kinds of intracellular signaling proteins, leading to the formation of large signaling complexes on the cytosolic tail of the RTK. One of the key members of these signaling complexes is **Ras**—a small GTP-binding protein that is bound by a lipid tail to the cytoplasmic face of the plasma membrane. Virtually all RTKs activate Ras, including platelet-derived growth factor (PDGF) receptors, which mediate cell proliferation in wound healing, and nerve growth factor (NGF) receptors, which play an important part in the development of certain vertebrate neurons.

The Ras protein is a member of a large family of small GTP-binding proteins, often called **monomeric GTPases** to distinguish them from the trimeric G proteins that we encountered earlier. Ras resembles the α subunit of a G protein and functions as a molecular switch in much the same way. It cycles between two distinct conformational states—active when GTP is bound and inactive when GDP is bound. Interaction with an activating protein called Ras-GEF encourages Ras to exchange its GDP for GTP, thus switching Ras to its activated state (**Figure 16–33**); after a delay, Ras is switched off by a GAP protein called Ras-GAP (see Figure 16–16), which promotes the hydrolysis of its bound GTP to GDP (Movie 16.6).

In its active state, Ras initiates a phosphorylation cascade in which a series of serine/threonine protein kinases phosphorylate and activate one another in sequence, like an intracellular game of dominoes. This relay system, which carries the signal from the plasma membrane to the nucleus, includes a three-protein-kinase module called the **MAP-kinase signaling module**, in honor of the final kinase in the chain, the mitogen-activated protein kinase, or **MAP kinase**. (As we discuss in Chapter 18, *mitogens* are extracellular signal molecules that stimulate cell proliferation.) In this pathway, outlined in **Figure 16–34**, MAP kinase

Figure 16–33 RTKs activate Ras.
An adaptor protein docks on a particular phosphotyrosine on the activated receptor (the other signaling proteins that are shown bound to the receptor in Figure 16–32 are omitted for simplicity). The adaptor recruits a Ras guanine nucleotide exchange factor (Ras-GEF) that stimulates Ras to exchange its bound GDP for GTP. The activated Ras protein can now stimulate several downstream signaling pathways, one of which is shown in Figure 16–34. Note that the Ras protein contains a covalently attached lipid group (red) that helps anchor the protein to the inside of the plasma membrane.

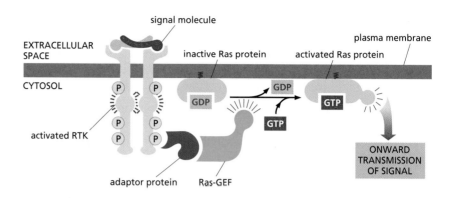

is phosphorylated and activated by an enzyme called, logically enough, MAP kinase kinase. And this protein is itself switched on by a MAP kinase kinase kinase (which is activated by Ras). At the end of the MAP-kinase cascade, MAP kinase phosphorylates various effector proteins, including certain transcription regulators, altering their ability to control gene transcription. This change in the pattern of gene expression may stimulate cell proliferation, promote cell survival, or induce cell differentiation: the precise outcome will depend on which other genes are active in the cell and what other signals the cell receives. How biologists unravel such complex signaling pathways is discussed in **How We Know**, pp. 556–557.

Before Ras was discovered in normal cells, a mutant form of it was found in human cancer cells; the mutation inactivated the GTPase activity of Ras, so that the protein could not shut itself off, promoting uncontrolled cell proliferation and the development of cancer. About 30% of human cancers contain such activating mutations in a *Ras* gene; of the cancers that do not, many have mutations in genes that encode proteins that function in the same signaling pathway as Ras. Many of the genes that encode normal intracellular signaling proteins were initially identified in the hunt for cancer-promoting *oncogenes* (discussed in Chapter 20).

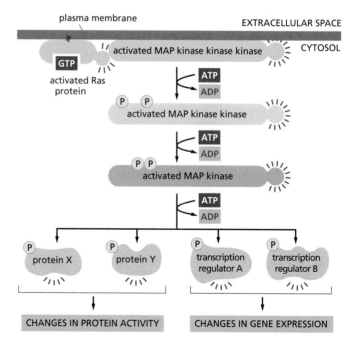

Figure 16–34 Ras activates a MAP-kinase signaling module. The Ras protein activated by the process shown in Figure 16–33 activates a three-kinase signaling module, which relays the signal. The final kinase in the module, MAP kinase, phosphorylates various downstream signaling or effector proteins.

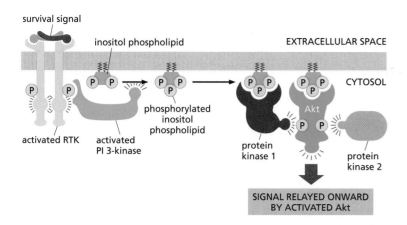

Figure 16–35 RTKs activate the PI-3-kinase–Akt signaling pathway. An extracellular survival signal, such as IGF, activates an RTK, which recruits and activates PI 3-kinase. PI 3-kinase then phosphorylates an inositol phospholipid that is embedded in the cytosolic side of the plasma membrane. The resulting phosphorylated inositol phospholipid then attracts intracellular signaling proteins that have a special domain that recognizes it. One of these signaling proteins, Akt, is a protein kinase that is activated at the membrane by phosphorylation mediated by two other protein kinases (here called protein kinases 1 and 2); protein kinase 1 is also recruited by the phosphorylated lipid docking sites. Once activated, Akt is released from the plasma membrane and phosphorylates various downstream proteins on specific serines and threonines (not shown).

RTKs Activate PI 3-Kinase to Produce Lipid Docking Sites in the Plasma Membrane

Many of the extracellular signal proteins that stimulate animal cells to survive and grow act through RTKs. These include signal proteins belonging to the insulin-like growth factor (IGF) family. One crucially important signaling pathway that these RTKs activate to promote cell growth and survival relies on the enzyme **phosphoinositide 3-kinase** (**PI 3-kinase**), which phosphorylates inositol phospholipids in the plasma membrane. These phosphorylated lipids then serve as docking sites for specific intracellular signaling proteins, which relocate from the cytosol to the plasma membrane, where they can activate one another. One of the most important of these relocated signaling proteins is the serine/threonine protein kinase *Akt* (**Figure 16–35**).

Akt, also called protein kinase B (PKB), promotes the growth and survival of many cell types, often by inactivating the signaling proteins it phosphorylates. For example, Akt phosphorylates and inactivates a cytosolic protein called Bad. In its active state, Bad encourages the cell to kill itself by indirectly activating a cell-suicide program called apoptosis (discussed in Chapter 18). Phosphorylation by Akt thus promotes cell survival by inactivating a protein that otherwise promotes cell death (**Figure 16–36**).

In addition to promoting cell survival, the *PI-3-kinase–Akt signaling pathway* also stimulates cells to grow in size. It does so by indirectly activating a large serine/threonine kinase called *Tor*. Tor stimulates cells to grow both by enhancing protein synthesis and by inhibiting protein

> ### QUESTION 16–9
>
> Would you expect to activate RTKs by exposing the exterior of cells to antibodies that bind to the respective proteins? Would your answer be different for GPCRs? (Hint: review Panel 4–3, on pp. 164–165, regarding the properties of antibody molecules.)

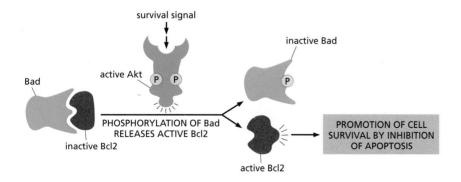

Figure 16–36 Activated Akt promotes cell survival. One way it does so is by phosphorylating and inactivating a protein called Bad. In its unphosphorylated state, Bad promotes apoptosis (a form of cell death) by binding to and inhibiting a protein, called Bcl2, which otherwise suppresses apoptosis. When Bad is phosphorylated by Akt, Bad releases Bcl2, which now blocks apoptosis, thereby promoting cell survival.

UNTANGLING CELL SIGNALING PATHWAYS

Intracellular signaling pathways are never mapped out in a single experiment. Although insulin was first isolated from dog pancreas in the early 1920s, the molecular chain of events that links the binding of insulin to its receptor with the activation of the transporter proteins that take up glucose has taken decades to untangle—and is still not completely understood.

Instead, investigators figure out, piece by piece, how all the links in the chain fit together—and how each contributes to the cell's response to an extracellular signal molecule such as the hormone insulin. Here, we discuss the kinds of experiments that allow scientists to identify individual links and, ultimately, to piece together complex signaling pathways.

Close encounters

Most signaling pathways depend on proteins that physically interact with one another. There are several ways to detect such direct contact. One involves using a protein as "bait." For example, to isolate the receptor that binds to insulin, one could attach insulin to a chromatography column. Cells that respond to the hormone are broken open and their contents poured over the column. Proteins that bind to insulin will stick to this column and can later be eluted and identified (see Figure 4–48).

Protein–protein interactions in a signaling pathway can also be identified by *co-immunoprecipitation*. For example, cells exposed to an extracellular signal molecule can be broken open, and antibodies can be used to grab the receptor protein known to recognize the signal molecule (see Panel 4–2, pp. 146–147, and Panel 4–3, pp. 164–165). If the receptor is strongly associated with other proteins, these will be captured as well. In this way, researchers can identify which proteins interact when an extracellular signal molecule stimulates cells.

Once two proteins are known to bind to each other, the experimenter can use recombinant DNA technology to pinpoint which parts of the proteins are required for the interaction. For example, to determine which phosphorylated tyrosine on a receptor tyrosine kinase (RTK) a certain intracellular signaling protein binds, a series of mutant receptors is constructed, each missing a different tyrosine from its cytoplasmic domain (**Figure 16–37**). In this way, the specific tyrosines required for binding can be determined. Similarly, one can determine whether this phosphotyrosine docking site is required for the receptor to transmit a signal to the cell.

Jamming the pathway

Ultimately, one wants to assess what part a particular protein plays in a signaling pathway. A first test may involve using recombinant DNA technology to introduce into cells a gene encoding a constantly active form of the protein, to see if this mimics the effect of the extracellular signal molecule. Consider Ras, for example. The mutant form of Ras involved in human

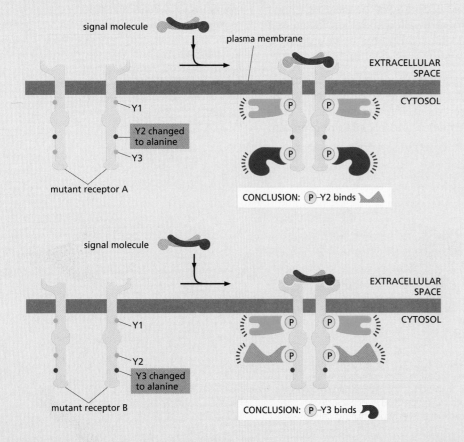

Figure 16–37 Mutant proteins can help to determine exactly where an intracellular signaling molecule binds. As shown in Figure 16–32, on binding their extracellular signal molecule, a pair of RTKs come together and phosphorylate specific tyrosines on each other's cytoplasmic tails. These phosphorylated tyrosines attract different intracellular signaling proteins, which then become activated and pass on the signal. To determine which tyrosine binds to a specific intracellular signaling protein, a series of mutant receptors is constructed. In the mutants shown, tyrosines Y2 and Y3 have been replaced, one at a time, by an alanine (*red*). As a result, the mutant receptors no longer bind to one of the intracellular signaling proteins shown in Figure 16–32. The effect on the cell's response to the signal can then be determined. It is important that the mutant receptor be tested in a cell that does not have its own normal receptors for the signal molecule.

cancers is constantly active because it has lost its ability to hydrolyze the bound GTP that keeps the Ras protein switched on. This continuously active form of Ras can stimulate some cells to proliferate, even in the absence of a proliferation signal.

Conversely, a specific signaling protein can be inactivated. In the case of Ras, for example, one could "knock down" the activity of the *Ras* gene in cells by RNA interference (see Figure 8–26). Such cells do not proliferate in response to extracellular mitogens, indicating the importance of normal Ras signaling in the proliferative response.

Making mutants

One powerful strategy that scientists use to identify proteins that participate in cell signaling involves screening tens of thousands of animals—fruit flies or nematode worms, for example (discussed in Chapter 19)—to search for mutants in which a signaling pathway is not functioning properly. By examining enough mutant animals,

many of the genes that encode the proteins involved in a signaling pathway can be identified.

Such classical genetic screens can also help determine the order in which intracellular signaling proteins act in a pathway. Suppose that a genetic screen uncovers a pair of new proteins, X and Y, involved in the Ras signaling pathway. To determine whether these proteins lie upstream or downstream of Ras, one could create cells that express an inactive, mutant form of each, and then ask whether these mutant cells can be "rescued" by the addition of a continuously active form of Ras. If the constantly active Ras overcomes the blockage created by the mutant protein, the protein must operate upstream of Ras in the pathway (**Figure 16–38A**). However, if Ras operates upstream of the protein, a constantly active Ras would be unable to transmit a signal past the obstruction caused by the disabled protein (**Figure 16–38B**). Through such experiments, even the most complex intracellular signaling pathways can be mapped out, one step at a time (**Figure 16–38C**).

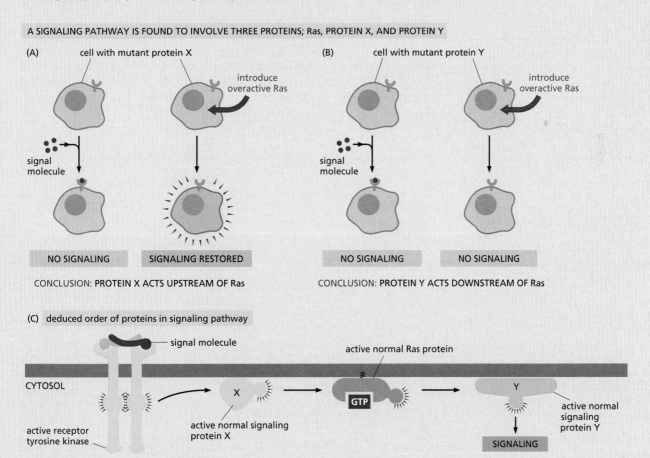

Figure 16–38 **The use of mutant cell lines and an overactive form of Ras can help dissect an intracellular signaling pathway.** In this hypothetical pathway, Ras, protein X, and protein Y are required for proper signaling. (A) In cells in which protein X has been inactivated, signaling does not occur. However, this signaling blockage can be overcome by the addition of an overactive form of Ras, such that the pathway is active even in the absence of the extracellular signal molecule. This result indicates that protein X acts upstream of Ras in the pathway. (B) Signaling is also disrupted in cells in which protein Y has been inactivated. In this case, introduction of an overactive Ras does not restore normal signaling, indicating that protein Y operates downstream of Ras. (C) Based on these results, the deduced order of the signaling pathway is shown.

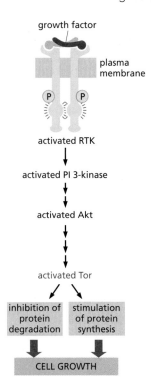

Figure 16–39 Akt stimulates cells to grow in size by activating the serine/threonine kinase Tor. The binding of a growth factor to an RTK activates the PI-3-kinase–Akt signaling pathway (as shown in Figure 16–35). Akt then indirectly activates Tor by phosphorylating and inhibiting a protein that helps to keep Tor shut down (not shown). Tor stimulates protein synthesis and inhibits protein degradation by phosphorylating key proteins in these processes (not shown). The anticancer drug rapamycin slows cell growth by inhibiting Tor. In fact, the Tor protein derives its name from the fact that it is a _t_arget _o_f _r_apamycin.

degradation (**Figure 16–39**). The anticancer drug rapamycin works by inactivating Tor, indicating the importance of this signaling pathway in regulating cell growth and survival—and the consequences of its disregulation in cancer.

Figure 16–40 summarizes the main intracellular signaling cascades activated by GPCRs and RTKs.

Some Receptors Activate a Fast Track to the Nucleus

Not all receptors trigger complex signaling cascades to carry a message to the nucleus. Some use a more direct route to control gene expression. One such receptor is the protein Notch.

Notch is a crucially important receptor in all animals, both during development and in adults. Among other things, it controls the development of neural cells in *Drosophila*, as mentioned earlier (see Figure 16–4). In this simple signaling pathway, the receptor itself acts as a transcription regulator. When activated by the binding of Delta, which is a transmembrane signal protein on the surface of a neighboring cell, the Notch receptor is cleaved. This cleavage releases the cytosolic tail of the receptor, which is then free to move to the nucleus where it helps to activate the appropriate set of Notch-responsive genes (**Figure 16–41**).

Figure 16–40 Both GPCRs and RTKs activate multiple intracellular signaling pathways. The figure reviews five of these pathways: two leading from GPCRs—through adenylyl cyclase and through phospholipase C—and three leading from RTKs—through phospholipase C, Ras, and PI 3-kinase. Each pathway differs from the others, yet they use some common components to transmit their signals. Because all five eventually activate protein kinases, it seems that each is capable in principle of regulating practically any process in the cell.

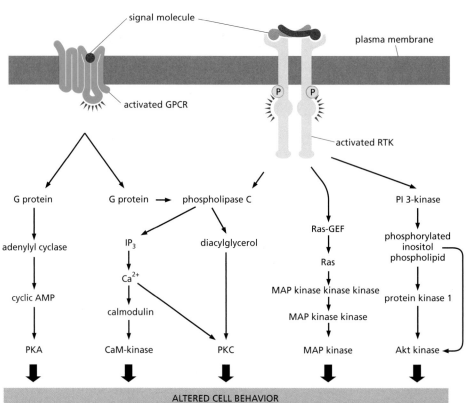

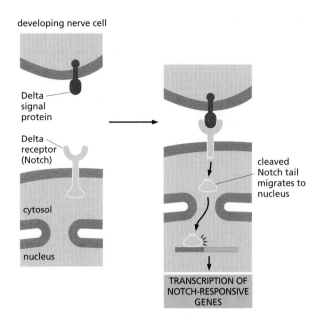

developing nerve cell

Delta signal protein

Delta receptor (Notch)

cytosol

nucleus

cleaved Notch tail migrates to nucleus

TRANSCRIPTION OF NOTCH-RESPONSIVE GENES

Figure 16–41 The Notch receptor itself is a transcription regulator. When the membrane-bound signal protein Delta binds to its receptor, Notch, on a neighboring cell, the receptor is cleaved. The released part of the cytosolic tail of Notch migrates to the nucleus, where it activates Notch-responsive genes. One consequence of this signaling process is shown in Figure 16–4.

Cell–Cell Communication Evolved Independently in Plants and Animals

Plants and animals have been evolving independently for more than a billion years, the last common ancestor being a single-celled eukaryote that most likely lived on its own. Because these kingdoms diverged so long ago—when it was still "every cell for itself"—each has evolved its own molecular solutions to the complex problem of becoming multicellular. Thus the mechanisms for cell–cell communication in plants and animals are in some ways quite different. At the same time, however, plants and animals started with a common set of eukaryotic genes—including some used by single-celled organisms to communicate among themselves—so their signaling systems also show some similarities.

Like animals, plants make extensive use of transmembrane cell-surface receptors—especially enzyme-coupled receptors. The spindly weed *Arabidopsis thaliana* (see Figure 1–32) has hundreds of genes encoding *receptor serine/threonine kinases*. These are, however, structurally distinct from the receptor serine/threonine kinases found in animal cells (which we do not discuss in this chapter). The plant receptors are thought to play an important part in a large variety of cell signaling processes, including those governing plant growth, development, and disease resistance. In contrast to animal cells, plant cells seem not to use RTKs, steroid-hormone-type nuclear receptors, or cyclic AMP, and they seem to use few GPCRs.

One of the best-studied signaling systems in plants mediates the response of cells to ethylene—a gaseous hormone that regulates a diverse array of developmental processes, including seed germination and fruit ripening. Tomato growers use ethylene to ripen their fruit, even after it has been picked. Although ethylene receptors are not evolutionarily related to any of the classes of receptor proteins that we have discussed so far, they function as enzyme-coupled receptors. Surprisingly, it is the empty receptor that is active: in the absence of ethylene, the empty receptor activates an associated protein kinase that ultimately shuts off the ethylene-responsive genes in the nucleus; when ethylene is present, the receptor and kinase are inactive, and the ethylene-responsive genes are transcribed (**Figure 16–42**). This strategy, whereby signals act to relieve transcriptional inhibition, is commonly used in plants.

Figure 16–42 The ethylene signaling pathway turns on genes by relieving inhibition. (A) In the absence of ethylene, the receptor directly activates an associated protein kinase, which then indirectly promotes the destruction of the transcription regulator that switches on ethylene-responsive genes. As a result, the genes remain turned off. (B) In the presence of ethylene, the receptor and kinase are both inactive, and the transcription regulator remains intact and stimulates the transcription of the ethylene-responsive genes. The kinase that ethylene receptors interact with is a serine/threonine kinase that is closely related to the MAP kinase kinase kinase found in animal cells (see Figure 16–34).

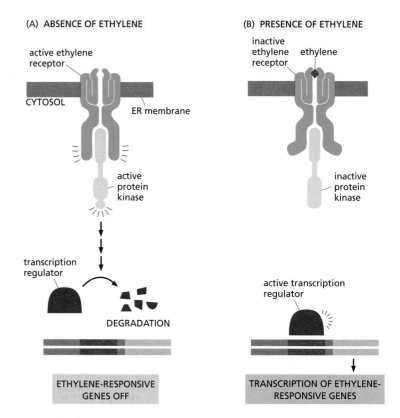

Protein Kinase Networks Integrate Information to Control Complex Cell Behaviors

Although the signaling pathways we have described thus far may seem dauntingly complex, the complexity of cell signaling is actually much greater than we have let on. First, we have not discussed all of the intracellular signaling pathways that operate in cells, even though many of these are critical for normal development. Second, although we depict these signaling pathways as being linear and self-contained, they do not work independently of one another. Instead, they are connected by interactions of many sorts. The most extensive links are those mediated by the protein kinases present in each pathway. These kinases often phosphorylate, and hence regulate, components in other signaling pathways, in addition to components in their own pathway. Thus a certain amount of cross-talk occurs between the different pathways. To give an idea of the scale of the complexity, genome sequencing studies suggest that about 2% of our ~21,000 protein-coding genes code for protein kinases; moreover, hundreds of distinct types of protein kinases are thought to be present in a single mammalian cell. How can we make sense of this tangled web of interacting signaling pathways, and what is the function of such complexity?

A cell receives messages from many sources, and it must integrate this information to generate an appropriate response: to live or die, to divide, to differentiate, to change shape, to move, to send out a chemical message of its own, and so on (see Figure 16–6 and Movies 16.7 and 16.8). Through the cross-talk between signaling pathways, the cell is able to bring together multiple bits of information and react to the combination. Thus some intracellular signaling proteins act as integrating devices, usually by having several potential phosphorylation sites, each of which can be phosphorylated by a different protein kinase. Information received from different sources can converge on such proteins, which then convert the input to a single outgoing signal (**Figure 16–43**, and see Figure 16–13). The integrating proteins in turn can deliver a signal to many downstream

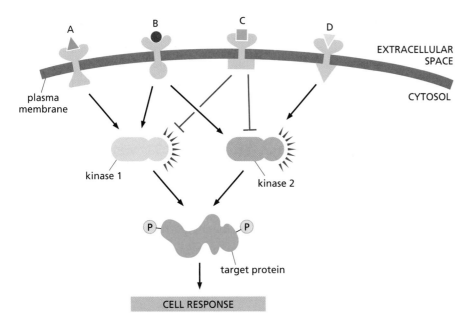

Figure 16–43 Intracellular signaling proteins serve to integrate incoming signals. Extracellular signals A, B, C, and D activate different receptors in the plasma membrane. The receptors act upon two protein kinases, which they either activate (arrowhead) or inhibit (crossbar). The kinases phosphorylate a same target protein and when it is fully phosphorylated, the target protein triggers a cellular response.
It can be seen that signal molecule B activates both protein kinases and therefore produces a strong output response. Signals A and D each activate a different kinase and therefore produce a response only if they are simultaneously present. Signal molecule C inhibits the cell response and will compete with the other signal molecules. The net outcome will depend both on the numbers of signaling molecules and the strengths of their connections. In a real cell these parameters would be determined by evolution.

targets. In this way, the intracellular signaling system may act like a network of nerve cells in the brain—or like a collection of microprocessors in a computer—interpreting complex information and generating complex responses.

Our understanding of these intricate networks is still evolving: we are still discovering new links in the chains, new signaling partners, new connections, and even new pathways. Unraveling the intracellular signaling pathways—in both animals and plants—is one of the most active areas of research in cell biology, and new discoveries are being made every day. Genome sequencing projects continue to provide long lists of components involved in signal transduction in a large variety of organisms. Even when we have identified all the components, however, it will remain a major challenge to figure out exactly how they work together to allow cells to integrate the diverse array of signals in their environment and respond in the appropriate manner.

In a way, learning how cells "think" is a problem akin to learning how we, as humans, think. Although we know, for example, how neurotransmitters activate certain neurons and how one neuron communicates with another, we are nowhere near understanding how all these components operate together to enable us to reason, converse, laugh, love, and attempt to uncover the fundamental features of life on Earth.

ESSENTIAL CONCEPTS

- Cells in multicellular organisms communicate through a large variety of extracellular chemical signals.

- In animals, hormones are carried in the blood to distant target cells, but most other extracellular signal molecules act over only a short distance. Neighboring cells often communicate through direct cell–cell contact.

- For an extracellular signal molecule to influence a target cell it must interact with a receptor protein on or in a target cell. Each receptor protein recognizes a particular signal molecule.

- Small, hydrophobic, extracellular signal molecules, such as steroid hormones and nitric oxide, can cross the plasma membrane and activate intracellular proteins, which are usually either transcription regulators or enzymes.

- Most extracellular signal molecules cannot pass through the plasma membrane; they bind to cell-surface receptor proteins that convert (transduce) the extracellular signal into different intracellular signals, which are usually organized into signaling pathways.

- There are three main classes of cell-surface receptors: (1) ion-channel-coupled receptors, (2) G-protein-coupled receptors (GPCRs), and (3) enzyme-coupled receptors.

- GPCRs and enzyme-coupled receptors respond to extracellular signals by activating one or more intracellular signaling pathways, which, in turn, activate effector proteins that alter the behavior of the cell.

- Turning off signaling pathways is as important as turning them on. Each activated component in a signaling pathway must be subsequently inactivated or removed for the pathway to function again.

- GPCRs activate trimeric GTP-binding proteins called G proteins; these act as molecular switches, transmitting the signal onward for a short period before switching themselves off by hydrolyzing their bound GTP to GDP.

- G proteins directly regulate ion channels or enzymes in the plasma membrane. Some directly activate (or inactivate) the enzyme adenylyl cyclase, which increases (or decreases) the intracellular concentration of the small messenger molecule cyclic AMP; others directly activate the enzyme phospholipase C, which generates the small messenger molecules inositol trisphosphate (IP_3) and diacylglycerol.

- IP_3 opens Ca^{2+} channels in the membrane of the endoplasmic reticulum, releasing a flood of free Ca^{2+} ions into the cytosol. The Ca^{2+} itself acts as a second messenger, altering the activity of a wide range of Ca^{2+}-responsive proteins. These include calmodulin, which activates various target proteins such as Ca^{2+}/calmodulin-dependent protein kinases (CaM-kinases) .

- A rise in cyclic AMP activates protein kinase A (PKA), while Ca^{2+} and diacylglycerol in combination activate protein kinase C (PKC).

- PKA, PKC, and CaM-kinases phosphorylate selected signaling and effector proteins on serines and threonines, thereby altering their activity. Different cell types contain different sets of signaling and effector proteins and are therefore affected in different ways.

- Enzyme-coupled receptors have intracellular protein domains that function as enzymes or are associated with intracellular enzymes. Many enzyme-coupled receptors are receptor tyrosine kinases (RTKs), which phosphorylate themselves and selected intracellular signaling proteins on tyrosines. The phosphotyrosines on RTKs then serve as docking sites for various intracellular signaling proteins.

- Most RTKs activate the monomeric GTPase Ras, which, in turn, activates a three-protein MAP-kinase signaling module that helps relay the signal from the plasma membrane to the nucleus.

- Ras mutations stimulate cell proliferation by keeping Ras (and, consequently, the Ras–MAP kinase signaling pathway) constantly active and are a common feature of many human cancers.

- Some RTKs stimulate cell growth and cell survival by activating PI 3-kinase, which phosphorylates specific inositol phospholipids in the cytosolic leaflet of the plasma membrane lipid bilayer. This inositol phosphorylation creates lipid docking sites that attract specific signaling proteins from the cytosol, including the protein kinase Akt, which becomes active and relays the signal onward.

- Other receptors, such as Notch, have a direct pathway to the nucleus. When activated, part of the receptor migrates from the plasma membrane to the nucleus, where it regulates the transcription of specific genes.

- Plants, like animals, use enzyme-coupled cell-surface receptors to recognize the extracellular signal molecules that control their growth and development; these receptors often act by relieving the transcriptional repression of specific genes.

- Different intracellular signaling pathways interact, enabling each cell type to produce the appropriate response to a combination of extracellular signals. In the absence of such signals, most animal cells have been programmed to kill themselves by undergoing apoptosis.

- We are far from understanding how a cell integrates all of the many extracellular signals that bombard it to generate an appropriate response.

KEY TERMS

adaptation	hormone	phospholipase C
adenylyl cyclase	inositol phospholipid	protein kinase
Ca^{2+}/calmodulin-dependent protein kinase (CaM-kinase)	inositol 1,4,5-trisphosphate (IP$_3$)	protein kinase C (PKC) protein phosphatase
calmodulin	intracellular signaling pathway	Ras
cell signaling	ion-channel-coupled receptor	receptor
cyclic AMP	local mediator	receptor serine/threonine kinase
cyclic-AMP-dependent protein kinase (PKA)	MAP kinase MAP-kinase signaling module	receptor tyrosine kinase (RTK) serine/threonine kinase
diacylglycerol (DAG)	molecular switch	signal transduction
enzyme-coupled receptor	monomeric GTPase	small intracellular signaling
extracellular signal molecule	neurotransmitter	molecule
G protein	nitric oxide (NO)	steroid hormone
G-protein-coupled receptor (GPCR)	nuclear receptor phosphoinositide 3-kinase	tyrosine kinase
GTP-binding protein	(PI 3-kinase)	

QUESTIONS

QUESTION 16–10

If some cell-surface receptors, including Notch, can rapidly signal to the nucleus by activating latent transcription regulators at the plasma membrane, why do most cell-surface receptors use long, indirect signaling cascades to influence gene transcription in the nucleus?

QUESTION 16–11

Which of the following statements are correct? Explain your answers.

A. The extracellular signal molecule acetylcholine has different effects on different cell types in an animal and often binds to different cell-surface receptor molecules on different cell types.

B. After acetylcholine is secreted from cells, it is long-lived, because it has to reach target cells all over the body.

C. Both the GTP-bound α subunits and nucleotide-free βγ complexes—but not GDP-bound, fully assembled G proteins—can activate other molecules downstream of GPCRs.

D. IP$_3$ is produced directly by cleavage of an inositol phospholipid without incorporation of an additional phosphate group.

E. Calmodulin regulates the intracellular Ca^{2+} concentration.

F. Different signals originating from the plasma membrane can be integrated by cross-talk between different signaling pathways inside the cell.

G. Tyrosine phosphorylation serves to build binding sites for other proteins to bind to RTKs.

QUESTION 16–12

The Ras protein functions as a molecular switch that is set to its "on" state by other proteins that cause it to expel its bound GDP and bind GTP. A GTPase-activating protein helps reset the switch to the "off" state by inducing Ras to hydrolyze its bound GTP to GDP much more rapidly than it would without this encouragement. Thus, Ras works like a light switch that one person turns on and another turns off. You are given a mutant cell that lacks the GTPase-activating protein. What abnormalities would you expect to find in the way in which Ras activity responds to extracellular signals?

QUESTION 16–13

A. Compare and contrast signaling by neurons, which secrete neurotransmitters at synapses, with signaling carried out by endocrine cells, which secrete hormones into the blood.

B. Discuss the relative advantages of the two mechanisms.

QUESTION 16–14

Two intracellular molecules, X and Y, are both normally synthesized at a constant rate of 1000 molecules per second per cell. Molecule X is broken down slowly: each molecule of X survives on average for 100 seconds. Molecule Y is broken down 10 times faster: each molecule of Y survives on average for 10 seconds.

A. Calculate how many molecules of X and Y the cell contains at any time.

B. If the rates of synthesis of both X and Y are suddenly increased tenfold to 10,000 molecules per second per cell—without any change in their degradation rates—how many molecules of X and Y will there be after one second?

C. Which molecule would be preferred for rapid signaling?

QUESTION 16–15

"One of the great kings of the past ruled an enormous kingdom that was more beautiful than anywhere else in the world. Every plant glistened as brilliantly as polished jade, and the softly rolling hills were as sleek as the waves of the summer sea. The wisdom of all of his decisions relied on a constant flow of information brought to him daily by messengers who told him about every detail of his kingdom so that he could take quick, appropriate actions when needed. Despite the beauty and efficiency, his people felt doomed living under his rule, for he had an adviser who had studied cell signal transduction and accordingly administered the king's Department of Information. The adviser had implemented the policy that all messengers will be immediately beheaded whenever spotted by the Royal Guard, because for rapid signaling the lifetime of messengers ought to be short. Their plea "Don't hurt me, I'm only the messenger!" was to no avail, and the people of the kingdom suffered terribly because of the rapid loss of their sons and daughters." Why is the analogy on which the king's adviser based his policies inappropriate? Briefly discuss the features that set cell signaling pathways apart from the human communication pathway described in the story.

QUESTION 16–16

In a series of experiments, genes that code for mutant forms of an RTK are introduced into cells. The cells also express their own normal form of the receptor from their normal gene, although the mutant genes are constructed so that the mutant RTK is expressed at considerably higher concentration than the normal RTK. What would be the consequences of introducing a mutant gene that codes for an RTK (A) lacking its extracellular domain, or (B) lacking its intracellular domain?

QUESTION 16–17

Discuss the following statement: "Membrane proteins that span the membrane many times can undergo a conformational change upon ligand binding that can be sensed on the other side of the membrane. Thus, individual protein molecules can transmit a signal across a membrane. In contrast, individual single-span membrane proteins cannot transmit a conformational change across the membrane but require oligomerization."

QUESTION 16–18

What are the similarities and differences between the reactions that lead to the activation of G proteins and the reactions that lead to the activation of Ras?

QUESTION 16–19

Why do you suppose cells use Ca^{2+} (which is kept by Ca^{2+} pumps at a cytosolic concentration of 10^{-7} M) for intracellular signaling and not another ion such as Na^+ (which is kept by the Na^+ pump at a cytosolic concentration of 10^{-3} M)?

QUESTION 16–20

It seems counterintuitive that a cell, having a perfectly abundant supply of nutrients available, would commit suicide if not constantly stimulated by signals from other cells (see Figure 16–6). What do you suppose might be the advantages of such regulation?

QUESTION 16–21

The contraction of the myosin–actin system in muscle cells is triggered by a rise in intracellular Ca^{2+}. Muscle cells have specialized Ca^{2+} channels—called ryanodine receptors because of their sensitivity to the drug ryanodine—that are embedded in the membrane of the sarcoplasmic reticulum, a specialized form of the endoplasmic reticulum. In contrast to the IP_3-gated Ca^{2+} channels in the endoplasmic reticulum shown in Figure 16–27, the signaling molecule that opens ryanodine receptors is Ca^{2+} itself. Discuss the consequences of ryanodine channels for muscle cell contraction.

QUESTION 16–22

Two protein kinases, K1 and K2, function sequentially in an intracellular signaling pathway. If either kinase contains a mutation that permanently inactivates its function, no response is seen in cells when an extracellular signal is received. A different mutation in K1 makes it permanently active, so that in cells containing that mutation a response is observed even in the absence of an extracellular signal. You characterize a double-mutant cell that contains K2 with the inactivating mutation and K1 with the activating mutation. You observe that the response is seen even in the absence of an extracellular signal. In the normal signaling pathway, does K1 activate K2 or does K2 activate K1? Explain your answer.

QUESTION 16–23

A. Trace the steps of a long and indirect signaling pathway from a cell-surface receptor to a change in gene expression in the nucleus.

B. Compare this pathway with two short and direct pathways from the cell surface to the nucleus.

QUESTION 16–24

How does PI 3-kinase activate the Akt kinase after activation of RTK?

QUESTION 16–25

Animal cells and plant cells have some very different intracellular signaling mechanisms but also share some common mechanisms. Why do you think this is so?

Cytoskeleton

The ability of eukaryotic cells to adopt a variety of shapes, organize the many components in their interior, interact mechanically with the environment, and carry out coordinated movements depends on the **cytoskeleton**—an intricate network of protein filaments that extends throughout the cytoplasm (**Figure 17–1**). This filamentous architecture helps to support the large volume of cytoplasm, a function that is particularly important in animal cells, which have no cell walls. Although some cytoskeletal components are present in bacteria, the cytoskeleton is most prominent in the large and structurally complex eukaryotic cell.

Unlike our own bony skeleton, however, the cytoskeleton is a highly dynamic structure that is continuously reorganized as a cell changes shape, divides, and responds to its environment. The cytoskeleton is not only the "bones" of a cell but its "muscles" too, and it is directly responsible for large-scale movements, including the crawling of cells along a surface, the contraction of muscle cells, and the changes in cell shape that take place as an embryo develops. Without the cytoskeleton, wounds would never heal, muscles would not contract, and sperm would never reach the egg.

Like any factory making a complex product, the eukaryotic cell has a highly organized interior in which organelles that carry out specialized functions are concentrated in different areas and linked by transport systems (discussed in Chapter 15). The cytoskeleton controls the location of the organelles and provides the machinery for transport between them. It is also responsible for the segregation of chromosomes into two daughter cells at cell division and for pinching apart those two new cells, as we discuss in Chapter 18.

INTERMEDIATE FILAMENTS

MICROTUBULES

ACTIN FILAMENTS

MUSCLE CONTRACTION

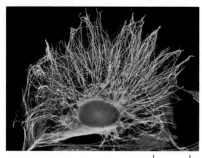

Figure 17–1 The cytoskeleton gives a cell its shape and allows the cell to organize its internal components and to move. An animal cell in culture has been labeled to show two of its major cytoskeletal systems, the microtubules (*green*) and the actin filaments (*red*). Where the two filaments overlap, they appear *yellow*. The DNA in the nucleus is labeled in *blue*. (Courtesy of Albert Tousson.)

The cytoskeleton is built on a framework of three types of protein filaments: *intermediate filaments*, microtubules, and *actin filaments*. Each type of filament has distinct mechanical properties and is formed from a different protein subunit. A family of fibrous proteins forms the intermediate filaments; globular *tubulin* subunits form microtubules; and globular *actin* subunits form actin filaments (**Figure 17–2**). In each case, thousands of subunits assemble into fine threads that sometimes extend across the entire cell.

In this chapter, we consider the structure and function of each of these protein filament networks. We begin with intermediate filaments, which provide cells with mechanical strength. We then see how microtubules organize the cytoplasm of eukaryotic cells and form the hairlike motile appendages that enable cells like protozoa and sperm to swim. We next consider how the actin cytoskeleton supports the cell surface and allows fibroblasts and other cells to crawl. Finally, we discuss how the actin cytoskeleton enables our muscles to contract.

Figure 17–2 The three types of protein filaments that form the cytoskeleton differ in their composition, mechanical properties, and roles inside the cell. They are shown here in epithelial cells, but they are all found in almost all animal cells.

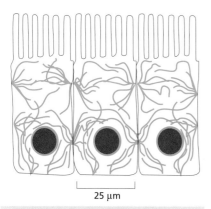

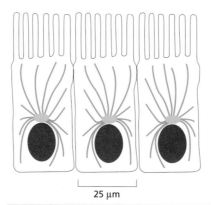

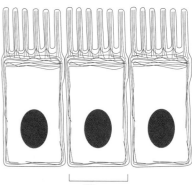

|— 25 μm —| |— 25 μm —| |— 25 μm —|

| INTERMEDIATE FILAMENTS | MICROTUBULES | ACTIN FILAMENTS |

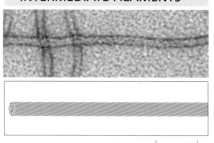

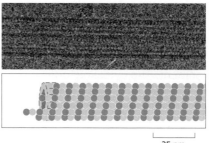

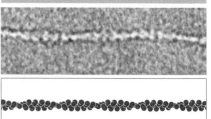

|— 25 nm —| |— 25 nm —| |— 25 nm —|

Intermediate filaments are ropelike fibers with a diameter of about 10 nm; they are made of fibrous intermediate filament proteins. One type of intermediate filament forms a meshwork called the nuclear lamina just beneath the inner nuclear membrane. Other types extend across the cytoplasm, giving cells mechanical strength and distributing the mechanical stresses in an epithelial tissue by spanning the cytoplasm from one cell–cell junction to another. Intermediate filaments are very flexible and have great tensile strength. They deform under stress but do not rupture. (Micrograph courtesy of Roy Quinlan.)

Microtubules are hollow cylinders made of the protein tubulin. They are long and straight and typically have one end attached to a single microtubule-organizing center called a *centrosome*. With an outer diameter of 25 nm, microtubules are more rigid than actin filaments or intermediate filaments, and they rupture when stretched. (Micrograph courtesy of Richard Wade.)

Actin filaments (also known as *microfilaments*) are helical polymers of the protein actin. They are flexible structures, with a diameter of about 7 nm, that are organized into a variety of linear bundles, two-dimensional networks, and three-dimensional gels. Although actin filaments are dispersed throughout the cell, they are most highly concentrated in the *cortex*, the layer of cytoplasm just beneath the plasma membrane. (Micrograph courtesy of Roger Craig.)

INTERMEDIATE FILAMENTS

Intermediate filaments have great tensile strength, and their main function is to enable cells to withstand the mechanical stress that occurs when cells are stretched. The filaments are called "intermediate" because, in the smooth muscle cells where they were first discovered, their diameter (about 10 nm) is between that of the thinner actin filaments and the thicker *myosin filaments*. Intermediate filaments are the toughest and most durable of the cytoskeletal filaments: when cells are treated with concentrated salt solutions and nonionic detergents, the intermediate filaments survive, while most of the rest of the cytoskeleton is destroyed.

Intermediate filaments are found in the cytoplasm of most animal cells. They typically form a network throughout the cytoplasm, surrounding the nucleus and extending out to the cell periphery. There they are often anchored to the plasma membrane at cell–cell junctions called *desmosomes* (discussed in Chapter 20), where the plasma membrane is connected to that of another cell (**Figure 17–3**). Intermediate filaments are also found within the nucleus of all eukaryotic cells. There they form a meshwork called the *nuclear lamina*, which underlies and strengthens the nuclear envelope. In this section, we see how the structure and assembly of intermediate filaments makes them particularly suited to strengthening cells and protecting them from tearing.

Intermediate Filaments Are Strong and Ropelike

An intermediate filament is like a rope in which many long strands are twisted together to provide tensile strength (**Movie 17.1**). The strands of this cable are made of intermediate filament proteins, fibrous subunits each containing a central elongated rod domain with distinct unstructured domains at either end (**Figure 17–4A**). The rod domain consists of an extended α-helical region that enables pairs of intermediate filament

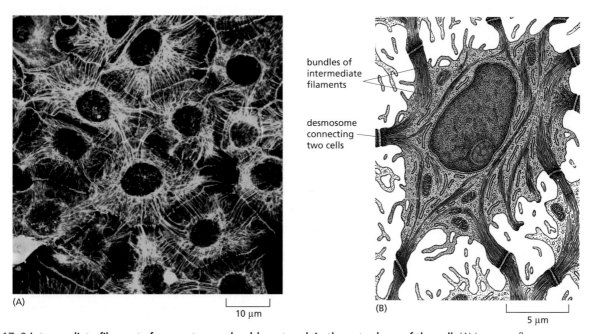

(A)

⌐——————————⌐
10 μm

bundles of
intermediate
filaments

desmosome
connecting
two cells

(B)

⌐——————⌐
5 μm

Figure 17–3 Intermediate filaments form a strong, durable network in the cytoplasm of the cell. (A) Immuno-fluorescence micrograph of a sheet of epithelial cells in culture stained to show the lacelike network of intermediate keratin filaments (*green*), which surround the nuclei and extend through the cytoplasm of the cells. The filaments in each cell are indirectly connected to those of neighboring cells through the desmosomes, establishing a continuous mechanical link from cell to cell throughout the epithelial sheet. A second protein (*blue*) has been stained to show the locations of the cell boundaries. (B) Drawing from an electron micrograph of a section of a skin cell showing the bundles of intermediate filaments that traverse the cytoplasm and are inserted at desmosomes. (A, courtesy of Kathleen Green and Evangeline Amargo; B, from R.V. Krstić, Ultrastructure of the Mammalian Cell: An Atlas. Berlin: Springer, 1979. With permission from Springer-Verlag.)

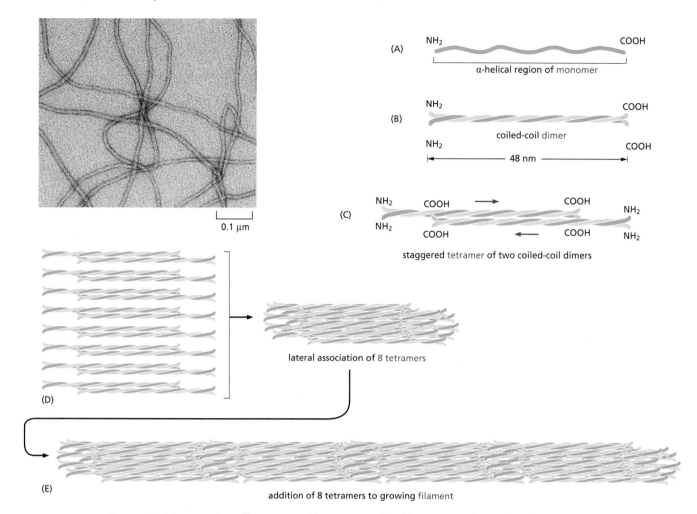

0.1 μm

(A) α-helical region of monomer

(B) coiled-coil dimer

48 nm

(C) staggered tetramer of two coiled-coil dimers

(D)

lateral association of 8 tetramers

(E)

addition of 8 tetramers to growing filament

Figure 17–4 Intermediate filaments are like ropes made of long, twisted strands of protein. The intermediate filament monomer consists of an α-helical central rod domain (A) with unstructured regions at either end (not shown). Pairs of monomers associate to form a dimer (B), and two dimers then line up to form a staggered, antiparallel tetramer (C). Tetramers can pack together into a helical array containing eight tetramer strands (D), which in turn assemble into the final ropelike intermediate filament (E). An electron micrograph of intermediate filaments is shown on the upper left. (Micrograph courtesy of Roy Quinlan.)

proteins to form stable dimers by wrapping around each other in a coiled-coil configuration (**Figure 17–4B**), as described in Chapter 4. Two of these coiled-coil dimers, running in opposite directions, associate to form a staggered tetramer (**Figure 17–4C**). These dimers and tetramers are the soluble subunits of intermediate filaments. The tetramers associate with each other side-by-side (**Figure 17–4D**) and then assemble to generate the final ropelike intermediate filament (**Figure 17–4E**).

Because the two dimers point in opposite directions, the two ends of the tetramer are the same, as are the two ends of assembled intermediate filaments; as we will see, this distinguishes these filaments from microtubules and actin filaments, whose structural polarity is crucial for their function. All the interactions between the intermediate filament proteins depend solely on noncovalent bonding; it is the combined strength of the overlapping lateral interactions along the length of the proteins that gives intermediate filaments their great tensile strength.

The central rod domains of different intermediate filament proteins are all similar in size and amino acid sequence, so that when they pack together they always form filaments of similar diameter and internal structure. By contrast, the terminal domains vary greatly in both size and amino acid sequence from one type of intermediate filament protein to another.

These unstructured domains are exposed on the surface of the filament, where they allow it to interact with specific components in the cytoplasm.

Intermediate Filaments Strengthen Cells Against Mechanical Stress

Intermediate filaments are particularly prominent in the cytoplasm of cells that are subject to mechanical stress. They are present in large numbers, for example, along the length of nerve cell axons, providing essential internal reinforcement to these extremely long and fine cell extensions. They are also abundant in muscle cells and in epithelial cells such as those of the skin. In all these cells, intermediate filaments distribute the effect of locally applied forces, thereby keeping cells and their membranes from tearing in response to mechanical shear. A similar principle is used to strengthen composite materials such as fiberglass or reinforced concrete, in which tension-bearing linear elements such as carbon fibers (in fiberglass) or steel bars (in concrete) are embedded in a space-filling matrix to give the material strength.

Intermediate filaments can be grouped into four classes: (1) *keratin filaments* in epithelial cells; (2) *vimentin* and *vimentin-related filaments* in connective-tissue cells, muscle cells, and supporting cells of the nervous system (glial cells); (3) *neurofilaments* in nerve cells; and (4) *nuclear lamins*, which strengthen the nuclear envelope. The first three filament types are found in the cytoplasm, whereas the fourth is found in the nucleus (**Figure 17–5**). Filaments of each class are formed by polymerization of their corresponding intermediate filament subunits.

The **keratin filaments** are the most diverse class of intermediate filament. Every kind of epithelium in the vertebrate body—whether in the tongue, the cornea, or the lining of the gut—has its own distinctive mixture of keratin proteins. Specialized keratins also occur in hair, feathers, and claws. In each case, the keratin filaments are formed from a mixture of different keratin subunits. Keratin filaments typically span the interiors of epithelial cells from one side of the cell to the other, and filaments in adjacent epithelial cells are indirectly connected through desmosomes (see Figure 17–3B). The ends of the keratin filaments are anchored to the desmosomes, and the filaments associate laterally with other cell components through the globular head and tail domains that project from their surface. This cabling of high tensile strength, formed by the filaments throughout the epithelial sheet, distributes the stress that occurs when the skin is stretched. The importance of this function is illustrated by the rare human genetic disease *epidermolysis bullosa simplex*, in which mutations in the keratin genes interfere with the formation of keratin filaments in the epidermis. As a result, the skin is highly vulnerable to mechanical injury, and even a gentle pressure can rupture its cells, causing the skin

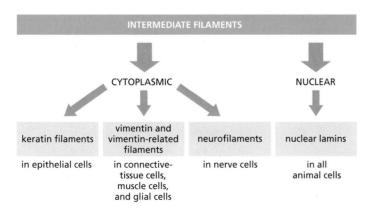

Figure 17–5 Intermediate filaments are divided into four major classes. These classes can include numerous subtypes. Humans, for example, have more than 50 keratin genes.

Figure 17–6 A mutant form of keratin makes skin more prone to blistering. A mutant gene encoding a truncated keratin protein was introduced into a mouse. The defective protein assembles with the normal keratins and thereby disrupts the keratin filament network in the skin. (A) Light micrograph of a cross section of normal skin, which is resistant to mechanical pressure. (B) Cross section of skin from mutant mouse shows the formation of a blister, which results from the rupturing of cells in the basal layer of the mutant epidermis (short *red* arrow). (From P.A. Coulombe et al., *J. Cell Biol.* 115:1661–1674, 1991. With permission from The Rockefeller University Press.)

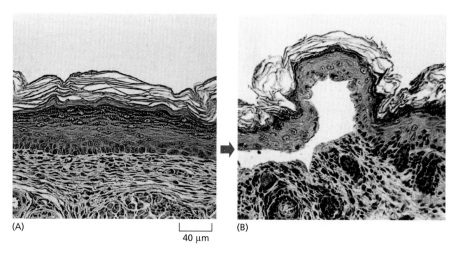

(A) (B)

40 μm

to blister. The disease can be reproduced in transgenic mice expressing a mutant *keratin* gene in their skin (**Figure 17–6**).

Many of the intermediate filaments are further stabilized and reinforced by accessory proteins, such as *plectin*, that cross-link the filaments into bundles and link them to microtubules, to actin filaments, and to adhesive structures in the desmosomes (**Figure 17–7**). Mutations in the gene for plectin cause a devastating human disease that combines features of epidermolysis bullosa simplex (caused by disruption of skin keratin), muscular dystrophy (caused by disruption of intermediate filaments in muscle), and neurodegeneration (caused by disruption of neurofilaments). Mice lacking a functional plectin gene die within a few days of birth, with blistered skin and abnormal skeletal and heart muscle. Thus although plectin may not be necessary for the initial formation of intermediate filaments, its cross-linking action is required to provide cells with the strength they need to withstand mechanical stress.

The Nuclear Envelope Is Supported by a Meshwork of Intermediate Filaments

Whereas cytoplasmic intermediate filaments form ropelike structures, the intermediate filaments lining and strengthening the inside surface of the inner nuclear membrane are organized as a two-dimensional meshwork (**Figure 17–8**). As mentioned earlier, the intermediate filaments that form this tough **nuclear lamina** are constructed from a class of intermediate filament proteins called *lamins* (not to be confused with laminin, which is an extracellular matrix protein). The nuclear lamina disassembles and re-forms at each cell division, when the nuclear envelope breaks down during mitosis and then re-forms in each daughter cell (discussed in Chapter 18). Cytoplasmic intermediate filaments also disassemble in mitosis.

The disassembly and reassembly of the nuclear lamina are controlled by the phosphorylation and dephosphorylation of the lamins. When the

---

QUESTION 17–1

Which of the following types of cells would you expect to contain a high density of intermediate filaments in their cytoplasm? Explain your answers.
A. *Amoeba proteus* (a free-living amoeba)
B. Skin epithelial cell
C. Smooth muscle cell in the digestive tract
D. *Escherichia coli*
E. Nerve cell in the spinal cord
F. Sperm cell
G. Plant cell

Figure 17–7 Plectin aids in the bundling of intermediate filaments and links these filaments to other cytoskeletal protein networks. In this scanning electron micrograph of the cytoskeletal protein network from cultured fibroblasts, the actin filaments have been removed, and the plectin, intermediate filaments, and microtubules have been artificially colored. Note how the plectin (*green*) links an intermediate filament (*blue*) to three microtubules (*red*). The *yellow dots* are gold particles linked to antibodies that recognize plectin. (From T.M. Svitkina and G.G. Borisy, *J. Cell Biol.* 135:991–1007, 1996. With permission from The Rockefeller University Press.)

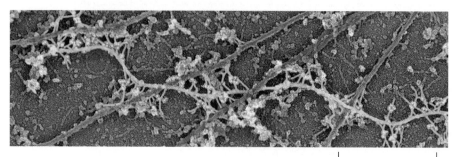

0.5 μm

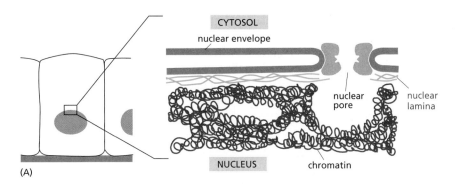

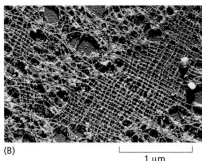

1 µm

Figure 17–8 Intermediate filaments support and strengthen the nuclear envelope. (A) Schematic cross section through the nuclear envelope. The intermediate filaments of the nuclear lamina line the inner face of the nuclear envelope and are thought to provide attachment sites for the chromosomes. (B) Electron micrograph of a portion of the nuclear lamina from a frog egg. The lamina is formed from a square lattice of intermediate filaments composed of lamins. (The nuclear lamina in other cell types is not always as regularly organized as the one shown here.) (B, courtesy of Ueli Aebi.)

lamins are phosphorylated by protein kinases (discussed in Chapter 4), the consequent conformational change weakens the binding between the lamin tetramers and causes the filaments to fall apart. Dephosphorylation by protein phosphatases at the end of mitosis causes the lamins to reassemble (see Figure 18–30).

Defects in a particular nuclear lamin are associated with certain types of *progeria*—rare disorders that cause affected individuals to age prematurely. Children with progeria have wrinkled skin, lose their teeth and hair, and often develop severe cardiovascular disease by the time they reach their teens (**Figure 17–9**). How the loss of a nuclear lamin could lead to this devastating condition is not yet known, but it may be that the resulting nuclear instability leads to impaired cell division, increased cell death, a diminished capacity for tissue repair, or some combination of these.

MICROTUBULES

Microtubules have a crucial organizing role in all eukaryotic cells. These long and relatively stiff hollow tubes of protein can rapidly disassemble in one location and reassemble in another. In a typical animal cell, microtubules grow out from a small structure near the center of the cell called the *centrosome* (**Figure 17–10A and B**). Extending out toward the cell periphery, they create a system of tracks within the cell, along which vesicles, organelles, and other cell components can be transported. These cytoplasmic microtubules are the part of the cytoskeleton mainly responsible for transporting and positioning membrane-enclosed organelles within the cell and for guiding the intracellular transport of various cytosolic macromolecules.

When a cell enters mitosis, the cytoplasmic microtubules disassemble and then reassemble into an intricate structure called the *mitotic spindle*. As we discuss in Chapter 18, the mitotic spindle provides the machinery that will segregate the chromosomes equally into the two daughter cells just before a cell divides (**Figure 17–10C**). Microtubules can also form stable structures, such as rhythmically beating *cilia* and *flagella* (**Figure 17–10D**). These hairlike structures extend from the surface of many eukaryotic cells, which use them either to swim or to sweep fluid over their surface. The core of a eukaryotic cilium or flagellum consists of a highly organized and stable bundle of microtubules. (Bacterial flagella have an entirely different structure and allow the cells to swim by a very different mechanism.)

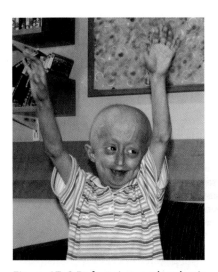

Figure 17–9 Defects in a nuclear lamin can cause a rare class of premature aging disorders called progeria. Children with progeria begin to show advanced features of aging around 18 to 24 months of age. (Courtesy of Progeria Research Foundation.)

In this section, we first consider the structure and assembly of microtubules. We then discuss their role in organizing the cytoplasm—an ability that depends on their association with accessory proteins, especially the *motor proteins* that propel organelles along cytoskeletal tracks. Finally, we discuss the structure and function of cilia and flagella, in which microtubules are stably associated with motor proteins that power the beating of these mobile appendages.

Microtubules Are Hollow Tubes with Structurally Distinct Ends

Microtubules are built from subunits—molecules of **tubulin**—each of which is itself a dimer composed of two very similar globular proteins called α-tubulin and β-tubulin, bound tightly together by noncovalent interactions. The tubulin dimers stack together, again by noncovalent bonding, to form the wall of the hollow cylindrical microtubule. This tubelike structure is made of 13 parallel protofilaments, each a linear chain of tubulin dimers with α- and β-tubulin alternating along its length (**Figure 17–11**). Each protofilament has a structural polarity,

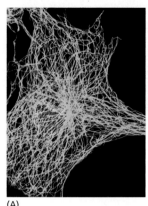

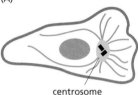

(A)

(B) NONDIVIDING CELL

centrosome

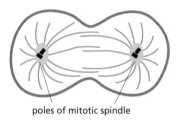

(C) DIVIDING CELL

poles of mitotic spindle

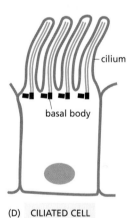

cilium

basal body

(D) CILIATED CELL

Figure 17–10 Microtubules usually grow out from an organizing center.
(A) Fluorescence micrograph of a cytoplasmic array of microtubules in a cultured fibroblast. Unlike intermediate filaments, microtubules (*dark green*) extend from organizing centers such as (B) a centrosome, (C) the two poles of a mitotic spindle, or (D) the basal body of a cilium. They can also grow from fragments of existing microtubules (not shown). (A, courtesy of Michael Davidson and The Florida State University Research Foundation.)

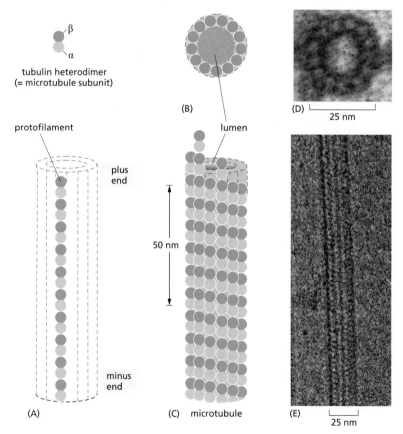

Figure 17–11 Microtubules are hollow tubes made of globular tubulin subunits.
(A) One tubulin subunit (an αβ dimer) and one protofilament are shown schematically, together with their location in the microtubule wall. Note that the tubulin dimers are all arranged in the protofilament with the same orientation. (B and C) Schematic diagrams of a microtubule, showing how tubulin dimers pack together in the microtubule wall. At the top, 13 β-tubulin molecules are shown in cross section. Below this, a side view of a short section of a microtubule shows how the dimers are aligned in the same orientation in all the protofilaments; thus, the microtubule has a definite structural polarity—with a designated plus and a minus end. (D) Electron micrograph of a cross section of a microtubule with its ring of 13 distinct subunits, each of which corresponds to a separate tubulin dimer. (E) Electron micrograph of a microtubule viewed lengthwise. (D, courtesy of Richard Linck; E, courtesy of Richard Wade.)

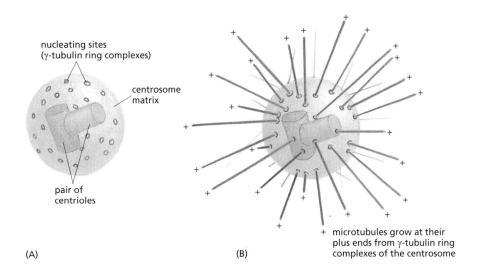

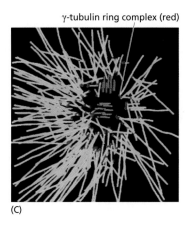

(C)

nucleating sites
(γ-tubulin ring complexes)

centrosome
matrix

pair of
centrioles

(A)

(B)

+ microtubules grow at their
plus ends from γ-tubulin ring
complexes of the centrosome

with α-tubulin exposed at one end and β-tubulin at the other, and this polarity—the directional arrow embodied in the structure—is the same for all the protofilaments, giving a structural polarity to the microtubule as a whole. One end of the microtubule, thought to be the β-tubulin end, is called its *plus end*, and the other, the α-tubulin end, its *minus end*.

In a concentrated solution of pure tubulin in a test tube, tubulin dimers will add to either end of a growing microtubule. However, they add more rapidly to the plus end than to the minus end, which is why the ends were originally named this way—not because they are electrically charged. The polarity of the microtubule—the fact that its structure has a definite direction, with the two ends being chemically and functionally distinct— is crucial, both for the assembly of microtubules and for their role once they are formed. If microtubules had no polarity, they could not guide intracellular transport, for example.

The Centrosome Is the Major Microtubule-organizing Center in Animal Cells

Inside cells, microtubules grow from specialized organizing centers that control the location, number, and orientation of the microtubules. In animal cells, for example, the **centrosome**—which is typically close to the cell nucleus when the cell is not in mitosis—organizes an array of microtubules that radiates outward through the cytoplasm (see Figure 17–10B). The centrosome consists of a pair of **centrioles**, surrounded by a matrix of proteins. The centrosome matrix includes hundreds of ring-shaped structures formed from a special type of tubulin, called *γ-tubulin*, and each *γ-tubulin ring complex* serves as the starting point, or *nucleation site*, for the growth of one microtubule (**Figure 17–12A**). The αβ-tubulin dimers add to each γ-tubulin ring complex in a specific orientation, with the result that the minus end of each microtubule is embedded in the centrosome, and growth occurs only at the plus end that extends into the cytoplasm (**Figure 17–12B and C**).

The paired centrioles at the center of an animal cell centrosome are curious structures; each centriole, sitting perpendicular to its partner, is made of a cylindrical array of short microtubules. Yet centrioles have no role in the nucleation of microtubules from the centrosome (the γ-tubulin ring complex alone is sufficient), and their function remains something of a mystery, especially as plant cells lack them. Centrioles do, however, act as the organizing centers for the microtubules in cilia and flagella, where they are called *basal bodies* (see Figure 17–10D), as we discuss later.

Figure 17–12 Tubulin polymerizes from nucleation sites on a centrosome.
(A) Schematic drawing showing that an animal cell centrosome consists of an amorphous matrix of various proteins, including the γ-tubulin rings (*red*) that nucleate microtubule growth, surrounding a pair of centrioles, oriented at right angles to each other. Each member of the centriole pair is made up of a cylindrical array of short microtubules. (B) Diagram of a centrosome with attached microtubules. The minus end of each microtubule is embedded in the centrosome, having grown from a γ-tubulin ring complex, whereas the plus end of each microtubule extends into the cytoplasm. (C) A reconstructed image of a centrosome of a *C. elegans* cell showing a dense thicket of microtubules emanating from γ-tubulin ring complexes. A pair of centrioles (*blue*) can be seen at the center. (C, from E.T. O'Toole et al., *J. Cell Biol.* 163:451–456, 2003. With permission from The Rockefeller University Press.)

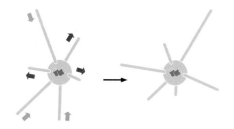

Figure 17–13 Each microtubule grows and shrinks independently of its neighbors. The array of microtubules anchored in a centrosome is continually changing, as new microtubules grow (*red arrows*) and old microtubules shrink (*blue arrows*).

Why do microtubules need nucleating sites such as those provided by the γ-tubulin rings in the centrosome? The answer is that it is much harder to start a new microtubule from scratch, by first assembling a ring of αβ-tubulin dimers, than it is to add such dimers to a preexisting γ-tubulin ring complex. Although purified αβ-tubulin dimers at a high concentration can polymerize into microtubules spontaneously *in vitro*, in a living cell, the concentration of free αβ-tubulin is too low to drive the difficult first step of assembling the initial ring of a new microtubule. By providing organizing centers at specific sites, and keeping the concentration of free αβ-tubulin dimers low, cells can control where microtubules form.

Growing Microtubules Display Dynamic Instability

Once a microtubule has been nucleated, it typically grows outward from the organizing center for many minutes by the addition of αβ-tubulin dimers to its plus end. Then, without warning, the microtubule can suddenly undergo a transition that causes it to shrink rapidly inward by losing tubulin dimers from its free plus end (Movie 17.2). It may shrink partially and then, no less suddenly, start growing again, or it may disappear completely, to be replaced by a new microtubule that grows from the same γ-tubulin ring complex (**Figure 17–13**).

This remarkable behavior—switching back and forth between polymerization and depolymerization—is known as **dynamic instability**. It allows microtubules to undergo rapid remodeling, and is crucial for their function. In a normal cell, the centrosome (or other organizing center) is continually shooting out new microtubules in different directions in an exploratory fashion, many of which then retract. A microtubule growing out from the centrosome can, however, be prevented from disassembling if its plus end is stabilized by attachment to another molecule or cell structure so as to prevent its depolymerization. If stabilized by attachment to a structure in a more distant region of the cell, the microtubule will establish a relatively stable link between that structure and the centrosome (**Figure 17–14**). The centrosome can be compared to a fisherman casting a line: if there is no bite at the end of the line, the line is quickly withdrawn, and a new cast is made; but, if a fish bites, the line remains in place, tethering the fish to the fisherman. This simple strategy of random exploration and selective stabilization enables the centrosome and other nucleating centers to set up a highly organized system of microtubules in selected parts of the cell. The same strategy is used to position organelles relative to one another.

Dynamic Instability is Driven by GTP Hydrolysis

The dynamic instability of microtubules stems from the intrinsic capacity of tubulin dimers to hydrolyze GTP. Each free tubulin dimer contains one GTP molecule tightly bound to β-tubulin, which hydrolyzes the GTP to GDP shortly after the dimer is added to a growing microtubule. This GDP remains tightly bound to the β-tubulin. When polymerization

Figure 17–14 The selective stabilization of microtubules can polarize a cell. A newly formed microtubule will persist only if both its ends are protected from depolymerization. In cells, the minus ends of microtubules are generally protected by the organizing centers from which the microtubules grow. The plus ends are initially free but can be stabilized by binding to specific proteins. Here, for example, a nonpolarized cell is depicted in (A), with new microtubules growing from a centrosome in many directions before shrinking back randomly. If a plus end happens to encounter a protein (capping protein) in a specific region of the cell cortex, it will be stabilized (B). Selective stabilization at one end of the cell will bias the orientation of the microtubule array (C) and, ultimately, will convert the cell to a strongly polarized form (D).

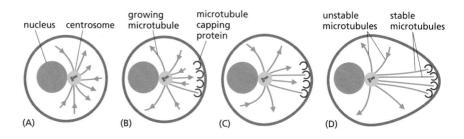

Figure 17–15 GTP hydrolysis controls the dynamic instability of microtubules. (A) Tubulin dimers carrying GTP (*red*) bind more tightly to one another than do tubulin dimers carrying GDP (*dark green*). Therefore, rapidly growing plus ends of microtubules, which have freshly added tubulin dimers with GTP bound, tend to keep growing. (B) From time to time, however, especially if microtubule growth is slow, the dimers in this GTP cap will hydrolyze their GTP to GDP before fresh dimers loaded with GTP have time to bind. The GTP cap is thereby lost. Because the GDP-carrying dimers are less tightly bound in the polymer, the protofilaments peel away from the plus end, and the dimers are released, causing the microtubule to shrink (Movie 17.3).

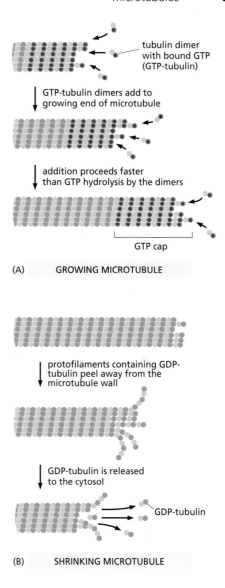

is proceeding rapidly, tubulin dimers add to the end of the microtubule faster than the GTP they carry is hydrolyzed. As a result, the end of a rapidly growing microtubule is composed entirely of GTP-tubulin dimers, which form a "GTP cap." GTP-associated dimers bind more strongly to their neighbors in the microtubule than do dimers that bear GDP, and they pack together more efficiently. Thus the microtubule will continue to grow (**Figure 17–15A**).

Because of the randomness of chemical processes, however, it will occasionally happen that the tubulin dimers at the free end of the microtubule will hydrolyze their GTP before the next dimers are added, so that the free ends of protofilaments are now composed of GDP-tubulin. These GDP-bearing dimers associate less tightly, tipping the balance in favor of disassembly (**Figure 17–15B**). Because the rest of the microtubule is composed of GDP-tubulin, once depolymerization has started, it will tend to continue; the microtubule starts to shrink rapidly and continuously and may even disappear.

The GDP-tubulin that is freed as the microtubule depolymerizes joins the pool of unpolymerized tubulin already in the cytosol. In a typical fibroblast, for example, at any one time about half of the tubulin in the cell is in microtubules, while the remainder is free in the cytosol, forming a pool of tubulin dimers available for microtubule growth. The tubulin dimers joining the pool rapidly exchange their bound GDP for GTP, thereby becoming competent again to add to another microtubule that is in a growth phase.

Microtubule Dynamics Can be Modified by Drugs

Drugs that prevent the polymerization or depolymerization of tubulin dimers can have a rapid and profound effect on the organization of microtubules—and thereby on the behavior of the cell. Consider the mitotic spindle, the microtubule-based apparatus that guides the chromosomes during mitosis (see Figure 17–10C). If a cell in mitosis is exposed to the drug *colchicine*, which binds tightly to free tubulin dimers and prevents their polymerization into microtubules, the mitotic spindle rapidly disappears, and the cell stalls in the middle of mitosis, unable to partition the chromosomes into two groups. This finding, and others like it, demonstrates that the mitotic spindle is normally maintained by a continuous balanced addition and loss of tubulin subunits: when tubulin addition is blocked by colchicine, tubulin loss continues until the spindle disappears.

The drug *Taxol* has the opposite effect. It binds tightly to microtubules and prevents them from losing subunits. Because new subunits can still be added, the microtubules can grow but cannot shrink. However, despite this difference in their mechanism of action, Taxol has the same overall effect as colchicine—arresting dividing cells in mitosis. These experiments show that for the mitotic spindle to function, microtubules must be able to assemble and disassemble. We discuss the behavior of the spindle in more detail in Chapter 18, when we consider mitosis.

QUESTION 17–2

Why do you suppose it is much easier to add tubulin to existing microtubules than to start a new microtubule from scratch? Explain how γ-tubulin in the centrosome helps to overcome this hurdle.

TABLE 17–1 DRUGS THAT AFFECT MICROTUBULES	
Microtubule-specific drugs	**Action**
Taxol	binds and stabilizes microtubules
Colchicine, colcemid	binds tubulin dimers and prevents their polymerization
Vinblastine, vincristine	binds tubulin dimers and prevents their polymerization

The inactivation or destruction of the mitotic spindle eventually kills dividing cells. Because cancer cells divide in a less controlled way than do normal cells of the body, they can sometimes be killed preferentially by microtubule-stabilizing or microtubule-destabilizing *antimitotic drugs*. These drugs include colchicine, Taxol, vincristine, and vinblastine—all of which are used in the treatment of human cancer (**Table 17–1**). As we discuss shortly, there are also drugs that stabilize or destabilize actin filaments.

Microtubules Organize the Cell Interior

Cells are able to modify the dynamic instability of their microtubules for particular purposes. As cells enter mitosis, for example, microtubules become initially more dynamic, switching between growing and shrinking much more frequently than cytoplasmic microtubules normally do. This change enables microtubules to disassemble rapidly and then reassemble into the mitotic spindle. On the other hand, when a cell has differentiated into a specialized cell type, the dynamic instability of its microtubules is often suppressed by proteins that bind to either the ends or the sides of the microtubules and stabilize them against disassembly. The stabilized microtubules then serve to maintain the organization of the differentiated cell.

Most differentiated animal cells are polarized; that is, one end of the cell is structurally or functionally different from the other. Nerve cells, for example, put out an axon from one end of the cell and dendrites from the other (see Figure 12–29). Cells specialized for secretion have their Golgi apparatus positioned toward the site of secretion, and so on. The cell's polarity is a reflection of the polarized systems of microtubules in its interior, which help to position organelles in their required location within the cell and to guide the streams of vesicular and macromolecular traffic moving between one part of the cell and another. In the nerve cell, for example, all the microtubules in the axon point in the same direction, with their plus ends toward the axon terminals; along these oriented tracks, the cell is able to transport organelles, membrane vesicles, and macromolecules—either from the cell body to the axon terminals or in the opposite direction (**Figure 17–16**).

Figure 17–16 Microtubules guide the transport of organelles, vesicles, and macromolecules in both directions along a nerve cell axon. All of the microtubules in the axon point in the same direction, with their plus ends toward the axon terminal. The oriented microtubules serve as tracks for the directional transport of materials synthesized in the cell body but required at the axon terminal. For an axon passing from your spinal cord to a muscle in your shoulder, say, the journey takes about two days. In addition to this outward traffic (*red* circles), which is driven by one set of motor proteins, there is traffic in the reverse direction (*blue* circles), which is driven by another set of motor proteins. The backward traffic includes worn-out mitochondria and materials ingested by the axon terminals.

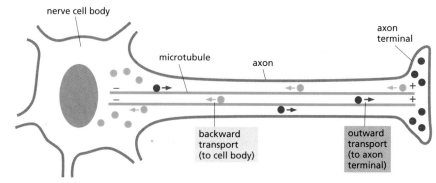

mitochondrion

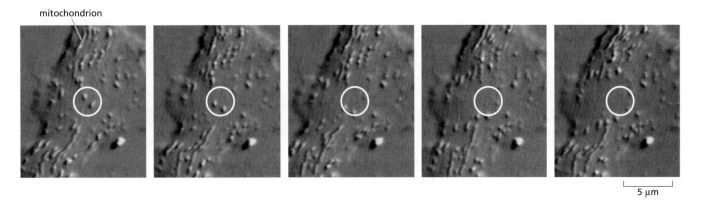

5 μm

Figure 17–17 **Organelles can move rapidly and unidirectionally in a nerve cell axon.** In this series of video-enhanced images of a flattened area of an invertebrate nerve axon, numerous membrane vesicles and mitochondria are present, many of which can be seen to move. The *white circle* provides a fixed frame of reference. These images were recorded at intervals of 400 milliseconds. The two vesicles in the circle are moving outward along microtubules, toward the axon terminal. (Courtesy of P. Forscher.)

Some of the traffic along axons travels at speeds in excess of 10 cm a day (**Figure 17–17**), which means that it could still take a week or more for materials to reach the end of a long axon in larger animals. Nonetheless, movement guided by microtubules is immeasurably faster and more efficient than movement driven by free diffusion. A protein molecule traveling by free diffusion could take years to reach the end of a long axon—if it arrived at all (see Question 17–12).

The microtubules in living cells do not act alone. Their activity, like those of other cytoskeletal filaments, depends on a large variety of accessory proteins that bind to them. Some of these **microtubule-associated proteins** stabilize microtubules against disassembly, for example, while others link microtubules to other cell components, including the other types of cytoskeletal filaments (see Figure 17–7). Still others are motor proteins that actively transport organelles, vesicles, and other macromolecules along microtubules.

Motor Proteins Drive Intracellular Transport

If a living cell is observed in a light microscope, its cytoplasm is seen to be in continual motion. Mitochondria and the smaller membrane-enclosed organelles and vesicles travel in small, jerky steps—moving for a short period, stopping, and then moving again. This *saltatory* movement is much more sustained and directional than the continual, small, Brownian movements caused by random thermal motions. Saltatory movements can occur along either microtubules or actin filaments. In both cases, the movements are driven by **motor proteins**, which use the energy derived from repeated cycles of ATP hydrolysis to travel steadily along the microtubule or actin filament in a single direction (see Figure 4–46). Because the motor proteins also attach to other cell components, they can transport this cargo along the filaments. There are dozens of different motor proteins; they differ in the type of filament they bind to, the direction in which they move along the filament, and the cargo they carry.

The motor proteins that move along cytoplasmic microtubules, such as those in the axon of a nerve cell, belong to two families: the **kinesins** generally move toward the plus end of a microtubule (outward from the cell body in Figure 17–16); the **dyneins** move toward the minus end (toward the cell body in Figure 17–16). Both kinesins and dyneins are generally

QUESTION 17–3

Dynamic instability causes microtubules either to grow or to shrink rapidly. Consider an individual microtubule that is in its shrinking phase.
A. What must happen at the end of the microtubule in order for it to stop shrinking and to start growing again?
B. How would a change in the tubulin concentration affect this switch?
C. What would happen if only GDP, but no GTP, were present in the solution?
D. What would happen if the solution contained an analog of GTP that cannot be hydrolyzed?

Figure 17–18 **Both kinesins and dyneins move along microtubules using their globular heads.** (A) Kinesins and cytoplasmic dyneins are microtubule motor proteins that generally move in opposite directions along a microtubule. Each of these proteins (drawn here roughly to scale) is a dimer composed of two identical subunits. Each dimer has two globular heads at one end, which bind and hydrolyze ATP and interact with microtubules, and a single tail at the other end, which interacts with cargo (not shown). (B) Schematic diagram of a generic motor protein "walking" along a filament; these proteins use the energy of ATP hydrolysis to move in one direction along the filament, as illustrated in Figure 4–46. (See also Figure 17–22B.)

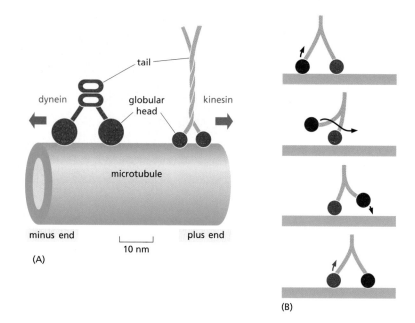

(A)

(B)

dimers that have two globular ATP-binding heads and a single tail (**Figure 17–18A**). The heads interact with microtubules in a stereospecific manner, so that the motor protein will attach to a microtubule in only one direction. The tail of a motor protein generally binds stably to some cell component, such as a vesicle or an organelle, and thereby determines the type of cargo that the motor protein can transport (**Figure 17–19**). The globular heads of kinesin and dynein are enzymes with ATP-hydrolyzing (ATPase) activity. This reaction provides the energy for driving a directed series of conformational changes in the head that enable it to move along the microtubule by a cycle of binding, release, and rebinding to the microtubule (see Figure 17–18B and Figure 4–46). For a discussion of the discovery and study of motor proteins, see **How We Know**, pp. 580–581.

Microtubules and Motor Proteins Position Organelles in the Cytoplasm

Microtubules and motor proteins play an important part in positioning organelles within a eukaryotic cell. In most animal cells, for example, the tubules of the endoplasmic reticulum (ER) reach almost to the edge of the cell (Movie 17.4), whereas the Golgi apparatus is located in the cell interior, near the centrosome (**Figure 17–20A**). The ER extends out from its points of connection with the nuclear envelope along microtubules, which reach from the centrally located centrosome out to the plasma membrane. As a cell grows, kinesins attached to the outside of the ER membrane (via receptor proteins) pull the ER outward along microtubules, stretching it like a net (**Figure 17–20B**). Cytoplasmic *dyneins* attached to

Figure 17–19 **Different motor proteins transport different types of cargo along microtubules.** Most kinesins move toward the plus end of a microtubule, whereas dyneins move toward the minus end (Movie 17.5). Both types of microtubule motor proteins exist in many forms, each of which is thought to transport a different type of cargo. The tail of the motor protein determines what cargo the protein transports.

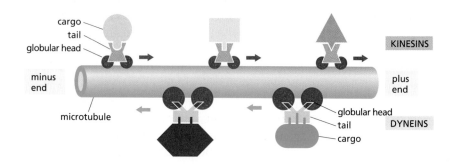

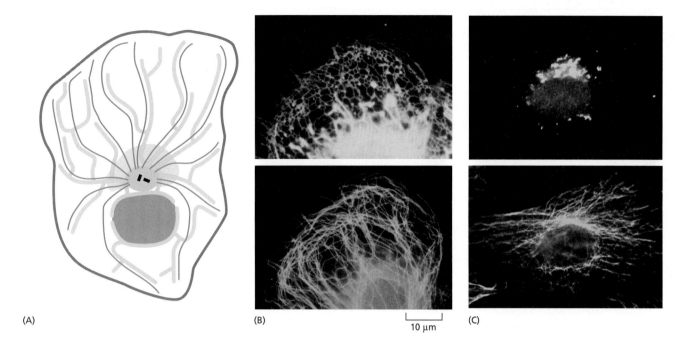

(A)   (B)   10 μm   (C)

Figure 17–20 Microtubules help position organelles in a eukaryotic cell. (A) Schematic diagram of a cell showing the typical arrangement of cytoplasmic microtubules (*dark green*), endoplasmic reticulum (*blue*), and Golgi apparatus (*yellow*). The nucleus is shown in *brown*, and the centrosome in *light green*. (B) One part of a cell in culture stained with antibodies to the endoplasmic reticulum (*blue, upper panel*) and to microtubules (*green, lower panel*). Kinesin motor proteins pull the endoplasmic reticulum outward along the microtubules. (C) A different cell in culture stained with antibodies to the Golgi apparatus (*yellow, upper panel*) and to microtubules (*green, lower panel*). In this case, cytoplasmic dyneins pull the Golgi apparatus inward along the microtubules to its position near the centrosome, which is not visible but is located on the Golgi side of the nucleus. (B, courtesy of Mark Terasaki, Lan Bo Chen, and Keigi Fujiwara; C, courtesy of Viki Allan and Thomas Kreis.)

the Golgi membranes pull the Golgi apparatus along microtubules in the opposite direction, inward toward the nucleus (**Figure 17–20C**). In this way, the regional differences in these internal membranes—crucial for their respective functions—are created and maintained.

When cells are treated with colchicine—a drug that causes microtubules to disassemble—both the ER and the Golgi apparatus change their location dramatically. The ER, which is connected to the nuclear envelope, collapses around the nucleus; the Golgi apparatus, which is not attached to any other organelle, fragments into small vesicles, which then disperse throughout the cytoplasm. When the colchicine is removed, the organelles return to their original positions, dragged by motor proteins moving along the re-formed microtubules.

Cilia and Flagella Contain Stable Microtubules Moved by Dynein

We mentioned earlier that many microtubules in cells are stabilized through their association with other proteins and therefore do not show dynamic instability. Cells use such stable microtubules as stiff supports in the construction of a variety of polarized structures, including motile cilia and flagella. **Cilia** are hairlike structures about 0.25 μm in diameter, covered by plasma membrane, that extend from the surface of many kinds of eukaryotic cells; each cilium contains a core of stable microtubules, arranged in a bundle, that grow from a cytoplasmic *basal body*, which serves as an organizing center (see Figure 17–10D).

Cilia beat in a whiplike fashion, either to move fluid over the surface of a cell or to propel single cells through a fluid. Some protozoa, for example,

PURSUING MICROTUBULE-ASSOCIATED MOTOR PROTEINS

The movement of organelles throughout the cell cytoplasm has been observed, measured, and speculated about since the middle of the nineteenth century. But it was not until the mid-1980s that biologists identified the molecules that drive this movement of organelles and vesicles from one part of the cell to another.

Why the lag between observation and understanding? The problem was in the proteins—or, more precisely, in the difficulty of studying them in isolation outside the cell. To investigate the activity of an enzyme, for example, biochemists first purify the polypeptide: they break open cells or tissues and separate the protein of interest from other molecular components (see Panels 4–4 and 4–5, pp. 166–167). They can then study the protein in a test tube (*in vitro*), controlling its exposure to substrates, inhibitors, ATP, and so on. Unfortunately, this approach did not seem to work for studies of the motile machinery that underlies intracellular transport. It is not possible to break open a cell and pull out an intact, fully active transport system, free of extraneous material, that continues to carry mitochondria and vesicles from place to place.

That problem was solved by technical advances in two separate fields. First, improvements in microscopy allowed biologists to see that an operational transport system (with extraneous material still attached) could be squeezed from the right kind of living cell. At the same time, biochemists realized that they could assemble a working transport system from scratch—using purified cytoskeletal filaments, motors, and cargo—outside the cell. One such breakthrough started with a squid.

Teeming cytoplasm

Neuroscientists interested in the electrical properties of nerve cell membranes have long studied the giant axon from squid (see How We Know, pp. 406–407). Because of its large size, researchers found that they could squeeze the cytoplasm from the axon like toothpaste, and then study how ions move back and forth through various channels in the empty, tubelike plasma membrane (see Figure 12–33). The physiologists simply discarded the cytoplasmic jelly, as it appeared to be inert (and thus uninteresting) when examined under a standard light microscope.

Then along came video-enhanced microscopy. This type of microscopy, developed by Shinya Inoué, Robert Allen, and others, allows one to detect structures that are smaller than the resolving power of standard light microscopes, which is only about 0.2 μm, or 200 nm (see Panel 1–1, pp. 10–11). Sample images are captured by a video camera and then enhanced by computer processing to reduce the background and heighten contrast. When researchers in the early 1980s applied this new technique to

preparations of squid axon cytoplasm (axoplasm), they observed, for the first time, the motion of vesicles and other organelles along cytoskeletal filaments.

Under the video-enhanced microscope, extruded axoplasm is seen to be teeming with tiny particles—from vesicles 30–50 nm in diameter to mitochondria some 5000 nm long, all moving to and fro along cytoskeletal filaments at speeds of up to 5 μm per second. If the axoplasm is spread thinly enough, individual filaments can be seen.

The movement continues for hours, allowing researchers to manipulate the preparation and study the effects. Ray Lasek and Scott Brady discovered, for example, that the organelle movement requires ATP. Substitution of ATP analogs, such as AMP-PNP, which resemble ATP but cannot be hydrolyzed (and thus provide no energy), inhibit the translocation.

Snaking tubes

More work was needed to identify the individual components that drive the transport system in squid axoplasm. What kind of filaments support this movement? What are the molecular motors that shuttle the vesicles and organelles along these filaments? Identifying the filaments was relatively easy: antibodies to tubulin revealed that they are microtubules. But what about the motor proteins? To find these, Ron Vale, Thomas Reese, and Michael Sheetz set up a system in which they could fish for proteins that power organelle movement.

Their strategy was simple yet elegant: add together microtubules and organelles and then look for molecules that induce motion. They used purified microtubules from squid brain, added organelles isolated from squid axons, and showed that organelle movement could be triggered by the addition of an extract from squid axoplasm. In this preparation, the researchers could either watch the organelles travel along the microtubules or watch the microtubules glide snakelike over the surface of a glass coverslip that had been coated with an axoplasm extract (see Question 17–18). Their challenge was to isolate the protein responsible for movement in this reconstituted system.

To do that, Vale and his colleagues took advantage of the earlier work with the ATP analog AMP-PNP. Although this analog inhibits the movement of vesicles along microtubules, it still allows organelles to attach to the microtubule filaments. So the researchers incubated the axoplasm extract with microtubules and organelles in the presence of AMP-PNP; they then pulled out the microtubules with what they hoped were the motor proteins still attached. Vale and his team then added ATP to

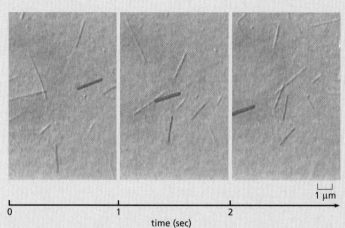

Figure 17–21 Kinesin causes microtubule gliding *in vitro*. In an *in vitro* motility assay, purified kinesin is mixed with microtubules in the presence of ATP. When a drop of the mixture is placed on a glass slide and examined by video-enhanced microscopy, individual microtubules can be seen gliding over the slide. They are driven by kinesin molecules, which attach to the glass slide by their tails. Images were recorded at 1-second intervals. The artificially colored microtubules moved at about 1–2 μm/sec. (Courtesy of Nick Carter and Rob Cross.)

release the attached proteins, and they found a 110-kilo-dalton polypeptide that could stimulate the gliding of microtubules along a glass coverslip (**Figure 17–21**). They dubbed the molecule kinesin (from the Greek *kinein*, "to move").

Similar *in vitro* motility assays have been instrumental in the study of other motor proteins—such as myosins, which move along actin filaments, as we discuss later. Subsequent studies showed that kinesin moves along microtubules from the minus end to the plus end; they also identified many other motor proteins of the kinesin family.

Lights, camera, action

Combining such assays with ever more refined microscopic techniques, researchers can now monitor the movement of individual motor proteins along single microtubules, even in living cells.

Observation of kinesin molecules coupled with green fluorescent protein (GFP) revealed that this motor protein marches along microtubules processively—that is, each molecule takes multiple "steps" along the filament (100 or so) before falling off. The length of each step is 8 nm, which corresponds to the spacing of individual tubulin dimers along the microtubule. Combining these observations with assays of ATP hydrolysis, researchers have confirmed that one molecule of ATP is hydrolyzed per step. Kinesin can move in a processive manner because it has two heads. This enables it to walk toward the plus end of the microtubule in a "hand-over-hand" fashion, each head repetitively binding and releasing the filament as it swings past the bound head in front (**Figure 17–22**). Such studies now allow us to follow the footsteps of these fascinating and industrious proteins—step by molecular step.

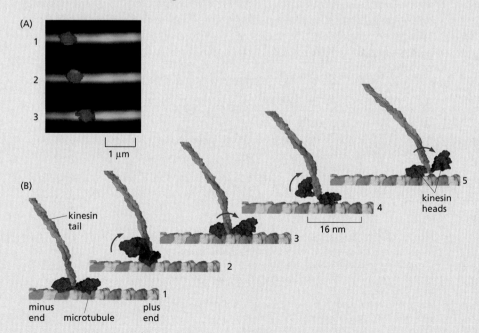

Figure 17–22 A single molecule of kinesin moves along a microtubule. (A) Three frames, separated by intervals of 1 second, record the movement of an individual kinesin-GFP molecule (*green*) along a microtubule (*red*); the labeled kinesin moves at a speed of 0.3 μm/sec. (B) A series of molecular models of the two heads of a kinesin molecule, showing how they are thought to walk processively along a microtubule in a series of 8-nm steps in which one head swings past the other (Movie 17.6). (A and B, courtesy of Ron Vale.)

Figure 17–23 Many hairlike cilia project from the surface of the epithelial cells that line the human respiratory tract. In this scanning electron micrograph, thick tufts of cilia can be seen extended from these ciliated cells, which are interspersed with the dome-shaped surfaces of nonciliated epithelial cells. (Reproduced from R.G. Kessel and R.H. Karden, Tissues and Organs. San Francisco: W.H. Freeman & Co., 1979.)

5 μm

use cilia to collect food particles, and others use them for locomotion. On the epithelial cells lining the human respiratory tract (**Figure 17–23**), huge numbers of beating cilia (more than a billion per square centimeter) sweep layers of mucus containing trapped dust particles and dead cells up toward the throat, to be swallowed and eventually eliminated from the body. Similarly, beating cilia on the cells of the oviduct wall create a current that helps to carry eggs along the oviduct. Each cilium acts as a small oar, moving in a repeated cycle that generates the movement of fluid over the cell surface (**Figure 17–24**).

The **flagella** (singular flagellum) that propel sperm and many protozoa are much like cilia in their internal structure but are usually very much longer. They are designed to move the entire cell, rather than moving fluid across the cell surface. Flagella propagate regular waves along their length, propelling the attached cell along (**Figure 17–25**).

The microtubules in cilia and flagella are slightly different from cytoplasmic microtubules; they are arranged in a curious and distinctive pattern, which was one of the most striking revelations of early electron microscopy. A cross section through a cilium shows nine doublet microtubules arranged in a ring around a pair of single microtubules (**Figure 17–26A**). This "9 + 2" array is characteristic of almost all eukaryotic cilia and flagella—from those of protozoa to those in humans.

The movement of a cilium or a flagellum is produced by the bending of its core as the microtubules slide against each other. The microtubules are associated with numerous accessory proteins (**Figure 17–26B**), which project at regular positions along the length of the microtubule bundle. Some of these proteins serve as cross-links to hold the bundle of microtubules together; others generate the force that causes the cilium to bend.

The most important of the accessory proteins is the motor protein *ciliary dynein*, which generates the bending motion of the core. It closely

power stroke

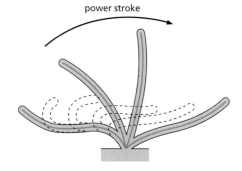

Figure 17–24 A cilium beats by performing a repetitive cycle of movements, consisting of a power stroke followed by a recovery stroke. In the fast power stroke, the cilium is fully extended and fluid is driven over the surface of the cell; in the slower recovery stroke, the cilium curls back into position with minimal disturbance to the surrounding fluid. Each cycle typically requires 0.1–0.2 second and generates a force parallel to the cell surface.

Figures 17–25 Flagella propel a cell through fluid using repetitive wavelike motion. The movement of a single flagellum on an invertebrate sperm is seen in a series of images captured by stroboscopic illumination at 400 flashes per second. (Courtesy of Charles J. Brokaw.)

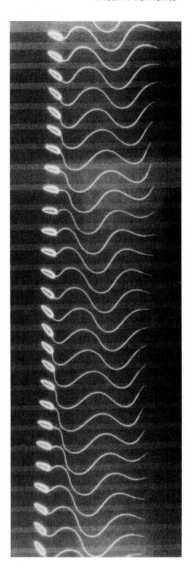

resembles cytoplasmic dynein and functions in much the same way. Ciliary dynein is attached by its tail to one microtubule, while its two heads interact with an adjacent microtubule to generate a sliding force between the two microtubules. Because of the multiple links that hold the adjacent microtubule doublets together, the sliding force between adjacent microtubules is converted to a bending motion in the cilium (**Figure 17–27**). In humans, hereditary defects in ciliary dynein cause Kartagener's syndrome. Men with this disorder are infertile because their sperm are nonmotile, and they have an increased susceptibility to bronchial infections because the cilia that line their respiratory tract are paralyzed and thus unable to clear bacteria and debris from the lungs.

Many animal cells that lack beating cilia contain a single, nonmotile *primary cilium*. This appendage is much shorter than a beating cilium and functions as an antenna for sensing certain extracellular signal molecules.

ACTIN FILAMENTS

Actin filaments, polymers of the protein **actin**, are present in all eukaryotic cells and are essential for many of the cell's movements, especially those involving the cell surface. Without actin filaments, for example, an animal cell could not crawl along a surface, engulf a large particle by phagocytosis, or divide in two. Like microtubules, many actin filaments are unstable, but by associating with other proteins they can also form stable structures in cells, such as the contractile apparatus of muscle cells. Actin filaments interact with a large number of *actin-binding proteins* that enable the filaments to serve a variety of functions in cells. Depending on which of these proteins they associate with, actin filaments can form stiff and stable structures, such as the *microvilli* on the epithelial cells lining

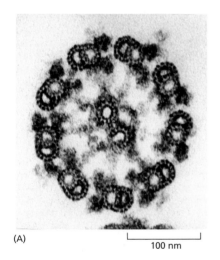

(A) |———— 100 nm ————|

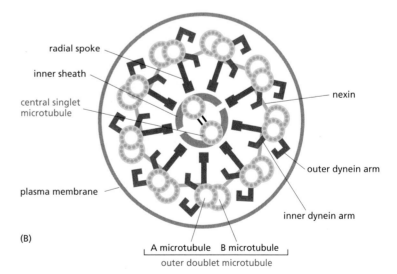

radial spoke
inner sheath
central singlet microtubule
plasma membrane
nexin
outer dynein arm
inner dynein arm

A microtubule   B microtubule
outer doublet microtubule

(B)

Figure 17–26 Microtubules in a cilium or flagellum are arranged in a "9 + 2" array. (A) Electron micrograph of a flagellum of the unicellular alga *Chlamydomonas* shown in cross section, illustrating the distinctive 9 + 2 arrangement of microtubules. (B) Diagram of the flagellum in cross section. The nine outer microtubules (each a special paired structure) carry two rows of dynein molecules. The heads of each dynein molecule appear in this view like arms reaching toward the adjacent doublet microtubule. In a living cilium, these dynein heads periodically make contact with the adjacent doublet microtubule and move along it, thereby producing the force for ciliary beating. The various other links and projections shown are proteins that serve to hold the bundle of microtubules together and to convert the sliding force produced by dyneins into bending, as illustrated in Figure 17–27. (A, courtesy of Lewis Tilney.)

Figure 17–27 **The movement of dynein causes the flagellum to bend.** (A) If the outer doublet microtubules and their associated dynein molecules are freed from other components of a sperm flagellum and then exposed to ATP, the doublets slide against each other, telescope-fashion, due to the repetitive action of their associated dyneins. (B) In an intact flagellum, however, the doublets are tied to each other by flexible protein links so that the action of the system produces bending rather than sliding.

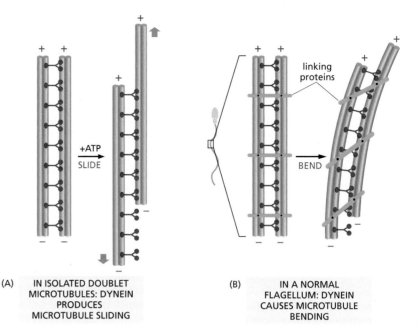

(A) IN ISOLATED DOUBLET MICROTUBULES: DYNEIN PRODUCES MICROTUBULE SLIDING

(B) IN A NORMAL FLAGELLUM: DYNEIN CAUSES MICROTUBULE BENDING

QUESTION 17–4

Dynein arms in a cilium are arranged so that, when activated, the heads push their neighboring outer doublet outward toward the tip of the cilium. Consider a cross section of a cilium (see Figure 17–26). Why would no bending motion of the cilium result if all dynein molecules were active at the same time? What pattern of dynein activity can account for the bending of a cilium in one direction?

the intestine (**Figure 17–28A**) or the small *contractile bundles* that can contract and act like tiny muscles in most animal cells (**Figure 17–28B**). They can also form temporary structures, such as the dynamic protrusions formed at the leading edge of a crawling cell (**Figure 17–28C**) or the *contractile ring* that pinches the cytoplasm in two when an animal cell divides (**Figure 17–28D**). Actin-dependent movements usually require actin's association with a motor protein called *myosin*.

In this section, we see how the arrangements of actin filaments in a cell depend on the types of actin-binding proteins present. Even though actin filaments and microtubules are formed from unrelated types of subunit proteins, we will see that the principles by which they assemble and disassemble, control cell structure, and work with motor proteins to bring about movement are strikingly similar.

Actin Filaments Are Thin and Flexible

Actin filaments appear in electron micrographs as threads about 7 nm in diameter. Each filament is a twisted chain of identical globular actin monomers, all of which "point" in the same direction along the axis of the chain. Like a microtubule, therefore, an actin filament has a structural polarity, with a plus end and a minus end (**Figure 17–29**).

Actin filaments are thinner, more flexible, and usually shorter than microtubules. There are, however, many more of them, so that the total length of all the actin filaments in a cell is generally many times greater than the total length of all of the microtubules. Unlike intermediate filaments and

Figure 17–28 **Actin filaments allow animal cells to adopt a variety of shapes and perform a variety of functions.** The actin filaments in four different structures are shown here in *red*: (A) microvilli; (B) contractile bundles in the cytoplasm; (C) fingerlike *filopodia* protruding from the leading edge of a moving cell; (D) contractile ring during cell division.

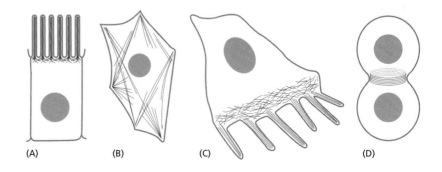

(A)   (B)   (C)   (D)

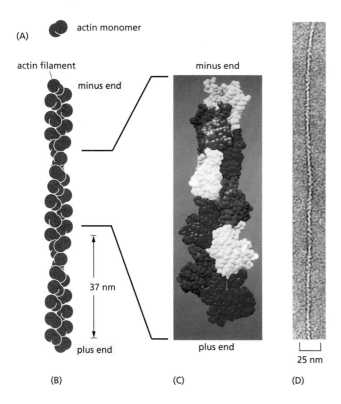

(A) actin monomer

actin filament

minus end

minus end

37 nm

plus end

(B)

minus end

plus end

(C)

25 nm

(D)

Figure 17–29 Actin filaments are thin, flexible protein threads. (A) The subunit of each actin filament is an actin monomer. A cleft in the monomer provides a binding site for ATP or ADP. (B) Arrangement of actin monomers in an actin filament. Each filament may be thought of as a two-stranded helix with a twist repeating every 37 nm. Multiple, lateral interactions between the two strands prevent the strands from separating. (C) Close-up view showing the identical subunits of an actin filament in different colors to emphasize the close interactions between each actin molecule and its four nearest neighbors. (D) Electron micrograph of a negatively stained actin filament. (C, from K.C. Holmes et al., *Nature* 347:44–49, 1990. With permission from Macmillan Publishers Ltd; D, courtesy of Roger Craig.)

microtubules, actin filaments rarely occur in isolation in the cell; they are generally found in cross-linked bundles and networks, which are much stronger than the individual filaments.

Actin and Tubulin Polymerize by Similar Mechanisms

Although actin filaments can grow by the addition of actin monomers at either end, like microtubules, their rate of growth is faster at the plus end than at the minus end. A naked actin filament, like a microtubule without associated proteins, is inherently unstable, and it can disassemble from both ends. In living cells, free actin monomers carry a tightly bound nucleoside triphosphate, in this case ATP. The actin monomer hydrolyzes its bound ATP to ADP soon after it is incorporated into the filament. As with the hydrolysis of GTP to GDP in a microtubule, hydrolysis of ATP to ADP in an actin filament reduces the strength of binding between the monomers, thereby decreasing the stability of the polymer. Thus in both cases, nucleotide hydrolysis promotes depolymerization, helping the cell to disassemble its microtubules and actin filaments after they have formed.

If the concentration of free actin monomers is very high, an actin filament will grow rapidly, adding monomers at both ends. At intermediate concentrations of free actin, however, something interesting takes place. Actin monomers add to the plus end at a rate faster than the bound ATP can be hydrolyzed, so the plus end grows. At the minus end, by contrast, ATP is hydrolyzed faster than new monomers can be added; because ADP-actin destabilizes the structure, the filament loses subunits from its minus end at the same time as it adds them to the plus end (**Figure 17–30**). Inasmuch as an individual monomer moves through the filament from the plus to the minus end, this behavior is called *treadmilling*.

Both the treadmilling of actin filaments and the dynamic instability of microtubules rely on the hydrolysis of a bound nucleoside triphosphate to regulate the length of the polymer. But the result is usually different.

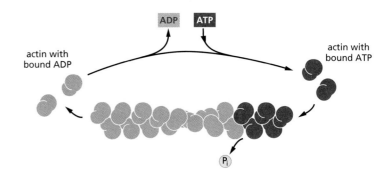

Figure 17–30 ATP hydrolysis decreases the stability of the actin polymer. Actin monomers in the cytosol carry ATP, which is hydrolyzed to ADP soon after assembly into a growing filament. The ADP molecules remain trapped within the actin filament, unable to exchange with ATP until the actin monomer that carries them dissociates from the filament.

Treadmilling involves a simultaneous gain of monomers at the plus end of an actin filament and loss at the minus end: when the rates of addition and loss are equal, the filament remains the same size (**Figure 17–31A**). Dynamic instability, on the other hand, involves a rapid switch from growth to shrinkage (or from shrinkage to growth) at only the plus end of the microtubule. As a result, microtubules tend to undergo more drastic changes in length than do actin filaments—either growing rapidly or collapsing rapidly (**Figure 17–31B**).

Actin filament function can be perturbed experimentally by certain toxins produced by fungi or marine sponges. Some, such as *cytochalasin* and *latrunculin*, prevent actin polymerization; others, such as *phalloidin*, stabilize actin filaments against depolymerization (**Table 17–2**). Addition of these toxins to the medium bathing cells or tissues, even in low concentrations, instantaneously freezes cell movements such as cell locomotion. Thus as with microtubules, many of the functions of actin filaments depend on the ability of the filament to assemble and disassemble, the rates of which depend on the dynamic equilibrium between the actin filaments, the pool of actin monomers, and various actin-binding proteins.

Many Proteins Bind to Actin and Modify Its Properties

About 5% of the total protein in a typical animal cell is actin; about half of this actin is assembled into filaments, and the other half remains as actin monomers in the cytosol. Thus unlike the situation for tubulin dimers, the concentration of actin monomers is high—much higher than the concentration required for purified actin monomers to polymerize spontaneously in a test tube. What, then, keeps the actin monomers in cells from polymerizing totally into filaments? The answer is that cells contain

QUESTION 17–5

The formation of actin filaments in the cytosol is controlled by actin-binding proteins. Some actin-binding proteins significantly increase the rate at which the formation of an actin filament is initiated. Suggest a mechanism by which they might do this.

Figure 17–31 Treadmilling of actin filaments and dynamic instability of microtubules regulate polymer length in different ways. (A) Treadmilling occurs when ATP-actin adds to the plus end of an actin filament at the same time that ADP-actin is lost from the minus end. When the rates of addition and loss are equal, the filament stays the same length—although individual actin monomers (three of which are numbered) move through the filament from the plus to the minus end. (B) In dynamic instability, GTP-tubulin adds to the plus end of a growing microtubule. As discussed earlier, when GTP-tubulin addition is faster than GTP hydrolysis, a GTP cap forms at that end; when the rate of addition slows, the GTP cap is lost, and the filament experiences catastrophic shrinkage via the loss of GDP-tubulin from the same end. The microtubule will shrink until the GTP cap is regained—or until the microtubule disappears (see Figure 17–15).

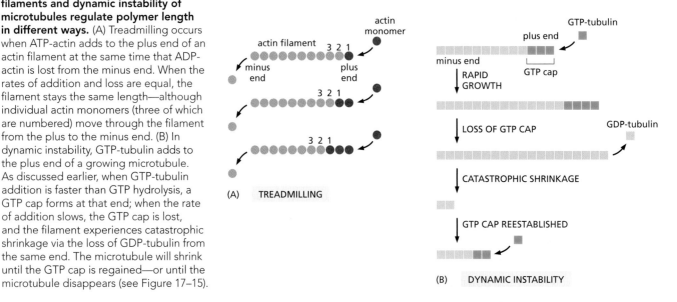

TABLE 17–2 DRUGS THAT AFFECT ACTIN FILAMENTS	
Actin-specific drugs	
Phalloidin	binds and stabilizes filaments
Cytochalasin	caps filament plus ends, preventing polymerization there
Latrunculin	binds actin monomers and prevents their polymerization

small proteins, such as *thymosin* and *profilin*, that bind to actin monomers in the cytosol, preventing them from adding to the ends of actin filaments. By keeping actin monomers in reserve until they are required, these proteins play a crucial role in regulating actin polymerization. When actin filaments are needed, other actin-binding proteins such as *formins* and *actin-related proteins* (*ARPs*) promote actin polymerization.

There are a great many **actin-binding proteins** in cells. Most of these bind to assembled actin filaments rather than to actin monomers and control the behavior of the intact filaments (**Figure 17–32**). Actin-bundling proteins, for example, hold actin filaments together in parallel bundles in microvilli; others cross-link actin filaments together in a gel-like meshwork within the *cell cortex*—the specialized layer of actin-filament-rich

Figure 17–32 Actin-binding proteins control the behavior of actin filaments in vertebrate cells. Actin is shown in *red*, and the actin-binding proteins are shown in *green*.

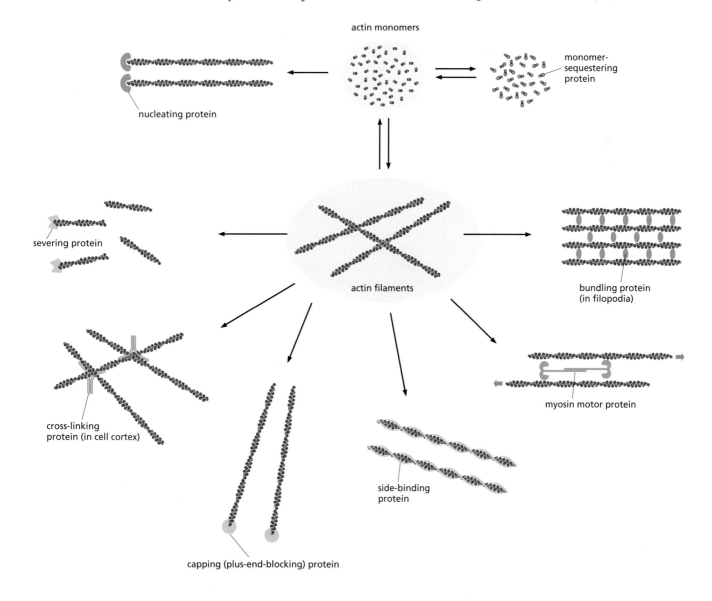

cytoplasm just beneath the plasma membrane. Filament-severing proteins fragment actin filaments into shorter lengths and thus can convert an actin gel to a more fluid state. Actin filaments can also associate with myosin motor proteins to form contractile bundles, as in muscle cells. And they often form tracks along which myosin motor proteins transport organelles, a function that is especially conspicuous in plant cells.

In the remainder of this chapter, we consider some characteristic structures that actin filaments can form, and we discuss how different types of actin-binding proteins are involved in their assembly. We begin with the cell cortex and its role in cell locomotion, and we conclude with the contractile apparatus of muscle cells.

A Cortex Rich in Actin Filaments Underlies the Plasma Membrane of Most Eukaryotic Cells

Although actin is found throughout the cytoplasm of a eukaryotic cell, in most cells it is highly concentrated in a layer just beneath the plasma membrane. In this region, called the **cell cortex**, actin filaments are linked by actin-binding proteins into a meshwork that supports the plasma membrane and gives it mechanical strength. In human red blood cells, a simple and regular network of fibrous proteins—including actin and spectrin filaments—attaches to the plasma membrane, providing the support necessary for the cells to maintain their simple discoid shape (see Figure 11–29). The cell cortex of other animal cells, however, is thicker and more complex, and it supports a far richer repertoire of cell shapes and cell-surface movements. Like the cortex in the red blood cell, the cortex in other cells contains spectrin; however, it also includes a much denser network of actin filaments. These filaments become cross-linked into a three-dimensional meshwork, which governs cell shape and the mechanical properties of the plasma membrane: the rearrangements of actin filaments within the cortex provide much of the molecular basis for changes in both cell shape and cell locomotion.

Cell Crawling Depends on Cortical Actin

Many eukaryotic cells move by crawling over surfaces, rather than by swimming by means of beating cilia or flagella. Carnivorous amoebae crawl continually, in search of food. The advancing tip of a developing axon migrates in response to growth factors, following a path of chemical signals to its eventual synaptic target cells. White blood cells known as *neutrophils* migrate out of the blood into infected tissues when they "smell" small molecules released by bacteria, which the neutrophils seek out and destroy. For these hunters, such chemotactic molecules binding to receptors on the cell surface trigger changes in actin filament assembly that help direct the cells toward their prey (see Movie 17.7).

The molecular mechanisms of these and other forms of cell crawling entail coordinated changes of many molecules in different regions of the cell, and no single, easily identifiable locomotory organelle, such as a flagellum, is responsible. In broad terms, however, three interrelated processes are known to be essential: (1) the cell pushes out protrusions at its "front," or leading edge; (2) these protrusions adhere to the surface over which the cell is crawling; and (3) the rest of the cell drags itself forward by traction on these anchorage points (**Figure 17–33**).

All three processes involve actin, but in different ways. The first step, the pushing forward of the cell surface, is driven by actin polymerization. The leading edge of a crawling fibroblast in culture regularly extends thin, sheetlike **lamellipodia**, which contain a dense meshwork of actin filaments, oriented so that most of the filaments have their plus ends close

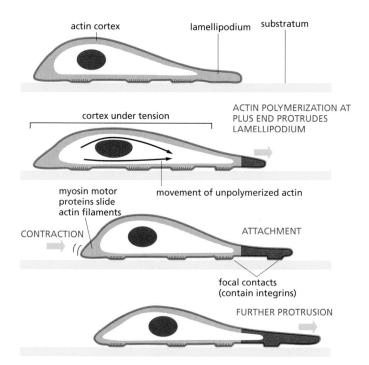

actin cortex   lamellipodium   substratum

cortex under tension

ACTIN POLYMERIZATION AT
PLUS END PROTRUDES
LAMELLIPODIUM

myosin motor
proteins slide
actin filaments

movement of unpolymerized actin

CONTRACTION   ATTACHMENT

focal contacts
(contain integrins)

FURTHER PROTRUSION

Figure 17–33 Forces generated in the actin-filament-rich cortex help move a cell forward. Actin polymerization at the leading edge of the cell *pushes* the plasma membrane forward (protrusion) and forms new regions of actin cortex, shown here in *red*. New points of anchorage are made between the bottom of the cell and the surface (substratum) on which the cell is crawling (attachment). Contraction at the rear of the cell—mediated by myosin motor proteins moving along actin filaments—then draws the body of the cell forward. New anchorage points are established at the front, and old ones are released at the back, as the cell crawls forward. The same cycle is repeated over and over again, moving the cell forward in a stepwise fashion.

to the plasma membrane. Many cells also extend thin, stiff protrusions called **filopodia**, both at the leading edge and elsewhere on their surface (**Figure 17–34**). These are about 0.1 μm wide and 5–10 μm long, and each contains a loose bundle of 10–20 actin filaments (see Figure 17–28C), again oriented with their plus ends pointing outward. The advancing tip (*growth cone*) of a developing nerve cell axon extends even longer filopodia, up to 50 μm long, which help it to probe its environment and find the correct path to its target cell. Both lamellipodia and filopodia are exploratory, motile structures that form and retract with great speed, moving at around 1 μm per second. Both are thought to be generated by the rapid local growth of actin filaments, which assemble close to the plasma membrane and elongate by the addition of actin monomers at their plus ends. In this way, the filaments push out the membrane without tearing it.

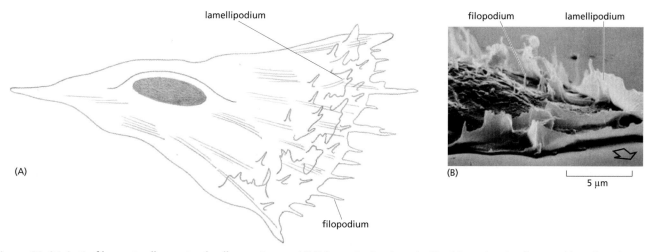

lamellipodium

filopodium   lamellipodium

(A)   (B)

5 μm

filopodium

Figure 17–34 Actin filaments allow animal cells to migrate. (A) Schematic drawing of a fibroblast, showing flattened lamellipodia and fine filopodia projecting from its surface, especially in the regions of the leading edge. (B) Scanning electron micrograph showing lamellipodia and filopodia at the leading edge of a human fibroblast migrating in culture; the arrow shows the direction of cell movement. As the cell moves forward, the lamellipodia that fail to attach to the substratum are swept backward over the upper surface of the cell— a movement referred to as ruffling. (B, courtesy of Julian Heath.)

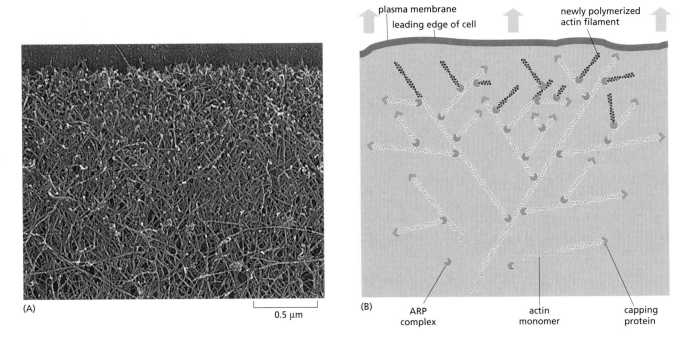

plasma membrane
leading edge of cell
newly polymerized
actin filament

(A)

0.5 μm

(B)   ARP
         complex

actin
monomer

capping
protein

Figure 17–35 A web of polymerizing actin filaments pushes the leading edge of a lamellipodium forward.
(A) A highly motile keratocyte from frog skin was fixed, dried, and shadowed with platinum, and examined in
an electron microscope. Actin filaments form a dense network, with the fast-growing plus ends of the filaments
terminating at the leading margin of the lamellipodium (top of figure). (B) Drawing showing how the nucleation of
new actin filaments (red) is mediated by ARP complexes (brown) attached to the sides of preexisting actin filaments.
The resulting branching structure pushes the plasma membrane forward. The plus ends of the actin filaments
become protected from depolymerizing by capping proteins (blue), while the minus ends of actin filaments nearer
the center of the cell continually disassemble through the action of depolymerizing proteins (not shown). The web
of actin as a whole thereby undergoes a continual rearward movement due to the assembly of filaments at the front
and their disassembly at the rear. This actin network is drawn to a different scale than the network shown in (A).
(A, courtesy of Tatyana Svitkina and Gary Borisy.)

The formation and growth of actin filaments at the leading edge of a
cell are assisted by various actin-binding proteins. The actin-related proteins—or ARPs—mentioned earlier promote the formation of a web of
branched actin filaments in lamellipodia. ARPs form complexes that bind
to the sides of existing actin filaments and nucleate the formation of new
filaments, which grow out at an angle to produce side branches. With the
aid of additional actin-binding proteins, this web undergoes continual
assembly at the leading edge and disassembly further back, pushing the
lamellipodium forward (**Figure 17–35**).

The other kind of cell protrusion, the filopodium, depends on *formins*, a
nucleating protein that attaches to the growing plus ends of actin filaments and promotes the addition of new monomers to form straight,
unbranched filaments. Formins are also used elsewhere to assemble
unbranched filaments, as in the contractile ring that pinches a dividing
animal cell in two.

When the lamellipodia and filopodia touch down on a favorable surface,
they stick: transmembrane proteins in their plasma membrane, known
as *integrins* (discussed in Chapter 20), adhere to molecules either in the
extracellular matrix or on the surface of a neighboring cell over which
the moving cell is crawling. Meanwhile, on the intracellular face of the
crawling cell's plasma membrane, integrins capture actin filaments in the
cortex, thereby creating a robust anchorage for the crawling cell (see
Figures 17–33 and 20–15C). To use this anchorage to drag its body forward, the cell calls on the help of myosin motor proteins, as we now
discuss.

QUESTION 17–6

Suppose that the actin molecules
in a cultured skin cell have been
randomly labeled in such a way
that 1 in 10,000 molecules carries
a fluorescent marker. What would
you expect to see if you examined
the lamellipodium (leading edge)
of this cell through a fluorescence
microscope? Assume that your
microscope is sensitive enough to
detect single fluorescent molecules.

Actin Associates with Myosin to Form Contractile Structures

All actin-dependent motor proteins belong to the **myosin** family. They bind to and hydrolyze ATP, which provides the energy for their movement along actin filaments toward the plus end. Myosin, along with actin, was first discovered in skeletal muscle, and much of what we know about the interaction of these two proteins was learned from studies of muscle. There are various types of myosins in cells, of which the myosin-I and myosin-II subfamilies are the most abundant. Myosin-I is present in all types of cells, whereas muscle cells make use of a specialized form of myosin-II. Because myosin-I is simpler in structure and mechanism of action, we discuss it first.

Myosin-I molecules have a head domain and a tail (**Figure 17–36A**). The head domain binds to an actin filament and has the ATP-hydrolyzing motor activity that enables it to move along the filament in a repetitive cycle of binding, detachment, and rebinding (Movie 17.8). The tail varies among the different types of myosin-I and determines what type of cargo the myosin drags along. For example, the tail may bind to a particular type of vesicle and propel it through the cell along actin filament tracks (**Figure 17–36B**), or it may bind to the plasma membrane and pull it into a different shape (**Figure 17–36C**).

Extracellular Signals Can Alter the Arrangement of Actin Filaments

We have seen that myosin and other actin-binding proteins can regulate the location, organization, and behavior of actin filaments. But the activities of these proteins are, in turn, controlled by extracellular signals, allowing the cell to rearrange its actin cytoskeleton in response to the environment.

The extracellular signal molecules that regulate the actin cytoskeleton activate a variety of cell-surface receptor proteins, which in turn activate various intracellular signaling pathways. These pathways often converge on a group of closely related *monomeric GTPase* proteins called the **Rho protein family**. As discussed in Chapter 16, monomeric GTPases behave as molecular switches that control intracellular processes by

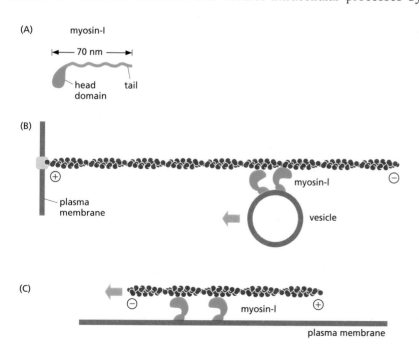

Figure 17–36 Myosin-I is the simplest myosin. (A) Myosin-I has a single globular head that attaches to an actin filament and a tail that attaches to another molecule or organelle in the cell. (B) This arrangement allows the head domain to move a vesicle relative to an actin filament, which in this case is anchored to the plasma membrane. (C) Myosin-I can also bind to an actin filament in the cell cortex, ultimately pulling the plasma membrane into a new shape. Note that the head group always walks toward the plus end of the actin filament.

Figure 17–37 Activation of Rho family GTPases can have a dramatic effect on the organization of actin filaments in fibroblasts. In these micrographs, actin is stained with fluorescently labeled phalloidin, a molecule that binds specifically to actin filaments (see Table 17–2, p. 587). (A) Unstimulated fibroblasts have actin filaments primarily in the cortex. (B) Microinjection of an activated form of Rho promotes the rapid assembly of bundles of long, unbranched actin filaments; because myosin is associated with these bundles, they are contractile. (C) Microinjection of an activated form of Rac, a GTP-binding protein similar to Rho, causes the formation of an enormous lamellipodium that extends from the entire circumference of the cell. (D) Microinjection of an activated form of Cdc42, another Rho family member, stimulates the protrusion of many long filopodia at the cell periphery. (From A. Hall, *Science* 279:509–514, 1998. With permission from AAAS.)

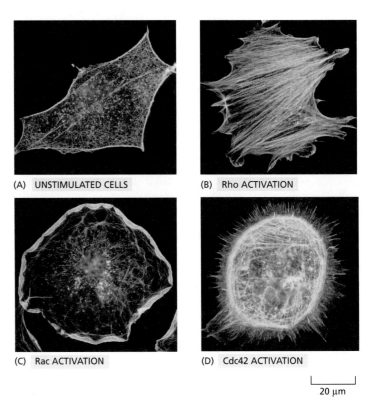

(A) UNSTIMULATED CELLS

(B) Rho ACTIVATION

(C) Rac ACTIVATION

(D) Cdc42 ACTIVATION

20 μm

QUESTION 17–7

At the leading edge of a crawling cell, the plus ends of actin filaments are located close to the plasma membrane, and actin monomers are added at these ends, pushing the membrane outward to form lamellipodia or filopodia. What do you suppose holds the filaments at their other ends to prevent them from just being pushed into the cell's interior?

---

cycling between an active GTP-bound state and an inactive GDP-bound state (see Figure 16–15B). In the case of the actin cytoskeleton, activation of different members of the Rho family affects the organization of actin filaments in different ways. For example, activation of one Rho family member triggers actin polymerization and filament bundling to form filopodia; activation of another promotes lamellipodia formation and ruffling; and activation of Rho itself drives the bundling of actin filaments with myosin-II and the clustering of cell-surface integrins, thus promoting cell crawling (**Figure 17–37**).

These dramatic and complex structural changes occur because the Rho family GTP-binding proteins, together with the protein kinases and accessory proteins with which they interact, act like a computational network to control actin organization and dynamics. This network receives external signals from nutrients, growth factors, and contacts with neighboring cells and the extracellular matrix, along with intracellular information about the cell's metabolic state and readiness for division. The Rho network then processes these inputs and activates intracellular signaling pathways that shape the actin cytoskeleton—for example, by activating the formin proteins that promote the formation of filopodia or by stimulating ARP complexes at the leading edge of the cell to generate large lamellipodia.

One of the most rapid rearrangements of cytoskeletal elements occurs when a muscle fiber contracts in response to a signal from a motor nerve, as we now discuss.

MUSCLE CONTRACTION

Muscle contraction is the most familiar and best understood of animal cell movements. In vertebrates, running, walking, swimming, and flying all depend on the ability of *skeletal muscle* to contract strongly and move various bones. Involuntary movements such as heart pumping and gut peristalsis depend on *cardiac muscle* and *smooth muscle*, respectively,

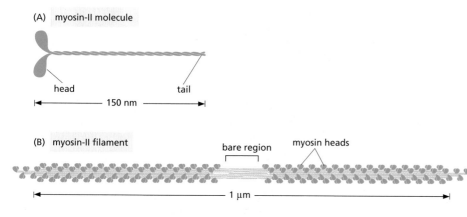

(A) myosin-II molecule

head   tail

|← 150 nm →|

(B) myosin-II filament

bare region   myosin heads

|← 1 μm →|

Figure 17–38 Myosin-II molecules can associate with one another to form myosin filaments. (A) A molecule of myosin-II contains two identical heavy chains, each with a globular head and an extended tail. (It also contains two light chains bound to each head, but these are not shown.) The tails of the two heavy chains form a single coiled-coil tail. (B) The coiled-coil tails of myosin-II molecules associate with one another to form a bipolar myosin filament in which the heads project outward from the middle in opposite directions. The bare region in the middle of the filament consists of tails only.

which are formed from muscle cells that differ in structure from skeletal muscle but use actin and myosin in a similar way to contract. Although muscle cells are highly specialized, many cell movements—from the locomotion of whole cells down to the motion of some components inside cells—also depend on the interaction of actin and myosin. Much of our understanding of the mechanisms of cell movement originated from studies of muscle cell contraction. In this section, we discuss how actin and myosin interact to produce this contraction.

Muscle Contraction Depends on Interacting Filaments of Actin and Myosin

Muscle myosin belongs to the **myosin-II** subfamily of myosins, all of which are dimers, with two globular ATPase heads at one end and a single coiled-coil tail at the other (**Figure 17–38A**). Clusters of myosin-II molecules bind to each other through their coiled-coil tails, forming a bipolar **myosin filament** from which the heads project (**Figure 17–38B**).

The myosin filament is like a double-headed arrow, with the two sets of myosin heads pointing in opposite directions, away from the middle. One set binds to actin filaments in one orientation and moves the filaments one way; the other set binds to other actin filaments in the opposite orientation and moves the filaments in the opposite direction. As a result, a myosin filament slides sets of oppositely oriented actin filaments past one another (**Figure 17–39**). We can see how, therefore, if actin filaments and myosin filaments are organized together in a bundle, the bundle can generate a strong contractile force. This is seen most clearly in muscle contraction, but it also occurs in the much smaller *contractile bundles* of actin filaments and myosin-II filaments (see Figure 17–28B) that assemble transiently in nonmuscle cells, and in the *contractile ring* that pinches a dividing cell in two by contracting and pulling inward on the plasma membrane (see Figure 17–28D).

QUESTION 17–8
If both the actin and myosin filaments of muscle are made up of subunits held together by weak noncovalent bonds, how is it possible for a human being to lift heavy objects?

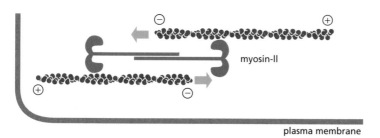

myosin-II

plasma membrane

Figure 17–39 A small, bipolar myosin-II filament can slide two actin filaments of opposite orientation past each other. This sliding movement mediates the contraction of interacting actin and myosin-II filaments in both muscle and nonmuscle cells. As with myosin-I, a myosin-II head group walks toward the plus end of the actin filament with which it interacts.

(A)

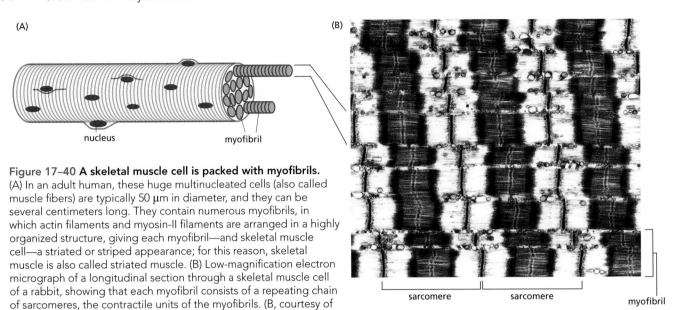

nucleus

myofibril

Figure 17–40 A skeletal muscle cell is packed with myofibrils. (A) In an adult human, these huge multinucleated cells (also called muscle fibers) are typically 50 μm in diameter, and they can be several centimeters long. They contain numerous myofibrils, in which actin filaments and myosin-II filaments are arranged in a highly organized structure, giving each myofibril—and skeletal muscle cell—a striated or striped appearance; for this reason, skeletal muscle is also called striated muscle. (B) Low-magnification electron micrograph of a longitudinal section through a skeletal muscle cell of a rabbit, showing that each myofibril consists of a repeating chain of sarcomeres, the contractile units of the myofibrils. (B, courtesy of Roger Craig.)

(B)

sarcomere   sarcomere

myofibril

Actin Filaments Slide Against Myosin Filaments During Muscle Contraction

Skeletal muscle fibers are huge, multinucleated individual cells formed by the fusion of many separate smaller cells. The nuclei of the contributing cells are retained in the muscle fiber and lie just beneath the plasma membrane. The bulk of the cytoplasm is made up of **myofibrils**, the contractile elements of the muscle cell. These cylindrical structures are 1–2 μm in diameter and may be as long as the muscle cell itself (**Figure 17–40A**).

A myofibril consists of a chain of identical tiny contractile units, or **sarcomeres**. Each sarcomere is about 2.5 μm long, and the repeating pattern of sarcomeres gives the vertebrate myofibril a striped appearance (**Figure 17–40B**). Sarcomeres are highly organized assemblies of two types of filaments—actin filaments and myosin filaments composed of a muscle-specific form of myosin-II. The myosin filaments (the *thick filaments*) are centrally positioned in each sarcomere, whereas the more slender actin filaments (the *thin filaments*) extend inward from each end of the sarcomere, where they are anchored by their plus ends to a structure known as the *Z disc*. The minus ends of the actin filaments overlap the ends of the myosin filaments (**Figure 17–41**).

The contraction of a muscle cell is caused by a simultaneous shortening of all the cell's sarcomeres, which is caused by the actin filaments sliding past the myosin filaments, with no change in the length of either type of filament (**Figure 17–42**). The sliding motion is generated by myosin heads that project from the sides of the myosin filament and interact with adjacent actin filaments (see Figure 17–39). When a muscle is stimulated

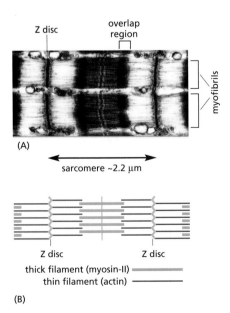

Z disc   overlap region

myofibrils

(A)

sarcomere ~2.2 μm

Z disc   Z disc

thick filament (myosin-II)
thin filament (actin)

(B)

Figure 17–41 Sarcomeres are the contractile units of muscle. (A) Detail of the electron micrograph from Figure 17–40 showing two myofibrils; the length of one sarcomere and the region where the actin and myosin filaments overlap are indicated. (B) Schematic diagram of a single sarcomere showing the origin of the light and dark bands seen in the microscope. Z discs at either end of the sarcomere are attachment points for the plus ends of actin filaments. The centrally located thick filaments (*green*) are each composed of many myosin-II molecules. The thin vertical line running down the center of the thick filament bundle in (A) corresponds to the bare regions of the myosin filaments, as seen in Figure 17–38B. (A, courtesy of Roger Craig.)

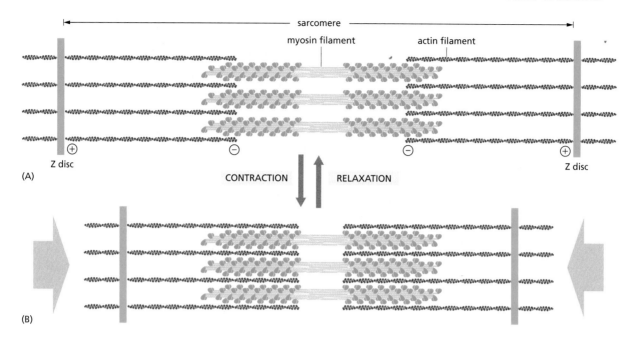

Figure 17–42 Muscles contract by a sliding-filament mechanism. (A) The myosin and actin filaments of a sarcomere overlap with the same relative polarity on either side of the midline. Recall that actin filaments are anchored by their plus ends to the Z disc and that myosin filaments are bipolar. (B) During contraction, the actin and myosin filaments slide past each other without shortening. The sliding motion is driven by the myosin heads walking toward the plus end of the adjacent actin filaments (Movie 17.9).

to contract, the myosin heads start to walk along the actin filament in repeated cycles of attachment and detachment. During each cycle, a myosin head binds and hydrolyzes one molecule of ATP. This causes a series of conformational changes that move the tip of the head by about 5 nm along the actin filament toward the plus end. This movement, repeated with each round of ATP hydrolysis, propels the myosin molecule unidirectionally along the actin filament (**Figure 17–43**). In so doing, the myosin heads pull against the actin filament, causing it to slide against the myosin filament. The concerted action of many myosin heads pulling the actin and myosin filaments past each other causes the sarcomere to contract. After a contraction is completed, the myosin heads all lose contact with the actin filaments, and the muscle relaxes.

A myosin filament has about 300 myosin heads. Each myosin head can attach and detach from actin about five times per second, allowing the myosin and actin filaments to slide past one another at speeds of up to 15 μm per second. This speed is sufficient to take a sarcomere from a fully extended state (3 μm) to a fully contracted state (2 μm) in less than one-tenth of a second. All of the sarcomeres of a muscle are coupled together and are triggered simultaneously by the signaling system we describe next, so the entire muscle contracts almost instantaneously.

Muscle Contraction Is Triggered by a Sudden Rise in Cytosolic Ca^{2+}

The force-generating molecular interaction between myosin and actin filaments takes place only when the skeletal muscle receives a signal from a motor nerve. The neurotransmitter released from the nerve terminal triggers an action potential (discussed in Chapter 12) in the muscle cell plasma membrane. This electrical excitation spreads in a matter of milliseconds into a series of membranous tubes, called *transverse* (or *T*) *tubules*, that extend inward from the plasma membrane around each myofibril. The electrical signal is then relayed to the *sarcoplasmic reticulum*, an adjacent sheath of interconnected flattened vesicles that surrounds each myofibril like a net stocking (**Figure 17–44**).

The sarcoplasmic reticulum is a specialized region of the endoplasmic reticulum in muscle cells. It contains a very high concentration of Ca^{2+}, and in response to the incoming electrical excitation, much of this Ca^{2+} is released into the cytosol through a specialized set of ion channels that

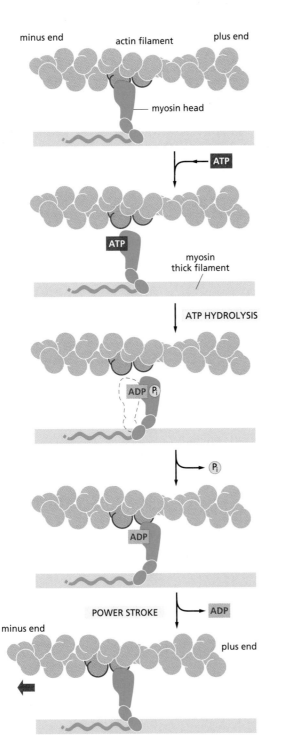

ATTACHED At the start of the cycle shown in this figure, a myosin head lacking a bound ATP or ADP is attached tightly to an actin filament in a *rigor* configuration (so named because it is responsible for *rigor mortis*, the rigidity of death). In an actively contracting muscle, this state is very short-lived, being rapidly terminated by the binding of a molecule of ATP to the myosin head.

RELEASED A molecule of ATP binds to the large cleft on the "back" of the myosin head (that is, on the side furthest from the actin filament) and immediately causes a slight change in the conformation of the domains that make up the actin-binding site. This reduces the affinity of the head for actin and allows it to move along the filament. (The space drawn here between the head and actin emphasizes this change, although in reality the head probably remains very close to the actin.)

COCKED The cleft closes like a clam shell around the ATP molecule, triggering a large shape change that causes the head to be displaced along the actin filament by a distance of about 5 nm. Hydrolysis of ATP occurs, but the ADP and inorganic phosphate (P_i) produced remain tightly bound to the myosin head.

FORCE-GENERATING A weak binding of the myosin head to a new site on the actin filament causes release of the inorganic phosphate produced by ATP hydrolysis, concomitantly with the tight binding of the head to actin. This release triggers the power stroke—the force-generating change in shape during which the head regains its original conformation. In the course of the power stroke, the head loses its bound ADP, thereby returning to the start of a new cycle.

ATTACHED At the end of the cycle, the myosin head is again bound tightly to the actin filament in a rigor configuration. Note that the head has moved to a new position on the actin filament, which has slid to the left along the myosin filament.

Figure 17–43 The head of a myosin-II molecule walks along an actin filament through an ATP-dependent cycle of conformational changes. Two actin monomers are highlighted to make the movement of the actin filament easier to see. Movie 17.10 shows actin and myosin in action. (Based on I. Rayment et al., *Science* 261:50–58, 1993. With permission from AAAS.)

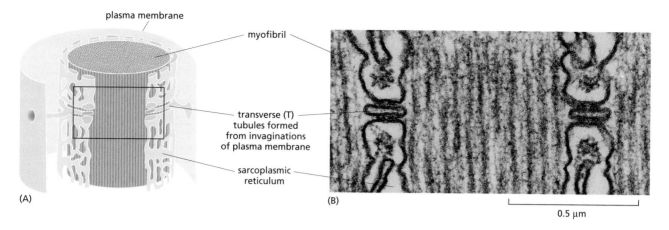

plasma membrane

myofibril

transverse (T) tubules formed from invaginations of plasma membrane

sarcoplasmic reticulum

(A)

(B)

0.5 μm

Figure 17–44 T tubules and the sarcoplasmic reticulum surround each myofibril. (A) Drawing of the two membrane systems that relay the signal to contract from the muscle cell plasma membrane to all of the myofibrils in the muscle cell. (B) Electron micrograph showing a cross section of two T tubules and their adjacent sarcoplasmic reticulum compartments. (B, courtesy of Clara Franzini-Armstrong.)

open in the sarcoplasmic reticulum membrane in response to the change in voltage across the plasma membrane and T tubules (**Figure 17–45**). As discussed in Chapter 16, Ca^{2+} is widely used as a small intracellular signal to relay a message from the exterior to the interior of cells. In muscle, the rise in cytosolic Ca^{2+} concentration activates a molecular switch made of specialized accessory proteins closely associated with the actin filaments (**Figure 17–46A**). One of these proteins is *tropomyosin*, a rigid, rod-shaped molecule that binds in the groove of the actin helix, where it prevents the myosin heads from associating with the actin filament. The other is *troponin*, a protein complex that includes a Ca^{2+}-sensitive protein associated with the end of a tropomyosin molecule. When the concentration of Ca^{2+} rises in the cytosol, Ca^{2+} binds to troponin and induces a change in the shape of the troponin complex. This in turn causes the tropomyosin molecules to shift their positions slightly, allowing myosin heads to bind to the actin filaments, initiating contraction (**Figure 17–46B**).

> **QUESTION 17–9**
>
> Compare the structure of intermediate filaments with that of the myosin-II filaments in skeletal muscle cells. What are the major similarities? What are the major differences? How do the differences in structure relate to their function?

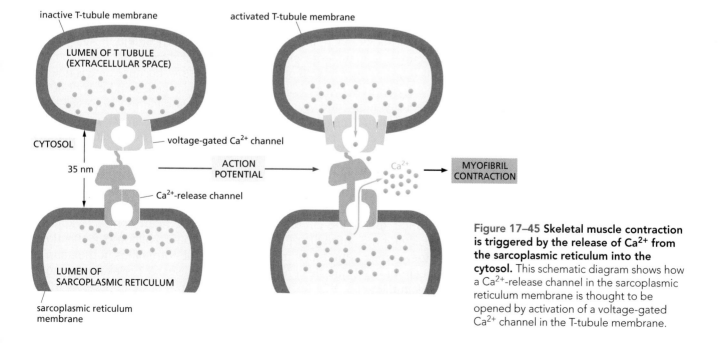

inactive T-tubule membrane

activated T-tubule membrane

LUMEN OF T TUBULE (EXTRACELLULAR SPACE)

CYTOSOL

voltage-gated Ca^{2+} channel

35 nm

Ca^{2+}-release channel

ACTION POTENTIAL

Ca^{2+}

MYOFIBRIL CONTRACTION

LUMEN OF SARCOPLASMIC RETICULUM

sarcoplasmic reticulum membrane

Figure 17–45 Skeletal muscle contraction is triggered by the release of Ca^{2+} from the sarcoplasmic reticulum into the cytosol. This schematic diagram shows how a Ca^{2+}-release channel in the sarcoplasmic reticulum membrane is thought to be opened by activation of a voltage-gated Ca^{2+} channel in the T-tubule membrane.

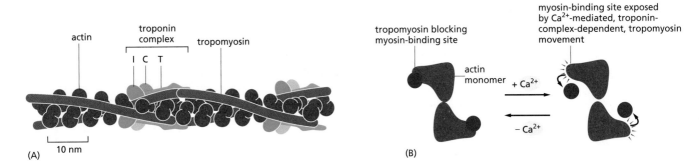

Figure 17–46 Skeletal muscle contraction is controlled by tropomyosin and troponin complexes. (A) An actin filament in muscle showing the positions of tropomyosin and troponin complexes along the filament. Every tropomyosin molecule has seven evenly spaced regions of similar amino acid sequence, each of which is thought to bind to an actin monomer in the filament. (B) When Ca^{2+} binds to a troponin complex, the complex moves the tropomyosin, which otherwise blocks the interaction of actin with the myosin heads. Here, the actin filament from (A) is shown end-on.

> ## QUESTION 17–10
>
> A. Note that in Figure 17–46, troponin molecules are evenly spaced along an actin filament, with one troponin found every seventh actin molecule. How do you suppose troponin molecules can be positioned this regularly? What does this tell you about the binding of troponin to actin filaments?
>
> B. What do you suppose would happen if you mixed actin filaments with (i) troponin alone, (ii) tropomyosin alone, or (iii) troponin plus tropomyosin, and then added myosin? Would the effects be dependent on Ca^{2+}?

Because the signal from the plasma membrane is passed within milliseconds (via the transverse tubules and sarcoplasmic reticulum) to every sarcomere in the cell, all the myofibrils in the cell contract at the same time. The increase in Ca^{2+} in the cytosol is transient because, when the nerve signal terminates, the Ca^{2+} is rapidly pumped back into the sarcoplasmic reticulum by abundant Ca^{2+}-pumps in its membrane (discussed in Chapter 12). As soon as the Ca^{2+} concentration returns to the resting level, troponin and tropomyosin molecules move back to their original positions. This reconfiguration once again blocks myosin binding to actin filaments, thereby ending the contraction.

Different Types of Muscle Cells Perform Different Functions

The highly specialized contractile machinery in muscle cells is thought to have evolved from the simpler contractile bundles of myosin and actin filaments found in all eukaryotic cells. The myosin-II in nonmuscle cells is also activated by a rise in cytosolic Ca^{2+}, but the mechanism of activation is different from that of the muscle-specific myosin-II. An increase in Ca^{2+} leads to the phosphorylation of nonmuscle myosin-II, which alters the myosin conformation and enables it to interact with actin. A similar activation mechanism operates in *smooth muscle*, which is present in the walls of the stomach, intestine, uterus, and arteries, and in many other structures that undergo slow and sustained involuntary contractions. This mode of myosin activation is relatively slow, because time is needed for enzyme molecules to diffuse to the myosin heads and carry out the phosphorylation and subsequent dephosphorylation. However, this mechanism has the advantage that—unlike the mechanism used by skeletal muscle cells—it can be activated by a variety of extracellular signals: thus smooth muscle, for example, is triggered to contract by adrenaline, serotonin, prostaglandins, and several other signal molecules.

In addition to skeletal and smooth muscle, other forms of muscle each perform a specific mechanical function. Heart—or *cardiac*—muscle, for instance, drives the circulation of blood. The heart contracts autonomously for the entire life of the organism—some 3 billion (3×10^9) times in an average human lifetime. Even subtle abnormalities in the actin or myosin of heart muscle can lead to serious disease. For example, mutations in the genes that encode cardiac myosin-II or other proteins in the sarcomere cause familial hypertrophic cardiomyopathy, a hereditary disorder responsible for sudden death in young athletes.

The contraction of muscle cells represents a highly specialized use of the basic components of the eukaryotic cytoskeleton. In the following chapter, we discuss the crucial roles the cytoskeleton has in perhaps the most fundamental cell movements of all: the segregation of newly replicated chromosomes and the formation of two daughter cells during the process of cell division.

ESSENTIAL CONCEPTS

- The cytoplasm of a eukaryotic cell is supported and organized by a cytoskeleton of intermediate filaments, microtubules, and actin filaments.

- Intermediate filaments are stable, ropelike polymers—built from fibrous protein subunits—that give cells mechanical strength. Some intermediate filaments form the nuclear lamina that supports and strengthens the nuclear envelope; others are distributed throughout the cytoplasm.

- Microtubules are stiff, hollow tubes formed by globular tubulin dimers. They are polarized structures, with a slow-growing minus end and a fast-growing plus end.

- Microtubules grow out from organizing centers such as the centrosome, in which the minus ends remain embedded.

- Many microtubules display dynamic instability, alternating rapidly between growth and shrinkage. Shrinkage is promoted by the hydrolysis of the GTP that is tightly bound to tubulin dimers, reducing the affinity of the dimers for their neighbors and thereby promoting microtubule disassembly.

- Microtubules can be stabilized by localized proteins that capture the plus ends, thereby helping to position the microtubules and harness them for specific functions.

- Kinesins and dyneins are microtubule-associated motor proteins that use the energy of ATP hydrolysis to move unidirectionally along microtubules. They carry specific organelles, vesicles, and other types of cargo to particular locations in the cell.

- Eukaryotic cilia and flagella contain a bundle of stable microtubules. Their rhythmic beating is caused by bending of the microtubules, driven by the ciliary dynein motor protein.

- Actin filaments are helical polymers of globular actin monomers. They are more flexible than microtubules and are generally found in bundles or networks.

- Like microtubules, actin filaments are polarized, with a fast-growing plus end and a slow-growing minus end. Their assembly and disassembly are controlled by the hydrolysis of ATP tightly bound to each actin monomer and by various actin-binding proteins.

- The varied arrangements and functions of actin filaments in cells stem from the diversity of actin-binding proteins, which can control actin polymerization, cross-link actin filaments into loose networks or stiff bundles, attach actin filaments to membranes, or move two adjacent filaments relative to each other.

- A concentrated network of actin filaments underneath the plasma membrane forms the bulk of the cell cortex, which is responsible for the shape and movement of the cell surface, including the movements involved when a cell crawls along a surface.

- Myosins are motor proteins that use the energy of ATP hydrolysis to move along actin filaments. In nonmuscle cells, myosin-I can carry organelles or vesicles along actin-filament tracks, and myosin-II can cause adjacent actin filaments to slide past each other in contractile bundles.

- In skeletal muscle cells, repeating arrays of overlapping filaments of actin and myosin-II form highly ordered myofibrils, which contract as these filaments slide past each other.

- Muscle contraction is initiated by a sudden rise in cytosolic Ca^{2+}, which delivers a signal to the myofibrils via Ca^{2+}-binding proteins associated with the actin filaments.

KEY TERMS

actin-binding protein	kinesin
actin filament	lamellipodium
cell cortex	microtubule
centriole	microtubule-associated protein
centrosome	motor protein
cilium	myofibril
cytoskeleton	myosin
dynamic instability	myosin filament
dynein	nuclear lamina
filopodium	polarity
flagellum	Rho protein family
intermediate filament	sarcomere
keratin filament	tubulin

QUESTIONS

QUESTION 17–11

Which of the following statements are correct? Explain your answers.

A. Kinesin moves endoplasmic reticulum membranes along microtubules so that the network of ER tubules becomes stretched throughout the cell.

B. Without actin, cells can form a functional mitotic spindle and pull their chromosomes apart but cannot divide.

C. Lamellipodia and filopodia are "feelers" that a cell extends to find anchor points on the substratum that it will then crawl over.

D. GTP is hydrolyzed by tubulin to cause the bending of flagella.

E. Cells having an intermediate-filament network that cannot be depolymerized would die.

F. The plus ends of microtubules grow faster because they have a larger GTP cap.

G. The transverse tubules in muscle cells are an extension of the plasma membrane, with which they are continuous; similarly, the sarcoplasmic reticulum is an extension of the endoplasmic reticulum.

H. Activation of myosin movement on actin filaments is triggered by the phosphorylation of troponin in some situations and by Ca^{2+} binding to troponin in others.

QUESTION 17–12

The average time taken for a molecule or an organelle to diffuse a distance of x cm is given by the formul

$$t = x^2/2D$$

where t is the time in seconds and D is a constant called the diffusion coefficient for the molecule or particle. Using the above formula, calculate the time it would take for a small molecule, a protein, and a membrane vesicle to diffuse from one side to another of a cell 10 μm across. Typical diffusion coefficients in units of cm^2/sec are: small molecule, 5×10^{-6}; protein molecule, 5×10^{-7}; vesicle, 5×10^{-8}. How long would a membrane vesicle take to reach the end of an axon 10 cm long by free diffusion? How long would it take if it was transported along microtubules at 1 μm/sec?

QUESTION 17–13

Why do eukaryotic cells, and especially animal cells, have such large and complex cytoskeletons? List the differences between animal cells and bacteria that depend on the eukaryotic cytoskeleton.

QUESTION 17–14

Examine the structure of an intermediate filament shown in Figure 17–4. Does the filament have a unique polarity—that is, could you distinguish one end from the other by chemical or other means? Explain your answer.

QUESTION 17–15

There are no known motor proteins that move on intermediate filaments. Suggest an explanation for this.

QUESTION 17–16

When cells enter mitosis, their existing array of cytoplasmic microtubules has to be rapidly broken down and replaced with the mitotic spindle that forms to pull the chromosomes into the daughter cells. The enzyme katanin, named after Japanese samurai swords, is activated during the onset of mitosis, and chops microtubules into short pieces. What do you suppose is the fate of the microtubule fragments created by katanin? Explain your answer.

QUESTION 17–17

The drug Taxol, extracted from the bark of yew trees, has an opposite effect to the drug colchicine, an alkaloid from autumn crocus. Taxol binds tightly to microtubules and stabilizes them; when added to cells, it causes much of the free tubulin to assemble into microtubules. In contrast, colchicine prevents microtubule formation. Taxol is just as pernicious to dividing cells as colchicine, and both are used as anticancer drugs. Based on your knowledge of microtubule dynamics, suggest why both drugs are toxic to dividing cells despite their opposite actions.

QUESTION 17–18

A useful technique for studying microtubule motors is to attach them by their tails to a glass coverslip (which can be accomplished quite easily because the tails stick avidly to a clean glass surface) and then allow them to settle. The microtubules may then be viewed in a light microscope as they are propelled over the surface of the coverslip by the heads of the motor proteins. Because the motor proteins attach at random orientations to the coverslip, however, how can they generate coordinated movement of individual microtubules rather than engaging in a tug-of-war? In which direction will microtubules crawl on a 'bed' of kinesin molecules (i.e., will they move plus-end first, or minus-end first)?

QUESTION 17–19

A typical time course of polymerization of purified tubulin to form microtubules is shown in **Figure Q17–19**.

A. Explain the different parts of the curve (labeled A, B, and C). Draw a diagram that shows the behavior of tubulin molecules in each of the three phases.

B. How would the curve in the figure change if centrosomes were added at the outset?

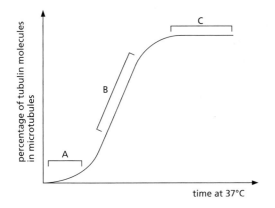

Figure Q17–19

QUESTION 17–20

The electron micrographs shown in **Figure Q17–20A** were obtained from a population of microtubules that were growing rapidly. **Figure Q17–20B** was obtained from microtubules undergoing "catastrophic" shrinking. Comment on any differences between A and B, and suggest likely explanations for the differences that you observe.

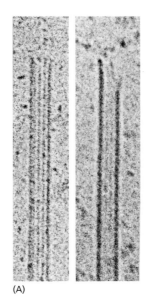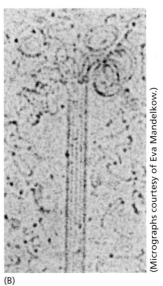

(A)   (B)

(Micrographs courtesy of Eva Mandelkow.)

Figure Q17–20

QUESTION 17–21

The locomotion of fibroblasts in culture is immediately halted by the drug cytochalasin, whereas colchicine causes fibroblasts to cease to move directionally and to begin extending lamellipodia in seemingly random directions. Injection of fibroblasts with antibodies to vimentin has no discernible effect on their migration. What do these observations suggest to you about the involvement of the three different cytoskeletal filaments in fibroblast locomotion?

QUESTION 17–22

Complete the following sentence accurately, explaining your reason for accepting or rejecting each of the four phrases (more than one can be correct). The role of calcium in muscle contraction is:

A. To detach myosin heads from actin.

B. To spread the action potential from the plasma membrane to the contractile machinery.

C. To bind to troponin, cause it to move tropomyosin, and thereby expose actin filaments to myosin heads.

D. To maintain the structure of the myosin filament.

QUESTION 17–23

Which of the following changes takes place when a skeletal muscle contracts?

A. Z discs move farther apart.

B. Actin filaments contract.

C. Myosin filaments contract.

D. Sarcomeres become shorter.

CHAPTER **EIGHTEEN**

18

The Cell-Division Cycle

"Where a cell arises, there must be a previous cell, just as animals can only arise from animals and plants from plants." This statement, which appears in a book written by German pathologist Rudolf Virchow in 1858, carries with it a profound message for the continuity of life. If every cell comes from a previous cell, all living organisms—from a unicellular bacterium to a multicellular mammal—are products of repeated rounds of cell growth and division that stretch back to the beginnings of life more than 3 billion years ago.

A cell reproduces by carrying out an orderly sequence of events in which it duplicates its contents and then divides in two. This cycle of duplication and division, known as the **cell cycle**, is the essential mechanism by which all living things reproduce. The details of the cell cycle vary from organism to organism and at different times in an individual organism's life. In unicellular organisms, such as bacteria and yeasts, each cell division produces a complete new organism, whereas many rounds of cell division are required to make a new multicellular organism from a fertilized egg. Certain features of the cell cycle, however, are universal, as they allow every cell to perform the fundamental task of copying and passing on its genetic information to the next generation of cells.

To explain how cells reproduce, we have to consider three major questions: (1) How do cells duplicate their contents—including the chromosomes, which carry the genetic information? (2) How do they partition the duplicated contents and split in two? (3) How do they coordinate all the steps and machinery required for these two processes? The first question is considered elsewhere in this book: in Chapter 6, we discuss how DNA is replicated, and in Chapters 7, 11, 15, and 17, we describe how the eukaryotic cell manufactures other components, such as proteins, membranes, organelles, and cytoskeletal filaments. In this chapter, we tackle the second and third questions: how a eukaryotic cell distributes—or *segregates*—its duplicated contents to produce two genetically identical daughter cells, and how it coordinates the various steps of this reproductive cycle.

We begin with an overview of the events that take place during a typical cell cycle. We then describe the complex system of regulatory proteins called the *cell-cycle control system*, which orders and coordinates these events to ensure that they occur in the correct sequence. We next discuss in detail the major stages of the cell cycle, in which the chromosomes are duplicated and then segregated into the two daughter cells. At the end of the chapter, we consider how animals use extracellular signals to control the survival, growth, and division of their cells. These signaling systems allow an animal to regulate the size and number of its cells—and, ultimately, the size and form of the organism itself.

OVERVIEW OF THE CELL CYCLE

The most basic function of the cell cycle is to duplicate accurately the vast amount of DNA in the chromosomes and then to segregate the DNA into genetically identical daughter cells such that each cell receives a complete copy of the entire genome (**Figure 18–1**). In most cases, a cell also duplicates its other macromolecules and organelles and doubles in size before it divides; otherwise, each time a cell split it would get smaller and smaller. Thus, to maintain their size, dividing cells coordinate their growth with their division. We return to the topic of cell-size control later in the chapter; here, we focus on cell division.

The duration of the cell cycle varies greatly from one cell type to another. In an early frog embryo, cells divide every 30 minutes, whereas a mammalian fibroblast in culture divides about once a day (**Table 18–1**). In this section, we describe briefly the sequence of events that occur in fairly rapidly dividing (proliferating) mammalian cells. We then introduce the cell-cycle control system that ensures that the various events of the cycle take place in the correct sequence and at the correct time.

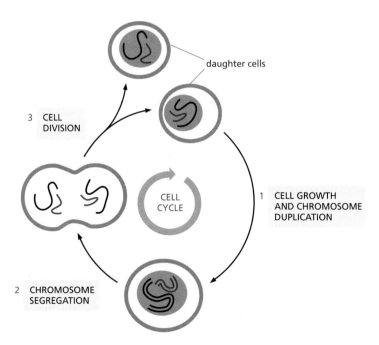

Figure 18–1 Cells reproduce by duplicating their contents and dividing in two, a process called the cell cycle. For simplicity, we use a hypothetical eukaryotic cell—with only one copy each of two different chromosomes—to illustrate how each cell cycle produces two genetically identical daughter cells. Each daughter cell can divide again by going through another cell cycle, and so on for generation after generation.

TABLE 18–1 SOME EUKARYOTIC CELL-CYCLE TIMES	
Cell Type	**Cell-Cycle Times**
Early frog embryo cells	30 minutes
Yeast cells	1.5 hours
Mammalian intestinal epithelial cells	~12 hours
Mammalian fibroblasts in culture	~20 hours

The Eukaryotic Cell Cycle Usually Includes Four Phases

Seen in a microscope, the two most dramatic events in the cell cycle are when the nucleus divides, a process called *mitosis,* and when the cell later splits in two, a process called *cytokinesis.* These two processes together constitute the **M phase** of the cycle. In a typical mammalian cell, the whole of M phase takes about an hour, which is only a small fraction of the total cell-cycle time (see Table 18–1).

The period between one M phase and the next is called **interphase**. Viewed with a microscope, it appears, deceptively, as an uneventful interlude during which the cell simply increases in size. Interphase, however, is a very busy time for a proliferating cell, and it encompasses the remaining three phases of the cell cycle. During **S phase** (S = synthesis), the cell replicates its DNA. S phase is flanked by two "gap" phases—called **G_1** and **G_2**—during which the cell continues to grow (**Figure 18–2**). During these gap phases, the cell monitors both its internal state and external environment. This monitoring ensures that conditions are suitable for reproduction and that preparations are complete before the cell commits to the major upheavals of S phase (which follows G_1) and mitosis (following G_2). At particular points in G_1 and G_2, the cell decides whether to proceed to the next phase or pause to allow more time to prepare.

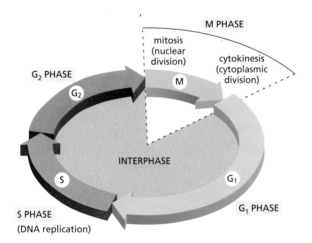

Figure 18–2 The eukaryotic cell cycle usually occurs in four phases. The cell grows continuously in interphase, which consists of three phases: G_1, S, and G_2. DNA replication is confined to S phase. G_1 is the gap between M phase and S phase, and G_2 is the gap between S phase and M phase. During M phase, the nucleus divides in a process called mitosis; then the cytoplasm divides, in a process called cytokinesis. In this figure—and in subsequent figures in the chapter—the lengths of the various phases are not drawn to scale: M phase, for example, is typically much shorter and G_1 much longer than shown.

QUESTION 18–2

A population of proliferating cells is stained with a dye that becomes fluorescent when it binds to DNA, so that the amount of fluorescence is directly proportional to the amount of DNA in each cell. To measure the amount of DNA in each cell, the cells are then passed through a flow cytometer, an instrument that measures the amount of fluorescence in individual cells. The number of cells with a given DNA content is plotted on the graph below.

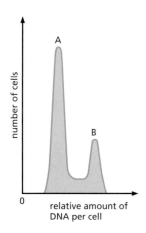

Indicate on the graph where you would expect to find cells that are in G_1, S, G_2, and mitosis. Which is the longest phase of the cell cycle in this population of cells?

During all of interphase, a cell generally continues to transcribe genes, synthesize proteins, and grow in mass. Together with S phase, G_1 and G_2 provide the time needed for the cell to grow and duplicate its cytoplasmic organelles. If interphase lasted only long enough for DNA replication, the cell would not have time to double its mass before it divided and would consequently shrink with each division. Indeed, in some special circumstances that is just what happens. In an early frog embryo, for example, the first cell divisions after fertilization (called *cleavage divisions*) serve to subdivide the giant egg cell into many smaller cells as quickly as possible (see Table 18–1). In such embryonic cell cycles, the G_1 and G_2 phases are drastically shortened, and the cells do not grow before they divide.

A Cell-Cycle Control System Triggers the Major Processes of the Cell Cycle

To ensure that they replicate all their DNA and organelles, and divide in an orderly manner, eukaryotic cells possess a complex network of regulatory proteins known as the *cell-cycle control system*. This system guarantees that the events of the cell cycle—DNA replication, mitosis, and so on—occur in a set sequence and that each process has been completed before the next one begins. To accomplish this, the control system is itself regulated at certain critical points of the cycle by feedback from the process currently being performed. Without such feedback, an interruption or a delay in any of the processes could be disastrous. All of the nuclear DNA, for example, must be replicated before the nucleus begins to divide, which means that a complete S phase must precede M phase. If DNA synthesis is slowed down or stalled, mitosis and cell division must also be delayed. Similarly, if DNA is damaged, the cycle must arrest in G_1, S, or G_2 so that the cell can repair the damage, either before DNA replication is started or completed or before the cell enters M phase. The cell-cycle control system achieves all of this by employing molecular brakes, sometimes called *checkpoints*, to pause the cycle at certain transition points. In this way, the control system does not trigger the next step in the cycle unless the cell is properly prepared.

The cell-cycle control system regulates progression through the cell cycle at three main transition points (**Figure 18–3**). At the transition from G_1 to S phase, the control system confirms that the environment is favorable for proliferation before committing to DNA replication. Cell proliferation in animals requires both sufficient nutrients and specific signal molecules

Figure 18–3 **The cell-cycle control system ensures that key processes in the cycle occur in the proper sequence.** The cell-cycle control system is shown as a controller arm that rotates clockwise, triggering essential processes when it reaches particular transition points on the outer dial. These processes include DNA replication in S phase and the segregation of duplicated chromosomes in mitosis. The control system can transiently halt the cycle at specific transition points—in G_1, G_2, and M phase—if extracellular or intracellular conditions are unfavorable.

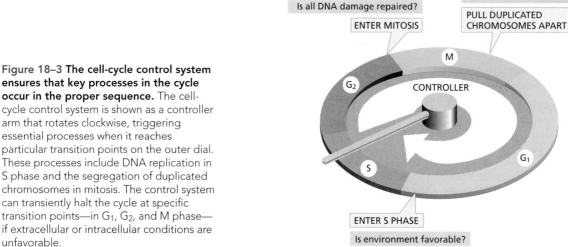

in the extracellular environment; if these extracellular conditions are unfavorable, cells can delay progress through G_1 and may even enter a specialized resting state known as G_0 (G zero). At the transition from G_2 to M phase, the control system confirms that the DNA is undamaged and fully replicated, ensuring that the cell does not enter mitosis unless its DNA is intact. Finally, during mitosis, the cell-cycle control machinery ensures that the duplicated chromosomes are properly attached to a cytoskeletal machine, called the *mitotic spindle*, before the spindle pulls the chromosomes apart and segregates them into the two daughter cells.

In animals, the transition from G_1 to S phase is especially important as a point in the cell cycle where the control system is regulated. Signals from other cells stimulate cell proliferation when more cells are needed—and block it when they are not. The cell-cycle control system therefore plays a central part in the regulation of cell numbers in the tissues of the body; if the control system malfunctions such that cell division is excessive, cancer can result. We discuss later how extracellular signals influence the decisions made at the G_1 to S transition.

Cell-Cycle Control is Similar in All Eukaryotes

Some features of the cell cycle, including the time required to complete certain events, vary greatly from one cell type to another, even within the same organism. The basic organization of the cycle, however, is essentially the same in all eukaryotic cells, and all eukaryotes appear to use similar machinery and control mechanisms to drive and regulate cell-cycle events. The proteins of the cell-cycle control system first appeared more than a billion years ago, and they have been so well conserved over the course of evolution that many of them function perfectly when transferred from a human cell to a yeast (see How We Know, pp. 609–610).

Because of this similarity, biologists can study the cell cycle and its regulation in a variety of organisms and use the findings from all of them to assemble a unified picture of how the cycle works. Many discoveries about the cell cycle have come from a systematic search for mutations that inactivate essential components of the cell-cycle control system in yeasts. Likewise, studies of both cultured mammalian cells and the embryos of frogs and sea urchins have been critical for examining the molecular mechanisms that underlie the cycle and its control in multicellular organisms like ourselves.

THE CELL-CYCLE CONTROL SYSTEM

Two types of machinery are involved in cell division: one manufactures the new components of the growing cell, and another hauls the components into their correct places and partitions them appropriately when the cell divides in two. The **cell-cycle control system** switches all this machinery on and off at the correct times, thereby coordinating the various steps of the cycle. The core of the cell-cycle control system is a series of molecular switches that operate in a defined sequence and orchestrate the main events of the cycle, including DNA replication and the segregation of duplicated chromosomes. In this section, we review the protein components of the control system and discuss how they work together to trigger the different phases of the cycle.

The Cell-Cycle Control System Depends on Cyclically Activated Protein Kinases called Cdks

The cell-cycle control system governs the cell-cycle machinery by cyclically activating and then inactivating the key proteins and protein

cyclin

cyclin-dependent
kinase (Cdk)

Figure 18–4 Progression through the cell cycle depends on cyclin-dependent protein kinases (Cdks). A Cdk must bind a regulatory protein called a cyclin before it can become enzymatically active. This activation also requires an activating phosphorylation of the Cdk (not shown, but see Movie 18.1). Once activated, a cyclin–Cdk complex phosphorylates key proteins in the cell that are required to initiate particular steps in the cell cycle. The cyclin also helps direct the Cdk to the target proteins that the Cdk phosphorylates.

complexes that initiate or regulate DNA replication, mitosis, and cytokinesis. This regulation is carried out largely through the phosphorylation and dephosphorylation of proteins involved in these essential processes.

As discussed in Chapter 4, phosphorylation followed by dephosphorylation is one of the most common ways by which cells switch the activity of a protein on and off (see Figure 4–42), and the cell-cycle control system uses this mechanism extensively and repeatedly. The phosphorylation reactions that control the cell cycle are carried out by a specific set of protein kinases, while dephosphorylation is performed by a set of protein phosphatases.

The protein kinases at the core of the cell-cycle control system are present in proliferating cells throughout the cell cycle. They are activated, however, only at appropriate times in the cycle, after which they are quickly inactivated. Thus, the activity of each of these kinases rises and falls in a cyclical fashion. Some of these protein kinases, for example, become active toward the end of G_1 phase and are responsible for driving the cell into S phase; another kinase becomes active just before M phase and drives the cell into mitosis.

Switching these kinases on and off at the appropriate times is partly the responsibility of another set of proteins in the control system—the **cyclins**. Cyclins have no enzymatic activity themselves, but they need to bind to the cell-cycle kinases before the kinases can become enzymatically active. The kinases of the cell-cycle control system are therefore known as **cyclin-dependent protein kinases**, or **Cdks** (**Figure 18–4**). Cyclins are so-named because, unlike the Cdks, their concentrations vary in a cyclical fashion during the cell cycle. The cyclical changes in cyclin concentrations help drive the cyclic assembly and activation of the cyclin–Cdk complexes. Once activated, cyclin–Cdk complexes help trigger various cell-cycle events, such as entry into S phase or M phase (**Figure 18–5**). We discuss how the Cdks and cyclins were discovered in **How We Know**, pp. 609–610.

Different Cyclin–Cdk Complexes Trigger Different Steps in the Cell Cycle

There are several types of cyclins and, in most eukaryotes, several types of Cdks involved in cell-cycle control. Different cyclin–Cdk complexes trigger different steps of the cell cycle. As shown in Figure 18–5, the cyclin that acts in G_2 to trigger entry into M phase is called **M cyclin**, and the active complex it forms with its Cdk is called **M-Cdk**. Other cyclins, called **S cyclins** and **G_1/S cyclins**, bind to a distinct Cdk protein late in G_1 to form **S-Cdk** and **G_1/S-Cdk**, respectively; these cyclin-Cdk complexes help launch S phase. The actions of S-Cdk and M-Cdk are indicated in

Figure 18–5 The accumulation of cyclins helps regulate the activity of Cdks. The formation of active cyclin–Cdk complexes drives various cell-cycle events, including entry into S phase or M phase. The figure shows the changes in cyclin concentration and Cdk protein kinase activity responsible for controlling entry into M phase. Increasing concentration of the relevant cyclin (called M cyclin) helps direct the formation of the active cyclin–Cdk complex (M–Cdk) that drives entry into M phase. Although the enzymatic activity of each type of cyclin-Cdk complex rises and falls during the course of the cell cycle, the concentration of the Cdk component does not (not shown).

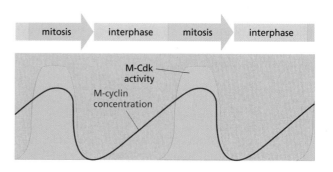

mitosis | interphase | mitosis | interphase

M-Cdk
activity

M-cyclin
concentration

DISCOVERY OF CYCLINS AND CDKS

For many years, cell biologists watched the "puppet show" of DNA synthesis, mitosis, and cytokinesis but had no idea what was behind the curtain, controlling these events. The cell-cycle control system was simply a "black box" inside the cell. It was not even clear whether there was a separate control system, or whether the cell-cycle machinery somehow controlled itself. A breakthrough came with the identification of the key proteins of the control system and the realization that they are distinct from the components of the cell-cycle machinery—the enzymes and other proteins that perform the essential processes of DNA replication, chromosome segregation, and so on.

The first components of the cell-cycle control system to be discovered were the cyclins and cyclin-dependent kinases (Cdks) that drive cells into M phase. They were found in studies of cell division conducted on animal eggs.

Back to the egg

The fertilized eggs of many animals are especially suitable for biochemical studies of the cell cycle because they are exceptionally large and divide rapidly. An egg of the frog *Xenopus*, for example, is just over 1 mm in diameter (**Figure 18–6**). After fertilization, it divides rapidly to partition the egg into many smaller cells. These rapid cell cycles consist mainly of repeated S and M phases, with very short or no G_1 or G_2 phases between them. There is no new gene transcription: all of the mRNAs and most of the proteins required for this early stage of embryonic development are already packed into the very large egg during its development as an oocyte in the ovary of the mother. In these early division cycles (*cleavage divisions*), no cell growth occurs, and all the cells of the embryo divide synchronously, growing smaller and smaller with each division (see Movie 18.2).

Because of the synchrony, it is possible to prepare an extract from frog eggs that is representative of the cell-cycle stage at which the extract is made. The biological activity of such an extract can then be tested by injecting it into a *Xenopus* oocyte (the immature precursor of the unfertilized egg) and observing, microscopically, its effects on cell-cycle behavior. The *Xenopus* oocyte is an especially convenient test system for detecting an activity that drives cells into M phase, because of its large size, and because it has completed DNA replication and is arrested at a stage in the meiotic cell cycle (discussed in Chapter 19) that is equivalent to the G_2 phase of a mitotic cell cycle.

Give us an M

In such experiments, Kazuo Matsui and colleagues found that an extract from an M-phase egg instantly drives the oocyte into M phase, whereas cytoplasm from a cleaving egg at other phases of the cycle does not. When they first made this discovery, they did not know the molecules or the mechanism responsible, so they referred to the unidentified agent as *maturation promoting factor*, or MPF (**Figure 18–7**). By testing cytoplasm from different stages of the cell cycle, Matsui and colleagues found that MPF activity oscillates dramatically during the course of each cell cycle: it increased rapidly just before the start of mitosis and fell rapidly to zero toward the end of mitosis (see Figure 18–5). This oscillation made MPF a strong candidate for a component involved in cell-cycle control.

When MPF was finally purified, it was found to contain a protein kinase that was required for its activity. But the kinase portion of MPF did not act alone. It had to have a specific protein (now known to be M cyclin) bound to it in order to function. M cyclin was discovered in a different type of experiment, involving clam eggs.

0.5 mm

Figure 18–6 **A mature *Xenopus* egg provides a convenient system for studying the cell cycle.** (Courtesy of Tony Mills.)

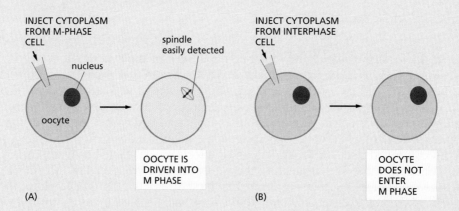

Figure 18–7 MPF activity was discovered by injecting *Xenopus* egg cytoplasm into *Xenopus* oocytes. (A) A *Xenopus* oocyte is injected with cytoplasm taken from a *Xenopus* egg in M phase. The cell extract drives the oocyte into M phase of the first meiotic division (a process called maturation), causing the large nucleus to break down and a spindle to form. (B) When the cytoplasm is instead taken from a cleaving egg in interphase, it does not cause the oocyte to enter M phase. Thus, the extract in (A) must contain some activity—a maturation promoting factor (MPF)—that triggers entry into M phase.

Fishing in clams

M cyclin was initially identified by Tim Hunt as a protein whose concentration rose gradually during interphase and then fell rapidly to zero as cleaving clam eggs went through M phase (see Figure 18–5). The protein repeated this performance in each cell cycle. Its role in cell-cycle control, however, was initially obscure. The breakthrough occurred when cyclin was found to be a component of MPF and to be required for MPF activity. Thus, MPF, which we now call M-Cdk, is a protein complex containing two subunits—a regulatory subunit, M cyclin, and a catalytic subunit, the mitotic Cdk. After the components of M-Cdk were identified, other types of cyclins and Cdks were isolated, whose concentrations or activities, respectively, rose and fell at other stages in the cell cycle.

All in the family

While biochemists were identifying the proteins that regulate the cell cycles of frog and clam embryos, yeast geneticists—led by Lee Hartwell, studying baker's yeast (*S. cerevisiae*), and Paul Nurse, studying fission yeast

(*S. pombe*)—were taking a genetic approach to dissecting the cell-cycle control system. By studying mutants that get stuck or misbehave at specific points in the cell cycle, these researchers were able to identify many genes responsible for cell-cycle control. Some of these genes turned out to encode cyclin or Cdk proteins, which were unmistakably similar—in both amino acid sequence and function—to their counterparts in frogs and clams. Similar genes were soon identified in human cells.

Many of the cell-cycle control genes have changed so little during evolution that the human version of the gene will function perfectly well in a yeast cell. For example, Nurse and colleagues were the first to show that a yeast with a defective copy of the gene encoding its only Cdk fails to divide, but it divides normally if a copy of the appropriate human gene is artificially introduced into the defective cell. Surely, even Darwin would have been astonished at such clear evidence that humans and yeasts are cousins. Despite a billion years of divergent evolution, all eukaryotic cells—whether yeast, animal, or plant—use essentially the same molecules to control the events of their cell cycle.

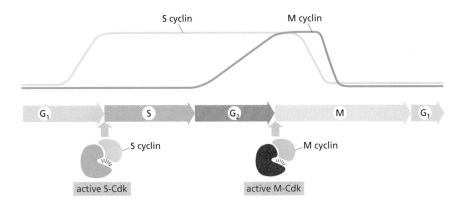

Figure 18–8 Distinct Cdks associate with different cyclins to trigger the different events of the cell cycle. For simplicity, only two types of cyclin–Cdk complexes are shown—one that triggers S phase and one that triggers M phase.

Figure 18–8. Another group of cyclins, called **G₁ cyclins**, act earlier in G₁ and bind to other Cdk proteins to form **G₁-Cdks**, which help drive the cell through G₁ toward S phase. We see later that the formation of these G₁-Cdks in animal cells usually depends on extracellular signal molecules that stimulate cells to divide. The names of the main cyclins and their Cdks are listed in **Table 18–2**.

Each of these cyclin–Cdk complexes phosphorylates a different set of target proteins in the cell. G₁-Cdks, for example, phosphorylate regulatory proteins that activate transcription of genes required for DNA replication. By activating different sets of target proteins, each type of complex triggers a different transition step in the cycle.

Cyclin Concentrations are Regulated by Transcription and by Proteolysis

As discussed in Chapter 7, the concentration of a given protein in the cell is determined by the rate at which the protein is synthesized and the rate at which it is degraded. Over the course of the cell cycle, the concentration of each type of cyclin rises gradually and then falls abruptly (see Figure 18–8). The gradual increase in cyclin concentration stems from increasing transcription of cyclin genes, whereas the rapid fall in cyclin concentration is precipitated by a full-scale targeted destruction of the protein.

The abrupt degradation of M and S cyclins part way through M phase depends on a large enzyme complex called—for reasons that will become clear later—**anaphase-promoting complex (APC)**. This complex tags these cyclins with a chain of ubiquitin. As discussed in Chapter 7, proteins marked in this way are directed to proteasomes where they are rapidly degraded (see Figure 7–40). The ubiquitylation and degradation of the cyclin returns its Cdk to an inactive state (**Figure 18–9**).

TABLE 18–2 THE MAJOR CYCLINS AND CDKS OF VERTEBRATES		
Cyclin–Cdk Complex	**Cyclin**	**Cdk Partner**
G₁-Cdk	cyclin D*	Cdk4, Cdk6
G₁/S-Cdk	cyclin E	Cdk2
S-Cdk	cyclin A	Cdk2
M-Cdk	cyclin B	Cdk1
*There are three D cyclins in mammals (cyclins D1, D2, and D3).		

Figure 18–9 The activity of some Cdks is regulated by cyclin degradation. Ubiquitylation of S or M cyclin by APC marks the protein for destruction in proteasomes (as discussed in Chapter 7). The loss of cyclin renders its Cdk partner inactive.

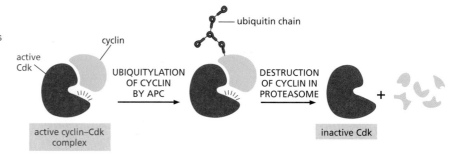

Cyclin destruction can help drive the transition from one phase of the cell cycle to the next. For example, M-cyclin degradation and the resulting inactivation of M-Cdk leads to the molecular events that take the cell out of mitosis.

The Activity of Cyclin–Cdk Complexes Depends on Phosphorylation and Dephosphorylation

The appearance and disappearance of cyclin proteins play an important part in regulating Cdk activity during the cell cycle, but there must be more to the story: although cyclin concentrations increase gradually, the activity of the associated cyclin–Cdk complexes tends to switch on abruptly at the appropriate time in the cell cycle (see Figure 18–5). What could trigger the abrupt activation of these complexes? It turns out that the cyclin–Cdk complex contains inhibitory phosphates, and to become active, the Cdk must be dephosphorylated by a specific protein phosphatase (**Figure 18–10**). Thus protein kinases and phosphatases regulate the activity of specific cyclin–Cdk complexes and help control progression through the cell cycle.

Cdk Activity Can be Blocked by Cdk Inhibitor Proteins

In addition to phosphorylation and dephosphorylation, the activity of Cdks can also be modulated by the binding of **Cdk inhibitor proteins**. The cell-cycle control system uses these inhibitors to block the assembly or activity of certain cyclin–Cdk complexes. Some Cdk inhibitor proteins, for example, help maintain Cdks in an inactive state during the G_1 phase of the cycle, thus delaying progression into S phase (**Figure 18–11**). Pausing at this transition point in G_1 gives the cell more time to grow, or allows it to wait until extracellular conditions are favorable for division.

The Cell-Cycle Control System Can Pause the Cycle in Various Ways

As mentioned earlier, the cell-cycle control system can transiently delay progress through the cycle at various transition points to ensure that the major events of the cycle occur in a specific order. At these transitions,

Figure 18–10 For M-Cdk to be active, inhibitory phosphates must be removed. As soon as the M cyclin–Cdk complex is formed, it is phosphorylated at two adjacent sites by an inhibitory protein kinase called Wee1. (For simplicity, only one inhibitory phosphate is shown.) This modification keeps M-Cdk in an inactive state until these phosphates are removed by an activating protein phosphatase called Cdc25. It is still not clear how the timing of the critical Cdc25 phosphatase triggering step shown here is controlled.

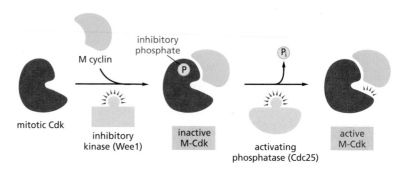

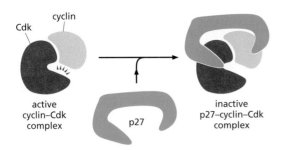

Figure 18–11 The activity of a Cdk can be blocked by the binding of a Cdk inhibitor. In this instance, the inhibitor protein (called p27) binds to an activated cyclin–Cdk complex. Its attachment prevents the Cdk from phosphorylating target proteins required for progress through G$_1$ into S phase.

the control system monitors the cell's internal state and the conditions in its environment, before allowing the cell to continue through the cycle. For example, it allows entry into S phase only if environmental conditions are appropriate; it triggers mitosis only after the DNA has been completely replicated; and it initiates chromosome segregation only after the duplicated chromosomes are correctly aligned on the mitotic spindle.

To accomplish these feats, the control system uses a combination of the mechanisms we have described. At the G$_1$-to-S transition, it uses Cdk inhibitors to keep cells from entering S phase and replicating their DNA (see Figure 18–11). At the G$_2$-to-M transition, it suppresses the activation of M-Cdk by inhibiting the phosphatase required to activate the Cdk (see Figure 18–10). And it can delay the exit from mitosis by inhibiting the activation of APC, thus preventing the degradation of M cyclin (see Figure 18–9).

These mechanisms, summarized in **Figure 18–12**, allow the cell to make "decisions" about whether to progress through the cell cycle. In the next section, we take a closer look at how the cell-cycle control system decides whether a cell in G$_1$ should commit to divide.

G$_1$ PHASE

In addition to being a bustling period of metabolic activity, cell growth, and repair, G$_1$ is an important point of decision-making for the cell. Based on intracellular signals that provide information about the size of the cell and extracellular signals reflecting the environment, the cell-cycle control machinery can either hold the cell transiently in G$_1$ (or in a more prolonged nonproliferative state, G$_0$), or allow it to prepare for entry into

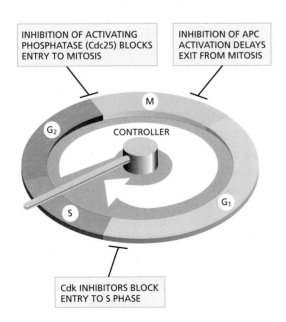

Figure 18–12 The cell-cycle control system uses various mechanisms to pause the cycle at specific transition points.

Figure 18–13 The transition from G$_1$ to S phase offers the cell a crossroad. The cell can commit to completing another cell cycle, pause temporarily until conditions are right, or withdraw from the cell cycle altogether—either temporarily in G$_0$, or permanently in the case of terminally differentiated cells.

QUESTION 18–3

Why do you suppose cells have evolved a special G$_0$ phase to exit from the cell cycle, rather than just stopping in G$_1$ and not moving on to S phase?

---

the S phase of another cell cycle. Once past this critical G$_1$-to-S transition, a cell usually continues all the way through the rest of the cell cycle quickly—typically within 12–24 hours in mammals. In yeasts, the G$_1$-to-S transition is therefore sometimes called *Start*, because passing it represents a commitment to complete a full cell cycle (**Figure 18–13**).

In this section, we consider how the cell-cycle control system decides between these options—and what it does once the decision is made. The molecular mechanisms involved are especially important, as defects in them can lead to unrestrained cell proliferation and cancer.

Cdks are Stably Inactivated in G$_1$

During M phase, when cells are actively dividing, the cell is awash with active cyclin–Cdk complexes. If those S-Cdks and M-Cdks are not disabled by the end of M phase, the cell will immediately replicate its DNA and initiate another round of division, without spending any significant time in the G$_1$ or G$_2$ phases. Such rapid cycling is typically seen in some early embryos, where the cells get smaller with each division, having little time to grow in between.

To usher a cell from the upheaval of M phase to the relative tranquility of G$_1$, the cell-cycle control machinery must inactivate its inventory of S-Cdk and M-Cdk. It does so by eliminating all of the existing cyclins, by blocking the synthesis of new ones, and by deploying Cdk inhibitor proteins to muffle the activity of any remaining cyclin–Cdk complexes. The use of multiple mechanisms makes this system of suppression robust, ensuring that essentially all Cdk activity is shut down. This wholesale inactivation resets the cell-cycle control system and generates a stable G$_1$ phase, during which the cell can grow and monitor its environment before committing to a new round of division.

Mitogens Promote the Production of the Cyclins that Stimulate Cell Division

As a general rule, mammalian cells will multiply only if they are stimulated to do so by extracellular signals, called *mitogens*, produced by other cells. If deprived of such signals, the cell cycle arrests in G$_1$; if the cell is deprived of mitogens for long enough, it will withdraw from the cell cycle and enter a nonproliferating state, in which the cell can remain for days or weeks, months, or even for the lifetime of the organism, as we discuss shortly.

Escape from cell-cycle arrest—or from certain nonproliferating states—requires the accumulation of cyclins. Mitogens act by switching on cell signaling pathways that stimulate the synthesis of G$_1$ cyclins, G$_1$/S cyclins, and other proteins involved in DNA synthesis and chromosome duplication. The buildup of these cyclins triggers a wave of G$_1$/S-Cdk activity, which ultimately relieves the negative controls that otherwise block progression from G$_1$ to S phase.

A crucial negative control is mediated by the *Retinoblastoma (Rb) protein*. Rb was initially identified from studies of a rare childhood eye tumor called retinoblastoma, in which the Rb protein is missing or defective. Rb is abundant in the nuclei of all vertebrate cells, where it binds to particular transcription regulators and prevents them from turning on the genes required for cell proliferation. Mitogens release the Rb brake by triggering the activation of G$_1$-Cdks and G$_1$/S-Cdks. These complexes phosphorylate the Rb protein, altering its conformation so that it releases its bound transcription regulators, which are then free to activate the genes required for cell proliferation (**Figure 18–14**).

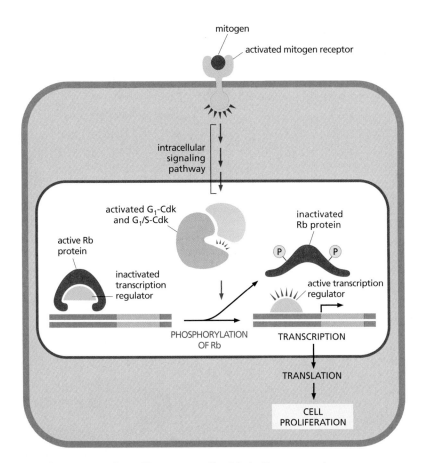

mitogen

activated mitogen receptor

intracellular signaling pathway

activated G₁-Cdk and G₁/S-Cdk

active Rb protein

inactivated Rb protein

inactivated transcription regulator

active transcription regulator

PHOSPHORYLATION OF Rb

TRANSCRIPTION

TRANSLATION

CELL PROLIFERATION

Figure 18–14 One way in which mitogens stimulate cell proliferation is by inhibiting the Rb protein. In the absence of mitogens, dephosphorylated Rb protein holds specific transcription regulators in an inactive state; these transcription regulators are required to stimulate the transcription of target genes that encode proteins needed for cell proliferation. Mitogens binding to cell-surface receptors activate intracellular signaling pathways that lead to the formation and activation of G₁-Cdk and G₁/S-Cdk complexes. These complexes phosphorylate, and thereby inactivate, the Rb protein, releasing the transcription regulators that activate the transcription of genes required for cell proliferation.

DNA Damage Can Temporarily Halt Progression Through G₁

The cell-cycle control system uses several distinct mechanisms to halt progress through the cell cycle if DNA is damaged. It can do so at various transition points from one phase of the cell cycle to the next. The mechanism that operates at the G₁-to-S transition, which prevents the cell from replicating damaged DNA, is especially well understood. DNA damage in G₁ causes an increase in both the concentration and activity of a protein called **p53**, which is a transcription regulator that activates the transcription of a gene encoding a Cdk inhibitor protein called p21. The p21 protein binds to G₁/S-Cdk and S-Cdk, preventing them from driving the cell into S phase (**Figure 18–15**). The arrest of the cell cycle in G₁ gives the cell time to repair the damaged DNA before replicating it. If the DNA damage is too severe to be repaired, p53 can induce the cell to kill itself by undergoing a form of programmed cell death called *apoptosis*, which we discuss later. If p53 is missing or defective, the unrestrained replication of damaged DNA leads to a high rate of mutation and the production of cells that tend to become cancerous. In fact, mutations in the *p53* gene are found in about half of all human cancers (Movie 18.3).

Cells Can Delay Division for Prolonged Periods by Entering Specialized Nondividing States

As mentioned earlier, cells can delay progress through the cell cycle at specific transition points, to wait for suitable conditions or to repair damaged DNA. They can also withdraw from the cell cycle for prolonged periods—either temporarily or permanently.

The most radical decision that the cell-cycle control system can make is to withdraw the cell from the cell cycle permanently. This decision has a special importance in multicellular organisms. Many cells in the

Figure 18–15 DNA damage can arrest the cell cycle in G$_1$. When DNA is damaged, specific protein kinases respond by both activating the p53 protein and halting its normal rapid degradation. Activated p53 protein thus accumulates and stimulates the transcription of the gene that encodes the Cdk inhibitor protein p21. The p21 protein binds to G$_1$/S-Cdk and S-Cdk and inactivates them, so that the cell cycle arrests in G$_1$.

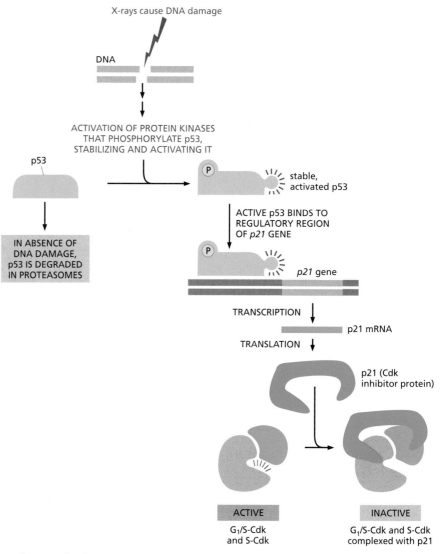

human body permanently stop dividing when they differentiate. In such *terminally differentiated* cells, such as nerve or muscle cells, the cell-cycle control system is dismantled completely and genes encoding the relevant cyclins and Cdks are irreversibly shut down.

In the absence of appropriate signals, other cell types withdraw from the cell cycle only temporarily, entering an arrested state called G$_0$. They retain the ability to reassemble the cell-cycle control system quickly and to divide again. Most liver cells, for example, are in G$_0$, but they can be stimulated to proliferate if the liver is damaged.

Most of the diversity in cell-division rates in the adult body lies in the variation in the time that cells spend in G$_0$ or in G$_1$. Some cell types, including liver cells, normally divide only once every year or two, whereas certain epithelial cells in the gut divide more than twice a day to renew the lining of the gut continually. Many of our cells fall somewhere in between: they can divide if the need arises but normally do so infrequently.

S PHASE

Before a cell divides, it must replicate its DNA. As we discuss in Chapter 6, this replication must occur with extreme accuracy to minimize the risk of mutations in the next cell generation. Of equal importance, every nucleotide in the genome must be copied once—and only once—to prevent the damaging effects of gene amplification. In this section, we consider the

---

QUESTION 18–4

What might be the consequences if a cell replicated damaged DNA before repairing it?

elegant molecular mechanisms by which the cell-cycle control system initiates DNA replication and, at the same time, prevents replication from happening more than once per cell cycle.

S-Cdk Initiates DNA Replication and Blocks Re-Replication

Like any monumental task, configuring chromosomes for replication requires a certain amount of preparation. For eukaryotic cells, this preparation begins early in G_1, when DNA is made replication-ready by the recruitment of proteins to the sites along each chromosome where replication will begin. These nucleotide sequences, called *origins of replication*, serve as landing pads for the proteins and protein complexes that control and carry out DNA synthesis.

One of these protein complexes, called the *origin recognition complex* (*ORC*), remains perched atop the replication origins throughout the cell cycle. In the first step of replication initiation, the ORC recruits a protein called Cdc6, whose concentration rises early in G_1. Together these proteins load the DNA helicases that will open up the double helix and ready the origin of replication. Once this *prereplicative complex* is in place, the replication origin is loaded and ready to "fire."

The signal to commence replication comes from S-Cdk, the cyclin–Cdk complex that triggers S phase. S-Cdk is assembled and activated at the end of G_1. During S phase, it activates the DNA helicases in the prereplicative complex and promotes the assembly of the rest of the proteins that form the *replication fork* (see Figure 6–19). In doing so, S-Cdk essentially "pulls the trigger" that initiates DNA replication (**Figure 18–16**).

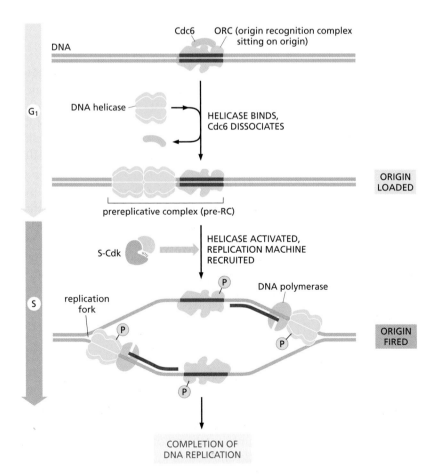

Figure 18–16 The initiation of DNA replication takes place in two steps. During G_1, Cdc6 binds to the ORC, and together these proteins load a pair of DNA helicases to form the prereplicative complex. At the start of S phase, S-Cdk triggers the firing of this loaded replication origin by guiding the assembly of the DNA polymerase (*green*) and other proteins (not shown) that initiate DNA synthesis at the replication fork (discussed in Chapter 6). S-Cdk also blocks re-replication by helping to phosphorylate Cdc6, which marks the protein for degradation (not shown).

S-Cdk not only triggers the initiation of DNA synthesis at a replication origin; it also helps prevent re-replication. It does so by helping phosphorylate Cdc6, which marks that protein for degradation. Eliminating Cdc6 helps ensure that DNA replication can not be reinitiated later in the same cell cycle.

Incomplete Replication Can Arrest the Cell Cycle in G$_2$

Earlier, we described how DNA damage can signal the cell-cycle control system to delay progress through the G$_1$-to-S transition, preventing the cell from replicating damaged DNA. But what if errors occur during DNA replication—or if replication is delayed? How does the cell keep from dividing with DNA that is incorrectly or incompletely replicated?

To address these issues, the cell-cycle control system uses a mechanism that can delay entry into M phase. As we saw in Figure 18–10, the activity of M-Cdk is inhibited by phosphorylation at particular sites. For the cell to progress into mitosis, these inhibitory phosphates must be removed by an activating protein phosphatase called Cdc25. When DNA is damaged or incompletely replicated, Cdc25 is itself inhibited, preventing the removal of the inhibitory phosphates. As a result, M-Cdk remains inactive and M phase is delayed until DNA replication is complete and any DNA damage is repaired.

Once a cell has successfully replicated its DNA in S phase, and progressed through G$_2$, it is ready to enter M phase. During this relatively brief period, the cell will divide its nucleus (mitosis) and then its cytoplasm (cytokinesis; see Figure 18–2). In the next three sections, we describe the events that occur during M phase. We first present a brief overview of M phase as a whole and then discuss, in sequence, the mechanics of mitosis and of cytokinesis, with a focus on animal cells.

M PHASE

Although M phase (mitosis plus cytokinesis) occurs over a relatively short amount of time—about one hour in a mammalian cell that divides once a day, or even once a year—it is by far the most dramatic phase of the cell cycle. During this brief period, the cell reorganizes virtually all of its components and distributes them equally into the two daughter cells. The earlier phases of the cell cycle, in effect, serve to set the stage for the drama of M phase.

The central problem for a cell in M phase is to accurately segregate the chromosomes that were duplicated in the preceding S phase, so that each new daughter cell receives an identical copy of the genome. With minor variations, all eukaryotes solve this problem in a similar way: they assemble two specialized cytoskeletal machines, one that pulls the duplicated chromosomes apart (during mitosis) and another that divides the cytoplasm into two halves (cytokinesis). We begin our discussion of M phase with an overview of how the cell sets the processes of M phase in motion.

M-Cdk Drives Entry Into M Phase and Mitosis

One of the most remarkable features of the cell-cycle control system is that a single protein complex, M-Cdk, brings about all the diverse and intricate rearrangements that occur in the early stages of mitosis. Among its many duties, M-Cdk helps prepare the duplicated chromosomes for segregation and induces the assembly of the mitotic spindle—the machinery that will pull the duplicated chromosomes apart.

M-Cdk complexes accumulate throughout G$_2$. But this stockpile is not activated until the end of G$_2$, when the activating phosphatase Cdc25 removes the inhibitory phosphates holding M-Cdk activity in check (see Figure 18–10). This activation is self-reinforcing: once activated, each M-Cdk complex can indirectly activate additional M-Cdk complexes—by phosphorylating and activating more Cdc25 (**Figure 18–17**). Activated M-Cdk also shuts down the inhibitory kinase Wee1 (see Figure 18–10), further promoting the production of activated M-Cdk. The overall consequence is that, once M-Cdk activation begins, it ignites an explosive increase in M-Cdk activity that drives the cell abruptly from G$_2$ into M phase.

Cohesins and Condensins Help Configure Duplicated Chromosomes for Separation

When the cell enters M phase, the duplicated chromosomes condense, becoming visible under the microscope as threadlike structures. Protein complexes, called **condensins**, help carry out this **chromosome condensation**, which reduces mitotic chromosomes to compact bodies that can be more easily segregated within the crowded confines of the dividing cell. The assembly of condensin complexes onto the DNA is triggered by the phosphorylation of condensins by M-Cdk.

Even before mitotic chromosomes become condensed, the replicated DNA is handled in a way that allows cells to keep track of the two copies. Immediately after a chromosome is duplicated during S phase, the two copies remain tightly bound together. These identical copies—called **sister chromatids**—each contain a single, double-stranded molecule of DNA, along with its associated proteins. The sisters are held together by protein complexes called **cohesins**, which assemble along the length of each chromatid as the DNA is replicated. This cohesion between sister chromatids is crucial for proper chromosome segregation, and it is broken completely only in late mitosis to allow the sisters to be pulled apart by the mitotic spindle. Defects in sister-chromatid cohesion lead to major errors in chromosome segregation. In humans, such mis-segregation can lead to abnormal numbers of chromosomes, as in individuals with Down Syndrome, who have three copies of chromosome 21.

Cohesins and condensins are structurally related, and both are thought to form ring structures around chromosomal DNA. But, whereas cohesins tie the two sister chromatids together (**Figure 18–18A**), condensins assemble on each individual sister chromatid at the start of M phase and help each of these double helices to coil up into a more compact form (**Figure 18–18B and C**). Together, these proteins help configure replicated chromosomes for mitosis.

Different Cytoskeletal Assemblies Carry Out Mitosis and Cytokinesis

After the duplicated chromosomes have condensed, two complex cytoskeletal machines assemble in sequence to carry out the two mechanical processes that occur in M phase. The mitotic spindle carries out nuclear division (mitosis), and, in animal cells and many unicellular eukaryotes, the *contractile ring* carries out cytoplasmic division (cytokinesis) (**Figure 18–19**). Both structures disassemble rapidly after they have performed their tasks.

The mitotic spindle is composed of microtubules and the various proteins that interact with them, including microtubule-associated motor proteins (discussed in Chapter 17). In all eukaryotic cells, the mitotic spindle is responsible for separating the duplicated chromosomes and allocating one copy of each chromosome to each daughter cell.

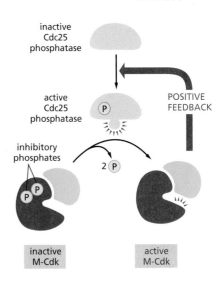

Figure 18–17 Activated M-Cdk indirectly activates more M-Cdk, creating a positive feedback loop. Once activated, M-Cdk phosphorylates, and thereby activates, more Cdk-activating phosphatase (Cdc25). This phosphatase can now activate more M-Cdk by removing the inhibitory phosphate groups from the Cdk subunit.

QUESTION 18–5

A small amount of cytoplasm isolated from a mitotic cell is injected into an unfertilized frog oocyte, causing the oocyte to enter M phase (see Figure 18–7A). A sample of the injected oocyte's cytoplasm is then taken and injected into a second oocyte, causing this cell also to enter M phase. The process is repeated many times until, essentially, none of the original protein sample remains, and yet, cytoplasm taken from the last in the series of injected oocytes is still able to trigger entry into M phase with undiminished efficiency. Explain this remarkable observation.

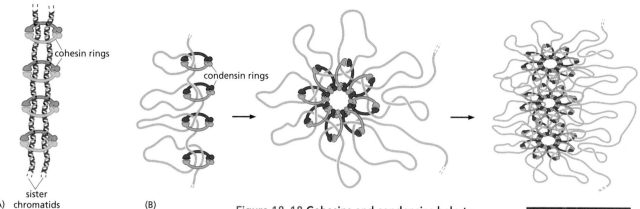

(A) sister chromatids

(B)

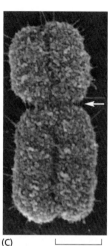

Figure 18–18 Cohesins and condensins help to configure duplicated chromosomes for segregation.
(A) Cohesins tie together the two adjacent sister chromatids in each duplicated chromosome. They are thought to form large protein rings that surround the sister chromatids, preventing them from coming apart, until the rings are broken late in mitosis.
(B) Condensins help coil each sister chromatid (in other words, each DNA double helix) into a smaller, more compact structure that can be more easily segregated during mitosis. (C) A scanning electron micrograph of a condensed, duplicated mitotic chromosome, consisting of two sister chromatids joined along their length. The constricted region (*arrow*) is the centromere, where each chromatid will attach to the mitotic spindle, which pulls the sister chromatids apart toward the end of mitosis. (C, courtesy of Terry D. Allen.)

(C)   1 μm

The *contractile ring* consists mainly of actin filaments and myosin filaments arranged in a ring around the equator of the cell (see Chapter 17). It starts to assemble just beneath the plasma membrane toward the end of mitosis. As the ring contracts, it pulls the membrane inward, thereby dividing the cell in two (see Figure 18–19). We discuss later how plant cells, which have a cell wall to contend with, divide their cytoplasm by a very different mechanism.

M Phase Occurs in Stages

Although M phase proceeds as a continuous sequence of events, it is traditionally divided into a series of stages. The first five stages of M phase—prophase, prometaphase, metaphase, anaphase, and telophase—constitute **mitosis**, which was originally defined as the period in which the chromosomes are visible (because they have become condensed). *Cytokinesis*, which constitutes the final stage of M phase, begins before mitosis ends. The stages of M phase are summarized in **Panel 18–1** (pp. 622–623). Together, they form a dynamic sequence in which many independent cycles—involving the chromosomes, cytoskeleton, and centrosomes—are coordinated to produce two genetically identical daughter cells (Movie 18.4 and Movie 18.5).

Figure 18–19 Two transient cytoskeletal structures mediate M phase in animal cells. The mitotic spindle assembles first to separate the duplicated chromosomes. Then, the contractile ring assembles to divide the cell in two. Whereas the mitotic spindle is based on microtubules, the contractile ring is based on actin and myosin filaments. Plant cells use a very different mechanism to divide the cytoplasm, as we discuss later.

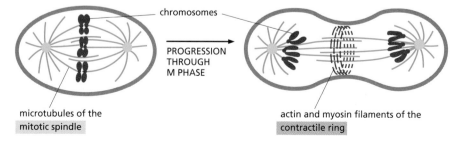

chromosomes

PROGRESSION THROUGH M PHASE

microtubules of the mitotic spindle

actin and myosin filaments of the contractile ring

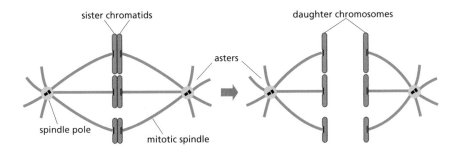

Figure 18–20 Sister chromatids separate at the beginning of anaphase. The mitotic spindle then pulls the resulting chromosomes to opposite poles of the cell.

MITOSIS

Before nuclear division, or mitosis, begins, each chromosome has been duplicated and consists of two identical sister chromatids, held together along their length by cohesin proteins (see Figure 18–18A). During mitosis, the cohesin proteins are cleaved, the sister chromatids split apart, and the resulting daughter chromosomes are pulled to opposite poles of the cell by the mitotic spindle (**Figure 18–20**). In this section, we examine how the mitotic spindle assembles and functions. We discuss how the dynamic instability of microtubules and the activity of microtubule-associated motor proteins contribute to both the assembly of the spindle and its ability to segregate the duplicated chromosomes. We then consider the mechanism that operates during mitosis to ensure the synchronous separation of these chromosomes. Finally we discuss how the daughter nuclei form.

Centrosomes Duplicate To Help Form the Two Poles of the Mitotic Spindle

Before M phase begins, two critical events must be completed: DNA must be fully replicated, and, in animal cells, the centrosome must be duplicated. The **centrosome** is the principal *microtubule-organizing center in animal cells.* It duplicates so that it can help form the two poles of the mitotic spindle and so that each daughter cell can receive its own centrosome.

Centrosome duplication begins at the same time as DNA replication. The process is triggered by the same Cdks—G_1/S-Cdk and S-Cdk—that initiate DNA replication. Initially, when the centrosome duplicates, both copies remain together as a single complex on one side of the nucleus. As mitosis begins, however, the two centrosomes separate, and each nucleates a radial array of microtubules called an **aster**. The two asters move to opposite sides of the nucleus to form the two poles of the mitotic spindle (**Figure 18–21**). The process of centrosome duplication and separation is known as the **centrosome cycle**.

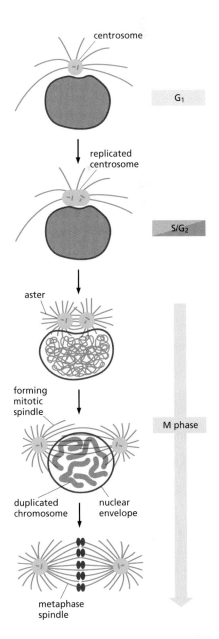

Figure 18–21 The centrosome in an interphase cell duplicates to form the two poles of a mitotic spindle. In most animal cells in interphase (G_1, S, and G_2), a centriole pair (shown here as a pair of *dark green bars*) is associated with the centrosome matrix (*light green*), which nucleates microtubule outgrowth. (The volume of the centrosome matrix is exaggerated in this diagram for clarity.) Centrosome duplication begins at the start of S phase and is complete by the end of G_2. Initially, the two centrosomes remain together, but, in early M phase, they separate, and each nucleates its own aster of microtubules. The centrosomes then move apart, and the microtubules that interact between the two asters elongate preferentially to form a bipolar mitotic spindle, with an aster at each pole. When the nuclear envelope breaks down, the spindle microtubules are able to interact with the duplicated chromosomes.

CELL DIVISION AND THE CELL CYCLE

INTERPHASE

S

G₁ G₂

CELL CYCLE

CYTOKINESIS

1 PROPHASE

2 PROMETAPHASE

5 TELOPHASE

4 ANAPHASE

3 METAPHASE

MITOSIS

M PHASE

The division of a cell into two daughters occurs in the M phase of the cell cycle. M phase consists of nuclear division, or mitosis, and cytoplasmic division, or cytokinesis. In this figure, M phase has been greatly expanded for clarity. Mitosis is itself divided into five stages, and these, together with cytokinesis, are described in this panel.

INTERPHASE

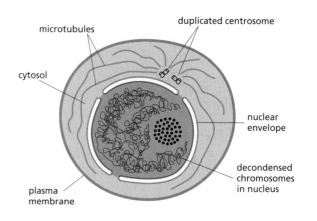

microtubules
duplicated centrosome
cytosol
nuclear envelope
plasma membrane
decondensed chromosomes in nucleus

During interphase, the cell increases in size. The DNA of the chromosomes is replicated, and the centrosome is duplicated.

In the light micrographs of dividing animal cells shown in this panel, chromosomes are stained *orange* and microtubules are *green*.

(Micrographs courtesy of Julie Canman and Ted Salmon; "Metaphase" from cover of *J. Cell. Sci.* 115(9), 2002. With permission from The Company of Biologists Ltd; "Telophase" from J.C. Canman et al., *Nature* 424:1074–1078, 2003. With permission from Macmillan Publishers Ltd.)

MITOSIS

1 PROPHASE

centrosome
intact nuclear envelope
forming mitotic spindle
kinetochore
condensing duplicated chromosome with two sister chromatids held together along their length

At prophase, the duplicated chromosomes, each consisting of two closely associated sister chromatids, condense. Outside the nucleus, the mitotic spindle assembles between the two centrosomes, which have begun to move apart. For simplicity, only three chromosomes are drawn.

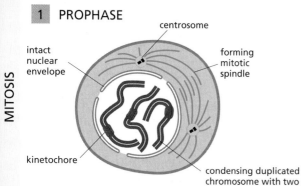

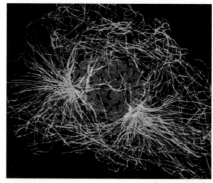

time = 0 min

MITOSIS

2 PROMETAPHASE

spindle pole
fragments of nuclear envelope
kinetochore microtubule
chromosome in motion

Prometaphase starts abruptly with the breakdown of the nuclear envelope. Chromosomes can now attach to spindle microtubules via their kinetochores and undergo active movement.

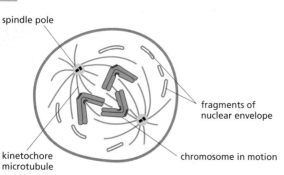

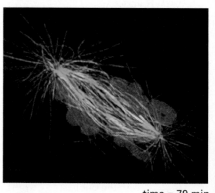

time = 79 min

3 METAPHASE

MITOSIS

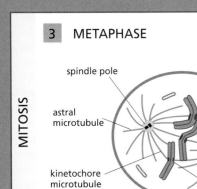

- spindle pole
- astral microtubule
- kinetochore microtubule
- spindle pole
- kinetochores of all chromosomes aligned in a plane midway between two spindle poles

At metaphase, the chromosomes are aligned at the equator of the spindle, midway between the spindle poles. The kinetochore microtubules on each sister chromatid attach to opposite poles of the spindle.

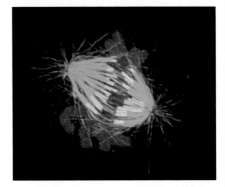

time = 250 min

4 ANAPHASE

MITOSIS

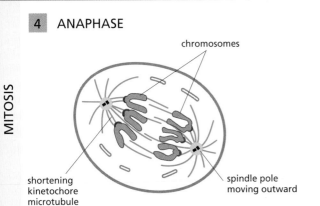

- chromosomes
- shortening kinetochore microtubule
- spindle pole moving outward

At anaphase, the sister chromatids synchronously separate and are pulled slowly toward the spindle pole to which they are attached. The kinetochore microtubules get shorter, and the spindle poles also move apart, both contributing to chromosome segregation.

time = 279 min

5 TELOPHASE

MITOSIS

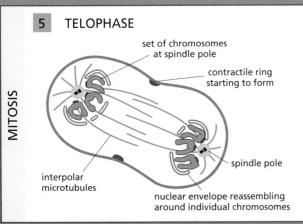

- set of chromosomes at spindle pole
- contractile ring starting to form
- spindle pole
- nuclear envelope reassembling around individual chromosomes
- interpolar microtubules

During telophase, the two sets of chromosomes arrive at the poles of the spindle. A new nuclear envelope reassembles around each set, completing the formation of two nuclei and marking the end of mitosis. The division of the cytoplasm begins with the assembly of the contractile ring.

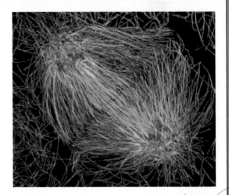

time = 315 min

CYTOKINESIS

CYTOKINESIS

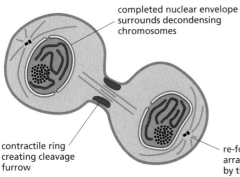

- completed nuclear envelope surrounds decondensing chromosomes
- contractile ring creating cleavage furrow
- re-formation of interphase array of microtubules nucleated by the centrosome

During cytokinesis of an animal cell, the cytoplasm is divided in two by a contractile ring of actin and myosin filaments, which pinches the cell into two daughters, each with one nucleus.

time = 362 min

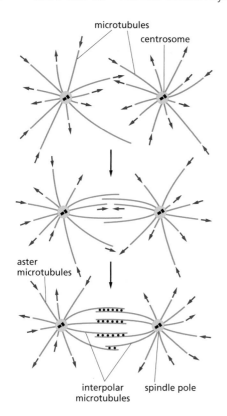

microtubules

centrosome

aster
microtubules

interpolar
microtubules   spindle pole

Figure 18–22 A bipolar mitotic spindle is formed by the selective stabilization of interacting microtubules. New microtubules grow out in random directions from the two centrosomes. The two ends of a microtubule (by convention, called the plus and the minus ends) have different properties, and it is the minus end that is anchored in the centrosome (discussed in Chapter 17). The free plus ends are dynamically unstable and switch suddenly from uniform growth (*outward-pointing red arrows*) to rapid shrinkage (*inward-pointing red arrows*). When two microtubules from opposite centrosomes interact in an overlap zone, motor proteins and other microtubule-associated proteins cross-link the microtubules together (*black dots*) in a way that stabilizes the plus ends by decreasing the probability of their depolymerization.

The Mitotic Spindle Starts to Assemble in Prophase

The mitotic spindle begins to form in **prophase**. This assembly of the highly dynamic spindle depends on the remarkable properties of microtubules. As discussed in Chapter 17, microtubules continuously polymerize and depolymerize by the addition and loss of their tubulin subunits, and individual filaments alternate between growing and shrinking—a process called *dynamic instability* (see Figure 17–13). At the start of mitosis, the stability of microtubules decreases—in part because M-Cdk phosphorylates microtubule-associated proteins that influence the stability of the microtubules. As a result, during prophase, rapidly growing and shrinking microtubules extend in all directions from the two centrosomes, exploring the interior of the cell.

Some of the microtubules growing from one centrosome interact with the microtubules from the other centrosome (see Figure 18–21). This interaction stabilizes the microtubules, preventing them from depolymerizing, and it joins the two sets of microtubules together to form the basic framework of the **mitotic spindle**, with its characteristic bipolar shape (**Movie 18.6**). The two centrosomes that give rise to these microtubules are now called **spindle poles**, and the interacting microtubules are called *interpolar microtubules* (**Figure 18–22**). The assembly of the spindle is driven, in part, by motor proteins associated with the interpolar microtubules that help to cross-link the two sets of microtubules.

In the next stage of mitosis, the duplicated chromosomes attach to the spindle microtubules in such a way that, when the sister chromatids separate, they will be drawn to opposite poles of the cell.

Chromosomes Attach to the Mitotic Spindle at Prometaphase

Prometaphase starts abruptly with the disassembly of the nuclear envelope, which breaks up into small membrane vesicles. This process is triggered by the phosphorylation and consequent disassembly of nuclear pore proteins and the intermediate filament proteins of the nuclear lamina, the network of fibrous proteins that underlies and stabilizes the nuclear envelope (see Figure 17–8). The spindle microtubules, which have been lying in wait outside the nucleus, now gain access to the duplicated chromosomes and capture them (see Panel 18–1, pp. 622–623).

The spindle microtubules grab hold of the chromosomes at **kinetochores**, protein complexes that assemble on the centromere of each condensed chromosome during late prophase (**Figure 18–23**). Each duplicated chromosome has two kinetochores—one on each sister chromatid—which face in opposite directions. Kinetochores recognize the special DNA sequence present at the centromere: if this sequence is altered, kinetochores fail to assemble and, consequently, the chromosomes fail to segregate properly during mitosis.

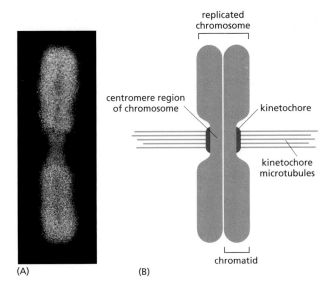

replicated chromosome

centromere region of chromosome

kinetochore

kinetochore microtubules

chromatid

(A)   (B)

Figure 18–23 Kinetochores attach chromosomes to the mitotic spindle. (A) A fluorescence micrograph of a duplicated mitotic chromosome. The DNA is stained with a fluorescent dye, and the kinetochores are stained *red* with fluorescent antibodies that recognize kinetochore proteins. These antibodies come from patients with scleroderma (a disease that causes progressive overproduction of connective tissue in skin and other organs), who, for unknown reasons, produce antibodies against their own kinetochore proteins. (B) Schematic drawing of a mitotic chromosome showing its two sister chromatids attached to kinetochore microtubules, which bind to the kinetochore by their plus ends. Each kinetochore forms a plaque on the surface of the centromere. (A, courtesy of B.R. Brinkley.)

Once the nuclear envelope has broken down, a randomly probing microtubule encountering a kinetochore will bind to it, thereby capturing the chromosome. This kinetochore microtubule links the chromosome to a spindle pole (see Figure 18–23 and Panel 18–1, pp. 622–623). Because kinetochores on sister chromatids face in opposite directions, they tend to attach to microtubules from opposite poles of the spindle, so that each duplicated chromosome becomes linked to both spindle poles. The attachment to opposite poles, called **bi-orientation**, generates tension on the kinetochores, which are being pulled in opposite directions. This tension signals to the sister kinetochores that they are attached correctly and are ready to be separated. The cell-cycle control system monitors this tension to ensure correct chromosome attachment (see Figure 18–3), a safeguard we discuss in detail shortly.

The number of microtubules attached to each kinetochore varies among species: each human kinetochore binds 20–40 microtubules, for example, whereas a yeast kinetochore binds just one. The three classes of microtubules that form the mitotic spindle are differently colored in **Figure 18–24**.

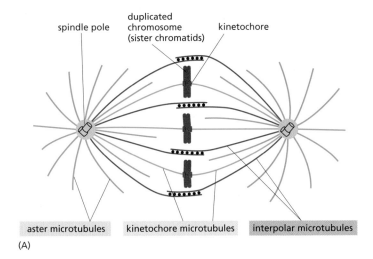

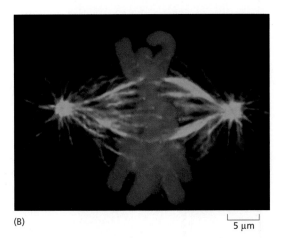

spindle pole

duplicated chromosome (sister chromatids)

kinetochore

aster microtubules   kinetochore microtubules   interpolar microtubules

(A)

(B)

5 μm

Figure 18–24 Three classes of microtubules make up the mitotic spindle. (A) Schematic drawing of a spindle with chromosomes attached, showing the three types of spindle microtubules: astral microtubules, kinetochore microtubules, and interpolar microtubules. In reality, the chromosomes are much larger than shown, and usually multiple microtubules are attached to each kinetochore. (B) Fluorescence micrograph of duplicated chromosomes at the metaphase plate of a real mitotic spindle. In this image, kinetochores are labeled in *red*, microtubules in *green*, and chromosomes in *blue*. (B, from A. Desai, *Curr. Biol.* 10:R508, 2000. With permission from Elsevier.)

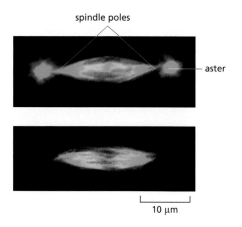

spindle poles

aster

10 μm

Figure 18–25 **Motor proteins and chromosomes can direct the assembly of a functional bipolar spindle in the absence of centrosomes.** In these fluorescence micrographs of embryos of the insect *Sciara*, the microtubules are stained *green* and the chromosomes *red*. The top micrograph shows a normal spindle formed with centrosomes in a normally fertilized embryo. The bottom micrograph shows a spindle formed without centrosomes in an embryo that initiated development without fertilization and thus lacks the centrosome normally provided by the sperm when it fertilizes the egg. Note that the spindle with centrosomes has an aster at each pole, whereas the spindle formed without centrosomes does not. As shown, both types of spindles are able to segregate the chromosomes. (From B. de Saint Phalle and W. Sullivan, *J. Cell Biol.* 141:1383–1391, 1998. With permission from The Rockefeller University Press.)

Chromosomes Assist in the Assembly of the Mitotic Spindle

Chromosomes are more than passive passengers in the process of spindle assembly: they can themselves stabilize and organize microtubules into functional mitotic spindles. In cells without centrosomes—including all plant cells and some animal cell types—the chromosomes nucleate microtubule assembly, and motor proteins then move and arrange the microtubules and chromosomes into a bipolar spindle. Even in animal cells that normally have centrosomes, a bipolar spindle can still be formed in this way if the centrosomes are removed (**Figure 18–25**). In cells with centrosomes, the chromosomes, motor proteins, and centrosomes work together to form the mitotic spindle.

Chromosomes Line Up at the Spindle Equator at Metaphase

During prometaphase, the duplicated chromosomes, now attached to the mitotic spindle, begin to move about, as if jerked first this way and then that. Eventually, they align at the equator of the spindle, halfway between the two spindle poles, thereby forming the *metaphase plate*. This event defines the beginning of **metaphase** (**Figure 18–26**). Although the forces that act to bring the chromosomes to the equator are not completely understood, both the continual growth and shrinkage of the microtubules and the action of microtubule motor proteins are required. A continuous balanced addition and loss of tubulin subunits is also required to maintain the metaphase spindle: when tubulin addition to the ends of microtubules is blocked by the drug colchicine, tubulin loss continues until the metaphase spindle disappears.

The chromosomes gathered at the equator of the metaphase spindle oscillate back and forth, continually adjusting their positions, indicating that the tug-of-war between the microtubules attached to opposite poles

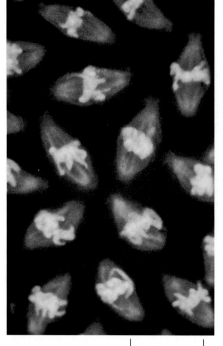

4 μm

Figure 18–26 **During metaphase, duplicated chromosomes gather halfway between the two spindle poles.** This fluorescence micrograph shows multiple mitotic spindles at metaphase in a fruit fly (*Drosophila*) embryo. The microtubules are stained *red*, and the chromosomes are stained *green*. At this stage of *Drosophila* development, there are multiple nuclei in one large cytoplasmic compartment, and all of the nuclei divide synchronously, which is why all of the nuclei shown here are at the same metaphase stage of the cell cycle (**Movie 18.7**). Metaphase spindles are usually pictured in two dimensions, as they are here; when viewed in three dimensions, however, the chromosomes are seen to be gathered at a platelike region at the equator of the spindle—the so-called metaphase plate. (Courtesy of William Sullivan.)

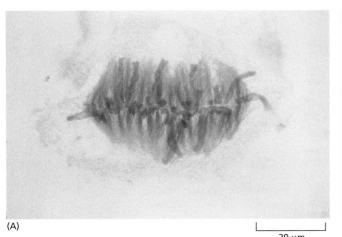
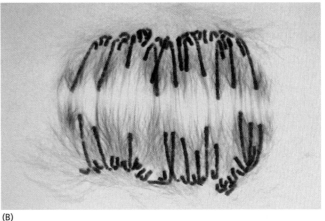

(A)

(B)

$\underline{}$
20 μm

Figure 18–27 Sister chromatids separate at anaphase. In the transition from metaphase (A) to anaphase (B), sister chromatids (stained *blue*) suddenly separate, allowing the resulting chromosomes to move toward opposite poles, as seen in these plant cells stained with gold-labeled antibodies to label the microtubules (*red*). Plant cells generally do not have centrosomes and therefore have less sharply defined spindle poles than animal cells (see Figure 18–34); nonetheless, spindle poles are present here at the top and bottom of each micrograph, although they cannot be seen. (Courtesy of Andrew Bajer.)

of the spindle continues to operate after the chromosomes are all aligned. If one of the pair of kinetochore attachments is artificially severed with a laser beam during metaphase, the entire duplicated chromosome immediately moves toward the pole to which it remains attached. Similarly, if the attachment between sister chromatids is cut, the two chromosomes separate and move toward opposite poles. These experiments show that the duplicated chromosomes are not simply deposited at the metaphase plate. They are suspended there under tension. In anaphase, that tension will pull the sister chromatids apart.

Proteolysis Triggers Sister-Chromatid Separation at Anaphase

Anaphase begins abruptly with the breakage of the cohesin linkages that hold sister chromatids together (see Figure 18–18A). This release allows each chromatid—now considered a chromosome—to be pulled to the spindle pole to which it is attached (**Figure 18–27**). This movement segregates the two identical sets of chromosomes to opposite ends of the spindle (see Panel 18–1, pp. 622–623).

The cohesin linkage is destroyed by a protease called *separase*. Before anaphase begins, this protease is held in an inactive state by an inhibitory protein called *securin*. At the beginning of anaphase, securin is targeted for destruction by APC—the same protein complex, discussed earlier, that marks M cyclin for degradation. Once securin has been removed, separase is then free to sever the cohesin linkages (**Figure 18–28**).

Chromosomes Segregate During Anaphase

Once the sister chromatids separate, the resulting chromosomes are pulled to the spindle pole to which they are attached. They all move at the same speed, which is typically about 1 μm per minute. The movement is the consequence of two independent processes that depend on different parts of the mitotic spindle. The two processes are called *anaphase A* and *anaphase B*, and they occur more or less simultaneously. In anaphase A, the kinetochore microtubules shorten and the attached chromosomes

QUESTION 18–6

If fine glass needles are used to manipulate a chromosome inside a living cell during early M phase, it is possible to trick the kinetochores on the two sister chromatids into attaching to the same spindle pole. This arrangement is normally unstable, but the attachments can be stabilized if the needle is used to gently pull the chromosome so that the microtubules attached to both kinetochores (via the same spindle pole) are under tension. What does this suggest to you about the mechanism by which kinetochores normally become attached and stay attached to microtubules from opposite spindle poles? Is the finding consistent with the possibility that a kinetochore is programmed to attach to microtubules from a particular spindle pole? Explain your answers.

Figure 18–28 APC triggers the separation of sister chromatids by promoting the destruction of cohesins. APC indirectly triggers the cleavage of the cohesins that hold sister chromatids together. It catalyzes the ubiquitylation and destruction of an inhibitory protein called securin. Securin inhibits the activity of a proteolytic enzyme called separase; when freed from securin, separase cleaves the cohesin complexes, allowing the mitotic spindle to pull the sister chromatids apart.

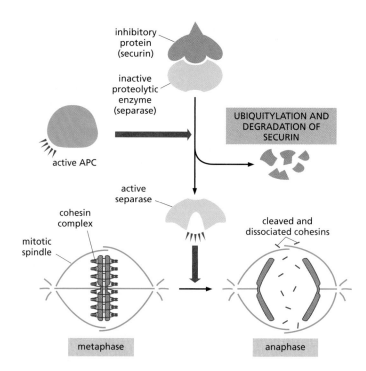

Figure 18–29 Two processes segregate daughter chromosomes at anaphase. In *anaphase A*, the daughter chromosomes are *pulled* toward opposite poles as the kinetochore microtubules depolymerize. The force driving this movement is generated mainly at the kinetochore. In *anaphase B*, the two spindle poles move apart as the result of two separate forces: (1) the elongation and sliding of the interpolar microtubules past one another *pushes* the two poles apart, and (2) forces exerted on the outward-pointing astral microtubules at each spindle pole *pull* the poles away from each other, toward the cell cortex. Both forces are thought to depend on the action of motor proteins associated with the microtubules.

move poleward. In anaphase B, the spindle poles themselves move apart, further segregating the two sets of chromosomes (**Figure 18–29**).

The driving force for the movements of anaphase A is thought to be provided mainly by the loss of tubulin subunits from both ends of the kinetochore microtubules. The driving forces in anaphase B are thought to be provided by two sets of motor proteins—members of the kinesin and dynein families—operating on different types of spindle microtubules (see Figure 17–21). Kinesin proteins act on the long, overlapping interpolar microtubules, sliding the microtubules from opposite poles past one another at the equator of the spindle and pushing the spindle poles apart. Dynein proteins, anchored to the cell cortex that underlies the plasma membrane, pull the poles apart (see Figure 18–29B).

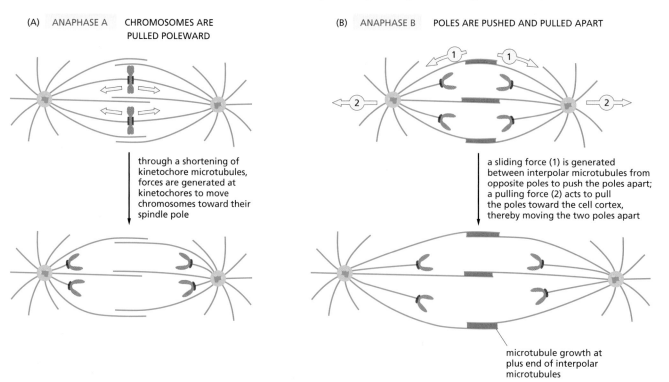

An Unattached Chromosome Will Prevent Sister-Chromatid Separation

If a dividing cell were to begin to segregate its chromosomes before all the chromosomes were properly attached to the spindle, one daughter cell would receive an incomplete set of chromosomes, while the other would receive a surplus. Both situations could be lethal. Thus, a dividing cell must ensure that every last chromosome is attached properly to the spindle before it completes mitosis. To monitor chromosome attachment, the cell makes use of a negative signal: unattached chromosomes send a "stop" signal to the cell-cycle control system. Although only some of the details are known, the signal inhibits further progress through mitosis by blocking the activation of the APC (see Figure 18–28). Without active APC, the sister chromatids remain glued together. Thus, none of the duplicated chromosomes can be pulled apart until every chromosome has been positioned correctly on the mitotic spindle. This so-called *spindle assembly checkpoint* thereby controls the onset of anaphase, as well as the exit from mitosis, as mentioned earlier (see Figure 18–12).

The Nuclear Envelope Re-forms at Telophase

By the end of anaphase, the daughter chromosomes have separated into two equal groups, one at each pole of the spindle. During **telophase**, the final stage of mitosis, the mitotic spindle disassembles, and a nuclear envelope reassembles around each group of chromosomes to form the two daughter nuclei (Movie 18.8). Vesicles of nuclear membrane first cluster around individual chromosomes and then fuse to re-form the nuclear envelope (see Panel 18–1, pp. 622–623). During this process, the nuclear pore proteins and nuclear lamins that were phosphorylated during prometaphase are now dephosphorylated, which allows them to reassemble and the nuclear envelope and lamina to re-form (**Figure 18–30**). Once the nuclear envelope has re-formed, the pores pump in nuclear proteins, the nucleus expands, and the condensed chromosomes decondense into their interphase state. As a consequence of this decondensation, gene transcription is able to resume. A new nucleus has been created, and mitosis is complete. All that remains is for the cell to complete its division into two separate daughter cells.

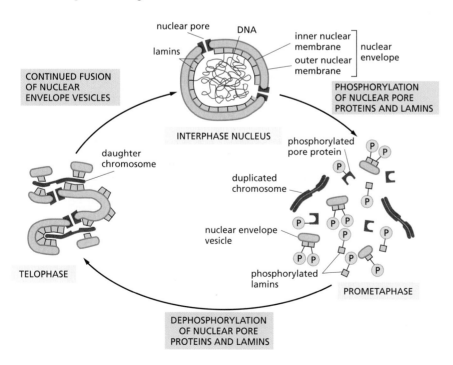

Figure 18–30 The nuclear envelope breaks down and re-forms during mitosis. The phosphorylation of nuclear pore proteins and lamins helps trigger the disassembly of the nuclear envelope at prometaphase. Dephosphorylation of these proteins at telophase helps reverse the process.

QUESTION 18–7

Consider the events that lead to the formation of the new nucleus at telophase. How do nuclear and cytosolic proteins become properly re-sorted so that the new nucleus contains nuclear proteins but not cytosolic proteins?

CYTOKINESIS

Cytokinesis, the process by which the cytoplasm is cleaved in two, completes M phase. It usually begins in anaphase but is not completed until the two daughter nuclei have formed in telophase. Whereas mitosis depends on a transient microtubule-based structure, the mitotic spindle, cytokinesis in animal cells depends on a transient structure based on actin and myosin filaments, the *contractile ring* (see Figure 18–19). Both the plane of cleavage and the timing of cytokinesis, however, are determined by the mitotic spindle.

The Mitotic Spindle Determines the Plane of Cytoplasmic Cleavage

The first visible sign of cytokinesis in animal cells is a puckering and furrowing of the plasma membrane that occurs during anaphase (**Figure 18–31**). The furrowing invariably occurs in a plane that runs perpendicular to the long axis of the mitotic spindle. This positioning ensures that the *cleavage furrow* cuts between the two groups of segregated chromosomes, so that each daughter cell receives an identical and complete set of chromosomes. If the mitotic spindle is deliberately displaced (using a fine glass needle) as soon as the furrow appears, the furrow disappears and a new one develops at a site corresponding to the new spindle location and orientation. Once the furrowing process is well under way, however, cleavage proceeds even if the mitotic spindle is artificially sucked out of the cell or depolymerized using the drug colchicine.

How does the mitotic spindle dictate the position of the cleavage furrow? The mechanism is still uncertain, but it appears that, during anaphase, the overlapping interpolar microtubules that form the *central spindle* recruit and activate proteins that signal to the cell cortex to initiate the assembly of the contractile ring at a position midway between the spindle poles. Because these signals originate in the anaphase spindle, this mechanism also contributes to the timing of cytokinesis in late mitosis.

When the mitotic spindle is located centrally in the cell—the usual situation in most dividing cells—the two daughter cells produced will be of equal size. During embryonic development, however, there are some instances in which the dividing cell moves its mitotic spindle to an asymmetrical position, and, consequently, the furrow creates two daughter cells that differ in size. In most of these *asymmetric divisions*, the daughters also differ in the molecules they inherit, and they usually develop into different cell types.

Figure 18–31 The cleavage furrow is formed by the action of the contractile ring underneath the plasma membrane. In these scanning electron micrographs of a dividing fertilized frog egg, the cleavage furrow is unusually well defined. (A) Low-magnification view of the egg surface. (B) A higher-magnification view of the cleavage furrow. (From H.W. Beams and R.G. Kessel, *Am. Sci.* 64:279–290, 1976. With permission of Sigma Xi.)

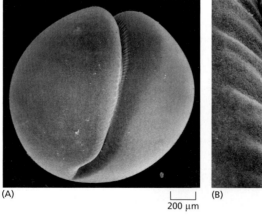

(A) 200 µm (B) 25 µm

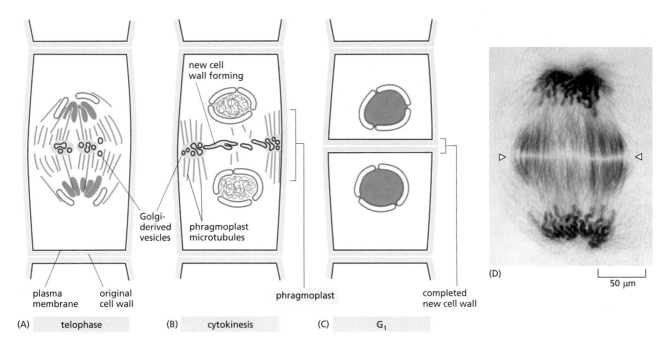

(A)   telophase   (B)   cytokinesis   (C)   G₁

(D)

50 µm

each cell cycle. The ER in interphase cells is continuous with the nuclear membrane and is organized by the microtubule cytoskeleton (see Figure 17–20A). Upon entry into M phase, the reorganization of the microtubules releases the ER; in most cells, the released ER remains intact during mitosis and is cut in two during cytokinesis. The Golgi apparatus fragments during mitosis; the fragments associate with the spindle microtubules via motor proteins, thereby hitching a ride into the daughter cells as the spindle elongates in anaphase. Other components of the cell—including the other membrane-enclosed organelles, ribosomes, and all of the soluble proteins—are inherited randomly when the cell divides.

Having discussed how cells divide, we now turn to the general problem of how the size of an animal or an organ is determined, which leads us to consider how cell number and cell size are controlled.

CONTROL OF CELL NUMBERS AND CELL SIZE

A fertilized mouse egg and a fertilized human egg are similar in size—about 100 µm in diameter. Yet an adult mouse is much smaller than an adult human. What are the differences between the control of cell behavior in humans and mice that generate such big differences in size? The same fundamental question can be asked about each organ and tissue in an individual's body. What adjustment of cell behavior explains the length of an elephant's trunk or the size of its brain or its liver? These questions are largely unanswered, but it is at least possible to say what the ingredients of an answer must be. Three fundamental processes largely determine organ and body size: cell growth, cell division, and cell death. Each of these processes, in turn, depends on programs intrinsic to the individual cell, regulated by signals from other cells in the body.

In this section, we first consider how organisms eliminate unwanted cells by a form of programmed cell death called *apoptosis*. We then discuss how extracellular signals balance cell death, cell growth, and cell division—thereby helping control the size of an animal and its organs. We conclude the section with a brief discussion of the extracellular signals that help keep these three processes in check.

Figure 18–34 Cytokinesis in a plant cell is guided by a specialized microtubule-based structure called the phragmoplast. At the beginning of telophase, after the chromosomes have segregated, a new cell wall starts to assemble inside the cell at the equator of the old spindle (A). The interpolar microtubules of the mitotic spindle remaining at telophase form the *phragmoplast* and guide vesicles, derived from the Golgi apparatus, toward the equator of the spindle. The vesicles, which are filled with cell-wall material, fuse to form the growing new cell wall that grows outward to reach the plasma membrane and original cell wall (B). The pre-existing plasma membrane and the membrane surrounding the new cell wall (both shown in red) then fuse, completely separating the two daughter cells (C). A light micrograph of a plant cell in telophase is shown in (D) at a stage corresponding to (A). The cell has been stained to show both the microtubules and the two sets of chromosomes segregated at the two poles of the spindle. The location of the growing new cell wall is indicated by the arrowheads. (D, courtesy of Andrew Bajer.)

QUESTION 18–9

The Golgi apparatus is thought to be partitioned into the daughter cells at cell division by a random distribution of fragments that are created at mitosis. Explain why random partitioning of chromosomes would not work.

Apoptosis Helps Regulate Animal Cell Numbers

The cells of a multicellular organism are members of a highly organized community. The number of cells in this community is tightly regulated—not simply by controlling the rate of cell division, but also by controlling the rate of cell death. If cells are no longer needed, they can commit suicide by activating an intracellular death program—a process called **programmed cell death**. In animals, the most common form of programmed cell death is called **apoptosis** (from a Greek word meaning "falling off," as leaves fall from a tree).

The amount of apoptosis that occurs in both developing and adult animal tissues can be astonishing. In the developing vertebrate nervous system, for example, more than half of some types of nerve cells normally die soon after they are formed. In a healthy adult human, billions of cells die in the bone marrow and intestine every hour. It seems remarkably wasteful for so many cells to die, especially as the vast majority are perfectly healthy at the time they kill themselves. What purposes does this massive cell suicide serve?

In some cases, the answers are clear. Mouse paws—and our own hands and feet—are sculpted by apoptosis during embryonic development: they start out as spadelike structures, and the individual fingers and toes separate because the cells between them die (**Figure 18–35**). In other cases, cells die when the structure they form is no longer needed. When a tadpole changes into a frog at metamorphosis, the cells in the tail die, and the tail, which is not needed in the frog, disappears (**Figure 18–36**). In these cases, the unneeded cells die largely through apoptosis.

In adult tissues, cell death usually exactly balances cell division, unless the tissue is growing or shrinking. If part of the liver is removed in an adult rat, for example, liver cells proliferate to make up the loss. Conversely, if a rat is treated with the drug phenobarbital, which stimulates liver cell division, the liver enlarges. However, when the phenobarbital treatment is stopped, apoptosis in the liver greatly increases until the organ has returned to its original size, usually within a week or so. Thus, the liver is kept at a constant size through regulation of both the cell death rate and the cell birth rate.

Apoptosis Is Mediated by an Intracellular Proteolytic Cascade

Cells that die as a result of acute injury typically swell and burst, spilling their contents all over their neighbors, a process called *cell necrosis* (**Figure 18–37A**). This eruption triggers a potentially damaging inflammatory response. By contrast, a cell that undergoes apoptosis dies neatly, without damaging its neighbors. A cell in the throes of apoptosis may

Figure 18–35 Apoptosis in the developing mouse paw sculpts the digits. (A) The paw in this mouse embryo has been stained with a dye that specifically labels cells that have undergone apoptosis. The apoptotic cells appear as *bright green dots* between the developing digits. (B) This cell death eliminates the tissue between the developing digits, as seen in the paw shown one day later. Here, few, if any, apoptotic cells can be seen—demonstrating how quickly apoptotic cells can be cleared from a tissue. (From W. Wood et al., *Development* 127:5245–5252, 2000. With permission from The Company of Biologists Ltd.)

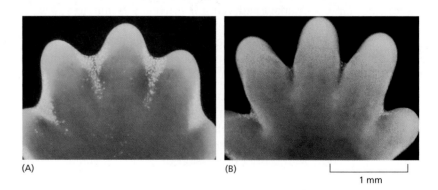

(A)　　　　(B)　　　　1 mm

Figure 18–36 As a tadpole changes into a frog, the cells in its tail are induced to undergo apoptosis. All of the changes that occur during metamorphosis, including the induction of apoptosis in the tadpole tail, are stimulated by an increase in thyroid hormone in the blood.

develop irregular bulges—or *blebs*—on its surface; but it then shrinks and condenses (**Figure 18–37B**). The cytoskeleton collapses, the nuclear envelope disassembles, and the nuclear DNA breaks up into fragments (Movie 18.9). Most importantly, the cell surface is altered in such a manner that it immediately attracts phagocytic cells, usually specialized phagocytic cells called macrophages (see Figure 15–32B). These cells engulf the apoptotic cell before it spills its contents (**Figure 18–37C**). This rapid removal of the dying cell avoids the damaging consequences of cell necrosis, and it also allows the organic components of the apoptotic cell to be recycled by the cell that ingests it.

The molecular machinery responsible for apoptosis, which seems to be similar in most animal cells, involves a family of proteases called **caspases**. These enzymes are made as inactive precursors, called *procaspases*, which are activated in response to signals that induce apoptosis (**Figure 18–38A**). Two types of caspases work together to take a cell apart. *Initiator caspases* cleave, and thereby activate, downstream *executioner caspases*. Some of these executioner caspases then activate additional executioners, kicking off an amplifying, proteolytic cascade;

> **QUESTION 18–10**
>
> Why do you think apoptosis occurs by a different mechanism from the cell death that occurs in cell necrosis? What might be the consequences if apoptosis were not achieved in so neat and orderly a fashion, whereby the cell destroys itself from within and avoids leakage of its contents into the extracellular space?

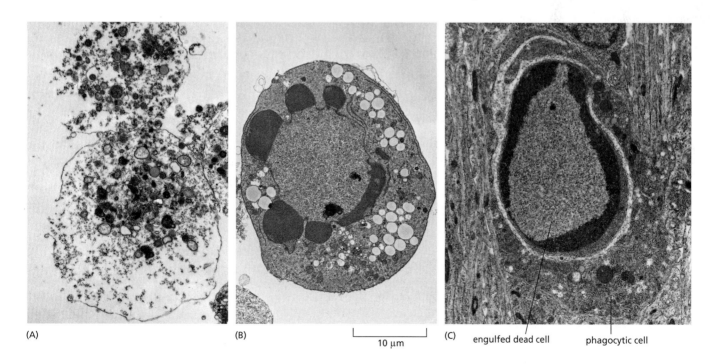

(A)   (B) |——————| 10 μm   (C)   engulfed dead cell   phagocytic cell

Figure 18–37 Cells undergoing apoptosis die quickly and cleanly. Electron micrographs showing cells that have died by necrosis (A) or by apoptosis (B and C). The cells in (A) and (B) died in a culture dish, whereas the cell in (C) died in a developing tissue and has been engulfed by a phagocytic cell. Note that the cell in (A) seems to have exploded, whereas those in (B) and (C) have condensed but seem relatively intact. The large vacuoles seen in the cytoplasm of the cell in (B) are a variable feature of apoptosis. (Courtesy of Julia Burne.)

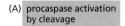

(A) procaspase activation by cleavage

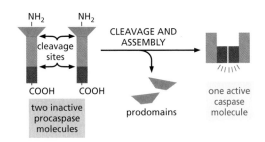

(B) amplifying caspase cascade

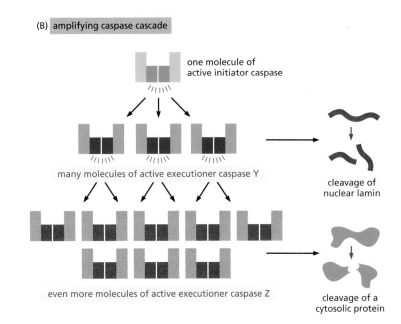

Figure 18–38 Apoptosis is mediated by an intracellular proteolytic cascade.
(A) Each suicide protease (caspase) is made as an inactive proenzyme, a procaspase, which is itself often activated by proteolytic cleavage by another member of the same protease family. Two cleaved fragments from each of two procaspase molecules associate to form an active caspase, which is formed from two small and two large subunits; the two prodomains are usually discarded.
(B) Each activated initiator caspase molecule can cleave many executioner procaspase molecules, thereby activating them, and these can activate even more procaspase molecules. In this way, an initial activation of a small number of initiator caspase molecules can lead, via an amplifying chain reaction (a cascade), to the explosive activation of a large number of executioner caspase molecules. Some of the activated executioner caspases break down key proteins in the cell, such as the nuclear lamins, leading to the controlled death of the cell. The proteolytic cascade begins when initiator procaspases are activated, as we discuss shortly.

others dismember other key proteins in the cell (**Figure 18–38B**). For example, one executioner caspase targets the lamin proteins that form the nuclear lamina underlying the nuclear envelope; this cleavage causes the irreversible breakdown of the nuclear lamina, which allows nucleases to enter the nucleus and break down the DNA. In this way, the cell dismantles itself quickly and cleanly, and its corpse is rapidly taken up and digested by another cell.

Activation of the apoptotic program, like entry into a new stage of the cell cycle, is usually triggered in an all-or-none fashion. The caspase cascade is not only destructive and self-amplifying but also irreversible; once a cell reaches a critical point along the path to destruction, it cannot turn back. Thus it is important that the decision to die be tightly controlled.

The Intrinsic Apoptotic Death Program Is Regulated by the Bcl2 Family of Intracellular Proteins

All nucleated animal cells contain the seeds of their own destruction: in these cells, inactive procaspases lie waiting for a signal to destroy the cell. It is therefore not surprising that caspase activity is tightly regulated inside the cell to ensure that the death program is held in check until it is needed—for example, to eliminate cells that are superfluous, mislocated, or badly damaged.

The main proteins that regulate the activation of caspases are members of the **Bcl2 family** of intracellular proteins. Some members of this protein family promote caspase activation and cell death, whereas others inhibit these processes. Two of the most important death-inducing family members are proteins called *Bax* and *Bak*. These proteins—which are activated in response to DNA damage or other insults—promote cell death by inducing the release of the electron-transport protein cytochrome *c* from mitochondria into the cytosol. Other members of the Bcl2 family (including Bcl2 itself) inhibit apoptosis by preventing Bax and Bak from releasing cytochrome *c*. The balance between the activities of pro-apoptotic and anti-apoptotic members of the Bcl2 family largely determines whether a mammalian cell lives or dies by apoptosis.

The cytochrome *c* molecules released from mitochondria activate initiator procaspases—and induce cell death—by promoting the assembly of a large, seven-armed, pinwheel-like protein complex called an *apoptosome*.

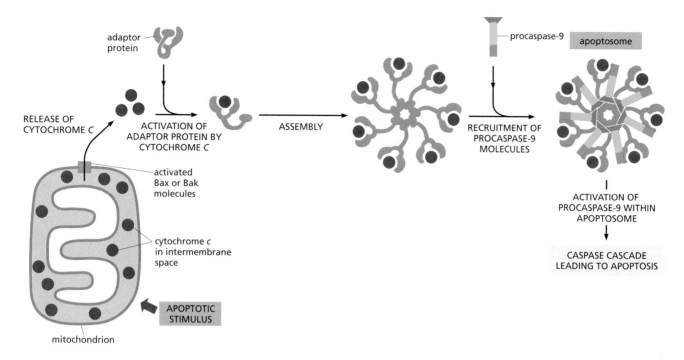

The apoptosome then recruits and activates a particular initiator procaspase, which then triggers a caspase cascade that leads to apoptosis (**Figure 18–39**).

Extracellular Signals Can Also Induce Apoptosis

Sometimes the signal to commit suicide is not generated internally, but instead comes from a neighboring cell. Some of these extracellular signals activate the cell death program by affecting the activity of members of the Bcl2 family of proteins. Others stimulate apoptosis more directly by activating a set of cell-surface receptor proteins known as *death receptors*.

One particularly well-understood death receptor, called *Fas*, is present on the surface of a variety of mammalian cell types. Fas is activated by a membrane-bound protein, called *Fas ligand*, present on the surface of specialized immune cells called *killer lymphocytes*. These killer cells help regulate immune responses by inducing apoptosis in other immune cells that are unwanted or are no longer needed—and activating Fas is one way they do so. The binding of Fas ligand to its receptor triggers the assembly of a death-inducing signaling complex, which includes specific initiator procaspases that, when activated, launch a caspase cascade that leads to cell death (**Figure 18–40**).

Animal Cells Require Extracellular Signals to Survive, Grow, and Divide

In a multicellular organism, the fate of individual cells is controlled by signals from other cells. For either tissue growth or cell replacement, cells must grow before they divide. Nutrients are not enough for an animal cell to survive, grow, or divide. It must also receive chemical signals from other cells, usually its neighbors. Such controls ensure that a cell survives only when it is needed and divides only when another cell is required, either to allow tissue growth or to replace cell loss.

Most of the extracellular signal molecules that influence cell survival, cell growth, and cell division are either soluble proteins secreted by other cells or proteins that are bound to the surface of other cells or to the

Figure 18–39 Bax and Bak are death-promoting members of the Bcl2 family of intracellular proteins that can trigger apoptosis by releasing cytochrome c from mitochondria. When Bak or Bax proteins are activated by an apoptotic stimulus, they aggregate in the outer mitochondrial membrane, leading to the release of cytochrome c by an unknown mechanism. The cytochrome c is released into the cytosol from the mitochondrial intermembrane space (along with other proteins in this space—not shown). Cytochrome c then binds to an adaptor protein, causing it to assemble into a seven-armed complex. This complex then recruits seven molecules of a specific initiator procaspase (procaspase-9) to form a structure called an apoptosome. The procaspase-9 proteins become activated within the apoptosome and then go on to activate executioner procaspases in the cytosol, leading to a caspase cascade and apoptosis.

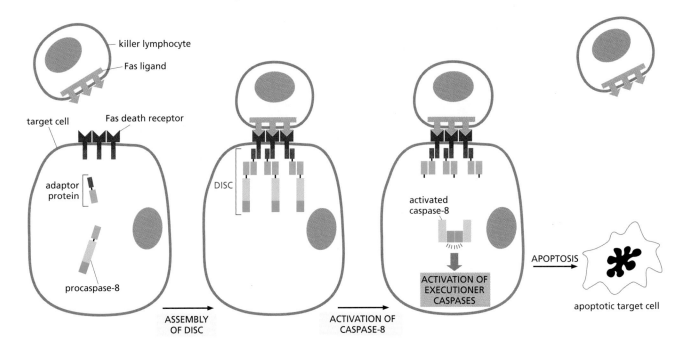

Figure 18–40 Activated death receptors initiate an intracellular signaling pathway that leads to apoptosis. Fas ligand on the surface of a killer lymphocyte activates Fas receptors on the surface of a target cell. This triggers the assembly of a collection of intracellular proteins to form a death-inducing signaling complex (DISC), which includes a specific initiator procaspase (procaspase-8 or 10). The procaspases cleave and activate one another, and the resulting active caspases then activate executioner procaspases in the cytosol, leading to a caspase proteolytic cascade and apoptosis.

extracellular matrix. Although most act positively to stimulate one or more of these cell processes, some act negatively to inhibit a particular process. The positively acting signal proteins can be classified, on the basis of their function, into three major categories:

1. **Survival factors** promote cell survival, largely by suppressing apoptosis.
2. **Mitogens** stimulate cell division, primarily by overcoming the intracellular braking mechanisms that tend to block progression through the cell cycle.
3. **Growth factors** stimulate cell growth (an increase in cell size and mass) by promoting the synthesis and inhibiting the degradation of proteins and other macromolecules.

These categories are not mutually exclusive, as many signal molecules have more than one of these functions. Unfortunately, the term "growth factor" is often used as a catch-all phrase to describe a protein with any of these functions. Indeed, the phrase "cell growth" is frequently used inappropriately to mean an increase in cell number, which is more correctly termed "cell proliferation."

In the following three sections, we examine each of these types of signal molecules in turn.

Survival Factors Suppress Apoptosis

Animal cells need signals from other cells just to survive. If deprived of such survival factors, cells activate a caspase-dependent intracellular suicide program and die by apoptosis. This requirement for signals from other cells helps ensure that cells survive only when and where they are needed. Many types of nerve cells, for example, are produced in excess in the developing nervous system and then compete for limited amounts of survival factors that are secreted by the target cells they contact. Those nerve cells that receive enough survival factor live, while the others die by apoptosis. In this way, the number of surviving nerve cells is automatically adjusted to match the number of cells with which they connect (**Figure 18–41**). A similar dependence on survival signals from

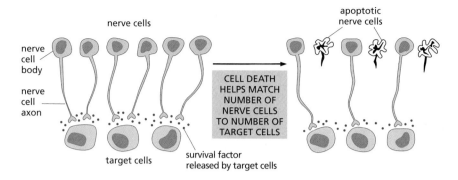

Figure 18–41 Cell death can help adjust the number of developing nerve cells to the number of target cells they contact. If more nerve cells are produced than can be supported by the limited amount of survival factor released by the target cells, some cells will receive insufficient amounts of survival factor to keep their suicide program suppressed and will undergo apoptosis. This strategy of overproduction followed by culling can help ensure that all target cells are contacted by nerve cells and that the "extra" nerve cells are automatically eliminated.

neighboring cells is thought to help control cell numbers in other tissues, both during development and in adulthood.

Survival factors usually act by activating cell-surface receptors. Once activated, the receptors turn on intracellular signaling pathways that keep the apoptotic death program suppressed, usually by regulating members of the Bcl2 family of proteins. Some survival factors, for example, increase the production of Bcl2, a protein that suppresses apoptosis (**Figure 18–42**).

Mitogens Stimulate Cell Division by Promoting Entry into S Phase

Most mitogens are secreted signal proteins that bind to cell-surface receptors. When activated by mitogen binding, these receptors initiate various intracellular signaling pathways (discussed in Chapter 16) that stimulate cell division. As we saw earlier, these signaling pathways act mainly by releasing the molecular brakes that block the transition from the G_1 phase of the cell cycle into S phase (see Figure 18–14).

Most mitogens have been identified and characterized by their effects on cells in culture. One of the first mitogens identified in this way was *platelet-derived growth factor*, or *PDGF*, the effects of which are typical of many others discovered since. When blood clots form (in a wound, for example), blood platelets incorporated in the clots are stimulated to release PDGF. PDGF then binds to receptor tyrosine kinases (discussed in Chapter 16) in surviving cells at the wound site, stimulating the cells to proliferate and help heal the wound. Similarly, if part of the liver is lost through surgery or acute injury, a mitogen called *hepatocyte growth factor* helps stimulate the surviving liver cells to proliferate.

Growth Factors Stimulate Cells to Grow

The growth of an organism or organ depends on cell growth as much as on cell division. If cells divided without growing, they would get progressively smaller, and there would be no increase in total cell mass. In single-celled organisms such as yeasts, both cell growth and cell division require only nutrients. In animals, by contrast, both cell growth and cell division depend on signals from other cells. Cell growth, unlike cell division, does not depend on the cell-cycle control system. Indeed, many

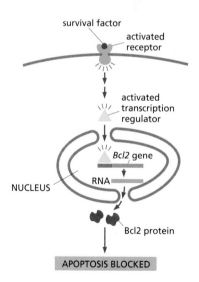

Figure 18–42 Survival factors often suppress apoptosis by regulating Bcl2 family members. In this case, the survival factor binds to cell-surface receptors that activate an intracellular signaling pathway, which in turn activates a transcription regulator in the cytosol. This protein moves to the nucleus, where it activates the gene encoding Bcl2, a protein that inhibits apoptosis.

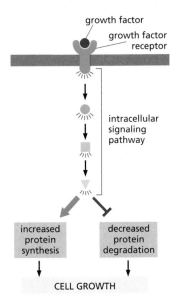

Figure 18–43 Extracellular growth factors increase the synthesis and decrease the degradation of macromolecules. This change leads to a net increase in macromolecules and thereby cell growth (see also Figure 16–39).

animal cells, including nerve cells and most muscle cells, do most of their growing after they have terminally differentiated and permanently stopped dividing.

Like most survival factors and mitogens, most extracellular growth factors bind to cell-surface receptors that activate intracellular signaling pathways. These pathways lead to the accumulation of proteins and other macromolecules. Growth factors both increase the rate of synthesis of these molecules and decrease their rate of degradation (**Figure 18–43**).

Some extracellular signal proteins, including PDGF, can act as both growth factors and mitogens, stimulating both cell growth and progression through the cell cycle. Such proteins help ensure that cells maintain their appropriate size as they proliferate.

Compared to cell division, there has been surprisingly little study of how cell size is controlled in animals. As a result, it remains a mystery how different cell types in the same animal grow to be so different in size (**Figure 18–44**).

Some Extracellular Signal Proteins Inhibit Cell Survival, Division, or Growth

The extracellular signal proteins that we have discussed so far—survival factors, mitogens, and growth factors—act positively to increase the size of organs and organisms. Some extracellular signal proteins, however, act to oppose these positive regulators and thereby inhibit tissue growth. *Myostatin*, for example, is a secreted signal protein that normally inhibits the growth and proliferation of the precursor cells (myoblasts) that fuse to form skeletal muscle cells during mammalian development. When the gene that encodes myostatin is deleted in mice, their muscles grow to be several times larger than normal, because both the number and the size of muscle cells is increased. Remarkably, two breeds of cattle that were bred for large muscles turned out to have mutations in the gene encoding myostatin (**Figure 18–45**).

Cancers are similarly the products of mutations that set cells free from the normal "social" controls operating on cell survival, growth, and proliferation. Because cancer cells are generally less dependent than normal cells on signals from other cells, they can out-survive, outgrow, and out-divide their normal neighbors, producing tumors that can kill their host (see Chapter 20).

In our discussions of cell division, we have thus far focused entirely on the ordinary divisions that produce two daughter cells, each with a full and identical complement of the parent cell's genetic material. There is, however, a different and highly specialized type of cell division called

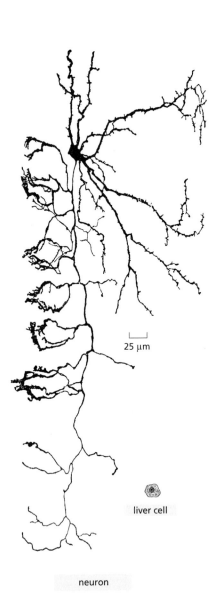

Figure 18–44 The cells in an animal can differ greatly in size. The neuron and liver cell shown here are drawn at the same scale and both contain the same amount of DNA. A neuron grows progressively larger after it has terminally differentiated and permanently stopped dividing. During this time, the ratio of cytoplasm to DNA increases enormously—by a factor of more than 10^5 for some neurons. (Neuron adapted from S. Ramón y Cajal, Histologie du Système Nerveux de l'Homme et de Vertébrés, 1909–1911. Paris: Maloine; reprinted, Madrid: C.S.I.C., 1972.)

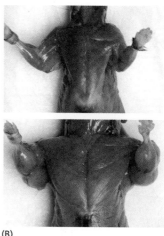

(A)

(B)

Figure 18–45 Mutation of the *myostatin* gene leads to a dramatic increase in muscle mass. (A) This Belgian Blue was produced by cattle breeders and was only later found to have a mutation in the *myostatin* gene. (B) Mice purposely made deficient in the same gene also have remarkably big muscles. A normal mouse is shown at the top for comparison with the muscular mutant shown at the bottom. (A, from H.L. Sweeney, *Sci. Am.* 291:62–69, 2004. With permission from Scientific American. B, from S.-J. Lee, *PLoS ONE* 2:e789, 2007.)

meiosis, which is required for sexual reproduction in eukaryotes. In the next chapter, we describe the special features of meiosis and how they underlie the genetic principles that define the laws of inheritance.

ESSENTIAL CONCEPTS

- The eukaryotic cell cycle consists of several distinct phases. In interphase, the cell grows and the nuclear DNA is replicated; in M phase, the nucleus divides (mitosis) followed by the cytoplasm (cytokinesis).

- In most cells, interphase consists of an S phase when DNA is duplicated, plus two gap phases—G_1 and G_2. These gap phases give proliferating cells more time to grow and prepare for S phase and M phase.

- The cell-cycle control system coordinates events of the cell cycle by sequentially and cyclically switching on and off the appropriate parts of the cell-cycle machinery.

- The cell-cycle control system depends on cyclin-dependent protein kinases (Cdks), which are cyclically activated by the binding of cyclin proteins and by phosphorylation and dephosphorylation; when activated, Cdks phosphorylate key proteins in the cell.

- Different cyclin–Cdk complexes trigger different steps of the cell cycle: M-Cdk drives the cell into mitosis; G_1-Cdk drives it through G_1; G_1/S-Cdk and S-Cdk drive it into S phase.

- The control system also uses protein complexes, such as APC, to trigger the destruction of specific cell-cycle regulators at particular stages of the cycle.

- The cell-cycle control system can halt the cycle at specific transition points to ensure that intracellular and extracellular conditions are favorable and that each step is completed before the next is started. Some of these control mechanisms rely on Cdk inhibitors that block the activity of one or more cyclin–Cdk complexes.

- S-Cdk initiates DNA replication during S phase and helps ensure that the genome is copied only once. The cell-cycle control system can delay cell-cycle progression during G_1 or S phase to prevent cells from replicating damaged DNA. It can also delay the start of M phase to ensure that DNA replication is complete.

- Centrosomes duplicate during S phase and separate during G_2. Some of the microtubules that grow out of the duplicated centrosomes interact to form the mitotic spindle.

- When the nuclear envelope breaks down, the spindle microtubules capture the duplicated chromosomes and pull them in opposite directions, positioning the chromosomes at the equator of the metaphase spindle.

- The sudden separation of sister chromatids at anaphase allows the chromosomes to be pulled to opposite poles; this movement is driven by the depolymerization of spindle microtubules and by microtubule-associated motor proteins.

- A nuclear envelope re-forms around the two sets of segregated chromosomes to form two new nuclei, thereby completing mitosis.

- In animal cells, cytokinesis is mediated by a contractile ring of actin filaments and myosin filaments, which assembles midway between the spindle poles; in plant cells, by contrast, a new cell wall forms inside the parent cell to divide the cytoplasm in two.

- In animals, extracellular signals regulate cell numbers by controlling cell survival, cell growth, and cell proliferation.

- Most animal cells require survival signals from other cells to avoid apoptosis—a form of cell suicide mediated by a proteolytic caspase cascade; this strategy helps ensure that cells survive only when and where they are needed.

- Animal cells proliferate only if stimulated by extracellular mitogens produced by other cells; mitogens release the normal intracellular brakes that block progression from G_1 or G_0 into S phase.

- For an organism or an organ to grow, cells must grow as well as divide; animal cell growth depends on extracellular growth factors that stimulate protein synthesis and inhibit protein degradation.

- Some extracellular signal molecules inhibit rather than promote cell survival, cell growth, or cell division.

- Cancer cells fail to obey these normal "social" controls on cell behavior and therefore outgrow, out-divide, and out-survive their normal neighbors.

KEY TERMS

anaphase	condensin	metaphase
anaphase-promoting complex (APC)	contractile ring	mitogen
apoptosis	cyclin	mitosis
aster	cytokinesis	mitotic spindle
Bcl2 family	G_1-Cdk	p53
bi-orientation	G_1 cyclin	phragmoplast
caspase	G_1 phase	programmed cell death
Cdk (cyclin-dependent protein kinase)	G_2 phase	prometaphase
Cdk inhibitor protein	G_1/S-Cdk	prophase
cell cycle	G_1/S cyclin	S-Cdk
cell-cycle control system	growth factor	S cyclin
centrosome	interphase	S phase
centrosome cycle	kinetochore	sister chromatid
chromosome condensation	M-Cdk	spindle pole
cohesin	M cyclin	survival factor
	M phase	telophase

QUESTIONS

QUESTION 18–11

Roughly, how long would it take a single fertilized human egg to make a cluster of cells weighing 70 kg by repeated divisions, if each cell weighs 1 nanogram just after cell division and each cell cycle takes 24 hours? Why does it take very much longer than this to make a 70-kg adult human?

QUESTION 18–12

The shortest eukaryotic cell cycles of all—shorter even than those of many bacteria—occur in many early animal embryos. These so-called cleavage divisions take place without any significant increase in the weight of the embryo. How can this be? Which phase of the cell cycle would you expect to be most reduced?

QUESTION 18–13

One important biological effect of a large dose of ionizing radiation is to halt cell division.

A. How does this occur?

B. What happens if a cell has a mutation that prevents it from halting cell division after being irradiated?

C. What might be the effects of such a mutation if the cell is not irradiated?

D. An adult human who has reached maturity will die within a few days of receiving a radiation dose large enough to stop cell division. What does that tell you (other than that one should avoid large doses of radiation)?

QUESTION 18–14

If cells are grown in a culture medium containing radioactive thymidine, the thymidine will be covalently incorporated into the cell's DNA during S phase. The radioactive DNA can be detected in the nuclei of individual cells by autoradiography (i.e., by placing a photographic emulsion over the cells, radioactive cells will activate the emulsion and be labeled by black dots when looked at under a microscope). Consider a simple experiment in which cells are radioactively labeled by this method for only a short period (about 30 minutes). The radioactive thymidine medium is then replaced with one containing unlabeled thymidine, and the cells are grown for some additional time. At different time points after replacement of the medium, cells are examined in a microscope. The fraction of cells in mitosis (which can be easily recognized because the cells have rounded up and their chromosomes are condensed) that have radioactive DNA in their nuclei is then determined and plotted as a function of time after the labeling with radioactive thymidine (**Figure Q18–14**).

A. Would all cells (including cells at all phases of the cell cycle) be expected to contain radioactive DNA after the labeling procedure?

B. Initially there are no mitotic cells that contain radioactive DNA (see Figure Q18–14). Why is this?

C. Explain the rise and fall and then rise again of the curve.

D. Estimate the length of the G_2 phase from this graph.

QUESTION 18–15

One of the functions of M-Cdk is to cause a precipitous drop in M-cyclin concentration halfway through M phase. Describe the consequences of this sudden decrease and suggest possible mechanisms by which it might occur.

QUESTION 18–16

Figure 18–5 shows the rise of cyclin concentration and the rise of M-Cdk activity in cells as they progress through the cell cycle. It is remarkable that the cyclin concentration rises slowly and steadily, whereas M-Cdk activity increases suddenly. How do you think this difference arises?

QUESTION 18–17

What is the order in which the following events occur during cell division:

A. anaphase

B. metaphase

C. prometaphase

D. telophase

E. lunar phase

F. mitosis

G. prophase

Where does cytokinesis fit in?

QUESTION 18–18

The lifetime of a microtubule in a mammalian cell, between its formation by polymerization and its spontaneous disappearance by depolymerization, varies with the stage of the cell cycle. For an actively proliferating cell, the average lifetime is 5 minutes in interphase and 15 seconds in mitosis. If the average length of a microtubule in interphase is 20 μm, how long will it be during mitosis, assuming that the rates of microtubule elongation due to the addition of tubulin subunits in the two phases are the same?

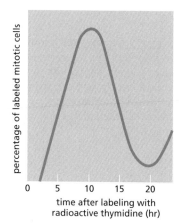

Figure Q18–14

QUESTION 18–19

The balance between plus-end-directed and minus-end-directed motor proteins that bind to interpolar microtubules in the overlap region of the mitotic spindle is thought to help determine the length of the spindle. How might each type of motor protein contribute to the determination of spindle length?

QUESTION 18–20

Sketch the principal stages of mitosis, using Panel 18–1 (pp. 622–623) as a guide. Color one sister chromatid and follow it through mitosis and cytokinesis. What event commits this chromatid to a particular daughter cell? Once initially committed, can its fate be reversed? What may influence this commitment?

QUESTION 18–21

The polar movement of chromosomes during anaphase A is associated with microtubule shortening. In particular, microtubules depolymerize at the ends at which they are attached to the kinetochores. Sketch a model that explains how a microtubule can shorten and generate force yet remain firmly attached to the chromosome.

QUESTION 18–22

Rarely, both sister chromatids of a replicated chromosome end up in one daughter cell. How might this happen? What could be the consequences of such a mitotic error?

QUESTION 18–23

Which of the following statements are correct? Explain your answers.

A. Centrosomes are replicated before M phase begins.

B. Two sister chromatids arise by replication of the DNA of the same chromosome and remain paired as they line up on the metaphase plate.

C. Interpolar microtubules attach end-to-end and are therefore continuous from one spindle pole to the other.

D. Microtubule polymerization and depolymerization and microtubule motor proteins are all required for DNA replication.

E. Microtubules nucleate at the centromeres and then connect to the kinetochores, which are structures at the centrosome regions of chromosomes.

QUESTION 18–24

An antibody that binds to myosin prevents the movement of myosin molecules along actin filaments (the interaction between actin and myosin is described in Chapter 17). How do you suppose the antibody exerts this effect? What might be the result of injecting this antibody into cells (A) on the movement of chromosomes at anaphase or (B) on cytokinesis? Explain your answers.

QUESTION 18–25

Look carefully at the electron micrographs in Figure 18–37. Describe the differences between the cell that died by necrosis and those that died by apoptosis.

How do the pictures confirm the differences between the two processes? Explain your answer.

QUESTION 18–26

Which of the following statements are correct? Explain your answers.

A. Cells do not pass from G_1 into M phase of the cell cycle unless there are sufficient nutrients to complete an entire cell cycle.

B. Apoptosis is mediated by special intracellular proteases, one of which cleaves nuclear lamins.

C. Developing neurons compete for limited amounts of survival factors.

D. Some vertebrate cell-cycle control proteins function when expressed in yeast cells.

E. The enzymatic activity of a Cdk protein is determined both by the presence of a bound cyclin and by the phosphorylation state of the Cdk.

QUESTION 18–27

Compare the rules of cell behavior in an animal with the rules that govern human behavior in society. What would happen to an animal if its cells behaved as people normally behave in our society? Could the rules that govern cell behavior be applied to human societies?

QUESTION 18–28

In his highly classified research laboratory, Dr. Lawrence M. is charged with the task of developing a strain of dog-sized rats to be deployed behind enemy lines. In your opinion, which of the following strategies should Dr. M. pursue to increase the size of rats?

A. Block all apoptosis.

B. Block p53 function.

C. Overproduce growth factors, mitogens, or survival factors.

D. Obtain a taxi driver's license and switch careers.

Explain the likely consequences of each option.

QUESTION 18–29

PDGF is encoded by a gene that can cause cancer when expressed inappropriately. Why do cancers not arise at wounds in which PDGF is released from platelets?

QUESTION 18–30

What do you suppose happens in mutant cells that

A. cannot degrade M-cyclin?

B. always express high levels of p21?

C. cannot phosphorylate Rb?

QUESTION 18–31

Liver cells proliferate excessively both in patients with chronic alcoholism and in patients with liver cancer. What are the differences in the mechanisms by which cell proliferation is induced in these diseases?

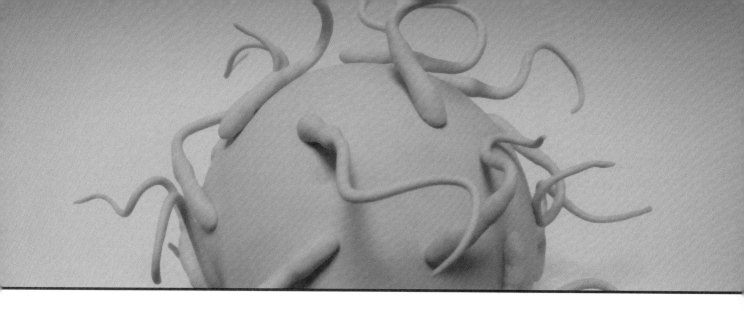

CHAPTER **NINETEEN**

19

Sexual Reproduction and the Power of Genetics

Individual cells reproduce by replicating their DNA and dividing in two. This basic process of cell proliferation occurs in all living species—in the cells of multicellular organisms and in free-living cells such as bacteria and yeasts—and it allows each cell to pass on its genetic information to future generations.

Yet reproduction in a multicellular organism—in a fish or a fly, a person or a plant—is a much more complicated affair. It entails elaborate developmental cycles, in which all of the organism's cells, tissues, and organs must be generated afresh from a single cell. This starter cell is no ordinary cell. It has a very peculiar origin: for most animal and plant species, it is produced by the union of a pair of cells that hail from two completely separate individuals—a mother and a father. As a result of this cell fusion—a central event in *sexual reproduction*—two genomes merge to form the genome of a new individual. The mechanisms that govern genetic inheritance in sexually reproducing organisms are therefore different, and more complex, than those that operate in organisms that pass on their genetic information asexually—by a straighforward cell division or by budding off a brand new individual.

In this chapter, we explore the cell biology of sexual reproduction. We discuss what organisms gain from sex, and we describe how they do it. We examine the reproductive cells produced by males and females, and we explore the specialized form of cell division, called *meiosis*, that generates them. We discuss how Gregor Mendel, a nineteenth-century Austrian monk, deduced the basic logic of genetic inheritance by studying the progeny of pea plants. Finally, we describe how scientists can exploit the genetics of sexual reproduction to gain insights into human biology, human origins, and the molecular underpinnings of human disease.

THE BENEFITS OF SEX

MEIOSIS AND FERTILIZATION

MENDEL AND THE LAWS OF INHERITANCE

GENETICS AS AN EXPERIMENTAL TOOL

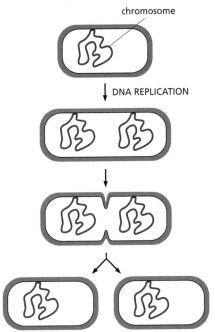

chromosome

DNA REPLICATION

Figure 19–1 Bacteria reproduce by simple cell division. The division of one bacterium into two takes 20–25 minutes under ideal growth conditions.

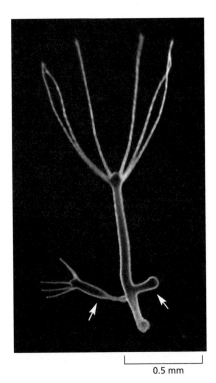

0.5 mm

Figure 19–2 A hydra reproduces by budding. This form of asexual reproduction involves the production of buds (*arrows*), which pinch off to form progeny that are genetically identical to their parent. Eventually, these buds will detach from their parent and live independently. (Courtesy of Amata Hornbruch.)

THE BENEFITS OF SEX

Most of the creatures we see around us reproduce sexually. However, many organisms, especially those invisible to the naked eye, can produce offspring without resorting to sex. Most bacteria and other single-celled organisms multiply by simple cell division (**Figure 19–1**). Many plants also reproduce asexually, forming multicellular offshoots that later detach from the parent to make new independent plants. Even in the animal kingdom, there are species that can procreate without sex. Hydra produce young by budding (**Figure 19–2**). Certain worms, when split in two, can regenerate the "missing halves" to form two complete individuals. And in some species of insects, lizards, and even birds, the females can lay eggs that develop *parthenogenetically*—without the help of males, sperm, or fertilization—into healthy daughters that can also reproduce the same way.

But while such forms of **asexual reproduction** are simple and direct, they give rise to offspring that are genetically identical to the parent organism. **Sexual reproduction**, on the other hand, involves the mixing of DNA from two individuals to produce offspring that are genetically distinct from one another and from both their parents. This mode of reproduction apparently has great advantages, as the vast majority of plants and animals have adopted it.

Sexual Reproduction Involves Both Diploid and Haploid Cells

Organisms that reproduce sexually are generally *diploid*: each cell contains two sets of chromosomes—one inherited from each parent. Because the two parents are members of the same species, the *maternal* chromosome set and the *paternal* chromosome set are very similar. The most notable difference between them is the *sex chromosomes*, which, in some species, distinguish males from females. With the exception of these sex chromosomes, the maternal and paternal versions of every chromosome—called the maternal and paternal **homologs**—carry the same set of genes. Each diploid cell, therefore, carries two copies of every gene (except for those found on the sex chromosomes, which may be present in only one copy).

Unlike the majority of cells in a diploid organism, however, the specialized cells that perform the central process in sexual reproduction—the **germ cells**, or **gametes**—are *haploid*: they each contain only one set of chromosomes. For most organisms, the males and females produce different types of gametes. In animals, one is large and nonmotile and is referred to as the *egg*; the other is small and motile and is referred to as the *sperm* (**Figure 19–3**). These two dissimilar haploid gametes join together to regenerate a diploid cell, called the fertilized egg, or *zygote*, which has chromosomes from both the mother and father. The zygote thus produced develops into a new individual with a diploid set of chromosomes that is distinct from that of either parent (**Figure 19–4**).

For almost all multicellular animals, including vertebrates, practically the whole life cycle is spent in the diploid state. The haploid cells exist only briefly and are highly specialized for their function as genetic ambassadors. These haploid gametes are generated from diploid precursor cells by a specialized form of reductive division called *meiosis*, a process we discuss shortly. This precursor cell lineage, which is dedicated solely to the production of germ cells, is called the **germ line**. The cells forming the rest of the animal's body—the **somatic cells**—ultimately leave no

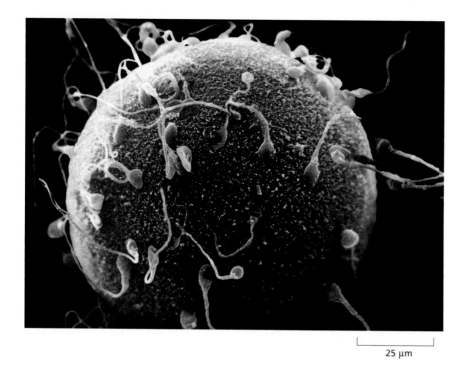

25 μm

Figure 19–3 Despite their tremendous difference in size, sperm and egg contribute equally to the genetic character of the offspring. This difference in size between male and female gametes (in which eggs contain a large quantity of cytoplasm, whereas sperm contain almost none) is consistent with the fact that the cytoplasm is not the basis of inheritance. If it were, the female's contribution to the makeup of the offspring would be much greater than the male's. Shown here is a scanning electron micrograph of an egg with human sperm bound to its surface. Although many sperm are bound to the egg, only one will fertilize it. (Courtesy of David M. Phillips/Photo Researchers, Inc.)

progeny of their own (**Figure 19–5** and see Figure 9–3). They exist, in effect, only to help the cells of the germ line survive and propagate.

The sexual reproductive cycle thus involves an alternation of haploid cells, each carrying a single set of chromosomes, with generations of diploid cells, each carrying two sets of chromosomes. One benefit of this arrangement is that it allows sexually reproducing organisms to produce offspring that are genetically diverse, as we discuss next.

Sexual Reproduction Generates Genetic Diversity

Sexual reproduction produces novel chromosome combinations. During meiosis, the maternal and paternal chromosome sets present in diploid germ-line cells are partitioned out into the single chromosome sets of the gametes. Each gamete will receive a mixture of maternal homologs and paternal homologs; when the genomes of two gametes combine during fertilization, they produce a zygote with a unique chromosomal complement.

Of course, if the maternal and paternal homologs carry the same genes, why should such chromosomal assortment matter? One answer is that although the set of genes on each homolog is the same, the paternal and maternal version of each gene is not. Genes occur in variant versions, called **alleles**. For any given gene, many different alleles may be present in the "gene pool" of a species. The existence of these variant alleles means that the two copies of any given gene in a particular individual are likely to be somewhat different from each other—and from those carried by other individuals. What makes individuals within a species genetically unique is the inheritance of different combinations of alleles. And with its cycles of diploidy, meiosis, haploidy, and cell fusion, sex breaks up old combinations of alleles and generates new ones.

Sexual reproduction also generates genetic diversity through a second mechanism—genetic recombination. We discuss this process, which scrambles the genetic information on each chromosome during meiosis, a bit later.

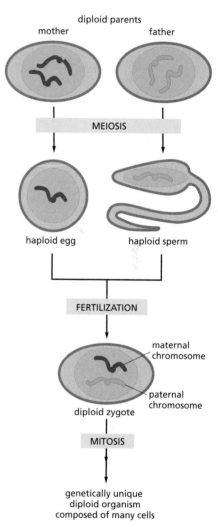

Figure 19–4 Sexual reproduction involves both haploid and diploid cells. Sperm and egg are produced by meiosis of diploid germ-line cells. During fertilization, a haploid egg and a haploid sperm fuse to form a diploid zygote. For simplicity, only one chromosome is shown for each gamete, and the sperm cell has been greatly enlarged. Human gametes have 23 chromosomes, and the egg is much larger than the sperm (see, for example, Figure 19–3).

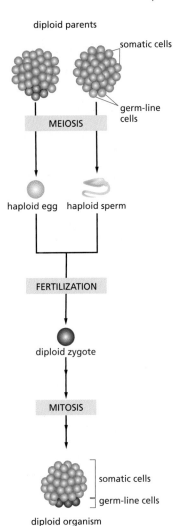

diploid parents

somatic cells

germ-line cells

MEIOSIS

haploid egg   haploid sperm

FERTILIZATION

diploid zygote

MITOSIS

somatic cells

germ-line cells

diploid organism

Figure 19–5 Germ-line cells and somatic cells carry out fundamentally different functions. In sexually reproducing animals, diploid germ-line cells, which are specified early in development, give rise to haploid gametes by meiosis. The gametes propagate genetic information into the next generation. Somatic cells (*gray*) form the body of the organism and are therefore necessary to support sexual reproduction, but they themselves leave no progeny.

Sexual Reproduction Gives Organisms a Competitive Advantage in a Changing Environment

The processes that generate genetic diversity during meiosis operate at random, as we will shortly discuss. That means that the alleles an individual receives from its parents are as likely to represent a change for the worse as they are a change for the better. Why, then, should the ability to try out new genetic combinations give organisms that reproduce sexually an evolutionary advantage over those that "breed true" through an asexual process? This question continues to perplex evolutionary geneticists, but one advantage seems to be that reshuffling genetic information through sexual reproduction can help a species survive in an unpredictably variable environment. If two parents produce many offspring with a wide variety of gene combinations, they increase the odds that at least one of their progeny will have a combination of features necessary for survival in a variety of environmental conditions. They are more likely, for example, to survive infections by bacteria, viruses, and parasites, which themselves continually change in a never-ending evolutionary battle. This genetic gamble may explain why even unicellular organisms, such as yeasts, intermittently indulge in a simple form of sexual reproduction. Typically, they switch on this behavior as an alternative to ordinary cell division when times are hard and starvation looms. Yeasts with a genetic defect that makes them unable to reproduce sexually show a reduced ability to evolve and adapt when they are subjected to harsh conditions.

Sexual reproduction may also be advantageous for another reason. In any population, new mutations continually occur, giving rise to new alleles—and many of these new mutations may be harmful. Sexual reproduction can speed up the elimination of these deleterious alleles and help to prevent them from accumulating in the population. By mating with only the fittest males, females select for good combinations of alleles and allow bad combinations to be lost from the population more efficiently than they would otherwise be. According to this theory, which is supported by some careful calculations of costs and benefits, sexual reproduction is favored because males can serve as a genetic filtering device: the males who succeed in mating allow the best, and only the best, collections of genes to be passed on, whereas males who fail to mate serve as a genetic "trash can"—a way of discarding bad collections of alleles from the population. Of course, for social organisms especially, it must be conceded that males may sometimes make themselves useful in other ways.

Whatever its advantages, sex has clearly been favored by evolution. In the following section, we review the central features of this popular form of reproduction, beginning with meiosis, the process by which gametes are formed.

MEIOSIS AND FERTILIZATION

Our modern understanding of the fundamental cycle of events involved in sexual reproduction grew out of discoveries reported in 1888, when Theodor Boveri noted that the fertilized egg of a parasitic roundworm contains four chromosomes, whereas the worm's gametes (sperm and egg) contain only two. This study was the first to demonstrate that gametes are **haploid**—they carry only a single set of chromosomes. All of the other cells of the body, including the germ-line cells that give rise to the gametes, are **diploid**—they carry two sets of chromosomes, one derived from the mother and the other from the father. Therefore, sperm and eggs must be produced by a special kind of "reductive" cell division in which the number of chromosomes is precisely halved (see Figure 19–4). The term **meiosis** was coined to describe this form of cell division; it comes from a Greek word meaning "diminution," or "lessening."

From Boveri's experiments on worms and other species, it became clear that the behavior of the chromosomes, which at that time were simply microscopic bodies of unknown function, matched the pattern of inheritance, in which the two parents make equal contributions to determining the character of the progeny despite the enormous difference in size between egg and sperm (see Figure 19–3). This was the first clue that chromosomes contain the material of heredity. The study of sexual reproduction and meiosis therefore has a central place in the history of cell biology.

In this section, we describe the cell biology of sexual reproduction from a modern point of view, focusing on the elaborate dance of chromosomes that occurs when a cell undertakes meiosis. We begin with an overview of how meiosis distributes chromosomes to the gametes. We then take a closer look at how chromosomes pair, recombine, and are segregated into germ cells during meiosis, thereby shuffling the maternal and paternal genes into novel combinations. We also discuss what happens when meiosis goes awry. Finally, we consider briefly the process of fertilization, through which gametes come together to form a new, genetically distinct individual.

Meiosis Involves One Round of DNA Replication Followed by Two Rounds of Cell Division

Before a diploid cell divides by mitosis, it duplicates its two sets of chromosomes. This duplication allows a full set of chromosomes—including a complete maternal set plus a complete paternal set—to be transmitted to each daughter cell (discussed in Chapter 18). Although meiosis ultimately halves this diploid chromosome complement—producing haploid gametes that carry only a single set of chromosomes—it, too, begins with a round of chromosome duplication. The subsequent reduction in chromosome number occurs because this single round of duplication is followed by two successive cell divisions without further DNA replication (**Figure 19–6**). One can imagine that meiosis might instead occur by a simple modification of mitotic cell division: if DNA replication (S phase) were omitted completely, a single round of cell division could produce two haploid cells directly. But, for reasons that are still unclear, this is not the way meiosis works.

Meiosis begins in specialized diploid germ-line cells that reside in the ovaries or testes. Like somatic cells, these germ-line cells are diploid; each contains two copies of every chromosome—a paternal homolog, inherited from the organism's father, and a maternal homolog, inherited from its mother. In the first step of meiosis, all of these chromosomes are duplicated, and the resulting copies remain closely attached to each other, as they would during an ordinary mitosis (see "Prophase" in Panel 18–1, pp. 622–623). The next phase of the process, however, is unique to meiosis. Each duplicated paternal chromosome first locates and then attaches itself to the corresponding duplicated maternal homolog, a process called *pairing*. Pairing ensures that the homologs will segregate properly during the two subsequent cell divisions and that each of the final gametes will receive a complete haploid set of chromosomes.

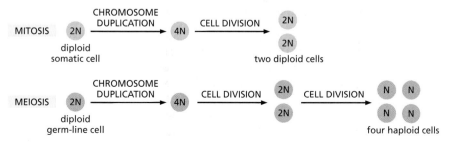

Figure 19–6 Mitosis and meiosis both begin with a round of chromosome duplication. In mitosis, this duplication is followed by a single round of cell division to yield two diploid cells. In meiosis, chromosome duplication in a diploid germ-line cell is followed by two rounds of cell division, without further DNA replication, to produce four haploid cells. *N* represents the number of chromosomes in the haploid cell.

Together, the two successive meiotic cell divisions, called *meiotic division I* (meiosis I) and *meiotic division II* (meiosis II), parcel out one complete set of chromosomes to each of the four haploid cells produced. Because the assignment of each homolog to the haploid daughter cells is random, each of the resulting gametes will receive a different mixture of maternal and paternal chromosomes.

Thus, meiosis produces four cells that are genetically dissimilar and that contain exactly half as many chromosomes as the original parent germ-line cell. Mitosis, in contrast, produces two genetically identical daughter cells. **Figure 19–7** summarizes the molecular events that distinguish

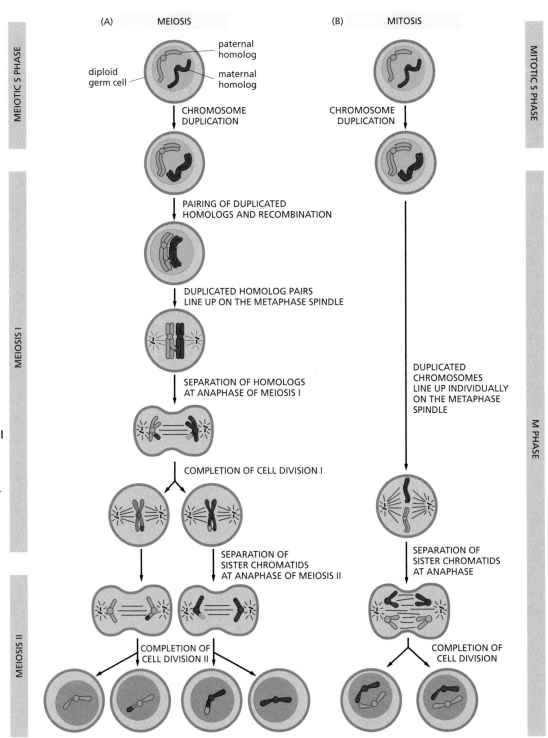

Figure 19–7 Meiosis generates four nonidentical haploid cells, whereas mitosis produces two identical diploid cells. As in Figure 19–4, only one pair of homologous chromosomes is shown. (A) In meiosis, two cell divisions are required after chromosome duplication to produce haploid cells. Each diploid cell that enters meiosis therefore produces four haploid cells, whereas (B) each diploid cell that divides by mitosis produces two diploid cells. Although mitosis and meiosis II are usually accomplished within hours, meiosis I can last days, months, or even years, because of the long time spent in prophase I.

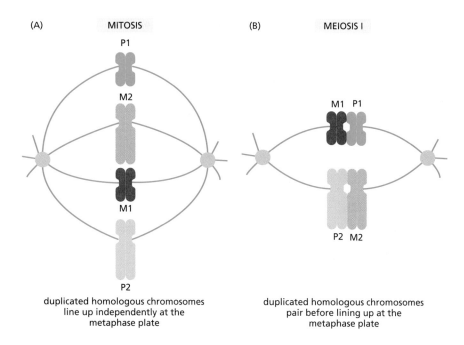

Figure 19–8 During meiosis, duplicated homologous chromosomes pair before lining up on the meiotic spindle. (A) In mitosis, the individual duplicated maternal (M) and paternal (P) chromosomes line up independently at the metaphase plate; each consists of a pair of sister chromatids, which will separate just before the cell divides. (B) By contrast, in division I of meiosis, duplicated maternal and paternal homologs pair long before lining up at the metaphase plate. The maternal and paternal homologs separate during the first meiotic division, and the sister chromatids separate during meiosis II. The mitotic and meiotic spindles are shown in *green*.

these two types of cell division—differences we now discuss in greater detail, beginning with the meiosis-specific pairing of maternal and paternal chromosomes.

Meiosis Requires the Pairing of Duplicated Homologous Chromosomes

As mentioned earlier, before a eukaryotic cell divides—by either meiosis or mitosis—it first duplicates all of its chromosomes. The twin copies of each duplicated chromosome, called **sister chromatids**, at first remain tightly linked along their length. The way these duplicated chromosomes are handled, however, differs between meiosis and mitosis. In mitosis, as we discuss in Chapter 18, the duplicated chromosomes line up, single file, at the metaphase plate (**Figure 19–8A**). As mitosis continues, the sister chromatids separate and are segregated into one or other of the two daughter cells.

In meiosis, however, the need to halve the number of chromosomes introduces an extra demand on the cell-division machinery. To ensure that each of the four haploid cells produced by meiosis will receive a single sister chromatid from each chromosome set, a germ-line cell must keep track of both the maternal and paternal **homologous chromosomes** (homologs). It does so by **pairing** the duplicated homologs before they line up at the metaphase plate (**Figure 19–8B**). Each pairing forms a structure called a **bivalent**, in which all four sister chromatids stick together until the cell is ready to divide (**Figure 19–9**). The maternal and paternal homologs will separate during meiotic division I, and the individual sister chromatids will separate during meiotic division II.

How the homologs (and the two sex chromosomes) recognize each other during pairing is still not fully understood. In many organisms, the initial association depends on an interaction between matching maternal and paternal DNA sequences at numerous sites that are widely dispersed along the homologous chromosomes. Once formed, bivalents are very stable: they remain associated throughout the long prophase of meiosis I, a stage that in some organisms can last for years.

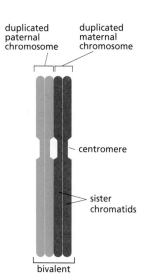

Figure 19–9 Duplicated maternal and paternal chromosomes pair during meiosis I to form bivalents. Each bivalent contains four sister chromatids and forms during prophase of meiosis I, well before attaching to the meiotic spindle.

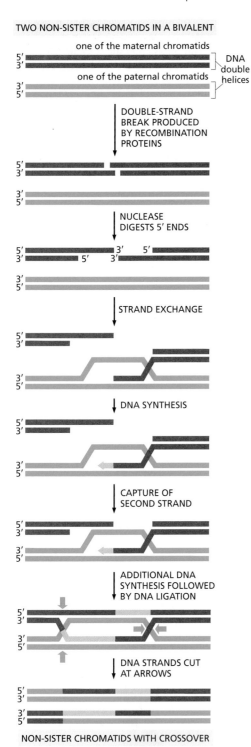

TWO NON-SISTER CHROMATIDS IN A BIVALENT

one of the maternal chromatids

one of the paternal chromatids

DNA double helices

DOUBLE-STRAND BREAK PRODUCED BY RECOMBINATION PROTEINS

NUCLEASE DIGESTS 5′ ENDS

STRAND EXCHANGE

DNA SYNTHESIS

CAPTURE OF SECOND STRAND

ADDITIONAL DNA SYNTHESIS FOLLOWED BY DNA LIGATION

DNA STRANDS CUT AT ARROWS

NON-SISTER CHROMATIDS WITH CROSSOVER

Figure 19–10 During meiosis I, non-sister chromatids in each bivalent swap segments of DNA. Here, only two of the four sister chromatids in the bivalent are shown, each drawn as a DNA double helix. During meiosis, the protein complexes that carry out this homologous recombination (not shown) first produce a double-strand break in the DNA of one of the chromatids (either the maternal or paternal chromatid) and then promote the formation of a cross-strand exchange with the other chromatid. When this exchange is resolved, each chromatid will contain a segment of DNA from the other. Many of the steps that produce chromosome crossovers during meiosis resemble those that guide the repair of DNA double-strand breaks in somatic cells (see Figure 6–30).

Crossing-Over Occurs Between the Duplicated Maternal and Paternal Chromosomes in Each Bivalent

The picture of meiotic division I we have just painted is greatly simplified, in that it leaves out a crucial feature. In sexually reproducing organisms, the pairing of the maternal and paternal chromosomes is accompanied by **homologous recombination**, a process in which two identical or very similar nucleotide sequences exchange genetic information. In Chapter 6, we discussed how homologous recombination is used to mend damaged chromosomes from which genetic information has been lost. This type of repair, uses information from an intact DNA double helix to restore the correct nucleotide sequence to a damaged, newly duplicated homolog (see Figure 6–30). A similar process takes place when homologous chromosomes pair during the long prophase of the first meiotic division. In meiosis, however, the recombination occurs between the non-sister chromatids in each bivalent, rather than between the identical sister chromatids within each duplicated chromosome. As a result, the maternal and paternal homologs end up physically swapping homologous chromosomal segments in a complex, multistep process called **crossing-over** (**Figure 19–10**).

Crossing-over is facilitated by the formation of a *synaptonemal complex.* As the duplicated homologs pair, this elaborate protein complex helps to hold the bivalent together and align the homologs so that strand exchange can readily occur between the non-sister chromatids. Each of the chromatids in a duplicated homolog (that is, each of these very long DNA double helices) can form a crossover with either (or both) of the chromatids from the other chromosome in the bivalent. The synaptonemal complex also helps space out the crossover events that take place along each chromosome.

By the time prophase I ends, the synaptonemal complex has disassembled, allowing the homologs to separate along most of their length. But each bivalent remains held together by at least one **chiasma** (plural **chiasmata**), a structure named after the Greek letter chi, χ, which is shaped like a cross. Each chiasma corresponds to a crossover between two non-sister chromatids (**Figure 19–11A**). Most bivalents contain more than one chiasma, indicating that multiple crossovers occur between homologous chromosomes (**Figure 19–11B and C**). In human oocytes—the cells that give rise to the egg—an average of two to three crossover events occur within each bivalent (**Figure 19–12**).

Crossovers that occur during meiosis are a major source of genetic variation in sexually reproducing species. By scrambling the genetic constitution of each of the chromosomes in the gamete, crossing-over helps to produce individuals with novel assortments of alleles. Crossing-over also has a second important role in meiosis. By holding homologous

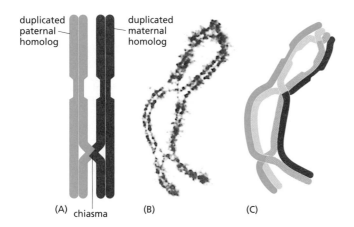

Figure 19–11 Crossover events create chiasmata between non-sister chromatids in each bivalent. (A) Schematic set of paired homologs in which one crossover event has occurred, creating a single chiasma. (B) Micrograph of a grasshopper bivalent with three chiasmata. (C) As the maternal and paternal homologs start to separate in meiosis I, chiasmata like those shown here help to hold the bivalent together. (B, courtesy of Bernard John.)

chromosomes together during prophase I, the chiasmata help ensure that the maternal and paternal homologs will segregate from one another correctly at the first meiotic division, as we discuss next.

Chromosome Pairing and Crossing-Over Ensure the Proper Segregation of Homologs

In most organisms, crossing-over during meiosis is required for the correct segregation of the two duplicated homologs into separate daughter nuclei. The chiasmata created by crossover events keep the maternal and paternal homologs bundled together until the spindle separates them during meiotic anaphase I. Before anaphase I, the two poles of the spindle pull on the duplicated homologs in opposite directions, and the chiasmata resist this pulling (**Figure 19–13A**). In so doing, the chiasmata help to position and stabilize bivalents at the metaphase plate.

In addition to the chiasmata, which hold the maternal and paternal homologs together, *cohesin* proteins (described in Chapter 18) keep the sister chromatids glued together along their entire length at meiosis I (see Figures 19–11B and 18–18). At the start of anaphase I, the cohesin proteins that hold the chromosome arms together are suddenly degraded. This release allows the arms to separate and the recombined homologs to be pulled apart (**Figure 19–13B**). This release is necessary because if the arms did not come apart, the duplicated maternal and paternal homologs would remain tethered to one another by the homologous DNA segments they had exchanged.

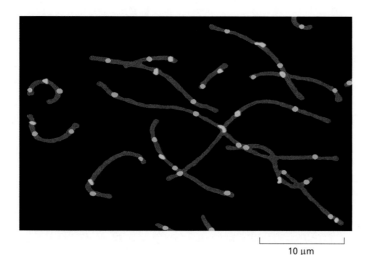

10 μm

Figure 19–12 Multiple crossovers can occur between the duplicated homologous chromosomes in a bivalent. Fluorescence micrograph shows a spread of chromosomes from a human oocyte (egg-cell precursor) at the stage where all four chromatids—both maternal and paternal homologs—are still tightly associated: each single long thread (stained *red*) is a bivalent, containing four DNA double helices. Sites of crossing-over are marked by the presence of a protein (stained *green*) that is a key component of the meiotic recombination machinery. *Blue* staining marks the position of centromeres (see Figure 19–9). (From C. Tease et al., *Am. J. Hum. Genet.* 70:1469–1479, 2002. With permission from Elsevier.)

Figure 19–13 Chiasmata help ensure proper segregation of duplicated homologs during the first meiotic division. (A) In metaphase of meiosis I, chiasmata created by crossing-over hold the maternal and paternal homologs together. At this stage, cohesin proteins (not shown) keep the sister chromatids glued together along their entire length. The kinetochores of sister chromatids function as a single unit in meiosis I, and microtubules that attach to them point toward the same spindle pole. (B) At anaphase of meiosis I, the cohesins holding the arms of the sister chromatids together are suddenly degraded, allowing the homologs to be separated. Cohesins in the centromere continue to hold the sister chromatids together as the homologs are pulled apart.

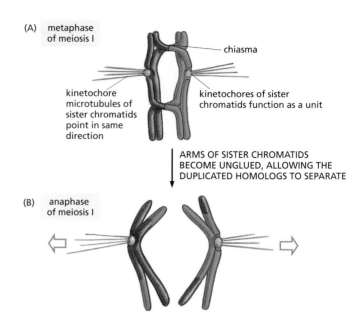

The Second Meiotic Division Produces Haploid Daughter Cells

To separate the sister chromatids and produce cells with a haploid amount of DNA, a second round of division, meiosis II, follows soon after the first—without further DNA replication or any significant interphase period. A meiotic spindle forms, and the kinetochores on each pair of sister chromatids now attach to kinetochore microtubules that point in opposite directions, as they would in an ordinary mitotic division. At anaphase of meiosis II, the remaining, meiosis-specific cohesins—located in the centromere—are degraded, and the sister chromatids are drawn into different daughter cells (**Figure 19–14**). The entire process is shown in Movie 19.1.

Haploid Gametes Contain Reassorted Genetic Information

Even though they share the same parents, no two siblings are genetically the same (unless they are identical twins). These genetic differences are initiated long before sperm meets egg, when meiosis I produces two kinds of randomizing genetic reassortment.

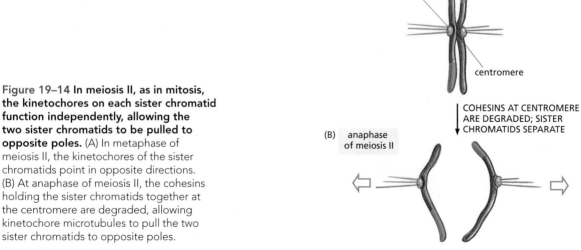

Figure 19–14 In meiosis II, as in mitosis, the kinetochores on each sister chromatid function independently, allowing the two sister chromatids to be pulled to opposite poles. (A) In metaphase of meiosis II, the kinetochores of the sister chromatids point in opposite directions. (B) At anaphase of meiosis II, the cohesins holding the sister chromatids together at the centromere are degraded, allowing kinetochore microtubules to pull the two sister chromatids to opposite poles.

First, as we have seen, the maternal and paternal chromosomes are shuffled and dealt out randomly during meiosis I. Although the chromosomes are carefully distributed so that each cell receives one and only one copy of each chromosome, the choice between the maternal or paternal homolog is made by chance, like the flip of a coin. Thus, each gamete contains the maternal versions of some chromosomes and the paternal versions of others (**Figure 19–15A**). This random assortment depends solely on the way each bivalent happens to be positioned when it lines up on the spindle during metaphase of meiosis I. Whether the maternal or paternal homolog is captured by the microtubules from one pole or the other depends on which way the bivalent is facing when the microtubules connect to its kinetochore (see Figure 19–13). Because the orientation of each bivalent at the moment of capture is completely random, the assortment of maternal and paternal chromosomes is random as well.

Thanks to this random reassortment of maternal and paternal homologs, an individual could in principle produce 2^n genetically different gametes, where n is the haploid number of chromosomes. With 23 chromosomes to choose from, each human, for example, could in theory produce 2^{23}— or 8.4×10^6—genetically distinct gametes. The actual number of different gametes each person can produce, however, is much greater than that, because the crossing-over that takes place during meiosis provides a second source of randomized genetic reassortment. Between two and three crossovers occur on average between each pair of human homologs, generating new chromosomes with novel assortments of maternal and paternal alleles. Because crossing-over occurs at more or less random sites along the length of a chromosome, each meiosis will produce four sets of entirely novel chromosomes (**Figure 19–15B**).

Taken together, the random reassortment of maternal and paternal chromosomes, coupled with the genetic mixing of crossing-over, provides a

(A)

(B)

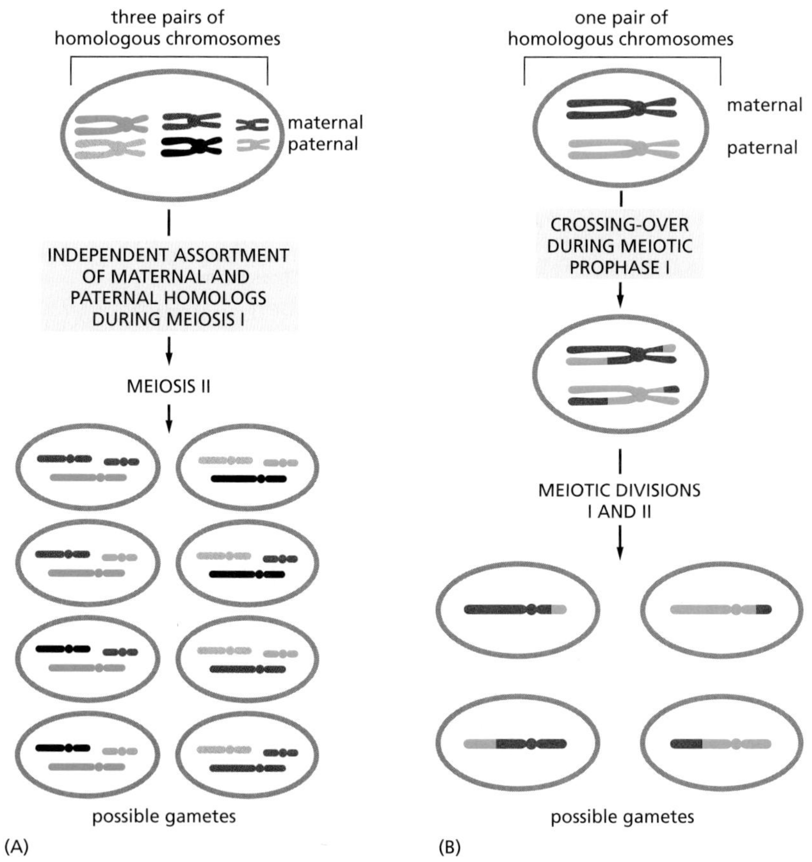

Figure 19–15 **Two kinds of genetic reassortment generate new chromosome combinations during meiosis.** (A) The independent assortment of the maternal and paternal homologs during meiosis produces 2^n different haploid gametes for an organism with n chromosomes. Here $n = 3$, and there are 2^3, or 8, different possible gametes. For simplicity, chromosome crossing-over is not shown here. (B) Crossing-over during meiotic prophase I exchanges segments of DNA between homologous chromosomes and thereby reassorts genes on each individual chromosome. For simplicity, only a single pair of homologous chromosomes is shown. Both independent chromosome assortment and crossing-over occur during every meiosis.

nearly limitless source of genetic variation in the gametes produced by a single individual. Considering that every person is formed by the fusion of such gametes, produced by two completely different individuals, the richness of human variation that we see around us, even within a single family, is not at all surprising.

Meiosis Is Not Flawless

The sorting of chromosomes that takes place during meiosis is a remarkable feat of molecular bookkeeping: in humans, each meiosis requires that the starting cell keep track of 92 chromosomes (23 pairs, each of which has duplicated), handing out one complete set to each gamete. Not surprisingly, mistakes can occur in the distribution of chromosomes during this elaborate process.

Occasionally, homologs fail to separate properly—a phenomenon known as *nondisjunction*. As a result, some of the haploid cells that are produced lack a particular chromosome, while others have more than one copy of it. Upon fertilization, such gametes form abnormal embryos, most of which die. Some survive, however. *Down syndrome*, for example—a disorder associated with cognitive disability and characteristic physical abnormalities—is caused by an extra copy of Chromosome 21. This error results from nondisjunction of a Chromosome 21 pair during meiosis I, giving rise to a gamete that contains two copies of that chromosome instead of one (**Figure 19–16**). When this abnormal gamete fuses with a normal gamete at fertilization, the resulting embryo contains three copies of Chromosome 21 instead of two. This chromosome imbalance produces an extra dose of the proteins encoded by Chromosome 21 and thereby interferes with the proper development of the embryo and normal functions in the adult.

The frequency of chromosome mis-segregation during the production of human gametes is remarkably high, particularly in females: nondisjunction occurs in about 10% of the meioses in human oocytes, giving rise to eggs that contain the wrong number of chromosomes (a condition called *aneuploidy*). Aneuploidy occurs less often in human sperm, perhaps because sperm development is subjected to more stringent quality

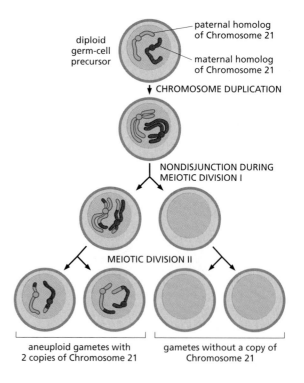

Figure 19–16 Errors in chromosome segregation during meiosis can result in gametes with incorrect numbers of chromosomes. In this example, the duplicated maternal and paternal copies of Chromosome 21 fail to separate normally during the first meiotic division. As a result, two of the gametes receive no copy of the chromosome, while the other two gametes receive two copies instead of the proper single copy. Gametes that receive an incorrect number of chromosomes are called *aneuploid* gametes. If one of them participates in the fertilization process, the resulting zygote will also have an abnormal number of chromosomes. A child that receives three copies of Chromosome 21 will have Down syndrome.

control than egg development. If meiosis goes wrong in male cells, a cell-cycle checkpoint mechanism is activated, arresting meiosis and leading to cell death by apoptosis. Regardless of whether the segregation error occurs in the sperm or the egg, nondisjunction is thought to be one reason for the high rate of miscarriages (spontaneous abortions) in early pregnancy in humans.

Fertilization Reconstitutes a Complete Diploid Genome

Having seen how chromosomes are parceled out during meiosis to form haploid germ cells, we now briefly consider how they are reunited in the process of **fertilization** to form a new zygote with a diploid set of chromosomes.

Of the 300 million human sperm ejaculated during coitus, only about 200 reach the site of fertilization in the oviduct. Sperm are attracted to an ovulated egg by chemical signals released by both the egg and the supporting cells that surround it. Once a sperm finds the egg, it must migrate through a protective layer of cells and then bind to, and tunnel through, the egg coat, called the *zona pellucida*. Finally, the sperm must bind to and fuse with the underlying egg plasma membrane (**Figure 19–17**). Although fertilization normally occurs by this process of sperm–egg fusion, it can also be achieved artificially by injecting the sperm directly into the egg cytoplasm; this is often done in infertility clinics when there is a problem with natural sperm–egg fusion.

Although many sperm may bind to an egg (see Figure 19–3), only one normally fuses with the egg plasma membrane and introduces its DNA into the egg cytoplasm. The control of this step is especially important because it ensures that the fertilized egg—also called a **zygote**—will contain two, and only two, sets of chromosomes. Several mechanisms prevent multiple sperm from entering an egg. In one mechanism, the first successful sperm triggers the release of a wave of Ca^{2+} ions in the egg cytoplasm. This flood of Ca^{2+} in turn triggers the secretion of enzymes that cause a "hardening" of the zona pellucida, which prevents "runner up" sperm from penetrating the egg. The Ca^{2+} wave also helps trigger the development of the egg. To watch a fertilization-induced calcium wave, see Movie 19.2.

The process of fertilization is not complete, however, until the two haploid nuclei (called *pronuclei*) come together and combine their chromosomes into a single diploid nucleus. Soon after the pronuclei fuse, the diploid cell begins to divide, forming a ball of cells that—through repeated rounds of cell division and differentiation—will give rise to an embryo and, eventually, an adult organism. Fertilization marks the beginning of one of the most remarkable phenomena in all of biology—the process by which a single-celled zygote initiates the developmental program that directs the formation of a new individual.

MENDEL AND THE LAWS OF INHERITANCE

In organisms that reproduce without sex, the genetic material of the parent is transmitted faithfully to its progeny. The resulting offspring are thus genetically identical to a single parent. Before Mendel started working with peas, some biologists suspected that inheritance might work that way in humans (**Figure 19–18**).

Although children resemble their parents, they are not carbon copies of either the mother or the father. Thanks to the mechanisms of meiosis just described, sex breaks up existing collections of genetic information, shuffles alleles into new combinations, and produces offspring that tend

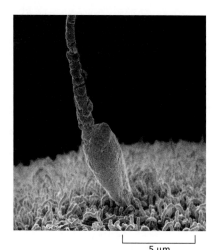

|—————— 5 μm ——————|

Figure 19–17 A sperm binds to the plasma membrane of an egg. Shown here is a scanning electron micrograph of a human sperm coming in contact with a hamster egg. The egg has been stripped of its zona pellucida, exposing the plasma membrane, which is covered in fingerlike microvilli. Such uncoated hamster eggs were sometimes used in infertility clinics to assess whether a man's sperm were capable of penetrating an egg. The zygotes resulting from this test do not develop. (Courtesy of David M. Phillips.)

Figure 19–18 One disproven theory of inheritance suggested that genetic traits are passed down solely from the father. In support of this particular theory of uniparental inheritance, some early microscopists fancied that they could detect a small, perfectly formed human crouched inside the head of each sperm.

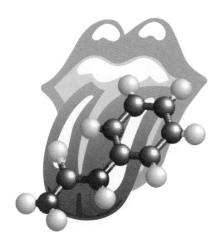

Figure 19–19 Some people taste it, some people don't. The ability to taste the chemical phenylthiocarbamide (PTC) is governed by a single gene. Although geneticists have known since the 1930s that the inability to taste PTC is inherited, it was not until 2003 that researchers identified the responsible gene, which encodes a bitter-taste receptor. Nontasters produce a PTC receptor protein that contains amino acid substitutions that are thought to reduce the receptor's activity.

to exhibit a mixture of traits derived from both parents, as well as novel ones. The ability to track characteristics that show some variation from one generation to the next enabled geneticists to begin to decipher the rules that govern heredity in sexually reproducing organisms.

The simplest traits to follow are those that are easy to see or to measure. In humans, these include the tendency to sneeze when exposed to bright sun, whether a person's earlobes are attached or pendulous, or the ability to detect certain odors or flavors (**Figure 19–19**). Of course, the laws of inheritance were not worked out by studying people with pendulous earlobes, but by following traits in organisms that are easy to breed and that produce large numbers of offspring. Gregor Mendel, the father of genetics, focused on peas. But similar breeding experiments can be performed in fruit flies, worms, dogs, cats, or any other plant or animal that possesses characteristics of interest, because the same basic laws of inheritance apply to all sexually reproducing organisms, from peas to people.

In this section, we describe the logic of genetic inheritance in sexually reproducing organisms. We see how the behavior of chromosomes during meiosis—their segregation into gametes that then unite at random to form genetically unique offspring—explains the experimentally derived laws of inheritance. But first, we discuss how Mendel, breeding peas in his monastery garden, discovered these laws more than 150 years ago.

Mendel Studied Traits That Are Inherited in a Discrete Fashion

Mendel chose to study pea plants because they are easy to cultivate in large numbers and could be raised in a small space—such as an abbey garden. He controlled which plants mated with which by removing sperm (pollen) from one plant and brushing it onto the female structures of another. This careful cross-pollination ensured that Mendel could be certain of the parentage of every pea plant he examined.

Perhaps more important for Mendel's purposes, pea plants were available in many varieties. For example, one variety has purple flowers, another has white. One variety produces seeds (peas) with smooth skin, another produces peas that are wrinkled. Mendel chose to follow seven traits—such as flower color and pea shape—that were distinct, easily observable, and, most importantly, inherited in a discrete fashion: for example, the plants have either purple flowers or white flowers—nothing in between (**Figure 19–20**).

Mendel Disproved the Alternative Theories of Inheritance

The breeding experiments that Mendel performed were straightforward. He started with stocks of genetically pure, "true-breeding" plants—those that produce offspring of the same variety when allowed to self-fertilize. If he followed pea color, for example, he used plants with yellow peas that always produced offspring with yellow peas, and plants with green peas that always produced offspring with green peas.

Mendel's predecessors had focused on organisms that varied in multiple traits. These investigators often wound up trying to characterize offspring whose appearance differed in such a complex way that they could not easily be compared with their parents. But Mendel took the unique approach of studying each trait one at a time. In a typical experiment, he would cross-pollinate two of his true-breeding varieties. He then recorded the inheritance of the chosen trait in the next generation. For example, Mendel crossed plants producing yellow peas with plants producing green peas and discovered that the resulting hybrid offspring,

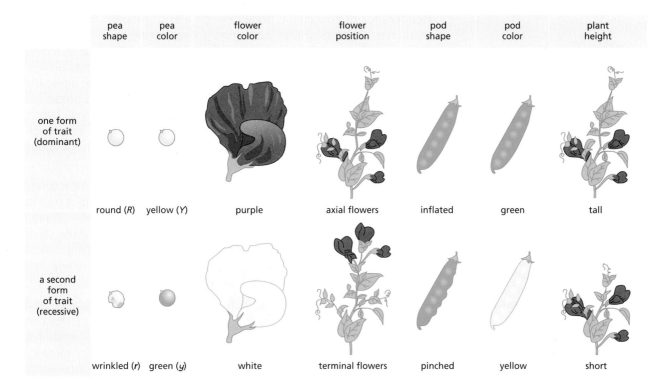

pea shape	pea color	flower color	flower position	pod shape	pod color	plant height

one form of trait (dominant)

round (*R*) — yellow (*Y*) — purple — axial flowers — inflated — green — tall

a second form of trait (recessive)

wrinkled (*r*) — green (*y*) — white — terminal flowers — pinched — yellow — short

called the first filial, or *F₁*, generation, all had yellow peas (**Figure 19–21**). He obtained a similar result for every trait he followed: the F₁ hybrids all resembled only one of their two parents.

Had Mendel stopped there—observing only the F₁ generation—he might have developed some mistaken ideas about the nature of heredity: these results appear to support the theory of uniparental inheritance, which states that the appearance of the offspring will match one parent or the other (see, for example, Figure 19–18). Fortunately, Mendel took his breeding experiments to the next step: he crossed the F₁ plants with one another (or allowed them to self-fertilize) and examined the results.

Figure 19–20 Mendel studied seven traits that are inherited in a discrete fashion. For each trait, the plants display either one variation or the other, with nothing in between. As we will see shortly, one form of each trait is dominant, whereas the other is recessive.

Mendel's Experiments Revealed the Existence of Dominant and Recessive Alleles

One look at the offspring of Mendel's initial cross-fertilization experiments, such as those shown in Figure 19–21, raises an obvious question: what happened to the traits that disappeared in the F₁ generation? Did the plants bearing green peas, for example, fail to make a genetic contribution to their offspring? To find out, Mendel allowed the F₁ plants to self-fertilize. If the trait for green peas had been lost, then the F₁ plants would produce only plants with yellow peas in the next, F₂, generation. Instead, he found that the "disappearing trait" returned: although three-quarters of the offspring in the F₂ generation had yellow peas, one-quarter had green peas (**Figure 19–22**). Mendel saw the same type of behavior for each of the other six traits he examined.

To account for these observations, Mendel proposed that the inheritance of traits is governed by hereditary factors (which we now call genes) and that these factors come in alternative versions that account for the variations seen in inherited characteristics. The gene dictating pea color, for example, exists in two "flavors"—one that directs the production of yellow peas and one that directs production of green peas. Such alternative versions of a gene are now called *alleles*, and the whole collection of alleles possessed by an individual—its genetic makeup—is called its **genotype**.

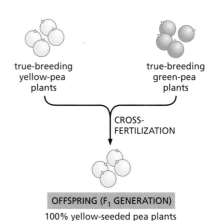

true-breeding yellow-pea plants

true-breeding green-pea plants

CROSS-FERTILIZATION

OFFSPRING (F₁ GENERATION)
100% yellow-seeded pea plants

Figure 19–21 True-breeding varieties, when cross-fertilized with each other, produce hybrid offspring that resemble one parent. In this case, true-breeding green-pea plants, crossed with true-breeding yellow-pea plants, always produce offspring with yellow peas.

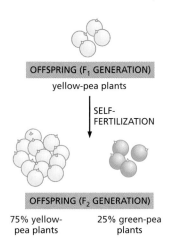

OFFSPRING (F₁ GENERATION)

yellow-pea plants

SELF-FERTILIZATION

OFFSPRING (F₂ GENERATION)

75% yellow-pea plants   25% green-pea plants

Figure 19–22 The appearance of the F₂ generation shows that an individual carries two alleles of each gene. When the F₁ plants in Figure 19–21 are allowed to self-fertilize (or are bred with each other), 25% of the progeny produce green peas.

Mendel's major conceptual breakthrough was to propose that for each characteristic, an organism must inherit two copies, or alleles, of each gene—one from its mother and one from its father. The true-breeding parental strains, he theorized, each possessed a pair of identical alleles—the yellow-pea plants possessed two alleles for yellow peas, the green-pea plant two alleles for green peas. An individual that possesses two identical alleles is said to be **homozygous** for that trait. The F₁ hybrid plants, on the other hand, had received two dissimilar alleles—one specifying yellow peas and the other green. These plants were **heterozygous** for the trait of interest.

The appearance, or **phenotype**, of the organism depends on which versions of each allele it inherits. To explain the disappearance of a trait in the F₁ generation—and its reappearance in the F₂ generation—Mendel supposed that for any pair of alleles, one allele is *dominant* and the other is *recessive*, or hidden. The dominant allele, whenever it is present, would dictate the plant's phenotype. In the case of pea color, the allele that specifies yellow peas is dominant; the green-pea allele is recessive.

One important consequence of heterozygosity, and of dominance and recessiveness, is that not all of the alleles that an individual carries can be detected in its phenotype. Humans have about 30,000 genes, and each of us is heterozygous for a very large number of these. Thus, we all carry a great deal of genetic information that remains hidden in our personal phenotype but that can turn up in future generations.

Each Gamete Carries a Single Allele for Each Character

Mendel's theory—that for every gene, an individual inherits one copy from its mother and one copy from its father—raised some logistical issues. If an organism has two copies of every gene, how does it pass only one copy of each to its progeny? And how do these gene sets come together again in the resulting offspring?

Mendel postulated that when sperm and eggs are formed, the two copies of each gene present in the parent separate so that each gamete receives only one allele for each trait. For his pea plants, each egg (ovum) and each sperm (pollen) receives only one allele for pea color (either yellow or green), one allele for pea shape (smooth or wrinkled), one allele for flower color (purple or white), and so on. During fertilization, sperm carrying one or other allele unites with an egg carrying one or other allele to produce a fertilized egg or zygote with two alleles. Which type of sperm unites with which type of egg at fertilization is entirely a matter of chance.

This principle of heredity is laid out in Mendel's first law, the **law of segregation**. It states that the two alleles for each trait separate (or segregate) during gamete formation and then unite at random—one from each parent—at fertilization. According to this law, the F₁ hybrid plants with yellow peas will produce two classes of gametes: half the gametes will get a yellow-pea allele and half will get a green-pea allele. When the hybrid plants self-pollinate, these two classes of gametes will unite at random. Thus, four different combinations of alleles can come together in the F₂ offspring (**Figure 19–23**). One-quarter of the F₂ plants will receive two alleles specifying green peas; these plants will obviously produce green peas. One-quarter of the plants will receive two yellow-pea alleles and will produce yellow peas. But one-half of the plants will inherit one allele for yellow peas and one allele for green. Because the yellow allele is dominant, these plants—like their heterozygous F₁ parents—will all produce yellow peas. All told, three-quarters of the offspring will have yellow peas and one-quarter will have green peas. Thus Mendel's law of segregation explains the 3:1 ratio that he observed in the F₂ generation.

Figure 19–23 Parent plants produce gametes that each contain one allele for each trait; the phenotype of the offspring depends on which combination of alleles it receives. Here we see both the genotype and phenotype of the pea plants that were bred in the experiments illustrated in Figures 19–21 and 19–22. The true-breeding yellow-pea plants produce only Y-bearing gametes; the true-breeding green plants produce only y gametes. The F_1 offspring of a cross between these parents all produce yellow peas, and they have the genotype Yy. When these hybrid plants are bred with each other, 75% of the offspring have yellow peas, 25% have green. The gray box at the bottom, called a Punnett square after a British mathematician who was a follower of Mendel, allows one to track the segregation of alleles during gamete formation and to predict the outcomes of breeding experiments like the one outlined in Figure 19–22. According to the system invented by Mendel, capital letters indicate a dominant allele and lowercase letters a recessive allele.

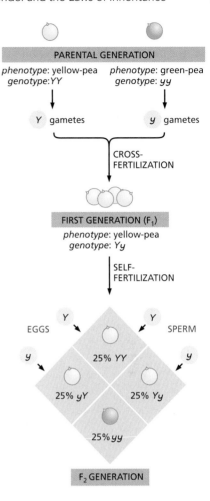

Mendel's Law of Segregation Applies to All Sexually Reproducing Organisms

Mendel's law of segregation explained the data for every trait he examined in pea plants, and he replicated his basic findings with corn and beans. But his rules governing inheritance are not limited to plants: they apply to all sexually reproducing organisms (**Figure 19–24**).

Consider a phenotype in humans that reflects the action of a single gene. The major form of *albinism*—Type II albinism—is a rare condition that is inherited in a recessive manner in many animals, including humans. Like the pea plants that produce green seeds, albinos are homozygous recessive: their genotype is *aa*. The dominant allele of the gene (denoted *A*) encodes an enzyme involved in making melanin, the pigment responsible for most of the brown and black color present in hair, skin, and the retina of the eye. Because the recessive allele codes for a version of this enzyme that is only weakly active or completely inactive, albinos have white hair, white skin, and pupils that look pink because a lack of melanin in the eye allows the red color of the hemoglobin in blood vessels in the retina to be visible.

The trait for albinism is inherited in the same manner as any other recessive trait, including Mendel's green peas. If a Type II albino man (genotype *aa*) has children with a Type II albino woman (also *aa*), all of their children will be albino (*aa*). However, if a nonalbino man (*AA*) marries and has children with an albino woman (*aa*), their children will all be heterozygous (*Aa*) and normally pigmented (**Figure 19–25**). If two nonalbino individuals with an *Aa* genotype start a family, each of their children would have a 25% chance of being an albino (*aa*).

Of course, humans generally don't have families large enough to guarantee accurate Mendelian ratios. (Mendel arrived at his ratios by breeding

Figure 19–24 Mendel's law of segregation applies to all sexually reproducing organisms. Dogs are bred specifically to enhance certain phenotypic traits, including a diverse range of body size, coat color, head shape, snout length, ear position, and fur patterns. Scientists have been conducting genetic analyses on scores of dog breeds to search for the alleles that underlie these common canine characteristics. A single growth factor gene has been linked to body size, and three additional genes have been shown to account for coat length, curliness, and the presence or absence of "furnishings"—bushy eyebrows and beards—in almost all dog breeds. (Courtesy of Ester Inbar.)

Figure 19–25 Recessive alleles all follow the same Mendelian laws of inheritance. Here, we trace the inheritance of Type II albinism, a recessive trait that is associated with a single gene in humans. Note that normally pigmented individuals can be either homozygous (*AA*) or heterozygous (*Aa*) for the dominant allele *A*.

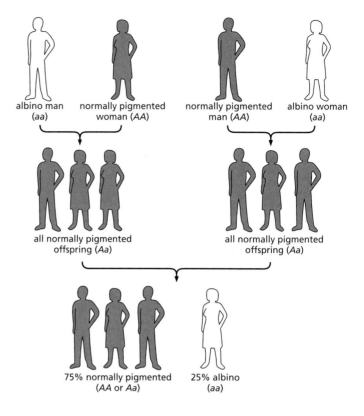

and counting thousands of pea plants for most of his crosses.) Geneticists that follow the inheritance of specific traits in humans get around this problem by working with large numbers of families—or with several generations of a few large families—and preparing **pedigrees** that show the phenotype of each family member for the relevant trait. **Figure 19–26** shows the pedigree for a family that harbors a recessive allele for deafness. It also illustrates an important practical consequence of Mendel's laws: first-cousin marriages create a greatly increased risk of producing children that are homozygous for a deleterious recessive mutation.

Alleles for Different Traits Segregate Independently

Mendel deliberately simplified the problem of heredity by starting with breeding experiments that focused on the inheritance of one trait at a time, called *monohybrid crosses*. He then turned his attention to multihybrid crosses, examining the simultaneous inheritance of two or more apparently unrelated traits.

In the simplest situation, a *dihybrid cross*, Mendel followed the inheritance of two traits at once: for example, pea color and pea shape. In the case of pea color, we have already seen that yellow is dominant over

Figure 19–26 A pedigree shows the risks of first-cousin marriages. Shown here is an actual pedigree for a family that harbors a rare recessive mutation causing deafness. According to convention, *squares* represent males, *circles* are females. Here, family members that show the deaf phenotype are indicated by a *blue* symbol, those that do not by a *gray* symbol. A *black* horizontal line connecting a male and female represents a mating between unrelated individuals, and a *red* horizontal line represents a mating between blood relatives. The offspring of each mating are shown underneath, in order of their birth from left to right.

   Individuals within each generation are labeled sequentially from left to right for purposes of identification. In the third generation in this pedigree, for example, individual 2, a man who is not deaf, marries his first cousin, individual 3, who is also not deaf. Three out of their five children (individuals 7, 8, and 9 in the fourth generation) are deaf. Meanwhile, individual 1, the brother of 2, also marries a first cousin, individual 4, the sister of 3. Two out of their five children are deaf. (Adapted from Z.M. Ahmed et al., *BMC Med. Genet.* 5:24, 2004. With permission from BMC Medical Genetics.)

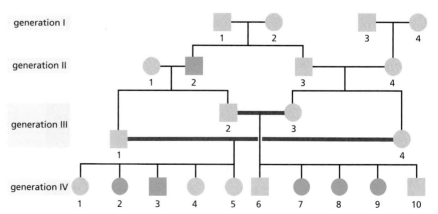

green; for pea shape, round is dominant over wrinkled (see Figure 19–20). What would happen when plants that differ in both of these characters are crossed? Again, Mendel started with true-breeding parental strains: the dominant strain produced yellow round peas (its genotype is *YYRR*), the recessive strain produced green wrinkled peas (*yyrr*). One possibility is that the two characters, pea color and shape, would be transmitted from parents to offspring as a linked package. In other words, plants would always produce either yellow round peas or green wrinkled ones. The other possibility is that pea color and shape would be inherited independently, which means that at some point plants that produce a novel mix of traits—yellow wrinkled peas or green round peas—would arise.

The F_1 generation of plants all showed the expected phenotype: each produced peas that were yellow and round. But this result would occur whether or not the parental alleles were linked. When the F_1 plants were then allowed to self-fertilize, the results were clear: the two alleles for seed color segregated independently from the two alleles for seed shape, producing four different pea phenotypes: yellow-round, yellow-wrinkled, green-round, and green-wrinkled (**Figure 19–27**). Mendel tried his seven pea characters in various pairwise combinations and always observed a characteristic 9:3:3:1 phenotypic ratio in the F_2 generation. The independent segregation of each pair of alleles during gamete formation is Mendel's second law—the **law of independent assortment**.

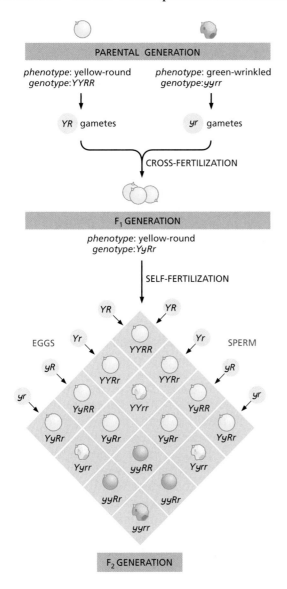

Figure 19–27 **A dihybrid (two trait) cross demonstrates that alleles can segregate independently.** Alleles that segregate independently are packaged into gametes in all possible combinations. So the *Y* allele is equally likely to be packaged with the *R* or *r* allele during gamete formation; and the same holds true for the *y* allele. Thus four classes of gametes are produced in roughly equal numbers: *YR*, *Yr*, *yR*, and *yr*. When these gametes are allowed to combine at random to produce the F_2 generation, the resulting pea phenotypes are yellow-round, yellow-wrinkled, green-round, and green-wrinkled in a ratio of 9:3:3:1.

The Behavior of Chromosomes During Meiosis Underlies Mendel's Laws of Inheritance

So far we have discussed alleles and genes as if they are disembodied entities. We now know that Mendel's "factors"—the things we call genes—are carried on chromosomes that are parceled out during the formation of gametes and then brought together in novel combinations in the zygote at fertilization. Chromosomes therefore provide the physical basis for Mendel's laws, and their behavior during meiosis and fertilization—which we discussed earlier—explains these laws perfectly.

During meiosis, the maternal and paternal homologs—and the genes that they contain—pair and then separate from each other as they are parceled out into gametes. These maternal and paternal chromosome copies will possess different variants—or alleles—of many of the genes they carry. Take, for example, a pea plant that is heterozygous for the yellow-pea gene (*Yy*). During meiosis, the chromosomes bearing the *Y* and *y* alleles will be separated, producing two types of haploid gametes—ones that contain a *Y* allele and others that contain a *y*. In a plant that self-fertilizes, these haploid gametes come together to produce the diploid individuals of the next generation—which may be *YY*, *Yy*, or *yy*. Together, the meiotic mechanisms that distribute the alleles into gametes and the combining of gametes at fertilization provide the physical foundation for Mendel's law of segregation.

But what about independent assortment of multiple traits? Because each pair of duplicated homologs attaches to the spindle and lines up at the metaphase plate independently during meiosis, each gamete will inherit a random mixture of paternal and maternal chromosomes (see Figure 19–15A). Thus the alleles of genes on different chromosomes will segregate independently.

Consider a pea plant that is heterozygous for both seed color (*Yy*) and seed shape (*Rr*). The homolog pair carrying the color alleles will attach to the meiotic spindle with a certain orientation: whether the *Y*-bearing homolog or its *y*-bearing counterpart is captured by the microtubules from one pole or the other depends on which way the bivalent happens to be facing at the moment of attachment (**Figure 19–28**). The same is true for the homolog pair carrying the alleles for seed shape. Thus, whether the final gamete receives the *YR*, *Yr*, *yR*, or *yr* allele combination depends entirely on which way the two homolog pairs were facing when they were captured by the meiotic spindle; each outcome has the same degree of randomness as the tossing of a coin.

Even Genes on the Same Chromosome Can Segregate Independently by Crossing-Over

Mendel studied seven traits, each of which is controlled by a separate gene. It turns out that most of these genes are located on different chromosomes, which readily explains the independent segregation he observed. But the independent segregation of different traits does not necessarily require that the responsible genes lie on different chromosomes. If two genes are far enough away from each other on the same chromosome, they will also sort independently, because of the crossing-over that occurs during meiosis. As we discussed earlier, when duplicated homologs pair to form bivalents, the maternal and paternal homologs always undergo crossing-over. This genetic exchange can separate alleles that were formerly together on the same chromosome, causing them to segregate into different gametes (**Figure 19–29**). We now know, for

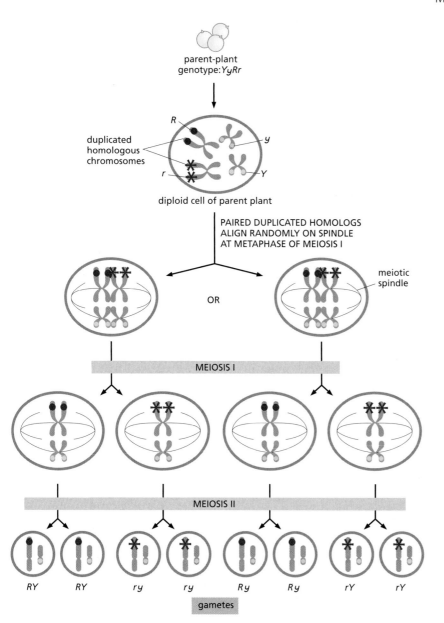

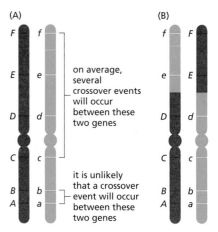

Figure 19–28 The separation of duplicated homologous chromosomes during meiosis explains Mendel's laws of segregation and independent assortment. Here we show independent assortment of the alleles for seed color, yellow (*Y*) and green (*y*), and for seed shape, round (*R*) and wrinkled (*r*), as an example of how two genes on different chromosomes segregate independently. Although crossovers are not shown, they would not affect the independent assortment of these traits, as the two genes lie on different chromosomes.

example, that the genes for pea shape and pod color that Mendel studied are located on the same chromosome, but because they are far apart they segregate independently.

Not all genes segregate independently as per Mendel's second law. If genes lie close together on a chromosome, they are likely to be inherited as a unit. For example, human genes associated with red–green color-blindness and hemophilia are typically inherited together for this reason. By measuring how frequently genes are co-inherited, geneticists can determine whether they reside on the same chromosome and, if so, how far apart they are. These measurements of *genetic linkage* have been used to map the relative positions of the genes on each chromosome of many organisms. Such **genetic maps** have been crucial for isolating and characterizing mutant genes responsible for human genetic diseases such as cystic fibrosis.

Mutations in Genes Can Cause a Loss of Function or a Gain of Function

Mutations produce heritable changes in DNA sequence. They can arise in various ways (discussed in Chapter 6) and can be classified by the effect

Figure 19–29 Genes that lie far enough apart on the same chromosome will segregate independently. (A) Because several crossover events occur randomly along each chromosome during prophase of meiosis I, two genes on the same chromosome will obey Mendel's law of independent assortment if they are far enough apart. Thus, for example, there is a high probability of crossovers occurring in the long region between *C/c* and *F/f*, meaning that a gamete carrying the *F* allele will wind up with the *c* allele as often as it will the *C* allele. In contrast, the *A/a* and *B/b* genes are close together, so there is only a small chance of crossing-over between them: thus the *A* allele is likely to be co-inherited with the *B* allele, and the *a* allele with the *b* allele. From the frequency of recombination, one can estimate the distances between the genes. (B) An example of a crossover that has separated the *C/c* and *F/f* alleles, but not the *A/a* and *B/b* alleles.

Figure 19–30 Mutations in protein-coding genes can affect the protein product in a variety of ways. (A) In this example, the normal or "wild-type" protein has a specific function denoted by the *red rays*. (B) Various loss-of-function mutations decrease or eliminate this activity. (C) Gain-of-function mutations boost this activity, as shown, or lead to an increase in the amount of the normal protein (not shown).

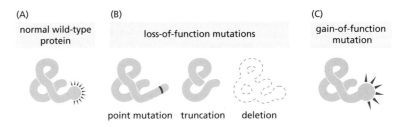

(A) normal wild-type protein

(B) loss-of-function mutations

point mutation   truncation   deletion

(C) gain-of-function mutation

they have on gene function. Mutations that reduce or eliminate the activity of a gene are called **loss-of-function mutations** (**Figure 19–30**). An organism in which both alleles of a gene bear loss-of-function mutations will generally display an abnormal phenotype—one that differs from the most commonly occurring phenotype (although the difference may sometimes be subtle and hard to detect). By contrast, the heterozygote, which possesses one mutant allele and one normal, "wild-type" allele, generally makes enough active gene product to function normally and retain a normal phenotype. Thus loss-of-function mutations are usually recessive, because—for most genes—decreasing the normal amount of gene product by 50% has little impact.

In the case of Mendel's peas, the gene that dictates seed shape codes for an enzyme that helps convert sugars into branched starch molecules. The dominant, wild-type allele, *R*, produces an active enzyme; the recessive, mutant allele, *r*, does not. Because they lack this enzyme, plants that are homozygous for the *r* allele contain more sugar and less starch than plants that possess the dominant *R* allele, which gives their peas a wrinkled appearance (see Figure 19–20). The sweet peas available in the supermarket are often wrinkled mutants of the same type that Mendel studied.

Mutations that increase the activity of a gene or its product, or result in the gene being expressed in inappropriate circumstances, are called **gain-of-function mutations** (see Figure 19–30). Such mutations are usually dominant. For example, as we saw in Chapter 16, certain mutations in the *Ras* gene generate a form of the protein that is always active. Because the normal Ras protein is involved in controlling cell proliferation, the mutant protein drives cells to multiply inappropriately, even in the absence of signals that are normally required to stimulate cell division—thereby promoting the development of cancer. About 30% of all human cancers contain such dominant, gain-of-function mutations in the *Ras* gene.

Each of Us Carries Many Potentially Harmful Recessive Mutations

As we saw in Chapter 9, mutations provide the fodder for evolution. They can alter the fitness of an organism, making it either less or more likely for the individual to survive and leave progeny. Natural selection determines whether these mutations are preserved: those that confer a selective advantage on an organism tend to be perpetuated, while those that compromise an organism's fitness or ability to procreate tend to be lost.

The great majority of chance mutations are either neutral, with no effect on phenotype, or deleterious. A deleterious mutation that is dominant—one that exerts its negative effects when present even in a single copy—will be eliminated almost as soon as it arises. In an extreme case, if a mutant organism is unable to reproduce, the mutation that causes that failure will be lost from the population when the mutant individual dies. For deleterious mutations that are recessive, things are a little more complicated. When such a mutation first arises, it will generally be present in only a single copy. The organism that caries the mutation can produce just as many progeny as other individuals; the majority of these progeny will inherit a single copy of the mutation, and they too will appear fit and healthy. But as they and their descendants begin to mate with one another, some individuals will inherit two copies of the mutant allele and display an abnormal phenotype.

If such a homozygous individual fails to reproduce, two copies of the mutant allele will be lost from the population. Eventually, an equilibrium is reached, where the rate at which new mutations occur in the gene balances the rate at which these mutant alleles are lost through matings that yield abnormal homozygous mutant individuals. As a consequence, many deleterious recessive mutations are present in heterozygous individuals at a surprisingly high frequency, even though homozygous individuals showing the deleterious phenotype are rare. Thus the most common form of hereditary deafness (due to mutations in a gene that encodes a gap-junction protein; see Figure 20–29) occurs in about one in 4000 births, but about one in 30 of us are carriers of a loss-of-function mutant allele of the gene.

GENETICS AS AN EXPERIMENTAL TOOL

Our understanding of how chromosomes shuttle genetic information from one generation to the next did more than demystify the basis of inheritance: it united the science of genetics with other life sciences, from cell biology and biochemistry to physiology and medicine. **Genetics** provides a powerful way to discover what specific genes do and how variations in those genes underlie the differences between one species and another or between individuals within a species. Such knowledge also has practical benefits, as understanding the genetic and biological basis of diseases can help us to better diagnose, treat, and prevent them.

In this section, we outline the *classical genetic approach* to identifying genes and determining how they influence the phenotype of an experimental organism such as yeast or flies. The process begins with the generation of a very large number of mutants and the identification of those rare individuals that show a phenotype of interest. By analyzing these rare mutant individuals and their progeny, we can track down the genes responsible and work out what these genes normally do—and how mutations that alter their activity affect how an organism looks and behaves.

Modern technologies—particularly new methods for obtaining and comparing genome sequences—have made it possible to analyze the genotypes of large numbers of individuals, including humans. In the final part of this section, we discuss how analyses of DNA collected from human families and populations all over the world are providing clues about our evolutionary history and about the genes that influence our susceptibility to disease.

The Classical Genetic Approach Begins with Random Mutagenesis

Before the advent of recombinant DNA technology (discussed in Chapter 10), most genes were identified and characterized by observing the processes disrupted when the gene was mutated. This type of analysis begins with the isolation of mutants that have an interesting or unusual phenotype: fruit flies that have white eyes or curly wings or that become paralyzed when exposed to high temperatures, for example. Working backward from the abnormal phenotype, one then determines the change

> **QUESTION 19–3**
>
> Imagine that each chromosome undergoes one and only one crossover event on each chromatid during each meiosis. How would the co-inheritance of traits that are determined by genes at opposite ends of the same chromosome compare with the co-inheritance observed for genes on two different chromosomes? How does this compare with the actual situation?

Figure 19–31 **Mutations come in various forms.** Different mutagens tend to produce different types of changes. Some common types of mutation are shown here. Other examples include changes in larger segments of DNA, including deletions, duplications, and chromosomal rearrangements (not shown).

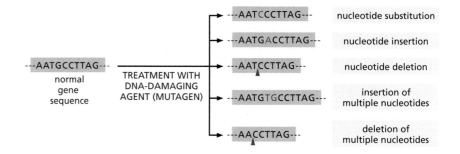

in DNA that is responsible. This **classical genetic approach**—searching for mutant phenotypes and then isolating the responsible genes—is most easily performed in "model organisms" that reproduce rapidly and are amenable to genetic manipulation, such as bacteria, yeasts, nematode worms, zebrafish, and fruit flies. A brief review of this classical approach is presented in **Panel 19–1** (p. 669).

Although spontaneous mutants with interesting phenotypes can be found by combing through a collection of thousands or millions of organisms, the process can be made much more efficient by generating mutations artificially with agents that damage DNA, called *mutagens*. Different mutagens generate different types of DNA mutations (**Figure 19–31**). Not all mutations will lead to a noticeable change in phenotype. But by treating large numbers of organisms with mutagens, collections of mutants can be generated quickly, increasing the odds of finding an interesting phenotype, as we discuss next.

Genetic Screens Identify Mutants Deficient in Specific Cell Processes

A **genetic screen** typically involves examining many thousands of mutagenized individuals to find the few that show a specific altered phenotype of interest. To search for genes involved in cell metabolism, for example, one might screen mutagenized bacterial or yeast cells to pick out those that have lost the ability to grow in the absence of a particular amino acid or other nutrient.

Even genes involved in complex phenotypes, such as social behavior, can be identified by genetic screens in multicellular organisms. For example, by screening for worms that feed alone rather than in clusters as do wild-type individuals, scientists identified and isolated a gene that affects this "social behavior" (**Figure 19–32**).

Advances in modern technologies have made it possible to carry out genome-wide, high-throughput genetic screens on collections of

Figure 19–32 **Genetic screens can be used to identify mutations that affect an animal's behavior.** (A) Wild-type *C. elegans* engage in social feeding. The worms swim around until they encounter their neighbors and only then settle down to feed. (B) Mutant worms dine alone. (Courtesy of Cornelia Bargmann, cover of *Cell* 94, 1998. With permission from Elsevier.)

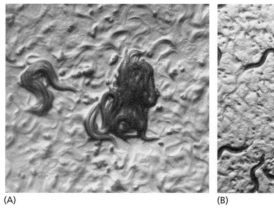
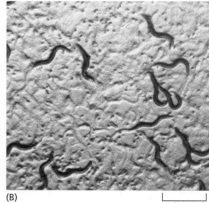

(A)

(B)

1 mm

GENES AND PHENOTYPES

Gene: a functional unit of inheritance, corresponding to the segment of DNA coding for a protein or noncoding RNA molecule.

Genome: all of an organism's DNA sequences.

locus: the site of a gene in the genome

alleles: alternative forms of a gene

Wild type: the normal, naturally occurring type

Mutant: differing from the wild type because of a genetic change (a mutation)

GENOTYPE: the specific set of alleles forming the genome of an individual

PHENOTYPE: the visible or functional characteristics of the individual

homozygous *A/A*   heterozygous *a/A*   homozygous *a/a*

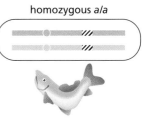

allele *A* is dominant (relative to *a*); allele *a* is recessive (relative to *A*)

In the example above, the phenotype of the heterozygote is the same as that of one of the homozygotes; in cases where it is different from both homozygotes, the two alleles are said to be co-dominant.

MEIOSIS AND GENETIC MAPPING

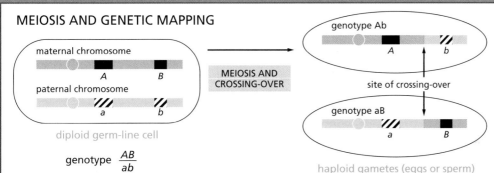

maternal chromosome

A   *B*

paternal chromosome

a   *b*

diploid germ-line cell

genotype $\dfrac{AB}{ab}$

MEIOSIS AND CROSSING-OVER

genotype Ab

A   *b*

site of crossing-over

genotype aB

a   *B*

haploid gametes (eggs or sperm)

The greater the distance between two loci on a single chromosome, the greater is the chance that they will be separated by crossing-over occurring at a site between them. If two genes are thus reassorted in x% of gametes, they are said to be separated on a chromosome by a genetic map distance of x map units (or x centimorgans).

TWO GENES OR ONE?

Given two mutations that produce the same phenotype, how can we tell whether they are mutations in the same gene? If the mutations are recessive (as they most often are), the answer can be found by a complementation test.

In the simplest type of complementation test, an individual who is homozygous for one mutation is mated with an individual who is homozygous for the other. The phenotype of the offspring gives the answer to the question.

| COMPLEMENTATION: MUTATIONS IN TWO DIFFERENT GENES | NONCOMPLEMENTATION: TWO INDEPENDENT MUTATIONS IN THE SAME GENE |

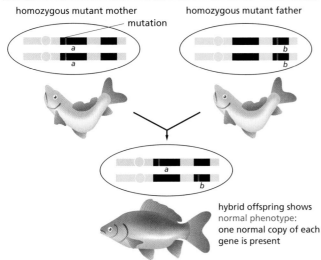

homozygous mutant mother   homozygous mutant father

mutation

a   *b*
a   *b*

hybrid offspring shows normal phenotype: one normal copy of each gene is present

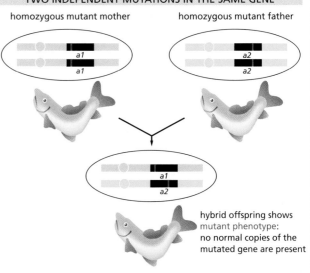

homozygous mutant mother   homozygous mutant father

a1   *a2*
a1   *a2*

hybrid offspring shows mutant phenotype: no normal copies of the mutated gene are present

Figure 19–33 **RNA interference provides a convenient method for conducting genome-wide genetic screens.** In this experiment, each well in this 96-well plate is filled with *E. coli* that produce a different double-stranded (ds), interfering RNA. *E. coli* are a standard diet for *C. elegans* raised in the laboratory. Each interfering RNA matches the nucleotide sequence of a single *C. elegans* gene, thereby inactivating it. About 10 worms are added to each well, where they ingest the genetically modified bacteria. The plate is incubated for several days, which gives the RNAs time to inactivate their target genes—and the worms time to grow, mate, and produce offspring. The plate is then examined in a microscope, which can be controlled robotically, to screen for genes that affect the worms' ability to survive, reproduce, develop, and behave. Shown here are wild-type worms alongside a mutant that shows an impaired ability to reproduce. (From B. Lehner et al., *Nat. Genet.* 38:896–903, 2006. With permission from Macmillan Publishers Ltd.)

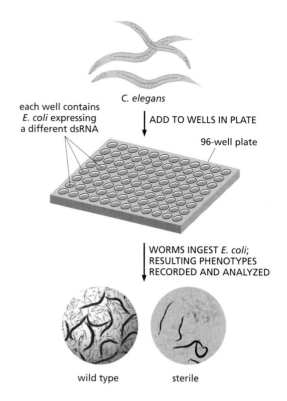

individuals in which nearly all of the protein-coding genes have been individually inactivated. Moreover, such mutant collections can often be screened using automated robots. For example, investigators have made use of RNA interference (explained in Figure 10–34) to generate nematode worms in which the activity of every protein-coding gene has been disrupted, with each worm being deficient in just one gene. These collections can be rapidly screened for dramatic changes in phenotype, such as stunted growth, uncoordinated movement, decreased fertility, or impaired embryonic development (**Figure 19–33**). Using this strategy, the genes needed for a particular characteristic can be identified.

Conditional Mutants Permit the Study of Lethal Mutations

Genetic screens are a powerful approach for isolating and characterizing mutations that are compatible with life—those that change the appearance or behavior of an organism without killing it. A problem arises, however, if we wish to study essential genes—those that are absolutely required for fundamental cell processes, such as RNA synthesis or cell division. Defects in these genes are usually lethal, which means that special strategies are needed to isolate and propagate such mutants: if the mutants cannot be bred, their genes cannot be studied.

If the organism is diploid—a mouse or a pea plant, say—and the mutant phenotype is recessive, there is a simple solution. Individuals that are heterozygous for the mutation will have a normal phenotype and can be propagated. When they are mated with one another, 25% of the progeny will be homozygous mutants and will show the lethal mutant phenotype; 50% will be heterozygous carriers of the mutation like their parents and can maintain the breeding stock.

But what if the organism is haploid, as is the case for many yeast and bacteria? One way to study lethal mutations in such organisms makes use of *conditional mutants*, in which the protein product of the mutant gene is only defective under certain conditions. For example, in mutants that are *temperature-sensitive*, the protein functions normally within a

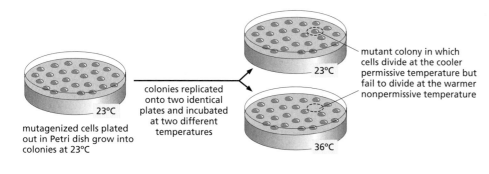

mutant colony in which cells divide at the cooler permissive temperature but fail to divide at the warmer nonpermissive temperature

colonies replicated onto two identical plates and incubated at two different temperatures

mutagenized cells plated out in Petri dish grow into colonies at 23°C

Figure 19–34 Temperature-sensitive mutants are valuable for identifying the genes and proteins involved in essential cell processes. In this example, yeast cells are treated with a mutagen, spread on a culture plate at a relatively cool temperature, and allowed to proliferate to form colonies. The colonies are then transferred to two identical Petri plates using a technique called replica plating. One of these plates is incubated at a cool temperature, the other at a warmer temperature. Those cells that contain a temperature-sensitive mutation in a gene essential for proliferation can be readily identified, because they divide at the cooler permissive temperature—but fail to divide at the warmer non-permissive temperature.

certain range of temperatures (called the *permissive* temperature) but can be inactivated by a shift to a *nonpermissive temperature* outside this range. Thus the abnormal phenotype can be switched on and off simply by changing the temperature. A cell containing a temperature-sensitive mutation in an essential gene can be propagated at the permissive temperature and then be driven to display its mutant phenotype by a shift to a nonpermissive temperature (**Figure 19–34**).

Many temperature-sensitive bacterial mutants were isolated to identify the genes that encode the bacterial proteins required for DNA replication; investigators treated large populations of bacteria with mutagens and then screened for cells that stopped making DNA when they were warmed from 30°C to 42°C. Similarly, temperature-sensitive yeast mutants were used to identify many of the proteins involved in regulating the cell cycle (see How We Know, pp. 30–31) and in transporting proteins through the secretory pathway (see Figure 15–28).

A Complementation Test Reveals Whether Two Mutations Are in the Same Gene

A large-scale genetic screen can turn up many mutant organisms with the same phenotype. These mutations might affect the same gene or they might affect different genes that function in the same process. How can we distinguish between the two? If the mutations are recessive and cause a loss of function, a **complementation test** can reveal whether they affect the same or different genes.

In the simplest type of complementation test, an individual that is homozygous for one recessive mutation is mated with an individual that is homozygous for the other mutation. If the two mutations affect the same gene, the offspring will show the mutant phenotype, because they carry only defective copies of the gene in question. If, in contrast, the mutations affect different genes, the resulting offspring will show the normal, wild-type phenotype, because they will have one normal copy (and one mutant copy) of each gene (see Panel 19–1, p. 669).

Whenever the normal phenotype is restored in such a test, the alleles inherited from the two parents are said to complement each other (**Figure 19–35**). For example, complementation tests on mutants identified during genetic screens have revealed that 5 genes are required for yeast cells

Figure 19–35 A complementation test can reveal that mutations in two different genes are responsible for the same abnormal phenotype. When an albino (white) bird from one strain is bred with an albino from a different strain, the resulting offspring have normal coloration. This restoration of the wild-type plumage implies that the two white breeds lack color because of recessive mutations in different genes. (From W. Bateson, *Mendel's Principles of Heredity*, 1st ed. Cambridge, UK: Cambridge University Press, 1913. With permission from Cambridge University Press.)

to digest the sugar galactose, that 20 genes are needed for *E. coli* to build a functional flagellum, and many hundreds are essential for the normal development of an adult nematode worm from a fertilized egg.

Rapid and Cheap DNA Sequencing Has Revolutionized Human Genetic Studies

Genetic screens in model experimental organisms have been spectacularly successful in identifying genes and relating them to various phenotypes, including many that are conserved between these organisms and humans. But the same approach cannot be used in humans. Unlike flies, worms, yeast, and bacteria, humans do not reproduce rapidly, and intentional mutagenesis in humans is out of the question. Moreover, an individual with a serious defect in an essential process such as DNA replication would die long before birth—before we can assess the phenotype.

Nonetheless, humans are becoming increasingly attractive subjects for genetic studies. Because the human population is so large, spontaneous, nonlethal mutations have arisen in all human genes—many times over. A substantial proportion of these remain in the genomes of present-day humans. The most deleterious of these mutations are discovered when the mutant individuals call attention to themselves by seeking medical help—a uniquely human behavior.

With the recent advances that have enabled the sequencing of entire human genomes cheaply and quickly, we can now identify such mutations and study their evolution and inheritance in ways that were impossible even a few years ago. By comparing the sequences of thousands of human genomes from all around the world, we can now identify directly the DNA differences that distinguish one individual from another. These differences hold clues to our evolutionary origins and can be used to explore the roots of disease.

Linked Blocks of Polymorphisms Have Been Passed Down from Our Ancestors

When we compare the sequences of multiple human genomes, we find that any two individuals will differ in about 1 nucleotide pair in 1000. Most of these variations are common and relatively harmless. When two sequence variants coexist in the population and are both common, the variants are called **polymorphisms**. The majority of polymorphisms are due to the substitution of a single nucleotide, called **single-nucleotide polymorphisms** or **SNPs** (Figure 19–36). The rest are due largely to insertions or deletions—called *indels* when the change is small, or *copy number variants* (*CNVs*) when it is large.

Although these common variants can be found throughout the genome, they are not scattered randomly—or even independently. Instead, they tend to travel in groups called **haplotype blocks**—combinations of polymorphisms or other DNA markers that are inherited as a unit.

QUESTION 19–4

When two individuals from different isolated, inbred subpopulations of a species come together and mate, their offspring often show "hybrid vigor": that is, they appear more robust, healthy, and fertile than either parent. Can you suggest a possible explanation for this phenomenon?

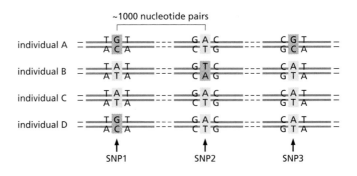

Figure 19–36 Single-nucleotide polymorphisms (SNPs) are sites in the genome where two or more alternative choices of a nucleotide are common in the population. Most such variations in the human genome occur at locations where they do not significantly affect a gene's function.

To understand why such haplotype blocks exist, we need to consider our evolutionary history. It is thought that modern humans expanded from a relatively small population—perhaps around 10,000 individuals—that existed in Africa about 60,000 years ago. Among that small group of our ancestors, some individuals will have carried one set of genetic variants, others a different set. The chromosomes of a present-day human represent a shuffled combination of chromosome segments from different members of this small ancestral group of people. Because only about two thousand generations separate us from them, large segments of these ancestral chromosomes have passed from parent to child, unbroken by the crossover events that occur during meiosis. (Remember, only a few crossovers occur between each set of homologous chromosomes (see Figure 19–12).)

As a result, certain sets of DNA sequences—and their associated polymorphisms—have been inherited in linked groups, with little genetic rearrangement across the generations. These are the haplotype blocks. Like genes that exist in different allelic forms, haplotype blocks also come in a limited number of variants that are common in the human population, each representing a combination of DNA polymorphisms passed down from a particular ancestor long ago.

Our Genome Sequences Provide Clues to our Evolutionary History

A detailed examination of haplotype blocks has provided intriguing insights into the history of human populations. New alleles of genes are continually being generated by mutation; many of these variants will be neutral, in that they will not affect the reproductive success of the individual. These have a chance of becoming common in the population. The more time that has elapsed since the origin of a relatively common allele like a SNP, the smaller should be the haplotype block that surrounds it: over the course of many generations, crossover events will have had many chances to separate an ancient allele from other polymorphisms nearby. Thus by comparing the sizes of haplotype blocks from different human populations, it is possible to estimate how many generations have elapsed since the origin of a specific neutral mutation. Combining such genetic comparisons with archaeological findings, scientists have traced our history from that small set of ancestors and deduce the most probable routes our ancestors took when they left Africa (**Figure 19–37**).

More recent studies, comparing the genome sequences of living humans with those of Neanderthals and another extinct relation from southern Siberia, suggest that our exit from Africa was a bit more convoluted. Some of us share a few percent of our nucleotide sequences with these

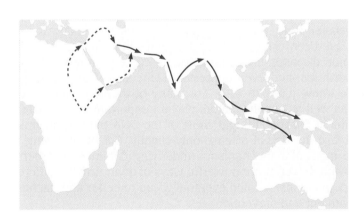

Figure 19–37 The human populations that are now dispersed around the world originated in Africa about 60,000 to 80,000 years ago. The map shows the routes of the earliest successful human migrations. Dotted lines indicate two alternative routes that our ancestors seem to have taken out of Africa. Studies of the size of haplotype blocks suggest that modern Europeans are descended from a small ancestral population that existed about 30,000 to 50,000 years ago. Haplotype blocks in a Nigerian population are significantly smaller, indicating that the Nigerian population was established before the European. These findings agree with archaeological findings, which suggest that the ancestors of modern native Australians (*solid red* arrows)—and of modern European and Middle Eastern populations (migration routes not shown)—reached their destinations about 45,000 years ago. (Modified from P. Forster and S. Matsumura, *Science* 308:965–966, 2005. With permission from AAAS.)

archaic humans, suggesting that a number of our ancestors interbred with their neighbors as they made their way across the globe.

Genome analyses can also be used to estimate when and where humans acquired mutations that have conferred an evolutionary benefit, such as resistance to infection. Such favorable mutations will rapidly accumulate in the population because individuals that carry it will be more likely to survive an epidemic and pass the mutation on to their offspring. A haplotype analysis can be used to "date" the appearance of such a favorable mutation. If it cropped up in the population relatively recently, there will have been fewer opportunities for recombination to break up the DNA sequence around it, so the surrounding haplotype block will be large.

Such is the case for two alleles that confer resistance to malaria. These alleles are widespread in Africa, where malaria is rife. They are embedded in unusually large haplotype blocks, suggesting that they arose recently in the African gene pool—probably about 2500 years ago for one of them and about 6500 years ago for the other. In this way, analyses of modern human genomes can highlight important events in ancient human history, including our initial exposures to specific infections.

Polymorphisms Can Aid the Search for Mutations Associated with Disease

The study of polymorphisms may also have more practical relevance to human health. CNVs, indels, and SNPs can be used as markers for building human genetic maps or for conducting searches for mutations that predispose individuals to a specific disease. Mutations that give rise, in a reproducible way, to rare but clearly defined abnormalities, such as albinism or congenital deafness, can often be identified by studies of affected families. Such single-gene, or monogenic, disorders are often referred to as *Mendelian* because their pattern of inheritance is as easy to track as the wrinkled peas and purple flowers that were studied by Mendel. But for many common diseases, the genetic roots are more complex. Instead of a single allele of a single gene, such disorders stem from a combination of contributions from multiple genes. For these *multigenic* conditions, such as diabetes or arthritis, population studies are often helpful in tracking down the genes that increase the risk of getting the disease.

In population studies, investigators collect DNA samples from a large number of people who have the disease and compare them to samples from a group of people who do not have the disease. They look for variants—SNPs, for example—that are more common among the people who have the disease. Because DNA sequences that are close together on a chromosome tend to be inherited together, the presence of such SNPs could indicate that an allele that increases the risk of the disease might lie nearby (**Figure 19–38**). Although, in principle, the disease could be caused by the SNP itself, the culprit is much more likely to be a change that is merely linked to the SNP.

Such *genome-wide association studies*—which initially focused on SNPs—have been used to search for genes that predispose individuals to common diseases, including diabetes, coronary artery disease, rheumatoid arthritis, and even depression. One such study is described in **How We Know** (pp. 676–677). For many of these conditions, environmental as well as genetic factors play an important part in determining susceptibility. Disappointingly, most of the DNA polymorphisms identified increase the risk of disease only slightly. But by providing insights into the molecular mechanisms underlying common disorders, these results—and newer, more powerful ways of analyzing the differences in human populations—should eventually lead to better approaches to disease treatment and prevention.

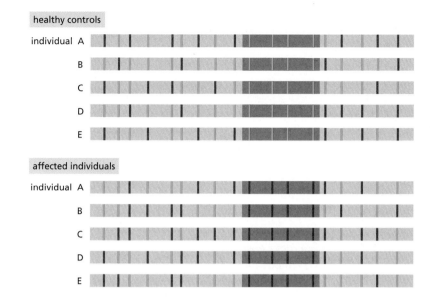

healthy controls

individual A
B
C
D
E

affected individuals

individual A
B
C
D
E

Figure 19–38 Genes that affect the risk of developing a common disease can often be tracked down through linkage to SNPs. Here, the patterns of SNPs are compared between two sets of individuals—a set of healthy controls and a set affected by a particular common disease. A segment of a typical chromosome is shown. For most polymorphic sites in this segment, it is a random matter whether an individual has one SNP variant (*red* vertical bars) or another (*blue* vertical bars); and the same randomness is seen both for the control group and for the affected individuals. However, in the part of the chromosome that is shaded in *darker gray*, a bias is seen, such that most normal individuals have the *blue* SNP variants, whereas most affected individuals have the *red* SNP variants. This suggests that this region contains, or is close to, a gene that is genetically linked to these red SNP variants and that predisposes to the disease. Using carefully selected controls and thousands of affected individuals, this approach can help track down disease-related genes, even when they confer only a slight increase in the risk of developing the disease.

Genomics Is Accelerating the Discovery of Rare Mutations that Predispose Us to Serious Disease

The genetic variants that have thus far allowed us to track our ancestors and identify some of the genes that increase our risk of disease are common ones. They arose long ago in our evolutionary past and are now present, in one form or another, in a substantial portion (1% or more) of the population. Such polymorphisms are thought to account for about 90% of the differences between one person's genome and another. But when we try to tie these common variants to differences in disease susceptibility or other heritable traits, such as height, we find that they do not have as much predictive power as we had anticipated: thus, for example, most confer relatively small increases—less than twofold—in the risk of developing a common disease.

In contrast to polymorphisms, rare DNA variants—those much less frequent in humans than SNPs—can have large effects on the risk of developing some common diseases. For example, a number of different loss-of-function mutations, each individually rare, have been found to increase greatly the predisposition to autism and schizophrenia. Many of these are *de novo* mutations, which arose spontaneously in the germ-line cells of one or other parent. The fact that these mutations arise spontaneously with some frequency could help explain why these common disorders—each observed in about 1% of the population—remain with us, even though the affected individuals leave few or no descendants. These rare mutations, which may arise in any one of hundreds of different genes, can greatly increase the risk of autism and schizophrenia—and could explain much of their clinical variability. Because they are kept rare by natural selection, most such variants with a large effect on risk would be missed by genome-wide association studies.

Now that the price of DNA sequencing has plummeted, the most efficient and cost-effective way to identify these rare, large-effect mutations is by sequencing all the exons (the *exome*)—or even the whole genomes—of affected individuals, along with those of their parents and siblings as controls. Although exome sequencing will miss the noncoding variations that affect gene regulation, the majority of rare, large-effect mutations have thus far been found to lie within exons; common, small-effect variations, by contrast, have been found mainly in noncoding sequences.

QUESTION 19–5

In a recent automated analysis, thousands of SNPs across the genome were analyzed in pooled DNA samples from humans who had been sorted into groups according to their age. For the vast majority of these sites, there was no change in the relative frequencies of different variants as these humans aged. Sometimes, albeit rarely, a particular variant at one position was found to decrease in frequency progressively for people over 50 years old. Which of the possible explanations seems most likely?
A. The nucleotide in that SNP at that position is unstable, and mutates with age.
B. Those people born more than 50 years ago came from a population that tended to lack the disappearing SNP variant.
C. The SNP variant alters an important gene product in a way that shortens the human life-span, or is linked to a neighboring allele that has this effect.

USING SNPs TO GET A HANDLE ON HUMAN DISEASE

For diseases that have their roots in genetics, finding the gene or genes responsible can be the first step toward improved diagnosis, treatment, and even prevention. The task is not simple, but having access to polymorphisms such as SNPs can help. In 1999, an international group of scientists set out to collect and catalog 300,000 SNPs—the single-nucleotide polymorphisms that are common in the human population (see Figure 19–36). Today, the database has grown to include a catalog of more than 17 million SNPs. These SNPs not only help to define the differences between one individual and another; for geneticists, they also serve as signposts that can point the way toward the genes involved in common human disorders, such as diabetes, obesity, asthma, arthritis, and even gallstones and restless leg syndrome.

Making a map

One way that SNPs have facilitated the search for alleles that predispose to disease is by providing the physical markers needed to construct detailed genetic linkage maps. A genetic linkage map displays the relative locations of a set of genes. Such maps are based on the frequency with which two alleles are co-inherited—something we can discover by seeing how often the phenotypic traits associated with those alleles show up together in an individual. Genes that lie close to one another on the same chromosome will be inherited together much more frequently than those that lie farther apart. By determining how often crossing-over separates two genes, the relative distance between them can be calculated (see Panel 19–1, p. 669).

The same sort of analysis can be used to discover linkage between a SNP and an allele. One simply looks for co-inheritance of the SNP with a certain phenotype, such as an inherited disease. Finding such a linkage indicates that the mutation responsible for the phenotype is either the SNP itself or, more likely, that lies close to the SNP (**Figure 19–39**). And because we know the exact location in the human genome sequence of every SNP we examine, the linkage tells us the neighborhood in which the causative mutation resides. A more detailed analysis of the DNA in that region—to look for deletions, insertions, or other functionally significant abnormalities in the DNA sequence of affected individuals—can then lead to a precise identification of the critical gene.

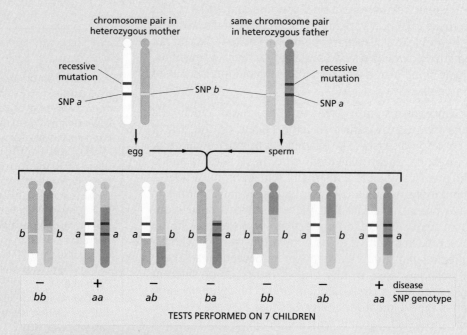

Disease is seen only in progeny with SNP genotype *aa*.
CONCLUSION: Recessive mutation causing the disease is co-inherited with SNP *a*. If this same correlation is observed in other families that have been examined, the mutation causing the disease must lie close to SNP *a*.

Figure 19–39 SNP analysis can pin down the location of a mutation that causes a genetic disease. In this approach, one studies the co-inheritance of a specific human phenotype (here a genetic disease) with a particular set of SNPs. The figure shows the logic for the common case of a family in which both parents are carriers of a recessive mutation. If individuals with the disease, and only such individuals, are homozygous for a particular SNP, then the SNP and the recessive mutation that causes the disease are likely to be close together on the same chromosome, as shown here. To prove that an apparent linkage is statistically significant, a few dozen individuals from such families may need to be examined. With more individuals and using more SNPs, it is possible to locate the mutation more precisely. These days it can be just as fast and cheap to use whole-genome sequencing to find the mutation.

Such linkage analyses are usually carried out in families that are particularly prone to a disorder—the larger the family, the better. And the method works best where there is a simple cause-and-effect relationship, such that a particular mutant gene directly and reliably causes the disorder—as is the case, for example, for the mutant gene that causes cystic fibrosis. But most common disorders are not like this. Instead, many factors affect the disease risk—some genetic, some environmental, some just a matter of chance. For such conditions, a different approach is needed to identify risk genes.

Making associations

Genome-wide association studies allow us to discover common genetic variants that affect the risk for a common disease, even if each variant alters susceptibility only slightly. Because mutations that destroy the activity of a key gene are likely to have a disastrous effect on the fitness of the mutant individual, they tend to be eliminated from the population by natural selection and so are rarely seen. Genetic variants that alter a gene's function only slightly, on the other hand, are much more common. By tracking down these common variants, or polymorphisms, we can sniff out some of the genes that contribute to the biology of common diseases.

Genome-wide association studies use genetic markers, such as SNPs, that are located throughout the genome to compare directly the DNA sequences of two populations: individuals who have a particular disease and those who do not. The approach identifies SNPs that are present in the people who have the disease more often than would be expected by chance.

Consider the case of *age-related macular degeneration (AMD)*, a degenerative disorder of the retina that is a leading cause of blindness in the elderly. To search for genetic variations that are associated with AMD, researchers looked at a panel of just over 100,000 SNPs that spanned the genome. They determined the nucleotide sequence at each of these SNPs in 96 people who had AMD, and 50 who did not. Among the 100,000 SNPs, they discovered that one particular SNP was present significantly more often in the individuals who had the disease (**Figure 19–40**).

The SNP is located in a gene called *Cfh* (*complement factor H*). But it falls within one of the gene's introns and appears unlikely to have any effect on the protein product. This SNP itself, therefore, did not seem likely to be the cause of the increase in susceptibility to AMD. But it focused the researchers' attention on the Cfh gene. So they resequenced the region to look for additional polymorphisms that might also be inherited more often by people with AMD, along with the SNP that they had

already identified. They discovered three variants that affect the amino acid sequence of the Cfh protein. One substitutes a histidine for a tyrosine at one particular place in the protein, and it was strongly associated with the disease (and almost always coupled with the original SNP that had put the researchers on the track of the *Cfh* gene). Individuals who carried two copies of this risky allele were five to seven times more likely to develop AMD than those who harbored a different allele of the *Cfh* gene.

Several other research teams, using a similar genetic association approach, have also pointed to *Cfh* variants as increasing the likelihood of developing AMD, making it almost certain that the *Cfh* gene has something to do with the biology of the disease. The Cfh protein is part of the complement system; it helps prevent the system from becoming overactive, a condition that can lead to inflammation and tissue damage. Interestingly, the environmental risk factors associated with the disease—smoking, obesity, and age—also affect inflammation and the activity of the complement system. Thus, whatever the detailed mechanism by which the *Cfh* gene influences the risk of AMD, the finding that complement is critical could lead to new tests for the early diagnosis of the disorder, as well as potential new avenues for treatment.

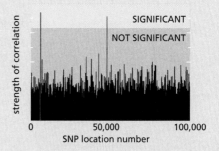

Figure 19–40 Genome-wide association studies identify DNA variations that are significantly more frequent in people with a given disease. In this study, scientists examined more than 100,000 SNPs in each of 146 people. The x-axis of the graph shows the relative position of each SNP in the genome, starting at the left with the SNPs on Chromosome 1. The y-axis shows the strength of each SNP's observed correlation with AMD. The *blue* region indicates a cutoff level for statistical significance, corresponding to a probability of less than 5% of finding that strength of correlation by pure chance anywhere among the whole set of 100,000 tested SNPs. The SNP marked in *red* is the one that led the way to the relevant gene, *Cfh*. The initial association of the other prominent SNP (*black*) with the disease disappeared when additional sequencing at that site was performed. (Adapted from R.J. Klein et al., *Science* 308:385–389, 2005. With permission from AAAS.)

Exome and genome sequencing efforts are turning up many previously unreported genetic variants—in both disease and apparently healthy populations. One recent study suggests that we all harbor about 100 loss-of-function mutations in protein-coding genes—20 of which eliminate the activity of both gene copies—indicating that humans do not actually need all of our genes to develop and function as "normal."

ESSENTIAL CONCEPTS

- Sexual reproduction involves the cyclic alternation of diploid and haploid states: diploid germ-line cells divide by meiosis to form haploid gametes, and the haploid gametes from two individuals fuse at fertilization to form a new diploid cell—the zygote.

- During meiosis, the maternal and paternal homologs are parceled out to gametes such that each gamete receives one copy of each chromosome. Because the segregation of these homologs occurs randomly, and crossing-over occurs between them, many genetically different gametes can be produced from a single individual.

- In addition to enhancing genetic mixing, crossing-over helps ensure the proper segregation of chromosomes during meiosis.

- Although most of the mechanical features of meiosis are similar to those of mitosis, the behavior of the chromosomes is different: meiosis produces four genetically distinct haploid cells by two consecutive cell divisions, whereas mitosis produces two genetically identical diploid cells by a single cell division.

- Mendel unraveled the laws of heredity by studying the inheritance of a handful of discrete traits in pea plants.

- Mendel's law of segregation states that the maternal and paternal alleles for each trait separate from one another during gamete formation and then reunite randomly during fertilization.

- Mendel's law of independent assortment states that, during gamete formation, different pairs of alleles segregate independently of one another.

- The behavior of chromosomes during meiosis explains both of Mendel's laws.

- If two genes are close to each other on a chromosome, they tend to be inherited as a unit; if they are far apart, they will typically be separated by crossing-over. The frequency with which two genes are separated by crossovers can be used to construct a genetic map that shows their order on a chromosome.

- Mutant alleles can be either dominant or recessive. If a single copy of the mutant allele alters the phenotype of an individual that also possesses a wild-type allele, the mutant allele is dominant; if not, it is recessive.

- Complementation tests reveal whether two mutations that produce the same phenotype affect the same gene or different genes.

- Mutant organisms can be generated by treating animals with mutagens, which damage DNA. Such mutants can then be screened to identify phenotypes of interest and, ultimately, to isolate the responsible genes.

- With the possible exception of identical twins, no two human genomes are alike. Each of us carries a unique set of polymorphisms—variations in nucleotide sequence that in some cases contribute to our individual phenotypes.

- Some of the common polymorphisms—including SNPs, indels, and CNVs—provide useful markers for genetic mapping.

- The human genome consists of large haplotype blocks—segments of nucleotide sequence that have been passed down intact from our distant ancestors and, in most individuals, have not yet been broken up by crossovers. The relative sizes of haplotype blocks can give us clues to our evolutionary history.

- DNA sequencing studies are identifying an increasing number of rare mutations that can greatly increase the risk of developing the most common human disorders.

KEY TERMS

allele	homolog
asexual reproduction	homologous chromosome
bivalent	homologous recombination
chiasma (plural chiasmata)	homozygous
classical genetic approach	law of independent assortment
complementation test	law of segregation
crossing-over	loss-of-function mutation
diploid	meiosis
fertilization	pairing
gain-of-function mutation	pedigree
gamete	phenotype
genetic map	polymorphism
genetic screen	recombination
genetics	segregation
genotype	sexual reproduction
germ cell	sister chromatid
germ line	SNP (single-nucleotide
haploid	polymorphism)
haplotype block	somatic cell
heterozygous	zygote

QUESTIONS

QUESTION 19–6

It is easy to see how deleterious mutations in bacteria, which have a single copy of each gene, are eliminated by natural selection: the affected bacteria die and the mutation is thereby lost from the population. Eukaryotes, however, have two copies of most genes—that is, they are diploid. Often an individual with two normal copies of the gene (homozygous normal) is indistinguishable in phenotype from an individual with one normal copy and one defective copy of the gene (heterozygous). In such cases, natural selection can operate only against an individual with two copies of the defective gene (homozygous defective). Consider the situation in which a defective form of the gene is lethal when homozygous, but without effect when heterozygous. Can such a mutation ever be eliminated from the population by natural selection? Why or why not?

QUESTION 19–7

Which of the following statements are correct? Explain your answers.

A. The egg and sperm cells of animals contain haploid genomes.

B. During meiosis, chromosomes are allocated so that each germ cell obtains one and only one copy of each of the different chromosomes.

C. Mutations that arise during meiosis are not transmitted to the next generation.

QUESTION 19–8

What might cause chromosome nondisjunction, where two copies of the same chromosome end up in the same daughter cell? What could be the consequences of this event occurring (a) in mitosis and (b) in meiosis?

QUESTION 19–9

Why do sister chromatids have to remain paired in division I of meiosis? Does the answer suggest a strategy for washing your socks?

QUESTION 19–10

Distinguish between the following genetic terms:

A. Gene and allele.

B. Homozygous and heterozygous.

C. Genotype and phenotype.

D. Dominant and recessive.

QUESTION 19–11

You have been given three wrinkled peas, which we shall call A, B, and C, each of which you plant to produce a mature pea plant. Each of these three plants, once self-pollinated, produces only wrinkled peas.

A. Given that you know that the wrinkled-pea phenotype is recessive, as a result of a loss-of-function mutation, what can you say about the genotype of each plant?

B. Can you safely conclude that each of the three plants carries a mutation in the same gene?

C. If not, how could you rule out the possibility that each plant carries a mutation in a different gene, each of which gives the wrinkled-pea phenotype?

QUESTION 19–12

Susan's grandfather was deaf, and passed down a hereditary form of deafness within Susan's family as shown in **Figure Q19–12**.

A. Is this mutation most likely to be dominant or recessive?

B. Is it carried on an autosome or a sex chromosome? Why?

C. A complete SNP analysis has been done for all of the 11 grandchildren (4 affected, and 7 unaffected). In comparing these 11 SNP results, how long a haplotype block would you expect to find around the critical gene? How might you detect it?

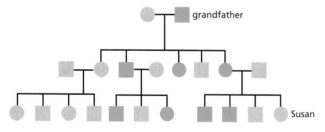

Figure Q19–12

QUESTION 19–13

Given that the mutation causing deafness in the family shown in Figure 19–26 is very rare, what is the most probable genotype of each of the four children in generation II?

QUESTION 19–14

In the pedigree shown in **Figure Q19–14**, the first born in each of three generations is the only person affected by a dominant genetically inherited disease, D. Your friend concludes that the first child born has a greater chance of inheriting the mutant *D* allele than do later children.

A. According to Mendel's laws, is this conclusion plausible?

B. What is the probability of obtaining this result by chance?

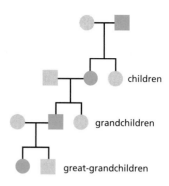

children

grandchildren

great-grandchildren

Figure Q19–14

C. What kind of additional data would be needed to test your friend's idea?

D. Is there any way in which your friend's hypothesis might turn out to be right?

QUESTION 19–15

Suppose one person in 100 is a carrier of a fatal recessive mutation, such that babies homozygous for the mutation die soon after birth. In a population where there are 1,000,000 births per year, how many babies per year will be born with the lethal homozygous condition?

QUESTION 19–16

Certain mutations are called *dominant-negative mutations*. What do you think this means and how do you suppose these mutations act? Explain the difference between a dominant-negative mutation and a gain-of-function mutation.

QUESTION 19–17

Discuss the following statement: "We would have no idea today of the importance of insulin as a regulatory hormone if its absence were not associated with the devastating human disease diabetes. It is the dramatic consequences of its absence that focused early efforts on the identification of insulin and the study of its normal role in physiology."

QUESTION 19–18

Early genetic studies in *Drosophila* laid the foundation for our current understanding of genes. *Drosophila* geneticists were able to generate mutant flies with a variety of easily observable phenotypic changes. Alterations from the fly's normal brick-red eye color have a venerable history because the very first mutant found by Thomas Hunt Morgan was a white-eyed fly (**Figure Q19–18**). Since that time, a large number of mutant flies with intermediate eye colors have been isolated and given names that challenge your color sense: garnet, ruby, vermilion, cherry, coral, apricot, buff, and carnation. The mutations responsible for these eye-color phenotypes are all recessive. To determine whether the mutations affected the same or different genes, homozygous flies for each mutation were bred to one another in pairs and the eye colors of their progeny were noted. In **Table Q19–18**, a + or a – indicates the phenotype of the progeny flies produced by mating the fly listed at the top of the column with the fly listed to the left of the row;

TABLE Q19–18 COMPLEMENTATION ANALYSIS OF *DROSOPHILA* EYE-COLOR MUTATIONS

Mutation	white	garnet	ruby	vermilion	cherry	coral	apricot	buff	carnation
white	–	+	+	+	–	–	–	–	+
garnet		–	+	+	+	+	+	+	+
ruby			–	+	+	+	+	+	+
vermilion				–	+	+	+	+	+
cherry					–	–	–	–	+
coral						–	–	–	+
apricot							–	–	+
buff								–	+
carnation									–

+ indicates that progeny of a cross between individuals showing the indicated eye colors are phenotypically normal; – indicates that the eye color of the progeny is abnormal.

brick-red

flies with other eye colors

white

Figure Q19–18

brick-red wild-type eyes are shown as + and other colors are indicated as –.

A. How is it that flies with two different eye colors—ruby and white, for example—can give rise to progeny that all have brick-red eyes?

B. Which mutations are alleles of the same gene and which affect different genes?

C. How can different alleles of the same gene give different eye colors?

QUESTION 19–19

What are single-nucleotide polymorphisms (SNPs), and how can they be used to locate a mutant gene by linkage analysis?

CHAPTER **TWENTY**

20

Cell Communities: Tissues, Stem Cells, and Cancer

Cells are the building blocks of multicellular organisms. This seems a simple statement, but it raises deep problems. Cells are not like bricks: they are small and squishy. How can they be used to construct a giraffe or a giant redwood tree? Each cell is enclosed in a flimsy membrane less than a hundred-thousandth of a millimeter thick, and it depends on the integrity of this membrane for its survival. How, then, can cells be joined together robustly to form muscles that can lift an elephant's weight? Most mysterious of all, if cells are the building blocks, where is the builder and where are the architect's plans? How are all the different cell types in a plant or an animal produced, with each in its proper place in an elaborate pattern (**Figure 20–1**)?

Most of the cells in multicellular organisms are organized into cooperative assemblies called **tissues**, such as the nervous, muscle, epithelial, and connective tissues found in vertebrates (**Figure 20–2**). In this chapter, we begin by discussing the architecture of tissues from a mechanical point of view. We will see that tissues are composed not only of cells, with their internal framework of cytoskeletal filaments (discussed in Chapter 17), but also of **extracellular matrix**, which cells secrete around themselves; it is this matrix that gives supportive tissues such as bone or wood their strength. The matrix provides one way to bind cells together, but cells can also attach to one another directly. Thus, we also discuss the *cell junctions* that link cells together in the flexible, mobile tissues of animals. These junctions transmit forces either from the cytoskeleton of one cell to that of the next, or from the cytoskeleton of a cell to the extracellular matrix.

EXTRACELLULAR MATRIX AND
CONNECTIVE TISSUES

EPITHELIAL SHEETS AND CELL
JUNCTIONS

TISSUE MAINTENANCE AND
RENEWAL

CANCER

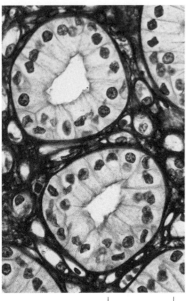

50 μm

Figure 20–1 Multicellular organisms are built from organized collections of cells. This section of cells in the urine-collecting ducts of the kidney was stained with a combination of dyes, hematoxylin and eosin, commonly used in histology. Each duct is made of closely packed "principal" cells (with nuclei stained *red*), which form an epithelial tube, seen here in cross section as a ring. The ducts are embedded in an extracellular matrix, stained *purple* and populated by other types of cells. (From P.R. Wheater et al., Functional Histology, 2nd ed. London: Churchill Livingstone, 1987.)

But there is more to the organization of tissues than mechanics. Just as a building needs plumbing, telephone lines, and other fittings, so an animal tissue requires blood vessels, nerves, and other components formed from a variety of specialized cell types. All the tissue components have to be appropriately organized and coordinated, and many of them require continual maintenance and renewal. Cells die and have to be replaced with new cells of the right type, in the right places, and in the right numbers. In the third section of this chapter, we discuss how these processes are organized, as well as the crucial role that *stem cells*, self-renewing undifferentiated cells, play in the renewal and repair of some tissues.

Disorders of tissue renewal are a major medical concern, and those due to the misbehavior of mutant cells underlie the development of *cancer*. We discuss cancer in the final section of this chapter and of the book as a whole. Its study requires a synthesis of knowledge of cells and tissues at every level, from the molecular biology of DNA repair to the principles of natural selection and the social organization of cells in tissues. Many fundamental advances in cell biology have been driven by cancer research, and basic cell biology in return continues to deepen our understanding of the disease and provide us with renewed optimism about its treatment.

EXTRACELLULAR MATRIX AND CONNECTIVE TISSUES

Plants and animals have evolved their multicellular organization independently, and their tissues are constructed on different principles. Animals prey on other living things—and often are preyed on by other animals—and for this they must be strong and agile: they must possess tissues capable of rapid movement, and the cells that form those tissues must be able to generate and transmit forces and to change shape quickly. Plants, by contrast, are sedentary: their tissues are more or less rigid, although their cells are weak and fragile if isolated from the supporting matrix that surrounds them.

Figure 20–2 Cells are organized into tissues. Simplified drawing of a cross section through part of the wall of the intestine of a mammal. This long, tubelike organ is constructed from epithelial tissues (*red*), connective tissues (*green*), and muscle tissues (*yellow*). Each tissue is an organized assembly of cells, held together by cell–cell adhesions, extracellular matrix, or both.

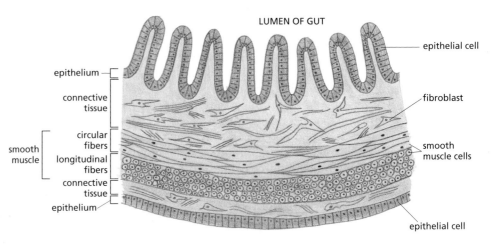

LUMEN OF GUT

epithelium

connective tissue

circular fibers

longitudinal fibers

connective tissue

epithelium

smooth muscle

epithelial cell

fibroblast

smooth muscle cells

epithelial cell

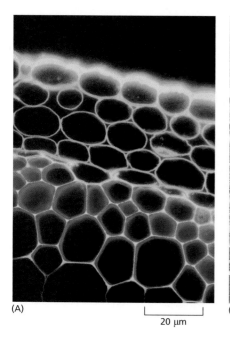

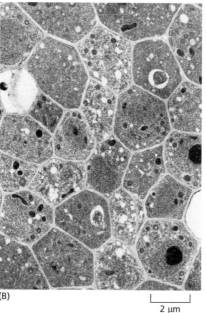

(A)

20 μm

(B)

2 μm

Figure 20–3 Plant tissues are strengthened by cell walls. (A) A cross section of part of the stem of the flowering plant *Arabidopsis* is shown, stained with fluorescent dyes that label two different cell wall polysaccharides—cellulose in *blue*, and pectin in *green*. The cells themselves are unstained and invisible in this preparation. Regions rich in both cellulose and pectin appear white. Pectin predominates in the outer layers of cells, which have only primary cell walls (deposited while the cell is still growing). Cellulose is more plentiful in the inner layers, which have thicker, more rigid secondary cell walls (deposited after cell growth has ceased). (B) Cells and their walls are clearly seen in this electron micrograph of the young cells in the root of the same plant. These cells are much smaller than those in the stem, as can be seen by the different scale bars in the two micrographs. (Courtesy of Paul Linstead.)

In plants, the supportive matrix is called the **cell wall**, a boxlike structure that encloses, protects, and constrains the shape of each of its cells (**Figure 20–3**). Plant cells themselves synthesize, secrete, and control the composition of this extracellular matrix: a cell wall can be thick and hard, as in wood, or thin and flexible, as in a leaf. But the principle of tissue construction is the same in either case: many tiny boxes are cemented together, with a delicate cell living inside each one. Indeed, as we noted in Chapter 1, it was the close-packed mass of microscopic chambers that Robert Hooke saw in a slice of cork three centuries ago that inspired the term "cell."

Animal tissues are more diverse. Like plant tissues, they consist of both cells and extracellular matrix, but these components are organized in many different ways. In some tissues, such as bone or tendon, extracellular matrix is plentiful and mechanically all-important; in others, such as muscle or epidermis, extracellular matrix is scanty, and the cytoskeletons of the cells themselves carry the mechanical load. We begin with a brief discussion of plant cells and tissues, before considering those of animals.

Plant Cells Have Tough External Walls

A naked plant cell, artificially stripped of its wall, is a delicate and vulnerable thing. With care, it can be kept alive in culture; but it is easily ruptured, and even a small decrease in the osmotic strength of the culture medium can cause the cell to swell and burst. Its cytoskeleton lacks the tension-bearing intermediate filaments found in animal cells, and as a result, it has virtually no tensile strength. An external wall, therefore, is essential.

The plant cell wall has to be tough, but it does not necessarily have to be rigid. Osmotic swelling of the cell, limited by the resistance of the cell wall, can keep the chamber distended, and a mass of such swollen chambers cemented together forms a semirigid tissue. Such is the state of a crisp lettuce leaf (**Figure 20–4**). If water is lacking so that the cells shrink, the leaf wilts.

Most newly formed cells in a multicellular plant initially make relatively thin *primary cell walls*, which can slowly expand to accommodate cell growth (see Figure 20–3B). The driving force for cell growth is the same

100 μm

Figure 20–4 A scanning electron micrograph shows the cells in a crisp lettuce leaf. The cells, swollen by osmotic pressure, are stuck together via their walls. (Courtesy of Kim Findlay.)

Figure 20–5 A scale model shows a portion of a primary plant cell wall. The *green* bars represent cellulose microfibrils, which provide tensile strength. Other polysaccharides (*red* strands) cross-link the cellulose microfibrils, while the polysaccharide pectin (*blue* strands) fills the spaces between the microfibrils, providing resistance to compression. The middle lamella (*yellow*) is rich in pectin and is the layer that cements one cell wall to another.

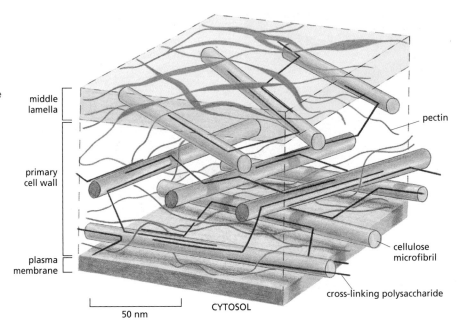

as that keeping the lettuce leaf crisp—a swelling pressure, called the *turgor pressure*, that develops as the result of an osmotic imbalance between the interior of the plant cell and its surroundings. Once cell growth stops and the wall no longer needs to expand, a more rigid *secondary cell wall* is often produced (see Figure 20–3A)—either by thickening of the primary wall or by deposition of new layers with a different composition underneath the old ones. When plant cells become specialized, they generally produce specially adapted types of walls: waxy, waterproof walls for the surface epidermal cells of a leaf; hard, thick, woody walls for the xylem cells of the stem; and so on.

Cellulose Microfibrils Give the Plant Cell Wall Its Tensile Strength

Like all extracellular matrices, plant cell walls derive their tensile strength from long fibers oriented along the lines of stress. In higher plants, the long fibers are generally made from the polysaccharide *cellulose*, the most abundant organic macromolecule on Earth. These **cellulose microfibrils** are interwoven with other polysaccharides and some structural proteins, all bonded together to form a complex structure that resists both compression and tension (**Figure 20–5**). In woody tissue, a highly cross-linked network of *lignin* (a complex polymer built from aromatic alcohol groups) is deposited within this matrix to make it more rigid and waterproof.

For a plant cell to grow or change its shape, the cell wall has to stretch or deform. Because the cellulose microfibrils resist stretching, their orientation governs the direction in which the growing cell enlarges: if, for example, they are arranged circumferentially as a corset, the cell will grow more readily in length than in girth (**Figure 20–6**). By controlling the

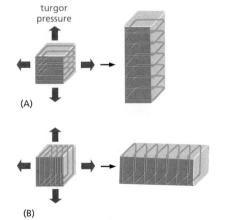

Figure 20–6 The orientation of cellulose microfibrils within the plant cell wall influences the direction in which the cell elongates. The cells in (A) and (B) start off with identical shapes (shown here as *cubes*) but with different orientations of cellulose microfibrils in their walls. Although turgor pressure is uniform in all directions, each cell tends to elongate in a direction perpendicular to the orientation of the microfibrils, which have great tensile strength. The final shape of an organ, such as a shoot, is determined by the direction in which its cells expand.

way that it lays down its wall, the plant cell consequently controls its own shape and thus the direction of growth of the tissue to which it belongs.

Cellulose is produced in a radically different way from most other extracellular macromolecules. Instead of being made inside the cell and then exported by exocytosis (discussed in Chapter 15), it is synthesized on the outer surface of the cell by enzyme complexes embedded in the plasma membrane. These complexes transport sugar monomers across the plasma membrane and incorporate them into a set of growing polymer chains at their points of membrane attachment. Each set of chains assembles to form a cellulose microfibril. The enzyme complexes move in the membrane, spinning out new polymers and laying down a trail of oriented cellulose microfibrils behind them (**Figure 20–7A**).

The paths followed by the enzyme complexes dictate the orientation in which cellulose is deposited in the cell wall; but what guides the enzyme complexes? Just underneath the plasma membrane, microtubules are aligned exactly with the cellulose microfibrils outside the cell (**Figure 20–7B**). These microtubules serve as tracks that help guide the movement of the enzyme complexes (**Figure 20–7C**). In this curiously indirect way, the cytoskeleton controls the shape of the plant cell and the modeling of the plant tissues. We will see that animal cells use their cytoskeleton to control tissue architecture in a much more direct manner.

> ## QUESTION 20–1
>
> Cells in the stem of a seedling that is grown in the dark orient their microtubules horizontally. How would you expect this to affect the growth of the plant?

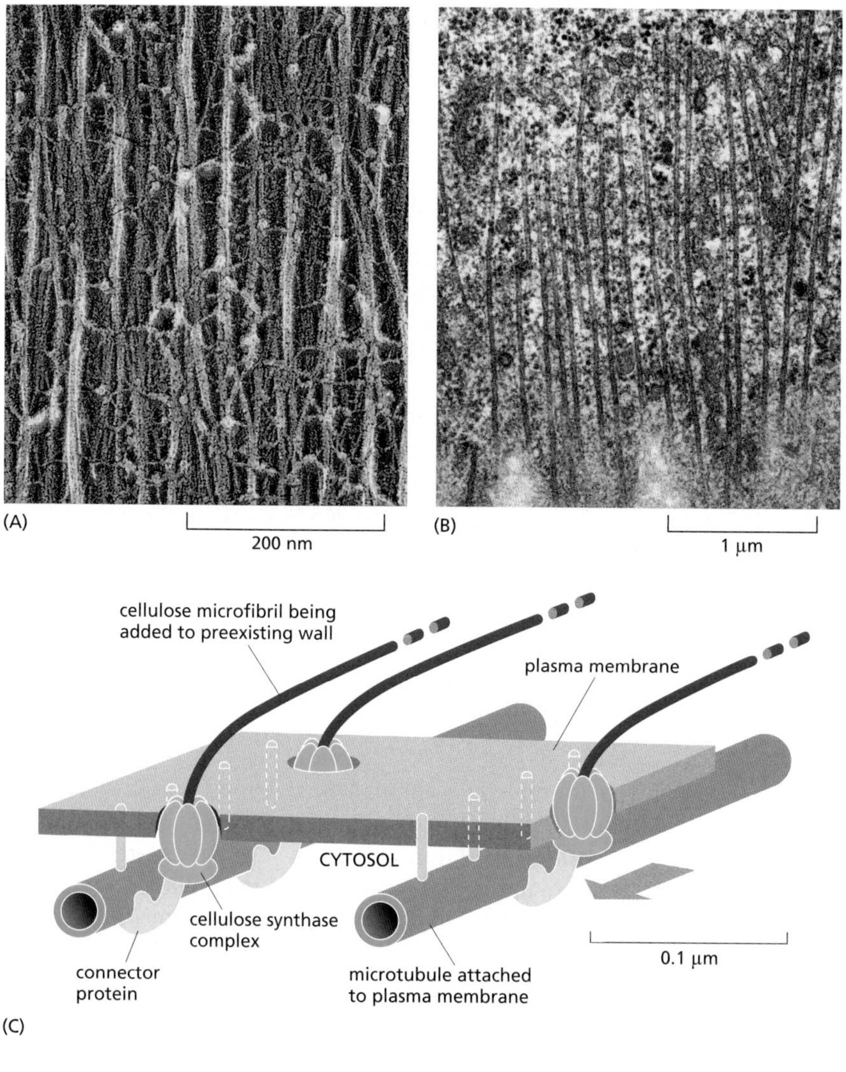

(A) ⊢————⊣ 200 nm

(B) ⊢————⊣ 1 µm

(C)

cellulose microfibril being added to preexisting wall

plasma membrane

CYTOSOL

cellulose synthase complex

connector protein

microtubule attached to plasma membrane

⊢————⊣ 0.1 µm

Figure 20–7 Microtubules help direct the deposition of cellulose in the plant cell wall. Electron micrographs show (A) oriented cellulose microfibrils in a plant cell wall and (B) microtubules just beneath a plant cell's plasma membrane. (C) The orientation of the newly deposited extracellular cellulose microfibrils (*dark blue*) is determined by the orientation of the underlying intracellular microtubules (*dark green*). The large *cellulose synthase* enzyme complexes are integral membrane proteins that continuously synthesize cellulose microfibrils on the outer face of the plasma membrane. The distal ends of the stiff microfibrils become integrated into the texture of the cell wall, and their elongation at the other end pushes the synthase complex along in the plane of the plasma membrane (*red arrow*). The cortical array of microtubules attached to the plasma membrane by transmembrane proteins (*green* vertical bars) helps determine the direction in which the microfibrils are laid down. (A, courtesy of Brian Wells and Keith Roberts; B, courtesy of Brian Gunning.)

Figure 20–8 Extracellular matrix is plentiful in connective tissue such as bone. In this micrograph, cells in a cross section of bone appear as small, dark, antlike objects embedded in the bone matrix, which occupies most of the volume of the tissue and provides all its mechanical strength. The alternating light and dark bands are layers of matrix containing oriented collagen fibrils (made visible with the help of polarized light). Calcium phosphate crystals (not visible) filling the interstices between the collagen fibrils make bone matrix resistant to both compression and tension, like reinforced concrete.

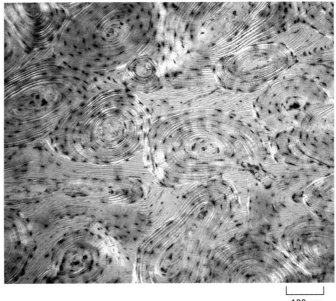

100 μm

Animal Connective Tissues Consist Largely of Extracellular Matrix

It is traditional to distinguish four major types of tissues in animals: connective, epithelial, nervous, and muscular. But the basic architectural distinction is between connective tissues and the rest. In **connective tissues**, extracellular matrix is plentiful and carries the mechanical load. In other tissues, such as epithelia, extracellular matrix is scanty, and the cells are directly joined to one another and carry the mechanical load themselves. We discuss connective tissues first.

Animal connective tissues are enormously varied. They can be tough and flexible like tendons or the dermis of the skin; hard and dense like bone; resilient and shock-absorbing like cartilage; or soft and transparent like the jelly that fills the interior of the eye. In all these examples, the bulk of the tissue is occupied by extracellular matrix, and the cells that produce the matrix are scattered within it like raisins in a pudding (**Figure 20–8**). In all of these tissues, the tensile strength—whether great or small—is chiefly provided not by a polysaccharide, as in plants, but by a fibrous protein: collagen. The various types of connective tissues owe their specific characters to the type of collagen that they contain, to its quantity, and, most importantly, to the other molecules that are interwoven with it in varying proportions. These include the rubbery protein *elastin*, which gives the walls of arteries their resilience as blood pulses through them, as well as a host of specialized polysaccharide molecules, which we discuss shortly.

Collagen Provides Tensile Strength in Animal Connective Tissues

Collagen is a protein found in all animals, and it comes in many varieties. Mammals have about 20 different collagen genes, coding for the variant forms of collagen required in different tissues. Collagens are the chief proteins in bone, tendon, and skin (leather is pickled collagen), and they constitute 25% of the total protein mass in a mammal—more than any other type of protein.

The characteristic feature of a typical collagen molecule is its long, stiff, triple-stranded helical structure, in which three collagen polypeptide

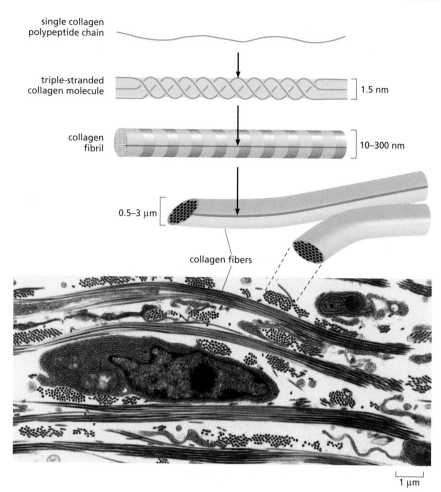

single collagen
polypeptide chain

triple-stranded
collagen molecule ⟩ 1.5 nm

collagen
fibril ⟩ 10–300 nm

0.5–3 μm

collagen fibers

1 μm

Figure 20–9 Collagen fibrils are organized into bundles. The drawings show the steps of collagen assembly, from individual polypeptide chains to triple-stranded collagen molecules, then to fibrils and, finally, fibers. The electron micrograph shows fully assembled collagen in the connective tissue of embryonic chick skin. The fibrils are organized into bundles (fibers), some running in the plane of the section, others approximately at right angles to it. The cell in the micrograph is a fibroblast, which secretes collagen and other extracellular matrix components. (Photograph from C. Ploetz et al., *J. Struct. Biol.* 106:73–81, 1991. With permission from Elsevier.)

chains are wound around one another in a ropelike superhelix (see Figure 4–29A). Some types of collagen molecules in turn assemble into ordered polymers called *collagen fibrils*, which are thin cables 10–300 nm in diameter and many micrometers long; these can pack together into still thicker *collagen fibers* (**Figure 20–9**). Other types of collagen molecules decorate the surface of collagen fibrils and link the fibrils to one another and to other components in the extracellular matrix.

The connective-tissue cells that manufacture and inhabit the extracellular matrix go by various names according to the tissue: in skin, tendon, and many other connective tissues, they are called **fibroblasts** (**Figure 20–10** and see Figure 20–9); in bone, they are called *osteoblasts*. They make both the collagen and the other macromolecules of the matrix. Almost all of these molecules are synthesized intracellularly and then secreted in the standard way, by exocytosis (discussed in Chapter 15). Outside the cell, they assemble into huge, cohesive aggregates. If assembly were to occur prematurely, before secretion, the cell would become choked with its own products. In the case of collagen, the cells avoid this catastrophe by secreting collagen molecules in a precursor form, called *procollagen*, with additional peptide extensions at each end that obstruct premature assembly into collagen fibrils. Extracellular enzymes—called procollagen proteinases—cut off these terminal extensions to allow assembly only after the molecules have emerged into the extracellular space (**Figure 20–11**).

Some people have a genetic defect in one of these proteinases, or in procollagen itself, so that their collagen fibrils do not assemble correctly.

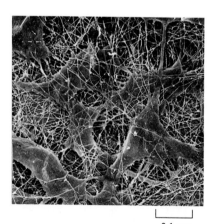

0.1 μm

Figure 20–10 Fibroblasts produce the extracellular matrix of some connective tissues. A scanning electron micrograph showing fibroblasts and collagen fibers in connective tissue from the cornea of a rat. Other components that normally form a hydrated gel filling the spaces between the collagen fibrils and fibers have been removed by enzyme and acid treatment. (From T. Nishida et al., *Invest. Ophthalmol. Vis. Sci.* 29:1887–1890, 1988. With permission from ARVO.)

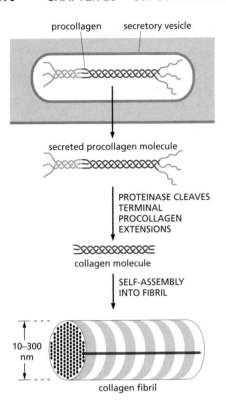

procollagen   secretory vesicle

secreted procollagen molecule

PROTEINASE CLEAVES
TERMINAL
PROCOLLAGEN
EXTENSIONS

collagen molecule

SELF-ASSEMBLY
INTO FIBRIL

10–300
nm

collagen fibril

Figure 20–11 Procollagen precursors are cleaved to form mature collagen outside the cell. Collagen is synthesized as a procollagen molecule that has unstructured peptides at either end. These peptides prevent collagen from assembling into a fibril inside the fibroblast. When the procollagen is secreted, an extracellular enzyme removes its terminal peptides, producing mature collagen molecules. These molecules can then self-assemble into ordered collagen fibrils (see also Figure 20–9).

As a result, their connective tissues have a lower tensile strength and are extraordinarily stretchable (**Figure 20–12**).

Cells in tissues have to be able to degrade matrix as well as make it. This ability is essential for tissue growth, repair, and renewal; it is also important where migratory cells, such as macrophages, need to burrow through the thicket of collagen and other extracellular matrix polymers. Matrix proteases that cleave extracellular proteins play a part in many disease processes, ranging from arthritis, where they contribute to the breakdown of cartilage in affected joints, to cancer, where they help cancer cells invade normal tissue.

Cells Organize the Collagen That They Secrete

To do their job, collagen fibrils must be correctly aligned. In skin, for example, they are woven in a wickerwork pattern, or in alternating layers with different orientations so as to resist tensile stress in multiple directions (**Figure 20–13**). In tendons, which attach muscles to bone, they are aligned in parallel bundles along the major axis of tension.

The connective-tissue cells that produce collagen control this orientation, first by depositing the collagen in an oriented fashion and then by rearranging it. During development of the tissue, fibroblasts work on the collagen they have secreted, crawling over it and pulling on it—helping to compact it into sheets and draw it out into cables. This mechanical role of fibroblasts in shaping collagen matrices has been demonstrated dramatically in cell culture. When fibroblasts are mixed with a meshwork of randomly oriented collagen fibrils that form a gel in a culture dish, the fibroblasts tug on the meshwork, drawing in collagen from their surroundings and compacting it. If two small pieces of embryonic tissue containing fibroblasts are placed far apart on a collagen gel, the intervening collagen becomes organized into a dense band of aligned fibers that connect the two explants (**Figure 20–14**). The fibroblasts migrate out from the explants along the aligned collagen fibers. Thus, the fibroblasts influence the alignment of the collagen fibers, and the collagen fibers in turn affect the distribution of the fibroblasts. Fibroblasts presumably play a similar role in generating long-range order in the extracellular matrix inside the body—in helping to create tendons, for example, and the tough, dense layers of connective tissue that ensheathe and bind together most organs. Fibroblast migration is also important for healing wounds (Movie 20.1).

Figure 20–12 Incorrect collagen assembly can cause the skin to be hyperextensible. James Morris, "the elastic skin man," from a photograph taken in about 1890. Abnormally stretchable skin is part of a genetic syndrome that results from a defect in collagen assembly. In some individuals, this condition arises from a lack of an enzyme that converts procollagen to collagen.

Figure 20–13 Collagen fibrils in skin are arranged in a plywoodlike pattern. The electron micrograph shows a cross section of tadpole skin. Successive layers of fibrils are laid down nearly at right angles to each other (see also Figure 20–9). This arrangement is also found in mature bone and in the cornea. (Courtesy of Jerome Gross.)

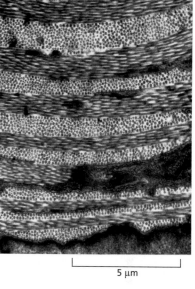

5 μm

Integrins Couple the Matrix Outside a Cell to the Cytoskeleton Inside It

If cells are to pull on the matrix and crawl over it, they must be able to attach to it. Cells do not attach well to bare collagen. Another extracellular matrix protein, **fibronectin**, provides a linkage: part of the fibronectin molecule binds to collagen, while another part forms an attachment site for a cell (**Figure 20–15A and B**).

A cell attaches itself to fibronectin by means of a receptor protein, called an **integrin**, which spans the cell's plasma membrane. When the extracellular domain of the integrin binds to fibronectin, the intracellular domain binds (through a set of adaptor molecules) to an actin filament inside the cell (**Figure 20–15C**). Without this internal anchorage to the cytoskeleton, integrins would be ripped out of the flimsy lipid bilayer of the plasma membrane as the cell attempted to pull itself along the matrix.

It is the formation and breakage of the attachments on either end of an integrin molecule that allows a cell to crawl through a tissue, grabbing hold of the matrix at its front end and releasing its grip at the rear (see Figure 17–33). Integrins coordinate these "catch-and-release" manoeuvers by undergoing remarkable conformational changes. Binding to a molecule on one side of the plasma membrane causes the integrin molecule to stretch out into an extended, activated state so that it can then latch onto a different molecule on the opposite side—an effect that operates in either direction across the membrane (**Figure 20–16**). Thus, an intracellular signaling molecule can activate the integrin from the cytosolic side, causing it to reach out and grab hold of an extracellular structure. Similarly, binding to an external structure can switch on intracellular signaling pathways by activating protein kinases that associate with the intracellular end of the integrin. In this way, a cell's external attachments help regulate whether it lives or dies, and—if it does survive—whether it grows, divides, or differentiates.

Humans make at least 24 kinds of integrins, each of which recognizes distinct extracellular molecules and has distinct functions, depending on

migrating fibroblasts

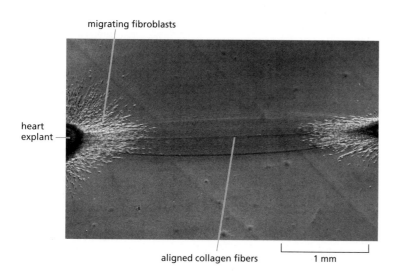

heart explant

aligned collagen fibers   1 mm

Figure 20–14 Fibroblasts influence the alignment of collagen fibers. This micrograph shows a region between two pieces of embryonic chick heart (rich in fibroblasts, as well as heart muscle cells) that have grown in culture on a collagen gel for four days. A dense tract of aligned collagen fibers has formed between the explants, presumably as a result of the fibroblasts, which have migrated out from the explants, tugging on the collagen. Elsewhere in the culture dish, the collagen remains disorganized and unaligned, so that it appears uniformly gray. (From D. Stopak and A.K. Harris, *Dev. Biol.* 90:383–398, 1982. With permission from Elsevier.)

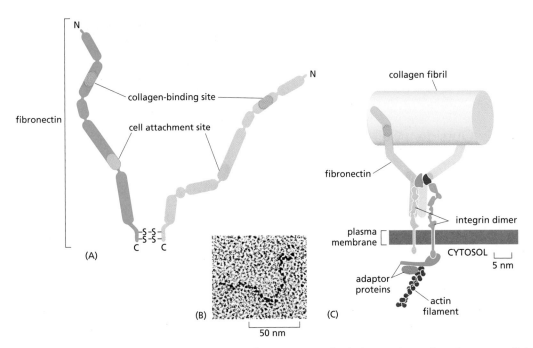

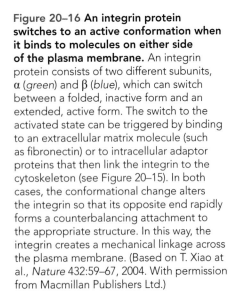

Figure 20–15 **Fibronectin and integrin proteins help attach a cell to the extracellular matrix.** Fibronectin molecules outside the cell bind to collagen fibrils. Integrins in the plasma membrane bind to the fibronectin and tether it to the cytoskeleton inside the cell. (A) Diagram and (B) electron micrograph of a molecule of fibronectin. (C) The transmembrane linkage mediated by an integrin protein (*blue* and *green* dimer). The integrin molecule transmits tension across the plasma membrane: it is anchored inside the cell via adaptor proteins to the actin cytoskeleton and externally via fibronectin to other extracellular matrix proteins, such as the collagen fibril shown. The integrin shown here links fibronectin to an actin filament inside the cell. Other integrins can connect different extracellular proteins to the cytoskeleton (usually to actin filaments, but sometimes to intermediate filaments). (B, from J. Engel et al., *J. Mol. Biol.* 150:97–120, 1981. With permission from Elsevier.)

the cell type in which it resides. For example, the integrins on white blood cells help the cells crawl out of blood vessels at sites of infection so as to deal with marauding microbes. People who lack this type of integrin develop a disease called *leucocyte adhesion deficiency* and suffer from repeated bacterial infections. A different form of integrin is found on blood platelets, and individuals who lack this integrin bleed excessively because their platelets cannot bind to the necessary clotting factor in the extracellular matrix.

Gels of Polysaccharides and Proteins Fill Spaces and Resist Compression

While collagen provides tensile strength to resist stretching, a completely different group of macromolecules in the extracellular matrix of

Figure 20–16 **An integrin protein switches to an active conformation when it binds to molecules on either side of the plasma membrane.** An integrin protein consists of two different subunits, α (*green*) and β (*blue*), which can switch between a folded, inactive form and an extended, active form. The switch to the activated state can be triggered by binding to an extracellular matrix molecule (such as fibronectin) or to intracellular adaptor proteins that then link the integrin to the cytoskeleton (see Figure 20–15). In both cases, the conformational change alters the integrin so that its opposite end rapidly forms a counterbalancing attachment to the appropriate structure. In this way, the integrin creates a mechanical linkage across the plasma membrane. (Based on T. Xiao at al., *Nature* 432:59–67, 2004. With permission from Macmillan Publishers Ltd.)

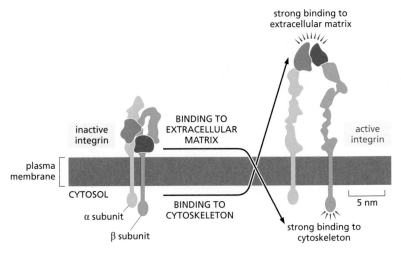

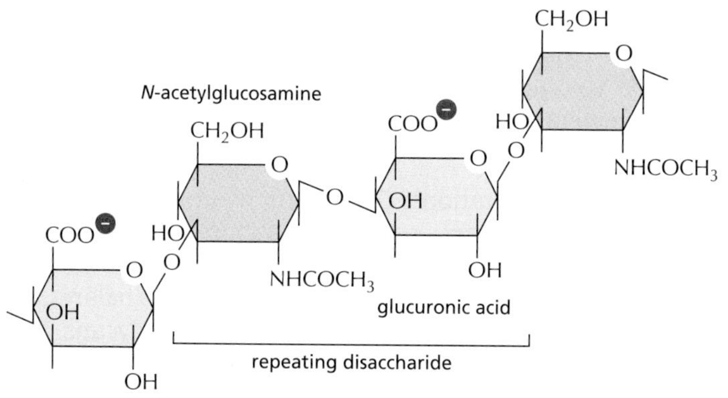
N-acetylglucosamine
glucuronic acid
repeating disaccharide

Figure 20–17 Glycosaminoglycans (GAGs) are built from repeating disaccharide units. Hyaluronan, a relatively simple GAG, is shown here. It consists of a single long chain of up to 25,000 repeated disaccharide units, each carrying a negative charge (*red*). As in other GAGs, one of the sugar monomers (*green*) in each disaccharide unit is an amino sugar. Many GAGs have additional negatively charges, often from sulfate groups (not shown).

animal tissues provides the complementary function, resisting compression. These are the **glycosaminoglycans** (**GAGs**), negatively charged polysaccharide chains made of repeating disaccharide units (**Figure 20–17**). GAGs are usually covalently linked to core proteins to form **proteoglycans**, which are extremely diverse in size, shape, and chemistry. Typically, many GAG chains are attached to a single core protein, which may in turn be linked at one end to another GAG, creating an enormous aggregate resembling a bottlebrush, with a molecular weight in the millions (**Figure 20–18**).

In dense, compact connective tissues such as tendon and bone, the proportion of GAGs is small, and the matrix consists almost entirely of collagen (or, in the case of bone, of collagen plus calcium phosphate crystals). At the other extreme, the jellylike substance in the interior of the eye consists almost entirely of one particular type of GAG, plus water,

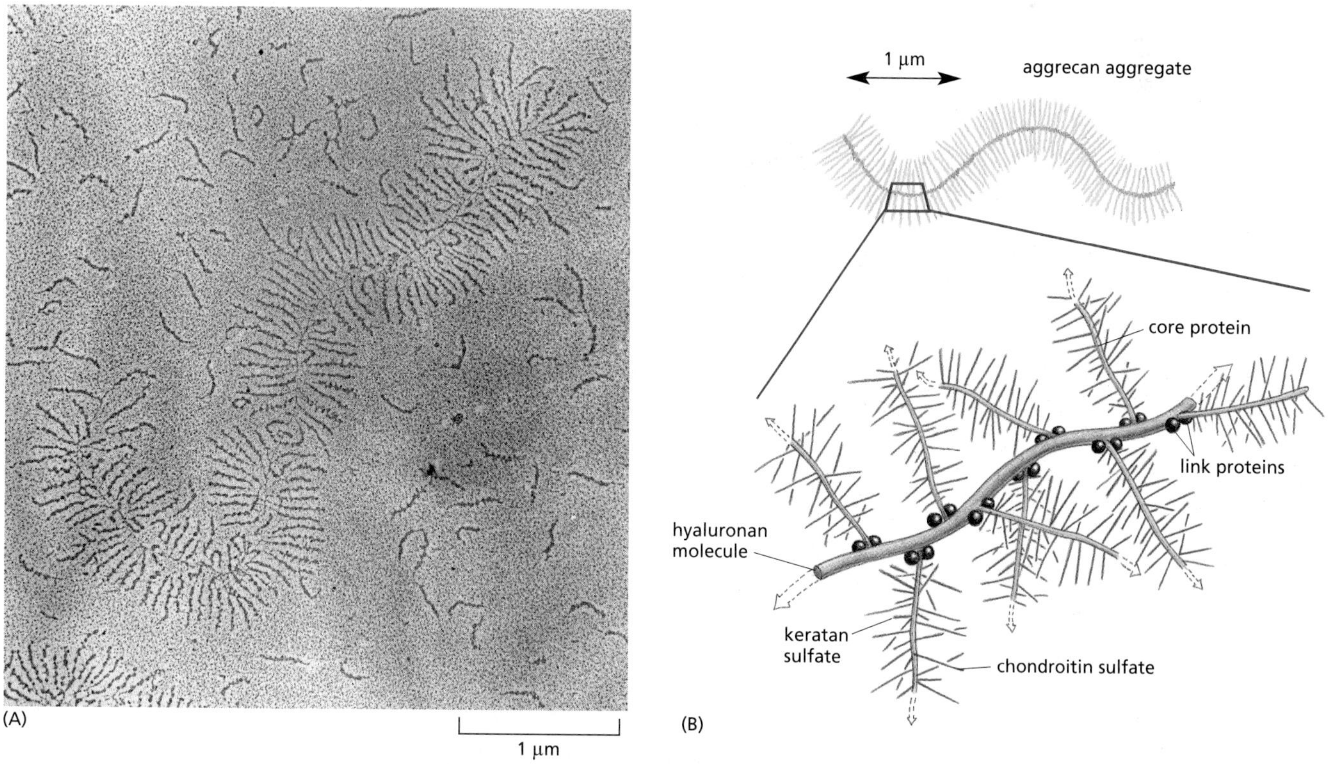
(A)
1 μm
(B)
1 μm
aggrecan aggregate
core protein
link proteins
hyaluronan molecule
keratan sulfate
chondroitin sulfate

Figure 20–18 Proteoglycans and GAGs can form large aggregates. (A) Electron micrograph of an aggregate from cartilage, spread out on a flat surface. Many free subunits—themselves large proteoglycan molecules—can also be seen. (B) Schematic drawing of the giant aggregate illustrated in (A), showing how it is built up from GAGs (*red* and *blue*) and proteins (*green* and *black*). The mass of such a complex can be 10^8 daltons or more, and it occupies a volume equivalent to that of a bacterium, which is about 2×10^{-12} cm^3. (A, courtesy of Lawrence Rosenberg.)

with only a small amount of collagen. In general, GAGs are strongly hydrophilic and tend to adopt highly extended conformations, which occupy a huge volume relative to their mass (see Figure 20–18). Thus GAGs act as effective "space fillers" in the extracellular matrix of connective tissues.

Even at very low concentrations, GAGs form hydrophilic gels: their multiple negative charges attract a cloud of cations, such as Na^+, that are osmotically active, causing large amounts of water to be sucked into the matrix. This gives rise to a swelling pressure, which is balanced by tension in the collagen fibers interwoven with the proteoglycans. When the matrix is rich in collagen and large quantities of GAGs are trapped in its meshes, both the swelling pressure and the counterbalancing tension are enormous. Such a matrix is tough, resilient, and resistant to compression. The cartilage matrix that lines the knee joint, for example, has this character: it can support pressures of hundreds of kilograms per square centimeter.

Proteoglycans perform many sophisticated functions in addition to providing hydrated space around cells. They can form gels of varying pore size and charge density that act as filters to regulate the passage of molecules through the extracellular medium. They can bind secreted growth factors and other proteins that serve as extracellular signals for cells. They can block, encourage, or guide cell migration through the matrix. In all these ways, the matrix components influence the behavior of cells, often the same cells that make the matrix—a reciprocal interaction that has important effects on cell differentiation and the arrangement of cells in a tissue. Much remains to be learned about how cells weave the tapestry of matrix molecules and how the chemical messages they deposit in this intricate biochemical fabric are organized and act.

EPITHELIAL SHEETS AND CELL JUNCTIONS

There are more than 200 visibly different cell types in the body of a vertebrate. The majority of these are organized into **epithelia** (singular **epithelium**)—multicellular sheets in which the cells are joined together, side to side. In some cases, the sheet is many cells thick, or *stratified*, as in the epidermis (the outer layer of the skin); in other cases, it is a *simple epithelium*, only one cell thick, as in the lining of the gut. Epithelial cells can take many forms. They can be tall and columnar, squat and cuboidal, or flat and squamous (**Figure 20–19**). Within a given sheet, the cells may be all the same type or a mixture of different types. Some epithelia, like the epidermis, act mainly just as a protective barrier; others have complex biochemical functions. Some secrete specialized products such as hormones, milk, or tears; others, such as the epithelium lining the gut, absorb nutrients; yet others detect signals, such as light, sensed by

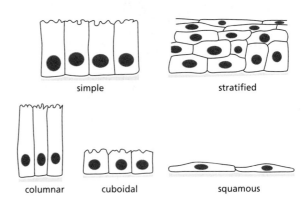

Figure 20–19 Cells can be packed together in different ways to form an epithelial sheet. Five basic types of epithelia are shown.

simple

stratified

columnar

cuboidal

squamous

the layer of photoreceptors in the retina of the eye, or sound, sensed by the epithelium containing the auditory hair cells in the ear (see Figure 12–27). Despite these and many other variations, one can recognize a standard set of structural features that virtually all animal epithelia share. The arrangement of cells into epithelia is so commonplace that one easily takes it for granted; yet it requires a collection of specialized devices, as we will see, and these are common to a wide variety of epithelial cell types.

Epithelia cover the external surface of the body and line all its internal cavities, and they must have been an early feature in the evolution of animals. Cells joined together into an epithelial sheet create a barrier, which has the same significance for the multicellular organism that the plasma membrane has for a single cell. It keeps some molecules in, and others out; it takes up nutrients and exports wastes; it contains receptors for environmental signals; and it protects the interior of the organism from invading microorganisms and fluid loss.

Epithelial Sheets Are Polarized and Rest on a Basal Lamina

An epithelial sheet has two faces: the **apical** surface is free and exposed to the air or to a watery fluid; the **basal** surface is attached to a sheet of connective tissue called the basal lamina (**Figure 20–20**). The **basal lamina** consists of a thin, tough sheet of extracellular matrix, composed mainly of a specialized type of collagen (type IV collagen) and a protein called *laminin* (**Figure 20–21**). Laminin provides adhesive sites for integrin molecules in the basal plasma membranes of epithelial cells, and it thus serves a linking role like that of fibronectin in other connective tissues.

The apical and basal faces of an epithelium are chemically different, reflecting the polarized organization of the individual epithelial cells: each has a top and a bottom, with different properties and functions. This polarity is crucial for epithelial function. Consider, for example, the simple columnar epithelium that lines the small intestine of a mammal. It mainly consists of two intermingled cell types: absorptive cells, which

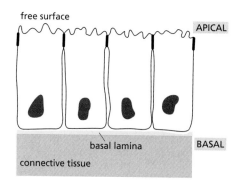

Figure 20–20 A sheet of epithelial cells has an apical and a basal surface. The basal surface sits on a specialized sheet of extracellular matrix called the basal lamina, while the apical surface is free.

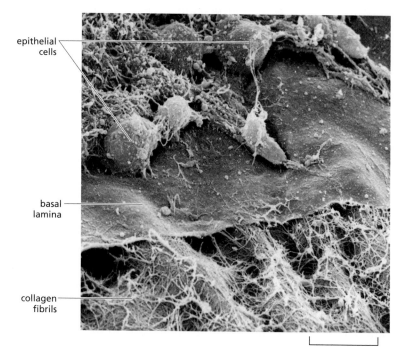

10 µm

Figure 20–21 The basal lamina supports a sheet of epithelial cells. Scanning electron micrograph of a basal lamina in the cornea of a chick embryo. Some of the epithelial cells have been removed to expose the upper surface of the matlike basal lamina, which is woven from type IV collagen and laminin proteins. A network of other collagen fibrils in the underlying connective tissue interacts with the lower face of the lamina. (Courtesy of Robert Trelstad.)

Figure 20–22 Functionally polarized cell types line the intestine. Absorptive cells, which take up nutrients from the intestine, are mingled in the gut epithelium with goblet cells (*brown*), which secrete mucus into the gut lumen. The absorptive cells are often called *brush-border cells*, because of the brushlike mass of microvilli on their apical surface; the microvilli serve to increase the area of apical plasma membrane for the transport of small molecules into the cell. The goblet cells owe their gobletlike shape to the mass of secretory vesicles that distends the cytoplasm in their apical region. (Adapted from R. Krstić, Human Microscopic Anatomy. Berlin: Springer, 1991. With permission from Springer-Verlag.)

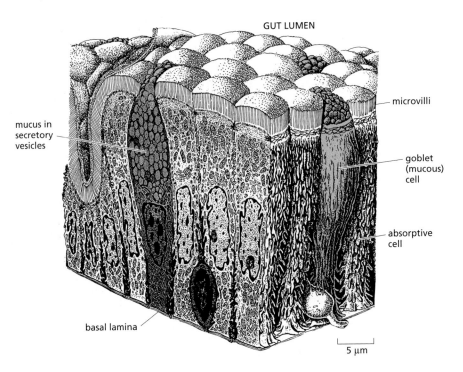

GUT LUMEN

mucus in secretory vesicles

microvilli

goblet (mucous) cell

absorptive cell

basal lamina

5 μm

take up nutrients, and goblet cells (so called because of their shape), which secrete the mucus that protects and lubricates the gut lining (**Figure 20–22**). Both cell types are polarized. The absorptive cells import food molecules from the gut lumen through their apical surface and export these molecules from their basal surface into the underlying tissues. To do this, absorptive cells require different sets of membrane transport proteins in their apical and basal plasma membranes (see Figure 12–14). The goblet cells also have to be polarized, but in a different way: they have to synthesize mucus and then discharge it from their apical end only (see Figure 20–22); their Golgi apparatus, secretory vesicles, and cytoskeleton are all polarized so as to bring this about. For both types of epithelial cells polarity depends on the junctions that the cells form with one another and with the basal lamina. These in turn control the arrangement of an elaborate system of membrane-associated intracellular proteins that create the polarized organization of the cytoplasm.

Tight Junctions Make an Epithelium Leak-proof and Separate Its Apical and Basal Surfaces

Epithelial **cell junctions** can be classified according to their function. Some provide a tight seal to prevent the leakage of molecules across the epithelium through the gaps between its cells; some provide strong mechanical attachments; and some provide for a special type of intimate chemical communication. In most epithelia, all these types of junctions are present. As we will see, each type of junction is characterized by its own class of membrane proteins.

The sealing function is served (in vertebrates) by **tight junctions**. These junctions seal neighboring cells together so that water-soluble molecules cannot easily leak between them. If a tracer molecule is added to one side of an epithelial cell sheet, it will usually not pass beyond the tight junction (**Figure 20–23A and B**). The tight junction is formed from proteins called *claudins* and *occludins*, which are arranged in strands along the lines of the junction to create the seal (**Figure 20–23C**). Without tight

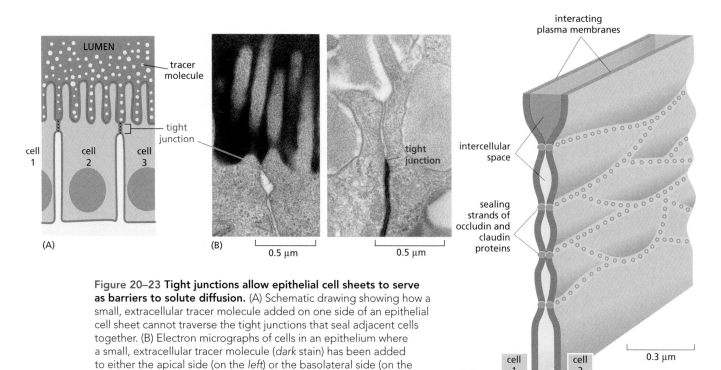

Figure 20–23 Tight junctions allow epithelial cell sheets to serve as barriers to solute diffusion. (A) Schematic drawing showing how a small, extracellular tracer molecule added on one side of an epithelial cell sheet cannot traverse the tight junctions that seal adjacent cells together. (B) Electron micrographs of cells in an epithelium where a small, extracellular tracer molecule (*dark* stain) has been added to either the apical side (on the *left*) or the basolateral side (on the *right*); in both cases the tracer is stopped by the tight junction. (C) A model of the structure of a tight junction, showing how the cells are sealed together by branching strands of transmembrane proteins, called claudins and occludins (*green*), in the plasma membranes of the interacting cells. Each type of protein binds to the same type in the apposed membrane (not shown). (B, courtesy of Daniel Friend.)

junctions to prevent leakage, the pumping activities of absorptive cells such as those in the gut would be futile, and the composition of the extracellular fluid would become the same on both sides of the epithelium. Tight junctions also play a key part in maintaining the polarity of the individual epithelial cells in two ways. First, the tight junctions around the apical region of each cell prevents diffusion of proteins within the plasma membrane and so keeps the apical domain of the plasma membrane different from the basal (or basolateral) domain (see Figure 11–32). Second, in many epithelia, the tight junctions are sites of assembly for the complexes of intracellular proteins that govern the apico-basal polarity of the cell interior.

Cytoskeleton-linked Junctions Bind Epithelial Cells Robustly to One Another and to the Basal Lamina

The cell junctions that hold an epithelium together by forming mechanical attachments are of three main types. *Adherens junctions* and *desmosomes* bind one epithelial cell to another, while *hemidesmosomes* bind epithelial cells to the basal lamina. All of these junctions provide mechanical strength by the same strategy: the proteins that form the cell adhesion span the plasma membrane and are linked inside the cell to cytoskeletal filaments. In this way, the cytoskeletal filaments are tied into a network that extends from cell to cell across the whole expanse of the epithelial sheet.

Adherens junctions and desmosomes are both built around transmembrane proteins that belong to the **cadherin** family: a cadherin molecule in the plasma membrane of one cell binds directly to an identical cadherin molecule in the plasma membrane of its neighbor (**Figure 20–24**).

Figure 20–24 Cadherin molecules mediate mechanical attachment of one cell to another. Identical cadherin molecules in the plasma membranes of adjacent cells bind to each other extracellularly; inside the cell, they are attached, via linker proteins, to cytoskeletal filaments—either actin filaments or keratin intermediate filaments. As cells touch one another, their cadherins become concentrated at the point of attachment (Movie 20.2).

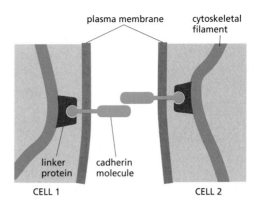

Such binding of like-to-like is called *homophilic* binding. In the case of cadherins, binding also requires that Ca^{2+} be present in the extracellular medium—hence the name.

At an **adherens junction**, each cadherin molecule is tethered inside its cell, via several linker proteins, to actin filaments. Often, the adherens junctions form a continuous adhesion belt around each of the interacting epithelial cells; this belt is located near the apical end of the cell, just below the tight junctions (**Figure 20–25**). Bundles of actin filaments are thus connected from cell to cell across the epithelium. This network of actin filaments can contract, giving the epithelial sheet the capacity to develop tension and to change its shape in remarkable ways. By shrinking the apical surface of an epithelial sheet along one axis, the sheet can roll itself up into a tube (**Figure 20–26A and B**). Alternatively, by shrinking its apical surface locally along all axes at once, the sheet can invaginate into a cup and eventually create a spherical vesicle by pinching off from the rest of the epithelium (**Figure 20–26C**). Epithelial movements such as these are important in embryonic development, where they create structures such as the neural tube (see Figure 20–26B), which gives rise to the central nervous system, and the lens vesicle, which develops into the lens of the eye (see Figure 20–26C).

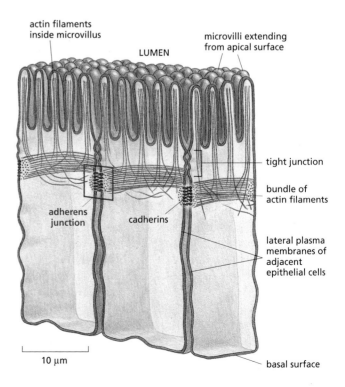

Figure 20–25 Adherens junctions form adhesion belts around epithelial cells in the small intestine. A contractile bundle of actin filaments runs along the cytoplasmic surface of the plasma membrane near the apex of each cell. These bundles are linked to those in adjacent cells via transmembrane cadherin molecules (see Figure 20–24).

At a **desmosome**, a different set of cadherin molecules connects to *keratin filaments*—the intermediate filaments found specifically in epithelial cells (see Figure 17–5). Bundles of ropelike keratin filaments criss-cross the cytoplasm and are "spot-welded" via desmosome junctions to the bundles of keratin filaments in adjacent cells (**Figure 20–27**). This arrangement confers great tensile strength on the epithelial sheet and is characteristic of tough, exposed epithelia such as the epidermis of the skin.

Blisters are a painful reminder that it is not enough for epidermal cells to be firmly attached to one another: they must also be anchored to the underlying connective tissue. As we noted earlier, the anchorage is mediated by integrins in the cells' basal plasma membranes. The extracellular domains of these integrins bind to laminin in the basal lamina; inside the cell, the integrin tails are linked to keratin filaments, creating a structure that looks superficially like half a desmosome. These attachments

Figure 20–26 Epithelial sheets can bend to form a tube or a vesicle. Contraction of apical bundles of actin filaments linked from cell to cell via adherens junctions causes the epithelial cells to narrow at their apex. Depending on whether the contraction of the epithelial sheet is oriented along one axis, or is equal in all directions, the epithelium will either roll up into a tube or invaginate to form a vesicle. (A) Diagram showing how an apical contraction along one axis of an epithelial sheet can cause the sheet to form a tube. (B) Scanning electron micrograph of a cross section through the trunk of a two-day chick embryo, showing the formation of the neural tube by the process shown in (A). Part of the epithelial sheet that covers the surface of the embryo has thickened and rolled up by apical contraction; the opposing folds are about to fuse, after which the structure will pinch off to form the neural tube. (C) Scanning electron micrograph of a chick embryo showing the formation of the eye cup and lens. A patch of surface epithelium overlying the forming eye cup has become concave and has pinched off as a separate vesicle—the lens vesicle—within the eye cup. This process is driven by an apical narrowing of epithelial cells in all directions. (B, courtesy of Jean-Paul Revel; C, courtesy of K.W. Tosney.)

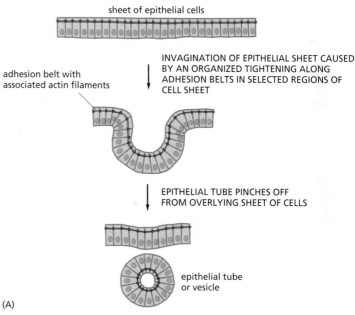

sheet of epithelial cells

adhesion belt with associated actin filaments

INVAGINATION OF EPITHELIAL SHEET CAUSED BY AN ORGANIZED TIGHTENING ALONG ADHESION BELTS IN SELECTED REGIONS OF CELL SHEET

EPITHELIAL TUBE PINCHES OFF FROM OVERLYING SHEET OF CELLS

epithelial tube or vesicle

(A)

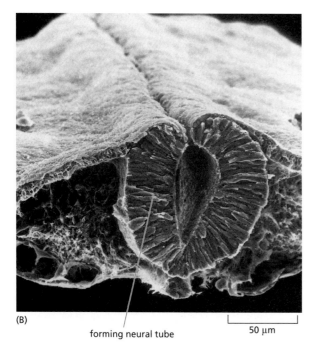

(B)

forming neural tube

50 μm

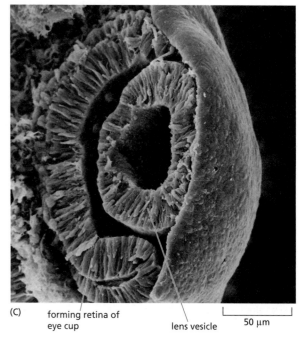

(C)

forming retina of eye cup

lens vesicle

50 μm

Figure 20–27 Desmosomes link the keratin intermediate filaments of one epithelial cell to those of another. (A) An electron micrograph of a desmosome joining two cells in the epidermis of newt skin, showing the attachment of keratin filaments. (B) Schematic drawing of a desmosome. On the cytoplasmic surface of each interacting plasma membrane is a dense plaque composed of a mixture of intracellular linker proteins. A bundle of keratin filaments is attached to the surface of each plaque. The cytoplasmic tails of transmembrane cadherin proteins bind to the outer face of each plaque; their extracellular domains interact to hold the cells together. (A, from D.E. Kelly, *J. Cell Biol.* 28:51–72, 1966. With permission from The Rockefeller University Press.)

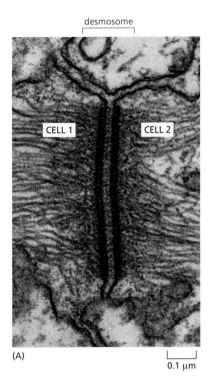

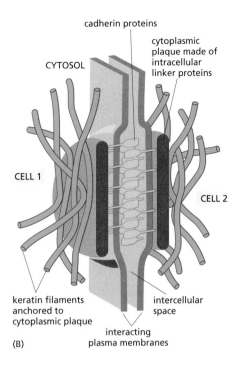

desmosome

CELL 1   CELL 2

(A)

0.1 µm

cadherin proteins

cytoplasmic plaque made of intracellular linker proteins

CYTOSOL

CELL 1

CELL 2

keratin filaments anchored to cytoplasmic plaque

intercellular space

interacting plasma membranes

(B)

of epithelial cells to basal lamina beneath them are therefore called **hemidesmosomes** (**Figure 20–28**).

Gap Junctions Allow Cytosolic Inorganic Ions and Small Molecules to Pass from Cell to Cell

The final type of epithelial cell junction, found in virtually all epithelia and in many other types of animal tissues, serves a totally different purpose. In the electron microscope, this **gap junction** appears as a region where the membranes of two cells lie close together and exactly parallel, with a very narrow gap of 2–4 nm between them. The gap, however, is not entirely empty; it is spanned by the protruding ends of many identical, transmembrane protein complexes that lie in the plasma membranes of the two apposed cells. These complexes, called *connexons*, are aligned end-to-end to form narrow, water-filled channels across the two plasma membranes (**Figure 20–29A**). The channels allow inorganic ions and small, water-soluble molecules (up to a molecular mass of about 1000 daltons) to move directly from the cytosol of one cell to the cytosol of the other. This creates an electrical and a metabolic coupling between the cells. Gap junctions between cardiac muscle cells, for example, provide

QUESTION 20–4

Analogs of hemidesmosomes are the focal contact sites described in Chapter 17, which are also sites where the cell attaches to the extracellular matrix. These junctions are prevalent in fibroblasts but largely absent in epithelial cells. On the other hand, hemidesmosomes are prevalent in epithelial cells but absent in fibroblasts. In focal contact sites, intracellular connections are made to actin filaments, whereas in hemidesmosomes connections are made to intermediate filaments. Why do you suppose these two different cell types attach differently to the extracellular matrix?

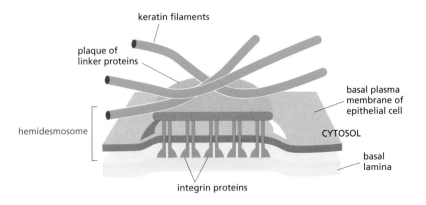

keratin filaments

plaque of linker proteins

hemidesmosome

basal plasma membrane of epithelial cell

CYTOSOL

basal lamina

integrin proteins

Figure 20–28 Hemidesmosomes anchor the keratin filaments in an epithelial cell to the basal lamina. The linkage is mediated by transmembrane integrins.

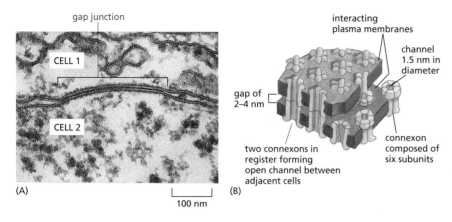

gap junction

CELL 1

CELL 2

(A)

100 nm

interacting
plasma membranes

channel
1.5 nm in
diameter

gap of
2–4 nm

two connexons in
register forming
open channel between
adjacent cells

connexon
composed of
six subunits

(B)

Figure 20–29 Gap junctions provide neighboring cells with a direct channel of communication. (A) Thin-section electron micrograph of a gap junction between two cells in culture. (B) A model of a gap junction. The drawing shows the interacting plasma membranes of two adjacent cells. The apposed membranes are penetrated by protein assemblies called *connexons* (*green*), each of which is formed from six identical protein subunits. Two connexons join across the intercellular gap to form an aqueous channel connecting the cytosols of the two cells. (A, from N.B. Gilula, in Cell Communication [R.P. Cox, ed.], pp. 1–29. New York: Wiley, 1974. With permission from John Wiley & Sons, Inc.)

the electrical coupling that allows electrical waves of excitation to spread synchronously through the heart, triggering the coordinated contraction of the cells that produces each heart beat.

Gap junctions in many tissues can be opened or closed in response to extracellular or intracellular signals. The neurotransmitter dopamine, for example, reduces gap-junction communication within a class of neurons in the retina in response to an increase in light intensity (**Figure 20–30**). This reduction in gap-junction permeability alters the pattern of electrical signaling and helps the retina switch from using rod photoreceptors, which are good detectors of low light, to cone photoreceptors, which detect color and fine detail in bright light. The function of gap junctions—and of the other junctions found in animal cells—are summarized in **Figure 20–31**.

Plant tissues lack all the types of cell junctions we have discussed so far, as their cells are held together by their cell walls. Curiously, however, they have a functional counterpart of the gap junction. The cytoplasms of adjacent plant cells are connected via minute communicating channels called **plasmodesmata**, which span the intervening cell walls. In contrast to gap-junction channels, plasmodesmata are cytoplasmic channels lined with plasma membrane (**Figure 20–32**). Thus in plants, the cytoplasm is, in principle, continuous from one cell to the next. Inorganic small molecules, and even macromolecules—including some proteins and regulatory RNAs—can pass through plasmodesmata. The controlled traffic of transcription regulators and regulatory RNAs from one cell to another is important in plant development.

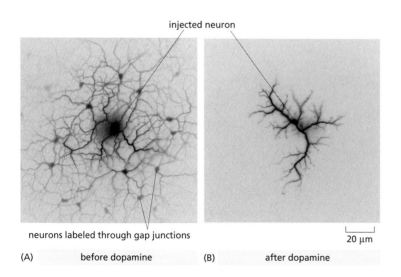

injected neuron

neurons labeled through gap junctions

20 µm

(A)   before dopamine

(B)   after dopamine

Figure 20–30 Extracellular signals can regulate the permeability of gap junctions. (A) A neuron in a rabbit retina (center) was injected with a dye that passes readily through gap junctions. The dye diffuses rapidly from the injected cell to label the surrounding neurons, which are connected by gap junctions. (B) Treatment of the retina with the neurotransmitter dopamine prior to dye injection decreases the permeability of the gap junctions and hampers the spread of the dye. (Courtesy of David Vaney.)

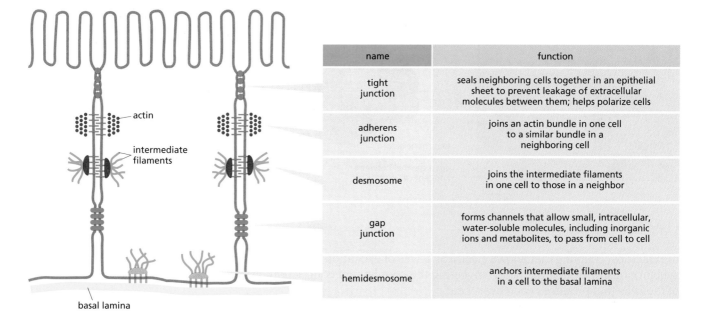

name	function
tight junction	seals neighboring cells together in an epithelial sheet to prevent leakage of extracellular molecules between them; helps polarize cells
adherens junction	joins an actin bundle in one cell to a similar bundle in a neighboring cell
desmosome	joins the intermediate filaments in one cell to those in a neighbor
gap junction	forms channels that allow small, intracellular, water-soluble molecules, including inorganic ions and metabolites, to pass from cell to cell
hemidesmosome	anchors intermediate filaments in a cell to the basal lamina

Figure 20–31 Several types of cell junctions are found in epithelia in animals. Whereas tight junctions are peculiar to epithelia, the other types also occur, in modified forms, in various nonepithelial tissues.

TISSUE MAINTENANCE AND RENEWAL

One cannot contemplate the organization of tissues without wondering how these astonishingly patterned structures come into being. This question raises an even more challenging one—a puzzle that is one of the most ancient and fundamental in all of biology: how is a complex multicellular organism generated from a single fertilized egg?

In the process of development, the fertilized egg cell divides repeatedly to give a clone of cells—about 10,000,000,000,000 for a human—essentially all containing the same genome but specialized in different ways. This clone has a structure. It may take the form of a daisy or an oak tree, a sea urchin, a whale, or a mouse (**Figure 20–33**). The structure is determined by the genome that the fertilized egg contains. The linear sequence of A, G, C, and T nucleotides in the DNA directs the production of a variety of distinct cell types, each expressing different sets of genes and arranged in a precise, intricate, three-dimensional pattern, which self-assembles during development.

Although the final structure of an animal's body may be enormously complex, it is generated by a limited repertoire of cell activities. Examples of all these activities have been discussed in earlier pages of this book. Cells grow, divide, migrate, and die. They form mechanical attachments and generate forces that bind epithelial sheets. They differentiate by switching on or off the production of specific sets of proteins and regulatory

Figure 20–32 Plant cells are connected via plasmodesmata. (A) The cytoplasmic channels of plasmodesmata pierce the plant cell wall and connect the interiors of all cells in a plant. (B) Each plasmodesma is lined with plasma membrane common to the two connected cells. It usually also contains a fine tubular structure, the desmotubule, derived from smooth endoplasmic reticulum.

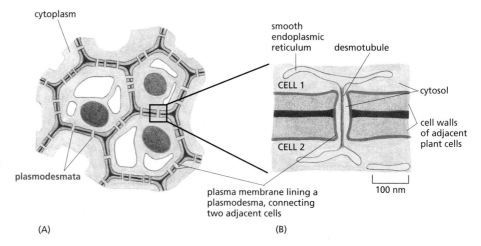

(A)
100 μm

(B)

(C)
50 μm

(D)

Figure 20–33 The genome of the fertilized egg determines the ultimate structure of the clone of cells that will develop from it. (A and B) A sea-urchin egg gives rise to a sea urchin; (C and D) a mouse egg gives rise to a mouse. (A, courtesy of David McClay; B, courtesy of Alaska Department of Fish and Game; C, courtesy of Patricia Calarco, from G. Martin, *Science* 209:768–776, 1980, with permission from AAAS; D, courtesy of US Department of Agriculture, Agricultural Research Service.)

RNAs. They produce molecular signals to influence neighboring cells, and they respond to signals that neighboring cells deliver to them. They remember the effects of previous signals they have received, and so progressively become more and more specialized in the characteristics they adopt. The genome, identical in virtually every cell, defines the rules by which these various possible cell activities are called into play. Through its operation in each cell individually, the genome guides the whole intricate process by which a multicellular organism is generated from a fertilized egg. Movies 1.1, 20.3, and 20.4 offer some visual examples of how development unfurls for the embryos of a frog, a fruit fly, and a zebrafish, respectively.

For developmental biologists, the challenge is to explain how genes orchestrate the entire sequence of interlocking events that lead from the egg to the adult organism. We will not attempt to set out an answer to this problem here: we do not have space to do it justice, even though a great deal of the genetic and cell biological basis of development is now understood. But the same basic activities that combine to create the organism during development continue even in the adult body, where fresh cells are continually generated in precisely controlled patterns. It is this more limited topic that we discuss in this section, focusing on the organization and maintenance of the tissues of adult vertebrates.

Tissues Are Organized Mixtures of Many Cell Types

Although the specialized tissues in our body differ in many ways, they all have certain basic requirements, usually provided for by a mixture of cell types, as illustrated for the skin in **Figure 20–34**. As discussed earlier, all tissues need mechanical strength, which is often supplied by a supporting framework of connective tissue inhabited by fibroblasts. In this connective tissue, blood vessels lined with endothelial cells satisfy the need for oxygen, nutrients, and waste disposal. Likewise, most tissues

Figure 20–34 Mammalian skin is made of a mixture of cell types. Schematic diagrams showing the cellular architecture of the main layers of thick skin. Skin can be viewed as a large organ composed of two main tissues: epithelial tissue (the *epidermis*) on the outside, and connective tissue on the inside. The outermost layer of the epidermis consists of flat dead cells, whose intracellular organelles have disappeared (see Figure 20–37). The connective tissue consists of the tough *dermis* (from which leather is made) and the underlying fatty *hypodermis*. The dermis and hypodermis are richly supplied with blood vessels and nerves; some of the nerves extend into the epidermis, as shown.

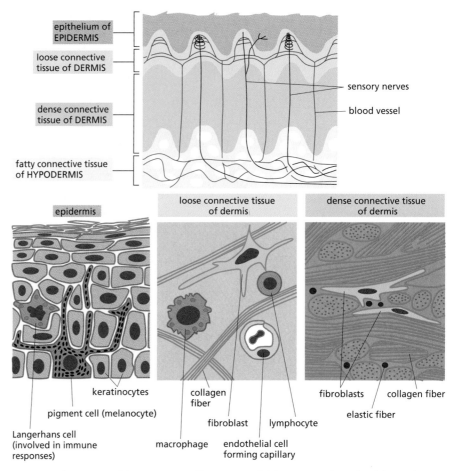

are innervated by nerve-cell axons, which are ensheathed by Schwann cells that can wrap around the axons to provide electrical insulation. Macrophages dispose of dead and damaged cells and other unwanted debris, and lymphocytes and other white blood cells combat infection. Most of these cell types originate outside the tissue and invade it, either early in the course of its development (endothelial cells, nerve-cell axons, and Schwann cells) or continually during life (macrophages and other white blood cells).

A similar supporting apparatus is required to maintain the principal specialized cells of many tissues: the contractile cells of muscle, the secretory cells of glands, or the blood-forming cells of bone marrow, for example. Almost every tissue is therefore an intricate mixture of many cell types that must remain different from one another while coexisting in the same environment. Moreover, in almost all adult tissues, cells are continually dying and being replaced; throughout this hurly-burly of cell replacement and tissue renewal, the organization of the tissue must be preserved.

Three main factors contribute to this stability.

1. *Cell communication*: each type of specialized cell continually monitors its environment for signals from other cells and adjusts its behavior accordingly; in fact, the very survival of most cells depends on such social signals (discussed in Chapter 16). This communication ensures that new cells are produced and survive only when and where they are required.

2. *Selective cell adhesion*: because different cell types have different cadherins and other cell adhesion molecules in their plasma membrane, they tend to stick selectively, by homophilic binding, to other cells of the same type. They may also form selective attachments to certain other cell types and to specific extracellular matrix components. The selectivity of these cell adhesions prevents the different cell types in a tissue from becoming chaotically mixed.

3. *Cell memory*: as discussed in Chapter 8, specialized patterns of gene expression, evoked by signals that acted during embryonic development, are afterward stably maintained, so that cells autonomously preserve their distinctive character and pass it on to their progeny. A fibroblast divides to produce more fibroblasts, an endothelial cell divides to produce more endothelial cells, and so on.

Different Tissues Are Renewed at Different Rates

Tissues vary enormously in their rate and pattern of *cell turnover*. At one extreme is nervous tissue, in which most of the nerve cells last a lifetime without replacement. At the other extreme is the intestinal epithelium, in which cells are replaced every three to six days. Between these extremes there is a spectrum of different rates and styles of cell turnover and tissue renewal. Bone (see Figure 20–8) has a turnover time of about ten years in humans, involving renewal of the matrix as well as of cells: old bone matrix is slowly eaten away by a set of cells called *osteoclasts*, akin to macrophages, while new matrix is deposited by another set of cells, *osteoblasts*, akin to fibroblasts. New red blood cells in humans are generated continually by blood-forming precursor cells in the bone marrow; they are released into the bloodstream, where they recirculate continually for about 120 days before being removed and destroyed in the liver and spleen. In the skin, the outer layers of the epidermis are continually flaking off and being replaced from below, so that the epidermis is renewed with a turnover time of about two months. And so on.

Our life depends on these renewal processes. A large dose of ionizing radiation blocks cell division and thus halts renewal: within a few days, the lining of the intestine, for example, becomes denuded of cells, leading to the devastating diarrhea and water loss characteristic of acute radiation sickness.

Clearly, there have to be elaborate control mechanisms to keep cell production and cell loss in balance in the normal, healthy adult body. Cancers originate through violation of these controls, allowing cells in the self-renewing tissues to survive and proliferate to excess. To understand cancer, therefore, it is important to understand the normal social controls on cell turnover that cancer perverts.

Stem Cells Generate a Continuous Supply of Terminally Differentiated Cells

Most of the specialized, or **differentiated**, cells that need continual replacement are themselves unable to divide. Red blood cells, the epidermal cells on the skin surface, and the absorptive and goblet cells of the gut epithelium are all examples of this type. Such cells are referred to as *terminally differentiated*: they lie at the dead end of their developmental pathway.

The cells that replace the terminally differentiated cells that are lost are generated from a stock of proliferating *precursor cells*, which themselves usually derive from a much smaller number of self-renewing **stem cells**. Both stem cells and proliferating precursor cells are retained in the corresponding tissues along with the differentiated cells. Stem cells are not differentiated and can divide without limit (or at least for the lifetime of the animal). When a stem cell divides, though, each daughter has a choice: either it can remain a stem cell, or it can embark on a course leading to terminal differentiation, usually via a series of precursor cell divisions (**Figure 20–35**). The job of the stem cells and precursor cells, therefore, is not to carry out the specialized function of the differentiated cells, but

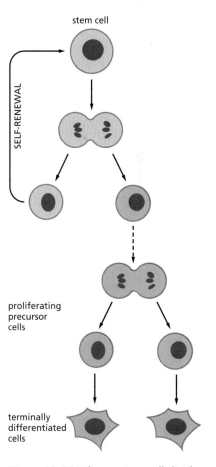

Figure 20–35 When a stem cell divides, each daughter can either remain a stem cell or go on to become terminally differentiated. The terminally differentiated cells usually develop from precursor cells that divide a limited number of times before they terminally differentiate. Stem-cell divisions can also produce two stem cells or two precursor cells, as long as the pool of stem cells is maintained.

> ### QUESTION 20–6
>
> Why does ionizing radiation stop cell division?

rather to produce cells that will. Stem cells are usually present in small numbers and often have a nondescript appearance, making them difficult to identify. Although stem cells and precursor cells are not differentiated, they are nonetheless developmentally restricted: under normal conditions, they stably express sets of transcription regulators that ensure that their differentiated progeny will be of the appropriate cell types.

The pattern of cell replacement varies from one stem-cell-based tissue to another. In the lining of the small intestine, for example, the absorptive and secretory cells are arranged as a single-layered, simple epithelium covering the surfaces of the fingerlike villi that project into the gut lumen. This epithelium is continuous with the epithelium lining the *crypts*, which descend into the underlying connective tissue (**Figure 20–36A**). The stem cells lie near the bottom of the crypts, where they give rise mostly to proliferating precursor cells, which move upward in the plane of the epithelial sheet. As they move upward, the precursor cells terminally differentiate into absorptive or secretory cells, which are shed into the gut lumen and die when they reach the tips of the villi (**Figure 20–36B**).

A contrasting example is the epidermis, a stratified epithelium. In the epidermis, proliferating stem cells and precursor cells are confined to the basal layer, adhering to the basal lamina. The differentiating cells travel

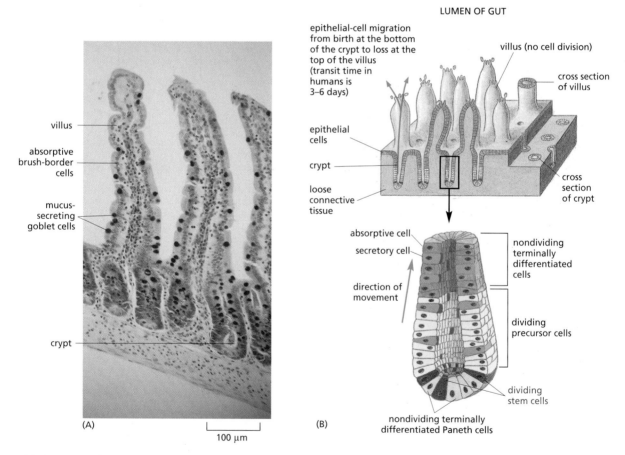

Figure 20–36 Renewal occurs continuously in the epithelial lining of the adult mammalian intestine. (A) Micrograph of a section of part of the lining of the small intestine, showing the villi and crypts. Mucus-secreting goblet cells (stained *purple*) are interspersed among the absorptive brush-border cells in the epithelium covering the villi. Smaller numbers of two other secretory cell types— enteroendocrine cells (not visible here), which secrete gut hormones, and Paneth cells, which secrete antibacterial proteins—are also present and derive from the same stem cells. (B) Cartoon showing the pattern of cell turnover and the proliferation of stem cells and precursor cells. The stem cells give rise mainly to proliferating precursor cells that slide continuously upward and terminally differentiate into secretory or absorptive cells, which are shed from the tip of the villus. The stem cells also give rise directly to terminally differentiated Paneth cells, which remain at the bottom of the crypt.

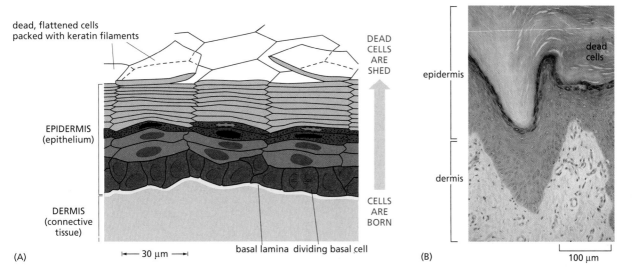

dead, flattened cells
packed with keratin filaments

DEAD
CELLS
ARE
SHED

EPIDERMIS
(epithelium)

DERMIS
(connective
tissue)

CELLS
ARE
BORN

(A)   ⊢—— 30 μm ——⊣   basal lamina dividing basal cell

epidermis

dead
cells

dermis

(B)   ⊢—— 100 μm ——⊣

Figure 20–37 The epidermis of the skin is renewed from stem cells in its basal layer. (A) The basal layer contains a mixture of stem cells and dividing precursor cells that are produced from the stem cells. On emerging from the basal layer, the precursor cells stop dividing and move outward, differentiating as they go. Eventually, the cells undergo a special form of cell death: the nucleus and other organelles disintegrate, and the cell shrinks to the form of a flattened scale, packed with keratin filaments. The scales are ultimately shed from the skin surface. (B) Light micrograph of a cross section through the sole of a human foot, stained with hematoxylin and eosin.

outward from their site of origin in a direction perpendicular to the plane of the cell sheet; terminally differentiated cells and their corpses are eventually shed from the skin surface (**Figure 20–37**).

Often, a single type of stem cell gives rise to several types of differentiated progeny: the stem cells of the intestine, for example, produce absorptive cells, goblet cells, and several other secretory cell types. The process of blood-cell formation, or *hemopoiesis*, provides an extreme example of this phenomenon. All of the different cell types in the blood—both the red blood cells that carry oxygen and the many types of white blood cells that fight infection (**Figure 20–38**)—ultimately derive from a shared *hemopoietic stem cell* found in the bone marrow (**Figure 20–39**).

Specific Signals Maintain Stem-Cell Populations

Every stem-cell system requires control mechanisms to ensure that new cells are generated in the appropriate places and in the right numbers. The controls depend on extracellular signals exchanged between the stem cells, their progeny, and other cell types in the area. These signals, and the intracellular signaling pathways they activate, fall into a surprisingly small number of families, corresponding to half-a-dozen basic signaling mechanisms, some of which are discussed in Chapter 16. These few mechanisms are used again and again—in different combinations, evoking different responses in different contexts, in both the embryo and the adult.

Almost all these signaling mechanisms contribute to the task of maintaining the complex organization of a stem-cell system such as that of the intestine. Thus, a class of signal molecules known as the **Wnt proteins** serves to promote the proliferation of the stem cells and precursor cells at the base of each intestinal crypt (**Figure 20–40**). Cells in the crypt produce, in addition, other signals that act at longer range to prevent activation of the Wnt pathway outside the crypts. They also exchange yet other signals to control their diversification, so that some differentiate into secretory cells while others become absorptive cells.

QUESTION 20–7

Why do you suppose epithelial cells lining the gut are renewed frequently, whereas most neurons last for the lifetime of the organism?

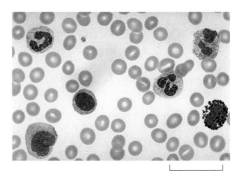

⊢—— 20 μm ——⊣

Figure 20–38 Blood contains many circulating cell types, all derived from a single type of stem cell. A sample of blood is smeared onto a glass coverslip, chemically fixed (see Panel 1–1, p.10), and stained with hematoxylin, which reacts with nucleic acids. Microscopic examination reveals numerous small erythrocytes (red blood cells), which lack DNA. Larger, purple-stained cells are different types of white blood cell: lymphocytes, eosinophils, basophils, neutrophils, and monocytes. Blood smears of this kind are routinely used as a clinical test in hospitals. (Courtesy of Peter Takizawa.)

Figure 20–39 **A hemopoietic stem cell divides to generate more stem cells, as well as precursor cells (not shown) that proliferate and differentiate into the mature blood cell types found in the circulation.** The macrophages found in many tissues of the body and the osteoclasts that eat away bone matrix originate from the same precursor cells, as do a number of other cell types not shown in this scheme. Megakaryocytes give rise to blood platelets by shedding cell fragments (Movie 20.5). A large number of extracellular signal molecules are known to act at various points in this cell lineage to control the production of each cell type and to maintain appropriate numbers of precursor cells and stem cells.

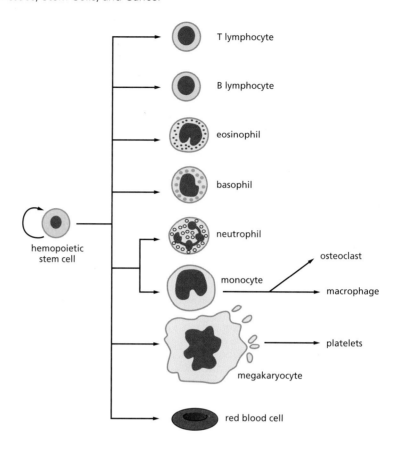

Disorders of these signaling mechanisms disrupt the structure of the gut lining. In particular, as we see later, defects in the regulation of Wnt signaling underlie the commonest forms of human intestinal cancer.

Stem Cells Can Be Used to Repair Lost or Damaged Tissues

Because stem cells can proliferate indefinitely and produce progeny that differentiate, they provide for both continual renewal of normal tissue and repair of tissue lost through injury. For example, by transfusing a few hemopoietic stem cells into a mouse whose own blood stem cells have been destroyed by irradiation, it is possible to fully repopulate the animal with new blood cells and ultimately rescue it from death by anemia, infection, or both. A similar approach is used in the treatment of human leukemia with irradiation (or cytotoxic drugs) followed by bone marrow transplantation.

Although stem cells taken directly from adult tissues such as bone marrow have already proven their clinical value, another type of stem cell, first identified through experiments in mice, may have even greater

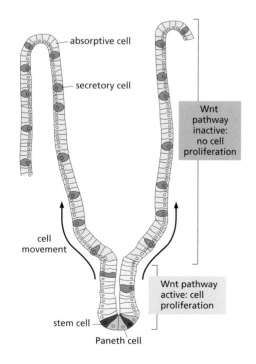

Figure 20–40 **The Wnt signaling pathway helps to control the production of differentiated cells from stem cells in the intestinal crypt.** Wnt signaling maintains proliferation in the crypt. The Wnt proteins are secreted by cells in and around the crypt base, especially the Paneth cells—a subclass of terminally differentiated secretory cells that are generated from the gut stem cells, but that move down to the crypt bottom instead of up to the tip of the villus. Paneth cells have a dual function: they secrete antimicrobial peptides to keep infection at bay, and at the same time they provide the signals to sustain the stem-cell population from which they themselves derive.

potential—both for treating and understanding human disease. It is possible, through cell culture, to derive from early mouse embryos an extraordinary class of stem cells called **embryonic stem cells**, or **ES cells**. Under appropriate conditions, these cells can be kept proliferating indefinitely in culture and yet retain unrestricted developmental potential and are thus said to be **pluripotent**: if the cells from the culture dish are put back into an early embryo, they can give rise to all the tissues and cell types in the body, including germ cells. Their descendants in the embryo are able to integrate perfectly into whatever site they come to occupy, adopting the character and behavior that normal cells would show at that site. These cells can also be induced to differentiate in culture into a large variety of cell types (**Figure 20–41**).

Cells with properties similar to those of mouse ES cells can now be derived from early human embryos, creating a potentially inexhaustible supply of cells that might be used for the replacement or repair of mature human tissues that are damaged. For example, experiments in mice suggest that it should be possible to use ES cells to replace the skeletal muscle fibers that degenerate in victims of muscular dystrophy, the nerve cells that die in patients with Parkinson's disease, the insulin-secreting cells that are destroyed in type 1 diabetics, and the cardiac muscle cells that die during a heart attack. Perhaps one day it might even become possible to grow entire organs from ES cells by a recapitulation of embryonic development (**Figure 20–42**).

There are, however, many hurdles to be cleared before such dreams can become reality. One major problem concerns immune rejection: if the transplanted cells are genetically different from the cells of the patient into whom they are grafted, they are likely to be rejected and destroyed by the immune system. Beyond the practical scientific difficulties, there have been ethical concerns about the use of human embryos and the purposes to which human ES cells might be put. One anxiety, for example, has centered on the possibility of using ES cells for human "cloning." But what exactly does this mean?

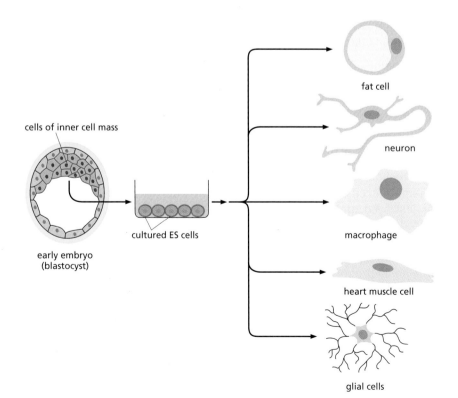

Figure 20–41 **ES cells derived from an embryo can give rise to all of the tissues and cell types of the body.** ES cells are harvested from the inner cell mass of an early embryo and can be maintained indefinitely as pluripotent stem cells in culture. If they are put back into an embryo, they will integrate perfectly and differentiate to suit whatever environment they are placed in. Alternatively, these cells can be induced to differentiate into specific cell types in culture when provided with the appropriate extracellular signal molecules (Movie 20.6). (Based on data from E. Fuchs and J.A. Segré, *Cell* 100:143–155, 2000. With permission from Elsevier.)

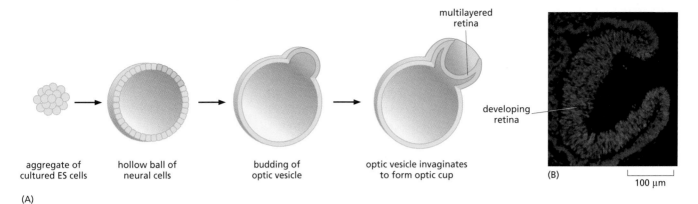

Figure 20–42 Cultured ES cells can give rise to a three-dimensional organ. (A) Remarkably, under appropriate conditions, mouse ES cells in culture can proliferate, differentiate, and interact to form a three-dimensional, eye-like structure, which includes a multilayered retina similar in organization to the one that forms *in vivo*. (B) Fluorescent micrograph of an optic cup formed by ES cells in culture. The structure includes a developing retina, containing multiple layers of neural cells, which produce a protein (*pink*) that serves as a marker for retinal tissue. (A, Adapted from M. Eiraku and Y. Sasai, *Curr. Opin. Neurobiol.* 22: 768–777, 2012; B, from M. Eiraku et al., *Nature* 472:51–56, 2011. With permission from Macmillan Publishers Ltd.)

Therapeutic Cloning and Reproductive Cloning Are Very Different Enterprises

The term "cloning" has been used in confusing ways as a shorthand term for several quite distinct types of procedure, particularly in public debates about the ethics of stem-cell research. It is important to understand the distinctions.

As biologists define the term, a clone is simply a set of cells or individuals that are essentially genetically identical, by virtue of their descent from a single ancestor cell. The simplest type of cloning is the cloning of cells in a culture dish. For instance, one can take a single epidermal stem cell from the skin and let it proliferate in culture to obtain a large clone of genetically identical epidermal cells. Such cells could be used to help reconstruct the skin of a badly burned patient. This kind of cloning is no more than an extension by artificial means of the processes of cell proliferation and differentiation that occur in a normal human body.

The cloning of entire multicellular animals, called **reproductive cloning**, is a very different enterprise, involving a far more radical departure from the ordinary course of nature. As we discuss in Chapter 19, each individual animal normally has both a mother and a father and is not genetically identical to either of them. In reproductive cloning, the need for two parents and sexual union is bypassed. For mammals, this difficult feat has been achieved in mice and sheep and a variety of other domestic animals by a technique called *nuclear transplantation*. The procedure begins with an unfertilized egg cell. The nucleus of this haploid gamete is sucked out or destroyed, and in its place a nucleus from a regular diploid cell is introduced. The diploid donor cell can, for example, be taken from a tissue of an adult individual. The hybrid cell, consisting of a diploid donor nucleus in a host egg cytoplasm, is allowed to develop for a few days in culture. In a small proportion of cases, this cell will give rise to an early embryo (a blastocyst) containing about 200 cells, which is then transferred into the uterus of a foster mother (**Figure 20–43**). If the experimenter is lucky, development continues as it would in a normal embryo, eventually giving rise to a whole new animal. An individual produced in this way, should be genetically identical to the adult individual who donated the diploid cell (except for the small amount of genetic information in mitochondria, which are inherited with the egg cytoplasm).

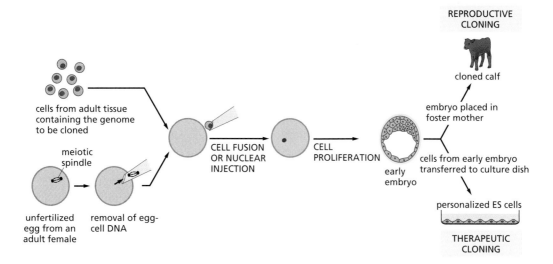

Figure 20–43 Nuclear transplantation can be used for "cloning" in two quite different senses of the word. In reproductive cloning, a whole new multicellular individual is generated; in therapeutic cloning, only cells (personalized ES cells) are produced. Both procedures begin with nuclear transplantation, in which a nucleus taken from an adult cell is transferred into the cytoplasm of an enucleated egg cell, so as to create a cell that has an embryonic character but carries the genes of the adult cell.

A different procedure, called **therapeutic cloning**, uses the technique of nuclear transplantation to produce cultured ES cells, rather than a cloned animal (see Figure 20–43). This approach is an elaborate method for generating *personalized ES cells*, with the aim of generating various cell types that can be used for tissue repair or to study disease mechanisms. Because the cells obtained are genetically almost identical to the original donor cell, they can be grafted back into the adult from whom the donor nucleus was taken, thereby minimizing immunological rejection. Nuclear transplantation, however, is technically very difficult, and it has only recently been possible to use it to produce personalized human ES cells. Moreover, the procedure requires a supply of human egg cells, which raises ethical issues. Indeed, nuclear transplantation into human egg cells is outlawed in some countries.

Induced Pluripotent Stem Cells Provide a Convenient Source of Human ES-like Cells

The problems associated with making personalized ES cells by nuclear transplantation can now be bypassed by an alternative approach, in which cells are taken from an adult tissue, grown in culture, and reprogrammed into an ES-like state by artificially driving the expression of a set of three transcription regulators called Oct3/4, Sox2, and Klf4. This treatment is sufficient to convert fibroblasts into cells with practically all the properties of ES cells, including the ability to proliferate indefinitely and differentiate in diverse ways and to contribute to any tissue (**Figure 20–44**). These ES-like cells are called **induced pluripotent stem cells** (**iPS cells**). The conversion rate is low, however—only a tiny proportion of the fibroblasts make the switch—and there are serious worries about the safety of implanting into humans derivatives of cells with such an abnormal developmental history. Much work remains to be done to allow this approach to be used to treat human diseases.

Figure 20–44 Induced pluripotent stem cells (iPS cells) can be generated by transformation of cultured cells isolated from adult tissues. In the example shown, genes that encode several transcription regulators normally expressed in ES cells are introduced into cultured fibroblasts using genetically manipulated viruses as vectors. After a few weeks in culture, a small proportion of the fibroblasts have transformed into cells that look and behave like ES cells and have the same ability as ES cells to differentiate into any of the cell types in the body.

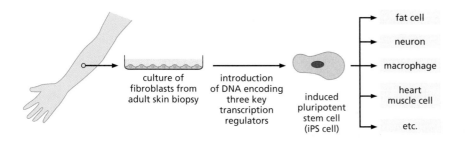

Meanwhile, however, human ES cells and especially human iPS cells are proving to be valuable in other ways. They can be used to generate large, homogeneous populations of differentiated human cells of a specific type in culture; these can be used to test for potential toxic or beneficial effects of candidate drugs on specific human cell types. Moreover, it is possible to create iPS cells containing the genomes of patients who suffer from a genetic disease, and to use these patient-specific stem cells to study the disease mechanism and to search for drugs that might be useful in the treatment of that disease. An example is Timothy syndrome, a rare genetic disease caused by mutations in a gene that encodes a specific type of Ca^{2+} channel. The defective channel fails to close properly after opening, leading to abnormalities in heart rhythm and, in some individuals, to autism. The iPS cells produced from such individuals have been coaxed to differentiate in culture into neurons and heart muscle cells, which are now being used to study the physiological consequences of the Ca^{2+} channel abnormality and to hunt for drugs that can correct the defects.

In addition, experiments on the pluripotent stem cells themselves are providing insights into some of the many unsolved mysteries of developmental and stem-cell biology, including the mechanisms that make the specialized characters of most cells in adult tissues so remarkably stable under normal circumstances.

CANCER

Humans pay a price for having tissues that can renew and repair themselves. The delicately adjusted mechanisms that control these processes can go wrong, leading to catastrophic disruption of tissue structure. Foremost among the diseases of tissue renewal is **cancer**, which stands alongside infectious illness, malnutrition, war, and heart disease as a major cause of death in human populations. In Europe and North America, for example, one in five of us will die of cancer.

Cancer arises from violations of the basic rules of social cell behavior. To make sense of the origins and progression of the disease, and to devise treatments, we have to draw upon almost every part of our knowledge of how cells work and interact in tissues. Conversely, much of what we know about cell and tissue biology has been discovered as a by-product of cancer research. In this section, we examine the causes and mechanisms of cancer, the types of cell misbehavior that contribute to its progress, and the ways in which we hope to use our understanding to defeat these misbehaving cells and, hence, the disease. Although there are many types of cancer, each with distinct properties, we will refer to them collectively by the umbrella term "cancer," as they are united by certain common principles.

Cancer Cells Proliferate, Invade, and Metastasize

As tissues grow and renew themselves, each individual cell must adjust its behavior according to the needs of the organism as a whole. The cell must divide only when new cells of that type are needed, and refrain from dividing when they are not; it must live as long as it is needed, and kill itself when it is not; it must maintain its specialized character; and it must occupy its proper place and not stray into inappropriate territories.

In a large organism, no significant harm is done if an occasional single cell misbehaves. But a potentially devastating breakdown of order occurs when a single cell suffers a genetic alteration that allows it to survive and divide when it should not, producing daughter cells that behave in the same antisocial way. Such a relentlessly expanding clone of abnormal

normal liver tissue   secondary tumors (metastases)

cancer cells   normal liver cells

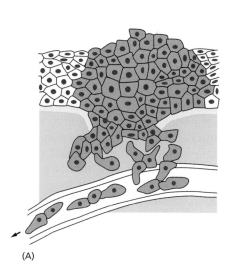

(A)

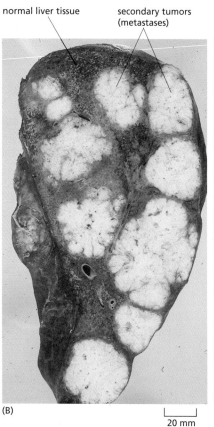

(B)   20 mm

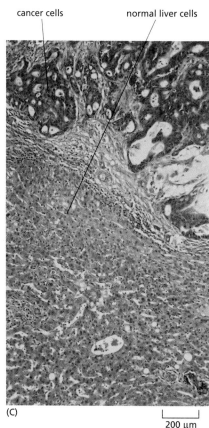

(C)   200 μm

cells can disrupt the organization of the tissue, and eventually that of the body as a whole. It is this catastrophe that happens in cancer.

Cancer cells are defined by two heritable properties: they and their progeny (1) proliferate in defiance of the normal constraints and (2) invade and colonize territories normally reserved for other cells (Movie 20.7). It is the combination of these socially deviant features that creates the lethal danger. Cells that have the first property but not the second proliferate excessively but remain clustered together in a single mass, forming a tumor. But the tumor in this case is said to be *benign*, and it can usually be removed cleanly and completely by surgery. A tumor is cancerous only if its cells have the ability to invade surrounding tissue, in which case the tumor is said to be *malignant*. Malignant tumor cells with this invasive property often break loose from the primary tumor and enter the bloodstream or lymphatic vessels, where they form secondary tumors, or **metastases**, at other sites in the body (**Figure 20–45**). The more widely the cancer spreads, the harder it is to eradicate.

Epidemiological Studies Identify Preventable Causes of Cancer

Prevention is always better than cure, but to prevent cancer we need to know what causes it. Do factors in our environment or features of our way of life trigger the disease and help it to progress? If so, what are they? Answers to these questions come mainly from *epidemiology*—the statistical analysis of human populations, looking for factors that correlate with disease incidence. This approach has provided strong evidence that the environment plays an important part in the causation of most cases of cancer. The types of cancers that are common, for example, vary from country to country, and studies of migrants show that it is usually where people live, rather than where they were born, that governs their cancer risk.

Figure 20–45 Cancers invade surrounding tissues and often metastasize to distant sites. (A) To give rise to a colony in a new site—called a secondary tumor or metastasis—the cells of a primary tumor in an epithelium must typically cross the basal lamina, migrate through connective tissue, and get into either blood or lymphatic vessels. They then have to exit from the bloodstream or lymph and settle, survive, and proliferate in a new location. (B) Secondary tumors in a human liver, originating from a primary tumor in the colon. (C) Higher-magnification view of one of the secondary tumors, stained differently to show the contrast between the normal liver cells and the cancer cells. (B and C, courtesy of Peter Isaacson.)

Although it is still hard to discover which specific factors in the environment or lifestyle are significant, and many remain unknown, some have been precisely identified. For example, it was noted long ago that cervical cancer, which arises in the epithelium lining the cervix (neck) of the uterus, was much more common in women who were sexually experienced than in those who were not, suggesting a cause related to sexual activity. We now know, through modern epidemiological studies, that most cases of cervical cancer depend on infection of the cervical epithelium with certain subtypes of a common virus, called *human papillomavirus*. This virus is transmitted through sexual intercourse and can sometimes, if one is unlucky, provoke uncontrolled proliferation of the infected cells. Knowing this, we can attempt to prevent the cancer by preventing the infection—for example, by vaccination against papillomavirus. Such a vaccine is now available, conferring a high level of protection if given to young people before they become sexually active.

In the great majority of human cancers, however, viruses do not appear to play a part: as we will see, cancer is not an infectious disease. But epidemiology reveals that other factors increase the risk of cancer. Obesity is one such factor. Smoking tobacco is another: tobacco smoke is not only responsible for almost all cases of lung cancer, but it also raises the incidence of several other cancers, such as those of the bladder. By stopping the use of tobacco, we could prevent about 30% of all cancer deaths. No other single policy or treatment is known that would have such a dramatic impact on the cancer death rate.

As we will explain, although environmental factors affect the incidence of cancer and are critical for some forms of the disease, it would be wrong to conclude that they are the only cause of cancers. No matter how hard we try to prevent cancer by healthy living, we will never be able to eradicate this disease. To devise effective treatments, we need to derive a deep understanding of the biology of cancer cells and the mechanisms that underlie the growth and spread of tumors.

Cancers Develop by an Accumulation of Mutations

Cancer is fundamentally a genetic disease: it arises as a consequence of pathological changes in the information carried by DNA. It differs from other genetic diseases in that the mutations underlying cancer are mainly somatic mutations—those that occur in individual somatic cells of the body—as opposed to germ-line mutations, which are handed down via the germ cells from which the entire multicellular organism develops.

Most of the identified agents known to contribute to the causation of cancer, including ionizing radiation and most chemical carcinogens, are mutagens: they cause changes in the nucleotide sequence of DNA. But even in an environment that is free of tobacco smoke, radioactivity, and all the other external mutagens that worry us, mutations will occur spontaneously as a result of fundamental limitations on the accuracy of DNA replication and DNA repair (discussed in Chapter 6). In fact, environmental carcinogens other than tobacco smoke probably account for only a small fraction of the mutations responsible for cancer, and elimination of all these external risk factors would still leave us prone to the disease.

Although DNA is replicated and repaired with great accuracy, an average of one mistake slips by for every 10^9 or 10^{10} nucleotides copied, as we discuss in Chapter 6. This means that spontaneous mutations occur at an estimated rate of about 10^{-6} or 10^{-7} mutations per gene per cell division, even without encouragement by external mutagens. About 10^{16} cell divisions take place in a human body in the course of an average lifetime; thus, every single gene is likely to have acquired a mutation on more than 10^9 separate occasions in any individual. From this point of view, the problem of cancer seems to be not why it occurs, but why it occurs so infrequently.

The explanation is that it takes more than a single mutation to turn a normal cell into a cancer cell. Precisely how many are required is still a matter of debate, but for most full blown cancers it could be at least 10—and, as we will see, they have to affect the right type of gene. These mutations do not all occur at once, but sequentially, usually over a period of many years.

Cancer, therefore, is most often a disease of old age, because it takes a long time for an individual clone of cells—those derived from a common founder—to accumulate a large number of mutations (see Figure 6–32). In fact, most human cancer cells not only contain many mutations, but they are also genetically unstable. This **genetic instability** results from mutations that interfere with the accurate replication and maintenance of the genome and thereby increase the mutation rate itself. Sometimes, the increased mutation rate may result from a defect in one of the many proteins needed to repair damaged DNA or to correct errors in DNA replication. Sometimes, there may be a defect in the cell-cycle checkpoint mechanisms that normally prevent a cell with damaged DNA from attempting to divide before it has completed the repair (discussed in Chapter 18). Sometimes, there may be a fault in the machinery of mitosis, which can lead to chromosomal damage, loss, or gain. These potential sources of genetic instability are summarized in **Table 20–1**.

Genetic instability can generate extra chromosomes, as well as chromosome breaks and rearrangements—gross abnormalities that can be seen in a karyotype (**Figure 20–46**). It can also help drive the evolution of cancer, as we now discuss.

Cancer Cells Evolve, Giving Them an Increasingly Competitive Advantage

The mutations that lead to cancer do not cripple the mutant cells. On the contrary, they give these cells a competitive advantage over their neighbors. It is this advantage enjoyed by the mutant cells that leads to disaster for the organism as a whole. As an initial population of mutant cells grows, it slowly evolves: new chance mutations occur, some of which are favored by natural selection because they enhance cell proliferation and

TABLE 20–1 A VARIETY OF FACTORS CAN CONTRIBUTE TO GENETIC INSTABILITY
Defects in DNA replication
Defects in DNA repair
Defects in cell-cycle checkpoint mechanisms
Mistakes in mitosis
Abnormal chromosome numbers

(A)

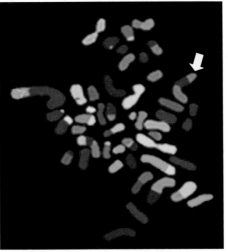

(B)

Figure 20–46 Cancer cells often have highly abnormal chromosomes, reflecting genetic instability. In the example shown here, chromosomes were prepared from a breast cancer cell in metaphase, spread on a glass slide, and stained with (A) a general DNA stain or (B) a combination of fluorescent stains that give a different color for each human chromosome. The staining (displayed in false color) shows multiple translocations, including one chromosome (*white* arrow) that has undergone two translocations, so that it is now made up of two pieces of chromosome 8 (*olive*) and a piece of chromosome 17 (*purple*). The karyotype also contains 48 chromosomes, instead of the normal 46. Such abnormalities in chromosome number can further cause chromosome-segregation errors when the cell divides, so that the degree of genetic disruption goes from bad to worse (see Table 20–1). (Courtesy of Joanne Davidson and Paul Edwards.)

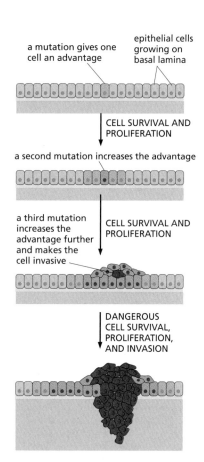

a mutation gives one cell an advantage

epithelial cells growing on basal lamina

CELL SURVIVAL AND PROLIFERATION

a second mutation increases the advantage

a third mutation increases the advantage further and makes the cell invasive

CELL SURVIVAL AND PROLIFERATION

DANGEROUS CELL SURVIVAL, PROLIFERATION, AND INVASION

Figure 20–47 Tumors evolve by repeated rounds of mutation, proliferation, and natural selection. The final outcome is a fully malignant tumor. At each step, a single cell undergoes a mutation that enhances its ability to proliferate, or survive, or both, so that its progeny become a dominant clone in the tumor. Proliferation of this clone then hastens occurrence of the next step of tumor progression by increasing the size of the cell population at risk of undergoing an additional mutation. Some cancers contain multiple malignant clones, each with its own collection of mutations, in addition to a common set of mutations that reflect the tumor's origin from a founding mutant cell (not shown).

cell survival. This process of random mutation followed by selection culminates in the genesis of cancer cells that run riot within the population of cells that form the body, upsetting its regular structure (**Figure 20–47**).

Non-mutagenic environmental or lifestyle factors such as obesity may favor the development of cancer by altering the selection pressures that operate in tissues. A glut of circulating nutrients, or abnormal increases in hormones, mitogens, or growth factors, for example, may help cells with dangerous mutations survive, grow, and proliferate. Eventually, cells emerge that have all the abnormalities required for full-blown cancer.

To be successful, a cancer cell must acquire a whole range of abnormal properties—a collection of subversive behaviors. A proliferating precursor cell in the epithelial lining of the gut, for example, must undergo changes that permit it to carry on dividing when it would normally stop (see Figure 20–36). That cell and its progeny must also be able to avoid cell death, displace their normal neighbors, and attract a blood supply to nourish continued tumor growth. For the tumor cells to then become invasive, they must be able to detach from the epithelial sheet and digest their way through the basal lamina into the underlying connective tissue. To spread to other organs and form *metastases*, they must be able to get in, and then out, of blood or lymph vessels and settle, survive, and proliferate in new sites (see Figure 20–45).

Different cancers require different combinations of properties. Nevertheless, we can draw up a general list of characteristics that distinguish cancer cells from normal cells.

1. Cancer cells have a reduced dependence on signals from other cells for their survival, growth, and division. Often, this is because they contain mutations in components of the cell signaling pathways that normally respond to such stimuli. An activating mutation in a *Ras* gene (discussed in Chapter 16), for example, can cause an intracellular signal for proliferation even in the absence of the extracellular cue that would normally be needed to turn Ras on, like a faulty doorbell that rings even when nobody is pressing the button.

2. Cancer cells can survive levels of stress and internal derangement that would cause normal cells to kill themselves by apoptosis. This avoidance of cell suicide is often the result of mutations in genes that regulate the intracellular death program responsible for apoptosis (discussed in Chapter 18). For example, about 50% of all human cancers have an inactivating mutation in the *p53* gene. The p53 protein normally acts as part of a DNA damage response that causes cells with DNA damage to either cease dividing (see Figure 18–15) or die by apoptosis. Chromosome breakage, for example, if not repaired, will generally cause a cell to commit suicide; but if the cell is defective in p53, it may survive and divide, creating highly abnormal daughter cells that have the potential for further mischief.

3. Unlike most normal human cells, cancer cells can often proliferate indefinitely. Most normal human somatic cells will only divide a limited number of times in culture, after which they permanently stop; this is at least partly because they have lost the ability to produce the enzyme *telomerase*, so the telomeres at the ends of their chromosomes become progressively shorter with each cell division (see page 210). Cancer cells typically break through this proliferation barrier by reactivating production of telomerase, enabling them to maintain telomere length indefinitely.

4. Most cancer cells are genetically unstable, with a greatly increased mutation rate and an abnormal number of chromosomes.

5. Cancer cells are abnormally invasive, at least partly because they often lack certain cell adhesion molecules, such as cadherins, that help hold normal cells in their proper place.

6. Cancer cells have an abnormal metabolism that makes them avid for nutrients, which they use to fuel their biosynthesis and growth, rather than for energy generation by oxidative phosphorylation.

7. Cancer cells can survive and proliferate in abnormal locations, whereas most normal cells die when misplaced. This colonization of unfamiliar territory may result from the ability of cancer cells to produce their own extracellular survival signals and to suppress their apoptosis program (as described in #2, above).

To understand the molecular biology of cancer, we have to identify the mutations responsible for these abnormal properties.

Two Main Classes of Genes Are Critical for Cancer: Oncogenes and Tumor Suppressor Genes

Investigators have made use of a variety of approaches to track down the genes and mutations that are critical for cancer—from studying viruses that cause cancer in chickens to following families in which a particular cancer occurs unusually often. Though many of the most important of these genes have been identified, the hunt for others continues.

For many cancer-critical genes, the dangerous mutations are ones that render the encoded protein hyperactive. These *gain-of-function mutations* have a dominant effect: only one gene copy needs to be mutated to cause trouble. The resulting mutant gene is called an **oncogene**, and the corresponding normal form of the gene is called a **proto-oncogene (Figure 20–48A)**. **Figure 20–49** shows a variety of ways in which a proto-oncogene can be converted into its corresponding oncogene.

For other genes, the danger lies in mutations that destroy their activity. These *loss-of-function mutations* are generally recessive: both copies of the gene must be lost or inactivated before an effect is seen; the normal

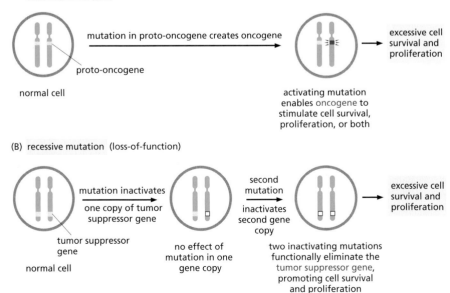

(A) dominant mutation (gain-of-function)

normal cell — proto-oncogene — mutation in proto-oncogene creates oncogene — activating mutation enables oncogene to stimulate cell survival, proliferation, or both → excessive cell survival and proliferation

(B) recessive mutation (loss-of-function)

normal cell — tumor suppressor gene — mutation inactivates one copy of tumor suppressor gene — no effect of mutation in one gene copy — second mutation inactivates second gene copy — two inactivating mutations functionally eliminate the tumor suppressor gene, promoting cell survival and proliferation → excessive cell survival and proliferation

Figure 20–48 Genes that are critical for cancer are classified as proto-oncogenes or tumor suppressor genes, according to whether the dangerous mutations are dominant or recessive. (A) Oncogenes act in a dominant manner: a gain-of-function mutation in a single copy of the proto-oncogene can drive a cell toward cancer. (B) Loss-of-function mutations in tumor suppressor genes generally act in a recessive manner: the function of both copies of the gene must be lost to drive a cell toward cancer. In this diagram, normal genes are represented by *light blue* squares, activating mutations by *red* rays, and inactivating mutations by hollow *red* rectangles.

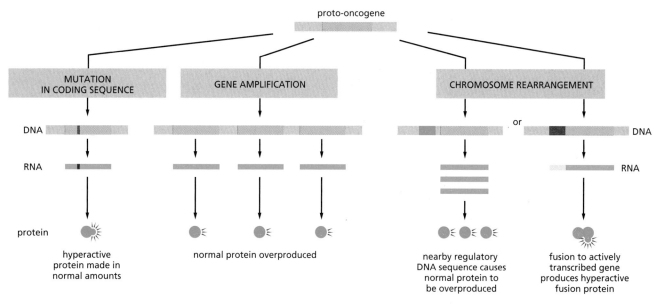

Figure 20–49 Several kinds of genetic change can convert a proto-oncogene into an oncogene. In each case, the change leads to an increase in the gene's function—that is, it is a gain-of-function mutation.

gene is called a **tumor suppressor gene** (Figure 20–48B). In addition to such genetic alterations, tumor suppressor genes can also be silenced by *epigenetic changes*, which alter gene expression without changing the gene's nucleotide sequence (as discussed in Chapter 8). Epigenetic changes are thought to silence some tumor suppressor genes in most human cancers. **Figure 20–50** highlights a few of the ways in which the activity of a tumor suppressor gene can be lost.

The variety of proto-oncogenes and tumor suppressor genes code for proteins of many different types, corresponding to the many kinds of misbehavior that cancer cells display. Some of these proteins are involved in signaling pathways that regulate cell survival, cell growth, or cell division. Others take part in DNA repair, mediate the DNA damage response, modify chromatin, or help regulate the cell cycle or apoptosis. Still others (such as cadherins) are involved in cell adhesion or other properties critical for metastasis, or have roles that we do not yet properly understand.

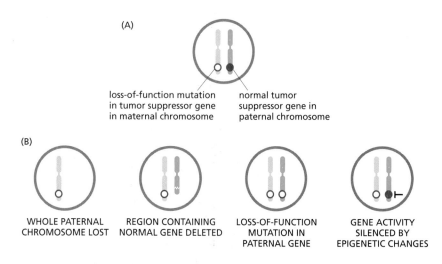

Figure 20–50 Several kinds of genetic events can eliminate the activity of a tumor suppressor gene. Note that both copies of such a gene must be lost to eliminate its function. (A) A cell in which the maternal copy of the suppressor gene is inactive because of loss-of-function mutation. (B) The same cell in which the paternal copy of the gene is inactivated in different ways, as shown.

Cancer-causing Mutations Cluster in a Few Fundamental Pathways

From the point of view of a cancer cell, oncogenes and tumor suppressor genes—and the mutations that affect them—are flip sides of the same coin. Activation of an oncogene and inactivation of a tumor suppressor gene can both promote the development of cancer. And both types of mutations are called into play in most cancers. In classifying cancer-critical genes, it seems that the type of mutation—gain-of-function or loss-of-function—matters less than the pathway in which it acts.

Rapid, low-cost DNA sequencing is now providing an unprecedented amount of information about the mutations that drive a variety of cancers. We can now compare the complete genome sequences of the cancer cells from a patient's tumor to the genome sequence of the noncancerous cells in the same individual—or of cancer cells that have spread to another location in the body. By putting together such data from many different patients, we can begin to draw up exhaustive lists of the genes that are critical for specific classes of cancer; and by analysis of data from a single patient, we can deduce the "family tree" of his or her cancer cells, showing how the progeny of the original founder cell have evolved and diversified as they multiplied and metastasized to different sites.

One remarkable finding has been that many of the genes mutated in individual tumors fall into a small number of key regulatory pathways: those that govern the initiation of cell proliferation, control cell growth, and regulate the cell's response to DNA damage and stress. For example, in almost every case of glioblastoma—the most common type of human brain tumor—mutations disrupt all three of these fundamental pathways, and the same pathways are subverted, in one way or another, in almost all human cancers (**Figure 20–51**). In any given patient, only a single gene tends to be mutated in each pathway, but not always the same gene: it is the under- or overactivity of the pathway that matters for cancer development, not the way in which this malfunction is achieved. Because the same three fundamental control systems are subverted in so wide a variety of cancers, it seems that their misregulation must be key to most cancers' success.

Colorectal Cancer Illustrates How Loss of a Tumor Suppressor Gene Can Lead to Cancer

Colorectal cancer provides one well-studied example of how a tumor suppressor can be identified and its role in tumor growth determined. Colorectal cancer arises from the epithelium lining the colon and rectum; most cases are seen in old people and do not have any discernible hereditary cause. A small proportion of cases, however, occur in families that are exceptionally prone to the disease and show an unusually early onset. In one set of such "predisposed" families, the affected individuals develop colorectal cancer in early adult life, and the onset of their disease is foreshadowed by the development of hundreds or thousands of little tumors, called polyps, in the epithelial lining of the colon and rectum.

By studying these families, investigators traced the development of the polyps to a deletion or inactivation of a tumor suppressor gene called *APC*—for *Adenomatous Polyposis Coli*. (Note that the protein encoded by this gene is different from the anaphase-promoting complex, also abbreviated APC, discussed in Chapter 18.) Affected individuals inherit one mutant copy of the gene and one normal copy. Although one normal gene copy is enough for normal cell behavior, all the cells of these individuals are only one mutational step away from total loss of the gene's function (as compared to two steps away for a person who inherits two

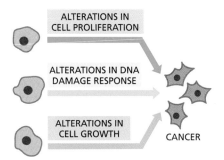

Figure 20–51 Three key regulatory pathways are perturbed in almost all human cancers. These pathways regulate cell proliferation, cell growth, and the cell's response to DNA damage or stress.

normal copies of the gene). The individual tumors arise from cells that have undergone a somatic mutation that inactivates the remaining good copy. Because the number of new mutations required is smaller, the disease strikes these individuals at an earlier age.

But what about the great majority of colorectal cancer patients, who have inherited two good copies of *APC* and do not have the hereditary condition or any significant family history of cancer? When their tumors are analyzed, it turns out that in more than 60% of cases, although both copies of *APC* are present in the adjacent normal tissue, the tumor cells themselves have lost or inactivated both copies of this gene, presumably through two independent somatic mutations.

All these findings clearly identify *APC* as a tumor suppressor gene and, knowing its sequence and mutant phenotype, one can begin to decipher how its loss helps to initiate the development of cancer. As explained in **How We Know** (pp. 722–723), the *APC* gene was found to encode an inhibitory protein that normally restricts the activation of the Wnt signaling pathway, which is involved in stimulating cell proliferation in the crypts of the gut lining, as described earlier (see Figure 20–40). When *APC* is lost, the pathway is hyperactive and epithelial cells proliferate to excess, generating a polyp (**Figure 20–52**). Within this growing mass of tissue, further mutations occur, sometimes resulting in invasive cancer (**Figure 20–53**).

An Understanding of Cancer Cell Biology Opens the Way to New Treatments

The better we understand the tricks that cancer cells use to survive, proliferate, and spread, the better are our chances of finding ways to defeat them. The task is made more challenging because cancer cells are highly mutable and, like weeds or parasites, rapidly evolve resistance to treatments used to exterminate them. Moreover, because mutations arise randomly, every case of cancer is likely to have its own unique combination of genes mutated. Even within an individual patient, tumor cells do not all contain the same genetic lesions. Thus, no single treatment is likely to work in every patient, or even for every cancer cell within the same patient. And the fact that cancers generally are not detected until the primary tumor has reached a diameter of 1 cm or more—by

Figure 20–52 Colorectal cancer often begins with loss of the tumor suppressor gene APC, leading to growth of a polyp. (A) Thousands of small polyps, and a few much larger ones, are seen in the lining of the colon of a patient with an inherited *APC* mutation (whereas individuals without an *APC* mutation might have one or two polyps). Through further mutations, some of these polyps will progress to become invasive cancers, unless the tissue is removed surgically. (B) Cross section of one such polyp; note the excessive quantities of deeply infolded epithelium, corresponding to crypts full of abnormal, proliferating cells. (A, courtesy of John Northover and Cancer Research UK; B, courtesy of Anne Campbell.)

(A)

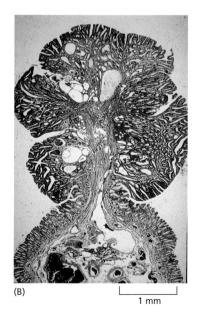
(B)

1 mm

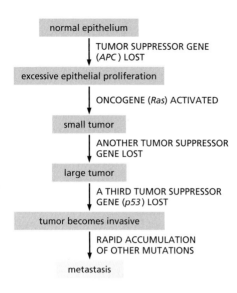

Figure 20–53 A polyp in the epithelial lining of the colon or rectum, caused by loss of the *APC* gene, can progress to cancer by accumulation of further mutations. The diagram shows a sequence of mutations that might underlie a typical case of colorectal cancer. After the initial mutation, all subsequent mutations occur randomly in a single cell that had already acquired the previous mutations. A sequence of events such as that shown here would usually be spread over 10 to 20 years or more. Though most colorectal cancers are thought to begin with loss of the *APC* tumor suppressor gene, the subsequent sequence of mutations is quite variable; indeed, many polyps never progress to cancer.

which time it consists of hundreds of millions of cells that are already genetically diverse and often have already begun to metastasize (**Figure 20–54**)—makes treatment even harder still.

Yet, in spite of these difficulties, an increasing number of cancers can be treated effectively. Surgery remains a highly effective tactic, and surgical techniques are continually improving: in many cases, if a cancer has not spread far, it can often be cured by simply cutting it out. Where surgery fails, therapies based on the intrinsic peculiarities of cancer cells can be used. Lack of normal cell-cycle control mechanisms, for example, may help make cancer cells particularly vulnerable to DNA damage: whereas a normal cell will halt its proliferation until such damage is repaired, a cancer cell may charge ahead regardless, producing daughter cells that may die because they inherit too many unrepaired breakages in their chromosomes. Presumably for this reason, cancer cells can often be killed by doses of radiotherapy or DNA-damaging chemotherapy that leave normal cells relatively unharmed.

Surgery, radiation, and chemotherapy are long-established treatments, but many novel approaches are also being enthusiastically pursued. In some cases, as with loss of a normal response to DNA damage, the very feature that helps to make the cancer cell dangerous also makes it vulnerable, enabling doctors to kill it with a properly targeted treatment. Some cancers of the breast and ovary, for example, owe their genetic instability to the lack of a protein (Brca1 or Brca2) needed for accurate repair of double-strand breaks in DNA (discussed in Chapter 6); the cancer cells survive by relying on alternative types of DNA repair mechanisms. Drugs that inhibit one of these alternative DNA repair mechanisms kill the cancer cells by raising their genetic instability to such a level that the cells die from chromosome fragmentation when they attempt to divide. Normal cells, which have an intact double-strand break repair mechanism, are relatively unaffected, and the drugs seem to have few side effects.

Another set of strategies aims to use the immune system to kill the tumor cells, taking advantage of tumor-specific cell-surface molecules to target the attack. Antibodies that recognize these tumor molecules can be produced *in vitro* and injected into the patient to mark the tumor cells for destruction. Other antibodies, aimed at the immune cells, can promote the elimination of cancer cells by neutralizing the inhibitory cell-surface molecules that keep the immune system's killer cells in check. The latter antibodies have been remarkably effective in clinical trials and, in principle, should be useful for treating a variety of different cancers.

In some cancers, it is becoming possible to target the products of specific oncogenes directly so as to block their action, causing the cancer cells to die. In chronic myeloid leukemia (CML), the misbehavior of the cancer cells depends on a mutant intracellular signaling protein (a tyrosine kinase) that causes the cells to proliferate when they should not. A small drug molecule, called imatinib (trade name Gleevec), blocks the activity

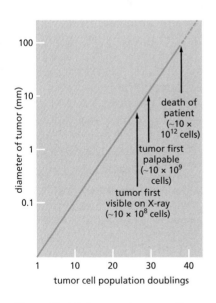

Figure 20–54 A tumor is generally not diagnosed until it has grown to contain hundreds of millions of cells. Here, the growth of a typical tumor is plotted on a logarithmic scale. Years may elapse before the tumor becomes noticeable. The doubling time of a typical breast tumor, for example, is about 100 days.

MAKING SENSE OF THE GENES THAT ARE CRITICAL FOR CANCER

The search for genes that are critical for cancer sometimes begins with a family that shows an inherited predisposition to a particular form of the disease. *APC*—a tumor suppressor gene that is frequently deleted or inactivated in colorectal cancer—was tracked down by searching for genetic defects in such families prone to the disease. But identifying the gene is only half the battle. The next step is determining what the normal gene does in a normal cell—and why alterations in the gene promote cancer.

Guilt by association

Determining what a gene—or its encoded product—does inside a cell is not a simple task. Imagine isolating an uncharacterized protein and being told that it acts as a protein kinase. That information does not reveal how the protein functions in the context of a living cell. What proteins does the kinase phosphorylate? In which tissues is it active? What role does it have in the growth, development, or physiology of the organism? A great deal of additional information is required to understand the biological context in which the kinase acts.

Most proteins do not function in isolation: they interact with other proteins in the cell. Thus one way to begin to decipher a protein's biological role is to identify its binding partners. If an uncharacterized protein interacts with a protein whose role in the cell is understood, the function of the unknown protein is likely to be in some way related. The simplest method for identifying proteins that bind to one another tightly is co-immunoprecipitation (see Panel 4–3, pp. 164–165). In this technique, an antibody is used to capture and precipitate a specific target protein from an extract prepared by breaking open cells; if this target protein is associated tightly with another protein, the partner protein will precipitate as well. This is the approach that was taken to characterize the Adenomatous Polyposis Coli gene product, APC.

Two groups of researchers used antibodies against APC to isolate the protein from extracts prepared from cultured human cells. The antibodies captured APC along with a second protein. When the researchers examined the amino acid sequence of this partner, they recognized the protein as β-catenin.

The discovery that APC interacts with β-catenin initially led to some wrong guesses about the role of APC in colorectal cancer. In mammals, β-catenin was known primarily for its role at adherens junctions, where it serves as a linker to connect membrane-spanning cadherin proteins to the intracellular actin cytoskeleton (see, for example, Figure 20–24). Thus, for some time, scientists thought that APC might be involved in cell adhesion. But within a few years, it emerged that β-catenin also has another completely different function. It is this unexpected function that turned out to be the one that is relevant for understanding APC's role in cancer.

Wingless flies

Not long before the discovery that APC binds to β-catenin, developmental biologists working on the fruit fly *Drosophila* had noticed that the human β-catenin protein is very similar in amino acid sequence to a *Drosophila* protein called Armadillo. Armadillo was known to be a key protein in a signaling pathway that has an important role in normal development in flies. The pathway is activated by the Wnt family of extracellular signal proteins, the founding member of which was called *Wingless*, after its mutant phenotype in flies. Wnt proteins bind to receptors on the surface of a cell, switching on an intracellular signaling pathway that ultimately leads to the activation of a set of genes that influence cell growth, division, and differentiation. Mutations in any of the proteins in this pathway lead to developmental errors that disrupt the basic body plan of the fly. The least devastating mutations cause flies to develop without wings; most mutations, however, result in the death of the embryo. In either case, the damage is done through effects on gene expression. This strongly suggested that Armadillo, and hence its vertebrate homolog β-catenin, were not just involved in cell adhesion, but somehow mediated the control of gene expression through the Wnt signaling pathway.

Although the Wnt pathway was discovered and studied intensively in fruit flies, it was later found to control many aspects of development in vertebrates, including mice and humans. Indeed, some of the proteins in the Wnt pathway function almost interchangeably in *Drosophila* and vertebrates. The direct link between β-catenin and gene expression became clear from work in mammalian cells. Just as APC could be used as "bait" to catch its partner β-catenin by immunoprecipitation, so β-catenin could be used as bait to catch the next protein in the signaling pathway. This was found to be a transcription regulator called LEF-1/TCF, or TCF for short. It too was found to have a *Drosophila* counterpart in the Wnt pathway, and a combination of *Drosophila* genetics and mammalian cell biology revealed how the gene control mechanism works.

Wnt transmits its signal by promoting the accumulation of "free" β-catenin (or, in flies, Armadillo)—that is, of β-catenin that is not locked up in cell junctions. This free protein migrates from the cytoplasm into the nucleus. There it binds to the TCF transcription regulator, creating a complex that activates transcription of various

Wnt-responsive genes, including genes whose products stimulate cell proliferation (**Figure 20–55**).

It turns out that APC regulates the activity of this pathway by facilitating degradation of β-catenin and thereby preventing it from activating TCF in cells where no Wnt signal has been received (see Figure 20–55A). Loss of APC allows the concentration of β-catenin to rise, so that TCF is activated and Wnt-responsive genes are turned on even in the absence of a Wnt signal. But how does this promote the development of colorectal cancer? To find out, researchers turned to mice that lack TCF4, a member of the TCF gene family that is specifically expressed in the gut epithelial lining.

Tales from the crypt

Although it may seem counterintuitive, one of the most direct ways of finding out what a gene normally does is to see what happens to the organism when that gene is missing. If one can pinpoint the processes that are disrupted or compromised, one can begin to decipher the gene's function.

With this in mind, researchers generated "knockout" mice in which the gene encoding TCF4 was disrupted. The mutation is lethal: mice lacking TCF4 die shortly after birth. But the animals showed an interesting abnormality in their intestines. The intestinal crypts, which contain the stem cells responsible for the renewal of the gut lining (see Figure 20–36), completely failed to develop. The researchers concluded that TCF4 is normally required for maintaining the pool of proliferating gut stem cells.

When APC is missing, we see the other side of the coin: without APC to promote its degradation, β-catenin accumulates in excessive quantities, binds to the TCF4 transcription regulator, and thereby overactivates the TCF4-responsive genes. This drives the formation of polyps by promoting the inappropriate proliferation of gut stem cells. Differentiated progeny cells continue to be produced and discarded into the gut lumen, but the crypt cell population grows too fast for this disposal mechanism to keep pace. The result is crypt enlargement and a steady increase in the number of crypts. The growing mass of tissue bulges out into the gut lumen as a polyp (see Figure 20–52 and Movie 20.8). A number of additional mutations are needed, however, to convert this primary tumor into an invasive cancer.

More than 60% of human colorectal tumors harbor mutations in the *APC* gene. Interestingly, among the minority class of tumors that retain functional APC, about a quarter have activating mutations in β-catenin instead. These mutations tend to make the β-catenin protein more resistant to degradation and thus produce the same effect as loss of APC. In fact, mutations that enhance the activity of β-catenin have been found in a wide variety of other tumor types, including melanomas, stomach cancers, and liver cancers. Thus, the genes that encode proteins that act in the Wnt signaling pathway provide multiple targets for mutations that can spur the development of cancer.

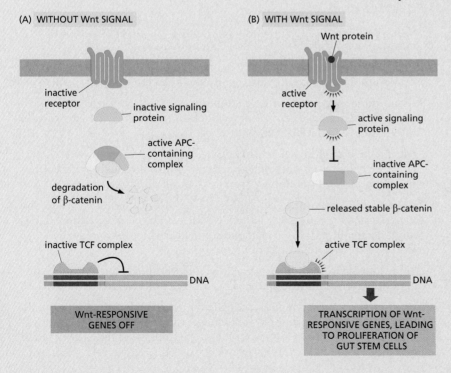

(A) WITHOUT Wnt SIGNAL

inactive receptor

inactive signaling protein

active APC-containing complex

degradation of β-catenin

inactive TCF complex

DNA

Wnt-RESPONSIVE GENES OFF

(B) WITH Wnt SIGNAL

Wnt protein

active receptor

active signaling protein

inactive APC-containing complex

released stable β-catenin

active TCF complex

DNA

TRANSCRIPTION OF Wnt-RESPONSIVE GENES, LEADING TO PROLIFERATION OF GUT STEM CELLS

Figure 20–55 The APC protein keeps the Wnt signaling pathway inactive when the cell is not exposed to Wnt protein. It does this by promoting degradation of the signaling molecule β-catenin. In the presence of Wnt, or in the absence of active APC, free β-catenin becomes plentiful and combines with the transcription regulator TCF to drive transcription of Wnt-responsive genes and, ultimately, the proliferation of stem cells in the intestinal crypt (see Figure 20–40). In the colon, mutations that inactivate APC initiate tumors by causing excessive activation of the Wnt signaling pathway.

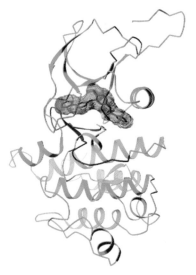

Figure 20–56 The drug Gleevec (imatinib) blocks the activity of a hyperactive oncogenic protein, thus inhibiting the growth of cancers that depend on that protein. The structure of a complex of Gleevec (solid *blue*) with the tyrosine kinase domain of the Abl protein (ribbon diagram), as determined by X-ray crystallography. (From T. Schindler et al., *Science* 289:1938–1942, 2000. With permission from AAAS.)

of this hyperactive mutant kinase (**Figure 20–56**). The results have been a dramatic success: in many patients, the abnormal proliferation and survival of the leukemic cells are strongly inhibited, providing many years of symptom-free survival. The same drug is also effective in some other cancers that depend on similar oncogenes.

With these examples before us, we can hope that our modern understanding of the molecular biology of cancer will soon allow us to devise effective rational treatments for even more forms of cancer. At the same time, cancer research has taught us many important lessons about basic cell biology. The applications of that knowledge go far beyond the treatment of cancer, giving us insight into the way the whole living world works.

ESSENTIAL CONCEPTS

- Tissues are composed of cells and extracellular matrix.

- In plants, each cell surrounds itself with extracellular matrix in the form of a cell wall, which is made chiefly of cellulose and other polysaccharides.

- An osmotic swelling pressure on plant cell walls keeps plant tissue turgid.

- Cellulose microfibrils in the plant cell wall confer tensile strength, while other polysaccharide components resist compression.

- The orientation in which the cellulose microfibrils are deposited controls the orientation of plant cell growth.

- Animal connective tissues provide mechanical support; these tissues consist mainly of extracellular matrix, which is secreted by a sparse scattering of embedded cells.

- In the extracellular matrix of animals, tensile strength is provided by the fibrous protein collagen, while glycosaminoglycans (GAGs), covalently linked to proteins to form proteoglycans, act as space-fillers and provide resistance to compression.

- Transmembrane integrin proteins link extracellular matrix proteins such as collagen and fibronectin to the intracellular cytoskeleton of cells that contact the matrix.

- Cells are connected via cell junctions in epithelial sheets that line all external and internal surfaces of the animal body.

- Proteins of the cadherin family span the epithelial cell plasma membrane and bind to identical cadherins in adjacent epithelial cells.

- At an adherens junction, the cadherins are linked intracellularly to actin filaments; at a desmosome junction, they are linked to keratin intermediate filaments.

- During development, the actin bundles at the adherens junctions that connect cells in an epithelial sheet can contract, causing the epithelium to bend and pinch off, forming an epithelial tube or vesicle.

- Hemidesmosomes attach the basal face of an epithelial cell to the basal lamina, a specialized sheet of extracellular matrix; the attachment is mediated by transmembrane integrin proteins, which are linked to intracellular keratin filaments.

- Tight junctions seal one epithelial cell to the next, barring the diffusion of water-soluble molecules across the epithelium.

- Gap junctions form channels that allow the direct passage of inorganic ions and small, hydrophilic molecules from cell to cell; plasmodesmata in plants form a different type of channel that allows both small and large molecules to pass from cell to cell.

- Most tissues in vertebrates are complex mixtures of cell types that are subject to continual turnover.

- The tissues of an adult animal are maintained and renewed by the same basic processes that generated them in the embryo: cell proliferation, cell movement, and cell differentiation. As in the embryo, these processes are controlled by intercellular communication, selective cell–cell adhesion, and cell memory.

- In many tissues, nondividing, terminally differentiated cells are generated from stem cells, usually via the proliferation of precursor cells.

- Embryonic stem cells (ES cells) can proliferate indefinitely in culture and remain capable of differentiating into any cell type in the body—that is, they are pluripotent.

- Induced pluripotent stem cells (iPS cells), which resemble ES cells, can be generated from cells of adult human tissues through the artificial expression of a small set of transcription regulators.

- Cancer cells fail to obey the social constraints that normally ensure that cells survive and proliferate only when and where they should, and do not invade regions where they do not belong.

- Cancers arise from the accumulation of many mutations in a single somatic cell lineage; they are genetically unstable, having increased mutation rates and, often, chromosomal abnormalities.

- Unlike most normal human cells, cancer cells typically express telomerase, enabling them to proliferate indefinitely without losing DNA at their chromosome ends.

- Most human cancer cells harbor mutations in the *p53* gene, allowing them to survive and divide even when their DNA is damaged.

- The mutations that promote cancer can do so either by converting proto-oncogenes into hyperactive oncogenes or by inactivating tumor suppressor genes.

- Sequencing of cancer genomes reveals that most cancers have mutations that subvert the same three key pathways, controlling cell proliferation, cell growth, and the response to DNA damage and stress. In different cases of cancer, these pathways are subverted in different ways.

- Knowing the molecular abnormalities that underlie a particular cancer, one can begin to design specifically targeted treatments.

KEY TERMS

adherens junction	glycosaminoglycan (GAG)
apical	hemidesmosome
basal	induced pluripotent stem (iPS) cell
basal lamina	integrin
cadherin	metastasis
cancer	oncogene
cell junction	plasmodesma
cell wall	(plural plasmodesmata)
cellulose microfibril	pluripotent
collagen	proteoglycan
connective tissue	proto-oncogene
desmosome	reproductive cloning
embryonic stem (ES) cell	stem cell
epithelium (plural epithelia)	therapeutic cloning
extracellular matrix	tight junction
fibroblast	tissue
fibronectin	tumor suppressor gene
gap junction	Wnt protein
genetic instability	

QUESTIONS

QUESTION 20–9

Which of the following statements are correct? Explain your answers.

A. Gap junctions connect the cytoskeleton of one cell to that of a neighboring cell or to the extracellular matrix.

B. A wilted plant leaf can be likened to a deflated bicycle tire.

C. Because of their rigid structure, proteoglycans can withstand a large amount of compressive force.

D. The basal lamina is a specialized layer of extracellular matrix to which sheets of epithelial cells are attached.

E. Skin cells are continually shed and are renewed every few weeks; for a permanent tattoo, it is therefore necessary to deposit pigment below the epidermis.

F. Although stem cells are not differentiated, they are specialized and therefore give rise only to specific cell types.

QUESTION 20–10

Which of the following substances would you expect to spread from one cell to the next through (a) gap junctions and (b) plasmodesmata: glutamic acid, mRNA, cyclic AMP, Ca^{2+}, G proteins, and plasma membrane phospholipids?

QUESTION 20–11

Discuss the following statement: "If plant cells contained intermediate filaments to provide the cells with tensile strength, their cell walls would be dispensable."

QUESTION 20–12

Through the exchange of small metabolites and ions, gap junctions provide metabolic and electrical coupling between cells. Why, then, do you suppose that neurons communicate primarily through synapses rather than through gap junctions?

QUESTION 20–13

Gelatin is primarily composed of collagen, which is responsible for the remarkable tensile strength of connective tissue. It is the basic ingredient of jello; yet, as you probably experienced many times yourself while consuming the strawberry-flavored variety, jello has virtually no tensile strength. Why?

QUESTION 20–14

"The structure of an organism is determined by the genome that the egg contains." What is the evidence on which this statement is based? Indeed, a friend challenges you and suggests that you replace the DNA of a stork's egg with human DNA to see if a human baby results. How would you answer him?

QUESTION 20–15

Leukemias—that is, cancers arising through mutations that cause excessive production of white blood cells—have an earlier average age of onset than other cancers. Propose an explanation for why this might be the case.

QUESTION 20–16

Carefully consider the graph in **Figure Q20–16**, showing the number of cases of colon cancer diagnosed per 100,000 women per year as a function of age. Why is this graph so steep and curved, if mutations occur with a similar frequency throughout a person's life-span?

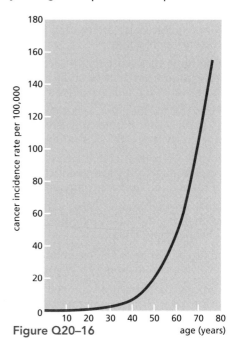

Figure Q20–16

QUESTION 20–17

Heavy smokers or industrial workers exposed for a limited time to a chemical carcinogen that induces mutations in DNA do not usually begin to develop cancers characteristic of their habit or occupation until 10, 20, or even more years after the exposure. Suggest an explanation for this long delay.

QUESTION 20–18

High levels of the female sex hormone estrogen increase the incidence of some forms of cancer. Thus, some early types of contraceptive pills containing high concentrations of estrogen were eventually withdrawn from use because this was found to increase the risk of cancer of the lining of the uterus. Male transsexuals who use estrogen preparations to give themselves a female appearance have an increased risk of breast cancer. High levels of androgens (male sex hormones) increase the risk of some other forms of cancer, such as cancer of the prostate. Can one infer that estrogens and androgens are mutagenic?

QUESTION 20–19

Is cancer hereditary?

Answers

Chapter 1

ANSWER 1–1   Trying to define life in terms of properties is an elusive business, as suggested by this scoring exercise (Table A1–1). Vacuum cleaners are highly organized objects, and take matter and energy from the environment and transform the energy into motion, responding to stimuli from the operator as they do so. On the other hand, they cannot reproduce themselves, or grow and develop—but then neither can old animals. Potatoes are not particularly responsive to stimuli, and so on. It is curious that standard definitions of life usually do not mention that living organisms on Earth are largely made of organic molecules, that life is carbon based. As we now know, the key types of "informational macromolecules"—DNA, RNA, and protein—are the same in every living species.

TABLE A1–1 PLAUSIBLE "LIFE" SCORES FOR A VACUUM CLEANER, A POTATO, AND A HUMAN			
Characteristic	Vacuum cleaner	Potato	Human
1. Organization	Yes	Yes	Yes
2. Homeostasis	Yes	Yes	Yes
3. Reproduction	No	Yes	Yes
4. Development	No	Yes	Yes
5. Energy	Yes	Yes	Yes
6. Responsiveness	Yes	No	Yes
7. Adaptation	No	Yes	Yes

ANSWER 1–2   Most random changes to the shoe design would result in objectionable defects: shoes with multiple heels, with no soles, or with awkward sizes would obviously not sell and would therefore be selected against by market forces. Other changes would be neutral, such as minor variations in color or in size. A minority of changes, however, might result in more desirable shoes: deep scratches in a previously flat sole, for example, might create shoes that would perform better in wet conditions; the loss of high heels might produce shoes that are more comfortable. The example illustrates that random changes can lead to significant improvements if the number of trials is large enough and selective pressures are imposed.

ANSWER 1–3   It is extremely unlikely that you created a new organism in this experiment. Far more probably, a spore from the air landed in your broth, germinated, and gave rise to the cells you observed. In the middle of the nineteenth century, Louis Pasteur invented a clever apparatus to disprove the then widely accepted belief that life could arise spontaneously. He showed that sealed flasks never grew anything if properly heat-sterilized first. He overcame the objections of those who pointed out the lack of oxygen or who suggested that his heat sterilization killed the life-generating principle, by using a special flask with a slender "swan's neck," which was designed to prevent spores carried in the air from contaminating the culture (Figure A1–3). The cultures in these flasks never showed any signs of life; however, they were capable of supporting life, as could be demonstrated by washing some of the "dust" from the neck into the culture.

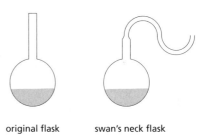

original flask   swan's neck flask

Figure A1–3

ANSWER 1–4 6×10^{39} (= 6×10^{27} g/10^{-12} g) bacteria would have the same mass as the Earth. And $6 \times 10^{39} = 2^{t/20}$, according to the equation describing exponential growth. Solving this equation for t results in $t = 2642$ minutes (or 44 hours). This represents only 132 generation times(!), whereas 5×10^{14} bacterial generation times have passed during the last 3.5 billion years. Obviously, the total mass of bacteria on this planet is nowhere close to the mass of the Earth. This illustrates that exponential growth can occur only for very few generations, i.e., for minuscule periods of time compared with evolution. In any realistic scenario, food supplies very quickly become limiting.

This simple calculation shows us that the ability to grow and divide quickly when food is ample is only one factor in the survival of a species. Food is generally scarce, and individuals of the same species have to compete with one another for the limited resources. Natural selection favors mutants that either win the competition or find ways to exploit food sources that their neighbors are unable to use.

ANSWER 1–5   By engulfing substances, such as food particles, eukaryotic cells can sequester them and feed on them efficiently. Bacteria, in contrast, have no way of capturing lumps of food; they can export substances that help break down food substances in the environment, but the products of this labor must then be shared with other cells in the same neighborhood.

ANSWER 1–6   Conventional light microscopy is much easier to use and requires much simpler instruments. Objects that are 1 μm in size can easily be resolved; the lower limit of resolution is 0.2 μm, which is a theoretical limit imposed by the wavelength of visible light. Visible light is nondestructive and passes readily through water, making it possible to observe living cells. Electron microscopy, on the other hand, is much more complicated, both in the preparation of the sample (which needs to be extremely thinly sliced, stained with electron-dense heavy metal, and completely dehydrated) and in the nature of the instrument. Living cells cannot be observed in an electron microscope. The resolution of electron microscopy is much higher, however, and biological objects as small as 1 nm can be resolved. To see any structural detail, microtubules, mitochondria, and bacteria would need to be analyzed by electron microscopy. It is possible, however, to stain them with specific dyes and then determine their location by light microscopy; if the dye is fluorescent, the stained objects can be seen with high resolution in a fluorescence microscope.

ANSWER 1–7   Because the basic workings of cells are so similar, a great deal has been learned from studying model systems. Brewer's yeast is a good model system, because yeast cells are much simpler than human cancer cells. We can grow cells inexpensively and in vast quantities, and we can manipulate them genetically and biochemically much more easily than human cells. This allows us to use yeast to decipher the ground rules governing how cells divide and grow. Cancer cells divide when they should not (and therefore give rise to tumors), and a basic understanding of how cell division is normally controlled is therefore directly relevant to the cancer problem. Indeed, the National Cancer Institute, the American Cancer Society, and many other institutions that are devoted to finding a cure for cancer strongly support basic research on various aspects of cell division in different model systems, including yeast.

ANSWER 1–8   Check your answers using the Glossary and Panel 1–2 (p. 25).

ANSWER 1–9

A. False. The hereditary information is encoded in the cell's DNA, which in turn specifies its proteins (via RNA).

B. True. Bacteria do not have a nucleus.

C. False. Plants are composed of eukaryotic cells that contain chloroplasts as cytoplasmic organelles. The chloroplasts are thought to be evolutionarily derived from prokaryotic cells.

D. True. The number of chromosomes varies from one organism to another, but is constant in all cells (except germ cells) of the same organism.

E. False. The cytosol is the cytoplasm excluding all membrane-enclosed organelles.

F. True. The nuclear envelope is a double membrane, and mitochondria are surrounded by both an inner and an outer membrane.

G. False. Protozoans are single-celled organisms and therefore do not have different tissues or cell types.

They have a complex structure, however, that has highly specialized parts.

H. Somewhat true. Peroxisomes and lysosomes contain enzymes that catalyze the breakdown of substances produced in the cytosol or taken up by the cell. One can argue, however, that many of these substances are degraded to generate food molecules, and as such are certainly not "unwanted."

ANSWER 1–10   One average brain cell weighs 10^{-9} g (= 1000 g/10^{12}). Because 1 g of water occupies 1 ml = 1 cm^3 (= 10^{-6} m^3), the volume of one cell is 10^{-15} m^3 (= 10^{-9} g × 10^{-6} m^3/g). Taking the cube root yields a side length of 10^{-5} m, or 10 μm (10^6 μm = 1 m) for each cell. The page of the book has a surface of 0.057 m^2 (= 21 cm × 27.5 cm), and each cell has a footprint of 10^{-10} m^2 (10^{-5} m × 10^{-5} m). Therefore, 57 × 10^7 (= 0.057 m^2/10^{-10} m^2) cells fit on this page when spread out as a single layer. Thus, 10^{12} cells would occupy 1750 pages (= 10^{12}/[57 × 10^7]).

ANSWER 1–11   In this plant cell, A is the nucleus, B is a vacuole, C is the cell wall, and D is a chloroplast. The scale bar is about 10 μm, the width of the nucleus.

ANSWER 1–12   The three major filaments are actin filaments, intermediate filaments, and microtubules. Actin filaments are involved in rapid cell movement, and are the most abundant filaments in a muscle cell; intermediate filaments provide mechanical stability and are the most abundant filaments in epidermal cells of the skin; and microtubules function as "railroad tracks" for intracellular movements, and are responsible for the separation of chromosomes during cell division. Other functions of all these filaments are discussed in Chapter 17.

ANSWER 1–13   It takes only 20 hours, i.e., less than a day, before mutant cells become more abundant in the culture. Using the equation provided in the question, we see that the number of the original ("wild-type") bacterial cells at time t minutes after the mutation occurred is 10^6 × $2^{t/20}$. The number of mutant cells at time t is 1 × $2^{t/15}$. To find out when the mutant cells "overtake" the wild-type cells, we simply have to make these two numbers equal to each other (i.e., 10^6 × $2^{t/20}$ = $2^{t/15}$). Taking the logarithm to base 10 of both sides of this equation and solving it for t results in t = 1200 minutes (or 20 hours). At this time, the culture contains 2 × 10^{24} cells (10^6 × 2^{60} + 1 × 2^{80}). Incidentally, 2 × 10^{24} bacterial cells, each weighing 10^{-12} g, would weigh 2 × 10^{12} g (= 2 × 10^9 kg, or 2 million tons!). This can only have been a thought experiment.

ANSWER 1–14   Bacteria continually acquire mutations in their DNA. In the population of cells exposed to the poison, one or a few cells may harbor a mutation that makes them resistant to the action of the drug. Antibiotics that are poisonous to bacteria because they bind to certain bacterial proteins, for example, would not work if the proteins have a slightly changed surface so that binding occurs more weakly or not at all. These mutant bacteria would continue dividing rapidly while their cousins are slowed down. The antibiotic-resistant bacteria would soon become the predominant species in the culture.

ANSWER 1–15 10^{13} = $2^{(t/1)}$. Therefore, it would take only 43 days [t = 13/log(2)]. This explains why some cancers can progress extremely rapidly. Many cancer cells divide much more slowly, however, or die because of their internal abnormalities or because they do not have sufficient blood supply, and the actual progression of cancer is therefore usually slower.

ANSWER 1–16   Living cells evolved from nonliving matter, but grow and replicate. Like the material they originated from, they are governed by the laws of physics, thermodynamics, and chemistry. Thus, for example, they cannot create energy *de novo* or build ordered structures without the expenditure of free energy. We can understand virtually all cellular events, such as metabolism, catalysis, membrane assembly, and DNA replication, as complicated chemical reactions that can be experimentally reproduced, manipulated, and studied in test tubes.

Despite this fundamental reducibility, a living cell is more than the sum of its parts. We cannot randomly mix proteins, nucleic acids, and other chemicals together in a test tube, for example, and make a cell. The cell functions by virtue of its organized structure, and this is a product of its evolutionary history. Cells always come from preexisting cells, and the division of a mother cell passes both chemical constituents and structures to its daughters. The plasma membrane, for example, never has to form *de novo*, but grows by expansion of a preexisting membrane; there will always be a ribosome, in part made up of proteins whose function it is to make more proteins, including those that build more ribosomes.

ANSWER 1–17   In a multicellular organism, different cells take on specialized functions and cooperate with one another, so that any one cell type does not have to perform all activities for itself. Through such divisions of labor, multicellular organisms are able to exploit food sources that are inaccessible to single-celled organisms. A plant, for example, can reach the soil with its roots to take up water and nutrients, while at the same time, its leaves above ground can harvest light energy and CO_2 from the air. By protecting its reproductive cells with other specialized cells, the multicellular organism can develop new ways to survive in harsh environments or to fight off predators. When food runs out, it may be able to preserve its reproductive cells by allowing them to draw upon resources stored by their companions—or even to cannibalize relatives (a common process, in fact).

ANSWER 1–18   The volume and the surface area are 5.24×10^{-19} m^3 and 3.14×10^{-12} m^2 for the bacterial cell, and 1.77×10^{-15} m^3 and 7.07×10^{-10} m^2 for the animal cell, respectively. From these numbers, the surface-to-volume ratios are 6×10^6 m^{-1} and 4×10^5 m^{-1}, respectively. In other words, although the animal cell has a 3375-fold larger volume, its membrane surface is increased only 225-fold. If internal membranes are included in the calculation, however, the surface-to-volume ratios of both cells are about equal. Thus, because of their internal membranes, eukaryotic cells can grow bigger and still maintain a sufficiently large membrane area, which—as we shall discuss in more detail in later chapters—is required for many essential functions.

ANSWER 1–19   There are many lines of evidence for a common ancestor. Analyses of modern-day living cells show an amazing degree of similarity in the basic components that make up the inner workings of otherwise vastly different cells. Many metabolic pathways, for example, are conserved from one cell to another, and the compounds that make up nucleic acids and proteins are the same in all living cells, even though it is easy to imagine that a different choice of compounds (e.g., amino acids with different side chains) would have worked just as well. Similarly, it is not uncommon to find that important proteins have closely similar detailed structures in prokaryotic and eukaryotic cells. Theoretically, there would be many different ways to build proteins that could perform the same functions. The evidence overwhelmingly shows that most important processes were "invented" only once and then became fine-tuned during evolution to suit the particular needs of specialized cells and specific organisms.

It seems highly unlikely, however, that the first cell survived to become the primordial founder cell of today's living world. As evolution is not a directed process with a purposeful progression, it is more likely that there were a vast number of unsuccessful trial cells that replicated for a while and then became extinct because they could not adapt to changes in the environment or could not survive in competition with other types of cells. We can therefore speculate that the primordial ancestor cell was a "lucky" cell that ended up in a relatively stable environment in which it had a chance to replicate and evolve.

ANSWER 1–20   See Figure A1–20.

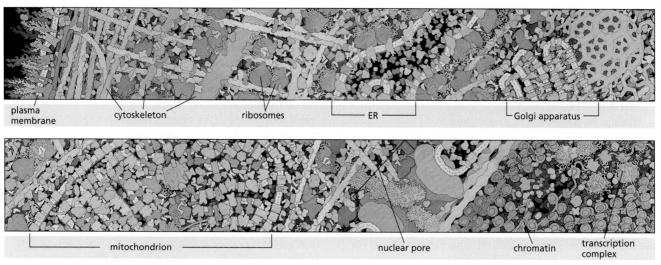

plasma membrane   cytoskeleton   ribosomes   ER   Golgi apparatus

mitochondrion   nuclear pore   chromatin   transcription complex

Figure A1–20 Courtesy of D. Goodsell.

ANSWER 1–21   A quick inspection might reveal beating cilia on the cell surface; their presence would tell you that the cell was eukaryotic. If you don't see them—and you are quite likely not to—you will have to look for other distinguishing features. If you are lucky, you might see the cell divide. Watch it then with the right optics, and you might be able to see condensed mitotic chromosomes, which again would tell you that it was a eukaryote. Fix the cell and stain it with a dye for DNA: if this is contained in a well-defined nucleus, the cell is a eukaryote; if you cannot see a well-defined nucleus, the cell may be a prokaryote. Alternatively, stain it with fluorescent antibodies that bind actin or tubulin (proteins that are highly conserved in eukaryotes but absent in bacteria). Embed it, section it, and look with an electron microscope: can you see organelles such as mitochondria inside your cell? Try staining it with the Gram stain, which is specific for molecules in the cell wall of some classes of bacteria. But all these tests might fail, leaving you still uncertain. For a definitive answer, you could attempt to analyze the sequences of the DNA and RNA molecules that it contains, using the sophisticated methods described later in this book. The sequences of highly conserved molecules, such as those that form the core components of the ribosome, provide a molecular signature that can tell you whether your cell is a eukaryote, a bacterium, or an archaeon. If you can't detect any RNA, you are probably looking not at a cell but at a piece of dirt.

Chapter 2

ANSWER 2–1   The chances are excellent because of the enormous size of Avogadro's number. The original cup contained one mole of water, or 6×10^{23} molecules, and the volume of the world's oceans, converted to cubic centimeters, is 1.5×10^{24} cm^3. After mixing, there should be on average 0.4 of an ancient water molecule per cm^3 ($6 \times 10^{23}/1.5 \times 10^{24}$), or 7.2 molecules in 18 g of Pacific Ocean.

ANSWER 2–2
A. The atomic number is 6; the atomic weight is 12 (= 6 protons + 6 neutrons).
B. The number of electrons is six (= the number of protons).
C. The first shell can accommodate two and the second shell eight electrons. Carbon therefore needs four additional electrons (or would have to give up four electrons) to obtain a full outermost shell. Carbon is most stable when it shares four additional electrons with other atoms (including other carbon atoms) by forming four covalent bonds.
D. Carbon 14 has two additional neutrons in its nucleus. Because the chemical properties of an atom are determined by its electrons, the chemical behavior of carbon 14 is identical to that of carbon 12.

ANSWER 2–3   The statement is correct. Both ionic and covalent bonds are based on the same principles: electrons can be shared equally between two interacting atoms, forming a nonpolar covalent bond; electrons can be shared unequally between two interacting atoms, forming a polar covalent bond; or electrons can be completely lost from one atom and gained by the other, forming an ionic bond. There are bonds of every conceivable intermediate state, and for borderline cases it becomes arbitrary whether a bond is described as a very polar covalent bond or an ionic bond.

ANSWER 2–4   The statement is correct. The hydrogen–oxygen bond in water molecules is polar, so that the oxygen atom carries a more negative charge than the hydrogen atoms. These partial negative charges are attracted to the positively charged sodium ions but are repelled from the negatively charged chloride ions.

ANSWER 2–5
A. Hydronium (H_3O^+) ions result from water dissociating into protons and hydroxyl ions, each proton binding to a water molecule to form a hydronium ion ($2H_2O \rightarrow H_2O + H^+ + OH^- \rightarrow H_3O^+ + OH^-$). At neutral pH, i.e., in the absence of an acid providing more H_3O^+ ions or a base providing more OH^- ions, the concentrations of H_3O^+ ions and OH^- ions are equal. We know that at neutrality the pH = 7.0, and therefore, the H^+ concentration is 10^{-7} M. The H^+ concentration equals the H_3O^+ concentration.
B. To calculate the ratio of H_3O^+ ions to H_2O molecules, we need to know the concentration of water molecules. The molecular weight of water is 18 (i.e., 18 g/mole), and 1 liter of water weighs 1 kg. Therefore, the concentration of water is 55.6 M (= 1000 [g/l]/[18 g/mole]), and the ratio of H_3O^+ ions to H_2O molecules is 1.8×10^{-9} (= $10^{-7}/55.6$); i.e., fewer than two water molecules in a billion are dissociated at neutral pH.

ANSWER 2–6   The synthesis of a macromolecule with a unique structure requires that in each amino acid position only one stereoisomer is used. Changing one amino acid from its L- to its D-form would result in a different protein. Thus, if for each amino acid a random mixture of the D- and L-forms were used to build a protein, its amino acid sequence could not specify a single structure, but many different structures (2^N different structures would be formed, where N is the number of amino acids in the protein).
   Why L-amino acids were selected in evolution as the exclusive building blocks of proteins is a mystery; we could easily imagine a cell in which certain (or even all) amino acids were used in the D-forms to build proteins, as long as these particular stereoisomers were used exclusively.

ANSWER 2–7   The term "polarity" can be used in two ways. In one meaning, it refers to directional asymmetry—for example, in linear polymers such as polypeptides (which have an N-terminus and a C-terminus) or nucleic acids (which have a 3' and a 5' end). Because the covalent bonds that link the subunits together form only between the amino and the carboxyl groups of the amino acids in a polypeptide, and between the 3' and the 5' ends of nucleotides in a nucleic acid, polypeptides and nucleic acids always have two different ends, which give the chain a defined chemical polarity.
   In the other meaning, polarity refers to a separation of electric charge in a bond or molecule. This kind of polarity promotes hydrogen-bonding to water molecules, and because the water solubility, or hydrophilicity, of a molecule depends upon its being polar in this sense, the term "polar" also indicates water solubility.

ANSWER 2–8   A major advantage of condensation reactions is that they are readily reversible by hydrolysis (and water is readily available in the cell). This allows cells to break down their macromolecules (or macromolecules of other organisms that were ingested as food) and to recover the subunits intact so that they can be "recycled," i.e., used to build new macromolecules.

ANSWER 2–9   Many of the functions that macromolecules perform rely on their ability to associate with and dissociate from other molecules readily. This allows cells, for example, to remodel their interior when they move or divide, and to transport components from one organelle to another. Covalent bonds would be too stable for such a purpose, requiring a specific enzyme to break each kind of bond.

ANSWER 2–10
A. True. All nuclei are made of positively charged protons and uncharged neutrons; the only exception is the hydrogen nucleus, which consists of only one proton.
B. False. Atoms are electrically neutral. The number of positively charged protons is always balanced by an equal number of negatively charged electrons.
C. True—but only for the cell nucleus (see Chapter 1), and not for the atomic nucleus discussed in this chapter.
D. False. Elements can have different isotopes, which differ only in their number of neutrons.
E. True. In certain isotopes, the large number of neutrons destabilizes the nucleus, which decomposes in a process called radioactive decay.
F. True. Examples include granules of glycogen, a polymer of glucose, found in liver cells; and fat droplets, made of aggregated triacylglycerols, found in fat cells.
G. True. Individually, these bonds are weak and readily broken by thermal motion, but because interactions between two macromolecules involve a large number of such bonds, the overall binding can be quite strong, and because hydrogen bonds form only between correctly positioned groups on the interacting macromolecules, they are very specific.

ANSWER 2–11
A. One cellulose molecule has a molecular weight of $n \times (12[C] + 2 \times 1[H] + 16[O])$. We do not know n, but we can determine the ratio with which the individual elements contribute to the weight of cellulose. The contribution of carbon atoms is 40% [= 12/(12 + 2 + 16) × 100%]. Therefore, 2 g (40% of 5 g) of carbon atoms are contained in the cellulose that makes up this page. The atomic weight of carbon is 12 g/mole, and there are 6×10^{23} atoms or molecules in a mole. Therefore, 10^{23} carbon atoms [= (2 g/12 [g/mole]) × 6 × 10²³ (molecules/mole)] make up this page.
B. The volume of the page is 4×10^{-6} m³ (= 21.2 cm × 27.6 cm × 0.07 mm), which is the same as the volume of a cube with a side length of 1.6 cm (= $\sqrt[3]{4 \times 10^{-6}\ m^2}$). Because we know from part A that the page contains 10^{23} carbon atoms, geometry tells us that there could be about 4.6×10^7 carbon atoms (= $\sqrt[3]{10^{23}}$) lined up along each side of this cube. Therefore, in cellulose, about 200,000 carbon atoms (= $4.6 \times 10^7 \times 0.07 \times 10^{-3}$ m/1.6 $\times 10^{-2}$ m) span the thickness of the page.
C. If tightly stacked, 350,000 carbon atoms with a 0.2-nm diameter would span the 0.07-mm thickness of the page.
D. There are two reasons for the 1.7-fold difference in the two calculations: (1) carbon is not the only atom in cellulose; and (2) paper is not an atomic lattice of precisely arranged cellulose molecules (as a diamond would be for precisely arranged carbon atoms), but a random meshwork of fibers.

ANSWER 2–12
A. The occupancies of the three electron shells, from the nucleus outward, are 2, 8, 8.
B. helium   already has full level
   oxygen   gain 2
   carbon   gain 4 or lose 4
   sodium   lose 1
   chlorine   gain 1
C. Helium with its fully occupied electron shell is chemically unreactive. Sodium and chlorine, on the other hand, are extremely reactive and readily form Na^+ and Cl^- ions, which can form ionic bonds to produce NaCl (table salt).

ANSWER 2–13   Whether a substance is a liquid or gas at a given temperature depends on the attractive forces between its molecules. H_2S is a gas at room temperature and H_2O is a liquid because the hydrogen bonds that hold H_2O molecules together do not form between H_2S molecules. A sulfur atom is much larger than an oxygen atom, and because of its larger size, the outermost electrons are not as strongly attracted to the nucleus of the sulfur atom as they are in an oxygen atom. Consequently, the hydrogen–sulfur bond is much less polar than the hydrogen–oxygen bond. Because of the reduced polarity, the sulfur in a H_2S molecule is not strongly attracted to the hydrogen atoms in an adjacent H_2S molecule, and hydrogen bonds, which are so predominant in water, do not form.

ANSWER 2–14   The reactions are diagrammed in Figure A2–14, where R_1 and R_2 are amino acid side chains.

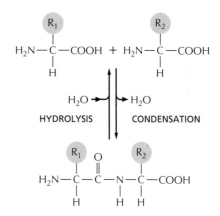

Figure A2–14

ANSWER 2–15
A. False. The properties of a protein depend on both the amino acids it contains and the order in which they are linked together. The diversity of proteins is due to the almost unlimited number of ways in which 20 different amino acids can be combined in a linear sequence.
B. False. Phospholipids assemble into bilayers in an aqueous environment by noncovalent forces. Lipid bilayers are therefore not macromolecules.
C. True. The backbone of nucleic acids is made up of alternating ribose (or deoxyribose in DNA) and phosphate groups. Ribose and deoxyribose are sugars.
D. True. About half of the 20 naturally occurring amino acids have hydrophobic side chains. In folded proteins, many of these side chains face toward the inside of a folded-up globular protein, because they are repelled from water.

E. True. Hydrophobic hydrocarbon tails contain only nonpolar bonds. Thus, they cannot participate in hydrogen-bonding and are repelled from water. We consider the underlying principles in more detail in Chapter 11.

F. False. RNA contains the four listed bases, but DNA contains T instead of U. T and U are very much alike, however, and differ only by a single methyl group.

ANSWER 2–16

A. (a) 400 (= 20^2); (b) 8000 (= 20^3); (c) 160,000 (= 20^4).

B. A protein with a molecular weight of 4800 daltons is made of about 40 amino acids; thus there are 1.1×10^{52} (= 20^{40}) different ways to make such a protein. Each individual protein molecule weighs 8×10^{-21} g (= $4800/6 \times 10^{23}$); thus a mixture of one molecule of each weighs 9×10^{31} g (= 8×10^{-21} g $\times 1.1 \times 10^{52}$), which is 15,000 times the total weight of the planet Earth, weighing 6×10^{24} kg. You would need a quite large container, indeed.

C. Given that most cell proteins are even larger than the one used in this example, it is clear that only a minuscule fraction of the total possible amino acid sequences are used in living cells.

ANSWER 2–17   Because all living cells are made up of chemicals and because all chemical reactions (whether in living cells or in test tubes) follow the same rules, an understanding of basic chemical principles is fundamentally important to the understanding of cell biology. For this reason, in later chapters, we will frequently refer back to these principles, on which all of the more complicated pathways and reactions that occur in cells are based.

ANSWER 2–18

A. Hydrogen bonds form between two specific chemical groups; one is always a hydrogen atom linked in a polar covalent bond to an oxygen or a nitrogen atom, and the other is usually a nitrogen or an oxygen atom. Van der Waals attractions are weaker and occur between any two atoms that are in close enough proximity. Both hydrogen bonds and van der Waals attractions are short-range interactions that come into play only when two molecules are already in close proximity. Both types of bonds can therefore be thought of as means of "fine-tuning" an interaction, i.e., helping to position two molecules correctly with respect to each other once they have been brought together by diffusion.

B. Van der Waals attractions would occur in all three examples. Hydrogen bonds would form only in (c).

ANSWER 2–19   Noncovalent bonds form between the covalently linked subunits of a macromolecule such as a polypeptide or RNA chain causing the chain to fold into a unique shape. These noncovalent bonds include hydrogen bonds, ionic interactions, van der Waals attractions, and hydrophobic interactions. Because these interactions are weak, they can be broken with relative ease; thus, most macromolecules can be unfolded by heating, which increases thermal motion.

ANSWER 2–20   Amphipathic molecules have both a hydrophilic and a hydrophobic end. Their hydrophilic end can hydrogen-bond to water, but their hydrophobic end is repelled from water because it interferes with the water structure. Consequently, the hydrophobic ends of amphipathic molecules tend to be exposed to air at air–water interfaces, or, in the interior of an aqueous solution, they will always cluster together to minimize their contact with water molecules. (See **Figure A2–20**.)

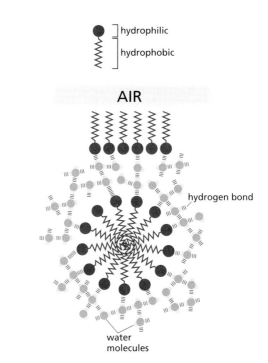

Figure A2–20

ANSWER 2–21

A,B. (A) and (B) are both correct formulas of the amino acid phenylalanine. In formula (B), phenylalanine is shown in the ionized form that exists in an aqueous solution, where the basic amino group is protonated and the acidic carboxylic group is deprotonated.

C. Incorrect. This structure of a peptide bond is missing a hydrogen atom bound to the nitrogen.

D. Incorrect. This formula of an adenine base features one double bond too many, creating a five-valent carbon atom and a four-valent nitrogen atom.

E. Incorrect. In this formula of a nucleoside triphosphate, there should be two additional oxygen atoms, one between each of the phosphorus atoms.

F. This is the correct formula of ethanol.

G. Incorrect. Water does not hydrogen-bond to hydrogens bonded to carbon. The lack of the capacity to hydrogen-bond makes hydrocarbon chains hydrophobic, i.e., water-hating.

H. Incorrect. Na and Cl form an ionic bond, Na^+Cl^-, but a covalent bond is drawn.

I. Incorrect. The oxygen atom attracts electrons more than the carbon atom; the polarity of the two bonds should therefore be reversed.

J. This structure of glucose is correct.

K. Almost correct. It is more accurate to show that only one hydrogen is lost from the –NH$_2$ group, and the –OH group is lost from the –COOH group.

Chapter 3

ANSWER 3–1   The equation represents the "bottom line" of photosynthesis, which occurs as a large set of individual reactions that are catalyzed by many individual enzymes. Because sugars are more complicated molecules than CO_2 and H_2O, the reaction generates a more ordered state inside the cell. As demanded by the second law of thermodynamics, this increase in order must be accompanied by a greater increase in disorder, which occurs because heat is generated at many steps on the long pathway leading to the products summarized in this equation.

ANSWER 3–2   Oxidation is defined as removal of electrons, and reduction represents a gain of electrons. Therefore, (A) is an oxidation, and (B) is a reduction. The *red* carbon atom in (C) remains largely unchanged; the neighboring carbon atom, however, loses a hydrogen atom (i.e., an electron and a proton) and hence becomes oxidized. The *red* carbon atom in (D) becomes oxidized because it loses a hydrogen atom, whereas the *red* carbon atom in (E) becomes reduced because it gains a hydrogen atom.

ANSWER 3–3
A. Both states of the coin, H and T, have an equal probability. There is therefore no driving force, i.e., no energy difference, that would favor H turning to T or vice versa. Therefore, $\Delta G° = 0$ for this reaction. However, a reaction proceeds if H and T coins are not present in the box in equal numbers. In this case, the concentration difference between H and T creates a driving force and $\Delta G \neq 0$; when the reaction reaches equilibrium—i.e., when there are equal numbers of H and T—$\Delta G = 0$.
B. The amount of shaking corresponds to the temperature, as it results in the "thermal" motion of the coins. The activation energy of the reaction is the energy that needs to be expended to flip the coin, i.e., to stand it on its rim, from where it can fall back facing either side up. Jigglase would speed up the flipping by lowering the energy required for this; it could, for example, be a magnet that is suspended above the box and helps lift the coins. Jigglase would not affect where the equilibrium lies (at an equal number of H and T), but it would speed up the process of reaching the equilibrium, because in the presence of jigglase more coins would flip back and forth.

ANSWER 3–4   See Figure A3–4. Note that $\Delta G°_{X \rightarrow Y}$ is positive, whereas $\Delta G°_{Y \rightarrow Z}$ and $\Delta G°_{X \rightarrow Z}$ are negative. The graph also shows that $\Delta G°_{X \rightarrow Z} = \Delta G°_{X \rightarrow Y} + \Delta G°_{Y \rightarrow Z}$. We do not know from the information given in Figure 3–12 how high the activation energy barriers are; they are therefore drawn to an arbitrary height (solid lines). The activation energies would be lowered by enzymes that catalyze these reactions, thereby speeding up the reaction rates (dotted lines), but the enzymes would not change the $\Delta G°$ values.

ANSWER 3–5   The reaction rates might be limited by (1) the concentration of the substrate, i.e., how often a molecule of CO_2 collides with the active site on the enzyme; (2) how many of these collisions are energetic enough to lead to a reaction; and (3) how fast the enzyme can release

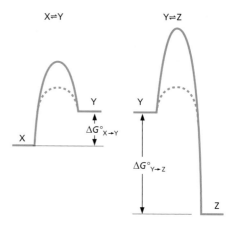

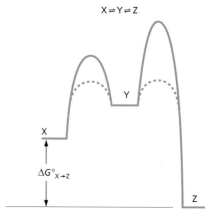

Figure A3–4

the products of the reaction and therefore be free to bind more CO_2. The diagram in **Figure A3–5** shows that the enzyme lowers the activation energy barrier, so that more CO_2 molecules have sufficient energy to undergo the reaction. The area under the curve from point A to infinite energy or from point B to infinite energy indicates the total number of molecules that will react without or with the enzyme, respectively. Although not drawn to scale, the ratio of these two areas should be 10^7.

ANSWER 3–6   All reactions are reversible. If the compound AB can dissociate to produce A and B, then it must also be possible for A and B to associate to form AB. Which of the two reactions predominates depends on the equilibrium constant of the reaction and the concentration of A, B, and AB (as discussed in Figure 3–19). Presumably, when this enzyme was isolated its activity was detected by

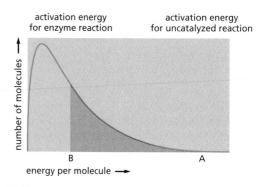

Figure A3–5

supplying A and B in relatively large amounts and measuring the amount of AB generated. But suppose, however, that in the cell there is a large concentration of AB, in which case the enzyme would actually catalyze AB → A + B. (This question is based on an actual example in which an enzyme was isolated and named according to the reaction in one direction, but was later shown to catalyze the reverse reaction in living cells.)

ANSWER 3–7

A. The rocks in Figure 3–30B provide the energy to lift the bucket of water. In the reaction X + ATP → Y + ADP + P_i, ATP hydrolysis is driving the reaction; thus ATP corresponds to the rocks on top of the cliff. The broken debris in Figure 3–30B corresponds to ADP and P_i, the products of ATP hydrolysis. In the reaction, ATP hydrolysis is coupled to the conversion of X to Y. X, therefore, is the starting material, the bucket on the ground, which is converted to Y, the bucket at its highest point.

B. (i) The rock hitting the ground would be the futile hydrolysis of ATP—for example, in the absence of an enzyme that uses the energy released by the ATP hydrolysis to drive an otherwise unfavorable reaction; in this case, the energy stored in the phosphoanhydride bond of ATP would be lost as heat. (ii) The energy stored in Y could be used to drive another reaction. If Y represented the activated form of amino acid X, for example, it could undergo a condensation reaction to form a peptide bond during protein synthesis.

ANSWER 3–8   The free energy ΔG derived from ATP hydrolysis depends on both the ΔG° and the concentrations of the substrate and products. For example, for a particular set of concentrations, one might have

$$\Delta G = -12 \text{ kcal/mole} = -7.3 \text{ kcal/mole} + 0.616 \ln \frac{[ADP] \times [P_i]}{[ATP]}$$

ΔG is smaller than ΔG°, largely because the ATP concentration in cells is high (in the millimolar range) and the ADP concentration is low (in the 10 μM range). The concentration term of this equation is therefore smaller than 1 and its logarithm is a negative number.

ΔG° is a constant for the reaction and will not vary with reaction conditions. ΔG, in contrast, depends on the concentrations of ATP, ADP, and phosphate, which can be somewhat different between cells.

ANSWER 3–9   Reactions B, D, and E all require coupling to other, energetically favorable reactions. In each case, higher-order structures are formed that are more complicated and have higher-energy bonds than the starting materials. In contrast, reaction A is a catabolic reaction that leads to compounds in a lower energy state and will occur spontaneously. The nucleoside triphosphates in reaction C contain enough energy to drive DNA synthesis (see Figure 3–41).

ANSWER 3–10

A. Nearly true, but strictly speaking, false. Because enzymes enhance the rate but do not change the equilibrium point of a reaction, a reaction will always occur in the absence of the enzyme, though often at a minuscule rate. Moreover, competing reactions may use up the substrate more quickly, thus further impeding

the desired reaction. Thus, in practical terms, without an enzyme, some reactions may never occur to an appreciable extent.

B. False. High-energy electrons are more easily transferred, i.e., more loosely bound to the donor molecule. This does not mean that they move any faster.

C. True. Hydrolysis of an ATP molecule to form AMP also produces a pyrophosphate (PP_i) molecule, which in turn is hydrolyzed into two phosphate molecules. This second reaction releases almost the same amount of energy as the initial hydrolysis of ATP, thereby approximately doubling the energy total yield.

D. True. Oxidation is the removal of electrons, which reduces the diameter of the carbon atom.

E. True. ATP, for example, can donate both chemical-bond energy and a phosphate group.

F. False. Living cells have a particular kind of chemistry in which most oxidations are energy-releasing events; under different conditions, however, such as in a hydrogen-containing atmosphere, reductions would be energy-releasing events.

G. False. All cells, including those of cold- and warm-blooded animals, radiate comparable amounts of heat as a consequence of their metabolic reactions. For bacterial cells, for example, this becomes apparent when a compost pile heats up.

H. False. The equilibrium constant of the reaction X ↔ Y remains unchanged. If Y is removed by a second reaction, more X is converted to Y so that the ratio of X to Y remains constant.

ANSWER 3–11   The free-energy difference (ΔG°) between Y and X due to three hydrogen bonds is –3 kcal/mole. (Note that the free energy of Y is lower than that of X, because energy would need to be expended to break the bonds to convert Y to X. The value for ΔG° for the transition X → Y is therefore negative.) The equilibrium constant for the reaction is therefore about 100 (from Table 3–1, p. 98); i.e., there are about 100 times more molecules of Y than of X at equilibrium. An additional three hydrogen bonds would increase ΔG° to –6 kcal/mole and increase the equilibrium constant about another 100-fold to 10^4. Thus, relatively small differences in energy can have a major effect on equilibria.

ANSWER 3–12

A. The equilibrium constant is defined as K = [AB]/([A] × [B]). The square brackets indicate the concentration. Thus, if A, B, and AB are each 1 μM (10^{-6} M), K will be 10^6 liters/mole [= $10^{-6}/(10^{-6} \times 10^{-6})$].

B. Similarly, if A, B, and AB are each 1 nM (10^{-9} M), then K will be 10^9 liters/mole.

C. This example illustrates that interacting proteins that are present in cells in lower concentrations need to bind to each other with higher affinities so that a significant fraction of the molecules are bound at equilibrium. In this particular case, lowering the concentration by 1000-fold (from μM to nM) requires an increase in the equilibrium constant by 1000-fold to maintain the AB protein complex in the same proportion (corresponding to –4.3 kcal of free energy; see Table 3–1). This corresponds to about four or five extra hydrogen bonds.

ANSWER 3–13   The statement is correct. The criterion for whether a reaction proceeds spontaneously is ΔG, not ΔG°, and takes the concentrations of the reacting components into account. A reaction with a negative $\Delta G°$, for example, would not proceed spontaneously under conditions where there is a large enough excess of products, i.e., more than at equilibrium. Conversely, a reaction with a positive $\Delta G°$ might spontaneously go forward under conditions where there is a huge excess of substrate.

ANSWER 3–14

A. A maximum of 57 ATP molecules (= 686/12) corresponds to the total energy released by the complete oxidation of glucose to CO_2 and H_2O.

B. The overall efficiency of ATP production would be about 53%, calculated as the ratio of actually produced ATP molecules (30) divided by the number of ATP molecules that could be obtained if all the energy stored in a glucose molecule could be harvested as chemical energy in ATP (57).

C. During the oxidation of 1 mole of glucose, 322 kcal (the remaining 47% of the available 686 kcal in one mole of glucose that is not stored as chemical energy in ATP) would be released as heat. This amount of energy would heat your body by 4.3°C (= 322 kcal/75 kg). This is a significant amount of heat, considering that 4°C of elevated temperature would be a quite incapacitating fever and that 1 mole (180 g) of glucose is no more than two cups of sugar.

D. If the energy yield were only 20%, then instead of 47% in the example above, 80% of the available energy would be released as heat and would need to be dissipated by your body. The heat production would be more than 1.7-fold higher than normal, and your body would certainly overheat.

E. The chemical formula of ATP is $C_{10}H_{12}O_{13}N_5P_3$, and its molecular weight is therefore 503 g/mole. Your resting body therefore hydrolyzes about 80 moles (= 40 kg/0.503 kg/mole) of ATP in 24 hours (this corresponds to about 1000 kcal of liberated chemical energy). Because every mole of glucose yields 30 moles of ATP, this amount of energy could be produced by oxidation of 480 g glucose (= 180 g/mole × 80 moles/30).

ANSWER 3–15

This scientist is definitely a fake. The 57 ATP molecules would store 684 kcal (= 57 × 12 kcal) of chemical energy, which implies that the efficiency of ATP production from glucose would have been greater than 99%. This impossible degree of efficiency would leave virtually no energy to be released as heat, and this release is required according to the laws of thermodynamics.

ANSWER 3–16

A. From Table 3–1 (p. 98) we know that a free-energy difference of 4.3 kcal/mole corresponds to an equilibrium constant of 10^{-3}, i.e., [A*]/[A] = 10^{-3}. The concentration of A* is therefore 1000-fold lower than that of A at equilibrium.

B. The ratio of A to A* would be unchanged. Lowering the activation energy barrier with an enzyme would accelerate the rate of the reaction, i.e., it would allow more molecules in a given time period to convert from A $\rightarrow$ A* and from A* $\rightarrow$ A, but it would not affect the ratio of A $\rightarrow$ A* at equilibrium.

ANSWER 3–17

A. The mutant mushroom would probably be safe to eat. ATP hydrolysis can provide approximately –12 kcal/mole of energy. This amount of energy shifts the equilibrium point of a reaction by an enormous factor: about 10^8-fold (from Table 3–1, p. 98, we see that –5.7 kcal/mole corresponds to an equilibrium constant of 10^4; thus, –12 kcal/mole corresponds to about 10^8. Note that, for coupled reactions, energies are additive, whereas equilibrium constants are multiplied). Therefore, if the energy of ATP hydrolysis cannot be utilized by the enzyme, 10^8-fold less poison is made. This example illustrates that coupling a reaction to the hydrolysis of an activated carrier molecule can shift the equilibrium point drastically.

B. It would be risky to consume this mutant mushroom. Slowing down the reaction rate would not affect its equilibrium point, and if the reaction were allowed to proceed for a long enough time, the mushroom would likely be loaded with poison. It is possible that the reaction would not reach equilibrium, but it would not be advisable to take a chance.

ANSWER 3–18

Enzyme A is beneficial. It allows the interconversion of two energy-carrier molecules, both of which are required as the triphosphate form for many metabolic reactions. Any ADP that is formed is quickly converted to ATP, and thus the cell maintains a high ATP/ADP ratio. Because of enzyme A, called nucleotide phosphokinase, some of the ATP is used to keep the GTP/GDP ratio similarly high.

Enzyme B would be highly detrimental to the cell. Cells use NAD^+ as an electron acceptor in catabolic reactions and must maintain high concentrations of this form of the carrier as it is used in reactions that break down glucose to make ATP. In contrast, NADPH is used as an electron donor in biosynthetic reactions and is kept at high concentration in the cells so as to allow the synthesis of nucleotides, fatty acids, and other essential molecules. Since enzyme B would deplete the cell's reserves of both NAD^+ and NADPH, it would decrease the rates of both catabolic and biosynthetic reactions.

ANSWER 3–19

Because enzymes are catalysts, enzyme reactions have to be thermodynamically feasible; the enzyme only lowers the activation energy barrier that otherwise slows the rate with which the reaction occurs. Heat confers more kinetic energy to substrates so that a higher fraction of them can surmount the normal activation energy barrier. Many substrates, however, have many different ways in which they could react, and all of these potential pathways will be enhanced by heat. An enzyme, by contrast, acts selectively to facilitate only one particular pathway that, in evolution, was selected to be useful for the cell. Heat, therefore, cannot substitute for enzyme function, and chicken soup must exert its claimed beneficial effects by other mechanisms, which remain to be discovered.

ANSWER 3–20

A. When [S] << K_M, the term ([S] + K_M) approaches K_M. Therefore, the equation is simplified to rate = V_{max}[S]/K_M. Therefore, the rate is proportional to [S].

B. When [S] = K_M, the term [S]/([S] + K_M) equals ½. Therefore, the reaction rate is half of the maximal rate V_{max}.

C. If [S] >> K_M, the term ([S] + K_M) approaches [S]. Therefore, [S]/([S] + K_M) equals 1 and the reaction occurs at its maximal rate V_{max}.

ANSWER 3–21   The substrate concentration is 1 mM. This value can be obtained by substituting values in the equation, but it is simpler to note that the desired rate (50 μmole/sec) is exactly half of the maximum rate, V_{max}, where the substrate concentration is typically equal to the K_M. The two plots requested are shown in **Figure A3–21**. A plot of 1/rate versus 1/[S] is a straight line because rearranging the standard equation yields the equation listed in Question 3–23B.

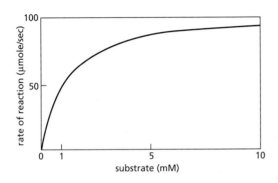

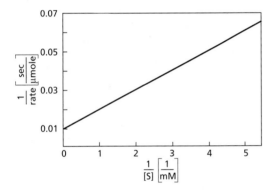

Figure A3–21

ANSWER 3–22   If [S] is very much smaller than K_M, the active site of the enzyme is mostly unoccupied. If [S] is very much greater than K_M, the reaction rate is limited by the enzyme concentration (because most of the catalytic sites are fully occupied).

ANSWER 3–23

A,B. The data in the boxes have been used to plot the red curve and red line in **Figure A3–23**. From the plotted data, the K_M is 1 μM and the V_{max} is 2 μmole/min. Note that the data are much easier to interpret in the linear plot, because the curve in (A) approaches, but never reaches, V_{max}.

C. It is important that only a small quantity of product is made, because otherwise the rate of reaction would decrease as the substrate was depleted and product accumulated. Thus the measured rates would be lower than they should be.

D. If the K_M increases, then the concentration of substrate needed to give a half-maximal rate is increased. As more substrate is needed to produce the same rate, the enzyme-catalyzed reaction has been inhibited by the phosphorylation. The expected data plots for the

phosphorylated enzyme are the green curve and the green line in Figure A3–23.

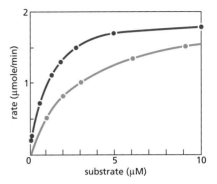

DATA FOR A AND B

[S] (μM)	$\frac{1}{[S]}\left[\frac{1}{μM}\right]$	rate (μmole/min)	$\frac{1}{rate}\left[\frac{min}{μmole}\right]$
0.08	12.50	0.15	6.7
0.12	8.30	0.21	4.8
0.54	1.85	0.70	1.4
1.23	0.81	1.1	0.91
1.82	0.56	1.3	0.77
2.72	0.37	1.5	0.67
4.94	0.20	1.7	0.59
10.00	0.10	1.8	0.56

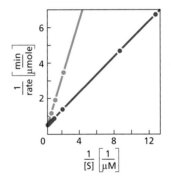

Figure A3–23

Chapter 4

ANSWER 4–1   Urea is a very small organic molecule that functions both as an efficient hydrogen-bond donor (through its –NH_2 groups) and as an efficient hydrogen-bond acceptor (through its –C=O group). As such, it can squeeze between hydrogen bonds that stabilize protein molecules and thus destabilize protein structures. In addition, the nonpolar side chains of a protein are held together in the interior of the folded structure because they would disrupt the structure of water if they were exposed. At high concentrations of urea, the hydrogen-bonded network of water molecules becomes disrupted so that these hydrophobic forces are significantly diminished. Proteins unfold in urea as a consequence of its effect on these two forces.

ANSWER 4–2   The amino acid sequence consists of alternating nonpolar and charged or polar amino acids. The resulting strand in a β sheet would therefore be polar on one side and hydrophobic on the other. Such a strand would probably be surrounded on either side by similar strands that together form a β sheet with a hydrophobic face and a polar face. In a protein, such a β sheet (called "amphipathic," from the Greek *amphi*, "of both kinds," and *pathos*, "passion," because of its two surfaces with such different properties) would be positioned so that the hydrophobic side would face the protein's interior and the polar side would be on its surface, exposed to the water outside.

ANSWER 4–3   Mutations that are beneficial to an organism are selected in evolution because they confer a reproductive or survival advantage to the organism. Examples might be the better utilization of a food source, enhanced resistance to environmental insults, or an enhanced ability to attract a mate for sexual reproduction. In contrast, useless proteins are detrimental to organisms, as the metabolic energy required to make them is a wasted cost. If such mutant proteins were made in excess, the synthesis of normal proteins would suffer because the synthetic capacity of the cell is limited. In more severe cases, a mutant protein could interfere with the normal workings of the cell; a mutant enzyme that still binds an activated carrier molecule but does not catalyze a reaction, for example, may compete for a limited amount of this carrier and therefore inhibit normal processes. Natural selection therefore provides a strong driving force that eliminates both useless and harmful proteins.

ANSWER 4–4   Strong reducing agents that break all of the S–S bonds would cause all of the keratin filaments to separate. Individual hairs would be weakened and fragment. Indeed, strong reducing agents are used commercially in hair-removal creams sold by your local pharmacist. However, mild reducing agents are used in treatments that either straighten or curl hair, the latter requiring hair curlers. (See Figure A4–4.)

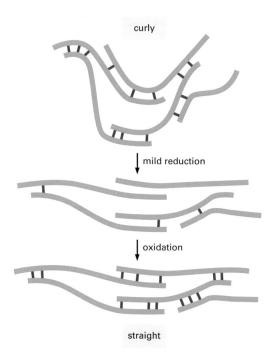

Figure A4–4

ANSWER 4–5   See Figure A4–5.

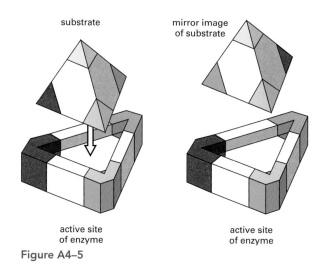

Figure A4–5

ANSWER 4–6
A. Feedback inhibition from Z that affects the reaction B → C would increase the flow through the B → X → Y → Z pathway, because the conversion of B to C is inhibited. Thus, the more Z there is, the more production of Z would be stimulated. This is likely to result in an uncontrolled "runaway" amplification of this pathway.
B. Feedback inhibition from Z affecting Y → Z would only inhibit the production of Z. In this scheme, however, X and Y would still be made at normal rates, even though both of these intermediates are no longer needed at this level. This pathway is therefore less efficient than the one shown in Figure 4–38.
C. If Z is a positive regulator of the step B → X, then the more Z there is, the more B will be converted to X and therefore shunted into the pathway producing more Z. This would result in a runaway amplification similar to that described for (A).
D. If Z is a positive regulator of the step B → C, then accumulation of Z leads to a redirection of the pathway to make more C. This is a second possible way, in addition to that shown in the figure, to balance the distribution of compounds into the two branches of the pathway.

ANSWER 4–7   Both nucleotide binding and phosphorylation can induce allosteric changes in proteins. These can have a multitude of consequences, such as altered enzyme activity, drastic shape changes, and changes in affinity for other proteins or small molecules. Both mechanisms are quite versatile. An advantage of nucleotide binding is the fast rate with which a small nucleotide can diffuse to the protein; the shape changes that accompany the function of motor proteins, for example, require quick nucleotide replenishment. If the different conformational states of a motor protein were controlled by phosphorylation, for example, a protein kinase would either need to diffuse into position at each step, a much slower process, or be associated permanently with each motor protein. One advantage of phosphorylation is that it requires only a single amino acid on the protein's surface, rather than a specific binding site. Phosphates can therefore be added to many different side chains on the same protein (as long as protein kinases with the proper specificities exist), thereby vastly increasing the complexity of regulation that can be achieved for a single protein.

ANSWER 4–8   In working together in a complex, all three proteins contribute to the specificity (by binding to the safe and key directly). They help position one another correctly, and provide the mechanical bracing that allows them to perform a task that they could not perform individually (the key is grasped by two of the proteins, for example). Moreover, their functions are generally coordinated in time (for instance, the binding of ATP to one subunit is likely to require that ATP has already been hydrolyzed to ADP by another).

ANSWER 4–9   The α helix is right-handed. The three strands that form the large β sheet are antiparallel. There are no knots in the polypeptide chain, presumably because a knot would interfere with the folding of the protein into its three-dimensional conformation after protein synthesis.

ANSWER 4–10
A. True. Only a few amino acid side chains contribute to the active site. The rest of the protein is required to maintain the polypeptide chain in the correct conformation, provide additional binding sites for regulatory purposes, and localize the protein in the cell.
B. True. Some enzymes form covalent intermediates with their substrates (see middle panels of Figure 4–35); however, in all cases the enzyme is restored to its original structure after the reaction.
C. False. β sheets can, in principle, contain any number of strands because the two strands that form the rims of the sheet are available for hydrogen-bonding to other strands. (β sheets in known proteins contain from 2 to 16 strands.)
D. False. It is true that the specificity of an antibody molecule is exclusively contained in polypeptide loops on its surface; however, these loops are contributed by both the folded light and heavy chains (see Figure 4–33).
E. False. The possible linear arrangements of amino acids that lead to a stably folded protein domain are so few that most new proteins evolve by alteration of old ones.
F. True. Allosteric enzymes generally bind one or more molecules that function as regulators at sites that are distinct from the active site.
G. False. Although single noncovalent bonds are weak, many such bonds acting together are a major contributor to the three-dimensional structure of macromolecules.
H. False. Affinity chromatography separates specific macromolecules because of their interactions with specific ligands, not because of their charge.
I. False. The larger an organelle is, the more centrifugal force it experiences and the faster it sediments, despite an increased frictional resistance from the fluid through which it moves.

ANSWER 4–11   In an α helix and in the central strands of a β sheet, all of the N–H and C=O groups in the polypeptide backbone are engaged in hydrogen bonds. This gives considerable stability to these secondary structural elements, and it allows them to form in many different proteins.

ANSWER 4–12   No. It would not have the same or even a similar structure, because the peptide bond has a polarity. Looking at two sequential amino acids in a polypeptide chain, the amino acid that is closer to the N-terminal end contributes the carboxyl group and the other amino acid contributes the amino group to the peptide bond that links the two amino acids. Changing their order would put the side chains into different positions with respect to the peptide backbone and therefore change the way the polypeptide folds.

ANSWER 4–13   As it takes 3.6 amino acid residues to complete a turn of an α helix, this sequence of 14 amino acids would make close to 4 full turns. It is remarkable because its polar and hydrophobic amino acids are spaced so that all the polar ones are on one side of the α helix and all the hydrophobic residues are on the other. It is therefore likely that such an amphipathic α helix is exposed on the protein surface with its hydrophobic side facing the protein's interior. In addition, two such helices might wrap around each other as shown in Figure 4–16.

ANSWER 4–14
A. ES represents the enzyme–substrate complex.
B. Enzyme and substrate are in equilibrium between their free and bound states; once bound to the enzyme, a substrate molecule may either dissociate again (hence the bidirectional arrows) or be converted to product. As the substrate is converted to product (with the concomitant release of free energy), however, a reaction usually proceeds strongly in the forward direction, as indicated by the unidirectional arrow.
C. The enzyme is a catalyst and is therefore liberated in an unchanged form after the reaction; thus, E appears at both ends of the equation.
D. Often, the product of a reaction resembles the substrate sufficiently that it can also bind to the enzyme. Any enzyme molecules that are bound to the product (i.e., are part of the EP complex) are unavailable for catalysis; excess P therefore inhibits the reaction by lowering the concentration of free E.
E. Compound X would act as an inhibitor of the reaction and work similarly by forming an EX complex. However, since P has to be made before it can inhibit the reaction, it takes longer to act than X, which is present from the beginning of the reaction.

ANSWER 4–15   The polar amino acids Ser, Ser-P, Lys, Gln, His, and Glu are more likely to be found on a protein's surface, and the hydrophobic amino acids Leu, Phe, Val, Ile, and Met are more likely to be found in its interior. The oxidation of two cysteine residues to form a disulfide bond eliminates their potential to form hydrogen bonds and therefore makes them even more hydrophobic; thus disulfide bonds are usually found in the interior of proteins. Irrespective of the nature of their side chains, the most N-terminal amino acid and the most C-terminal amino acid each contain a charged group (the amino and carboxyl groups, respectively, that mark the ends of the polypeptide chain) and hence are usually found on the protein's surface.

ANSWER 4–16   Many secondary structural elements are not stable in isolation but are stabilized by other parts of the polypeptide chain. Hydrophobic regions of fragments, which would normally be hidden in the inside of a folded protein, would be exposed to water in an aqueous solution; such fragments would tend to aggregate nonspecifically, and not have a defined structure, and they would be inactive for ligand binding, even if they contained all of the amino acids that would normally contribute to the ligand-binding site. A protein domain, in contrast, is considered a folding unit, and fragments of a polypeptide chain that correspond to intact domains are often able to fold correctly. Thus, separated protein domains often retain their activities, such as ligand binding, if the binding site is contained entirely within the domain. Thus the most likely place in which the polypeptide chain of the protein in Figure 4–19 could be severed to give rise to stable fragments is at the boundary between the two domains (i.e., at the loop between the two α helices at the bottom right of the structure shown).

ANSWER 4–17   Because of the lack of secondary structure, the C-terminal region of neurofilament proteins undergoes continual Brownian motion. The high density of negatively charged phosphate groups means that the C-terminals also experience repulsive interactions, which cause them to stand out from the surface of the neurofilament like the bristles of a brush. In electron micrographs of a cross section of an axon, the region occupied by the extended C-terminals appears as a clear zone around each neurofilament, from which organelles and other neurofilaments are excluded.

ANSWER 4–18   The heat-inactivation of the enzyme suggests that the mutation causes the enzyme to have a less stable structure. For example, a hydrogen bond that is normally formed between two amino acid side chains might no longer be formed because the mutation replaces one of these amino acids with a different one that cannot participate in the bond. Lacking such a bond that normally helps to keep the polypeptide chain folded properly, the protein partially or completely unfolds at a temperature at which it would normally be stable. Polypeptide chains that denature when the temperature is raised often aggregate, and they rarely refold into active proteins when the temperature is decreased.

ANSWER 4–19   The motor protein in the illustration can move just as easily to the left as to the right and so will not move steadily in one direction. However, if just one of the steps is coupled to ATP hydrolysis (for example, by making detachment of one foot dependent on binding of ATP and coupling the reattachment to hydrolysis of the bound ATP), then the protein will show unidirectional movement that requires the continued consumption of ATP. Note that, in principle, it does not matter which step is coupled to ATP hydrolysis (Figure A4–19).

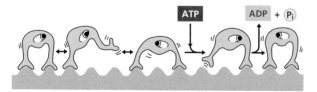

Figure A4–19

ANSWER 4–20   The slower migration of small molecules through a gel-filtration column occurs because smaller molecules have access to many more spaces in the porous beads that are packed into the column than do larger molecules. However, it is important that the flow rate through the column is slow enough to give the smaller molecules sufficient time to diffuse into the spaces inside the beads. At very rapid flow rates, all molecules will move rapidly around the beads, so that large and small molecules will now tend to exit together from the column.

ANSWER 4–21   The α helix in the figure is right-handed, whereas the coiled-coil is left-handed. The reversal occurs because of the staggered positions of hydrophobic side chains in the α helix.

ANSWER 4–22   The atoms at the binding sites of proteins must be precisely located to fit the molecules that they bind. Their location in turn requires the precise positioning of many of the amino acids and their side chains in the core of the protein, distant from the binding site itself. Thus, even a small change in this core can disrupt protein function by altering the conformation at a binding site far away.

Chapter 5

ANSWER 5–1
A. False. The polarity of a DNA strand commonly refers to the orientation of its sugar–phosphate backbone, one end of which contains a phosphate group and the other a hydroxyl group.
B. True. G-C base pairs are held together by three hydrogen bonds, whereas A-T base pairs are held together by only two.

ANSWER 5–2   Histone octamers occupy about 9% of the volume of the nucleus. The volume of the nucleus is

$$V = 4/3 \times 3.14 \times (3 \times 10^3 \text{ nm})^3$$

$$V = 1.13 \times 10^{11} \text{ nm}^3$$

The volume of the histone octamers is

$$V = 3.14 \times (4.5 \text{ nm})^2 \times (5 \text{ nm}) \times (32 \times 10^6)$$

$$V = 1.02 \times 10^{10} \text{ nm}^3$$

The ratio of the volume of histone octamers to the nuclear volume is 0.09; thus, histone octamers occupy about 9% of the nuclear volume. Because the DNA also occupies about 9% of the nuclear volume, together they occupy about 18% of the volume of the nucleus.

ANSWER 5–3   In contrast to most proteins, which accumulate amino acid changes over evolutionary time, the functions of histone proteins must involve nearly all of their amino acids, so that a change in any position would be deleterious to the cell.

ANSWER 5–4   Men have only one copy of the X chromosome in their cells; a defective gene carried on it therefore has no backup copy. Women, on the other hand, have two copies of the X chromosome in their cells, one inherited from each parent, so a defective copy of the gene on one X chromosome can generally be compensated for by a normal copy on the other chromosome. This is the case with regard to the gene that causes color blindness. However, during female development, one X chromosome in each cell is inactivated by compaction into heterochromatin, shutting down gene expression from that chromosome (see Figure 5–30). This occurs at random in each cell to one or the other of the two X chromosomes, and therefore some cells of the woman will express the mutant copy of the gene, whereas others will express the normal copy. This results in a retina in which on average only every other cone cell is color sensitive, and women carrying the mutant gene on one X chromosome therefore see colored objects with reduced resolution.

A woman who is color-blind must have two defective copies of this gene, one inherited from each parent. Her father must therefore carry the mutation on his X chromosome; because this is his only copy of the gene, he would be color-blind. Her mother could carry the defective gene on either or both of her X chromosomes: if she carried it on both, she would be color-blind; if she carried it on one, she would have color vision but reduced resolution, as described above. Several different types of inherited color blindness are found in the human population; this question applies to only one type.

ANSWER 5–5

A. The complementary strand reads 5′-TGATTGTGGACAAAAATCC-3′. Paired DNA strands have opposite polarity, and the convention is to write a single-stranded DNA sequence in the 5′-to-3′ direction.

B. The DNA is made of four nucleotides (100% = 13% A + x% T + y% G + z% C). Because A pairs with T, the two nucleotides are represented in equimolar proportions in DNA. Therefore, the bacterial DNA in question contains 13% thymidine. This leaves 74% [= 100% – (13% + 13%)] for G and C, which also form base pairs and hence are equimolar. Thus $y = z = 74/2 = 37$.

C. A single-stranded DNA molecule that is N nucleotides long can have any one of 4^N possible sequences, but the number of possible double-stranded DNA molecules is more difficult to calculate. Many of the 4^N single-stranded sequences will be the complement of another possible sequence in the list; for example, 5′-AGTCC-3′ and 5′-GGACT-3′ form the same double-stranded DNA molecule and therefore count as a single, double-stranded possibility. If N is an odd number, then every single-stranded sequence will complement another sequence in the list so that the number of double-stranded sequences will be 0.5×4^N. If N is an even number, then there will be slightly more than this, since some sequences will be self-complementary (such as 5′-ACTAGT-3′) and the actual value can be calculated to be $0.5 \times 4^N + 0.5 \times 4^{N/2}$.

D. To specify a unique sequence which is N nucleotides long, 4^N has to be larger than 3×10^6. Thus, $4^N > 3 \times 10^6$, solved for N, gives $N > \ln(3 \times 10^6)/\ln(4) = 10.7$. Thus, on average, a sequence of only 11 nucleotides in length is unique in the genome. Performing the same calculation for the genome size of an animal cell yields a minimal stretch of 16 nucleotides. This shows that a relatively short sequence can mark a unique position in the genome and is sufficient, for example, to serve as an identity tag for one specific gene.

ANSWER 5–6   If the wrong bases were frequently incorporated during DNA replication, genetic information could not be inherited accurately. Life, as we know it, could

not exist. Although the bases can form hydrogen-bonded pairs as indicated, these do not fit into the structure of the double helix. The angle with which the A residue is attached to the sugar–phosphate backbone is vastly different in the A-C pair, and the spacing between the two sugar–phosphate strands is increased in the A-G pair, where two large purine rings interact. Consequently, it is energetically unfavorable to incorporate a wrong base in DNA, and such errors occur only very rarely.

ANSWER 5–7

A. The bases V, W, X, and Y can form a DNA-like double-helical molecule with virtually identical properties to those of bona fide DNA. V would always pair with X, and W with Y. Therefore, the macromolecules could be derived from a living organism that uses the same principles to replicate its genome as those used by organisms on Earth. In principle, different bases, such as V, W, X, and Y, could have been selected during evolution on Earth as building blocks for DNA. (Similarly, there are many more conceivable amino acid side chains than the set of 20 selected in evolution that make up all proteins.)

B. None of the bases V, W, X, or Y can replace A, T, G, or C. To preserve the distance between the two sugar–phosphate strands in a double helix, a pyrimidine always has to pair with a purine (see, for example, Figure 5–6). Thus, the eight possible combinations would be V-A, V-G, W-A, W-G, X-C, X-T, Y-C, and Y-T. Because of the positions of hydrogen-bond acceptors and hydrogen-bond donor groups, however, no stable base pairs would form in any of these combinations, as shown for the pairing of V and A in Figure A5–7, where only a single hydrogen bond could form.

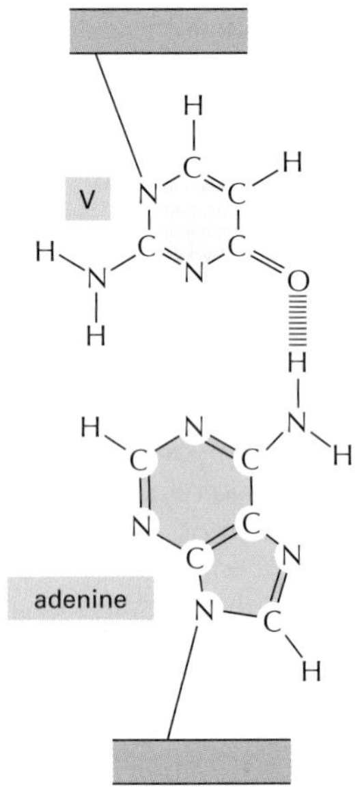

Figure A5–7

ANSWER 5–8   As the two strands are held together by hydrogen bonds between the bases, the stability of a DNA double helix is largely dependent on the number of hydrogen bonds that can be formed. Thus two parameters determine the stability: the number of nucleotide pairs and the number of hydrogen bonds that each nucleotide pair contributes. As shown in Figure 5–6, an A-T pair contributes two hydrogen bonds, whereas a G-C pair contributes three hydrogen bonds. Therefore, helix C (containing a total of 34 hydrogen bonds) would melt at the lowest temperature, helix B (containing a total of 65 hydrogen bonds) would melt next, and helix A (containing a total of 78 hydrogen bonds) would melt last. Helix A is the most stable, largely owing to its high GC content. Indeed, the DNA of organisms that grow in extreme temperature environments, such as certain prokaryotes that grow in geothermal vents, has an unusually high GC content.

ANSWER 5–9   The DNA would be enlarged by a factor of 2.5×10^6 ($= 5 \times 10^{-3}/2 \times 10^{-9}$ m). Thus the extension cord would be 2500 km long. This is approximately the distance from London to Istanbul, San Francisco to Kansas City, Tokyo to the southern tip of Taiwan, and Melbourne to Cairns. Adjacent nucleotides would be about 0.85 mm apart (which is only about the thickness of a stack of 12 pages of this book). A gene that is 1000 nucleotide pairs long would be about 85 cm in length.

ANSWER 5–10

A. It takes two bits to specify each nucleotide pair (for example, 00, 01, 10, and 11 would be the binary codes for the four different nucleotides, each paired with its appropriate partner).
B. The entire human genome (3×10^9 nucleotide pairs) could be stored on two CDs ($3 \times 10^9 \times 2$ bits/4.8×10^9 bits).

ANSWER 5–11

A. True.
B. False. Nucleosome core particles are approximately 11 nm in diameter.

ANSWER 5–12   The definitions of the terms can be found in the Glossary. DNA assembles with specialized proteins to form *chromatin*. At a first level of packing, *histones* form the core of *nucleosomes*. In a nucleosome, the DNA is wrapped almost twice around this core. Between nuclear divisions—that is, in interphase—the *chromatin* of the *interphase chromosomes* is in a relatively extended form in the nucleus, although some regions of it, the *heterochromatin*, remain densely packed and are transcriptionally inactive. During nuclear division—that is, in mitosis—replicated chromosomes become condensed into *mitotic chromosomes*, which are transcriptionally inactive and are designed to be readily distributed between the two daughter cells.

ANSWER 5–13   Colonies are clumps of cells that originate from a single founder cell and grow outward as the cells divide again and again. In the lower colony of Figure Q5–13, the *Ade2* gene is inactivated when placed near a telomere, but apparently it can become spontaneously activated in a few cells, which then turn white. Once activated in a cell, the *Ade2* gene continues to be active in the descendants of that cell, resulting in clumps of white cells (the white sectors) in the colony. This result shows both that the inactivation of a gene positioned close to a telomere can be reversed and that this change is passed on to further generations. This change in *Ade2* expression probably results from a spontaneous decondensation of the chromatin structure around the gene.

ANSWER 5–14   In the electron micrographs, one can detect chromatin regions of two different densities; the densely stained regions correspond to heterochromatin, while less condensed chromatin is more lightly stained. The chromatin in A is mostly in the form of condensed, transcriptionally inactive heterochromatin, whereas most of the chromatin in B is decondensed and therefore potentially transcriptionally active. The nucleus in A is from a reticulocyte, a red blood cell precursor, which is largely devoted to making a single protein, hemoglobin. The nucleus in B is from a lymphocyte, which is active in transcribing many different genes.

ANSWER 5–15   Helix A is right-handed. Helix C is left-handed. Helix B has one right-handed strand and one left-handed strand. There are several ways to tell the handedness of a helix. For a vertically oriented helix, like the ones in Figure Q5–15, if the strands in front point up to the right, the helix is right-handed; if they point up to the left, the helix is left-handed. Once you are comfortable identifying the handedness of a helix, you will be amused to note that nearly 50% of the "DNA" helices shown in advertisements are left-handed, as are a surprisingly high number of the ones shown in books. Amazingly, a version of Helix B was used in advertisements for a prominent international conference, celebrating the 30-year anniversary of the discovery of the DNA helix.

ANSWER 5–16   The packing ratio within a nucleosome core is 4.5 [(147 bp × 0.34 nm/bp)/(11 nm) = 4.5]. If there is an additional 54 bp of linker DNA, then the packing ratio for "beads-on-a-string" DNA is 2.3 [(201 bp × 0.34 nm/bp)/(11 nm + {54 bp × 0.34 nm/bp}) = 2.3]. This first level of packing represents only 0.023% (2.3/10,000) of the total condensation that occurs at mitosis.

Chapter 6

ANSWER 6–1

A. The distance between replication forks 4 and 5 is about 280 nm, corresponding to 824 nucleotides (= 280/0.34). These two replication forks would collide in about 8 seconds. Forks 7 and 8 move away from each other and would therefore never collide.
B. The total length of DNA shown in the electron micrograph is about 1.5 μm, corresponding to 4400 nucleotides. This is only about 0.002% [= (4400/1.8 × 10^8) × 100%] of the total DNA in a fly cell.

ANSWER 6–2   Although the process may seem wasteful, it is not possible to proofread during the initial stages of primer synthesis. To start a new primer on a piece of single-stranded DNA, one nucleotide needs to be put in place and then linked to a second and then to a third, and so on. Even if these first nucleotides were perfectly matched to the template strand, they would bind with very low affinity, and it would consequently be difficult to distinguish the correct from incorrect bases by a hypothetical primase with proofreading activity; the enzyme would therefore stall. The task of the primase is to "just polymerize nucleotides that bind reasonably well to the template without worrying too much about accuracy." Later, these sequences are removed and replaced by DNA polymerase, which uses newly synthesized (and therefore proofread) DNA as its primer.

ANSWER 6–3

A. Without DNA polymerase, no replication can take place at all. RNA primers will be laid down at the origin of replication.

B. DNA ligase links the DNA fragments that are produced on the lagging strand. In the absence of ligase, the newly replicated DNA strands will remain as fragments, but no nucleotides will be missing.

C. Without the sliding clamp, the DNA polymerase will frequently fall off the DNA template. In principle, it can rebind and continue, but the continual falling off and rebinding will be time-consuming and will greatly slow down DNA replication.

D. In the absence of RNA-excision enzymes, the RNA fragments will remain covalently attached to the newly replicated DNA fragments. No ligation will take place, because the DNA ligase will not link DNA to RNA. The lagging strand will therefore consist of fragments composed of both RNA and DNA.

E. Without DNA helicase, the DNA polymerase will stall because it cannot separate the strands of the template DNA ahead of it. Little or no new DNA will be synthesized.

F. In the absence of primase, RNA primers cannot begin on either the leading or the lagging strand. DNA replication therefore cannot begin.

ANSWER 6–4

DNA damage by deamination and depurination reactions occurs spontaneously. This type of damage is not the result of replication errors and is therefore equally likely to occur on either strand. If DNA repair enzymes recognized such damage only on newly synthesized DNA strands, half of the defects would go uncorrected. The statement is therefore incorrect.

ANSWER 6–5

If the old strand were "repaired" using the new strand that contains a replication error as the template, then the error would become a permanent mutation in the genome. The old information would be erased in the process. Therefore, if repair enzymes did not distinguish between the two strands, there would be only a 50% chance that any given replication error would be corrected.

ANSWER 6–6

The argument is severely flawed. You cannot transform one species into another simply by introducing random changes into the DNA. It is exceedingly unlikely that the 5000 mutations that would accumulate every day in the absence of the DNA repair enzyme would be in the very positions where human and chimpanzee DNA sequences are different. It is very likely that, at such a high mutation frequency, many essential genes would be inactivated, leading to cell death. Furthermore, your body is made up of about 10^{13} cells. For you to turn into an ape, not just one but many of these cells would need to be changed. And even then, many of these changes would have to occur during development to effect changes in your body plan (making your arms longer than your legs, for example).

ANSWER 6–7

A. False. Identical DNA polymerase molecules catalyze DNA synthesis on the leading and lagging strands of a bacterial replication fork. The replication fork is asymmetrical because the lagging strand is synthesized in pieces that are then stitched together.

B. False. Only the RNA primers are removed by an RNA nuclease; Okazaki fragments are pieces of newly synthesized DNA on the lagging strand that are eventually joined together by DNA ligase.

C. True. With proofreading, DNA polymerase has an error rate of one mistake in 10^7 nucleotides polymerized; 99% of its errors are corrected by DNA mismatch repair enzymes, bringing the final error rate to one in 10^9.

D. True. Mutations would accumulate rapidly, inactivating many genes.

E. True. If a damaged nucleotide also occurred naturally in DNA, the repair enzyme would have no way of identifying the damage. It would therefore have only a 50% chance of fixing the right strand.

F. True. Usually, multiple mutations of specific types need to accumulate in a somatic cell lineage to produce a cancer. A mutation in a gene that codes for a DNA repair enzyme can make a cell more liable to accumulate further mutations, thereby accelerating the onset of cancer.

ANSWER 6–8

With a single origin of replication, which launches two DNA polymerases in opposite directions on the DNA, each moving at 100 nucleotides per second, the number of nucleotides replicated in 24 hours will be 1.73×10^7 ($= 2 \times 100 \times 24 \times 60 \times 60$). To replicate all the 6×10^9 nucleotides of DNA in the cell in this time, therefore, will require at least 348 ($= 6 \times 10^9/1.73 \times 10^7$) origins of replication. The estimated 10,000 origins of replication in the human genome are therefore more than enough to satisfy this minimum requirement.

ANSWER 6–9

A. Dideoxycytosine triphosphate (ddCTP) is identical to dCTP, except it lacks the 3′-hydroxyl group on the sugar ring. ddCTP is recognized by DNA polymerase as dCTP and becomes incorporated into DNA; because it lacks the crucial 3′-hydroxyl group, however, its addition to a growing DNA strand creates a dead end to which no further nucleotides can be added. Thus, if ddCTP is added in large excess, new DNA strands will be synthesized until the first G (the nucleotide complementary to C) is encountered on the template strand. ddCTP will then be incorporated instead of C, and the extension of this strand will be terminated.

B. If ddCTP is added at about 10% of the concentration of the available dCTP, there is a 1 in 10 chance of its being incorporated whenever a G is encountered on the template strand. Thus a population of DNA fragments will be synthesized, and from their lengths one can deduce where the G residues are located on the template strand. This strategy forms the basis of methods used to determine the sequence of nucleotides in a stretch of DNA (discussed in Chapter 10).

The same chemical phenomenon is exploited by a drug, 3′-azido-3′-deoxythymidine (AZT), that is now commonly used in HIV-infected patients to treat AIDS. AZT is converted in cells to the triphosphate form and is incorporated into the growing viral DNA. Because the drug lacks a 3′-OH group, it blocks DNA synthesis and replication of the virus. AZT inhibits viral replication preferentially because reverse transcriptase has a higher affinity for the drug than for thymidine triphosphate; human cellular DNA polymerases do not show this preference.

C. Dideoxycytosine monophosphate (ddCMP) lacks the 5'-triphosphate group as well as the 3'-hydroxyl group of the sugar ring. It therefore cannot provide the energy that drives the polymerization reaction of nucleotides into DNA and therefore will not be incorporated into the replicating DNA. Addition of this compound should not affect DNA replication.

ANSWER 6–10   See Figure A6–10.

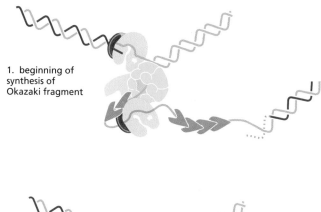

1. beginning of synthesis of Okazaki fragment

2. midpoint of synthesis of Okazaki fragment

Figure A6–10

ANSWER 6–11   Both strands of the bacterial chromosome contain 6×10^6 nucleotides. During the polymerization of nucleoside triphosphates into DNA, two phosphoanhydride bonds are broken for each nucleotide added: the nucleoside triphosphate is hydrolyzed to produce the nucleoside monophosphate added to the growing DNA strand, and the released pyrophosphate is hydrolyzed to phosphate. Therefore, 1.2×10^7 high-energy bonds are hydrolyzed during each round of bacterial DNA replication. This requires 4×10^5 $(= 1.2 \times 10^7/30)$ glucose molecules, which weigh 1.2×10^{-16} g $(= 4 \times 10^5$ molecules $\times 180$ g/mole/ 6×10^{23} molecules/mole), which is 0.01% of the total weight of the cell.

ANSWER 6–12   The statement is correct. If the DNA in somatic cells is not sufficiently stable (that is, if it accumulates mutations too rapidly), the organism dies (of cancer, for example), and because this may often happen before the organism can reproduce, the species will die out. If the DNA in reproductive cells is not sufficiently stable, many mutations will accumulate and be passed on to future generations, so that the species will not be maintained.

ANSWER 6–13   As shown in Figure A6–13, thymine and uracil lack amino groups and therefore cannot be deaminated. Deamination of adenine and guanine produces

Figure A6–13

purine rings that are not found in conventional nucleic acids. In contrast, deamination of cytosine produces uracil. Therefore, if uracil were a naturally occurring base in DNA (as it is in RNA), repair enzymes could not distinguish whether a uracil is the appropriate base or whether it arose through spontaneous deamination of cytosine. This dilemma is not encountered, however, because thymine, rather than uracil, is used in DNA. Therefore, if a uracil base is found in DNA, it can be automatically recognized as a damaged base and then excised and replaced by cytosine.

ANSWER 6–14

A. Because DNA polymerase requires a 3'-OH to synthesize DNA, without telomeres and telomerase, the ends of linear chromosomes would shrink during each round of DNA replication (Figure A6–14). For bacterial chromosomes, which have no ends, the problem does not arise; there will always be a 3'-OH group available to prime the DNA polymerase that replaces the RNA primer with DNA. Telomeres and telomerase prevent the shrinking of chromosomes because they extend the 3' end of a DNA strand (see Figure 6–22). This extension of the lagging-strand template provides the "space" to begin the final Okazaki fragments.

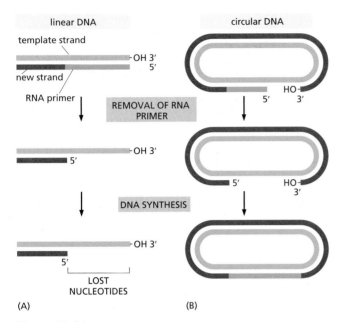

Figure A6–14

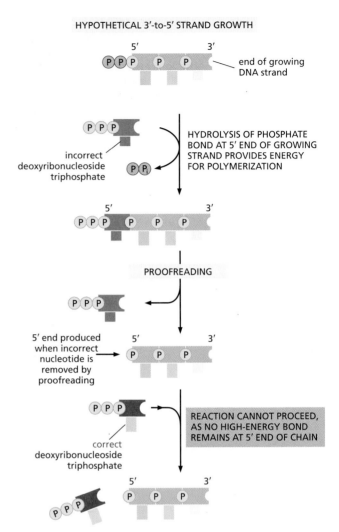

Figure A6–16

B. As shown in Figure A6–14, telomeres and telomerase are still needed even if the last fragment of the lagging strand were initiated by primase at the very 3′ end of chromosomal DNA, inasmuch as the RNA primer must be removed.

ANSWER 6–15

A. If the single origin of replication were located exactly in the center of the chromosome, it would take more than 8 days to replicate the DNA [= 75×10^6 nucleotides/(100 nucleotides/sec)]. The rate of replication would therefore severely limit the rate of cell division. If the origin were located at one end, the time required to replicate the chromosome would be approximately double this.

B. A chromosome end that is not "capped" with a telomere would lose nucleotides during each round of DNA replication and would gradually shrink. Eventually, essential genes would be lost, and the chromosome's ends might be recognized by the DNA damage-response mechanisms, which would stop cell division or induce cell death.

C. Without centromeres, which attach mitotic chromosomes to the mitotic spindle, the two new chromosomes that result from chromosome duplication would not be partitioned accurately between the two daughter cells. Therefore, many daughter cells would die, because they would not receive a full set of chromosomes.

ANSWER 6–16 The addition of each nucleotide by a hypothetical polymerase that synthesized DNA in the reverse 3′-to-5′ direction would require the energy provided by hydrolysis of the high-energy phosphate bond at the 5′ end of the growing chain—rather than at the 5′ end of the incoming nucleotide, as do the actual DNA polymerases. If an incorrectly incorporated nucleotide were removed from such a growing chain, DNA synthesis would grind to a halt, as there would be no high-energy bonds remaining at the 5′ end of the chain to fuel further polymerization (see Figure A6–16).

Chapter 7

ANSWER 7–1   Perhaps the best answer was given by Francis Crick himself, who coined the term in the mid-1950s: "I called this idea the central dogma for two reasons, I suspect. I had already used the obvious word hypothesis in the sequence hypothesis, which proposes that genetic information is encoded in the sequence of the DNA bases, and in addition I wanted to suggest that this new assumption was more central and more powerful…. As it turned out, the use of the word dogma caused more trouble than it was worth. Many years later Jacques Monod pointed out to me that I did not appear to understand the correct use of the word dogma, which is a belief that cannot be doubted. I did appreciate this in a vague sort of way but since I thought that all religious beliefs were without serious foundation, I used the word in the way I myself thought about it, not as the world does, and simply applied it to a grand hypothesis that, however plausible, had little direct experimental support at the time." (Francis Crick, *What Mad Pursuit: A Personal View of Scientific Discovery*. Basic Books, 1988.)

ANSWER 7–2   Actually, the RNA polymerases are not moving at all in the micrograph, because they have been fixed and coated with metal to prepare the sample for viewing in the electron microscope. However, before they were fixed, they were moving from left to right, as indicated by the gradual lengthening of the RNA transcripts.

The RNA transcripts are shorter because they begin to fold up (i.e., to acquire a three-dimensional structure) as they are synthesized (see, for example, Figure 7–5), whereas the DNA is an extended double helix.

ANSWER 7–3   At first glance, the catalytic activities of an RNA polymerase used for transcription could replace the DNA primase. Upon further reflection, however, there are some serious problems. (1) The RNA polymerase used to make primers would need to initiate every few hundred bases, which is much more often than promoters are spaced on the DNA. Initiation would therefore need to occur in a promoter-independent fashion or many more promoters would have to be present in the DNA, both of which would be problematic for the control of transcription. (2) Similarly, the RNA primers used in DNA replication are much shorter than mRNAs. The RNA polymerase would therefore need to terminate much more frequently than during transcription. Termination would need to occur spontaneously, i.e., without requiring a terminator sequence in the DNA, or many more terminators would need to be present. Again, both of these scenarios would be problematic for the control of transcription.

Although it might be possible to overcome this problem if special control proteins became attached to RNA polymerase during replication, the problem has been solved during evolution by using separate enzymes with specialized properties. Some small DNA viruses, however, do utilize the host RNA polymerase to make DNA primers for their replication.

ANSWER 7–4   This experiment demonstrates that the ribosome does not check the amino acid that is attached to a tRNA. Once an amino acid has been coupled to a tRNA, the ribosome will "blindly" incorporate that amino acid into the position according to the match between the codon and anticodon. We can therefore conclude that a significant part of the correct reading of the genetic code, i.e., the matching of a codon in an mRNA with the correct amino acid, is performed by the synthetase enzymes that correctly match tRNAs and amino acids.

ANSWER 7–5   The mRNA will have a 5′-to-3′ polarity, opposite to that of the DNA strand that serves as the template. Thus the mRNA sequence will read 5′-GAAAAAAGCCGUUAA-3′. The N-terminal amino acid coded for by GAA is glutamic acid. UAA specifies a stop codon, so the C-terminal amino acid is coded for by CGU and is an arginine. Note that the convention in describing the sequence of a gene is to give the sequence of the DNA strand that is *not* used as a template for RNA synthesis; this sequence is the same as that of the RNA transcript, with T written in place of U.

ANSWER 7–6   The first statement is probably correct: RNA is thought to have been the first self-replicating catalyst and, in modern cells, is no longer self-replicating. We can debate, however, whether this represents a "loss." RNA now serves many roles in the cell: as messengers, as adaptors for protein synthesis, as primers for DNA replication, and as catalysts for some of the most fundamental reactions, especially RNA splicing and protein synthesis.

ANSWER 7–7

A. False. Ribosomes can make any protein that is specified by the particular mRNA that they are translating. After translation, ribosomes are released from the mRNA and can then start translating a different mRNA. It is true, however, that a ribosome can only make one type of protein at a time.

B. False. mRNAs are translated as linear polymers; there is no requirement that they have any particular folded structure. In fact, such structures that are formed by mRNA can inhibit its translation, because the ribosome has to unfold the mRNA in order to read the message it contains.

C. False. Ribosomal subunits exchange partners after each round of translation. After a ribosome is released from an mRNA, its two subunits dissociate and enter a pool of free small and large subunits from which new ribosomes assemble around a new mRNA.

D. False. Ribosomes are cytoplasmic organelles, but they are not individually enclosed in a membrane.

E. False. The position of the promoter determines the direction in which transcription proceeds and therefore which DNA strand is used as the template. Transcription in the opposite direction would produce an mRNA with a completely different (and probably meaningless) sequence.

F. False. RNA contains uracil but not thymine.

G. False. The level of a protein depends on its rate of synthesis and degradation but not on its catalytic activity.

ANSWER 7–8   Because the deletion in the Lacheinmal mRNA is internal, it is likely that the deletion arises from an mRNA splicing defect. The simplest interpretation is that the *Lacheinmal* gene contains a 173-nucleotide-long exon (labeled "E2" in Figure A7–8), and that this exon is lost during the processing of the mutant precursor mRNA (pre-mRNA). This could occur, for example, if the mutation changed the 3′ splice site in the preceding intron ("I1") so that it was no longer recognized by the splicing machinery (a change in the CAG sequence shown in Figure 7–19 could do this). The snRNP would search for the next available 3′ splice site, which is found at the 3′ end of the next intron ("I2"), and the splicing reaction would therefore remove E2 together with I1 and I2, resulting in a shortened mRNA. The mRNA is then translated into a defective protein, resulting in the Lacheinmal deficiency.

Because 173 nucleotides do not amount to an integral number of codons, the lack of this exon in the mRNA will shift the reading frame at the splice junction. Therefore, the Lacheinmal protein would be made correctly only through exon E1. As the ribosome begins translating sequences in exon E3, it will be in a different reading frame and therefore will produce a protein sequence that is unrelated to the Lacheinmal sequence normally encoded by exon E3. Most likely, the ribosome will soon encounter a stop codon, which in RNA sequences that do not code for protein would be expected to occur on average about once in every 21 codons (there are 3 stop codons in the 64 codons of the genetic code).

ANSWER 7–9   Sequence 1 and sequence 4 both code for the peptide Arg-Gly-Asp. Because the genetic code is redundant, different nucleotide sequences can encode the same amino acid sequence.

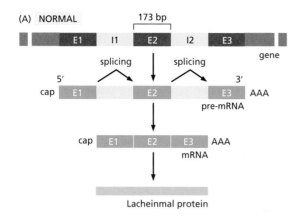

(A) NORMAL

173 bp

splicing   splicing

gene

5′
cap

3′
AAA

pre-mRNA

cap   E1   E2   E3   AAA

mRNA

Lacheinmal protein

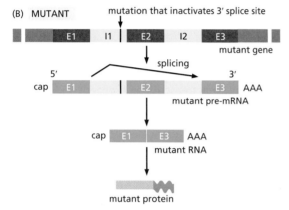

(B) MUTANT

mutation that inactivates 3′ splice site

mutant gene

splicing

5′
cap

3′
AAA

mutant pre-mRNA

cap   E1   E3   AAA

mutant RNA

mutant protein

Figure A7–8

ANSWER 7–10

A. Incorrect. The bonds are not covalent, and their formation does not require input of energy.

B. Correct. The aminoacyl-tRNA enters the ribosome at the A site and forms hydrogen bonds with the codon in the mRNA.

C. Correct. As the ribosome moves along the mRNA, the tRNAs that have donated their amino acid to the growing polypeptide chain are ejected from the ribosome and the mRNA. The ejection takes place two cycles after the tRNA first enters the ribosome (see Figure 7–34).

ANSWER 7–11
Replication. Dictionary definition: the creation of an exact copy; molecular biology definition: the act of duplicating DNA. Transcription. Dictionary definition: the act of writing out a copy, especially from one physical form to another; molecular biology definition: the act of copying the information stored in DNA into RNA. Translation. Dictionary definition: the act of putting words into a different language; molecular biology definition: the act of polymerizing amino acids into a defined linear sequence using the information provided by the linear sequence of nucleotides in mRNA. (Note that "translation" is also used in a quite different sense, both in ordinary language and in scientific contexts, to mean a movement from one place to another.)

ANSWER 7–12
With four different nucleotides to choose from, a code of two nucleotides could specify 16 different amino acids (= 4^2), and a triplet code in which the position of the nucleotides is not important could specify 20 different amino acids (= 4 possibilities of 3 of the same bases +

12 possibilities of 2 bases the same and one different + 4 possibilities of 3 different bases). In both cases, these maximal amino acid numbers would need to be reduced by at least 1, because of the need to specify translation stop codons. It is relatively easy to envision how a doublet code could be translated by a mechanism similar to that used in our world by providing tRNAs with only two relevant bases in the anticodon loop. It is more difficult to envision how the nucleotide composition of a stretch of three nucleotides could be translated without regard to their order, because base-pairing can then no longer be used: AUG, for example, will not base-pair with the same anticodon as UGA.

ANSWER 7–13
It is likely that in early cells the matching between codons and amino acids was less accurate than it is in present-day cells. The feature of the genetic code described in the question may have allowed early cells to tolerate this inaccuracy by allowing a blurred relationship between sets of roughly similar codons and roughly similar amino acids. One can easily imagine how the matching between codons and amino acids could have become more accurate, step by step, as the translation machinery evolved into that found in modern cells.

ANSWER 7–14
The codon for Trp is 5′-UGG-3′. Thus a normal Trp-tRNA contains the sequence 5′-CCA-3′ as its anticodon (see Figure 7–30). If this tRNA contains a mutation so that its anticodon is changed to UCA, it will recognize a UGA codon and lead to the incorporation of a tryptophan residue instead of causing translation to stop. Many other protein-encoding sequences, however, contain UGA codons as their natural stop sites, and these stops would also be affected by the mutant tRNA. Depending on the competition between the altered tRNA and the normal translation release factors (Figure 7–38), some of these proteins would be made with additional amino acids at their C-terminal end. The additional lengths would depend on the number of codons before the ribosomes encounter a non-UGA stop codon in the mRNA in the reading frame in which the protein is translated.

ANSWER 7–15
One effective way of driving a reaction to completion is to remove one of the products, so that the reverse reaction cannot occur. ATP contains two high-energy bonds that link the three phosphate groups. In the reaction shown, PP_i is released, consisting of two phosphate groups linked by one of these high-energy bonds. Thus PP_i can be hydrolyzed with a considerable gain of free energy, and thereby can be efficiently removed. This happens rapidly in cells, and reactions that produce and further hydrolyze PP_i are therefore virtually irreversible (see Figure 3–40).

ANSWER 7–16

A. A titin molecule is made of 25,000 (3,000,000/120) amino acids. It therefore takes about 3.5 hours [(25,000/2)× (1/60) × (1/60)] to synthesize a single molecule of titin in muscle cells.

B. Because of its large size, the probability of making a titin molecule without any mistakes is only 0.08 [= $(1 - 10^{-4})^{25,000}$]; i.e., only 8 in 100 titin molecules synthesized are free of mistakes. In contrast, over 97% of newly synthesized proteins of average size are made correctly.

C. The error rate limits the sizes of proteins that can be synthesized accurately. Similarly, if a eukaryotic ribosomal protein were synthesized as a single molecule, a large portion (87%) of this hypothetical giant ribosomal protein would be expected to contain at least one mistake. It is therefore more advantageous to make ribosomal proteins individually, because in this way only a small proportion of each type of protein will be defective, and these few bad molecules can be individually eliminated by proteolysis to ensure that there are no defects in the ribosome as a whole.

D. To calculate the time it takes to transcribe a titin mRNA, you would need to know the size of its gene, which is likely to contain many introns. Transcription of the exons alone (25,000 × 3 = 75,000 nucleotides) requires about 42 minutes [(75,000/30) × (1/60)]. Because introns can be quite large, the time required to transcribe the entire gene is likely to be considerably longer.

ANSWER 7–17   Mutations of the type described in (B) and (D) are often the most harmful. In both cases, the reading frame would be changed, and because this frameshift occurs near the beginning or in the middle of the coding sequence, much of the protein will contain a nonsensical and/or truncated sequence of amino acids. In contrast, a reading-frame shift that occurs toward the end of the coding sequence, as described in (A), will result in a largely correct protein that may be functional. Deletion of three consecutive nucleotides, as described in (C), leads to the deletion of an amino acid but does not alter the reading frame. The deleted amino acid may or may not be important for the folding or activity of the protein; in many cases such mutations are silent, i.e., have no or only minor consequences for the organism. Substitution of one nucleotide for another, as in (E), is often completely harmless. In some cases, it will not change the amino acid sequence of the protein; in other cases it will change a single amino acid; at worst, it may create a new stop codon, giving rise to a truncated protein.

Chapter 8

ANSWER 8–1

A. Transcription of the tryptophan operon would no longer be regulated by the absence or presence of tryptophan; the enzymes would be permanently turned on in scenarios (1) and (2) and permanently shut off in scenario (3).

B. In scenarios (1) and (2), the normal tryptophan repressor molecules would completely restore the regulation of the tryptophan biosynthesis enzymes. In contrast, expression of the normal protein would have no effect in scenario (3), because the tryptophan operator would remain occupied by the mutant protein, even in the presence of tryptophan.

ANSWER 8–2   Contacts can form between the protein and the edges of the base pairs that are exposed in the major groove of the DNA (**Figure A8–2**). These sequence-specific contacts can include hydrogen bonds with the highlighted oxygen, nitrogen, and hydrogen atoms, as well as hydrophobic interactions with the methyl group on thymine (*yellow*). Note that the arrangement of hydrogen-bond donors (*blue*) and hydrogen-bond acceptors (*red*) of a T-A pair is different from that of a C-G pair. Similarly, the

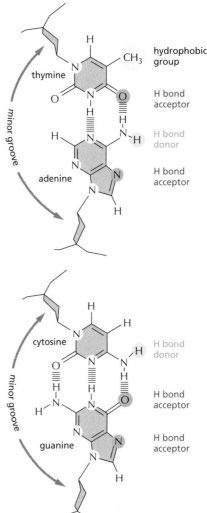

Figure A8–2

arrangement of hydrogen-bond donors and hydrogen-bond acceptors of A-T and G-C pairs would be different from one another and from the two pairs shown in the figure. These differences allow recognition of specific DNA sequences via the major groove. In addition to the contacts shown in the figure, electrostatic attractions between the positively charged amino acid side chains of the protein and the negatively charged phosphate groups in the DNA backbone usually stabilize DNA–protein interactions.

ANSWER 8–3
Bending proteins can help to bring distant DNA regions together that normally would contact each other only inefficiently (**Figure A8–3**). Such proteins are found in both prokaryotes and eukaryotes and are involved in many examples of transcriptional regulation.

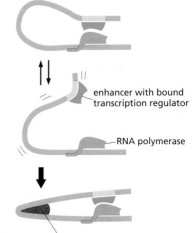

Figure A8–3

ANSWER 8–4

A. UV light throws the switch from the prophage to the lytic state: when cI protein is destroyed, Cro is made and turns off the further production of cI. The virus produces coat proteins, and new virus particles are made.

B. When the UV light is switched off, the virus remains in the lytic state. Thus, cI and Cro form a gene regulatory switch that "memorizes" its previous setting.

C. This switch makes sense in the viral life cycle: UV light tends to damage the bacterial DNA (see Figure 6–24), thereby rendering the bacterium an unreliable host for the virus. A prophage will therefore switch to the lytic state and leave the "sinking ship" in search for new host cells to infect.

ANSWER 8–5

A. True. Prokaryotic mRNAs are often transcripts of entire operons. Ribosomes can initiate translation at internal AUG start sites of these "polycistronic" mRNAs (see Figures 7–36 and 8–6).

B. True. The major groove of double-stranded DNA is sufficiently wide to allow a protein surface, such as one face of an α helix, access to the base pairs. The sequence of H-bond donors and acceptors in the major groove can then be "read" by the protein to determine the sequence and orientation of the DNA.

C. True. It is advantageous to exert control at the earliest possible point in a pathway. This conserves metabolic energy because unnecessary products are not made in the first place.

ANSWER 8–6   From our knowledge of enhancers, one would expect their function to be relatively independent of their distance from the promoter—possibly weakening as this distance increases. The surprising feature of the data (which have been adapted from an actual experiment) is the periodicity: the enhancer is maximally active at certain distances from the promoter (50, 60, or 70 nucleotides), but almost inactive at intermediate distances (55 or 65 nucleotides). The periodicity of 10 suggests that the mystery can be explained by considering the structure of double-helical DNA, which has 10 base pairs per turn. Thus, placing an enhancer on the side of the DNA opposite to that of the promoter (**Figure A8–6**) would make it more difficult for the

enhancer with bound transcription regulator

50 bp   RNA polymerase

55 bp

60 bp   **Figure A8–6**

activator that binds to it to interact with the proteins bound at the promoter. At longer distances, there is more DNA to absorb the twist, and the effect is diminished.

ANSWER 8–7   The affinity of the dimeric λ repressor for its binding site is the sum of the interactions made by each of the two DNA-binding domains. A single DNA-binding domain can make only half the contacts and provide just half the binding energy compared with the dimer. Thus, although the concentration of binding domains is unchanged, they are no longer coupled, and their individual affinities for DNA are sufficiently weak that they cannot remain bound. As a result, the genes for lytic growth are turned on.

ANSWER 8–8   The function of these *Arg* genes is to synthesize arginine. When arginine is abundant, expression of the biosynthetic genes should be turned off. If ArgR acts as a gene repressor (which it does in reality), then binding of arginine should increase its affinity for its regulatory sites, allowing it to bind and shut off gene expression. If ArgR acted as a gene activator instead, then the binding of arginine would be predicted to reduce its affinity for its regulatory DNA, preventing its binding and thereby shutting off expression of the *Arg* genes.

ANSWER 8–9   The results of this experiment favor DNA looping, which should not be affected by the protein bridge (so long as it allowed the DNA to bend, which it does). The scanning or entry-site model, however, is predicted to be affected by the nature of the linkage between the enhancer and the promoter. If the proteins enter at the enhancer and scan to the promoter, they would have to traverse the protein linkage. If such proteins are geared to scan on DNA, they would likely have difficulty scanning across such a barrier.

ANSWER 8–10   The most definitive result is one showing that a single differentiated cell taken from a specialized tissue can re-create a whole organism. This proves that the cell must contain all the information required to produce a whole organism, including all of its specialized cell types. Experiments of this type are summarized in Figure 8–2.

ANSWER 8–11   In principle, you could create 16 different cell types with 4 different transcription regulators (all the 8 cell types shown in Figure 8–17, plus another 8 created by adding an additional transcription regulator). MyoD by itself is sufficient to induce muscle-specific gene expression only in certain cell types, such as some kinds of fibroblasts. The action of MyoD is therefore consistent with the model shown in Figure 8–17: if muscle cells were specified, for example, by the combination of transcription regulators 1, 3, and MyoD, then the addition of MyoD would convert only two of the cell types of Figure 8–17 (cells F and H) to muscle.

ANSWER 8–12   The induction of a transcriptional activator protein that stimulates its own synthesis can create a positive feedback loop that can produce cell memory. The continued self-stimulated synthesis of activator A can in principle last for many cell generations, serving as a constant reminder of an event that took place in the past. By contrast, the induction of a transcriptional repressor that inhibits its own synthesis creates a negative feedback loop which ensures that the response to the transient stimulus will be similarly transient. Because repressor R shuts off its own synthesis, the cell will quickly return to the state that existed before the signal.

ANSWER 8–13   Many transcription regulators are continually made in the cell; that is, their expression is constitutive and the activity of the protein is controlled by signals from inside or outside the cell (e.g., the availability of nutrients, as for the tryptophan repressor, or by hormones, as for the glucocorticoid receptor), thereby adjusting the transcriptional program to the physiological needs of the cell. Moreover, a given transcription regulator usually controls the expression of many different genes. Transcription regulators are often used in various combinations and can affect each other's activity, thereby further increasing the possibilities for regulation with a limited set of proteins. Nevertheless, most cells devote a large fraction of their genomes to the control of transcription: about 10% of genes in eukaryotic cells code for transcription regulators.

Chapter 9

ANSWER 9–1   When it comes to genetic information, a balance must be struck between stability and change. If the mutation rate were too high, a species would eventually die out because all its individuals would accumulate mutations in genes essential for survival. And for a species to be successful—in evolutionary terms—individual members must have a good genetic memory; that is, there must be high fidelity in DNA replication. At the same time, occasional changes are needed if the species is to adapt to changing conditions. If the change leads to an improvement, it will persist by selection; if it is neutral, it may or may not accumulate; but if the change proves disastrous, the individual organism that was the unfortunate subject of nature's experiment will die, but the species will survive.

ANSWER 9–2   In single-celled organisms, the genome is the germ line and any modification is passed on to the next generation. By contrast, in multicellular organisms, most of the cells are somatic cells and make no contribution to the next generation; thus, modification of those cells by horizontal gene transfer would have no consequence for the next generation. The germ-line cells are usually sequestered in the interior of multicellular organisms, minimizing their contact with foreign cells, viruses, and DNA, thus insulating the species from the effects of horizontal gene transfer. Nevertheless, horizontal gene transfer is possible for multicellular organisms. For example, the genomes of some insect species contain DNA that was horizontally transferred from bacteria that infect them.

ANSWER 9–3   It is unlikely that any gene came into existence perfectly optimized for its function. Ribosomal RNA genes presumably varied a great deal when they first appeared on Earth. But this would have been at the very early stage of a common ancestral cell (see Figure 9–23). Since then there has been much less leeway for change since ribosomal RNA (and other highly conserved genes) play such a fundamental role in living processes. Nonetheless, the environment an organism finds itself in is changeable, so no gene can be optimal indefinitely. Thus we find there are indeed significant differences in ribosomal RNAs among species.

ANSWER 9–4   Each time another copy of a transposon is inserted into a chromosome, the change can be either neutral, beneficial, or detrimental for the organism. Because individuals that accumulate detrimental insertions would be selected against, the proliferation of transposons is controlled by natural selection. If a transposon arose that proliferated uncontrollably, it is unlikely that a viable host organism could be maintained. For this reason, most transposons have evolved to transpose only rarely. Many transposons, for example, synthesize only infrequent bursts of very small amounts of the transposase that is required for their movement.

ANSWER 9–5   Viruses cannot exist as free-living organisms: they have no metabolism, do not communicate with other viruses, and cannot reproduce themselves. They thus have none of the attributes that one normally associates with life. Indeed, they can even be crystallized. Only inside cells can they redirect normal cellular biosynthetic activities to the task of making more copies of themselves. Thus, the only aspect of "living" that viruses display is their capacity to direct their own reproduction once inside a cell.

ANSWER 9–6   Mobile genetic elements could provide opportunities for homologous recombination events, thereby causing genomic rearrangements. They could insert into genes, possibly obliterating splicing signals and thereby changing the protein produced by the gene. They could also insert into the regulatory region of a gene, where insertion between an enhancer and a transcription start site could block the function of the enhancer and therefore reduce the level of expression of a gene. In addition, the mobile genetic element could itself contain an enhancer and thereby change the time and place in the organism where the gene is expressed.

ANSWER 9–7   With their ability to facilitate genetic recombination, mobile genetic elements have almost certainly played an important part in the evolution of modern-day organisms. They can facilitate gene duplication and the creation of new genes via exon shuffling, and they can change the way in which existing genes are expressed. Although the transposition of a mobile genetic element can be harmful for an individual organism—if, for example, it disrupts the activity of a critical gene—these agents of genetic change may well be beneficial to the species as a whole.

ANSWER 9–8   About 7.6% of each gene is converted to mRNA [(5.4 exons/gene × 266 nucleotide pairs/exon)/ (19,000 nucleotide pairs/gene) = 7.6%]. Protein-coding genes occupy about 28% of Chromosome 22 [(700 genes × 19,000 nucleotide pairs/gene)/(48 × 10^6 nucleotide pairs) = 27.7%]. However, over 90% of this DNA is made of introns.

ANSWER 9–9   This statement is probably true. For example, nearly half our DNA is composed of defunct mobile genetic elements. And only about 9% of the human genome appears to be under positive selection. However, it is possible that future research will uncover a function for some portion of our seemingly unimportant DNA.

ANSWER 9–10   The HoxD cluster is packed with complex and extensive regulatory sequences that direct each of its genes to be expressed at the correct time and place during development. Insertion of mobile genetic elements into the

(A) POSITIONS OF HUMAN β-GLOBIN EXONS

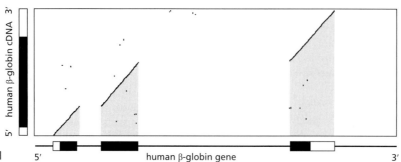

(B) HOMOLOGY BETWEEN MOUSE AND HUMAN GENES

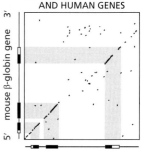

Figure A9–11

HoxD cluster is thought to be selected against because it would disrupt proper regulation of its resident genes.

ANSWER 9–11

A. The exons in the human β-globin gene correspond to the positions of sequence similarity (in this case identity) with the cDNA, which is a direct copy of the mRNA and thus contains no introns. The introns correspond to the regions between the exons. The positions of the introns and exons in the human β-globin gene are indicated in **Figure A9–11A.** Also shown (in open bars) are sequences present in the mature β-globin mRNA (and in the gene) that are not translated into protein.

B. From the positions of the exons, as defined in Figure A9–11A, it is clear that the first two exons of the human β-globin gene have counterparts, with similar sequence, in the mouse β-globin gene (Figure A9–11B). However, only the first half of the third exon of the human β-globin gene is similar to the mouse β-globin gene. The similar portion of the third exon contains sequences that encode protein, whereas the portion that is different represents the 3' untranslated region of the gene. Because this portion of the gene does not encode protein (nor does it contain extensive regulatory sequences), its sequence is not constrained and the mouse and human sequences have drifted apart.

C. The human and mouse β-globin genes are also similar at their 5' ends, as indicated by the cluster of points along the same diagonal as the first exon (Figure A9–11B). These sequences correspond to the regulatory regions upstream of the start sites for transcription. Functional sequences, which are under selective pressure, diverge much more slowly than sequences without function.

D. The diagon plot shows that the first intron is nearly the same length in the human and mouse genes, but the length of the second intron is noticeably different (Figure A9–11B). If the introns were the same length, the line segments that represent sequence similarity would fall on the same diagonal. The easiest way to test for the colinearity of the line segments is to tilt the page and sight along the diagonal. It is impossible to tell from this comparison if the change in length is due to a shortening of the mouse intron or to a lengthening of the human intron, or some combination of those possibilities.

ANSWER 9–12

Computer algorithms that search for exons are complex affairs, as you might imagine. To identify unknown genes, these programs combine statistical information derived from known genes, such as:

1. An exon that encodes protein will have an open reading frame. If the amino acid sequence specified by this open reading frame matches a protein sequence in any database, there is a high likelihood that it is an authentic exon.

2. The reading frames of adjacent exons in the same gene will match up when the intron sequences are omitted.

3. Internal exons (excluding the first and the last) will have splicing signals at each end; most of the time (98.1%) these will be AG at the 5' ends of the exons and GT at the 3' ends.

4. The multiple codons for most individual amino acids are not used with equal frequency. This so-called coding bias can be factored in to aid in the recognition of true exons.

5. Exons and introns have characteristic length distributions. The median length of exons in human genes is about 120 nucleotide pairs. Introns tend to be much larger: a median length of about 2 kb in genomic regions of 30–40% GC content, and a median length of about 500 nucleotide pairs in regions above 50% GC.

6. The initiation codon for protein synthesis (nearly always an ATG) has a statistical association with adjacent nucleotides that seem to enhance its recognition by translation factors.

7. The terminal exon will have a signal (most commonly AATAAA) for cleavage and polyadenylation close to its 3' end.

The statistical nature of these features, coupled with the low frequency of coding information in the genome (2–3%) and the frequency of alternative splicing (an estimated 95% of human genes), makes it especially impressive that current algorithms can identify more than 70% of individual exons in the human genome. As shown in Figure 9–37, bioinformatic approaches are usually coupled with direct experimental data, such as those obtained from RNA Seq.

ANSWER 9–13

It is not a simple matter to determine the function of a gene from scratch, nor is there a universal recipe for doing so. Nevertheless, there are a variety of standard questions whose answers help to narrow down the possibilities. Below we list some of these questions.

In what tissues is the gene expressed? If the gene is expressed in all tissues, it is likely to have a general function. If it is expressed in one or a few tissues, its function is likely to be more specialized, perhaps related to the specialized functions of the tissues. If the gene is expressed in the embryo but not the adult, it probably functions in development.

In what compartment of the cell is the protein found? Knowing the subcellular localization of the protein—nucleus, plasma membrane, mitochondria, etc.—can also help to suggest categories of potential function. For example, a protein that is localized to the plasma membrane is likely to be a transporter, a receptor or other component of a signaling pathway, a cell-adhesion molecule, etc.

What are the effects of mutations in the gene? Mutations that eliminate or modify the function of the gene product can also provide clues to function. For example, if the gene product is critical at a certain time during development, mutant embryos will often die at that stage or develop obvious abnormalities. Unless the abnormality is highly specific, it is usually difficult to deduce its function. And often the links are indirect, becoming apparent only after the gene's function is known.

With what other proteins does the encoded protein interact? In carrying out their function, proteins often interact with other proteins involved in the same or closely related processes. If an interacting protein can be identified, and if its function is already known (through previous research or through the searching of databases), the range of possible functions can be narrowed dramatically.

Can mutations in other genes alter effects of mutation in the unknown gene? Searching for such mutations can be a very powerful approach to investigating gene function, especially in organisms such as bacteria and yeast, which have simple genetic systems. Although much more difficult to perform in the mouse, this type of approach can nonetheless be used. The rationale for this strategy is analogous to that of looking for interacting proteins: genes that interact genetically—so that the double mutant phenotype is more selective than either of the individual mutants—are often involved in the same process or in closely related processes. Identification of such an interacting gene (and knowledge of its function) would provide an important clue to the function of the unknown gene.

Addressing each of these questions requires specialized experimental expertise and a substantial time commitment from the investigator. It is no wonder that progress is made much more rapidly when a clue to a gene's function can be found simply by identifying a similar gene of known function in the database. As more and more genes are studied, this strategy will become increasingly successful.

ANSWER 9–14   In a very long, random sequence of DNA, each of the 64 different codons will occur with equal frequency. Because 3 of the 64 are stop codons, they will be expected to occur on average every 21 codons (64/3 = 21.3).

ANSWER 9–15   On the surface, its resistance to mutation suggests that the genetic code was shaped by forces of natural selection. An underlying assumption, which seems reasonable, is that resistance to mutation is a valuable feature of a genetic code. This reasoning suggests that it would have been a lucky accident indeed—roughly a one-in-a-million chance—to stumble on a code as error-proof as our own.

But all is not so simple. If resistance to mutation is an essential feature of any code that can support complex organisms such as ourselves, then the only codes we *could* observe are ones that are error-resistant. A less favorable frozen accident, giving rise to a more error-prone code, might have limited the complexity of life to organisms too simple to contemplate their own genetic code. This is akin to the anthropic principle of cosmology: many universes

may be possible, but few are compatible with life that can ponder the nature of the universe.

Beyond these considerations, there is ample evidence that the code is not static, and thus could respond to the forces of natural selection. Altered versions of the standard genetic code have been identified in the mitochondrial and nuclear genomes of several organisms. In each case, one or a few codons have taken on a new meaning.

ANSWER 9–16   All of these mechanisms contribute to the evolution of new protein-coding genes. A, B, C, and E were discussed in the text. Recent studies indicate that certain short protein-coding genes arose from previously untranslated regions of genomes, so choice D is also correct.

ANSWER 9–17
A. Because synonymous changes do not alter the amino acid sequence of the protein, they usually do not affect the overall fitness of the organism and are therefore not selected against. By contrast, nonsynonymous changes, which substitute a new amino acid in place of the original one, can alter the function of the encoded protein and change the fitness of the organism. Since most amino acid substitutions probably harm the protein, they tend to be selected against.
B. Virtually all amino acid substitutions in the histone H3 protein are deleterious and are therefore selected against. The extreme conservation of histone H3 argues that its function is very tightly constrained, probably because of extensive interactions with other proteins and with DNA.
C. Histone H3 is clearly not in a "privileged" site in the genome because it undergoes synonymous nucleotide changes at about the same rate as other genes.

ANSWER 9–18
A. The data embodied in the phylogenetic tree (Figure Q9–18) refutes the hypothesis that plant hemoglobin genes were acquired by horizontal transfer. Looking at the more familiar parts of the tree, we see that the hemoglobins of vertebrates (fish to human) have approximately the same phylogenetic relationships as do the species themselves. Plant hemoglobins also form a distinct group that displays accepted evolutionary relationships, with barley, a monocot, diverging before bean, alfalfa, and lotus, which are all dicots (and legumes). The basic hemoglobin gene, therefore, was in place long ago in evolution. The phylogenetic tree of Figure Q9–18 indicates that the hemoglobin genes in modern plant and animal species were inherited from a common ancestor.
B. Had the plant hemoglobin genes arisen by horizontal transfer from a nematode, then the plant sequences would have clustered with the nematode sequences in the phylogenetic tree in Figure Q9–18.

ANSWER 9–19   In each human lineage, new mutations will be introduced at a rate of 10^{-10} alterations per nucleotide per cell generation, and the differences between two human lineages will accumulate at twice this rate. To accumulate 10^{-3} differences per nucleotide will thus take $10^{-3}/(2 \times 10^{-10})$ cell generations, corresponding to $(1/200) \times 10^{-3}/(2 \times 10^{-10})$ = 25,000 human generations, or 750,000 years. In reality, we are not descended from one pair of genetically identical ancestral humans; rather, it is likely that we are descended from a relatively small founder population of humans who were already genetically diverse. More sophisticated analysis suggests that this founder population existed about 150,000 years ago.

ANSWER 9–20   The AIDS virus (the human immunodeficiency virus, HIV) is a retrovirus, and thus synthesizes DNA from an RNA template using reverse transcriptase. This leads to frequent mutation of the viral genome. In fact, AIDS patients often carry many different genetic variants of HIV that are distinct from the original virus that infected them. This poses great problems in treating the infection: drugs that block essential viral enzymes work only temporarily, because new strains of the virus resistant to these drugs arise rapidly by mutation.

RNA replicases (enzymes that synthesize RNA using RNA as a template) do not proofread either. Thus, RNA viruses that replicate their RNA genomes directly (that is, without using DNA as an intermediate) also mutate frequently. In such a virus, this tends to produce changes in the coat proteins that cause the mutated virus to appear "new" to our immune systems; the virus is therefore not suppressed by immunity that has arisen to the previous version. This is part of the explanation for the new strains of the influenza (flu) virus and the common cold virus that regularly appear.

Chapter 10

ANSWER 10–1   The presence of a mutation in a gene does not necessarily mean that the protein expressed from it is defective. For example, the mutation could change one codon into another that still specifies the same amino acid, and so does not change the amino acid sequence of the protein. Or, the mutation may cause a change from one amino acid to another in the protein, but in a position that is not important for the folding or function of the protein. In assessing the likelihood that such a mutation might cause a defective protein, information on the known β-globin mutations that are found in humans is essential. You would therefore want to know the precise nucleotide change in your mutant gene, and whether this change has any known or predictable consequences for the function of the encoded protein. If your mate has two normal copies of the globin gene, 50% of your children would be carriers of your mutant gene.

ANSWER 10–2
A. Digestion with EcoRI produces two products:
5'-AAGAATTGCGG   AATTCGGGCCTTAAGCGCCGCGTCGAGGCCTTAAA-3'
3'-TTCTTAACGCCTTAA   GCCCGGAATTCGCGGCGCAGCTCCGGAATTT-5'

B. Digestion with HaeIII produces three products:
5'-AAGAATTGCGGAATTCGGG   CCTTAAGCGCCGCGTCGAGG   CCTTAAA-3'
3'-TTCTTAACGCCTTAAGCCC   GGAATTCGCGGCGCAGCTCC   GGAATTT-5'

C. The sequence lacks a HindIII cleavage site.

D. Digestion with all three enzymes therefore produces:
5'-AAGAATTGCGG   AATTCGGG   CCTTAAGCGCCGCGTCGAGG   CCTTAAA-3'
3'-TTCTTAACGCCTTAA   GCCC   GGAATTCGCGGCGCAGCTCC   GGAATTT-5'

ANSWER 10–3   Protein biochemistry is still very important because it provides the link between the amino acid sequence (which can be deduced from DNA sequences) and the functional properties of the protein. We are still not able to infallibly predict the folding of a polypeptide chain from its amino acid sequence, and in most cases information regarding the function of the protein, such as its catalytic

activity, cannot be deduced from the gene sequence alone. Instead, such information must be obtained experimentally by analyzing the properties of proteins biochemically. Furthermore, the structural information that can be deduced from DNA sequences is necessarily incomplete. We cannot, for example, accurately predict covalent modifications of the protein, proteolytic processing, the presence of tightly bound small molecules, or the association of the protein with other subunits. Moreover, we cannot accurately predict the effects these modifications might have on the activity of the protein.

ANSWER 10–4
A. After an additional round of amplification there will be 2 gray, 4 green, 4 red, and 22 yellow-outlined fragments; after a second additional round there will be 2 gray, 5 green, 5 red, and 52 yellow-outlined fragments. Thus the DNA fragments outlined in yellow increase exponentially and will eventually overrun the other reaction products. Their length is determined by the DNA sequence that spans the distance between the two primers plus the length of the primers.

B. The mass of one DNA molecule 500 nucleotide pairs long is 5.5×10^{-19} g [$= 2 \times 500 \times 330$ (g/mole)$/6 \times 10^{23}$ (molecules/mole)]. Ignoring the complexities of the first few steps of the amplification reaction (which produce longer products that eventually make an insignificant contribution to the total DNA amplified), this amount of product approximately doubles for every amplification step. Therefore, 100×10^{-9} g $= 2^N \times 5.5 \times 10^{-19}$ g, where N is the number of amplification steps of the reaction. Solving this equation for $N = \log(1.81 \times 10^{11})/\log(2)$ gives $N = 37.4$. Thus, only about 40 cycles of PCR amplification are sufficient to amplify DNA from a single molecule to a quantity that can be readily handled and analyzed biochemically. This whole procedure is automated and takes only a few hours in the laboratory.

ANSWER 10–5   If the ratio of dideoxyribonucleoside triphosphates to deoxyribonucleoside triphosphates is increased, DNA polymerization will be terminated more frequently and thus shorter DNA strands will be produced. Such conditions are favorable for determining nucleotide sequences that are close to the DNA primer used in the reaction. In contrast, decreasing the ratio of dideoxyribonucleoside triphosphates to deoxyribonucleoside triphosphates will produce longer DNA fragments, thus allowing one to determine nucleotide sequences more distant from the primer.

ANSWER 10–6   Although several explanations are possible, the simplest is that the DNA probe has hybridized predominantly with its corresponding mRNA, which is typically present in many more copies per cell than is the gene. The different extents of hybridization probably reflect different levels of gene expression. Perhaps each of the different cell types that make up the tissue expresses the gene at a different level.

ANSWER 10–7   Like the vast majority of mammalian genes, the attractase gene likely contains introns. Bacteria do not have the splicing machinery required to remove introns, and therefore the correct protein would not be expressed from the gene. For expression of most mammalian genes in bacterial cells, a cDNA version of the gene must be used.

ANSWER 10–8

A. False. Restriction sites are found at random throughout the genome, within as well as between genes.

B. True. DNA bears a negative charge at each phosphate, giving DNA an overall negative charge.

C. False. Clones isolated from cDNA libraries do not contain promoter sequences. These sequences are not transcribed and are therefore not part of the mRNAs that are used as the templates to make cDNAs.

D. True. Each polymerization reaction produces double-stranded DNA that must, at each cycle, be denatured to allow new primers to hybridize so that the DNA strand can be copied again.

E. False. Digestion of genomic DNA with restriction nucleases that recognize four-nucleotide sequences produces fragments that are *on average* 256 nucleotides long. However, the actual lengths of the fragments produced will vary considerably on both sides of the average.

F. True. Reverse transcriptase is first needed to copy the mRNA into single-stranded DNA, and DNA polymerase is then required to make the second DNA strand.

G. True. Using a sufficient number of STRs, individuals can be uniquely "fingerprinted" (see Figure 10–18).

H. True. If cells of the tissue do not transcribe the gene of interest, it will not be represented in a cDNA library prepared from this tissue. However, it will be represented in a genomic library prepared from the same tissue.

ANSWER 10–9

A. The DNA sequence, from its 5′ end to its 3′ end, is read starting from the bottom of the gel, where the smallest DNA fragments migrate. Each band results from the incorporation of the appropriate dideoxyribonucleoside triphosphate, and as expected there are no two bands that have the same mobility. This allows one to determine the DNA sequence by reading off the bands in strict order, proceeding upward from the bottom of the gel, and assigning the correct nucleotide according to which lane the band is in.

　　The nucleotide sequence of the top strand (Figure A10–9A) was obtained directly from the data of Figure Q10–9, and the bottom strand was deduced from the complementary base-pairing rules.

B. The DNA sequence can then be translated into an amino acid sequence using the genetic code. However, there are two strands of DNA that could be transcribed into RNA and three possible reading frames for each strand. Thus there are six amino acid sequences that can in principle be encoded by this stretch of DNA. Of the three reading frames possible from the top strand, only one is not interrupted by a stop codon (*yellow* blocks in Figure A10–9B).

　　From the bottom strand, two of the three reading frames also have stop codons (not shown). The third frame gives the following sequence:

`SerAlaLeuGlySerSerGluAsnArgProArgThrProAlaArg`
`ThrGlyCysProValTyr`

　　It is not possible from the information given to tell which of the two open reading frames corresponds to the actual protein encoded by this stretch of DNA. What additional experiment could distinguish between these two possibilities?

ANSWER 10–10

A. Cleavage of human genomic DNA with HaeIII would generate about 11×10^6 different fragments [$= 3 \times 10^9/4^4$] and with EcoRI about 730,000 different fragments [$= 3 \times 10^9/4^6$]. There will also be some additional fragments generated because the maternal and paternal chromosomes are very similar but not identical in DNA sequence.

B. A set of overlapping DNA fragments will be generated. Libraries constructed from sets of overlapping fragments are valuable because they can be used to order cloned sequences in relation to their original order in the genome and thus obtain the DNA sequence of a long stretch of DNA (see Figure 10–26).

ANSWER 10–11 By comparison with the positions of the size markers, we find that EcoRI treatment gives two fragments of 4 kb and 6 kb; HindIII treatment gives one fragment of 10 kb; and treatment with EcoRI + HindIII gives three fragments of 6 kb, 3 kb, and 1 kb. This gives a total length of 10 kb calculated as the sum of the fragments in each lane. Thus the original DNA molecule must be 10 kb (10,000 nucleotide pairs) long. Because treatment with HindIII gives a fragment 10 kb long it could be that the original DNA is a linear molecule with no cutting site for HindIII. But we can rule that out by the results of the EcoRI + HindIII digestion. We know that EcoRI cleavage alone produces two fragments of 6 kb and 4 kb, and in the double digest this 4-kb fragment is further cleaved by HindIII into a 3-kb and a 1-kb fragment. The DNA therefore contains a single HindIII cleavage site, and thus it must be circular, as a single fragment of 10 kb is produced when it is cut with HindIII alone. Arranging the cutting sites on a circular DNA to give the appropriate sizes of fragments produces the map illustrated in Figure A10–11.

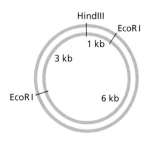

Figure A10–11

(A)　5′ -TATAAACTGGACAACCAGTTCGAGCTGGTGTTCGTGGTCGGTTTTCAGAAGATCCTAACGCTGACG-3′
　　　3′ -ATATTTGACCTGTTGGTCAAGCTCGACCACAAGCACCAGCCAAAAGTCTTCTAGGATTGCGACTGC-5′

(B)　　　5′　top strand of DNA　　　　　　　　　　　　　　　　　　　　　　　　　3′
　　　TATAAACTGGACAACCAGTTCGAGCTGGTG TTCGTGGTCGGTTTTCAGAAGATCCTAACGCTGACG
　　1　TyrLysLeuAspAsnGlnPheGluLeuVal PheValValGlyPheGlnLysIleLeuThrLeuThr
　　2　IleAsnTrpThrThrSerSerSerTrpCys SerTrpSerValPheArgArgSer　　Arg　　Ar
　　3　　　ThrGlyGlnProValArgAlaGlyVal ArgGlyArgPheSerGluAspProAsnAlaAsp

Figure A10–9

ANSWER 10–12

A. The genetic code is degenerate, and there is more than one possible codon for each amino acid, with the exception of tryptophan and methionine. Therefore, to detect the nucleotide sequence that codes for the amino acid sequence of the protein, many DNA molecules must be made and pooled to ensure that the mixture will contain the one that exactly matches the DNA sequence of the gene. For the three peptide sequences given in this question, the following probes need to be made (alternative bases at the same position are given in parentheses):

Peptide 1:

`5'-TGGATGCA(C,T)CA(C,T)AA(A,G)-3'`

Because of the three twofold degeneracies, you would need eight (= 2^3) different DNA sequences in the mixture.

Peptide 2:

`5'(T,C)T(G,A,T,C)(A,T)(G,C)(G,A,T,C)(A,C)`
`G(G,A,T,C)(T,C)T(G,A,T,C)(A,C)G(G,A,T,C)-3'`

The mixture representing peptide sequence #2 is much more complicated. Leu, Ser, and Arg are each encoded by six different codons; you would therefore need to synthesize a mixture of 7776 (= 6^5) different DNA molecules.

Peptide 3:

`5'-TA(C,T)TT(C,T)GG(G,A,T,C)ATGCA(A,G)3'`

Because of three twofold and one fourfold degeneracies, you would need 32 (= $2^3 \times 4$) different sequences in the mixture.

You would probably first use probe #1 to screen your library by hybridization. Because there are only eight possible DNA sequences, the ratio of the one correct sequence to the incorrect ones is highest, giving you the best chance of finding a matching clone. Probe #2 is practically useless, because only 1/7776 of the DNA in the mixture would perfectly hybridize to your gene of interest. You could use probe #3 to verify that the clone you obtained is correct. Any library clones that hybridize to probes #1 and #3 are very likely to contain the gene of interest.

B. Knowing that peptide sequence #3 contains the last amino acid of the protein is valuable information because it tells you that the other two peptide sequences must precede it, that is, they must be located farther toward the N-terminal end of the protein. Knowing this order is important, because DNA primers can be extended by DNA polymerases only from their 3' ends; thus, the 3' ends of two primers need to "face" each other during a PCR amplification reaction (see Figure 10–14). A PCR primer based on peptide sequence #3 must therefore be the complementary sequence of probe #3 (so that its 3' end corresponds to the first nucleotide of the sequence complementary to the Trp codon):

`5'-(TC)TGCAT(G,A,T,C)CC(G,A)AA(G,A)TA-3'`

As before, this "primer" would contain 32 different DNA sequences, only one of which will perfectly match the gene. Probe #1 could be your choice for the second primer. Probe #2, again because of its high degeneracy, would be a much less suitable choice.

C. The ends of the final amplification product are derived from the primers, which are each 15 nucleotides long. Therefore, a 270-nucleotide segment of the cDNA of the gene has been amplified. This will encode 90 amino acids; adding the amino acids encoded by the primers gives you a protein-coding sequence of 100 amino acids. This is unlikely to represent the whole gene. To your satisfaction, however, you note that CTATCACGCTTAGG encodes peptide sequence #2. This information therefore confirms that your PCR product indeed encodes a fragment of the protein you originally isolated.

ANSWER 10–13

The products will comprise a large number of different single-stranded DNA molecules, one for each nucleotide in the sequence. However, each DNA molecule will be one of four colors, depending on which of the four dideoxyribonucleotides terminated the polymerization reaction of that chain. Separation by gel electrophoresis will generate a ladder of bands, each one nucleotide apart, and the sequence can be read from the order of colors (**Figure A10–13**). The method described here forms the basis for the DNA sequencing strategy used in most automated DNA sequencing machines (see Figure 10–21).

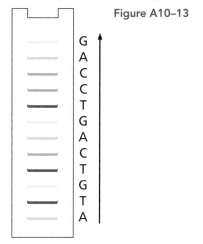

Figure A10–13

ANSWER 10–14

A. cDNA clones could not be used because there is no overlap between cDNA clones from adjacent genes.

B. Such repetitive DNA sequences can confuse chromosome walks, because the walk would appear to branch off in many different directions at once. The general strategy for avoiding these problems is to use genomic clones that are sufficiently long to span beyond the repetitive DNA sequences.

ANSWER 10–15

A. Infants 2 and 8 have identical STR patterns and therefore must be identical twins. Infants 3 and 6 also have identical STR patterns and must also be identical twins. The other two sets of twins must be fraternal twins because their STR patterns are not identical. Fraternal twins, like any pair of siblings born to the same parents, will have roughly half their genome in common. Thus, roughly half the STR polymorphisms in fraternal twins will be identical. Using this criterion, you can identify infants 1 and 7 as fraternal twins and infants 4 and 5 as fraternal twins.

B. You can match infants to their parents by using the same sort of analysis of STR polymorphisms. Every band present in the analysis of an infant should have a matching band in one or the other of the parents, and, on average, each infant will share half of its polymorphisms with each parent. Thus, the degree of match between each child and each parent will be approximately the same as that between fraternal twins.

ANSWER 10–16   Mutant bacteria that do not produce ice-protein have probably arisen many times in nature. However, bacteria that produce ice-protein have a slight growth advantage over bacteria that do not, so it would be difficult to find such mutants in the wild. Recombinant DNA technology makes these mutants much easier to obtain. In this case, the consequences, both advantageous and disadvantageous, of using a genetically modified organism are therefore nearly indistinguishable from those of a natural mutant. Indeed, bacterial and yeast strains have been selected for centuries for desirable genetic traits that make them suitable for industrial-scale applications such as cheese and wine production. The possibilities of recombinant DNA technology are endless, however, and as with any technology, there is a finite risk of unforeseen consequences. Recombinant DNA experimentation, therefore, is regulated, and the risks of individual projects are carefully assessed by review panels before permissions are granted. The state of our knowledge is sufficiently advanced that the consequences of some changes, such as the disruption of a bacterial gene in the example above, can be predicted with reasonable certainty. Other applications, such as germ-line gene therapy to correct human disease, may have far more complex outcomes, and it will take many more years of research and ethical debate to determine whether such treatments will eventually be used.

Chapter 11

ANSWER 11–1   Water is a liquid, and thus hydrogen bonds between water molecules are not static; they are continually formed and broken again by thermal motion. When a water molecule happens to be next to a hydrophobic molecule, it is more restricted in motion and has fewer neighbors with which it can interact because it cannot form any hydrogen bonds in the direction of the hydrophobic molecule. It will therefore form hydrogen bonds to the more limited number of water molecules in its proximity. Bonding to fewer partners results in a more ordered water structure, which represents the cagelike structure in Figure 11–9. This structure has been likened to ice, although it is a more transient, less organized, and less extensive network than even a tiny ice crystal. The formation of any ordered structure decreases the entropy of the system and is thus energetically unfavorable (discussed in Chapter 3).

ANSWER 11–2   (B) is the correct analogy for lipid bilayer assembly because exclusion from water rather than attractive forces between the lipid molecules is involved. If the lipid molecules formed bonds with one another, the bilayer would be less fluid, and might even become rigid, depending on the strength of the interaction.

ANSWER 11–3   The fluidity of the bilayer is strictly confined to one plane: lipid molecules can diffuse laterally in their own monolayer but do not readily flip from one monolayer to the other. Specific types of lipid molecules inserted into one monolayer therefore remain in it unless they are actively transferred by an enzyme—called a flippase.

ANSWER 11–4   In both an α helix and a β barrel the polar peptide bonds of the polypeptide backbone can be completely shielded from the hydrophobic environment of the lipid bilayer by the hydrophobic amino acid side chains. Internal hydrogen bonds between the peptide bonds stabilize the α helix and β barrel.

ANSWER 11–5   The sulfate group in SDS is charged and therefore hydrophilic. The OH group and the C–O–C groups in Triton X-100 are polar; they can form hydrogen bonds with water molecules and are therefore hydrophilic. In contrast, the blue portions of the detergents are either hydrocarbon chains or aromatic rings, neither of which has polar groups that could form hydrogen bonds with water molecules; they are therefore hydrophobic. (See Figure A11–5.)

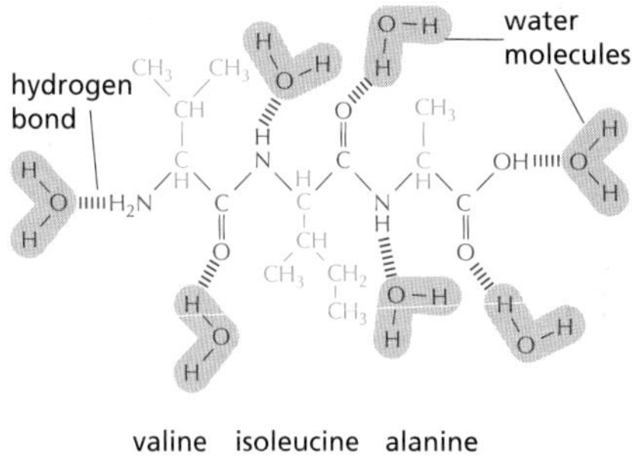

Figure A11–5

ANSWER 11–6   Some of the molecules of the two different transmembrane proteins are anchored to the spectrin filaments of the cell cortex. These molecules are not free to rotate or diffuse within the plane of the membrane. There is an excess of transmembrane proteins over the available attachment sites in the cortex, however, so that some of the transmembrane protein molecules are not anchored and are free to rotate and diffuse within the plane of the membrane. Indeed, measurements of protein mobility show that there are two populations of each transmembrane protein, corresponding to those proteins that are anchored and those that are not.

ANSWER 11–7   The different ways in which membrane proteins can be restricted to different regions of the membrane are summarized in Figure 11–31. The mobility of the membrane proteins is drastically reduced if they are bound to other proteins such as those of the cell cortex or the extracellular matrix. Some membrane proteins are confined to membrane domains by barriers, such as tight junctions. The fluidity of the lipid bilayer is not significantly affected by the anchoring of membrane proteins; the sea of lipid molecules flows around anchored membrane proteins like water around the posts of a pier.

ANSWER 11–8   All of the statements are correct.
A, B, C, D. The lipid bilayer is fluid because its lipid molecules can undergo these motions.
E. Glycolipids are mostly restricted to the monolayer of membranes that faces away from the cytosol. Some special glycolipids, such as phosphatidylinositol (discussed in Chapter 16), are found specifically in the cytosolic monolayer.
F. The reduction of double bonds (by hydrogenation) allows the resulting saturated lipid molecules to pack

more tightly against one another and therefore increases viscosity—that is, it turns oil into margarine.

G. Examples include the many membrane enzymes involved in cell signaling (discussed in Chapter 16).

H. Polysaccharides are the main constituents of mucus and slime; the carbohydrate coat of a cell, which is made up of polysaccharides and oligosaccharides, is a very important lubricant—for cells that line blood vessels or circulate in the bloodstream, for example.

ANSWER 11–9 In a two-dimensional fluid, the molecules are free to move only in one plane; the molecules in a normal fluid, in contrast, can move in three dimensions.

ANSWER 11–10

A. You would have a detergent. The diameter of the lipid head would be much larger than that of the hydrocarbon tail, so that the shape of the molecule would be a cone rather than a cylinder, and the molecules would aggregate to form micelles rather than bilayers.

B. Lipid bilayers formed would be much more fluid, as the tails would have less tendency to interact with one another. The bilayers would also be less stable, as the shorter hydrocarbon tails would be less hydrophobic, so the forces that drive the formation of the bilayer would be reduced.

C. The lipid bilayers formed would be much less fluid. Whereas a normal lipid bilayer has the viscosity of olive oil, a bilayer made of the same lipids but with saturated hydrocarbon tails would have the consistency of bacon fat.

D. The lipid bilayers formed would be much more fluid. Also, because the lipids would pack together less well, there would be more gaps and the bilayer would be more permeable to small, water-soluble molecules.

E. If we assume that the lipid molecules are completely intermixed, the fluidity of the membrane would be unchanged. In such bilayers, however, the saturated lipid molecules would tend to aggregate with one another because they can pack so much more tightly and would therefore form patches of much-reduced fluidity. The bilayer would not, therefore, have uniform properties over its surface. Because, normally, one saturated and one unsaturated hydrocarbon tail are linked to the same hydrophilic head in membrane phospholipid molecules, such segregation does not occur in cell membranes.

F. The lipid bilayers formed would have virtually unchanged properties. Each lipid molecule would now span the entire membrane, with one of its two head groups exposed at each surface. Such lipid molecules are found in the membranes of thermophilic bacteria, which can live at temperatures approaching boiling water. Their bilayers do not come apart at elevated temperatures, as usual bilayers do, because the original two monolayers are now covalently linked into a single structure.

ANSWER 11–11 Phospholipid molecules are approximately cylindrical in shape. Detergent molecules, by contrast, are conical or wedge-shaped. A phospholipid molecule with only one hydrocarbon tail, for example, would be a detergent. To make a phospholipid molecule into a detergent, you would have to make its hydrophilic head larger or remove one of its tails so that it could form a micelle. Detergent molecules also usually have shorter hydrocarbon tails than phospholipid molecules. This makes

them slightly water-soluble, so that detergent molecules leave and re-enter micelles frequently in aqueous solution. Because of this, some monomeric detergent molecules are always present in aqueous solution and therefore can enter the lipid bilayer of a cell membrane to solubilize the proteins (see Figure 11–26).

ANSWER 11–12

A. There are about 4000 lipid molecules, each 0.5 nm wide, between one end of the bacterial cell and the other. So if a lipid molecule at one end moved directly in a straight line it would require only 4×10^{-4} sec ($= 4000 \times 10^{-7}$) to reach the other end. In reality, however, the lipid molecule would move in a random path so that it would take considerably longer. We can calculate the approximate time required from the equation: $t = x^2/2D$ where x is the average distance moved, t is the time taken, and D is a constant called the diffusion coefficient. Inserting step values $x = 0.5$ nm and $t = 10^{-7}$ sec we obtain $D = 1.25 \times 10^{-7}$ cm^2/sec. Using this value in the same equation but with distance $x = 2 \times 10^{-4}$ cm ($= 2$ μm) gives the time taken $t = 1.6$ seconds.

B. Similarly, if a 4-cm ping-pong ball exchanged partners every 10^{-7} seconds and moved in a linear fashion it would reach the opposite wall in 1.5×10^{-5} sec (traveling at 1,440,000 km/hr. But a random walk would take longer. Using the equation above, we calculate the constant D in this case to be 8×10^7 cm^2/sec and the time required to travel 6 m about 2 msec ($= 600^2/(1.6 \times 10^8)$).

ANSWER 11–13 Transmembrane proteins anchor the plasma membrane to the underlying cell cortex, strengthening the membrane so that it can withstand the forces on it when the red blood cell is pumped through small blood vessels. Transmembrane proteins also transport nutrients and ions across the plasma membrane.

ANSWER 11–14 The hydrophilic faces of the five membrane-spanning α helices, each contributed by a different subunit, are thought to come together to form a pore across the lipid bilayer that is lined with the hydrophilic amino acid side chains (Figure A11–14). Ions can pass through this hydrophilic pore without coming into contact with the hydrophobic lipid tails of the bilayer. The hydrophobic side chains on the opposite face of the α helices interact with the hydrophobic lipid tails.

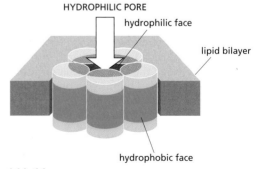

HYDROPHILIC PORE
hydrophilic face
lipid bilayer
hydrophobic face

Figure A11–14

ANSWER 11–15 There are about 100 lipid molecules (i.e., phospholipid + cholesterol) for every protein molecule in the membrane [= (2/50,000)/(1/800 + 1/386)]. A similar protein/lipid ratio is seen in many cell membranes.

ANSWER 11–16 Membrane fusion does not alter the orientation of the membrane proteins with their attached color tags: the portion of each transmembrane protein that is exposed to the cytosol always remains exposed to the cytosol, and the portion exposed to the outside always remains exposed to the outside despite diffusional mixing (Figure A11–16). At 0°C, the fluidity of the membrane is reduced, and the mixing of the membrane proteins is significantly slowed.

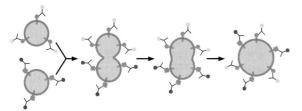

Figure A11–16

ANSWER 11–17 The exposure of hydrophobic amino acid side chains to water is energetically unfavorable. There are two ways that such side chains can be sequestered away from water to achieve an energetically more favorable state. First, they can form transmembrane segments that span a lipid bilayer. This requires about 20 of them to be located sequentially in the polypeptide chain. Second, the hydrophobic amino acid side chains can be sequestered in the interior of the folded polypeptide chain. This is one of the major forces that lock the polypeptide chain into a unique three-dimensional structure. In either case, the hydrophobic forces in the lipid bilayer or in the interior of a protein are based on the same principles.

ANSWER 11–18 (A) Antarctic fish live at subzero temperatures and are cold-blooded. To keep their membranes fluid at these temperatures, they have an unusually high percentage of unsaturated phospholipids.

ANSWER 11–19 Sequence B is most likely to form a transmembrane helix. It is composed primarily of hydrophobic amino acids, and therefore can be stably integrated into a lipid bilayer. In contrast, sequence A contains many polar amino acids (S, T, N, Q), and sequence C contains many charged amino acids (K, R, H, E, D), which would be energetically disfavored in the hydrophobic interior of the lipid bilayer.

Chapter 12

ANSWER 12–1

A. The movement of a solute mediated by a transporter can be described by a strictly analogous equation:
equation 1: $T + S \leftrightarrow TS \rightarrow T + S^*$
where S is the solute, S* is the solute on the other side of the membrane (i.e., although it is still the same molecule, it is now located in a different environment), and T is the transporter.

B. This equation is useful because it describes a binding step, followed by a delivery step. The mathematical treatment of this equation would be very similar to that described for enzymes (see Figure 3–24); thus

transporters are characterized by a K_M value that describes their affinity for a solute and a V_{max} value that describes their maximal rate of transfer.

To be more accurate, one could include the conformational change of the transporter in the reaction scheme:
equation 2: $T + S \leftrightarrow TS \leftrightarrow T^*S^* \rightarrow T^* + S^*$
equation 3: $T \leftrightarrow T^*$
where T* is the transporter after the conformational change that exposes its solute-binding site on the other side of the membrane. This account requires a second equation (3) that allows the transporter to return to its starting conformation.

C. The equations do not describe the behavior of channels because solutes passing through channels do not bind to them in the way that a substrate binds to an enzyme.

ANSWER 12–2   If the Na⁺ pump is not working at full capacity because it is partially inhibited by ouabain or digitalis, the electrochemical gradient of Na⁺ that the pump generates is less steep than that in untreated cells. Consequently, the Ca²⁺-Na⁺ antiport works less efficiently, and Ca²⁺ is removed from the cell more slowly. When the next cycle of muscle contraction begins, there is still an elevated level of Ca²⁺ left in the cytosol. The entry of the same number of Ca²⁺ ions into the cell therefore leads to a higher Ca²⁺ concentration than in untreated cells, which in turn leads to a stronger and longer-lasting muscle contraction. Because the Na⁺ pump fulfills essential functions in all animal cells, both to maintain osmotic balance and to generate the Na⁺ gradient used to power many transporters, the drugs are deadly poisons if too much is taken.

ANSWER 12–3

A. The properties define a transporter acting as a symport.

B. No additional properties need to be specified. The important feature that provides the coupling of the two solutes is that the protein cannot switch its conformation if only one of the two solutes is bound. Solute B, which is driving the transport of solute A, is in excess on the side of the membrane from which transport initiates and therefore occupies its binding site most of the time. In this state, the transporter is prevented from switching its conformation until a solute A molecule binds, which it will occasionally. With both binding sites occupied, the transporter switches conformation. Now exposed to the other side of the membrane, the binding site for solute B is mostly empty because there is little of it in the solution on this side of the membrane. Although the binding site for A is now more frequently occupied, the transporter can switch back only after solute A is unloaded as well.

C. An antiport could be similarly constructed with a transmembrane protein with the following properties. It has two binding sites, one for solute A and one for solute B. The protein can undergo a conformational change to switch between two states: either both binding sites are exposed exclusively on one side of the membrane or both are exposed exclusively on the other side. The protein can switch between the two conformational states only if one binding site is occupied, but not if both binding sites are either occupied or empty.

Note that these rules described in B and C provide an alternative model to that shown in Figure 12–14.

Thus, in principle, there are two possible ways to couple the transport of two solutes: (1) provide cooperative solute-binding sites and allow the transporter to switch between the two states randomly as shown in Figure 12–14 or (2) allow independent binding of both solutes and make the switch between the two states conditional on the occupancy of the binding sites. As the structure of a coupled transporter has not yet been determined, we do not know which of the two mechanisms such transporters use.

ANSWER 12–4

A. Each of the rectangular peaks corresponds to the opening of a single channel that allows a small current to pass. You note from the recording that the channels present in the patch of membrane open and close frequently. Each channel remains open for a very short, somewhat variable time, averaging about 10 milliseconds. When open, the channels allow a small current with a unique amplitude (4 pA; one picoampere $= 10^{-12}$ A) to pass. In one instance, the current doubles, indicating that two channels in the same membrane patch opened simultaneously.

B. If acetylcholine is omitted or is added to the solution outside the pipette, you would measure only the baseline current. Acetylcholine must bind to the extracellular portion of the acetylcholine receptor in the membrane patch to allow the channel to open frequently enough to detect; in the membrane patch shown in Figure 12–24, only, the cytoplasmic side of the receptor is exposed to the solution outside the microelectrode.

ANSWER 12–5   The equilibrium potential of K^+ is –90 mV [= 62 mV $\log_{10}$ (5 mM/140 mM)], and that of Na^+ is +72 mV [= 62 mV $\log_{10}$ (145 mM/10 mM)]. The K^+ leak channels are the main ion channels open in the plasma membrane of a resting cell, and they allow K^+ to come to equilibrium; the membrane potential of the cell is therefore close to –90 mV. When Na^+ channels open, Na^+ rushes in, and, as a result, the membrane potential reverses its polarity to a value nearer to +72 mV, the equilibrium value for Na^+. Upon closure of the Na^+ channels, the K^+ leak channels allow K^+, now no longer at equilibrium, to exit from the cell until the membrane potential is restored to the equilibrium value for K^+, about –90 mV.

ANSWER 12–6   When the resting membrane potential of an axon (inside negative) rises to a threshold value, voltage-gated Na^+ channels in the immediate neighborhood open and allow an influx of Na^+. This depolarizes the membrane further, causing more voltage-gated Na^+ channels to open, including those in the adjacent plasma membrane. This creates a wave of depolarization that spreads rapidly along

the axon, called the action potential. Because Na^+ channels become inactivated soon after they open, the outward flow of K^+ through voltage-gated K^+ channels and K^+ leak channels is rapidly able to restore the original resting membrane potential. (96 words)

ANSWER 12–7   If the number of functional acetylcholine receptors is reduced by the antibodies, the neurotransmitter (acetylcholine) that is released from the nerve terminals cannot (or can only weakly) stimulate the muscle to contract.

ANSWER 12–8   Although the concentration of Cl^- outside cells is much higher than inside, when transmitter-gated Cl^- channels open in the plasma membrane of a postsynaptic neuron in response to an inhibitory neurotransmitter, very little Cl^- enters the cell. This is because the driving force for the influx of Cl^- across the membrane is close to zero at the resting membrane potential, which opposes the influx. If, however, the excitatory neurotransmitter opens Na^+ channels in the postsynaptic membrane at the same time that an inhibitory neurotransmitter opens Cl^- channels, the resulting depolarization caused by the Na^+ influx will cause Cl^- to move into the cell through the open Cl^- channels, neutralizing the effect of the Na^+ influx. In this way, inhibitory neurotransmitters suppress the production of an action potential by making the target cell membrane much harder to depolarize.

ANSWER 12–9   By analogy to the Na^+ pump shown in Figure 12–9, ATP might be hydrolyzed and donate a phosphate group to the transporter when—and only when—it has the solute bound on the cytosolic face of the membrane (step 1 → 2). The attachment of the phosphate would trigger an immediate conformational change (step 2 → 3), thereby capturing the solute and exposing it to the other side of the membrane. The phosphate would be removed from the protein when—and only when—the solute had dissociated, and the now empty, nonphosphorylated transporter would switch back to the starting conformation (step 3 → 4) (Figure A12–9).

ANSWER 12–10

A. False. The plasma membrane contains transport proteins that confer selective permeability to many but not all charged molecules. In contrast, a pure lipid bilayer lacking proteins is highly impermeable to all charged molecules.

B. False. Channels do not have binding pockets for the solute that passes through them. Selectivity of a channel is achieved by the size of the internal pore and by charged regions at the entrance of the pore that attract or repel ions of the appropriate charge.

C. False. Transporters are slower. They have enzymelike properties, i.e., they bind solutes and need to undergo

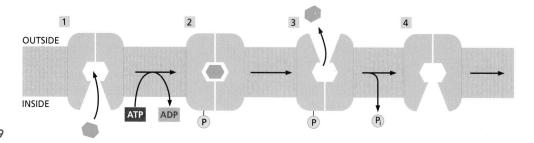

Figure A12–9

conformational changes during their functional cycle. This limits the maximal rate of transport to about 1000 solute molecules per second, whereas channels can pass up to 1,000,000 solute molecules per second.

D. True. The bacteriorhodopsin of some photosynthetic bacteria pumps H^+ out of the cell, using energy captured from visible light.

E. True. Most animal cells contain K^+ leak channels in their plasma membrane that are predominantly open. The K^+ concentration inside the cell still remains higher than outside, because the membrane potential is negative and therefore inhibits the positively charged K^+ from leaking out. K^+ is also continually pumped into the cell by the Na^+ pump.

F. False. A symport binds two different solutes on the same side of the membrane. Turning it around would not change it into an antiport, which must also bind to different solutes, but on opposing sides of the membrane.

G. False. The peak of an action potential corresponds to a transient shift of the membrane potential from a negative to a positive value. The influx of Na^+ causes the membrane potential first to move toward zero and then to reverse, rendering the cell positively charged on its inside. Eventually, the resting potential is restored by an efflux of K^+ through voltage-gated K^+ channels and K^+ leak channels.

ANSWER 12–11 The permeabilities are N_2 (small and nonpolar) > ethanol (small and slightly polar) > water (small and polar) > glucose (large and polar) > Ca^{2+} (small and charged) > RNA (very large and charged).

ANSWER 12–12
A. Both couple the movement of two different solutes across a cell membrane. Symports transport both solutes in the same direction, whereas antiports transport the solutes in opposite directions.

B. Both are mediated by membrane transport proteins. Passive transport of a solute occurs downhill, in the direction of its concentration or electrochemical gradient, whereas active transport occurs uphill and therefore needs an energy source. Active transport can be mediated by transporters but not by channels, whereas passive transport can be mediated by either.

C. Both terms describe gradients across a membrane. The membrane potential refers to the voltage gradient; the electrochemical gradient is a composite of the voltage gradient and the concentration gradient of a specific charged solute (ion). The membrane potential is defined independently of the solute of interest, whereas an electrochemical gradient refers to the particular solute.

D. A pump is a specialized transporter that uses energy to transport a solute uphill—against an electrochemical gradient for a charged solute or a concentration for an uncharged solute.

E. Both transmit electrical signals by means of electrons in wires and ion movements across the plasma membrane in axons. Wires are made of copper, axons are not. The signal passing down an axon does not diminish in strength, because it is self-amplifying, whereas the signal in a wire decreases over distance (by leakage of current across the insulating sheath).

F. Both affect the osmotic pressure in a cell. An ion is a solute that bears a charge.

ANSWER 12–13 A bridge allows vehicles to pass over water in a steady stream; the entrance can be designed to exclude, for example, oversized trucks, and it can be intermittently closed to traffic by a gate. By analogy, gated channels allow ions to pass across a cell membrane, imposing size and charge restrictions.

A ferry, in contrast, loads vehicles on one side of the body of water, crosses, and unloads on the other side—a slower process. During loading, particular vehicles could be selected from the waiting line because they fit particularly well on the car deck. By analogy, transporters bind solutes on one side of the membrane and then, after a conformational movement, release them on the other side. Specific binding selects the molecules to be transported. As in the case of coupled transport, sometimes you have to wait until the ferry is full before you can go.

ANSWER 12–14 Acetylcholine is being transported into the vesicles by an H^+–acetylcholine antiport in the vesicle membrane. The H^+ gradient that drives the uptake is generated by an ATP-driven H^+ pump in the vesicle membrane, which pumps H^+ into the vesicle (hence the dependence of the reaction on ATP). Raising the pH of the solution surrounding the vesicles decreases the H^+ concentration of the solution, thereby increasing the outward gradient across the vesicle membrane, explaining the enhanced rate of acetylcholine uptake.

ANSWER 12–15 The voltage gradient across the membrane is about 150,000 V/cm (70×10^{-3} V/4.5×10^{-7} cm). This extremely powerful electric field is close to the limit at which insulating materials—such as the lipid bilayer—break down and cease to act as insulators. The large field indicates what a large amount of energy can be stored in electrical gradients across the membrane, as well as the extreme electrical forces that proteins can experience in a membrane. A voltage of 150,000 V would instantly discharge in an arc across a 1-cm-wide gap (that is, air would be an insufficient insulator for this strength of field).

ANSWER 12–16
A. Nothing. You require ATP to drive the Na^+ pump.

B. The ATP becomes hydrolyzed, and Na^+ is pumped into the vesicles, generating a concentration gradient of Na^+ across the membrane. At the same time, K^+ is pumped out of the vesicles, generating a concentration gradient of K^+ of opposite polarity. When all the K^+ is pumped out of the vesicle or the ATP runs out, the pump would stop.

C. The pump would initiate a transport cycle and then cease. Because all reaction steps must occur strictly sequentially, dephosphorylation and the accompanying conformational switch cannot occur in the absence of K^+. The Na^+ pump will therefore become stuck in the phosphorylated state, waiting indefinitely for a potassium ion. The number of sodium ions transported would be minuscule, because each pump molecule would have functioned only a single time.

Similar experiments, leaving out individual ions and analyzing the consequences, were used to determine the sequence of steps by which the Na^+ pump works.

D. ATP would become hydrolyzed, and Na^+ and K^+ would be pumped across the membrane as described in (B). However, the pump molecules that sit in the membrane in the reverse orientation would be completely inactive (i.e., they would not—as one might have erroneously assumed—pump ions in the opposite direction), because ATP would not have access to the site on these molecules where phosphorylation occurs, which is normally exposed to the cytosol. ATP is highly charged and cannot cross membranes without the help of specific transporters.

E. ATP becomes hydrolyzed, and Na^+ and K^+ are pumped across the membrane, as described in (B). K^+, however, immediately flows back into the vesicles through the K^+ leak channels. K^+ moves down the K^+ concentration gradient formed by the action of the Na^+ pump. With each K^+ that moves into the vesicle through a leak channel, a positive charge is moved across the membrane, generating a membrane potential that is positive on the inside of the vesicles. Eventually, K^+ will stop flowing through the leak channels when the membrane potential balances the K^+ concentration gradient. The scenario described here is a slight oversimplification: the Na^+ pump in mammalian cells actually moves three sodium ions out of cells for each two potassium ions that it pumps, thereby driving an electric current across the membrane and making a small additional contribution to the resting membrane potential (which therefore corresponds only approximately to a state of equilibrium for K^+ moving via K^+ leak channels).

ANSWER 12–17 Ion channels can be ligand-gated, voltage-gated, or mechanically (stress) gated.

ANSWER 12–18 The cell has a volume of 10^{-12} liters (= 10^{-15} m^3) and thus contains 6×10^4 calcium ions (= 6×10^{23} molecules/mole $\times 100 \times 10^{-9}$ moles/liter $\times 10^{-12}$ liters). Therefore, to raise the intracellular Ca^{2+} concentration fiftyfold, another 2,940,000 calcium ions have to enter the cell (note that at 5 µM concentration there are 3×10^6 ions in the cell, of which 60,000 are already present before the channels are opened). Because each of the 1000 channels allows 10^6 ions to pass per second, each channel has to stay open for only 3 milliseconds.

ANSWER 12–19 Animal cells drive most transport processes across the plasma membrane with the electrochemical gradient of Na^+. ATP is needed to fuel the Na^+ pump to maintain the Na^+ gradient.

ANSWER 12–20
A. If H^+ is pumped across the membrane into the endosomes, an electrochemical gradient of H^+ results—composed of both an H^+ concentration gradient and a membrane potential, with the interior of the vesicle positive. Both of these components add to the energy that is stored in the gradient and that must be supplied to generate it. The electrochemical gradient will limit the transfer of more H^+. If, however, the membrane also contains Cl^- channels, the negatively charged Cl^- in the cytosol will flow into the endosomes and diminish their membrane potential. It therefore becomes energetically less expensive to pump more H^+ across the membrane, and the interior of the endosomes can become more acidic.

B. No. As explained in (A), some acidification would still occur in their absence.

ANSWER 12–21
A. See Figure A12–21A.
B. The transport rates of compound A are proportional to its concentration, indicating that compound A can diffuse through membranes on its own. Compound A is likely to be ethanol, because it is a small and relatively nonpolar molecule that can diffuse readily through the lipid bilayer (see Figure 12–2).

In contrast, the transport rates of compound B saturate at high concentrations, indicating that compound B is transported across the membrane by some sort of membrane transport protein. Transport rates cannot increase beyond a maximal rate at which this protein can function. Compound B is likely to be acetate, because it is a charged molecule that could not cross the membrane without the help of a membrane transport protein.

C. For ethanol, the graph shows a linear relationship between concentration and transport rate. Thus, at 0.5 mM the transport rate would be 10 µmol/min, and at 100 mM the transport rate would be 2000 µmol/min (2 mmol/min).

For the transport-protein-mediated movement of acetate, the relationship between concentration, S, and transport rate can be described by the Michaelis–Menten equation, which describes simple enzyme reactions:

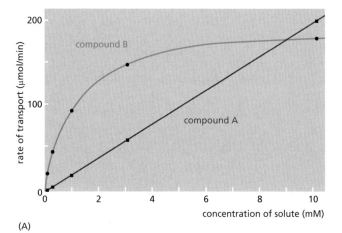

(A)

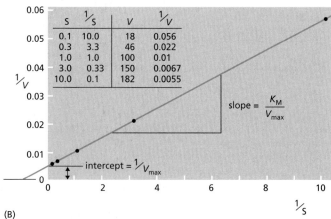

(B)

Figure A12–21

equation 1: transport rate = $V_{max} \times S/[K_M + S]$

Recall from Chapter 3 (see Question 3–20, p. 118) that to determine the V_{max} and K_M, a trick is used in which the Michaelis–Menten equation is transformed so that it is possible to plot the data as a straight line. A simple transformation yields

equation 2: **1/rate = (K_M/V_{max}) (1/S) + $1/V_{max}$**
(i.e., an equation of the form **y = ax + b**)

Calculation of 1/rate and 1/S for the given data and plotting them in a new graph as in **Figure A12–21B** gives a straight line. The K_M (= 1.0 mM) and V_{max} (= 200 μmol/min) are determined from the intercept of the line with the y axis ($1/V_{max}$) and from its slope (K_M/V_{max}). Knowing the values for K_M and V_{max} allows you to calculate the transport rates for 0.5 mM and 100 mM acetate using equation (1). The results are 67 μmol/min and 198 μmol/min, respectively.

ANSWER 12–22 The membrane potential and the steep extracellular Na^+ concentration provide a large inward electrochemical driving force and a large reservoir of Na^+ ions, so that mostly Na^+ ions enter the cell as acetylcholine receptors open. Ca^{2+} ions will also enter the cell, but their influx is much more limited because of their lower extracellular concentration. (Most of the Ca^{2+} that enters the cytosol to stimulate muscle contraction is released from intracellular stores, as we discuss in Chapter 17). Because of the high intracellular K^+ concentration and the opposing direction of the membrane potential, there will be little if any movement of K^+ ions upon opening of a cation channel.

ANSWER 12–23 The diversity of neurotransmitter-gated ion channels is a good thing for the industry, as it raises the possibility of developing new drugs specific for each channel type. Each of the diverse subtypes of these channels is expressed in a narrow subset of neurons. This narrow range of expression should make it possible, in principle, to discover or design drugs that affect particular receptor subtypes present in a selected set of neurons, thus to target particular brain functions with greater specificity.

Chapter 13

ANSWER 13–1 To keep glycolysis going, cells need to regenerate NAD^+ from NADH. There is no efficient way to do this without fermentation. In the absence of regenerated NAD^+, step 6 of glycolysis (the oxidation of glyceraldehyde 3-phosphate to 1,3-bisphosphoglycerate (Panel 13–1, pp. 428–429) could not occur and the product glyceraldehyde 3-phosphate would accumulate. The same thing would happen in cells unable to make either lactate or ethanol: neither would be able to regenerate NAD^+, and so glycolysis would be blocked at the same step.

ANSWER 13–2 Arsenate instead of phosphate becomes attached in step 6 of glycolysis to form 1-arseno-3-phosphoglycerate (**Figure A13–2**). Because of its sensitivity to hydrolysis in water, the high-energy bond is destroyed before the molecule that contains it can diffuse to reach the next enzyme. The product of the hydrolysis, 3-phosphoglycerate, is the same product normally formed in step 7 by the action of phosphoglycerate kinase. But because hydrolysis occurs nonenzymatically, the energy liberated by breaking the high-energy bond cannot be

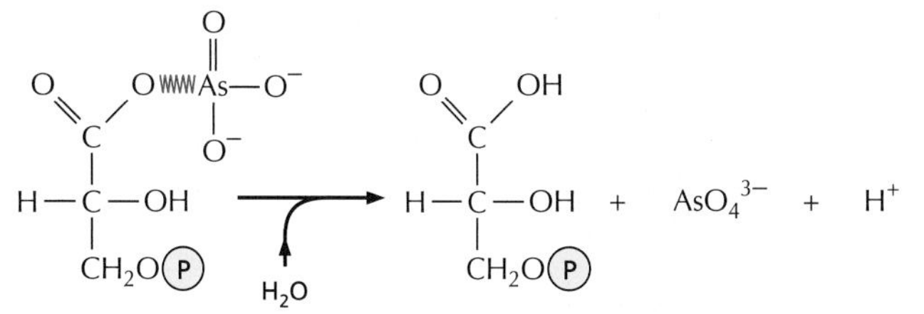

Figure A13–2

captured to generate ATP. In Figure 13–7, therefore, the reaction corresponding to the downward-pointing arrow would still occur, but the wheel that provides the coupling to ATP synthesis is missing. Arsenate wastes metabolic energy by uncoupling many phosphotransfer reactions by the same mechanism, which is why it is so poisonous.

ANSWER 13–3 The oxidation of fatty acids breaks the carbon chain into two-carbon units (acetyl groups) that become attached to CoA. Conversely, during biogenesis, fatty acids are constructed by linking together acetyl groups. Most fatty acids therefore have an even number of carbon atoms.

ANSWER 13–4 Because the function of the citric acid cycle is to harvest the energy released during the oxidation, it is advantageous to break the overall reaction into as many steps as possible (see Figure 13–1). Using a two-carbon compound, the available chemistry would be much more limited, and it would be impossible to generate as many intermediates.

ANSWER 13–5 It is true that oxygen atoms are returned as part of CO_2 to the atmosphere. The CO_2 released from the cells, however, does not contain those specific oxygen atoms that were consumed as part of the oxidative phosphorylation process and converted into water. One can show this directly by incubating living cells in an atmosphere that includes molecular oxygen containing the ^{18}O isotope of oxygen instead of the naturally abundant isotope, ^{16}O. In such an experiment, one finds that all the CO_2 released from cells contains only ^{16}O. Therefore, the oxygen atoms in the released CO_2 molecules do not come directly from the atmosphere but from organic molecules that the cell has first made and then oxidized as fuel (see top of first page of Panel 13–2, pp. 434–435).

ANSWER 13–6 The cycle continues because intermediates are replenished as necessary by reactions leading into the citric acid cycle (instead of away from it). One of the most important reactions of this kind is the conversion of pyruvate to oxaloacetate by the enzyme pyruvate carboxylase:

$$\text{pyruvate} + CO_2 + \text{ATP} + H_2O \rightarrow$$
$$\text{oxaloacetate} + \text{ADP} + P_i + 2H^+$$

This is one of the many examples of how metabolic pathways are carefully coordinated to work together to maintain appropriate concentrations of all metabolites required by the cell (see Figure A13–6).

ANSWER 13–7 The carbon atoms in sugar molecules are already partially oxidized, in contrast to all but the very first carbon atoms in the acyl chains of fatty acids. Thus,

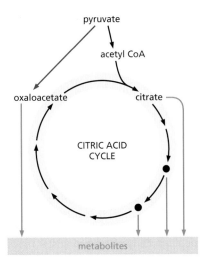

Figure A13–6

two carbon atoms from glucose are lost as CO_2 during the conversion of pyruvate to acetyl CoA, and only four of the six carbon atoms of the sugar molecule are recovered and can enter the citric acid cycle, where most of the energy is captured. In contrast, all carbon atoms of a fatty acid are converted into acetyl CoA.

ANSWER 13–8
A. False. If this were the case, then the reaction would be useless for the cell. No chemical energy would be harvested in a useful form (e.g., ATP) to be used for metabolic processes. (Cells would be nice and warm, though!)
B. False. No energy-conversion process can be 100% efficient. Recall that entropy in the universe always has to increase, and for most reactions this is accomplished by releasing heat.
C. True. The carbon atoms in glucose are in a reduced state compared with those in CO_2, in which they are fully oxidized.
D. False. The reaction does indeed produce some water, but water is so abundant in the biosphere that this is no more than "a drop in the ocean."
E. True. If it had occurred in only one step, then all the energy would be released at once and it would be impossible to harness it efficiently to drive other reactions, such as the synthesis of ATP.
F. False. Molecular oxygen (O_2) is used only in the very last step of the reaction.
G. True. Plants convert CO_2 into sugars by harvesting the energy of light in photosynthesis. O_2 is produced in the process and released into the atmosphere by plant cells.
H. True. Anaerobically growing cells use glycolysis to oxidize sugars to pyruvate. Animal cells convert pyruvate to lactate, and no CO_2 is produced; yeast cells, however, convert pyruvate to ethanol and CO_2. It is this CO_2 gas, released from yeast cells during fermentation, that makes bread dough rise and that carbonates beer and champagne.

ANSWER 13–9   Darwin exhaled the carbon atom, which therefore must be the carbon atom of a CO_2 molecule. After spending some time in the atmosphere, the CO_2 molecule must have entered a plant cell, where it became "fixed" by photosynthesis and converted into part of a

sugar molecule. While it is certain that these early steps must have happened this way, there are many different paths from there that the carbon atom could have taken. The sugar could have been broken down by the plant cell into pyruvate or acetyl CoA, for example, which then could have entered biosynthetic reactions to build an amino acid. The amino acid might have been incorporated into a plant protein, maybe an enzyme or a protein that builds the cell wall. You might have eaten the delicious leaves of the plant in your salad, and digested the protein in your gut to produce amino acids again. After circulating in your bloodstream, the amino acid might have been taken up by a developing red blood cell to make its own protein, such as the hemoglobin in question. If we wish, of course, we can make our food chain scenario more complicated. The plant, for example, might have been eaten by an animal that in turn was consumed by you during lunch break. Moreover, because Darwin died more than 100 years ago, the carbon atom could have traveled such a route many times. In each round, however, it would have started again as fully oxidized CO_2 gas and entered the living world following its reduction during photosynthesis.

ANSWER 13–10 Yeast cells grow much better aerobically. Under anaerobic conditions they cannot perform oxidative phosphorylation and therefore have to produce all their ATP by glycolysis, which is less efficient. Whereas one glucose molecule yields a net gain of two ATP molecules by glycolysis, the additional use of the citric acid cycle and oxidative phosphorylation boosts the energy yield up to about 30 ATP molecules.

ANSWER 13–11 The amount of free energy stored in the phosphate bond in creatine phosphate is larger than that of the anhydride bonds in ATP. Hydrolysis of creatine phosphate can therefore be directly coupled to the production of ATP.

creatine phosphate + ADP → creatine + ATP

The $\Delta G°$ for this reaction is –3 kcal/mole, indicating that it proceeds rapidly to the right, as written.

ANSWER 13–12 The extreme conservation of glycolysis is some of the evidence that all present cells are derived from a single founder cell as discussed in Chapter 1. The elegant reactions of glycolysis would therefore have evolved only once, and then they would have been inherited as cells evolved. The later invention of oxidative phosphorylation allowed follow-up reactions to capture 15 times more energy than is possible by glycolysis alone. This remarkable efficiency is close to the theoretical limit and hence virtually eliminates the opportunity for further improvements. Thus, the generation of alternative pathways would result in no obvious reproductive advantage that would have been selected in evolution.

ANSWER 13–13 If one glucose produces 30 ATPs, then to generate 10^9 ATP molecules will require $1 \times 10^9/30 = 3.3 \times 10^7$ glucose molecules and $6 \times 3.3 \times 10^7 = 2 \times 10^8$ molecules of oxygen. Thus in one minute the cell will consume $2 \times 10^8/(6 \times 10^{23})$ or 3.3×10^{-16} moles of oxygen, which would occupy $3.3 \times 10^{-16} \times 22.4 = 7.4 \times 10^{-15}$ liters in gaseous form. The volume of the cell is 10^{-15} cubic meters (= $(10^{-5})^3$), which is 10^{-12} liter. The cell therefore consumes about 0.7% of its volume of O_2 gas every minute, or its own volume of O_2 gas in 2 hours and 15 minutes.

ANSWER 13–14 The reactions each have negative ΔG values and are therefore energetically favorable (see **Figure A13–14** for energy diagrams).

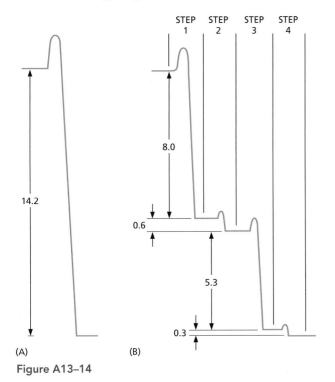

Figure A13–14

ANSWER 13–15

A. Pyruvate is converted to acetyl CoA, and the labeled ^{14}C atom is released as $^{14}CO_2$ gas (see Figure 13–10A).

B. By following the ^{14}C-labeled atom through every reaction in the cycle, shown in Panel 13–2 (pp. 434–435), you find that the added ^{14}C label would be quantitatively recovered in oxaloacetate. The analysis also reveals, however, that it is no longer in the keto group but in the methylene group of oxaloacetate (**Figure A13–15**).

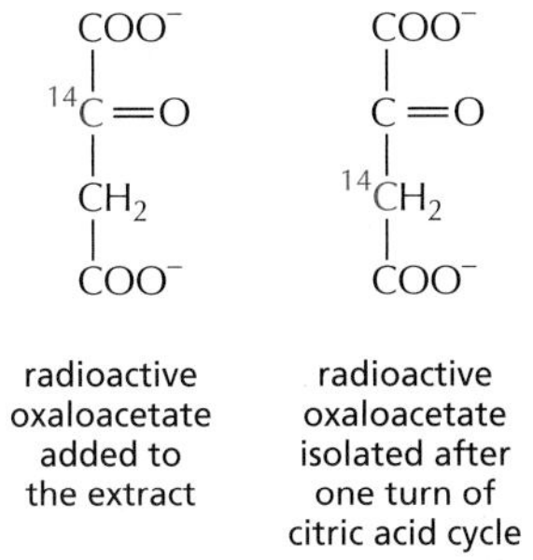

radioactive oxaloacetate added to the extract

radioactive oxaloacetate isolated after one turn of citric acid cycle

Figure A13–15

ANSWER 13–16 In the presence of molecular oxygen, oxidative phosphorylation converts most of the cellular NADH to NAD^+. Since fermentation requires NADH, it is severely inhibited by the availability of oxygen gas.

Chapter 14

ANSWER 14–1 By making membranes permeable to protons, DNP collapses—or at very small concentrations diminishes—the proton gradient across the inner mitochondrial membrane. Cells continue to oxidize food molecules to feed high-energy electrons into the electron-transport chain, but H^+ ions pumped across the membrane flow back into the mitochondria in a futile cycle. As a result, the energy of the electrons cannot be tapped to drive ATP synthesis, and instead is released as heat. Patients who have been given small doses of DNP lose weight because their fat reserves are used more rapidly to feed the electron-transport chain, and the whole process simply "wastes" energy as heat.

A similar mechanism of heat production is used naturally in a specialized tissue composed of brown fat cells, which is abundant in newborn humans and in hibernating animals. These cells are packed with mitochondria that leak part of their H^+ gradient futilely back across the membrane for the sole purpose of warming up the organism. These cells are brown because they are packed with mitochondria, which contain high concentrations of pigmented proteins, such as cytochromes.

ANSWER 14–2 The inner mitochondrial membrane is the site of oxidative phosphorylation, and it produces most of the cell's ATP. Cristae are portions of the mitochondrial inner membrane that are folded inward. Mitochondria that have a higher density of cristae have a larger area of inner membrane and therefore a greater capacity to carry out oxidative phosphorylation. Heart muscle expends a lot of energy during its continuous contractions, whereas skin cells have a smaller energy demand. An increased density of cristae therefore increases the ATP-production capacity of the heart muscle cell. This is a remarkable example of how cells adjust the abundance of their individual components according to need.

ANSWER 14–3

A. The DNP collapses the electrochemical proton gradient completely. H^+ ions that are pumped to one side of the membrane flow back freely, and therefore no energy to drive ATP synthesis can be stored across the membrane.

B. An electrochemical gradient is made up of two components: a concentration gradient and an electrical potential. If the membrane is made permeable to K^+ with nigericin, K^+ will be driven into the matrix by the electrical potential of the inner membrane (negative inside, positive outside). The influx of positively charged K^+ will abolish the membrane's electrical potential. In contrast, the concentration component of the H^+ gradient (the pH difference) is unaffected by nigericin. Therefore, only part of the driving force that makes it energetically favorable for H^+ ions to flow back into the matrix is lost.

ANSWER 14–4

A. Such a turbine running in reverse is an electrically driven water pump, which is analogous to what the ATP synthase becomes when it uses the energy of ATP hydrolysis to pump protons against their electrochemical gradient across the inner mitochondrial membrane.

B. The ATP synthase should stall when the energy that it can draw from the proton gradient is just equal to the ΔG required to make ATP; at this equilibrium point there will be neither net ATP synthesis nor net ATP consumption.

C. As the cell uses up ATP, the ATP/ADP ratio in the matrix

falls below the equilibrium point just described, and ATP synthase uses the energy stored in the proton gradient to synthesize ATP in order to restore the original ATP/ADP ratio. Conversely, when the electrochemical proton gradient drops below that at the equilibrium point, ATP synthase uses ATP in the matrix to restore this gradient.

ANSWER 14–5 An electron pair causes 10 H^+ to be pumped across the membrane when passing from NADH to O_2 through the three respiratory complexes. Four H^+ are needed to make each ATP: three for synthesis from ADP and one for ATP export to the cytosol. Therefore, 2.5 ATP molecules are synthesized from each NADH molecule.

ANSWER 14–6 One can describe four essential roles for the proteins in the process. First, the chemical environment provided by a protein's amino acid side chains sets the redox potential of each Fe ion such that electrons can be passed in a defined order from one component to the next, giving up their energy in small steps and becoming more firmly bound as they proceed. Second, the proteins position the Fe ions so that the electrons can move efficiently between them. Third, the proteins prevent electrons from skipping an intermediate step; thus, as we have learned for other enzymes (discussed in Chapter 4), they channel the electron flow along a defined path. Fourth, the proteins couple the movement of the electrons down their energy ladder to the pumping of protons across the membrane, thereby harnessing the energy that is released and storing it in a proton gradient that is then used for ATP production.

ANSWER 14–7 It would not be productive to use the same carrier in two steps. If ubiquinone, for example, could transfer electrons directly to the cytochrome c oxidase, the cytochrome c reductase complex would often be skipped when electrons are collected from NADH dehydrogenase. Given the large difference in redox potential between ubiquinone and cytochrome c oxidase, a large amount of energy would be released as heat and thus be wasted. Electron transfer directly between NADH dehydrogenase and cytochrome c would similarly allow the cytochrome c reductase complex to be bypassed.

ANSWER 14–8 Protons pumped across the inner mitochondrial membrane into the intermembrane space equilibrate with the cytosol, which functions as a huge H^+ sink. Both the mitochondrial matrix and the cytosol support many metabolic reactions that require a pH around neutrality. The H^+ concentration difference, ΔpH, that can be achieved between the mitochondrial matrix and the cytosol is therefore relatively small (less than one pH unit). Much of the energy stored in the mitochondrial electrochemical proton gradient is instead due to the membrane potential (see Figure 14–15).

In contrast, chloroplasts have a smaller, dedicated compartment into which H^+ ions are pumped. Much higher concentration differences can be achieved (up to a thousandfold, or 3 pH units), and much of the energy stored in the thylakoid H^+ gradient is due to the H^+ concentration difference between the thylakoid space and the stroma.

ANSWER 14–9 NADH and NADPH differ by the presence of a single phosphate group. That phosphate gives NADPH a slightly different shape from NADH, which allows these molecules to be recognized by different enzymes, and thus to deliver their electrons to different destinations.

Such a division of labor is useful because NADPH tends to be involved in biosynthetic reactions, where high-energy electrons are used to produce energy-rich biological molecules. NADH, on the other hand, is involved in reactions that oxidize energy-rich food molecules to produce ATP. Inside the cell, the ratio of NAD^+ to NADH is kept high, whereas the ratio of $NADP^+$ to NADPH is kept low. This provides plenty of NAD^+ to act as an oxidizing agent and plenty of NADPH to act as a reducing agent—as required for their special roles in catabolism and anabolism, respectively.

ANSWER 14–10
A. Photosynthesis produces sugars, most importantly sucrose, that are transported from the photosynthetic cells through the sap to root cells. There, the sugars are oxidized by glycolysis in the root cell cytoplasm and by oxidative phosphorylation in the root cell mitochondria to produce ATP, as well as being used as the building blocks for many other metabolites.
B. Mitochondria are required even during daylight hours in chloroplast-containing cells to supply the cell with ATP derived by oxidative phosphorylation. Glyceraldehyde 3-phosphate made by photosynthesis in chloroplasts moves to the cytosol and is eventually used as a source of energy to drive ATP production in mitochondria.

ANSWER 14–11 All statements are correct.
A. This is a necessary condition. If it were not true, electrons could not be removed from water and the reaction that splits water molecules ($H_2O \rightarrow 2H^+ + \frac{1}{2}O_2 + 2e^-$) would not occur.
B. Only when excited by light energy does chlorophyll have a low enough affinity for an electron to pass it to an electron carrier with a low electron affinity. This transfer allows the energy of the photon to be harnessed as energy that can be utilized in chemical conversions.
C. It can be argued that this is one of the most important obstacles that had to be overcome during the evolution of photosynthesis: partially reduced oxygen molecules, such as the superoxide radical O_2^-, are dangerously reactive and will attack and destroy almost any biologically active molecule. These intermediates therefore have to remain tightly bound to the metals in the active site of the enzyme until all four electrons have been removed from two water molecules. This requires the sequential capture of four photons by the same reaction center.

ANSWER 14–12
A. True. NAD^+ and quinones are examples of compounds that do not have metal ions but can participate in electron transfer.
B. False. The potential is due to protons (H^+) that are pumped across the membrane from the matrix to the intermembrane space. Electrons remain bound to electron carriers in the inner mitochondrial membrane.
C. True. Both components add to the driving force that makes it energetically favorable for H^+ to flow back into the matrix.
D. True. Both move rapidly in the plane of the membrane.
E. False. Not only do plants need mitochondria to make ATP in cells that do not have chloroplasts, such as root cells, but mitochondria make most of the cytosolic ATP in all plant cells.

F. True. Chlorophyll's physiological function requires it to absorb light; heme just happens to be a colored compound from which blood derives its red color.

G. False. Chlorophyll absorbs light and transfers energy in the form of an energized electron, whereas the iron in heme is a simple electron carrier.

H. False. Most of the dry weight of a tree comes from carbon derived from the CO_2 that has been fixed during photosynthesis.

ANSWER 14–13 It takes three protons. The precise value of the ΔG for ATP synthesis depends on the concentrations of ATP, ADP, and P_i (as described in Chapter 3). The higher the ratio of the concentration of ATP to ADP, the more energy it takes to make additional ATP. The lower value of 11 kcal/mole therefore applies to conditions where cells have expended a lot of energy and have therefore decreased the normal ATP/ADP ratio.

ANSWER 14–14 If no O_2 is available, all components of the mitochondrial electron-transport chain will accumulate in their *reduced* form. This is the case because electrons derived from NADH enter the chain but cannot be transferred to O_2. The electron-transport chain therefore stalls with all of its components in the reduced form. If O_2 is suddenly added again, the electron carriers in cytochrome *c* oxidase will become *oxidized before* those in NADH dehydrogenase. This is true because, after O_2 addition, cytochrome *c* oxidase will donate its electrons directly to O_2, thereby becoming oxidized. A wave of increasing oxidation then passes backward with time from cytochrome *c* oxidase through the components of the electron-transport chain, as each component regains the opportunity to pass on its electrons to downstream components.

ANSWER 14–15 As oxidized ubiquinone becomes reduced, it picks up two electrons but also two protons from water (Figure 14–23). Upon oxidation, these protons are released. If reduction occurs on one side of the membrane and oxidation at the other side, a proton is pumped across the membrane for each electron transported. Electron transport by ubiquinone thereby contributes directly to the generation of the H^+ gradient.

ANSWER 14–16 Photosynthetic bacteria and plant cells use the electrons derived in the reaction $2H_2O \rightarrow 4e^- + 4H^+ + O_2$ to reduce $NADP^+$ to NADPH, which is then used to produce useful metabolites. If the electrons were used instead to produce H_2 in addition to O_2, the cells would lose any benefit they derive from carrying out the reaction, because the electrons could not take part in metabolically useful reactions.

ANSWER 14–17

A. The switch in solutions creates a pH gradient across the thylakoid membrane. The flow of H^+ ions down its electrochemical potential drives ATP synthase, which converts ADP to ATP.

B. No light is needed, because the H^+ gradient is established artificially without a need for the light-driven electron-transport chain.

C. Nothing. The H^+ gradient would be in the wrong direction; ATP synthase would not work.

D. The experiment provided early supporting evidence for the chemiosmotic model by showing that an H^+ gradient alone is sufficient to drive ATP synthesis.

ANSWER 14–18

A. When the vesicles are exposed to light, H^+ ions (derived from H_2O) pumped into the vesicles by the bacteriorhodopsin flow back out through the ATP synthase, causing ATP to be made in the solution surrounding the vesicles in response to light.

B. If the vesicles are leaky, no H^+ gradient can form and thus ATP synthase cannot work.

C. Using components from widely divergent organisms can be a very powerful experimental tool. Because the two proteins come from such different sources, it is very unlikely that they form a direct functional interaction. The experiment therefore strongly suggests that electron transport and ATP synthesis are separate events. This approach is therefore a valid one.

ANSWER 14–19 The redox potential of $FADH_2$ is too low to transfer electrons to the NADH dehydrogenase complex, but high enough to transfer electrons to ubiquinone (Figure 14–24). Therefore, electrons from $FADH_2$ can enter the electron-transport chain only at this step (**Figure A14–19**). Because the NADH dehydrogenase complex is bypassed, fewer H^+ ions are pumped across the membrane and less ATP is made. This example shows the versatility of the electron-transport chain. The ability to use vastly different sources of electrons from the environment to feed electron transport is thought to have been an essential feature in the early evolution of life.

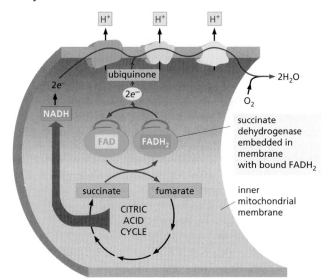

Figure A14–19

ANSWER 14–20 If these bacteria used a proton gradient to make their ATP in a fashion analogous to that in other bacteria (that is, fewer protons inside than outside), they would need to raise their cytoplasmic pH even higher than that of their environment (pH 10). Cells with a cytoplasmic pH greater than 10 would not be viable. These bacteria must therefore use gradients of ions other than H^+, such as Na^+ gradients, in the chemiosmotic coupling between electron transport and an ATP synthase.

ANSWER 14–21 Statements A and B are accurate. Statement C is incorrect, because the chemical reactions that are carried out in each cycle are completely different, even though the net effect is the same as that expected for simple reversal.

ANSWER 14–22 This experiment would suggest a two-step model for ATP synthase function. According to this model, the flow of protons through the base of the synthase drives rotation of the head, which in turn causes ATP synthesis. In their experiment, the authors have succeeded in uncoupling these two steps. If rotating the head mechanically is sufficient to produce ATP in the absence of any applied proton gradient, the ATP synthase is a protein machine that indeed functions like a "molecular turbine." This would be a very exciting experiment indeed, because it would directly demonstrate the relationship between mechanical movement and enzymatic activity. There is no doubt that it should be published and that it would become a "classic."

ANSWER 14–23 Only under condition (E) is electron transfer observed, with cytochrome *c* becoming reduced. A portion of the electron-transport chain has been reconstituted in this mixture, so that electrons can flow in the energetically favored direction from reduced ubiquinone to the cytochrome *c* reductase complex to cytochrome *c*. Although energetically favorable, the transfer in (A) cannot occur spontaneously in the absence of the cytochrome *c* reductase complex to catalyze this reaction. No electron flow occurs in the other experiments, whether the cytochrome *c* reductase complex is present or not: in experiments (B) and (F) both ubiquinone and cytochrome *c* are oxidized; in experiments (C) and (G) both are reduced; and in experiments (D) and (H) electron flow is energetically disfavored because an electron in reduced cytochrome *c* has a lower free energy than an electron added to oxidized ubiquinone.

Chapter 15

ANSWER 15–1 Although the nuclear envelope forms one continuous membrane, it has specialized regions that contain special proteins and have a characteristic appearance. One such specialized region is the inner nuclear membrane. Membrane proteins can indeed diffuse between the inner and outer nuclear membranes, at the connections formed around the nuclear pores. Those proteins with particular functions in the inner membrane, however, are usually anchored there by their interaction with other components such as chromosomes and the nuclear lamina (a protein meshwork underlying the inner nuclear membrane that helps give structural integrity to the nuclear envelope).

ANSWER 15–2 Eukaryotic gene expression is more complicated than prokaryotic gene expression. In particular, prokaryotic cells do not have introns that interrupt the coding sequences of their genes, so that an mRNA can be translated immediately after it is transcribed, without a need for further processing (discussed in Chapter 7). In fact, in prokaryotic cells, ribosomes start translating most mRNAs before transcription is finished. This would have disastrous consequences in eukaryotic cells, because most RNA transcripts have to be spliced before they can be translated. The nuclear envelope separates the transcription and translation processes in space and time: a primary RNA transcript is held in the nucleus until it is properly processed to form an mRNA, and only then is it allowed to leave the nucleus so that ribosomes can translate it.

ANSWER 15–3 An mRNA molecule is attached to the ER membrane by the ribosomes translating it. This ribosome population, however, is not static; the mRNA is continuously moved through the ribosome. Those ribosomes that have finished translation dissociate from the 3′ end of the mRNA and from the ER membrane, but the mRNA itself remains bound by other ribosomes, newly recruited from the cytosolic pool, that have attached to the 5′ end of the mRNA and are still translating the mRNA. Depending on its length, there are about 10–20 ribosomes attached to each membrane-bound mRNA molecule.

ANSWER 15–4

A. The internal signal sequence functions as a membrane anchor, as shown in Figure 15–17. Because there is no stop-transfer sequence, however, the C-terminal end of the protein continues to be translocated into the ER lumen. The resulting protein therefore has its N-terminal domain in the cytosol, followed by a single transmembrane segment, and a C-terminal domain in the ER lumen (Figure A15–4A).

B. The N-terminal signal sequence initiates translocation of the N-terminal domain of the protein until translocation is stopped by the stop-transfer sequence. A cytosolic domain is synthesized until the start-transfer sequence initiates translocation again. The situation now resembles that described in (A), and the C-terminal domain of the protein is translocated into the lumen of the ER. The resulting protein therefore spans the membrane twice. Both its N-terminal and C-terminal domains are in the ER lumen, and a loop domain between the two transmembrane regions is exposed in the cytosol (Figure A15–4B).

C. It would need a cleaved signal sequence, followed by an internal stop-transfer sequence, followed by pairs of start- and stop-transfer sequences (Figure A15–4C).

These examples demonstrate that complex protein topologies can be achieved by simple variations and combinations of the two basic mechanisms shown in Figures 15–16 and 15–17.

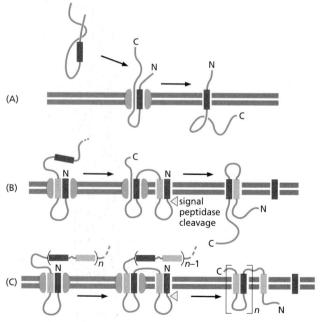

Figure A15–4

ANSWER 15–5

A. Clathrin coats cannot assemble in the absence of

adaptins that link the clathrin to the membrane. At high clathrin concentrations and under the appropriate ionic conditions, clathrin cages assemble in solution, but they are empty shells, lacking other proteins, and they contain no membrane. This shows that the information to form clathrin baskets is contained in the clathrin molecules themselves, which are therefore able to self-assemble.

B. Without clathrin, adaptins still bind to receptors in the membrane, but no clathrin coat can form and thus no clathrin-coated pits or vesicles are produced.

C. Deeply invaginated clathrin-coated pits form on the membrane, but they do not pinch off to form closed vesicles (see Figure A15–13).

D. Prokaryotic cells do not perform endocytosis. A prokaryotic cell therefore does not contain any receptors with appropriate cytosolic tails that could mediate adaptin binding. Therefore, no clathrin can bind and no clathrin coats can assemble.

ANSWER 15–6   The preassembled sugar chain allows better quality control. The assembled oligosaccharide chains can be checked for accuracy before they are added to the protein; if a mistake were made in adding sugars individually to the protein, the whole protein would have to be discarded. Because far more energy is used in building a protein than in building a short oligosaccharide chain, this is a much more economical strategy. This difficulty becomes apparent as the protein moves to the cell surface: although sugar chains are continually modified by enzymes in various compartments of the secretory pathway, these modifications are often incomplete and result in considerable heterogeneity of the glycoproteins that leave the cell. This heterogeneity is largely due to the restricted access that the enzymes have to the sugar trees attached to the surface of proteins. The heterogeneity also explains why glycoproteins are more difficult to study and purify than nonglycosylated proteins.

ANSWER 15–7   Aggregates of the secretory proteins would form in the ER, just as they do in the *trans* Golgi network. As the aggregation is specific for secretory proteins, ER proteins would be excluded from the aggregates. The aggregates would eventually be degraded.

ANSWER 15–8   Transferrin without Fe bound does not interact with its receptor and circulates in the bloodstream until it catches an Fe ion. Once iron is bound, the iron–transferrin complex can bind to the transferrin receptor on the surface of a cell and be endocytosed. Under the acidic conditions of the endosome, the transferrin releases its iron, but the transferrin remains bound to the transferrin receptor, which is recycled back to the cell surface, where it encounters the neutral pH environment of the blood. The neutral pH causes the receptor to release the transferrin into the circulation, where it can pick up another Fe ion to repeat the cycle. The iron released in the endosome, like the LDL in Figure 15–33, moves on to lysosomes, from where it is transported into the cytosol.

The system allows cells to take up iron efficiently even though the concentration of iron in the blood is extremely low. The iron bound to transferrin is concentrated at the cell surface by binding to transferrin receptors; it becomes further concentrated in clathrin-coated pits, which collect

the transferrin receptors. In this way, transferrin cycles between the blood and endosomes, delivering the iron that cells need to grow.

ANSWER 15–9
A. True.
B. False. The signal sequences that direct proteins to the ER contain a core of eight or more hydrophobic amino acids. The sequence shown here contains many hydrophilic amino acid side chains, including the charged amino acids His, Arg, Asp, and Lys, and the uncharged hydrophilic amino acids Gln and Ser.
C. True. Otherwise they could not dock at the correct target membrane or recruit a fusion complex to a docking site.
D. True.
E. True. Lysosomal proteins are selected in the *trans* Golgi network and packaged into transport vesicles that deliver them to the late endosome. If not selected, they would enter by default into transport vesicles that move constitutively to the cell surface.
F. False. Lysosomes also digest internal organelles by autophagy.
G. False. Mitochondria do not participate in vesicular transport, and therefore *N*-linked glycoproteins, which are exclusively assembled in the ER, cannot be transported to mitochondria.

ANSWER 15–10 They must contain a nuclear localization signal as well. Proteins with nuclear export signals shuttle between the nucleus and the cytosol. An example is the A1 protein, which binds to mRNAs in the nucleus and guides them through the nuclear pores. Once in the cytosol, a nuclear localization signal ensures that the A1 protein is re-imported so that it can participate in the export of further mRNAs.

ANSWER 15–11 Influenza virus enters cells by endocytosis and is delivered to endosomes, where it encounters an acidic pH that activates its fusion protein. The viral membrane then fuses with the membrane of the endosome, releasing the viral genome into the cytosol (Figure A15–11). NH_3 is a small molecule that readily penetrates membranes. Thus, it can enter all intracellular compartments, including endosomes, by diffusion. Once in a compartment that has an acidic pH, NH_3 binds H^+ to form NH_4^+, which is a charged ion and therefore cannot cross the membrane by diffusion. NH_4^+ ions therefore accumulate in acidic compartments, raising their pH. When the pH of the endosome is raised, viruses are still endocytosed, but

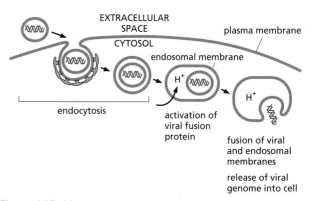

Figure A15–11

because the viral fusion protein cannot be activated, the virus cannot enter the cytosol. Remember this the next time you have the flu and have access to a stable.

ANSWER 15–12
A. The problem is that vesicles having two different kinds of v-SNAREs in their membrane could dock on either of two different membranes.
B. The answer to this puzzle is currently not known, but we can predict that cells must have ways of turning the docking ability of SNAREs on and off. This may be achieved through other proteins that are, for example, co-packaged in the ER with SNAREs into transport vesicles and facilitate the interactions of the correct v-SNARE with the t-SNARE in the *cis* Golgi network.

ANSWER 15–13 Synaptic transmission involves the release of neurotransmitters by exocytosis. During this event, the membrane of the synaptic vesicle fuses with the plasma membrane of the nerve terminals. To make new synaptic vesicles, membrane must be retrieved from the plasma membrane by endocytosis. This endocytosis step is blocked if dynamin is defective, as the protein is required to pinch off the clathrin-coated endocytic vesicles. The first clue to deciphering the role of dynamin came from electron micrographs of synapses of the mutant flies (Figure A15–13). Note that there are many flasklike invaginations of the plasma membrane, representing deeply invaginated clathrin-coated pits that cannot pinch off. The collars visible around the necks of these invaginations are made of mutant dynamin.

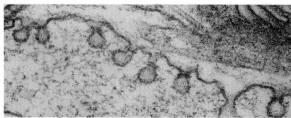

From J.H. Koenig and K. Ikeda, *J. Neurosci.* 9:3844–3860, 1989.
With permission from The Society for Neuroscience.

Figure A15–13

ANSWER 15–14 The first two sentences are correct. The third is not. It should read: "Because the contents of the lumen of the ER or any other compartment in the secretory or endocytic pathways never mix with the cytosol, proteins that enter these pathways will never need to be imported again."

ANSWER 15–15 The protein is translocated into the ER. Its ER signal sequence is recognized as soon as it emerges from the ribosome. The ribosome then becomes bound to the ER membrane, and the growing polypeptide chain is transferred through the ER translocation channel. The nuclear localization sequence is therefore never exposed to the cytosol. It will never encounter nuclear import receptors, and the protein will not enter the nucleus.

ANSWER 15–16 (1) Proteins are imported into the nucleus after they have been synthesized, folded, and, if appropriate, assembled into complexes. In contrast, unfolded polypeptide chains are translocated into the ER as they are being made by the ribosomes. Ribosomes are assembled in the nucleus yet function in the cytosol, and

the enzyme complexes that catalyze RNA transcription and splicing are assembled in the cytosol yet function in the nucleus. Thus, both ribosomes and these enzyme complexes need to be transported through the nuclear pores intact. (2) Nuclear pores are gates, which are always open to small molecules; in contrast, translocation channels in the ER membrane are normally closed, and open only after the ribosome has attached to the membrane and the translocating polypeptide chain has sealed the channel from the cytosol. It is important that the ER membrane remain impermeable to small molecules during the translocation process, as the ER is a major store for Ca^{2+} in the cell, and Ca^{2+} release into the cytosol must be tightly controlled (discussed in Chapter 16). (3) Nuclear localization signals are not cleaved off after protein import into the nucleus; in contrast, ER signal peptides are usually cleaved off. Nuclear localization signals are needed to repeatedly re-import nuclear proteins after they have been released into the cytosol during mitosis, when the nuclear envelope breaks down.

ANSWER 15–17 The transient intermixing of nuclear and cytosolic contents during mitosis supports the idea that the nuclear interior and the cytosol are indeed evolutionarily related. In fact, one can consider the nucleus as a subcompartment of the cytosol that has become surrounded by the nuclear envelope, with access only through the nuclear pores.

ANSWER 15–18 The actual explanation is that the single amino acid change causes the protein to misfold slightly so that, although it is still active as a protease inhibitor, it is prevented by chaperone proteins in the ER from exiting this organelle. It therefore accumulates in the ER lumen and is eventually degraded. Alternative interpretations might have been that (1) the mutation affects the stability of the protein in the bloodstream so that it is degraded much faster in the blood than the normal protein, or (2) the mutation inactivates the ER signal sequence and prevents the protein from entering the ER. (3) Another explanation could have been that the mutation altered the sequence to create an ER retention signal, which would have retained the mutant protein in the ER. One could distinguish between these possibilities by using fluorescently tagged antibodies against the protein or by expressing the protein as a fusion with GFP to follow its transport in the cells (see How We Know, pp. 512–513).

ANSWER 15–19 Critique: "Dr. Outonalimb proposes to study the biosynthesis of forgettin, a protein of significant interest. The main hypothesis on which this proposal is based, however, requires further support. In particular, it is questionable whether forgettin is indeed a secreted protein, as proposed. ER signal sequences are normally found at the N-terminus. C-terminal hydrophobic sequences will be exposed outside the ribosome only after protein synthesis has already terminated and can therefore not be recognized by an SRP during translation. It is therefore unlikely that forgettin will be translocated by an SRP-dependent mechanism; it is more likely that it will remain in the cytosol. Dr. Outonalimb should take these considerations into account when submitting a revised application."

ANSWER 15–20 The Golgi apparatus may have evolved from specialized patches of ER membrane. These regions of the ER might have pinched off, forming a new compartment

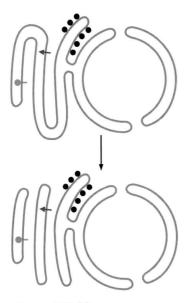

Figure A15–20

(Figure A15–20), which still communicates with the ER by vesicular transport. For the newly evolved Golgi compartment to be useful, transport vesicles would also have to have evolved.

ANSWER 15–21 This is a chicken-and-egg question. In fact, the situation never arises in present-day cells, although it must have posed a considerable problem for the first cells that evolved. New cell membranes are made by expansion of existing membranes, and the ER is never made *de novo*. There will always be an existing piece of ER with translocation channels to integrate new translocation channels. Inheritance is therefore not limited to the propagation of the genome; a cell's organelles must also be passed from generation to generation. In fact, the ER translocation channels can be traced back to structurally related translocation channels in the prokaryotic plasma membrane.

ANSWER 15–22
A. Extracellular space
B. Cytosol
C. Plasma membrane
D. Clathrin coat
E. Membrane of deeply invaginated, clathrin-coated pit
F. Captured cargo particles
G. Lumen of deeply invaginated, clathrin-coated pit

ANSWER 15–23 A single, incomplete round of nuclear import would occur. Because nuclear transport is fueled by GTP hydrolysis, under conditions of insufficient energy, GTP would be used up and no Ran-GTP would be available to unload the cargo protein from its nuclear import receptor upon arrival in the nucleus (see Figure 15–10). Unable to release its cargo, the nuclear import receptor would be stuck at the nuclear pore and not return to the cytosol. Because the nuclear cargo protein is not released, it would not be functional, and no further import could occur.

Chapter 16

ANSWER 16–1 Most paracrine signaling molecules are very short-lived after they are released from a signaling cell: they are either degraded by extracellular enzymes or are rapidly taken up by neighboring target cells. In addition, some become attached to the extracellular matrix and are thus prevented from diffusing too far.

ANSWER 16–2 Polar groups are hydrophilic, so cholesterol, with only one polar –OH group, would be too hydrophobic to be an effective hormone by itself. Because it is virtually insoluble in water, it could not move readily as a messenger from one cell to another via the extracellular fluid, unless carried by specific proteins.

ANSWER 16–3 The protein could be an enzyme that produces a large number of small intracellular signaling molecules such as cyclic AMP or cyclic GMP. Or, it could be an enzyme that modifies a large number of intracellular target proteins—for example, by phosphorylation.

ANSWER 16–4 In the case of the steroid-hormone receptor, a one-to-one complex of steroid and receptor binds to DNA to activate or inactivate gene transcription; there is thus no amplification between ligand binding and transcriptional regulation. Amplification occurs later, because transcription of a gene gives rise to many mRNAs, each of which is translated to give many copies of the protein it encodes (discussed in Chapter 7). For the ion-channel-coupled receptors, a single ion channel will let through thousands of ions in the time it remains open; this serves as the amplification step in this type of signaling system.

ANSWER 16–5 The mutant G protein would be almost continuously activated, because GDP would dissociate spontaneously, allowing GTP to bind even in the absence of an activated GPCR. The consequences for the cell would therefore be similar to those caused by cholera toxin, which modifies the α subunit of G_s so that it cannot hydrolyze GTP to shut itself off. In contrast to the cholera toxin case, however, the mutant G protein would not stay permanently activated: it would switch itself off normally, but then it would instantly become activated again as the GDP dissociated and GTP re-bound.

ANSWER 16–6 Rapid breakdown keeps the intracellular cyclic AMP concentrations low. The lower the cAMP levels are, the larger and faster the increase achieved upon activation of adenylyl cyclase, which makes new cyclic AMP. If you have $100 in the bank and you deposit another $100, you have doubled your wealth; if you have only $10 to start with and you deposit $100, you have increased your wealth tenfold, a much larger proportional increase resulting from the same deposit.

ANSWER 16–7 Recall that the plasma membrane constitutes a rather small area compared with the total membrane surfaces in a cell (discussed in Chapter 15). The endoplasmic reticulum is especially abundant and spans the entire volume of the cell as a vast network of membrane tubes and sheets. The Ca^{2+} stored in the endoplasmic reticulum can therefore be released throughout the cytosol. This is important because the rapid clearing of Ca^{2+} ions from the cytosol by Ca^{2+} pumps prevents Ca^{2+} from diffusing any significant distance in the cytosol.

ANSWER 16–8 Each reaction involved in the amplification scheme must be turned off to reset the signaling pathway to a resting level. Each of these off switches is equally important.

ANSWER 16–9 Because each antibody has two antigen-binding sites, it can cross-link the receptors and cause them to cluster on the cell surface. This clustering is likely to activate RTKs, which are usually activated by dimerization. For RTKs, clustering allows the individual kinase domains of the receptors to phosphorylate adjacent receptors in the cluster. The activation of GPCRs is more complicated, because the ligand has to induce a particular conformational change; only very special antibodies mimic receptor ligands sufficiently well to induce the conformational change that activates a GPCR.

ANSWER 16–10 The more steps there are in an intracellular signaling pathway, the more places the cell has to regulate the pathway, amplify the signal, integrate signals from different pathways, and spread the signal along divergent paths (see Figure 16–13).

ANSWER 16–11

A. True. Acetylcholine, for example, slows the beating of heart muscle cells by binding to a GPCR and stimulates the contraction of skeletal muscle cells by binding to a different acetylcholine receptor, which is an ion-channel-coupled receptor.

B. False. Acetylcholine is short-lived and exerts its effects locally. Indeed, the consequences of prolonging its lifetime can be disastrous. Compounds that inhibit the enzyme acetylcholinesterase, which normally breaks down acetylcholine at a nerve–muscle synapse, are extremely toxic: for example, the nerve gas sarin, used in chemical warfare, is an acetylcholinesterase inhibitor.

C. True. Nucleotide-free $\beta\gamma$ complexes can activate ion channels, and GTP-bound α subunits can activate enzymes. The GDP-bound form of trimeric G proteins is the inactive state.

D. True. The inositol phospholipid that is cleaved to produce IP$_3$ contains three phosphate groups, one of which links the sugar to the diacylglycerol lipid. IP$_3$ is generated by a simple hydrolysis reaction (see Figure 16–27).

E. False. Calmodulin senses but does not regulate intracellular Ca^{2+} levels.

F. True. See Figure 16–40.

G. True. See Figure 16–32.

ANSWER 16–12

1. You would expect a high background level of Ras activity, because Ras cannot be turned off efficiently.

2. Because some Ras molecules are already GTP-bound, Ras activity in response to an extracellular signal would be greater than normal, but this activity would be liable to saturate when all Ras molecules are converted to the GTP-bound form.

3. The response to a signal would be much less rapid, because the signal-dependent increase in GTP-bound Ras would occur over an elevated background of preexisting GTP-bound Ras (see Question 16–6).

4. The increase in Ras activity in response to a signal would also be prolonged compared to the response in normal cells.

ANSWER 16–13

A. Both types of signaling can occur over a long range: neurons can send action potentials along very long axons (think of the axons in the neck of a giraffe, for example), and hormones are carried via the bloodstream throughout the organism. Because neurons secrete large amounts of neurotransmitters at a synapse, a small, well-defined space between two cells, the concentrations of these signal molecules are high; neurotransmitter receptors, therefore, need to bind to neurotransmitters with only low affinity. Hormones, in contrast, are vastly diluted in the bloodstream, where they circulate at often minuscule concentrations; hormone receptors therefore generally bind their hormone with extremely high affinity.

B. Whereas neuronal signaling is a private affair, with one neuron talking to a select group of target cells through specific synaptic connections, endocrine signaling is a public announcement, with any target cell with appropriate receptors able to respond to the hormone in the blood. Neuronal signaling is very fast, limited only by the speed of propagation of the action potential and the workings of the synapse, whereas endocrine signaling is slower, limited by blood flow and diffusion over larger distances.

ANSWER 16–14

A. There are 100,000 molecules of X and 10,000 molecules of Y in the cell (= rate of synthesis × average lifetime).

B. After one second, the concentration of X will have increased by 10,000 molecules. The concentration of X, therefore, one second after its synthesis is increased, is about 110,000 molecules per cell—which is a 10% increase over the concentration of X before the boost of its synthesis. The concentration of Y will also increase by 10,000 molecules, which, in contrast to X, represents a full twofold increase in its concentration (for simplicity, we can neglect the breakdown in this estimation because X and Y are relatively stable during the one-second stimulation).

C. Because of its larger proportional increase, Y is the preferred signaling molecule. This calculation illustrates the surprising but important principle that the time it takes to switch a signal on is determined by the lifetime of the signaling molecule.

ANSWER 16–15 The information transmitted by a cell signaling pathway is contained in the *concentration* of the messenger, be it a small molecule or a phosphorylated protein. Therefore, to allow the detection of a change in concentration, the original messenger has to be both rapidly destroyed and rapidly resynthesized. The shorter the average lifetime of the messenger population, the faster the system can respond to changes. Human communication relies on messages that are delivered only once and that are generally not interpreted by their abundance but by their *content*. So it is a mistake to kill the human messengers; they can be used more than once.

ANSWER 16–16

A. The mutant RTK lacking its extracellular ligand-binding domain is inactive. It cannot bind extracellular signals, and its presence has no consequences for the function of the normal RTK (Figure A16–16A).

B. The mutant RTK lacking its intracellular domain is also inactive, but its presence will block signaling by the normal receptors. When a signal molecule binds to either receptor, it will induce their dimerization. Two

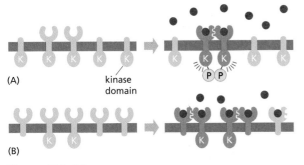

(A)   kinase domain

(B)

Figure A16–16

normal receptors have to come together to activate each other by phosphorylation. In the presence of an excess of mutant receptors, however, normal receptors will usually form mixed dimers, in which their intracellular domain cannot be activated because their partner is a mutant and lacks a kinase domain (Figure A16–16B).

ANSWER 16–17 The statement is correct. Upon ligand binding, transmembrane helices of multispanning receptors, like the GPCRs, shift and rearrange with respect to one another (Figure A16–17A). This conformational change is sensed on the cytosolic side of the membrane because of a change in the arrangement of the cytoplasmic loops. A single transmembrane segment is not sufficient to transmit a signal across the membrane directly; no rearrangements in the membrane are possible upon ligand binding. Upon ligand binding, single-span receptors such as most RTKs

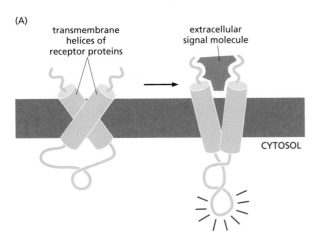

(A)

transmembrane helices of receptor proteins

extracellular signal molecule

CYTOSOL

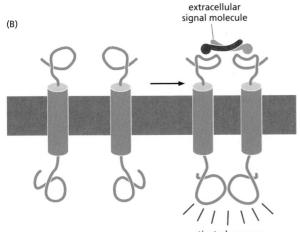

(B)

extracellular signal molecule

activated enzyme domain of receptors

Figure A16–17

tend to dimerize, thereby bringing their intracellular kinase domains into proximity so that they can cross-phosphorylate and activate each other (Figure A16–17B).

ANSWER 16–18 Activation in both cases depends on proteins that catalyze GDP–GTP exchange on the G protein or Ras protein. Whereas activated GPCRs perform this function directly for G proteins, enzyme-linked receptors assemble multiple signaling proteins into a signaling complex when the receptors are activated by phosphorylation; one of these is an adaptor protein that recruits a guanine nucleotide exchange factor that fulfills this function for Ras.

ANSWER 16–19 Because the cytosolic concentration of Ca^{2+} is so low, an influx of relatively few Ca^{2+} ions leads to large changes in its cytosolic concentration. Thus, a tenfold increase in cytosolic Ca^{2+} can be achieved by raising its concentration into the micromolar range, which would require far fewer ions than would be required to change significantly the cytosolic concentration of a more abundant ion such as Na^+. In muscle, a greater than tenfold change in cytosolic Ca^{2+} concentration can be achieved in microseconds by releasing Ca^{2+} from the sarcoplasmic reticulum, a task that would be difficult to accomplish if changes in the millimolar range were required.

ANSWER 16–20 In a multicellular organism such as an animal, it is important that cells survive only when and where they are needed. Having cells depend on signals from other cells may be a simple way of ensuring this. A misplaced cell, for example, would probably fail to get the survival signals it needs (as its neighbors would be inappropriate) and would therefore kill itself. This strategy can also help regulate cell numbers: if cell type A depends on a survival signal from cell type B, the number of B cells could control the number of A cells by making a limited amount of the survival signal, so that only a certain number of A cells could survive. There is indeed evidence that such a mechanism does operate to help regulate cell numbers—in both developing and adult tissues (see Figure 18–41).

ANSWER 16–21 Ca^{2+}-activated Ca^{2+} channels create a positive feedback loop: the more Ca^{2+} that is released, the more Ca^{2+} channels open. The Ca^{2+} signal in the cytosol is therefore propagated explosively throughout the entire muscle cell, thereby ensuring that all myosin–actin filaments contract almost synchronously.

ANSWER 16–22 K2 activates K1. If K1 is permanently activated, a response is observed regardless of the status of K2. If the order were reversed, K1 would need to activate K2, which cannot occur because in our example K2 contains an inactivating mutation.

ANSWER 16–23
A. Three examples of extended signaling pathways to the nucleus are (1) extracellular signal → RTK → adaptor protein → Ras-activating protein → MAP kinase kinase kinase → MAP kinase kinase → MAP kinase → transcription regulator; (2) extracellular signal → GPCR → G protein → phospholipase C → IP_3 → Ca^{2+} → calmodulin → CaM-kinase → transcription regulator; (3) extracellular signal → GPCR → G protein → adenylyl cyclase → cyclic AMP → PKA → transcription regulator.
B. An example of a direct signaling pathway to the nucleus is Delta → Notch → cleaved Notch tail → transcription.

ANSWER 16–24 When PI 3-kinase is activated by an activated RTK, it phosphorylates a specific inositol phospholipid in the plasma membrane. The resulting phosphorylated inositol phospholipid then recruits to the plasma membrane both Akt and another protein kinase that helps phosphorylate and activate Akt. A third kinase that is permanently associated with the membrane also helps activate Akt (see Figure 16–35).

ANSWER 16–25 Animals and plants are thought to have evolved multicellularity independently and therefore will be expected to have evolved some distinct signaling mechanisms for their cells to communicate with one another. On the other hand, animal and plant cells are thought to have evolved from a common eukaryotic ancestor cell, and so plants and animals would be expected to share some intracellular signaling mechanisms that the common ancestor cell used to respond to its environment.

Chapter 17

ANSWER 17–1   Cells that migrate rapidly from one place to another, such as amoebae (A) and sperm cells (F), do not in general need intermediate filaments in their cytoplasm, since they do not develop or sustain large tensile forces. Plant cells (G) are pushed and pulled by the forces of wind and water, but they resist these forces by means of their rigid cell walls rather than by their cytoskeleton. Epithelial cells (B), smooth muscle cells (C), and the long axons of nerve cells (E) are all rich in cytoplasmic intermediate filaments, which prevent them from rupturing as they are stretched and compressed by the movements of their surrounding tissues.

All of the above eukaryotic cells possess at least intermediate filaments in their nuclear lamina. Bacteria, such as *Escherichia coli* (D), have none whatsoever.

ANSWER 17–2   Two tubulin dimers have a lower affinity for each other (because of a more limited number of interaction sites) than a tubulin dimer has for the end of a microtubule (where there are multiple possible interaction sites, both end-to-end of tubulin dimers adding to a protofilament and side-to-side of the tubulin dimers interacting with tubulin subunits in adjacent protofilaments forming the ringlike cross section). Thus, to initiate a microtubule from scratch, enough tubulin dimers have to come together and remain bound to one another for long enough for other tubulin molecules to add to them. Only when a number of tubulin dimers have already assembled will the binding of the next subunit be favored. The formation of these initial "nucleating sites" is therefore rare and will not occur spontaneously at cellular concentrations of tubulin.

Centrosomes contain preassembled rings of γ-tubulin (in which the γ-tubulin subunits are held together in much tighter side-to-side interactions than αβ-tubulin can form) to which αβ-tubulin dimers can bind. The binding conditions of αβ-tubulin dimers resemble those of adding to the end of an assembled microtubule. The γ-tubulin rings in the centrosome can therefore be thought of as permanently preassembled nucleation sites.

ANSWER 17–3

A. The microtubule is shrinking because it has lost its GTP cap, i.e., the tubulin subunits at its end are all in their GDP-bound form. GTP-loaded tubulin subunits from solution will still add to this end, but they will be short-lived—either because they hydrolyze their GTP or because they fall off as the microtubule rim around them disassembles. If, however, enough GTP-loaded subunits are added quickly enough to cover up the GDP-containing tubulin subunits at the microtubule end, a new GTP cap can form and regrowth is favored.

B. The rate of addition of GTP-tubulin will be greater at higher tubulin concentrations. The frequency with which shrinking microtubules switch to the growing mode will therefore increase with increasing tubulin concentration. The consequence of this regulation is that the system is self-balancing: the more microtubules shrink (resulting in a higher concentration of free tubulin), the more frequently microtubules will start to grow again. Conversely, the more microtubules grow, the lower the concentration of free tubulin will become and the rate of GTP-tubulin addition will slow down; at some point GTP hydrolysis will catch up with new GTP-tubulin addition, the GTP cap will be destroyed, and the microtubule will switch to the shrinking mode.

C. If only GDP were present, microtubules would continue to shrink and eventually disappear, because tubulin dimers with GDP have very low affinity for each other and will not add stably to microtubules.

D. If GTP is present but cannot be hydrolyzed, microtubules will continue to grow until all free tubulin subunits have been used up.

ANSWER 17–4   If all the dynein arms were equally active, there could be no significant relative motion of one microtubule to the other as required for bending (think of a circle of nine weightlifters, each trying to lift his neighbor off the ground: if they all succeeded, the group would levitate!). Thus, a few ciliary dynein molecules must be activated selectively on one side of the cilium. As they move their neighboring microtubules toward the tip of the cilium, the cilium bends away from the side containing the activated dyneins.

ANSWER 17–5   Any actin-binding protein that stabilizes complexes of two or more actin monomers without blocking the ends required for filament growth will facilitate the initiation of a new filament (nucleation).

ANSWER 17–6   Only fluorescent actin molecules assembled into filaments are visible, because unpolymerized actin molecules diffuse so rapidly they produce a dim uniform background. Since, in your experiment, so few actin molecules are labeled (1:10,000), there should be at most one labeled actin monomer per filament (see Figure 17–29). The lamellipodium as a whole has many actin filaments, some of which overlap and therefore show a random speckled pattern of actin molecules, each marking a different filament.

This technique (called "speckle fluorescence") can be used to follow the movement of polymerized actin in a migrating cell. If you watch this pattern with time, you will see that individual fluorescent spots move steadily back from the leading edge toward the interior of the cell, a movement that occurs whether or not the cell is actually migrating. Rearward movement takes place because actin monomers are added to filaments at the plus end and are lost from the minus end (where they are depolymerized) (see Figure

17–35B). In effect, actin monomers "move through" the actin filaments, a phenomenon termed "treadmilling." Treadmilling has been demonstrated to occur in isolated actin filaments in solution and also in dynamic microtubules, such as those within a mitotic spindle.

ANSWER 17–7 Cells contain actin-binding proteins that bundle and cross-link actin filaments (see Figure 17–32). The filaments extending from lamellipodia and filopodia become firmly connected to the filamentous meshwork of the cell cortex, thus providing the mechanical anchorage required for the growing rodlike filaments to deform the cell membrane.

ANSWER 17–8 Although the subunits are indeed held together by noncovalent bonds that are individually weak, there are a very large number of them, distributed among a very large number of filaments. As a result, the stress a human being exerts by lifting a heavy object is dispersed over so many subunits that their interaction strength is not exceeded. By analogy, a single thread of silk is not nearly strong enough to hold a human, but a rope woven of such fibers is.

ANSWER 17–9 Both filaments are composed of subunits in the form of protein dimers that are held together by coiled-coil interactions. Moreover, in both cases, the dimers polymerize through their coiled-coil domains into filaments. Whereas intermediate filament dimers assemble head-to-head, however, and thereby create a filament that has no polarity, all myosin molecules in the same half of the myosin filament are oriented with their heads pointing in the same direction. This polarity is necessary for them to be able to develop a contractile force in muscle.

ANSWER 17–10

A. Successive actin molecules in an actin filament are identical in position and conformation. After a first protein (such as troponin) had bound to the actin filament, there would be no way in which a second protein could recognize every seventh monomer in a naked actin filament. Tropomyosin, however, binds along the length of an actin filament, spanning precisely seven monomers, and thus provides a molecular "ruler" that measures the length of seven actin monomers. Troponin becomes localized by binding to the evenly spaced ends of tropomyosin molecules.

B. Calcium ions influence force generation in the actin–myosin system only if both troponin (to bind the calcium ions) and tropomyosin (to transmit the information that troponin has bound calcium to the actin filament) are present. (i) Troponin cannot bind to actin without tropomyosin. The actin filament would be permanently exposed to the myosin, and the system would be continuously active, independently of whether calcium ions were present or not (a muscle cell would therefore be continuously contracted with no possibility of regulation). (ii) Tropomyosin would bind to actin and block binding of myosin completely; the system would be permanently inactive, no matter whether calcium ions were present, because tropomyosin is not affected by calcium. (iii) The system will contract in response to calcium ions.

ANSWER 17–11

A. True. A continual outward movement of ER is required; in the absence of microtubules, the ER collapses toward the center of the cell.

B. True. Actin is needed to make the contractile ring that causes the physical cleavage between the two daughter cells, whereas the mitotic spindle that partitions the chromosomes is composed of microtubules.

C. True. Both extensions are associated with transmembrane proteins that protrude from the plasma membrane and enable the cell to form new anchor points on the substratum.

D. False. To cause bending, ATP is hydrolyzed by the dynein motor proteins that are attached to the outer microtubules in the flagellum.

E. False. Cells could not divide without rearranging their intermediate filaments, but many terminally differentiated and long-lived cells, such as nerve cells, have stable intermediate filaments that are not known to depolymerize.

F. False. The rate of growth is independent of the size of the GTP cap. The plus and minus ends have different growth rates because they have physically distinct binding sites for the incoming tubulin subunits; the rate of addition of tubulin subunits differs at the two ends.

G. True. Both are nice examples of how the same membrane can have regions that are highly specialized for a particular function.

H. False. Myosin movement is activated by the phosphorylation of myosin, or by calcium binding to troponin.

ANSWER 17–12 The average time taken for a small molecule (such as ATP) to diffuse a distance of 10 μm is given by the calculation

$$(10^{-3})^2 / (2 \times 5 \times 10^{-6}) = 0.1 \text{ seconds}$$

Similarly, a protein takes 1 second and a vesicle 10 seconds on average to travel 10 μm. A vesicle would require on average 10^9 seconds, or more than 30 years, to diffuse to the end of a 10-cm axon. This calculation makes it clear why kinesin and other motor proteins evolved to carry molecules and organelles along microtubules.

ANSWER 17–13 (1) Animal cells are much larger and more diversely shaped, and do not have a cell wall. Cytoskeletal elements are required to provide mechanical strength and shape in the absence of a cell wall. (2) Animal cells, and all other eukaryotic cells, have a nucleus that is shaped and held in place in the cell by intermediate filaments; the nuclear lamins attached to the inner nuclear membrane support and shape the nuclear membrane, and a meshwork of intermediate filaments surrounds the nucleus and spans the cytosol. (3) Animal cells can move by a process that requires a change in cell shape. Actin filaments and myosin motor proteins are required for these activities. (4) Animal cells have a much larger genome than bacteria; this genome is fragmented into many chromosomes. For cell division, chromosomes need to be accurately distributed to the daughter cells, requiring the function of the microtubules that form the mitotic spindle. (5) Animal cells have internal organelles. Their localization in the cell is dependent on motor proteins that move them along microtubules. A remarkable example is the long-distance travel of membrane-enclosed vesicles (organelles) along microtubules in an axon that can be up to 1 m (≈3 ft) long in the case of the nerve cells that extend from your spinal cord to your feet.

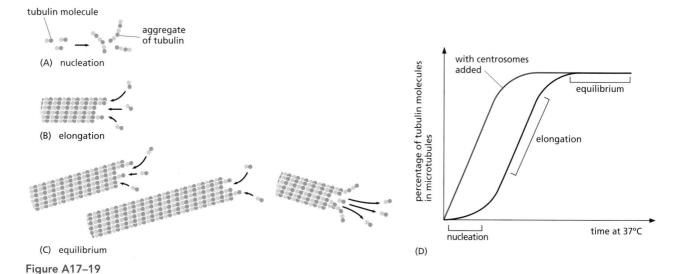

Figure A17–19

ANSWER 17–14 The ends of an intermediate filament are indistinguishable from each other, because the filaments are built by the assembly of symmetrical tetramers made from two coiled-coil dimers. In contrast to microtubules and actin filaments, intermediate filaments therefore have no polarity.

ANSWER 17–15 Intermediate filaments have no polarity; their ends are chemically indistinguishable. It would therefore be difficult to envision how a hypothetical motor protein that bound to the middle of the filament could sense a defined direction. Such a motor protein would be equally likely to attach to the filament facing one end or the other.

ANSWER 17–16 Katanin breaks microtubules along their length, and at positions remote from their GTP caps. The fragments that form therefore contain GDP-tubulin at their exposed ends and rapidly depolymerize. Katanin thus provides a very quick means of destroying existing microtubules.

ANSWER 17–17 Cell division depends on the ability of microtubules both to polymerize and to depolymerize. This is most obvious when one considers that the formation of the mitotic spindle requires the prior depolymerization of other cellular microtubules to free up the tubulin required to build the spindle. This rearrangement is not possible in Taxol-treated cells, whereas in colchicine-treated cells, division is blocked because a spindle cannot be assembled. On a more subtle but no less important level, both drugs block the dynamic instability of microtubules and would therefore interfere with the workings of the mitotic spindle, even if one could be properly assembled.

ANSWER 17–18 Motor proteins are unidirectional in their action; kinesin always moves toward the plus end of a microtubule and dynein toward the minus end. Thus if kinesin molecules are attached to glass, only those individual motors that have the correct orientation in relation to the microtubule that settles on them can attach to the microtubule and exert force on it to propel it forward. Since kinesin moves toward the plus end of the microtubule, the microtubule will always crawl minus-end first over the coverslip.

ANSWER 17–19

A. Phase A corresponds to a lag phase, during which tubulin molecules assemble to form nucleation centers (Figure A17–19A). Nucleation is followed by a rapid rise (phase B) to a plateau value as tubulin dimers add to the ends of the elongating microtubules (Figure A17–19B). At phase C, equilibrium is reached with some microtubules in the population growing while others are rapidly shrinking (Figure A17–19C). The concentration of free tubulin is constant at this point, because polymerization and depolymerization are balanced (see also Question 17–3, p. 577).

B. The addition of centrosomes introduces nucleation sites that eliminate the lag phase A as shown by the red curve in Figure A17–19D. The rate of microtubule growth (i.e., the slope of the curve in the elongation phase B) and the equilibrium level of free tubulin remain unchanged, because the presence of centrosomes does not affect the rates of polymerization and depolymerization.

ANSWER 17–20 The ends of the shrinking microtubule are visibly frayed, and the individual protofilaments appear to come apart and curl as the end depolymerizes. This micrograph therefore suggests that the GTP cap (which is lost from shrinking microtubules) holds the protofilaments properly aligned with each other, perhaps by strengthening the side-to-side interactions between αβ-tubulin subunits when they are in their GTP-bound form.

ANSWER 17–21 Cytochalasin interferes with actin filament formation, and its effect on the cell demonstrates the importance of actin to cell locomotion. The experiment with colchicine shows that microtubules are required to give a cell a polarity that then determines which end becomes the leading edge (see Figure 17–14). In the absence of microtubules, cells still go through the motions normally associated with cell movement, such as the extension of lamellipodia, but in the absence of cell polarity these are futile exercises because they happen indiscriminately in all directions.

Antibodies bind tightly to the antigen (in this case vimentin) to which they were raised (see Panel 4–2, pp. 146–147). When bound, an antibody can interfere with the function of the antigen by preventing it from interacting properly with other cell components. The antibody injection experiment therefore suggests that intermediate filaments are not required for the maintenance of cell polarity or for the motile machinery.

ANSWER 17–22 Either (B) or (C) would complete the sentence correctly. The direct result of the action potential in the plasma membrane is the release of Ca^{2+} into the cytosol from the sarcoplasmic reticulum; muscle cells are triggered to contract by this rapid rise in cytosolic Ca^{2+}. Calcium ions at high concentrations bind to troponin, which in turn causes tropomyosin to move to expose myosin-binding sites on the actin filaments. (A) and (D) would be wrong because Ca^{2+} has no effect on the detachment of the myosin head from actin, which is the result of ATP hydrolysis. Nor does it have any role in maintaining the structure of the myosin filament.

ANSWER 17–23 Only (D) is correct. Upon contraction, the Z discs move closer together, and neither actin nor myosin filaments contract (see Figures 17–41 and 17–42).

Chapter 18

ANSWER 18–1 Because all cells arise by division of another cell, this statement is correct, assuming that "first cell division" refers to the division of the successful founder cell from which all life as we know it has derived. There were probably many other unsuccessful attempts to start the chain of life.

ANSWER 18–2 Cells in peak B contain twice as much DNA as those in peak A, indicating that they contain replicated DNA, whereas the cells in peak A contain unreplicated DNA. Peak A therefore contains cells that are in G_1, and peak B contains cells that are in G_2 and mitosis. Cells in S phase have begun but not finished DNA synthesis; they therefore have various intermediate amounts of DNA and are found in the region between the two peaks. Most cells are in G_1, indicating that it is the longest phase of the cell cycle (see Figure 18–2).

ANSWER 18–3 For multicellular organisms, the control of cell division is extremely important. Individual cells must not proliferate unless it is to the benefit of the whole organism. The G_0 state offers protection from aberrant activation of cell division, because the cell-cycle control system is largely dismantled. If, on the other hand, a cell just paused in G_1, it would still contain all of the cell-cycle control system and could readily be induced to divide. The cell would also have to remake the "decision" not to divide almost continuously. To re-enter the cell cycle from G_0, a cell has to resynthesize all of the components that have disappeared.

ANSWER 18–4 The cell would replicate its damaged DNA and therefore would introduce mutations to the two daughter cells when the cell divides. Such mutations could increase the chances that the progeny of the affected daughter cells would eventually become cancer cells.

ANSWER 18–5 Before injection, the frog oocytes must contain inactive M-Cdk. Upon injection of the M-phase cytoplasm, the small amount of the active M-Cdk in the injected cytoplasm activates the inactive M-Cdk by switching on the activating phosphatase (Cdc25), which

removes the inhibitory phosphate groups from the inactive M-Cdk (see Figure 18–17). An extract of the second oocyte, now in M phase itself, will therefore contain as much active M-Cdk as the original cytoplasmic extract, and so on.

ANSWER 18–6 The experiment shows that kinetochores are not preassigned to one or other spindle pole; microtubules attach to the kinetochores that they are able to reach. For the chromosomes to remain attached to a microtubule, however, tension has to be exerted. Tension is normally achieved by the opposing pulling forces from opposite spindle poles. The requirement for such tension ensures that if two sister kinetochores ever become attached to the same spindle pole, so that tension is not generated, one or both of the connections would be lost, and microtubules from the opposing spindle pole would have another chance to attach properly.

ANSWER 18–7 Recall from Figure 18–30 that the new nuclear envelope reassembles on the surface of the chromosomes. The close apposition of the envelope to the chromosomes prevents cytosolic proteins from being trapped between the chromosomes and the envelope. Nuclear proteins are then selectively imported through the nuclear pores, causing the nucleus to expand while maintaining its characteristic protein composition.

ANSWER 18–8 The membranes of the Golgi vesicles fuse to form part of the plasma membranes of the two daughter cells. The interiors of the vesicles, which are filled with cell-wall material, become the new cell-wall matrix separating the two daughter cells. Proteins in the membranes of the Golgi vesicles thus become plasma membrane proteins. Those parts of the proteins that were exposed to the lumen of the Golgi vesicle will end up exposed to the new cell wall (**Figure A18–8**).

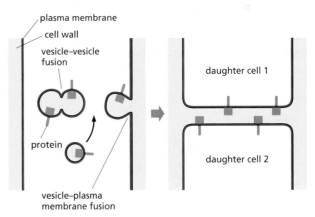

Figure A18–8

ANSWER 18–9 In a eukaryotic organism, the genetic information that the organism needs to survive and reproduce is distributed between multiple chromosomes. It is therefore crucial that each daughter cell receives a copy of each chromosome when a cell divides; if a daughter cell receives too few or too many chromosomes, the effects are usually deleterious or even lethal. Only two copies of each chromosome are produced by chromosome replication in mitosis. If the cell were to randomly distribute the chromosomes when it divided, it would be very unlikely that each daughter cell would receive precisely one copy of each chromosome. In contrast, the Golgi apparatus fragments into tiny vesicles that are all alike, and by random distribution it is very likely that each daughter cell will receive an approximately equal number of them.

ANSWER 18–10 As apoptosis occurs on a large scale in both developing and adult tissues, it must not trigger alarm reactions that are normally associated with cell injury. Tissue injury, for example, leads to the release of signal molecules that stimulate the proliferation of surrounding cells so that the wound heals. It also causes the release of signals that can cause a destructive inflammatory reaction. Moreover, the release of intracellular contents could elicit an immune response against molecules that are normally not encountered by the immune system. Such reactions would be self-defeating if they occurred in response to the massive cell death that occurs in normal development.

ANSWER 18–11 Because the cell population is increasing exponentially, doubling its weight at every cell division, the weight of the cell cluster after N cell divisions is $2^N \times 10^{-9}$ g. Therefore, 70 kg (70×10^3 g) = $2^N \times 10^{-9}$ g, or $2^N = 7 \times 10^{13}$. Taking the logarithm of both sides allows you to solve the equation for N. Therefore, $N = \ln (7 \times 10^{13}) / \ln 2 = 46$; i.e., it would take only 46 days if cells proliferated exponentially. Cell division in animals is tightly controlled, however, and most cells in the human body stop dividing when they become highly specialized. The example demonstrates that exponential cell proliferation occurs only for very brief periods, even during embryonic development.

ANSWER 18–12 The egg cells of many animals are big and contain stores of enough cell components to last for many cell divisions. The daughter cells that form during the first cell divisions after fertilization are progressively smaller in size and thus can be formed without a need for new protein or RNA synthesis. Whereas normally dividing cells would grow continuously in G_1, G_2, and S phases, until their size doubled, there is no cell growth in these early cleavage divisions, and both G_1 and G_2 are virtually absent. As G_1 is usually longer than G_2 and S phase, G_1 is the most drastically reduced in these divisions.

ANSWER 18–13

A. Radiation leads to DNA damage, which activates a checkpoint mechanism (mediated by p53 and p21; see Figure 18–15) that arrests the cell cycle until the DNA has been repaired.

B. The cell will replicate damaged DNA and thereby introduce mutations in the daughter cells when the cell divides.

C. The cell will be able to divide normally, but it will be prone to mutations, because some DNA damage always occurs as the result of natural irradiation caused, for example, by cosmic rays. The checkpoint mechanism mediated by p53 is mainly required as a safeguard against the devastating effects of accumulating DNA damage, but not for the natural progression of the cell cycle in undamaged cells.

D. Cell division in humans is an ongoing process that does not cease upon reaching maturity, and it is required for survival. Blood cells and epithelial cells in the skin or lining the gut, for example, are being constantly produced by cell division to meet the body's needs;

each day, your body produces about 10^{11} new red blood cells alone.

ANSWER 18–14

A. Only the cells that were in the S phase of their cell cycle (i.e., those cells making DNA) during the 30-minute labeling period contain any radioactive DNA.

B. Initially, mitotic cells contain no radioactive DNA because these cells were not engaged in DNA synthesis during the labeling period. Indeed, it takes about two hours before the first labeled mitotic cells appear.

C. The initial rise of the curve corresponds to cells that were just finishing DNA replication when the radioactive thymidine was added. The curve rises as more labeled cells enter mitosis; the peak corresponds to those cells that had just started S phase when the radioactive thymidine was added. The labeled cells then exit from mitosis, and are replaced by unlabeled mitotic cells, which were not yet in S phase during the labeling period. After 20 hours the curve starts rising again, because the labeled cells enter their second round of mitosis.

D. The intial two-hour lag before any labeled mitotic cells appear corresponds to the G_2 phase, which is the time between the end of S phase and the beginning of mitosis. The first labeled cells seen in mitosis were those that were just finishing S phase (DNA synthesis) when the radioactive thymidine was added.

ANSWER 18–15 Loss of M cyclin leads to inactivation of M-Cdk. As a result, the M-Cdk target proteins become dephosphorylated by phosphatases, and the cells exit from mitosis: they disassemble the mitotic spindle, reassemble the nuclear envelope, decondense their chromosomes, and so on. The M cyclin is degraded by ubiquitin-dependent destruction in proteasomes, and the activation of M-Cdk leads to the activation of the APC, which ubiquitylates the cyclin, but with a substantial delay. As discussed in Chapter 7, ubiquitylation tags proteins for degradation in proteasomes.

ANSWER 18–16 M cyclin accumulates gradually as it is steadily synthesized. As it accumulates, it will tend to form complexes with the mitotic Cdk molecules that are present. After a certain threshold level has been reached, a sufficient amount of M-Cdk has been formed so that it is activated by the appropriate kinases and phosphatases that phosphorylate and dephosphorylate it. Once activated, M-Cdk acts to enhance the activity of the activating phosphatase; this positive feedback leads to the explosive activation of M-Cdk (see Figure 18–17). Thus, M-cyclin accumulation acts like a slow-burning fuse, which eventually helps trigger the explosive self-activation of M-Cdk. The precipitous destruction of M cyclin terminates M-Cdk activity, and a new round of M-cyclin accumulation begins.

ANSWER 18–17 The order is G, C, B, A, D. Together, these five steps are referred to as mitosis (F). No step in mitosis is influenced by the phases of the moon (E). Cytokinesis is the last step in M phase, which overlaps with anaphase and telophase. Mitosis and cytokinesis are both part of M phase.

ANSWER 18–18 If the growth rate of microtubules is the same in mitotic and in interphase cells, their length is proportional to their lifetime. Thus, the average length of microtubules in mitosis is 1 μm (= 20 μm × 15 s/300 s).

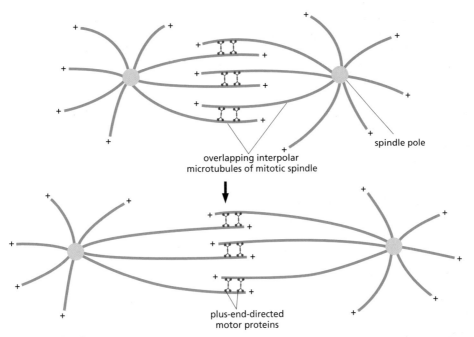

Figure A18–19

ANSWER 18–19 As shown in Figure A18–19, the overlapping interpolar microtubules from opposite poles of the spindle have their plus ends pointing in opposite directions. Plus-end-directed motor proteins cross-link adjacent antiparallel microtubules together and tend to move the microtubules in the direction that will push the two poles of the spindle apart, as shown in the figure. Minus-end-directed motor proteins also cross-link adjacent antiparallel microtubules together but move in the opposite direction, tending to pull the spindle poles together (not shown).

ANSWER 18–20 The sister chromatid becomes committed when a microtubule from one of the spindle poles attaches to the kinetochore of the chromatid. Microtubule attachment is still reversible until a second microtubule from the other spindle pole attaches to the kinetochore of its partner sister chromatid so that the duplicated chromosome is now put under mechanical tension by pulling forces from both poles. The tension ensures that both microtubules remain attached to the chromosome. The position of a chromatid in the cell at the time that the nuclear envelope breaks down will influence which spindle pole it will be pulled to, as its kinetochore is most likely to become attached to the spindle pole toward which it is facing.

ANSWER 18–21 It is still not certain what drives the poleward movement of chromosomes during anaphase. In principle, two possible models could explain it (Figure A18–21). In the model shown in (A), microtubule motor proteins associated with the kinetochore dash toward the minus end of the depolymerizing microtubule, dragging the chromosome toward the pole. Although this model is appealingly simple, there is little evidence that motor proteins are required for chromosome movement during anaphase. Instead, current experimental evidence greatly supports the model outlined in (B). In this model, chromosome movement is driven by kinetochore proteins

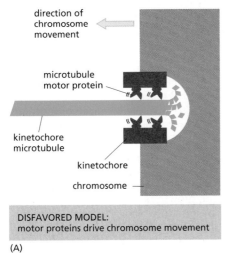

(A)

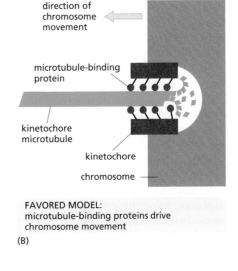

(B)

Figure A18–21

that cling to the sides of the depolymerizing microtubule. These proteins frequently detach from—and reattach to—the kinetochore microtubule. As tubulin subunits continue to dissociate, the kinetochore must slide poleward to maintain its grip on the retreating end of the shrinking microtubule.

ANSWER 18–22 Both sister chromatids could end up in the same daughter cell for any of a number of reasons. (1) If the microtubules or their connections with a kinetochore were to break during anaphase, both sister chromatids could be drawn to the same pole, and hence into the same daughter cell. (2) If microtubules from the same spindle pole attached to both kinetochores, the chromosome would be pulled to the same pole. (3) If the cohesins that link sister chromatids were not degraded, the pair of chromatids might be pulled to the same pole. (4) If a duplicated chromosome never engaged microtubules and was left out of the spindle, it would also end up in one daughter cell.

Some of these errors in the mitotic process would be expected to activate a checkpoint mechanism that delays the onset of anaphase until all chromosomes are attached properly to both poles of the spindle. This "spindle assembly checkpoint" mechanism should allow most chromosome attachment errors to be corrected, which is one reason why such errors are rare.

The consequences of both sister chromatids ending up in one daughter cell are usually dire. One daughter cell would contain only one copy of all the genes carried on that chromosome and the other daughter cell would contain three copies. The altered gene dosage, leading to correspondingly changed amounts of the mRNAs and proteins produced, is often detrimental to the cell. In addition, there is the possibility that the single copy of the chromosome may contain a defective gene with a critical function, which would normally be taken care of by the good copy of the gene on the other chromosome that is now missing.

ANSWER 18–23
A. True. Centrosomes replicate during interphase, before M phase begins.
B. True. Sister chromatids separate completely only at the start of anaphase.
C. False. The ends of interpolar microtubules overlap and attach to one another via proteins (including motor proteins) that bridge between the microtubules.
D. False. Microtubules and their motor proteins play no role in DNA replication.
E. False. To be a correct statement, the terms "centromere" and "centrosome" must be switched.

ANSWER 18–24 Antibodies bind tightly to the antigen (in this case myosin) to which they were raised. When bound, an antibody can interfere with the function of the antigen by preventing it from interacting properly with other cell components. (A) The movement of chromosomes at anaphase depends on microtubules and their motor proteins and does not depend on actin or myosin. Injection of an anti-myosin antibody into a cell will therefore have no effect on chromosome movement during anaphase. (B) Cytokinesis, on the other hand, depends on the assembly and contraction of a ring of actin and myosin filaments, which forms the cleavage furrow that splits the cell in two. Injection of anti-myosin antibody will therefore block cytokinesis.

ANSWER 18–25 The plasma membrane of the cell that died by necrosis in Figure 18–37A is ruptured; a clear break is visible, for example, at a position corresponding to the 12 o'clock mark on a watch. The cell's contents, mostly membranous and cytoskeletal debris, are seen spilling into the surroundings through these breaks. The cytosol stains lightly, because most soluble cell components were lost before the cell was fixed. In contrast, the cell that underwent apoptosis in Figure 18–37B is surrounded by an intact membrane, and its cytosol is densely stained, indicating a normal concentration of cell components. The cell's interior is remarkably different from a normal cell, however. Particularly characteristic are the large "blobs" that extrude from the nucleus, probably as the result of the breakdown of the nuclear lamina. The cytosol also contains many large, round, membrane-enclosed vesicles of unknown origin, which are not normally seen in healthy cells. The pictures visually confirm the notion that necrosis involves cell lysis, whereas cells undergoing apoptosis remain relatively intact until they are phagocytosed and digested by another cell.

ANSWER 18–26
A. False. There is no G_1 to M phase transition. The statement is correct, however, for the G_1 to S phase transition, in which cells commit themselves to a division cycle.
B. True. Apoptosis is an active process carried out by special proteases (caspases).
C. True. This mechanism is thought to adjust the number of neurons to the number of specific target cells to which the neurons connect.
D. True. An amazing evolutionary conservation!
E. True. Association of a Cdk protein with a cyclin is required for its activity (hence its name <u>c</u>yclin-<u>d</u>ependent <u>k</u>inase). Furthermore, phosphorylation at a specific site and dephosphorylation at other sites on the Cdk protein are required for the cyclin–Cdk complex to be active.

ANSWER 18–27 Cells in an animal must behave for the good of the organism as a whole—to a much greater extent than people generally act for the good of society as a whole. In the context of an organism, unsocial behavior would lead to a loss of organization and to cancer. Many of the rules that cells have to obey would be unacceptable in a human society. Most people, for example, would be reluctant to kill themselves for the good of society, yet our cells do it all the time.

ANSWER 18–28 The most likely approach to success (if that is what the goal should be called) is plan C, which should result in an increase in cell numbers. The problem is, of course, that cell numbers of each tissue must be increased similarly to maintain balanced proportions in the organism, yet different cells respond to different growth factors. As shown in **Figure A18–28**, however, the approach has indeed met with limited success. A mouse producing very large quantities of growth hormone (*left*)—which acts to stimulate the production of a secreted protein that acts as a survival factor, growth factor, or mitogen, depending on the cell type—grows to almost twice the size of a normal mouse (*right*). To achieve this twofold change in size, however, growth hormone was massively overproduced (about fiftyfold). And note that the mouse did not even attain the size of a rat, let alone a dog.

Figure A18–28   Courtesy of Ralph Brinster

The other approaches have conceptual problems:
A. Blocking all apoptosis would lead to defects in development, as rat development requires the selective death of many cells. It is unlikely that a viable animal would be obtained.
B. Blocking p53 function would eliminate an important checkpoint of the cell cycle that detects DNA damage and stops the cycle so that the cell can repair the damage; removing p53 would increase mutation rates and lead to cancer. Indeed, mice without p53 usually develop normally but die of cancer at a young age.
D. Given the circumstances, switching careers might not be a bad option.

ANSWER 18–29 The on-demand, limited release of PDGF at a wound site triggers cell division of neighboring cells for a limited amount of time, until the PDGF is degraded. This is different from the continuous release of PDGF from mutant cells, where PDGF is made in an uncontrolled way at high levels. Moreover, the mutant cells that make PDGF often express their own PDGF receptor inappropriately, so that they can stimulate their own proliferation, thereby promoting the development of cancer.

ANSWER 18–30 All three types of mutant cells would be unable to divide. The cells
A. would enter mitosis but would not be able to exit mitosis.
B. would arrest permanently in G_1 because the cyclin–Cdk complexes that act in G_1 would be inactivated.
C. would not be able to activate the transcription of genes required for cell division because the required transcription regulators would be constantly inhibited by unphosphorylated Rb.

ANSWER 18–31 In alcoholism, liver cells proliferate because the organ is overburdened and becomes damaged by the large amounts of alcohol that have to be metabolized. This need for more liver cells activates the control mechanisms that normally regulate cell proliferation. Unless badly damaged and full of scar tissue, the liver will usually shrink back to a normal size after the patient stops drinking excessively. In liver cancer, in contrast, mutations abolish normal cell proliferation control and, as a result, cells divide and keep on dividing in an uncontrolled manner, which is usually fatal.

Chapter 19

ANSWER 19–1 Although each daughter cell ends up with a diploid amount of DNA after the first meiotic division, each cell has effectively only a haploid set of chromosomes (albeit in two copies), representing only one or other homolog of each type of chromosome (although some mixing will have occurred during crossing-over). Because the maternal and paternal chromosomes of a pair will carry different versions of many of the genes, these daughter cells will not be genetically identical; each one will, however, have lost either the paternal or the maternal version of each chromosome. In contrast, somatic cells dividing by mitosis inherit a diploid set of chromosomes, and all daughter cells are genetically identical and inherit both maternal and paternal gene copies. The role of gametes produced by meiosis is to mix and reassort gene pools during sexual reproduction, and thus it is a definite advantage for each of them to have a slightly different genetic constitution. The role of somatic cells on the other hand is to build an organism that contains the same genes in all its cells and retains in each cell both maternal and paternal genetic information.

ANSWER 19–2 A typical human female produces fewer than 1000 mature eggs in her lifetime (12 per year over about 40 years); this is less than one-tenth of a percent of the possible gametes, excluding the effects of meiotic crossing-over. A typical human male produces billions of sperm during a lifetime, so in principle, every possible chromosome combination is sampled many times.

ANSWER 19–3 For simplicity, consider the situation where a father carries genes for two dominant traits, M and N, on one of his two copies of human Chromosome 1. If these two genes were located at opposite ends of this chromosome, and there were one and only one crossover event per chromosome as postulated in the question, half of his children would express trait M and the other half would express trait N—with no child resembling the father in carrying both traits. This is very different from the actual situation, where there are multiple crossover events per chromosome, causing the traits M and N to be inherited as if they were on separate chromosomes. By constructing a Punnett square like that in Figure 19–27, one can see that in this latter, more realistic case, we would actually expect one-fourth of the children of this father to inherit both traits, one-fourth to inherit trait M only, one-fourth to inherit trait N only, and one-fourth to inherit neither trait.

ANSWER 19–4 Inbreeding tends to give rise to individuals who are homozygous for many genes. To see why, consider the extreme case where inbreeding takes the form of brother–sister matings (as among the Pharaohs of ancient Egypt): because the parents are closely related, there is a high probability that the maternal and paternal alleles inherited by the offspring will be the same. Inbreeding continued over many generations gives rise to individuals who are all alike and homozygous for almost every gene. Because of the randomness of the mechanism of inheritance, there is a large chance that some deleterious alleles will become prevalent in the population in this way, giving all individuals a reduced fitness. In another, separate inbred population, the same thing will happen, but chances are a different set of deleterious alleles will become prevalent. When individuals from the two separate inbred populations mate, their offspring will inherit deleterious alleles of genes *A*, *B*, and *C* for example, from the mother, but good alleles of those genes from the father; conversely, they will inherit deleterious alleles of genes *D*, *E*, and *F* from the father, but good alleles of those genes from the mother. Most deleterious mutations are recessive. The hybrid offspring, because they are heterozygous for these genes, will thus escape the deleterious effects seen in the parents.

ANSWER 19–5   Although any one of the three explanations could in principle account for the observed result, A and B can be ruled out as being implausible.

A. There is no precedent for any instability in DNA so great as to be detectable in such a SNP analysis; in any case, the hypothesis would predict a steady decrease in the frequency of the SNP with age, not a drop in frequency that begins only at age 50.

B. Human genes change only very slowly over time (unless a massive population migration brings an influx of individuals who are genetically different). People born 50 years ago will be, on average, virtually the same genetically as the population being born today.

C. This hypothesis is correct. A SNP with these properties has been used to discover a gene that appears to cause a substantial increase in the probability of death from cardiac abnormalities.

ANSWER 19–6   Natural selection alone is not sufficient to eliminate recessive lethal genes from the population. Consider the following line of reasoning. Homozygous defective individuals can arise only as the offspring of a mating between two heterozygous individuals. By the rules of Mendelian genetics, offspring of such a mating will be in the ratio of 1 homozygous normal: 2 heterozygous: 1 homozygous defective. Thus, statistically, heterozygous individuals should always be more numerous than the homozygous, defective individuals. And although natural selection effectively eliminates the defective genes in homozygous individuals through death, it cannot act to eliminate the defective genes in heterozygous individuals because they do not affect the phenotype. Natural selection will keep the frequency of the defective gene low in the population, but, in the absence of any other effect, there will always be a reservoir of defective genes in the heterozygous individuals.

At low frequencies of the defective gene, another important factor—chance—comes into play. Chance variation can increase or decrease the frequency of heterozygous individuals (and thereby the frequency of the defective gene). By chance, the offspring of a mating between heterozygotes could all be normal, which would eliminate the defective gene from that lineage. Increases in the frequency of a deleterious gene are opposed by natural selection; however, decreases are unopposed and can, by chance, lead to elimination of the defective gene from the population. On the other hand, new mutations are continually occurring, albeit at a low rate, creating fresh copies of the deleterious recessive allele. In a large population, a balance will be struck between the creation of new copies of the allele in this way, and their elimination through the death of homozygotes.

ANSWER 19–7
A. True.

B. True.
C. False. Mutations that occur during meiosis can be propagated, unless they give rise to nonviable gametes.

ANSWER 19–8   Two copies of the same chromosome can end up in the same daughter cell if one of the microtubule connections breaks before sister chromatids are separated. Alternatively, microtubules from the same spindle pole could attach to both kinetochores of the chromosome. As a consequence, one daughter cell would receive only one copy of all the genes carried on that chromosome, and the other daughter cell would receive three copies. The imbalance of the genes on this chromosome compared with the genes on all the other chromosomes would produce imbalanced levels of protein which, in most cases, is detrimental to the cell. If the mistake happens during meiosis, in the process of gamete formation, it will be propagated in all cells of the organism. A form of mental retardation called Down syndrome, for example, is due to the presence of three copies of Chromosome 21 in all of the nucleated cells in the body.

ANSWER 19–9   Meiosis begins with DNA replication, producing a tetraploid cell containing four copies of each chromosome. These four copies have to be distributed equally during the two sequential meiotic divisions into four haploid cells. Sister chromatids remain paired so that (1) the cells resulting from the first division receive two complete sets of chromosomes and (2) the chromosomes can be evenly distributed again in the second meiotic division. If the sister chromatids did not remain paired, it would not be possible in the second division to distinguish which chromatids belong together, and it would therefore be difficult to ensure that precisely one copy of each chromatid is pulled into each daughter cell. Keeping two sister chromatids paired in the first meiotic division is therefore an easy way to keep track of which chromatids belong together.

This biological principle suggests that you might consider clamping your socks together in matching pairs before putting them into the laundry. In this way, the cumbersome process of sorting them out afterward—and the seemingly inevitable mistakes that occur during that process—could be avoided.

ANSWER 19–10
A. A gene is a stretch of DNA that codes for a protein or functional RNA. An allele is an alternative form of a gene. Within the population, there are often several "normal" alleles, whose functions are indistinguishable. In addition, there may be many rare alleles that are defective to varying degrees. An individual, however, normally carries a maximum of two alleles of each gene.

B. An individual is said to be homozygous if the two alleles of a gene are the same. An individual is said to be heterozygous if the two alleles of a gene are different. An individual can be heterozygous for gene A and homozygous for gene B.

C. The genotype is the specific set of alleles present in the genome of an individual. In practice, for organisms studied in a laboratory, the genotype is usually specified as a list of the known differences between the individual and the wild type, which is the standard, naturally occurring type. The phenotype is a description of the visible characteristics of the individual. In practice, the phenotype is usually a list of the differences in visible characteristics between the individual and the wild type.

D. An allele A is dominant (relative to a second allele a) if the presence of even a single copy of A is enough to affect the phenotype; that is, if heterozygotes (with genotype Aa) appear different from aa homozygotes. An allele a is recessive (relative to a second allele A) if the presence of a single copy makes no difference to the phenotype, so that Aa individuals look just like AA individuals. If the phenotype of the heterozygous individual differs from the phenotypes of individuals that are homozygous for either allele, the alleles are said to be co-dominant.

ANSWER 19–11

A. Since the pea plant is diploid, any true-breeding plant must carry two mutant copies of the same gene—both of which have lost their function.

B. No, the same phenotype can be produced by mutations in different genes.

C. If each plant carries a mutation in a different gene, this will be revealed by complementation tests (see Panel 19–1, p. 669). When plant A is crossed with plant B, all of the F$_1$ plants will produce only round peas. And the same result will be obtained when plant B is crossed with plant C, or when plant A is crossed with plant C. In contrast, a cross between any two true-breeding plants that carry loss-of-function mutations in the same gene should produce only plants with wrinkled peas. This is true if the mutations themselves lie in different parts of the gene.

ANSWER 19–12

A. The mutation is likely to be dominant, because roughly half of the progeny born to an affected parent—in each of three marriages to hearing partners—are deaf, and it is unlikely that all these hearing partners were heterozygous carriers of the mutation.

B. The mutation is present on an autosome. If it were instead carried on a sex chromosome, either only the female progeny should be affected (expected if the mutation arose in a gene on the grandfather's X chromosome), or only the male progeny should be affected (expected if the mutation arose in a gene on the grandfather's Y chromosome). In fact, the pedigree reveals that both males and females have inherited the mutant form of the gene.

C. Suppose that the mutation was present on one of the two copies of the grandfather's Chromosome 12. Each of these copies of Chromosome 12 would be expected to carry a different pattern of SNPs, since one of them was inherited from his father and the other was inherited from his mother. Each of the copies of Chromosome 12 that was passed to his grandchildren will have gone through two meioses—one meiosis per generation.

Because two or three crossover events occur per chromosome during a meiosis, each chromosome inherited by a grandchild will have been subjected to about five crossovers since it left the grandfather, dividing it into six segments. An identical pattern of SNPs should surround whatever gene causes the deafness in each of the four affected grandchildren; moreover, this SNP pattern should be clearly different from that surrounding the same gene in each of the seven grandchildren who are normal. These SNPs would form an unusually long haplotype block—one that extends for about one-sixth of the length of Chromosome 12. (One-quarter of the DNA of each grandchild will have been inherited from the grandfather, in roughly 70 segments of this length scattered among the grandchild's 46 chromosomes.)

ANSWER 19–13 Individual 1 might be either heterozygous (+/–) or homozygous for the normal allele (+/+). Individual 2 must be homozygous for the recessive deafness allele (–/–). (Both his parents must have been heterozygous because they produced a deaf son.) Individual 3 is almost certainly heterozygous (+/–) and responsible for transmitting the mutant allele to his children and grandchildren. Given that the mutant allele is rare, individual 4 is most probably homozygous for the normal allele (+/+).

ANSWER 19–14 Your friend is wrong.

A. Mendel's laws, and the clear understanding that we now have concerning the mechanisms that produce them, rule out many false ideas concerning human heredity. One of them is that a first-born child has a different chance of inheriting particular traits from its parents than its siblings.

B. The probability of this type of pedigree arising by chance is one-fourth for each generation, or one in 64 for the three generations shown.

C. Data from an enlarged sampling of family members, or from more generations, would quickly reveal that the regular pattern observed in this particular pedigree arose by chance.

D. An opposite result, if it had strong statistical significance, would suggest that some process of selection was involved: for example, parents who had had a first child that was affected might regularly opt for screening of subsequent pregnancies and selectively terminate those pregnancies in which the fetus was found to be affected. Fewer second children would then be born with the abnormality.

ANSWER 19–15 Each carrier is a heterozygote, and 50% of his sperm or her eggs will carry the lethal allele. When two carriers marry, there is therefore a 25% chance that any baby will inherit the lethal allele from both parents and so will show the fatal phenotype. Because one person in 100 is a carrier, one partnership in 10,000 (100 × 100) will be a partnership of carriers (assuming that people choose their partners at random). Other things being equal, one baby in 40,000 will then be born with the defect, or 25 babies per year out of a total of a million babies born.

ANSWER 19–16 A dominant-negative mutation gives rise to a mutant gene product that interferes with the function of the normal gene product, causing a loss-of-function phenotype even in the presence of a normal copy of the gene. For example, if a protein forms a hexamer, and the mutant protein can interact with the normal subunits and inhibit the function of the hexamer, the mutation will be dominant. This ability of a single defective allele to determine the phenotype is the reason why such an allele is dominant. A gain-of-function mutation increases the activity of the gene or makes it active in inappropriate circumstances. The change in activity often has a phenotypic consequence, which is why such mutations are usually dominant.

ANSWER 19–17 This statement is largely true. Diabetes is one of the oldest diseases described by humans, dating at least back to the time of the ancient Greeks. Diabetes itself comes from the Greek word for siphon, which was used to describe the main symptoms: "The disease was called diabetes, as though it were a siphon, because it converts the human body into a pipe for the transflux of liquid humors" (in other words, untreated patients have constant thirst, balanced by high output of urine). If there were no human disease, the role of insulin would not have come to our attention in so demanding a way. We would have ultimately understood its role—and by now may have. Yet it is difficult to overstate the case for the role of disease in focusing our efforts toward a molecular understanding. Even today, the quest to understand and alleviate human disease is a principal driving force in biomedical research.

ANSWER 19–18

A. As outlined in **Figure A19–18**, if flies that are defective in different genes mate, their progeny will have one normal gene at each locus. In the case of a mating between a ruby-eyed fly and a white-eyed fly, every progeny fly will inherit one functional copy of the white gene from one parent and one functional copy of the ruby gene from the other parent. Note that the normal white allele produces brick-red eyes and the mutated form of the gene produces white eyes. Because each of the mutant alleles is recessive to the corresponding wild-type allele, the progeny will have the wild-type phenotype—brick-red eyes.

B. Garnet, ruby, vermilion, and carnation complement one another and the various alleles of the white gene (that is, when these mutant flies are mated with each other, they produce flies with a normal eye color); thus each of these mutants defines a separate gene. In contrast, white, cherry, coral, apricot, and buff do not complement each other; thus, they must be alleles of the same gene,

which has been named the white gene. Thus, these nine different eye-color mutants define five different genes.

C. Different alleles of the same gene, like the five alleles of the white gene, often have different phenotypes. Different mutations compromise the function of the gene product to different extents, depending on the location of the mutation. Alleles that do not produce any functional product (null alleles), even if they result from different DNA sequence changes, do have the same phenotype.

ANSWER 19–19 SNPs are single-nucleotide differences between individuals for which two or more variants are each found at high frequency in the population. In the human population, SNPs occur roughly once per 1000 nucleotides of sequence. Many have been identified and mapped in various organisms, including several million in the human genome. SNPs, which can be detected by sequencing or oligonucleotide hybridization, serve as physical markers whose genomic locations are known. By tracking a mutant gene through different matings, and correlating the presence of the gene with the co-inheritance of particular SNP variants, one can narrow down the potential location of a gene to a chromosomal region that may contain only a few genes. These candidate genes can then be tested for the presence of a mutation that could account for the original mutant phenotype (see Figure 19–38).

Chapter 20

ANSWER 20–1 The horizontal orientation of the microtubules will be associated with a horizontal orientation of cellulose microfibrils deposited in the cell walls. The growth of the cells will therefore be in a vertical direction, expanding the distance between the cellulose microfibrils without stretching them. In this way, the stem will rapidly

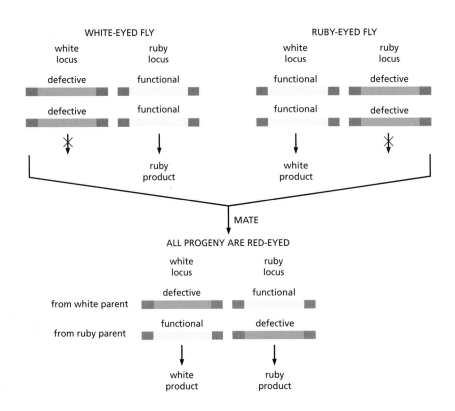

Figure A19–18

elongate; in a typical natural environment, this will hasten emergence from darkness into light.

ANSWER 20–2

A. As three collagen chains have to come together to form the triple helix, a defective molecule will impair assembly, even if normal collagen chains are present at the same time. Collagen mutations are therefore dominant; that is, they have a deleterious effect even in the presence of a normal copy of the gene.

B. The different severity of the mutations results from a polarity in the assembly process. Collagen monomers assemble into the triple-helical rod starting from their amino-terminal ends. A mutation in an "early" glycine therefore allows only short rods to form, whereas a mutation farther downstream allows for longer, more normal rods.

ANSWER 20–3   The remarkable ability to swell and thus occupy a large volume of space depends on the negative charges. These attract a cloud of positive ions, chiefly Na^+, which by osmosis draw in large amounts of water, thus giving proteoglycans their unique properties. Uncharged polysaccharides such as cellulose, starch, and glycogen, by contrast, are easily compacted into fibers or granules.

ANSWER 20–4   Focal contact sites are common in connective tissue, where fibroblasts exert traction forces on the extracellular matrix, and in cell culture, where cell crawling is observed. The forces for pulling on matrix or for driving crawling movement are generated by the actin cytoskeleton. In mature epithelium, focal contact sites are presumably less prominent because the cells are largely fixed in place and have no need to crawl over the basal lamina or actively pull on it.

ANSWER 20–5   Suppose a cell is damaged so that its plasma membrane becomes leaky. Ions present in high concentration in the extracellular fluid, such as Na^+ and Ca^{2+}, then rush into the cell, and valuable metabolites leak out. If the cell were to remain connected to its healthy neighbors, these too would suffer from the damage. But the influx of Ca^{2+} into the sick cell causes its gap junctions to close immediately, effectively isolating the cell and preventing damage from spreading in this way.

ANSWER 20–6   Ionizing (high-energy) radiation tears through matter, knocking electrons out of their orbits and breaking chemical bonds. In particular, it creates breaks and other damage in DNA, and thus causes cells to arrest in the cell cycle (discussed in Chapter 18). If the damage is so severe that it cannot be repaired, cells become permanently arrested and undergo apoptosis; that is, they activate a suicide program.

ANSWER 20–7   Cells in the gut epithelium are exposed to a quite hostile environment, containing digestive enzymes and many other substances that vary drastically from day to day depending on the food intake of the organism. The epithelial cells also form a first line of defense against potentially hazardous compounds and mutagens that are ubiquitous in our environment. The rapid turnover protects the organism from harmful consequences, as wounded and sick cells are discarded. If an epithelial cell started to divide inappropriately as the result of a mutation, for example,

it and its unwanted progeny would most often simply be discarded by natural disposal from the tip of a villus: even though such mutations must occur often, they rarely give rise to a cancer.

A neuron, on the other hand, lives in a very protected environment, insulated from the outside world. Its function depends on a complex system of connections with other neurons—a system that is created during development and is not easy to reconstruct if the neuron subsequently dies.

ANSWER 20–8   Every cell division generates one additional cell; so if the cells were never lost or discarded from the body, the number of cells in the body should equal the number of divisions plus one. The number of divisions is 1000-fold greater than the number of cells because, in the course of a lifetime, 1000 cells are discarded by mechanisms such as apoptosis for every cell that is retained in the body.

ANSWER 20–9

A. False. Gap junctions are not connected to the cytoskeleton; their role is to provide cell–cell communication by allowing small molecules to pass from one cell to another.

B. True. Upon wilting, the turgor pressure in the plant cell is reduced, and consequently the cell walls, having tensile but little compressive strength, like a rubber tire, no longer provide rigidity.

C. False. Proteoglycans can withstand a large amount of compressive force but do not have a rigid structure. Their space-filling properties result from their tendency to absorb large amounts of water.

D. True.

E. True.

F. True. Stem cells stably express control genes that ensure that their daughter cells will be of the appropriate differentiated cell types.

ANSWER 20–10 Small cytosolic molecules, such as glutamic acid, cyclic AMP, and Ca^{2+} ions, pass readily through both gap junctions and plasmodesmata, whereas large cytosolic macromolecules, such as mRNA and G proteins, are excluded. Plasma membrane phospholipids diffuse in the plane of the membrane through plasmodesmata because the plasma membranes of adjacent cells are continuous through these junctions. This traffic is not possible through gap junctions, because the membranes of the connected cells remain separate.

ANSWER 20–11 Plants are exposed to extreme changes in the environment, which often are accompanied by huge fluctuations in the osmotic properties of their surroundings. An intermediate filament network as we know it from animal cells would not be able to provide full osmotic support for cells: the sparse, rivetlike attachment points would not be able to prevent the membrane from bursting in response to a huge osmotic pressure applied from the inside of the cell.

ANSWER 20–12 Action potentials can, in fact, be passed from cell to cell through gap junctions. Indeed, heart muscle cells are connected in this way, which ensures that they contract synchronously when stimulated. This mechanism of passing the signal from cell to cell is rather limited, however. As we discuss in Chapter 12, synapses are far more sophisticated and allow signals to be modulated and to be integrated with other signals received by the cell. Thus, gap junctions are like simple soldered joints between electrical components, while synapses are like complex relay devices, enabling systems of neurons to perform computations.

ANSWER 20–13 To make jello, gelatin is boiled in water, which denatures the collagen fibers. Upon cooling, the disordered fibers form a tangled mess that solidifies into a gel. This gel actually resembles the collagen as it is initially secreted by fibroblasts. It is not until the fibers have been aligned, bundled, and cross-linked that they acquire their ability to resist tensile forces.

ANSWER 20–14 The evidence that DNA is the blueprint that specifies all the structural characteristics of an organism is based on observations that small changes in the DNA by mutation result in changes in the organism. Although DNA provides the plans that specify structure, these plans need to be executed during development. This requires a suitable environment (a human baby would not fit into a stork's egg shell), suitable nourishment, suitable tools (such as the appropriate transcription regulators required for early development), suitable spatial organization (such as the asymmetries in the egg cell required to allow for appropriate cell differentiation during the early cell divisions), and so on. Thus inheritance is not restricted to the passing on of the organism's DNA, because development requires appropriate conditions to be set up by the parent. Nevertheless, when all these conditions are met, the plans that are archived in the genome will determine the structure of the organism to be built.

ANSWER 20–15 White blood cells circulate in the bloodstream and migrate into and out of tissues in performance of their normal function of defending the body against infection: they are naturally invasive. Once mutations have occurred to upset the normal controls on production of these cells, there is no need for additional mutations to enable the cells to spread through the body. Thus, the number of mutations that have to be accumulated in order to give rise to leukemia is smaller than for other types of cancer.

ANSWER 20–16 The shape of the curve reflects the need for multiple mutations to accumulate in a cell before a cancer results. If a single mutation were sufficient, the graph would be a straight horizontal line: the likelihood of occurrence of a particular mutation, and therefore of cancer, would be the same at any age. If two specific mutations were required, the graph would be a straight line sloping upward from the origin: the second mutation has an equal chance of occurring at any time, but will tip the cell into cancerous behavior only if the first mutation has already occurred in the same cell lineage; and the likelihood that the first mutation has already occurred will be proportional to the age of the individual. The steeply curved graph shown in the figure goes up approximately as the fifth power of the age, and this indicates that far more than two mutations have to be accumulated before cancer sets in. It is not easy to say precisely how many, because of the complex ways in which cancers develop. Successive mutations can alter cell numbers and cell behavior, and thereby change both the probability of subsequent mutations and the selection pressures that drive the evolution of cancer.

ANSWER 20–17 During exposure to the carcinogen, mutations are induced, but the number of relevant mutations in any one cell is usually not enough to convert it directly into a cancer cell. Over the years, the cells that have become predisposed to cancer through the induced mutations accumulate progressively more mutations. Eventually, one of them will turn into a cancer cell. The long delay between exposure and cancer has made it extremely difficult to hold cigarette manufacturers or producers of industrial carcinogens legally responsible for the damage that is caused by their products.

ANSWER 20–18 By definition, a carcinogen is any substance that promotes the occurrence of one or more types of cancer. The sex hormones can therefore be classified as naturally occurring carcinogens. Although most carcinogens act by directly causing mutations, carcinogenic effects are also often exerted in other ways. The sex hormones increase both the rate of cell division and the numbers of cells in hormone-sensitive organs such as breast, uterus, and prostate. The first effect increases the mutation rate per cell, because mutations, regardless of environmental factors, are spontaneously generated in the course of DNA replication and chromosome segregation; the second effect increases the number of cells at risk. In these and possibly other ways, the hormones can favor the development of cancer, even though they do not directly cause mutations.

ANSWER 20–19 The short answer is no—cancer in general is not a hereditary disease. It arises from new mutations occurring in our own somatic cells, rather than mutations we inherit from our parents. In some rare types of cancer, however, there is a strong heritable risk factor, so that parents and their children both show the same predisposition to a specific form of the disease. This occurs, for example, in families carrying a mutation that knocks out one of the two copies of the tumor suppressor gene *APC*; the children then inherit a propensity to colorectal cancer. Much weaker heritable tendencies are also seen in several other cancers, including breast cancer, but the genes responsible for these effects are still mostly unknown.

Glossary

acetyl CoA (acetyl coenzyme A)
Activated carrier that donates the carbon atoms in its readily transferable acetyl group to many metabolic reactions, including the citric acid cycle and fatty acid biosynthesis; the acetyl group is linked to coenzyme A (CoA) by a thioester bond that releases a large amount of energy when hydrolyzed.

acetyl group
Chemical group derived from acetic acid.

acid
A molecule that releases a proton when dissolved in water; this dissociation generates hydronium (H_3O^+) ions, thereby lowering the pH.

actin filament
Thin, flexible protein filament  made from a chain of globular actin molecules; a major constituent of all eukaryotic cells, this cytoskeletal element is essential for cell movement and for the contraction of muscle cells.

actin-binding protein
Protein that interacts with actin monomers or filaments to control the assembly, structure, and behavior of actin filaments and networks.

action potential
Traveling wave of electrical excitation caused by rapid, transient, self-propagating depolarization of the plasma membrane in a neuron or other excitable cell; also called a nerve impulse.

activated carrier
A small molecule that stores energy or chemical groups in a form that can be donated to many different metabolic reactions. Examples include ATP, acetyl CoA, and NADH.

activation energy
The energy that must be acquired by a molecule to undergo a chemical reaction.

activator
A protein that binds to a specific regulatory region of DNA to permit transcription of an adjacent gene.

active site
Region on the surface of an enzyme that binds to a substrate molecule and catalyzes its chemical transformation.

active transport
The movement of a solute across a membrane against its electrochemical gradient; requires an input of energy, such as that provided by ATP hydrolysis.

acyl group
Functional group derived from a carboxylic acid.

adaptation
Adjustment of sensitivity following repeated stimulation; allows a cell or organism to register small changes in a signal despite a high background level of stimulation.

adenylyl cyclase
Enzyme that catalyzes the formation of cyclic AMP from ATP; an important component in some intracellular signaling pathways.

adherens junction
Cell junction that helps hold together epithelial cells in a sheet of epithelium; actin filaments inside the cell attach to its cytoplasmic face.

ADP (adenosine 5′-diphosphate)
Nucleoside diphosphate produced by hydrolysis of the terminal phosphate of ATP. (*See* Figure 3–31.)

alcohol
Organic compound containing a hydroxyl group (–OH) bound to a saturated carbon atom, for example, ethanol. (*See* Panel 2–1, pp. 66–67.)

aldehyde
Reactive organic compound that contains the HC=O group, for example, glyceraldehyde. (*See* Panel 2–1, pp. 66–67.)

alkyl group
Functional group consisting solely of single-bonded carbon and hydrogen atoms, such as methyl (–CH_3) or ethyl (–CH_2CH_3) groups.

allele
An alternative form of a gene; for a given gene, many alleles may exist in the gene pool of the species.

allosteric
Describes a protein that can exist in multiple conformations depending on the binding of a molecule (ligand) at a site other than the catalytic site; changes from one conformation to another often alter the protein's activity or ligand affinity.

alpha helix (α helix)
Folding pattern, common in many proteins, in which a single polypeptide chain twists around itself to form a rigid cylinder stabilized by hydrogen bonds between every fourth amino acid.

alternative splicing
The production of different mRNAs (and proteins) from the same gene by splicing its RNA transcripts in different ways.

***Alu* sequence**
Family of mobile genetic elements that comprises about 10% of the human genome; this short, repetitive sequence is no longer mobile on its own, but requires enzymes encoded by other elements to transpose.

amide
Molecule containing the functional group –CONH₂. (*See* Panel 2–1, pp. 66–67.)

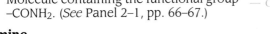

amine
Molecule containing an amino group (–NH₂). (*See* Panel 2–1, pp. 66–67.)

amino acid
Small organic molecule containing both an amino group and a carboxyl group; it serves as the building block of proteins. (*See* Panel 2–5, pp. 74–75.)

amino acid sequence
The order of the amino acid subunits in a protein chain. Sometimes called the primary structure of a protein.

amino group
Functional group (–NH₂) derived from ammonia. Can accept a proton and carry a positive charge in aqueous solution. (*See* Panel 2–1, pp. 66–67.)

amino terminus—*see* **N-terminus**

aminoacyl-tRNA synthetase
During protein synthesis, an enzyme that attaches the correct amino acid to a tRNA molecule to form a "charged" aminoacyl-tRNA.

AMP (adenosine 5′ monophosphate)
Nucleotide produced by the energetically favorable hydrolysis of the final two phosphate groups from ATP, a reaction that drives the synthesis of DNA and RNA. (*See* Figure 3–40.)

amphipathic
Having both hydrophobic and hydrophilic regions, as in a phospholipid or a detergent molecule.

anabolic pathway
Series of enzyme-catalyzed reactions by which large biological molecules are synthesized from smaller subunits; usually requires an input of energy.

anabolism
Set of metabolic pathways by which large molecules are made from smaller ones.

anaerobic
Describes a cell, organism, or metabolic process that operates in the absence of air or, more precisely, in the absence of molecular oxygen.

anaphase
Stage of mitosis during which the two sets of chromosomes separate and are pulled toward opposite ends of the dividing cell.

anaphase-promoting complex (APC)
A protein complex that triggers the separation of sister chromatids and orchestrates the carefully timed destruction of proteins that control progress through the cell cycle; the complex catalyzes the ubiquitylation of its targets.

anion
Negatively charged ion, such as Cl⁻ or CH₃COO⁻.

antenna complex
In chloroplasts and photosynthetic bacteria, the part of the membrane-bound photosystem that captures energy from sunlight; contains an array of proteins that bind hundreds of chlorophyll molecules and other photosensitive pigments.

antibody
Protein produced by B lymphocytes in response to a foreign molecule or invading organism. Binds to the foreign molecule or cell extremely tightly, thereby inactivating it or marking it for destruction.

anticodon
Set of three consecutive nucleotides in a transfer RNA molecule that recognizes, through base-pairing, the three-nucleotide codon on a messenger RNA molecule; this interaction helps to deliver the correct amino acid to a growing polypeptide chain.

antigen
Molecule or fragment of a molecule that is recognized by an antibody.

antiparallel
Describes two similar structures arranged in opposite orientations, such as the two strands of a DNA double helix.

antiport
Type of coupled transporter that transfers two different ions or small molecules across a membrane in opposite directions, either simultaneously or in sequence.

APC—*see* **anaphase-promoting complex**

apical
Describes the top or the tip of a cell, structure, or organ; in an epithelial cell, for example, this surface is opposite the base, or basal surface.

apoptosis
A tightly controlled form of programmed cell death that allows cells that are unneeded or unwanted to be eliminated from an adult or developing organism.

archaea
One of the two divisions of prokaryotes, often found in hostile environments such as hot springs or concentrated brine. (*See also* **bacteria**.)

asexual reproduction
Mode of reproduction in which offspring arise from a single parent, producing an individual genetically identical to that parent; includes budding, binary fission, and parthenogenesis.

aster
Star-shaped array of microtubules emanating from a centrosome or from a pole of a mitotic spindle.

atom
The smallest particle of an element that still retains its distinctive chemical properties; consists of a positively charged nucleus surrounded by a cloud of negatively charged electrons.

atomic mass
The mass of an atom expressed in daltons, the atomic mass unit that closely approximates the mass of a hydrogen atom.

ATP (adenosine 5′-triphosphate)
Molecule that serves as the principal carrier of energy in cells; this nucleoside triphosphate is composed of adenine, ribose, and three phosphate groups. (*See* Figure 2–24.)

ATP synthase
Membrane-associated enzyme complex that catalyzes the formation of ATP from ADP and inorganic phosphate during oxidative phosphorylation and photosynthesis.

autophagy
Mechanism by which a cell "eats itself," digesting molecules and organelles that are damaged or obsolete.

Avogadro's number
The number of molecules in a mole, the quantity of a substance equal to its molecular weight in grams; approximately 6×10^{23}.

axon
Long, thin extension that conducts electrical signals away from a nerve cell body toward remote target cells.

bacteria (singular bacterium)
One of the two divisions of prokaryotes; some species cause disease. The term is sometimes used to refer to any prokaryotic microorganism, although the world of prokaryotes also includes archaea, which are only distantly related. (*See also* **archaea**.)

bacteriorhodopsin
Pigmented protein found in abundance in the plasma membrane of the salt-loving archaeon *Halobacterium halobium*; pumps protons out of the cell in response to light.

basal
Situated near the base; opposite of apical.

basal body—*see* **centriole**

basal lamina
Thin mat of extracellular matrix, secreted by epithelial cells, upon which the cells sit.

base
Molecule that accepts a proton when dissolved in water; also used to refer to the nitrogen-containing purines or pyrimidines in DNA and RNA.

base pair
Two complementary nucleotides in an RNA or a DNA molecule that are held together by hydrogen bonds— for example, G with C, and A with T or U.

Bcl2 family
Related group of intracellular proteins that regulates apoptosis; some family members promote cell death, others inhibit it.

beta sheet
(β sheet)

Folding pattern found in many proteins in which neighboring regions of the polypeptide chain associate side by side with each other through hydrogen bonds to give a rigid, flattened structure.

bi-orientation
The symmetrical attachment of a sister chromatid pair on the mitotic spindle, such that one chromatid in the duplicated chromosome is attached to one spindle pole and the other is attached to the opposite pole.

binding site
Region on the surface of a protein, typically a cavity or groove, that interacts with another molecule (a ligand) through the formation of multiple noncovalent bonds.

biosynthesis
An enzyme-catalyzed process by which complex molecules are formed from simple substances by living cells; also called anabolism.

bivalent
Structure formed when a duplicated chromosome pairs with its homolog at the beginning of meiosis; contains four sister chromatids.

bond—*see* **chemical bond**

bond energy
The strength of the chemical linkage between two atoms, measured by the energy in kilocalories needed to break it.

bond length
Average distance between two interacting atoms in a molecule, usually those linked covalently.

buffer
Mixture of weak acids and bases that maintains the pH of a solution by releasing and taking up protons.

C-terminus (carboxyl terminus)
The end of a polypeptide chain that carries a free carboxyl group (–COOH).

Ca²⁺ pump
An active transporter that uses energy supplied by ATP hydrolysis to actively expel Ca^{2+} from the cell cytosol.

Ca²⁺/calmodulin-dependent protein kinase (CaM kinase)
Enzyme that phosphorylates target proteins in response to an increase in Ca^{2+} ion concentration through its interaction with the Ca^{2+}-binding protein calmodulin.

cadherin
A member of a family of Ca^{2+}-dependent proteins that mediates the attachment of one cell to another in animal tissues.

calmodulin (CaM)
Small Ca^{2+}-binding protein that modifies the activity of many target proteins in response to changes in Ca^{2+} concentration.

calorie
Unit of heat. Equal to the amount of heat needed to raise the temperature of 1 gram of water by 1°C.

CaM—*see* **calmodulin**

cancer
Disease caused by abnormal and uncontrolled cell proliferation, followed by invasion and colonization of body sites normally reserved for other cells.

carbohydrate
General term for sugars and related compounds with the general formula $(CH_2O)_n$. (*See* Panel 2–3, pp. 70–71.)

carbohydrate layer
Protective layer of sugar residues, including the polysaccharide portions of proteoglycans and oligosaccharides attached to protein or lipid molecules, on the outer surface of a cell. Also called the glycocalyx.

carbon fixation
Process by which green plants and other photosynthetic organisms incorporate carbon atoms from atmospheric carbon dioxide into sugars. The second stage of photosynthesis.

carbonyl group
Carbon atom linked to an oxygen atom by a double bond. (*See* Panel 2–1, pp. 66–67.)

carboxyl group
Carbon atom linked to an oxygen atom by a double bond and to a hydroxyl group (–COOH). In aqueous solution, acts as a weak acid. (*See* Panel 2–1, pp. 66–67.)

carboxyl terminus—*see* **C-terminus**

cascade—*see* **signaling cascade**

caspase
A family of proteases that, when activated, mediates the destruction of the cell by apoptosis.

catabolism
Set of enzyme-catalyzed reactions by which complex molecules are degraded to simpler ones with release of energy; intermediates in these reactions are sometimes called catabolites.

catalysis
The acceleration of a chemical reaction brought about by the action of a catalyst; virtually all reactions in a cell require such assistance to occur under conditions present in living organisms.

catalyst
Substance that accelerates a chemical reaction by lowering its activation energy; enzymes perform this role in cells.

cation
Positively charged ion, such as Na^+ or $CH_3NH_3^+$.

Cdk inhibitor protein
Regulatory protein that blocks the assembly or activity of cyclin–Cdk complexes, delaying progression primarily through the G_1 and S phases of the cell cycle.

cDNA library
Collection of DNA fragments synthesized using all of the mRNAs present in a particular type of cell as a template.

cell
The basic unit from which a living organism is made; consists of an aqueous solution of organic molecules enclosed by a membrane.

cell cortex
Specialized layer of cytoplasm on the inner face of the plasma membrane. In animal cells, it is rich in the actin filaments that govern cell shape and drive cell movement.

cell cycle
The orderly sequence of events by which a cell duplicates its contents and divides into two.

cell division
Separation of a cell into two daughter cells. In eukaryotic cells, entails the splitting of the nucleus (mitosis) closely followed by cleavage of the cytoplasm (cytokinesis).

cell junction
Specialized region of connection between two cells or between a cell and the extracellular matrix.

cell line
Population of cells derived from a plant or animal capable of dividing indefinitely in culture.

cell locomotion
Active movement of a cell from one location to another.

cell memory
The ability of differentiated cells and their descendants to maintain their identity.

cell respiration
Process by which cells harvest the energy stored in food molecules; usually accompanied by the uptake of O_2 and the release of CO_2.

cell signaling
The molecular mechanisms by which cells detect and respond to external stimuli and send messages to other cells.

cell wall
Mechanically strong fibrous layer deposited by a cell outside its plasma membrane. Prominent in most plants, bacteria, algae, and fungi, but not present in most animal cells.

cell-cycle control system
Network of regulatory proteins that govern the orderly progression of a eukaryotic cell through the stages of cell division.

cellulose
Structural polysaccharide consisting of long chains of covalently linked glucose units. Provides tensile strength in plant cell walls.

cellulose microfibril
Long, thin strand of cellulose that helps strengthen plant cell walls.

central dogma
The principle that genetic information flows from DNA to RNA to protein.

centriole
Cylindrical array of microtubules usually found in pairs at the center of a centrosome in animal cells. Also found at the base of cilia and flagella, where they are called basal bodies.

centromere
Specialized DNA sequence that allows duplicated chromosomes to be separated during M phase; can be seen as the constricted region of a mitotic chromosome.

centrosome
Microtubule-organizing center that sits near the nucleus in an animal cell; during the cell cycle, this structure duplicates to form the two poles of the mitotic spindle.

centrosome cycle
Process by which the centrosome duplicates (during interphase) and the two new centrosomes separate (at the beginning of mitosis) to form the poles of the mitotic spindle.

channel
A protein that forms a hydrophilic pore across a membrane, through which selected small molecules or ions can passively diffuse.

chaperone protein
Molecule that steers proteins along productive folding pathways, helping them to fold correctly and preventing them from forming aggregates inside the cell.

checkpoint
Mechanism by which the cell-cycle control system can regulate progression through the cycle, ensuring that conditions are favorable and each process has been completed before proceeding to the next stage.

chemical bond
An exchange of electrons that holds two atoms together. Types found in living cells include ionic bonds, covalent bonds, and hydrogen bonds.

chemical group
Combination of atoms, such as a hydroxyl group ($-OH$) or an amino group ($-NH_2$), with distinct chemical and physical properties that influences the behavior of the molecule in which it resides.

chemiosmotic coupling
Mechanism that uses the energy stored in a transmembrane proton gradient to drive an energy-requiring process, such as the synthesis of ATP or the transport of a molecule across a membrane.

chiasma (plural chiasmata)
X-shaped connection between paired homologous chromosomes during meiosis; represents a site of crossing-over between two non-sister chromatids.

chlorophyll
Light-absorbing green pigment that plays a central part in photosynthesis.

chloroplast
Specialized organelle in algae and plants that contains chlorophyll and serves as the site in which photosynthesis takes place.

cholesterol
Short, rigid lipid molecule present in large amounts in the plasma membranes of animal cells, where it makes the lipid bilayer less flexible.

chromatid—_see_ **sister chromatid**

chromatin
Complex of DNA and proteins that makes up the chromosomes in a eukaryotic cell.

chromatin-remodeling complex
Enzyme (typically multisubunit) that uses the energy of ATP hydrolysis to alter the arrangement of nucleosomes in eukaryotic chromosomes, changing the accessibility of the underlying DNA to other proteins, including those involved in transcription.

chromatography
Technique used to separate the individual molecules in a complex mixture on the basis of their size, charge, or their ability to bind to a particular chemical group. In a common form of the technique, the mixture is run through a column filled with a material that either binds or lets through the desired molecule.

chromosome
Long, threadlike structure composed of DNA and proteins that carries the genetic information of an organism; becomes visible as a distinct entity when a plant or animal cell prepares to divide.

chromosome condensation
Process by which a duplicated chromosome becomes packed into a more compact structure prior to cell division.

cilium (plural cilia)
Hairlike structure made of microtubules found on the surface of many eukaryotic cells; when present in large numbers, its rhythmic beating can drive the movement of fluid over the cell surface, as in the epithelium of the lungs.

cis
On the same side as.

cis Golgi network
Section of the Golgi apparatus that receives materials from the endoplasmic reticulum.

citric acid cycle
Series of reactions that generates large amounts of NADH by oxidizing acetyl groups derived from food molecules to CO_2. In eukaryotic cells, this central metabolic pathway takes place in the mitochondrial matrix.

classical genetic approach
Experimental techniques used to isolate the genes responsible for an interesting phenotype.

clathrin
Protein that makes up the coat of a type of transport

vesicle that buds from either the Golgi apparatus (on the outward secretory pathway) or from the plasma membrane (on the inward endocytic pathway).

coated vesicle
Small membrane-enclosed sac that wears a distinctive layer of proteins on its cytosolic surface. It is formed by pinching-off of a protein-coated region of cell membrane.

codon
Group of three consecutive nucleotides that specifies a particular amino acid or that starts or stops protein synthesis; applies to the nucleotides in an mRNA or in a coding sequence of DNA.

coenzyme A
Small molecule used to carry and transfer acetyl groups needed for a variety of metabolic reactions, such as the synthesis of fatty acids. (*See also* **acetyl CoA** and Figure 3–36.)

cohesin
Protein complex that holds sister chromatids together after DNA has been replicated in the cell cycle.

coiled-coil
Stable, rodlike protein structure formed when two or more α helices twist around each other.

collagen
Triple-stranded, fibrous protein that is a major component of the extracellular matrix and connective tissues; it is the main protein in animal tissues, and different forms can be found in skin, tendon, bone, cartilage, and blood vessels.

combinatorial control
Describes the way in which groups of transcription regulators work together to regulate the expression of a single gene.

complementary
Describes two molecular surfaces that fit together closely and form noncovalent bonds with each other. Examples include complementary base pairs, such as A and T, and the two complementary strands of a DNA molecule.

complementary DNA (cDNA)
DNA molecule synthesized from an mRNA molecule and therefore lacking the introns that are present in genomic DNA.

complementation test
Genetic experiment that determines whether two mutations that are associated with the same phenotype lie in the same gene or in different genes.

complex
A collection of macromolecules that are bound to each other by noncovalent bonds to form a large structure with a specific function.

complex trait
A heritable characteristic whose transmission to progeny does not appear to obey Mendel's laws. Such characteristics, for example height, usually result from the interaction of multiple genes.

condensation—*see* chromosome condensation

condensation reaction
Chemical reaction in which a covalent bond is formed between two molecules as water is expelled; used to build polymers, such as proteins, polysaccharides, and nucleic acids.

condensin
Protein complex that helps configure duplicated chromosomes for segregation by making them more compact.

conformation
Precise, three-dimensional shape of a protein or other macromolecule, based on the spatial location of its atoms in relation to one another.

connective tissue
Tissues such as bone, tendons, and the dermis of the skin, in which extracellular matrix makes up the bulk of the tissue and carries the mechanical load.

conserved synteny
The preservation of gene order and location in the genomes of different species.

contractile ring
Structure made of actin and myosin filaments that forms a belt around a dividing cell, pinching it in two.

copy-number variation (CNV)
Large segment of DNA, 1000 nucleotide pairs or greater, that has been duplicated or lost in an individual genome (compared to the "reference" genome sequence).

coupled pump
Active transporter that uses the movement of one solute down its electrochemical gradient to drive the uphill transport of another solute across the same membrane.

coupled reaction
Linked pair of chemical reactions in which free energy released by one reaction serves to drive the other reaction.

covalent bond
Stable chemical link between two atoms produced by sharing one or more pairs of electrons.

crossing-over
Process whereby two homologous chromosomes break at corresponding sites and rejoin to produce two recombined chromosomes that have physically exchanged segments of DNA.

cyclic AMP (cAMP)
Small intracellular signaling molecule generated from ATP in response to hormonal stimulation of cell-surface receptors.

cyclic-AMP-dependent protein kinase (protein kinase A, PKA)
Enzyme that phosphorylates target proteins in response to a rise in intracellular cyclic AMP concentration.

cyclin
Regulatory protein whose concentration rises and falls at specific times during the eukaryotic cell cycle;

cyclins help control progression from one stage of the cell cycle to the next by binding to cyclin-dependent protein kinases (Cdks).

cyclin-dependent protein kinase (Cdk)
Enzyme that, when complexed with a regulatory cyclin protein, can trigger various events in the cell-division cycle by phosphorylating specific target proteins.

cytochrome
Membrane-bound, colored, heme-containing protein that transfers electrons during cellular respiration and photosynthesis.

cytochrome *c* oxidase
Protein complex that serves as the final electron carrier in the respiratory chain; removes electrons from cytochrome *c* and passes them to O_2 to produce H_2O.

cytokine
Small signaling molecule, made and secreted by cells, that acts on neighboring cells to alter their behavior. Usually a protein, polypeptide, or glycoprotein.

cytokinesis
Process by which the cytoplasm of a plant or animal cell divides in two to form individual daughter cells.

cytoplasm
Contents of a cell that are contained within its plasma membrane but, in the case of eukaryotic cells, contained outside the nucleus.

cytoskeleton
System of protein filaments in the cytoplasm of a eukaryotic cell that gives the cell shape and the capacity for directed movement. Its most abundant components are actin filaments, microtubules, and intermediate filaments.

cytosol
Contents of the main compartment of the cytoplasm, excluding membrane-enclosed organelles such as endoplasmic reticulum and mitochondria. The cell fraction remaining after membranes, cytoskeletal components, and other organelles have been removed.

dalton
Unit of molecular mass. Defined as one-twelfth the mass of an atom of carbon 12 (1.66×10^{-24} g); approximately equal to the mass of a hydrogen atom.

dark reactions
In photosynthesis, the set of reactions that produce sugars from CO_2; these reactions, also called carbon fixation, can occur in the absence of sunlight.

denature
To cause a dramatic change in the structure of a macromolecule by exposing it to extreme conditions, such as high heat or harsh chemicals. Usually results in the loss of biological function.

dendrite
Short, branching structure that extends from the surface of a nerve cell and receives signals from other neurons.

deoxyribonucleic acid—*see* **DNA**

depolarization
A shift in the membrane potential, making it less negative.

desmosome
Specialized cell–cell junction, usually formed between two epithelial cells, that serves to connect the ropelike keratin filaments of the adjoining cells, providing tensile strength.

detergent
Soapy substance used to solubilize membrane proteins.

diacylglycerol (DAG)
Small messenger molecule produced by the cleavage of membrane inositol phospholipids in response to extracellular signals. Helps activate protein kinase C.

dideoxy sequencing or Sanger sequencing
The standard method of determining the nucleotide sequence of DNA; utilizes DNA polymerase and a set of chain-terminating nucleotides.

differentiation
Process by which a cell undergoes a progressive, coordinated change to a more specialized cell type, brought about by large-scale changes in gene expression.

diffusion
Process by which molecules and small particles move from one location to another by random, thermally driven motion.

dimer
A molecule composed of two structurally similar subunits.

diploid
Describes a cell or organism containing two sets of homologous chromosomes, one inherited from each parent. (*See also* **haploid**.)

disulfide bond
Covalent cross-link formed between the sulfhydryl groups on two cysteine side chains; often used to reinforce a secreted protein's structure or to join two different proteins together.

divergence
Differences in sequence that accumulate over time in DNA segments derived from a common ancestral sequence.

DNA (deoxyribonucleic acid)
Double-stranded polynucleotide formed from two separate chains of covalently linked deoxyribonucleotide units. It serves as the cell's store of genetic information that is transmitted from generation to generation.

DNA cloning
Production of many identical copies of a DNA sequence.

DNA library
Collection of cloned DNA molecules, representing either an entire genome (genomic library) or copies of the mRNA produced by a cell (cDNA library).

DNA ligase
Enzyme that reseals nicks that arise in the backbone of a DNA molecule; in the laboratory, can be used to join together two DNA fragments.

DNA methylation
The enzymatic addition of methyl groups to cytosine bases in DNA; this covalent modification generally turns off genes by attracting proteins that block gene expression.

DNA microarray
A surface on which a large number of short DNA molecules (typically in the tens of thousands) have been immobilized in an orderly pattern. Each of these DNA fragments acts as a probe for a specific gene, allowing the activities of thousands of genes to be monitored at the same time.

DNA repair
Collective term for the enzymatic processes that correct deleterious changes affecting the continuity or sequence of a DNA molecule.

DNA replication
The process by which a copy of a DNA molecule is made.

DNA transcription—*see* **transcription**

domain
Small discrete region of a structure; in a protein, a segment that folds into a compact and stable structure. In a membrane, a region of the bilayer with a characteristic lipid and protein composition.

double bond
Chemical linkage formed when two atoms share four electrons.

double helix
The typical structure of a DNA molecule in which the two complementary polynucleotide strands are wound around each other with base-pairing between the strands.

dynamic instability
The rapid switching between growth and shrinkage shown by microtubules.

dynein
Motor protein that uses the energy of ATP hydrolysis to move toward the minus end of a microtubule. One form of the protein is responsible for the bending of cilia.

electrochemical gradient
Driving force that determines which way an ion will move across a membrane; consists of the combined influence of the ion's concentration gradient and the membrane potential.

electron
Negatively charged subatomic particle that occupies space around an atomic nucleus (e^-).

electron acceptor
Atom or molecule that readily takes up electrons, thereby becoming reduced.

electron carrier
Molecule capable of picking up an electron from a molecule with weak electron affinity and transferring it to a molecule with a higher electron affinity.

electron donor
Molecule that easily gives up an electron, thereby becoming oxidized.

electron microscope
Instrument that illuminates a specimen using beams of electrons to reveal and magnify the structures of very small objects, such as organelles and large molecules.

electron-transport chain
A series of membrane-embedded electron carrier molecules that facilitate the movement of electrons from a higher to a lower energy level, as in oxidative phosphorylation and photosynthesis.

electrophoresis
Technique for separating a mixture of proteins or DNA fragments by placing them on a polymer gel and subjecting them to an electric field. The molecules migrate through the gel at different speeds depending on their size and net charge.

electrostatic attraction
Force that draws together oppositely charged atoms. Examples include ionic bonds and the attractions between molecules containing polar covalent bonds.

element
Substance that cannot be broken down to any other chemical form; composed of a single type of atom.

embryonic stem cell (**ES cell**)
An undifferentiated cell type derived from the inner cell mass of an early mammalian embryo and capable of differentiating to give rise to any of the specialized cell types in the adult body.

endocytosis
Process by which cells take in materials through an invagination of the plasma membrane, which surrounds the ingested material in a membrane-enclosed vesicle. (*See also* **pinocytosis** and **phagocytosis**.)

endomembrane system
Interconnected network of membrane-enclosed organelles in a eukaryotic cell; includes the endoplasmic reticulum, Golgi apparatus, lysosomes, peroxisomes, and endosomes.

endoplasmic reticulum (**ER**)
Labyrinthine membrane-enclosed compartment in the cytoplasm of eukaryotic cells where lipids and proteins are made.

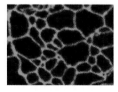

endosome
Membrane-enclosed compartment of a eukaryotic cell through which material ingested by endocytosis passes on its way to lysosomes.

enhancer
Regulatory DNA sequence to which transcription regulators bind, influencing the rate of transcription of a gene that may be many thousands of base pairs away.

entropy
Thermodynamic quantity that measures the degree of disorder in a system.

enzyme
A protein that catalyzes a specific chemical reaction.

enzyme-coupled receptor
Transmembrane protein that, when stimulated by the binding of a ligand, activates an intracellular enzyme (either a separate enzyme or part of the receptor itself).

epigenetic inheritance
The transmission of a heritable pattern of gene expression from one cell to its progeny that does not involve altering the nucleotide sequence of the DNA.

epithelium (plural epithelia)
Sheet of cells covering an external surface or lining an internal body cavity.

equilibrium
State in which the forward and reverse rates of a chemical reaction are equal so that no net chemical change occurs.

equilibrium constant (K)
For a reversible chemical reaction, the ratio of substrate to product when the rates of the forward and reverse reactions are equal. (*See* Table 3–1, p. 98.)

***Escherichia coli* (*E. coli*)**
Rodlike bacterium normally found in the colon of humans and other mammals and widely used in biomedical research.

eubacteria
The proper term for the bacteria of common occurrence, used to distinguish them from archaea.

euchromatin
One of the two main states in which chromatin exists within an interphase cell. Prevalent in gene-rich areas, its less compact structure allows access for proteins involved in transcription. (*See also* **heterochromatin**.)

eukaryote
An organism whose cells have a distinct nucleus and cytoplasm.

evolution
Process of gradual modification and adaptation that occurs in living organisms over generations.

exocytosis
Process by which most molecules are secreted from a eukaryotic cell. These molecules are packaged in membrane-enclosed vesicles that fuse with the plasma membrane, releasing their contents to the outside.

exon
Segment of a eukaryotic gene that is transcribed into RNA and dictates the amino acid sequence of part of a protein.

exon shuffling
Mechanism for the evolution of new genes; in the process, coding sequences from different genes are brought together to generate a protein with a novel combination of domains.

extracellular matrix
Complex network of polysaccharides (such as glycosaminoglycans or cellulose) and proteins (such as collagen) secreted by cells. A structural component of tissues that also influences their development and physiology.

extracellular signal molecule
Any molecule present outside the cell that can elicit a response inside the cell when the molecule binds to a receptor.

FAD—*see* **FADH$_2$**

FADH$_2$ (reduced flavin adenine dinucleotide)
A high-energy electron carrier produced by reduction of FAD during the breakdown of molecules derived from food, including fatty acids and acetyl CoA.

fat
Type of lipid used by living cells to store metabolic energy. Mainly composed of triacylglycerols. (*See* Panel 2–4, pp. 72–73.)

fatty acid

Molecule that consists of a carboxylic acid attached to a long hydrocarbon chain. Used as a major source of energy during metabolism and as a starting point for the synthesis of phospholipids. (*See* Panel 2–4, pp. 72–73.)

feedback inhibition
A form of metabolic control in which the end product of a chain of enzymatic reactions reduces the activity of an enzyme early in the pathway.

fermentation
The breakdown of organic molecules without the involvement of molecular oxygen. This form of oxidation yields less energy than aerobic cell respiration.

fertilization
The fusion of two gametes—sperm and egg—to produce a new individual organism.

fibroblast
Cell type that produces the collagen-rich extracellular matrix in connective tissues such as skin and tendon. Proliferates readily in wounded tissue and in tissue culture.

fibronectin
Extracellular matrix protein that helps cells attach to the matrix by acting as a "linker" that binds to a cell-surface integrin molecule on one end and to a matrix component, such as collagen, on the other.

fibrous protein
A protein with an elongated, rodlike shape, such as collagen or a keratin filament.

filopodium (plural filopodia)
Long, thin, actin-containing extension on the surface of an animal cell. Sometimes has an exploratory function, as in a growth cone.

flagellum (plural flagella)
Long, whiplike structure capable of propelling a cell through a fluid medium with its rhythmic beating.

Eukaryotic flagella are longer versions of cilia; bacterial flagella are completely different, being smaller and simpler in construction.

fluorescence microscope
Instrument used to visualize a specimen that has been labeled with a fluorescent dye; samples are illuminated with a wavelength of light that excites the dye, causing it to fluoresce.

free energy (G)
Energy that can be harnessed to do work, such as driving a chemical reaction.

free-energy change (ΔG)
"Delta G": in a chemical reaction, the difference in free energy between reactant and product molecules. A large negative value of ΔG indicates that the reaction has a strong tendency to occur. The standard free-energy change ($\Delta G°$) is the free-energy change measured at defined concentration, temperature, and pressure.

G, ΔG, ΔG°—*see* free energy, free-energy change

G protein
A membrane-bound GTP-binding protein involved in intracellular signaling; composed of three subunits, this intermediary is usually activated by the binding of a hormone or other ligand to a transmembrane receptor.

G-protein-coupled receptor (GPCR)
Cell-surface receptor that associates with an intracellular trimeric GTP-binding protein (G protein) after activation by an extracellular ligand. These receptors are embedded in the membrane by seven transmembrane α helices.

G_1 cyclin
Regulatory protein that helps drive a cell through the first gap phase of the cell cycle and toward S phase.

G_1 phase
Gap 1 phase of the eukaryotic cell cycle; falls between the end of cytokinesis and the start of DNA synthesis.

G_1-Cdk
Protein complex whose activity drives the cell through the first gap phase of the cell cycle; consists of a G_1 cyclin plus a cyclin-dependent protein kinase (Cdk).

G_1/S cyclin
Regulatory protein that helps to launch the S phase of the cell cycle.

G_1/S-Cdk
Protein complex whose activity triggers entry into S phase of the cell cycle; consists of a G_1/S cyclin plus a cyclin-dependent protein kinase (Cdk).

G_2 phase
Gap 2 phase of the eukaryotic cell cycle; falls between the end of DNA synthesis and the beginning of mitosis.

gain-of-function mutation
Genetic change that increases the activity of a gene or makes it active in inappropriate circumstances; such mutations are usually dominant.

gamete
Cell type in a diploid organism that carries only one set of chromosomes and is specialized for sexual reproduction. A sperm or an egg; also called germ cell.

gap junction
In animal tissues, specialized connection between juxtaposed cells through which ions and small molecules can pass from one cell to the other.

GDP (guanosine 5'-diphosphate)
Nucleotide that is produced by the hydrolysis of the terminal phosphate of GTP, a reaction that also produces inorganic phosphate.

gene
Unit of heredity containing the instructions that dictate the characteristics or phenotype of an organism; in molecular terms, a segment of DNA that directs the production of a protein or functional RNA molecule.

gene duplication and divergence
A process by which new genes can form; involves the accidental generation of an additional copy of a stretch of DNA containing one or more genes, followed by an accumulation of mutations that over time can alter the function of either the original or its copy.

gene expression
The process by which a gene makes a product that is useful to the cell or organism by directing the synthesis of a protein or an RNA molecule with a characteristic activity.

gene family
A set of related genes that has arisen through a process of gene duplication and divergence.

gene knockout
A genetically engineered animal in which a specific gene has been inactivated.

gene replacement
Technique that substitutes a mutant form of a gene for its normal counterpart to investigate the gene's function.

general transcription factors
Proteins that assemble on the promoters of many eukaryotic genes near the start site of transcription and load the RNA polymerase in the correct position.

genetic code
Set of rules by which the information contained in the nucleotide sequence of a gene and its corresponding RNA molecule is translated into the amino acid sequence in a protein.

genetic engineering—*see* recombinant DNA technology

genetic instability
An increased rate of mutation often caused by defects

in the systems that govern the accurate replication and maintenance of the genome; the resulting mutations sometimes drive the evolution of cancer.

genetic map
A graphic representation of the order of genes in chromosomes spaced according to the amount of recombination that occurs between them.

genetic screen
Experimental technique used to search through a collection of mutants for a particular phenotype.

genetics
The study of genes, heredity, and the variation that gives rise to differences between one living organism and another.

genome
The total genetic information carried by all the chromosomes of a cell or organism.

genomic DNA library
Collection of cloned DNA molecules that represents the entire genome of a cell.

genotype
The genetic makeup of a cell or organism, including which alleles (gene variants) it carries.

germ cell
Cell type in a diploid organism that carries only one set of chromosomes and is specialized for sexual reproduction. A sperm or an egg; also called gamete.

germ line
The lineage of reproductive cells that contributes to the formation of a new generation of organisms, as distinct from somatic cells, which form the body and leave no descendants in the next generation.

globular protein
Any protein in which the polypeptide chain folds into a compact, rounded shape. Includes most enzymes.

gluconeogenesis
Set of enzyme-catalyzed reactions by which glucose is synthesized from small organic molecules such as pyruvate, lactate, or amino acids; in effect, the reverse of glycolysis.

glucose
Six-carbon sugar that plays a major role in the metabolism of living cells. Stored in polymeric form as glycogen in animal cells and as starch in plant cells. (*See* Panel 2–3, pp. 70–71.)

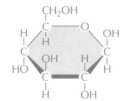

glycocalyx
Protective layer of carbohydrates on the outside surface of the plasma membrane formed by the sugar residues of membrane glycoproteins, proteoglycans, and glycolipids.

glycogen
Branched polymer composed exclusively of glucose units used to store energy in animal cells. Granules of this material are especially abundant in liver and muscle cells.

glycolipid
Membrane lipid molecule that has a short carbohydrate chain attached to its hydrophilic head.

glycolysis
Series of enzyme-catalyzed reactions in which sugars are partially degraded and their energy captured by the activated carriers ATP and NADH. (Literally, "sugar splitting.")

glycoprotein
Any protein with one or more covalently linked oligosaccharide chains. Includes most secreted proteins and most proteins exposed on the outer surface of the plasma membrane.

glycosaminoglycan (GAG)
Polysaccharide chain that can form a gel that acts as a "space filler" in the extracellular matrix of connective tissues; helps animal tissues resist compression.

Golgi apparatus
Membrane-enclosed organelle in eukaryotic cells that modifies the proteins and lipids made in the endoplasmic reticulum and sorts them for transport to other sites.

green fluorescent protein (GFP)
Fluorescent protein, isolated from a jellyfish, that is used experimentally as a marker for monitoring the location and movement of proteins in living cells.

group—*see* **chemical group**

growth factor
Extracellular signaling molecule that stimulates a cell to increase in size and mass. Examples include epidermal growth factor (EGF) and platelet-derived growth factor (PDGF).

GTP (guanosine 5′-triphosphate)
Nucleoside triphosphate used in the synthesis of RNA and DNA. Like the closely related ATP, serves as an activated carrier in some energy-transfer reactions. Also has a special role in microtubule assembly, protein synthesis, and cell signaling.

GTP-binding protein
Intracellular signaling protein whose activity is determined by its association with either GTP or GDP. Includes both trimeric G proteins and monomeric GTPases, such as Ras.

haploid
Describes a cell or organism with only one set of chromosomes, such as a sperm cell or a bacterium. (*See also* **diploid**.)

haplotype block
A combination of alleles or other DNA markers that has been inherited as a unit, undisturbed by genetic recombination, across many generations.

helix
An elongated structure whose subunits twist in a regular fashion around a central axis, like a spiral staircase.

hemidesmosome
Structure that anchors epithelial cells to the basal lamina beneath them.

heredity
The genetic transmission of traits from parents to offspring.

heterochromatin
Highly condensed region of an interphase chromosome; generally gene-poor and transcriptionally inactive. (*See also* **euchromatin**.)

heterozygous
Possessing dissimilar alleles for a given gene.

high-energy bond
Covalent bond whose hydrolysis releases an unusually large amount of free energy under the conditions existing in a cell. Examples include the phosphodiester bonds in ATP and the thioester linkage in acetyl CoA.

histone
One of a group of abundant highly conserved proteins around which DNA wraps to form nucleosomes, structures that represent the most fundamental level of chromatin packing.

histone deacetylase
Enzyme that removes acetyl groups from lysines present in histones; its action often allows chromatin to pack more tightly.

homolog
A gene, chromosome, or any structure that has a close similarity to another as a result of common ancestry. (*See also* **homologous chromosome**.)

homologous
Describes genes, chromosomes, or any structures that are similar because of their common evolutionary origin. Can also refer to similarities between protein sequences or nucleic acid sequences.

homologous chromosome
In a diploid cell, one of the two copies of a particular chromosome, one of which comes from the father and the other from the mother.

homologous gene—*see* **homologous**

homologous recombination
Mechanism by which double-strand breaks in a DNA molecule can be repaired flawlessly; uses an undamaged, duplicated, or homologous chromosome to guide the repair. During meiosis, the mechanism results in an exchange of genetic information between the maternal and paternal homologs.

homozygous
Possessing identical alleles for a given gene.

horizontal gene transfer
Process by which DNA is passed from the genome of one organism to that of another, even to an individual from another species. This contrasts with "vertical" gene transfer, which refers to the transfer of genetic information from parent to progeny.

hormone
Extracellular signal molecule that is secreted and transported via the bloodstream (in animals) or the sap (in plants) to target tissues on which it exerts a specific effect.

hybridization
Experimental technique in which two complementary nucleic acid strands come together and form hydrogen bonds to produce a double helix; used to detect specific nucleotide sequences in either DNA or RNA.

hydrogen bond
A weak noncovalent interaction between a positively charged hydrogen atom in one molecule and a negatively charged atom, such as nitrogen or oxygen, in another; these interactions are key to the structure and properties of water.

hydrogen ion
Positively charged ion generated by the removal of an electron from a hydrogen atom; often used to refer to a proton (H^+) in aqueous solution. Its presence is the basis of acidity. (*See* Panel 2–2, pp. 68–69.)

hydrolysis
Chemical reaction that involves cleavage of a covalent bond with the accompanying consumption of water (its –H being added to one product of the cleavage and its –OH to the other); the reverse of condensation.

hydronium ion (H_3O^+)
The form taken by a proton (H^+) in aqueous solution.

hydrophilic
Molecule or part of a molecule that readily forms hydrogen bonds with water, allowing it to dissolve; literally, "water loving."

hydrophobic
Nonpolar, uncharged molecule or part of a molecule that forms few or no hydrogen bonds with water molecules and therefore does not dissolve; literally, "water fearing."

hydrophobic interaction
Type of noncovalent bond that forces together the hydrophobic portions of dissolved molecules to minimize their disruption of the hydrogen-bonded network of water; helps push together membrane phospholipids and fold proteins into a compact, globular shape.

hydroxyl (–OH)
Chemical group consisting of a hydrogen atom linked to an oxygen, as in an alcohol. (*See* Panel 2–1, pp. 66–67.)

***in situ* hybridization**
Technique in which a single-stranded RNA or DNA probe is used to locate a complementary nucleotide sequence in a chromosome, cell, or tissue; used to diagnose genetic disorders or to track gene expression.

in vitro
Term used by biochemists to describe a process that takes place in an isolated cell-free extract. Also used by cell biologists to refer to cells growing in culture, as opposed to in an organism.

in vivo
In an intact cell or organism. (Latin for "in life.")

induced pluripotent stem cell (iPS cell)
Somatic cell that has been reprogrammed to resemble and behave like a pluripotent embryonic stem (ES) cell through the artificial introduction of a set of genes encoding particular transcription regulators.

initiation factor
Protein that promotes the proper association of ribosomes with mRNA and is required for the initiation of protein synthesis.

initiator tRNA
Special tRNA that initiates the translation of an mRNA in a ribosome. It always carries the amino acid methionine.

inorganic
Not composed of carbon and hydrogen.

inositol
Sugar molecule with six hydroxyl groups that forms the structural basis for inositol phospholipids, which can act as membrane-bound signaling molecules.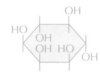

inositol 1,4,5-trisphosphate (IP₃)
Small intracellular signaling molecule that triggers the release of Ca^{2+} from the endoplasmic reticulum into the cytosol; produced when a signal molecule activates a membrane-bound protein called phospholipase C.

inositol phospholipid
Minor lipid component of plasma membranes that plays a part in signal transduction in eukaryotic cells; cleavage yields two small messenger molecules, IP₃ and diacylglycerol.

integrin
Family of transmembrane proteins present on cell surfaces that enable cells to make and break attachments to the extracellular matrix, allowing them to crawl through a tissue.

intermediate filament
Fibrous cytoskeletal element, about 10 nm in diameter, that forms ropelike networks in animal cells; helps cells resist tension applied from outside.

interphase
Long period of the cell cycle between one mitosis and the next. Includes G_1 phase, S phase, and G_2 phase.

interphase chromosome
State in which a eukaryotic chromosome exists when the cell is between divisions; more extended and transcriptionally active than mitotic chromosomes.

intracellular signaling molecule
Molecule that is part of the mechanism for transducing and transmitting signals inside a cell.

intracellular signaling pathway
A set of proteins and small-molecule second messengers that interact with each other to relay a signal from the cell membrane to its final destination in the cytoplasm or nucleus.

intrinsically disordered sequence
Region in a polypeptide chain that lacks a definite structure.

intron
Noncoding sequence within a eukaryotic gene that is transcribed into an RNA molecule but is then excised by RNA splicing to produce an mRNA.

ion
An atom carrying an electrical charge, either positive or negative.

ion channel
Transmembrane protein that forms a pore across the lipid bilayer through which specific inorganic ions can diffuse down their electrochemical gradients.

ion-channel-coupled receptor
Transmembrane receptor protein or protein complex that opens in response to the binding of a ligand to its external face, allowing the passage of a specific inorganic ion.

ionic bond
Interaction formed when one atom donates electrons to another; this transfer of electrons causes both atoms to become electrically charged.

iron–sulfur center
Metal complex found in electron carriers that operate early in the electron-transport chain; has a relatively weak affinity for electrons.

isomer (stereoisomer)
One of two or more substances that contains the same atoms and has the same molecular formula (such as $C_6H_{12}O_6$) as the other, but differs from the other in the spatial arrangement of these atoms. Optical isomers are mirror images of each other.

isotope
A variant of an element that has the same number of protons but a different atomic weight. Some are radioactive.

K⁺
Potassium ion—the most abundant positively charged ion in living cells.

K⁺ leak channel
Ion channel permeable to K⁺ that randomly flickers between an open and closed state; largely responsible for the resting membrane potential in animal cells.

karyotype
An ordered display of the full set of chromosomes of a cell arranged with respect to size, shape, and number.

keratin filament
Class of intermediate filament abundant in epithelial cells, where it provides tensile strength; main structural component of hair, feathers, and claws.

kilocalorie (kcal)
Unit of heat equal to 1000 calories. Often used to express the energy content of food or molecules: bond strengths, for example, are measured in kcal/mole. An alternative unit in wide use is the kilojoule.

kilojoule (kJ)
Standard unit of energy equal to 0.239 kilocalories.

kinase—*see* **protein kinase**

kinesin
A large family of motor proteins that uses the energy

of ATP hydrolysis to move toward the plus end of a microtubule.

kinetochore
Protein complex that assembles on the centromere of a condensed mitotic chromosome; the site to which spindle microtubules attach.

K_M
The concentration of substrate at which an enzyme works at half its maximum rate. Large values of K_M usually indicate that the enzyme binds to its substrate with relatively low affinity.

knockout mouse
Genetically engineered mouse in which a specific gene has been inactivated, for example, by introducing a deletion in its DNA.

***L1* element**
Type of retrotransposon that constitutes 15% of the human genome; also called *LINE-1*.

lagging strand
At a replication fork, the DNA strand that is made discontinuously in short separate fragments that are later joined together to form one continuous new strand.

lamellipodium (plural lamellipodia)
Dynamic sheetlike extension on the surface of an animal cell, especially one migrating over a surface.

law of independent assortment
Principle that, during gamete formation, the alleles for different traits segregate independently of one another; Mendel's second law of inheritance.

law of segregation
Principle that the maternal and paternal alleles for a trait separate from one another during gamete formation and then reunite during fertilization; Mendel's first law of inheritance.

leading strand
At a replication fork, the DNA strand that is made by continuous synthesis in the 5'-to-3' direction.

ligand
General term for a molecule that binds to a specific site on a protein.

ligand-gated channel
An ion channel that is stimulated to open by the binding of a small molecule such as a neurotransmitter.

ligase
Enzyme that reseals nicks that arise in the backbone of a DNA molecule; in the laboratory, can be used to join together two DNA fragments.

light reactions
In photosynthesis, the set of reactions that converts the energy of sunlight into chemical energy in the form of ATP and NADPH.

lipid
Organic molecule that is insoluble in water but dissolves readily in nonpolar organic solvents; typically contains long hydrocarbon chains or multiple rings. One class, the phospholipids, forms the structural basis of biological membranes.

lipid bilayer
Thin pair of closely juxtaposed sheets, composed mainly of phospholipid molecules, that forms the structural basis for all cell membranes.

local mediator
Secreted signal molecule that acts at a short range on adjacent cells.

long noncoding RNA
Class of RNA molecules more than 200 nucleotides in length that does not encode proteins.

loss-of-function mutation
A genetic alteration that reduces or eliminates the activity of a gene. Such mutations are usually recessive: the organism can function normally as long as it retains at least one normal copy of the affected gene.

lumen
The space inside a hollow or tubular structure; can refer to the cavity in a tissue or within an organelle.

lymphocyte
White blood cell that mediates the immune response to foreign molecules (antigens). Can be an antibody-secreting B cell type or the T cell type that recognizes and ultimately eliminates infected cells.

lysosome
Membrane-enclosed organelle that breaks down worn-out proteins and organelles and other waste materials, as well as molecules taken up by endocytosis; contains digestive enzymes that are typically most active at the acid pH found inside these organelles.

lysozyme
Enzyme that severs the polysaccharide chains that form the cell walls of bacteria; found in many secretions including saliva and tears.

M cyclin
Regulatory protein that binds to mitotic Cdk to form M-Cdk, the protein complex that triggers the M phase of the cell cycle.

M phase
Period of the eukaryotic cell cycle during which the nucleus and cytoplasm divide.

M-Cdk
Protein complex that triggers the M phase of the cell cycle; consists of an M cyclin plus a mitotic cyclin-dependent protein kinase (Cdk).

macromolecule
Polymer built from covalently linked subunits; includes proteins, nucleic acids, and polysaccharides with a molecular mass greater than a few thousand daltons.

macrophage
Cell found in animal tissues that defends against infections by ingesting invading microbes by a process of phagocytosis; derived from a type of white blood cell.

MAP kinase
Mitogen-activated protein kinase. Signaling molecule that is the final kinase in a three-kinase sequence called the MAP-kinase signaling module.

MAP-kinase signaling module
Set of three functionally interlinked protein kinases that allows cells to respond to extracellular signal molecules that stimulate proliferation; includes a mitogen-activated protein kinase (MAP kinase), a MAP kinase kinase, and a MAP kinase kinase kinase.

mass spectrometry
Technique for determining the exact mass of every peptide present in a sample of purified protein or protein mixture.

matrix
Large internal compartment within a mitochondrion.

mechanically gated channel
An ion channel that allows the passage of select ions across a membrane in response to a physical perturbation.

meiosis
Specialized type of cell division by which eggs and sperm cells are made. Two successive nuclear divisions with only one round of DNA replication generate four haploid cells from an initial diploid cell.

membrane
Thin sheet of lipid molecules and associated proteins that encloses all cells and forms the boundaries of many eukaryotic organelles.

membrane domain
Functionally and structurally specialized region in the membrane of a cell or organelle; typically characterized by the presence of specific proteins.

membrane potential
Voltage difference across a membrane due to a slight excess of positive ions on one side and of negative ions on the other.

membrane protein
A protein associated with the lipid bilayer of a cell membrane.

membrane transport protein
Any transmembrane protein that provides a passageway for the movement of select substances across a cell membrane.

membrane-enclosed organelle
Any organelle in a eukaryotic cell that is surrounded by a lipid bilayer, for example, the endoplasmic reticulum, Golgi apparatus, and lysosome.

membrane-enclosed organelle
Any organelle in the eukaryotic cell that is surrounded by a lipid bilayer; for example, the endoplasmic reticulum, Golgi apparatus, and lysosome.

messenger RNA (mRNA)
RNA molecule that specifies the amino acid sequence of a protein.

metabolic pathway
Interconnected sequence of enzymatic reactions in which the product of one reaction is the substrate of the next.

metabolism
The sum total of the chemical reactions that take place in the cells of a living organism.

metaphase
Stage of mitosis in which chromosomes are firmly attached to the mitotic spindle at its equator but have not yet segregated toward opposite poles.

metastasis
The spread of cancer cells from the initial site of the tumor to form secondary tumors at other sites in the body.

methyl (–CH$_3$) group
Hydrophobic chemical group derived from methane (CH$_4$). (*See* Panel 2–1, pp. 66–67.)

Michaelis constant (K_M)
Concentration of substrate at which an enzyme works at half its maximum velocity; serves as a measure of how tightly the substrate is bound.

micro-
In the metric system, prefix denoting 10^{-6}.

micrograph
Any photograph or digital image taken through a microscope. Can be a light micrograph or an electron micrograph, depending on the type of microscope used.

micrometer
Unit of length equal to one millionth (10^{-6}) of a meter or 10^{-4} centimeter.

microRNA (miRNA)
Small noncoding RNA that controls gene expression by base-pairing with a specific mRNA to regulate its stability and its translation.

microscope
Instrument for viewing extremely small objects. A light microscope utilizes a focused beam of visible light and is used to examine cells and organelles. An electron microscope utilizes a beam of electrons and can be used to examine objects as small as individual molecules.

microtubule
Long, stiff, cylindrical structure composed of the 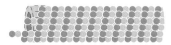 protein tubulin. Used by eukaryotic cells to organize their cytoplasm and guide the intracellular transport of macromolecules and organelles.

microtubule-associated protein
Accessory protein that binds to microtubules; can stabilize microtubule filaments, link them to other cell structures, or transport various components along their length.

milli-
In the metric system, prefix denoting 10^{-3}.

mismatch repair
Mechanism for recognizing and correcting incorrectly paired nucleotides—those that are noncomplementary.

mitochondrion (plural mitochondria)
Membrane-enclosed organelle, about the size of a

bacterium, that carries out oxidative phosphorylation and produces most of the ATP in eukaryotic cells.

mitogen
An extracellular signal molecule that stimulates cell proliferation.

mitosis
Division of the nucleus of a eukaryotic cell.

mitotic chromosome
Highly condensed duplicated chromosome in which the two new chromosomes (also called sister chromatids) are still held together at the centromere. The structure chromosomes adopt during mitosis.

mitotic spindle
Array of microtubules and associated molecules that forms between the opposite poles of a eukaryotic cell during mitosis and pulls duplicated chromosome sets apart.

mobile genetic element
Short segment of DNA that can move, sometimes through an RNA intermediate, from one location in a genome to another; an important source of genetic variation in most genomes. Also called a transposon.

model organism
A living thing selected for intensive study as a representative of a large group of species. Examples include the mouse (representing mammals), the yeast *Saccharomyces cerevisiae* (representing a unicellular eukaryote), and *Escherichia coli* (representing bacteria).

mole
The amount of a substance, in grams, that is equal to its molecular weight; this quantity will contain 6×10^{23} molecules of the substance.

molecular mass
The weight of a molecule expressed in daltons, the atomic mass unit that closely approximates the mass of a hydrogen atom.

molecular switch
Intracellular signaling protein that toggles between an active and inactive state in response to receiving a signal.

molecular weight
Sum of the atomic weights of the atoms in a molecule; as a ratio of molecular masses, it is a number without units.

molecule
Group of atoms joined together by covalent bonds.

monomer
Small molecule that can be linked to others of a similar type to form a larger molecule (polymer).

monomeric GTPase
Small, single-subunit GTP-binding protein. Proteins of this family, such as Ras and Rho, are part of many different signaling pathways.

motor protein
Protein such as myosin or kinesin that uses energy derived from ATP hydrolysis to propel itself along a protein filament or polymeric molecule.

mutation
A randomly produced, permanent change in the nucleotide sequence of DNA.

myofibril
Long, cylindrical structure that constitutes the contractile element of a muscle cell; constructed of arrays of highly organized bundles of actin, myosin, and other accessory proteins.

myosin
Type of motor protein that uses ATP to drive movements along actin filaments. One subtype interacts with actin to form the thick contractile bundles of skeletal muscle.

myosin filament
Polymer composed of interacting molecules of myosin-II; interaction with actin promotes contraction in muscle and nonmuscle cells.

myosin-I
Simplest type of myosin, present in all cells; consists of a single actin-binding head and a tail that can attach to other molecules or organelles.

myosin-II
Type of myosin that exists as a dimer with two actin-binding heads and a coiled-coil tail; can associate to form long myosin filaments.

N-terminus (amino terminus)
The end of a polypeptide chain that carries a free α-amino group.

Na⁺
Sodium ion—a positively charged ion that is a major constituent of living cells.

Na⁺ pump (sodium pump)
Transporter found in the plasma membrane of most animal cells that actively pumps Na^+ out of the cell and K^+ in using the energy derived from ATP hydrolysis.

NAD⁺ (nicotine adenine dinucleotide)
Activated carrier that accepts a hydride ion (H^-) from a donor molecule, thereby producing NADH. Widely used in the energy-producing breakdown of sugar molecules. (*See* Figure 3–34.)

NADH
Activated carrier widely used in the energy-producing breakdown of sugar molecules. (*See* Figure 3–34.)

NADPH (nicotine adenine dinucleotide phosphate)
Activated carrier closely related to NADH and used as an electron donor in biosynthetic pathways. In the process it is oxidized to $NADP^+$. (*See* Figure 3–35.)

nanometer
Unit of length that represents 10^{-9} (one billionth of a) meter; commonly used to measure molecules and organelles.

Nernst equation
An equation that relates the concentrations of an inorganic ion on the two sides of a permeable membrane to the membrane potential at which there would be no net movement of the ion across the membrane.

nerve terminal
Structure at the end of an axon that signals to another neuron or target cell.

neuron
An electrically excitable cell that integrates and transmits information as part of the nervous system; a nerve cell.

neurotransmitter
Small signaling molecule secreted by a nerve cell at a synapse to transmit information to a postsynaptic cell. Examples include acetylcholine, glutamate, GABA, and glycine.

nitric oxide (NO)
Locally acting gaseous signal molecule that diffuses across cell membranes to affect the activity of intracellular proteins.

nitrogen fixation
Conversion of nitrogen gas from the atmosphere into nitrogen-containing molecules by soil bacteria and cyanobacteria.

noncovalent bond
Chemical association that does not involve the sharing of electrons; singly are relatively weak, but can sum together to produce strong, highly specific interactions between molecules. Examples are hydrogen bonds and van der Waals attractions.

nonhomologous end joining
A quick-and-dirty mechanism for repairing double-strand breaks in DNA that involves quickly bringing together, trimming, and rejoining the two broken ends; results in a loss of information at the site of repair.

nonpolar
Describes a molecule that lacks a local accumulation of positive or negative charge; generally insoluble in water.

nuclear envelope
Double membrane surrounding the nucleus. Consists of outer and inner membranes, perforated by nuclear pores.

nuclear lamina
Fibrous layer on the inner surface of the inner nuclear membrane formed as a network of intermediate filaments made from nuclear lamins.

nuclear magnetic resonance (NMR) spectroscopy
Technique used for determining the three-dimensional structure of a protein in solution.

nuclear pore
Channel through which selected large molecules move between the nucleus and the cytoplasm.

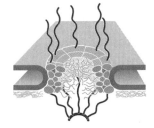

nuclear receptor
Protein inside a eukaryotic cell that, on binding to a signal molecule, enters the nucleus and regulates transcription.

nucleic acid
Macromolecule that consists of a chain of nucleotides joined together by phosphodiester bonds; RNA or DNA.

nucleolus
Large structure within the nucleus where ribosomal RNA is transcribed and ribosomal subunits are assembled.

nucleoside
Molecule made of a nitrogen-containing ring compound attached to a sugar, either ribose (in RNA) or deoxyribose (in DNA).

nucleosome
Beadlike structural unit of a eukaryotic chromosome composed of a short length of DNA wrapped around a core of histone proteins; includes a nucleosomal core particle (DNA plus histone protein) along with a segment of linker DNA that ties the core particles together.

nucleotide
Basic building block of the nucleic acids, DNA and RNA; includes a nucleoside with a series of one or more phosphate groups linked to its sugar.

nucleus
In biology, refers to the prominent, rounded structure that contains the DNA of a eukaryotic cell. In chemistry, refers to the dense, positively charged center of an atom.

Okazaki fragment
Short length of DNA produced on the lagging strand during DNA replication. Adjacent fragments are rapidly joined together by DNA ligase to form a continuous DNA strand.

oligo-
Prefix that denotes a short polymer (oligomer). May be made of amino acids (oligopeptide), sugars (oligosaccharide), or nucleotides (oligonucleotide).

oncogene
A gene that, when activated, can potentially make a cell cancerous. Typically a mutant form of a normal gene (proto-oncogene) involved in the control of cell growth or division.

open reading frame (ORF)
Long sequence of nucleotides that contains no stop codon; used to identify potential protein-coding sequences in DNA.

optogenetics
Technique that uses light to control the activity of neurons into which light-gated ion channels have been artificially introduced.

organelle
A discrete structure or subcompartment of a eukaryotic cell that is specialized to carry out a particular function. Examples include mitochondria and the Golgi apparatus.

organic chemistry
The branch of chemistry concerned with compounds made of carbon. Includes essentially all of the molecules from which living cells are made, apart from water and metal ions such as Na^+.

organic molecule
Chemical compound that contains carbon and hydrogen.

origin recognition complex (ORC)
Assembly of proteins that is bound to the DNA at origins of replication in eukaryotic chromosomes throughout the cell cycle.

osmosis
Passive movement of water across a cell membrane from a region where the concentration of water is high (because the concentration of solutes is low) to a region where the concentration of water is low (and the concentration of solutes is high).

oxidation
Removal of electrons from an atom, as occurs during the addition of oxygen to a carbon atom or when a hydrogen is removed from a carbon atom. The opposite of reduction. (*See* Figure 3–11.)

oxidative phosphorylation
Process in bacteria and mitochondria in which ATP formation is driven by the transfer of electrons from food molecules to molecular oxygen.

p53
Transcription regulator that controls the cell's response to DNA damage, preventing the cell from entering S phase until the damage has been repaired or inducing the cell to commit suicide if the damage is too extensive; mutations in the gene encoding this protein are found in many human cancers.

pairing
In meiosis, the process by which a pair of duplicated homologous chromosomes attach to one another to form a structure containing four sister chromatids.

passive transport
The spontaneous movement of a solute down its concentration gradient across a cell membrane via a membrane transport protein, such as a channel or a transporter.

patch-clamp recording
Technique used to monitor the activity of ion channels in a membrane; involves the formation of a tight seal between the tip of a glass electrode and a small region of cell membrane, and manipulation of the membrane potential by varying the concentrations of ions in the electrode.

pedigree
Chart showing the line of descent, or ancestry, of an individual organism.

peptide bond
Chemical bond between the carbonyl group of one amino acid and the amino group of a second amino acid. (*See* Panel 2–5, pp. 74–75.)

peroxisome
Small membrane-enclosed organelle that contains enzymes that degrade lipids and destroy toxins.

pH scale
Concentration of hydrogen ions in a solution, expressed as a logarithm. Thus, an acidic solution with pH 3 will contain 10^{-3} M hydrogen ions.

phagocytic cell
A cell such as a macrophage or neutrophil that is specialized to take up particles and microorganisms by phagocytosis.

phagocytosis
The process by which particulate material is engulfed ("eaten") by a cell. Prominent in predatory cells, such as *Amoeba proteus*, and in cells of the vertebrate immune system such as macrophages.

phenotype
The observable characteristics of a cell or organism.

phosphatidylcholine
Common phospholipid present in abundance in most cell membranes; uses choline attached to a phosphate as its head group.

phosphodiester bond
Strong covalent bond that forms the backbone of DNA and RNA molecules; links the 3′ carbon of one sugar to the 5′ carbon of another. (*See* Figure 2–26.)

phosphoinositide 3-kinase (PI 3-kinase)
Enzyme that phosphorylates inositol phospholipids in the plasma membrane, which generates docking sites for intracellular signaling proteins that promote cell growth and survival.

phospholipase C
Enzyme associated with the plasma membrane that generates two small messenger molecules in response to activation.

phospholipid
A major type of lipid molecule in many cell membranes. Generally composed of two fatty acid tails linked to one of a variety of phosphate-containing polar groups.

phosphorylation—see **protein phosphorylation**

photosynthesis
The process by which plants, algae, and some bacteria use the energy of sunlight to drive the synthesis of organic molecules from carbon dioxide and water.

photosystem
Large multiprotein complex containing chlorophyll that captures light energy and converts it into chemical energy; consists of a set of antenna complexes and a reaction center.

phragmoplast
In a dividing plant cell, structure made of microtubules and membrane vesicles that guides the formation of a new cell wall.

phylogenetic tree
Diagram or "family tree" showing the evolutionary history of a group of organisms or proteins.

pinocytosis
Type of endocytosis in which soluble materials are taken up from the environment and incorporated into vesicles for digestion. (Literally, "cell drinking.")

plasma membrane
The protein-containing lipid bilayer that surrounds a living cell.

plasmid
Small circular DNA molecule that replicates

independently of the genome. Used extensively as a vector for DNA cloning.

plasmodesma (plural plasmodesmata)
Cell–cell junction that connects one plant cell to the next; consists of a channel of cytoplasm lined by membrane.

pluripotent
Capable of giving rise to any type of cell or tissue.

pluripotent stem cell
Cell capable of giving rise to any of the specialized cell types in the body.

point mutation
Change in a single nucleotide pair in a DNA sequence.

polar
In chemistry, describes a molecule or bond in which electrons are distributed unevenly.

polarity
An inherent asymmetry that allows one end of an object to be distinguished from another; can refer to a molecule, a polymer (such as an actin filament), or even a cell (for example, an epithelial cell that lines the mammalian small intestine).

polyadenylation
The addition of multiple adenine nucleotides to the 3′ end of a newly synthesized mRNA molecule.

polymer
Long molecule made by covalently linking multiple identical or similar subunits (monomers).

polymerase
General term for an enzyme that catalyzes addition of subunits to a nucleic acid polymer. DNA polymerase, for example, makes DNA, and RNA polymerase makes RNA.

polymerase chain reaction (**PCR**)
Technique for amplifying selected regions of DNA by multiple cycles of DNA synthesis; can produce billions of copies of a given sequence in a matter of hours.

polymorphism
DNA sequence for which two or more variants are present at high frequency in the general population.

polynucleotide
A molecular chain of nucleotides chemically linked by a series of phosphodiester bonds. A strand of RNA or DNA.

polypeptide backbone
Repeating sequence of atoms (–N–C–C–) that forms the core of a protein molecule and to which the amino acid side chains are attached.

polypeptide; polypeptide chain
Linear polymer composed of multiple amino acids. Proteins are composed of one or more long polypeptide chains.

polyribosome
Messenger RNA molecule to which multiple ribosomes are attached and engaged in protein synthesis.

polysaccharide
Linear or branched polymer composed of sugars. Examples are glycogen, hyaluronic acid, and cellulose.

positive feedback loop
An important form of regulation in which the end product of a reaction or pathway stimulates continued activity; controls a variety of biological processes, including enzyme activity, cell signaling, and gene expression.

post-transcriptional control
Regulation of gene expression that occurs after transcription of the gene has begun; examples include RNA splicing and RNA interference.

primary structure
The amino acid sequence of a protein.

primary transcript—*see* **transcription**

primase
An RNA polymerase that uses DNA as a template to produce an RNA fragment that serves as a primer for DNA synthesis.

primer
In DNA replication, a short length of RNA made at the beginning of the synthesis of each DNA fragment; these RNA segments are subsequently removed and filled in with DNA.

processive
Describes an ability to catalyze consecutive reactions or undergo multiple conformational changes without releasing a substrate. Examples include replication by DNA polymerase or the movement of motor proteins involved in transport, such as kinesin.

programmed cell death
A tightly controlled form of cell suicide that allows cells that are unneeded or unwanted to be eliminated from an adult or developing organism; also called apoptosis.

prokaryote
Major category of living cells distinguished by the absence of a nucleus. Prokaryotes include the archaea and the eubacteria (commonly called bacteria).

prometaphase
Stage of mitosis in which the nuclear envelope breaks down and duplicated chromosomes are captured by the spindle microtubules; precedes metaphase.

promoter
DNA sequence that initiates gene transcription; includes sequences recognized by RNA polymerase.

proofreading
The process by which DNA polymerase corrects its own errors as it moves along DNA.

prophase
First stage of mitosis, during which the duplicated chromosomes condense and the mitotic spindle forms.

protease
Enzyme that degrades proteins by hydrolyzing their peptide bonds.

proteasome

Large protein machine that degrades proteins that are damaged, misfolded, or no longer needed by the cell; its target proteins are marked for destruction primarily by the attachment of a short chain of ubiquitin.

protein

Polymer built from amino acids that provides cells with their shape and structure and performs most of their activities.

protein complex—*see* complex

protein domain

Segment of a polypeptide chain that can fold into a compact stable structure and that usually carries out a specific function.

protein family

A group of polypeptides that shares a similar amino acid sequence or three-dimensional structure, reflecting a common evolutionary origin. Individual members often have related but distinct functions, such as kinases that phosphorylate different target proteins.

protein kinase

Enzyme that catalyzes the transfer of a phosphate group from ATP to a specific amino acid side chain on a target protein.

protein kinase C (PKC)

Enzyme that phosphorylates target proteins in response to a rise in diacylglycerol and Ca^{2+} ions.

protein machine

Large assembly of protein molecules that operates as a unit to perform a complex series of biological activities, such as replicating DNA.

protein phosphatase

Enzyme that catalyzes the removal of a phosphate group from a protein, often with high specificity for the phosphorylated site.

protein phosphorylation

The covalent addition of a phosphate group to a side chain of a protein, catalyzed by a protein kinase; serves as a form of regulation that usually alters the activity or properties of the target protein.

proteoglycan

Molecule consisting of one or more glycosaminoglycan chains attached to a core protein; these aggregates can form gels that regulate the passage of molecules through the extracellular medium and guide cell migration.

proteolysis

Degradation of a protein by means of a protease.

proteomics

The large-scale study of the structure and function of proteins.

proto-oncogene

Gene that when mutated or overexpressed can transform a normal cell into a cancerous one.

proton

Positively charged particle found in the nucleus of every atom; also, another name for a hydrogen ion (H^+).

proton (H^+) pump

A transporter that actively moves H^+ across a cell membrane, thereby generating a gradient that can be used by the cell, for example, to import other solutes.

protozoan (plural protozoa)

A free-living, nonphotosynthetic, single-celled, motile eukaryote.

pump

Transporter that uses a source of energy, such as ATP hydrolysis or sunlight, to actively move a solute across a membrane against its electrochemical gradient.

purifying selection

Preservation of a specific nucleotide sequence driven by the elimination of individuals carrying mutations that interfere with its functions.

purine

A double-ringed, nitrogen-containing compound found in DNA and RNA. Examples are adenine and guanine. (*See* Panel 2–6, pp. 76–77.)

pyrimidine

A nitrogen-containing, six-membered ring compound found in DNA and RNA. Examples are thymine, cytosine, and uracil. (*See* Panel 2–6, pp. 76–77.)

pyruvate

Three-carbon metabolite that is the end product of the glycolytic breakdown of glucose; provides a crucial link to the citric acid cycle and many biosynthetic pathways.

quaternary structure

Complete structure formed by multiple, interacting polypeptide chains within a protein molecule.

quinone

Small, lipid-soluble, mobile electron carrier molecule found in the respiratory and photosynthetic electron-transport chains. (*See* Figure 14–23.)

Rab protein

A family of small GTP-binding proteins present on the surfaces of transport vesicles and organelles that serves as a molecular marker to help ensure that transport vesicles fuse only with the correct membrane.

Ras

One of a large family of small GTP-binding proteins (the monomeric GTPases) that helps relay signals from cell-surface receptors to the nucleus. Many human cancers contain an overactive mutant form of the protein.

reaction center

In photosynthetic membranes, a protein complex that contains a specialized pair of chlorophyll molecules that performs photochemical reactions to convert the energy of photons (light) into high-energy electrons for transport down the photosynthetic electron-transport chain.

reading frame

One of the three possible ways in which a set of successive nucleotide triplets can be translated into protein, depending on which nucleotide serves as the starting point.

receptor
Protein that recognizes and responds to a specific signal molecule.

receptor serine/threonine kinase
Enzyme-coupled receptor that phosphorylates target proteins on serine or threonine.

receptor tyrosine kinase (RTK)

Enzyme-coupled receptor in which the intracellular domain has a tyrosine kinase activity, which is activated by ligand binding to the receptor's extracellular domain.

receptor-mediated endocytosis
Mechanism of selective uptake of material by animal cells in which a macromolecule binds to a receptor in the plasma membrane and enters the cell in a clathrin-coated vesicle.

recombinant DNA molecule
A DNA molecule that is composed of DNA sequences from different sources.

recombinant DNA technology
The collection of techniques by which DNA segments from different sources are combined to make new DNA. Recombinant DNAs are widely used in the cloning of genes, in the genetic modification of organisms, and in molecular biology generally.

recombination
Process in which an exchange of genetic information occurs between two chromosomes or DNA molecules. Enzyme-mediated recombination can occur naturally in living cells or in a test tube using purified DNA and enzymes that break and re-ligate DNA strands.

redox pair
Two molecules that can be interconverted by the gain or loss of an electron; for example, NADH and NAD^+.

redox potential
A measure of the tendency of a given redox pair to donate or accept electrons.

redox reaction
A reaction in which electrons are transferred from one chemical species to another. An oxidation–reduction reaction.

reduction
Addition of electrons to an atom, as occurs during the addition of hydrogen to a carbon atom or the removal of oxygen from it. The opposite of oxidation. (*See* Figure 3–11.)

regulatory DNA sequence
DNA sequence to which a transcription regulator binds to determine when, where, and in what quantities a gene is to be transcribed into RNA.

regulatory RNA
RNA molecule that plays a role in controlling gene expression.

replication fork
Y-shaped junction that forms at the site where DNA is being replicated.

replication origin
Nucleotide sequence at which DNA replication is initiated.

reporter gene
Gene encoding a protein whose activity is easy to monitor experimentally; used to study the expression pattern of a target gene or the localization of its protein product.

repressor
A protein that binds to a specific regulatory region of DNA to prevent transcription of an adjacent gene.

reproductive cloning
The artificial production of genetically identical copies of an animal by, for example, the transplantation of a somatic cell nucleus into an enucleated fertilized egg.

respiration
General term for any process in a cell in which the uptake of molecular oxygen (O_2) is coupled to the production of CO_2.

respiratory enzyme complex
Set of proteins in the inner mitochondrial membrane that facilitates the transfer of high-energy electrons from NADH to water while pumping protons into the intermembrane space.

resting membrane potential
Voltage difference across the plasma membrane when a cell is not stimulated.

restriction nuclease

Enzyme that can cleave a DNA molecule at a specific, short sequence of nucleotides. Extensively used in recombinant DNA technology.

retrotransposon
Type of mobile genetic element that moves by being first transcribed into an RNA copy that is reconverted to DNA by reverse transcriptase and inserted elsewhere in the chromosomes.

retrovirus
RNA-containing virus that replicates in a cell by first making a double-stranded DNA intermediate that becomes integrated into the cell's chromosome.

reverse transcriptase
Enzyme that makes a double-stranded DNA copy from a single-stranded RNA template molecule. Present in retroviruses and as part of the transposition machinery of retrotransposons.

Rho protein family
Family of small, monomeric GTPases that controls the organization of the actin cytoskeleton.

ribosomal RNA (rRNA)
RNA molecule that forms the structural and catalytic core of the ribosome.

ribosome
Large macromolecular complex, composed of ribosomal RNAs and ribosomal proteins, that translates messenger RNA into protein.

ribozyme
An RNA molecule with catalytic activity.

RNA (ribonucleic acid)
Molecule produced by the transcription of DNA; usually single-stranded, it is a polynucleotide composed of covalently linked ribonucleotide subunits. Serves a variety of structural, catalytic, and regulatory functions in cells.

RNA capping
The modification of the 5′ end of a maturing RNA transcript by the addition of an atypical nucleotide.

RNA interference (RNAi)
Cellular mechanism activated by double-stranded RNA molecules that results in the destruction of RNAs containing a similar nucleotide sequence. It is widely exploited as an experimental tool for preventing the expression of selected genes (gene silencing).

RNA polymerase
Enzyme that catalyzes the synthesis of an RNA molecule from a DNA template using nucleoside triphosphate precursors.

RNA primer—*see* **primer**

RNA processing
Broad term for the modifications that a precursor mRNA undergoes as it matures into an mRNA. It typically includes 5′ capping, RNA splicing, and 3′ polyadenylation.

RNA splicing
Process in which intron sequences are excised from RNA molecules in the nucleus during the formation of a mature messenger RNA.

RNA transcript
RNA molecule produced by transcription that is complementary to one strand of DNA.

RNA world
Hypothetical period in Earth's early history in which life-forms were thought to use RNA both to store genetic information and to catalyze chemical reactions.

RNA-Seq
Sequencing technique used to determine directly the nucleotide sequence of a collection of RNAs.

rough endoplasmic reticulum
Region of the endoplasmic reticulum associated with ribosomes and involved in the synthesis of secreted and membrane-bound proteins.

S cyclin
Regulatory protein that helps to launch the S phase of the cell cycle.

S phase
Period during a eukaryotic cell cycle in which DNA is synthesized.

S-Cdk
Protein complex whose activity initiates DNA replication; consists of an S cyclin plus a cyclin-dependent protein kinase (Cdk).

sarcomere
Highly organized assembly of actin and myosin filaments that serves as the contractile unit of a myofibril in a muscle cell.

saturated
Describes an organic molecule that contains a full complement of hydrogen; in other words, no double or triple carbon–carbon bonds.

second messenger
Small intracellular signaling molecule generated or released in response to an extracellular signal. Examples include cAMP, IP$_3$, and Ca^{2+}.

secondary structure
Regular local folding pattern of a polymeric molecule. In proteins, it refers to α helices and β sheets.

secretion
Production and release of a substance from a cell.

secretory vesicle
Membrane-enclosed organelle in which molecules destined for secretion are stored prior to release. Sometimes called a secretory granule because darkly staining contents make the organelle visible as a small solid object.

segregation
During cell division, the process by which duplicated chromosomes are organized and then separated into the chromosome sets that will be inherited by each of the daughter cells.

sequence
The linear order of monomers in a large molecule, for example amino acids in a protein or nucleotides in DNA; encodes information that specifies a macromolecule's precise biological function.

serine/threonine kinase
Enzyme that phosphorylates target proteins on serines or threonines.

sex chromosome
Type of chromosome that determines the sex of an individual and directs the development of sexual characteristics. In mammals, the X and Y chromosomes.

sexual reproduction
Mode of reproduction in which the genomes of two individuals are mixed to produce an individual that is genetically distinct from its parents.

side chain
Portion of an amino acid not involved in forming peptide bonds; its chemical identity gives each amino acid its unique properties.

signal sequence
Amino acid sequence that directs a protein to a specific location in the cell, such as the nucleus or mitochondria.

signal transduction
Conversion of an impulse or stimulus from one physical or chemical form to another. In cell biology, the process by which a cell responds to an extracellular signal.

signaling cascade
Sequence of linked reactions, often including phosphorylation and dephosphorylation, that carries information within a cell, often amplifying an initial signal.

single-nucleotide polymorphism (SNP)
Form of genetic variation in which one portion of the population differs from another in terms of which nucleotide is found at a particular position in the genome.

sister chromatid
Copy of a chromosome, produced by DNA replication, that remains bound to the other copy.

site-directed mutagenesis
Technique by which a mutation can be made at a particular site in DNA.

site-specific recombination
Type of genetic exchange in which one segment of DNA is inserted into another at a particular nucleotide sequence; does not require extensive similarity between the two participating DNA sequences, which can be on different DNA molecules or within a single DNA molecule.

small interfering RNA (siRNA)
Short length of RNA produced from double-stranded RNA during the process of RNA interference. It base-pairs with identical sequences in other RNAs, leading to the inactivation or destruction of the target RNA.

small intracellular signaling molecule
Nucleotide, lipid, ion, or other small molecule generated or released in response to an extracellular signal. Examples include cAMP, IP$_3$, and Ca^{2+}. Also called second messengers.

small messenger—*see* **second messenger**

small nuclear ribonucleoprotein (snRNP)
Complex made of RNA and protein that recognizes RNA splice sites and participates in the chemistry of splicing; together these complexes form the core of the spliceosome.

small nuclear RNA (snRNA)
RNA molecule of around 200 nucleotides that participates in RNA splicing.

smooth endoplasmic reticulum (SER)
Region of the endoplasmic reticulum not associated with ribosomes; involved in the synthesis of lipids.

SNARE
One of a family of membrane proteins responsible for the selective fusion of vesicles with a target membrane inside the cell.

sodium pump—*see* **Na$^+$ pump**

solute
Any substance that is dissolved in a liquid. The liquid is called the solvent.

somatic cell
Any cell that forms part of the body of a plant or animal that is not a germ cell or germ-line precursor.

specificity
Selective affinity of one molecule for another that permits the two to bind or react, even in the presence of a vast excess of unrelated molecular species.

spindle pole
Centrosome from which microtubules radiate to form the mitotic spindle.

spliceosome
Large assembly of RNA and protein molecules that splices introns out of pre-mRNA in the nucleus of eukaryotic cells.

starch
Polysaccharide composed exclusively of glucose units, used as an energy store in plant cells.

stem cell
Relatively undifferentiated, self-renewing cell that produces daughter cells that can either differentiate into more specialized cell types or can retain the developmental potential of the parent cell.

steroid hormone
Hydrophobic signal molecule related to cholesterol; can pass through the plasma membrane to interact with intracellular receptors that affect gene expression in the target cell. Examples include estrogen and testosterone.

stroma
In a chloroplast, the large interior space that contains the enzymes needed to incorporate CO$_2$ into sugars during the carbon-fixation stage of photosynthesis; equivalent to the matrix of a mitochondrion.

substrate
A molecule on which an enzyme acts.

substratum
Solid surface to which a cell adheres.

subunit
A monomer that forms part of a larger molecule, such as an amino acid residue in a protein or a nucleotide residue in a nucleic acid. Can also refer to a complete molecule that forms part of a larger molecule. Many proteins, for example, are composed of multiple polypeptide chains, each of which is called a protein subunit.

sugar
A substance made of carbon, hydrogen, and oxygen with the general formula $(CH_2O)_n$. A carbohydrate or saccharide. The "sugar" of everyday use is sucrose, a sweet-tasting disaccharide made of glucose and fructose.

sulfhydryl group (–SH, thiol)
Chemical group containing sulfur and hydrogen found in the amino acid cysteine and other molecules. Two can join together to produce a disulfide bond.

survival factor
Extracellular signal molecule that must be present to suppress apoptosis.

symbiosis
Intimate association between two organisms of different species from which both derive a long-term selective advantage.

symport
A transporter that transfers two different solutes across a cell membrane in the same direction.

synapse

Specialized junction where a nerve cell communicates with another cell (such as a nerve cell, muscle cell, or gland cell), usually via a neurotransmitter secreted by the nerve cell.

synaptic plasticity

The ability of a synapse to adjust its strength for a prolonged period, either up or down, depending on its use; thought to play an important role in learning and memory.

synaptic vesicle

Small membrane-enclosed sac filled with neurotransmitter that releases its contents by exocytosis at a synapse.

telomerase

Enzyme that elongates telomeres, synthesizing the repetitive nucleotide sequences found at the ends of eukaryotic chromosomes.

telomere

Repetitive nucleotide sequence that caps the ends of linear chromosomes. Counteracts the tendency of the chromosome otherwise to shorten with each round of replication.

telophase

Final stage of mitosis in which the two sets of separated chromosomes decondense and become enclosed by a nuclear envelope.

template

A molecular structure that serves as a pattern for the production of other molecules. For example, one strand of DNA directs the synthesis of the complementary DNA strand.

tertiary structure

Complete three-dimensional structure of a fully folded protein.

therapeutic cloning

Procedure that uses nuclear transplantation to generate cells for tissue repair and other such purposes, as opposed to producing whole multicellular individuals.

thioester bond

High-energy bond formed by a condensation reaction between an acid (acyl) group and a thiol group (–SH); seen, for example, in acetyl CoA and in many enzyme–substrate complexes.

thylakoid

In a chloroplast, the flattened disclike sac whose membranes contain the proteins and pigments that convert light energy into chemical energy during photosynthesis.

tight junction

Cell–cell junction that seals adjacent epithelial cells together, preventing the passage of most dissolved molecules from one side of the epithelial sheet to the other.

tissue

Cooperative assembly of cells and matrix woven together to form a distinctive multicellular fabric with a specific function.

trans

Beyond, or on the other side.

trans **Golgi network (TGN)**

Portion of the Golgi apparatus furthest from the endoplasmic reticulum and from which proteins and lipids leave for lysosomes, secretory vesicles, or the cell surface.

transcription

Process in which RNA polymerase uses one strand of DNA as a template to synthesize a complementary RNA sequence.

transcription factor

Term loosely applied to any protein required to initiate or regulate transcription in eukaryotes. Includes transcription regulators as well as the general transcription factors.

transcription regulator

Protein that binds specifically to a regulatory DNA sequence and is involved in controlling whether a gene is switched on or off.

transcriptional activator

A protein that binds to a specific regulatory region of DNA to permit transcription of an adjacent gene.

transcriptional repressor

A protein that binds to a specific regulatory region of DNA to prevent transcription of an adjacent gene.

transfer RNA (tRNA)

Small RNA molecule that serves as an adaptor that "reads" a codon in mRNA and adds the correct amino acid to the growing polypeptide chain.

transformation

Process by which cells take up DNA molecules from their surroundings and then express genes on that DNA.

transgenic organism

A plant or animal that has stably incorporated into its genome one or more genes derived from another cell or organism.

transition state

Structure that forms transiently during the course of a chemical reaction; in this configuration, a molecule has the highest free energy, and is no longer a substrate, but is not yet a product.

translation

Process by which the sequence of nucleotides in a messenger RNA molecule directs the incorporation of amino acids into protein.

translation initiation factor

Protein that promotes the proper association of ribosomes with mRNA and is required for the initiation of protein synthesis.

transmitter-gated ion channel

Transmembrane receptor protein or protein complex that opens in response to the binding of a neurotransmitter, allowing the passage of a specific inorganic ion; its activation can trigger an action potential in a postsynaptic cell.

transport vesicle

Membrane vesicle that carries proteins from one

intracellular compartment to another, for example, from the endoplasmic reticulum to the Golgi apparatus.

transporter
Membrane transport protein that moves a solute across a cell membrane by undergoing a series of conformational changes.

transposon
General name for short segments of DNA that can move from one location to another in the genome. Also known as mobile genetic elements.

triacylglycerol
Compound made of three fatty acid tails covalently attached to glycerol. A storage form of fat, the main constituent of fat droplets in animal tissues (in which the fatty acids are saturated) and of vegetable oil from plants (in which the fatty acids are mainly unsaturated).

tryptophan repressor
In bacteria, a transcription regulator that, in the presence of tryptophan, shuts off production of the tryptophan biosynthetic enzymes by binding to the promoter region that controls expression of those genes.

tubulin
Protein from which microtubules are made.

γ-tubulin ring
Protein complex in centrosomes from which microtubules grow.

tumor suppressor gene
A gene that in a normal tissue cell inhibits cancerous behavior. Loss or inactivation of both copies of such a gene from a diploid cell can cause it to behave as a cancer cell.

turgor pressure
Force exerted on a plant cell wall when water enters the cell by osmosis; keeps plant from wilting.

turnover number
The number of substrate molecules an enzyme can convert into product per second.

tyrosine kinase
Enzyme that phosphorylates target proteins on tyrosines.

unfolded protein response (UPR)
Molecular program triggered by the accumulation of misfolded proteins in the endoplasmic reticulum. Allows cells to expand the endoplasmic reticulum and produce more of the molecular machinery needed to restore proper protein folding and processing.

unsaturated
Describes an organic molecule that contains one or more double or triple bonds between its carbon atoms.

valence
The number of electrons an atom must gain or lose (either by electron sharing or electron transfer) to achieve a filled outer shell. For example, Na must lose one electron, and Cl must gain one electron. This number is also equal to the number of single bonds that the atom can form.

van der Waals attraction
Weak noncovalent interaction, due to fluctuating electrical charges, that comes into play between two atoms within a short distance of each other.

vector
DNA molecule that is used as a vehicle to carry a fragment of DNA into a recipient cell for the purpose of gene cloning; examples include plasmids, engineered viruses, and artificial chromosomes.

vesicle
Small, membrane-enclosed, spherical sac in the cytoplasm of a eukaryotic cell.

vesicular transport
Movement of material between organelles in the eukaryotic cell via membrane-enclosed vesicles.

virus
Particle consisting of nucleic acid (RNA or DNA) enclosed in a protein coat and capable of replicating within a host cell and spreading from cell to cell. Often the cause of disease.

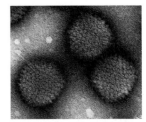

V_{max}
The maximum rate of an enzymatic reaction, reached when the active sites of the enzyme molecules in a sample are fully occupied by substrate.

voltage-gated channel
Channel protein that permits the passage of selected ions, such as Na+, across a membrane in response to changes in the membrane potential. Found primarily in electrically excitable cells such as nerve and muscle.

voltage-gated ion channel
Protein that selectively allows particular ions to cross a membrane in response to a change in membrane potential. Found mainly in electrically excitable cells such as nerve and muscle cells.

voltage-gated Na+ channel
Protein in the plasma membrane of electrically excitable cells that opens in response to membrane depolarization, allowing Na+ to enter the cell. It is responsible for action potentials in these cells.

wild type
Typical non-mutant form of a species, gene, or characteristic as it occurs in nature.

Wnt protein
Member of a family of extracellular signal molecules that regulates cell proliferation and migration during embryonic development and that maintains stem cells in a proliferative state.

X chromosome
Larger of the two sex chromosomes in mammals. The cells of males contain one, and females possess two.

X-ray crystallography
Technique used to determine the three-dimensional structure of a protein molecule by analyzing the pattern produced when a beam of X-rays is passed through an ordered array of the protein.

Y chromosome
Smaller of the two sex chromosomes of mammals. Present in a single copy only in the cells of males, contains genes that direct the development of male sex organs and characteristics.

yeast
Common term for several families of eukaryotic unicellular fungi used as model organisms. Includes species used for brewing beer and making bread, as well as species that cause disease.

zygote
Diploid cell produced by fusion of a male and a female gamete. A fertilized egg.

Index

Note: The suffixes F and T after a page number indicate a relevant Figure or Table appearing on a page separated from the main text discussion.

Note: The suffixes F and T after a page number indicate a relevant Figure or Table appearing on a page separated from the main text discussion.

Note: The suffixes F and T after a page number indicate a relevant Figure or Table appearing on a page separated from the main text discussion.

Note: The suffixes F and T after a page number indicate a relevant Figure or Table appearing on a page separated from the main text discussion.

Note: The suffixes F and T after a page number indicate a relevant Figure or Table appearing on a page separated from the main text discussion.

Note: The suffixes F and T after a page number indicate a relevant Figure or Table appearing on a page separated from the main text discussion.

Note: The suffixes F and T after a page number indicate a relevant Figure or Table appearing on a page separated from the main text discussion.

Note: The suffixes F and T after a page number indicate a relevant Figure or Table appearing on a page separated from the main text discussion.

Note: The suffixes F and T after a page number indicate a relevant Figure or Table appearing on a page separated from the main text discussion.

Note: The suffixes F and T after a page number indicate a relevant Figure or Table appearing on a page separated from the main text discussion.

Note: The suffixes F and T after a page number indicate a relevant Figure or Table appearing on a page separated from the main text discussion.

Note: The suffixes F and T after a page number indicate a relevant Figure or Table appearing on a page separated from the main text discussion.

Note: The suffixes F and T after a page number indicate a relevant Figure or Table appearing on a page separated from the main text discussion.

Note: The suffixes F and T after a page number indicate a relevant Figure or Table appearing on a page separated from the main text discussion.